Geomorphology of Desert Environments

Second Edition

Geomorphology of Desert Environments

Second Edition

Edited by

Anthony J. Parsons
University of Sheffield, UK

and

Athol D. Abrahams
State University of New York at Buffalo, USA

Editors
Anthony J. Parsons
University of Sheffield
Dept. Geography
Winter Street
Sheffield
United Kingdom S10 2TN
a.j.parsons@sheffield.ac.uk

Athol D. Abrahams
State University of New York
Dept. Geography
Buffalo NY 14261
USA
abrahams@buffalo.edu

This is a second revised and enlarged edition of the first edition published by Chapman and Hall, 1994.

100652281\

ISBN 978-1-4020-5718-2   e-ISBN 978-1-4020-5719-9

DOI 10.1007/978-1-4020-5719-9

Library of Congress Control Number: 2008939014

Cover design based on 'Space between Mesas' by Ed Mell, whom we thank for allowing us to use his work.

Printed on acid-free paper

9 8 7 6 5 4 3 2 1

springer.com

Preface to the First Edition

About one-third of the Earth's land surface experiences a hyperarid, arid, or semi-arid climate, and this area supports approximately 15% of the planet's population. This percentage continues to grow and with this growth comes the need to learn more about the desert environment. Geomorphology is only one aspect of this environment, but an important one, as geomorphic phenomena such as salt weathering, debris flows, flash flooding, and dune encroachment pose major problems to desert settlement and transportation.

The geomorphology of deserts has been the subject of scientific enquiry for more than a century, but desert geomorphology did not emerge as an identifiable sub-discipline in geomorphology until the 1970s when the first textbooks on the subject appeared, namely *Geomorphology in deserts* in 1973 and *Desert landforms* in 1977. Also, in 1977 the Eighth Annual (Binghamton) Geomorphology Symposium was devoted to the theme 'Geomorphology in Arid Lands' and the proceedings of the symposium were published in the same year. The 1980s have seen the appearance of titles dealing with particular topics within desert geomorphology, the most notable of these being *Urban geomorphology in drylands* and *Dryland rivers*. As we enter the 1990s, a new generation of textbooks on desert geomorphology has reached the bookstores. *Arid zone geomorphology* and *Desert geomorphology* incorporate the advances in knowledge that have occurred during the past 20 years but are primarily written for the college student. By contrast, the present volume assumes that the reader already has some knowledge of desert geomorphology. It is pitched at a level somewhat higher than the standard text and is intended to serve mainly as a reference book.

To achieve this goal we sought out authors who were active researchers and authorities in their fields. We asked each to write an up-to-date review of an assigned topic related to their speciality. These reviews are assembled in this book and together represent a comprehensive treatment of the state of knowledge in desert geomorphology. The treatment, perhaps inevitably, contains a geographical bias, in that 14 of the 22 authors are based in North America. Although most of them have experience in deserts on other continents, their discussions and the examples they draw upon are lop-sidedly American. The bias was perhaps inevitable (despite our best efforts to avoid it) because modern research in desert geomorphology published in English is dominated by investigations conducted in the Deserts of the American South-west. Faced with this geographical bias, we specifically requested authors to include research conducted outside North America. Different authors have succeeded in this regard in varying degrees. Thus in spite of the bias, we believe this book will have appeal and relevance beyond North America and will be useful to geomorphologists working in deserts around the globe.

The idea for this volume emerged during an informal field trip of desert geomorphologists through the Mojave Desert and Death Valley prior to the Annual Meeting of the Association of American Geographers in Phoenix in April 1988. One author submitted his chapter in September, even before we had found a publisher! As we write this Preface, almost four years later, the final chapter has just arrived. Assembling within a limited time frame 26 chapters from 22 authors, all of whom have busy schedules and other commitments and obligations, is a daunting task. Those who submitted their chapters early or on time have waited patiently for those less prompt, while those running late have had to sustain regular badgering by the editors. Finally, however, the book is complete. As is generally the case with edited volumes, the quality of the product depends very heavily on the quality of the individual chapters, and the quality of the chapters depends on the authors. Recognizing this, we would like to thank the authors for their efforts in writing this book. We are pleased with the final product, and we hope they are too.

Athol D. Abrahams   Anthony J. Parsons
Buffalo, USA   Keele, UK

Preface to the Second Edition

When we were approached by Springer to consider a second edition of *Geomorphology of Desert Environments*, our initial inclination was to say no. Before doing so, however, we contacted the authors who contributed chapters to the first edition and asked them if they would be prepared to update their chapters. To our considerable surprise, their response was overwhelmingly positive. With very few exceptions, those still active in the field expressed enthusiasm for the idea. The appearance of this volume is, therefore, more a credit to the contributors to the first edition than it is to the editors! We are grateful to them for their support of this new edition, and to those new contributors, some of whom have filled in the gaps, but the majority of whom have provided chapters additional to those in the first edition. It may be invidious to single out a single contributor, but we should specifically acknowledge Dorothy Sack who not only revised her own chapter from the first edition but offered to take on the revision of the two chapters that the late Don Currey had contributed.

In the decade and a half since the preparation of the first edition, progress in the multitude of subjects that comprise the field of desert geomorphology has varied greatly, and this variation has had a profound effect on the character of the field. Some subjects (for example, dust) have burgeoned over the period to merit a chapter in their own right. Others have seen significant changes, particularly those in which the advances in dating techniques have had an impact. Yet other areas of research have seen relatively little progress and appear to have fallen from fashion. In the course of revising *Geomorphology of Desert Environments*, we therefore made an effort to adjust the coverage of the various subjects to reflect the changes that have occurred in these subjects since the printing of the first edition. Thus the *raison d'être* for the second edition is to provide a balanced and up-to-date synthesis of the geomorphic processes that operate in desert environments and the landforms they produce.

Sheffield, UK

Buffalo, USA

Anthony J. Parsons

Athol D. Abrahams

Acknowledgments

We would like to thank the following publishers, organizations, and individuals for permission to reproduce the following figures.

Figure 1.2   Thomas, DSG 1997. Dating of desert sequences. In *Arid Zone* Geomorphology, D.S.G. Thomas (ed.), 577–605. Reprinted by permission of John Wiley & Sons, Ltd.

Figure 8.9   Selby, M.J. 1987. Rock slopes. In *Slope stability: geotechnical engineering and geomorphology*, M.G. Anderson and K.S. Richards (eds.), 475–504. Reprinted by permission of John Wiley & Sons, Ltd.

Figure 8.22   Donald O. Doehring

Figure 8.29   van Nostrand Reinhold

Figure 8.33   Zeitschrift fur Geomorphologie

Figure 8.34   Arthur L. Lange

Figure 9.9   Academic Press

Figures 9.11 and 9.13   Elsevier Science Publishers

Figure 9.15   Bunte, K. and Poesen, J. 1994. Effects of rock fragment size and cover on overland flow hydraulics, local turbulence and sediment yield on an erodible soil surface. *Earth Surface Processes and Landforms*, **19**, 115–35. Copyright 1994 John Wiley & Sons Ltd. Reproduced with permission.

Figure 9.16   Parsons, A.J., J. Wainwright and A.D. Abrahams 1996. Runoff and erosion on semi-arid hillslopes. In *Advances in Hillslope Processes*, M.J. Anderson and S.M. Brooks (eds.), 1061–1078. Copyright 1994 John Wiley & Sons Ltd. Reproduced with permission.

Figure 9.17   Parsons, A.J., J. Wainwright and A.D. Abrahams 1996. Runoff and erosion on semi-arid hillslopes. In *Advances in Hillslope Processes*, M.J. Anderson and S.M. Brooks (eds.), 1061–1078. Copyright 1994 John Wiley & Sons Ltd. Reproduced with permission.

Figure 9.18	Abrahams, A.D., G. Li and A.J. Parsons. 1996. Rill hydraulics on a semiarid hillslope, southern Arizona. *Earth Surface Processes and Landforms*, **21**, 35–47. Copyright 1996 John Wiley & Sons Ltd. Reproduced with permission.
Figure 9.21a	Elsevier Science Publishers
Figure 9.25	Kirkby, M.J. 1969. Erosion by water on hillslopes. In *Water, earth and man*, R.J. Chorley (ed.), 229–38. Reproduced by permission of Methuen & Co.
Figure 9.26	Catena Verlag
Figure 9.27	Parsons, A.J. and J. Wainwright 2006. Depth distribution of overland flow and the formation of rills. *Hydrological Processes*, **20**, 1511–23. Copyright 2006 John Wiley & Sons Ltd. Reproduced with permission.
Figure 10.7	Elsevier Science Publishers
Figure 10.8	Catena Verlag
Figure 23.1	M. Servant
Figure 23.2	M. R. Talbot
Figure 23.3	Frostick, L.E. and I. Reid 1989. Is structure the main control of river drainage and sedimentation in rifts? *Journal of African Earth Sciences* **8**, 165–82. Reprinted with permission of Pergamon Press PLC.
Figures 23.4 and 23.6	D.A. Adamson
Figures 23.9 and 23.10	Maizels, J.K. 1987. Plio-Pleistocene raised channel systems of the western Sharqiya (Wahiba), Oman. In *Desert sediments: ancient and modern*, L.E. Frostick and I. Reid (eds.), 35–50. Reproduced by permission of the Geological Society and J.K. Maizels.
Figures 23.11 and 23.12	Baker, V.R. 1978. Adjustment of fluvial systems to climate and source terrain in tropical and subtropical environments. In *Fluvial sedimentology*, A.D. Miall (ed.), 211–30. Reproduced with permission of the Canadian Society of Petroleum Geologists.
Figures 23.13 and 23.15a,b	W. L. Graf
Figure 23.15c	Graf, W.L. 1979. The development of montane arroyos and gullies. *Earth Surface Processes* **4**, 1–14. Copyright 1979 John Wiley & Sons, Ltd. Reprinted by permission of John Wiley & Sons, Ltd.
Figure 23.16	Schumm, S.A. and R.F. Hadley 1957. Arroyos and the semi-arid cycle of erosion. *American Journal of Science* **255**, 161–74. Reprinted by permission of American Journal of Science and S.A. Schumm.
Figure 23.18	M.R. Talbot
Figure 23.20	Grossman, S. and R. Gerson 1987. Fluviatile deposists and morphology of alluvial surfaces as indicators of Quaternary environmental changes in the southern Negev, Israel. In *Desert sediments: ancient and modern*, L.E. Frostick and I. Reid (eds.), 17–29. Reproduced by permission of the Geological Society and S. Grossman.

Figure 23.21   Maizels, J.K. 1987. Plio-Pleistocene raised channel systems of the western Sharqiya (Wahiba), Oman. In *Desert sediments: ancient and modern*, L.E. Frostick and I. Reid (eds.), 31–50. Reproduced by permission of the Geological Society and J.K. Maizels.

Figure 23.22   D. Adamson

Figure 25.9   Elsevier Science Publishers and D.R. Currey

Figure 26.1   Tchakerian, V.P. 1999. Dune palaeoenvironments. In *Aeolian environments, sediments and landforms*, A.S. Goudie, I. Livingstone and S. Stokes (eds.), 261–292. Copyright 2006 John Wiley & Sons Ltd. Reproduced with permission.

Figure 26.4   Tchakerian, V.P. 1999. Dune palaeoenvironments. In *Aeolian environments, sediments and landforms*, A.S. Goudie, I. Livingstone and S. Stokes (eds.), 261–292. Copyright 2006 John Wiley & Sons Ltd. Reproduced with permission.

Figure 26.5   Tchakerian, V.P. 1999. Dune palaeoenvironments. In *Aeolian environments, sediments and landforms*, A.S. Goudie, I. Livingstone and S. Stokes (eds.), 261–292. Copyright 2006 John Wiley & Sons Ltd. Reproduced with permission.

Figure 26.6   Tchakerian, V.P. 1999. Dune palaeoenvironments. In *Aeolian environments, sediments and landforms*, A.S. Goudie, I. Livingstone and S. Stokes (eds.), 261–292. Copyright 2006 John Wiley & Sons Ltd. Reproduced with permission.

Figure 28.2   Williams, M.A.J., P.I. Abell and B.W. Sparks 1987. Quaternary landforms, sediments, and depositional environments and gastropod isotope ratios at Adrar Bous, Tenere Desert of Niger, south-central Sahara. In *Desert sediments: ancient and modern*, L.E. Frostick and I. Reid (eds.), 105–25. Reproduced by permission of the Geological Society and I. Reid.

Contents

Part IV   Rivers

Part V   Piedmonts

Part VI   Lake Basins

Part VII   Aeolian Surfaces

Part VIII   Climatic Change

Contributors

Athol D. Abrahams Department of Geography, State University of New York at Buffalo, Buffalo, NY 14261, USA, abrahams@buffalo.edu

Terence C. Blair Blair & Associates LLC, 1949 Hardscrabble Place, Boulder, CO 80305, USA, tcblair@aol.com

Joanna E. Bullard Department of Geography, Loughborough University of Technology, Loughborough, LE11 3TU, UK, J.E.Bullard@lboro.ac.uk

Donald R. Currey (deceased) Department of Geography, University of Utah, Salt Lake City, UT 84112, USA

John C. Dixon Department of Geosciences, University of Arkansas, Fayetteville, AR 72701, USA, jcdixon@uark.edu

John C. Dohrenwend Southwest Satellite Imaging, PO Box 1467, Moab, UT 84532, USA, dohrenwend@scinternet.net

Ronald I. Dorn School of Geographical Sciences, Arizona State University, Tempe, AZ 85287, USA, ronald.dorn@asu.edu

Andrew S. Goudie School of Geography, Oxford University, South Parks Road, Oxford, OX1 3QY, UK, andrew.goudie@stx.ox.ac.uk

Alan D. Howard Department of Environmental Sciences, University of Virginia, Charlottesville, VA 22904-4123, USA, ah6p@cms.mail.virginia.edu

Julie E. Laity Department of Geography, California State University, Northridge, CA 91330, USA, julie.laity@csun.edu

Nicholas Lancaster Desert Research Institute, 2215 Raggio Parkway, Reno, NV 89512-1095, USA, nick.lancaster@dri.edu

Ian Livingstone School of Applied Sciences, University of Northampton, Northampton NN2 7AL, UK, ian.livingstone@northampton.ac.uk

Cheryl Mckenna Neuman Department of Geography, Trent University Peterborough, Ontario, Canada, K9J 7B8, cmckneuman@trentu.ca

Sue J. Mclaren Department of Geography, University of Leicester, University Road, Leicester LE1 7RH, UK, sjm11@le.ac.uk

John G. Mcpherson ExxonMobil Exploration Company, 12 Riverside Quay, Southbank, Victoria, Australia 3006, john.mcpherson@exxonmobil.com

William G. Nickling Wind Erosion Laboratory, Department of Geography, University of Guelph, Guelph, Ontario N1G 2W1, Canada, nickling@uoguelph.ca

Anthony J. Parsons Sheffield Centre for International Drylands Research, Department of Geography, University of Sheffield, Sheffield S10 2TN, UK, a.j.parsons@sheffield.ac.uk

D. Mark Powell Department of Geography, University of Leicester, University Road, Leicester LE1 7RH, UK, dmp6@le.ac.uk

Ian Reid Department of Geography, Loughborough University, Loughborough, LE11 3TU, UK, ian.reid@lboro.ac.uk

Tim Reynolds School of Continuing Education, Birkbeck, University of London, 26 Russell Square, London WC1B 5DQ, UK, te.reynolds@bbk.ac.uk

Dorothy Sack Department of Geography, Ohio University, Athens, OH 45701, USA, sack@ohio.edu

Karl-Heinz Schmidt Department of Geoscience, Universität Halle, Von-Seckendorff-Platz 4, 06120 Halle, Germany, karl-heinz.schmidt@geo.uni-halle.de

Michael J. Selby Department of Earth Sciences, University of Waikato, Private Bag 3105, Hamilton, New Zealand

B.J. Smith School of Geography, Archaeology and Palaeoecology, Queen's University of Belfast, Belfast BT7 1NN, UK, b.smith@qub.ac.uk

Vatche P. Tchakerian Department of Geography, Texas A & M University, College Station, TX 77843, USA, v-tchakerian@tamu.edu

John B. Thornes (deceased) Department of Geography, King's College London, Strand, London WC2R 2LS, UK

John Wainwright Sheffield Centre for International Drylands Research, Department of Geography, University of Sheffield, Sheffield S10 2TN, UK, j.wainwright@sheffield.ac.uk

M.A.J. Williams Geographical and Environmental Studies, University of Adelaide, Adelaide, SA 5005, Australia, martin.williams@adelaide.edu.au

Part I
Introduction

Chapter 1

Geomorphology of Desert Environments

Anthony J. Parsons and Athol D. Abrahams

The Concept of Desert Geomorphology

The notion that the desert areas of the world possess a distinct geomorphology has a long history and, in many ways, is informed by the popular concept of deserts as places that are different. Not surprisingly, early explorers in deserts, particularly Europeans travelling in the Sahara from the late 18th century onwards, were impressed by, and reported on, the unusual features of these areas. Rock pedestals, sand dunes, and bare-rock hills rising almost vertically from near-horizontal, gravel-covered plains all contributed to the impression of a unique landscape. This spirit of exploration in a totally alien landscape continued into the 20th century, so that as late as 1935 R.A. Bagnold wrote of his travels in North Africa during the preceding decade under the title *Libyan sands: travels in a dead world* (Bagnold, 1935). Emphasis on the unusual and remarkable landforms of desert areas and a coincident emphasis on the hot tropical deserts had a profound impact on attempts to explain the geomorphology of deserts.

Of particular influence in shaping a view of the uniqueness of desert geomorphology, due in large measure to his influence in shaping geomorphology overall, was W.M. Davis who was sufficiently persuaded of the distinctiveness of desert landscapes that in 1905 he published his cycle of erosion in arid climates. Davis held the opinion that, notwithstanding the infrequency of rainfall in desert areas, the landforms resulted primarily from fluvial processes. Only towards the end of

his cycle of erosion did aeolian processes come to play a dominant role. Subsequently, there was substantial debate on the relative importance of fluvial and aeolian processes in desert landform evolution, and only in recent times has there been a recognition of, and attention paid to, the links that exist between aeolian and fluvial processes (e.g. Bullard and Livingstone, 2002) and the extent to which desert landforms owe their character to these two sets of processes acting in concert (e.g. Parsons et al., 2003). However, whether through agencies of wind and/or water, the essence of Davis's viewpoint, namely that arid areas are subject to a unique cycle of erosion, was maintained for much of the 20th century in the work of, for example, Cotton (1947) and, in a wider context, in the many writings that stem from the concept of climatic geomorphology (e.g. Birot, 1960; Tricart and Cailleux, 1969; Budel 1963).

As the emphasis in geomorphology moved, in the latter part of the twentieth century, away from cycles of erosion and morphogenesis within specific areas towards the study of geomorphological processes, the distinctiveness of desert geomophology was undermined. Thus, in his study of the anabranching of Red Creek in arid Wyoming (mean annual precipitation of 165 mm) Schumann (1989) drew a parallel between the flashy regime of this river and that of the Yallahs River studied by Gupta (1975) in Jamaica, where the mean annual rainfall exceeds 2000 mm. Likewise, Abrahams and Parsons (1991) compared their finding that resistance to overland flow is related to the concentration of gravel on hillslope surfaces in southern Arizona (mean annual precipitation of 288 mm) to similar findings by Roels (1984) in the Ardeche basin, France (mean annual rainfall of 1036 mm).

In the minds of many (e.g. Young, 1978 p.78) emphasis on short-term, small-scale processes was

A.J. Parsons (✉)
Department of Geography, Sheffield Centre for International Drylands Research, University of Sheffield, Sheffield S10 2TN, UK
e-mail: a.j.parsons@sheffield.ac.uk

A.J. Parsons, A.D. Abrahams (eds.), *Geomorphology of Desert Environments*, 2nd ed., DOI 10.1007/978-1-4020-5719-9_1, © Springer Science+Business Media B.V. 2009

no more than a stepping stone in the history of geomorphology towards an improved understanding of landscapes. However, making the link back from the greater understanding of geomorphological processes that has been achieved in the past half century to a more informed and quantitatively based understanding of landscape evolution has proven to be more complex than at first envisaged (Sugden et al., 1997). Consequently, although geomorphology has showed renewed and increasing interest in long-term landscape evolution (Summerfield, 2005), particularly in response to the development of techniques to date landscape surfaces and deposits, progress in tying such quantitative information on rates of landscape change to process mechanisms has been both limited, often focused within the confines of individual process domains, and poorly linked to the growing record of climatic oscillations.

Central to the concept of desert landforms and landform evolution is the assumption that similarities of climate throughout desert areas outweigh differences that may arise from other influences and similarities (such as those that arise from tectonic history or character of the substrate) that transcend climatic setting. This assumption may be challenged not only from the perspective of the relative importance of other influences and similarities (see Mabbutt, 1977) but also from an assessment of the geomorphological significance of the supposed similarity of desert climates.

Desert Climates

In scientific terms, deserts are usually defined in terms of aridity. However, providing a universally acceptable definition of aridity upon which to base a definition of desert areas has not been straightforward. Several attempts based upon a variety of geomorphic, climatic, and/or vegetational indices of aridity have been made to identify the world distribution of deserts. The UNEP World Atlas of Desertification (UNEP, 1997) classifies deserts on the basis of an Aridity Index. This index is derived from monthly data on temperature and precipitation (P) over the period 1951–1980 for a worldwide network of meteorological stations. From the temperature data, together with monthly data on daylight hours, potential evapotranspiration (PET) is calculated. The aridity index is simply the value of P/PET. For purposes of mapping (Fig. 1.1) the Aridity Index is classified into four:

Hyperarid regions – P/PET < 0.05
Arid regions – 0.05 < P/PET < 0.2
Semi-arid regions – 0.2 < P/PET < 0.5
Dry-subhumid regions – 0.5 < P/PET < 0.65

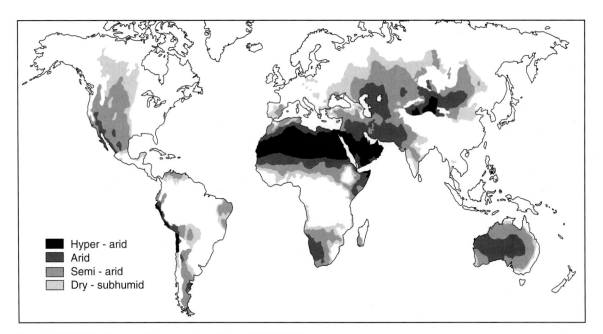

Fig. 1.1 World distribution of deserts (adapted from UNEP 1997)

Table 1.1 Land areas in each of the four Aridity Classes defined by UNEP (1997)

Aridity Class	World Land Area (%)
Hyperarid	7.5
Arid	12.1
Semi-arid	17.7
Dry subhumid	9.9

Global land area in each of these four aridity classes is given in Table 1.1.

To what extent, however, are these aridity zones geomorphologically meaningful? As the subsequent chapters of this book will show, it is not aridity *per se* that is of significance for geomorphological processes in deserts. Rather it is the availability of moisture and the timescales of that availability that matter: directly so in the case of water-driven processes, and indirectly so in the case of aeolian processes through the effects of water availability on vegetation cover.

Similarly, the lack of any simple relationship between current aridity and present-day geomorphological processes raises questions about the inferences that may be drawn from palaeoclimatic information for the geomorphological inheritance of deserts. While it has been recognised that the world's deserts have very different climatic histories (Thomas, 1997; Fig. 1.2), the broad geomorphological implications of these different histories, couched as they are in terms of varying aridity, are far from obvious and almost certainly not straighforward. Indeed questions must arise about the data upon which climatic histories are based. Where the data are drawn from evidence based upon geomorphological processes, then their interpretation in terms of simple aridty may be suspect. On the other hand, where the data come from other climatic sources or proxies, their value in explaining the suite of landforms extant today is dubious.

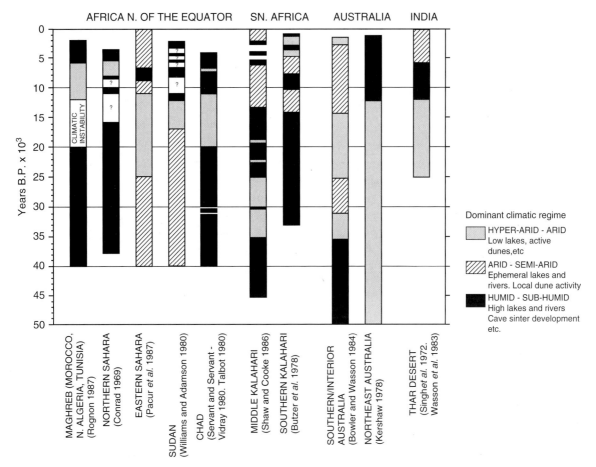

Fig. 1.2 Late Quaternary climatic changes in the world's desert areas (after Thomas 1997)

Is There a Geomorphology of Deserts?

If general scientific notions of aridity are insufficient to characterize a geomorphology of deserts, then what is? Two arguments may be made. The first is that employed by practitioners of geomorphology. A number of geomorphologists focus on the geomorphology of deserts. Whether these geomorphologists are interested in rivers, sand dunes or weathering processes, the environmental context – that is, the totality of desert geomorphology – will be pertinent to their study. There is a geomorphology of deserts because those who study component aspects of it need the totality to exist. The second argument is that which derives from the landscape itself. Notwithstanding all the problems that may be encountered in defining a set of unique and characteristic landforms for the world's arid lands, the fact remains that along transects, either equatorward from temperate areas or poleward from the wet tropics, there are progressive climatic and vegetational changes. Along these transects (i) rainfall diminishes in amount and becomes less frequent, and more sporadic, (ii) vegetation becomes smaller and patchy, and (iii) bare ground becomes more common. Desert geomorphology can effectively be defined as the geomorphic consequences of these climatic and vegetational changes. Under this definition, as in Fig. 1.1, the term desert is used in this volume broadly to include all hot, warm, and temperate arid and semi-arid parts of the world.

However, neither argument creates a watertight definition. Practitioners often extend their expertise outside deserts, and landforms common in deserts are seldom unique to them. Consequently, although many of the luminescence studies conducted by Bateman, for example, focus on environmental change in deserts (e.g. Bateman et al., 2003), others address comparable aeolian processes in quite different environments (e.g. Bateman and van Huissteden, 1999). Understanding the geomorphology of desert environments draws upon knowledge gained in other settings. Likewise, our understanding of deserts is frequently helpful in understanding landforms outside the desert realm.

Organization of the Book

This book focuses on the geomorphic processes that operate in desert environments and the landforms they produce. The effects of most processes are spatially limited so that it is possible to identify within any landscape a set of process domains within which particular processes dominate. The book is mainly organized around these process domains. Because different domains dominate different deserts, a first consideration needs to be the distribution of these domains across the deserts of the world. In the second chapter of the introductory section, therefore, the world's deserts are compared from the point of view of these process domains. Because all deserts are characterised by patchy vegetation and all geomorphological processes are influenced by this vegetation, chapter three of the introductory session considers the nature and geomorphological significance of vegetation in desert environments.

Some processes, particularly weathering and soil formation, are less constrained into specific process domains than others. Because of their widespread effects across all desert terrain types, these processes are considered in the second section of the book. The next five sections examine the processes of the five main process domains of deserts: hillslopes, rivers, piedmonts, lake basins, and aeolian surfaces. In the final section of the book, we step outside the present spatial pattern of processes and process domains, which are no more than a short-term expression of the contemporary climate, to examine how the processes and process domains of deserts respond to and are able to provide information about climatic change.

References

Abrahams, A.D. and A.J. Parsons 1991. Resistance to overland flow on desert pavement and its implications for sediment transport modeling. *Water Resources Research* **27**, 1827–36.

Bagnold, R.A. 1935. *Libyan Sands: Travels in a Dead World.* London: The Travel Book Club.

Bateman, M.D. and van Huissteden, J. (1999). The timing of Last Glacial periglacial and aeolian events, twente, Eastern Netherlands. *Journal of Quaternary Science*, **14**, 277–283.

Bateman, M.D., Thomas, D.S.G. and Singhvi, A.K. (2003). Extending the aridity record of the Southwest Kalahari: current problems and future perspectives. *Quaternary International*, **111**, 37–49.

Bullard, J.E. and I. Livingstone 2002. Interactions between aeolian and fluvial systems in dryland environments. *Area* **34**, 8–16.

Birot, P. 1960. *Le Cycle d'erosion sous les Differents Climats.* Rio de Janeiro: University of Brazil. English translation by C.I. Jackson and K.M. Clayton, London: Batsford, 1968.

Budel, J. 1963. Klima-genetische Geomorphologie. *Geographische Rundschau* **15**, 269–85.

Cotton, C.A. 1947. *Climatic Accidents in Landscape-Making.* Christchurch: Whitcombe & Tombs.

Davis, W.M. 1905. The Geographical Cycle in an arid climate. *Journal of Geology* **13**, 381–407.

Gupta, A. 1975. Stream characteristics in eastern Jamaica, an environment of seasonal flow and large floods. *American Journal of Science* **275**, 825–47.

Mabbutt, J.A. 1977. *Desert landforms.* Canberra: ANU Press, 340pp.

Parsons, A.J., Wainwright, J., Schlesinger, W.H. & Abrahams A.D. 2003 The role of overland flow in sediment and nitrogen budgets of mesquite dunefields, southern New Mexico. *Journal of Arid Environments*, **53**, 61–71.

Roels, J.M. 1984. Flow resistance in concentrated overland flow on rough slope surfaces. *Earth Surface Processes and Landforms* **9**, 541–51.

Schumann, R.R. 1989. Morphology of Red Creek, Wyoming, an arid-region anastomosing channel system. *Earth Surface Processes and Landforms* **14**, 277–88.

Sugden, D.E., M.A. Summerfield and T.P. Burt. 1997. Linking short-term geomorphic processes to landscape evolution. *Earth Surface Processes and Landforms*, **22**, 193–194.

Summerfield, M.A. 2005. A tale of two scales, or the two geomorphologies. *Institute of British Geographers: Transactions* **30**: 402–415.

Tricart, J. and A. Cailleux 1969. *Traite de geomorphology IV: le modele des regions seches.* Paris: Societe d'edition d'enseignement superieur.

Thomas, D.S.G. 1997. *Arid Zone Geomorphology (2nd Edn).* Chichester: John Wiley & Sons.

UNEP 1997. *World Atlas of Desertification (2nd Edn).* London: Arnold.

Young, A. 1978. Slopes: 1970–1975. In: Embleton, C., Brunsden, D. and Jones, D.K.C. (eds) *Geomorphology: Present Problems and Future Prospects.* Oxford: University Press. 73–83.

Chapter 2

Global Deserts and Their Geomorphological Diversity

Andrew S. Goudie

Introduction

The world's deserts show great diversity in terms of both their landscapes and their geomorphological processes (Goudie, 2002). Climate is one major control of their character. Thus aridity determines the extent to which different types of salt can accumulate, but above all it determines the nature of the vegetation cover, which in turn controls the rate of operation of slope, fluvial and aeolian processes. For example, dunes will not for the most part move if is there is a substantial vegetation cover, nor will dust storms be generated. Deserts such as the Atacama, Libyan and Namib are hyper-arid, whereas those of the Thar, Kalahari and Australia are considerably moister. Some deserts are high energy wind environments, while others are not, and this helps to explain variations in dune forms, and the presence or absence of wind erosion features such as *yardangs*. Some have unidirectional wind regimes, whereas others are more variable. Some deserts, especially coastal ones such as the Namib (Fig. 2.1) and Atacama, are foggy and this may influence rates of weathering. Other deserts are cold in winter and to may be subject to frost processes. Climatic history is also important. Some deserts are very ancient, but others are less so. Some, because of continental drift show the imprint of having travelled through zones of different climates (e.g. Australia).

The other great control of desert character is tectonic history. Deserts on active plate margins (e.g. the Atacama) are different from those on passive

Fig. 2.1 The Namib Desert of southern Africa is a coastal, foggy desert, where fog precipitation may be greater than that provided by rainfall. This plays a major role in rock weathering

margins (e.g. the Namib), while those on old shields (e.g. Australia) are very different from those where orogeny is active (e.g. Iran). Some deserts occur in areas of ongoing erosion and uplift, while others occur in areas of sediment accumulation and subsidence (e.g. the Kalahari).

Let us now consider these general propositions by examining what it is that creates the distinctive nature of a selection of eight of the world's deserts. Descriptions of other deserts are given in Petrov (1976), while thorough treatments of particular deserts not covered here include Busche (1998) on the Sahara and Edgell (2006) on the Arabian deserts.

The Libyan Desert

The Libyan Desert (which is called the Western Desert in Egypt) forms part of the eastern Sahara and is the

A.S. Goudie (✉)
School of Geography, Oxford University, South Parks Road, Oxford, OX1 3QY, UK
e-mail: andrew.goudie@stx.ox.ac.uk

A.J. Parsons, A.D. Abrahams (eds.), *Geomorphology of Desert Environments*, 2nd ed., DOI 10.1007/978-1-4020-5719-9_2, © Springer Science+Business Media B.V. 2009

largest expanse of profound aridity on the face of
the Earth. It has been used as an analogue for Mars.
For the most part it is rather flat and only limited
areas reach altitudes more than a few hundred meters
above sea-level. Much of it is underlain by relatively
gently-dipping limestones, shales and sandstones that
create low escarpments and gently sloping plateaus.
Higher land only tends to occur in the south west of
the region, where the Gilf Kebir forms a flat plateau of
sandstone attaining heights of more than 100 m above
sea-level, and where the granitic Gebel Uweinat rises
to over 1900 m. The erodible sedimentary rocks that
characterise most of the region, however, have been
excavated to produce some great closed depressions –
Fayum, Qattara, Farafra, Bahariya, Dakhla (Fig. 2.2),
Kurkur, Kharga and Siwa – places where the underground aquifers approach to or attain the surface, so
producing oases. The Qattara has been excavated to
−133 m below sea-level. There has been considerable
debate about the origin of these depressions, and they
may owe some of their form to excavation by Eocene
karstic processes or to incision by now defunct river
systems, but wind action has certainly played a highly
significant role, aided and abetted by salt attack (Aref
et al., 2002). Indeed, because large areas only have a
few mm of precipitation per year, and because they
are subjected to the persistent northerly trade winds,
aeolian processes are evident in the form of dune fields
(Fig. 2.3) and wind fluted terrain (Embabi, 2004). It has
been a classic area for dune research, (Bagnold, 1941.)
However, the closed depressions of the Libyan Desert
have been much affected by past humid climates in the
mid Holocene and portions of the Pleistocene. Large

Fig. 2.3 The Libyan Desert is a classic area for aeolian research
and contains large expanses of classic barchans and linear dunes.
These barchanic forms are in the Kharga depression

freshwater lakes existed as did active rivers such as
Wadi Howar (Pachur and Kröpelin, 1987) and Hoelzmann et al. (2001), have established the existence of
what they term the 'West Nubian Palaeolake.' This
covered as much as 7000 km² between 9500 and 4000
years BP. The moist phases are also represented by
widespread spring deposits and carbonate tufas, by
large landslips in shales, and by groundwater-sapped
cliffs. However, some of the distinctiveness of the
Libyan Desert is created by the existence of The Nile,
both in terms of its present and its former courses. The
Nile as we see it today is a young river in geological
terms. Its course has been affected by the retreat of the
Tethys Ocean and the desiccation of the Mediterranean
basin around 6 million years ago, and the plate splitting that led to the uplift of the Red Sea Hills and the
mountains of Ethiopia (Issawi and McCauley, 1992;
Goudie, 2005). For example, at the end of the
Oligocene (c 24 Ma) a river, (the Gilf system) flowed
westward from the newly uplifting Red Sea Hills
through Aswan and Dakhla to Siwa, whereas in the
middle Miocene (c 16 Ma) drainage in the area (the
Qena system) was essentially south westwards from
the Red Sea Hills towards the Chad Basin. Around
6 Ma, at a time of very low sea level in the Mediterranean (the Messinian Salinity Crisis), a precursor
of the present Nile (the Eonile) cut back southwards
along a great canyon to capture the Qena system.

The Namib

The Namib is a very dry desert which extends for
2000 km along the South Atlantic coastline of southern

Fig. 2.2 The Dakhla oasis in the Libyan Desert of Egypt is
formed by aeolian excavation into limestones and shales, and
contains Holocene lake beds which have been excavated by late
Holocene wind activity

Africa, and occupies portions of South Africa, Namibia and Angola. It is, however, narrow (only 120–200 km wide), being bounded to its east by The Great Escarpment. It is hyper-arid (rainfall at the coast is often only 10–20 mm per annum), but is characterised by frequent, wetting fogs (Olivier, 1995).

Its landscape demonstrates the importance of tectonic setting. The Great Escarpment, the sloping plains of the Namib itself, and its major inselbergs, can be explained by the opening of the South Atlantic in the Early Cretaceous, the separation of southern Africa from South America, and the development of a major hot-spot track associated offshore with the Walvis Ridge and the Tristan and Gough islands (Goudie and Eckardt, 1999). Igneous extrusive and intrusive activity occurred, leading to the formation of large spreads of lava (the Etendeka lavas) (Fig. 2.4) and the development of some large plutons and associated inselbergs (e.g. Erongo, Brandberg and Spitzkoppje) (Fig. 2.5). The Great Escarpment formed as a result of uplift and incision following the break up of Gondwanaland, and is comparable to that of other passive margin settings. Deeply incised into it is the Fish River canyon, one of the world's largest examples of this type of feature (Fig. 2.6).

The Namib is also an ancient desert and this also must have been controlled to a considerable extent by its plate tectonic history, which influenced the opening up of the seaways of the Southern Ocean, the location of Antarctica with respect to the South Pole, and the subsequent initiation of the cold, offshore Benguela Current. The date of the onset of aridity is the subject of debate, but it could date back to the early Cretaceous,

Fig. 2.5 Spitzkoppje is a granite mass that was intruded into the Central Namib in the Early Cretaceous, and which has been exhumed by subsequent erosion

Fig. 2.6 The uplift of the western passive margin of southern Africa produced a Great escarpment into which the Fish River has become deeply incised

Fig. 2.4 The opening of the South Atlantic in the early Cretaceous caused the eruption of large volumes of lava (the Etendeka lavas) in the Skeleton Coast

for dune beds are found inter-digitated with Etendeka lavas (Jerram et al., 2000). Ward et al. (1983) believe that the Namib has not experienced climates significantly more humid than semi-arid at any time during the last 80 million years. The present Namib sand sea is underlain by a lithified erg composed of the Tsondab Sandstone, and this dates back to at least the lower Miocene (Senut et al., 1994.) By the late Miocene, offshore dust inputs were increasing and river inputs were decreasing (Kastanja et al., 2006).

The Namib today has a wide diversity of landforms that includes wind fluted terrain (*yardangs*), especially in northern Namibia (just to the south of the Cunene river) (Goudie, 2007) and in the southern Namib near Luderitz. There are also four major *ergs* or sand seas (which from north to south are the Baia dos Tigres erg

Fig. 2.7 The Namib Sand Sea of the Central Namib, seen here at Sossus Vlei, contains some of the world's highest linear and star dunes

in Angola, the Cunene Erg, the Skeleton Coast erg and the Namib Sand Sea) (Fig. 2.7). The wind has also created pans, wind streaks and dust storms, and these have been created particularly by high velocity winds blowing out from the interior plateau (*Berg* winds) (Eckardt et al., 2001).

The coastal portions of the Namib Desert, because of the prevalence of fog and large quantities of salts, including gypsum (Eckardt and Spiro, 1999), are sites of very rapid salt weathering. As in the Atacama, this may explain the presence of extensive, featureless plains (Goudie et al., 1997). The salts themselves are derived from wind-blown aerosols that have accumulated since the initiation of the desert and which have been redistributed by wind action and sporadic surface runoff (Eckardt et al., 2001).

The Kalahari

In the interior of southern Africa, much of it in Botswana, lies the Kalahari (Thomas and Shaw, 1991). Most of it is not a true desert but an extensively wooded 'thirstland'. Over enormous distances the relief is highly subdued and the landscape monotonous. In the extreme south west on the borders of Botswana, Namibia and South Africa, the rainfall (<200 mm per annum) is just sufficient to allow present day dune activity (Fig. 2.8), but to the north the Kalahari is largely a relict sand desert, which extends into Angola, Zambia, Zimbabwe and the Congo (Shaw and Goudie, 2002), and has mean annual rainfall levels that exceed 800 mm.

Fig. 2.8 In the south western Kalahari, there is sufficiently low rainfall for linear dunes to be partially active, but many of the dunes of the mega-Kalahari are relicts of previously more extensive dry conditions

The Kalahari owes its gross form and subdued morphology to the fact that following the break up of Gondwanaland it became an area of down-warping that was bounded on the west by the highlands of Namibia and Angola, and on the east by mountains such as the Drakensberg and Lubombo. It became a basin of sedimentation and this largely accounts for its flatness. The Kalahari Beds that fill this basin are often over 100 m in thickness and in parts of the Etosha region of northern Namibia they are over 300 m thick. They consist of dune sands, alluvial deposits, calcretes and marls. Faulting within the basin has led to the formation of the great Okavango Delta (McCarthy et al., 1988).

Apart from its relict dunes (Thomas, 1984), the Kalahari contains large numbers of pans and associated lunettes (Goudie and Thomas, 1985), together with two large closed depressions – Etosha Pan and the Mkgadikgadi Depression. Today these are major

Fig. 2.9 The Molopo River, on the border between Botswana and South Africa is probably a testament to former wetter conditions. It is incised into the thick calcretes that are a feature of the Kalahari basin

Fig. 2.10 In the north of Pakistan is the Hunza Valley. At Pasu there are the huge peaks of the Karakoram Mountains

sources of dust plumes (Washington et al., 2003). During pluvials, and perhaps because of tectonically controlled water inputs form the Zambezi, the latter was occupied by a huge lake (Grove, 1968), which covered over 120,000 km². Waters may also have been contributed in the past by a series of fossil valleys – *mekgacha* (Shaw et al., 1992).

Another characteristic of the Kalahari is the excellent development that has occurred of calcrete, silcrete and combinations of the two (Watts, 1980; Nash et al., 1994). The reason why calcretes in particular are so well developed probably relates to the Kalahari's long history of gentle sedimentation, though the carbonates making up the calcretes are largely derived from the expanses of limestones and dolomites that occur on its margins. Ancient river systems, such as the Molopo, are incised into the calcrete (Fig. 2.9).

The Kalahari then provides a fine contrast to the Namib, because of its relatively high rainfall and because of its basinal form.

The Thar

The Thar desert of India (Allchin et al., 1977), like the Kalahari, is not an area of profound aridity and very little of the area has less than 100 mm of mean annual rainfall. The Indian arid zone, however, shows greater diversity than the Kalahari, for there is a striking contrast in relief between the arid foothills and valleys of the Karakorams (Fig. 2.10) and Ladakh in the north, the enormous alluvial plain and delta of the

Indus River, the salty sabkha of the Rann of Kutch in the south, and the ancient mountain stumps of the Aravallis in the east.

The Indus, which derives its waters from the high mountains of Asia (Shroder, 1993), is a dominant influence on the desert, but the mountains also provided in the past the discharge of a whole series of 'lost rivers' (Wilhelmy, 1969) that are a feature of the Punjab. Other rivers, such as the Luni, flow from the Aravallis, which are notable because they are one of the oldest mountains systems, still maintaining some relief, in the world (Spate, 1957). Indeed, four orogenic events have been identified, ranging in age from 3000 Ma to 750 Ma ago (Mishra et al., 2000).

The Thar is a relatively moist and low velocity wind environment, but it has large expanses of dunes, the sand for which comes from a wide range of sources: the coastline of the Arabian Sea, the large alluvial plains and the weathering of extensive areas of sandstones and granites. Uniquely in the world, many of the dunes are rake-like parabolics (Kar, 1993), which have formed transverse to the dominant early summer south-westerly monsoon winds. Dunes were, however, much more extensive under past more arid conditions (Goudie et al., 1973: Allchin et al., 1977), and this includes the highly lithified aeolianites (*miliolites*) of Saurashtra (Fig. 2.11) (Sperling and Goudie, 1975: Goudie and Sperling, 1977). The Thar contains some lake basins, created in part by aeolian disruption of drainage lines, and these provide evidence for former wetter conditions (Wasson et al., 1984: Singh et al., 1990), not least in early to mid-Holocene times (Fig. 2.12). Likewise there have been alternations of fluvial and aeolian accumulation in the southern Thar

Fig. 2.11 On the coast of Saurashtra (Kathiawar) in northwest India, as at Junagadh Hill, there are lithified ancient dunes, called *aeolianite* (miliolite). The cross bedding structures are well displayed

Fig. 2.12 At Pushkar, in Rajasthan, northwest India, there is a small lake basin, surrounded by dunes. Such basins were occupied by larger freshwater lakes in the early Holocene

during the late Pleistocene (Juyal et al., 2006). These reflect fluctuations in the nature of the south west monsoon.

Atacama and Altiplano

To the west of the Andean cordillera between latitudes 5 and 30° S lies the largest west coast desert in the world (Bowman, 1926). It is also the world's driest desert and Quillagua (mean average rainfall 0.05 mm) can lay claim to the driest place on Earth (Middleton, 2001). There are fogs (the *garuá* of Peru and the *camanchaca* of Chile), and there are occasional high rainfall years associated with El Niño conditions that cause great floods (Magilligan and Goldstein, 2001), but aridity is intense. It has also persisted for a long time, and like the Namib, the Atacama has a very extended history that goes back to at least the late

Eocene and possibly to the Triassic (Alpers and Brimhall, 1988; Clarke, 2006). One consequence of long continued intense aridity, is that the Atacama contains the most famous and important *caliche* (sodium nitrate) deposits in the world. Nitrate is highly soluble and can only accumulate under very dry conditions (Fig. 2.13). The nitrates mantle the landscape, break up the underlying bedrock and are largely derived from atmospheric sources that have provided material to old, desert surfaces (Ericksen, 1981; Searl and Rankin, 1993; Bohlke et al., 1997). Precipitation seems to have plummeted between 19 and 13 Ma (from >200 mm per annum, to <20 mm) as the uplift of the Andes blocked the ingress of the South American summer monsoon into the Atacama. Nitrate accumulation may have begun at that time (i.e. in the middle Miocene) (Rech et al., 2006). The combination of fogs and salt at altitudes below c 1100 m create an aggressive environment for salt weathering (Goudie et al., 2002).

Indeed, a major influence on the geomorphology of the Atacama has been the growth and presence of the Andes (Fig. 2.14). Tectonic uplift and eastward migration of the Andes volcanic arc associated with the subduction of the Nazca oceanic plate beneath the South American continental plate have created some of the greatest altitudinal contrasts to be found on Earth. Over a horizontal distance of no more than 300 km one moves from the Peru-Chile French (at some 7600 m below sea-level) to Andean peaks that rise up to over 6000 m above sea level. Thus, whereas the Namib is on a passive margin, the Atacama is on an active margin. It has much evidence of volcanic activity, folding and faulting, high mountain development and, in

Fig. 2.13 In the Atacama of Chile, inland from Iquique, there are large expanses of salt deposits, which include the famous nitrate accumulations called *caliche*

Fig. 2.14 Near Putre in the Atacama Desert of northern Chile, the snow capped Andes form an impressive backdrop to the world's driest desert

the Altiplano, basin and range topography containing large depressions (Lamb et al., 1997). The grain of the land runs approximately north to south with a very narrow or non-existent coastal plain, a coastal range (Cordillera de la Costa), a longitudinal Central Valley and then, to the east, the higher level Andes and Altiplano. The Altiplano is a high plateau composed of the sedimentary infill of a series of intermontane tectonic trenches. It is characterised by some large basins (*salars*) which have in the past contained large bodies of water (Rauchy et al., 1996; Placzek et al., 2001). One of these the Salar de Uyuni in Bolivia, is now the major source of dust in South America (Washington et al., 2003).

Taklamakan and Tarim

The deserts of China and its neighbours cover a wide range of geomorphological and tectonic settings from the Turfan (Turpan) Depression (−150 m) to the high mountains of the Kunlun and Karakorams, where altitudes exceed 5000 m over extensive tracts. The Taklamakan, 'the place from which there is no return', is the largest desert in China and is very dry, with mean annual precipitation dropping to as low as 10 mm in its driest parts. It occurs within the Tarim Basin, which, with an area of 530,000 km^2 is one of the largest closed basins on Earth. The subsidence that produced it was initiated in the Oligocene, and there are huge thicknesses (up to 3300 m) of Pliocene and Pleistocene sediments underlying it. The largest part in the basin is 'the wandering lake of Lop Nor', at only 780 m above sea level (Zhao and Xia, 1984).

The Tarim Basin is bounded on the south by the Kunlun Mountains and on the north by the Tian Shan. Both ranges produce alluvial fans and gravel aprons and generally feed the basin with sediment. It is for this reason the Taklamakan can lay claim to have the most positive budget of any sand sea in the world (Mainguet and Chemin, 1986). At 337,600 km^2 it is indeed a huge sand sea with a diverse range of dune types, many of which are 80–200 m in height (Zhu, 1984).

It is likely that the winnowing of fine sediment from the Tarim Basin has been a major source of material for dust storms and for the great areas of aeolian silt (loess) that reach their ultimate development in the Loess Plateau downwind to the east. Indeed, the Taklamakan is one of the dustiest places on earth (Zhang et al., 1998; Kes and Fedorovich, 1976), because of its aridity, its plentiful supply of mountain-derived sediment, and it topographically funnelled winds (Washington et al., 2003). Dust from it is not only transported to other parts of China, but also to Korea, Japan and even North America. In addition, the area is the classic location for yardang formation and this attests to the importance of wind action (Hedin, 1903; Halimov and Fezer, 1989).

Aridity in the area may be of some antiquity. The uplift of Tibet took place in the Miocene, with a rapid rise at about 8 m.y. ago (Molnar et al., 1993). This caused a major shift in climate and a transformation in the nature of the monsoonal system at that time (Fluteau et al., 1999). Wang et al. (1999), on the basis of the study of sediments in the Qaidam Basin, argue that the Tibetan Plateau must have reached a threshold elevation in the latest Miocene that caused a drying in central Asia and the intensification of the East Asian monsoon. The Pliocene Red Clay Formation (PRCF) of China, which is in part a product of aeolian dust accumulation and has loessic characteristics, has been dated to around 7.2–8.35 million years ago (Qiang et al., 2001; Ding and Yang, 2000), though dust derived from the Tibetan Plateau and the Gobi is evident in ocean core deposits going back to at least 11 million years (Pettke et al., 2000).

Australia

Australia is the world's second driest continent, but aridity is not especially intense and nowhere does mean annual rainfall drop below 100–125 mm. This reflects the inland setting of the arid zone, the absence of very high relief barriers against the inland penetration of moist air (particularly of tropical air from the north) and the lack of a definite cold inshore oceanic current along the west coast (contrast this with the hyperarid west coast deserts of the Namib and Atacama.) Australia is also for the most part at low altitude, with about 40% of its area standing less than about 200 m above sea-level. It is also dominated by large plainlands, associated with such typically Australian phenomena as stone mantles (*gibbers*) and duricrusts. Australia is also an ancient continent with extensive venerable shield areas and land surfaces that have been exposed to sub-aerial processes for hundreds of millions for years (Twidale, 2000). As Oberlander (1994, p. 26) observed, 'The erosional flattening of Australia is so thorough that any sharp protruberance constitutes a major landmark'. It is geomorphologically comatose and a museum of relict features, with some of the lowest denudation rates of any land surface in the world (Gale, 1992).

Australia is a fragment of Gondwanaland and its landscapes and its climates have been affected by continental drift that has been ongoing since the Middle Jurassic. Over that time it has moved northwards, and continues to do so. In the process it has moved through different climate belts. During the Tertiary it moved from being in a high-latitude near-polar climatic zone, through a mid-latitude humid zone, into a zone of tropical and sub-tropical climates (including desert). Ancient, broad, infilled valley systems, now dismembered and containing strings of salt lakes (Fig. 2.15), especially in Western Australia, may have been beheaded about 75 Ma by the rifting that initiated separation of Australia from Antarctica. Deep weathering profiles and etchplains, often associated with a range of duricrust types (particularly ferricretes and silcretes), are among the geomorphological phenomena that date back to the early Tertiary and before.

That said, other phenomena near witness to more recent climate changes, including a massive anticlockwise whirl of sand deserts (Wasson et al., 1988), composed very largely of linear dunes; great networks of anastomosing and anabranching rivers created by intense tropical storms (Tooth and Nanson, 1999; Bourke and Pickup, 1999); large numbers of salt lakes that were filled by large water bodies (Harrison and Dodson, 1993) at various times in the Pleistocene; and

Fig. 2.15 In Western Australia, there are large, ancient valley systems that now contain strings of salt lakes which have been moulded by Aeolian activity

clay and sandy lunettes developed on the lee sides of many ephemeral basins.

North American Deserts

Two main physiographical provinces – The Basin and Range and The Colorado Plateau – contain the most important of the North American deserts, but there are marked differences between them. Within the former lies the Sonoran, Chihuahuan, Mojave and Great Basin deserts (Tchakerian, 1997). These are characterised by block-faulted, more or less north to sound trending mountain ranges and basins (Morrison 1991; Peterson, 1981), which started to develop in the late Oligocene as a response to crustal extension. The juxtaposition of topographic highs and

Fig. 2.16 During the Pleistocene the Basin and Range province of America was occupied by many large lakes. This figure shows the shorelines of Lake Bonneville in Utah

lows provides a situation where there are many alluvial fans, extensive pediments, active runoff, and the formation of large numbers of closed basins, which in pluvial times contained large lake basins (Tchakerian and Lancaster, 2002). Notable among these were lakes Bonneville (Fig. 2.16) and Lahontan.

The Colorado Plateau is very different. It exhibits sedimentary strata, which on account of their very limited dips give rise to extensive mesa and scarp landscapes, sandstone canyons, and intricately dissected fluvial landscapes. Indeed, in the nineteenth century it became the epicentre for innovative fluvial geomorphology. The Colorado Canyon's initiation is relatively recent, having started around 5.5 million years ago. Incision has taken place at 470–800 m per million years (Patton et al., 1991). Rapid denudation has also created the great cliffs of the Plateau, with their magnificent escarpments, mesas, natural arches, box canyons and ground water-sapped alcoves (Laity and Malin, 1985.)

In the early Tertiary, the Colorado Plateau, made famous by the pioneer researches of John Wesley Powell and Clarence Dutton, was a low-lying basin. Uplift started in the Mid-Eocene, and lasted into the late Miocene, causing the plateau to become a high region into which drainage became incised. Tectonic activity also created some major igneous landforms, including the classic intrusions of the Henry Mountains and elsewhere (Gilbert, 1877).

Although these two provinces contain some dune fields (Algodones, White Sands etc.) and generate some dust from desiccated lakes, they are not areas where aeolian processes and phenomena are generally dominant. Fluvial and slope processes, driven by water

and gravity, give the North American deserts their distinctiveness.

Conclusions

There is a great diversity of landscapes both between and within deserts. One can, for example, distinguish at a gross scale between shield deserts and mountain-and-basin deserts, or, between deserts on active plate margins (e.g. the Atacama) and those on passive plate margins (e.g. The Namib), but within such deserts it is possible to subdivide landscapes into sand deserts, stony deserts, clay plains, riverine deserts etc. Equally there is a great diversity of climatic conditions in different desert areas and this serves to control the relative importance of different geomorphological processes. The range of desert landscape types is also diversified because of the near ubiquitous occurrence of relict landforms produced under a range of former climatic conditions. Each of the eight deserts discussed in this chapter has particular features that render it distinctive.

Plainly plate tectonic history has been crucial in many of them. In South America, orogeny and the eastward migration of the Andean volcanic arc, associated with the subduction of the oceanic Nazca plate beneath the South American continental plate, have created great contrasts in altitude, abundant volcanism and the many closed depressions of the Altiplano. In North America the Basin and Range Province is a classic area of crustal extension, volcanism and faulting. In northern Africa, the Atlas Mountains, the highlands of the central Sahara (e.g. Tibesti and Hoggar), the uplift of the Red Sea Hills and the evolution of the Nile have all been affected by plate tectonic processes and are major controls and features of the area's topography. In southern Africa the geomorphology of the Namib owes much to the opening of the South Atlantic by sea-floor spreading in the late Jurassic and early Cretaceous and the presence of the great hot-spot associated with the Walvis Ridge. In addition, uplift of the Great Escarpment and the basinal form of the Kalahari are associated with passive-margin evolution. In the Middle East, features such as the Red Sea rift, the Dead Sea fault, the Zagros Mountains and the ophiolite ranges of Oman and the United Arab Emirates are a response to the area being at a crossroads of major plate boundaries. In Asia the uplift of the Himalayas and the Tibetan Plateau radically modified climatic conditions and accounts for the development of major mountain ranges in close proximity to enormous closed basins (such as Tarim). By contrast, Australia is a low, dry, ancient continent with a long history of comparative orogenic stability over large areas, so that many of the present landscape features are inherited from a great variety of climates that may go back to the Jurassic or earlier.

References

Allchin B, Goudie A.S. and Hegde K.T.M. (1977). *The Palaeogeography and Prehistory of the Great Indian Desert*. London: Academic Press.

Alpers C.N. and Brimhall G.H. (1988). Middle Miocene climatic change in the Atacama Desert, northern Chile: evidence from supergene mineralization at La Escondida. *Bulletin Geological Society of America* 100, 1640–56.

Aref M.A., El-Khoriby E. and Hamdan M.A. (2002). The role of salt weathering in the origin of the Qattara Depression, Western Desert, Egypt. *Geomorphology* 45, 403–414.

Bagnold R.A. (1941). *The Physics of Blown Sand and Desert Dunes*. London, Methuen.

Bohlke J.K., Ericksen G.E. and Revesz K. (1997). Stable isotope evidence for an atmospheric origin of desert nitrate deposits in northern Chile and southern California, USA. *Chemical Geology* 136, 135–152.

Bourke M.C. and Pickup G. (1999). Fluvial form variability in arid central Australia. In: A.J. Miller and A. Gupta (eds.) *Varieties of Fluvial Form*. Chichester: Wiley, pp. 249–271.

Bowman I. (1926). *Desert Trails of Atacama*. New York: American Geographical Society.

Busche D. (1998). *Die Zentrale Sahara*. Gotha: Justus Perthes verlag.

Clarke J.D.A. (2006). Antiquity of aridity in the Chilean Atacama Desert. *Geomorphology* 73, 101–114.

Ding Z.L. and Yang S.L. (2000). C3/C4 vegetation evolution over the last 7.0 Myr in the Chinese Loess Plateau: evidence from pedogenic carbonate δ13C. *Palaeogeography, Palaeoclimatology, Palaeoecology* 160, 292–299.

Eckardt F.D. and Spiro B. (1999). The origin of sulphur in gypsum and dissolved sulphate in the Central Namib Desert, Namibia. *Sedimentary Geology* 123, 255–273.

Eckardt F.D., Drake N., Goudie A.S., White K. and Viles H. (2001). The role of playas in pedogenic gypsum crust formation in the Central Namib Desert: a theoretical model. *Earth Surface Processes and Landforms* 26, 1177–1193.

Edgell H.S. (2006). *Arabian Deserts. Nature, origin and evolution*. Dordrecht: Springer.

Embabi N.S. (2004). *The geomorphology of Egypt. Vol.1, The Nile Valley and the Western Desert*. Cairo: Egyptian Geographical Society.

Ericksen G.E. (1981). Geology and origin of the Chilean nitrate deposits. *United States Geological Survey Professional Paper* 1188, 37pp.

Fluteau F., Ramstein G. and Besse J. (1999). Simulating the evolution of the Asian and African monsoons during the past 30 Myr using an atmospheric general circulation model. *Journal of Geophysical Research* 104 (D10), 11995–12018.

Gale S.J. (1992). Long-term landscape evolution in Australia. *Earth Surface Processes and Landforms* 17, 323–43.

Gilbert G.K. (1877). *Report on the Geology of the Henry Mountains*. US Geographical and Geological Survey.

Goudie A.S. (2002). *Great Warm Deserts of the World. Landscapes and evolution*. Oxford: Oxford University Press.

Goudie A.S. (2005). The drainage of Africa since the Cretaceous. *Geomorphology* 67, 437–456.

Goudie A.S. (2007). Mega-yardangs: a global analysis. *Geography Compass* 1, 65–81.

Goudie A.S. and Eckardt F. (1999). The evolution of the morphological framework of the Central Namib Desert, Namibia, since the early Cretaceous. *Geografiska Annaler* 81A, 443–458.

Goudie A.S. and Sperling C.H.B. (1977). Long distance transport of Foraminiferal tests by wind in the Thar Desert, northwest India. *Journal of Sedimentary Petrology* 47, 630–33.

Goudie A.S. and Thomas D.S.G. (1985). Pans in southern Africa with particular reference to South Africa and Zimbabwe. *Zeitschrift fur Geomorphologie* 29, 1–19.

Goudie A.S., Allchin B. and Hegde K.T.M. (1973). The former extensions of the Great Indian Sand Desert. *Geographical Journal* 139, 243–57.

Goudie A.S., Viles H.A. and Parker A.G. (1997). Monitoring of rapid salt weathering in the central Namib using limestone blocks. *Journal of Arid Environments* 37, 581–598.

Goudie A.S., Wright E. and Viles H.A. (2002). The roles of salt (sodium nitrate) and fog in weathering: a loaboratory simulation of conditions in the northern Atacama Desert, Chile. *Catena* 48, 255–266.

Grove A.T. (1968). Landforms and climatic change in the Kalahari and Ngamiland. *Geographical Journal* 135, 191–212.

Halimov M. and Fezer, F. (1989). Eight yardang types in Central asia. *Zeitschrift für Geomorphologie* 33, 205–217.

Harrison S.P. and Dodson J. (1993). Climates of Australia and New Guinea since 18,000 yr B.P., in H.E. Wright, J.E. Kutzbach, T. Webb, W.F. Ruddiman, F.A. Street-Perrott and P.J. Bartlein (eds.) *Global Climates Since the Last Glacial Maximum*. Minneapolis: University of Minnesota Press, pp. 265–293.

Hedin S. (1903). *Central Asia and Tibet*. New York: Scribners.

Hoelzmann P., Keding B., Berke H., Kröpelin S. and Kruse H.-J. (2001). Environmental change and archaeology: lake evolution and human occupation in the Eastern Sahara during the Holocene. *Palaeogeography, Palaeoclimatology, Palaeoecology* 169, 193–217.

Issawi B. and McCauley J.F. (1992). The Cenozoic Rivers of Egypt: the Nile problem. In: *The Followers of Horus* (eds. B. Adams and R. Friedman), pp. 1–18. Oxford: Oxbow Press.

Jerram D.A., Mountney N.P., Howell J.A., Long D. and Stollhofen, H. (2000). Death of a sand sea: an active aeolian erg systematically buried by the Etendeka flood basalts of NW Namibia. *Journal of the Geological Society, London*, 157, 513–516.

Kar A. (1993). Aeolian processes and bedforms in the Thar Desert. *Journal of Arid Environments* 25, 83–96.

Kastanja, M.M., Diekmann, B. and Henrich, R. (2006). Controls on terrigenous deposition in the incipient Benguela upwelling system during the middle to the late Miocene (ODP Sites 1085 and 1087). *Palaeogeography, Palaeoclimatology, Palaeoecology* 241, 515–530.

Kes A.S. and Fedorovich B.A. (1976). Process of forming of aeolian dust in space and time. *23rd International Geographical Congress*, Section 1, 174–77.

Juyal N., Chamyal, L.S., Bhandari, S., Bhusan, R. and Singhvi, A.K. (2006). Continental record of the southwest monsoon during the last 130 ka: evidence from the southern margin of the Thar desert, India. *Quaternary Science Reviews* 25, 2632–2650.

Laity J.E. and Malin M.C. (1985). Sapping processes and the development of theater-headed valley networks on the Colorado Plateau. *Bulletin Geological Society of America* 96, 203–217.

Lamb S., Hoke L., Kenna L. and Dewey J. (1997). Cenozoic evolution of the Central Andes in Bolivia and northern Chile. *Geological Society of London Special Publication* 121, 237–64.

Magilligan F.J. and Goldstein P.S. (2001). El Niño floods and culture change: a late Holocene flood history for the Rio Moquegua, southern Peru. *Geology* 29, 431–434.

Mainguet M. and Chemin M.C. (1986). Wind system and sand dunes in the Taklamakan Desert (People's Republic of China). *Paper Presented at the Twentieth International Symposium on Remote Sensing of Environment, Nairobi, Kenya*, 827–833.

McCarthy T.S., Stannistreat I.G., Cairncross B., Ellery W.N., Ellery K., Oelofse R. and Grobicki T.S.A. (1988). Incremental aggradation on the Okavango Delta-fan, Botswana. *Geomorphology* 1, 267–78.

Middleton N.J. (2001). *Going to extremes; mud, sweat and frozen tears*. London: Channel 4 books.

Mishra D.C., Singh B., Tiwari V.M., Gupta S.B. and Rao M.B.S.V. (2000). Two cases of continental collisions and related tectonics during the Proterozonic period in India – insights from gravity modelling constrained by seismic and magnetotelluric studies. *Precambrian Research* 99, 149–169.

Molnar P., England P. and Martinod J. (1993). Mantle dynamics, uplift of the Tibetan Plateau and its margins. *Review of Geophysics* 31, 357–396.

Morrison R.B. (1991). Quaternary geology of the Southern Basin and Range province. In Vol. K-2. *Quaternary Nonglacial Geology: Conterminous U.S.* Boulder: Geological Society of America, pp. 353–371.

Nash D.J., Shaw P.A. and Thomas D.S.G. (1994). Duricrust development and valley evolution: Process landform links in the Kalahari. *Earth Surface Processes and Landforms* 19, 299–317.

Oberlander T.M. (1994). Global deserts: a geomorphic comparison. In A.D.Abrhams and A.J. Parsons (eds.), *Geomorphology of Desert Environments*. London: Chapman and Hall, pp. 13–35.

Olivier J. (1995). Spatial distribution of fog in the Namib. *Journal of Arid Environments* 29, 129–38.

Pachur H.J. and Kröpelin S. (1987). Wadi Howar: Paleoclimatic evidence from an extinct river system in the southeastern Sahara. *Science* 237, 298–300.

Patton P.C., Biggar N., Condit C.D., Gillam M.L., Love D.W., Machette M.N., Mayer L., Morrison R.B. and Rosholt J.N. (1991). Quaternary geology of the Colorado Plateau. In Vol.

K-2, *Quaternary Nonglacial Geology: Conterminous U.S.*, Boulder: Geological Society of America, pp. 363–406.

Peterson F.F. (1981). Landforms of the Basin and Range Province defined for soil survey. *Nevada Agricultural Experiment Station, Technical Bulletin* 28, 52pp.

Petrov, M.P. (1976). *Deserts of the World*. New York: Wiley.

Pettke T., Halliday A.N., Hall, C.M. and Rea, D.K. (2000). Dust production and deposition in Asia and the north Pacific Ocean over the past 12 Myr. *Earth and Planetary Science Letters* 178, 397–413.

Placzek, C., Quade J. and Betancourt J.L. (2001). Holocene lake-level fluctuations of Lake Aricota, southern Peru. *Quaternary Research* 56, 181–190.

Qiang X.K., Li, Z.X., Powell C.McA. and Zheng H.B. (2001). Magnetostratigraphic record of the Late Miocene onset of the East Asian monsoon, and Pliocene uplift of northern Tibet. *Earth and Planetary Science Letters* 187, 83–93.

Rauchy J.M., Servant, J.M., Fournier M. and Causse C. (1996). Extensive carbonate algal bioherms in upper Pleistocene saline lakes of the central Altiplano of Bolivia. *Sedimentology* 43, 973–993.

Rech J.A., Currie B.S., Michalski G. and Cowan, A.M. (2006). Neogene climate change and uplift of the Atacama Desert, Chile. *Geology* 34, 761–764.

Searl A. and Rankin S. (1993). A preliminary petrographic study of the Chilean nitrates. *Geological Magazine* 130, 319–333.

Senut B., Pickford M. and Ward J. (1994). Biostratigraphie de éolianites néogènes du sud de la Sperrgebiet (Désert de Namib, Namibie). *Comptes Rendus de l'Academie des Sciences* 318, 1001–7.

Shaw A. and Goudie A.S. (2002). Geomorphological evidence for the extension of the mega-Kalahari into south-central Angola. *South African Geographical Journal* 84, 182–194.

Shaw P.A., Thomas D.S.G. and Nash D.J. (1992). Late Quaternary fluvial activity in the dry valleys (megacha) of the Middle and Southern Kalahari, southern Africa. *Journal of Quaternary Science* 7, 273–281.

Shroder J.F. (1993). *Himalaya to the Sea*. London: Routledge.

Singh G., Wasson R.J. and Agrawal D.P. (1990). Vegetational and seasonal climatic changes since the last full glacial in the Thar Desert, northwest India. *Review of Palaeobotany and Palynology* 64, 351–358.

Spate O.H.K. (1957). *India and Pakistan: A General and Regional Geography* (2nd edition). London: Methuen.

Sperling C.H.B. and Goudie A.S. (1975). The miliolite of western India: a discussion of the aeolian and marine hypotheses. *Sedimentary Geology* 13, 71–5.

Tchakerian V.P. (1997) North America. In D.S.G. Thomas (ed.) *Arid Zone Geomorphology: Process Form and Change in Drylands* (2nd edition). Chichester: Wiley, pp. 523–541.

Tchakerian V.P. and Lancaster N. (2002). Late Quaternary arid/humid cycles in the Mojave Desert and Western Great Basin of North America. *Quaternary Science Reviews* 21, 799–810.

Thomas D.S.G. (1984). Ancient ergs of the former arid zones of Zimbabwe, Zambia and Angola. *Transactions of the Institute of British Geographers* NS 9, 75–88.

Thomas D.S.G. and Shaw P. (1991). *The Kalahari Environment*. Cambridge: Cambridge University Press.

Tooth S. and Nanson G.C. (1999). Anabranching rivers on the Northern Plains of arid central Australia. *Geomorphology* 29, 211–233.

Twidale C.R. (2000). Early Mesozoic (?Triassic) landscapes in Australia: evidence, argument and implications. *Journal of Geology* 108, 537–552.

Wang J., Wang Y.J., Liu Z.C., Li J.Q. and Xi P. (1999). Cenozoic environmental evolution of the Qaidam Basin and its implications for the uplift of the Tibetan Plateau and the drying of Central Asia. *Palaeogeography, Palaeoclimatology, Palaeoecology* 152, 37–47.

Ward J.D., Seely, M.K. and Lancaster N. (1983). On the antiquity of the Namib. *South African Journal of Science* 79, 175–183.

Washington R., Todd, M., Middleton, N.J. and Goudie, A.S. (2003). Dust-storm source areas determined by the Total Ozone Monitoring Spectrometer and surface observations. *Annals of the Association of American Geographers* 93, 297–313.

Wasson R.J., Smith G.I. and Agrawal D.P. (1984). Late Quaternary sediments, minerals and inferred geochemical history of Didwana Lake, Thar Desert, India. *Palaeogeography, Palaeoclimatology, Palaeoecology* 46, 345–72.

Wasson R.J., Fitchett K., Mackey B. and Hyde R. (1988). Large-scale patterns of dune type, spacing and orientation in the Australian continental dunefield. *Australian Geographer* 19, 89–104.

Watts N.L. (1980). Quaternary pedogenic calcretes from the Kalahari (southern Africa): mineralogy, genesis and diagenesis. *Sedimentology* 27, 661–86.

Wilhelmy H. (1969). Das Urstromtal am Ostrand der Indusbene und dar Sarasvati Problem. *Zeitschrift für Geomorphologie Supplementband* 8, 76–93.

Zhang X.Y., Arimoto R., Zhu, G.H. and Zhang G.Y. (1998). Concentration, size-distribution and deposition of mineral aerosol over Chinese desert regions. *Tellus* 50B, 317–330.

Zhao S. and Xia X. (1984). Evolution of the Lop Desert and the Lop Nor. *Geographical Journal* 150, 311–21.

Zhu Z. (1984). Aeolian landforms in the Taklimakan Desert. In: F. El-Baz (ed.) *Deserts and Arid Lands*. The Hague: Nijhoff, pp. 133–144.

Chapter 3

Desert Ecogeomorphology

John Wainwright

Introduction

Previous reviews have suggested that the rôle of vegetation has often been given scant regard in the understanding of dryland geomorphology (Francis, 1994; Bullard, 1997). Bullard (1997) emphasized the landmark collections of papers in Viles (1988), Thornes (1990a) and Millington and Pye (1994) as reflecting a turning point in geomorphological perspectives, that is further emphasized by the 118 papers recorded in the ISI database since 1990 (but none before) up to mid-2007 which deal explicitly with the topic in some way. While it is untrue to suggest that work on the subject was not carried out before the 1980s – for example, Bryan (1928), Cooke and Reeves (1976), Hadley (1961), Huntington (1914), Melton (1965), Rempel (1936) and White (1969) – what has changed is the framework in which such research is carried out in dryland environments. This change is two-fold. First, geomorphologists have more explicitly recognized the need to incorporate a consideration of vegetation and, more broadly, ecosystems into their research designs. Secondly, ecologists have equally perceived the need for a more explicit evaluation of geomorphic and related hydrologic processes in order to be able to understand vegetation and ecosystem patterning. There have been parallel developments in ecology, hydrology and geomorphology alike that suggest that there is a need for producing understanding across different

spatial and temporal scales. Likewise, patterns and process change depending on the scale of observation. This trans-scale understanding is particularly critical in relation to considerations of system dynamics and the move away from narrow equilibrium perceptions (e.g. De Angelis and Waterhouse, 1987; Perry, 2002; Sullivan and Rohde, 2002; Bracken and Wainwright, 2006). Furthermore, understanding the development of spatial patterns in both vegetation and landforms requires a move away from small-scale and highly reductionist foci on local processes. To provide these understandings, both inter- and trans-disciplinarity work within an integrated framework are critical. This framework requires openness in both methodology (e.g. Balsiger, 2004) and discourse and dialogue, especially when applied to human aspects of ecosystems and environments (e.g. Wear, 1999; Tress et al., 2007; MacMynowski, 2007).

Research Frameworks

While systems approaches have been common in geomorphology since the 1960s (e.g. Stoddart, 1967; Chorley and Kennedy, 1971), the shift from a rather abstract "biosphere" system to an (explicit) ecosystemic approach has been somewhat slower to develop. In part, this sloth may relate to the relative detachment of biogeography from process-based and quantitative developments in physical geography from the 1960s (Gregory, 2000). The work of Thornes (1985, 1988, 1990b) in developing the population-modelling approach of May (1975) and applying it to understanding vegetation-erosion dynamics stands out as an early exception. This work notwithstanding, few of the

J. Wainwright (✉)
Department of Geography, Sheffield Centre for International Drylands Research, University of Sheffield, Sheffield S10 2TN, UK
e-mail: j.wainwright@sheffield.ac.uk

A.J. Parsons, A.D. Abrahams (eds.), *Geomorphology of Desert Environments*, 2nd ed., DOI 10.1007/978-1-4020-5719-9_3, © Springer Science+Business Media B.V. 2009

papers in the collection edited by Thornes (1990a) go beyond broadly empirical contributions. Despite Stoddart's (1967) early recognition of the importance of the ecosystems concept to the discipline, it was not until the 1980s that a more widespread acceptance developed (Simmons, 1980; but note that only three papers before 1992 in the *Journal of Biogeography* considered ecosystems important enough to mention them in the abstract). The more rapid advances since the 1990s have in part been due to a cross-fertilization of expertise, with the number of (especially landscape) ecologists finding homes in geography departments and carrying out collaborative research increasing significantly.

At the same time, there have been developments leading to the definition of the new field of ecohydrology. Early definitions of ecohydrology (see reviews in Baird, 1999; Kundzewicz, 2002) tended to focus on wetland systems, but from the later 1990s, the utility of linking ecology and hydrology was recognized more broadly. One of the problems with the establishment of ecohydrology has been the difficulty of definition (e.g. the debate in Kundzewicz, 2002; Zalewski, 2002; Nuttle, 2002; Bonell, 2002), which may reflect its immaturity as a scientific approach. Definitions vary from the utilitarian "science which seeks to describe the hydrologic mechanisms that underlie ecological patterns and processes" (Rodriguez-Iturbe, 2000: 3) via the evolutionary engineering approach of Eagleson (2002) to "a more general or "universal" understanding about how environmental systems work . . . combining Newtonian principles of simplification, ideal systems, and predictive understanding (often, but not solely embraced by hydrologists) with Darwinian principles of complexity, contingency, and interdependence (often, but not solely embraced by ecologists)" (Newman et al., 2006: 2). More extreme proponents suggest taking on Gaian principles by considering catchments as "superorganisms" (Zalewski, 2002: 827). Often this work is also framed within an applied (or even [green] engineering) framework, especially in the literature using terms such as hydro-ecology (i.e. the impact of flows on stream ecosystems: e.g. Acreman, 2001) and hydromorphology (which occurs from the late 1980s, predominantly in the applied water management or ecological literatures). This framing may produce problems in the move from more descriptive approaches to ones that are clearly founded in the need to *understand* complex systems and their behaviour. What is clear is that ecohydrology is yet to emerge as an approach that

is more than the sum of its parts, but nevertheless, progress will not be made without more explicit linkages at conceptual, methodological and interpretive stages of research.

Conversely, there has been little effort to develop corresponding ideas that might be called "ecogeomorphology". While the collection of papers edited by Viles (1988) is introduced in relation to ecological principles it is clearly largely focused on the more empirical interaction between plants, animals and landforms "to assess what is known about the *biological* component of geomorphology" (*ibid.*, p. 3, emphasis added), and ecosystems get nary mention in the main text. For this reason, Viles' term "biogeomorphology" does seem appropriate. The fact that more than just hydrological processes drive landform development in drylands points to the notion that a broader approach than just ecohydrology is necessary. The uneven development of the necessary integrated approaches towards an overarching ecogeomorphology, of which ecohydrology would be part, reflects the more general lack of involvement in interdisciplinary developments such as Earth-System Science (Wainwright, 2009).

Conceptual Framework

Within this chapter, an approach will be taken that integrates both ecohydrological and ecogeomorphological perspectives. It does so in the knowledge that a fully integrated approach is a long way from being developed. In the first part, it concentrates on mechanisms of vegetation adaptations to the extremes of desert environments. The principal purpose here is essentially to answer the question *why* are plants found in deserts? Despite the recent increase in investigations discussed above, there is still a poor level of understanding as to the frequency with which specially adapted plants are found in drylands. The chapter then moves on to the understanding of processes and process interactions with a range of Earth-surface (sub-)systems, in order to evaluate *how* plants are important to the evolution of dryland landforms. At a third level, the chapter evaluates the patterns that emerge both in terms of spatial and temporal distributions of vegetation and in terms of the interrelated distributions of landforms. The aim here is to demonstrate *what* patterns develop and *when*, and in what ways they are important to understanding landscape evolution.

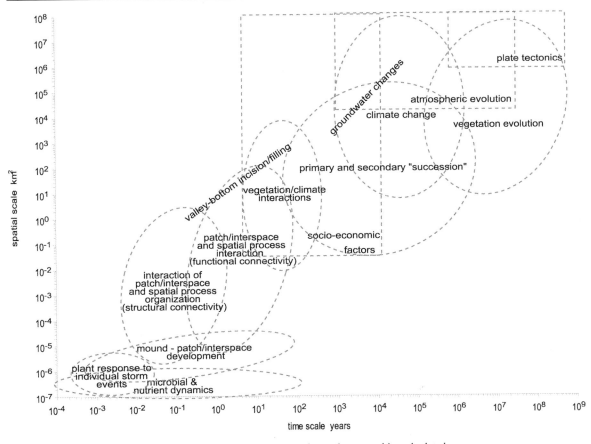

Fig. 3.1 Spatial and temporal scales of interaction of desert vegetation and geomorphic and related processes

Across all these questions, the issue of scale is fundamental. The set of processes driving different aspects of vegetation dynamics operates across many orders of magnitude of both spatial and temporal scale (Fig. 3.1). At the smallest spatial scale, microbial communities have developed and may persist beneath individual stones on the desert surface; much has been made of these lifeforms as potential analogues in the search for life on other planets. At progressively larger scales, it is necessary to understand how individual plants and groups of plants grow within harsh desert environments, and how interactions between them and their environment lead to the development of spatial patterns from hillslope to landscape scales. The evolution of different plant types, with notable stages in the development of land plants and the more recent development and dispersal of grass species, occurs at the largest spatial scale.

Timescales of adaptation are generally positively correlated with spatial scales, albeit with a great deal of fuzziness and overlap. Processes of evolution and climate change are strongly controlled by plate tectonics and the Earth's orbital variation. Climate variability on a range of progressively shorter timescales may affect the patterning on a range of spatial scales, generally with significant connectivity across different scales. For example, extreme storm events may last for a period of tens of minutes or a few hours in many dryland areas, but may significantly disrupt riparian vegetation with spatial effects potentially over hundreds of kilometres and temporal effects lasting for decades. Similarly, disturbance of biological crusts may take decades to recover, with significant feedbacks on dust emissions, atmospheric variability and nutrient cycles. The development of such complex behaviour and responses is an inherent and highly significant component of desert vegetation and thus desert environments as a whole. Explaining the complexity of desert geomorphology within an ecohydrological/ecogeomorphological framework as considered above is thus the aim of this chapter.

Types and Evolution of Desert Plants

Plant Functions and Stresses in Extreme Environments

Given the extremes of both temperature and moisture in desert environments (Noy-Meir, 1973; Cooke et al., 1993; D'Odorico and Porporato, 2006; Thomas, 1997; Wainwright, 2006a; Wainwright et al., 1999a), plants need to develop appropriate coping strategies. Plants interact with their environment in order to extract energy, carbon, water and mineral resources (Fitter and Hay, 1987). Photosynthesis converts solar energy to chemical energy. The radiation-use efficiency (proportion of radiation energy with wavelengths of 400–700 nm that can be stored as chemical energy) of photosynthesis is about 8–10% under ambient CO_2 conditions (Larcher, 1995). To allow CO_2 from the atmosphere to enter the plant so that it can be used in photosynthesis, stomata on the surface of the leaf must open. As water is an important component of the capture of CO_2 by dissolution and subsequent photosynthetic processes, the mesophyll cells immediately below a stoma are typically saturated. This saturation creates a vapour-pressure deficit between the plant and the atmosphere, especially in deserts, where the air is rarely saturated. Water molecules thus diffuse out of the mesophyll cells and into the atmosphere through the open stoma in the process of transpiration (Fitter and Hay, 1987). Transpiration is thus a significant potential problem in drylands because the rate of transpiration is proportional to the vapour-pressure deficit, which is generally high. The loss of water through transpiration also sets up a potential gradient through the plant, and it is this gradient that allows water to be drawn up through the roots, and with it nutrients from the soil. However, given that desert soils are generally dry – in other words at high soil-moisture suctions – the water-potential gradient in the plant must drop sufficiently to allow it to be less than the water potential in the soil. The danger to the plant in so doing is that it can lead to unsustainable water loss, and the development of high internal pressures. In order to avoid the former, roots send chemical signals to the stomata to remain closed when soil-moisture conditions are unfavourable (Tyree, 1999), although this mechanism may not be sufficiently rapid in more extreme conditions, so that some plants can respond to changes in relative humidity surrounding the leaf, or to

leaf biochemical processes directly. Transpiration thus causes problems for desert plants in inducing water stress, but it is a necessary evil in that the same process drives the movement of water through the plant, which in turn leads to the transport of photosynthesized sucrose around the plant and thus to growth.

Water stress is thus almost an omnipresent problem for desert plants. According to Fitter and Hay (1987) the consequences of water stress are multiple. As the cell water potential reduces, plant turgor – and thus efficiency of water transport within the plant – decreases until ultimately leaves wilt – and thus become inefficient for transpiration and thus photosynthesis. Extreme and persistent drought can lead to permanent cell damage. Increasing water stress also affects the plant biochemistry and metabolic processes, again making their growth increasingly inefficient.

In hot deserts, plants are also subjected to frequent temperature stress. As well as the heating effect of the ambient temperature, leaves absorb energy during photosynthesis, with pigments such as chlorophyll and xanthophyll absorbing the visible light and leaf water absorbing mid-infrared light. This incoming energy must be balanced by outgoing energy (as long-wave radiation), or the plant will progressively heat up. The resulting temperature stress leads to the breakdown of metabolic processes of photosynthesis, respiration and enzyme activity at temperatures above 45–55°C, even for durations as short as 30 min (Fitter and Hay, 1987). While transpiration is an effective regulator of leaf temperature by changing the microclimate, it is inactive during periods of stomatal closure when plants are water stressed. Radiation is not an effective means of cooling as it is controlled by the ambient temperature, so desert plants must cool themselves either by convection or by metabolic adaptation. Temperature stress can also occur in relation to periods of cold, with freezing conditions occurring often at night during winter months (e.g. Wainwright, 2005), and the limits of a number of desert plants are often discussed in terms of their tolerance of these periods (e.g. Nobel, 1980; Pockman and Sperry 1997; Loik and Redar, 2003).

Plant Responses to Extreme Environments

To survive in deserts, plants must find strategies for adjusting to the consequences of water and/or temperature stress. These strategies can be considered broadly as relating to *avoidance* of the stress or *adaptation* to the stress. Stress avoidance generally entails either spatial or temporal adaptations. Many desert plants have a high concentration of their biomass in the roots, so that it is minimally affected by stress at the surface and can allow the plant to respond rapidly to changes in resource availability. Between 60 and 90% of biomass is concentrated in the roots of desert plants (Fitter and Hay, 1987). The pattern of these roots varies with species, with grasses typically concentrated nearer the surface to take rapid advantage of infrequent rainfall, while shrubs have deeper roots to allow them to take up moisture stored on an interannual basis. Some species such as mesquite and types of acacia have very deep tap roots, with some examples noted as extending several tens of metres below the surface (Gibbens et al., 2005; Fig. 3.2). Plants growing on coastal dunes often increase root-biomass production during periods of burial by sand as an adaptation mechanism (e.g. Maun, 1994; Zhang, 1996; Bach, 1998) and there is some evidence of dryland plants reacting in the same way (e.g. Brown, 1997; Shi et al., 2004; Zhao et al., 2007).

Other direct spatial adaptations relate to the location within the landscape. The riparian zones of channels provide less variable supplies of water than elsewhere, and biomass is often concentrated here. Species such as honey mesquite (*Prosopis glandulosa*), which occur as shrubs elsewhere in the landscape, will commonly exhibit tree phenotypes in these locations. Species such as desert willow (*Chilopsis linearis*) are confined to growth in the riparian zone. Local hollows or depositional environments may also allow plants to avoid water stress by concentrating moisture (and usually other resources). For example, Wainwright et al. (2002) demonstrated that depositional splays or "beads" along small channels in New Mexico were foci for enhanced grass and shrub growth (see also Bull, 1997). Hollows may also occur in joints in rock faces, and the concentration of root biomass often allows desert plants to extract the moisture from relatively deep within them. Rock faces also provide opportunities for shading to avoid heat stress, but there are issues here for C_4 plants (see below), which are adapted to tolerate high temperatures (Table 3.1) but cannot tolerate more than 25% shading. Temporal adaptations may also occur on a range of different scales. Dormancy of seeds is a commonly adopted mechanism, but as Fitter and Hay (1987) point out, this approach can be un-

reliable in drylands unless the plants also have some means of establishing whether appropriate conditions will be maintained. Some seeds require prolonged hydration while others produce both dormant and nondormant seeds as a way of overcoming this problem. Venable (2007) has demonstrated that Sonoran winter annuals with the highest degree of dormancy have the highest long-term reproductive success. Short-lived rains in most deserts produce a rapid response of annuals. Senescence is also a useful avoidance mechanism, where a plant drops leaf, and in progressively more extreme conditions shoot, biomass. Some plants may take this approach to the extreme that they appear to have died totally, but they have simply reduced transpirational needs to the absolute minimum, and grow back again from the ample root system once conditions become favourable. To maximize efficiency, dryland plants have developed a range of mechanisms for resorbing nutrients in leaves, and may resorb about 50% of nitrogen and phosphorus before excision (Killingbeck and Whitford, 2001). Some evergreen species are drought-deciduous in a similar way.

Adaptation to stress can occur in a number of ways. Changes to leaf morphology can help this adaptation by a number of mechanisms. Many dryland plants have very small leaves as a means of maximizing convection by reducing boundary-layer resistance and thus minimizing temperature stress. This effect is further emphasized in plants with sparse canopies with nonoverlapping leaves. Similarly, thick leaves – as in many succulents – may be an adaptation to temperature stress because the high specific heat of the high leaf-water content minimizes temperature fluctuations (Fitter and Hay, 1987). There may be a further advantage in this approach as it allows the plant to produce more photosynthetic mesophyll for a corresponding transpiring leaf area, and thus be more photosynthetically efficient in limited moisture conditions. Some species increase the cuticular thickness to prevent transcuticular diffusion of moisture, and this can be seen (and felt) in their surface morphology with thick layers of cutin and wax. Others produce hairs on the leaf to increase the boundary-layer thickness and thus increase resistance to loss of moisture. The clustered pattern of leaf hairs in some species acts to restrict gas exchange through stomata. Leaf hairs have been shown to increase plant albedo (Ehleringer, 1980), for example in brittlebush (*Encelia farinosa*) and bursage (*Ambrosia dumosa*), thus reducing temperature stress. In some cases, plants

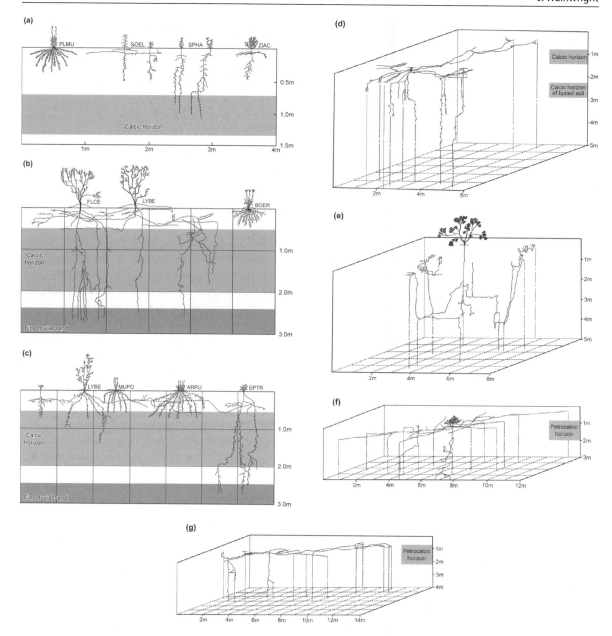

Fig. 3.2 Examples of root distributions for a range of grass, forb and shrub species in the Chihuahuan Desert, New Mexico, USA: (**a**) tobosa grass (PLMU: *Pleuraphis mutica*), and the forbs silverleaf nightshade (SOEL: *Solanum elaeagnifolium*), wrinkled globemallow (SPHA: *Sphaeralcea hastulata*) and desert zinnia (ZIAC: *Zinnia acerosa*) in a fine-loamy, Typic Calciargid; (**b**) tarbush (FLCE: *Flourensia cernua*) and Berlandier's wolfberry (LYBE, *Lycium berlandieri*) shrubs, and black grama grass (BOER, *Bouteloua eriopoda*) growing in a fine-loamy Typic Calciorthid; (**c**) Berlandier's wolfberry shrub (LYBE: *Lycium berlandieri*), bush muhly (MUPO: *Muhlenbergia porteri*) and red threeawn (ARPU: *Aristida purpurea*) bunch grasses, and a longleaf ephedra shrub (EPTR: *Ephedra trifurca*) growing in a fine-loamy, Typic Calciorthid; (**d**) two creosotebush (*Larrea tridentata*) shrubs growing in a fine-loamy, Typic Calciargid (above-ground part of plant not shown); (**e**) large crucifixion thorn (*Koberlinia spinosa*) shrub growing in a fine-silty Ustic Haplocalcid; (**f**) small mesquite (*Prosopis glandulosa*) shrub growing in a coarse-loamy, Argic Petrocalcid; and *g*. four-wing saltbush (*Atriplex canescens*) shrub growing in a coarse, loamy, Argic Petrocalcid. Where the roots end in arrows, they descend deeper but were not followed further (Gibbens and Lenz, 2001)

Table 3.1 Plant characteristics according to different photosynthetic pathways (based on Fitter and Hay, 1987; Larcher, 1999; Long, 1999; Whitford, 2002)

Characteristic	C_3	C_4	CAM
Initial CO_2-fixing enzyme	RuBP carboxylase	PEP carboxylase	RuBP (in light) PEP (in dark)
First product of photosynthesis	C_3 acids (PGA)	C_4 acids (oxaloacetate, malate, aspartate)	PGA (in light) Malate (in dark)
Operating internal CO_2 concentration (ppm)	220–260	100–150	
Photosynthetic rate (mg CO_2 dm^{-2} leaf^{-1} h^{-1})	25	60	3
Water-loss rate (g H_2O g^{-1} C fixed)	650–800	250–350	~50
Nitrogen-use efficiency (g C fixed g^{-1} N)	53–81	66–130	?
Maximum light-use efficiency (μmol CO_2 mol^{-1})	$53.8 - 1.3\,T + 0.099\,C$[†]	50–70	Similar to C_3[‡]
Optimal temperature (°C)	~25	~45	~35
Light saturation	1/4 full sunlight	>full sunlight	Fixes CO_2 at night
Redistribution of assimilation products	Slow	Rapid	Variable
Dry matter production	Medium	High	Low
Carbon-isotope ratio in photosynthates (δ^{13}C, ‰)	−20 to −40	−10 to −20	−10 to −35

[†] Based on model of Ehleringer et al. (1997): T is daytime growing-season temperature (°C) and C is atmospheric CO_2 concentration (ppmV).

[‡] Drennan and Nobel (2000).

adapt the growth of different types of leaves at different periods. For example, brittlebush has hairy leaves in the dry part of the year to minimize stresses and less hairy leaves during the rainy season to maximize transpiration and thus potential growth (Ehleringer et al., 1976; Sandquist and Ehleringer, 1998). Creosotebush produces larger leaves in the summer than in response to spring rainfall (Barker et al., 2006), probably relating to the more likely continued availability of moisture through the summer monsoon. Some species such as crucifixion thorn (*Canotia holacantha*) and Mormon tea (*Ephedra spp.*) have evolved to have their photosynthetic tissue in the stems. As well as changes to leaf morphology, leaf angle in the canopy can be an important adaptation. High leaf angles minimize radiation interception and thus the potential for temperature (and thus moisture) stress. In some cases, dynamic changes occur to the leaf shape, particularly the rolling of the leaf to shade stomata in the hotter parts of the day. Leaf angle may also be adjusted dynamically. Paraheliotropic sun tracking involves maintaining the leaves parallel to the sun through the day in order to minimize incident radiation (Ehleringer and Forseth, 1989). Paradoxically, some desert plants exhibit diaheliotropic sun tracking (maintenance of the leaf at right angles to the sun), which has been interpreted as a means of maximizing growth during the short time periods when sufficient moisture is available. Smith et al. (1998) suggest that the ability of plants to track the sun has evolved in parallel with changes in the leaf morphology, so that dryland plants will exhibit a number of the adaptive traits described here (see also Sayed, 1996). Some plants – for example the saltbush (*Atriplex hymenelytra*) – secrete salt onto the leaf surface to increase albedo and thus reflectance of radiation. This mechanism is likely to be more present in halophytes and in particular C_4 plants. Certain plants have modified leaf cells to allow them to absorb moisture from dew or fog, which is a particularly important mechanism in the coastal deserts of southern America and southern Africa, where much of the precipitation occurs in this form. A direct adaptation to moisture stress is to store moisture directly in plant tissues. In some species, this approach is carried out within seeds or tubers, while in others it occurs within the plant as in succulents, cacti and some thick-trunked trees. Succulence is also thought to be a means of adaptation to the high salt conditions found in many drylands (Greenway and Munns, 1980; Gul et al., 2001). Other adaptations to elevated salinity levels are leaf glands that secrete salts and selective uptake of different ions by plant roots (e.g. Arndt et al., 2004).

Photosynthetic Pathways

A more fundamental adaptation of some plants to desert conditions may relate to the photosynthetic pathway used. Until the mid-1960s, it was thought that all plants used the Calvin-Benson cycle to produce carbon from the photosynthetic process (Fig. 3.3a). The output of the Calvin-Benson cycle are molecules with three carbon atoms, hence plants that employ this process alone are called C_3 plants. The C_3 approach to photosynthesis is most effective in relatively high atmospheric CO_2 concentrations and lower temperatures and can tolerate shaded conditions (Table 3.1). However, they are relatively inefficient in their use of water and nutrients, both of which are typically problematic in dryland environments.

An alternative approach to photosynthesis is to divide its operation spatially within the leaf. Plants employing this approach convert incoming CO_2 into aspartate or malate within the mesophyll cell (Fig. 3.3b). These molecules have four carbon atoms, so plants using this pathway are termed C_4 plants. The aspartate or malate is then passed into bundle sheath cells, where the Calvin cycle produces sucrose that can be used by the plant. At least three mechanisms

have evolved in different species to accomplish this process (Sage, 2004). Plants with C_4 photosynthesis perform better under conditions of lower CO_2 and higher temperature, and are relatively efficient users of water and nutrients (Table 3.1). Recent confounding results where some C_4 plants perform better under elevated current atmospheric CO_2 may be explained by the elevated CO_2 causing moisture stress to be less significant (Körner, 2006), so the simple explanation that C_3 persistence relates to a climate feedback (Gill et al., 2002; Polley et al., 2002, 2003; Morgan et al., 2004) must be employed with care (see also Hanson et al., 1993; Archer et al., 1995). C_4 plants tolerate shade conditions less well, however. They have also been found to require sodium as a micronutrient (Brownell and Crossland, 1972; Grof et al., 1989), which may be one reason why they are more salt-adapted or tolerate conditions of higher salinity typical of many dryland environments (Fitter and Hay, 1987).

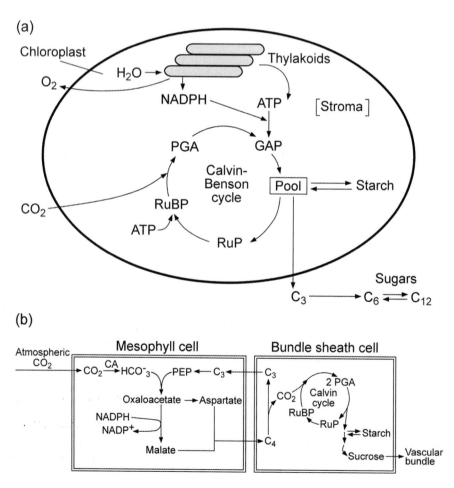

Fig. 3.3 Photosynthetic pathways in plants: (**a**) the Calvin-Benson cycle used by C_3 plants; and (**b**) the two-part approach used by C_4 plants (after Larcher, 1995)

A third photosynthetic pathway is that of crassulacean acid metabolism, or CAM, so-called because it was first observed in plants of the Crassulaceae family. CAM uses a temporal separation of the stages in photosynthesis. With their stomata closed at nighttime to minimize water loss, the process is similar to the C_4 photosynthetic pathway, producing CO_2, which is stored within the leaves of the plant. During the day, this CO_2 is converted to sucrose by means of the Calvin cycle, with a feedback in the process producing the phosphoenolpyruvate required in the nighttime C_4 process. Although the CAM pathway is much more water-efficient than the other two pathways, the complexity of the process means that the photosynthetic rate is comparatively much lower. They have similar temperature and atmospheric CO_2 preferences to C_3 plants (Table 3.1), and indeed many CAM plants revert to C_3 photosynthesis when they are not moisture-stressed.

One further consequence of the different photosynthetic pathways is that they produce different isotopic fractionation. The rubisco enzyme that controls C_3 photosynthesis strongly favours the lighter ^{12}C rather than the ^{13}C isotope, so that it is possible to use $\delta^{13}C$ values to estimate presence of different types of vegetation. C_3 plants have significantly lower values of $\delta^{13}C$ than C_4 plants, with CAM plants having intermediate values (Table 3.1). Thus, analysis of carbon isotopes in soils and fossil materials can be used to evaluate the relative balance of plants with different photosynthetic pathways in the landscape. This approach has been used to document the first appearance of C_4 plants in the Oligocene, and their rapid expansion in many drylands in the later Miocene (8–5 Ma) (Sage, 2004; Osborne and Beerling, 2006; Tipple and Pagani, 2007).

Although plants using CAM are the stereotypical desert plants of cacti and other succulents, the division is not so straightforward, and examples of C_3 and C_4 plants are commonly found in drylands (and some CAM plants are adapted to wetland conditions). Typical examples of C_3 plants include shrub and tree species such as creosotebush (*Larrea tridentata* or *L. divaricata*), mesquite (*Prosopis glandulosa*) in the North American deserts; holm and kermès oak (*Quercus ilex* and *Q. coccifera*), Aleppo pine (*Pinus halepensis*), retama (*Retama spaerocarpa*) and oleander (*Nerium oleander*) in the Mediterranean; acacia (*Acacia spp.*), bushwillow (*Combretum spp.*) and guiera (*Guiera senegalensis*) in Africa; tamarisk

(*Tamarix spp.*) in Asia (and subsequently introduced into North America where it has expanded widely); and mulga (*A. aneura*) and poplar box (*Eucalyptus populnea*) in Australia (Fig. 3.4). C_4 plants on the other hand are dominated by grasses such as *Aristida*, *Bouteloua*, *Andropogon* and *Panicum*, as well as herbaceous and shrubby plants (eudicots) such as saltbush (*Atriplex*), hogweed (*Boerhavia*), amaranths, samphires (*Halosarcia*), *Bienertia*, *Blepharis*, *Aerva* and *Zygophyllum*. Sage (2004) defines at least four centres – Mexican, South American, African and Central Asian – where C_4 plants evolved separately, with the possibility of a fifth centre in Australia.

Investigation of the presence of C_4 plants in the geological record has demonstrated that they are a relatively recent adaptation. Although undisputed fossil evidence only extends back to about 12.5 Ma in a sample from California, isotopic evidence (see above) and "molecular clock" techniques suggest that they first emerged between 32 and 25 Ma, probably related to declining global atmospheric CO_2 values (Osborne and Beerling, 2006). It was initially thought that their rapid expansion in the late Miocene (8–5 Ma) was also explained by declining atmospheric CO_2 values, but more recent reviews have demonstrated that this hypothesis is unlikely, not least because CO_2 values slightly increased during this period. Osborne and Beerling (2006) emphasize the importance of disturbance régimes such as herbivory and fire, as well as seasonal drought such as the development of monsoonal systems, with the relative importance of these factors being different in different locations where C_4 plants came to dominate. The development of savannah landscapes in semi-arid areas is intimately related to the development of these climatic and disturbance régimes. It should be noted, therefore, that the relatively recent appearance of C_4 plants means that some caution is necessary in the investigation of some modern desert environments in order to understand the functioning and sedimentology of past (pre-Miocene or in places pre-Oligocene) desert environments. The uniformitarian assessments in so doing will be flawed as the conditions are not equivalent. Similarly, conditions further back in time may vary due to the presence of different plant structures, and pre-Ordovician deserts would have had no land plants (and possibly not until much later, as the earliest land plants have affinities with wet-adapted species: Wellman et al., 2003; although Belnap, 2003, points

Fig. 3.4 (continued)

Fig. 3.4 Examples of dryland plants: (**a**) C₃ trees and shrubs (i. is creosotebush [*Larrea tridentata*]; ii. is detail of the leaves and seeds of a creosotebush; iii. is holm oak [*Quercus ilex*] with a mixed shrub understorey including kermès oak [*Q. coccifera*] in Catalunya; iv. are phreatophytes growing in Tunisia; and v. shows the use of leaf-curling as a stress-avoidance mechanisms in the Sahel); (**b**) C₄ grasses (i. is black grama [*Bouteloua eri-* *opoda*] grassland at Sevilleta, New Mexico in the Chihuahuan Desert; and ii. is a bunchgrass growing on the Kelso Dunes, Mojave Desert); and (**c**) cacti and succulents using the CAM photosynthetic pathway (i. is Mojave yucca or Spanish dagger [*Yucca schidigera*]; ii. is soaptree yucca [*Yucca elata*]; and iii. is prickly pear [*Opuntia spp.*])

to the evidence for microphytes as early as 1,200 Ma, which would probably have had some stabilizing effects – see below).

Plant Interactions

Of course, an ecosystem perspective requires a consideration not only of single plants, but also of interactions between plants. Interactions may include processes of *competition* and of *facilitation*. Competition relates to the ability of different plants (which may be of the same species) to access water, nutrient or light resources, and corresponding strategies that have been developed to adapt to spatial and temporal variability in resources. The conventional view follows Walter (1971), whose two-layer hypothesis suggests shallow-rooting (e.g. <50 cm) grasses and annuals are able to compete more effectively for water following short-lived storm events, whereas shrubs and trees with deeper roots are able to access deeper reserves and thus more reliable water sources in the longer term (see also Reynolds et al., 1999; TM Scanlon et al., 2005). This binary split may not be present in all circumstances, however. Rodriguez et al. (2007) demonstrated that roots of both creosotebush (*Larrea divaricata*) and *Stipa tenuis* grass could occupy the top 50 cm of the soil profile, and suggested the two-layer hypothesis may relate to drylands where rainfall is not seasonally variable. Spatial competition for resources may mean that shrub roots also radiate over larger areas than the canopy (Brisson and Reynolds, 1994), and some authors have suggested that vertical distributions of roots of plants of the same species will often appear very different if the plants are adjacent (Fitter and Hay, 1987). Novoplansky and Goldberg (2001) suggest that processes of competition are complex and poorly understood in general, not least because studies have often focussed on too short a time scale to evaluate their presence or absence.

The "islands of fertility" hypothesis (Garcia-Moya and McKell, 1970; Charley and West, 1975) has been used to explain patchy distributions of desert shrubs (Schlesinger et al., 1990). In this hypothesis, shrub canopies focus water and nutrient resources at their base, which results in a positive feedback for more shrub growth. Away from the shrub canopy, erosion creates conditions where less water infiltrates and nutrients are stripped, thus making the inter-plant loca-

tions more extreme and less likely to be colonized. C_4 grasses and annuals are less likely to compete under the canopy of the shrubs due to their relative intolerance to shading. More controversially, some authors have argued for the existence of allelopathy, or the production of chemicals that inhibit or prevent the growth of other plants in the neighbourhood, and in extreme cases kill them. Various phenolic and turpene substances produced by plants have been suggested as playing this rôle. Fitter and Hay (1987) have discussed the difficulties in demonstrating the existence of allelopathy, and in particular that experimental approaches have used plant extracts that are not directly produced by the plant in question. They cite studies of Californian chapparal as being probably the best-documented example (Muller, 1965; 1966), with turpenes produced by shrubs inhibiting the growth of grasses, and thus being a potential explanation for bare areas around the shrubs. Hyder et al. (2002) have suggested that phenolics in all parts of creosotebush plants may be a way of inhibiting herbivory, which has been suggested as being one way in which these C_3 plants can outcompete C_4 plants in the same location (Knipe and Herbel, 1966).

Facilitation between plants may also occur in the "islands of fertility" because of the availability of nutrients and differentiation in root zones. Indeed, taller shrubs have been suggested as having an important "nurse-maid" rôle in the propagation of younger plants, by providing protection from harsh conditions – either from solar radiation or from predators. A similar mechanism for propagation in tiger bush has been proposed (e.g. Lefever and Lejeune, 1997; Lejeune et al., 1999). Haase (2001; Haase et al., 1997) has argued that in Mediterranean environments, plants growing near to *Artemisia*, especially *Anthyllis*, enjoy a nurse-maid effect because *Artemisia* is strongly aromatic in order to discourage grazers. Maestre et al. (2003) have demonstrated that Aleppo pine (*Pinus halepensis*) in Mediterranean settings tends to produce microclimates that favour the germination of a perennial grass understorey, but not that of shrubs – which might subsequently compete with the pine for resources. Recent work has also demonstrated that certain mycorrhizal fungi may have a symbiotic rôle in the propagation and survival of arid region plants. Such fungi are present in both grasses (Barrow et al., 2007) and shrubs (Barrow et al., 1997; Barrow and Aaltonen, 2001). One suggested benefit for the plant is the ability of the fungi to transfer water into the root cells when soil conditions are so dry that the plant would be unable to exert sufficient suction to extract moisture.

The discussion above has tended to focus on individual plants or groups of individuals. In reality, most vegetation communities in drylands contain a mixture of different species, often with different growth forms. In ecosystem terms, these different plants inhabit different niches and provide different functions and habitats within the landscape. There is a tendency to use terms such as "shrubland" or "grassland", which suggest homogeneity within the landscape. Usually, most shrublands will contain grasses and annuals, and conversely most grasslands will contain scattered shrubs. Only in relatively extreme conditions will monospecific stands be found – e.g. the *Thymnus*-dominated matorral in parts of southern Spain. As well as variability in niches and habitats, most desert vegetation is found in distinctive spatial patterns. The types and mechanisms of formation of these patterns will be discussed in more detail below following a consideration of the impacts of vegetation on geomorphic and related processes within the landscape.

Process Interactions

Climate and Microclimate

Charney (1975) suggested that the presence of vegetation in drylands is strongly coupled to precipitation via feedback with surface albedo. Bare surfaces have a relatively high albedo, producing higher reflection of energy which reinforces the sinking circulation of dry air masses over the continental areas. These dry air masses produce little or no precipitation, and thus surfaces typically support less vegetation. Conversely, vegetation has a higher albedo, producing higher surface heating and thus stronger land-ocean temperature gradients, which in turn enhance monsoonal circulation in the tropics. Enhanced monsoons and rising air masses produce more precipitation, and thus typically more vegetation. Charney thus argued that precipitation-albedo feedbacks operating via the removal of vegetation by grazing would control the pattern of vegetated and unvegetated areas in the Sahel. Numerical modelling

of coupled vegetation-atmosphere conditions, with varying degrees of complexity, has tended to support Charney's hypothesis (Xue and Shukla, 1993; Claussen, 1997; Zeng et al., 1999; Zhou et al., 2007). However, there has been some debate as to whether there are strong correlations between satellite-derived albedo measurements and rainfall over the Sahel, or whether observed albedo changes are as high as those suggested theoretically by Charney (Jackson and Idso, 1975; Wendler and Eaton, 1983). Some authors have used similar hypotheses to posit significant past climate changes as a function of human action. Reale and Dirmeyer (2000; Reale and Shukla, 2000) argued for significant aridification of the northern African climate following Roman deforestation. However, their study implies that there was more extensive forest in the region before this period than was actually the case (see review in Wainwright and Thornes, 2003), and thus probably over-emphasizes the impact.

A further process in the feedback between vegetation and climate that may explain this disparity is that of soil moisture. Entekhabi et al. (1992) suggest that soil moisture as affected by vegetation cover (see below) is likely to have strong feedbacks on regional climate in three ways. First, the inverse relationship between soil-moisture content and albedo changes the radiative régime. Secondly, wetter soils typically have higher values of thermal diffusivity, thermal conductivity and heat capacity, and so more energy is transferred into the soils. Thirdly, soil moisture provides a direct supply of moisture to the atmosphere by evaporation, or an indirect supply via transpiration. Thus there should be a feedback between vegetation cover and precipitation. Entekhabi et al. argue that the longer "memory" of soil-moisture changes would mean that the link between albedo and precipitation would not be as distinct as proposed by critics of the Charney hypothesis. They used a simple atmosphere-hydrology model to demonstrate that patterns of persistent drought could result following large perturbations to the system because of these feedbacks. Scheffer et al. (2005) and Dekker et al. (2007) have taken these studies further by suggesting very local scale feedbacks with vegetation and infiltration, and again supported them with model simulations. Given the speed with which evaporation from bare surfaces in drylands occurs, it is more likely that the deep channelling and reuse of water from depth by transpiration is the source of the longer memory posited by Entekhabi et al. (1992). This effect has been demonstrated experimentally using weighing lysimeters in the Mojave (BR Scanlon et al., 2005a). While such a simple point-feedback with infiltration may be considered to be reasonable, there are a number of reasons as discussed in the following sections why it may not be as straightforward as it first seems.

As well as regional scale feedbacks, vegetation is also important in creating microclimates. Humidity will be elevated around transpiring leaves, but the extent to which this moisture remains local rather than contributing to some of the feedbacks noted above is controlled largely by the dynamics of wind flow and their interaction with the canopy structure, as discussed below. Temperatures under the canopy are lower than elsewhere, and Whitford (2002) notes a moderation of 10–12°C of peak temperatures under surface litter, which is concentrated under the canopy by a variety of processes. Ambient temperatures may be reached between 15–45 cm below the surface, making the area under the canopy a very attractive habitat for animals. Breshears et al. (1998) considered temperature and soil-moisture differences between piñon-juniper canopy and intercanopy areas dominated by blue grama (*Bouteloua gracilis*). They found that canopy temperatures were more moderate, providing protection at the surface both from cold in winter and heat in summer. Evaporation rates were also significantly reduced with the strongest differences seen for the driest initial conditions. Breshears et al. concluded that these microclimatic effects were a strong feedback on germination processes, and particularly enhanced the ability of C_3 plants to germinate in the canopy areas. Similar feedbacks have been observed beneath *Retama* canopies in southern Spain (Moro et al., 1997), for cacti beneath shrubs in the Sonoran Desert (Franco and Nobel, 1989), for savanna in Kenya (Belsky et al., 1989) and in tiger bush (Lefever and Lejeune, 1997) suggesting that it is a widespread feature of dryland systems.

Canopy

Canopy-related processes can thus be considered to be fundamental in the ways that vegetation interacts with climate and thus geomorphic processes in deserts. There is still a tendency to consider aeolian and plu-

vial/fluvial processes separately, often resulting in different conceptual frameworks. However, this separation belies a number of similarities.

Both wind and water flows can be considered using the same fluid-dynamics framework. The interaction between vegetation and fluid flows produces in general terms a deceleration of the fluid. The extent of this deceleration is called the effect of vegetation roughness on the fluid flow. Vegetation is one of several surface components that make up the total roughness of a surface-flow interaction. The vertical velocity profile of a fluid flow is described by the Prandtl-von Kármán equation, which is also often called the "law of the wall":

$$u\left(z\right) = \frac{u_*}{\kappa} \ln\left(\frac{z}{z_0}\right) \tag{3.1}$$

where $u(z)$ is the mean fluid velocity $[\text{m s}^{-1}]$ at height z [m] above the surface, u_* $[\text{m s}^{-1}]$ is shear velocity, κ is the von Kármán coefficient [dimensionless, usually with a value of 0.35–0.45 and thus often incorrectly called a constant], and z_0 [m] is the roughness height. Where the roughness elements are dense, they can absorb the entire momentum of the flow and move the entire wind profile upwards:

$$u\left(z\right) = \frac{u_*}{\kappa} \ln\left(\frac{z - d}{z_0}\right) \tag{3.2}$$

where d [m] is the displacement height. In theory, the definition of the effects of vegetation on the values of z_0 and d should therefore completely describe the interaction of vegetation and wind or water flow. For wind flows, it is common to estimate the parameters directly by the empirical fitting of vertical wind profiles. Factors affecting the values of the parameters include canopy density (or conversely, porosity), height and volume, and flexibility of stems and branches (Table 3.2).

In practice, there are a number of complications. The work of Wolfe and Nickling (1993), Judd et al. (1996) and Leenders et al. (2007) suggests that for wind flows, these equations are applicable at large scales. At the resolution of a single plant, they break down because of the more complex interactions between the porous vegetation canopy and the wind flow. These interactions can be divided into six zones (Fig. 3.5). In the approach zone, the flow decelerates

Table 3.2 Vegetation controls on wind profiles. Effects of simple and complicated vegetated surfaces on the displacement height (z_0) of the logarithmic wind profile (Equation 3.1) with comparative data for representative, unvegetated desert surfaces

Site description	Mean z_0 (m)	Range of z_0 (m)	Mean vegetation height (m)	Mean vegetation width (m)	Mean spacing between plants (m)	Source
annuals, nebkha, rough interspace	0.0279	0.0166–0.046	0.2–0.45	0.3–0.5	0.5–5	MacKinnon et al. (2004)
creosotebush, annuals, rough interspace	0.0194	0.0089–0.042	0.1–1.7	0.1–2.3	0.5–9.7	MacKinnon et al. (2004)
creosotebush, annuals, rough interspace	0.0054	0.0026–0.0113	0.09–1.3	0.1–2.3	0.75–12	MacKinnon et al. (2004)
succulents, annuals, nebkha	0.0310	0.0194–0.0496	0.4–0.7	0.5–0.71	2.5–2.8	MacKinnon et al. (2004)
creosotebush and borage	0.0266	0.0157–0.0449	0.2–1.3	0.1–1.4	0.2–6.6	MacKinnon et al. (2004)
grass and nebkha	0.0559	0.0335–0.0933	0.5–0.71	0.77–0.77	1.9–1.9	MacKinnon et al. (2004)
creosotebush, bursage, pavement	0.0148	0.0091–0.024	0.01–1.4	0.01–1.9	0.05–9.3	MacKinnon et al. (2004)
shrub, nebkha, salt push-up structures	0.00014	0.000053–0.00039	0.02–0.8	0.1–1.1	0.5–30	MacKinnon et al. (2004)
shrub, nebkha, annuals	0.0178	0.0116–0.0273	0.2–0.55	0.1–0.96	0.5–5	MacKinnon et al. (2004)
creosotebush, other shrubs and nebkha	0.071	0.0468–0.1078	0.6–1.7	0.8–2.6	3–7.5	MacKinnon et al. (2004)
shrubs and nebkha	0.0071	0.00194–0.0115	0.26–0.83	1.5–1.7	5–6.1	MacKinnon et al. (2004)
sparse vegetation	0.0037				(0.04% cover)	Wolfe (1993)
sparse vegetation	0.054				(8% cover)	Wolfe (1993)
sparse vegetation	0.068				(10.3% cover)	Wolfe (1993)
sparse vegetation	0.072				(13.5% cover)	Wolfe (1993)
sparse vegetation	0.083				(26% cover)	Wolfe (1993)
grassland	0.00088					Gillette et al. (1980)
alluvial fan surface	0.00175	0.00076–0.0031				Blumberg and Greeley (1993)
playa		0.000077–0.000166				Blumberg and Greeley (1993)
desert pavement	0.00133					Gillette et al. (1980)
alluvial fan	0.00088					Gillette et al. (1980)
crusted playa	0.00059					Gillette et al. (1980)
playa	0.00083					Gillette et al. (1980)
sand dune	0.00007					Gillette et al. (1980)
bare agricultural field	0.00031					Gillette et al. (1980)

Fig. 3.5 Wind flow around a
single vegetation element
(based on Wolfe and
Nickling, 1993, Judd
et al., 1996, and Leenders
et al., 2007)

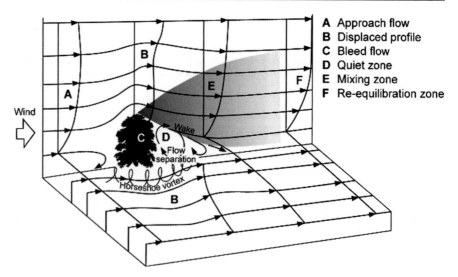

A Approach flow
B Displaced profile
C Bleed flow
D Quiet zone
E Mixing zone
F Re-equilibration zone

and diverges around the plant. Flow above and to the sides of the plant will accelerate because of streamline compression, while the part of the flow that continues through the plant will decelerate as a function of the canopy density. Downwind, there is a low-velocity zone behind the plant and a mixing zone above, before the flow profile reestablishes itself. Understanding these local variations are thus important in drylands, where vegetation patterns and canopies are sparse.

Leenders et al. (2007) used sonic anemometers positioned around a number of shrub and tree species in Burkina Faso to evaluate the variations. They found that the shrubs had a significant effect in reducing flow velocities near ground level, while single-trunked trees caused acceleration (Fig. 3.6). Deceleration was displaced upwards to heights of >2 m, relating to the location of the main canopy. Gillette et al. (2006) measured variations in roughness height, z_0, and displacement height, d, around mesquite bushes on nebkhas in New Mexico. They found that d was non-zero for distances between up to five and ten times the vegetation/nebkha height. Beyond this, $d = 0$ m and $z_0 < 0.06$ m, so that the vegetation had minimal impact on the wind flow. Within the vegetation canopy, $d > 0.4$ m and $z_0 > 0.06$ m, while in the transitional zone they found three different types of intermediate behaviour, depending on whether measurements were taken upwind or downwind, with interference from other plants important in certain wind directions. The roughness parameters thus vary spatially over short scales, and will vary temporally as a function of wind direction.

Morris (1955) defined three categories of flow that allow a distinction to be drawn as to whether

vegetation-windflow interaction should be considered at the level of single plants rather than as compound elements. Isolated-roughness flow tends to occur with sparse vegetation cover ($<16\%$) and where the spaces between elements is more than 3.5 times their height (Lee and Soliman, 1977). Wake-interference flow occurs in intermediate conditions so that only the tails of the mixing zones are affected. Skimming flow, where the entire wind profile is displaced upwards, occurs in covers of more than 40%, and where the spacing between roughness elements is less than 2.25 times their height (Fig. 3.7). The greatest protection against wind erosion (see Chapter 17) occurs where skimming flow is developed. Other feedbacks in relation to point-scale changes in aeolian entrainment and the development of fetch effects through areas of interconnected, unvegetated space are dealt with in Chapter 17.

There is a further feedback with climate, in that as vegetation roughness decreases, faster, more turbulent flows occur closer to the surface, which will act to increase evapotranspiration and thus the upward moisture flux. This feedback may be one reason why the vegetation-albedo feedback may not be as strong as first anticipated.

For water flows, the main complication is the fact that except for rare occasions within channels and riparian zones, the vegetation will not be completely inundated by the flow. Water flowing across the surface will interact in the same way with single plants as the wind-flow interaction described above, except for the compression of streamlines above the plant in non-inundated situations. The law of the

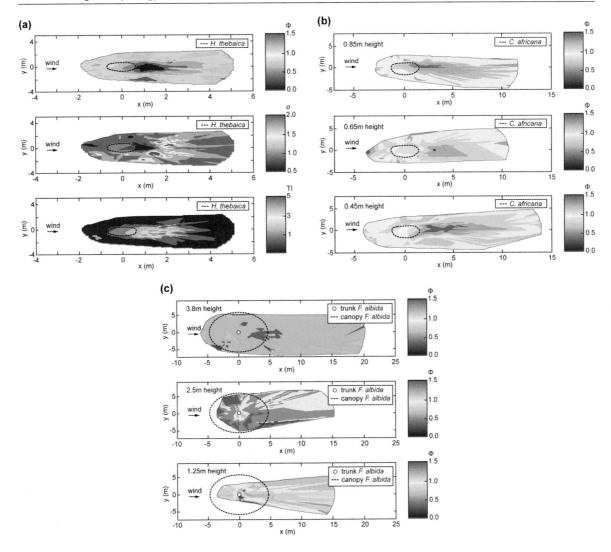

Fig. 3.6 Interactions between wind speeds and vegetation elements for: (**a**) a 0.6-m tall *Hyphaene thebaica* shrub; (**b**) a 1.9-m tall *Commiphora africana* shrub; and (**c**) a 11.5-m tall *Faidherbia albida* tree in Burkina Faso (Leenders et al., 2007)

wall parameters tend not to be fitted in studies of resistance to water flow, not least because of the practical difficulties in measuring velocity profiles that are only a few millimetres deep in unconcentrated overland flows (e.g. Abrahams and Parsons, 1990, 1991; Parsons et al., 1994) to some tens of millimetres in concentrated rill flows (e.g. Parsons and Wainwright, 2006); or due to the equipment damage that may occur in highly turbid, deeper gully and ephemeral channel flows. As an alternative, research has tended to focus on the estimation of flow resistance using the dimensionless Darcy-Weisbach friction factor *ff*:

$$ff = \frac{8\,g\,h\,S}{v^2} \qquad (3.3)$$

where g is acceleration due to gravity [m s^{-2}], h is flow depth [m], S is surface slope [-] and v is mean flow velocity [m s^{-1}] – or dimensionally inaccurate but equivalent equations using Manning's n or Chézy's C parameters as the roughness terms. Abrahams et al. (1988, 1994) demonstrated that *ff* might be a complex function of dimensionless flow discharge (Reynolds number) in shrublands in Arizona due to its interaction with a variety of microtopographic features including vegetation and the mounds on which the

Fig. 3.7 Effects of multiple roughness (vegetation) elements on wind profiles (Wolfe and Nickling, 1993)

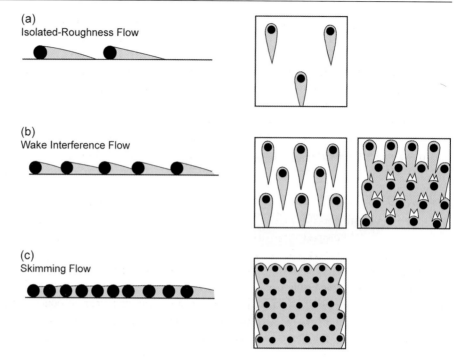

vegetation stands. Significantly higher roughness values were found for desert grasslands in the same setting (Weltz et al., 1992; Parsons et al., 1994). This effect is opposite to the one found for resistance to wind flow. In part, this result occurs because the grass is more fully inundated so that it has a greater proportional effect, but also because of other feedbacks imparted by the vegetation. Grasses often cause the formation of tread-and-riser topographies because the higher resistance causes flow deceleration and significant local sediment deposition. These topographies increase the apparent resistance by increasing the flow path length and decreasing the local slope (Parsons et al., 1997). Similar topographies have been described in matorral in Spain (Boer and Puigdefábregas, 2005). Shrubs tend to promote the formation of mounds (see below), which divert flow around them but tend to channel it into deeper, more hydraulically efficient threads. Surfaces dominated by biological crusts (see below) tend to have very low roughness (Belnap et al., 2005), except in cases where the soil surface freezes.

The effects of vegetation on roughness and thus resistance to flow are not independent of other local factors, as noted above. Some authors have attempted to separate the different effects, using the roughness partitioning approach originally due to Schlichtling (1936), which states that different roughness elements have an additive effect to total roughness. For example, in the approach of Weltz et al. (1992):

$$ff_e = ff_{rs} + ff_{rr} + ff_{gc} + ff_{pb} \qquad (3.4)$$

where the subscript e denotes the effective total value, and the other subscripts relate to grain roughness, microtopography, surface cover and standing vegetation. This approach is incorporated in the WEPP soil-erosion model. Abrahams et al. (1992) have also developed a similar analysis for overland flows, and it has been widely used in wind-erosion studies (e.g. Marshall, 1971; Gillette and Stockton, 1989; Lancaster and Baas, 1998; Crawley and Nickling, 2003). In the Weltz et al. approach, vegetation would have a direct effect on resistance through the ff_{pb} term, and indirect effects through ff_{rs} by the effects on grain size (e.g. relating to differential erosion processes as discussed by Parsons et al., 1992), ff_{rr} by changing the local topography (e.g. the tread-and-riser or mound forms described above) and ff_{gc} by changing stone cover (e.g. the inverse relationship described by Scoging et al., 1992 resulting in pavement formation as discussed by Wainwright et al., 1995, 1999b) and concentrations of surface litter. An implication of the Schlichtling approach is that it suggests shear stress should also be linearly separable in the same way. A complication that this relationship produces is that it implies that shear velocity, u_*, is not independent of roughness height, z_0, in equations (3.1) and (3.2). Other implications of the approach are discussed by Crawley and Nickling (2003).

The canopy also interacts with the environment by processes of interception. In the case of wind flow, interception is of dust and other particles (e.g. pollen and more recently, anthropic pollutants) that are deposited on the canopy. Dust may contain important nutrients that can be transported to the soil by leafdrip or stemflow or fixed by symbiotic fungi, and thus become available to plants. Dust deposition usually provides a net input of nitrogen to the desert ecosystem in this way (Schlesinger et al., 2000; Baez et al., 2007), and the deposited dust is a significant soil-forming material in a number of environments (e.g. McFadden et al., 1987; Yaalon, 1997). Dust may travel for significant distances while gradually settling through the atmosphere (Chapter 20). Vegetation is likely to be a significant location for dust deposition, because the flow resistance slows the wind flow and thus increases the rate of settling, especially in the slower-flowing areas immediately upwind and in the lee of individual plants. In general, the greater the resistance, the more deposition will occur, and this may be enhanced by the waxy nature of the leaves, and conditions where the leaves have been wetted by prior rainfall, dew or fog, or in conditions where salts are deposited on the surface – either by the plant itself, or due to its location (Grantz et al., 2003). Dust deposition on the plant surface can also cause negative effects, for example by reducing photosynthesis, by causing abrasion and damaging plant tissue, and by increasing the albedo of the surface. Recent studies have also demonstrated the potential effects of nitrogen toxicity in areas of elevated deposition, for example downwind of industrial areas (Clark and Tilman, 2008) and the strong co-limitation of productivity by nitrogen and water availability (Hooper and Johnson, 1999).

Interception of water occurs during rainfall events as water lands on leaves and branches during storm events. Martinez-Meza and Whitford (1996) estimated that interception by creosotebush was 44% of rainfall on average, compared to 42–47% for tarbush and 36–38% for mesquite, albeit with significant seasonal variability. The remaining rain passes through the canopy as throughfall and hits the ground directly. Some intercepted water remains on the canopy and is subsequently lost by evaporation. Experiments by Abrahams et al. (2003) suggest that canopy storage is small in desert shrubs, and even mature creosotebush (1.29–1.9-m tall with diameters of 1.37–2.50 m) may only store <5 mm of rain in this way. A similar amount was found for juniper (*Juniperus ashei*) in Texas (Owens et al., 2005). The remaining water reaches the ground by one of two pathways. First, water accumulating on leaves produces drops that are big enough to exceed the storage capacity of the leaf and/or associated surface tension, and falls to the ground (or is reintercepted in more dense canopies). Depending on the shape and structure of the leaves, these drops may be larger than the raindrops formed even in intense storms, and their effect on impact will be a function of their velocity as controlled by fall height. Brandt (1989) demonstrated that tall canopies can produce drops approaching terminal velocity, which are thus more erosive than the smaller raindrops. In drylands, this situation may occur in savannah where there are isolated trees with little understorey. For shrubs, Wainwright et al. (1999c) demonstrated that despite the sparse canopies of creosotebush, there was a 30% reduction in total kinetic energy beneath the plant, although this figure is dominated by throughfall drops that do not interact with the canopy. The kinetic energy of leafdrip was estimated as only 6% of the incoming rainfall energy. Leafdrip had a smaller drop size than the rainfall, which is probably a function of the small leaf size in creosotebush.

The second pathway is by stemflow. It has typically been assumed that water flowing along stems and infiltrating into the crowns of shrubs is a significant water supply. Martinez-Meza and Whitford (1996) estimated values of 10% of rainfall in creosotebush and tarbush and 5% in mesquite entered the ground in this way. However, Abrahams et al. (2003) subsequently demonstrated that the value for creosotebush was probably only 6.7% on average, because some of the water runs off across the surface, especially given the high rates of throughfall also reaching the ground. Runoff generated from stemflow was only moderated or eliminated where there was a significant understorey (usually of muhly grass [*Muhlenbergia porteri*]). Owens et al. (2006) found only 5% of rainfall on juniper in Texas was converted into stemflow. Stemflow may be more important in some other environments. For example values as high as 42% have been observed in acacia and eucalyptus shrubs in Australia (Pressland, 1973; Nulsen et al., 1986).

As well as interception of rainfall, interception of dew and fog by desert plants can also have a significant impact on the water budget. In hyperarid areas such as the Atacama, the presence of fog is directly related to that of lichens, in areas otherwise devoid of vegetation (Warren-Rhodes et al., 2007). Lange et al. (2006) found that maximum rates of photosynthesis in epilithic lichens in the Namib Desert coincided with peaks in fog formation in spring. Epiphytic lichens also occur in the Namib, taking in moisture at night and having short bursts of photosynthetic activity shortly after sunrise (Lange et al., 2007). There seems to be little information about whether these lichens modify the microclimate sufficiently to benefit the host plant. Days with fog in the Namib can produce 0.5–2.3 mm day^{-1}, which can be an important source where 60–200 days per year are foggy (Shanyengana et al., 2002). Kidron (2005)

demonstrated that interception of both fog and dew in the Negev was inversely related to the angle of the receiving surface. Dew was measured as producing 0.12–0.28 mm day^{-1} on average in the Negev (Kidron, 2000). Malek et al. (1999) measured dew deposition of 13 mm a^{-1} in semi-arid shrubland in the Great Basin. The recent review of Agam and Berliner (2006) suggests that adsorption of soil-water vapour may be a more important process than fog or dew interception in these extreme environments. Ramírez et al. (2007) have also suggested the existence of this process in Mediterranean Spain in an area with a mean annual rainfall of 291 mm.

As noted several times already, one indirect consequence of the presence of the vegetation canopy is the development of mounds. In areas with actively blowing sand, the vegetation canopy decelerates the wind to cause deposition in streamlined forms known as nebkha (pl. nebkhat), which may be up to 25-m long and 5-m high (Fig. 3.8). Although a variety of

Fig. 3.8 Nebkha dunes in: (**a**) Tunisia (looking upwind – the nebkha in the foreground is approximately 1-m tall (cf. Fig. 3.9); (**b**) the Chihuahuan Desert around mesquite plants (the nebkha in the centre is about 1.5-m tall and 3.5-m in diameter – note also the evidence of fluvial activity in this landscape in the form of the rill running to the left of this nebkha and towards the bottom left of the image); and (**c**) miniature nebkhas formed behind grass clumps in the Mojave (lens cap is 55 mm in diameter)

names has appeared in the literature, such as coppice dune, phytogenic mound, kthib, rebdou or rehboub, convergence on the widely used "nebkha" seems preferred (see Cooke et al., 1993: 356 for yet further alternatives). Tengberg and Chen (1998) studied fields of nebkhat around *Ziziphus lotus* in Tunisia and *Acacia sp.* and *Balanites aegyptiaca* in Burkino Faso and found that dune height correlated well with dune length up to a threshold length then, for further increases in length, the relationship broke down (Fig. 3.9). This pattern was interpreted as being the result of three phases of development. In the growth phase, the dune height can keep growing until it reaches the threshold height defined by the point at which the wind profile is sufficiently fast to reentrain sediment. At this point, the dune enters a stabilization phase, where it may grow in length by trapping progressively

more sediment, but cannot increase in height. The third, degradation phase occurs when sediment supply drops sufficiently for the wind to restart entrainment, and may occur because of changes in the surrounding supply area or due to interactions between the nebkhat themselves. Nebkha dunes have been recorded in a wide range of other desert settings in the US (Gibbens et al., 1983; Langford, 2000), Africa (Nickling and Wolfe, 1994; Dougill and Thomas, 2002), the Middle East (El Bana et al. 2003; Pease and Tchakerian, 2002; Saqqa and Altallah, 2004), China (Wang et al., 2006) and Australia (Hesp, 1979, 1989 cited in Cooke et al., 1993; Ash and Wasson, 1983). Parsons et al. (2003) noted that nebkhat provide a further strong feedback controlling runoff and fluvial erosion processes in mesquite nebkhat in the US southwest.

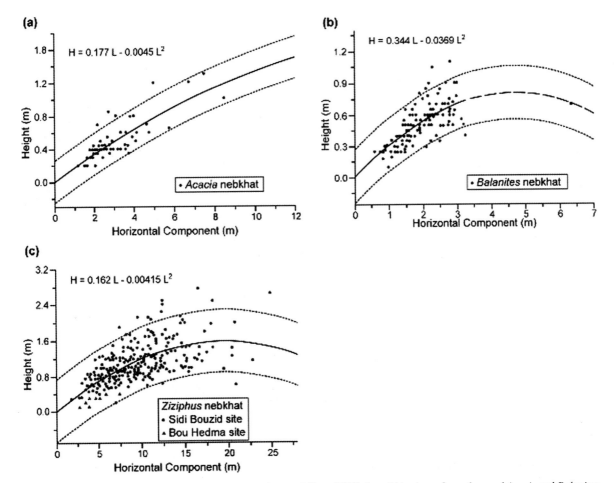

Fig. 3.9 Length-height relationships measured by Tengberg and Chen (1998) for nebkha dunes formed around *Acacia* and *Balanites* shrubs in Tunisia and Burkina Faso

As well as a response to aeolian deposition of sediment (e.g. Cooke et al., 1993), there are a number of hypotheses as to how mounds form in relation to the presence of vegetation. Rostagno and del Valle (1988) suggested that the mounds are upstanding remnants following erosion of the intershrub areas by overland-flow erosion. In the Negev, Shachack and Lovett (1998) found no significant difference in atmospheric dust deposition between shrub mounds and adjacent crust areas. Thus dust is part of the sediment source, but the mounds are due to subsequent redistribution of this material. Parsons et al. (1992) demonstrated a similar mechanism of erosion and deposition based on differential splash (Carson and Kirkby, 1972: 189), while Wainwright et al. (1995) supported this mechanism further, as well as its interaction with erosion related to un-concentrated overland flow. Pelletier et al. (2007) have more recently criticized this explanation for the formation of desert pavements and associated shrub mounds in the US southwest, preferring the aeolian deposition hypothesis for the formation of pavements. However, their hypothesis is unable to explain the presence of mounds or the development of pavements where dust deposition is shown to be negligible (Wainwright et al., 1999b), and lacks an independent demonstration of the process in operation. Biot (1990) also suggested that mounds could form in areas where plant roots and rates of termite digging are high, while Neave and Abrahams (2001) also emphasize the rôle of other animals. It is therefore likely that the interaction of several processes is likely to be important in the formation of mounds beneath vegetation.

Vegetation mounds have become central to the understanding of the functioning of dryland environments over the last three decades as part of the concept of "islands of fertility"; the vegetation canopy is critical in modifying processes to mitigate stresses on plant and animal life in dryland environments. A further interaction with the canopy and the development of islands of fertility is by the production of leaf and other litter. Whitford (2002) notes that litter-decomposition rates tend to be high in drylands due to the high temperatures and ultraviolet radiation present. Intermittent, often intense, rainfall can also contribute by the mechanical breakdown of litter fragments. Annual rates of litter mass loss may vary from 31% to 93%.

The canopy is also critical in a further process in most drylands – that of wildfire. As noted previously, it is thought that the evolution of C_4 grasses was favoured by drying climates during the Miocene. Many drylands – especially the savannah and grassland landscapes – would not exist in their present form without the presence of fire. In the US southwest, McPherson (1999) has suggested that woody plants tend not to occur in areas with annual fires, while they may occur in a scattered way if burning has not occurred for a period of 5–10 years. Only where fire has not recurred for more than 20 years are shrub communities able to persist. Drewa and Havstad (2001), however, suggested that the pattern was not necessarily so straightforward, and that it depends on the interactions between fire and other processes such as drought and grazing by large herbivores. Other drylands have C_3 species that have specially adapted, for example by resprouting from the trunk or having fire-adapted seeds, seen in Mediterranean trees such as holm oak (*Quercus ilex*) or aleppo pine (*Pinus halepensis*), respectively. Fire can have important impacts in changing the surface hydrology, in nutrient budgets, and on erosion and sedimentation patterns (see review in Wainwright and Thornes, 2003). The exact nature of fire occurrence in a particular location will depend on the vegetation state, especially the amount of combustible material, its moisture content and the connectivity of the canopy, as well as the weather (wind speed, temperature and relative humidity) and topography (local aspect and slope) (McPherson, 1995). Fire frequencies tend to demonstrate power-law scaling (Malamud et al., 2005), so that small fires are relatively frequent and very large fires infrequent, with important feedbacks for the connectivity of the system (Fig. 3.10).

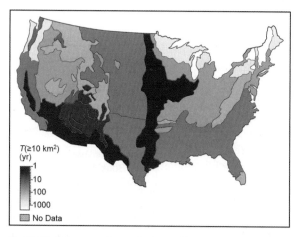

$T(\geq 10 \text{ km}^2)$
(yr)
1
10
100
1000
No Data

Fig. 3.10 Recurrence intervals mapped by ecoregion for the coterminous United States (Malamud et al., 2005)

Roots

Martinez-Meza and Whitford (1996) used dye-tracing experiments on mesquite, tarbush and creosotebush to demonstrate that water from stemflow is typically channelled along the roots of the shrub to depth, where it is available in periods of drought. The same process has been observed by Wang et al. (2007) for *Artemisia ordosica* and *Caragana korshinskii* shrubs in China, and in *Artemisia tridentata* (big sagebrush) in the US by Ryel et al. (2003; 2004). Decaying and decayed roots can also increase infiltration amount and depth. Devitt and Smith (2002) demonstrated using dye-tracing that plots where creosotebush plants had previously been located produced decaying root pathways that allowed water to reach a depth of more than 40 cm, compared to plots without these pathways where infiltration did not reach a depth of 20 cm. These preferential pathways for water flow may thus persist for significant periods, even once vegetation has been removed or died off, and thus may explain sources of variability in observed infiltration rates and consequently why surface properties are not always a good indicator of variability in infiltration (Wainwright et al., 2000). As well as leading to the development of such macropores, root decomposition also affects soil texture and porosity. Whitford et al. (1988) have demonstrated that roots in desert ecosystems tend to decay more rapidly than in other environments. The rate of decay was found to be highly correlated to the extent of termite colonization, and seems relatively independent of additional moisture supplies. The presence of termites will also enhance the porosity and macroporosity of the soil, providing a further feedback to higher infiltration rates. Decaying organic matter will also provide binding materials leading to the development of soil aggregates with similar consequences. Decomposition does also produce a range of non-water-soluble compounds, and these may be significant in dryland vegetation. The presence of such phenolic, turpene and related compounds may be to increase soil hydrophobicity, and thus lead to a decrease in infiltration rates. It should be remembered that these processes relating to decomposition may occur even if the main plant still survives, as parts of the plant biomass may senesce as a response to drought conditions, or die off due to predation by animals. Whitford (2002) points out that root decom-

position is the principal carbon source in desert soils. However, the link between this decomposition and reconstruction of past environments using soil-carbon isotopic composition (see above) is probably complicated by other processes including erosion (Turnbull et al., 2008a).

Roots also have a very high strength and can often penetrate along narrow fissures in bedrock to exploit water resources. Opening up of the fissures in this way provides a further feedback to infiltration, possibly to depth, and in the formation of soils. However, most literature suggests that roots have a very limited ability to penetrate through calcic and petrocalcic horizons (Chapters 5 and 6), resulting in limitations to water availability and consequently plant growth. These limitations affect shrubs (Cunningham and Burk, 1973; Hamerlynck et al., 2002), grasses (McAuliffe, 1995) and succulants (Escoto-Rodríguez and Bullock, 2002) alike (Gibbens and Lenz, 2001; Gibbens et al., 2005). Plant-available water contents have been measured to be higher in petrocalcic horizons than the overlying soils (Duniway et al., 2007) and the structure of the horizons loses strength when wetted (Larsen, *pers. comm.*), so it may be that the explanation for this lack of penetration is more a function of soil chemistry. Buxbaum and Vanderbilt (2007) have related the limitations of vegetation growth to the limitations imposed on the plant due to the high osmotic potential of $CaCO_3$-saturated water. However, this explanation cannot be complete because, as Buxbaum and Vanderbilt themselves point out, creosotebush is more tolerant to these higher potentials than grasses and yet creosotebushes are commonly constrained to a fraction of their potential size above shallow petrocalcic horizons.

Roots affect the vertical flow of water towards the surface as well as away from it. Hydraulic lift occurs when there is a sufficiently high hydraulic conductivity between the soil and root pores to enable flow into plant roots. The phenomenon can only occur if there is sufficient water stored in the soil to produce a hydraulic gradient from the soil to the root during conditions when the plant is not transpiring (and thus generating the gradient by this means). Water then flows upwards through the roots until it reaches levels where the surrounding soil is sufficiently dry to cause a reverse hydraulic gradient, and the water flows out into this soil, where it is available to this and other plants during normal transpiration-driven metabolism.

The process was first observed in big sagebrush shrubs (Richards and Caldwell, 1987), and has been subsequently documented in species of all photosynthetic pathways. Yoder and Nowak (1999) demonstrated its occurrence in C_3 plants with deep (creosotebush), intermediate (*Ephedra nevadensis* [Mormon tea]) and shallow (*Ambrosia dumosa* [white bursage]) roots, as well as the CAM *Yucca schidigera* and the C_4 perennial grass *Achnatherum hymenoides*. While hydraulic lift occurs during the night for the C_3 and C_4 plants, it occurs during the day for the CAM plants in relation to the relative timing of transpiration. However, there is also increasing evidence that some transpiration occurs at night in both C_3 and C_4 plants due to incomplete stomatal closure (and thus conductance of water from the plant) (Caird et al., 2007) and that the conductance mechanism occurs in parallel with hydraulic lift. Snyder et al. (2008) have suggested that night-time transpiration may be a response to nutrient limitation.

Deeper water flow into the vadose zone is heavily restricted if not absent in drylands. Although roots act to channel water to depth, most of this water does not travel beyond the root zone (Walvoord and Phillips, 2004) and is indeed used by plants at a later stage by the mechanisms outlined above. Sandvig and Phillips (2006) cored sites across a vegetation transect in the Sevilleta National Wildlife Reserve in order to evaluate potential water flux. The transect covered creosotebush (1,470–1,590 m asl, mean annual temperature 12.8–13.3°C, mean annual precipitation 230–235 mm), mixed grassland (*Bouteloua gracilis*, *B. eriopoda*, *Muhlenbergia porteri* and *Hilaria mutica* at 1,560–1,900 m asl, 12.0–12.8°C, 230–306 mm), juniper woodland (*Juniperus monosperma*, 1,930–2,050 m asl, 11.5–12.1°C, 306–316 mm) and ponderosa pine forest (2,300–2,380 m asl, 9.2–9.5°C, 327–336 mm). Preferential flow paths were suggested for the juniper and pine sites, usually below a depth of several metres. Limited preferential flow paths were found in the grassland, but none was apparent below about 0.5 m in creosotebush. The chloride mass-balance approach was used to estimate recharge rates. No recharge was found to have occurred beneath the creosotebush sites since 22.0 ± 2.9 ka. Complete flushing of the grassland site soil moisture has probably not occurred for 10.0 ± 1.5 ka, although a slow recharge rate of 0.069 ± 0.020 mm a^{-1} was measured. Turnover of soil moisture under juniper took place since 5.9 ± 1.9 ka with estimated recharge of

0.439 ± 0.430 mm a^{-1}, while under ponderosa pine the turnover varied between 300 a and 3.6 ka and estimated recharge was 2.26 ± 2.89 mm a^{-1}. While some caution must be expressed in interpreting these differences (the creosotebush is likely to have been much less extensive prior to European settlement; there are large uncertainties due to small sample sizes), the overall recharge rates under present conditions are probably <1% of precipitation, with most of this recharge coming from upland, wooded areas. Studies using weighing lysimeters with creosotebush vegetation and unvegetated lysimeters in the Mojave Desert found that even during wet periods (El Niño-related winter rainfalls) the creosotebush was very effective in using water that had percolated to depth (1.7 m) (BR Scanlon et al., 2005a). The evaporation from the unvegetated sites was unable on its own to remove all the water at this depth, and thus some deeper percolation could be expected from bare surfaces. Such bare surfaces would be limited for example to cultivated areas in reality, as the root distributions often extend well beyond the canopy into areas with a bare surface, as noted previously. Scanlon et al. interpreted these results as being consistent with chloride-based estimates that no recharge has occurred for the last 10–15 ka. A similar change was demonstrated by Wang et al. (2004) using vegetated and unvegetated weighing lysimeters under *Caragana korshinskii* shrubland in the Tengger Desert in China.

When indigenous vegetation was replaced by irrigation agriculture in the Amargosa Desert and High Plains, recharge occurred once the combined precipitation and applied irrigation water exceeded about 800 mm (BR Scanlon et al., 2005b). Such rates are however clearly unsustainable over broad areas. Seyfried and Wilcox (2006) reviewed the suggestions that removing woody (shrub) vegetation in drylands would lead to increased recharge and streamflow, given that the current literature has broadly varying interpretations of the consequences. By monitoring soil moisture post-fire at Reynolds Creek in Idaho (550 mm annual precipitation), they found that water uptake was much lower beyond depths of about 1 m, suggesting higher potential for recharge under the burnt areas. They suggest that the critical factor leading to a change in recharge under these circumstances is the extent to which the profile can store plant-available moisture. The change in availability is likely to be a complex function of local conditions. In more arid cases, the ability of precipitation to wet the whole profile will also limit the conditions for recharge to occur. Seyfried et al. (2005) carried out a review a model analysis of recharge in the US deserts since the late Pleistocene. They suggest that slow recharge may still be taking place below depths of about 20 m, representing water that infiltrated over 10,000 years ago. Above this zone is a general area of net upward flux (Fig. 3.11).

While recharge from much of the surface area of drylands is negligible at best, the same is not necessarily true of channels and the riparian zone. Riparian vegetation is generally well developed in drylands as the area surrounding the channel is well supplied by water, if only intermittently. The high porosity of many channel-bed materials also enhances infiltration through transmission losses, often by an order of magnitude compared to surrounding areas (see Chapter 11). Atchley et al. (1999) investigated differences in soil water and nutrients and photosynthetic and transpiration rates in *Fallugia paradoxa* (Apache plume), *Prosopis glandulosa* (honey mesquite) both of which grow in the riparian zone and elsewhere, with *Chilopsis linearis* (desert willow), which is an obligate riparian shrub. They found that patterns were highly variable both between and within sites, not least because of the highly variable spatial response

to convective storm events. In some locations, the multi-stemmed structure of the Apache plume was able to trap more nutrients and thus not be nutrient-limited. The relative location of the vegetation to rapid flow (straight channel sections) seemed to control the ability of the plant to act in this way via a feedback in the resistance to flow imparted by the vegetation canopy. Mesquite is able to exploit the moisture stored beneath the channel more effectively, and thus transpires more rapidly than mesquite growing away from the arroyos. The desert willow adjusted its transpiration in order to minimize water needs in periods when water was sparse. De Soyza et al. (2004) found this response to be the case even in conditions of extreme drought. At the end of drought conditions, all of the species observed responded quickly. The relative importance of recharge along arroyos will thus relate to the extent to which the additional vegetation and rapid response can increase transpiration rates relative to water supply. Scott et al. (2006) investigated transpiration rates of grassland, shrubland and wooded habitats on terraces of the San Pedro river in Arizona. They found that over a single growing season where precipitation was 233 mm, total evapotranspiration was 407, 639 and 450 mm, respectively (Fig. 3.12). In other words, groundwater use was 227, 473 and 265 mm, respectively, which was clearly observed in the water-table depth (2.6, 6.4 and 9.8 m, respectively) which responded rapidly to biomass growth. These results imply that riparian vegetation can limit or prevent enhanced recharge around channels. Wilcox (2002) reached similar conclusions in relation to a review of attempts to enhance streamflow by mesquite removal. The pattern in juniper rangelands was more equivocal, although Wilcox interpreted these results as being more due to the relatively shallow soils and sediments in the uplands where the juniper is commonly found.

The results of these studies imply that the response of riparian and other phreatophytic vegetation in drylands will be sensitive to changing ground-water levels. However, Naumburg et al. (2005) demonstrate that the consequences of change are not straightforward. Declining water levels may be beneficial in providing more root space and may also provide more beneficial conditions in saline soils. More often they may be detrimental, as the plant needs to exert higher leaf-pressure potentials to raise water from deeper in the soil profile. Threshold potentials before cavitation occurs in the leaves may be anything from –4 to –12 Mpa, so

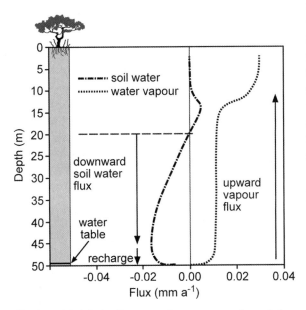

Fig. 3.11 Modelled soil-water and water-vapour fluxes below the root zone 10 ka after the establishment of xeric vegetation, which are compatible with rates observed using tracer studies (Seyfried et al., 2005)

Fig. 3.12 Weekly
evapotranspiration and
precipitation measured by
Scott et al. (2006) for the San
Pedro River, Arizona, in
relation to different types of
riparian vegetation

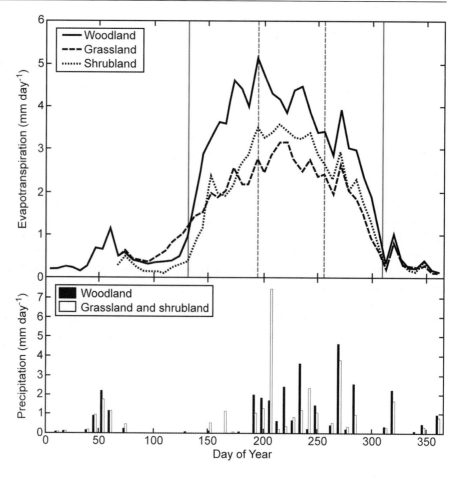

that responses will be species-specific. The impact on plants will also be a function of the rate of descent, as root growth may be able to keep pace with the drop – measured rates of root growth are between 1–13 mm day^{-1} in riparian species such as *Populus*, *Salix* and *Tamarix* – so that rates of groundwater fall up to 40 mm day^{-1} have been tolerated up to the maximum rooting depth of the species. Some species also maintain shallow roots and thus switch to more readily accessible near-surface moisture under these conditions. To minimize the resistance in the roots and thus maximize the ability to raise water from depth, deeper roots will tend to have a greater diameter, but the ultimate limitation on following declining groundwater will be the ability of plants to produce sufficient biomass to create new roots. Rising water levels also paradoxically affect phreatophytic vegetation. Waterlogging produces anoxic conditions, causing root die-off unless the plant can translate sufficient oxygen to the roots. The relative rate of impact will thus again be species-specific.

Root death will also reduce the ability of the plant to extract moisture to supply the above-ground biomass, and so produces a feedback to the ability of the plant to adapt. Plants will usually adapt to rising water tables by growing new roots higher in the profile, or by becoming dormant. These relative effects were evaluated using a numerical model by Naumberg et al. (2005), demonstrating that for an application to Owens Valley, shrubs were less responsive (more resilient) than grasses although the particular response was also a function of soil type. Biomass is typically lower in changing or fluctuating conditions than it is with steady water tables near the surface. Seyfried et al. (2005) suggested that riparian vegetation would have been especially sensitive to changes at the start of the Holocene, but that further studies are required in terms of the carbon costs that the plant incurs in adapting to a declining water table, unknown interactions between species (for example in the presence of hydraulic lift), the link with upland processes and the understanding community dynamics, particularly in defining critical thresholds. For example, Stromberg et al. (1996) demonstrated that well defined species associations along the San Pedro River in Arizona had clear thresholds of tolerance of distance to the water table. The species composition changed dramatically on passing 0.25 m, and then again at 1, 3 and 8 m. These thresholds are further complicated by the variability in the water table through the year (Stromberg et al., 2007). Increased rates of abstraction for agriculture, drinking water and industry in drylands will thus accentuate the sensitivity of riparian vegetation. Anthropic impacts are not limited to extraction; they are also a function of introduced species. Pataki et al. (2006) demonstrate the sensitivity of Fremont cottonwood (*Populus fremontii*) along the Colorado to invasion by salt cedar (*Tamarix ramosissima*). The ability of the latter species to tolerate more highly saline conditions allowed it to transpire more and thus outcompete the cottonwood. More saline conditions will again be a typical response to human intervention.

Roots also have a significant impact in providing a structure to soils and sediments. There are two specific effects. First, this structure imparts cohesion to soils and thus reduces their erodibility, in relation to splash and overland-flow erosion. Such cohesion will also reduce erodibility with respect to wind erosion, although it is likely that the modification of the velocity profile by the vegetation structure as discussed above is more important in this case. Secondly, roots typically have a high tensile strength and thus act as reinforcing elements, thereby minimizing the likelihood of slope or channel-bank failure. Tensile strength is inversely proportional to the root diameter, so that the overall effect is a function of the distribution of roots of different sizes, which is often poorly known, especially in the case of dryland plants (Pollen and Simon, 2005). Graf (1981, 1983a) has demonstrated the importance of vegetation in maintaining stability along the Gila River in Arizona, with thresholds due to extreme events leading to the erosion of stabilizing vegetation and corresponding subsequent increase of active-channel width. Subsequent recolonization and restabilization is a much slower process. Stromberg (1997) investigated a 25-year recurrence-interval flood event on the Hassayampa River in Arizona in January 1993 which created a 50-m-wide expansion of the active channel zone. She found that colonization of the expanded channel occurred very rapidly, with Fremont cottonwood germinating in March-April (i.e. two-three months after the flood event), Goodding willow in April-May, salt cedar in May-September, and arrow weed and seep willow in July-September. Different species were found to colonize different areas, with Fremont cottonwood preferring dry surface sediments, salt cedar and seep willow preferring saturated surface sediments and the other species found in both types of sediment. A second major event in 1995 removed a significant number of the colonizing plants, and reset the revegetation process. However, the more depositional nature of this flood in the study area tended to favour recolonization on the second occasion by Freemont cottonwood. Thus, the relative timing and nature of flood events will strongly interact with the types of riparian vegetation present and thus the likelihood that any particular event will be erosive or be resisted by the effects of vegetation. Diversity of vegetation type will be less in channels with increasingly ephemeral flows – including those due to human modifications – due to the removal of wetter microhabitats (Stromberg, 2001; Stromberg et al., 2007). Therefore, it is likely that the more ephemeral a channel is, the more likely it is to present a major response to extreme flows because of the sparseness and spatial patchiness of riparian vegetation. A further significant interaction between vegetation and channel flow is in the initiation of gullies. Graf (1983b, 1988) analyzed the relationship between vegetation cover and shear stresses generated to initiate gullies in the Henry Mountains in Utah. He found a strongly hysteretic relationship that was well described by a cusp-catastrophe model.

Surface and Near Surface

Vegetation provides a further control on infiltration by reducing the energy and rate of water arrival at the surface. As noted above, Wainwright et al. (1999c) identified a significant reduction in rainfall energy arriving at the surface below creosotebush, and Brandt (1989) measured the same phenomenon under a number of shrub plants. The effect of grass – especially species with a clumped and/or low growth form – can be considered to exacerbate this effect, although direct measurement is difficult. However, such vegetation controls are not universal and depend on the interaction with other surface features. For example, Descroix et al. (2001) demonstrated that on grass slopes in the Sierra Madre of Mexico, runoff and erosion was significantly higher than elsewhere in the area. They attributed this difference to higher proportions of embedded stones and surface crusting than found in association with other vegetation types.

Most discussion thus far has been of macrophytic vegetation. However, microphytes are also an important component of the desert ecosystem. Lichens, mosses, and cyanobacterial and chlorophytal algae have all been observed (Lange et al., 1992; 1994). They may be early colonizers and, in conditions where moisture from rain is sparse, they may be the only form of vegetation present in deserts. Growing at or near the surface because of their need to photosynthesize, they can produce dense mats at the surface, and reinforce the subsurface by the production of filamentous growths. Together, these reinforcing mechanisms produce what are commonly known as biological crusts. The filaments can often be seen in cross section and are a useful way of distinguishing between biological and mechanical crusts. These crusts tend to have low infiltration rates and fewer propensities for runon infiltration due to low roughness values (Belnap et al., 2005), except in cases where frost action is important and rough surfaces are created thus tending to lead to higher infiltration rates (Belnap, 2003). Conversely, Williams et al. (1999) found no measured difference between unsaturated hydraulic conductivity of biological crusts compared to other areas, which may be due to their use of a tension infiltrometer for measurement rather than considering the effect of rainfall. Belnap (2006) summarizes the difference in the literature regarding differential infiltration rates. She suggests that smooth cyanobacterial crusts decrease infiltration, while rugose crusts or lichen or moss surfaces will tend to increase infiltration, suggesting that different crust types in different environments will have significantly different effects on infiltration and runoff processes (Fig. 3.13). Microphytic crusts respond rapidly to pulses of rainfall (Cable and Huxman, 2004; Belnap et al., 2005; Bell et al., 2008; Loik, 2006). For example, Cable and Huxman (2004) found that crusts began photosynthesis within 100 h of a rainfall event and contributed a significant component of carbon flux into soils, especially for small rainfall events. Bell et al. (2008) found a positive correlation between soil moisture and photosynthetic activity. Crust growth may therefore be rapid and able to exploit water sources that are too sparse for

other plant activity. Thus, microphytic crusts may be significant in the rapid stabilization of the surface following disturbance due to erosion events or human activity. Belnap and Eldridge (2001) noted a 35-fold decrease in erosion on a well-developed crust surface, and Kidron (2001) and Neave and Rayburg (2007) also highlight the importance of these crusts in reducing erosion. However, once disturbance has taken place, full recovery of microphytic crust function may take decades or centuries (Belnap, 2003).

Microphytic crusts can also provide a significant nutrient input to the soil, with corresponding benefits to other plant life. Belnap (2003; see also Chapter 9) notes that this effect may be indirect in that the polysaccharide sheaths of bacteria are "sticky" and thus trap dust, an effect that may be emphasized where the crust surface is rugose. Bell et al. (2008) measured significant NH_4-N and NO_3-N production by microphytes in soils at Big Bend National Park, Texas. Nitrogen fixation by lichen may be as high as $10\,kg\,ha^{-1}\,a^{-1}$, and that by cyanobacteria as high as $1\,kg\,ha^{-1}\,a^{-1}$ (Belnap, 2003).

Certain microphytes inhabit very extreme environments. Schlesinger et al. (2003) note the presence of a range of hypolithic cyanobacteria beneath quartz pebbles in the Mojave Desert. Although most frequent beneath pebbles of 9–10 mm thick, some examples were found beneath 25-mm thick pebbles where light was approximately 0.08% that of ambient. Such cyanobacteria have been found to tolerate temperatures as high as 90°C. It has been suggested that these environments are analogues for where life may be found on other planets, or for the early evolution of plant life in desert systems. Hypolithic cyanobacteria may also increase the stability of pavement surfaces, although detailed research has yet to be carried out on this aspect of their function.

Habitat

A further rôle of vegetation in affecting geomorphological processes comes through its interaction with fauna. Beyond its source as a food supply (e.g. Noy-Meir, 1974; Tivy, 1990; Stafford Smith, 1996; Whitford, 2002) and the resultant redistribution of nutrients across the landscape as faecal matter, the vegetation structure also provides a habitat for animals. Birds nest and roost, often in upper branches away from potential predators, where there is also

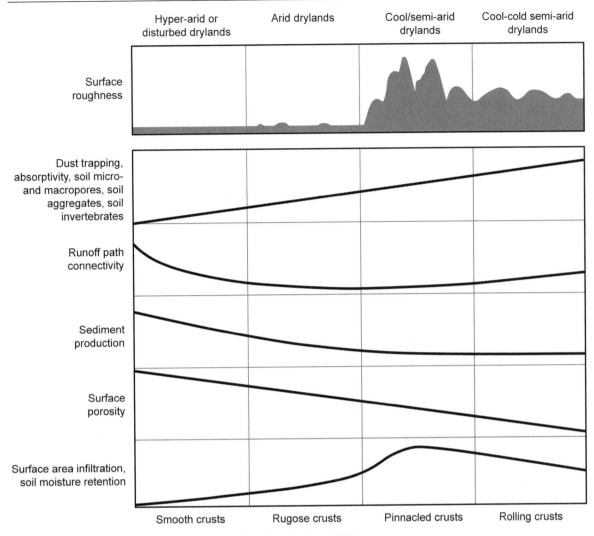

Fig. 3.13 Effects of different crust types on surface hydrology (Belnap, 2006)

the advantage of cooler microclimates. Reptiles may use branches for basking, while amphibians and a range of small mammals burrow beneath vegetation (Whitford, 2002). The vegetation provides a way for the animals to avoid the extremes of the environment, as well as providing protection from predation. One consequence of the burrowing and other digging activity (e.g. for food) is that it disturbs the surface below and around plant canopies. Neave and Abrahams (2001) demonstrated a positive relationship between the extent of such disturbance activity and subsequent erosion in overland flow. The disturbance creates a significant sediment source that is less readily exhausted during runoff events, so that erosion is higher and more prolonged than on undisturbed sites. Concentration of small animals in and around shrub canopies may contribute to the "islands of fertility" phenomenon by recycling of plant material and redeposition of it as faecal matter beneath and adjacent to plants. In contrast, large herbivores may have a more dispersive effect.

Pattern

The discussion of process-interactions above suggests that work on understanding the interactions between desert vegetation and geomorphic processes has focussed on two extremes of scale. The first is the regional scale of thousands of square kilometres or more, where there is reasonable evidence to suggest large scale feedbacks between vegetation and climate. The second is at the scale of individual plants or some component of their structure. This latter approach has been highly influential, not least since the development of the "islands of fertility" concept by Garcia-Moya and McKell (1970) and Charley and West (1975), and used as a basis for understanding dryland degradation by Schlesinger et al. (1990). This oft-cited paper has been used to support a large body of research that has informed our understanding of desert vegetation at the scale of individual plants, and because of the sparse distribution of plants, at the scale of the interspaces between the plants. Research at this scale is frequently carried out, not least because of the relative ease of carrying out experimental approaches at this scale. An implication of this research focus has been an often implicit conceptualization that once the plant-interspace scale has been fully investigated, desert ecosystems and their interaction with the landscape will be fully understood. Four considerations suggest that this logical leap is unfounded.

First, empirical considerations suggest that patterns are not quite so straightforward. When the Schlesinger et al. (1990) paper was published, it contained little in the way of direct evidence to support the concept of islands of fertility; it was more a manifesto for research that would be required to test the hypothesis. Schlesinger et al. (1996) compared patterns of soil N, P and K with distributions of black grama grass plants, and of creosotebush and mesquite shrubs at sites in the Chihuahuan, Mojave and Great Basin Deserts. Geostatistical analyses of 8×12-m areas suggested that nutrients were autocorrelated at shorter distances in the grasslands than in the shrubs, and that these patterns could be related to the average distance between plants in the different cases. This analysis was extended by Schlesinger and Pilmanis (1998) who noted changes in spatial autocorrelation as shrubs invade grassland, and that when shrubs are cleared, they tend to be more readily reestablished if the soils are not homogenized. In contrast, Müller et al. (2008) used geostatistical analyses over areas of 90×90 m to investigate spatial patterns of nutrients, soil moisture, infiltration and other soil properties in grassland and creosotebush, mesquite and tarbush shrubland. They found that although soil moisture and infiltration was closely related to vegetation size in all cases except for mesquite, other parameters were distributed in a much more complex way (Fig. 3.14). This complexity was attributed to processes

that transfer materials over longer distances such as overland flow in concentrated flow paths and rills. Indeed, these results are not incompatible with those of Schlesinger et al. (1996), who noted for example that only 35–76% of variance in grassland N and 35–51% in shrubland N was explained by the pattern of vegetation. More often than not, therefore, factors at larger scales than vegetation variability must explain the variance in N in their results also.

Secondly, and not least, it has long been known that dryland vegetation exhibits distinct distribution patterns and distributions at different scales. On moderate-angled slopes, the plant-interspace patchiness contains features at larger scales, such as concentrations of vegetation along rills and larger channels, and concentrations in splay areas that typically receive high amounts of runon infiltration (termed "beads" by Wainwright et al., 2002). The exact scale of these patterns is a function of local soil and climate conditions, but may induce variability on a scale of tens of metres to several kilometres (Fig. 3.15). At these larger scales, progressive sorting and the development of soil catenas and other evolutionary sequences control the structure of water availability and thus the larger scale patterns of vegetation distribution, including transitions between different dominant vegetation types (e.g. Phillips and MacMahon, 1978; McAuliffe, 1994). Larger scale vegetation patterns have also been observed in environments where wind is considered to be the dominant vector of sediment transport. Okin and Gillette (2001; Okin et al., 2006) have described elongated bare patches parallel to the dominant wind pattern in areas with mesquite nebkhas, so that there is a distinct anisotropy in the vegetation pattern here too. These elongated patches have been termed "streets". In reality, in all but the more hyperarid areas, there are likely to be significant interactions between water and wind movement of sediment along these streets. In areas dominated by more gentle slopes (c.1% is typically considered the threshold, but again the exact value will depend on local conditions), vegetation bands parallel to the slope contour have been described in deserts worldwide (Dunkerley and Brown, 1995). These areas of banded vegetation are known as mulga groves in Australia and vegetation arcs and tiger bush (*brousse tigrée* in Francophone areas) in Africa. The bands are usually associated with a stepped microtopography of about 10 cm. Bands occur dominantly in shrublands but

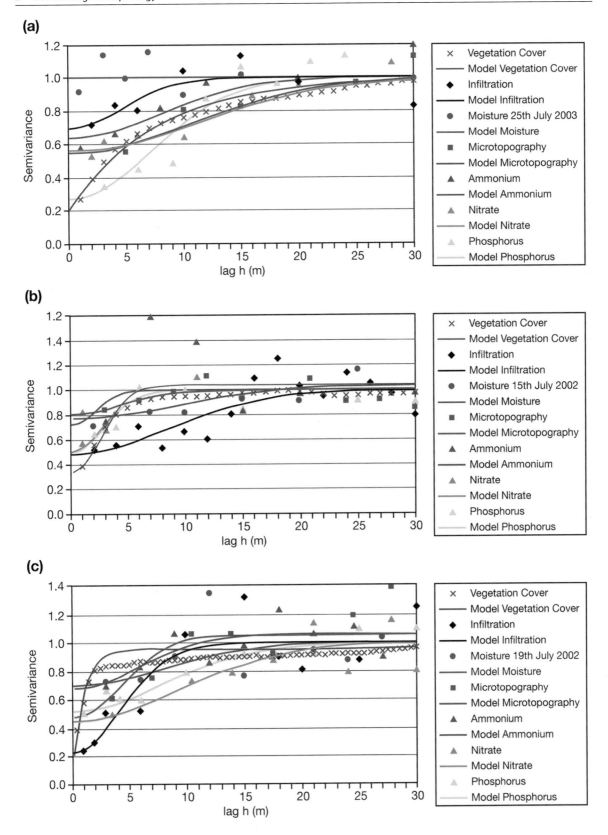

Fig. 3.14 Semivariograms showing different spatial relationships of vegetation and surface characteristics for a range of shrub and grass vegetation types, Jornada Experimental Range, New Mexico (Müller, 2004; see also Müller et al., 2008)

Fig. 3.15 (caption on next page)

have also been observed in grasslands (Worrall, 1959; Montaña, 1992); grasslands may also exhibit similar patterns at smaller scales, with arcuate patches and tread-and-riser topographies. Common explanations of banded vegetation are not dissimilar to those of "islands of fertility" in patchy vegetation. It is argued that runoff from the bare areas transports sediment, nutrients and propagules downslope. Erosion of sediment in this way creates the stepped microtopography and makes the interband areas less likely to be able to support vegetation. When the runoff encounters the vegetation band, it starts to deposit water, sediment and nutrients so that vegetation on the upslope edge of the band receives more resource than vegetation elsewhere. The relative effect of this resource accumulation can be estimated by observing the relative widths of the bands and interbands. Tongway and Ludwig (1990) note that bands are typically 10–20-m wide, and that interbands are 20–50-m wide. Goudie (2002) gives examples of interbands that are 100–200 m in width and generalizes that they are often two to four times as wide as the bands. Propagules will also have a favourable location to start new vegetation growth. Thus, it is often argued that vegetation bands should migrate upslope (but see Dunkerley, 1997 for an opposing viewpoint based on hydrological mechanisms only). Dunkerley and Brown (2002) observed banded vegetation on slopes as low as 0.4%, and there is nothing theoretically to prevent banding on less steep slopes, as long as a mechanism for building a pressure head exists. However, much uncertainty exists in the explanation of banded vegetation, inasmuch as different mechanisms may be in operation in different places. For example, Montaña (1992) demonstrated that banded mesquite-tobosa grass vegetation in the Chihuahuan Desert in Mexico migrated upslope as a result of colonization processes, while Dunkerley and Brown (2002) did not observe this process in relatively stable bands of mixed chenopod shrubs and tussock grasses. The latter study also demonstrated

the existence of bands at orientations of 45–70° to the slope direction, which are difficult to explain using the same mechanisms. A further consequence of vegetation patchiness is its effect on biomass productivity. A number of authors (Humphrey, 1958; Buffington and Herbel, 1965; Barbier et al., 2006) have suggested that patchy vegetation produces a greater biomass than uniform vegetation. Aguiar and Sala (1999) have suggested that the reason for this difference is that patchy vegetation is more easily able to exploit all precipitation rather than lose some "ineffective precipitation" below the biomass-production threshold. This theory implies that sufficient ineffective precipitation can be accessed by plants, which may not always be the case for low intensity events (e.g. runoff-runon mechanisms do no occur). The lack of a significant difference in biomass between grasslands and shrublands in the same location (Huenneke and Schlesinger, 2006) may imply that the patchiness at the scale at which it occurs in grasslands is also able to exploit this mechanism, and it is the relative scale of plant to interspace that is important. However, the analysis of Aguiar and Sala (1999) also implies that patchy vegetation is the result of dominant wind and animal vectors of movement, which is clearly contrary to observations elsewhere that show water vectors are highly sigificant in the development of patches. Oksanen (1990) has also suggested that herbivorous animals play a significant rôle in the development of patches in semi-arid environments, but that their impact is moderated in arid conditions.

Thirdly, and often as a response to the difficulties of field-based explanations of pattern, there has been a rash of related modelling studies. These studies either focus just on banded vegetation (e.g. Lefever and Lejeune, 1997; Lejeune and Tlidi, 1999; Klausmeier, 1999; Esteban and Fairén, 2006; Sherratt and Lord, 2007) or attempt to explain pattern in a generic framework, which usually consists of a classification into "patchy"/"spotty"/"leopard", "labyrinthine" and

Fig. 3.15 Spatial patterns of desert vegetation: (**a**) patchy low matorral in Almería, Spain; (**b**) patches, labyrinths and tiger bush near Niamey (bare ground in the tiger bush is 75–100-m in width), Niger; (**c**) patches in mesquite nebkhas (cf. Fig. 3.8b: this figure is in the same area of the M-NORT site of Okin and Gillette, 2001, with evidence of "streets" aligned SW-NE following dominant wind directions; note also the patches of denser vegetation in local topographic hollows); (**d**) patchy vegetation at various scales relating to the presence of riparian vegetation and discontinuous flows in the Sonoran Desert (cf. also Wainwright et al., 2002); and (**e**) vegetation following an altitudinal-catena gradient in the Chihuahuan Desert (the mountain has sparse juniper trees, the base of the mountain has remnant black grama-yucca grassland, which merges into creosotebush and then tarbush shrubland, and finally into a mixed tarbush-tobosa grass) "banded"/"tiger" (e.g. HilleRisLambers et al., 2001; Rietkerk et al., 2002). As with other classifications, its utility is probably more by demonstration of proof by exception. A common feature of this modelling approach is to use the concept of emergence of larger scale properties (the vegetation patterning) as a response to local scale interactions. Many of these studies are conducted implicitly or explicitly within a Turing-instability framework. Stewart et al. (in press) point to the conceptual and practical inadequacies of these approaches and present an alternative modelling framework that incorporates processes that operate at a range of scales within the landscape that produce patterns at a range of spatial scales.

Fourthly, strategies for mitigation of land degradation that have been based on the plant-interspace concept have generally proven inadequate. Rango et al. (2005) demonstrated that areas where shrubs had invaded grassland and where shrubs were subsequently removed were able to revert back to shrubland if no other measures were taken (see also Rango et al., 2006). One aspect of this inadequacy is the failure to incorporate aspects of historical legacies (Foster et al., 2003) on understanding the evolution of systems by affecting the boundary and initial conditions of the problem when observed in isolation. A similar problem is the impact of contingency both in terms of stochastic events and historical development. Allison and Hobbs (2004) have considered the effects of long-term (and large-scale) economic cycles and demonstrated their importance for understanding problems in the management of agricultural systems in arid Western Australia.

Thus, as noted previously, dryland landscapes can be seen to interact with vegetation on a range of spatial and temporal scales. For example, mound formation may be the result of plant-interspace processes at least initially in areas where rainsplash is the dominant mechanism. The presence of mounds starts to concentrate fluvial and aeolian erosion processes that not only concentrate differences at the plant-interspace scale, but also contribute to more advective patterns of redistribution. However, splash-related processes will operate more frequently because of the magnitude of most rainfall events, while advective transfers require larger magnitude events. Vegetation patterns thus evolve at a range of spatial scales on timescales ranging from years to centuries. Cooke and Reeves (1976) have demonstrated how changing

hydrological régimes produced during these vegetation changes can cause both deepening and widening of arroyos, which in turn affects and is affected by the presence of riparian vegetation. Consequently, these changes may have impacts on regional groundwater recharge on millennial timescales (see also Wilcox et al., 2006). Graf (1983c) has also demonstrated decadal consequences on patterns of incision and sedimentation, that are consistent with previous episodes in the Holocene (Hall, 1977). All of these changes will produce decadal to centennial variations by feedbacks in the climate. At the longest timescales, there will be feedbacks between landforms and tectonic processes. While this interaction is usually considered unidirectional, in the sense of deserts created in the lee of mountain chains (e.g. Chapter 28) or of structural controls (e.g. Campos-Enriquez et al., 1999; Schlemon and Riefner, 2006; Dill et al., 2006), Willett (1999) has demonstrated that there is a dynamic feedback between pattern of climate (e.g. pro- versus retro-wedge aridity in convergence zones) and uplift via isostatic adjustment following erosion. There are thus likely to be complex feedbacks in the presence, evolution and pattern of dryland vegetation at timescales of millions of years.

Implications

Understanding the complexity of the interaction of desert vegetation with landforms is not a trivial exercise. As noted by Newman et al. (2006), it requires both top-down (systems-based) and bottom-up (complexity-theory and evolutionary) approaches. A range of conceptual models has been proposed to try to explain different aspects of these interactions, and for the most part they are undergoing evaluation and refinement. The two-layer hypothesis of Walter (1971) that attempts to explain the coexistence of grasses and shrubs or trees in the same apparent niche has been demonstrated to be too simplistic, and temporal as well as spatial variability needs to be accounted for. This variability was central to the classic overview by Noy-Meir (1973). The trigger–transfer–reserve–pulse (TTRP) model of Ludwig et al. (2005) provides one means of accounting for both sources of variability (Fig. 3.16), with examples relating to runoff-runon dynamics and patterned vegetation in Australia and

Fig. 3.16 The trigger-transfer-reserve-pulse (TTRP) conceptual model of Ludwig et al. (2005). Direct consequences of a trigger event are shown as solid arrows, feedbacks as dashed lines and flows out of the (local) system as dotted lines

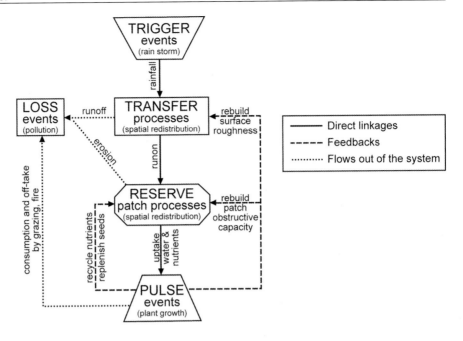

the USA. Loik et al. (2004) have used a similar pulse-based framework to consider vegetation-climate linkages from local to regional spatial scales and from timescales of individual storm events to decadal variability of climate (principally series of drought cycles). One emergent property of landscapes that undergo these processes with those driven principally by aeolian processes is that of connectivity, or in other words, the extent to which spatial patterns emerge that allow the effective transfer of resources across the landscape (Stewart et al., in press; Okin et al., 2009, and more general reviews in Bracken and Croke, 2007). The extent to which structural and functional connectivity link the dynamics of process with form in desert environments seems to be a useful concept allowing the understanding of changes from plant-interspace to landscape scales (Turnbull et al., 2008b).

One apparent paradox with desert vegetation is that it can both exhibit highly resilient characteristics and undergo very dramatic, rapid changes. The resilience of an ecosystem can be defined as its ability to withstand major disturbances without significant change (Holling, 1973). At the plant level, desert vegetation is resilient because it has had to develop mechanisms for adapting to extreme and highly variable environmental conditions, as discussed above, and in some conditions, the interaction between plants

in producing favourable environments for each other can lead to the operation of this resilience at larger spatial and temporal scales. Catastrophic changes are also well known (DeMenocal et al., 2000; Scheffer et al., 2001; Liu et al., 2007) and can be interpreted in terms of the non-linear and threshold behaviour of vegetation-vegetation and vegetation-landscape interactions. Models using catastrophe theory to explain different aspects of these interactions have been relatively widely employed (e.g. Graf, 1983b, 1988; Thornes, 1980; Lockwood and Lockwood, 1993; Rietkerk et al., 1996; Scheffer and Carpenter, 2003), and Turnbull et al. (2008b) have suggested that these conceptual approaches can be combined with those relating to connectivity as discussed above. Given that vegetation is a strong control on the types and rates of landscape evolution that occur in drylands, it seems reasonable to hypothesize that a TTRP model for geomorphic processes is appropriate, and explains some of the current difficulties in understanding the complexities of dryland landscape evolution.

As noted in the introduction, conceptual models that assume (simple) equilibrium conditions are particularly inappropriate to dryland environments, and there has been much debate about the existence of multiple stable states in savannah ecosystems, for example (Scoones et al., 1993; Sullivan and Rohde, 2002). Gillson (2004) has demonstrated that such conditions are likely to have persisted in Kenya for at least the last 1,400 years. It is thus unlikely that ecohydrological approaches that are based on assumptions of tendency towards equilibrium, such as those of Eagleson (2002), will provide strong explanatory mechanisms for understanding the evolution of dryland vegetation (see also the discussions on optimality from an ecological perspective in Hatton et al., 1997, and Kerkoff et al., 2004). Adding geomorphic processes increases the number of degrees of freedom in the system and the subsequent difficulty of demonstrating equilibrium conditions (Bracken and Wainwright, 2006), even if they do exist. Multiple equilibria or non-equilibrium systems are one reason why semi-arid geomorphic systems exhibit complex responses both spatially (Schumm, 1973) and in time (Wainwright, 2006b). They also help to understand why explanations of certain phenomena may fall foul of the problem of equifinality (e.g. Cooke and Reeves, 1976). Overall, the interactions of dryland vegetation and geomorphic processes, with their multifarious scales, delays, emergence, thresholds, catastrophic changes, nonlinearities and non- or multi-equilibrium states, are one reason why the reconstruction of environmental change in drylands (Chapter 28) is problematic.

Conclusions

Progress in understanding the interactions of desert vegetation with geomorphic processes over the last two decades or so has been considerable. In part, this progress has been due to a shift in perspective away from empirically based studies to more conceptually sound frameworks that often take on multidisciplinary perspectives. However, this shift needs to continue and to take on a more holistic perspective that allows the integration of process understanding and employ models of a range of complexities.

At the same time, it is important not to lose sight of the wood (as it were) from the trees. Although advances are being made, most dryland environments remain poorly understood, and the variability in understanding of different environments and systems is considerable. It is fundamental that new empirical studies are carried out in parallel with the development of new models and theories, in order to evaluate their general usefulness.

In so doing, it is important to build on the recent advances in research in dryland ecogeomorphology. Underpinning much of this work are various concepts of scale and scaling. In particular, there is a convergence in perspectives from ecological moves from individual leaf or plant scales to a focus on landscape, and geomorphological moves from small-scale processes again to a landscape perspective. Work on the "islands of fertility" concept over the last two decades has demonstrated that while the idea is useful for understanding behaviour at individual plant scale, it does not offer sufficient explanatory power for the emergence of vegetation patchiness and patterns at hillslope to landscape scales, nor for the longer term behaviour of landscapes at decadal and longer timescales. More work is required to elucidate how vegetation "islands" interconnect with other landscape elements; indeed, given their leakiness, it may be appropriate to stop using the island metaphor in the sense that it may impede further developments in understanding integrated vegetation-landscape systems. At present, the change in scale has tended to be driven from the bottom up, i.e. from a perspective of employing small-scale process understandings to interpret landscape-scale pattern. However, as shown above (Fig. 3.1), ecogeomorphic processes operate over multiple time and space scales, and so there is great scope for driving understanding in the opposite direction, from geological or Quaternary science perspectives for example. At a global scale, deserts produce significant feedbacks on climate systems, and there is also a greater need for understanding the rôle of these feedbacks within broader debates on climate change and Earth-System Science (Wainwright, 2009). A number of empirical observations and modelling results also suggest that dryland ecogeomorphology is particularly susceptible to initial conditions and to past trajectories, again emphasizing the need for integration with work carried out at longer timescales. This integration needs to operate in both directions, with a recognition that reconstructing past climate or vegetation patterns is not straightforward, but needs to be informed by ongoing process-based work. A significant aspect of contingency and path-dependence in the system is the impact of human activity, and new methodologies are being developed for the analysis and interpretation of these impacts (Wainwright, 2008; Wainwright and Millington, in press). However, there is also a need for historical and archaeological work to underpin these approaches with data to assess the integrated human-landscape interactions and their evolution. Along with human activity, the impacts of other animals and of fire, all of which have only been considered briefly within the scope of this chapter, need to be fully incorporated into studies of dryland ecogeomorphology. As more holistic or whole-system approaches are developed, there needs to be an evaluation of these processes as interrnal to the system, rather than as being external "disturbances", a perspective that tends to draw the focus away from their importance. By so doing, significant advances may also be made in understanding one of the key questions of dryland ecogeomorphology, namely why individual plants and plant components can exhibit such degrees of resilience while plant assemblages and landscapes are highly sensitivite. The development of these holistic approaches also requires a clearer recognition of the interactions between different processes, and in particular the removal of barriers imposed by the imposition of narrow geomorphic process domains. There are many advantages to interpreting aeolian and fluvial processes within the same framework, for example, and in all but the more hyper-arid or temperate extremes of drylands, both sets of process tend to operate to varying degrees. There is also the rather false split between hillslope and fluvial geomorphology, which has tended to hinder the understanding of whole catchment behaviour. Recent management as well as scientific developments will hopefully begin to remove this division. There needs to be a continued development of new techniques for understanding process, and for evaluating patterns over large areas. Continued improvements in Earth-Observation techniques are required in this respect, and there needs to be an ongoing evaluation of how such observations fit into a continuum of measurement from the point scale upwards. Given the extent to which desert vegetation concentrates resources below ground, there is a major need for improving techniques that can evaluate the subsurface, preferably non-destructively. Such improvements are vital to make advances from current techniques that base estimates on surface proxies. Finally, as we have previously noted (Wainwright et al., 2000), all of the above developments will require a move away from empiricism to general, conceptually underpinned approaches that recognize the need to move away from plant-/plot-based methodologies, and the need for better framing of individual case studies to relate their specific context to the bigger scientific picture.

Whither desert ecogeomorphology? Conceptual developments over the last decade suggest that the topic is a prime area for understanding non-equilibrium dynamics, complexity theory and spatio-temporal connectivity of process and form in whole-systems perspectives. Taking these conceptual models forwards employing integrated modelling and field approaches within an inter- and multidisciplinary framework is likely to provide fertile ground for future discoveries.

References

Abrahams, AD and AJ Parsons 1990 'Determining the mean depth of overland flow in field studies of flow hydraulics', *Water Resources Research* **26**, 501–503.

Abrahams, AD and AJ Parsons 1991 'Resistance to overland flow on desert pavement and its implications for sediment transport modeling', *Water Resources Research* **27**, 1827–1836.

Abrahams, AD, AJ Parsons and P Hirsch 1992 'Field and laboratory studies of resistance to interrill overland flow on semiarid hillslopes, southern Arizona', in AJ Parsons and AD Abrahams (eds) *Overland Flow: Hydraulics and Erosion Mechanics*, 1–23, UCL Press, London.

Abrahams, AD, AJ Parsons and S-H Luk 1988 'Resistance to overland flow on desert hillslopes', *Journal of Hydrology* **88**, 343–363.

Abrahams, AD, AJ Parsons and J Wainwright 1994 'Resistance to overland flow on semiarid grassland and shrubland hillslopes, Walnut Gulch, Southern Arizona', *Journal of Hydrology* **156**, 431–446.

Abrahams, AD, AJ Parsons and J Wainwright 2003 'Disposition of stemflow under creosotebush', *Hydrological Processes* **17**, 2555–2566.

Acreman, MC (ed.) 2001 *Hydro-ecology: Linking Hydrology and Aquatic Ecology*, IAHS Publication no. 266, Wallingford.

Agam, N and PR Berliner 2006 'Dew formation and water vapor adsorption in semi-arid environments – a review', *Journal of Arid Environments* **65**, 572–590.

Aguiar, MR and OE Sala 1999 'Patch structure, dynamics and implications for the functioning of arid ecosystems', *Trends in Ecosystems and Evolution* **14**, 273–277.

Allison, HE and RJ Hobbs 2004 'Resilience, adaptive capacity, and the "lock-in trap" of the Western Australian agricultural region', *Ecology and Society* **9(1)**, art. no. 3.

Archer, S, DS Schimel and EA Holland 1995 'Mechanisms of shrubland expansion: land use, climate or CO_2?', *Climatic Change* **29**, 91–99.

Arndt, SK, C Arampatsis, A Foetzki, X Li, F Zeng and X Zhang 2004 'Contrasting patterns of leaf solute accumulation and salt adaptation in four phreatophytic desert plants in a hyper-arid desert with saline groundwater', *Journal of Arid Environments* **59**, 259–270.

Ash, JE and RJ Wasson 1983 'Vegetation and sand mobility in the Australian desert dunefield', *Zeitschrift für Geomorphologie Supplementband* **45**, 7–25.

Atchley, MC, AG de Soyza and WG Whitford 1999 'Arroyo water storage and soil nutrients and their effects on gas-exchange of shrub species in the northern Chihuahuan Desert', *Journal of Arid Environments* **43**, 21–33.

Bach, CE 1998 'Interactive effects of herbivory and sand burial on growth of a tropical dune plant, *Ipomoea pes-caprae*', *Ecological Entomology* **23**, 238–245.

Baez, S, J Fargione, DI Moore, SL Collins and JR Gosz 2007 'Atmospheric nitrogen deposition in the northern Chihuahuan desert: Temporal trends and potential consequences', *Journal of Arid Environments* **68**, 640–651.

Baird, AJ 1999 'Introduction', in AJ Baird and RL Wilby (eds) *Eco-hydrology: Plants and Water in Terrestrial and Aquatic Environments*, 1–10, Routledge, London.

Balsiger, PW 2004 'Supradisciplinary research practices: history, objectives and rationale', *Futures* **36**, 407–421.

Barbier, N, P Couteron, J Lejoly, V Deblauwe and O Lejeune 2006 'Self-organized vegetation patterning as a fingerprint of climate and human impact on semi-arid ecosystems', *Journal of Ecology* **94**, 537–547.

Barker, DH, C Vanier, E Naumburg, TN Charlet, KM Nielsen, BA Newingham and SD Smith 2006 'Enhanced monsoon precipitation and nitrogen deposition affect leaf traits and photosynthesis differently in spring and summer in the desert shrub *Larrea tridentata*', *New Phytologist* **169**, 799–808.

Barrow JR and RE Aaltonen 2001 'Evaluation of the internal colonization of *Atriplex canescens* (Pursh) Nutt. roots by dark septate fungi and the influence of host physiological activity', *Mycorrhiza* **11**, 199–205.

Barrow, JR, KM Havstad and BD McCaslin 1997 'Fungal root endophytes in fourwing saltbush, *Atriplex canescens*, on arid rangelands of southwestern USA', *Arid Soil Research and Rehabilitation* **11**, 177–185.

Barrow J, M Lucero, I Reyes-Vera and K Havstad 2007 'Endosymbiotic fungi structurally integrated with leaves reveals a lichenous condition of C4 grasses', *In Vitro Cellular and Developmental Biology – Plant* **43**, 65–70.

Bell, C, N McIntyre, S Cox, D Tissue and J Zak 2008 'Soil microbial responses to temporal variations of moisture and temperature in a Chihuahuan Desert grassland', *Microbial Ecology* DOI 10.1007/s00248-007-9333-z.

Belnap, J 2003 'The world at your feet: desert biological soil crusts', *Frontiers in Ecology and Environment* **1**, 181–189.

Belnap, J 2006 'The potential roles of biological soil crusts in dryland hydrologic cycles', *Hydrological Processes* **20**, 3159–3178.

Belnap, J and D Eldridge 2001 'Disturbance and recovery of biological soil crusts', in J Belnap and OL Lange (eds) *Biological Soil Crusts: Structure, Function and Management*, 363–383, Springer-Verlag, Berlin.

Belnap, J, JR Welter, NB Grimm, N Barger and JA Ludwig 2005 'Linkages between microbial and hydrologic processes in arid and semiarid watersheds', *Ecology* **86**, 298–307.

Belsky, AJ, RG Amundson, JM Duxbry, SJ Riha, AR Ali and SM Mwonga 1989 'The effects of tree on their physical, chemical, and biological environments in a semi-arid savanna in Kenya', *Journal of Soil Science* **37**, 345–350.

Biot, Y 1990 'The use of tree mounds as benchmarks of previous land surfaces in a semi-arid tree savanna, Botswana, in JB Thornes (ed.) *Vegetation and Erosion*, 437–450, John Wiley and Sons, Chichester.

Blumberg, DG and R Greeley 1993 'Field studies of aerodynamic roughness length', *Journal of Arid Environments* **25**, 39–48.

Boer, M and J Puigdefábregas 2005 'Effects of spatially structured vegetation patterns on hillslope erosion in a semiarid Mediterranean environment: a simulation study', *Earth Surface Processes and Landforms* **30**,149–167.

Bonell, M 2002 'Ecohydrology – a completely new idea?', *Hydrological Sciences–Journal–des Sciences Hydrologiques* **47**, 809–810.

Bracken LJ and J Croke 2007 'The concept of hydrological connectivity and its contribution to understanding runoff-dominated geomorphic systems', *Hydrological Processes* **21**, 1749–1763.

Bracken, LJ and J Wainwright 2006 'Geomorphological equilibrium: myth and metaphor?', *Transactions of the Institute of British Geographers NS* **31**, 167–178.

Brandt, CJ 1989 'The size distribution of throughfall drops under vegetation canopies', *Catena* **16**, 507–524.

Breshears, DD, JW Nyhan, CE Heil and BP Wilcox 1998 'Effects of woody plants on microclimate in a semiarid woodland: soil temperature and evaporation in canopy and intercanopy patches', *International Journal of Plant Sciences* **159**, 1010–1017.

Brisson, J and JF Reynolds 1994 'The effect of neighbors on root distribution in a creosotebush (*Larrea tridentata*) population', *Ecology* **75**, 1693–1702.

Brown, JF 1997 'Effects of experimental burial on survival, growth, and resource allocation of three species of dune plants', *Journal of Ecology* **85**, 151–158.

Brownell, PF and CJ Crossland 1972 'Requirement for sodium as a micronutrient by species having C4 dicarboxylic photosynthetic pathway', *Plant Physiology* **49**, 794–797

Bryan, K 1928 'Historic evidence on changes in the channel of the Rio Puerco, a tributary of the Rio Grande in New Mexico', *Journal of Geology* **36**, 265–282.

Buffington, LC and CH Herbel 1965 'Vegetational changes on a semidesert grassland range from 1858 to 1963', *Ecological Monographs* **35**,139–164.

Bull, WB 1997 'Discontinuous ephemeral streams', *Geomorphology* **19**, 227–276.

Bullard, JE 1997 'Vegetation and dryland geomorphology', in DSG Thomas (ed.) *Arid Zone Geomorphology: Process, Form and Change in Drylands (2nd ed.)*, 109–131, John Wiley and Sons, Chichester.

Buxbaum, CAZ and K Vanderbilt 2007 'Soil heterogeneity and the distribution of desert and steppe plant species across a desert-grassland ecotone', *Journal of Arid Environments* **69**, 617–632.

Cable, JM and TE Huxman 2004 'Precipitation pulse size effects on Sonoran Desert soil microbial crusts', *Oecologia* **141**, 317–324.

Caird, MA, JH Richards and LA Donovan 2007 'Nighttime stomatal conductance and transpiration in C3 and C4 plants', *Plant Physiology* **143**, 4–10.

Campos-Enriquez, JO, J Ortega-Ramírez, D Alatriste-Vilchis, R Cruz-Gática and E Cabral-Cano 1999 'Relationship between extensional tectonic style and the paleoclimatic elements at Laguna El Fresnal, Chihuahua Desert, Mexico', *Geomorphology* **28**, 75–94.

Carson, MA and MJ Kirkby 1972 *Hillslope Form and Process.* Cambridge University Press, Cambridge.

Charley, JL and NE West 1975 'Plant-induced soil chemical patterns in some shrub-dominated semi-desert ecosystems of Utah', *Journal of Ecology* **63**, 945–963.

Charney, JG 1975 'Dynamics of deserts and drought in Sahel', *Quarterly Journal of the Royal Meteorological Society* **101**, 193–202.

Chorley, RJ and BA Kennedy 1971 *Physical Geography: A Systems Approach.* Prentice-Hall, London.

Clark, CM and D Tilman 2008 'Loss of plant species after chronic low-level nitrogen deposition to prairie grasslands', *Nature* **451**, 712–715.

Claussen, M 1997 'Modelling bio-geophysical feedback in the African and Indian monsoon region', *Climate Dynamics* **13**, 247–257.

Cooke, RU and RW Reeves 1976 *Arroyos and Environmental Change in the American South-West.* Clarendon Press, Oxford.

Cooke, RU, A Warren and A Goudie 1993 *Desert Geomorphology.* UCL Press, London

Crawley, DM and WG Nickling 2003 'Drag partition for regularly-arrayed rough surfaces', *Boundary-Layer Meteorology* **107**, 445–468.

Cunningham, GL and JH Burk 1973 'The effect of carbonate deposition layers ("caliche") on the water status of *Larrea divaricata*', *American Midland Naturalist* **90**, 474–480.

De Soyza, AG, KT Killingbeck and WG Whitford 2004 'Plant water relations and photosynthesis during and after drought in a Chihuahuan desert arroyo', *Journal of Arid Environments* **59**, 27–39.

DeAngelis DL and JC Waterhouse 1987 'Equilibrium and nonequilibrium concepts in ecological models', *Ecological Monographs* **57**, 1–21

Dekker, SC, M Rietkirk and MFP Bierkens 2007 'Coupling microscale vegetation-soil water and macroscale vegetation-precipitation feedbacks in semiarid systems', *Global Change Biology* **13**, 671–678.

deMenocal, P, J Ortiz, T Guilderson, J Adkins, M Sarnthein, L Baker and M Yarusinsky 2000 'Abrupt onset and termination of the African Humid Period: rapid climate responses to gradual insolation forcing', *Quaternary Science Reviews* **19**, 347–361.

Descroix L, D Viramontes, M Vauclin, JLG Barrios and M Esteves 2001 'Influence of soil surface features and vegetation on runoff and erosion in the Western Sierra Madre (Durango, Northwest Mexico)', *Catena* **43**, 115–135.

Devitt, DA and SD Smith 2002 'Root channel macropores enhance downward movement of water in a Mojave Desert ecosystem', *Journal of Arid Environments* **50**, 99–108.

Dill, HG, S Khishigsuren, Y Majigsuren, S Myagniarsuren and J Bulgamaa 2006 'Geomorphological studies along a transect from the taiga to the desert in Central Mongolia – evolution of landforms in the mid-latitude continental interior as a function of climate and vegetation', *Journal of Asian Earth Sciences* **27**, 241–264.

D'Odorico, P and A Porporato 2006 'Ecohydrology of arid and semiarid ecosystems: an introduction', in P D'Odorico and A Porporato (eds) *Dryland Ecohydrology*, 1–10, Springer, Berlin.

Dougill, AJ and AD Thomas 2002 'Nebkha dunes in the Molopo Basin, South Africa and Botswana: formation controls and their validity as indicators of soil degradation', *Journal of Arid Environments* **50**, 413–428.

Drennan, PM and PS Nobel 2000 'Responses of CAM species to increasing atmospheric CO_2 concentrations', *Plant Cell and Environment* **23**, 767–781.

Drewa, PB and KM Havstad 2001 'Effects of fire, grazing, and the presence of shrubs on Chihuahuan desert grasslands', *Journal of Arid Environments* **48**, 429–443.

Duniway, MC, JE Herrick and HC Monger 2007 'The high water-holding capacity of petrocalcic horizons', *Soil Science Society of America Journal* **71**, 812–819.

Dunkerley, DL 1997 'Banded vegetation: survival under drought and grazing pressure based on a simple cellular automaton model', *Journal of Arid Environments* **35**, 419–428.

Dunkerley, DL and KJ Brown 1995 'Runoff and runon areas in a patterned chenopod shrubland, arid western New South Wales, Australia: characteristics and origin', *Journal of Arid Environments* **30**, 41–55.

Dunkerley, DL and KJ Brown 2002 'Oblique vegetation banding in the Australian arid zone: implications for theories of pattern evolution and maintenance', *Journal of Arid Environments* **51**, 163–181.

Eagleson, PS 2002 *Ecohydrology: Darwinian Expression of Vegetation Form and Function.* Cambridge University Press, Cambridge.

Ehleringer, JR 1980 'Leaf morphology and reflectance in relation to water and temperature stress', in NC Turner and PJ Kramer (eds) *Adaptations of Plants to Water and High Temperature Stress*, 295–308, John Wiley and Sons, Chichester.

Ehleringer, JR and IN Forseth 1989 'Diurnal leaf movements and productivity in canopies', in G Russel, B Marshall and PG Marshall (eds) *Plant Canopies: Their Growth, Form and Function*, 129–142, Cambridge University Press, Cambridge.

Ehleringer, J, O Björkman and HA Mooney 1976 'Leaf pubescence: effects on absorptance and photosynthesis in a desert shrub', *Science* **192**, 376–377.

Ehleringer, JR, TE Cerling and BR Helliker 1997 'C_4 photosynthesis, atmospheric CO_2 and climate', *Oecologia* **112**, 285–299.

El Bana, MI, I Nijs and AHA Khedr 2003 'The importance of phytogenic mounds (nebkhas) for restoration of arid degraded rangelands in northern Sinai', *Restoration Ecology* **11**, 317–324.

Entekhabi, D, I Rodriguez-Iturbe and RL Bras 1992 'Variability in large-scale water-balance with land surface atmosphere interaction', *Journal of Climate* **5**, 798–813.

Escoto-Rodríguez, M and SH Bullock 2002 'Long-term growth rates of cirio (*Fouquieria columnaris*), a giant succulent of the Sonoran Desert in Baja California', *Journal of Arid Environments* **50**, 593–611.

Esteban, J and V Fairén 2006 'Self-organized formation of banded vegetation patterns in semi-arid regions: a model', *Ecological Complexity* **3**, 109–118.

Fitter, AH and RKM Hay 1987 *Environmental Physiology of Plants (2nd ed.).* Academic Press, London.

Foster, D, F Swanson, J Aber, I Burke, N Brokaw, D Tilman and A Knapp 2003 'The importance of land-use legacies to ecology and conservation', *BioScience* **53**, 77–88.

Francis, CF 1994 'Plants on desert hillslopes', in AD Abrahams and AJ Parsons (eds) *Geomorphology of Desert Environments (1st ed.)*, 243–254, Chapman and Hall, London.

Franco, AC and PS Nobel 1989 'Effect of nurse plants on the microhabitat and growth of cacti', *Journal of Ecology* **77**, 870–886.

Garcia-Moya, E and CM McKell 1970 'Contribution of shrubs to the nitrogen economy of a desert-wash plant community', *Ecology* **51**, 81–88.

Gibbens, RP and JM Lenz 2001 'Root systems of some Chihuahuan Desert plants', *Journal of Arid Environments* **49**, 221–263.

Gibbens, RP, JM Tromble, JL Hennessy and M Cardenas 1983 'Soil movement in mesquite dunelands and former grasslands of southern New Mexico from 1933 to 1980', *Journal of Range Management* **36**, 145–148.

Gibbens RP, RP McNeely, KM Havstad, RF Beck and B Nolen 2005 'Vegetation changes in the Jornada Basin from 1858 to 1998', *Journal of Arid Environments* **61**, 651–668.

Gill, RA, HW Polley, HB Johnson, LJ Anderson, H Maherali and RB Jackson 2002 'Nonlinear grassland responses to past and future atmospheric CO_2', *Nature* **417**, 279–282.

Gillette, DA, J Adams, E Endo and D Smith 1980 'Threshold velocities for input of soil particles into the air by desert soils', *Journal of Geophysical Research* **85(C10)**, 5621–5630.

Gillette, DA and PH Stockton 1989 'The effect of nonerodible particles on wind erosion of erodible surfaces', *Journal of Geophysical Research–Atmospheres* **94**, 12885–12893.

Gillette, DA, JE Herrick and GA Herbert 2006 'Wind characteristics of mesquite streets in the Northern Chihuahuan Desert, New Mexico, USA', *Environmental Fluid Mechanics* **6**, 241–275.

Gillson, L 2004 'Testing non-equilibrium theories in savannas: 1400 years of vegetation change in Tsavo National Park, Kenya', *Ecological Complexity* **1**, 281–298.

Goudie, AS 2002 *Great Warm Deserts of the World: Landscape and Evolution*. Oxford University Press, Oxford.

Graf, WL 1981 'Channel instability in a sand-river bed', *Water Resources Research* **17**, 1087–1094.

Graf, WL 1983a 'Flood-related channel change in an arid-region river', *Earth Surface Processes and Landforms* **8**, 125–139.

Graf, WL 1983b 'Downstream changes in stream power in the Henry Mountains, Utah', *Annals of the Association of American Geographers* **73**, 373–387.

Graf, WL 1983c 'Variability of sediment removal in a semi-arid watershed', *Water Resources Research* **19**, 643–652.

Graf, WL 1988 'Applications of catastrophe theory in fluvial geomorphology', in MG Anderson (ed) *Modelling Geomorphological Systems*, 33–48, John Wiley and Sons, Chichester.

Grantz DA, JHB Garner and DW Johnson 2003 'Ecological effects of particulate matter', *Environment International* **29**, 213–239.

Greenway, H and R Munns 1980 'Mechanisms of salt tolerance in nonhalophytes', *Annual Review of Plant Physiology* **31**, 149–190.

Gregory, KJ 2000 *The Changing Nature of Physical Geography*. Arnold, London.

Grof, CPL, M Johnston and PF Brownell 1989 'Effect of sodium nutrition on the ultrastructure of chloroplasts of C_4 plants', *Plant Physiology* **89**, 539–543.

Gul, B, DJ Weber and MA Khan 2001 'Growth, ionic and osmotic relations of an *Allenrolfea occidentalis* population in an inland salt playa of the Great Basin Desert', *Journal of Arid Environments* **48**, 445–460.

Haase, P 2001 'Can isotropy vs. anisotropy in the spatial association of plant species reveal physical vs. biotic facilitation?', *Journal of Vegetation Science* **12**, 127–136.

Haase, P, FI Pugnaire, SC Clark and LD Incoll 1997 'Spatial pattern in *Anthyllis cytisoides* shrubland on abandoned land in southeastern Spain', *Journal of Vegetation Science* **8**, 627–634.

Hadley, RF 1961 'Influence of riparian vegetation on channel shape, nort-eastern Arizona', *US Geological Survey Professional Paper* **424-C**, 30–31, Washington DC.

Hall, SA 1977 'Late Quaternary sedimentation and paleoecologic history of Chaco Canyon, New Mexico', *Geological Society of America Bulletin* **88**, 1593–1618.

Hamerlynck, EP, JR McAuliffe, EV McDonald and SD Smith 2002 'Ecological responses of two Mojave desert shrubs to soil horizon development and soil water dynamics', *Ecology* **83**, 768–779.

Hanson, JD, BB Baker and RM Bourdon 1993 'Comparison of the effects of different climate change scenarios on rangeland livestock production', *Agricultural Systems* **41**, 487–502.

Hatton, TJ, GD Salvucci and HI Wu 1997 'Eagleson's optimality theory of an ecohydrological equilibrium: *quo vadis?*', *Functional Ecology* **11**, 665–674.

HilleRisLambers, R., M Rietkerk, F van den Bosch, HHT Prins and H de Croon 2001 'Vegetation pattern formation in semi-arid grazing systems', *Ecology* **82**, 50–61.

Holling, CS 1973 'Resilience and stability of ecological systems', *Annual Review of Ecology and Systematics* **4**, 1–23.

Hooper, DU and L Johnson 1999 'Nitrogen limitation in dryland ecosystems: responses to geographical and temporal variation in precipitation', *Biogeochemistry* **46**, 247–293.

Huenneke, LF and WH Schlesinger 2006 'Patterns of net primary production in Chihuahuan Desert ecosystems', in KM Havstad, WH Schlesinger and LF Huenneke (eds) *Structure and Function of a Chihuahuan Desert Ecosystem: The Jornada LTER*, 232–246, Oxford University Press, Oxford.

Humphrey, RR 1958 'The desert grassland: a history of vegetational change and an analysis of causes', *Botanical Review* **24**, 164–193.

Huntington, E 1914 *The Climatic Factor as Illustrated in Arid America*. Carnegie Institute of Washington Publication n°. 192, Washington DC.

Hyder PW, EL Fredrickson, RE Estell, M Tellez and RP Gibbens 2002 'Distribution and concentration of total phenolics, condensed tannins, and nordihydroguaiaretic acid (NDGA) in creosotebush (*Larrea tridentata*)', *Biochemical Systematics and Ecology* **30**, 905–912.

Jackson, RD and SB Idso 1975 'Surface albedo and desertification', *Science* **189**, 1012–1013.

Judd, MJ, MR Raupach and JJ Finnigan 1996 'A wind tunnel study of turbulent flow around single and multiple windbreaks. Part I: velocity fields', *Boundary Layer Meteorology* **80**, 127–165.

Kerkhoff, AJ, SN Martens and BT Milne 2004 'An ecological evaluation of Eagleson's optimality hypotheses', *Functional Ecology* **18**, 404–413.

Kidron, GJ 2000 'Analysis of dew precipitation in three habitats within a small arid drainage basin, Negev Highlands, Israel', *Atmospheric Research* **55**, 257–270.

Kidron, GJ 2005 'Angle and aspect dependent dew and fog precipitation in the Negev desert' *Journal of Hydrology* **301**, 66–74.

Killingbeck, KT and WG Whitford 2001 'Nutrient resorption in shrubs growing by design, and by default in Chihuahuan Desert arroyos', *Oecologia* **128**, 351–359.

Klausmeier, CA 1999 'Regular and irregular patterns in semiarid vegetation', *Science* **284**, 1826–1828.

Knipe, D and CH Herbel 1966 'Germination and growth of some semidesert grassland species treated with aqueous extract from creosotebush', *Ecology* **47**, pp. 775–781.

Körner, C 2006 'Plant CO_2 responses: an issue of definition, time and resource supply', *New Phytologist* **172**, 393–411.

Kundzewicz, ZW 2002 'Ecohydrology – seeking consensus on interpretation of the notion', *Hydrological Sciences–Journal–des Sciences Hydrologiques* **47**, 799–804.

Lancaster, N and A Baas 1998 'Influence of vegetation cover on sand transport by wind: Field studies at Owens Lake, California', *Earth Surface Processes and Landforms* **23**, 69–82.

Lange, OL, GJ Kidron, B Budel, A Meyer, E Kilian and A Abeliovich 1992 'Taxonomic composition and photosynthetic characteristics of the "biological soil crusts" covering sand dunes in the Western Negev Desert', *Functional Ecology* **6**, 519–527.

Lange, OL, A Meyer, H Zellner and U Heber 1994 'Photosynthesis and water relations of lichen soil crusts: field measurements in the coastal fog zone of the Namib Desert', *Functional Ecology* **8**, 253–264.

Lange, OL, TGA Green, B Melzer, A Meyer and H Zellner 2006 'Water relations and CO_2 exchange of the terrestrial lichen *Teloschistes capensis* in the Namib fog desert: measurements during two seasons in the field and under controlled conditions', *Flora – Morphology, Distribution, Functional Ecology of Plants* **201**, 268–280.

Lange, OL, TGA Green, A Meyer and H Zellner 2007 'Water relations and carbon dioxide exchange of epiphytic lichens in the Namib fog desert', *Flora – Morphology, Distribution, Functional Ecology of Plants* **202**, 479–487.

Langford, RP 2000 'Nabkha (coppice dune) fields of south-central New Mexico, U.S.A.', *Journal of Arid Environments* **46**, 25–41.

Larcher, W 1995 *Physiological Plant Ecology. Ecophysiology and Stress Physiology of Functional Groups (3rd ed.).* Springer, Berlin.

Lee, BE and BF Soliman 1977 'An investigation of the forces on three dimensional bluff bodies in rough wall turbulent boundary layers', *Transactions of the ASME, Journal of Fluids Engineering* **99**, 503–510.

Leenders, JK, JH van Boxel and G Sterk 2007 'The effect of single vegetation elements on wind speed and sediment transport in the Sahelian Zone of Burkina Faso', *Earth Surface Processes and Landforms* **32**, 1454–1474.

Lefever, R and O Lejeune 1997 'On the origin of tiger bush', *Bulletin of Mathematical Biology* **59**, 263–294.

Lejeune, O and M Tlidi 1999 'A model for the explanation of vegetation strips (tiger bush)', *Journal of Vegetation Science* **10**, 201–208.

Lejeune, O, P Couteron and R Lefever 1999 'Short range co-operativity competing with long range inhibition explains vegetation patterns', *Acta Oecologica* **20**, 171–183.

Liu, Z, Y Wang, R Gallimore, F Gasse, T Johnson, P deMenocal, J Adkins, M Notaro, IC Prentice, J Kutzbach, R Jacob, P Behling, L Wang and E Ong 2007 'Simulating the transient evolution and abrupt change of Northern Africa atmosphere–ocean–terrestrial ecosystem in the Holocene', *Quaternary Science Reviews* **26**, 1818–1837.

Lockwood JA and DR Lockwood 1993 'Catastrophe theory: a unified paradigm for rangeland ecosystem dynamics', *Journal of Range Management* **46**, 282–288.

Loik, ME 2006 'Sensitivity of water relations and photosynthesis to summer precipitation pulses for *Artemisia tridentata* and *Purshia tridentata*', *Plant Ecology* **191**, 95–108.

Loik, ME and SP Redar 2003 'Microclimate, freezing tolerance, and cold acclimation along an elevation gradient for seedlings of the Great Basin Desert shrub, *Artemisia tridentata*', *Journal of Arid Environments* **54**, 769–782.

Loik, ME, DD Breshears, WK Lauenroth and J Belnap 2004 'A multi-scale perspective of water pulses in dryland ecosystems: climatology and ecohydrology of the western USA', *Oecologia* **141**, 269–281.

Long, SP 1999 'Environmental responses', in RF Sage and RK Monson (eds) C_4 *Plant Biology*, 215–250, Academic Press, London.

Ludwig, JA, BP Wilcox, DD Breshears, DJ Tongway and AC Imeson 2005 'Vegetation patches and runoff–erosion as interacting ecohydrological processes in semiarid landscapes', *Ecology* **86**, 288–297.

MacMynowski, DP 2007 'Pausing at the brink of interdisciplinarity: power and knowledge at the meeting of social and biophysical science', *Ecology and Society* **12(1)**, paper 20.

McAuliffe, JR 1994 'Landscape evolution, soil formation, and ecological patterns and processes in Sonoran Desert bajadas', *Ecological Monographs* **64**, 111–148.

McAuliffe, JR 1995 'Landscape evolution, soil formation, and Arizona's desert grasslands', in MP McClaran and TR van Devender (eds) *The Desert Grassland*, 100–129, University of Arizona Press, Tucson, AZ.

McFadden LD, SG Wells and MJ Jercinovich 1987 'Influences of eolian and pedogenic processes on the origin and evolution of desert pavements', *Geology* **15**, 504–508.

McPherson, GR 1995 'The role of fire in the desert grasslands', in MP McLaran and TR Van Devender (eds) *The Desert Grassland*, 130–151, University of Arizona Press, Tucson.

MacKinnon, DJ, GD Clow, RK Tigges, RL Reynolds and PS Chavez Jr 2004 'Comparison of aerodynamically and model-derived roughness lengths (z_o) over diverse surfaces, central Mojave Desert, California, USA', *Geomorphology* **63**, 103–113.

Maestre, FT, J Cortina, S Bautista and J Bellot 2003 'Does Pinus halepensis facilitate the establishment of shrubs in Mediterranean semi-arid afforestations?' *Forest Ecology and Management* **176**, 147–160.

Malamud, BD, JDA Millington and GLW Perry 2005 'Characterizing wildfire regimes in the United States', *Proceedings of the National Academy of Sciences of the United States of America* **102**, 4694–4699.

Malek, E, G McCurdy and B Giles 1999 'Dew contribution to the annual water balances in semi-arid desert valleys', *Journal of Arid Environments* **42**, 71–80.

Marshall, JK 1971 'Drag measurements in roughness arrays of varying density and distribution', *Agricultural Meteorology* **8**, 269–292.

Martinez Meza, E and WG Whitford 1996 'Stemflow, throughfall and channelization of stemflow by roots in three Chihuahuan desert shrubs', *Journal of Arid Environments* **32**, 271–287.

Maun, MA 1994 'Adaptations enhancing survival and establishment of seedlings on coastal dune systems', *Vegetatio* **111**, 59–70.

May, RM 1975 'Deterministic models with chaotic dynamics', *Nature* **256**, 165–166.

Melton, MA 1965 'The geomorphic and paleoclimatic significance of alluvial deposits in southern Arizona', *Journal of Geology* **73**, 1–38.

Millington, AC and K Pye (eds) 1994 *Effects of Environmental Change on Drylands*, John Wiley and Sons, Chichester.

Montaña, C 1992 'The colonization of bare areas in two-phase mosaics of an arid ecosystem', *Journal of Ecology* **80**, 315–327.

Morgan, JA, DE Pataki, C Körner, H Clark, SJ Grosso, JM Grünzweig, AK Knapp, AR Mosier, PCD Newton, PA Niklaus, JB Nippert, RS Nowak, WJ Parton, HW Polley and MR Shaw 2004 'Water relations in grassland and desert ecosystems exposed to elevated atmospheric CO_2', *Oecologia* **140**, 11–25.

Moro, MJ FI Pugnaire, P Haase and J Puigdefábregas 1997 'Mechanisms of interaction between a leguminous shrub and its understorey in a semi-arid environment', *Ecography* **20**, 175–184.

Morris, HM 1955 'Flow in rough conduits', *Transactions of the American Society of Agricultural Engineers* **120**, 373–398.

Müller, EN 2004 *Scaling Approaches to the Modelling of Water, Sediment and Nutrient Flows within Semi-Arid Landscapes, Jornada Basin, New Mexico*. PhD Thesis, University of London.

Müller, EN, J Wainwright and AJ Parsons 2008 'Spatial variability of soil and nutrient characteristics of semi-arid grasslands and shrublands, Jornada Basin, New Mexico', *Ecohydrology* **1**, 3–12.

Muller, CH 1966 'The role of chemical inhibition (allelopathy) in vegetational composition', *Bulletin of the Torrey Botanical Club* **93**, 332–351.

Muller, WH 1965 'Volatile materials produced by *Salvia leucophylla*: effects on seedling growth and soil bacteria', *Botanical Gazette* **126**, 195–200.

Naumburg, E, R Mata-Gonzalez, RG Hunter, T Mclendon and DW Martin 2005 'Phreatophytic vegetation and groundwater fluctuations: A review of current research and application of ecosystem response modeling with an emphasis on Great Basin vegetation', *Environmental Management* **35**, 726–740.

Neave M and AD Abrahams 2001 'Impact of small mammal disturbances on sediment yield from grassland and shrubland ecosystems in the Chihuahuan Desert', *Catena* **44**, 285–303.

Neave, M and S Rayburg 2007 'A field investigation into the effects of progressive rainfall-induced soil seal and crust development on runoff and erosion rates: the impact of surface cover', *Geomorphology* **87**, 378–390.

Newman, BD, BP Wilcox, SR Archer, DD Breshears, CN Dahm, CJ Duffy, NG McDowell, FM Phillips, BR Scanlon and ER Vivoni 2006 'Ecology of water-limited environments: a scientific vision', *Water Resources Research* **42**, W06302, doi: 10.1029/2005WR004141.

Nickling, WG and SA Wolfe 1994 'The morphology and origin of nabkha, region of Mopti, Mali, West Africa', *Journal of Arid Environments* **28**, 13–30.

Nobel, PS 1980 'Influences of minimum stem temperatures on ranges of cacti in southwestern United States and central Chile', *Oecologia* **47**, 10–15.

Novoplansky, A and D Goldberg 2001 'Interactions between neighbour environments and drought resistance', *Journal of Arid Environments* **47**, 11–32.

Noy-Meir, I 1973 'Desert ecosystems: environment and producers', *Annual Review of Ecology and Systematics* **4**, 25–41.

Noy-Meir, I 1974 'Desert ecosystems: higher trophic levels', *Annual Review of Ecology and Systematics* **5**, 195–214.

Nulsen, RA, KJ Bligh, IN Baxter, EJ Solin and DH Imrie 1986 'The fate of rainfall in a malle and heath vegetated catchment in southern Western Australia', *Australian Journal of Ecology* **11**, 361–371.

Nuttle, WK 2002 'Is ecohydrology one idea or many?', *Hydrological Sciences–Journal–des Sciences Hydrologiques* **47**, 805–807.

Okin, GS and DA Gillette 2001 'Distribution of vegetation in wind-dominated landscapes: Implications for wind erosion modeling and landscape processes', *Journal of Geophysical Research-Atmospheres* **106**, 9673–9683.

Okin GS, DA Gillette and JE Herrick 2006 'Multi-scale controls on and consequences of aeolian processes in landscape change in arid and semi-arid environments', *Journal of Arid Environments* **65**, 253–275.

Okin, GS, AJ Parsons, J Wainwright, JE Herrick, BT Bestelmeyer, DPC Peters and EL Fredrickson *submitted* 'Does connectivity explain desertification?', *BioScience*.

Oksanen, T 1990 'Exploitation systems in heterogeneous habitat complexes', *Evolutionary Ecology* **4**, 220–234.

Osborne, CP and DJ Beerling 2006 'Nature's green revolution: the remarkable evolutionary rise of C_4 plants', *Philosophical Transactions of the Royal Society B* **361**, 173–194.

Owens, MK, RK Lyons and CL Alejandro 2006 'Rainfall partitioning within semiarid juniper communities: effects of event size and canopy cover', *Hydrological Processes* **20**, 3179–3189.

Parsons, AJ and J Wainwright 2006 'Depth distribution of interrill overland flow and the formation of rills', *Hydrological Processes* **20**, 1511–1523.

Parsons, AJ, AD Abrahams and JR Simanton 1992 'Microtopography and soil-surface materials on semi-arid piedmont hillslopes, southern Arizona', *Journal of Arid Environments* **22**, 107–115.

Parsons, AJ, AD Abrahams and J Wainwright 1994 'On determining resistance to interrill overland flow', *Water Resources Research* **30**, 3515–3521.

Parsons, AJ, J Wainwright, AD Abrahams and JR Simanton 1997 'Distributed dynamic modelling of interrill overland flow', *Hydrological Processes* **11**, 1833–1859.

Parsons, AJ, J Wainwright, WH Schlesinger and AD Abrahams 2003 'Sediment and nutrient transport by overland flow in mesquite nabkha, southern New Mexico', *Journal of Arid Environments* **53**, 61–71.

Pataki, DE, SE Bush, P Gardner, DK Solomon and JR Ehleringer 2005 'Ecohydrology in a Colorado River riparian forest: Implications for the decline of *Populus fremontii*', *Ecological Applications* **15**, 1009–1018.

Pease, PP and VP Tchakerian 2002 'Composition and sources of sand in the Wahiba Sand Sea, Sultanate of Oman', *Annals of the Association of American Geographers* **92**, 416–434.

Pelletier, JD, M Cline and SB DeLong 2007 'Desert pavement dynamics: numerical modeling and field-based calibration', *Earth Surface Processes and Landforms* **32**, 1913–1927.

Perry, GLW 2002 'Landscapes, space and equilibrium: shifting viewpoints', *Progress in Physical Geography* **26**, 339–359.

Phillips, DL and JA MacMahon 1978 'Gradient analysis of a Sonoran Desert bajada', *Southwestern Naturalist* **23**, 669–680.

Pockman, WT and JS Sperry 1997 'Freezing-induced xylem cavitation and the northern limit of *Larrea tridentata*', *Oecologia* **109**, 19–27.

Pollen, N and A Simon 2005 'Estimating the mechanical effects of riparian vegetation on stream bank stability using a fiber bundle model', *Water Resources Research* **41**, W07025, doi:10.1029/2004WR003801.

Polley, HW, HB Johnson and JD Derner 2002 'Soil- and plant-water dynamics in a C_3/C_4 grassland exposed to a subambient to superambient CO_2 gradient', *Global Change Biology* **8**,1118–1129.

Polley, HW, HB Johnson and CR Tischler 2003 'Woody invasion of grasslands: evidence that CO_2 enrichment indirectly promotes establishment of Prosopis glandulosa', *Plant Ecology* **164**, 85–94.

Pressland, AJ 1973 'Rainfall partitioning by an arid woodland (*Acacia aneura* F. Meull.) in south-western Queensland', *Australian Journal of Botany* **21**, 235–245.

Ramírez, DA, J Bellot, F Domingo and A Blasco 2007 'Can water responses in *Stipa tenacissima* L. during the summer season be promoted by non-rainfall water gains in soil?', *Plant and Soil* **291**, 67–79.

Rango, A, L Huenneke, M Buonopane, JE Herrick and KM Havstad 2005 'Using historic data to assess effectiveness of shrub removal in southern New Mexico', *Journal of Arid Environments* **62**, 75–91.

Rango, A, S Tartowski, A Laliberte, J Wainwright and AJ Parsons 2006 'Islands of hydrologically enhanced biotic productivity in natural and managed arid ecosystems', *Journal of Arid Environments* **65**, 235–252.

Reale, O and P Dirmeyer 2000 'Modeling the effects of vegetation on Mediterranean climate during the Roman Classical Period, part I: climate history and model sensitivity', *Global and Planetary Change* **25**, 163–184.

Reale, O and J Shukla 2000 'Modeling the effects of vegetation on Mediterranean climate during the Roman Classical Period, part II: model simulation', *Global and Planetary Change* 25, 185–214.

Rempel, PJ 1936 'The crescentic dunes of the Salton Sea and their relation to the vegetation', *Ecology* **17**, 347–358.

Reynolds, JF, RA Virginia, PR Kemp, AG de Soyza and DC Tremmel 1999 'Impact of drought on desert shrubs: effects of seasonality and degree of resource island development', *Ecological Monographs* **69**, 69–106.

Richards, JH and MM Caldwell 1987 'Hydraulic lift: substantial nocturnal water transport between soil layers by *Artemisia tridentata* roots', *Oecologia* **73**, 486–489.

Rietkerk, M, P Ketner, L Stroosnijder and HHT Prins 1996 'Sahelian rangeland development: a catastrophe?', *Journal of Range Management* **49**, 512–519.

Rietkerk M, MC Boerlijst, F van Langevelde, R HilleRisLambers, J van de Koppel, L Kumar, HHTPrins and AM de Roos 2002 'Self-organization of vegetation in arid ecosystems', *American Naturalist* **160**, 524–530.

Rodriguez, MV, MB Bertiller and A Bisigato 2007 'Are fine roots of both shrubs and perennial grasses able to occupy the upper soil layer? A case study in the arid Patagonian Monte with non-seasonal precipitation', *Plant and Soil* **300**, 281–288.

Rodriguez-Iturbe, I 2000 'Ecohydrology: a hydrologic perspective of climate-soil-vegetation dynamics', *Water Resources Research* **36**, 3–9.

Rostagno, CM and HF del Valle 1988 'Mounds associated with shrubs in aridic soils of northeastern Patagonia: characteristics and probable genesis', *Catena* **15**, 347–359.

Ryel, RJ, MM Caldwell, AJ Leffler and CK Yoder 2003 'Rapid soil moisture recharge to depth by roots in a stand of *Artemisia tridentata*', *Ecology* **84**, 757–764.

Ryel RJ, AJ Leffler, MS Peek, CY Ivans and MM Caldwell 2004 'Water conservation in *Artemisia tridentata* through redistribution of precipitation', *Oecologia* **141**, 335–345.

Sage, RF 2004 'The evolution of C_4 photosynthesis', *New Phytologist* **161**, 341–370.

Sandquist, DR and JR Ehleringer 1998 'Intraspecific variation of drought adaptation in brittlebush: leaf pubescence and timing of leaf loss vary with rainfall', *Oecologia* **113**, 162–169.

Sandvig, RM and FM Phillips 2006 'Ecohydrological controls on soil moisture fluxes in arid to semiarid vadose zones', *Water Resources Research* **42**, W08422, doi:10.1029/2005WR004644.

Saqqa, W and M Altallah 2004 'Characterization of the aeolian terrain facies in Wadi Araba Desert, southwestern Jordan', *Geomorphology* **62**, 63–87.

Sayed, OH 1996 'Adaptational responses of *Zygophyllum qatarense* Hadidi to stress conditions in a desert environment', *Journal of Arid Environments* **32**, 445–452.

Scanlon, BR, DG Levitt, RC Reedy, KE Keese and MJ Sully 2005a 'Ecological controls on water-cycle response to climate variability in deserts', *Proceedings of the National Academy of Sciences of the United States of America* **102**, 6033–6038.

Scanlon, BR, RC Reedy, DA Stonestrom, DE Prudic and KF Dennehy 2005b 'Impact of land use and land cover change on groundwater recharge and quality in the southwestern US', *Global Change Biology* **11**, 1577–1593.

Scanlon, TM, KK Caylor, S Manfread, SA Levin and I Rodriguez-Iturbe 2005 'Dyamic response of grass cover to rainfall variability: implications for the function and persistence of savanna ecosystems', *Advances in Water Resources* **28**, 291–302.

Scheffer, M and SR Carpenter 2003 'Catastrophic regime shifts in ecosystems: linking theory to observation', *TRENDS in Ecology and Evolution* **18**, 648–656.

Scheffer, M, S Carpenter, JA Foley, C Folke and B Walker 2001 'Catastrophic shifts in ecosystems', *Nature* **413**, 591–596.

Scheffer, M, M Holmgren, V Brovkin and M Claussen 2005 'Synergy between small- and large-scale feedbacks of vegetation on the water cycle', *Global Change Biology* **11**, 1003–1012.

Schlemon, R and RE Riefner 2006 'The role of tectonic processes in the interaction between geology and ecosystems', in IS Zektser, B Marker, J Ridgway, L Rogachevskaya and G Vartanyan (eds) *Geology and Ecosystems*, 49–60, Springer Verlag, Berlin.

Schlesinger, WH and AM Pilmanis 1998 'Plant-soil interactions in deserts', *Biogeochemistry* **42**, 169–187.

Schlesinger, WH, TJ Ward and J Anderson 2000 'Nutrient losses in runoff from grassland and shrubland habitats in southern New Mexico: II. Field plots', *Biogeochemistry* **49**, 69–86.

Schlesinger WH, JS Pippen, MD Wallenstein, KS Hofmockel, DM Klepeis and BE Mahall 2003 'Community composition and photosynthesis by photoautotrophs under quartz pebbles, southern Mojave Desert', *Ecology* **84**, 3222–3231.

Schlesinger, WH, JA Raikes, AE Hartley and AF Cross 1996 'On the spatial pattern of soil nutrients in desert ecosystems', *Ecology* **77**, 364–374.

Schlesinger, WH, JF Reynolds, GL Cunningham, LF Huenneke, WM Jarrell, RA Virginia and WG Whitford 1990 'Biological feedbacks in global desertification', *Science* **247**, 1043–1048.

Schlichting, H 1936 'Experimentelle Untersuchungen zum Rauhigkeitsproblem', *Ingenieur Archiv* **7**, 1–34.

Schumm, SA 1973 'Geomorphic thresholds and complex response of drainage systems', in M Morisawa (ed.) *Fluvial Geomorphology, Proceedings of the 4th Annual Geomorphology Symposia Series, Binghamton*, 299–311, Allen and Unwin, London.

Scoging, H, AJ Parsons and AD Abrahams 1992 'Application of a dynamic overland-flow hydraulic model to a semi-arid hillslope, Walnut Gulch, Arizona', in AJ Parsons and AD Abrahams (eds) *Overland Flow: Hydraulics and Erosion Mechanics*, 105–145, UCL Press, London.

Scoones, I, R Behnke and C Kerven (eds) 1993 *Range Ecology at Disequilibrium. New Models of Natural Variability and Pastoral Adaptation in African Savannas*, Overseas Development Institute, London.

Scott, RL, TE Huxman, DG Williams and DC Goodrich 2006 'Ecohydrological impacts of woody-plant encroachment: seasonal patterns of water and carbon dioxide exchange within a semiarid riparian environment', *Global Change Biology* **12**, 311–324.

Seyfried, MS and BP Wilcox 2006 'Soil water storage and rooting depth: key factors controlling recharge on rangelands', *Hydrological Processes* **20**, 3261–3275.

Seyfried, MS, S Schwinning, MA Walvoord, WT Pockman, BD Newman, RB Jackson and EM Phillips 2005 'Ecohydrological control of deep drainage in arid and semiarid regions', *Ecology* **86**, 277–287.

Shachak, M and GM Lovett 1998 'Atmospheric deposition to a desert ecosystem and its implications for management', *Ecological Applications* **8**, 455–463.

Shanyengana, ES, JR Henschel, MK Seely and RD Sanderson 2002 'Exploring fog as a supplementary water source in Namibia', *Atmospheric Research* **64**, 251–259.

Sherratt, JA and GJ Lord 2007 'Nonlinear dynamics and pattern bifurcations in a model for vegetation stripes in semi-arid environments', *Theoretical Population Biology* **71**, 1–11.

Simmons, IG 1980 'Biogeography', in EH Brown (ed.) *Geography Yesterday and Tomorrow*, 146–166, Oxford University Press, Oxford.

Smith, WK, DT Bell and KA Shepherd 1998 'Associations between leaf structure, orientation, and sunlight exposure in five Western Australian communities', *American Journal of Botany* **85**, 56–63.

Snyder, KA, JJ James, JH Richards and LA Donovan 2008 'Does hydraulic lift or nighttime transpiration facilitate nitrogen acquisition?', *Plant and Soil* (in press) doi: 10.1007/s11104-008-9567-7.

Stafford Smith, M 1996 'Management of rangelands: paradigms at their limits', in J Hodgson and AW Illius (eds) *The Ecology and Management of Grazing Systems*, 325–357, CAB International, Wallingford.

Stewart, J, AJ Parsons, J Wainwright, GS Okin, B Bestelmeyer, E Fredrickson and WH Schlesinger in press 'Modelling emergent patterns of dynamic desert ecosystems as a function of changing landscape connectivity: part one – theoretical framework', *Ecological Modelling*.

Stoddart, DR 1967 'Organism and ecosystem as geographical models', in RJ Chorley and P Haggett (eds) *Models in Geography*, 511–548, Methuen, London.

Stromberg, JC 1997 'Growth and survivorship of Fremont cottonwood, Goodding willow, and salt cedar seedlings after large floods in central Arizona', *Great Basin Naturalist* **57**:198–208.

Stromberg, JC 2001 'Restoration of riparian vegetation in the south-western United States: importance of flow regimes and fluvial dynamism', *Journal of Arid Environments* **49**:17–34.

Stromberg, JC, VB Beauchamp, MD Dixon, SJ Lite and C Paradzick 2007 'Importance of low-flow and high-flow characteristics to restoration of riparian vegetation along rivers in arid south-western United States', *Freshwater Biology* **52**, 651–679.

Stromberg, JC, R Tiller and B Richter 1996 'Effects of groundwater decline on riparian vegetation of semiarid regions: the San Pedro, Arizona', *Ecological Applications* **6**, 113–131.

Sullivan, S and R Rohde 2002 'On non-equilibrium in arid and semi-arid grazing systems', *Journal of Biogeography* **29**, 1595–1618.

Tengberg, A and D Chen 1998 'A comparative analysis of nebkhas in central Tunisia and northern Burkina Faso', *Geomorphology* **22**, 181–192.

Thomas, DSG 1997 'Arid environments: their nature and extent', in DSG Thomas (ed.) *Arid Zone Geomorphology: Process, Form and Change in Drylands*, 3–12, John Wiley and Sons, Chichester.

Thornes JB. 1980 'Structural instability and ephemeral channel behaviour', *Zeitschrift für Geomorphologie Supplementband* **36**, 233–244.

Thornes, JB 1985 'The ecology of erosion', *Geography* **70**, 222–236.

Thornes, JB 1988 'Erosional equilibria under grazing', in J Bintliff, D Davidson and E Grant (eds) *Conceptual Issues in Environmental Archaeology*, 193–210, Edinburgh University Press, Edinburgh.

Thornes, JB (ed.) 1990a *Vegetation and Erosion*. John Wiley and Sons, Chichester.

Thornes, JB 1990b 'The interaction of erosional and vegetational dynamics in land degradation: spatial outcomes', in JB Thornes (ed.) *Vegetation and Erosion*, 41–53, John Wiley and Sons, Chichester.

Tipple, BJ and M Pagani 2007 'The early origins of terrestrial C_4 photosynthesis', *Annual Review of Earth and Planetary Sciences* **35**, 435–461.

Tivy, J 1990 *Agricultural Ecology*. Longman, Harlow.

Tongway, DJ and JA Ludwig 1990 'Vegetation and soil patterning in semiarid mulga lands of Eastern Australia', *Australian Journal of Ecology* **15**, 23–34.

Tress, G, B Tress and G Fry 2007 'Analysis of the barriers to integration in landscape research projects', *Land Use Policy* **24**, 374–385.

Turnbull, L, RE Brazier, J Wainwright, E Dixon and R Bol 2008a 'Use of carbon isotope analysis to understand soil erosion dynamics and long-term semi-arid land degradation', *Rapid Communications in Mass Spectrometry* **22**, 1697–1702.

Turnbull, L, J Wainwright and RE Brazier 2008b 'A conceptual framework for understanding semi-arid land degradation: ecohydrological interactions across multiple-space and time scales', *Ecohydrology* **1**, 23–34, DOI: 10.1002/eco.4.

Tyree, MT 1999 'Water relations of plants', in AJ Baird and RL Wilby (eds) *Eco-hydrology*, 11–38, Routledge, London.

Venable, DL 2007 'Bet hedging in a guild of desert annuals', *Ecology* **88**, 1086–1090.

Viles, H (ed.) 1988 *Biogeomorphology*. Blackwell, Oxford.

Wainwright, J 2006a 'Climate and climatological variations', in KM Havstad, WH Schlesinger and LF Huenneke (eds) *Structure and Function of a Chihuahuan Desert Ecosystem: The Jornada LTER*, 44–80, Oxford University Press, Oxford.

Wainwright, J 2006b 'Degrees of separation: hillslope-channel coupling and the limits of palaeohydrological reconstruction', *Catena* **66**, 93–106.

Wainwright, J 2008 'Can modelling enable us to understand the rôle of humans in landscape evolution?', *Geoforum* **39**, 659–674, doi: 10.1016/j.geoforum.2006.09.011.

Wainwright, J 2009 'Earth-system science', in N Castree, D Liverman, B Rhoads and D Demeritt (eds) *Blackwell Companion to Environmental Geography*, 145–167, Blackwell, Oxford.

Wainwright, J and JDA Millington in press 'Mind, the gap in landscape-evolution modelling', *Earth Surface Processes and Landforms*.

Wainwright, J and JB Thornes 2003 *Environmental Issues in the Mediterranean: Processes and Perspectives from the Past and Present*. Routledge, London.

Wainwright, J, M Mulligan and JB Thornes 1999a 'Plants and water in drylands', in AJ Baird and RL Wilby (eds) *Eco-hydrology*, 78–126, Routledge, London.

Wainwright, J, AJ Parsons and AD Abrahams 1995 'Simulation of raindrop erosion and the development of desert pavements', *Earth Surface Processes and Landforms* **20**, 277–291.

Wainwright, J, AJ Parsons and AD Abrahams 1999b 'Field and computer simulation experiments on the formation of desert pavement', *Earth Surface Processes and Landforms* **24**, 1025–1037.

Wainwright, J, AJ Parsons and AD Abrahams 1999c 'Rainfall energy under creosotebush', *Journal of Arid Environments* **43**, 111–120.

Wainwright, J, AJ Parsons and AD Abrahams 2000 'Plot-scale studies of vegetation, overland flow and erosion interactions: case studies from Arizona and New Mexico', *Hydrological Processes* **14**, 2921–2943.

Wainwright, J, AJ Parsons, WH Schlesinger and AD Abrahams 2002 'Hydrology–vegetation interactions in areas of discontinuous flow on a semi-arid bajada, southern New Mexico', *Journal of Arid Environments* **51**, 319–330.

Walter, H 1971 *Ecology of Tropical and Subtropical Vegetation*. Oliver and Boyd, Edinburgh.

Walvoord, MA and FM Phillips 2004 'Identifying areas of basin-floor recharge in the trans-Pecos region and the link to vegetation', *Journal of Hydrology* **292**, 59– 74.

Wang, XP, R Berndtsson, XR Li and ES Kang 2004 'Water balance change for a re-vegetated xerophyte shrub area', *Hydrological Sciences Journal-Journal des Sciences Hydrologiques* **49(2)**, 283–295.

Wang, XP, T Wang, Z Dong, X Liu and G Qian 2006 'Nebkha development and its significance to wind erosion and land degradation in semi-arid northern China', *Journal of Arid Environments* **65**, 129–141.

Wang, XP, XR Wang, HL Xiao, R Berndtsson and YX Pan 2007 'Effects of surface characteristics on infiltration patterns in an arid shrub desert', *Hydrological Processes* 72–79.

Warren-Rhodes, K, S Weinstein, JL Piatek, J Dohm, A Hock, E Minkley, D Pane, LA Ernst, G Fisher, S Emani, AS Waggoner, NA Cabrol, DS Wettergreen, E Grin, P Coppin, C Diaz, J Moersch, GG Oril, T Smith, K Stubbs, G Thomas, M Wagner, M Wyatt and LN Boyle 2007 'Robotic ecological mapping: habitats and the search for life in the Atacama Desert', *Journal of Geophysical Research-Biogeosciences* **112(G4)**, Art. No. G04S06.

Wear, DN 1999 'Challenges to interdisciplinary discourse', *Ecosystems* **2**, 299–301.

Wellman CH, PL Osterloff and U Mohiuddin 2003 'Fragments of the earliest land plants', *Nature* **425**, 282–285.

Weltz, MA, BA Awadis and LJ Lane 1992 'Hydraulic roughness coefficients for native rangelands', *Journal of Irrigation and Drainage Engineering* **118**, 776–790.

Wendler, G and F Eaton 1983 'On the desertification of the Sahel zone; 1. ground observations climatic change', *Climatic Change* **5**, 365–380.

White, LP 1969 'Vegetation arcs in Jordan', *Journal of Ecology* **57**, 461–464.

Whitford, WG 2002 *Ecology of Desert Systems*. Academic Press, London.

Whitford, WG, K Stinnett and J Anderson 1988 'Decomposition of roots in a Chihuahan desert ecosystem', *Oecologia* **75**, 8–11.

Wilcox, BP 2002 'Shrub control and streamflow on rangelands: a process based viewpoint', *Journal of Range Management* **55**, 318–326.

Wilcox, BP, MK Owens, WA Dugas, DN Ueckert and CR Hart 2006 'Shrubs, streamflow, and the paradox of scale', *Hydrological Processes* **20**, 3245–3259.

Willett, SD 1999 'Orogeny and orography: The effects of erosion on the structure of mountain belts', *Journal Of Geophysical Research–Solid Earth* **104**, 28957–28981.

Williams, JD, JP Dobrowolski and NE West 1999 'Microbiotic crust influence on unsaturated hydraulic conductivity', *Arid Soil Research and Rehabilitation* **13**, 145–154.

Wolfe, SA 1993 *Sparse Vegetation as a Control on Wind Erosion.* PhD thesis, University of Guelph, Guelph, Ontario.

Wolfe, SA and WG Nickling 1993 'The protective role of sparse vegetation in wind erosion', *Progress in Physical Geography* **17**, 50–68.

Wolfe SA and WG Nickling 1996 'Shear stress partitioning in sparsely vegetated desert canopies', *Earth Surface Processes and Landforms* **21**, 607–619.

Worrall, GA 1959 'The Butana grass patterns', *Journal of Soil Science* **10**, 34–53.

Xue, YK and J Shukla 1993 'The influence of land-surface properties on Sahel climate; 1: desertification', *Journal of Climate* **6**, 2232–2245.

Yaalon, DH 1997 'Soils in the Mediterranean region: what makes them different?', *Catena* **28**, 157–169.

Yoder, CK and RS Nowak 1999 'Hydraulic lift among native plant species in the Mojave Desert', *Plant and Soil* **215**, 93–102.

Zalewski, M 2002 'Ecohydrology – the use of ecological and hydrological processes for sustainable management on water resources', *Hydrological Sciences–Journal–des Sciences Hydrologiques* **47**, 823–832.

Zeng, N, JD Neelin, KM Lau and CJ Tucker 'Enhancement of interdecadal climate variability in the Sahel by vegetation interaction', *Science* **286**, 1537–1540.

Zhang, J 1996 'Interactive effects of soil nutrients, moisture and sand burial on the development, physiology, biomass and fitness of *Cakile edentula*', *Annals of Botany* **78**, 591–598.

Zhao, WZ, Z Zhang and Q Li 2007 'Growth and reproduction of *Sophora moorcroftiana* responding to altitude and sand burial in the middle Tibet', *Environmental Geology* **53**, 11–17.

Zhou, LM, RE Dickinson, YH Tian, RS Vose and YJ Dai 2007 'Impact of vegetation removal and soil aridation on diurnal temperature range in a semiarid region: Application to the Sahel', *Proceedings of the National Academy of Sciences of the United States of America* **104**, 17937–17942.

Part II
Weathering

Chapter 4

Weathering Processes and Forms

B.J. Smith

'See how that huge boulder has been split by the sun and wind.'
Commentary by tour guide, Namib Desert, July 2006.

Introduction

Previous reviews of weathering in deserts (e.g. Cooke et al. 1993, Goudie 1997) have been excellent at identifying the mechanisms considered to operate and the landforms with which they are generally associated. Invariably, however, such reviews – especially if orientated towards students – deal primarily with perceived certainties. In reality, weathering studies continue to be characterized more by uncertainties and gaps in knowledge – especially in deserts. This chapter will therefore attempt to concentrate upon the ongoing development of ideas. The aim is not to be exhaustive or comprehensive, but by focusing on a limited number of underlying themes it hopefully questions some traditionally held views and could stimulate future research.

Background to Weathering Studies in Deserts

The harshness and difficulty of access to many desert environments has meant that, compared to many other climate zones, much research is still at the exploratory stage; the nature of desert environments is incompletely understood and many features have yet to be fully described. Because of a comparative paucity of

B.J. Smith (✉)
School of Geography, Archaeology and Palaeoecology, Queen's University of Belfast, Belfast BT7 1NN, UK
e-mail: b.smith@qub.ac.uk

rigorous fieldwork, many geomorphological studies have also relied heavily upon the uncorroborated reports of early explorers and data collected for other purposes. This is particularly the case for weathering studies where, for example, views on temperature and moisture regimes have been strongly influenced by accounts of 'baking hot days', and the sounds of rocks spontaneously splitting during 'freezing nights'. Whilst such reports were no doubt successful in selling travelogues, they have had the unfortunate side effect of creating a popular view of deserts as places of uniformly extreme diurnal temperature regimes, and the total absence of moisture apart from the very infrequent flash flood. This has in turn influenced perceptions of how processes operate and encouraged a concentration upon the role of extreme events, be they the highest ever rock temperature or the longest period between rainfall. Nowhere is this better illustrated than in the still widespread popular belief in, and the numerous theories that have grown up around the concept of insolation weathering. Thus, as pointed out by Gómez-Heras et al. (2006 p, 237), 'insolation weathering has became 'a dogma that is solidly accepted' (Blackwelder 1933), 'a matter of speculation' (Cooke and Warren 1973), or an 'academic abstraction' (Camuffo 1998). This is despite, as Smith and Warke (1997) identified, a growing acknowledgement amongst researchers that, whilst temperature is a major driver of desert weathering, there are many other environmental agents operating with the means, motive and opportunity to weather rocks in deserts and the high probability that they will 'collude' to bring this about. An area where

A.J. Parsons, A.D. Abrahams (eds.), *Geomorphology of Desert Environments*, 2nd ed.,
DOI 10.1007/978-1-4020-5719-9_4, © Springer Science+Business Media B.V. 2009

this preoccupation with the effects of environmental extremes has had a particularly significant impact is in the design of weathering simulations, which, in the absence of field studies, have formed a cornerstone of much weathering research.

It must be acknowledged, however, that the last few decades, and even the period since the first edition of this book, have seen a significant increase in the number and objectivity of field studies into weathering in deserts. Coincident with this – at least in the English-speaking world – has been a tendency to concentrate on the investigation of weathering process and their reduction to the basic physical and chemical mechanisms that control their operation. In the specific context of desert weathering studies this can, however, lead to difficulties. First, the majority of *in situ* observations of weathering mechanisms tend to be short-term and of restricted spatial extent. This creates problems as to their long-term significance in regions where few areas have been mapped in detail, and where little is known about medium- and long-term climatic variability. Secondly, process studies focus attention upon features that exhibit measurable change, especially within the one or two years over which the fieldwork for most research projects is carried out. Consequently, features such as cavernous hollows (tafoni) have received much attention, but, until relatively recently (e.g. Migon et al. 2005) few studies have examined the intervening cliff faces or the debris slopes below them. Thirdly, the reductionist tendency has fostered a segregationist approach, wherein specific mechanisms such as insolation weathering, salt crystallization, or freeze–thaw are examined in isolation. This approach has been prevalent in laboratory-based simulation studies, where it is their aim – and their virtue – that they can examine a particular mechanism under controlled conditions. In doing this, however, environmental conditions may be unrealistically simplified and possible synergisms between weathering mechanisms and other geomorphic processes discounted. Finally, in common with other areas of geomorphology, we are only now beginning to question whether our improved understanding of weathering mechanisms can allow us to cross scale barriers and allow the explanation of either the present configuration of most desert landscapes or their long-term evolution. Process studies are only a limited end in themselves. If we are to justify our preoccupation with them we must ultimately breach the spatial and temporal boundaries that separate individual, short-term studies at the microscale from their cumulative impact on the landscape.

Despite the above observations our understanding of weathering processes and environments and the landforms they produce has steadily progressed. Many of the early misconceptions concerning weathering have been dismissed or modified and we are beginning to understand aspects of the fundamental controls upon rock breakdown and some of the key questions that need to be answered. For example, Cooke et al. (1993) listed six observations that they considered should guide our understanding of weathering in deserts.

(a) Weathering processes are likely to be distinctive because of distinctive diurnal and seasonal temperature and relative humidity regimes.
(b) Contrary to popular belief, moisture for weathering is widely available, from rainfall, dew, and fogs.
(c) Relative humidity is often high at night.
(d) Physical processes are probably significantly more important than elsewhere, but the role of chemical processes should not be ignored.
(e) Some of the debris and weathering features seen today may well be inherited from different climatic conditions.
(f) Despite a common emphasis upon weathering landforms, the most important role of weathering is to provide debris for fluvial and aeolian systems.

Perceptive as these observations are, like all generalizations, they are open to question and qualification. It is these qualifications that identify directions for future research.

The Role of Temperature

The Search for Extremes and a New-Found Variability

Discussions concerning the role of temperature in desert weathering have largely concentrated upon the extremes of diurnal variability – normally under summer conditions – and an apparent preoccupation with the search for maximum ground-surface temperatures (Goudie 1989, McGreevy and Smith 1982). These observations have in turn played an essential role in the design of laboratory experiments for the simulation of, for example, salt weathering. The selection of a realistic and appropriate temperature regime for an experimental simulation is, however, vital in ensuring, for example, that salt weathering and not thermoclasty is being investigated (McGreevy and Smith 1982). The use of extreme diurnal temperature regimes therefore raises a number of questions concerning the extent to which they represent weathering environments and, indeed, the realism of using one temperature regime to characterize a complex climatic zone (Jenkins and Smith 1990).

With regard to these points, 'arguably the most important change in our understanding of desert rock temperatures over the last 20 years has been the appreciation of their considerable spatial and temporal variability' (Smith et al. 2005, p.214). This point was examined initially in studies such as that by Smith (1977) in Morocco, which highlighted seasonal variations, and the fact that most deserts have winters. This study also demonstrated the importance of aspect and shading in controlling the timing, duration, and intensity of diurnal heating and cooling patterns. It has, for example, been long known that desert weathering concentrates in so-called 'shadow areas' (e.g. Evans 1970) to produce features such as cavernous hollows (tafoni), honeycombs, and pedestal rocks (Fig. 4.1). Rock surface temperatures and diurnal ranges associated with these features are reduced compared with those on surrounding exposed surfaces (Dragovich 1967, 1981, Rögner 1987, Smith and

Fig. 4.1 Cavernously weathered boulder, Skeleton Coast, Namibia showing accentuated weathering of the shadow zone, itself created by the cavernous weathering

McAlister 1986, Turkington 1998), and within shaded areas moisture availability appears to be the critical control upon rock breakdown rather than extreme temperatures. In particular, a rapid, early morning temperature should inhibit absorption of any moisture deposited overnight on exposed surfaces and hence restrict the operation of processes such as hydration, solution, and hydrolysis. High rock temperatures may therefore not only be unnecessary but undesirable for many desert weathering processes.

There is also debate concerning the representativeness of many of the very high rock surface temperatures recorded, even for exposed surfaces. This was reviewed in detail by McGreevy and Smith (1982, 1983), where it was noted that many early measurements, often in excess of 70°C, were made with less sophisticated equipment than now used. Peel (1974), for example, used thermocouples without a 0°C datum, and Williams (1923) used a black bulb thermometer. With the advent of reliable thermistors and, more recently, infrared thermometers, many of the maxima recorded from a wide range of desert conditions are below 60°C (Table 4.1). Where 60°C is exceeded there may also be exceptional circumstance, such as the bitumen-coated road surfaces measured by Potocki (1978). Unfortunately, very high absolute temperatures and temperature ranges have been used in, for example, many salt weathering simulations. Data from similar tests used for assessing building-stone durability have shown that rates of breakdown are closely related to the maximum temperatures to which samples are heated. Minty (1965) showed that when samples of dolerite were cycled in sea water between 105 and 110°C the rate of disaggregation was approximately 40 times greater than when similar samples were cycled between 48.9 and 65.6°C. Other workers have also noted that the type of damage caused during salt crystallization durability tests varies with different heating cycles. Marschner (1978) found that damage was restricted to surface layers in samples heated to 60°C but that deeper cracking occurred in samples heated to 105°C. The critical nature of the 60–70°C range was also demonstrated by Yong and Wang (1980), who found that microcracking could be initiated in granite only after samples were heated above 60–70°C, and by Goudie and Viles (2000) who produced microfracturing in marble by repeatedly heating and cooling it between 25 and 80°C. Thus, as pointed out by McGreevy and Smith (1983), the use

Table 4.1 Selected rock surface temperature (°C) measurements from desert environments

Surface temperature °C	Material	Location (altitude)	Time of year	Source
74.4	Black bulb	Egypt	August	Williams (1923)
62	Limestone	Egypt	August	Sutton (1945)
49	Quartz Monzanite	California	August	Roth (1965)
72.5	Rock	Sudan	September	Cloudsley-Thompson and Chadwick (1969)
78.5	Basalt	Tibesti	August	Peel (1974)
79.3	Dark Sandstone	Tibesti	August	Peel (1974)
78.8	Light Sandstone	Tibesti	August	Peel (1974)
57	Basalt	Tibesti	March	Jäkel and Dronia (1976)
46	Granite	Tibesti	March	Jäkel and Dronia (1976)
48	Sandstone	Tibesti	March	Jäkel and Dronia (1976)
48.1	Limestone	Morocco	August	Smith (1977)
21.1	Limestone	Morocco	January	Smith (1977)
62	Asphalt	Abu Dhabi	Summer	Potocki (1978)
73	Asphalt	Abu Dhabi	Summer	Potocki (1978)
41.0	Basalt	Karakoram Mountains	July/August	Whalley et al. (1984)
33.5	Basalt	Karakoram Mountains	July/August	Whalley et al. (1984)
46.3	Desert Varnish	Karakoram Mountains	July/August	Whalley et al. (1984)
54.0	Sandstone	Karakoram Mountains	July/August	Whalley et al. (1984)
50.8	Limestone	Negev	August	Rögner (1987)
56	Rock	Algeria	June	George (1986)
52.2	Sandstone	Tenerife (2070 m)	June	Jenkins and Smith (1990)
42.5	Sandstone	Tenerife (2070 m)	January	Jenkins and Smith (1990)
35.0	Sandstone	Tenerife (950 m)	June	Jenkins and Smith (1990)
29.0	Sandstone	Tenerife (950 m)	January	Jenkins and Smith (1990)
50.0	Sandstone	Tenerife (50 m)	June	Jenkins and Smith (1990)
41.0	Sandstone	Tenerife (50 m)	January	Jenkins and Smith (1990)

of very high rock temperatures and rapid temperature change in simulation experiments of salt crystallization under 'desert conditions' may enhance breakdown by increasing crystallization pressures, ensuring complete crystallization from solution and perhaps better crystal development, and by instigating effects such as the purely thermal expansion of salts and/or thermal fracturing of the rock itself.

By subjecting rocks to unnaturally high temperatures, especially for salt weathering environments, it may not be possible to limit or identify the mechanisms responsible for any breakdown. There may also 'be the danger of introducing thermal effects which would not be encountered in nature' (McGreevy and Smith 1983, p. 300). Laboratory studies that use extreme temperature cycles to produce insolation weathering (e.g. Rodriguez-Rey and Montoto 1978) must therefore be approached with considerable caution. This precautionary approach has found its way into an increasing number of studies in recent years. and is reflected in a 'less dogmatic view toward the appropriateness of a single temperature/humidity

regime and a willingness to tailor regimes to specific environmental conditions' (Smith et al. 2005 p. 215). As they point out, this is demonstrated in the paper by Goudie (1993) in which he used six different diurnal cycles ranging from the extreme 'Wady Digla' regime of Williams (1923) to the 'Negev Cycle'. The latter oscillates between 10 and 40°C and 20 and 100% relative humidity and is considered to more closely represent conditions associated with the regular overnight deposition of dew experienced in this desert (Goudie and Viles 1997).

The use of diurnally based thermal cycles in laboratory weathering simulations does, however, presuppose that this is the primary variable influencing physical breakdown. Yet, as noted by Cooke et al. (1993), there can be significant seasonal variations in temperature. Under winter conditions freezing temperatures may be frequently experienced at rock surfaces and freeze–thaw activity becomes a possibility. Both seasonal and diurnal variability are clearly systematic, but superimposed upon these are additional layers of spatial and temporal variability which, although far less predictable, may be of considerable significance. Reduced air and rock temperature maxima within caverns have already been noted, but to these must be added other aspect-related differences. This applies not only to obvious variations between north- and south-facing slopes, but also between east- and west-facing surfaces. In low latitudes, for example, any overnight moisture should be retained for longer on surfaces facing west. This moisture is then quickly driven off as rocks come out of shadow and are rapidly heated by a sun already high above the horizon (Smith 1977). This observation is corroborated by recent studies in Petra (Paradise 2002) that identified greatest weathering on east- and west-facing walls.

Added to these effects is the amelioration of rock temperature regimes resulting from maritime influences in coastal deserts, and a variety of effects associated with increasing altitude. Unfortunately, there are relatively few studies of rock temperature variations in high altitude desert areas, but those that have been made record surface maxima similar to those observed in low altitude environments (e.g. Whalley et al. 1984). These high surface temperatures frequently occur in conjunction with low air temperatures, and any interruption of incident radiation can result in a rapid drop in surface temperature. This phenomenon was illustrated by Jenkins and Smith (1990) in a study of altitudinal and seasonal variability in daytime temperatures on the island of Tenerife. By continuous measurement of surface temperatures on a standard sandstone block moved between three sites at different altitude, they showed that at an altitude of 2070 m there were numerous short-term fluctuations of 3–15 min duration related to shading by a light cloud cover and wind speed variations. During these fluctuations surface temperatures could drop up to 15°C with temperature gradients sometimes in excess of 2°C per minute. At mid-altitude (950 m) and coastal (50 m) sites, where cloud cover was greater, additional fluctuations of 1–2 h duration were noted which effectively destroyed the daytime element of any generalized diurnal temperature curve. The frequency and intensity of these short-term variations suggests that they could play an important role in disruptive processes such as granular disintegration and the formation of thin surface flakes, not least because recent studies using microthermistors have identified that the most important thermal gradients are produced within the upper few millimetres (Gómez-Heras et al. 2004)

rather than the upper few centimetres of exposed stone (Peel 1974).

Previously, short-term fluctuations have either failed to register when desert temperatures were recorded at set intervals (every 15, 30 min, etc.) or have been disregarded in insolation and salt weathering simulations based upon diurnal cycles of heating/cooling and wetting/drying. It is interesting to note, however, that in one of the few sets of experiments to register detectable alteration (using reflectance and microhardness changes) resulting from thermal fatigue, the heating and cooling cycle was of only 15 min duration (Aires-Barros et al. 1975, Aires-Barros 1977, 1978). The efficacy of short-term temperature cycling in producing near-surface salt weathering has also been demonstrated experimentally by Warke and Smith (1994) and by Smith et al. (2005). In the latter experiment blocks of sandstone and limestone, some impregnated with sodium chloride, were subjected to 24,000 heating and cooling cycles of 30 min duration under dry conditions. Only those samples impregnated with salt produced any debris (through surface disaggregation), and whilst the amounts were very small they are possibly more representatative of actual rates of decay under non-accelerated conditions and could indicate the viability of differential thermal expansion as a salt weathering mechanism.

The growth of interest in the localised action of weathering processes has paralled a growing appreciation of the importance of microclimatic controls. Warke (2000), for example, has stressed the importance of the boundary-layer climate and changes at the rock/environment interface in controling weathering, including feedbacks in which intrinsic and extrinsic rock properties influence the conditions experienced. In support of this contention Viles (2005a) has demonstrated, through the use of exposure trials, the importance of microclimate in determing both patterns and rates of weathering in Namibia. This recognition has also stimulated a renewed interest in the role of fatigue failure in generating surface breakdown, in as much as rapid near-surface temperature cycling dramatically multiplies the number of possible 'fatigue events' to which exposed rock could be subject. Hall and his co-workers have, for example, recently revisited and championed the possible role of fatigue failure under cold arid conditions in a number of papers. They used high frequency monitoring of temperature to confirm both the existence of repeated, rapid surface and immediate sub-surface temperature change (Hall and Andre 2001), and their importance in generating stress at the microscale that could lead to fatigue failure (Hall 1999). Under extreme conditions of very low air temperatures Hall and Hall (1991) also raised the possibility of any interruption to incoming insolation creating a 'thermal shock' in which the stresses generated occur more rapidly than can be accomodated by the required internal deformation (Yatsu 1988). They similarly drew attention to the possible importance of stresses generated between adjacent crystals with different coefficients of thermal expansion and/or crystallographic allignments (Hall and Andre 2003). This theme was further developed experimentally by Gómez-Heras et al. (2006), who identified a number of papers that have calculated crack resistance energy for grain boundaries as only some 0.4 times the crack resistance energy for bulk elements (Yang et al. 1990, Sridhar et al. 1994; Zimmermann et al. 2001, Weiss et al. 2002). In their experiments they demonstrated measurable temperature differences and rates of temperature change between adjacent crystals/grains for a range of rock types when heated in a controlled ambient environment using infra-red lamps. Albedo was shown to be the main control on individual and overall maximum temperatures, but crystal size was shown to be the major factor determining temperature differences between adjacent crystals/grains. Large differences in crystal/grain size appear to magnify stresses resulting from differential thermal expansion.

Thermal Stress and Rock Fracture

In light of the previous discussion, it is clear that the patterns of thermal stress experienced by desert rock surfaces are exceedingly complex. Recognition of the range of stresses to which rocks are subjected suggests that rarely, if ever, is rock breakdown a function of a single mechanism acting in isolation. Invariably it is a product of two or more mechanisms acting together or in alternation (Jenkins and Smith 1990). The most obvious example of this is insolation weathering itself. It is difficult to envisage any situation in which desert rocks are subjected to temperature fluctuations in isolation. It has already been noted that moisture is invariably available in some form within hot deserts (see

also next section). Additionally, all rocks exposed at desert surfaces have a unique stress history which will probably leave them weakened to a greater or lesser extent, and more or less susceptible to either mechanical breakdown or chemical decay.

Sources of pre-stressing in desert rocks are numerous. They include chemical alteration, either under present conditions of limited, but assured moisture availability or inherited from former periods of moister climate; dilatation acting at a range of scales; and previous exposure to processes such a salt and frost weathering. Frost weathering is particularly relevant when one considers that mountains consistently constitute the dominant terrain type in the world's major desert areas. A study by Fookes (1976) identified, for example, 43% of the Sahara desert, 39% of the Libyan Desert, 47% of Arabia, and 38.1% of the deserts in the south-western United States as mountainous. Other studies of desert terrain (e.g. Cooke et al. 1982) also identified a range of mountain–plain models as the most representative of the world's desert landscapes. Within these models most debris mantling alluvial fans and plains is derived initially from adjacent mountain catchments, where moisture may be more readily available and temperatures can frequently fall below 0°C. Debris produced under these conditions is likely to carry with it a memory of inbuilt stresses that can be exploited by other weathering processes more characteristic of hot desert environments *per se* (Warke 2007).

Potentially exploitable weaknesses in the form of microfractures can also be created within the rock mass prior to exposure. These will then be carried over into any debris derived from these rocks. The range of microfractures found within near-surface rocks has been discussed by Nur and Simmons (1970) and Simmons and Richter (1976), and was summarized by Whalley et al. (1982a) as consisting of (a) cracks at grain boundaries produced during ascent to the Earth's surface; (b) stress-induced cracks produced by the principle of non-hydrostatic stress; (c) radial and concentric cracks about grains enclosed by material with different volumetric properties; (d) tube cracks produced by magmatic fluid solution, dislocation, etching, etc.; (e) cracks induced by thermal shocks and gradients; and (f) cleavage cracks. To these crack-opening processes should be added potential weathering lines comprising, for example, certain mineralogical concentrations. The role of crack propogation in determining patterns of weathering has been demonstrated by Kane (1999 – reported in Smith et al. 2005), who also highlighted the accentuation of thermally driven expansion where rocks are bounded or 'buttressed' within a larger rock mass (Merrill 1906). She did this by studying the combined effects of confinement and compression on granite test blocks that were compressed between two metal plates before being subjected to repeated salt weathering cycles in a climatic cabinet. Results showed that the decay of unconfined test blocks progressed steadily through gradual surface disaggregation as salt crystallisation stresses are accommodated by the elastic expansion and contraction of the rock to form a network of near-surface microfractures. In contrast, blocks initially compressed at $900 \, kNm^2$ showed little initial loss. This is because, under compression any microfractures perpendicular to the direction of compression tend to close (Batzle et al. 1980) and inhibit salt penetration. There was some slight, initial loss of material as surface parallell cracks were exploited to produce limited surface flaking, but the rate of loss was no greater than that for unconfined blocks. However, the repeated low-magnitude stressing of the salt weathering cycles eventually propagated a microfracture network into the now inelastic granite through a combination of stress concentration at crack tips (Griffith 1921) and salt crystallisation through ever-widening cracks. The interaction of adjacent fractures can effectively create a localised failure plane (Janach 1977) that is typically at an oblique angle to the direction of compression, and it was the breaking away of angular fragments defined by these planes that eventually triggered the accelerated breakdown of the confined block. To test this interpretation, Kane took one further block and doubled the compressive loading. As predicted this had the effect of delaying even further any noticeable surface loss – through more effectively closing existing fractures into the block – but once it began, breakdown was noticeably more rapid and produced larger fragments than the block under the lower compression.

Understandably, microfractures and lines of potential weathering are normally associated with igneous and metamorphic rocks. Stress relief does, however, occur within sedimentary rocks, where unequal release of confining pressures by erosion of once deeply buried rocks can lead to exfoliation and splitting (e.g. Bradley 1963). Similarly, sedimentary and metamorphic rocks can contain a wide range of different lineations associated with stylolites, cleavage and bedding planes and alignments of grain boundaries and/or voids – even in apparently unbedded sandstones (e.g. Smith and McGreevy 1988).

Weathering susceptibility is also influenced by alteration skins, rock varnishes (see Chapter 6) and case-hardened layers (e.g. Conca and Rossman 1982) which frequently cover desert rock surfaces. Indeed it is not unknown for the presence of a rock varnish, and the consequent reduction in albedo, to be used as a justification for the high rock-surface temperatures recorded in deserts (e.g. Peel 1974). The importance of albedo has been demonstrated by Carter and Viles (2004) who showed, for example, how the growth of black lichen raised both surface and subsurface temperatures. Although, presumably the growth of a light-coloured lichen on a dark substrate would have the effect of increasing albedo (Fig. 4.2, Goudie and Viles 1999). Warke et al. (1996) also showed under experimental conditions how the artificial soiling of stone surfaces markedly increased both maximum surface temperatures and rates of temperature change under simulated insolation – especially in rocks with an initially high albedo. Ironically, however, most weathering simulations that have argued for the importance of albedo, and its role in facilitating extreme temperature variations, have subsequently done little to explore its significance. This is because most have utilised climatic cabinets in which rocks with different thermal characteristics are forced through the same heating and cooling cycle.

Fig. 4.2 Light-coloured lichen covering the upper surfaces of dark-coloured boulders and increasing their albedo, Skeleton Coast, Namibia

In doing this, they negate the very differences in thermal response that are known to occur in nature and undoubtedly contribute to the differential weathering observed in deserts (Warke and Smith 1998). Changes in albedo are also frequently associated with a physical and/or chemical modification of the outer layer of the rock. It would be reasonable therefore to explore the possibility of insolation weathering occurring in the presence of such crusts. Work on other materials such as concrete has shown that durability problems frequently occur because of expansion in materials with different surface layers. This can lead to patterns of rupture redolent of those found on desert rocks (Fig. 4.3).

Albedo is not, however, the only rock property that inflences response to weathering, and other studies have identified the significance of physical and chemical variability as a catalyst to breakdown. Experiments using 'coupled' samples of two rock types exposed to simulated salt weathering conditions have, for example, shown how moisture can be retained for longer at boundaries between different stone types. This can

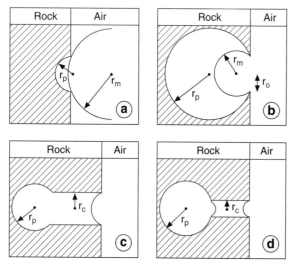

Fig. 4.3 Typical failure patterns in an elastic outer layer on a rigid base due to forces concentrated at the interface. (**a–d**) illustrate force concentrated along the interface, (**e–f**) illustrate force exerted through expansion at a point. (**a**), parallel separation of rigid plate; (**b**), peeling from base; (**c**), bending – plastic failure; (**d**), bending – elastic sudden failure; (**e**), cleavage – sudden failure; (**f–h**), punching or pop-out failure – pattern of failure depends upon brittleness of outer layer. At high brittleness numbers failure occurs suddenly (**f**). At low brittleness material behaves in a ductile fashion (**g**). At intermediate brittleness failure occurs through stable crack growth (**h**) (adapted from Bache 1985)

lead ultimately to a concentration of crystallized salt, if available, and enhanced disaggregation (Haneef et al. 1990a, b). Other salt weathering simulations (McGreevy and Smith 1984) have shown how the presence of swelling clays in sandstone samples, coupled with increases in microporosity, can enhance breakdown through flaking. Exploitation of potential cracks comprising aligned grain boundaries and pores in sandstones may be a factor in salt-induced contour scaling of sandstones (Smith and McGreevy 1988). Breakdown following the cyclical wetting and drying of basalts has similarly been attributed to the swelling of hydrothermally derived smectite clays (McGreevy 1982a). Systematic studies of the physical rock properties that influence mechanical breakdown have identified water-absorption capacity, porosity and microporosity, saturation coefficient, and tensile strength as amongst the most important properties (Cooke 1979). Compressive strength, specific gravity, compactness, hardness texture, shape, anisotropy, microfractures, and cleavage can also play a role in determining breakdown (Fookes et al. 1988). Additional factors which are of specific relevance to the role of temperature in weathering are variations in specific heat capacity, and thermal conductivity. These not only influence rock surface temperatures and internal temperature gradients (Kerr et al. 1984, McGreevy 1985), but also the establishment of other internal stresses as rock constituents with varying thermal properties, including additions such as clays and salts, expand and contract differentially.

The presence of physical variability, pre-stressing, and chemical alteration within rocks is one possible explanation why insolation weathering continues to find support from field studies (Ollier 1963, 1984) through observations of split and cracked bouulders (Fig. 4.4). The fact that laboratory simulations using unconfined, freshly cut and polished blocks have historically discounted thermal fatigue may find an explanation in that they simply did not continue for long enough, or that the weathering that did take place was not detected (Goudie 1989). Within surface rocks a wide range of potential weaknesses thus exists that can be exploited during thermally induced surface expansion and contraction. Folk and Begle-Patton (1982) and Rice (1976) have pointed out that the early experiments of Blackwelder (1925, 1933) and Griggs (1936) could have been more successful in producing failure if they had used larger blocks that were in some way confined.

Fig. 4.4 Parallel stone cracking of boulder, Skeleton Coast, Namibia. Such features have commonly been attributed to 'insolation weathering', although such stones are, or have been partially buried and this could contribute to stress concentrations that result in the characteristic fracture pattern

Similarly, such experiments may have produced a more realistic assessment of weathering in deserts if they had used samples characteristic of those exposed by erosion and transport at desert surfaces; rather than freshly quarried blocks selected for their purity, freedom from imperfections and absence of recognizable weathering. Possibly the major difficulty with these experiments is, however, the way in which they attempted to reproduce fatigue failure by subjecting rocks to what were, in effect, repeated thermal shocks.

The Role of Moisture

Variable Availability

The role of moisture in shaping desert environments has for many years appeared anomalous. On the one hand, deserts are perceived as areas of little if any moisture, while on the other we are faced with landscapes in which many elements are of patently fluvial origin. For many years explorers denied these origins by invoking other processes including: 'sculpture due to solar heat shattering the rocks and wind removing the pulverised residue' (Peel 1975, p. 110). But as understanding of fluvial processes progressed through the early part of the last century, alternative explanations became necessary. Many fluvial features were thus interpreted either as legacies from some former climate of more assured rainfall, or as products of infrequent but intense rainfall operating over long timespans. It is only relatively recently that alternative viewpoints have been explored in which moisture is seen as being more readily available in many hot deserts than previously surmised (e.g. Peel 1975, Goudie 1997). In his paper on water action in deserts, for example, Peel referenced Slatyer and Mabbutt (1964) who found that intense, 'catastrophic' rainfall of the kind traditionally associated with deserts is much more common around the margins of deserts. In a survey of various deserts, they found that 50% of the rain that falls usually occurs in gentle showers of moderate intensity. Similarly, Peel drew on observations by Dubief (1965) who showed that mean rainfall intensities over the whole of the Sahara are little higher than those in France. Rainfall of lower intensities is less likely to generate overland flow, more likely to be absorbed by rock and soil, and will most probably play a more important role in promoting a range of weathering processes. In his review of desert weathering, Goudie (1997) tabulated data from a range of world deserts which demonstrate that even in what are perceived to be dry locations there may be an appreciable number of days in which measurable quantities of rain are received. Frequent wetting enhances the efficacy of both mechanical and chemical weathering processes.

Rainfall can also concentrate locally and might lead to enhanced scope for weathering in certain situations. Apart from obvious orographic controls, it seems that there may be a spatial organization to desert storms (e.g. Sharon 1981) related to phenomena such as Rayleigh cells within the atmosphere, although there is no significant long-term tendency for storms to occur at preferred locations. However, local rainfall is not randomly scattered and there is evidence that it can concentrate in valleys which also receive rainfall deflected away from exposed interfluves by higher winds (Sharon 1970). Windward slopes will likewise receive substantially more rainfall than adjacent leeward slopes. Yair et al. (1978) suggested that windward slopes of 20° with incident rainfall of 40° would receive almost twice as much rainfall as opposing valley sides. Similar observations were made in Sinai by Schattner (1961) who noted that granular disintegration of granites was more rapid and penetrated more deeply on rocks that were not exposed to the 'strongest and largest insolation' (p. 254). Instead, disintegration is most intensive on surfaces facing north and west into the dominant rain-bearing winds. South-facing slopes, he found, rarely exhibited intensive disintegration.

The degree to which water is concentrated into certain areas is also dependent upon surface controls, such as the extent of soil and rock cover. Where bare rock dominates, surface runoff is more frequent and extensive, and infiltration concentrates in certain parts of catchments, such as slope foot zones of colluvial accumulation (Yair and Berkowicz 1989). Because of between-catchment differences in hydrological response, Yair and Berkowicz (1989) suggested that climatic parameters such as temperature and rainfall alone provide insufficient indication of ground surface aridity.

The Importance of Direct Precipitation of Moisture

Both Peel (1975) and Goudie (1997) drew attention to the importance of moisture from fogs. Goudie tabulated the relative contributions of rainfall and fog as moisture sources at Gobabeb in Namibia over the period 1963–1984. He found that rainfall averaged 24.5 mm and fog 31.7 mm per annum. More universally appreciated is the contribution that direct precipitation in the form of dew can make to the moisture budget of desert areas (Verheye 1976). Little is known in any detail of dewfall from many of the world's deserts, and most opinions concerning its significance have been based upon pioneering work by Duvdevani (e.g. 1947, 1953) and Evenari et al. (1971) in Israel, especially at the ancient city of Avdat in the Negev Desert. On occasions they reported that dewfall can exceed annual rainfall (28.4 mm as opposed to 25.6 mm in 1962–1963), but over the four-year period 1963–66 mean dewfall at the ground surface was 33 mm from an average of 195 dew nights and, at a

height of 100 cm, 28 mm from 172 dew nights. This compares with an average annual rainfall at Avdat of 83 mm for the period 1960–67 (Evenari et al. 1971, p. 30). Clearly these figures suggest that although dewfall is frequent, the amounts per night are small and are a product of low ground surface temperatures and low moisture contents of the contact zone rather than high levels of air moisture (Verheye 1976). A further demonstration of the importance of altitude and aspect in controlling deposition from dew and is to be found in the detailed studies by Kidron, again in the Negev (e.g. Kidron 2000). These show a 2–3 fold increase in average daily dew and fog amounts from sea level to 1000 m above m.s.l., and an increase in the percentage of days with heavy dew and fog (Kidron 1999), as well as demonstrating the importance of surface radiation input and temperature in controlling dew deposition (Kidron et al. 2000a). Moreover, continuous monitoring of dew deposition has shown that in sun-shaded habitats dew condensation can continue even after sunrise and may explain the higher overall dew values and its longer duration in these environments (Kidron 1999).

It is not surprising therefore, that direct precipitation has been invoked as a major moisture source and key component in the accentuated weathering associated with features such as tafoni and other cavernous hollows (e.g. Smith 1978, Rögner 1988a) and in low-lying areas (e.g. Schattner 1961). Recent studies (e.g. Goudie et al. 1997, Viles and Goudie 2007) have also demonstrated the efficacy of direct deposition from fog in promoting rapid rock breakdown in coastal deserts by salt weathering, as well as investigating its impact through linked laboratory simulations (Goudie and Parker 1998, Goudie et al. 2002a). The weathering significance of direct precipitation is further enhanced by its chemical composition. This is largely a function of the aerosols and dissolved salts that it can scavenge from the atmosphere as well as any dry fallout previously deposited on to rock surfaces and mobilized during condensation. The greatest potential for this mobilization is normally associated with polluted urban environments (e.g. Wisniewski 1982, Ashton and Sereda 1982). What observations there are from arid to semi-arid areas have, however, also shown markedly enhanced concentrations of calcium, bicarbonate, and sulphate ions in dewfall compared with rainfall (Yaalon and Ganor 1968). The most likely source for these ions is sea salts, but chemical analysis of airborne dusts blowing from desert areas suggests that they can contain elevated concentrations of numerous minerals. These include a range of clay minerals and soil nutrients (e.g. Wilke et al. 1984) as well as salts (Pye 1987, Table 6.3, p. 130).

Before ascribing too much importance to dewfall, certain other factors must be considered. First, the information we do have is drawn from meteorological observations which are not necessarily relevant to conditions experienced at rock surfaces. Secondly, we have little knowledge of the controls that humidity conditions exert upon weathering (McGreevy and Smith 1982). Is, for example, all moisture that condenses at a rock surface absorbed and, if so, what role does it play in mobilizing salts?

Meteorological observations are normally collected to reflect standard conditions, with figures invariably averaged over a given time period. The figures for dewfall at Avdat (only 80 km from the Mediterranean) are, for example, given in Evenari et al. (1971) as annual averages showing higher deposition at ground level than at 100 cm. Duvdevani (1953) showed, however, that under dry summer conditions dew deposition at various sites in Israel increases from ground level upwards – his so-called 'arid gradient'. It is only during moist winter months, when moisture can be drawn up from below, that dew deposition is greatest nearer the ground.

The mechanism of moisture deposition can itself be complicated. Under clear-sky conditions with rapid radiative heat loss at night, surface temperatures can fall below ambient air temperature. As they do so, condensation will occur at relative humidities progressively below 100%, especially if hygroscopic salts are present (Arnold 1982). Moisture flux from atmosphere to rock surface may, however, take place first in the vapour phase within pores before dew point is reached and before spontaneous condensation of liquid water.

In his review of condensation–evaporation cycles in capillary systems, Camuffo (1984) suggested that condensation begins with an adsorbed film of molecular moisture. This can form at relative humidities below 100% because 'the binding forces between the water molecules and the solid surface are larger than the binding forces between the first adsorbed layer and newly arriving vapour' (p. 152). As relative humidity in the environment increases, further molecular layers can accumulate which may eventually collapse under gravity into larger pores and take on 'bulk-liquid struc-

ture'. Such patterns of deposition have been observed within buildings (e.g. Camuffo 1983) under conditions that are otherwise effectively arid. Advantages of this mechanism are that it can promote deep moisture penetration, carrying with it gases scavenged from the atmosphere, and that moisture absorption occurs very frequently (Camuffo 1983). Such relatively small quantities of moisture may also be important in fractured rocks where, if it can form near crack tips, it may generate stresses through the formation of an electrical double layer (Ravina and Zaslavsky 1974).

As rock surface temperatures fall overnight and the relative humidity of air in contact increases, eventually a point is reached where surface condensation occurs. Whether such moisture is absorbed by the rock will depend on pore size and geometry. In the case of a small, hemispheric cavity facing the external environment (Fig. 4.5a) it will remain empty except for the monolayer of moisture until the radius of the meniscus of condensed moisture r_m equals the pore radius r_p, after which the cavity will progressively fill with moisture. This process works in reverse when relative humidity falls and the cavity progressively empties. In the case of large 'spheric pores' which open to the atmosphere via a small opening (Fig. 4.5b), the cavity remains empty until relative humidity reaches a critical level where r_m equals r_p. Thereafter the cavity will completely fill with no increase in relative humidity. If relative humidity subsequently falls, however, the pore will remain filled until a critical relative humidity is reached where r_m equals r_o, the radius of the opening, which then triggers evaporation of moisture. Where pores are connected to the atmosphere via capillaries, the pattern of filling and emptying depends upon the ratio of the pore radius r_p to that of the capillary r_c. Where $r_c > r_p/2$ (Fig. 4.5c) condensation occurs first in the pore and then in the capillary. On the other hand, where $r_c < r_p/2$ (Fig. 4.5d), condensation will occur within the capillary, trapping air within the pore. This trapped air can act as an effective block against further moisture ingress even though relative humidity may rise above 100%. Camuffo (1983) pointed out in fact that in rocks with pores showing this 'ink-bottle effect' the only way to fill inner cavities with moisture would be to submerge them in liquid water, although some moisture may be 'sucked in' if temperature falls and trapped air contracts. Conversely, if temperature were to rise, trapped air could expand and expel water without evaporation.

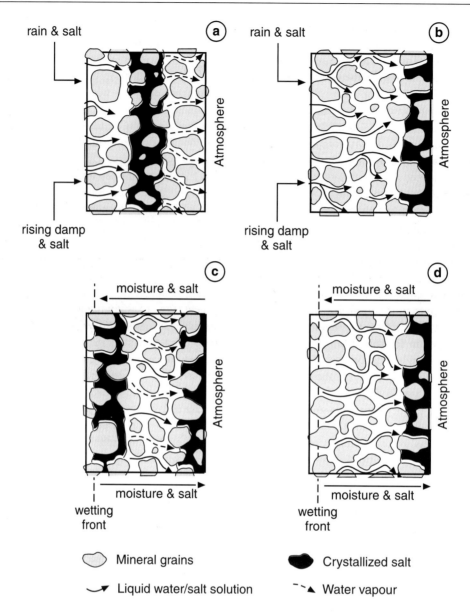

Fig. 4.5 Influences of pore and capillary characteristics upon moisture deposition and condensation at a rock–air interface. r_p = pore radius, r_m = radius of meniscus of liquid water (which increases with relative humidity such that $r_m = \infty$ at RH = 100%), r_c = capillary radius. (**a**) Hemispherical cav- ity facing large pore or external environment; (**b**) spheric cavity open to the environment through a small hole; (**c**) pore connected to environment by wide capillary ($r_c > r_p/2$); (**d**) pore connected to environment by narrow capillary ($r_c < r_p/2$) (adapted from Camuffo 1984)

Moisture and its Significance for Salt Weathering

Although it is not theoretically required, the norm is for drying to involve evaporation and, as Rodriguez-Navarro and Doehne (1999) have clearly shown,

evaporation rate can have a major impact on the crys- tallization of any salt that is present in solution within the rock. This occurs in two stages (Hall et al. 1984). In the first stage, drying is strongly influenced by both temperature and wind speed. A 10°C rise in temperature will approximately double the drying rate, as will a quadrupling of wind speed. The drying rate in the first stage is also directly proportional to relative humidity. A second phase will eventually be reached when the drying rate declines substantially. During this phase drying is not significantly affected by humidity. Towards the later stages of drying in a porous body 'liquid phase capillary continuity is lost and vapour phase diffusion no doubt becomes the only transport mechanism within the pores' (Hall et al. 1984, p. 18). A similar situation can be reached earlier under conditions of either very rapid surface moisture loss or where the rate of evaporation is greater than the rate of replenishment of capillary moisture from within the rock (Amoroso and Fassina 1983, Lewin 1981). Under these conditions work on building stones has shown that the liquid-to-vapour transition will occur at some depth below the rock surface. This depth can have very significant implications if the moisture contains dissolved salts, which will begin to crystallize out of solution at this depth as a subflorescence (Fig. 4.6a). Growth of crystals at this wet–dry interface has been found experimentally by Lewin and Charola (1979) to cause the outer surface of stonework to separate or crumble (also reported in Amoroso and Fassina 1983, pp. 33–34). A similar suggestion that scaling could be associated with wetting depths was made by Dragovich (1969) working on granites in Australia. If the rate of drying is less than that of moisture replenishment from within the stone (Fig. 4.6b), then evaporation will take place from the surface and an efflorescence of crystallized salt will develop. Under these conditions observations on the behaviour of building materials suggest that disruption will be reduced compared with subsurface crystallization (Amoroso and Fassina 1983, p. 29). There is, however, some evidence from studies of sandstone used in construction that, beneath an outer surface characterised by salt-induced flaking and granular disaggregation, a zone of structural weakening can occur which is rapidly exploited once the outer surface is lost (Warke and Smith 2000). They also observed that, in a mixed salt environment salts within a rock can become segregated on the basis of their solubility, and that it is possible for complex distributions with depth to occur that represent combinations of the scenarios described in Fig. 4.6.

Under desert conditions, salt-rich moisture migrating from within a rock is most likely to derive from rising groundwater. Indeed, groundwater is now recognized by many as a major source of moisture

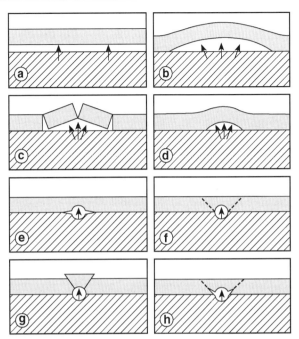

Fig. 4.6 Patterns of salt crystallization within a porous rock: (**a**) subsurface deposition where there is a high rate of surface evaporation and salt solution drawn from within the rock; (**b**) surface deposition where the rate of evaporation is less than the potential rate of outward salt solution migration; (**c**) surface and subsurface salt deposition after evaporation of surface-wetted rock (sodium sulphate, magnesium sulphate); (**d**) surface salt deposition after evaporation of surface-wetted rock (sodium chloride)

and salts in numerous low-lying and/or coastal desert areas. It has, for example, been identified as a major cause of building decay in these areas as well as being instrumental in the formation of many natural salt-weathering phenomena such as tafoni and honeycombs (see Cooke et al. 1982, Chapter 5 for a review). In urban areas from other environments, rising damp is seen to produce characteristic weathering zones above ground level (e.g. Arnold 1982), with maximum deterioration in a periodically dried zone some distance above the ground. Experimental work has also demonstrated the rapid effectiveness of salts derived from rising capillary moisture in causing rock breakdown (e.g. Goudie 1986, Benavente et al. 2001), in tests lasting as short as two weeks and with drying conditions of only 20°C and 60% relative humidity (Lewin 1981), although salt crystallization pressures are likey to be influenced by a range of variables, including pore size and salt-solution interfacial tension (La Iglesia et al. 1997). Under conditions of prolonged saturation it is also possible that salts could migrate through rock via ionic diffusion rather than in solution (Pel et al. 2003). This process could lead to complex, three dimensional distributions of salts throughout a rock mass and also facilitate chemical reactions with rock constituents (Turkington and Smith 2000). Salts deposited from groundwater are often complex mixtures and occur in high concentrations (e.g. Goudie and Cooke 1984). As a consequence, these salts can exploit a wide range of environmental conditions to exert stresses within rocks through repeated expansion and contraction. Complete pore-filling by salts could enhance the efficacy of differential thermal expansion, whereas mixtures of salts acting singly or synergistically hydrate and dehydrate or dissolve and crystallize across a wide range of temperature and humidity conditions.

Most desert environments are not, however, characterized either by an abundance of salts or of moisture. Instead, as already argued, moisture is frequently derived in small amounts from direct precipitation overnight at or near an exposed surface from where it is lost during the following day. Under these conditions large quantities of salt might even prevent ingress of small quantities of moisture by acting in effect as a 'passive pore-filler' (Smith and McAlister 1986). Smith and Kennedy (1999) have, for example, shown under experimental conditions how just one cycle of wetting and drying with salt solution can significantly modify subsequent patterns of infiltration and drying. Experimental work has also shown that, where sandstone blocks are sprayed with limited daily applications of weak salt solutions to one face, the subsequent salt distribution depends not just on the drying regime, but is also particularly sensitive to the detailed pore characteristics of the rock (Smith et al. 2005) and the solubility of the salt used (Smith and McGreevy 1988). So that in sandstone blocks treated with 10% solutions of sodium sulphate and magnesium sulphate, salt accumulated at or near the approximate diurnal wetting depth and just below the surface (see Fig. 4.6c) and eventually resulted in surface scaling. In contrast, in a block treated with more soluble sodium chloride, salt only accumulated in a narrow zone at and beneath the stone surface resulting in limited granular disintegration (see Fig. 4.6d). It must be remembered, however, that as salt weathering progresses it can significantly alter the porosity (Benavente et al. 1999, Nicholson 2001) and other physical characteristics of fresh rock. Any cor-

relation between weathering performance and original conditions must therefore be interpreted with care.

Clearly, the effectiveness of any salts precipitated out of solution in causing rock breakdown is dependent on where they crystallize. It must, however, also depend on the manner of their crystallization and the stresses that they can exert from within pores and capillaries on surrounding mineral grains (Scherer 1999), which, in turn can depend on factors such as the solution supersaturation ratio (Rodriguez-Navarro and Doehne 1999, Flatt 2002). The process of crystal growth out of solution itself is considered by many to be the most important mechanism by which stresses are exerted and rock fabric disrupted (Cooke et al. 1993). A situation is frequently envisaged where crystals bridge across a pore and continue to grow against confining pressures provided there is a film of solution at the salt–rock interface (Correns 1949, La Rijniers et al. 2005). Although, as Scherer (1999) has pointed out, the stress generated within a single pore is insufficient to cause failure as it acts on too small a volume. Instead, salts are required to propogate throughout a a zone within the porous medium so that sufficient stress is generated to cause failure by interacting with the larger flaws that ultimately control material strength. Observations of crystallized salts within pores from a range of environments suggest that pore-filling begins with a layer of salts crystallizing around the boundary of the pore. Bernardi et al. (1985), for example, identify authigenic calcite and gypsum growing around ooliths in limestones from the Ducal Palace in Urbino. These salts come into contact first either in capillaries or the narrow necks connecting pores. By concentrating stresses here they could cause disruption without the necessity either for pore-bridging or pore-filling. Indeed, although there are now numerous descriptions of salt efflorescences from many environments and of subfluorescences exposed when blisters are removed, there are few, if any, examples of pore-bridging within sound rock. Long, acicular gypsum crystals have been frequently described growing out of carbonaceous nucleii (e.g. Del Monte and Sabbioni 1984, Del Monte et al. 1984, Del Monte and Vittori 1985). But these are observed growing on the surface of polluted stonework or subsequently under moist laboratory conditions from sources such as fly-ash. More detail on the mechanics of salt weathering can be found in the excellent reviews by Goudie and Viles (1997) and Doehne (2002).

As an alternative to partial pore-filling by clearly defined salt crystals, relatively rapid evaporation under laboratory conditions can fill pores with microcrystalline salts (e.g. Smith and McGreevy 1988). Complete pore-filling of this kind may cause breakdown by crystallization if salts continue to migrate towards the pore, but will also result in stresses due to differential thermal expansion and/or hydration if environmental conditions and salt types permit it (Stambolov 1976, Sperling and Cooke 1980a, b).

The areal distribution of salt accumulation and ultimately salt-induced decay is very dependent upon the wetting and drying characteristics of stone surfaces. Varying pore size and shape characteristics will, as already discussed, lead to differing patterns of moisture condensation, absorption, and loss as well as affecting susceptibility to various salt weathering mechanisms. Variability of this kind has often been used to explain the location of so-called sidewall tafoni (Fig. 4.7) on cliff faces (e.g. Bryan 1925, Mustoe 1983, Turkington 1998). In contrast, basal tafoni are normally ascribed to greater moisture availability in cliff foot zones from either groundwater seepage or enhanced direct precipitation near ground level (Dragovich 1967, Smith 1978, Migon et al. 2005). Once created, any hollows protected from rainfall become sites of preferential salt accumulation (Fig. 4.8) and, it has been proposed, sites of accelerated weathering related to higher average levels of relative humidity and protection from direct insolation (see previous section). This proposition has, however, always required that at some stage moisture is evaporated from within the rock. This may be achieved through insolation reaching the interior of the hollow later in the day or through a general rise in

Fig. 4.8 Salt concentration at the back of a tafoni, Twyffelfontein, namibia

air temperatures and fall in relative humidity. Work on building stone decay has suggested that drying might be accelerated in honeycomb features by turbulent airflow (Torraca 1981, Rodriguez-Navarro et al. 1999), although Huinink et al. (2004) are unconvinced that the same effect is experienced in larger tafoni. Simulations of the environmental conditions experienced in the caverns created as building blocks retreat (Turkington et al. 2002), combined with field and laboratory analyses of the weathered blocks themselves, suggest however, that within caverns environmental cycles tend to be restricted to the outer few millimetres of the stone surface at a scale that is commensurate with the multiple flakes that are characteristic of many cavern interiors (Smith et al. 2002a).

It would seem, therefore, that moisture is available for weathering in deserts from a variety of sources. Often this moisture is spatially concentrated, but although it may occur frequently (e.g. as dewfall), it is normally available in limited quantities. The concentration of weathering in shadow zones where moisture is retained, rather than in areas experiencing the highest rock surface temperatures suggests that moisture availability is a critical control on processes such as salt weathering in deserts.

Chemical and Biochemical Weathering

Much debate about chemical weathering in deserts concerns the question of how much can be attributed to present-day climatic conditions and how much is

Fig. 4.7 Sidewall tafoni, Valley of Fire, Nevada

inherited from conditions of greater moisture availability. The best-documented use of the latter argument is the work of climatic geomorphologists such as Büdel, who proposed a climatogenetic interpretation of desert landscapes which combined them with tropical savanna and rainforest areas as a zone of Tropical Planation (e.g. Büdel 1982, pp. 120–85). Of paramount importance in this planation process is prior deep weathering, which he envisaged as having occurred at some stage across the whole zone. According to Büdel, most of the deeply weathered material has been removed from present-day desert areas under a geomorphic regime in which long-term fluvial erosion and aeolian deflation exceeds debris supply through weathering. Because of this, much of the evidence of former deep weathering consists not of weathered profiles but of the gross similarity between, for example, the inselberg and plain landscape of tropical West Africa and that of the Sahara to the north. Morphological evidence of this kind must be approached with caution, especially in light of the great antiquity of the land surfaces in these areas and the likelihood of morphological convergence towards landforms and landscapes conditioned by geological controls. In addition, there are few studies and a relatively poor understanding of what constitutes chemical weathering under hot desert conditions and the extent to which chemical processes are currently active in desert environments.

Contemporary Chemical Weathering

The clearest evidence that chemical weathering occurs in hot deserts is the already stated argument that moisture is readily available. This is supported by the widely held belief that salt weathering constitutes the principal mode of rock breakdown (e.g. Cooke and Smalley 1968, Evans 1970, Goudie 1985), which presupposes a major role for moisture absorption by rocks. Indeed, it is arguable whether certain salt weathering mechanisms are not themselves partly chemical in nature. Thus Winkler (1987a), writing about the distinction between chemical and physical weathering, makes the point that: 'on both natural outcrops and urban buildings the process (weathering), turns out to be more complex, and not readily separable into chemical and physical components. Where can we draw

the line between crystallization and hydration pressure, as these are physiochemical processes? There is also evidence of chemical reaction between salts in the masonry and the masonry substance (Arnold 1981) leading to new compounds. Today we still do not understand the physical behaviour and chemical effect of water trapped in capillaries under pressure in most types of stone' (p. 85).

Despite the presumption that desert environments are not conducive to chemical weathering, we can no longer assume that where deeply weathered rock is found it is inherited from moister conditions. In a study of saprolites on igneous rocks in the Negev, Singer (1984) has shown that the climate during their formation has never been moister than semi-arid. Nonetheless, clay contents of 45%, 53%, 11.3%, and 37% were found on andesitic basalt, Na–basanite, microgranite, and microdiorite. Moreover, a wide range of alteration products was identified (Table 4.2), which runs counter to traditional associations of arid conditions with limited alteration of silicate rocks only to smectite clays except under exceptional local circumstances (Barshad 1966, Tardy et al. 1973, Singer 1980). Crystallized iron oxides were also found in the clay mineral fraction of saprolites from basic rocks and argillation seemed to affect mafic minerals preferentially. Significantly, the alteration that Singer observed was not associated with, and thus not dependent upon, significant leaching of the weathered profiles. The presence of supergene mineralisation in the very ancient Atacama Desert has also been interpreted by Clarke (2006) as indicative of long-term weathering under arid conditions, without the possibility of previous periods of more humid climate.

The formation of iron oxides and their movement through rocks by capillary action has also been reported from other deserts. Selby (1977) described red stainings, principally of limonite, but also possibly

Table 4.2 Alteration products of deep weathering in the Negev Desert (Singer 1984)

Parent material	Alteration product
Na–basanite	Dioctahedral smectite
Andesite basalt	Hydroxy: interlayered, high charge smectite
Microgranite	Principally smectite and chlorite with moderate amounts of kaolinite
Microdiorite	Mica: after sericitization of feldspars and osmectite

of goethite, from granites in Namibia associated with scaling features. The staining could sometimes be seen emanating from biotite crystals and a few feldspar crystals showed evidence of moderate kaolinization and sericitization. This, together with some possible chloride formation, was taken to indicate slight chemical weathering. Selby also noted some strong leaching and decomposition of biotites in granites experiencing granular disintegration in the central Sahara. Conca and Rossman (1985) similarly noted haematite coatings on tonalite boulders in California derived from the leaching of biotite, while Osborn and Duford (1981) described streaks of iron oxide running down the sides of sandstone inselbergs in south-western Jordan. They also noted iron-rich weathering rinds up to 1 cm thick on many boulders. The rinds were found to reduce the degree of disintegration when boulders were dropped from cliffs and to offer a degree of protection to the underlying sandstone. A more recent study of these coatings by Goudie et al. (2002b) attributed the case hardening of sections of cliff face to 'iron and other solutions' precipitated by rainwash. Interestingly, they also identified an algal layer just below the case-hardened surfaces. Subsequent detailed analysis of the crusts by Viles and Goudies (2004) found iron, manganese and calcium to be common constituents of the cements within the crusts and and associated them with the presence of cryptoendolithic biofilms containing cyanobacteria and fungi. They are, however, careful to acknowledge that whilst these microorganisms may assist in the formation of the crusts, they are not a necessary component of all such crusts. Similar protection has been assumed by those who see tafoni and other cavernous features as forming through the breaching of weathering rinds. Breaching is followed by excavation of rock behind the rind that was preferentially weakened during rind formation by weathering and outward migration of certain constituents (e.g. Wilhelmy 1964, Winkler 1979). In addition to the surface precipitation of iron compounds, case hardening is also produced by the outward migration in solution and precipitation of calcium carbonate, which may be derived from within the rock or in non-calcareous rocks from lime-rich ground water (e.g. Bryan 1926). Calcite deposition can alternatively weaken rock structure when, for example, it crystallizes within the pores of a quartz sandstone (Laity 1983). The constituents of case hardened coatings can also originate as external

additions of wind-blown dusts which are subsequently mobilized by surface moisture. Conca and Rossman (1982) have described thin case-hardened layers on quartz sandstone in Nevada which contain calcite, hydrated calcium barite, and raised levels of kaolinite of possibly aeolian origin.

Behind these crusts the interior of boulders may be weakened by a process termed core softening by Conca and Rossman (1985). They describe how tonalite boulders in Baja California have undergone preferential decay beneath iron-rich surface layers. This decay consists of kaolinization of feldspars (normally along cleavage planes and grain boundaries), a loss of iron from biotite, and the formation of authigenic silica and traces of calcite and gypsum. They proposed that the weathering is accomplished by migrating capillary moisture. This may have an elevated chemical reactivity due to increased dissolved CO_2 and organic acids if it had previously flowed through the upper soil horizons within which the boulders are partly buried. Weakening of rock structure may similarly be accomplished in quartz sandstone by selective silica dissolution under saline conditions, revealed as solutional etching on grain surfaces (e.g. Young 1987). In light of these observations, Young proposed that this dissolution can lead to the development of cavernous hollows. Mustoe (1983) also made the point that such cavities can be produced by chemical weathering in addition to the normally assumed salt weathering. This weathering is possible given that the cycles of hydration and dehydration required for most salt weathering will also encourage a range of chemical reactions leading, for example, to the decay of feldspars. Silica solubility is also enhanced under saline conditions, and Winkler (1987b) described silica coatings on tuffs in Texas, which he attributed to the 'high solubility of silica in a dry and hot west Texas... enhanced by probably alkalic waters' (p. 975).

The increase in solubility of both quartz and amorphous silica with increasing pH is well established (Goudie 1997) and is reflected in the formation of extensive silicate deposits within the evaporite sequences of alkaline lakes (e.g. Eugster 1967, 1969, Eugster and Jones 1968). The corollary of this increase in solubility is that quartz and silicate rocks will be corroded if they come into contact with alkaline lake water. This is illustrated by Butler and Mount (1986) who described corroded cobbles along former lacustrine strandlines in Death Valley. This corrosion manifests itself in the form of pits and honeycombs, and it derives principally from the dissolution of quartz, feldspars, and phyllosilicates in a variety of rock types. Corrosion occurs preferentially along grain boundaries, laminations, cross beds, and foliations and can enlarge vesicles in extrusive igneous rocks. This process can lead to development of a secondary porosity, which increases moisture penetration and encourages further corrosion but can also increase susceptibility to salt weathering if the water is high in total dissolved solids.

Arid Karst and the Role of Biological Activity

In their study of alkaline corrosion, Butler and Mount (1986) observed little corrosion of limestone or dolomite. Again, it would seem that environmental factors, such as high temperatures and paucity of soil and vegetation giving low dissolved CO_2 contents in desert moisture, generally combine to produce little karstic weathering in deserts. Thus the traditional view is that espoused by Jennings (1983) who concluded that carbonate karst decreases with precipitation. This does not mean that karstic phenomena are absent; simply that they are invariably explained away as relics of former climates (Table 4.3). As such they are described as being destroyed by arid processes including aeolian abrasion (Krason 1961), mechanical weathering (Smith 1978), and fluvial incision (Marker 1972). Some features do, however, manage to retain a degree of development under arid conditions. Thus, Hunt

Table 4.3 Karstic features described from desert environments

Principal features	Location	Source
Dayas (dolines)	Algeria, Morocco	Conrad (1959, 1969), Clark et al. (1974), Castellani and Dragoni (1977)
Caves	Algeria	Conrad et al. (1967)
Solution hollows	Algeria	Quinif (1983)
Caves/hollows/karren	Morocco	Smith (1987)
Cone karst	Egypt	El Aref et al. (1986, 1987)
Caves	Australia	Grodzicki (1985)
Caves	Morocco	Castellani and Dragoni (1987)

et al. (1985) reported active karst on gypsum and anhydrite deposits of the Tripolitanian pre-desert in Libya, and Castellani and Dragoni (1987) described active erosion of solution pipes in south-eastern Morocco. The moisture for the latter erosion condenses as dew and is blown along the limestone surface. The ability of rivulets of coalesced dew to effect solution is further evidenced by studies of building stone, where they can cause serious damage (e.g. Camuffo et al. 1986). It is not likely that dew could initiate features such as extensive solution piping. Instead, the pipes probably formed under conditions of more assured rainfall, but continue to develop slowly at the present time.

Dewfall does, however, appear to play significant direct and indirect roles in the formation of microkarstic features found in deserts (Smith 1988). Wind-driven dew can form microrills or rillensteine (Laudermilk and Woodford 1932) and plays an important part in the solution facetting of limestone pebbles (Bryan 1929) and micropitting of pebbles buried within the vesicular layer beneath stone pavements (Smith 1988). Other active karst described from arid areas have included rillenkarren (Sweeting and Lancaster 1982) and solution pans (Smith 1987).

In addition to highlighting the role of direct precipitation, karst studies have stressed the predominance of features such as pans and pits (Fig. 4.9) which trap moisture (Smith et al. 2000) and the role of organisms in promoting solution (Smith 1988). The general role of microorganisms in weathering and particularly the erosion of limestone has been reviewed by Viles (1988, 1995) Naylor et al. (2002) and their specific role in desert weathering by Cooke et al. (1993). In deserts emphasis has been placed on endolithic cyanobacteria and blue-green algae in, for example, the formation of weathering pits (Danin et al. 1982, Danin and Garty 1983); on endolithic algae in the 'solubilizing' of minerals, including silicates (Friedman 1971); and on lichen in exfoliation (e.g. Friedman 1982). Similar observations were made by Smith et al. (2000) on limstones in southern Tunisa, who identified the active formation of a microkarst associated with endolithic and epilithic algae responsible for algal boring, plucking and etching (Fig. 4.10), and stressed the importance of microclimatic controls on the weathering environment and the interaction of biological processes with active salt weathering. The effectiveness of this synergistic

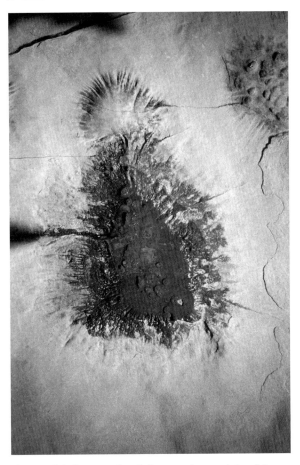

Fig. 4.9 Solution pan, Grand Canyon, characteristic of the active karst and pseudokarst of arid regions that persist because of their ability to trap and concentrate the limited moisture that is available

association has been identified elsewhere where it has contributed to accelerated urban stone decay (Papida et al. 2000). Smith et al. (2000) also drew attention to the occurrence of patches of iron amd manganese

Fig. 4.10 Small scale solution litting of a limestone in southern Tunisia. Each pit supports an epilithic algal community

rock coatings on the limestones. These have been studied in greater detail by Drake et al. (1993), are only found on stable surfaces and are considered to be of possible biogenic origin. Fungi and bacteria occur on both iron and manganese crusts, but rare microcolonial fungi are only found on the latter. On more unstable substrates, microbes were generally found living within, rather than on the surface of the rock. There are also observations which suggest that biological agents can, at least in the short-term, help to stabilize rock surfaces in ways other than facilitating the precipitation of mineralogic crusts (Carter and Viles 2005). For example, Kurtz and Netoff (2001) working in Utah proposed that sandstone surfaces could be temporarily stabilized through the deposition of extracellular polymeric substances and filamentous growths that retarded erosion by wind and water. There is also evidence that, through the retention of moisture, epilithic lichen can reduce the surface thermal stress experienced by rocks such as limestone (Carter and Viles 2003).

The widespread occurrence of microorganisms, especially in deserts such as the Negev, is generally attributed to the high frequency and regularity of dewfalls (Kappen et al. 1979) as a source of essential moisture. Because of the crucial role of dew, especially surface moisture duration (Kidron et al. 2000b), in determining the distribution of microbiotic crusts, the distributions of these organisms show a marked variability with aspect. Lichen, for example, were found by Kappen et al. (1980) to be much more productive on shaded north-facing slopes in the Negev than on drier east-facing slopes exposed to strong insolation from the rising sun. Danin and Garty (1983) similarly found that north-facing slopes were characterized by epilithic lichen, whereas south-facing slopes were inhabited by endolithic cyanobacteria which are more capable of surviving drought conditions (see also Viles 1988). These variations translate through to biological weathering effects, and Stretch and Viles (2002) showed on Lanzarote that lichen grew preferentially on N and NE facing surfaces and that this resulted in a rate of weathering rind formation that was approximately sixteen times that on lichen-free lavas. Such aspect-related variations can also operate at the micro-scale and, for example, Kidron (2002) noted that the the tops of cobbles in the Negev Desert could receive approximately twice as much dew as was precipitated on the sides. Care must, however, be taken in attributing all micro-solutional phenomena to present-day biotic weathering. Danin et al. (1982) found, for example, that although lichen and blue-green algae can produce characteristic jigsaw-like patterns of microgrooves at the boundaries between lichen colonies in Israel, they are only active in Mediterranean areas. Where these grooves occur in arid areas they are interpreted as indicating more humid conditions in the past.

From this very brief overview of chemical weathering, it would seem that there are very few chemical processes that cease completely or are totally excluded under desert conditions. Their rates of action, mode of operation, and frequency of occurrence do change significantly. Limestone solution, for example, does not cease, but no longer produces clearly defined macro- and mesoscale karstic phenomena. Instead it is responsible for a complex variety of microscale features and, it would seem, an undifferentiated slow loss of limestone in solution (Smith 1988).

Weathering of Debris in Deserts

As noted in Cooke et al. (1993), weathering of debris in deserts has tended to be neglected in favour of investigations into rock masses or, at best, boulders of a size that can support features such as cavernous hollows. Yet weathering does not cease with the production of coarse debris, but continues as long as the debris remains at or near the desert surface (Smith 1988). Indeed, weathering of debris exercises an important control upon the overall rate of erosion of many landscapes mantled and protected by stone pavements. Even when such debris is fractured or disaggregated, further comminution may be required before it can be removed by either low energy fluvial erosion or aeolian deflation. A crucial stage in this process is the reduction of sand-sized or coarser debris into silt-sized and finer fragments that are the principal constituents of many wind-borne dusts that blow out of the world's deserts.

Coarse Debris

Clearly, exceptions to the above generalization exist and many experimental studies of salt weathering have used 'debris-sized' blocks of stone (e.g. Goudie et al. 1970, Goudie 1974, Fahey 1986, Cooke 1979). Unfortunately, these studies have rarely been used to explain debris behaviour *per se*. Instead, results have invariably been applied to generalized questions such as relative stone durability, the effectiveness of various salts and salt combinations in causing weight loss, or the nature of weathering mechanisms. Little attention has been paid to patterns of decay or, with some exceptions (e.g. Smith and McGreevy 1983, 1988), to the reproduction of weathering features similar to those observed under field conditions. The effectiveness of salt weathering in causing debris breakdown has been further demonstrated through studies of natural debris (e.g. Goudie and Day 1980) and cut blocks of stone (e.g. Goudie and Cooke 1984, Viles and Goudie 2006, Warke 2007) placed in salt-rich desert environments. Progress has been made despite the problems envisaged earlier in causing mechanical breakdown in small, unconstrained blocks which can themselves expand and contract in response to temperature variations.

Salt weathering is not, however, the only weathering process to operate in deserts, and rock debris should be subject to the full range of weathering processes active in these environments. Such debris could originate in a high mountain environment where it is released from the rock mass by frost-related processes exploiting intrinsic microfractures. It might then experience abrasion and high energy collisions during subsequent fluvial and debris-flow transport and finally be exposed to extreme temperature variations and partial or complete burial in a salt-rich environment such as an alluvial fan. As a result, the debris found mantling many desert surfaces exhibits a wide range of different stress histories, which in turn may explain the variety of weathering patterns observed and the comparative ease with which some debris is seen to breakdown in response, for example, to temperature cycling. The stress history described represents a form of climatic change brought about by spatial relocation. While there has been some work on the combined effects of different weathering regimes on rock breakdown (e.g. frost and salt weathering, Williams and Robinson 1981, McGreevy 1982b, Robinson and Jerwood 1987, Jerwood et al. 1990), only recently has there been any systematic examination of the sequential exposure of rocks to different weathering mechanisms. This is despite the recognition that many weathering mechanisms, such as salt crystallization and wedging by expanding lattice clays, are exploitative in character and only operate in many originally non-porous rocks once fractures or microcracks have been initiated by other mechanisms (e.g. Smith and Magee 1990). Thus, Warke et al. (2006) were able to show how sandstone blocks subjected to alternating combinations of simulated frost and salt weathering underwent a period of internal structural weakening before rapid surface loss was triggerred. The length of this preparatory stage was dependent on the precise combination of frost and salt events and also on the texture/porosity of the sandstone.

A wide range of weathering patterns has been observed on desert debris (see Table 4.4), but it is possible to identify a number of underlying characteristics. First, enhanced moisture and salt availability (via upward migration in low-lying alluvial and lacustrine environments) can produce rapid and thorough disintegration. Secondly, processes such as insolation and salt weathering are enhanced where debris is partially buried within finer material. Partial burial can produce accentuated internal temperature gradients around the exposure and burial boundary, constraint upon the expansion of the buried portion, and possibly enhanced chemical weathering within the subsurface environment. Thirdly, many recent studies have stressed the role of biological agencies in

Table 4.4 Weathering features associated with debris on stone pavements

Weathering phenomena	Sources
Parallel stone cracking	Yaalon (1970), Cooke (1970)
Desert varnish	Verheye (1986), Bowman (1982), Dorn (1983)
Solution pitting	Amit and Gerson (1986), Rögner (1988a)
Vesicular weathering	Smith (1988)
Solution facetting	Bryan (1929)
Solution pinnacles	Joly (1962)
Microrills	Laudermilk and Woodford (1932)
Case hardening	Smith (1988)
Shattered rocks	Ollier (1963), Rögner (1988b)
Polygonal cracking	Soleilhavoup (1977), Bertouille et al. (1979)
Dirt cracking	Ollier (1965)
Granular disintegration	Cooke (1970), Goudie and Day (1980)
Honeycombing	Butler and Mount (1986)
Split/cleaved boulders	Whitaker (1974), Ollier (1984)
Weathering rinds	Osborn and Duford (1981)

facilitating the solution of limestone and the fixation of iron and manganese in desert varnishes. Finally, contrary to normal expectations, weathering need not be dominated by physical breakdown. On limestone debris, for example, it was noted in the previous section that a range of microsolutional features can be observed which are the most likely representatives of a truly arid karst.

Fine Material and the Question of Desert Loess

Comminution of boulder and cobble-sized debris on desert pavements produces a variety of material ranging from rock fragments to individual mineral grains and, less frequently, fragmented grains that can progressively form a complete surface cover (Al-Farraj and Harvey 2000). This material is in turn subject to further weathering and/or attrition which can, under suitable conditions, reduce material to fine sand or silt. This is especially the case in low-lying areas such as the Qattara Depression (Aref et al. 2002) and coastal pans (Viles and Goudie 2007) where salts are available in high concentrations and from where large quantities of the 'dust' produced can potentially be removed by deflation. Much of this material can be transported great distances with wide-ranging consequences (see, for example, Goudie and Middleton 2001 and Tafuro et al. 2006), but of particular interest within this material are the silt-sized particles that ultimately make up the loess (e.g. Yaalon and Dan 1974, Caudé-Gaussen et al. 1982, Caudé-Gaussen 1984) and loess-like deposits of desert marginal areas (e.g. Smith and Whalley 1981, McTainsh 1984). It is generally considered that the transformation of quartz sand into silt requires either the precise application of disruptive stresses or the general application of very large stresses. Thus glacial grinding has been invoked to explain the loess deposits of high latitudes. Indeed, the apparent pre-eminence of glacial theories for loess-silt formation has meant that, even where silt is blown out of deserts such as the Gobi, it has been proposed that it must have originated initially from surrounding mountains (e.g. Smalley and Krinsley 1978).

There is, however, increasing evidence that a number of weathering and other processes that operate in deserts are capable of producing quartz silt of loess size

Table 4.5 Possible origins of loess-sized quartz silt in desert environments

Process	Sources
Salt weathering of loose granular material	Goudie et al. (1979), Pye and Sperling (1983), Peterson (1980)
Salt weathering of sandstone	Goudie (1986), Smith et al. (1987)
Frost weathering	Moss et al. (1981), Lautridou and Ozouf (1982)
Biotite weathering in granitic profiles	Pye (1985)
Inherited tropical deep weathering	Boulet (1973), Leprun (1979), Nahon and Trompette (1982)
Inherited silica dissolution of dune sands	Little et al. (1978), Pye (1983)
Inherited weathering from temperate soils	Pye (1987)
Silica dissolution in saline environments	Young (1987), Magee et al. (1988)
Inherited glacial grinding	Smalley and Vita-Finzi (1968), Boulton (1978)
Aeolian abrasion	Whalley et al. (1982b, 1988)
Fluviatile abrasion	Moss (1972), Moss et al. (1973)
Fracturing and/or dissolution in duricrusts	Nahon and Trompette (1982)
Microdilation in granitic profiles	Nur and Simmons (1970), Moss and Green (1975), White (1976), Power et al. (1990)

(20–60 μm). These processes are reviewed in Table 4.5 and also by Pye (1987), McTainsh (1987), and Smith et al. (2002b). Some of the proposed processes, including salt weathering and aeolian and fluviatile abrasion, clearly act under hot desert conditions. Others, such as frost shattering, are most likely to be effective in either desert marginal areas or 'climatic islands' within deserts induced by altitude. A further and important group comprises processes which operate under moist tropical or temperate conditions, the effects of which may represent climatic inheritance when encountered in present-day deserts. It is a characteristic of many of the mechanisms listed in Table 4.5 that they operate through exploiting flaws in sand-sized and larger grains. These flaws could be inherited microfractures (Moss and Green 1975), cleavage planes (Wilson 1979), or dislocations formed by tectonic deformation (White 1976). Such flaws are not always necessary, and within sandstones fractures could be initiated by compressive loading of sand grains when salts expand within adjacent pores (Smith et al. 1987). It is difficult to envisage, however, how other processes such as salt weathering of loose grains could be effective in the absence of pre-existing flaws or microfractures.

In keeping with larger debris, therefore, sand-sized and finer material exposed at desert surfaces is invariably the product of a complex stress history. This history may represent sequential exposure to different stress mechanisms through climatic fluctuations, long-distance transport, and/or release from parent rock. Grains may additionally be subject to processes such as frost and salt weathering acting alternatively or in conjunction. The end products of exposure to these stresses are either comparatively flawless core grains of the type found in many dune sands or silt-sized fragments of a calibre prone to deflation and removal from the desert environment.

Conclusions and Future Directions

There are many aspects of desert weathering that require further investigation and refinement. In particular, it is clear that the weathering environments of hot deserts are numerous and varied. We need much more information on the range of environmental conditions experienced before rock responses can be fully understood. Moreover, the distinction must be made between conditions as measured for meteorological purposes and the temperature and moisture regimes experienced at the rock–atmosphere interface. The overall significance of weathering achieved within these environments cannot be appreciated until (a) we improve the precision with which weathering damage is recognized and assessed, and (b) we begin to investigate the overall contribution of weathering not only in the development of individual landforms but also in the long-term evolution of desert landscapes as a whole.

Damage Recognition and Assessment

A criticism levelled at many early simulation studies of rock weathering is that they dismissed weathering mechanisms such as insolation-related fatigue failure, simply because they employed techniques of damage assessment insufficiently sensitive to register changes that might have been expected over the duration of the experiment (see Rice 1976). Similarly, many weathering changes are threshold phenomena, in that specimens may exhibit little surface change while experiencing internal changes that can lead to sudden and catastrophic damage as strength thresholds are breached and, for example, large contour scales break away. As a result, damage assessment by weight loss alone can lead to misleading estimates of relative durability if simulation experiments are of limited duration. Alternatively, there is a temptation to increase the aggressivity of such tests to the point that, while measurable damage is achieved, conditions no longer relate to those encountered in nature. There is a need, therefore, for more precise and consistent methods of assessing surface and internal changes in rock properties consequent upon weathering. Ideally such methods should be applicable under both field and laboratory conditions to permit work to be integrated between the two situations.

Considerable progress has been made in recent years in terms of damage assessment, including the development of a rigorous terminology for defining weathering damage. A major driving force in providing this sytematic information has been 'crossover' research into the decay and conservation of stone monuments within arid enviroments. The leading proponents of this have been Fitzner and co-workers at Aachen who have devised a detailed and extremely comprehensive classification of weathering forms (e.g. Fitzner et al. 1992, 1995). They have in turn applied this to historic monuments in Cairo (Fitzner et al. 2002), in which they tailored an interpretation of weathering form and intensity to the petrological characteristics of the limestones used as well as environmental conditions. Other applications of this approach have included the deterioration of sanstone monuments in Petra (Heinrichs and Fitzner 2000), the weathering of Pharaonic monuments in Luxor (Fitzner et al. 2003a,b) and the use of damage assessment linked to diagnosis to identify a weathering progression at Petra (Heinrichs 2002). Petra has also been used by other researchers as a 'natural environmental laboratory' for weathering studies (e.g. Paradise 2005). The regularity of construction makes such cities, and individual buildings, ideal for the isolation of controls such as aspect on weathering processes (e.g. Paradise 2002). Their known age greatly facilitates studies of weathering rates under different environmental conditions and for different stone types (Paradise 1998) as well as the identification of decay thresholds (Paradise 1995).

Elsewhere, weathering rates and surface change over time have been established using a variety of strategies, including the analyses of previous conservation interventions and records (Selwitz 1990, Wüst 2000, Wüst and Mclane 2000), the protrusion of lead lettering (Klein 1984) and a movable frame for plotting surface contours (Sancho et al. 2003). Typically, such studies have also been associated with a forensic analysis of the immediate and underlying causes of decay as the first step towards the selection of appropriate conservation strategies. This includes not just external influences such as the origins and impact of soluble salts (e.g. Wüst and Mclane 2000), but also an emphasis on the importance of geological controls (e.g. Rodriguez-Navarro et al. 1997) that is sometimes absent from or underplayed in geomorphological studies.

The Need for Integration

There has long been a tendency in studies of desert weathering to isolate individual weathering mechanisms and to seek an understanding of weathering phenomena through segregation rather than integration. This tendency has, as indicated in the introduction, reflected a general trend in geomorphological investigations which have become increasingly process orientated and reductionist in character. Like these wider studies, the understanding of desert weathering mechanisms is not an end in itself; ultimately such mechanisms must be set within wide temporal and spatial contexts to understand the landscapes within which weathering occurs (Smith 1987, Turkington and Paradise 2005, Turkington et al. 2005). In doing this, it must be recognized that weathering mechanisms do not operate in isolation from one another and that integration must take place at a variety of levels. There is a need, for example, to integrate across the scale boundaries that separate the detailed understanding of weathering at the nano- and microscales from their role in the evolution of complex landforms and landscapes, especially through its control on debris character and rates of debris supply.

In relation to the up-scaling of weathering process and effects to the landform scale and beyond, Viles (2001) saw the identification of the spatio-temporal scales of weathering phenomena as an essential first step. In pursuit of this she tabulates weathering features in terms of their scale of operation, together with the factors that control weathering and how these change across a range of scales. This work is also an acknowledgement that many of the potential approaches to the crossing of scale boundaries are rooted in the interpretation of geomorphic phenomena as non-linear dynamical systems. This interpretation builds upon the work of Goudie and Viles (1999) in which they demonstrated the applicability of the magnitude/frequency concept to understanding the operation of a wide range of weathering processses. They concluded in turn that the magnitude and frequency of weathering events is largely controlled by climate, but that climate itself comprises different magnitude and frequency distributions including cyclical and non-cyclical events. This theme was later developed (Viles and Goudie 2003) through a specific examination of the links between climatic variability at different scales and geomorphic activity in general.

The non-linearity of the behaviour of weathering systems has been further investigated by Phillips (2005), through an analysis of their stability and instability at four different scales concerned with weathering processes, the allocation of weathering products, the interrelation of weathering and denudation and topographic and isostatic responses to weathering. Following a detailed analysis of these systems, he was led to conclude that instability is prevalent at local spatial scales, except over the longest timescales. At intermedite spatial scales stability is contingent upon the strength of the feedbacks between weathering and erosion – with stability more likely at shorter than longer timescales. Over the largest spatial scales, instability is likely, except possibly over intermediate timescales if weathering and erosion feedbacks are weak. As he points out, the significance of these analyses lies in the tendency for stable systems to experience convergent evolution, whereas instability favours divergent evolution in which original heterogeneities in the landscape tend to be accentuated over time. An understanding of systems behaviour at this level is essential if we are going to progress desert weathering beyond its exploration and descovery phase to the point where the interactions between process, form, material and environment can be modelled and future change predicted.

One area where the challenge of modelling system behaviour has been taken up in recent years is the evolution of cavernous hollows (honecombs and tafoni). As Viles (2005b) has pointed out, there have been a number of recent studies that have sought to model cavern development and raise the question of whether there are common underlying controls that cut across scale, lithological and broad environmental boundaries. For many years it has been proposed that caverns develop to optimise internal environmental conditions that favour the operation of processes such as salt weathering (e.g. Smith 1978), but more recently, Turkington and Phillips (2004) have formalised this concept by proposing that caverns develop as a self-organised response to dynamical instability within the weathering system. Although, as Phillips (1999) notes, self-organisation is a broad umbrella which encompasses concepts that range from internal adjustment towards the maintenance of a steady state; through to the tendency for natural systems to evolve towards a critical threshold. The potential for complexity in the application of this paradigm is demonstrated in their paper by observation of the disparity between caverns, which tend to minimise surface area through development towards a spherical interior, and the tendency for the outcrops on which they occur to progress towards maximum exposed surface as caverns develop. A similar invocation of self-organisation was presented by McBride and Picard (2004), although this time in relation to the initial distribution of salt-rich water within the rock and its potential role in controlling the location of caverns. Lastly, Huinink et al. (2004) use a detailed analysis of salt migration to model the development of a single cavern. As Viles (2005b) observes, the appearance of these three studies in the same journal volume highlights the opportunities and need for integration, not only across scales but also between field and laboratory, between different data sets and, possibly most important, across the mindsets of researchers coming from different backgrounds.

References

Aires-Barros, L. 1977. Experiments on thermal fatigue of non-igneous rocks. *Engineering Geology* **11**, 227–38.

Aires-Barros, L. 1978. Comparative study between rates of experimental laboratory weathering of rocks and their natural environmental weathering decay. *Bulletin of the International Association of Engineering Geology* **18**, 169–74.

Aires-Barros, L., R.C. Graça and A. Velez 1975. Dry and wet laboratory tests and thermal fatigue of rocks. *Engineering Geology* **9**, 249–65.

Al-Farraj, A. and A.M. Harvey 2000. Desert pavement characteristics on Wadi terrace and alluvial fan surfaces: Wadi Al-Bih, U.A.E. and Oman. *Geomorphology* **35**, 279–297.

Amit, R. and R. Gerson 1986. The evolution of Holocene reg (gravelly) soils in deserts – an example from the Dead Sea region. *Catena* **13**, 59–79.

Amoroso, G.G. and V. Fassina 1983. *Stone decay and conservation*. Amsterdam: Elsevier.

Aref, M.A.M., E. El-Khoriby and M.A. Hamdan 2002. The role of salt weathering in the origin of the Qattara Depression, Western desertm Egypt. *Geomorphology* **45**, 181–195.

Arnold, A. 1981. Nature and reactions of saline minerals in walls. In *The conservation of stone II*, 13–23. Bologna: Centro Cesare Gnudi.

Arnold, A. 1982. Rising damp and saline minerals. In *Proceedings of the 4th International Congress on Deterioration and Preservation of Stone Objects*, K.L. Gauri and J.A. Gwinn (eds) 11–28. Louisville, KY: University of Louisville.

Ashton, H.E. and P.J. Sereda 1982. Environment, microenvironment and durability of building materials. *Durability of Building Materials* **1**, 49–65.

Bache, H.H. 1985. Durability of concrete fracture mechanical aspects. *Nordic Concrete Research Publication* **4**, 7–25.

Barshad, I. 1966. The effects of variation of precipitation on the nature of clay minerals formed in soils from acid and basic igneous rocks. *Proceedings of International Clay Conference, Jerusalem* **1**, 167–73.

Batzle, M.L., G. Simmons and R.W. Siegfried 1980. Micro–crack closure in rocks under stress: direct observation. *Journal of Geophysical Research* **185**, 7072–7079.

Benavente, D., M.A. García del Cura, A. Bernabéu and S. Ordóñez 2001. Quantification of salt weathering in porous stones using an experimental continuous partial immersion method. *Engineering Geology* **59**, 313–325.

Benavente, D., M.A. García del Cura, R. Fort and S. Ordóñez 1999. Thermodynamic modelling of changes induced by salt pressure crystallization in porous media of stone. *Journal of Crystal growth* **204**, 168–178.

Bernardi, A., D. Camuffo and C. Sabbioni 1985. Microclimate and weathering of a historical building: the Ducal Palace in Urbino. *The Science of the Total Environment* **46**, 243–60.

Bertouille, H., J.P. Coutard, F. Soleilhavoup and J. Pellerin 1979. Les galets fissures: étude de la fissuration des galets Sahariens et comparaison avec issus de diverses zones climatiques. *Révue de Géomorphologie* **28**, 33–48.

Blackwelder, E. 1925. Exfoliations as a phase of rock weathering. *Journal of Geology* **33**, 793–806.

Blackwelder, E. 1933. The insolation hypothesis of rock weathering. *American Journal of Science* **266**, 97–113.

Boulet, R. 1973. Toposéquence de sols tropicaux en Haute Volta. Equilibre et deséquilibre pédobioclimatique. *Memoires ORSTOM* **85**, 1–272.

Boulton, G.S. 1978. Boulder shapes and grain size distribution of debris as indicators of transport paths through a glacier and till genesis. *Sedimentology* **25**, 773–99.

Bowman, D. 1982. Iron coating in recent terrace sequences under extremely arid conditions. *Catena* **9**, 353–9.

Bradley, W.C. 1963. Large-scale exfoliation in massive sandstones of the Colorado Plateau. *Bulletin of the Geological Society of America* **74**, 519–28.

Bryan, K. 1925. The Papago Country, Arizona. *U.S. Geological Survey Water-Supply Paper* 499.

Bryan, K. 1926. Niches and other cavities in sandstone at Chaco Canyon, New Mexico. *Zeitschrift für Geomorphologie* **3**, 125–40.

Bryan, K. 1929. Solution facetted limestone pebbles. *American Journal of Science* **18**, 193–208.

Büdel, J. 1982. *Climatic geomorphology*, Princeton, NJ: Princeton University Press.

Butler, P.R. and J.F. Mount 1986. Corroded cobbles in southern Death Valley: their relationship to honeycomb weathering and lake shorelines. *Earth Surface Processes and Landforms* **11**, 377–87.

Camuffo, D. 1983. Indoor dynamic climatology: investigations on the interactions between walls and indoor environment. *Atmospheric Environment* **17**, 1803–9.

Camuffo, D. 1984. Condensation–evaporation cycles in pore and capillary systems according to the Kelvin model. *Water, Air and Soil Pollution* **21**, 151–9.

Camuffo, D. 1998. *Microclimate for Cultural Heritage*. Developments in atmospheric science, **23**. Amsterdam: Elsevier.

Camuffo, D., A. Bernardi and A. Ongaro 1986. The challenges of the microclimate and the conservation of works of art. *2emes Rencontres Internationals pour la Protection du Patrimoine Cultural, Avignon, Nov. 1986*, 215–28.

Carter, N.E.A. and H.A. Viles 2003. Experimental investigations into the interactions between moiture, rock surface temperatures and an epilithic lichen cover in the bioprotection of limestone. *Building and Environment* **38**, 1225–1234.

Carter, N.E.A. and H.A. Viles 2004. Lichen hotspots: raised temperature beneath *Verrucaria nigrescens* on limestone. *Geomorphology* **62**, 1–16.

Carter, N.E.A. and H.A. Viles 2005. Bioprotection explored: the story of a little known earth surface process. *Geomorphology* **67**, 273–281.

Castellani, V. and W. Dragoni 1977. Surface Karst landforms in the Moroccan Hamada of Guir. *Proceedings of the 7th Congress of Speleology*, 98–101.

Castellani, V. and W. Dragoni 1987. Some considerations regarding Karstic evolution of desert limestone plateaus. In *International geomorphology*, V. Gardiner (ed.), 1199–206. Chichester: Wiley.

Caudé-Gaussen, G. 1984. Le cycle des poussieres éoliennes désertiques actuelles et la sédimentation des peridésertiques Quaternaires. *Bulletin des Centres de Récherches Exploration-Production Elf Aquitaine* **8**, 167–82.

Caudé-Gaussen, G., C. Mosser, P. Rognon and J. Torenq 1982. Une accumulation de loess Pleistocene superient dans le sud-Tunisien: la coupe de Techine. *Bulletin de la Société Géologique de France* **24**, 283–92.

Clark, D.M., C.W. Mitchell and J.A. Varley 1974. Geomorphic evolution of sediment-filled solution hollows in some arid regions. *Zeitschrift für Geomorphologie Supplement Band* **20**, 193–208.

Clarke, J.D.A. 2006. Antiquity of aridity in the Chilean Atacama Desert. *Geomorphology* **73**, 101–114.

Cloudsley-Thompson, J.L. and M.J. Chadwick 1969. *Life in deserts*. London: Foulis.

Conca, J.L. and G.R. Rossman 1982. Case hardening of sandstone. *Geology* **10**, 520–3.

Conca, J.L. and G.R. Rossman 1985. Core softening of cavernously weathered tonalite. *Journal of Geology* **93**, 59–73.

Conrad, G. 1959. Observations preliminaires sur la sedimentation dans les daias de la Hamada du Guir. *Compte Rendu Société Géologique Français* **7**, 156–62.

Conrad, G. 1969. *L'évolution continentale post-Hercynienne du Sahara Algerien*. Paris: Centre National du Récherche Scientifique.

Conrad, G., B. Geze and Y. Paloc 1967. Phénomenes Karstiques et pseudokarstiques du Sahara. *Révue de Géographie Physique et Géologie Dynamique* **9**, 357–69.

Cooke, R.U. 1970. Stone pavements in deserts. *Annals of the Association of American Geographers* **4**, 560–77.

Cooke, R.U. 1979. Laboratory simulation of salt weathering processes in arid environments. *Earth Surface Processes and Landforms* **4**, 347–59.

Cooke, R.U. and I.J. Smalley 1968. Salt weathering in deserts. *Nature* **220**, 1226–7.

Cooke, R.U. and A. Warren 1973. *Geomorphology in deserts*. London: Batsford.

Cooke, R.U., D. Brunsden, J.C. Doornkamp and D.K.C. Jones 1982. *Urban geomorphology in drylands*. Oxford: Oxford University Press.

Cooke, R.U., A. Warren and A.S. Goudie 1993. *Desert geomorphology*. London: UCL Press.

Correns, C.W. 1949. Growth and dissolution of crystals under linear pressure. *Discussions of the Faraday Society* **5**, 267–71.

Danin, A. and J. Garty 1983. Distribution of cyanobacteria and lichens on hillsides of the Negev highlands and their impact on biogenic weathering. *Zeitschrift für Geomorphologie* **27**, 423–44.

Danin, A., R. Gerson, K. Marton and J. Garty 1982. Patterns of limestone and dolomite weathering by lichens and blue-green algae and their palaeoclimatic significance. *Palaeogeography, Palaeoclimatology, Palaeoecology* **37**, 221–33.

Del Monte, M. and C. Sabbioni 1984. Gypsum crusts and fly ash particles on carbonate outcrops. *Archives for Meteorology, Geophysics and Bioclimatology Series B* **35**, 105–11.

Del Monte, M. and O. Vittori 1985. Air pollution and stone decay: the case of Venice. *Endeavour, New Series* **9**, 117–22.

Del Monte, M., C. Sabbioni, A. Ventura and G. Zappia 1984. Crystal growth from carbonaceous particles. *The Science of the Total Environment* **36**, 247–54.

Doehne, E. 2002. Salt weathering: a selective review. *Geological Society of London Special Publication* **205**, 51–64.

Dorn, R.I. 1983. Cation ratio dating: a new rock varnish age-determination technique. *Quaternary Research* **20**, 49–73.

Dragovich, D. 1967. Flaking, a weathering process operating on cavernous rock surfaces. *Bulletin of the Geological Society of America* **78**, 801–4.

Dragovich, D. 1969. The origin of cavernous surfaces (tafoni) in granitic rocks of Southern Australia. *Zeitschrift für Geomorphologie* **13**, 163–81.

Dragovich, D. 1981. Cavern microclimates in relation to preservation of rock art. *Studies in Conservation* **26**, 143–9.

Drake, N.J., M.T. Heydeman and K.H. White 1993. Distribution and formation of rock varnish in Southern Tunisia. *Earth Surface Processes and Landforms* **18**, 31–41.

Dubief, J. 1965. Le probleme de l'eau superficielle au Sahara. *La Méteorologie* **77**, 3–32.

Duvdevani, S. 1947. Dew in Palestine. *Nature* **159**, 4041.

Duvdevani, S. 1953. Dew gradients in relation to climate, soil and topography. *Desert Research Symposium, Jerusalem*, 136–52.

El Aref, M.M., F. Awadalah and S. Ahmed 1986. Karst landform development and related sediments in the Miocene rocks of the Red Sea coastal zone. *Geologie Rundschau* **76**, 781–90.

El Aref, M.M., A.M. Abou Khadrah and Z.H. Lofty 1987. Karst topography and karstification processes in the Eocene limestone plateau of El Bhaharia Oasis, Western Desert, Egypt. *Zeitschrift für Geomorphologie* **31**, 45–64.

Eugster, H.P. 1967. Hydrous sodium silicates from Lake Magadi, Kenya: precursors of bedded chert. *Science* **157**, 1177–80.

Eugster, H.P. 1969. Inorganic bedded cherts from the Magadi area, Kenya. *Contributions to Mineralogy and Petrology*, **22**, 1–31.

Eugster, H.P. and B.F. Jones 1968. Gels composed of sodium aluminium silicate, Lake Magadi, Kenya. *Science* **162**, 160–4.

Evans, I.S. 1970. Salt crystallization and rock weathering. *Revue de Géomorphologie Dynamique* **19**, 153–77.

Evenari, M., L. Shanan and N. Tadmor 1971. *The Negev*. Cambridge, MA: Harvard University Press.

Fahey, B.D. 1986. A comparative laboratory study of salt crystallisation and salt hydration as potential weathering agents in deserts. *Geografiska Annaler* **68A**, 107–11.

Fitzner, B. Heinrichs, K. and La Bouchardiere, D. 2003a. Limestone weathering of historical monuments in Cairo, Egypt. *Geological Society of London Special Publication* **205**, 217–239.

Fitzner, B., K. Heinrichs and D. La Bouchardiere 2003b. Weathering damage on Pharaonic sandstone monuments in Luxor-Egypt. *Building and Environment* **38**, 1089–1103.

Fitzner, B., K. Heinrichs and R. Kownatzki 1992. Classification and mapping of weathering forms. In *Proceedings of the 7th International Congress on Deterioration and Conservation of Stone*, J. Delgado-Rodriguez (ed.), 15–18. Lisbon: Laboratorio Nacional de Enghenaria Civil.

Fitzner, B., K. Heinrichs and R. Kownatzki 1995. Weathering forms–classification and mapping. In *Verwitterungsformen-Klassifizierung und Kartierung. Denkmalpflege und Naturwissenschaft, Natursteinkonservierung 1*, 41–88. Berlin: Verlag Ernst & Sohn.

Flatt, R.J. 2002. Salt damage in porous materials: how high supersaturation is generated. *Journal of Crystal Growth* **242**, 435–454.

Folk, R.L. and E. Begle-Patton 1982. Buttressed expansion of granite and development of grus in Central Texas. *Zeitschrift für Geomorphologie* **26**, 17–32.

Fookes, P.G. 1976. Middle East – inherent ground problems. In *Geological Society, Conference on Engineering Problems, November 1976*, 7–23.

Fookes, P.G., C.S. Gourlay and C. Ohikere 1988. Rock weathering in engineering time. *Quarterly Journal of Engineering Geology* **21**, 33–57.

Friedman, E.I. 1971. Light and scanning electron microscopy of endolithic desert algal habitat. *Phycologia* **10**, 411–28.

Friedman, E.I. 1982. Endolithic microorganisms in the Antarctic cold desert. *Science* **215**, 1045–53.

George, V. 1986. The thermal niche: desert sand and desert rock. *Journal of Arid Environments* **10**, 213–24.

Gómez-Heras, M., R. Fort González and B.J. Smith 2004. Problemas de escala en la interpretación del deterioro térmico por insolación de piedra de construcción. *Geo–temas* **6**, 263–266.

Gómez-Heras, M., B.J. Smith and R. Fort 2006. Surface temperature differences between minerals in crystalline rocks: implications for granular disaggregation of granites through thermal fatigue. *Geomorphology* **78**, 236–249.

Goudie, A.S. 1974. Further experimental investigation of rock weathering by salt and other mechanical processes. *Zeitschrift für Geomorphologie Supplement Band* **21**, 1–12.

Goudie, A.S. 1985. Salt weathering. *Oxford School of Geography Research Paper* **8**.

Goudie, A.S. 1986. Laboratory simulation of the 'wick effect' in salt weathering of rock. *Earth Surface Processes and Landforms* **11**, 275–85.

Goudie, A.S. 1989. Weathering Processes. In *Arid Zone Geomorphology*, D.S.G. Thomas (ed.), 11–24. London: Belhaven Press.

Goudie, A.S. 1997. Weathering Processes. In *Arid zone geomorphology (2nd Edition)*, D.S.G. Thomas (ed.), **25–39**. Chichester: Wiley.

Goudie, A.S. 1993. Salt weathering simulation using a single–immersion technique. *Earth Surface Processes and Landforms* **18**, 369–76.

Goudie, A.S. and R.U. Cooke 1984. Salt efflorescences and saline lakes; a distributional analysis. *Geoforum* **15**, 563–82.

Goudie, A.S. and M.J. Day 1980. Disintegration of fan sediments in Death Valley, California by salt weathering. *Physical Geography* **1**, 126–37.

Goudie, A.S. and N.J. Middleton 2001. Saharan dust storms: nature and consequences. *Earth Science Reviews* **56**, 179–204.

Goudie, A.S. and A.G. Parker 1998. Experimental simulation of rapid rock block disintegration by sodium chloride in a foggy coastal desert. *Journal of Arid Environments* **40**, 347–355.

Goudie, A.S. and A. Watson 1984. Rock block monitoring of rapid salt weathering in southern Tunisia. *Earth Surface Processes and Landforms* **9**, 95–8.

Goudie, A.S. and H.A. Viles 1997. Salt weathering. Chichester, UK: Wiley, 241pp.

Goudie, A.S. and H.A. Viles 1999. The frequency and magnitude concept in relation to rock weathering. *Zeitschrift für Geomorphologie Supplement* **115**, 175–89.

Goudie, A.S. and H.A. Viles 2000. The thermal degradation of marble. *Acta Universitatis Carolinae, Geographica* XXXV, 7–16.

Goudie, A.S., R.U. Cooke and I. Evans 1970. Experimental investigation of rock weathering by salts *Area* **4**, 42–8.

Goudie, A.S., R.U. Cooke and J.C. Doornkamp 1979. The formation of silt from quartz dune sand by salt processes in deserts. *Journal of Arid Environments* **2**, 105–12.

Goudie, A.S., H.A. Viles and A.G. Parker 1997. Monitoring of rapid salt weathering in the central Namib Desert using limestone blocks. *Journal of Arid Environments* **37**, 581–598.

Goudie, A.S., Wright, E. and Viles, H.A. 2002a. The roles of salt (sodium nitrate) and fog in weathering: A laboratory simulation of conditions in the northern Atacama desert, Chile. *Catena* **48**, 255–266.

Goudie, A.S., P. Migon, R.J. Allison and N. Rosser 2002b. Sandstone geomorphology of the Al-Quwayra area of south Jordan. *Zeitschrift für Geomorphologie* **46**, 165–190.

Griffith, A.A. 1921. The phenomenon of rupture and flow in solids. *Philosophical Transactions of the Royal Society* **A221**, 163–198.

Griggs, D.T. 1936. The factor of fatigue in rock exfoliation. *Journal of Geology* **74**, 733–96.

Grodzicki, J. 1985. Genesis of Nullarbor caves in southern Australia. *Zeitschrift für Geomorphologie* **29**, 37–49.

Hall, K. 1999. The role of thermal stress fatigue in the breakdown of rock in cold regions. *Geomorphology* **31**, 47–63.

Hall, K. and M–F. Andre 2001. New insights into rock weathering from high–frequency rock temperature data: An Antarctic study of weathering by thermal stress. *Geomorphology* **41**, 23–35.

Hall, K. and M–F. Andre 2003. Rock thermal data at the grain scale: applicability to granular disintegration in cold environments. *Earth Surface Processes and Landforms* **28**, 823–836.

Hall, K. and A. Hall 1991. Thermal gradients and rock weathering at low temperatures: Some simulation data. *Permafrost and Periglacial Processes* **2**, 103–112.

Hall, C., W.D. Hoff and M.R. Nixon 1984. Water movement in porous building materials VI. Evaporation and drying in brick and block materials. *Building and Environment* **19**, 13–20.

Haneef, S.J., C. Dickinson, J.B. Johnson, G.E. Thompson 1990a. Degradation of coupled stones by artificial acid rain solution. In *Science technology and the european cultural heritage*, N.S. Baer, C. Sabbioni and A.I. Sors (eds), 469–73, Oxford: Butterworth-Heinemann.

Haneef, S.J., C. Dickinson, J.B. Johnson, G.E. Thompson 1990b. Laboratory-based testing and simulation of dry and wet deposition of pollutants and their interaction with building stone. *European Cultural Heritage Newsletter* **4**, 9–16.

Heinrichs, F. and B. Fitzner 2000. Deterioration of rock monuments in Petra/Jordan. In *Proceedings, 9th International Congress on Deterioration and Conservation of Stone*, V. Fassina (ed.), 53–61. Amsterdam: Elsevier.

Heinrichs, K. 2002. Weathering progression on rock-cut monuments in Petra, Jordan. In *Science and technology in Archaeological Coservation*, T.S Akasheh (ed.), 255–279. Granada: Fundación El Legado Andalusia.

Hunt, C.O., S.J. Gale and D.D. Gilbertson 1985. The UNESCO Libyan valleys survey IX: anhydrite and limestone karst in the Tripolitanian pre-desert. *Libyan Studies* **16**, 1–13.

Huinink, H.P., Pel, L. and Kopinga, K. 2004. Simulating the growth of tafoni. *Earth Surface Processes and Landforms* **29**, 1225–1233.

Jäkel, D. and H. Dronia 1976. Ergebrisse von boden- und Gesteintemperatur-messungen in der Sahara. *Berliner Geographische Abhandlungen* **24**, 55–64.

Janach, W. 1977. Failure of granite under compression. *International Journal of Rock Mechanics and Mining Science and Geomechanics Abstracts* **14**, 209–215.

Jenkins, K.A. and B.J. Smith 1990. Daytime rock surface temperature variability and its implications for mechanical rock weathering: Tenerife, Canary Islands. *Catena* **17**, 449–59.

Jennings, J.N. 1983. The disregarded Karst of the arid and semi-arid domain. *Karstologia* **1**, 61–73.

Jerwood, L.C., D.A. Robinson and R.B.G. Williams 1990. Experimental frost and salt weathering of Chalk – 1. *Earth Surface Processes and Landforms* **15**, 611–24.

Joly, F. 1962. Etude sur le relief du sud-est Morocain. *Institute Scientifique Chérifien, Travaux* 10.

Kane, H.L. 1999. *Background Controls on the Weathering of Granite in Polluted Environments.* Unpublished PhD. Thesis, Queen's University, Belfast, UK, 348pp.

Kappen, L. O.L. Lange, E.-D. Schulze, M. Evenari 1979. Ecophysiological investigations on lichens of the Negev Desert. VI. Annual course of the photosynthetic production. *Flora* **168**, 85–108.

Kappen, L., O.L. Lange, E.-D. Schulze, U. Buschbom 1980. Ecophysiological investigations on lichens of the Negev Desert VII. The influence of the habitat exposure on dew inhibition and photosynthetic productivity. *Flora* **169**, 216–29.

Kerr, A., B.J. Smith, W.B. Whalley and J.P. McGreevy 1984. Rock temperature measurements from southeast Morocco and their significance for experimental rock weather studies. *Geology* **12**, 306–9.

Kidron, G.J. 1999. Altitude dependent dew and fog in the Negev Desert, Israel. *Agricultural and Forest Meteorology* **96**, 1–8.

Kidron, G.J. 2000. Analysis of dew precipitation in three habitats within a small arid drainage basin, Negev Highlands, Israel. *Atmospheric Research* **55**, 257–270.

Kidron, G.J. 2002. Causes of two patterns of lichen zonation on cobbles in the Negev Desert, Israel. *Lichenologist* **34**, 71–80.

Kidron, G.J., A. Yair and A. Danin 2000a. Dew variability within a small arid drainage basin in the Negev Highlands, Israel. *Quarterly Journal of the Royal Meteorological Society* **126**, 63–80.

Kidron, G.J., E. Barzilay and E. Sachs 2000b. Microclimatic control upon sand microbiotic crusts, western Negev Desert, Israel. *Geomorphology* **36**, 1–18.

Klein, M. 1984. Weathering rates of limestone tombstones measured in Haifa, Israel. *Zeitschrift für Geomorphologie* **28**, 105–11.

Krason, J. 1961. The caves in Maestrichtian limestone in the Arabic Desert. *Hohle* **12**, 57.

Kurtz, H.D.Jr. and D.I. Netoff 2001. Stabilization of friable sandstone surfaces in a desiccating, wind-abraded environment of south-central utah by rock surface microorganisms. *Journal of Arid Environments* **48**, 89–100.

La Iglesia, A., V. González, V. López-Acevedo and C. Viedma 1997. Salt crystallization in porous construction materials I Estimation of crystallization pressures. *Journal of Crystal Growth* **177**, 111–8.

Laity, J.E. 1983. Diagenetic controls on groundwater sapping and valley formation, Colorado Plateau, revealed by optical and electron microscopy. *Physical Geography* **4**, 103–25.

Laudermilk, J.D. and A.O. Woodford 1932. Concerning rillensteine. *American Journal of Science* **23**, 135–54.

Lautridou, J.P. and J.C. Ozouf 1982. Experimental frost shattering: fifteen years of research at the Centre de Geomorphologie du CNRS. *Progress in Physical Geography* **6**, 215–32.

Leprun, J.C. 1979. *Les cuirasses ferrugineuses des pays cristallins de l'Afrique Occidentale séche: genese, transformations, degradation.* Unpublished Doctorate thesis. Université de Strasbourg.

Lewin, S.Z. 1981. The mechanism of masonry decay through crystallization. In *Conservation of historic stone buildings and monuments*, S.M. Barkin (ed.), 1–31. Washington, D.C.: National Academy of Sciences.

Lewin, S.Z. and A.E. Charola 1979. The physical chemistry of deteriorated brick and its impregnation techniques. In *Atti del Convegno Internazionale il Mattone di Venezia, Venice October 22–23*, 189–214 (cited by Amoroso and Fassina 1983).

Little, I.P., T.M. Armitage and R.J. Gilkes 1978. Weathering of quartz in dune sands under subtropical conditions in Eastern Australia. *Geoderma* **20**, 225–37.

Magee, A.W., P.A. Bull and A.S. Goudie 1988. Chemical textures on quartz grains. An experimental approach using salts. *Earth Surface Processes and Landforms* **13**, 665–76.

Marker, M.E. 1972. Karst landform analysis as evidence for climatic change in the Transvaal, South Africa. *South African Geographical Journal* **54**, 152–62.

Marschner, H. 1978. Application of salt crystallization test to impregnated stones. In *Deterioration and Protection of Stone Monuments.* UNESCO/RILEM Conference, Paris, Section 3.4.

McBride, E.F. and M.D. Picard 2004. Origin of honeycombs and re,ated weathering forms in Oligocene Macigno sandstone, Tuscan coast near Livorno,Italy. *Earth Surface Processes and Landforms* **29**, 713–735.

McGreevy, J.P. 1982a. Hydrothermal alteration and earth surface rock weathering: a basalt example. *Earth Surface Processes and Landforms* **7**, 189–95.

McGreevy, J.P. 1982b. Frost and salt weathering: Further experimental results. *Earth Surface Processes and Landforms* **7**, 475–88.

McGreevy, J.P. 1985. Thermal rock properties as controls on rock surface temperature maxima, and possible implications for rock weathering. *Earth Surface Processes and Landforms* **10**, 125–36.

McGreevy, J.P. and B.J. Smith 1982. Salt weathering in hot deserts: observations on the design of simulation experiments. *Geografiska Annaler* **64A**, 161–70.

McGreevy, J.P. and B.J. Smith 1983. Salt weathering in hot deserts: observations on the design of simulation experiments. A reply. *Geografiska Annaler* **65A**, 298–302.

McGreevy, J.P. and B.J. Smith 1984. The possible role of clay minerals in salt weathering. *Catena* **11**, 169–75.

McTainsh, G. 1984. The nature and origin of aeolian mantles in northern Nigeria. *Geoderma* **33**, 13–37.

McTainsh, G. 1987. Desert loess in northern Nigeria. *Zeitschrift für Geomorphologie* **31**, 145–65.

Merrill, G.P. 1906. *A treatise on rocks, rock weathering and soils.* London: Macmillan.

Migon, P., A.S. Goudie, R. Allison and Rosser, N. 2005. The origin and evolution of footslope ramps in the sandstone desert environment of south-west Jordan. *Journal of Arid Environments* **60**, 303–320.

Minty, E.J. 1965. Preliminary reports on an investigation into the influence of several factors on the sodium sulphate test for aggregate. *Australian Road Research* **22**, 49–52.

Moss, A.J. 1972. Initial fluviatile fragmentation of granitic quartz. *Journal of Sedimentary Petrology* **42**, 905–16.

Moss, A.J. and P. Green 1975. Sand and silt grains: predetermination of their formation and properties by microfractures in quartz. *Journal of the Geological Society of Australia* **22**, 485–95.

Moss, A.J., P.H. Walker and J. Hutka 1973. Fragmentation of granitic quartz in water. *Sedimentology* **20**, 1489–511.

Moss, A.J., P. Green and J. Hutka 1981. Static breakage of granitic detritus by ice and water in comparison with breakage by flowing water. *Sedimentology* **28**, 261–72.

Mustoe, G.E. 1983. Cavernous weathering in the Capitol Reef Desert, Utah. *Earth Surface Processes and Landforms* **8**, 517–26.

Nahon, D. and R. Trompette 1982. Origin of siltstones: glacial grinding versus weathering. *Sedimentology* **29**, 25–35.

Naylor, L.A., H.A. Viles and N.E.A. Carter 2002. Biogeomorphology revisited: lookingtowards the future. *Geomorphology* **47**, 3–14.

Nicholson, D.T. 2001. Pore properties as indicators of breakdown mechanisms in experimentally weathered limestones. *Earth Surface Processes and Landforms* **26**, 819–838.

Nur, A. and G. Simmons 1970. The origin of small cracks in igneous rocks. *International Journal of Rock Mechanics and Mineral Science* **7**, 307–14.

Ollier, C.D. 1963. Insolation weathering: examples from central Australia. *American Journal of Science* **261**, 376–87.

Ollier, C.D. 1984. *Weathering*, 2nd edn. London: Longman.

Ollier, C.D. 1965. Dirt cracking – a type of insolation weathering. *Journal of Geology* **73**, 454–68.

Osborn, G. and J.M. Duford 1981. Geomorphological processes in the inselberg region of south-west Jordan. *Palestine Exploration Quarterly* January–June, 1–17.

Papida, S., W. Murphy and E. May 2000. Enhancement of physical weathering of building stones by microbial populations. *International Biodeterioration and Biodegradation* **46**, 305–317.

Paradise, T.R. 1995. Sandstone weathering thresholds in Petra. *Physical Geography* **16**, 205–222.

Paradise, T.R. 1998. Limestone weathering and rate variability, Great temple, Amman. *Physical Geography* **19**, 133–146.

Paradise, T.R. 2002. Sandstone weathering and aspect in Petra, Jordan. *Zeitschrift für Geomorphologie* **46**, 1–17.

Paradise, T.R. 2005. Petra revisited: an examination of sandstone weathering research in Petra, Jordan. *Geological Society of America, Special Paper* **390**, 39–50.

Peel, R.F. 1974. Insolation weathering: some measurements of diurnal temperature changes in exposed rocks in the Tibesti region, central Sahara. *Zeitschrift für Geomorphologie Supplement Band* **21**, 19–28.

Peel, R.F. 1975. Water action in deserts. In *Processes in physical and human geography*, R.F. Peel, M. Chisholm and P. Haggett (eds), 110–29. London: Heinemann.

Pel. L., H. Huinink and K. Kopinga 2003. Salt transport and crystallization in porous building materials. *Magnetic Resonance Imaging* 21, 317–320.

Peterson, G.L. 1980. Sediment size reduction by salt weathering in arid environments. *Geological Society of America Abstracts with Programs* **12**, 301.

Phillips, J.D. 1999. Divergence, convergence and self-organization in landscapes. *Annals of the Associationof American Geographers* **89**, 466–483.

Phillips, J.D. 2005. Weathering instability and landscape evolution. *Geomorphology* **67**, 255–272.

Potocki, F.P. 1978. Road temperatures and climatological observations in the Emirate of Abu Dhabi. *U.K. Transport and Road Research Laboratory, Supplementary Report* 412.

Power, E.T., B.J. Smith and W.B. Whalley 1990. Fracture patterns and grain release in physically weathered granitic rocks. In *Soil micromorphology: a basic and applied science*, L.A. Douglas (ed.), 545–50. Amsterdam: Elsevier.

Pye, K. 1983. Formation of quartz silt during humid tropical weathering of dune sands. *Sedimentary Geology* **34**, 267–82.

Pye, K. 1985. Granular disintegration of gneiss and migmatites. *Catena* 12, 191–9.

Pye, K. 1987. *Aeolian dust and dust deposits*. London: Academic Press.

Pye, K. and H.B. Sperling 1983. Experimental investigation of silt formation by static breakage processes: the effect of temperature, moisture and salt on quartz dune sand and granitic regolith. *Sedimentology* **30**, 49–62.

Quinif, Y. 1983. La reculée et le réseau karstique de Bou Akous (Hammamet, Algérie de l'Est) Geomorphologie et aspects évolutifs. *Révue Belge de Geographie* **107**, 89–111.

Ravina, I. and D. Zaslavsky 1974. The electrical double layer as a possible factor in desert weathering. *Zeitschrift für Geomorphologie Supplement Band* **21**, 13–18.

Rice, A. 1976. Insolation warmed over. *Geology* **4**, 61–2.

Rijniers, L.A., H.P. Huinink, L. Pel and K. Kopinga 2005. Experimental evidence of crystallization pressure inside porous media. *Physical Review Letters* **94**, 075503.

Robinson, D.A. and L.C. Jerwood 1987. Frost and salt weathering of chalk shore platforms near Brighton, Sussex, U.K. *Transactions of the Institute of British Geographers* **12**, 217–26.

Rodriguez–Navarro, C. and E. Doehne 1999. Salt weathering: influence of evaporation rate, supersaturation and crystallization pattern. *Earth Surface Processes and Landforms* **24**, 191–209.

Rodriguez–Navarro, C., E. Doehne and E. Sebastian 1999. Origins of honeycomb weathering: the role of salts and wind. *Geological Society of America Bulletin* **111**, 1250–1255.

Rodriguez–Navarro, C., E. Hansen, E. Sebastian and W. Ginell 1997. The role of clays in the decay of ancient Egyptian limestone sculptures. *Journal of the American Institute for Conservation* **36**, 151–163.

Rodriguez-Rey, A. and M. Montoto 1978. Insolation weathering phenomena in crystalline rocks. In *Deterioration and Protection of Stone Monuments*. UNESCO/RILEM Conference, Paris, Section 2.7.

Rögner, K. 1987. Temperature measurements of rock surfaces in hot deserts (Negev Israel). In *International geomorphology*, V. Gardiner (ed.), 1271–87. Chichester: Wiley.

Rögner, K. 1988a. Measurements of cavernous weathering at Machtesh Hagadol (Negev Israel). A semi-quantitative study. *Catena Supplement* **12**, 67–76.

Rögner, K. 1988b. Der Verwitterungsgrad von Geröllen auf Terrassenoberflächen. *Die Erde* **119**, 31–45.

Roth, E.S. 1965. Temperature and water content as factors in desert weathering. *Journal of Geology* **73**, 454–68.

Sancho, C., R. Fort and A. Belmonte 2003. Weathjering rates of historic sandstone in semi-arid environments (ebro Basin, NE Spain). *Catena* **53**, 53–64.

Schattner, I. 1961. Weathering phenomena in the crystalline of the Sinai in the light of current notions. *The Bulletin of the Research Council of Israel* **10G**, 247–66.

Scherer, G.W. 1999. Crystallization in pores. *Cement and Concrete Research* **29**, 1347–58.

Selby, M.J. 1977. On the origin of sheeting and laminae in granitic rocks: evidence from Antarctica, the Namib Desert and the Central Sahara. *Madoqua* **10**, 171–9.

Selwitz, C. 1990. Deterioration of the Great Sphinx: an assessment of the literature. *Antiquity* **64**, 853–859.

Sharon, D. 1970. Topography-conditioned variations in rainfall related to the runoff-contributing areas in a small watershed. *Israel Journal of Earth Sciences* **19**, 85–9.

Sharon, D. 1981. The distribution in space of local rainfall in the Namib Desert. *Journal of Climatology* **1**, 69–75.

Simmons, G. and D. Richter 1976. Microcracks in rocks. In *The physics and chemistry of minerals and rocks*, R.G.J. Strens (ed.) 105–37. Chichester: Wiley.

Singer, A. 1980. The palaeoclimatic interpretation of clay minerals in sediments – a review. *Earth-Science Reviews* **21**, 251–93.

Singer, A. 1984. Clay formation in saprolites of igneous rocks under semiarid to arid conditions, Negev, southern Israel. *Soil Science* **137**, 332–40.

Slatyer, R.O. and J.A. Mabbutt 1964. Hydrology of arid and semi-arid regions. In *Handbook of applied hydrology*, V.T. Chow (ed.), New York: McGraw Hill.

Smalley, I.J. and D.H. Krinsley 1978. Loess deposits associated with deserts. *Catena* **5**, 53–66.

Smalley, I.J. and C. Vita-Finzi 1968. The formation of fine particles in sandy deserts and the nature of desert loess. *Journal of Sedimentary Petrology* **38**, 766–74.

Smith, B.J. 1977. Rock temperature measurements from the northwest Sahara and their implications for rock weathering. *Catena* **41**, 41–63.

Smith, B.J. 1978. The origin and geomorphic implications of cliff foot recesses and tafoni on limestone hamadas in the northwest Sahara. *Zeitschrift für Geomorphologie* **22**, 21–43.

Smith, B.J. 1987. An integrated approach to the weathering of limestone in an arid area and its role in landscape evolution: a case study from southeast Morocco. In *International geomorphology*, V. Gardiner (ed.), 637–57. Chichester: Wiley.

Smith, B.J. 1988. Weathering of superficial limestone debris in a hot desert environment. *Geomorphology* **1**, 355–67.

Smith, B.J. and Kennedy, E.M. 1999. Moisture loss from Stone influenced by salt accumulation. In *Aspects of Stone Weathering, Decay and Conservation*, M.S. Jones, and R.D. Wakefield (eds), 55–64. London: Imperial College Press.

Smith, B.J. and R.W. Magee 1990. Weathering of granite in an urban environment: a case study from Rio de Janeiro. *Journal of Tropical Geography*, **11**, 143–53.

Smith, B.J. and J.J. McAlister 1986. Observations on the occurrence and origins of salt weathering phenomena near Lake Magadi, southern Kenya. *Zeitschrift für Geomorphologie* **30**, 445–60.

Smith, B.J. and J.P. McGreevy 1983. A simulation study of salt weathering in hot deserts. *Geografiska Annaler* **64a**, 127–33.

Smith, B.J. and J.P. McGreevy 1988. Contour scaling of a sandstone by salt weathering under simulated hot desert conditions. *Earth Surface Processes and Landforms* **13**, 697–706.

Smith, B.J. and P.A. Warke 1997. Controls and uncertainties in the weathering environment. In *Arid Zone Geomorphology 2nd edition*, D.S.G. Thomas (ed), 41–54. Chichester: Wiley.

Smith, B.J. and W.B. Whalley 1981. Late Quaternary drift deposits of north-central Nigeria examined by scanning electron microscopy. *Catena* **8**, 345–68.

Smith, B.J., J.P. McGreevy and W.B. Whalley 1987. The production of silt-size quartz by experimental salt weathering of a sandstone. *Journal of Arid Environments* **12**, 199–214.

Smith, B.J., P.A. Warke and C.M. Moses 2000. Limestone weathering in a contemporary arid environment: A case study from southern Tunisia. *Earth Surface Processes and Landforms* **25**, 1343–1354.

Smith, B.J., J.S. Wright and W.B. Whalley 2002b. Sources of non-glacial loess-size quartz silt and the origins of 'desert loess'. *Earth Science Reviews* **59**, 1–26.

Smith, B.J., P.A. Warke, J.P. McGreevy and H.L. Kane 2005. Salt-weathering simulations under hot desert conditions: agents of enlightenment or perpetuators of preconceptions? *Earth Surface Processes and Landforms* **67**, 211–227.

Smith, B.J., A.V. Turkington, P.A. Warke, P.A.M. Basheer, J.J. McAlister, J. Meneely and J.M. Curran 2002a. Modelling the rapid retreat of building sandstones. A case study from a polluted maritime environment. *Geological Society of London Special Publication* **205**, 339–354.

Soleilhavoup, F. 1977. Les cailleux fissurés des regs Sahariens: étude descriptive et typologie. *Géologie Mediterraneé* **4**, 335–64.

Sperling, C.H.B. and R.U. Cooke 1980a. Salt weathering in arid environments. I Theoretical considerations. *Bedford College London Papers in Geography* **8**.

Sperling, C.H.B. and R.U. Cooke 1980b. Salt weathering in arid environments. II Laboratory studies. *Bedford College London Papers in Geography* **9**.

Sridhar, N., W.H. Yang, D.J. Srolovitz and E.R. Fuller (Jr) 1994. Microstructural mechanics model of anisotropic–thermal–expansion–induced microcracking. *Journal of the American Ceramic Society* **77**, 1123–38.

Stambolov, T. 1976. The corrosive action of salts. *Lithoclastia* **I**, 5–8.

Stretch, R.C. and H.A. Viles 2002. The nature and rate of weathering by lichens on lava flows on Lanzarote. *Geomorphology* **47**, 87–94.

Sutton, L.J. 1945. Meteorological conditions in caves and ancient tombs in Egypt. *Ministry of Public Works, Egypt, Physical Department Paper* 52.

Sweeting, M.M. and N. Lancaster 1982. Solutional and wind erosion forms on limestone in the central Namib Desert. *Zeitschrift für Geomorphologie* **26**, 197–207.

Tafuro, A.M., F. Barnaba, F. de Tomasi, M.R. Perrone and G.P. Gobbi 2006. Saharan dust particle properties over the central Mediterranean. *Atmospheric Research* **81**, 67–93.

Tardy, Y., C. Bocquir, H. Paquet and G. Millet 1973. Formation of clay from granite and its distribution in relation to climate and topography. *Geoderma* **10**, 271–84.

Torraca, G. 1981. *Porous Materials – Materials Science for Architectural Conservation*. Rome: ICCROM.

Turkington, A.V. 1998. Cavernous weathering in sandstones: lessons to be learned from natural exposure. *Quarterly Journal of the Geological Society* **31**, 375–383.

Turkington, A.V. and T.R. Paradise 2005. Sandstone weathering: a century of research and innovation. *Geomorphology* **67**, 229–253.

Turkington, A.V. and J.D. Phillips 2004. Cavernous weathering, dynamical instability and sely-organization. *Earth Surface Processes and Landforms* **29**, 665–675.

Turkington, A.V. and B.J. Smith 2000. Observations of three-dimensional salt distribution in building sandstone, *Earth Surface Processes and Landforms* **25**, 1317–1332.

Turkington, A.V., J.D. Phillips and S.W. Campbell 2005. Weathering and landscape evolution. Geomorphology **67**, 1–6.

Turkington, A.V., B.J. Smith and P.A.M. Basheer 2002. The effect of block retreat on sub–surface temperature and moisture conditions in sandstone. In *Understanding and Managing Stone Decay*, R. Prykryl and H.A. Viles (eds), 113–126. Prague: Karolinum Press.

Verheye, W. 1976. Nature and impact of temperature and moisture in arid weathering and soil forming processes. *Pédologie* **26**, 205–24.

Verheye, W. 1986. Observations on desert pavements and desert patinas. *Pedologie* **36**, 303–13.

Viles, H.A. 1988. Organisms and karst geomorphology. In *Biogeomorphology*, H.A. Viles (ed.) 319–50. Oxford: Basil Blackwell.

Viles, H.A. 1995. Ecological perspectives on rock surface weathering: Towards a conceptual model. *Geomorphology* **13**, 21–25.

Viles, H.A. 2001. Scale issues in weathering studies. *Geomorphology* **41**, 63–71.

Viles, H.A. 2005a. Microclimate and weathering in the central Namib desert. *Geomorphology* **67**, 189–209.

Viles, H.A. 2005b. Self-organized or disorganized? Towards a general explanation of cavernous weathering. *Earth Surface Processes and Landforms* **30**, 1471–73.

Viles, H.A. and A.S. Goudie 2003. Interannual, decadal and multidecadal scale climatic variability and geomorphology. *Earth Science Reviews* **61**, 105–131.

Viles, H.A. and A.S. Goudies 2004. Biofilms and case hardening on sandstones from Al-Quwayra, Jordan. *Earth Surface Processes and landforms* **29**, 1473–14885.

Viles, H.A. and A.S. Goudie 2007. Rapid salt weathering in the coastal Namib desert: implications for landscape development. *Geomorphology* **85**, 49–62.

Warke, P.A. 2007. Complex weathering in drylands: Implications of 'stress' history for rock debris breakdown and sediment release. *Geomorphology* **85**, 30–48.

Warke, P.A. 2000. Micro–environmental conditions and rock weathering in hot, arid regions. *Zeitschrift für Geomorphologie, Supplement* **120**, 83–95.

Warke, P.A. and B.J. Smith, 1994. Short–term rock temperature fluctuations under simulated hot desert conditions : Some preliminary data. In *Rock Weathering and Landform Evolution*, D.A. Robinson and R.B.G. Williams (eds), 57–70. Chichester: Wiley.

Warke, P.A. and B.J. Smith 1998. Effects of direct and indirect heating on the validity of rock weathering simulation studies and durability tests. *Geomorphology* **22**, 347–357.

Warke, P.A. and B.J. Smith 2000. Salt distribution in clay-rich weathered sandstone. *Earth Surface Processes and Landforms* **25**, 1333–1342.

Warke, P.A., J. McKinley and B.J. Smith 2006. Variable weathering response in sandstone: factors controlling decay sequences. *Earth Surface Processes and Landforms* **31**, 715–735.

Warke, P.A., B.J. Smith, and R.W. Magee 1996. Thermal response characteristics of stone: Implications for weathering of soiled surfaces in urban environments. *Earth Surface Processes and Landforms* **21**, 295–306.

Weiss, T., E.R. Fuller (Jr) and S. Siegesmund 2002. Thermal stresses in calcite and dolomite marbles quantified by finite element modelling. *Geological Society of London Special Publication* **205**, 84–94.

Whalley, W.B., G.R. Douglas and J.P. McGreevy 1982a. Crack propagation and associated weathering in igneous rocks. *Zeitschrift für Geomorphologie* **26**, 33–54.

Whalley, W.B., J.R. Marshall and B.J. Smith 1982b. Origin of desert loess from some experimental observations. *Nature* **300**, 433–5.

Whalley, W.B., J.P. McGreevy and R.I. Ferguson 1984. Rock temperature observations and chemical weathering in the Hunza region, Karakorum: preliminary data. In *Proceedings of the International Karakorum Project*, K.J. Miller (ed.), 616–33. Cambridge: Cambridge University Press.

Whalley, W.B., B.J. Smith, J.J. McAlister and A.J. Edwards 1988. Aeolian abrasion of quartz particles and the production of silt-size fragments. Preliminary results and some possible implications for loess and silcrete formation. In *Desert sediments: ancient and modern*, L. Frostick and I. Reid (eds), 129–38. Oxford: Blackwell Scientific, Geological Society of London Special Publication 135.

Whitaker, C.R. 1974. Split boulder. *Australian Geographer* **12**, 562–3.

White, S. 1976. The role of dislocation processes during tectonic deformation, with particular reference to quartz. In *The physics and chemistry of minerals and rocks*, R.G.J. Strens (ed.), 75–91. Chichester: Wiley.

Wilhelmy, H. 1964. Cavernous rock surfaces (tafoni) in semi-arid and arid climates. *Pakistan Geographical Review* **19**, 9–13.

Wilke, B.M., B.J. Duke and W.L.O. Jimoh 1984. Mineralogy and chemistry of harmattan dust in northern Nigeria. *Catena* **11**, 91–6.

Williams, C.B. 1923. A short bio-climatic study in the Egyptian Desert. *Ministry of Agriculture, Egypt Technical and Scientific Service Bulletin* 29.

Williams, R.B.G. and D.A. Robinson 1981. Weathering of a sandstone by the combined action of frost and salt. *Earth Surface Processes and Landforms* **6**, 1–9.

Wilson, P. 1979. Experimental investigation of etch pit formation on quartz sand grains. *Geological Magazine* **116**, 447–82.

Winkler, E.M. 1979. The role of salts in development of granitic tafoni, South Australia. A discussion. *Journal of Geology* **87**, 119–20.

Winkler, E.M. 1987a. Weathering and weathering rates of natural stone. *Environmental Geology and Water Science* **9**, 85–92.

Winkler, E.M. 1987b. Comment on 'Capillary moisture flow and the origin of cavernous weathering in dolerites of Bull Pass, Antarctica.' *Geology* **15**, 975.

Wisniewski, J. 1982. The potential acidity associated with dews, frosts and fogs. *Water Air and Soil Pollution* **17**, 361–77.

Wüst, R.A.J. 2000. The origin of soluble salts in rocks of the ancient Thebes Mountains, Egypt: the damage potential to ancient Egyptian wall art. *Journal of Archaelogical Science* **27**, 1161–1172.

Wüst, R.A.J. and J. Mclane 2000. Rock deterioration in the Royal Tomb of Seti I, Valley of the Kings, Luxor, Egypt. *Engineering Geology* **58**, 161–90.

Yaalon, D.H. 1970. Parallel stone cracking, a weathering process on desert surfaces. *Geological Institute Technical Economic Bulletin (Bucharest)* **18**, 107–10.

Yaalon, D.H. and J. Dan 1974. Accumulation and distribution of loess-derived deposits in the semi-arid desert fringe areas of Israel. *Zeitschrift für Geomorphologie Supplement Band* **20**, 27–32.

Yaalon, D.H. and E. Ganor 1968. Chemical composition of dew and dry fallout in Jerusalem, Israel. *Nature* **217**, 1139–40.

Yair, A. and S.M. Berkowicz 1989. Climatic and non-climatic controls of aridity: the case of the northern Negev of Israel. *Catena Supplement* **14**, 145–58.

Yair, A., D. Sharon and H. Lavee 1978. An instrumental watershed for the study of partial area contribution of runoff in the arid zone. *Zeitschrift für Geomorphologie Supplement Band* **29**, 71–82.

Yang, W.H., D.J. Srolovitz, G.N. Hassold and M.P. Anderson 1990. The effect of grain boundary cohesion on the fracture of brittle, polycrystalline materials. In *Simulation and Theory of Evolving Microstructures*, M.P. Anderson and A.D. Rollett (eds), 277–284. Warrendale, Pennsylvania: The metallurgical Society.

Yatsu, E. 1988. *The Nature of Weathering: an Introduction.* Tokyo: Sozosha.

Yong, C. and C.Y. Wang 1980. Thermally induced acoustic emission in Westerly granite. *Geophysical Research Letters* **7**, 1089–92.

Young, A.R.M. 1987. Salt as an agent in the development of cavernous weathering. *Geology* **15**, 962–6.

Zimmermann, A., W.C. Carter and E.R. Fuller (Jr) 2001. Damage evolution during microcracking of brittle solids. *Acta Materialia* **49**, 127–137.

Chapter 5

Aridic Soils, Patterned Ground, and Desert Pavements

John C. Dixon

Introduction

Pedogenic and geomorphic processes operating in deserts are inextricably linked. These linkages are particularly well expressed in the development of patterned ground and desert pavement. In addition, the nature and efficacy of hydraulic, gravitational, and aeolian processes on desert surfaces are strongly influenced by the physical and chemical characteristics of the underlying soils. As a result, the evolution of a diversity of desert landforms is either directly or indirectly linked to pedogenic processes.

The significant role played by pedogenic processes in desert landscape evolution is also strongly reflected in the development of duricrusts – indurated accumulations of calcium carbonate, gypsum, and silica. These essentially pedogenic materials impart considerable relief in the form of tablelands, mesas, and buttes to otherwise low-relief landscapes. Erosion of these topographic eminences in turn provides the coarse debris for the formation of patterned ground and desert pavement.

In this chapter the nature and genesis of soils in arid and semi-arid environments are examined. This discussion is followed by an examination of the major types of patterned ground and the processes responsible for its formation. Finally, the nature and origin of desert pavements are examined. The nature and origin of duricrusts are discussed in Chapter 6.

J.C. Dixon (✉)
Department of Geosciences, University of Arkansas,
Fayetteville, AR 72701, USA
e-mail: jcdixon@uark.edu

Aridic Soils

Distribution

Desert soils occupy approximately $46.1\,M\,km^2$, or 31.5% of the Earth's surface (Dregne 1976). They are most widespread in Australia, occupying some 82% of the continental surface (Stephens 1962), and are also the dominant soils of Africa, covering some 59% of the continent. Arid soils cover substantially smaller areas of the remaining continents: 33% of Asia, 18% of North America, 16% of South America, and just under 7% of Europe (Dregne 1976). Desert soils are dominated by five soil orders (U.S. Comprehensive Soil Classification): Entisols (41% of arid zone), Aridisols (36%), Mollisols (12%), Alfisols (7%), and Vertisols (4%) (Dregne 1976). These soil orders are subdivided into thirteen suborders discussed in greater detail later in this chapter.

Characteristics

Aridic soils display a variety of distinctive morphological features. Among these are the presence of gravel-covered surfaces, the development of surface organic and inorganic crusts, and the widespread occurrence of vesicular A horizons. In addition they commonly display the development of ochric and mollic epipedons and are characterized by the formation of a variety of diagnostic subsurface horizons including cambic, argillic, calcic, petrocalcic, gypsic, petrogypsic, natric, salic, and duripan horizons (Buol et al. 1997, Southard 2000).

A.J. Parsons, A.D. Abrahams (eds.), *Geomorphology of Desert Environments*, 2nd ed.,
DOI 10.1007/978-1-4020-5719-9_5, © Springer Science+Business Media B.V. 2009

Surface Crusts

Desert soils commomly display the presence of thin (10–20 mm) crusts on surfaces largely devoid of vegetation and gravel cover. The crusts are typically massive but may possess a platy structure in their upper part. The lower portion of the crusts is commonly vesicular where it is in contact with the underlying vesicular horizon. Crusts typically display low permeability and accompanying enhanced runoff (Buol et al. 1997, Schaetzl and Anderson 2005), however some studies have suggested that some crusts (especially organic-rich crusts) may in fact enhance infiltration (Gifford 1972, Blackburn 1975, Dunkerley and Brown 1997). Crusting is widely believed to be the result of repeated wetting and drying of predominantly loamy soils (Buol et al. 1997). These authors believe that during wetting, soil plasma moves to contact points between skeletal grains where upon drying it acts as a weak reversable cement. The source of the binding agents (the plasma) has long been elusive (Sharon 1962, Schaetzl and Anderson 2005). For a long time it was widely believed that crusting was the result of raindrop impact (McIntyre 1958a, b, Schaetzl and Anderson 2005) which both compacted soil particles and produced and transported clay size particles into voids. However crusting probably includes both inorganic and organic processes and components. Inorganic components include clays (Frenkel et al. 1978, Benhur et al. 1985, Dunkerley and Brown 1997) which serve as binding agents, exchangeable sodium (Painuli and Abrol 1986) which results in soil dispersion, and the presence of small quantities of salts and calcium carbonate (Sharon 1962, Benhur et al. 1985).

Desert soil crusts are now generally attributed to biological processes resulting in the production of thin microphytic layers on the soil surfaces (Dunkerley and Brown 1997, Schaetzl and Anderson 2005). Microphytic crusts are dominated by assemblages of algae, mold, cyanobacteria, (Schaetzl and Anderson 2005) as well as mosses, liverwarts, lichen, bacteria, and fungi (Dunkerley and Brown 1997). These organisms grow in the upper few millimeters of the soil surface after rain events, and form biomantles in which their mycelia bind soil partices together. The soil mantles are commonly enriched in C and N as well as silt and clay (Fletcher and Martin 1948).

Vesicular Horizons

Immediately beneath the gravel surface layer of many aridic soils a vesicular Av horizon (Fig. 5.1) commonly occurs and is often associated with a surface crust. Formation of the vesicular horizon has been widely attributed to the saturation of the fine grained soil surface horizon (Miller 1971). Nettleton and Peterson (1983) argued that in the saturated state the soil plasma of this upper horizon is free to move and in so doing traps air. With repeated episodes of saturation and air entrapment, the vesicles in the surface horizon increase in size. Repeated destruction of the vesicular horizon is thought to be caused by wetting and drying cycles (Springer 1958, Miller 1971). Nettleton and Peterson (1983) alternatively suggested that vesicular horizon destruction is primarily the result of soil trampling by animals. Upon destruction, the formation of the vesicular horizon begins anew as soil saturation episodes begin and air entrapment resumes.

A substantially different explanation for the origin of the vesicular horizon has been proposed by Wells et al. (1985), McFadden et al. (1986, 1987, 1998) and Blank et al. (1996). These workers attributed the origin of the fine-grained surface horizon and accompanying vesicular structure to aeolian addition of fine grained materials and associated soluble salts, carbonates, and iron oxides. The development of vesicles is attributed to entrapment of air by aeolian infall with subsequent expansion due to heating following summer rainfall events. This model follows that of Evenari et al. (1974) for vesicular horizons developed in soils in Israel.

Fig. 5.1 Soil developed beneath desert pavement showing vesicular A horizon (Av) and carbonate and gypsum enriched B horizon (Btky). Bedrock rubble is designated (R) (photo courtesy of L.D. McFadden)

Fig. 5.2 Columnar structure of vesicular A horizon resulting from high shrink–swell capacity (photo courtesy of L.D. McFadden)

Supporting evidence for an aeolian sediment source includes the predominantly silt and clay dominated texture of the Av horizon (McFadden et al. 1987, 1998, Anderson et al. 2002), the presence of abundant amounts of salts and carbonate which exceed those in parent materials (Reheis et al. 1995), and the presence of mineralogies which differ from those of the soil parent materials (Blank et al. 1996). The subsequent stability of the vesicular horizon is attributed largely to the formation of thin $CaCO_3$ coatings (Evenari et al. 1974) and clay coatings (Sullivan and Koppi 1991). While aeolian infall is envisaged as the primary source of the fine grained surface layer, McFadden et al. (1987) also suggested that some of the sand and silt fraction may be the result of mechanical weathering of the surface gravels. The presence, and successive accumulation, of clay in the aeolian mantle leads to the development of a distinctive columnar structure in the Av horizon due to enhanced shrink–swell capacity (McFadden et al. 1986) (Fig. 5.2). Sullivan and Koppi (1991), working on desert loams in Australia, have also proposed that the fine grained materials in the vesicular horizon are externally derived. They suggested a combination of aeolian, colluvial, and overland flow sources.

Cambic Horizons

Intimately associated with the vesicular, ochric epipedons are cambic horizons. These horizons are found below ochric epipedons and are typically reddish or brownish in colour. There is evidence of alteration in the horizon in the form of obliteration of original structures due to mixing, development of new structure, accumulation of clay, and carbonate translocation through and into the horizon (Nettleton and Peterson 1983, Buol et al. 1997). The source of the fine material and the carbonates in this horizon is widely attributed to aeolian infall (Gile 1970, 1975, Gerson et al. 1985, McFadden et al. 1986, Gerson and Amit 1987, Reheis 1987a, Wells et al. 1987, Birkeland 1990). McFadden et al. (1986) suggested that as the vesicular horizon thickens due to continued aeolian addition, plasma migration near the base of the horizon results in destruction of vesicles and a slowing of the migration of infiltrating waters. This slowing of infiltration results in ferrous iron alteration as well as the accumulation of authigenic ferric iron oxyhydroxides. Soil reddening results, and a cambic horizon slowly develops. Clay and carbonate added by aeolian infall also migrate to the cambic horizon.

Argillic Horizons

Many aridic soils are characterized by the occurrence of strongly developed argillic (clay-rich) horizons. This horizon differs markedly from the aforedescribed cambic horizon in that it is substantially more abundant in clay and commonly has a noncalcareous matrix. Research on the origin of the argillic horizon has focused on two contrasting sources of clay. Smith and Buol (1968) and Nettleton et al. (1975) have shown that the degree of mineral weathering in A horizons of some argids in the south-western United States is comparable to, or greater than, that in underlying Bt horizons. This finding therefore suggests that chemical weathering in A horizons is a possible source of illuvial clay. Further, these authors have pointed to the occurrence of clay skins on mineral grains and pebbles in the argillic horizon as evidence for clay illuviation. The frequent lack of clay skins on ped faces is attributed by Nettleton and Peterson (1983) to destruction by swell–shrink mechanisms, bioturbation, and calcium carbonate crystal growth. Nettleton et al. (1975) also argued for clay illuviation on the basis of patterns of distribution of fine and coarse clay and total iron and aluminium down the profile. Finally, Nettleton and Peterson (1983) reasoned that the occurrence of clay skins on the walls of deep pipes or as downward extensions of the argillic horizon support an illuvial origin for the clay. Chartres (1982), on the other hand, has stressed the importance of the aeolian addition of fine debris in the formation of argillic horizons, citing a relative lack of weathered soil particles in the A horizon as evidence against an in-profile source of clay. Blank et al. (1996), suggest a multi-stage model of argillic horizon formation involving aeolian addition/weathering/translocation. Based on chemical, mineralogical and micromorphological evidence, they suggest the neoformation of secondary clay minerals resulting from the weathering of aeolian–derived sediment followed by subsequent translocation of the neoformed clays to depth within the soil.

Carbonates are commonly associated with the argillic horizon. In some cases they may dominate the soil matrix (Gile 1967), but more commonly they occur as nodules which have displaced the previously deposited clays (Gile and Grossman 1968, McFadden et al. 1986). Accumulation of carbonates in argillic horizons is widely interpreted to be indicative of significant climate change from humid to arid climatic regimes with an accompanying reduction in depth of leaching (Nettleton and Peterson 1983, McFadden and Tinsley 1985, McFadden et al. 1986, Reheis 1987b, Wells et al. 1987). Thick argillic horizons in Aridisols are therefore widely regarded as being largely relics of the Pleistocene.

Natric and Salic Horizons

The accumulation of soluble salts is a further characteristic of aridic soils. Accumulation results in the formation of either natric (sodic) or salic (saline) horizons. Natric horizons are argillic horizons characterized by high ($\geq 15\%$) exchangeable sodium or a sodium adsorption ratio of 13 or more and by the development of prismatic or columnar soil structure (Soil Survey Staff 1975, Dregne 1976).Magnesium salts are also commonly present and on occasion may exceed the abundance of exchangeable sodium, but their presence is not mandatory for the classification of natric horizons (Dregne 1976). In Holocene-age soils these horizons are typically thin, while in Pleistocene soils they thicken considerably (Nettleton and Peterson 1983). These horizons typically occur

above or in a zone of calcium carbonate accumulation (Nettleton and Peterson 1983). Although there is sodium accumulation, there is insufficient for these horizons to be classified as salic. In order for a horizon to be classified as salic it must be at least 15 cm thick, be enriched in salts that are more soluble than gypsum, contain at least 2% soluble salt and have a product of salt content times horizon thickness greater than 60 (Soil Survey Staff 1975, Dregne 1976).

Salts in natric and salic horizons appear to be derived from several sources. Airborne addition of salts has been suggested for these horizons in aridic soils in the United States by Alexander and Nettleton (1977), Peterson (1980), Nettleton and Peterson (1983), and McFadden et al. (1991). An airborne source of salts has also been suggested for saline soils in Israel (Yaalon 1964, Yaalon and Ganor 1973, Dan et al. 1982, Amit and Gerson 1986). The source of salts in sodic soils in Australia has also been widely attributed to airborne sources, particularly for those soils within about 200 km of the coastal environment (Hutton and Leslie 1958, Wetselaar and Hutton 1963, Chartres 1983, Isbell et al. 1983) as well as from the mobilization of salts from playa surfaces in arid and semi-arid continental interiors.

A second source of the sodium is from the incursion of very shallow ground waters. Nettleton and Peterson (1983) suggested such an origin for some of the natric horizons in soils in the United States. These workers also proposed that this process is particularly important in the formation of salic horizons. This process appears to be especially significant for soils occurring in the Riverine Plain of south-eastern Australia (Northcote and Skene 1972, Isbell et al. 1983). Detailed discussion of salinization of soils in Western Australia by saline ground waters is provided by Conacher (1975). Secondary salinization as a result of groundwater pumping for irrigation is a growing problem in some arid Australian soils. Similarly, poor irrigation-agriculture practices has led to the development of sodic solis due to groundwater infiltration in Argentina, Peru, and Iraq (Dregne 1976).

The third explanation for the formation of natric or salic horizons is the inheritance of salts from parent materials. Three parent-material sources have been suggested by workers from Australia. Chartres (1983) suggested that sodic horizons developed well below the depth of contemporary infiltration may well have originated in older sodic soils which have subsequently been eroded. A number of studies of sodic soils from western Queensland strongly suggest that many have derived their sodium from the underlying bedrock (Gunn and Richardson 1979, Isbell et al. 1983). In particular, soils developed on marine shales and gypsum or anhydrite beds are highly susceptible to salt accumulation as are obviously those developed directly on pre-existing salt deposits (Dregne 1976). In a related fashion, some sodic soils derive their salts from underlying deeply weathered parent materials which have in turn been derived from salt-rich bedrock. Sodic soils developed on laterite in Western Australia have been reported by Bettenay et al. (1964) and Dimmock et al. (1974). The significance of the deep sodic horizons as sources of secondary salinization have been investigated by Peck and Hurle (1973) and Peck (1978). Sodic soils developed in deeply weathered materials have also been reported from Queensland by Hubble and Isbell (1958) and Isbell (1962). More recently, Gunn (1967) has suggested that deep weathering profiles are also the source of salts in solodic and solodized–solonetz soils in central Queensland.

Salts may also be derived from surface water runoff, especially where those waters accumulate in depressions in the landscape such as in salt pans, salinas, sebkhas etc. In such settings slow rates of infiltration coupled with high rates of evaporation result in salt accumulation. Salic and natric horizons may also develop in soils which are affected by interconnected surface and groundwater sources as is the case for soils in the Dasht-i-Kavir in Iran (Dregne 1976).

Finally, some saline horizons may form as a result of the accumulation of salts from decomposing vegetation. Such salts represent accumulations of recycled salts from deeper layers within the soil which pass through halophytes and accumulate in soil surface horizons. Such salt accumulations have been reported from the western United states by Wallace et al. (1973).

Calcic and Gypsic Horizons

Aridic soils are frequently dominated by calcium carbonate and/or gypsum. With progressive concentration of carbonate or gypsum and accompanying induration, petrocalcic and petrogypsic horizons develop. These horizons are collectively referred to as duricrusts, which are discussed fully in Chapter 6. Calcic and gypsic soils are characterized by the occurrence of non-indurated accumulations of calcite and/or dolomite in the former and gypsum in the latter. These materials occur in a variety of forms including powdery fillings, nodules, pendents, or crusts beneath pebbles and cobbles. Crusts in calcic and gypsic soils slake in water. In the United States, calcic soil horizons must be 15 cm or more thick, contain the equivalent of $\geq 15\%$ $CaCO_3$, and have a carbonate content $\geq 5\%$ greater than the C horizon. Gypsic soil horizons must also be 15 cm or more thick, contain at least 5% more gypsum than the underlying horizon, and have a product of horizon thickness in centimetres and percentage gypsum content not less than 150 (Soil Survey Staff 1994).

Descriptions of calcic and gypsic soils are common in the desert soils literature. It is important to point out that many aridic soils contain intergrades of calcic–gypsic horizons. A comprehensive discussion of gypsic soils (and intergrades) is provided by Boyadgiev and Verheye (1996) in which they identify the geographic extent, geomorphic and climatic settings, pedogenic characteristics and classification of gypsosols. Calcic soils in the United States have been extensively investigated by Bachman and Machette (1977), Shlemon (1978), Gile et al. (1981), Nettleton and Peterson (1983), Machette (1985), Weide (1985), Reheis et al. (1989), Harden et al. (1991), McFadden et al. (1991), Monger et al. (1991), Rabenhorst et al. (1991), and Nettleton et al. (1991). Studies of gypsic soils in the United States are notably fewer but include those ofNelson et al. (1978), Nettleton et al. (1982), and Harden et al. (1991). Detailed descriptions of gypsic soils in Wyoming have been provided by Reheis (1987a) Figs. 5.3, 5.4 and 5.5). Studies of calcic soils from Australia have been summarized by Stace et al. (1968) and Northcote and Skene (1972), Hutton and Dixon (1981), Milnes and Hutton (1983), Isbell et al. (1983), and Akpokodje (1984). Early descriptions of gypsic soils in Australia were provided by Prescott (1931), Jessop (1960) and Bettenay (1962). Calcic and gypsic soils have been reported from many other arid and semi-arid environments, some of which were summarized by Eswaran and Zi-Tong (1991). These authors reported the extensive occurrence of gypsic soils in China, India, Pakistan, the Middle East,

Fig. 5.3 Stage I gypsum (arrows) in soil developed on alluvial fan, Bighorn Basin, Wyoming (photo courtesy of M. Reheis (U.S. Geological Survey Bulletin 1590-C)

Fig. 5.5 Stage IV gypsum (g) in soil developed on alluvial fan, Bighorn Basin, Wyoming (photo courtesy of M. Reheis (U.S. Geological Survey Bulletin 1590-C))

Fig. 5.4 Complex gypsic soil developed on alluvial fan, Bighorn Basin, Wyoming. Stage IV gypsum 20–50 cm, Stage II/III gypsum 50–75 cm, Stage II gypsum 75+ cm (photo courtesy of M. Reheis (U.S. Geological Survey Bulletin 1590-C))

The source of the gypsum and carbonate in aridic soils is widely believed to be from atmospheric addition as either dust or carbonate dissolved in rainwater (Bull 1991, Dohrenwend et al. 1991, Gustavson et al. 1991). However, some studies such as those by Akpokodje (1984) suggest an *in situ* origin of the carbonate and gypsum. Boettinger and Southard (1991) have provided compelling evidence from Aridisols developed on pediments in the Mojave Desert of California that the source of carbonate deep within calcareous Haplargids is derived from the weathering of the granite and accompanying release of calcium from plagioclase. Hutton and Dixon (1981) provided evidence that at least some of the carbonates in southern south Australia were derived from pre-existing dolomite-rich laccustrine parent materials. Other sources include capilliary rise from shallow groundwater sources and from surface water runoff. Additionally, biological processes may play significant roles in calcic and gypsic horizon development. The mineralization of dead plant material has been suggested by Dregne (1976) to contribute significant quantities of carbonate and gypsum to soils. Amit and Harrison (1995) have shown that calcic horizons in Israeli soils are intimately associated with fungal hyphae.

Duripans

and Africa. More limited occurrences of gypsic soils were recorded in southern Europe. Dan (1983) and Dan et al. (1982) described gypsic soils from Israel.

Many arid soils possess a duripan. These pans are subsurface indurations of predominantly silica, but may grade to petrocalcic (calcrete) horizons (Soil

Survey Staff 1975, Nettleton and Peterson 1983). In gross morphology, chemistry, and mineralogy they are analogous to silcretes described in detail in Chapter 5. Duripans are generally platy in structure with individual plates ranging in thickness from 1 to 15 cm. Pores and surfaces of the plates are commonly coated with opal, chalcedony, and/or microcrystalline silica. For a soil horizon to be classified as a duripan, at least one of three criteria must be met. The soil must display (a) some vertical coatings of opal, (b) siliceous nodule development, or (c) the development of silica pendents on the undersides of coarse fragments. In addition, opal or other forms of silica must partly fill interstices and form bridges between sand grains (Soil Survey Staff 1975).

Duripans in the United States appear to be most strongly developed in soils formed on volcanic ash or other pyroclastic parent materials containing abundant silica (Flach et al. 1969, 1973, Brasher et al. 1976, Nettleton and Peterson 1983). Boettinger and Southard (1991) examined duripans from a Durorthid pedon developed in grus on a pediment in the Mojave Desert. These workers suggested that the source of the silica is not volcanic glass, which is essentially absent from the profile, but the weathering of feldspars. In Australia, Stace et al. (1968) reported hardpan soils commonly occurring on strongly weathered alluvial and colluvial deposits as well as on eroded laterites. Duripans are most strongly developed on older landscape surfaces where there has been sufficient time for prolonged silica dissolution, translocation, and deposition. They form at the average depth of wetting, which progressively diminishes with increasing aridity. Duripans commonly form below or in the lower part of argillic or natric horizons and have also been observed to be interlayered with illuvial clay (Nettleton and Peterson 1983). In Australia, hardpan soils have been reported from Western Australia by Litchfield and Mabbutt (1962), Stace et al. (1968), and Brewer et al. (1972), from South Australia by Stace et al. (1968), Wright (1983), and Milnes et al. (1991), and from New South Wales by Chartres (1985).

Classification

As indicated previously, the dominant soil order of the arid regions of the world are entisols. Within this order three suborders are widely identified in arid environments: Fluvents, Orthents, and Psaments. Fluvents occur on floodplains, deltas, and alluvial fans where drainage is not impeded. Orthents occur where diagnostic surface horizons have been eroded or occur on gypsiferous surfaces with deep water tables. Typically they are associated with rocky slopes (Dregne 1976). Psaments are Entisols which are sandy in all subhorizons below an Ap horizon or below a depth of 25 cm to a depth of 1 m. They contain less than 35% gravel; gravelly psaments are classified as Orthents. Psaments are typically associated with stabilized and unstabilized sand dunes, cover sands and sand sheets. Entisols occur extensively on landsurfaces of Holocene age in the southwestern United States (McFadden 1988, Bull 1991). The dominant zonal soil order of the arid lands are aridisols. This soil order is dominated by two suborders, Argids and Orthids. Argids are characterized by the development of argillic or natric horizons while Orthids lack an argillic or natric horizon.

Each of these suborders can be further divided into a number of great groups. The principal great groups of Argids are Durargids, Haplargids, Natrargids, Nadurargids, and Paleargids (Guthrie 1982). Durargids possess a duripan that underlies an argillic horizon and lack natric horizons. The top of the duripan is within a metre of the ground surface. Haplargids lack a natric horizon and do not have a duripan or petrocalcic horizon within a metre of the surface. They have weakly developed argillic horizons with less than 35% clay accumulation. Natrargids are characterized by the occurrence of a natric horizon but lack a duripan or petrocalcic horizon within a metre of the ground surface. Nadurargids are Argids that possess a natric horizon above a duripan, and the surface of the duripan is within a metre of the ground surface. Paleargids are Argids that develop on old land surfaces. They are characterized by either a petrocalcic horizon below an argillic horizon or an argillic horizon with greater than 35% clay and an abrupt upper boundary (Soil Survey Staff 1975).

Orthids are aridic soils with one or more pedogenic horizons. However, they lack a natric or argillic horizon. They commonly have an accumulic horizon of soluble salts and calcium carbonate. Some orthids possess salic, calcic, gypsic, petrocalcic, petrogypsic, cambic, and duripan horizons (Soil Survey Staff 1975). Orthids are divided into six great groups: Calciorthids, Camborthids, Durorthids, Gypsiorthids, Paleorthids, and Salorthids. Calciorthids contain abundant amounts of lime derived from either the parent material or added from aeolian dust. Camborthids are characterized by the development of a cambic horizon that results in a brownish to reddish brown soil of uniform texture. Durorthids possess a duripan within a metre of the surface and are commonly calcareous throughout. Gypsiorthids contain a gypsic or petrogypsic horizon with an upper boundary within a metre of the soil surface. Paleorthids contain a petrocalcic horizon within a metre of the ground surface. They commonly display evidence of calcium carbonate engulfing a pre-existing argillic horizon. Salorthids are very salty soils commonly associated with the accumulation of salts from capilliary rise of salty waters. They are characterized by a salic horizon (Soil Survey Staff 1975).

In one of the most comprehensive studies of arid lands soils in the United States, the Desert Project (Gile et al. 1981) recognized that large parts of The Desert Project study sites in New Mexico are also occupied by Mollisols, Alfisols, and some Vertisols. Two suborders of Mollisols are widely recognized in arid regions: Ustolls, and Xerolls. Ustolls occur in regions with semi-arid and subhumid climates. They are free draining and show no evidence of saturation during the year. They possess ustic or aridic moisture regimes. The presence of lime or gypsum indicates that they are dry for extended periods of time through the year. Xerolls occur under environments of contrasing seasonal precipitation. Mean annual soil temperature is less than 22°C and seasonal soil temperatures differ by at least 5°C at a depth of 50 cm. Internal drainage is sufficient to preclude the development of mottles and Fe-Mn nodules larger than 2 mm in diameter. Xerolls have a xeric or aridic moisture regime. Vertisols are predominantly Usterts which are characterized by the presence of cracks that remain open for 90 consecutive days or more, but are closed for at least 60 consecutive days when soil temperature is above 8°C at 50 cm., mean annual soil temperature is greater than 22°C, and displays a seasonal difference of less than 5°C. Alfisols are dominated by Ustalfs and Xeralfs. Ustalfs possess a ustic moisture regime and thermal regimes that are at least thermic, but more commonly mesic, or isomesic. Calcic horizons commonly occur below the argillic horizon. Xeralfs possess moisture regimes that are xeric with extended dry

summers and moist winters. Soil temperatures are typically thermic (Dregne 1976).

Aridic Soils and Landscape Development

Research on aridic soils in the United States in recent years has focused heavily on the relationship between soil development and landscape evolution. These studies have been concerned with the age of landscapes, correlation of Quaternary deposits, and climate change. The use of aridic soils as relative age indicators is well illustrated in studies from the western and south-western United States (Christianson and Purcell 1985, Harden et al. 1985, Machette 1985, Ponti 1985, Dohrenwend et al. 1991).

Detailed studies of soil development as an indicator of environmental change have been undertaken by many workers in the south-western United States. Major differences in aeolian dust addition to soils between the Holocene and Pleistocene are reported by Machette (1985), McFadden et al. (1984, 1986, 1987), Wells et al. (1987), Chadwick and Davis (1990), Reheis (1987a, b, 1990), and Bull (1991). Studies of carbonate accumulation amounts and rates, and depth of infiltration have been undertaken by Machette (1985), McFadden and Tinsley (1985), Reheis (1987a, b), and Bull (1991). These studies all point to maximum carbonate accumulation during the Holocene and accompanying reduction in the depth of carbonate infiltration. The polygenetic nature of many aridic soils is also highlighted in studies by Reheis (1987a, b) Nettleton et al. (1989), and Bull (1991). Variability in rates of soil formation depending on geomorphic setting and climate is stressed by Reheis et al. (1989) and Bull (1991). Recent more detailed studies by some of these same workers from California and Nevada (McFadden et al. 1998, Reheis et al. 1995) have established that accelerated dust deposition at the Pleistocene-Holocene boundary resulted in a period of active, rapid soil formation. McFadden et al. (1998) demonstrate further the intimate relationship between aeolian dust deposition, carbonate accumulation, vesicular soil horizon development and desert pavement formation.

Soils in arid environments in the western United States have also been widely used as stratigraphic markers and as indicators of periods of stability within Quaternary depositional systems. Such studies are well exemplified by the work of Gustavson and Winkler (1988), Gustavson et al. (1991), and Holliday (1985a, b, c, 1988, 1989, 1990) from the southern High Plains of Texas and New Mexico. Similar studies in aeolian environments in the southern Colorado Plateau by Wells et al. (1990) and California/Nevada (Reheis et al. 1995, McFadden et al. 1998) further demonstrates the importance of soil formation in understanding landscape evolution in desert environments.

Patterned Ground

Patterned ground is a common and widespread feature of desert surfaces. It includes a variety of forms including gilgai, surface cracking, microtopography related to surface crust formation, and vegetation patterning. At a variety of scales, all of these forms are related to patterns of water movement on landscape surfaces of diverse topography.

Gilgai

Patterned ground associated with both stony deserts and playa surfaces is a widespread phenomenon in warm deserts (Cooke and Warren 1973, pp. 129–49, Mabbutt 1977, pp. 130–4). Perhaps the most common and widespread type of patterned ground is gilgai. Gilgai is an Australian aboriginal word meaning small water hole, and while originally applied to small depressions that held water, it is now used to refer to a wide variety of soil patterned ground phenomena. Although the distribution and diversity of gilgai forms is perhaps greatest on the Australian continent, gilgai has also been reported from the Middle East by Harris (1958, 1959) and White and Law (1969), from South Dakota by White and Bonestall (1960), and from Death Valley, California, by Denny (1965, 1967) and Hunt and Washburn (1960). In addition to gilgai being frequently associated with soils of high swelling potential such as Vertisols, it is also a common feature in arid and semi-arid environments where strong textural contrasts exist within soils and where the climate is characterized by pronounced seasonality of precipitation. Gilgai morphology occurs in soils with annual rainfalls ranging from less than 150 mm to more than 1500 mm.

Gilgai Types

In early work on gilgai (Hallsworth and Robertson 1951, Hallsworth et al. 1955, Verger 1964, Stace et al. 1968, pp. 417 and 420, Hubble et al. 1983), six principal types of morphology were recognized (Fig. 5.6). (a) Normal gilgai is the most common form and is characterized by the development of randomly oriented mounds and shelves. The magnitude of these features ranges from imperceptible to as much as 3 m vertically with a wavelength of 15 m. If mounds are subcircular, they are referred to as puffy gilgai. (b) Melon-hole gilgai consists of large mounds separated by shelves of complex morphology. The shelves are commonly depressions with one or more sinkholes at the bottom; they are typically 1–3 m wide and 15–20 cm deep. (c) Stony gilgai, which most closely resembles the patterned ground found at high latitudes and high altitudes, has stone-covered mounds which are wide and flat in form. (d) Lattice gilgai is complex in morphology. It includes mounds that are discontinuous and oriented parallel to the direction of the slope as well as semi-continuous mounds that form networks of diverse orientation. (e) Linear or wavy gilgai develops on hillslopes ranging in gradient from 15" to 3°. Mounds and shelves are continuous and are arranged at right angles to the contour. These forms are 5–10 cm in height, and in the dry season they are very puffy in appearance. (f) Tank gilgai is large-scale gilgai that is usually rectangular in shape. Vertical dimensions range from 60 cm to 1.5 m, while depressions are 10–20 m long and 15–20 m wide. Verger's (1964) work recognizes essentially the same six basic forms, but he distinguishes between positive and negative relief groups and simultaneously between random and oriented distributions (Fig. 5.7).

The most widely developed type of gilgai in the Australian desert is stony gilgai. Ollier (1966) and Mabbutt (1977, p. 131, 1979) recognized two principal types: circular and stepped (Fig. 5.6c i and ii). Circular gilgai is characterized by the development of a relatively fine grained inner depression surrounded by a slightly raised stony rim. The depressions are commonly about 3 m in diameter, while the outer rims have diameters of approximately 8 m. This type of stony gilgai commonly forms on surfaces of low slope. The soil beneath the stony mound is generally clay-rich and has a silt crust with embedded pebbles. In contrast, the soil beneath the depression is typically sandier in the upper 30–50 cm but at depth resembles the mound soil and has abundant coarse fragments. Circular gilgai may occur either in random patterns or in networks. Stepped gilgai – or, as Mabbutt (1977, 1979) calls them, lattice systems – occur on steeper slopes with gradients of 0.5–6°. These forms are essentially distorted gilgai that become aligned across the slope. They are characterized by the development of stony risers upslope and downslope of practically stonefree treads. As with the circular gilgai forms, the risers are underlain by fine grained soils, while sandier soils underlie the treads. Treads commonly display sink holes at the base of the upslope riser (Ollier 1966, Cooke and Warren 1973, Mabbutt 1979).

Mabbutt (1979) recognized a third type of stony gilgai in the desert pavement-covered areas of the Northern Territory. He referred to this type as sorted stone polygons (Fig. 5.6c iii). These polygons are between 40 and 80 cm in diameter and are outlined by a rim of silcrete boulders sitting on a pavement of smaller pebbles. The silcrete boulders are absent from the interior depression, though the smaller pebbles are still present. Topsoil thicknesses are generally greater in the interior of the polygon (Mabbutt 1977).

Although gilgai in the arid areas of Australia is primarily associated with Desert Loams (Stace et al. 1968), it is also associated with a variety of other soils with strong textural contrasts in seasonally wet–dry environments. In particular, gilgai is strongly developed on grey cracking clays in west-central Queensland and in north-central and south-central New South Wales (Stace et al. 1968).

Detailed analyses of gilgai morphology have been undertaken by Paton (1974) and Knight (1980).

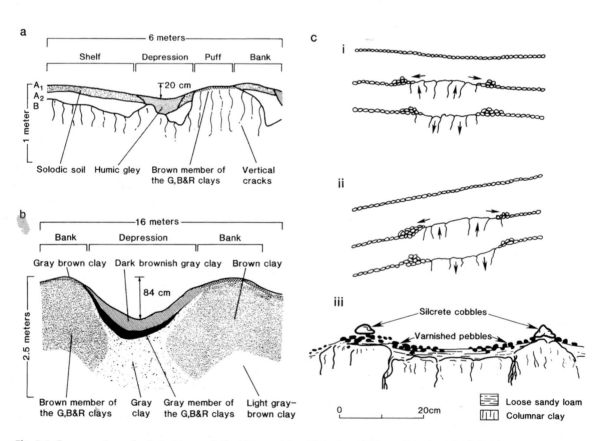

Fig. 5.6 Cross –sections of principal types of gilgai (**a**) normal; (**b**) melon-hole; (**c**) stony: (i) circular, (ii) stepped, (iii)polygonal; (d) lattice; (e) linear; (f) tank. (After Ollier 1966, Mabbutt 1977, Hubble et al. (1983)

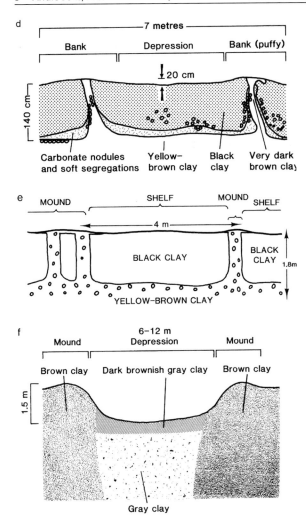

Fig. 5.6 (continued)

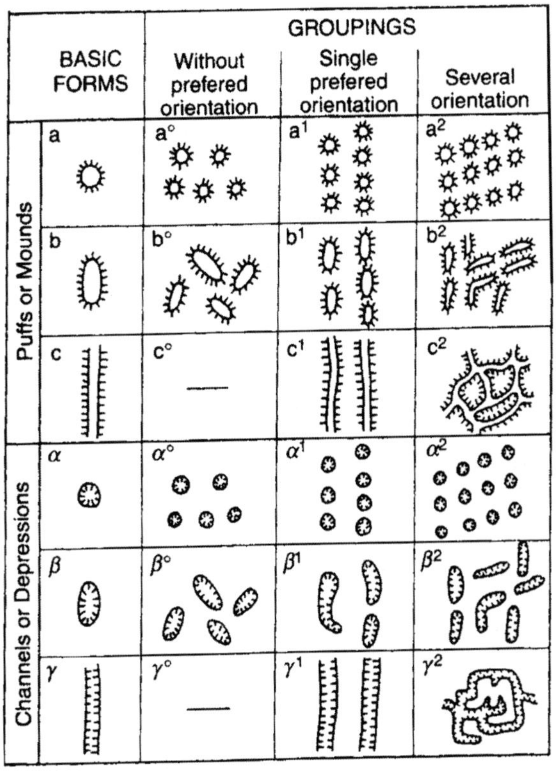

Fig. 5.7 Classification of gilgai morphology (After Verger 1964)

Knight analysed gilgai in south-eastern Australia using structural geological techniques. He recognized five recurring patterns based on the spatial arrangement of mounds, depressions, and shelves.

Gilgai Formation

A large literature exists on the origin of gilgai. Knight (1980) identified four types of gilgai-forming mechanisms: (a) heave between cracks, (b) heave over cracks, (c) contraction over cracks, and (d) heave due to loading (Fig. 5.8). Within the first type of mechanism he distinguished three subtypes (Fig. 5.8a). The first subtype involves soil compression, extrusion, and associated plastic flow, with resulting upward

movement of the mound. This mechanism was originally proposed by Leeper et al. (1936). The second subtype is perhaps one of the most widely accepted mechanisms of gilgai development and involves compression and block fracture (Hallsworth et al. 1955, Verger 1964). This mechanism envisages the falling or washing of surface materials into cracks with subsequent exertion of force on subsurface clays resulting in vertical upthrusting. The third subtype calls for compression and oblique slip. This mechanism was originally proposed by White and Bonestall (1960). However, these three subtypes are all regarded by Knight (1980) as being mechanically unsound.

Two subtypes of the heave-over-cracks mechanism were recognized by Knight (1980) (Fig. 5.8b). The first he classified as being due to cumulative internal vertical movements due to small oblique slips. The second subtype is that of vertical block movement due to heave and was proposed by Howard (1939). It is also the mechanism proposed by Ollier (1966) for stony gilgai.

The contraction-over-cracks mechanism is a hypothetical one proposed by McGarity (1953) (Fig. 5.8c). It envisages the formation of depressions in the vicinity of adjacent cracks and mounds between cracks. The depressions are the result of downward movement of soil due to drying.

The heave-due-to-loading mechanism can be divided into two subtypes (Fig. 5.8d). The first involves the upward flow of soil due to density contrasts between layers (Beckman 1966, Paton 1974), whereas the second involves the flow of fluidized soil up a crack through a solid layer (Hallsworth and Beckman 1969).

Maxwell (1994) argues that most of the pre-existing models of gilgai formation are inconsistent with observed patterns of moisture distribution on undulating surfaces and that they are also inconsistent with known patterns of stress within overconsolidated expansive clays. He suggests that gilgai forms are directly related to horizonal stresses occuring naturally within these clay-rich materials. Ground surface deformation occurs when relatively small horizontal stresses modify normal vertical stresses and differential vertical rebound occurs.

Surface Cracking

The second major type of patterned ground in warm deserts is related to surface cracking. This cracking is interpreted to be largely a desiccation feature caused by drying of the surface crust (Cooke and Warren 1973, pp. 139–40). Considerable literature exists on the nature and origin of cracks in sediments. However, there has been relatively little discussion of the origin of small-scale surface cracks in deserts. An exception is the work by Tucker (1978) in northern Iraq on the origin of patterned ground associated with gypsum crusts. Tucker examined non-orthogonal crack types (Lachenbruch 1962) which range in diameter from 50 cm to 2 m, with the cracks themselves being no wider than 5 cm. The cracks are commonly infilled with gypsum and stand as ridges above the general level of the surface crust. Polygons commonly are upturned at their edges. Tucker suggested that the development of cracks and resulting polygons is the result of changes in volume of the gypsum crust. These volume changes he attributed to diurnal and seasonal changes in temperature.

Another prominent type of patterned ground occurring in deserts is associated with playa lake surfaces and

is characterized by large scale dessication (Fig. 5.9). Such fissures have received considerable attention in the literature. In the United States extensive work on the nature and origin of these features has been undertaken in California by Neal (1965, 1968a, b, 1969), Neal and Motts (1967) and by Neal et al. (1968). These features are primarily the result of desiccation and subsequent cracking of crusts rich in swelling clays.

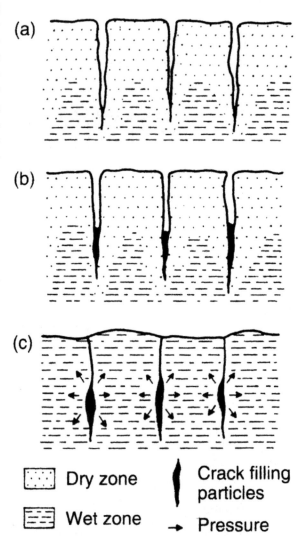

Fig. 5.8 Principal mechanisms of gilgai formation, (**a**) Heave between cracks: (i) by compression, extrusion and plastic flow, (ii) by compression and block fracture, (iii) by compression and oblique slip; (**b**) heave over cracks: (i) by cumulative internal oblique slip, (ii) by vertical block movement; (**c**) contraction over cracks due to downward movement associated with desiccation; (**d**) heave due to loading: (i) by upward flow due to density differences between layers, (ii) by upward flow of fluid soil through cracks in solid layer. (After Knight 1980)

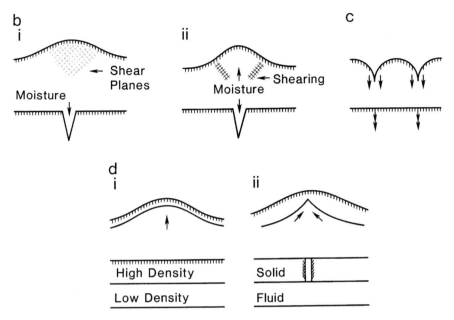

Fig. 5.8 (continued)

Fig. 5.9 Salt polygons, Death Valley Salt Pan, California

A great variety of patterned ground forms has been described from playa lakes in the western United States by Hunt and Washburn (1960, 1966). These forms are associated with different types of salt crusts. Smooth silt-rich surfaces with desiccation cracks and solution pits occur in the chloride zone beyond the limits of playa flooding. In the areas of gypsum crust, nets and polygons develop. In the carbonate zone near the edge of the salinas, sorted and non-sorted nets develop. The sorted nets form as a result of the washing of debris into the cracks. Sorted polygons develop where rock salt occurs beneath the surface. Coarse debris accumulates in depressions within silt overlying the salts. In addition, Hunt and Washburn (1966) have described a variety of patterned ground features developed on al-luvial fan surfaces, including sorted steps, sorted polygons, and stone stripes.

The various patterned ground phenomena described from playa lake basins are all essentially the result of the action of salt crystallization. In addition to the desiccation of salt-rich sediment crusts on and adjacent to playas, the expansion and contraction of salts as a result of heating and cooling as well as wetting and drying play an important role in the development of patterned ground features.

Patterned ground phenomena have been described from the Great Kavi of Iran. Krinsley (1968, 1970) reported a variety of morphologies (Fig. 5.10) including salt polygons, thrust polygons, desiccation cracks, and mud pinnacles. He attributed these features largely to the influx and episodic evaporation and subsequent desiccation of salt-rich ground waters. Patterned ground has also been described from the bed of Lake Eyre in central Australia, where its origin has been ascribed to the growth of salt crystals in unconsolidated muds and accompanying ground heave along desiccation cracks (Summerfield 1991, p. 147).

Microtopographic Patterned Ground

Microtopographic patterned ground forms are also common in desert environments. These forms are often

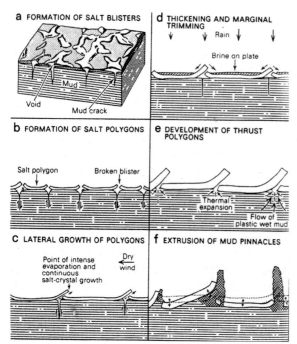

a FORMATION OF SALT BLISTERS

Mud

Void
Mud crack

b FORMATION OF SALT POLYGONS

Salt polygon   Broken blister

c LATERAL GROWTH OF POLYGONS

Point of intense
evaporation and   Dry
continuous   wind
salt-crystal growth

d THICKENING AND MARGINAL
TRIMMING
Rain

Brine on plate

e DEVELOPMENT OF THRUST
POLYGONS

Thermal
expansion
Flow of
plastic wet mud

f EXTRUSION OF MUD PINNACLES

Fig. 5.10 Development stages of polygons on salt crusts. (After Krinsley 1970)

not apparent to the casual observer but are highlighted by patterns of vegetation growth, such as vegetation arcs and water lanes recognized by MacFadyen (1950) and Hemming (1965) in Somalia. Water lanes have also been described from the Somaliland Plateau by Boaler and Hodge (1962). Similar patterns have been observed in Western Australia by Mabbutt (1963). He attributed the sorting of sediment to the combined action of wind and sheet flow. Large arcuate ripples have been described in Utah by Ives (1946). Similarly, microforms are reported from playa surfaces and other gypsum-rich environments (Gutiérrez 2006) and include hemispheric domes or tumuli which commonly display central collapse depressions. These forms are attributed to numerous processes including *in situ* dissolution and precipitation of surface gypsum which fills in existing pores of the deposit. Volume increase and associated crystallization pressures then lead to lateral gypsum expansion and dome building (Artieda 1993 in Gutiérrez 2006). More recently, Calaforra (1996 in Gutiérrez 2006) suggested several alternative hypotheses including compressive tectonic forces, and volumetric increases associated with the transformation of anhydrite to gypsum. Saline ramps composed of salt crystals and shaped as half-moons have been reported by Millington et al. (1995) from

playa surfaces in Tunisia. These forms are attributed to salt crystal redistribution by wind.

Desert Pavement

Morphology

Desert pavement is a stony surface generally composed of a layer of angular or subrounded gravels one or two stones thick sitting on a mantle of finer stone-free material (Mabbutt 1965, 1977, p. 119, Cooke and Warren 1973, p. 120) (Figs. 5.11 and 5.12). Desert pavement occurs widely in the warm deserts of the world (Cooke and Warren 1973, p. 121, Mabbutt 1977, p. 119). This surficial phenomenon is known by a

Fig. 5.11 Desert pavement, Cima volcanic field, California. Pavement underlain by a thick layer of aeolian-derived fine sediment (photo courtesy of L.D. McFadden)

Fig. 5.12 Desert pavement, Afton Canyon, California. Pavement underlain by a thick layer of aeolian-derived fine sediment

variety of names depending on the particular character of the pavement and the region of the world in which it occurs. Where the pavement is dominated by rock outcrops and boulders are relatively few, the landscape is referred to as hamada, an Arabic word meaning 'unfruitful'. Boulder hamada consists of extensive surfaces of large angular rock fragments and occurs extensively in deserts such as the Libyan Sahara (Mabbutt 1977, p. 121). Pavement dominated by smaller size gravel is referred to as reg in the old world and derives its name from the Arabic word meaning 'becoming smaller' (Dan et al. 1982, Amit and Gerson 1986). In the central Sahara, reg surfaces are referred to as serir (Mabbutt 1977). In Australia stone pavements are called gibber.

Genesis

Deflation

The development of desert pavement is generally attributed to one of five stone-concentrating processes. Perhaps the most commonly invoked process is that of deflation. The concentration of the coarse debris is believed to be the result of the removal of fine-grained material from the desert surface by wind, leaving the coarser debris behind as a lag deposit (e.g. Cooke 1970, Dan et al. 1981, 1982). Many workers, however, have questioned the ability of wind to transport fine-grained materials from desert surfaces. These materials are often incorporated into crusts which effectively preclude subsequent transportation (Cooke 1970). Several workers have also shown that as a gravel surface develops, the areas between the coarse fragments in fact become sheltered from the wind and fine-grained materials in these locations become progressively less likely to be transported (Pandastico and Ashaye 1956, Symmons and Hemming 1968). Finally, the fact that many desert pavements are underlain by a relatively thick layer of fine debris suggests the relative inefficiency of aeolian transport of fines from the desert surface. Some desert pavements, however, may in fact be the result of deflation, such as those observed in the Peruvian desert by Crolier et al. (1979). These workers noted that the cobbles forming the pavement are not supported by silts and clays but rather rest on sands and gravels.

Wash

A second mechanism for gravel concentration that has been widely proposed is the winnowing of fines by surface wash (Lowdermilk and Sundling 1950, Sharon 1962, Denny 1965, Cooke 1970, Dan et al. 1981, 1982 Parsons et al. 1992). Several of these workers have shown quantitatively that erosion of disturbed pavements yields considerable amounts of fine wash debris (Sharon 1962, Cooke 1970). McHargue (1981), working in the vicinity of the Aguila Mountains in Arizona, demonstrated that surface wash is a necessary component of pavement formation. He argued that incipient pavements form after the erosion of 1–3 cm of fine sediment and that stable pavements form after the removal of 3–15 cm of sediment. McHargue envisaged long-term stability of the pavement being related to subsequent addition of finegrained aeolian materials to the landscape. Similarly, Parsons et al. (1992) and Williams and Zimbelman (1994) emphasized the role of wash processes in the formation of desert pavements in southern Arizona, and Wainwright et al. (1999) demonstrated the crucial role played by wash in the reestablishment of pavement after its disturbance. However, the widespread occurrence of vesicular horizons beneath pavements has been argued to be incompatable with extensive wash and associated erosion (Williams and Zimbelman 1994)

Upward Migration of Stones

The third and perhaps most widely accepted explanation for the origin of desert pavements is the progressive upward migration of coarse particles through the underlying finer material. The mechanism most commonly invoked for this upward movement is alternate wetting and drying and associated swelling and shrinking of the fine-grained subpavement materials (Springer 1958, Jessop 1960, Cooke 1970, Dan et al. 1982). As the fine debris swells, coarse material is forced upward. When shrinkage occurs the coarse material fails to return to its former position which is occupied by fines. Repetition of this sequence of events causes the coarse material slowly to make its way to the ground surface. This process is clearly most pronounced in subpavement soils which are characterized by strong textural contrasts between the

A and B horizons. The presence of gypsic horizons beneath pavements may also impede the vertical migration of cobbles from the subsurface. Cooke and Warren (1973, p. 128) have suggested that freeze–thaw processes may also account for the formation of stone pavements in high altitude deserts such as those in South America and central Asia. The appeal of this mechanism is the fact that it offers an explanation for the fine-grained material found beneath the pavement. One of the limitations of this explanation is the fact that seldom are stones observed to be 'in transit' from below the fine-grained mantle to the ground surface (Mabbutt 1979). Further, adequate wetting depths for coarse fragment migration in deserts are difficult to invoke.

Cumulic Pedogenesis

Mabbutt (1977, 1979) proposed that perhaps some of the desert pavements in central Australia are the result of upward sorting above a fine textured mantle that is largely aeolian in origin. He suggested that aeolian dust is trapped by the rough surface of the pavement and that there is a consequent upward displacement of the pavement as dust accumulation proceeds. Sorting, if any, is limited to the uppermost part of the fine-grained aeolian mantle. More recently, a similar mechanism of pavement formation has been suggested by McFadden et al. (1987) Wells et al. (1995) and Haff and Werner (1996), who argued that the formation of desert pavement in the Cima volcanic field, California, is the result of the *in situ* mechanical weathering and accompanying downslope transport of basalt at the ground surface and the subsequent raising of the pavement as aeolian silts and clays accumulate below the gravels. In essence, the pavement is born at the earth's surface and, once formed, changes little through time. Soil development proceeds in the aeolian mantle beneath the pavement. McFadden et al. (1998) and Anderson et al. (2002) discuss the intimate relationship between vesicular horizon formation and pavements. They propose a three-stage model of pavement development involving (1) formation of a laminar crust adhered to the bottom of clasts. Crust formation is probably related to forces of adhesion and surface tension as water and sediment combine. (2) vesicular horizon development due to aeolian sediment and carbonate influx and associated clast lifting (3) formation of well developed soil structure permiting further development and thickening of the Av horizon.

Soil properties beneath desert pavements vary considerably depending on the precise nature of the pavement with respect to such characteristics as pavement coverage and clast arrangement. These characteristics have profound influences on soil properties and processes including soil texture, depth of leaching of fines and soluble salts (Wood et al. 2005). The effectiveness of cumulic pedogenesis appears to be also strongly influenced by clast size. Valentine and Harrington (2006) have demonstrated that if clasts are too small they do not form the interlocking pavement essential for silt capture and vesicular horizon formation. Support for this cumulic model of pavement formation has been provided by Wells et al. (1991) through cosmogenic dating of gravel clasts. These workers found that, using ^{3}He and ^{21}Ne, lava flow samples and clast samples have ages that are statistically inseparable from each other.

Subsurface Weathering

Mabbutt (1977) suggested a fifth process to account for the formation of desert pavement, namely differential weathering. He argued that moisture conditions are more favourable for rock weathering in the subsoil environment than at the desiccated surface, leading to the more rapid breakdown of coarse debris. The result of this enhanced weathering at depth is a layer of relatively fine materials with a lack of coarse debris. Mabbutt reasoned that this process may be particularly important in the generation of pavements in granitic terranes and that salt weathering may also be effective in the breakdown of coarse debris below the surface.

Desert pavements clearly result from a variety of processes. Some processes act independently, while others act in combination. The processes responsible for the formation of pavements vary from location to location depending on the climate, geomorphic setting, the nature of the clastic materials available, and the nature of the local soils (Bull 1991, p. 66).

Gravel Source

Central to an understanding of the origin of desert pavements is the source of the gravels. Two principal sources are generally recognized. Some pavements are composed of gravels that are fluvial in origin (Denny 1965, Cooke 1970, Mabbutt 1977, p. 122, Dan et al. 1982, Bull 1991, p. 65). These gravels are commonly rounded and occur in landscape settings in close proximity to stream channels. Some rounding of gravels can be accomplished by surface weathering processes, and so grain shape alone is not necessarily indicative of a fluvial origin. The second source is mechanical weathering of the local bedrock. This source has frequently been associated with the origin of hamadas (Cooke 1970, Mabbutt 1977, p. 121) but is now widely ascribed to finer grained pavement surfaces as well (Wells et al. 1985, McFadden et al. 1987).

Various processes have been postulated for the disintegration of bedrock and production of pavement gravels. Cooke (1970) suggested that pavement gravels in California and Chile might result from the combined influences of diurnal changes in insolation, the growth of frost needles and salt crystals, and a variety of chemical weathering processes. Peel (1974) proposed that gravels on pavements in the central Sahara are the result of insolation weathering, a view reiterated by Dan et al. (1982) for pavements in southern Israel and the Sinai. However, Dan et al. stress that cracking associated with salt crystal growth is a more important mechanism. In their detailed study of pavements in the Cima volcanic field, McFadden et al. (1987) suggested that much of the disintegration of the basalt flows is the result of mechanical disintegration processes. They proposed that salt- and clay-rich aeolian material deposited in fractures in the lava-flow surfaces experiences volumetric increase due to crystal growth and wetting and drying. These processes result in the vertical and lateral displacement of basalt clasts from the parent flows as well as subsequent breakdown of dislodged fragments. McFadden et al. suggested that some of the sand and silt in the fine-grained mantle may be derived from the mechanical breakdown of surface clasts. Some chemical alteration of the clasts has also occurred but appears to be limited to alteration of volcanic glass and iron oxides as well as olivine in the groundmass. Some authigenic clay formation is also suggested. However, much of the clay alteration seems to be related to chemical weathering of aeolian fines. A detailed discussion of weathering processes in arid environments is provided in Chapter 3.

Conclusions

Desert soils display a distinctive assemblage of physical chemical and biological characteristics. They are typically thin, gravelly, salt-dominated, and organic-poor. Their surfaces are characterized by the development of crusts and gravel lag deposits. The processes responsible for the development of these distinctive soils also result in the development of an assemblage of distinctive landform features.

The addition of considerable amounts of airborne clay to landscape surfaces in desert environments results in the development of soils with considerable swell–shrink capacity. Repeated swelling and shrinking of soils leads to the development of a variety of patterned ground forms.

Aeolian processes not only contribute significantly to soil formation and patterned ground development but they represent an important component of desert pavement formation. Silts and clays added to the desert landscape act to buoy the coarse gravel lag on the landscape surface. In addition, these fine particles as well as airborne salts serve as important agents of weathering, producing coarse debris that dominates some desert landscapes.

From the preceding discussion it is clear that a complete understanding of the geomorphic processes operating in desert environments cannot be obtained without a careful assessment of the nature of desert pedogenic processes.

References

Akpokodje, E.G. 1984. The influence of rock weathering on the genesis of gypsum and carbonates in some Australian arid zone soils. *Australian Journal of Soil Research* **22**, 243–51.

Alexander, E.B. and W.D. Nettleton 1977. Post-Mazama Natrargids in Dixie Valley, Nevada. *Soil Science Society of America Journal* **41**, 1210–12.

Amit, R. and R. Gerson 1986. The evolution of Holocene reg (gravelly) soils in deserts: an example from the Dead Sea region. *Catena* **13**, 59–79.

Amit, R. and J.B.J. Harrison 1995. Biogenic calcite horizon development under extremely arid conditions, Nizzana Sand Dunes, Israel. *Advances in Geoecology* **28**, 65–88.

Anderson, K., S.Wells and R. Graham 2002. Pedogenesis of vesicular horizons, Cima Volcanic Field, Mojave Desert, California. *Soil Science Society of America* **66**, 878–87.

Bachman, G.O. and M.N. Machette 1977. Calcic soils and calcretes in the southwestern United States. *U.S. Geological Survey Open File Report* 77–796.

Beckman, G.G. 1966. Gilgai phenomena as geologic structures: a theoretical discussion. *Proceedings of the Australian Conference of Soil Science* **3**, 1–2.

Benhur, M., I Shainberg, D Baker and R. Keren. 1985. Effect ofsoil texture and CaCO₃ content on water infiltration in crusted soil as related to water salinity. *Irrigation Science* **6**, 281–94.

Bettenay, E. 1962. The salt lake systems and their associated aeolian features in the semi-arid region of Western Australia. *Journal of Soil Science* **13**, 10–17.

Bettenay, E., A.V. Blackmore and F.J. Hingston 1964. Aspects of the hydrological cycle and related salinity in the Belka Valley, Western Australia. *Australian Journal of Soil Research* **2**, 187–210.

Birkeland, P.W. 1990 Soil-geomorphic research – a selected overview. *Geomorphology* **3**, 207–24.

Blackburn, W.H. 1975. Factors influencing infiltration infiltration rates and sediment production of semi-arid rangelands in Nevada. *Water Resources Research* **11**, 929–37.

Blank, R.R., J.A.Young and T.Lugaski 1996. Pedogenesis on talus slopes, the buckskin Range, Nevada. *Geoderma* **71**, 121–42.

Boaler, S.B. and C.A.H. Hodge 1962. Vegetation stripes in Somaliland. *Journal of Ecology* **50**, 465–76.

Boettinger, J.L. and R.J. Southard 1991. Silica and carbonate sources for aridisols on a granitic pediment, western Mojave Desert. *Soil Science Society of America Journal* **55**, 1057–67.

Boyadgiev, T.G. and W.H. Verheye 1996. Contribution to a utilitarian classification of gypsiferous soil. *Geoderma* **74**, 321–38.

Brasher, B.R., G. Borst and W.D. Nettleton 1976. Weak duripans in weathered rock in a Mediterranean climate. *Annual Meeting, American Society of Agronomy, Abstracts* 158.

Brewer, R., E. Bettenay and H.M. Churchward 1972. Some aspects of the origin and development of the red and brown hardpan soils of Bulloo Downs, W.A. *CSIRO (Australia) Division of Soils Technical Paper* 1B.

Bull, W.B. 1991. *Geomorphic responses to climate change.* New York: Oxford University Press.

Buol, S.W., F.D. Hole and R.J. McCracken 1997. *Soil genesis and classification, 4th edition.* Ames IA: Iowa State University Press.

Chadwick, O.A. and J.O. Davis 1990. Soil forming intervals caused by eolian sediment pulses in the Lahonton Basin, northwestern Nevada. *Geology* **18**, 243–6.

Chartres, C.J. 1982. The pedogenesis of desert loams in the Barrier Range, Western New South Wales, I. Soil parent materials. *Australian Journal of Soil Research* **20**, 269–81.

Chartres, C.J. 1983. The pedogenesis of desert loam soils in the Barrier Range, Western New South Wales. II. Weathering and soil formation. *Australian Journal of Soil Research* **21**, 1–13.

Chartres, C.J. 1985. A preliminary investigation of hardpan horizons in north-west New South Wales. *Australian Journal of Soil Research* **23**, 325–37.

Christianson, G.E. and C. Purcell 1985. Correlation and age of Quaternary alluvial fan sequences, Basin and Range province, southwestern United States. In *Soils and Quaternary geology of the southwstern United States*, D.L. Weide

(ed.), 115–22. Geological Society of America Special Paper 203.

Conacher, A.J. 1975. Throughflow as a mechanism responsible for excessive soil salinization in non-irrigated, previously arable lands in the Western Australia wheat belt: a field study. *Catena* **2**, 31–67.

Cooke, R.U. 1970. Stone pavements in deserts. *Annals of the Association of American Geographers* **60**, 560–77.

Cooke, R.U. and A. Warren 1973. *Geomorphology in deserts.* Berkeley: University of California Press.

Crolier, M.J., G.E. Ericksen, J.F. McCauley and P.C. Morris 1979. The desert landforms of Peru; a preliminary photographic atlas. *U.S. Geological Survey Interagency Report: Astrogeology* 57.

Dan, J. 1983. Soil chronosequences in Israel. *Catena* **10**, 287–399.

Dan, J., R. Gerson, H. Kogumdjisky and D.H. Yaalon 1981. *Arid Soils of Israel.* Agricultural Research Organization, Special Publication 190.

Dan, J., D.H. Yaalon, R. Moshe and S. Nissim 1982. Evolution of reg soils in southern Israel and Sinai. *Geoderma* **28**, 173–202.

Denny, C.S. 1965. Alluvial fans in the Death Valley region, California and Nevada. *U.S. Geological Survey, Professional Paper* 466.

Denny, C.S. 1967. Fans and pediments. *American Journal of Science* **265**, 81–105.

Dimmock, G.M., E. Bettenay and M.J. Mulcahy 1974. Salt content of laterite profiles in the Darling Range, Western Australia. *Australian Journal of Soil Research* **12**, 63–9.

Dohrenwend, J.C., W.B. Bull, L.D. McFadden, G.I. Smith 1991. Quaternary geology of the Basin and Range Province in California. In *Quaternary non-glacial geology: conterminous United States*, R.B. Morrison (ed.), 321–52. Geology of North America, Vol. K-2. Boulder, CO: Geological Society of America.

Dregne, H.E. 1976. *Soils of arid regions.* Amsterdam: Elsevier.

Dunkerley, D.L. and K.J. Brown 1997. Desert Soils. In *Arid Zone Geomorphology*, D.S.G. Thomas (ed.), 55–68. Chichester: John Wiley and Sons.

Eswaran, H. and G. Zi-Tong 1991. Properties, genesis, classification and distribution of soils with gypsum. In *Occurrence, characteristics, and genesis of carbonate, gypsum and silica accumulations in soils*, W.D. Nettleton (ed.), 89–119. Soil Science Society of America Special Publication 26.

Evenari, J., D.H. Yaalon and Y. Gutterman 1974. Note on soils with vesicular structures in deserts. *Zeitschrift für Geomorphologie* **18**, 162–72.

Flach, K.W., W.D. Nettleton, L.H. Gile and J.G. Cady 1969. Pedocementation: induration by silica, carbonates, and sesquioxides in the Quaternary. *Soil Science* **107**, 442–53.

Flach, K.W., W.D. Nettleton and R.E. Nelson 1973. The micromorphology of silica cemented soil horizons in western North America. In *Soil microscopy*, G.K. Rutherford (ed.) 714–29. Kingston: Queens University.

Fletcher, J.E. and W.P Martin 1948. some effects of algae and moulds in the raincrusts of desert soils. *Ecology* **29**, 95–100.

Frenkel, H., J.O. Goertzen and J.D. Rhoads 1978. Effects of clay type and content, exchangeable sodium percentage and electrolyte concentration on clay suspension and hydraulic conductivity. *Journal of the Soil Science Society of America* **42**, 32–9.

Gerson, R. and R. Amit 1987. Rates and modes of dust accretion and deposition in an arid region: the Negev, Israel. In *Desert sediments: ancient and modern*, L. Fostick and I. Reid (eds), 157–69. Geological Society Special Publication 35.

Gerson, R., R. Amit and S. Grossman 1985. *Dust availability in desert terrains, a study in the deserts of Israel, and the Sinai.* Jerusalem: Institute of Earth Sciences, the Hebrew University of Jerusalem.

Gifford, G.F. 1972. Infiltration rate and sediment production on a plowed big sagebrush site. *Journal of Range Management* **25**, 53–5.

Gile, L.H. 1967. Soils of an ancient basin floor near Las Cruces, New Mexico. *Soil Science* **103**, 265–76.

Gile, L.H. 1970. Soil of the Rio Grande Valley border in southern New Mexico. *Soil Science Society of America Journal* **34**, 465–72.

Gile, L.H. 1975. Holocene soils and soil geomorphic relations in an arid region of southern New Mexico. *Quaternary Research* **5**, 321–60.

Gile, L.H. and R.B. Grossman 1968. Morphology of the argillic horizon in desert soils of southern New Mexico. *Soil Science* **106**, 6–15.

Gile, L.H., J.W. Hawley and R.B. Grossman 1981. Soils and geomorphology in the Basin and Range area of Southern New Mexico: guidebook to the Desert Project, *New Mexico Bureau of Mines and Mineral Resources Memoir* **39**.

Gunn, R.H. 1967. A soil catena on denuded laterite profiles in Queensland. *Australian Journal of Soil Research* **5**, 117–32.

Gunn, R.H. and D.P. Richardson 1979. The nature and possible origins of soluble salts in deeply weathered landscapes of eastern Australia. *Australian Journal of Soil Research* **17**, 197–215.

Gustavson, T.C., R.W. Baumgardner Jr, S.C. Caran, V.T. Holliday 1991. Quaternary geology of the Southern Great Plains and an adjacent segment of the Rolling Plains. In *Quaternary non-glacial geology: conterminous United States*, R.B. Morrison (ed.), 477–501. Geology of North America, Vol. K-2. Boulder, CO: Geological Society of America.

Gustavson, T.C. and D.A. Winkler 1988. Depositional facies of the Miocene–Pliocene Ogallola Formation, northwestern Texas and eastern New Mexico. *Geology* **16**, 203–6.

Guthrie, P.L. 1982. Distribution of great groups of aridisols in the United States. *Catena Supplement* **1**, 29–36.

Gutiérrez M. 2006. *Climatic geomorphology*, Amsterdam: Elsevier.

Haff, P.K. and B.T. Werner 1996. Dynamical processes on desert pavements and the healing of surficial disturbances. *Quaternary Research* **45**, 38–46.

Hallsworth, E.G. and G.G. Beckman 1969. Gilgai in the quaternary. *Soil Science* **10**, 409–20.

Hallsworth, E.G. and G.K. Robertson 1951. The nature of gilgai and melonhole soils. *Australian Journal of Science* **13**, 181.

Hallsworth, E.G., G.K. Robertson and F.R. Gibbons 1955. Studies in pedogenesis in New South Wales. VII. The gilgai soils. *Journal of Soil Science* **6**, 1–31.

Harden, D.R., N.E. Biggar and M.L. Gillam 1985. Quaternary deposits and soils in and around Spanish Valley, Utah. In *Soils and Quaternary geology of the southwestern United States*, D.L. Weide (ed.), 43–64. Geological Society of America Special Paper 203.

Harden, J.W., E.M. Taylor, L.D. McFadden and M.C. Reheis 1991. Calcic, gypsic, and siliceous soil chronosequences in arid and semiarid environments. In *Occurrence, characteristics, and genesis of carbonate, gypsum and silica accumulations in soils*, W.D. Nettleton (ed.), 1–16. Soil Science Society of America. Special Publication 26.

Harris, S.A. 1958. Gilgaied and land structured soils of central Iraq. *Journal of Soil Science* **9**, 169–85.

Harris, S.A. 1959. The classification of gilgaied soils: some evidence from northern Iraq. *Journal of Soil Science* **10**, 27–33.

Hemming, C.F. 1965. Vegetation arcs in Somaliland. *Journal of Ecology* **53**, 57–67.

Holliday, V.T. 1985a. Holocene soil geomorphological relations in a semi-arid environment: the Southern High Plains of Texas. In *Soils and Quaternary landscape evolution*, J. Boardman (ed.), 325–57. New York: Wiley.

Holliday, V.T. 1985b. Morphology of late Holocene soils at the Lubbock Lake archeological site, Texas. *Soil Science Society of America Journal* **49**, 938–46.

Holliday, V.T. 1985c. Early and middle Holocene soils at the Lubbock Lake archeological site, Texas. *Catena* **12**, 61–78.

Holliday, V.T. 1988. Genesis of late Holocene soils at the Lubbock Lake archaeological site, Texas. *Annals of the Association of American Geographers* **78**, 594–610.

Holliday, V.T. 1989. The Blackwater Draw Formation (Quaternary). A 1.4 + m.y. record of eolian sedimentation and soil formation on the Southern High Plains. *Bulletin of the Geological Society of America* **101**, 1598–607.

Holliday, V.T. 1990. Soils and landscape evolution of eolian plains: the Southern High Plains of Texas and New Mexico. *Geomorphology* **3**, 489–515.

Howard, A. 1939. Crab-hole gilgai and self-mulching soils of the Murrumbidgee Irrigation Area. *Pedology* **8**, 14–8.

Hubble, G.D. and R.F. Isbell 1958. The occurrence of strongly acid clays beneath alkaline soils in Queensland. *Australian Journal of Science* **20**, 186–7.

Hubble, G.D., R.F. Isbell and K.H. Northcote 1983. Features of Australian soils. In *Soils: an Australian viewpoint*, 17–47. Melbourne: CSIRO/London: Academic Press.

Hunt, C.B. and A.L. Washburn 1960. Salt features that simulate ground patterns formed in cold climates. *U.S. Geological Survey Professional Paper* 400-B.

Hunt, C.B. and A.L. Washburn 1966. Patterned ground. *U.S. Geological Survey Professional Paper* 494-B, 104–33.

Hutton, J.T. and J.C. Dixon 1981. The chemistry and mineralogy of some South Australian calcretes and associated soft carbonates and their dolmitization. *Journal of the Geological Society of Australia* **28**, 71–9.

Hutton, J.T. and T.I. Leslie 1958. Accession of nonnitrogenous ions dissolved in rainwater to soils in Victoria. *Australian Journal of Agricultural Research* **14**, 319–29.

Isbell, R.F. 1962. Soils and vegetation of the brigalow lands, eastern Australia. *CSIRO (Australia) Division of Soils, Soils and Landuse Series* 43.

Isbell, R.F., R. Reeve and J.T. Hutton 1983. Salt and sodicity. In *Soils: an Australian viewpoint*, 107–17, Melbourne: CSIRO/London: Academic Press.

Ives, R.L. 1946. Desert ripples. *American Journal of Science* **244**, 492–501.

Jessop, R.W. 1960. The lateritic soils of the southeastern portion of the Australian Arid Zone. *Journal of Soil Science* **11**, 106–13.

Knight, M.J. 1980. Structural analysis and mechanical origins of gilgai at Boorook, Victoria, Australia. *Geoderma* **23**, 245–83.

Krinsley, D.B. 1968. Geomorphology of three kavirs in northern Iran. *Air Force Cambridge Research Laboratories Environmental Research Papers* 283.

Krinsley, D.B. 1970. A geomorphological and paleoclimatological study of the playas of Iran. *U.S. Geological Survey. Final Scientific Report Contract PROCP-70-800.*

Lachenbruch, A.H. 1962. Mechanics of thermal contraction cracks and ice-wedge polygons in permafrost. *Geological Society of America Special Paper* 70.

Leeper, G.W., A. Nicholls and S.M. Wadham 1936. Soil and pasture studies in the Mount Gellibrand area, Western District of Victoria. *Proceedings of the Royal Society of Victoria* **49**, 77–138.

Litchfield, W.H. and J.A. Mabbutt 1962. Hardpan soils of semiarid Western Australia. *Journal of Soil Science* **13**, 148–60.

Lowdermilk, W.C. and H.L. Sundling 1950. Erosion pavement, its formation and significance. *Transactions of the American Geophysical Union* **31**, 96–100.

Mabbutt, J.A. 1963. Wanderrie banks: micro-relief patterns in semiarid Western Australia. *Bulletin of the Geological Society of America* **74**, 529–40.

Mabbutt, J.A. 1965. Stone distribution in a stony tableland soil. *Australian Journal of Soil Research* **3**, 131–42.

Mabbutt, J.A. 1977. *Desert landforms*, Cambridge, Mass. MIT Press.

Mabbutt, J.A. 1979. Pavements and patterned ground in the Australian stony deserts. *Stuttgarter Geographische Studen* **93**, 107–23.

Machette, M.N. 1985. Calcic soils of the Southwestern United States. In *Soils and Quaternary geology of the southwestern United States*, D.L. Wide (ed.) 1–21. Geological Society of America Special Paper 203.

Maxwell, B. 1994. Influence of horizontal stresses on gilgai landforms. *Journal of Geotechnical Engineering* **120**, 1437–44.

McFadden, L.D. 1988. Climatic influences on rates and processes of soil development in Quaternary deposits of southern California. In *Paleosols and weathering through geologic time: principles and applications*, J. Reinhardt and W.R. Sigleo (eds), 153–77, Geological Society of America Special Paper 206.

McFadden, L.D. and J.C. Tinsley 1985. Rate and depth of pedogenic-carbonate accumulation in soils: formulation and testing of a comparintment model. In *Soils and Quaternary geology of the southwestern United States*, D.L. Weide (ed.), 23–41. Geological Society of America Special Paper 203.

McFadden, L.D., S.G. Wells, J.C. Dohrenwend and B.D. Turrin 1984. Cumulic soils formed in eolian parent materials on flows of the late Cenozoic Cima volcanic field, California. In *Surficial geology of the eastern Mojave Desert, California*, J.C. Dohrenwend (ed.), 134–49. Field Trip Guide, 97th Annual Meeting of the Geological Society of America.

McFadden, L.D., S.G. Wells and J.C. Dohrenwend 1986. Influences of Quaternary climate changes on processes of soil development on desert loess deposits of the Cima volcanic field; California. *Catena* **13**, 361–89.

McFadden, L.D., S.G. Wells and M.J. Jercinovich 1987. Influences of eolian and pedogenic processes on the origin and evolution of desert pavements. *Geology* **15**, 504–8.

McFadden, L.D., R.G. Amundson and O.A. Chadwick 1991. Numerical modeling, chemical, and isotopic studies of carbonate accumulation in soils of arid regions. In *Occurrence, characteristics, and genesis of carbonate, gypsum and silica accumulation in soils*, W.D. Nettleton (ed.), 17–35. Soil Science Society of America. Special Publication 26.

McFadden, L.D., E.V. McDonald, S.G. Wells, K.Anderson, J. Quade and S.L. Forman 1998. The vesicular layer and carbonate collars of desert soils and pavements: formation, age and relation to climate change. *Geomorphology* **24**, 101–45.

MacFadyen, W.A. 1950. Vegetation patterns in the semidesert plains of British Somalia. *Geographical Journal* **116**, 199–211.

McGarity, J.W. 1953. Melon hole formation in the Richmond River District of New South Wales. *Proceedings of the Australian Conference on Soil Science* **2**, 1–7.

McHargue, L.E. 1981. Late Quaternary deposition and pedogenesis on the Aquila Mountains piedmont, southeastern Arizona. M.S. thesis, University of Arizona, Tucson.

McIntyre, D.S. 1958a. Permeability measurements of soil crusts formed by raindrop impact. *Soil Science* **85**, 185–89.

McIntyre, D.S. 1958b. Soil splash and the formation of surface crusts by raindrop impact. *Soil Science* **85**, 261–66.

Millington, A.C., N.A. Drake, K. White and R.G. Bryant. 1995. Salt ramps: wind-induced depositional features on tunisian playas. *Earth Surface Processes and Landforms* **20**, 103–13.

Miller, D.E. 1971. Formation of vesicular structure in soil. *Soil Science Society of America Journal* **35**, 635–7.

Milnes, A.R. and J.T. Hutton 1983. Calcretes in Australia. In *Soils: an Australian viewpoint*, 119–62, Melbourne: CSIRO/London: Academic Press.

Milnes, A.R., M.J. Wright and M. Thiry 1991. Silica accumulations in saprolites and soils in South Australia. In *Occurrence, characteristics, and genesis of carbonate, gypsum and silica accumulations in soils*, W.D. Nettleton (ed.) 121–49. Soil Science Society of America Special Publication 26.

Monger, H.C., L.A. Daugherty and L.H. Gile 1991. A microscopic examination of pedogenic calcite in an aridisol of southern New Mexico. In *Occurrence, characteristics, and genesis of carbonate, gypsum and silica accumulations in soils*, W.D. Nettleton (ed.), 37–60. Soil Science Society of America. Special Publication 26.

Neal, J.T. 1965. Giant desiccation polygons of Great Basin playas. *Air Force Cambridge Research Laboratories Environmental Research Papers* 123, 1–30.

Neal, J.T. 1968a. Playa surface changes at Harper Lake California: 1962–1967. *Air Force Cambridge Research Laboratories Environmental Research Papers* 283.

Neal, J.T. (ed.) 1968b. Playa surface morphology: miscellaneous investigations. *Air Force Cambridge Research Laboratories Environmental Research Papers* 283.

Neal, J.T. 1969. Playa variations. In *Arid lands in perspective*, W.G. McGimies and B.J. Goldman (eds), 13–44. Tucson: University of Arizona Press.

Neal, J.T. and W.S. Motts 1967. Recent geomorphic changes in playas of the western United States. *Journal of Geology* **75**, 511–25.

Neal, J.T., A.M. Langer and P.F. Kerr 1968. Giant desiccation polygons of Great Basin playas. *Bulletin of the Geological Society of America* **79**, 69–90.

Nelson, R.E., L.C. Klameth and W.D. Nettleton 1978. Determining soil gypsum content and expressing properties of gypsiferous soils. *Soil Science Society of America Journal* **42**, 659–61.

Nettleton, W.D. and F.F. Peterson 1983. Aridisols. In *Pedogenesis and soil taxonomy II. The soil orders. Developments in Soil Science IIB*, L.P. Wilding, N.E. Smeck and G.F. Hall (eds), 165–215. Amsterdam: Elsevier.

Nettleton, W.D., J.G. Witty, R.E. Nelson and J.W. Hanley 1975. Genesis of argillic horizons in soils of desert areas of the southwestern. *U.S. Soil Science Society of America Journal* **39**, 919–26.

Nettleton, W.D., R.E. Nelson, B.R. Brasher and P.S. Deer 1982. Gypsiferous soils in the western United States. In *Acid sulfate weathering*, J.A. Kittrick, D.S. Fanning and L.R. Hossner (eds), 147–68. Soil Science Society of America. Special Publication 10.

Nettleton, W.D., E.E. Gamble, B.L. Allen, G. Borst 1989. Relief soils of subtropical regions of the United States. *Catena Supplement* **16**, 59–93.

Nettleton, W.D., B.R. Brasher and S.L. Baird 1991. Carbonate clay characterization by statistical methods. In *Occurrence, characteristics, and genesis of carbonate, gypsum and silica accumulations in soils* W.D. Nettleton (ed.), 75–88. Soil Science Society of America. Special Publication 25.

Northcote, K.H. and J.K.M. Skene 1972. Australian soils with saline and sodic properties. *CSIRO (Australia) Soil Publication* 27.

Ollier, C.D. 1966. Desert gilgai. *Nature* **212**, 581–3.

Painuli, D.K. and I.P. Abrol 1986. Effect of exchangeable sodium percent on surface scaling. *Agricultural Water Management* **11**, 247–256.

Pandastico, E.B. and T.I. Ashaye 1956. Demonstration of the effect of stone layers on soil transport and accretion. *In Experimental pedology*, E.G. Hallsworth and D.V. Crawford (eds), 384–90. London: Butterworth.

Parsons, A.J., A.D. Abrahams and J.R. Simanton 1992. Microtopography and soil-surface materials on semi-arid piedmont hillslopes, southern Arizona. *Journal of Arid Environments* **22**, 107–15.

Paton, T.A. 1974. Origin and terminology for gilgai in Australia. *Geoderma* **11**, 221–42.

Peck, A.J. 1978. Salinization of non-irrigated soils and associated streams: a review. *Australian Journal of Soil Research* **16**, 157–68.

Peck, A.J. and D.H. Hurle 1973. Chloride balance of some farmed and forested catchments in southwestern Australia. *Water Resource Research* **9**, 648–57.

Peel, R.F. 1974. Insolation weathering: some measurements of diurnal temperature change in exposed rocks in the Tibest region, central Sahara. *Zeitschrift für Geomorphologie Supplement Band* **21**, 19–28.

Peterson, F.F. 1980. Holocene desert soil formation under sodium salt influence in a playa-margin environment. *Quaternary Research* **13**, 172–86.

Ponti, D.J. 1985. The Quaternary alluvial sequence of the Antelope Valley, California. In *Soils and Quaternary geology of the southwestern United States*, D.L. Weide (ed.), 79–96. Geological Society of America Special Paper 203.

Prescott, J.A. 1931. The soils of Australia in relation to vegetation and climate. *CSIRO (Australia) Bulletin* **52**.

Rabenhorst, M.C., L.T. West and L.P. Wilding 1991. Genesis of calcic and petrocalcic horizons in soils over carbonate rocks. In *Occurrence, characteristics, and genesis of carbonate, gypsum and silica accumulations in soils*, W.D. Nettleton (ed.), 61–74, Soil Science Society of America. Special Publication 26.

Reheis, M.C. 1987a. Gypsic soils on the Kane alluvial fans, Big Horn Country, Wyoming. *U.S. Geological Survey Bulletin* **159-C**.

Reheis, M.C. 1987b. Climate implications of alternating clay and carbonate formation in semi arid soils of south-central Montana. *Quaternary Research* **29**, 270–82.

Reheis, M.C. 1990. Influence of climate and eolian dust on major-element chemistry and clay mineralogy of soils in the northern Bighorn Basin, U.S.A. *Catena* **17**, 219–48.

Reheis, M.C., J.W. Harden, L.D. McFadden and R.R. Shroba 1989. Development rates of late Quaternary soils: Silver Lake Playa, California. *Soil Science Society of America Journal* **53**, 1127–40.

Reheis, M.C., J.C. Goodmacher, J.W. Harden, L.D. McFadden, T.K. Rockwell, R.R. Shroba, J.M. Sowers and E.M. Taylor 1995. Quaternary soils and dust deposition in southern Nevada and California. *Geological Society of America Bulletin* **107**, 1003–22.

Schaetzl, R. and S. Anderson. 2005. *Soils: Genesis and Geomorphology*. Cambridge University Press: Cambridge.

Sharon, D. 1962. On the nature of hamadas in Israel. *Zeitschrift für Geomorphologie* **6**, 129–47.

Shlemon, R.J. 1978. Quaternary soil-geomorphic relationships, southeastern Mojave Desert, California and Arizona. In *Quaternary soils*, W.C. Mahany (ed.), 187–207, Norwich: Geo Books.

Smith, B.R. and S.W. Buol 1968. Genesis and relative weathering intensity studies in three semirarid soils. *Soil Science Society of America Journal* **32**, 261–5.

Soil Survey Staff 1975. Soil taxonomy: a basic system of soil classification for making and interpreting soil survey. *U.S. Department of Agriculture Agriculture Handbook* 436.

Soil Survey Staff 1994. *Keys to Soil Taxonomy 6th edition*. Pocahontas Press: Blacksburg VA.

Southard, R. 2000. Aridisols. In *Handbook of Soil Science*, M.E. Sumner (ed.), E-321- E 338, CRC Press: Boca Raton.

Springer, M.E. 1958. Desert pavement and vesicular layer in some soils of the Lahontan Basin, Nevada. *Soil Science Society of America Journal* **22**, 63–6.

Stace, H.C.T., G.D. Hubble, R. Brewer, K.H. Northcote 1968. *A handbook of Australian soils*, Glenside, South Australia: Rellim Technical Publications.

Stephens, C.G. 1962. *Manual of Australian soils*. 3rd edn. Melbourne: CSIRO.

Sullivan, L.A. and A.J. Koppi 1991. Morphology and genesis of silt and clay coatings in the vesicular layer of a desert loam soil. *Australian Journal of Soil Research* **29**, 579–86.

Summerfield, M.A. 1991. *Global geomorphology*, New York: Wiley.

Symmons, P.M. and C.F. Hemming 1968. A note on wind-stable stone-mantles in the southern Sahara. *Geographical Journal* **134**, 60–4.

Tucker, M.E. 1978. Gypsum crusts (gypcrete) and patterned ground from northern Iraq. *Zeitschrift für Geomorphologie* **22**, 89–100.

Valentine, G.A. and Harrington, C.D. 2006. Clast size controls and longevity of pleistocene desert pavements at Lathrop Wells and Red Cone volcanoes, southern Nevada. *Geology* **34**, 533–36.

Verger, F. 1964. Mottereaux et gilgais. *Annales de Géographie* **73**, 413–30.

Wainwright, J., A.J. Parsons and A.D. Abrahams 1999. Field and computer simulation experiments on the formation of desert pavement. *Earth Surface Processes and Landforms* **24**, 1025–1037.

Weide, D.L. (ed.) 1985. Soils and Quaternary geology of the southwestern United States. *Geological Society of America Special Paper* 203.

Wallace, A., E.M. Romney and V.O. Hale 1973. Sodium relations in desert plants: 1. Cation contents of some desert plant species from the Mojave and Great Basin deserts. *Soil Science* **115**, 284–87.

Wells, S.G., J.C. Dohrenwend, L.D. McFadden, B.D. Turrin 1985. Late Cenozoic landscape evolution on lava flow surfaces of the Cima volcanic field, Mojave Desert, California *Bulletin of the Geological Society of America* **96**, 1518–29.

Wells, S.G., L.D. McFadden and J.C. Dohrenwend 1987. Influence of late Quaternary climatic changes on geomorphic and pedogenic processes on a desert piedmont eastern Mojave Desert, California. *Quaternary Research* **27**, 130–46.

Wells, S.G., L.D. McFadden and J.D. Schultz 1990. Eolian landscape evolution and soil formation in the Chaco dune field, southern Colorado Plateau, New Mexico. *Geomorphology* **3**, 517–46.

Wells, S.G., L.D. McFadden and C.T. Olinger 1991. Use of cosmogenic 3Ne and 21Ne to understand desert pavement formation. *Geological Society of America Abstracts with Programs* **23** (5), 206.

Wells, S.G., L.D. McFadden, J. Poths and C.T. Olinger 1995. Cosmogenic ^{3}He surface exposure dating of stone pavements: implications for landscape evolution in deserts. *Geology* **23**, 613–16.

Wetselaar, R. and J.T. Hutton 1963. The ionic composition of rainwater of Katherine, N.T. and its part in the cycling of plant nutrients. *Australian Journal of Agricultural Research* **14**, 319–29.

White, E.M. and R.G. Bonestall 1960. Some gilgaied soils of South Dakota. *Soil Science Society of America Journal* **24**, 305–9.

White, L.P. and R. Law 1969. Channeling of alluvial depression soils in Iraq and Sudan. *Journal of Soil Science* **20**, 84–90.

Williams, S.H. and J.R. Zimbelman 1994. Desert pavement evolution: an example of the role of sheetflood. *Journal of Geology* **102**, 243–48.

Wood, Y.A., R.C. Graham and S.G. Wells 2005. Surface control of desert pavement pedologic process and landscape function, Cima Volcanic field, Mojave Desert, California. *Catena* **59** 205–30.

Wright, M.J. 1983. Red-brown hardpans and associated soils in Australia. *Transactions of the Royal Society of South Australia* **107**, 252–4.

Yaalon, D.H. 1964. Airborne salt as an active agent in pedogenic processes. *Transactions of the 8th Congress of the International Soil Science Society* **5**, 994–1000.

Yaalon, D.H. and E. Ganor 1973. The influence of dust on soils during the Quaternary. *Soil Science* **116**, 146–55.

Chapter 6

Duricrusts

John C. Dixon and Sue J. McLaren

Introduction

Warm desert environments are characterized by the occurrence of a variety of surface and near-surface, chemically precipitated crusts, including calcrete, silcrete, and gypcrete, and their intergrades (Dixon 1994). Collectively, these are referred to as duricrusts (Woolnough 1930). This chapter focuses on the morphology, chemistry, mineralogy, and origin of the major types of crusts found in dryland environments. Particular attention is given to recent research from Australia, the United States, the Middle East and Africa.

Calcrete

Terminology

Calcrete was defined by Netterberg (1969, p. 88) as almost any terrestrial material which has been cemented and/or replaced by dominantly calcium carbonate. The mechanism of calcification is not restricted and may be of pedogenic or non-pedogenic origin or both. Goudie (1983, p. 94–5) offered a more comprehensive definition. He defined calcrete as terrestrial materials composed dominantly, but not exclusively, of calcium carbonate, which occurs in states ranging from powdery to highly indurated. These materials involve the cementation of, accumulation in, and/or

replacement of greater or lesser quantities of soil, rock, or weathered material primarily in the vadose zone. Goudie recognized some limitations of this definition, pointing out that not all calcretes are indurated nor are they necessarily all formed in the vadose zone.

The term calcrete was introduced into the geomorphological literature by Lamplugh (1902) in reference to lime cemented gravels from Dublin, Ireland. Five years later Lamplugh (1907) introduced the related term valley calcretes to refer to calcareous deposits in the valley of the Zambezi River. In the same year that Lamplugh introduced the term calcrete, Blake (1902) used the term caliche in the United States to refer to carbonate accumulations at or near the Earth's surface in Arizona and New Mexico. While these two terms are the most commonly used in the English-speaking literature, they are by no means the only ones. A detailed summary of the various terms used to refer to terrestrial carbonates is provided by Goudie (1973). Some of the more commonly used terms include nari in Israel (Dan 1977), kankar in India and croute calcaire in the French literature (Durand 1963).

Preference is growing for the use of the term calcrete because of inconsistencies in the use of terms such as caliche and travertine (Hay and Reeder 1978, Netterberg and Caiger 1983). In the United States the term caliche is becoming less commonly used and is being replaced by the terms calcrete when referring to indurated terrestrial deposits (Machette 1985) and calcified soils in the case of non-indurated calcic accumulations (Machette 1985). Disfavour with the term caliche has grown as a result of its misapplication to a variety of surficial crusts. In addition to being applied to carbonate deposits, the term has been applied to many varieties of duricrust (Price 1933). In South America the term caliche has been applied to

J.C. Dixon (✉)
Department of Geosciences, University of Arkansas,
Fayetteville, AR 72701, USA
e-mail: jcdixon@uark.edu

A.J. Parsons, A.D. Abrahams (eds.), *Geomorphology of Desert Environments*, 2nd ed.,
DOI 10.1007/978-1-4020-5719-9_6, © Springer Science+Business Media B.V. 2009

accumulations of sodium nitrate (Ericksen 1994). Hunt and Mabey (1966) used the term to refer to calcium sulphate cemented gravels in Death Valley, California. Attempts to re-define caliche in order to avoid these confusions were made by Bretz and Horberg (1949) and Brown (1956), who both stressed restricting the term caliche to calcareous deposits that had accumulated in place in the weathering zone. Although Goudie (2007) note that carbonate crusts are now more sensibly referred to as calcretes and that the term caliche, they argue, should be used for nitrate deposits.

In Australia, early geologists referred to the widespread surficial accumulations of carbonate as travertine (Woolnough 1930, Crocker 1946). In the 1960s the term travertine was largely replaced by the term kunkar which is derived from the hindi term kankar meaning gravel (Goudie 1973, p. 9). The term kunkar was widely applied to indurated calcareous masses present within soil profiles (Johns 1963). Firman (1963) used the term to refer to limey B horizons present in many soils of South Australia. Crawford (1965) used the term for complex calcrete profiles on Yorke Peninsula in southern South Australia. In the mid-1960s Firman (1966) proposed that the term calcrete be adopted.

Throughout the past few decades the term calcrete has been widely used in the Australian literature on terrestrial carbonates (Dixon 1978, Hutton and Dixon 1981, Milnes and Hutton 1983, Phillips and Self 1987, Phillips et al. 1987), as well as in Southern Africa (Netterberg 1969, 1975, Watts 1980, Nash and McLaren 2003).

Distribution

Calcrete covers approximately 13% of the Earth's land surface (Yaalon 1981) and mostly occurs in areas where the annual precipitation is between 400 and 600 mm (Goudie 1983), although Strong et al. (1992) identified calcrete textures in a Pleistocene glacial gravel in North Yorkshire, U.K. Calcrete is widely distributed throughout Africa, the Middle East, southern Europe (Goudie 1973, pp. 73–9, 1983), and the semi-arid parts of India (Patil 1991). In the United States it occurs widely in the south-western states of New Mexico, Nevada, Arizona, Texas, and California (Reeves 1976, Machette 1985). It is also found in southern and central South Australia (Milnes and

Hutton 1983), central Western Australia (Mann and Horwitz 1979), and the Northern Territory (Jacobson et al. 1988).

Morphology

Macromorphology

Calcrete morphology has been described from many locations. However, only a few systematic, non-genetic, systems of calcrete classification have been developed. Durand (1949) recognized eight different types of calcrete crust: croute zonaire, encroutement, racine petrifiee, calcaire pulverulent, efflorescences, nodules, vernis desertique, and tufa or travertine. Today, however, desert varnish and tufa are not regarded as calcrete types.

In the late 1960s, a non-genetic, evolutionary, morphological sequence of calcrete types was developed by Netterberg (1969) for South Africa. Netterberg distinguished six fundamental calcrete types. The least strongly developed calcrete is a calcified soil which is cemented by carbonate to a firm or stiff consistency. With progressive calcification the calcified soil becomes a powder calcrete (Fig. 6.1) which is composed

Fig. 6.1 Powder calcretedeveloped in alluvium near Sbeitla, central Tunisia (McLaren photo)

of loose carbonate silt or sand particles with few or no host soil particles or nodules present. Nodular or glaebular calcrete consists of silt- to gravel-sized particles of soil material cemented by carbonate (Fig. 6.2). Included within the nodular calcrete type are calcified pedotubules and crotovinas. As calcrete nodules increase in size due to the progressive accumulation of carbonate, they begin to coalesce to form honeycomb calcrete. In this type of calcrete the coalesced nodules are separated by host soil material. When the voids of the honeycomb calcrete become infilled and cemented with carbonate, hardpan calcrete develops. Hardpan calcretes are firm to very hard sheet-like masses of calcrete and represent the most advanced stage of calcrete development (Fig. 6.3). They commonly overlie less strongly indurated materials (Fig. 6.4). Hardpan calcretes are commonly capped with a thin layer of laminar calcrete (Fig. 6.4). As hardpan calcretes are weathered and eroded, they break down into rounded, subrounded, subangular,

Fig. 6.3 Hardpan calcrete caprock., High Plains, New Mexico (Dixon photo)

Fig. 6.2 Nodular and powdery calcrete northwest Botswana (McLaren photo)

and blocky masses which are referred to as boulder calcrete (Fig. 6.5). Netterberg's classification system has been adopted in modified form by several workers in Australia (e.g. Hutton and Dixon 1981, Milnes and Hutton 1983, Phillips and Milnes 1988). Based on work by Netterberg (1967, 1980), Goudie (1983), Esteban and Klappa (1983), Wright (1994) and Alonso-Zarza (2003), Wright (2007) provides a useful table of the main morphological types of calcrete horizons.

In the United States, the most widely used system of morphological description of pedogenic calcretes is that of Gile et al. (1965). These authors introduced the concept of the K horizon to refer to layers of accumulated carbonate. To qualify as a K horizon a carbonate layer must consist of more than 90% K fabric. Such fabric consists of continuous authigenic carbonate that has engulfed and cemented pebble-, sand-, and silt-sized grains. Gile et al. (1966) identified

Fig. 6.4 Calcrete profile from Murray River valley, South Australia, showing progressive decrease in the degree of carbonate induration with depth and development of thin laminar layer at top of profile (Dixon photo)

Fig. 6.5 Boulder calcrete horizon overlying weakly indurated nodular calcrete horizon, Yorke Peninsula, South Australia (Dixon photo)

a morphological sequence of carbonate accumulation in calcic soils based on the degree of calcification. They proposed a four-stage sequence based in part on the nature of the texture of the parent material (Table 6.1).

The morphological scheme of Gile et al. (1966) has been modified by the inclusion of two additional stages of carbonate development which more completely de-

scribe later phases of carbonate induration (Bachman and Machette 1977, Machette 1985) (Table 6.2).

Multiple calcrete profiles displaying a variety of stages of development may be exposed in a single geological unit (Fig. 6.6). Gustavson and Winkler (1988) reported the occurrence of multiple calcretes in the Pleistocene age Ogallala Formation of west Texas and eastern New Mexico. Holliday (1989) recorded similar

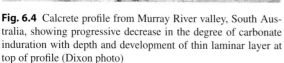

Table 6.1 Stages of the morphogenetic sequence of carbonate deposition in calcic soils[a]

| Stage | Diagnostic carbonate morphology | |
	Gravelly soils	Non-gravelly soils
I	Thin, discontinuous pebble coatings	Few filaments or faint coatings
II	Continuous pebble coatings, some interpebble fillings	Few to common nodules
III	Many interpebble fillings	Many nodules and internodular fillings
IV	Laminar horizon overlying plugged horizon	Increasing carbonate impregnation
	Thickened laminar and plugged horizonsf	Laminar horizon overlying plugged horizon

[a] Modified from Gile et al. (1966, p. 348, Table 1).

Table 6.2 Stages in the morphogenetic sequence of carbonate deposition used in this report[a]

Stage	Diagnostic carbonate morphology
I	Filaments or faint coatings. Thin, discontinuous coatings on lower surface of pebbles
II	Firm carbonate nodules few to common but isolated from one another. Matrix may include friable interstitial accumulations of carbonate. Pebble coatings continuous
III	Coalesced nodules in disseminated carbonate matrix
IV	Platy, massive indurated matrix. Relict nodules may be visible in places. Plugged. May have weak incipient laminae in upper surface. Case hardening common on vertical exposures
V	Platy to tabular, dense and firmly cemented. Well-developed laminar layer on upper surface. May have scattered incipient pisoliths in laminar zone. Case hardening common
VI	Massive, multilaminar and brecciated, with pisoliths common. Case hardening common

[a] From Bachman and Machette (1977, p. 40, Tables 2 and 3).

occurrences of multiple calcretes in the Holocene age Blackwater Draw Formation of west Texas.

Reeves (1976, pp. 42–65) recognized a diversity of calcrete morphologies largely related to the degree of induration of the carbonate mass. At the two extremes of induration he distinguished hard massive and soft powdery calcrete. Other macromorphologi-

Fig. 6.6 Multiple, buried calcrete profiles developed in the Ogalalla Formation, High Plains, west Texas (Dixon photo)

cal forms include breccia, conglomerate, concretions, glaebules (Brewer and Sleeman 1964), honeycomb calcrete, irregular masses of poorly indurated carbonate including cylindrical masses, laminar calcrete, nodules, pisolites, plates, and septaria. Indurated root mats can be common showing that calcification associated with roots is important (Klappa 1980, McLaren 1995, Kosir 2004). In Australia there is no widely agreed upon descriptive scheme of calcrete morphology. Early workers including Crocker (1946), Ward (1965, 1966), and Firman (1966) described field occurrences of calcrete as sheets, pans, or nodules. One of the earliest detailed descriptions of calcrete morphology was that of Crawford (1965) who recognized an upper unit of nodular calcrete overlying a unit of massive calcrete which was in turn underlain by a unit of calcrete nodules embedded in a matrix of marl. Similar terminologies had been used by Crocker (1946) in his brief descriptions of calcretes from southern South Australia.

Systematic detailed descriptions of the macromorphology of calcrete from Australia are relatively recent. Weatherby and Oades (1975) developed a scheme for calcrete descriptions based on texture, the amount and nature of the carbonate, and the nature of the horizon boundaries. Dixon (1978) and Hutton and Dixon (1981) adapted Netterberg's (1969) morphological scheme from South Africa to calcretes in the Murray River valley and on Yorke Peninsula, South Australia. Detailed macromorphological description of calcretes from the St Vincents Basin of southern South Australia has been undertaken (Phillips and Milnes 1988). These workers identified a wide variety of the calcrete morphologies, and to some they applied terms previously used by Netterberg, including carbonate silt, nodules, hardpan calcrete, and laminar zones. They recognized a variety of additional morphologies, including biscuits, pisoliths, coated clasts, wedges, stringers, and blotches. More recently, Hill et al. (1999) and Hill (2005) have presented a morphological classification scheme for regolith calcretes in southern Australia that more closely resembles that of Netterberg's (1969) scheme.

Micromorphology

Micromorphological analysis of calcretes has largely developed within the last few decades in an attempt to come to a more detailed and precise understanding of calcrete genesis. Recent reviews on calcrete microtextures can be found in Alonso-Zarza (2003) and Wright (2007). Khormali et al. (2006) have conducted a detailed analysis of the micromorphology of nodular calcrete in aridisols along a moisture gradient in Iran in an effort to better understand calcrete genesis. One of the earliest studies of calcrete micromorphology was undertaken in India by Sehgal and Stoops (1972). These workers identified seven distinct types of carbonate micromorphology: (a) microcrystalline interflorescences which consist of equant grains 2–3 μm in size; (b) neocalcitons which refer to calcite in the soil matrix surrounding pores; (c) spongy calcitic nodules comprised of irregular, diffuse nodules 100–200 μm in diameter; (d) compact calcitic soil nodules which are large dense nodules 250–500 μm in diameter; (e) crystal chambers filled with coarse calcite crystals; (f) needle-shaped calcite efflorescences consisting of calcite needles less than 50 μm in length; and (g) rounded calcite grains. Working in the West Indies at the same time, James (1972) recognized three distinct forms of calcite in calcretes: flower spar, needles, and micrite.

Wieder and Yaalon (1974) identified three distinct types of nodules within calcic soils in Israel: orthic, disorthic, and allothic nodules. This subdivision was based on the apparent origin of the grains. Orthic nodules have formed in place and are similar in appearance to the surrounding matrix; disorthic nodules show evidence of displacement; and allothic nodules are relict nodules related to some other soil.

Detailed studies of the micromorphology of calcretes were undertaken in Africa in the late 1970s. Hay and Reeder (1978) described calcretes from Olduvai Gorge in Tanzania, providing detailed micromorphological descriptions of laminar calcrete as well as of pisoliths above the laminar zone. N.A.

Watts (1978, 1980) working in the Kalahari desert provided detailed micromorphological descriptions of calcrete and associated diagenetic structures. Bachman and Machette (1977) and Monger et al. (1991a) described a diversity of micromorphological features from calcretes of the south-western United States, including micrite, microsparite, ooids, and pisoliths.

In the early 1980s attention shifted to detailed examination of the micromorphology of the various concentric forms of calcrete. Hay and Wiggins (1980), working on calcretes from a number of locations in the south-western United States, recognized three fundamental concentric forms of calcrete: pellets, ooids, and pisoliths. They distinguished among these forms on the basis of their internal structure and size. Similarly, Shafetz and Butler (1980) described the micromorphology of calcrete nodules and pisoliths from central Texas. A variety of other micromorphological forms have been recognized from the calcretes of central Texas, including calcans, neocalcans, needle efflorescences, concretions, crystal chambers, and intercalary crystals (Rabenhorst et al. 1984, Drees and Wilding 1987, West et al. 1988).

Wright and Tucker (1991) have attempted to classify calcretes on the basis of their microstructures, identifying two end members. Corresponding to the k-fabrics of Gile et al. (1965), they have termed non-biogenic calcretes that commonly have dense micritic to micro-sparitic textures with floating clasts alpha calcretes. Beta calcretes, on the other hand, show abundant evidence of biogenic fabrics.

Several workers have identified micromorphological features that have been interpreted to be the result of biological activity. Early work by James (1972) identified tubular structures in calcretes which he attributed to algal activity. Kahle (1977) suggested that calcretes might be formed by the combined influences of algae and fungi, while Knox (1977) suggested that bacteria also play an important role in calcrete formation. In a series of papers, Klappa (1978, 1979a, b) recognized a wide variety of micromorphological structures which he ascribed to biological activity. Klappa (1979a) recognized microborings, algal filaments, fungal hyphae, calcite spherules, spherical structures, and deci-micron calcite grains arranged in thin parallel layers in laminar calcrete from coastal Mediterranean sites and suggested that they were indicative of lichen stromatolites. Klappa (1979b), in a detailed discussion of calcified filamentous structures, including root hairs, algal filaments, Actinomycete, and fungal hyphae in calcretes from the Mediterranean, suggested that these features were indicative of organic templates for subsequent calcrete formation.

Detailed studies of the micromorphological characteristics of calcretes from South Australia reveal the widespread occurrence of needle fibre calcite and calcified filaments in voids within powder, nodular, platy, and hardpan calcrete (Phillips and Self 1987, Phillips et al. 1987). These workers have interpreted the needle fibre calcite as having two distinct biological origins, depending on the size of the needles. Micro-rods frequently associated with organic matter are interpreted as being calcified rod-shaped bacteria. The larger needles are interpreted as forming within mycelial strands from which they are later released. The calcified filaments are interpreted to be the remains of fungal hyphae. Verrecchia and Verrecchia (1994) found evidence that fungi play a role in the formation of needle fibre cements. Kosir (2004) has recently shown the distortion of plant cortical cells by the growth of calcite forms *Microcodium* structures.

Chemistry

Based on some 300 chemical analyses of calcrete from around the world, Goudie (1972) found the average composition of calcretes to consist of 79.28% $CaCO_3$, 12.30% SiO_2, 2.12% Al_2O_3, 2.03% Fe_2O_3, 3.05% MgO, and 42.62% CaO. There is, however, considerable variability in the chemistry of calcretes from one location to another. For example, Goudie (1973) pointed out that calcretes from India are on average less calcareous than the global average and that Australian calcretes are considerably more calcareous. Calcretes particularly abundant in magnesium have been reported from South Africa, Namibia, and Australia (Goudie 1973).

Aristarain (1970), working in the United States, cited numerous chemical analyses of calcretes from Oklahoma and New Mexico. These analyses reveal North American calcretes to be chemically different from the global average. Compilation of Aristarain's data, based on 22 analyses, reveals on average 34.93% CaO, 30.68% SiO_2, 1.81% MgO, 1.88 Al_2O_3, and 1.00% Fe_2O_3. He observed that calcium, carbon, oxygen, and hydrogen were added to the upper parts of the calcrete profile and decreased in abundance downward.

Magnesium, sulphur, and ferric iron were added in decreasing amounts from top to bottom of the profiles. Silicon and titanium increase slightly in the upper part of the profiles but appreciably in the middle. Potassium and phosphorus remain constant down the profiles. There are no clear profile trends in sodium and ferrous iron.

Chemical characteristics of calcretes from Nevada were reported by Gardner (1972). Unfortunately, these data were presented in parts per million concentrations and so are not comparable with other available data. More recently, a comprehensive survey of the major and minor element geochemistry of calcretes from the south-western United States has been presented by Bachman and Machette (1977).

Detailed chemical analyses of calcretes have been undertaken in Australia by numerous workers. Results of these investigations are summarized in a comprehensive discussion of Australian calcretes by Milnes and Hutton (1983). Investigations of the chemistry of calcretes in southern Australia have been undertaken by Dixon in the St Vincents Basin of South Australia which add to earlier work from the Murray Basin and Yorke Peninsula, South Australia (Dixon 1978, Hutton and Dixon 1981) and by Milnes (1992). Considerable variability in chemistry occurs in South Australian calcretes. Dixon (1978), Hutton and Dixon (1981) and Milnes (1992) found that this variability is related to both calcrete morphology and to profile position. Dixon (1978) showed that as the degree of induration of calcretes increases, so does their chemical purity, with hardpan calcretes being the most calcareous and powder calcretes the least. Hutton and Dixon (1981) found that there is a systematic decrease in CaO and an accompanying increase in MgO in calcretes with increasing depth. This trend was explained in terms of depth of leaching and associated degree of replacement of host materials by carbonate. McQueen et al. (1999) report similar geochemical trends in calcrete profiles from western New South Wales with high Ca:Mg ratios at the top of profiles and high Mg:Ca ratios at depth. Explanations of these trends are similar to those presented by Hutton and Dixon (1981) and coupled with changes in pH and CO_2 partial pressure (Hill et al. 1999).

Dixon has examined the role of topography in accounting for spatial variations in calcrete chemistry. He examined six calcrete profiles down a toposequence in the St Vincents Basin and found that there was a systematic decrease in calcium from top to bottom of the toposequence and an accompanying increase in magnesium, as well as systematic decreases in calcium with depth.

At a larger spatial scale, Milnes and Hutton (1983) recognized two distinct geochemical provinces of calcrete formation. The first province is associated with Pleistocene beach complexes and aeolianites and was given the name Bridgewater Formation province (Fig. 6.7). The second province is a largely non-calcareous continental province (Fig. 6.8) in which calcretes are associated with a diversity of bedrock types. It was named the Continental province. In the Bridgewater Formation province the chemistry of the calcretes is very similar to that of the calcareous parent materials. The chemistry of calcretes from the Continental province, on the other hand, is highly variable. Hutton et al. (1977) suggested that calcrete developed on norite (hypersthene gabbro) in the Murray Basin at Black Hill was in part derived from the weathering of this rock. Dixon (1978) also showed that calcrete chemistry varied markedly depending on the underlying lithology. Hutton and Dixon (1981)

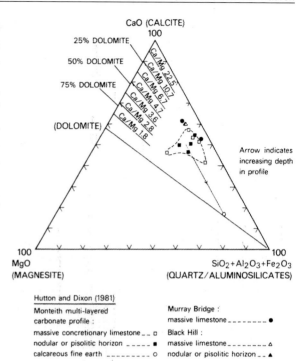

Fig. 6.8 Chemical and mineralogical composition of layered calcretes in continental locations Murray Valley, South Australia. (Ternary plot from Hutton and Milnes, 1993: Data from Dixon and Hutton, 1981)

established that the chemistry of the insoluble residue strongly reflected the chemistry of the underlying bedrock. Detailed discussion of the calcrete chemistry from each of the two provinces is provided by Milnes and Hutton (1983).

Attention has been given to the valuable metal contents of calcretes. Groundwater calcretes in particular have been the source of economic abundances of uranium and gold minerals. Uranium-rich calcretes have been reported from Western Australia by Butt et al. (1977) and Carlisle (1983). Gold-bearing calcretes are reported from southern Australia by McQueen et al. (1999) and Mumm and Reith (2007). Goudie (1983) also recorded uranium in calcretes from the Namib, Angola, Mauritania, and Somalia.

Chemical studies of calcrete in recent years have been focused on the isotopic compositions of these materials. Initially, these studies were interested in determining the source of the carbon and thereby ascertaining the origin of the calcretes but more recently $\delta^{18}O$ and $\delta^{13}C$ have been used to determine environmental and climatic conditions at the time of deposition. Over the last ten years there has been a strong focus

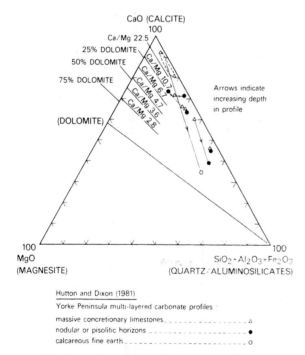

Fig. 6.7 Chemical and mineralogical composition of layered calcretes from Yorke Peninsula, South Australia (Ternary plot from Hutton and Milnes, 1993: Data from Hutton and Dixon, 1981)

on using isotopic data to try to elucidate palaeotemperatures, the importance of C_3 versus C_4 vegetation and atmospheric CO_2 (Andrews et al. 1998, Cerling 1999, Royer et al. 2001). However, such isotope studies are only of use if no secondary alteration or overprinting of the cements has occurred (Kelly et al. 2000). Early studies, such as Salomons et al. (1978) conducted extensive isotopic analyses of calcretes from Europe, Africa, and India. These workers found that despite a considerable spread in isotopic values for both carbon and oxygen, there was, in general, good agreement between values in calcrete and prevailing precipitation. From this observation they concluded that evaporation was not a significant factor in calcrete formation. Nor did they find any systematic difference in isotopic content with depth. Isotopic studies of calcretes in Africa have also been undertaken by Cerling and Hay (1986), who found that calcretes in Olduvai Gorge display progressive increases in ^{18}O and ^{13}C isotopes with time. From these data the researchers concluded that there had been a progressive drying of the climate and an accompanying increase in C_4 grasses.

Isotopic studies of calcretes in the United States are relatively limited. However, Gardner (1972) conducted an extensive analysis of the isotopic chemistry of calcretes from Nevada and New Mexico. More recently Amundson et al. (1988) examined the isotopic chemistry of calcic soils from the Mojave Desert. Following on from that study, Amundson et al. (1989) reported on oxygen and carbon isotopes in calcretes from Kyle Canyon in the Mojave Desert. Amundson et al. concluded that the isotopic composition of the calcretes is consistent with the prevailing climatic regime. These workers also determined the ages of calcrete lamelli using ^{14}C composition, but were unsure of the reliability of their dates due to possibility of contamination. The stable carbon and oxygen isotopic composition of calcretes from the Chihuahuan Desert was investigated by Monger et al. (1998) who identified stratigraphic discontinuities in both isotopes. These discontinuities were interpreted to reflect a shift from C_4–C_3 plants in the Holocene accompanying landscape erosion.

Early isotopic work on calcretes from Australia focused on ^{14}C concentrations in an attempt to determine the age of the calcretes (Williams and Polach 1969, 1971, Bowler and Polak 1971). These same workers also examined $^{12}C/^{13}C$ ratios in an attempt to come to some understanding of the origin of the carbonate in the calcrete. Milnes and Hutton (1983) examined

the ^{13}C content of calcretes from a variety of locations in southern South Australia. The significance of these data with respect to calcrete genesis is still poorly understood.

Mineralogy

The carbonate mineralogy of calcretes is dominated by calcite and dolomite, while the non-carbonate mineral fraction consists of quartz, opal, feldspars, and clay minerals (e.g. Goudie 1973, pp. 25–6, Reeves 1976, pp. 25–41, Milnes and Hutton 1983). The calcite is generally low-Mg calcite, though Watts (1980) reported high-Mg calcite from calcretes in southern Africa and McQueen et al. (1999) report high Mg calcite from calcrete profiles in western New South Wales, Australia. Aragonite is also occasionally found in calcretes (Watts 1980, Milnes and Hutton 1983). Minor non-carbonate minerals in calcretes include iron sulphide and glauconite (Goudie 1972, 1973). Gypsum has also been reported widely in calcretes (Reeves and Suggs 1964, Aristarain 1971, Goudie 1973, Rakshit and Sundaram 1998). Haematite, grossularite, magnetite, muscovite, rutile, tourmaline, and zircon have all been reported from calcretes in New Mexico by Aristarain (1971).

Mineralogical studies have focused primarily on the clay mineral fraction and have been largely concerned with the origin of the clays and the light they may throw on calcrete genesis. Clay mineral assemblages in calcretes appear to be characterized by a relatively small number of mineral species. In the United States, the dominant clay minerals appear to be palygorskite and sepiolite, and there are smaller abundances of illite, kaolinite, montmorillonite, interlayered illite–montmorillonite, and chlorite (Van den Heuvel 1966, Gile 1967, Aristarain 1970, 1971, Gardner 1972, Frye et al. 1974, Reeves 1976, pp. 37–9, Bachman and Machette 1977, Hay and Wiggins 1980). Many authors have ascribed the origin of the palygorskite and sepiolite to the alteration of montmorillonite and interlayered montmorillonite–illite. Hay and Wiggins (1980) believed that the sepiolite in calcretes from various locations in the south-western United States is neo-formational in origin. However, they argued for the formation of palygorskite from montmorillonite.

In Australia, the dominant clay minerals in calcretes are illite, kaolinite, and interstratified minerals (Milnes and Hutton 1983, Phillips and Milnes 1988). However, Dixon (1978) and Hutton and Dixon (1981) found that palygorskite and sepiolite are the dominant clay minerals in calcretes in selected areas of the Murray Basin of southern South Australia. These authors ascribed the origin of these minerals to neo-formation in an environment abundant in Mg, specifically a favourable underlying lithology that provides the necessary cations upon weathering. Clay mineralogy of calcretes appears to be strongly related to the provenance of the calcrete. Calcretes in the Bridgewater Formation province appear to be dominated by illite, kaolinite, and interlayered clays. In the Continental province the clay mineralogy is more diverse and includes sepiolite and palygorskite.

Clay mineralogical investigations of calcretes from Africa (Watts 1980) reveal mineral assemblages similar to those from the United States and Australia. Watts (1980) found that the calcretes from the Kalahari Desert were dominated by palygorskite and sepiolite. These clay minerals were interpreted to be primarily neo-formational in origin. However, Watts suggested that some of the palygorskite may be derived from the alteration of montmorillonite. The remainder of the clay mineral assemblage consists of illite, montmorillonite, mixed layer illite–montmorillonite, kaolinite, chlorite, and glauconite. Watts (1980) interpreted the minor clay mineral species present in the calcretes to be predominantly detrital. The mixed layer clays, however, appear to be weathering products derived from illite. Similar conclusions about clay mineral genesis in calcareous soils from south western Iran have recently been reached in a detailed study by Owliaie et al. (2006). Such clay mineral genesis studies provide important insights into calcrete genesis.

Detailed studies of dolomite in calcretes have been undertaken in both Australia and Africa (Dixon 1978, Watts 1980, Hutton and Dixon 1981, Phillips and Milnes 1988). Dixon (1978) and Hutton and Dixon (1981) identified abundant amounts of dolomite in calcretes from the southern Murray Basin and Yorke Peninsula, South Australia (Figs. 6.7 and 6.8). Dolomite abundances were greatest in the fine carbonate silt at the base of the calcrete profile and decreased progressively upward in the profiles as the degree of induration increased. Similar dolomite trends have been reported from calcrete profiles in the St Vincents Basin of southern South Australia (Phillips and Milnes 1988). These authors pointed out

that dolomite abundances are greatest in wedges and blotches developed in the calcretes. Detailed studies of regolith calcretes from western New South Wales by McQueen et al. (1999) reveal similar patterns of distribution of calcite and dolomite within calcrete profiles. However, these workers find the dolomite to be concentrated in the hardpan facies. In southern Africa, Watts (1980) reported generally similar trends with respect to the distribution of dolomite within the profiles. Dolomite is intimately associated with palygorskite and sepiolite in calcretes from both of these continents. Dolomite appears to be associated with a variety of calcrete morphologies, but most notably with depth within calcrete profiles.

Dixon has examined in detail the distribution of dolomite and calcite in calcrete profiles from the St Vincents Basin in South Australia. He conducted a study down a single hillslope developed on weakly indurated calcrete and found that there was a systematic decrease in calcite and an accompanying increase in dolomite. The calcrete profile at the top of the catena is characterized by a complete lack of dolomite through its entire thickness. The profile at the base of the catena contains only dolomite with a complete absence of calcite. The greater solubility of $MgCO_3$ compared with $CaCO_3$ accompanied by lateral transportation of Mg-rich waters may be responsible for this trend.

Quartz is often the dominant non-carbonate mineral in both bulk samples and clay size fractions of the calcretes (Reeves 1976, p. 30, Milnes and Hutton 1983, Phillips and Milnes 1988). Silica also occurs in various other forms, including opal and chalcedony. Opal has been widely reported from calcretes in west Texas (Reeves 1976, p. 28) as well as from the southwestern United States (Hay and Wiggins 1980). Watts reports the occurrence of opal in calcretes from southern Africa along with both length-slow and length-fast chalcedony. Much of the silica in calcretes occurs as cement. However, in some cases quartz is a replacement mineral of calcite.

Origin

Calcretes may be pedogenic or non-pedogenic in origin. However, most areally extensive calcretes appear to form in the soil environment. The specific processes responsible for calcrete formation in the soil environment are variable. Goudie (1973, pp. 141–4, 1983) suggested several models for the formation of pedogenic calcretes. These include the concentration of carbonate by downward translocation in percolating soil waters, the concentration of carbonate by capillary rise waters, *in situ* case hardening by carbonate, and cementation of calcrete detritus (McFadden et al. 1998). Of these models, the accumulation of carbonate in soils as a result of vertical translocation is the most widely accepted. Precipitation of the carbonate results from a variety of processes. Schlesinger (1985), working in the Mojave Desert, suggests that carbonate precipitation results from evaporation, increases in pH, and decreases in CO_2 partial pressure. Other mechanisms of carbonate precipitation include increases in temperature and accompanying loss of CO_2 (Barnes 1965), the common ion effect (Wigley 1973), and evaporation (Watts 1980, Watson 1989). Biological processes have also been shown to be important in calcite precipitation. Several authors have clearly established the significance of micro-organisms in calcrete formation (e.g. Klappa 1978, 1979a, b, Phillips and Self 1987, Phillips et al. 1987, Monger et al. 1991b, Wright and Tucker, 1991, Kosir 2004). In addition, the formation of calcrete around plant roots has been widely documented (Gill 1975, Klappa 1979a, b, Semeniuk and Meagher 1981, Warren 1983, Semeniuk and Searle 1985, McLaren 1995). In a recent study from South Australia, Mumm and Reith (2007) demonstrate the complex role played by plants and micro-organisms in the biomediation of calcrete as biological processes modify regolith geochemistry.

Central to the question of the origin of calcretes has been the origin of the calcium carbonate contained within them. In the United States there has been a long history of appealing to atmospheric sources for the carbonate. As early as 1956, Brown (1956) suggested that all of the carbonate necessary for the formation of calcretes could be derived from atmospheric dust and calcium dissolved in rainwater. Similar arguments have been presented by subsequent workers including Gile (1967), Ruhe (1967), Reeves (1970), Bachman and Machette (1977), Machette (1985), and McFadden and Tinsley (1983). Gardner (1972) provided quantitative support to the notion of an aeolian origin for the carbonate by showing that extreme thicknesses of overlying sediments would have to be weathered and leached in order to account for the thick calcrete

caprocks in Nevada. Similar conclusions had earlier been reached by several authors who had shown that there was little evidence for significant depletion of calcium from sediments overlying calcrete horizons (Gile et al. 1966, Van den Heuvel 1966, Aristarain 1970, McFadden 1982). More recently, $^{87}Sr/^{86}Sr$ and C have been used as isotopic tracers trying to identify the source of material for cements (local rocks versus atmospheric dust) (Quade et al. 1995). Studying carbonate soils from southeastern South Australia, Quade et al. (1995) found from both isotopes that the carbonates were derived from a combination of aeolian and *in situ* weathering sources. From the carbon isotopes it further becomes apparent that plant respiration and decay play a substantial role in calcrete genesis as previously suggested by micromorphological studies (Klappa 1979a and b).

In southern Australia, the carbonate mantle is also widely interpreted to be aeolian in origin. Early workers (Crocker 1946) interpreted the mantle to have been derived from the winnowing of the finer size fraction from coastal dune deposits. Studies by Milnes and Hutton (1983), Milnes et al. (1987) and Phillips and Milnes (1988) essentially concur but point out that the original sediments have been considerably modified since deposition and may also contain some inclusions of sediments from non-aeolian sources. Hutton and Dixon (1981) suggested that some of the carbonate, especially that in continental settings, was lacustrine in origin and had been modified in place by pedogenic processes. *In situ* formation of calcretes has occurred within calcareous coastal dunes and sand sheets on Yorke Peninsula (Dixon 1978, Hutton and Dixon 1981) as well as in similar deposits in Western Australia (Arakel 1985). In both locations cementation of host materials by percolating vadose waters is responsible for calcrete development. As indicated by isotope studies outlined above, (Quade et al. 1995) it is also apparent that when Ca-rich parent materials are present, their weathering facilitates the production of a souce of carbonate for calcrete formation.

Non-pedogenic calcretes have also long been recognized. As mentioned earlier, at approximately the same time that Lamplugh introduced the term calcrete he specifically recognized the existence of calcretes in association with river valleys. Considerable research in central and Western Australia has focused on the association of calcrete with present and palaeo-drainage ways. Much of this research has been undertaken in association with uranium exploration (e.g. Mann and Horwitz 1979, Arakel and Mc-Conchie 1982, Arakel 1986, Jacobson et al. 1988). In these calcretes the calcium carbonate is derived from groundwater sources and is subsequently precipitated in drainage channels as a result of low-precipitation and high-evaporation climatic regimes.

Non-pedogenic calcretes have been subdivided by Carlisle (1983) into superficial gravitational zone, groundwater, reconstituted and detrital types. Nash and McLaren (2003) have discussed the problems of trying to classify calcretes particularly considering that they form in a wide variety of hydrological and geomorphological settings. Nash and McLaren (2003) and McLaren (2004) have attempted to differentiate between alpha calcretes that form in different geomorphological settings in the hope of then being able to use the distinct features as tools in environmental reconstructions. Nash and McLaren (2003) have studied valley calcretes from the Kalahari. Mclaren (2004) has investigated channel calcretes from southern Jordan (McLaren et al. 2004) (Fig. 6.9) and Thomas et al.'s (2003) study involves analyses of Kalahari pan calcretes. To briefly summarise, the valley calcretes typically comprise quartzose sands cemented by fine often glaebular micrite (Nash and McLaren 2003). Grain coating and porefilling cements dominate and calcified roots are often present. These valley calcretes are thought to have formed in near surface environments with high rates of evaporation (Nash and McLaren 2003). The pan calcretes studied (Thomas et al. 2003) were similar in nature and origin to the valley calcretes, being relatively structureless with micrite cement and abundant glaebules and biogenic structures. The channel calcretes from Jordan varied in nature dependent upon the location within the former bedrock channel and age. Spar cements dominated at the base of the impermeable channels where ponded waters would allow slow crystallisation of large cement crystals (McLaren 2004). Near the top of the channels rapid evaporation resulted in micritic cements. In terms of age (height above the modern channel) there was an increase in secondary recrystallisation by porefilling cement over time (McLaren 2004).

Clay mineral assemblages present in the insoluble residue of calcrete can serve as keys to the origin of the sediments which have been calcified. The presence of clays such as illite and smectite typically are derived

Fig. 6.9 Calcretised conglomerate infilling a palaeochannel cut into bedrock, southern Jordan (McLaren photo)

from the weathering of feldspars and suggest that the calcrete parent material was a residual weathering profile derived from bedrock weathering. The presence of clay minerals such as sepiolite sugest calcification of a deposit rich in Mg and with restricted drainage such as a lake (Gurel and Kadir 2006).

Silcrete

Terminology

The term silcrete was introduced into the geomorphological literature at the same time as calcrete. Lamplugh (1902) originally used the term to refer to silicified sandstones and conglomerates and later (Lamplugh 1907) to indurated sandstones and quartzites cemented by chalcedony occurring in the Zambesi River basin. Today, silcrete is widely defined as a surface or near-surface deposit of saprolite, sediment, or soil that has been cemented by secondary silica (Milnes and Twidale 1983). Silicification has occurred at low temperatures and is not a product of volcanism, plutonism, metamorphism, or deep diagenesis (Summerfield 1983a).

Early work on silcretes linked them genetically to the deep weathering profile which frequently occurs beneath many massive silcrete caprocks (Woolnough 1930, Wopfner 1978) (Fig. 6.10). Silcrete in Australia and Africa has also commonly been related to surfaces of low relief (Woolnough 1927, Langford-Smith and Dury 1965, Wopfner and Twidale 1967, Twidale et al. 1970).

In Australia, silcretes have been reported from the states of Queensland (e.g. Gunn and Galloway 1978, Senior 1978, van Dijk and Beckman 1978), New South Wales (Dury 1966, S.H. Watts 1978b, Taylor 1978, Hill, 2005), and South Australia (Hutton et al. 1972, 1978, Wopfner 1978, 1983, Milnes and Twidale 1983, Milnes et al. 1991, Simon-Coinçon et al. 1996, Webb and Golding 1998, Thiry et al. 2006), Western Australia (Butt 1983, 1985, Lee and Gilkes, 2005) and central Australia (Brückner 1966).

Silcrete has not been widely reported from the United States. However, silicification of calcrete has been recognized in Texas (Price 1933, Reeves 1970, 1976, pp. 26–32). Where silicification of calcrete has been virtually complete, these deposits might be appropriately referred to as silcretes.

Fig. 6.10 Breakaway capped with silcrete and associated weathered pallid zone, northern New South Wales, Australia (photo courtesy of S.H. Watts)

Morphology

Macromorphology

Macroscopically, silcrete displays a diversity of forms, which reflects their mode of formation and the nature of the host material (Milnes and Thiry 1992). In a detailed study of the silcretes of western New South Wales, S.H. Watts (1978b) described a wide variety of silcrete morphologies both at the landform scale as well as at the outcrop scale. Watts recorded the common occurrence of massive, columnar-jointed silcretes capping upland surfaces (Fig. 6.11) that are

Distribution

Silcretes are most widely distributed in Australia and southern Africa but have also been reported from a variety of other locations. In Africa they have been reported from such diverse areas as the Sahara and its margins (Auzel and Cailleux 1949, Abdel-Wahab et al. 1998, McLaren et al. 2006), the Congo Basin, Zambesi Basin, Botswana, Namibia, South Africa, and some areas of east Africa (Summerfield 1982, Watson 1989, Shaw et al. 1990, Shaw and Nash 1998, Nash et al. 1994, 2004).

Fig. 6.11 Silcrete outcrop, showing typical columnar structure, Mount Stuart Station, New South Wales, Australia (photo courtesy of S.H. Watts)

often pedogenic in origin (Ballesteros et al. 1997). Individual columns often extend the complete thickness of the silcrete outcrop. Individual columns may be up to 5 m thick with widths of 2 m. In outcrop, silcrete surfaces frequently display concentric ring patterns, ropy surfaces, and nodular to concretionary weathering forms. Watts also described piping silcrete and accompanying rounded, lobe-like ridges and rings. Groundwater, pan and drainage-line silcretes tend to be less well jointed and more massive than pedogenic varieties (Thiry 1999).

In a study of silcrete in northern South Australia (Wopfner 1978) observed a similar diverse assemblage of macroscopic forms. He recognized polygonal prismatic or columnar silcrete as the most common morphological type. The columns or prisms are commonly 50–120 cm high with diameters of 25–80 cm. The columns are frequently draped with a thin skin of silica. Often associated with columnar silcrete is platy silcrete which occurs as large slabs from 8 to 15 cm thick. In detail, platy silcretes are either massive or pseudopebbly. Wopfner identified two other principal macromorphological types. Botryoidal silcrete is characteristically rough and jagged and occurs in thick profiles, whereas pillowy silcrete resembles the surface of lava. Like Watts (1978b), Wopfner described pipe silcrete as commonly occurring within outcroppings of botryoidal silcrete. On the basis of Wopfner's (1978) observations of silcrete macromorphology, Watts developed a comprehensive classification of silcretes. In addition to macroscopic characteristics, the classification includes microscopic parameters. The classification is based on the characteristics of the silcrete matrix mineralogy, macroscopic texture, host rock texture, and profile thickness and type.

A comprehensive summary of the macroscopic characteristics of silcretes in central Australia was provided by Milnes and Twidale (1983) and more recently Nash and Ullyott (2007). These workers recognized a variety of silcrete types occurring on mesa surfaces, including columnar, massive, bedded, and conglomeratic or nodular silcrete. Silcrete outcrop surfaces commonly display ropy, whorled, and custard-like forms as well as liesegang ring-like structures. Pipe-like structures are widespread, as are silcrete skins on bedrock outcrops (Twidale et al. 1970, Hutton et al. 1972, Milnes et al. 1991).

Micromorphology

Microscopically, silcretes reflect characteristics of both the host material and the silicification process (S.H. Watts 1978a, Summerfield 1983b, Milnes et al. 1991) and provide information relating to their formation and the diagenetic changes they have undergone. Primary characteristics of silcretes usually relate to such features as grain size, grain shape, mineralogy, and grain fabric. Secondary characteristics include cryptocrystalline and microcrystalline silcrete matrices, extensive embayment of quartz grains (Fig. 6.12), grain pitting and replacement by secondary silica, and textural modification of sediments from a grain-supported to a floating matrix texture (Watts 1978a). Summerfield (1982, 1983a) discussed a variety of microfabrics developed in silcretes from southern Africa. Detailed studies of the micromorphology of silcretes in Australia have been undertaken by Hutton et al. (1972), Milnes and Hutton (1974), Hutton et al. (1978), and Milnes and Twidale (1983).

A number of classification schemes based on micromorphological features have been developed, including those of Smale (1973), Watts (1978a), and Wopfner (1978, 1983). Perhaps the most comprehensive of these is that developed by Summerfield (1982, 1983a). He recognized four major fabric types with associated subtypes. The principal fabric types are GS (grain-supported) fabric, F (floating) fabric, M (matrix) fabric, and C (conglomeratic) fabric. Grain-supported fabrics are those in which skeletal grains of the silcrete are self-supporting. Skeletal grains may display over-

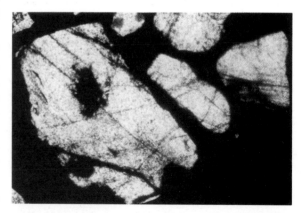

Fig. 6.12 Photomicrograph of silcrete, showing solutional embayments of original detrital quartz. Field of view is 1.8 mm. (photo courtesy of S.H. Watts)

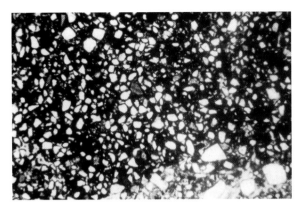

Fig. 6.13 Photomicrograph of silcrete from west Texas, showing floating (F) fabric with quartz grains separated by finer matrix materials. Field of view is 1.1 mm (Dixon photo)

growths of chalcedony or microquartz. The matrix of grain-supported silcrete fabrics is commonly characterized by microquartz and cryptocrystalline and opaline silica. Floating fabrics are those in which the skeletal grains float in a matrix of microcrystalline, cryptocrystalline, and opaline silica. The skeletal grain content of these fabrics is greater than 5%. Floating fabrics may be massive or glaebular (Fig. 6.13). Matrix fabrics are similar to F fabrics with the difference being they possess a skeletal grain content of less than 5%. As with F fabrics, M fabrics occur as massive and glaebular subtypes. Conglomeratic fabrics possess a discernible detrital fabric. The fabric is determined primarily on the basis of the size of the detrital grains which exceed 4 mm in diameter.

Lee and Gilkes (2005) make extensive use of micromorpholgy to characterize and interpret the origin of groundwater and pedogenic silcretes in southwestern Western Australia. These authors recognize three types of silcrete with distinctive micromorphologies reflecting their weathering histories. Micromorphology, coupled with petrofabric and mineraolgical analysis, has been demonstrated to be of fundamental importance in deciphering the history of silcification of silcretes in central Australia (Thiry et al. 2006).

Chemistry

The chemistry of silcretes is overwhelmingly dominated by silica, which may be as much as 99% by weight, with negligible amounts of other chemical constituents. Silcretes are characteristically poor in alkalis and alkali earths, though often they display enhanced concentrations of iron, titanium, zirconium, and aluminium. Titanium concentrations are associated with the presence of anatase and rutile (Hutton et al. 1972, Milnes and Twidale 1983). Aluminium concentrations are associated with clay minerals.

Chemical analyses of silcretes have been published by Goudie (1973, pp. 18–21) for Africa and by Summerfield (1982, 1983b, c) for Botswana and South Africa. The analyses of Summerfield were concerned with the chemical variability vertically down through the weathering profile. An examination of some 50 silcrete profiles reveals a mean silica abundance of 94.7%. The second most abundant constituent is titanium at 1.82%. Iron and aluminium are present in small and variable abundances. Silcrete profiles show enrichment in silica and often titanium but display depletion of aluminium, alkalis, and alkali earths. In some profiles there is also enrichment of iron where silcrete is associated with ferricrete and where late-stage surface weathering has occurred. On the basis of silcrete chemistry, distinction is made between weathering and non-weathering silcretes. Weathering silcretes are characterized by the enrichment of titanium as a result of the leaching of more soluble materials, while non-weathering silcretes are deficient in titanium (Summerfield 1983c).

Detailed chemical analyses of Australian silcretes are available from a number of studies. Hutton et al. (1972) examined the chemistry of silcretes and silcrete skins in South Australia. This study was significant in heightening awareness of concentrations of titanium and zirconium in silcretes. A comprehensive collection of papers edited by Langford-Smith (1978) provided a wide sampling of silcrete chemistries from a variety of areas in arid and semi-arid Australia (Gunn and Galloway 1978, Hutton et al. 1978, Langford-Smith and Watts 1978, Senior 1978, Wopfner 1978). Watts (1977) presented a detailed discussion of the chemistry of silcretes from western New South Wales. A summary of silcrete chemical analyses from a variety of sites in central and southern Australia has been presented by Milnes and Twidale (1983). Consistent with previous studies of silcrete chemistry these authors found the silcretes to be dominated by silica, enriched in titanium and zirconium, low in calcium and alkalis, and characterized by accumulations of iron oxides and aluminium. Work on pedogenic and groundwater silcretes from the opal fields of South Australia by Milnes et al. (1991) and Thiry and Milnes (1991) provided detailed chemical analyses of these diverse materials. These studies again revealed the trends in silcrete chemistry outlined above. An interesting finding of these studies was that as shale is transformed to opalite, there is a loss of aluminium, titanium, iron and water, and an accompanying accumulation of silica. Chemical analyses of groundwater calcrete from southwestern Western Australia is provided by Lee and Gilkes (2005). These workers make extensive use of Al:Si ratios to interpret the evolutionary history of the silcrete types.

Silicification of calcrete in the south-western United States is widespread and has been widely reported in the literature (Price 1933, Sidwell 1943, Brown 1956, Swineford and Franks 1959, Aristarain 1970, Reeves 1976). However, published chemical analyses are few. Field observations (Fig. 6.14) of

Fig. 6.14 Silcreted caprock calcrete, Ogalalla Formation, Palo Duro Canyon, west Texas (Dixon photo)

massive calcretes in the Ogallala Formation showed that many of them are extensively replaced by silica and, therefore, are more accurately referred to as silicicalcretes (Table 6.3). Considerable scope exists for the systematic study of silcretes in the south-western United States where these materials are undoubtedly more widespread than previously recognized (B.L. Allen, personal communication, 1991).

Mineralogy

Like its chemistry, the mineralogy of silcrete is relatively simple. However, the mineralogy is a reflection of both host materials as well as secondary alteration. Typically, silcretes are dominated by quartz (and it is the main detrital component of silcretes), with minor amounts of opaline silica, feldspars, and heavy minerals and clay minerals (Summerfield 1983a, c). The silcrete matrix can contain a range of silica polymorphs such as quartz (e.g. Ballesteros et al. 1997), chalcedony (e.g. Heaney 1993, 1995) and opal (e.g. Bustillo and Bustillo 2000).

Detailed mineralogical studies of silcretes from southern Africa (Summerfield 1982) revealed a mineralogy dominated by quartz with the presence of opal CT (cristobalite and tridymite) in the ground mass. Chalcedony was recognized in voids and vughs. The clay mineral fraction was dominated by glauconite–illite.

Similar mineralogies were reported from central and southern Australian silcretes by Milnes and Hutton (1983). These authors observed framework grains dominated by quartz with variable size lithic clasts. The micro- and cryptocrystalline matrix is a mixture of opal-C, opal-CT, and quartz. Overgrowths on framework grains consist of chalcedony and quartz. In some skin silcretes, in particular, the groundmass contains an abundance of anatase (Hutton et al. 1972, 1978). Other accessory minerals in Australian silcretes commonly include zircon and rutile. Mineralogical characteristics of silcretes from various locations in the Australian arid zone are reported by Gunn and Galloway 1978, Wopfner 1978, S.H. Watts 1978a). Groundwater silcretes from Western Australia are dominated by residual quartz opal-CT. Voids are commonly occupied by anatase, derived from the weathering of primary Ti-minerals. Oxidation of Fe-rich primary minerals has resulted in the coating

Table 6.3 Major element bulk chemistry of silicified calcretes. High Plains, west Texas

Sample	SiO$_2$	Al$_2$O$_3$	Fe$_2$O$_3$	MgO	CaO	Na$_2$O	K$_2$O	TiO$_2$	Loss on ignition	Total
S1A	65.7	2.2	0.83	1.4	14.4	0.28	0.45	0.13	15.1	100.5
S1B	65.5	2.2	0.79	1.3	14.5	0.29	0.46	0.12	15.1	100.3
S2A	71.1	2.7	0.97	1.4	10.8	0.35	0.66	0.14	11.8	99.9
S2B	70.8	2.7	0.95	1.4	10.8	0.35	0.67	0.14	11.8	99.9

of the silcrete by haematite and goethite. Secondary minerals associated with the silcrete include jarosite, ferrihydrite, and akagancite (Lee and Gilkes 2005). Plant root material in pedogenic silcretes in the same area contain quartz, halite, and gypsum, together with trace amounts of holloysite and opal-CT.

A variety of clay minerals have been identified in the matrix of the silcretes. van Dijk and Beckman (1978) found the clay size fraction of silcretes from south-eastern Queensland to be dominated by quartz and kaolinite, with smaller amounts of interstratified clay minerals, illite, and montmorillonite. Detailed mineralogical analyses of silcretes in the opal fields of South Australia (Milnes et al. 1991) revealed a diverse mineralogy dominated by kaolinite and smectite.

Origin

Silcrete formation involves the interaction of three elements: a silica source, silica transfer, and silica precipitation. Numerous sources of silica are possible for the formation of silcrete. Silica may be derived *in situ* from the dissolution of the host materials followed by reprecipitation. Evidence for this is seen in the frequent occurrence of embayed quartz and other aluminosilicate mineral grains within silcrete (Hutton et al. 1978, Watts 1978a).

Silcretes with accompanying weathering profiles or in close proximity to weathered sediments may derive the silica for their formation from the weathering of aluminosilicate minerals and/or dissolution of quartz (Senior 1978, Wopfner 1978, Summerfield 1982, 1983c, Milnes and Twidale 1983). Lee and Gilkes (2005) discuss silcrete formation in the kaolin-rich pallid zone of a granite weathering profile from Western Australia. The silcretes display a variety of forms, chemistry and mineralogy, but are all fundamentally the result of the weathering of the granite parent materials. The source of the silica is fundamentally the result of acid weathering of quartz and feldspar which has subsequently been deposited in the porous network of the kaolin. As indicated previously, anatase as well as opal-CT is also present in the silcrete and this mineral has been derived from the acid groundwater weathering of Ti-rich minerals. As the abundance of amorphous silica in the silcrete increases and the abundance of kaolin decreases there is a systematic decrease in the Al:Si ratio in the chemistry of the silcrete.

A long history exists in Australia of attempts to genetically link silcrete formation to the presence of basalt because of a common spatial association between the two. Many workers have suggested that silica released as a result of weathering is redeposited beneath the basalt as silcrete (Gunn and Galloway 1978, Ollier 1978). Taylor and Smith (1975), however, had previously argued against any genetic link between basalt and silcrete, suggesting that their frequent spatial association was merely fortuitous.

Silica sources for the formation of silcrete in southern Africa have been attributed to aeolian influx. Summerfield (1982, 1983a) has suggested that aeolian-transported quartz and aluminosilicate sediments are especially susceptible to dissolution because of the abrasion and associated disordering of their surfaces experienced during transport. In addition, grain size reduction during transport results in a greatly enhanced surface area to volume relationship, which increases their solubility.

Concentration of amorphous silica within their tissues is common amongst some grasses, reeds, palms, horsetails and hardwoods and represents another source of silica (Summerfield 1982, 1983a, Gardner and Hendry 1995). In lacustrine and fluvial environments, diatoms may also represent an important silica source (Summerfield 1982, Knauth 1994). Pedogenic silcretes in southwestern Western Australia display an abundance of silicified and ferruginized plant root material. Silcretes dominated by organic material display significantly greater abundances of secondary sulica compared to organic-poor silcretes. The plant root materials are almost pure silica suggesting a substantial organic influence in the genesis of these silcretes (Lee and Gilkes 2005).

Silica solution and transport takes place in normal aqueous systems as either silicic acid or as an aqueous sol (Milnes and Twidale 1983). Such solution and transport appears to occur at normal temperatures and under neutral to low pH conditions. However, it is widely known that the solubility and concentration of silica in solution in the soil environment is greatly increased in environments where pH is above 9 (Wilding et al. 1977) and silica availability may be greater in tropical environments.

Summerfield (1982, 1983a), based on his studies of silcrete genesis in southern Africa, distinguished between lateral transfer models and vertical transfer models of silcrete formation. Lateral transfer models, which include the formation of silcrete in lacustrine and fluvial models environments, have long been applied to the origin of silcretes in Australia and Africa. In some of the earliest work on the genesis of silcretes in Australia, Stephens (1964, 1971) suggested that the silcretes in central Australia were the result of the accumulation of silica leached and transported from the Eastern Highlands during the formation of lateritic iron crusts. Short-distance lateral transport models have been proposed by Milnes and Twidale (1983), Wopfner (1978), and Hutton et al. (1972, 1978). Summerfield (1982, 1983b), working in southern Africa, described silcretes from a variety of landscape settings whose origin can be best ascribed to the lateral transfer model. He reported the occurrence of silcretes in pan (lacustrine) settings in Botswana and the Northern Cape Province, South Africa, as well as on the floors of river valleys and in association with river terraces. Ambrose and Flint (1981) described silcretes associated with strandlines of a receding Miocene lake in northern South Australia.

Perhaps more widely applicable than the lateral transfer models are the vertical transfer models. The latter models include the *per ascendum, per descendum*, and groundwater models of Goudie (1973). Silcretes resulting from the vertical rise of soil waters (the *per ascendum* model) and subsequent deposition of silica as a result of evaporation and/or pH change were described in early investigations of silcrete genesis in Australia (Woolnough 1927) and Africa (Frankel and Kent 1938). Severe limitations with this model with respect to the distance of capillary rise through sands (Summerfield 1982) as well as

the efficiency of evaporative precipitation at depth (Summerfield 1983b) have cast serious doubts on its widespread applicability to silcrete genesis. The exceptions being in the upper parts of pedogenic silcrete profiles (Webb and Golding 1998) and in evaporite settings (Milnes et al. 1991).

Of broader applicability is the *per descendum* model in which silica-charged waters percolate to depth under the influence of gravity, eventually resulting in the deposition of silica. These models were championed early in the history of silcrete research by workers such as Whitehouse (1940) and more recently by Gunn and Galloway (1978) and Watts (1978a, b). Thiry and Milnes (1991), Simon-Coinçon et al. (1996) and Thiry et al. (2006) described pedogenic silcretes from the opal fields of South Australia in which they envisaged the accumulation of silica to be the result of the downward percolation of silica-rich waters derived from the weathering of overlying quartzite and dissolution of previous generations of silcrete.

Transport of silica in ground water (the groundwater model) and subsequent deposition with associated silcrete formation has received considerable attention in the literature. Groundwater silcretes have been described from arid central Australia by Ambrose and Flint (1981). Thiry and Milnes (1991), Simon-Coinçon et al. (1996) and Thiry et al. (2006) working in the opal fields of South Australia, proposed a model of silcrete formation in association with silica-laden ground waters. Successive episodes of meteoric water infiltration during wet periods react with Si-rich sediments producing unique phases of Si dissolution resulting in unique mineralogies, chemistries and micromorphologies of the silicified regolith. This model led them to conclude that multiple silcrete layers could be attributed to silicification associated with progressive water table lowering as landscape denudation proceeds.

Intergrade Duricrusts

Calcretes and silcretes exist as distinct entities but they also grade into one another and contain a mix of both calcium carbonate and silica cements (intergrade duricrusts Fig. 6.15). Nash and Shaw (1998) have suggested the terms calcrete-silcrete intergrade duricrust and silcrete-calcrete intergrade duricrust should be

Fig. 6.15 Cal-silcrete at the base of the section grading upwards into a calcrete, Kang Pan, central southern Botswana (McLaren photo)

used dependent upon the dominant cementing agent (until the deposit grades into silcretes or calcretes *sensu stricto* at either end of the spectrum). Intergrade duricrusts exist in a number of locations around the world including Southern Africa (Watts 1980, Summerfield 1984, Nash and Shaw 1998, Ringrose et al. 2002, Nash and McLaren 2003, Nash et al. 2004); North Africa (Thiry and Ben Brahim 1997, McLaren et al. 2006); Australia (Arakel et al. 1989); South America (King 1967); and North America (Vaniman et al. 1994). Intergrade duricrusts may show evidence of later alteration leading to secondary minerals partly replacing the primary cement. Alternatively secondary materials may have replaced pre-existing cements or precipitated in voids or pore spaces in the host material. In some situations both silica and carbonate cements may have occurred

in close succession or contemporaneously (Nash et al. 2004).

Silcretes also show evidence of intergrades in association with ferricrete. Lee and Gilkes (2005) describe iron cemented silcrete from southwestern Australia. From geochemical analyses and morphological observations they conclude that ferruginization occurred following the silicification of the regolth as a result of groundwater transport of both iron and aluminum through pores in the silcrete.

Gypsum Crusts

Occurrence

Gypsic crusts occur in warm desert environments receiving less than 200–250 mm of rainfall per annum (Watson 1983). However, in environments where precipitation levels are less than 25 mm per annum, more soluble halite crusts tend to form rather than gypcrete (Hartley and May 1998). Gypcretes have been reported from virtually every continent of the world (Watson 1983) where sulphate, calcium and evaporation of water lead to the accumulation of gypsum (Eckardt et al. 2001). In Australia, gypcretes have been described by Wopfner and Twidale (1967), Arakel and McConchie (1982), Warren (1982), Jacobson et al. (1988), Milnes et al. (1991) Chivas et al. (1991), Chen et al. (1991, 1995) and Chen (1997). In Africa, extensive areas of gypsum crusts have been recorded in Tunisia (Watson 1979, 1985, 1988, Drake et al. 2004), Algeria (Horta 1980), and the Namib Desert of southern Africa (Martin 1963, Scholz 1972, Watson 1983, 1985, 1988, Eckardt et al. 2001). Gypsic crusts have also been reported from the Middle East, specifically from Iraq (Tucker 1978), Israel (Dan et al. 1982, Amit and Gerson 1986), Egypt (Ali and West 1983), Saudi Arabia (Al Juaidi et al. 2003) and Jordan (Turner and Makhlouf 2005). In North America there have been limited reports of gypcretes from the southwestern United States (Nettleton et al. 1982) and the semi-arid Rocky Mountain states (Reider et al. 1974, Reheis 1987). Rech et al. (2003) have studied gypsum crusts from the Atacama Desert, Chile.

Morphology

Watson (1979, 1983, 1988) recognized three principal types of gypsum crust in Africa. (a) Evaporitic crust characteristically consists of packed microcrystalline gypsum strata. (b) True gypsum or *croute de nappe* occurs in two forms: either as lightly cemented gypsum crystals up to 1 mm in length commonly developed beneath non-gypsic sediment or as desert rose crust consisting of interlocking lenticular gypsum crystals ranging in size from a few millimetres to 20 cm. (c) Surface crust also occurs in two principal forms: gypsum powder which is composed of loose accumulations of small gypsum crystals, and indurated crust which is characteristically polygonal or columnar in appearance and consists of an outer zone of densely packed microcrystalline gypsum and a core of soft or puggy coarsely crystalline gypsum.

In a study of gypsic soils in the Rocky Mountain states, Reheis (1987) described gypsum accumulations in soils in terms of the morphologic sequence developed by Gile et al. (1966) for calcretes from the southwestern United States. Reheis (1987) identified four stages of gypcrete development. Stage I gypcrete is characterized by thin, discontinuous gypsum coatings on the undersides of stones. Stage II gypcrete has abundant gypsum pendents under stones and gypsum crystals scattered through the matrix or forming small, soft powdery nodules. Stage III gypcrete is distinguished by the presence of continuous gypsum through the soil matrix and large gypsum pendents beneath stones. Stage IV gypcrete consists of a continuous gypsum-plugged matrix, with stones and smaller debris floating in the gypsum matrix.

Hartley and May (1998) have identified two types of Miocene gypcrete from northern Chile. The first group (Type 1) are gypsum-cemented sandstones that comprise 'v-shaped' cracks infilled by clastic material, veins of fibrous gypsum, as well as alabastrine columns and nodules, which Hartley and May (1998) have argued represent weathered gypsic crusts. The Type 2 gypcretes are reddened, massive, mesocrystalline and poikilitic in nature with veins of fibrous gypsum and bedded clast-rich lenses. Hartley and May (1998) have proposed that the gypcretes are very similar to B horizons in Quaternary alluvial desert soils.

Descriptions of pedogenic gypcrete from central Australia have been made by Chen (1997), who describes them as white, powdery finely crystalline and massive deposits that commonly have a polygonal hardened surficial layer. Wopfner and Twidale (1967) recognized two main forms of gypsum. The first is an upper crystalline unit which these authors specifically referred to as gypcrete. This unit consists of coarsely crystalline double-twinned gypsum, together with fine to coarse clastic gypsum. The second lies below the crystalline unit and is a gypsum-rich sedimentary unit. These two units together constitute what is referred to as a gypsite profile. The underlying unit consists of finely crystalline gypsum dispersed throughout the clastic sediments. Gypsum prisms also occur within the matrix of the unit.

Gypcretes and gypsites have been described from Western Australia by Arakel and McConchie (1982) and the Northern Territory by Jacobson et al. (1988). The gypsites are morphologically diverse depending on whether they originate in the vadose or phreatic zones. Phreatic gypsite bodies are typically wedge or lensoid in shape. They consist of acicular and/or discoidal gypsum crystals with massive, tabular, or clotted micromorphologies. Vadose gypsites are typically massive in appearance and consist of loosely to tightly packed gypsum crystals, crystallites, or gypsic fragments. Grain sizes are variable and display a variety of micromorphological fabrics. Surface gypcretes are commonly thick and massive. They possess a variety of grain sizes and occasionally exhibit graded bedding. The gypcretes form benches adjacent to lake basins and occur as lenses in lake-bordering dunes. The gypcretes consist of weathered gypsum crystals that have undergone repeated solution and recrystallization (Jacobson et al. 1988). Pedogenic gypcretes from the Namib Desert have been found to comprise microcrystalline, lenticular and prismatic crystals with some lenses of microcrystalline alabastrine and fibrous gypsum (Eckardt et al. 2001).

Chemistry

A comprehensive summary of the characteristics of the chemistry of gypcretes from Tunisia and the Namib Desert was provided by Watson (1983, 1985). These analyses reflect the considerable diversity of gypcrete chemistries which vary with the degree of induration. Typically, the African gypcretes contain small abundances of calcium carbonate but are enriched in sodium, potassium, and magnesium. Iron and aluminium occur in smaller abundances than the soluble salts and probably reflect the mineralogy of the insoluble residue.

Comprehensive chemical analyses of gypcretes from North America are presented by Reheis (1987). The gypcretes are dominated by calcium which usually occurs in abundances of approximately 30%. Silicon is co-dominant in abundance with calcium and is indicative of the mineralogy of the insoluble residue. Magnesium is present in abundances of less than 10% and may be indicative of the presence of carbonates. As with the African gypcretes, iron and aluminium occur in substantially smaller abundances than the aforementioned elements. Like the silicon, these two elements are representative of the clay mineral fraction of the insoluble residue. Minor abundances of the other elements are present in the gypcretes. Sodium has been widely regarded as an important chemical constituent in gypcretes from Africa and linked to the role of ground water in its formation (Watson 1983). Chemical analyses of gypcretes from Wyoming, however, contain less than half a percent of sodium which may indicate minimal ground water effects.

Relatively little work has been undertaken on the mineralogy of gypcretes. Gypsum is the dominant mineral species. Analysis of about 150 samples from Tunisia and the Namib by Watson (1983) shows the mean abundance of this mineral to range from 61% in powdery gypcrete to greater than 77% in indurated surface crusts. Quartz is also an important constituent in the insoluble residue. Gypcretes from Wyoming (Reheis 1987) contain gypsum abundances of less than 10% in some stage I gypcretes to as much as 70% in stage IV materials. Chen (1997) reports that gypcretes from central Australia typically contain more than 80% gypsum with only minor amounts of carbonate, quartz, clays and heavy minerals.

Clay minerals that have been reported by Reheis (1987) from the insoluble residue of gypcretes from Wyoming, include kaolinite, smectite, and mica. These clay minerals commonly range in abundance from 10 to 30%, generally decreasing in abundance as the degree of crust induration increases (Reheis 1987). Minor quantities of palygorskite also occur in gypcretes from wyoming (Reheis 1987), but constitute the dominant secondary mineral in gypcretes from Africa (Watson 1989).

Origin

A diversity of models has been presented in the literature for the origin of different forms of gypcrete. These models fall basically into either pedogenic or non-pedogenic categories. It has been proposed that pedogenic gypcretes may result from the vertical rise and subsequent evaporation of gypsum-laden soil waters. This model, however, has been extensively criticized by Keen (1936) and more recently by Watson (1979, 1983, 1985). Keen argued that the structure of most desert soils precluded the capillary rise of waters, and Watson stressed the problem of the great quantities of water necessary to precipitate even thin gypsum crusts.

Chen (1997) has summarised pedogenic gypcrete genesis in central Australia as involving the dissolution, leaching and recrystallisation of primary gypsum sediments and has produced a useful table that provides examples of the key processes and changes that occur (Table 6.1 p. 41). Pedogenic gypcretes resulting from the downward movement of gypsum have been more widely embraced (Watson 1979, 1983, 1985, Chen 1997, Eckardt et al. 2001), the source of gypsum being generally regarded as aeolian (Page 1972, Watson 1979). Studies of gypsic soils in the Middle East have stressed the importance of wind-derived gypsum (Dan et al. 1982, Amit and Gerson 1986). This gypsum is then translocated to shallow depths and deposited by evaporation. Reheis (1987) and Chen et al. (1991) have argued that reworking of hydromorphic gypsum crust is also an important process. Where gypsum occurs deeper in the soil profile, wetter climatic regimes are implied. Reheis (1987) also ascribed the gypcretes in Wyoming to vertical translocation of aerosolically derived gypsum, the gypsum having been derived from nearby dunes and surrounding bedrock sources. Watson (1985) likewise recognized the significance of aeolian-derived gypsum as a source for surface and subsurface crusts in Tunisia and the Namib Desert. In Tunisia he considered the source to be deflation of seasonally flooded basins, while in the Namib he, along with Day (1993) and Heine and Walter (1996) emphasized the importance of fog in carrying dissolved salts. However, Eckardt and Schemenauer (1998) have shown that the ionic content of fog from the Namib Desert was very low and therefore unlikely to act as a major pathway of sulphur.

Recently, Drake et al. (2004) studied the sulphur isotope signatures of bedrock, groundwater, playa brines and sediments as well as gypsiferous crusts from southern Tunisia. Their data suggested that recycling of marine gypsum was the most likely source of sulphate in the groundwater, playa sediments and crusts. Drake et al. (2004) noted that although the ulimate gypsum source is from the bedrock, aeolian processes are thought to be the main transporting mechanism. Sea-salt sulphate and volatile biogenic sulphur have been identified as an important source of $\delta^{34}S$ in some coastal locations such as Western Australia (Chivas et al. (1991) and the Namib Desert (Eckardt and Spiro 1999).

Non-pedogenic models of gypcrete formation are related to deposition of gypsum in ground water or lacustrine environments. Extensive gypsite and gypcretes have recently been discovered in a series of playa lakes in central Australia (Jacobson et al. 1988). The gypsite and gypcretes result from the complex interaction of phreatic and vadose ground waters and infiltrating meteoric waters. Lacustrine gypcretes have also been found by Watson (1985, 1988) in the warm desert environments of Tunisia and the Namib. Watson described bedded gypsum crusts, which he attributed to evaporation of shallow lakes, and subsurface desert rose gypsum crusts, which he ascribed to the evaporation of shallow ground water.

Conclusions

From the foregoing discussion a number of general conclusions can be reached.

(a) Calcretes, silcretes, and gypcretes (and their intergrades) are widely distributed throughout the warm desert environments of Australia, Africa, North America, and the Middle East.

(b) Each of the principal types of duricrust displays considerable variability with respect to morphology, micromorphology, chemistry, and mineralogy. This variability reflects the diversity of environments of duricrust formation.

(c) The formation of duricrusts is the result of diverse processes which may be fundamentally divided into pedogenic and non-pedogenic categories. Complex interactions between formative processes also commonly occur.

(d) While the nature and origin of duricrusts have received considerable attention in the literature, considerable gaps exist geographically.

(e) There is a need for more work on the nature and origin of duricrusts in general in the American south-west. In Australia considerable opportunity exists for future research on the nature and origin of gypcretes.

(f) In all warm desert environments there is great scope for the assessment of the genetic relationship between different duricrust types.

References

Abdel-Wahab A., A.M.K. Salem and E.F. McBride 1998. Quartz cement of meteoric origin in silcrete and non-silcrete sandstones, Lower Carboniferous, western Sinai, Egypt. *Journal of African Earth Sciences* **27**, 277–290.

Ali Y.A. and I. West 1983. Relationships of modern gypsum nodules in sabkhas of loess to composition of brines and sediments in northern Egypt. *Journal of Sedimentary Petrology* **53**, 1151–68.

Al Juaidi F., A. Millington and S. McLaren 2003. Evaluating image fusion techniques for mapping geomorphological features on the eastern edge of the Arabian shield (Central Saudi Arabia). *Geographical Journal* **169**, 2, 117–131.

Alonso-Zarza A.M. 2003. Palaeoenvironmental significance of palustrine carbonates and calcretes in the geological record. *Earth Science Reviews* **60**, 261–298.

Ambrose G.J. and R. Flint 1981. A regressive Miocene lake system and silicified strand lines in northern South Australia: implications for regional stratigraphy and silcrete genesis. *Journal of the Geological Society of Australia* **28**, 81–94.

Amit R. and R. Gerson 1986. The evolution of Holocene reg (gravelly) soils in deserts – an example from the Dead Sea region. *Catena* **13**, 59–79.

Amundson R.G., O. Chadwick, J. Sowers and H. Donner 1988. The relationship between modern climate and vegetation and the stable isotope chemistry of Mojave Desert soils. *Quaternary Research* **29**, 245–54.

Amundson R.G., O. Chadwick, J. Sowers and H. Donner 1989. The stable isotope chemistry of pedogenic carbonate at Kyle Canyon, Nevada. *Soil Science Society of America Journal* **53**, 201–10.

Andrews J.E., A.K. Singhvi, A.J. Kailath, R. Kuhn, P.F. Dennis, S.K. Tandon and R.P. Dhir 1998. Do stable isotope data from calcrete record Late Pleistocene monsoonal climate variation in the Thar Desert of India? *Quaternary Research* **50**, 240–51.

Arakel A.V. 1985. Vadose diagenesis and multiple calcrete soil profile development in Hutt Lagoon area, Western Australia. *Revue de Geologie Dynamique et de Geographie Physique* **26**, 243–54.

Arakel A.V. 1986. Evolution of calcrete in palaeodranages of the Lake Napperby Area, central Australia. *Palaeogeography, Palaeoclimatology, Palaeoecology* **54**, 283–303.

Arakel A.V. and D. McConchie 1982. Classification and genesis of calcrete and gypsite lithofacies in paleodrainage systems of inland Australia and their relationship to carnotite mineralization. *Journal of Sedimentary Petrology* **52**, 1149–70.

Arakel A.V., G. Jacobsen, M. Salehi and C.M. Hill 1989. Silicification of calcrete in palaeodrainage basins of the Australian arid zone. *Australian Journal of Earth Sciences* **36**, 73–89.

Aristarain L.F. 1970. Chemical analysis of caliche profiles from High Plains, New Mexico. *Journal of Geology* **78**, 201–12.

Aristarain L.F. 1971. On the definition of caliche deposits. *Zeitschrift für Geomorphologie* **15**, 274–89.

Auzel M. and A. Cailleux 1949. Silicifications nord-saharieenes. *Bulletin de la Société géologique de France, Series* **519**, 553–559.

Bachman G.O. and M.N. Machette 1977. Calcic soils and calcretes in the southwestern United States. *U.S. Geological Survey Open-File Report* 77–794.

Ballesteros E.M., J.G. Talegón and M.A. Hernández 1997. Palaeoweathering profiles developed on the Iberian hercynian basement and their relationship to the oldest Tertiary surface in central and western Spain. In: Widdowson, M. (ed.) *Palaeosurfaces: Recognition, Reconstruction and Palaeoenvironmental Interpretation.* Special Publication, No. 120, Geological Society, London, pp. 175–185.

Barnes I. 1965. Geochemistry of Birch Creek, Inyo County, California, a travertine depositing creek in an arid environment. *Geochimica et Cosmochimica Acta* **29**, 85–112.

Blake W.P. 1902. The caliche of southern Arizona: an example of deposition by vadose circulation. *Transactions of the American Institute of Mining and Metalurgical Engineers* **31**, 220–6.

Bowler J.M. and H.A. Polak 1971. Radiocarbon analysis of soil carbonates: an evaluation from paleosols in southeastern Australia. In *Paleopedology – origin, nature, and dating of paleosols*, D. Yaalon (ed.), 97–108. Jerusalem: Israel University Press.

Bretz J.H. and L. Horberg 1949. Caliche in southeastern New Mexico. *Journal of Geology* **57**, 491–511.

Brewer R. and J.R. Sleeman 1964. Glaebules: their definition, classification and interpretation. *Journal of Soil Science* **15**, 66–78.

Brown C.N. 1956. The origin of caliche in the northeastern Llano Estacado, Texas. *Journal of Geology* **64**, 1–15.

Brückner W.D. 1966. Origin of silcretes in Central Australia. *Nature* **209**, 496–497.

Bustillo M.A. and M. Bustillo 2000. Miocene silcretes in argillaceous playa deposits, Madrid Basin, Spain: petrological and geochemical features. *Sedimentology* **47**, 1023–1037.

Butt C.R.M. 1983. Aluminosilicate cementation of saprolites, grits and silcretes in Western Australia. *Journal of the Geological Society of Australia* **30**, 179–86.

Butt C.R.M. 1985. Granite weathering and silcrete formation on the Yilgarn Block, Western Australia. *Australian Journal of Earth Sciences* **32**, 415–32.

Butt C.R.M., R.C. Horwitz and A.W. Mann 1977. Uranium occurrences in calcretes and associated sediments in Western Australia. *CSIRO Mineral Research Laboratories Division of Mineralogy. Report* FP 16.

Carlisle D. 1983. Concentration of uranium and vanadium in calcretes and gypcretes. In *Residual deposits: surface related weathering processes and materials*, R.C.L. Wilson (ed.) 185–95. Geological Society of London Special Publication 11.

Cerling T.E. and R.L. Hay 1986. An isotopic study of paleosol carbonates from Olduvai Gorge. *Quaternary Research* **25**, 63–78.

Cerling T.E. 1999. Stable isotopes in palaeosol carbonates In: Thiry, M. and Simon-Coincon, R. (eds.) *Palaeoweathering, Palaeosurfaces and Related Continental Deposits. International Association of Sedimentologists Special Publication* **27**, 43–60.

Chen X.Y., J.M. Bowler and J.W. Magee 1991. Gypsum ground: a new occurrence of gypsum sediment in playas of central Australia. *Sedimentary Geology* **72**, 79–95.

Chen X.Y. J. Chappell and A.S. Murray 1995. High (ground) water levels and dune development in central Australia: TL dates from gypsum and quartz dunes around Lake Lewis, Northern Territory. Geomorphology **11**, 311–322.

Chen X.Y. 1997. Pedogenic gypcrete formation in arid central Australia. Geoderma **77**, 39–61.

Chivas A.R. A.S. Andrew, W.B. Lyons, M.I. Bird and T.H. Donnelly 1991. Isotopic constraints on the origin of salts in Australian Playas, 1. Sulphur. *Palaeogeography, Palaeoclimatology, Palaeoecology* **84**, 309–332.

Crawford A.R. 1965. The geology of Yorke Peninsula. *Geological Survey of South Australia Bulletin* 39.

Crocker R.L. 1946. Post-Miocene climatic and geologic history and its significance in relation to the genesis of the major soil types of South Australia. *CSIRO Bulletin* 193.

Dan J. 1977. The distribution and origin of nari and other lime crusts in Israel. *Israel Journal of Earth Sciences* **26**, 68–83.

Dan J., D.H. Yaalon, R. Moshe and S. Nissim 1982. Evolution of reg soils in southern Israel and Sinai. *Geoderma* **28**, 173–202.

Day J.A. 1993. The major ion chemistry of some southern African saline systems. Hydrobiologia **267**, 37–59.

Dixon J.C. 1978. Morphology and genesis of calcrete in South Australia with special reference to the southern Murray Basin and Yorke Peninsula. M.A. thesis. University of Adelaide, South Australia.

Dixon J.C. 1994. Duricrusts In: Abrahams, A.D. and A.J. Parsons (eds.) *Geomorphology of Desert Environments*, pp 82–105, Chapman and Hall. London.

Drake N.A., F.D. Eckardt and K.H. White 2004. Sources of sulphur in gypsiferous sediments and crusts and pathways of gypsum redistribution in southern Tunisia. *Earth Surface Processes and Landforms* **29**, 1459–1473.

Drees L.R. and L.P. Wilding 1987. Micromorphic record and interpretation of carbonate forms in the Rolling Plains of Texas. *Geoderma* **40**, 157–76.

Durand G.H. 1949. Essai de nomenclature des croutes. *Bulletin Societe des Sciences Nautrelles de Tunisie* **3/4**, 141–2.

Durand J.H. 1963. Les croutes calcaires et gypseuses en Algerie: formation et age. *Bulletin Societe Geologique de France, Series 7* **6**, 959–68.

Dury G.H. 1966. Duricrusted residuals on the Barrier and Cobar pediplains of New South Wales. *Journal of the Geological Society of Australia* **13**, 299–307.

Eckardt F.D. and R.S. Schemenauer, 1998. Fogwater chemistry in the Namib Desert, Namibia. *Atmospheric Environment* **32**, 2595–2599.

Eckardt F.D. and B. Spiro 1999. The origin of sulphur in gypsum and dissolved sulphate in the Central Namib Desert, Namibia. *Sedimentary Geology* **123**, 255–73.

Eckardt F.D., N.A. Drake, A.S. Goudie, K. White and H. Viles 2001. The role of playas in the formation of the pedogenic gypsum crusts of the central Namib desert. *Earth Surface Processes and Landforms* **26**, 1177–93.

Ericksen G.E. 1994. Discussion of a petrographic study of the Chilean nitrates. *Geological Magazine* **131**, 849–52.

Esteban M. and C.F. Klappa 1983. Subaerial exposure environment. In: Scholle, P.A., Bebout, D.G. and Moore, C.H. (eds.) *Carbonate Depositional Environments. American Association of Petroleum Geologists Memoir 33*, 1–54.

Firman J.B. 1963. Quaternary geological events near Swan Reach in the Murray Basin, South Australia. *Quarterly Notes of the Geological Survey of South Australia* **5**, 2–4.

Firman J.B. 1966. Stratigraphy of the Chowilla area in the Murray Basin. *Quarterly Notes of the Geological Survey of South Australia* **20**, 3–7.

Frankel J.J. and L.E. Kent 1938. Grahamstown surface quartzites (silcretes). *Transactions of the Geological Society of South Africa* **15**, 1–42.

Frye J.C., H.D. Glass, A.B. Leonard and D. Coleman 1974. Caliche and clay mineral zonation of the Ogallala Formation, central-eastern New Mexico. *New Mexico Bureau of Mines and Mineral Resources Circular* 144.

Gardner L.R. 1972. Origin of the Mormon Mesa caliche, Clark County, Nevada. *Bulletin of the Geological Society of America* **83**, 143–56.

Gardner R.A.M. and D.A. Hendry 1995. Early silica diagenesis in aeolian sediments, south India. *Journal of the Geological Society of London*, **152**, 183–92.

Gile L.H. 1967. Soils of an ancient basin floor near Las Cruces. *Soil Science* **103**, 264–76.

Gile L.H., F.F. Peterson and R.B. Grossman 1965. The K horizon – a master soil horizon of carbonate accumulation. *Soil Science* **99**, 74–82.

Gile L.H., F.F. Peterson and R.B. Grossman 1966. Morphological and genetic sequences of carbonate accumulation in desert soils. *Soil Science* **101**, 347–60.

Gill E.D. 1975. Calcrete hardpans and rhizomorphs in western Victoria, Australia. *Pacific Geology* **9**, 1–16.

Goudie A. 1972. The chemistry of world calcretes. *Journal of Geology* **80**, 449–63.

Goudie A. 1973. *Duricrusts in Tropical and Subtropical Landscapes*. Oxford: Clarendon Press.

Goudie A. 1983. Calcrete. In: A.S. Goudie and K. Pye (eds.) *Chemical Sediments and Geomorphology: Precipitates and Residua in the Near Surface Environment*, 93–131. New York: Academic Press.

Goudie A.S. and Heslop, E. 2007. Sodium Nitrate Deposits and efflorescences. In: Nash, D.J. and McLaren, S.J. (eds.), 391–408. *Geochemical Sediments and Landscapes*. Blackwell, Oxford.

Gunn R.H. and R.W. Galloway 1978. Silcretes in south-central Queensland. In *Silcrete in Australia*, T. Langford-Smith (ed.), 51–71. Armidale: Department of Geography, University of New England.

Gurel A. and S. Kadir, 2006. Geology, mineralogy, and origin of clay minerals of the Pliocene fluvial lacustrine deposits in the Cappadocian volcanic province, central Anatolia, Turkey. Clays and Clay Minerals **54**, 555–570.

Gustavson T.C. and D.A. Winkler 1988. Depositional facies of the Miocene–Pliocene Ogallala Formation, northwestern Texas and eastern New Mexico. *Geology* **16**, 203–6.

Hartley A.J. and G. May 1998. Miocene gypcretes from the Calama Basin, northern Chile. *Sedimentology* **45**, 351–364.

Hay R.L. and R.J. Reeder 1978. Calcretes of Olduvai Gorge and the Ndolanya Beds of northern Tanzania. *Sedimentology* **25**, 649–73.

Hay R.L. and B. Wiggins 1980. Pellets, ooids, sepiolite and silica in three calcretes of the southwest United States. *Sedimentology* **27**, 559–76.

Heaney P.J. 1993. A proposed mechanism for the growth of chalcedony. *Contributions to Mineralogy and Petrology* **115**, 66–74.

Heaney P.J. 1995. Moganite as an indicator for vanished evaporites: a testament reborn? *Journal of Sedimentary Research* **A65**, 633–638.

Heine K. and R. Walter 1996. Gypcretes of the central Namib Desert, Namibia. *Palaeoecology of Africa* **24**, 173–201.

Hill S.M. 2005. Regolith and landscape evolution of far western New South Wales. In. R.R. Anand and P. de Broekert (eds.) *Regolith landscape Evolution Across Australia*. Bently, Western Australia. Cooperative research centre for landscape Evolution and Mineral Exploration. 130–145.

Hill S.M., K.G. McQueen, and K.A. Foster, 1999. Regolith carbonate accumulations in western and central NSW: characteristics and potential as an exploration sampling medium. In: Taylor, G.M. and C.F. Pain (eds.). *State of the Regolith*. Proceedings of the Regolith Conference, pp. 191–200. Canberra. Cooperative Centre for Landscape Evolution and Mineral Exploration.

Holliday V.T. 1989. The Blackwater Draw Formation (Quaternary): a 1.4-plus-m.y. record of eolian sedimentation and soil formation on the southern High Plains. *Bulletin of the Geological Society of America* **101**, 1598–607.

Horta O.S. 1980. Calcrete, gypcrete, and soil classification in Algeria. *Engineering Geology* **15**, 15–52.

Hunt C.B. and D.R. Mabey 1966. Stratigraphy and structure, Death Valley, California. *U.S. Geological Survey Professional Paper* 494A.

Hutton J.T. and J.C. Dixon, 1981. The chemistry and mineralogy of some South Australian calcretes and associated soft carbonates and their dolomitization. *Journal of the Geological Society of Australia* **28**, 71–9.

Hutton J.T., C.R. Twidale, A.R. Milnes and H. Rosser 1972. Composition and genesis of silcretes and silcrete skins from the Beda Valley, southern Arcoona Plateau, South Australia. *Journal of the Geological Society of Australia* **19**, 31–9.

Hutton J.T., D.S. Lindsay and C.R. Twidale. 1977. The weathering of norite at Black Hill. *Journal of the Geological Society of Australia* **24**, 37–50.

Hutton J.T., C.R. Twidale and A.R. Milnes 1978. Characteristics and origins of some Australian silcretes. In *Silcrete in Australia*, T. Langford-Smith (ed.), 19–40. Armidale: Department of Geography, University of New England.

Jacobson G., A.V. Arakel and C. Yijian 1988. The central Australian groundwater discharge zone: evolution of associated calcrete and gypcrete deposits. *Australian Journal of Earth Sciences* **35**, 549–65.

James N.P. 1972. Holocene and Pleistocene calcareous crust (caliche) profiles: criteria for subaerial exposure. *Journal of Sedimentary Petrology* **42**, 817–36.

Johns R.K. 1963. Limestone, dolomite, and magnesite resources of South Australia. *Geological Survey of South Australia Bulletin* 38.

Kahle C.E. 1977. Origin of subaerial Holocene calcareous crusts: role of algae, fungi, and sparmicritisation. *Sedimentology*, 24, 413–435.

Keen B.A. 1936. The circulation of water in the soil between the surface and the level of underground water. *Bulletin of the International Association of Scientific Hydrology* **22**, 328–31.

Kelly M., S. Black and J.S. Rowan 2000. A calcrete-based U/Th chronology for landform evolution in the Sorbas Basin, south east Spain. *Quaternary Science Reviews* **19**, 995–1010.

Khormali F., A. Abtahi and G. Steps, 2006. Micromorphology of calcic features in highly calcareously soils of Fars province, southern Iran. *Geoderma*, 132, 31–46.

King L.C. 1967. *The Morphology of the Earth*. Oliver and Boyd, Edinburgh.

Klappa C.F. 1978. Biolithogenesis of microcodium. *Sedimentology* **25**, 489–522.

Klappa C.F. 1979a. Lichen stromatolites: criterion for subaerial exposure and a mechanism for the formation of laminar calcretes (caliche). *Journal of Sedimentary Petrology* **49**, 387–400.

Klappa C.F. 1979b. Calcified filaments in Quaternary calcretes: organo-mineral interactions in the subaerial vadose environment. *Journal of Sedimentary Petrology* **49**, 955–68.

Klappa C.F. 1980. Rhizoliths in terrestrial carbonates: classification, recognition, genesis and significance. *Sedimentology* **27**, 613–629.

Knauth L.P. 1994. Petrogenesis of chert. In: Heaney, P.J., Prewitt, C.T. and Gibbs, G.V. (eds.) *Silica: Physical Behaviour, Geochemistry and Materials Applications*. Reviews in Mineralogy 29, Mineralogical Society of America, Washington, pp. 233–258.

Knox G.J. 1977. Caliche profile formation, Saldanha Bay (South Africa). *Sedimentology* **24**, 657–74.

Kosir A. 2004. Microcodium revisited: root calcification products of terrestrial plants on carbonate-rich substrates. *Journal of Sedimentary Research* **74**, 845–857.

Lamplugh G.H. 1907. The geology of the Zambezi Basin around the Batoka Gorge (Rhodesia). *Quarterly Journal of the Geological Society of London* **63**, 162–216.

Lamplugh G.W. 1902. 'Calcrete'. *Geological Magazine* **9**, 575.

Langford-Smith T. 1978. *Silcrete in Australia*. Armidale: Department of Geography, University of New England.

Langford-Smith T. and G.H. Dury 1965. Distribution, character and attitude of the duricrust in the northwest of New South Wales and adjacent areas of Queensland. *American Journal of Science* **263**, 170–90.

Langford-Smith T. and S.H. Watts 1978. The significance of co-existing siliceous and ferruginous weathering products at selected Australian localities. In *Silcrete in Australia*, T. Langford-Smith (ed.), 143–65. Armidale: Department of Geography, University of New England.

Machette M.N. 1985. Calcic soils of the southwestern United States. In *Soils and Quaternary geology of the southwestern*

United States, D. Weide (ed.) 1–21. Geological Society of America Special Paper 203.

Mann A.W. and R.C. Horwitz 1979. Groundwater calcretes in Australia: some observations from Western Australia. *Journal of the Geological Society of Australia* **26**, 293–303.

Martin H. 1963. A suggested theory for the origin and a brief description of some gypsum deposits of South West Africa. *Transactions of the Geological Society of South Africa* **55**, 345–50.

McFadden L.D. 1982. The impacts of temporal and spatial climatic changes on alluvial soil genesis in southern California. Ph.D. thesis. University of Arizona, Tucson.

McFadden L.D. and J.C. Tinsley 1983. Rate and depth of pedogenic–carbonate accumulation in soils: formulation and testing of a compartment model. In *Soils and Quaternary geology of the southwestern United States*, D. Weide (ed.), 23–41. Geological Society of America Special Paper 203.

McFadden L.D., E.V. McDonald, S.G. Wells, K. Anderson, J.Quade, and S.L. Forman 1998. The vesicular layer and carbonate collars of desert soils and pavements: formation, age and relation to climate change.*Geomorphology* **24**, 101–145.

McLaren S.J. 1995. Early diagenetic fabrics in the rhizosphere of late Pleistocene aeolian sediments. *Journal of the Geological Society of London* **152**, 173–181.

McLaren S., D. Gilbertson, J. Grattan, C. Hunt, G.A.T. Duller and G. Barker 2004. Quaternary palaeogeomorphologic evolution of the Wadi Faynan area, southern Jordan. *Palaeogeography, Palaeoclimatology, Palaeoecology* **205**, 1–2, 129–152.

McLaren S.J. 2004. Evolution and distribution of Quaternary channel calcretes, southern Jordan. *Earth Surface Processes and Landforms* **29**, 1487–1508.

McLaren S.J., N. Drake and K. White 2006. Late Quaternary environmental changes in the Fazzān, southern Libya: evidence from sediments and duricrusts. In: D. Mattingly, S. McLaren, E. Savage, Y. al-Fasatwi, K. Khadgood (eds.) *The Natural Resources and Cultural Heritage of Libya*. Socialist People's Libyan Arab Jamahariya The Society of Libyan Studies Monograph **6**, 157–166.

McQueen K.G., S.M. Hill and K.A. Foster. 1999. The nature and distribution of regolith carbonate accumulations in southeastern Australia and their potential as a sampling medium in geochemical exploration. *Journal of Geochemical Exploration* **67**, 67–82.

Milnes A.R. 1992. Calcrete. In: Martini, I.P. and Chesworth, W. (eds.) *Weathering, Soils and Palaeosols*. Developments in Earth Surface Processes 2. Elsevier, Amsterdam, pp. 349–377.

Milnes A.R. and J.T. Hutton 1974. The nature of microcryptocrystalline titania in 'silcrete' skins from the Beda Hill area of South Australia. *Search* **5**, 153–4.

Milnes A.R. and J.T. Hutton 1983. Calcretes in Australia. In *Soils: an Australian viewpoint*, 119–62. Melbourne: CSIRO/London: Academic Press.

Milnes A.R. and C.R. Twidale 1983. An overview of silicification in Cainozoic landscapes of arid central and southern Australia. *Australian Journal of Soil Research* **21**, 387–410.

Milnes A.R. and M. Thiry 1992. Silcretes. In: Martini, I.P. and Chesworth, W. (eds.) *Weathering, Soils and Palaeosols*. Developments in Earth Surface Processes 2. Elsevier, Amsterdam, pp. 349–377.

Milnes A.R., R.W. Kimber and S.E. Phillips 1987. Studies in eolian calcareous landscapes of southern Australia. In *Aspects of Loess Research*, L. Tungsheng (ed.), 130–9. Beijing: China Ocean Press.

Milnes A.R., M. Thiry and M.J. Wright 1991. Silica accumulations in saprolite and soils in South Australia. In *Occurrence, characteristics, and genesis of carbonate, gypsum and silica accumulation in soils*, W.D. Nettleton (ed.), 121–49. Soil Science Society of America, Special Publication 26.

Monger H.C., L.A. Daugherty, and L.H.Gile 1991a. A microscopic examination of pedogenic calcite in an aridisol of southern New Mexico. In *Occurrence,characteristics, and genesis of carbonate, gypsum, and silica accumulations in soils*. W.D. Nettleton (ed.). 37–60. Soil Science Society of America Special Publication 26.

Monger H.C., L.A. Daugherty, W.C. Lindemann and C.M. Liddell 1991b. Microbial precipitation of pedogenic calcite. *Geology* **19**, 997–1000.

Mumm A.S. and F.Reith 2007. Biomediation of calcrete at the gold anomaly of the Barns prospect, Gawler Craton, South Australia, *Journal of Geochemical Exploration* **92**, 13–33.

Nash D.J. and P.A. Shaw 1998. Silica and carbonate relationships in silcrete-calcrete intergrade duricrusts from the Kalahari Desert of Botswana and Namibia. *Journal of African Earth Sciences* **27**, 11–25.

Nash D.J. and S.J. McLaren 2003. Kalahari valley calcretes: their nature, origins, and environmental significance. *Quaternary International* **111**, 3–22.

Nash D.J., P.A. Shaw and D.S.G. Thomas 1994. Duricrust development and valley evolution: process-landform links in the Kalahari. *Earth Surface processes and Landforms* **19**, 299–317.

Nash D.J., S.J. McLaren and J.A. Webb 2004. Petrology, geochemistry and environmental significance of silcrete-calcrete intergrade duricrusts at Kang Pan and Tswaane, central Kalahari, Botswana. *Earth Surface Processes and Landforms* **29**, 1559–1586.

Nash D.J. and Ullyott, S. 2007. Silcretes. In: Nash, D.J. and McLaren, S.J. (eds.). *Geochemical Sediments and Landscapes*. Blackwell, Oxford.

Netterberg F. 1967. Some road making properties of South African calcretes. *Proceedings of the 4th Regional Conference of African Soil Mechanics and Foundation Engineers, Cape Town*, **1**, 77–81.

Netterberg F. 1969. Ages of calcretes in southern Africa. *South African Archeological Bulletin* **24**, 117–22.

Netterberg F. 1975. Dating and correlation of calcrete and other pedocretes. *Transactions of the Geological Society of South Africa* **81**, 379–391.

Netterberg F. 1980. Geology of South African calcretes I: Terminology, description macrofeatures and classification.*Transactions of the Geological Society of South Africa* **83**, 255–283.

Netterberg F. and J.H. Caiger 1983. A geotechnical classification of calcretes and other pedocretes. In *Residual deposits: surface and related weathering processes and materials*, R.C.L. Wilson (ed.), 235–43, Geological Society of London Special Publication 11.

Nettleton W.D., R.E. Nelson, B.R. Brasher and P.S. Derr. 1982. Gypsiferous soils of the western United States. In J.A. Kittrick, D.S. Fanning and L.R. Hossner (eds.) *Acid Sulphate Weathering*, 147–68. Soil Science Society of America Special Publication 10.

Ollier C.D. 1978. Silcrete and weathering. In *Silcrete in Australia*, T. Langford-Smith (ed.), 13–17. Armidale: Department of Geography, University of New England.

Owliaie H.R., A. Abtahi, and R.J. Heck, 2006. Pedogenesis and clay mineralogical investigation of soils formed on gypsiferous materials on a transect, southern Iran. Geoderma, 134, 62–81.

Page W.D. 1972. The geological setting of the archeological site at Oued al Akarit and the paleoenvironmental significance of gypsum soils, southern Tunisia. Ph.D. thesis. University of Colorado, Boulder.

Patil D.N. 1991. Basalt weathering and related chemical sediments and residuals from the Deccan volcanic province, Maharashtra, India. *Tropical Geomorphology Newsletter* **11**, 1–3.

Phillips S.E. and A.R. Milnes 1988. The Pleistocene terrestrial carbonate mantle on the southeastern margin of the St Vincent Basin, South Australia. *Australian Journal of Earth Sciences* **35**, 463–81.

Phillips S.E. and P.G. Self 1987. Morphology, crystallography and origin of needle-fibre calcite in Quaternary pedogenic calcretes of South Australia. *Australian Journal of Soil Research* **25**, 429–44.

Phillips S.E., A.R. Milnes and R.C. Foster 1987. Calcified filaments: an example of biological influences in the formation of calcrete in South Australia. *Australian Journal of Soil Research* **25**, 405–28.

Price W.A. 1933. The Reynosa problem of south Texas and the origin of caliche. *Bulletin of the Association of Petroleum Geologists* **17**, 488–522.

Quade J., A.R. Chivas and M.T. McCulloch 1995. Strontium and carbon isotope tracers and the origins of soil carbonate in South Australia and Victoria. *Palaeogeography, Palaeoclimatology, Palaeoecology* **113**, 103–117

Rabenhorst M.C., L.P. Wilding and C.L. Girdner 1984. Airborn dust in the Edwards Plateau region of Texas. *Soil Science Society of America Journal* **48**, 621–7.

Rakshit P. and R.M. Sundaram 1998. Calcrete and gypsum crusts of the Thar Desert, Rajasthan: their geomorphic locales and use as palaeoclimatic indicators. *Journal of the Geological Society of India* **51**, 249–255.

Rech J.A., J. Quade and W.S. Hart 2003. Isotopic evidence for the source of Ca and S in soil gypsum, anhydrite and calcite in the Atacama Desert, Chile. *Geochimica et Cosmochimica* **67**, 575–586.

Reeves C.C. 1970. Origin, classification, and geological history of caliche on the southern High Plains, Texas and eastern New Mexico. *Journal of Geology* **78**, 352–62.

Reeves C.C. 1976. *Caliche: Origin, Classification, Morphology and Uses*. Lubbock: Estacado Books.

Reeves C.C. and J.D. Suggs 1964. Caliche of central, and southern Llano Estacado, Texas. *Journal of Sedimentary Petrology* **34**, 669–72.

Reheis M.C. 1987. Gypsic soils on the Kane alluvial fans, Big Horn County, Wyoming. *U.S. Geological Survey Bulletin* 1590C.

Reider R.G., N.J. Kuniansky, D.M. Stiller and P.J. Uhl 1974. Preliminary investigation of comparative soil development on Pleistocene and Holocene geomorphic surfaces of the Laramie Basin, Wyoming. In *Applied Geology and Archeology: The Holocene History of Wyoming*, M. Wilson (ed.), 27–33. Geological Survey of Wyoming Report of Investigations 10.

Ringrose S., A.B. Kampunzu, B.W. Vink, W. Matheson and W.S. Downey 2002. Origin and palaeo-environments of calcareous sediments in the Moshaweng dry valley, southeast Botswana. *Earth Surface Processes and Landforms* **27**, 591–611.

Royer D.L., R.A. Berner and D.J. Beerling 2001. Phanerozoic atmospheric CO2 change: evaluating geochemical and paleobiological approaches. *Earth Science Reviews* **54**, 349–392.

Ruhe R.V. 1967. Geomorphic surfaces and surficial deposits in southern New Mexico. *New Mexico Bureau of Mines and Mineral Resources Memoir* 18.

Salomons W., A. Goudie and W.G. Mook 1978. Isotopic composition of calcrete deposits from Europe, Africa, and India. *Earth Surface Processes* **3**, 43–57.

Schlesinger W.H. 1985. The formation of caliche in soils of the Mojave Desert, California. *Geochimica et Cosmochimica Acta* **49**, 57–66.

Scholz H. 1972. The soils of the central Namib Desert with special consideration of the soils in the vicinity of Gobabeb. *Madoqua* **2**, 33–51.

Sehgal J.L. and G. Stoops 1972. Pedogenic calcite accumulations in arid and semi-arid regions of the Indo-Gangetic alluvial plain of erstwhile Punjab. *Geoderma* **8**, 59–72.

Semeniuk V. and T.D. Meagher 1981. Calcrete in Quaternary coastal dunes in southwestern Australia: a capillary rise phenomenon associated with plants. *Journal of Sedimentary Petrology* **51**, 47–68.

Semeniuk V. and D.J. Searle 1985. Distribution of calcrete in Holocene coastal sands in relationship to climate, southwestern Australia. *Journal of Sedimentary Petrology* **55**, 86–95.

Senior B.R. 1978. Silcrete and chemically weathered sediments in southwest Queensland. In *Silcrete in Australia*, T. Langford-Smith (ed.), 41–50. Armidale: Department of Geography, University of New England.

Shafetz H.S. and J.C. Butler 1980. Petrology of recent caliche pisolites, spherules and speleothems deposits from central Texas. *Sedimentology* **27**, 497–518.

Shaw P.A. and D.J. Nash 1998. Dual mechanisms for the formation of fluvial silcretes in the distal reaches of the Okavango Delta Fan, Botswana. *Earth Surface Processes and Landforms* **23**, 705–714.

Shaw P.A., H.J. Cooke and C.C. Perry 1990. Microbialitic silcretes in highly alkaline environments: some observations from Sua Pan, Botswana. *South African Journal of Geology* **93**, 803–808.

Sidwell R. 1943. Caliche deposits of the southern High Plains, Texas. *American Journal of Science* **241**, 257–61.

Simon-Coinçon R., A.R. Milnes, M. Thiry, and M.J. Wright 1996. Evolution of landscapes in northern South Australia in relation to the distribution and formation of silcretes. *Journal of the Geological Society, London* **153**, 467–480.

Smale D. 1973. Silcretes and associated silica diagenesis in southern Africa and Australia. *Journal of Sedimentary Petrology* **43**, 1077–89.

Stephens C.G. 1964. Silcretes of central Australia. *Nature* **203**, 1407.

Stephens C.G. 1971. Laterite and silcrete in Australia: a study of the genetic relationships of laterite and silcrete and their companion materials, and their collective significance in the weathered mantle, soils, relief, and drainage of the Australian continent. *Geoderma* **5**, 5–52.

Strong G.E., J.R.A. Giles and V.P. Wright 1992. A Holocene calcrete from North Yorkshire, England: implications for interpreting palaeoclimates using calcretes. *Sedimentology* **39**, 333–347.

Summerfield M.A. 1982. Distribution, nature and probable genesis of silcrete in arid and semi-arid southern Africa. *Catena Supplement* **1**, 37–65.

Summerfield M.A. 1983a. Silcrete. In: A.S. Goudie and K. Pye (eds.) *Chemical Sediments and Geomorphology: Precipitates and Residua in the Near Surface Environment*, 59–91. New York: Academic Press.

Summerfield M.A. 1983b. Petrology and diagenesis of silcrete from the Kalahari Basin and Cape coastal zone, southern Africa. *Journal of Sedimentary Petrology* **53**, 895–909.

Summerfield M.A. 1983c. Geochemistry of weathering profile silcretes, southern Cape Province, South Africa. In *Residual Deposits: Surface and Related Weathering Processes and Materials*, R.C.L. Wilson (ed.), 167–78. Geological Society of London Special Publication 11.

Summerfield M.A. 1984. Isovolumetric weathering and silcrete formation, Southern Cape Province, South Africa. *Earth Surface Processes and Landforms* **9**, 135–141.

Swineford A. and P.C. Franks 1959. Opal in the Ogalalla Formation in Kansas. *Society of Economic Paleontologists and Mineralogists Special Publication* 7.

Taylor G. 1978. Silcretes in the Walgett-Cumborah Region of New South Wales. In *Silcrete in Australia*, T. Langford-Smith (ed.), 187–93. Armidale: Department of Geography, University of New England.

Taylor G. and I.E. Smith 1975. The genesis of sub-basaltic silcretes from the Monaro, New South Wales. *Journal of the Geological Society of Australia* **22**, 377–85.

Thiry M. 1999. Diversity of continental silicification features: examples from the Cenozoic deposits in the Paris Basin and neighbouring basement. In: Thiry, M. and Simon-Coinçon, R. (eds.) *Palaeoweathering, Palaeosurfaces and Related Continental Deposits*. International Association of Sedimentologists, Special Publication No 27, Blackwell Science, Oxford, pp. 87–127.

Thiry M. and A.R. Milnes 1991. Pedogenic and groundwater silcretes at Sturt Creek opal field, South Australia. *Journal of Sedimentary Petrology* **61**, 111–27.

Thiry M. and M. Ben Brahim 1997. Ground-water silicifications in the calcareous facies of the Tertiary piedmont deposits of the Atlas Mountain (Hamada du Guir, Morocco). *Geodinamica Acta* **10**, 12–29.

Thiry M., A.R. Milnes, V. Rayot and R. Simon-Coinçon 2005. Interpretation of palaeoweathering features and successive silicification *Journal of the Geological Society, London* **163**, 723–36.

Thomas D.S.G., G. Brook, P.A. Shaw, M. Bateman, C. Appleton, D.J. Nash, S.J. McLaren and F. Davies 2003. Late Pleistocene wetting and drying in the NW Kalahari: an integrated study from the Tsodilo Hills, Botswana. *Quaternary International* **104**, 53–67.

Tucker M.E. 1978. Gypsum crusts (gypcrete) and patterned ground from northern Iraq. *Zeitschrift für Geomorphologie* **22**, 89–100.

Turner B.R. and I. Makhlouf. 2005. Quaternary sandstones, northeast Jordan: Age, depositional environments and climatic implications. *Palaeogeography, Palaeoclimatology, Palaeoecology* **229**, 230–50.

Twidale, C.R., J.A. Shepard and R.M. Thompson 1970. Geomorphology of the southern part of the Arcoona Plateau and the Tent Hill Region, west and north of Port Augusta, South Australia. *Transactions of the Royal Society of South Australia* **94**, 55–69.

Van den Heuvel R.C. 1966. The occurrence of sepiolite and attapulgite in the calcareous zone of a soil near Las Cruces, New Mexico. *Clays and Clay Minerals* **25**, 193–207.

van Dijk D.C. and G.G. Beckman 1978. The Yuleba hardpan and its relationship to soil geomorphic history in the Yuleba-Tora region, southeast Queensland. In *Silcrete in Australia*, T. Langford-Smith (ed.), 73–92. Armidale: Department of Geography, University of New England.

Vaniman D.T., S.J. Chipera and D.L. Bish 1994. Pedogenesis of siliceous calcretes at Yucca Mountain, Nevada. *Geoderma* **63**, 1–17.

Verrecchia E.P. and K.E. Verrecchia 1994. Needle-fiber calcite: a critical review and a proposed classification. *Journal of Sedimentary Research* **A64**, 650–664.

Ward W.T. 1965. Eustatic and climatic history of the Adelaide area, South Australia. *Journal of Geology* **73**, 592–602.

Ward W.T. 1966. Geology, geomorphology and soils of the southwestern part of County Adelaide. *CSIRO Soil Publication* 23.

Warren J.K. 1982. The hydrological setting, occurrence, and significance of gypsum in late Qaternary salt lakes in South Australia. *Sedimentology* **29**, 609–37.

Warren J.K. 1983. Pedogenic calcrete as it occurs in Quaternary calcareous dunes in coastal South Australia. *Journal of Sedimentary Petrology* **53**, 787–96.

Watson A. 1979. Gypsum crusts in deserts. *Journal of Arid Environments* **2**, 3–20.

Watson A. 1983. Gypsum crusts. In: A.S. Goudie and K. Pye (eds.) *Chemical Sediments and Geomorphology: Precipitates and Residua in the Near Surface Environment.* 133–6. New York: Academic Press.

Watson A. 1985. Structure, chemistry and origins of gypsum crusts in southern Tunisia and the central Namib Desert. *Sedimentology* **32**, 855–75.

Watson A. 1988. Desert gypsum crusts as palaeoenvironmental indicators: a micropetrographic study of crusts from southern Tunisia and the central Namib desert. *Journal of Arid Environments* **15**, 19–42.

Watson A. 1989. Desert crusts and varnishes. In *Arid zone geomorphology*, D.S.G. Thomas (ed.), 25–55. New York: Hallstead Press.

Watts N.L. 1978. Displacive calcite: evidence from recent and ancient calcretes. *Geology* **6**, 699–703.

Watts N.L. 1980. Quaternary pedogenic calcretes from the Kalahari (southern Africa): mineralogy, genesis and diagenesis. *Sedimentology* **27**, 661–86.

Watts S.H. 1977. Major element geochemistry of silcrete from a portion of inland Australia. *Geochimica et Cosmochimica Acta* **41**, 1164–7.

Watts S.H. 1978a. A petrographic study of silcrete from inland Australia. *Journal of Sedimentary Petrology* **48**, 987–94.

Watts S.H. 1978b. The nature and occurrence of silcrete in the Tibooburra area of northwestern New South Wales. In *Silcrete in Australia*, T. Langford-Smith (ed.), 167–85. Armidale: Department of Geography, University of New England.

Weatherby K.G. and J.M. Oades 1975. Classification of carbonate layers in highland soils of the northern Murray Valley, South Australia and their use in stratigraphic and land-use studies. *Australian Journal of Soil Research* **13**, 119–32.

Webb J.A. and S.D. Golding 1998. Geochemical mass-balance and oxygen-isotope constraints on silcrete formation and its palaeoclimatic implications in southern Australia. *Journal of Sedimentary Research* **68**, 981–993.

West, L.T., L.R. Drees, L.P. Wilding and M.C. Rabenhorst 1988. Differentiation of pedogenic and lithogenic carbonate forms in Texas. *Geoderma* **43**, 271–87.

Whitehouse F.W. 1940. *Studies in the Late Geological History of Queensland.* University of Queensland Papers in Geology 1.

Wieder M. and D. Yaalon 1974. Effects of matrix composition on carbonate nodule crystallization. *Geoderma* **11**, 95–121.

Wigley T.M.L. 1973. Chemical evolution of the system calcite–gypsum–water. *Canadian Journal of Earth Sciences* **10**, 306–15.

Wilding, L.P., N.E. Smeck and L.R. Drees 1977. Silica in soils: quartz, cristobalite, tridymite, and opal. In: J.B. Dixon and S.B. Weed (eds.) *Minerals in Soil Environments*, 471–522. Madison: Soil Science Society of America.

Williams G.E. and H.A. Polach 1969. The evaluation of C-14 ages for soil carbonate from the arid zone. *Earth and Planetary Science Letters* **7**, 240–2.

Williams G.E. and H.A. Polach 1971. Radiocarbon dating of arid zone calcareous paleosols. *Bulletin of the Geological Society of America* **82**, 3069–86.

Woolnough W.G. 1927. The duricrust of Australia. *Journal and Proceedings of the Royal Society of New South Wales* **61**, 24–53.

Woolnough W.G. 1930. Influence of climate and topography in the formation and distribution of products of weathering. *Geological Magazine* **67**, 123–32.

Wopfner H. 1978. Silcretes of northern South Australia and adjacent regions. In *Silcrete in Australia*, T. Langford-Smith (ed.), 93–141. Armidale: Department of Geography, University of New England.

Wopfner H. 1983. Environment of silcrete formation: A comparison of examples from Australia and the Cologne Embayment, West Germany. In *Residual Deposits: Surface and Related Weathering Processes and Materials*, R.C.L. Wilson (ed.), 151–8. Geological Society of London Special Publication 11.

Wopfner H. and C.R. Twidale 1967. Geomorphological history of the Lake Eyre Basin. In: J.N. Jennings and J.A. Mabbutt (eds.) *Landform Studies from Australia and New Guinea*, 118–43. Canberra: A.N.U. Press.

Wright V.P. 1994. Paleosols in shallow marine carbonate sequences. *Earth Science Reviews* **35**, 367–39.

Wright V.P. and M.E. Tucker (eds.) 1991. Calcretes: an Introduction. *International Association of Sedimentologists Reprint Series* **2**, 1–22.

Wright V.P. 2007. Calcretes. In: Nash D.J. and McLaren S.J. (eds.) *Geochemical Sediments and Landscapes*. Blackwell, Oxford

Yaalon D.H. 1981. Pedogenic carbonate in aridic soils: magnitude of the pool and annual fluxes. *Abstract, International Conference on Aridic Soils, Jerusalem.*

Chapter 7

Desert Rock Coatings

Ronald I. Dorn

Introduction

Desert landforms are characterized by an abundance of 'bare' rock and mineral surfaces. Mountains host widespread exposures of bedrock. Gravel desert pavements cap alluvial terraces and fans. Even sand dunes are themselves composed of rock fragments exposed to the atmosphere without substantive plant cover. This chapter focuses on an irony, that the supposed fundamental bare-rock nature of desert landforms stretches the truth.

In reality, rock coatings, even those thinner than $10\,\mu m$ (0.010 mm) substantially alter the appearance of almost all of the rock surfaces found in deserts (Fig. 7.1). Consider just a few of the icons of desert geomorphology. The main Petra tourist attraction of the Al-Khazneh Tomb façade is coated with a black manganese-rich varnish, allowing the carved portions of the elaborate burial chamber to stand out. The almost white colour of Ayers Rock is coated by mostly iron-clay orange accretions, facilitating photogenic displays. Dramatic sandstone escarpment faces of the Colorado Plateau in such places as Monument Valley are frequently coated by a reddish-brown silica glaze formed inside the unopened joint fracture and then exposed by block wasting.

The systematic study of rock coatings started tens of millennia ago, as prehistoric peoples targeted very specific rock coatings for petroglyph manufacturing as well as application of artificial rock coatings to create paintings (Whitley, 2001; Whitley, 2005). The scientific study of rock coatings started in 1799 (von Humboldt, 1812), when major differences in manganese composition between coatings and the underlying rock led to recognition of rock varnish's accretionary nature.

There are over a dozen different types of rock coatings (Table 7.1). Within each type, tremendous variety exists at spatial scales from kilometres to micrometres. For example, there are at least six different types of silica glazes (Dorn, 1998: 294–312). Interdigitation also exists between different types of rock coatings, resulting in complex microstratigraphic sequences. For example, lava flows in the arid Ashikule Basin of Tibet host carbonate skins, dust films, lithobiontic coatings, oxalate crusts, phosphate skins, rock varnish, silica glazes, and sulfate crusts. In another example, lithobionts like lichens are normally associated with rock weathering, but they can also play key roles in generating silica glazes (Lee and Parsons, 1999) and oxalate crusts (Beazley et al., 2002). Given the variety and complexity of rock coatings, it should be of no surprise that researchers not infrequently confuse different types of coatings in their data collection and analysis.

This chapter introduces the field of desert rock coatings. After introducing the paradigm perspective of landscape geochemistry that drives an overall conceptual understanding of rock coatings, I present current hypotheses of how different rock coatings form, followed by a summary of their use in desert geomorphology. This chapter ends by identifying future research needs.

R.I. Dorn (✉)
School of Geographical Sciences, Arizona State University, Tempe, AZ 85287, USA
e-mail: ronald.dorn@asu.edu

A.J. Parsons, A.D. Abrahams (eds.), *Geomorphology of Desert Environments*, 2nd ed., DOI 10.1007/978-1-4020-5719-9_7, © Springer Science+Business Media B.V. 2009

Fig. 7.1 Rock coatings change the appearance of bare rock landforms. *Upper Left*: A vertical face at Canyon de Chelly, Arizona, is streaked with heavy metal skins, iron films, lithobiont coatings, oxalate crusts, rock varnish, and silica glaze. *Upper Right*: Lava flows in the arid regions of Mauna Loa show a distinct color change within decades as a direct result of accumulation of silica glaze. The true color image, courtesy of NASA, has a length of 3.6 km with west at the *top*. The electron microscope images (backscatter detector) demonstrate that rock coatings are external accretions, exemplified by an oxalate crust from the arid Olary Province in South Australia in the *lower left* image that is about 500 μm thick. Rock varnish on the *lower right* from Kitt Peak in the Sonoran Desert is about 100 μm thick

Table 7.1 Major categories of rock coatings

Coating	Description	Related terms
Carbonate Skin	Composed primarily of carbonate, usually $CaCO_3$, but sometimes $MgCO_3$	Calcrete, travertine
Case Hardening	Addition of cementing agent to rock matrix material; the agent may be manganese, sulfate, carbonate, silica, iron, oxalate, organisms, or anthropogenic.	Sometimes called a particular type of rock coating
Dust Film	Light powder of clay- and silt-sized particles attached to rough surfaces and in rock fractures.	Clay skins; clay films; soiling
Heavy Metal Skins	Coatings of iron, manganese, copper, zinc, nickel, mercury, lead and other heavy metals on rocks in natural and human-altered settings.	Sometimes described by chemical composition
Iron Film	Composed primarily of iron oxides or oxyhydroxides; unlike orange rock varnish some films do not have clay as a major constituent.	Ferric oxide red staining, iron staining
Lithobiontic Coatings	Organisms forming rock coatings, for example lichens, moss, fungi, cyanobacteria, algae.	Organic mat, biofilms, biotic crust
Nitrate Crust	Potassium and calcium nitrate coatings on rocks, often in caves and rock shelters in limestone areas.	Saltpeter; niter; icing
Oxalate Crust	Mostly calcium oxalate and silica with variable concentrations of magnesium, aluminum, potassium, phosphorus, sulfur, barium, and manganese. Often found forming near or with lichens.	Oxalate patina, lichen-produced crusts, patina, scialbatura
Phosphate Skin	Various phosphate minerals (e.g. iron phosphates or apatite) sometimes mixed with clays and sometimes manganese.	Organophosphate film; epilithic biofilm
Pigment	Human-manufactured material placed on rock surfaces by people.	Pictograph, paint
Rock Varnish	Clay minerals, Mn and Fe oxides, and minor and trace elements; color ranges from orange to black in color produced by variable concentrations of different manganese and iron oxides	Desert varnish, patina, Wüstenlack
Salt Crust	Chloride precipitates formed on rock surfaces	Halite crust, efflorescence
Silica Glaze	Usually clear white to orange shiny lustre, but can be darker in appearance, composed primarily of amorphous silica and aluminum, but often with iron.	Desert glaze, turtle-skin patina, siliceous crusts, silica-alumina coating, silica skins
Sulfate Crust	Sulfates (e.g., barite, gypsum) on rocks; not gypsum crusts that are sedimentary deposits	Sulfate skin

Landscape Geochemistry Model of Rock Coating Development

Landscape geochemistry, as developed in Soviet geography (Polynov, 1937; Perel'man, 1961, 1966; Glazovskaya, 1968; Glazovskaya, 1973) and brought into the English literature (Fortescue, 1980), integrates studies of element abundance, element migration, geochemical flows, geochemical gradients, and geochemical barriers with the classification, interpretation, and spatial laws pertaining to geochemical landscapes. No other conceptual framework of looking at rock coatings explains the quantity or diversity of data, is as relevant to a variety of disciplines, or provides such a clear framework to analyze rock coatings (Dorn, 1998: 20–27). Furthermore, there is no alternative; landscape geochemistry currently provides the only larger theoretical framework for the study of desert rock coatings.

Viewed from the spectacles of landscape geochemistry, rock coatings occur where geochemical, biological, or physical barriers exist to the flow of coating constituents, so long as a rock surface is stable long enough to allow coatings to accrete. The geographical expression of these barriers can be extensive in area, such as the overall alkaline nature of desert rock surfaces facilitating the stability of iron (Fe) and manganese (Mn) in rock varnish (Fig. 7.2A). There

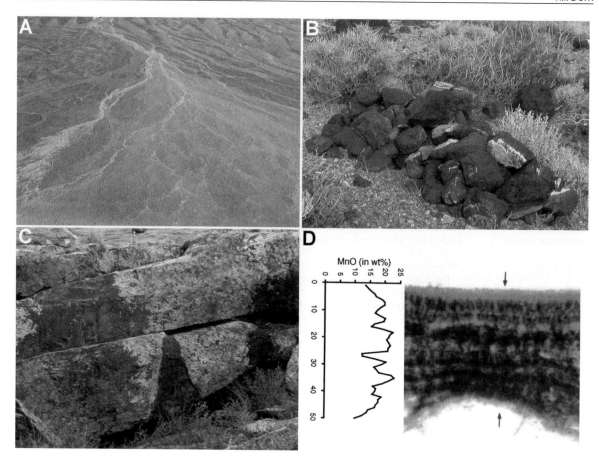

Fig. 7.2 Desert rock coatings accumulate at different types of geochemical barriers. (**A**) In southern Death Valley, California, the geochemical environment is alkaline enough to permit the accumulation of Fe and Mn on alluvial-fan cobbles. (**B**) The undersides of buried boulders on Panamint Valley shorelines, California, have accumulated carbonate, but in the case of this prehistoric cairn, the carbonate slowly dissolves from exposure to rain. (**C**) Lichen growth at Yunta Springs, South Australia, has dissolved most of previous iron film, rock varnish and silica glaze. (**D**) Rock varnish microlaminations of black Mn-rich layers record wetter climates, while orange Mn-low layers record drier conditions in the Zunggar Desert, western China (Zhou et al., 2000). The image is courtesy of Tanzhuo Liu (Liu, 2008)

are also linear barriers at chemical discontinuities, allowing such coatings as carbonate skins to accumulate on the undersides of desert boulders (Fig. 7.2B). These barriers shift in space and on timescales of 10^{-3} to 10^3 years, causing temporary coatings of ice to melt within a day (Hetu et al., 1994), silica glazes to form within decades (Fig. 7.1) carbonate skins to dissolve within centuries (Fig. 7.2C), or rock varnishes to accrete evidence of millennial-scale climatic change (Liu, 2003; Liu, 2008; Liu and Broecker, 2007; Liu and Broecker, 2008a, 2008b) (Fig. 7.2D).

Desert rock coatings can be interpreted by following a five-order hierarchical sequence of landscape geochemical controls (Dorn, 1998: 324–344).

First-Order Process: Geomorphic Controls

An obvious first control rests in the need for processes that generate the bare rock on which coatings accrete. Deserts are dominated by rock coatings, because deserts are weathering-limited landscapes where detachment and transport exceeds weathering (Gilbert, 1877). Rock coatings also dominate desert landforms because host rock surfaces remain stable long enough to accumulate coatings. Some surfaces can remain stable for 10^5 years (Nishiizumi et al., 1993; Liu and Broecker, 2000; Liu, 2003), although most rock surfaces do erode more rapidly

(Gordon and Dorn, 2005a). For example, biotite weathering promotes granite grussification that limits the development of rock coatings on granitic surfaces (Fig. 7.3A).

Second-Order Process: Inheritance from the Subsurface

Rock coatings can and do originate in soil (Engel and Sharp, 1958; Ha-mung, 1968; Hayden, 1976; Robinson and Williams, 1992) and in rock fissures (Dorn and Dragovich, 1990; Douglas et al., 1994; Mottershead and Pye, 1994; Frazier and Graham, 2000; Cerveny et al., 2006; Kim et al., 2006). Subaerial exposure occurs from soil erosion (Hunt and Wu, 2004) or spalling of the overlying block (Villa et al., 1995). These inherited rock coatings are extremely common, even though very few investigators note the possibility that the samples they collected may have originated underground. This general lack of appreciation may be because there are many possible trajectories for what happens to inherited coatings after exposure. The more common post-exposure sequences are: continued erosion exposes even more subsurface coatings; the rock under the inherited coatings erodes — effectively removing weathering rind and coating; inherited coatings erode, only to be replaced by a different sub-aerial coating; lithobionts grow on inherited coatings (Fig. 7.3B); or a similar or different coating grows on top of the inherited coating (Fig. 7.3C).

Third Order Process: Habitability for Lithobionts

Faster growing coatings such as lichens dominate over slower-growing accretions. Unless they are overcome by anthropogenic application of pigments (Lee et al., 1996; Wang and Hua, 1998; Saiz-Jimenez and Hermosin, 1999; Li et al., 2001), lithobiontic (Golubic et al., 1981) coatings such as lichens and fungi (Urzì et al., 1992; Souza-Egipsy et al., 2004) grow much faster than most natural inorganic coatings, effectively 'taking out' the competition (Fig. 7.3B). Some have even speculated that bacteria precipitate black Mn-rich coatings like rock varnish just to heat up

desert surfaces in order to reduce surface habitability for lichens. The latest research reveals that desert rock surfaces host an incredible variety of adventitious extreme organisms (Benzerara et al., 2006; Fajardo-Cavazos and Nicholson, 2006; Kuhlman et al., 2006). The speedy ones take possession of the surface.

Fourth- and fifth-order processes — typically the starting point for rock coating research — involve movement of coating constituents and their fixation on rock surfaces. However, these processes only influence the nature of desert rock coatings when: (1) rock faces are exposed by erosion; (2) exposed rock faces are stable enough to support rock coatings; and (3) lithobionts do not outcompete other coatings.

Fourth-Order Process: Transport Pathways

Transport of raw mineral ingredients involves two steps. The constituents must be present, and they must migrate to the rock face. Bird droppings (Arocena and Hall, 2003) or microorganisms (Konhauser et al., 1994), for example, generate the requisite material for a phosphate skin that is then mobilized and re-precipitated (Fig. 7.3D). Many oxalate-rich crusts found in deserts similarly rely on lithobionts to manufacture the oxalate that is then transported by water flow over rock surfaces (Beazley et al., 2002). Constituent availability alone, however, can be a factor in determining what type of rock coating grows. For example, water flows over sandstone cliff faces have an abundance of Mn and Fe precipitation, but these water-flow deposits often lack clay minerals; since clays are vital to the formation of rock varnish (Potter and Rossman, 1977; Potter, 1979), the net result is often the formation of a heavy metal skin (Dorn, 1998: ch 8) instead of a rock varnish. Similarly, the basalt flows of semi-arid Hawai'i lack the overwhelming aeolian deposition of clay minerals found in continental deserts, resulting in the dominance of silica glazes (Dorn, 1998: ch 13) over relatively rare Hawaiian rock varnish (Fig. 7.3E).

Fifth-Order Process: Barriers to Transport

Physical, geochemical, and biological barriers come into play only after all of above processes do not

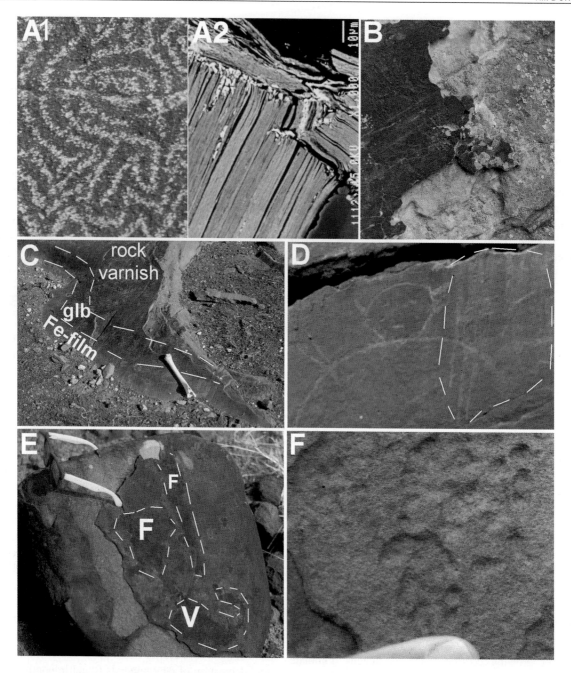

Fig. 7.3 Landscape geochemistry influences coating development. **A**. Petroglyph carved into varnished granodiorite, south-central Arizona (A1), where biotite oxidation and hydration from this site (A2) limits the accretion of rock varnish to the length of time it takes to erode a grus grain. **B**. Lichens, cyanobacteria, and fungi (*right side* of image) have almost completely eroded silica glaze from a sandstone joint face, Wyoming. The new lithobiont community now plays a key role in case hardening the surface, as has been found elsewhere (Viles and Goudie, 2004). **C**. Slow soil erosion at Karolta, South Australia, exposes two rock coatings. Erosion first exposed the ground-line band (glb), a very thin and shiny accretion of silica glaze and manganese that orig- inally forms at the soil-rock-atmosphere interface (Engel and Sharp, 1958). Then, continued erosion exposed iron film. Sub- aerial rock varnish then started to grow over both of these former subsurface coatings. **D**. Phosphate skin over sandstone, eastern Wyoming, where a bird droppings were mobilized then and pre- cipitated (inside *dashed area*) over mostly silica glaze. **E**. Basalt boulder on the rainshadow side of Kaho'olawe Island, Hawai'i, is mostly coated by silica glaze, but pockets of rock varnish (v) and fungi (f) also grow. **F**. Dust film deposited over sandstone in a Colorado Plateau alcove that is protected from rainsplash and water flow

rule out coating accretion. Dust coatings (Fig. 7.3F), for example, form at locales (Johnson et al., 2002) where the physical barrier of Van der Waals forces are *not* overcome by shear stresses imposed by water flow. Iron films, carbonate skins, and natural heavy metal skins are examples of coatings that may accrete at either geochemical (Krauskopf, 1957; Collins and Buol, 1970; Scheidegger et al., 1993; Huguen et al., 2001; Nanson et al., 2005) or biological barriers (Ha-mung, 1968; Chukhrov et al., 1973; Mustoe, 1981; Robbins et al., 1992; Robbins and Blackwelder, 1992). These geochemical barriers, in turn, can trap additional elements; for example, rock varnish forms at a barrier to Mn and Fe mobility where the Mn-Fe oxides in turn capture dozens of trace and rare elements (Wayne et al., 2006).

Formation

The processes by which different desert rock coatings form are best organized through the hierarchy discussed in the previous section. Since the first-order processes of exposing bare rock and the fourth-order processes of constituent transport do not in and of themselves produce accretions, the examples presented below focus on key steps making some of the more common desert rock coatings.

Second-Order Process: Fissuresol Coatings

A common cause of desert rock spalling is the gradual growth of fissuresols (Fig. 7.4A). Fissuresols are formed by the accumulation of rock coatings and sediment fill inside joints (Coudé-Gaussen et al., 1984; Villa et al., 1995; Frazier and Graham, 2000). As coatings and fill accrete, the fissure opens slowly until detachment occurs, exposing coatings originating inside of the crevice. Exposed fissuresols might only be a few centimetres across (Fig. 7.4B), or they might run completely through a 3-metre-diameter boulder (Fig. 7.4F).

Three types of rock coatings form inside fissures in drier deserts, and all three are exposed by rock spalling. Carbonate skins form in the deepest parts of the fracture (Fig. 7.4C, E, G), but this carbonate coating is not long-lived since it dissolves from exposure to carbonic acid in precipitation. The perimeter of a fracture develops a distinct zonation of an inner wider band of orange iron film and a narrow outer band of black rock varnish (Fig. 7.4B, E, F, G). These are not the Mn-Fe coatings found in saprolite fractures (Weaver, 1978). Rather, the iron films are clay-iron accretions similar to the orange coating found on the underside of desert pavement cobbles, and black coatings are manganese-rich rock varnish.

After exposure, if conditions are too xeric for fast-growing lithobionts (cf. Fig. 7.3B), then subaerial rock varnish grows on top of the fissuresol coatings (cf. Fig. 7.4D). The darkest rock varnishes seen on any given landform are usually those that start as a fissuresol, because there is foundation of a fairly complete coverage on the host rock. In contrast, those varnishes that start out on abraded clasts are not as dark and not as well coated. On rock surfaces exposed by abrasion processes (e.g., fluvial , glacial, littoral action), the rock varnish must first accumulate in nucleation sites in isolated microtopographic basins. Then, only after taking millennia to get this foothold does rock varnish grow together horizontally to form a complete coating. In contrast, the fissuresol acts like a paint primer that covers the whole surface, helping the black subaerial varnish to accrete a darker and more complete cover. This is why the sample selection criteria used by researchers (Harrington and Raymond, 1989; McFadden et al., 1989; Reneau, 1993) is based on an incomplete understanding of rock coating formation.

Fissuresol sequences (Fig. 7.4A) are the most common type of inherited rock coatings in drier deserts, but fractures in rocks in semi-arid environments typically contain silica glaze (Fig. 7.5D, E) that can be dense (Fig. 7.5F–H) or can be a more porous clay-rich fracture coating (Graham and Franco-Vı`zcaino, 1992; Thoma et al., 1992; Frazier and Graham, 2000).

Silica glaze in fractures is just the start. In the semi-arid sandstone cuestas of the Colorado Plateau and Wyoming basins, joint faces often accumulate black rock varnish after faces are exposed by erosion. Then, some of the iron and manganese in the varnish is leached (Dorn and Krinsley, 1991) to mix with silica glaze. The net effect is a case hardening of sandstone surfaces through a mixture of inherited silica glaze and iron and manganese leached out of varnish and washed into the rock (Fig. 7.5A, B, C).

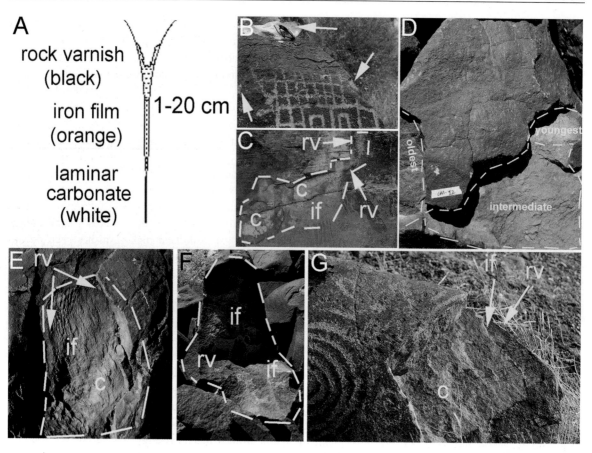

Fig. 7.4 Rock coatings formed within fractures in drier deserts. **A**: Generalized sequence of rock coatings found inside a still-closed desert rock fracture. **B–G**: Fissuresol sequences occur in all deserts and all rock types such as: **B**, granodiorite, southern Nevada; **C**. sandstone, southern Utah; **D** and **F**, basalt, eastern California; **E**, silicified dolomite, South Australia; **G**, hornfel, Sonoran Desert, Arizona. rv = rock varnish; if = iron film; c = carbonate skin. Horizonal photo dimensions are 0.7 m (B), 1.3 m (C), 0.4 m (D), 0.6 m (D), 2.5 m (F), and 0.6 m (G)

Third-Order Process: Lithobionts

Life coatings on rocks (lithobionts) include epiliths that live on the surface, euendoliths that bore tubes, chasmoendoliths occupying fissures, and cryptoendoliths that that live in weathering-rind pores (Golubic et al., 1981). Lithobionts can also be grouped into ≤1 mm biofilms, ~1–5 mm biorinds, and >5 mm biocrusts (Viles, 1995) that may be composed of bacteria, cyanobacteria, fungi, algae, and lichens. Although the general consensus in the past was that desert lithobiontic communities had low diversity, new methods reveal an astounding variety of organisms living on rock surfaces (Kuhlman et al., 2005; Benzerara et al., 2006; Fajardo-Cavazos and Nicholson, 2006). A single gram of rock varnish, for example, contains 10^8

microorganisms of *Proteobacteria*, *Actinobacteria*, eukaryota, and *Archaea* (Benzerara et al., 2006; Kuhlman et al., 2006).

A critical aspect of fast-growing lithobionts such as fungi and lichens is their capability of weathering inorganic rock coatings (Fig. 7.6A, B), as well as the underlying rock (Fig. 7.6B, C, D). By eroding the rock coating or its underlying substrate, lithobionts obtain possession of the surface.

Viles (1995: 32) modelled the weathering activity of lithobionts in conditions of varying moisture and rock hardness. A similar graphical presentation for rock varnish had considered growth also in terms of two simple factors, moisture and competition from lithobionts (Dorn and Oberlander, 1982). While these authors all acknowledge the simplicity of two variable perspectives, the interplay of lithobiont weathering

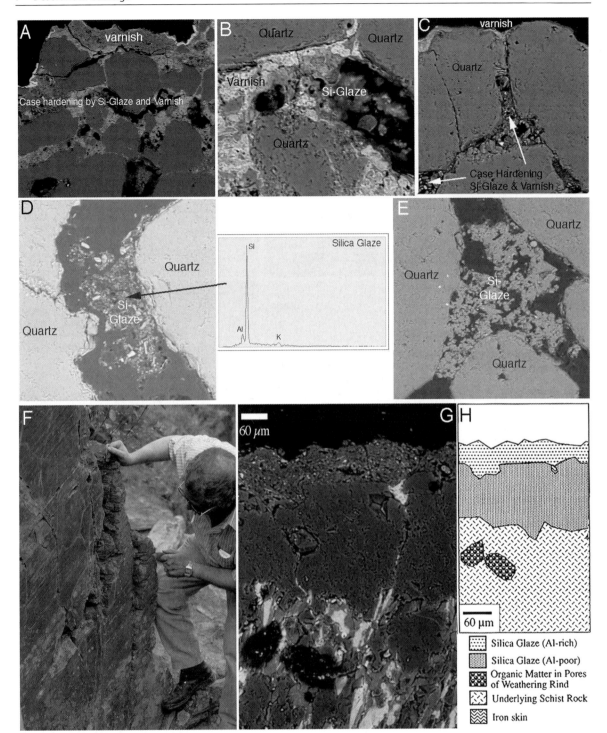

Fig. 7.5 Silica glaze rock coatings formed within fractures in semi-arid environments. Images **A–E** come from Wyoming sandstone cuestas and **F–G** from Portugal schist. Images **A–C** show how Mn and Fe leached from a very surface layer of rock varnish migrates into the sandstone, mixing with the pre-existing silica glaze to case harden the outer millimetres. Images **D** and **E** present silica glaze from an unopened joint face, collected ~40 cm up into a joint covered by an overlying sandstone block, but opened for sampling. A schist joint face in northern Portugal (**F**) hosts a fairly uniform layer of mostly silica under a clay-rich silica glaze (**G–H**). Images **A–E** and **G** are backscattered electron micrographs with image widths of: A (1800 μm); B (140 μm); C (140 μm); D (210 μm); E (260 μm)

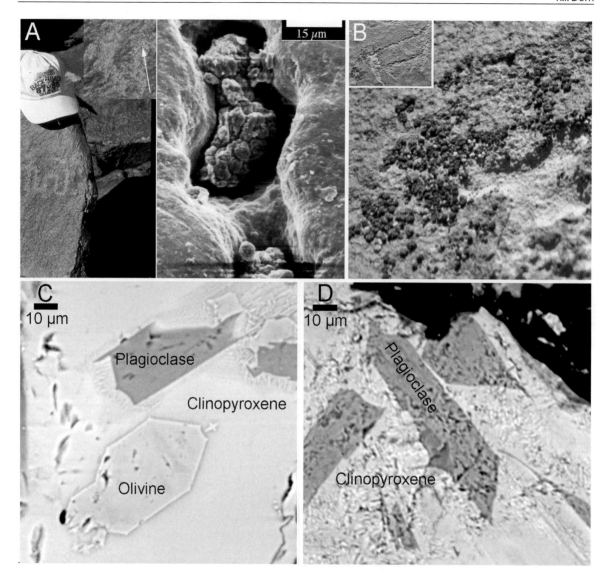

Fig. 7.6 Weathering activity of lithobionts. **A**. Varnished basalt boulder in the Mojave Desert with inset photo of petroglyph and showing location of the electron microscope image of a euendolith (tube boring) microcolonial fungi that is effectively dissolving rock varnish. **B**. Both silica glaze and rock varnish are being weathered and eroded by lichen growing on sandstone at Legend Rock, Wyoming. When the lichen is removed, the rock easily erodes, because the sandstone had become a silty powder under the lichen cover. **C** and **D** present backscattered electron microscopy imagery of weathering of lava flow f7dh7.9 on the desert side of Hualalai Volcano, Hawai'i. The less weathered sample (**C**) was collected away from lichens, and the sample with more porosity (dark 'holes' in **D**) was collected directly beneath *Stereocaulon vulcani* (image **D**)

and rock varnish growth can be understood at a general level by combining these two conceptualizations (Fig. 7.7).

Consider first the rock varnish perspective, viewed as the varnish rate deposition line in Fig. 7.7. In the drier deserts, varnish grows slowly, but subsurface lithobionts offer little competition. In conditions wetter than semi-arid South Australia, lichens and other epilithics reach a point that varnish erosion from the secretion of acids is faster than varnish formation.

Then, consider impact of lithobiont activity on rock varnish. The thinner lines in Fig. 7.7 represent the potential euendoliths that bore holes in varnish (Fig. 7.6A), biofilms of fungi, cyanobacteria, and

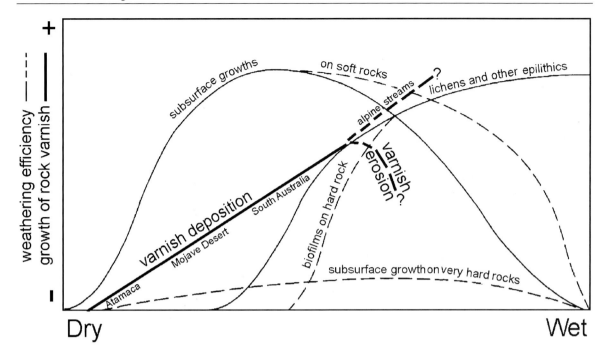

Fig. 7.7 Graphical conceptualization of how lithobionts and rock varnish interact together, adapted from Viles (1995) and Dorn and Oberlander (1982). The graph is a generalization of how moisture impacts the growth of rock varnish and the weathering efficiency of lithobionts. In the case of rock varnish, a second key factor other than moisture is competition from lithobionts. In the case of lithobionts, a second key factor is the hardness of the rock

algae that prevent varnish from growing, and lichen biorinds that secrete enough acid to completely dissolve away varnish (Figs. 7.3B and 7.6B). The efficiency of lithobionts in weathering both rock and varnish is highest on the wettest side of the graph, where varnishes are not found except under special conditions. The only circumstances where rock varnish can form in wetter environments are where epilithics have not yet colonized subglacial features, moraines and stream-side surfaces (Klute and Krasser, 1940; Whalley et al., 1990; Dorn, 1998).

Finally, examine the lithobiont's weathering efficiency in Fig. 7.7. Organisms impose three major 'styles' of biological weathering in their possession of rock surfaces: (1) epilithic biofilms and lichen-dominated surfaces; (2) endolithic-dominated; and (3) mixed biorinds (Viles, 1995). Epilithic biofilms and lichens dominate in the most mesic locales. Endolithic lithobionts can survive in extremely xeric settings. In the middle ground epilithic biofilms, epilithic lichens, and endolithic communities all mix together with varnish growth and varnish erosion. The hardness

or softness of the rock comes into play as a way of modelling how another factor can make life harder or easier on a lithobiont.

An important point must be made for those concerned about the impact of lithobionts on preservation of cultural resources such as rock art and stone monuments. The vast majority of the literature reveals overwhelming evidence of the weathering power of lithobionts (Jones and Wilson, 1985; Dragovich, 1986b; Cooks and Fourie, 1990; Viles, 1995; Banfield et al., 1999; Viles, 2001; Stretch and Viles, 2002; Souza-Egipsy et al., 2004; Gordon and Dorn, 2005b), and some may be tempted to remove this erosive force. However, it is often far better to leave lithobionts alone because of their ability to hold rock fragments in place (Gehrmann et al., 1988; Bjelland and Thorseth, 2002). Simply killing the lithobionts can end up 'releasing sediment behind a dam', allowing millimetres to centimetres of weathered fragments to erode in a short time (e.g. Fig. 7.6B). Another reason to why it is often better to leave lithobionts alone is because some have an ability to case harden surfaces (Viles and Goudie, 2004); others can help generate a protective coating of silica glaze, (Lee and Parsons, 1999); and still others help generate a protective coating of oxalate crust (Beazley et al., 2002; Souza-Egipsy et al., 2004). Thus, while it would be better for the stability of inorganic rock coatings and stone surfaces if lithobionts had never colonized, it is often far worse to intervene and use chemical or mechanical means to remove firmly-established lithobionts.

Fifth-Order Process: Rock Varnish

The most important aspect of understanding rock varnish formation was recognized more than two centuries ago, when the great enhancement of manganese (Mn) over iron (Fe) was first identified by Alexander von Humboldt. Mn is typically enhanced over Fe more than a factor of fifty above potential source materials such as dust, soils, water, and the underlying rock (von Humboldt, 1812; Lucas, 1905; Engel and Sharp, 1958; Dorn, 1998). The second most important varnish characteristic that must be explained is the dominance of clay minerals in rock varnish (Potter and Rossman, 1977; Potter and Rossman, 1979c; Dorn and Oberlander, 1982; Krinsley et al., 1990; Krinsley et al., 1995; Israel et al., 1997; Dorn, 1998; Krinsley, 1998; Diaz et al., 2002; Probst et al., 2002; Allen et al., 2004). Clay minerals make up about two-thirds of a typical subaerial varnish, Mn and Fe oxides a quarter, with several dozen minor and trace elements comprising the remainder of this black accretion that can grow to thicknesses exceeding 200 μm.

The composition of rock varnish must be tied to and explained by processes that fix Mn and Fe. Even though clay minerals abound in desert dust that falls on varnished rocks, it is the fixation of clays by Mn-minerals (Potter and Rossman, 1977; Potter and Rossman, 1979a; Potter and Rossman, 1979c) that explains why the dust remains cemented onto rock surfaces for millennia. Although minor and trace elements arrive already adsorbed to the desert dust, many are enhanced by the scavenging properties of Mn-Fe oxyhydroxides (Forbes et al., 1976), redistributed by wetting (Thiagarajan and Lee, 2004). Thus, all of the major, minor, and trace components of varnish depend on the biogeochemical barrier that fixes Mn and Fe.

Four general conceptual models have been proposed to explain varnish formation. The first model that saw general acceptance for almost a century invokes abiotic geochemical processes (Linck, 1901; Engel and Sharp, 1958; Hooke et al., 1969; Moore and Elvidge, 1982; Smith and Whalley, 1988) to increase Mn:Fe ratios two to three orders of magnitude above concentrations found in dust and rock material. Small pH/Eh fluctuations to more acid conditions dissolve Mn but not Fe (Krauskopf, 1957). The Mn released by slightly acidic precipitation is then fixed in clays after water evaporation or an increase in pH, as idealized in Fig. 7.8.

Although an abiotic geochemical barrier to Mn has not yet been falsified, there are a number of varnish characteristics that are incompatible with this model. First, varnishes are found in environments simply too wet and acidic to oxidize Mn (Dorn, 1998). Second, rock varnish is not very common in environments, such as coastal fog deserts or the rainshadows of Mauna Loa in Hawai'i, where repeated pH fluctuations would be at their maximum. Third, there is no extreme rate-limiting step in the abiotic model. Multiple dust deposition and carbonic acid wetting iterations take place annually, even in drought years. A bit of Mn leached from dust with each wetting event would generate varnish accretion hundreds to tens of thousands times faster than rates seen in typical varnishes (Dorn, 1998; Liu and Broecker, 2000; Liu and Broecker, 2007). While abiotic processes are involved in clay cementation and in trace-element enhancement, and while some abiotic oxidation may prove to be important in some locales, these and other characteristics of rock varnish are incompatible with an abiotic geochemical barrier to Mn transport.

The second general model holds that lithobionts or their organic remains produce and bind the constituents of varnish, including Mn. Lichens (Laudermilk, 1931; Krumbein, 1971), cyanobacteria (Scheffer et al., 1963; Krumbein, 1969), microcolonial fungi (Staley et al., 1982), pollen (White, 1924), peptides (Linck, 1928), refractory organic fragments (Staley et al., 1991), gram-negative bacteria (Drake et al., 1993; Sterflinger et al., 1999), gram-positive bacteria (Hungate et al., 1987), amino acids from gram-positive chemo-organotrophic bacteria (Nagy et al., 1991; Perry et al., 2004), fatty acid methyl esters (Schelble et al., 2005), a host of gram-negative Proteobacteria groups and Actinobacteria

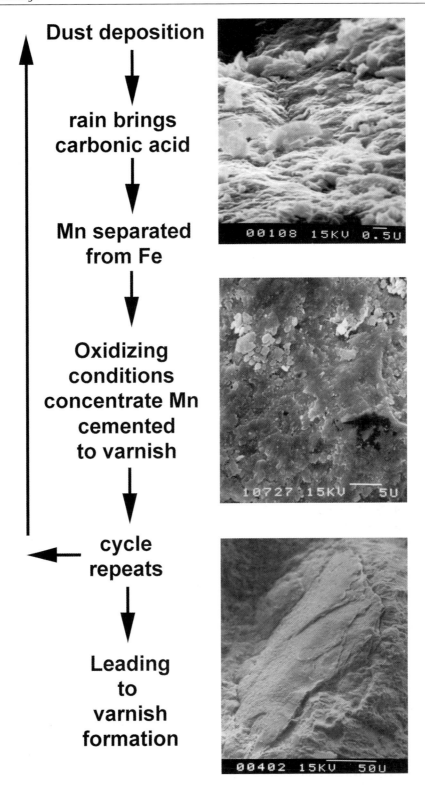

Fig. 7.8 The abiotic model of varnish formation postulates that acid solutions separate divalent Mn^{2+} from dust or tiny rock fragments that come to rest on surfaces. Then, oxidizing conditions trap the Mn^{4+} in varnish. This cycle would then repeat thousands of times to produce a 100 μm thick varnish

(Kuhlman et al., 2005), and a variety of microbial forms (Jones, 1991; Allen et al., 2004; Benzerara et al., 2006) are found on, in, or under rock varnish.

The vast majority of these lithobionts do not play a role in the fixation of Mn or Fe, and many of them actively erode varnish (Figs. 7.6 and 7.7). Because there are so very many adventitious organisms and organic remains associated with varnish, and because Mn-oxidizing organisms cultured from rock varnish (Krumbein and Jens, 1981; Palmer et al., 1985; Hungate et al., 1987) may not necessarily be those that form varnish, the only reliable evidence for a biotic origin of varnish must come from *in situ* observations of microbial forms coated with enhanced Mn and Fe (Dorn and Oberlander, 1981, 1982; Dorn et al., 1992; Dorn and Meek, 1995; Dorn, 1998; Krinsley, 1998). Such forms resemble budding bacteria that grow very slowly.

The most recent proposal to explain varnish formation is silica binding (Perry and Kolb, 2003; Perry et al., 2005; Perry et al., 2006) where silica is dissolved from dust and other mineral matter. The silica then gels and condenses, 'baking black opal in the desert sun' (Perry et al., 2006). This process, however, cannot produce rock varnish formation, since no aspect of the silica binding model explains either Mn enhancement in varnish or the birnessite-family minerals observed in varnish (Potter, 1979; Potter and Rossman, 1979a; Potter and Rossman, 1979b; McKeown and Post, 2001; Probst et al., 2001). The silica binding model also fails to explain slow rates of varnish accretion, since silica precipitation in silica glaze forms in years to decades, not millennia required by varnish. Other problems with silica binding include: not being able to explain the dominance of clay minerals; not accommodating the geography of rock coatings, being unable to answer the simple question of why would silica glazes dominate on Hawaiian lava flows (Curtiss et al., 1985; Gordon and Dorn, 2005b), but not rock varnish; and the 'baking' requirement fails to explain varnishes that grow in cold and dark places (Anderson and Sollid, 1971; Douglas, 1987; Dorn and Dragovich, 1990; Whalley et al., 1990; Dorn et al., 1992; Douglas et al., 1994; Villa et al., 1995; Dorn, 1998).

At the present time, a polygenetic model (Fig. 7.9) is the only proposed hypothesis that explains existing criteria (Table 7.2). As the polygenetic name suggests, rock varnish formation derives a combination of processes, where slow-growing bacteria fix Mn and Fe that is then abiotically cemented by clay minerals. Simply put, bacteria create the barrier to the movement of Mn. Bacteria concentrate Mn and Fe in fairly equal proportions in less alkaline times, but less Mn is enhanced in conditions of greater alkalinity. The geochemical barrier on bacterial casts breaks down over time, as acidic water slowly dissolves Mn and Fe. The resultant nanometre-scale granular fragments of Mn and Fe then move nanometres into the interstratified clay minerals deposited as dust on rock surfaces (Fig. 7.9). This process was predicted (Potter, 1979: 174–175) without benefit of the high resolution transmission electron microscopy imagery that showed the predicted steps of varnish formation (Dorn, 1998; Krinsley, 1998).

Fifth-Order Process: Silica Glaze

Hydrated silica (opal) accretes on the surfaces of rocks in all deserts (Stevenson, 1881; Hobbs, 1917; Jessup, 1951; Fisk, 1971; Haberland, 1975; Butzer et al., 1979; Farr and Adams, 1984; Bourman and Milnes, 1985; Watchman, 1985; Zhu et al., 1985; Smith and Whalley, 1988; Fullagar, 1991; Weed and Norton, 1991; Smoot, 1995). The mineralogy of the silica is most often x-ray amorphous (Curtiss et al., 1985), but some have noted silica minerals such as mogonite (Perry et al., 2006).

The general appearance varies quite a bit, ranging from almost transparent to opaque, from a charcoal black to ivory (Fig. 7.10A), and from dull to highly shiny (Fig. 7.10D). Thicknesses range from microns (Fig. 7.10E, D) to almost a millimetre, even on the same sample (Fig. 7.10F). Controls on thickness are not well understood, but they include the type of silica glaze, moisture conditions, and whether or not the silica glaze had experienced recent spalling along intra-glaze fractures. Over time, a silica glaze slow rock dissolution (Gordon and Dorn, 2005b) and silica movement into the underlying rock can case harden a weathering rind (Figs. 7.10B and 7.5A–C).

Silica glazes fall into six general categories (Dorn, 1998: 294–312), based on the abundance of non-silica constituents of iron, aluminum, and micron-sized bits of mineral detritus. Type I accretes as a fairly homogeneous and texturally uniform deposit of amorphous silica (Fig. 7.10B, C2; bottom glaze in Fig. 7.5G, H). Type II hosts a large amount of mineral detritus, where the silica acts as a glue for

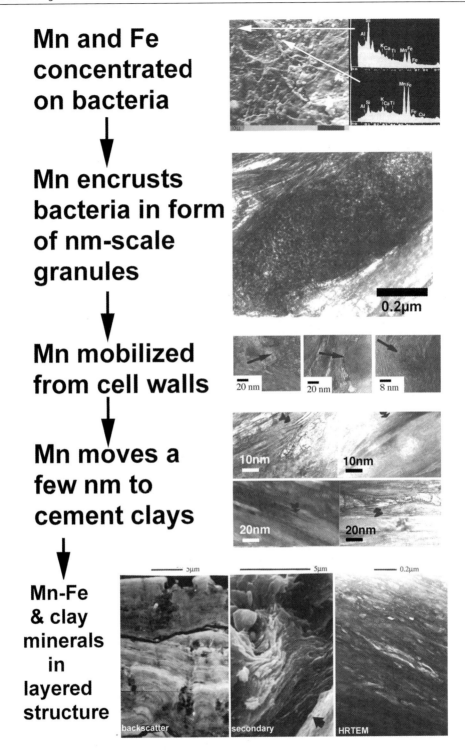

Fig. 7.9 The polygenetic model of rock varnish formation (Dorn, 1998; Krinsley, 1998) combines bacterial enhancement of Mn and Fe with abiotic fixation of the Mn by clay minerals. The process starts with bacteria fixing Mn on sheaths. Wetting events dissolve Mn, creating a granular nanometre-scale texture. The desert dust supplies interstratified clay minerals, and the nanometre-sized fragments of Mn-oxides fit into the weathered edges of these clay minerals, tightly cementing clays. The hexagonal arrangements of the oxygens in the tetrahedral or octahedral layers forms a template for the crystallization of layered birnessite, the Mn-mineral found most frequently in varnish. The net effect is highly layered texture at micrometre and nanometre scales imposed both by clay minerals and the cementing Mn-oxides

Table 7.2 Key criteria of rock varnish formation explained by polygenetic model

Criteria	The polygenetic model explains ...
Accretion Rate	... typical rates of accretion on the order of a 1-10 μm per millennia (Dorn, 1998; Liu and Broecker, 2000). Although faster-growing varnishes occur (Dorn and Meek, 1995), varnish accretion rates based on studies of over 10^5 microbasins (Liu, 2003; Liu, 2008; Liu and Broecker, 2007) demands the extreme rate-limiting step of budding bacteria concentrating Mn very slowly.
Clay Minerals	... the dominance of clay minerals in rock varnish (Potter and Rossman, 1977; Potter and Rossman, 1979c; Dorn and Oberlander, 1982; Krinsley et al., 1990; Krinsley et al., 1995; Israel et al., 1997; Dorn, 1998; Krinsley, 1998; Diaz et al., 2002; Probst et al., 2002; Allen et al., 2004), because the granular fragments of nanometre-Mn fragments from bacterial casts cement the clays to rock surfaces.
Fe behavior	... the differential enhancement of iron in different varnish layers (Liu et al., 2000; Broecker and Liu, 2001; Liu and Broecker, 2007) and different places (Adams et al., 1992; Dorn, 1998; Allen et al., 2004), because changes in alkalinity over time and space affect the ability of the bacteria to concentrate Mn.
Laboratory creation	... the creation of artificial varnish coatings by bacteria (Dorn and Oberlander, 1981; Krumbein and Jens, 1981; Dorn and Oberlander, 1982; Jones, 1991) may considered by some to be a vital criteria. However, given the extraordinary time scale jump between any laboratory experiment and natural varnish formation, and the extreme rate-limiting step involved in natural varnish formation, rigid application of this criteria may be problematic.
Lithobionts and organic remains	... the occurrence of different types of lithobionts and the nature of organic remains. The Mn-oxidizing bacteria actually making the varnish co-exist with these more abundant adventitious organisms, but the adventitious lithobionts are competitors.
Mn Enhancement	... the enhancement Mn typically more than a factor of fifty above potential source materials (von Humboldt, 1812; Lucas, 1905; Engel and Sharp, 1958), because the bacteria are seen *in situ* enhancing Mn.
Mn-mineralogy	... Mn-mineralogy characteristic of birnessite-family minerals (Potter, 1979; Potter and Rossman, 1979a; Potter and Rossman, 1979b; McKeown and Post, 2001; Probst et al., 2001), because the nanometre-scale fragments derived from bacteria fit well into the interstratified clays where they form layered phases such as birnessite.
Not just a few samples	... observations at sites around the world, because *in situ* enhancement of Mn-enhancing bacteria are seen globally (Dorn and Oberlander, 1982; Dorn et al., 1992; Dorn and Meek, 1995; Krinsley et al., 1995; Dorn, 1998; Krinsley, 1998; Spilde et al., 2002).
Paucity of microfossils	... the extremely infrequent occurrence of preserved microfossils, because the Mn-casts of bacteria are broken down by the varnish formation process Examination of 10^4 sedimentary microbasins (Liu, 2003; Liu, 2008; Liu and Broecker, 2007), and decades of research has generated only a few observations of microfossils (Dorn and Meek, 1995; Dorn, 1998; Krinsley, 1998; Flood et al., 2003), just what would be expected from the polygenetic model.
Rock Coating Geography	... why rock varnish grows in one place and other rock coatings elsewhere. Over a dozen major types of coatings form on terrestrial rock surfaces, a plethora of varieties for each rock coating. The polygenetic model explains this geography (Dorn, 1998).
Varnish Geography	... why different types of rock varnishes occur where they occur (Dorn, 1998).
Laminations (VML)	... the revolution in varnish microlamination (VML) understanding. Over ten thousand sedimentary microbasins analyzed by Liu (Liu, 1994; Liu and Broecker, 2000; Liu et al., 2000; Broecker and Liu, 2001; Liu, 2003; Liu, 2008; Liu and Broecker, 2007), a method subject to blind testing (Marston, 2003), reveals clear late Pleistocene and Holocene patterns in abundance of major varnish constituents connected to climate change. The characteristics of this single largest varnish data set are explained by the polygenetic model by changes in wetness altering the alkalinity of desert surfaces (Dorn, 1990; Drake et al., 1993; Diaz et al., 2002; Lee and Bland, 2003).

all of the bits and pieces of rock fragments coming to rest on a rock surface (Fig. 7.10C1; upper glaze in Fig. 7.5G, H). Silica is still dominant in Type III, but iron and aluminum concentrations are substantive, ranging from 5–40 oxide weight percent for FeO and 5–30% for Al_2O_3. Type III often has a dirty brown appearance and is found extensively in Australia (Watchman, 1992), in the Negev (Danin, 1985), in dryland Hawai'i (Curtiss et al., 1985), and in wetter climates (Matsukura et al., 1994; Mottershead and Pye, 1994). Types IV and V are about half silica, but aluminium is the only other dominant component in Type IV, while iron is the only other major constituent in Type V. Type VI glazes are dominated by Al_2O_3

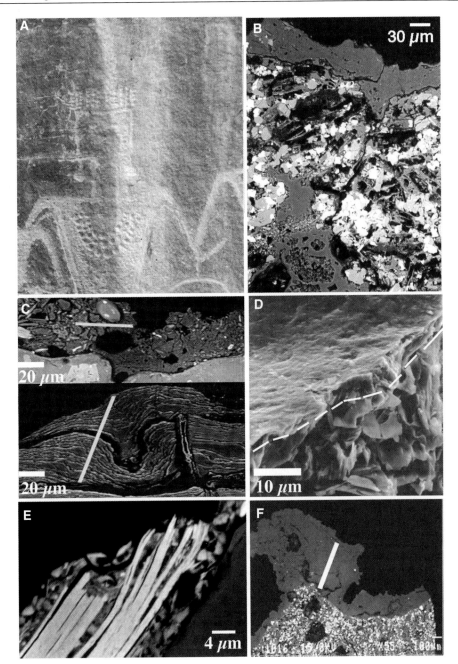

Fig. 7.10 (continued)

that sometimes reaches more than 50% by weight (Fig. 7.10F). These aluminium glazes are not well documented in the literature, and the reasons for the major enhancement of aluminium is not well understood.

Unlike rock varnish that typically accumulates only a few microns over a millennium, silica glazes can form very rapidly. Several different types have formed on historic lava flows in Hawai'i (Fig. 7.1) (Farr and Adams, 1984; Curtiss et al., 1985; Gordon and Dorn, 2005b). I have measured silica glaze formation within two years in the Owens Valley of eastern California (Fig. 7.10E) and within two decades on historic surfaces in the Mojave and Sonoran Deserts.

In addition to rapid formation rates, silica glaze also mechanically spalls along internal fractures. Thus, connecting thickness to age is extremely problematic.

Silica glazes form where there is a geochemical barrier to the migration of mobile silica. Since silica is ubiquitous in the host rock, dust, precipitation, groundwater seepage, and even opal phytoliths (Folk, 1978), there is no shortage of raw silica. Similarly, there is no great geochemical mystery in explaining silica movement to coated rock surfaces, since silica is easily dissolved in terrestrial weathering environments (Krauskopf, 1956).

Exactly how silica glaze is fixed to rock surfaces probably involves several different processes. The most common view is that monosilicic acid $(Si(OH)_4)$ precipitates as a gel (Krauskopf, 1956; Williams and Robinson, 1989), and experiments indicate that dissolved silica does precipitates as amorphous silica (Paraguassu, 1972; Whalley, 1978). Others argue for the importance of evaporation (Merrill, 1906; Fisk, 1971; Watchman, 1992) and complexing with organic matter (Watchman, 1992; Perry and Kolb, 2003; Perry et al., 2006). Still others have made arguments that different lithobionts can play a role in forming silica glazes (Fyfe et al., 1989; Urrutia and Beveridge, 1994; Lee and Parsons, 1999).

The above mechanisms are not mutually exclusive, and the great variety of silica glaze types (Dorn, 1998: Ch13) do argue for different processes creating different geochemical barriers to silica migration. Precipitation of silicic acid gels certainly makes sense for Type I silica glazes, for example, since the uniform texture and chemistry of these homogeneous glazes would be consistent with this simple silica-precipitation process.

Type II through Type VI silica glazes, however, call for processes able to explain variable concentrations of aluminium and iron. The explanation for abundant aluminium probably rests with soluble aluminium silicate complexes $(Al(OSi(OH)_3)^{2+})$. Soluble Al-Si complexes are ubiquitous at the water-rock interface (Lou and Huang, 1988; Browne and Driscoll, 1992), and they are easily released by weathering of phyllosilicate minerals (Robert and Tessier, 1992). The geochemical fixation of Al-Si complexes probably requires very gentle wetting events (Zorin et al., 1992) such as drizzle, as opposed to harsh convective storms. Once an initial silica glaze establishes itself, the silica acid or soluble Al-Si complex then more readily bonds to the pre-existing silica glaze (Casey et al., 1993). Other elements such as iron might be explained by strong adherence to silica surfaces through Fe-O-Si bonds (Scheidegger et al., 1993). The key to identifying processes responsible for the geochemical barrier rests in linking process to the type of silica glaze.

Fifth-Order Process: Other Coatings

The nature of this chapter does not permit a thorough explanation of the origin of every desert rock coating. Such information is presented in book form (Dorn, 1998). This section, however, summarizes explanations for the accumulation of four other desert rock coatings.

Iron films (Fig. 7.11A, D) are ubiquitous in and out of deserts, with at least three general categories. Type I iron films are mostly homogeneous iron with few other constituents (Fig. 7.11D). Type II iron films

---

Fig. 7.10 Silica glaze in different desert settings. **A.** This central Utah petroglyph carved into sandstone is covered by light and dark stripes of silica glaze. **B.** Backscattered electron microscope image of silica glaze on the boulder presented in Fig. 7.3E, collected from the arid rainshadow of Kaho'lawe Island, Hawai'i. Its composition of more than 90% silica moved through pore spaces into the underlying rock, filling pores and fractures. The net effect of case hardening is seen in Fig. 7.3E. **C.** Two backscatter images of silica glaze on a lava flow in the arid Ashikule basin, West Kunlun Mountains, Tibetan Plateau. Image C1 illustrates loess cemented by silica where the line indicates electron microprobe measurements of 50–60% SiO_2, 5–15% Al_2O_3, and 1–6% FeO by weight. Image C2 shows the layer-by-layer deposition of 80–85% SiO_2 with Al_2O_3 only ~5%. **D.** This flat surface, seen by secondary electrons, illustrates an ex-

tremely shiny, dark silica glaze collected from a Nasca trapezoid geoglyph, Peru. Its sheen derives from a combination of the smooth surface and from ~3% Mn. Silica glaze and manganese rock varnish often combine in the shiny black groundline band (Engel and Sharp, 1958) found on many desert pavement clasts. **E.** Silica glaze can form quite rapidly, in this case from the Owens Valley, eastern California within two years, freezing in place the weathered biotite mineral. **F.** This extremely thick accretion, collected from an opened joint face of Haleakala, Hawai'i, lava flow displays thicknesses ranging from <1 μm to almost a millimetre. The line on this backscattered image indicates the location of an electron microprobe transect along which Al_2O_3 values average 44%, but range from 19% all the way to 60% by weight (data table in Dorn, 1998: 311)

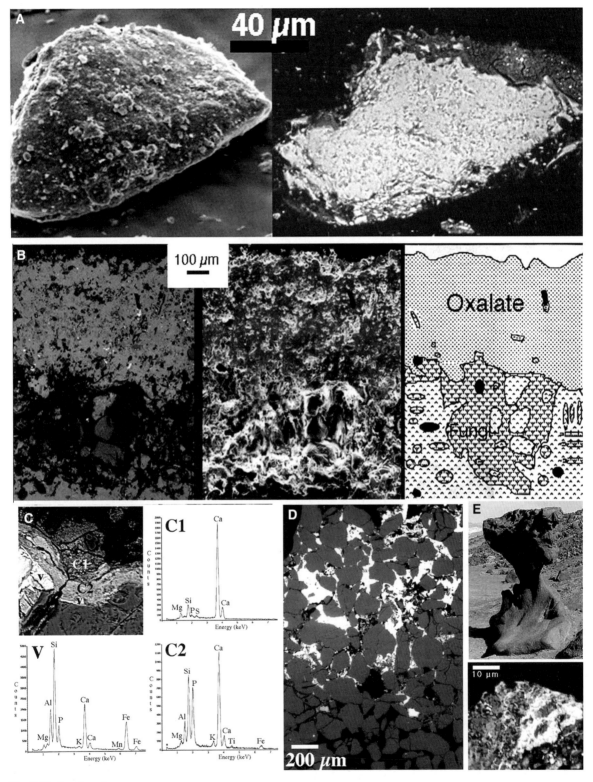

Fig. 7.11 (continued)

include aluminium and silicon as major elements, but still less abundant than iron. In Type III iron provides the orange to red coloration with concentrations of less than a third iron oxide by weight. The bulk of the coating comes from clay minerals (Fig. 7.11A).

The causes of iron films all involve biotic processes (Dorn and Oberlander, 1982; Adams et al., 1992; Konhauser et al., 1993; Schiavon, 1993; Konhauser et al., 1994; Dixon et al., 1995; Sterflinger et al., 1999; Fortin and Langley, 2005). Unlike rock varnish, however, no need exists for lithobionts to immobilize the iron. The barrier to iron movement on rock surfaces can simply be purely abiotic oxidation of iron, since the inorganic oxidation of Fe^{2+} to Fe^{3+} is rapid above a pH of 5 (Collins and Buol, 1970; Marshall, 1977; Holland, 1978). Desert rock surfaces typically have pH values well above 5 (Dorn, 1990). However, the processes behind iron film fixation are probably far more complex, perhaps involving the formation of chemical Fe-O-Si bonds in Type II iron films (Hazel et al., 1949; Scheidegger et al., 1993), perhaps involving photooxidation (McKnight et al., 1988) in Type I iron films, and likely involving interaction with interstratified clays in Type III iron films much like rock varnish (cf. Fig. 7.9).

Oxalate crusts (Figs. 7.1 and 7.11B) are far less common than any of the aforementioned rock coatings, and they are more frequently found in wetter microenvironments such as locales of water flow. Oxalate minerals include carbon, oxygen and a divalent cation, such as magnesium, calcium, or manganese. Whewellite ($CaC_2O_4 \bullet H_2O$) is the most common mineral. Oxalate crusts can vary considerably in appearance, including white, yellow, orange, red, red-brown, brown or black colours. Thickness also ranges considerably, from microns to a few millimetres. Al-

though the carbon can come from inorganic materials (Zák and Skála, 1993) and a host of plant and microbial sources (Lowenstam, 1981; Lapeyrie, 1988; Watchman, 1990; Cariati et al., 2000; Zhang et al., 2001), most of the oxalate found in deserts likely derives from the decay of lichens that synthesize the oxalate (Del Monte et al., 1987; Whitney and Arnott, 1987; Russ et al., 1996; Bjelland et al., 2002; Bjelland and Thorseth, 2002; Souza-Egipsy et al., 2004). After oxalate minerals crystallize, two additional steps are still required to form desert oxalate crusts. First, the oxalate must be mobilized away from the source, most typically by water flowing over a rock face away from lichens. Second, the oxalate must re-precipitate on desert rock surfaces; it is this last step that has eluded an explanation more detailed than evaporation.

Carbonate crusts (Figs. 7.2B and 7.11C) coat desert rocks in a variety of settings, including freshwater tufa, travertines and other carbonate deposits (Viles and Goudie, 1990; Carter et al., 2003), caves (Fyfe, 1996), lake shorelines (Benson, 1994), tropical beaches (Krumbein, 1979), subaerial rock faces mixed with clays and silica (Conca and Rossman, 1982; Conca, 1985; Dorn, 1998), and soils (Goudie, 1983). The mechanism of carbonate fixation varies greatly depending on environment, but both biotic (Krumbein, 1979; Viles and Goudie, 1990; Folk, 1993; Rodriguez-Navarro et al., 2003) and abiotic (Vardenoe, 1965; Gile et al., 1966; Bargar, 1978; Dandurand et al., 1982; Dunkerley, 1987; Benson, 1994) processes are invoked as key mechanisms in creating a geochemical barrier to carbonate movement.

Salt crusts (Fig. 7.11E) also appear as rock coatings in certain desert settings (Oberlander, 1988), but particularly associated with efflorescence (Goudie and Cooke, 1984) on porous rock surfaces (Smith, 1994;

Fig. 7.11 Other types of rock coatings. **A.** Iron film illustrated by secondary (*left*) and backscatter (*right*) electron microscope images of an orthoclase feldspar sand grain from the bright orange Parker Dune Field, western Arizona. Clay particles are cemented to sand grains by iron oxides. **B.** Oxalate crust illustrated by backscattered (*left*) and secondary (*middle*) electron microscope images of a sandstone face in the semi-arid Black Hills, Wyoming. Small pockets of rock varnish also occur, and a fibrous fungal mat has grown in the weathering rind underneath the rock coating. **C.** Carbonate crust (C1 and C2) that formed on top of a rock varnish (v), but only after a boulder had been flipped to form a geoglyph at least 8440±60 ^{14}C years ago, according

to a radiocarbon age on the more silica-rich (C2) inner carbonate crust (Cerveny et al., 2006). **D.** Iron film that impregnated sandstone at Petra, Jordan (Paradise, 2005). The iron film seen as bright white in this backscattered electron microscope image is a rock coating, but it also acts as a case hardening agent as the iron remobilizes and impregnates the host sandstone. **E.** At Mushroom Rock (E1), Death Valley, salt exists as efflorescent coatings seen on the surface, and as subflorescent deposits, as shown in the backscattered electron micrograph of barite (bright white) invading host basalt (mostly plagioclase) minerals (E1). Halite (NaCl) and celestite ($SrSO_4$) also coat surfaces on Mushroom Rock (Meek and Dorn, 2000)

Smith and Warke, 1996). Sulfates encrust desert surfaces (Goudie, 1972; Watson, 1988; Drake et al., 1993; White, 1993b). Nitrate can also coat rock surfaces (Mansfield and Boardman, 1932; Ericksen, 1981), as can phosphate skins (Trueman, 1965; Zanin, 1989; Konhauser et al., 1994; Arocena and Hall, 2003) (Fig. 7.3D).

Use of Rock Coatings as a Chronometric Tool

Desert geomorphologists and geoarchaeologists have long used rock coatings as indicators of the antiquity stone surfaces (Oberlander, 1994). Visual changes in such features as alluvial-fan sequences (Fig. 7.2B) (McFadden et al., 1989; Bull, 1991), the undersides of desert pavement clasts (Helms et al., 2003), inselberg debris slopes (Oberlander, 1989), glacial moraines (Staiger et al., 2006), and mass wasting (Moreiras, 2006) have all led to an intuitive belief that rock coatings can be used as way of obtaining minimum ages for erosive processes that 'wiped clean' prior rock coatings.

Very few investigators utilizing rock coatings as a chronometric tool, however, have written about the possibility that they may be including in their analyses 'inherited' rock coatings or that they may be sampling completely different types of coatings that look similar (McFadden et al., 1989; Harry et al., 1992; Reneau, 1993; Perry et al., 2006). I only present here chronometric tools that are clearly constrained by a landscape geochemistry perspective, grounded in hierarchical rock coating processes.

Rock Varnish

Rock varnish has been studied more extensively than any other rock coating, including more than a century of exploration on its use as a possible method to date desert landforms (Dorn, 1998: Ch 10). Several different dating methods have been proposed (Table 7.3). Yet up until only the last few years, all such proposed techniques have been highly experimental — tried in only a few selected circumstances and only rarely subjected to blind testing (Loendorf, 1991; Marston, 2003). To turn any dating method into a technique that can be practised widely requires the study of thousands of samples.

This level of extensive research has recently taken place only for the use of rock varnish microlaminations (VML) as a chronometric and palaeoclimatic research tool in desert geomorphology. The VML of orange (Mn-poor) and black (Mn-rich) layers accreted in subaerial varnishes (Perry and Adams, 1978; Dorn, 1990; Cremaschi, 1996; Leeder et al., 1998) can now be used with regularity and consistency. This revolution in varnish dating took detailed analyses of over ten thousand sedimentary microbasins studied through a decade of painstaking scholarship, laboratory work, and testing by Tanzhuo Liu (Liu, 1994; Liu and Dorn, 1996; Liu and Broecker, 2000; Liu et al., 2000; Zhou et al., 2000; Broecker and Liu, 2001; Liu, 2003; Marston, 2003; Liu, 2008; Liu and Broecker, 2007; Liu and Broecker, 2008a, 2008b). A complete discussion of the varnish microlaminations revolution for desert geomorphology is summarized in Chapter 21.

Some other rock varnish dating methods (Table 7.3) might also reach this stage of regular and consistent use. Yet, that next step would similarly require the level of funding and painstaking dedication achieved for the VML method.

Carbonate

Carbonate crusts are used extensively as a chronometric tool in deserts. Tufa crusts are used to radiocarbon date palaeoshorelines (Benson et al., 1995). Pedogenic carbonate crusts, with an awareness of confounding factors, can inform on palaeoclimate (Monger and Buck, 1999) and can date soils through radiocarbon, uranium-series, and its gradual accumulation (Gile et al., 1966; Machette, 1985; Chen and Polach, 1986; Bell et al., 1991; Amundson et al., 1994). Pedogenic carbonate can also serve as a vessel for accumulating cosmogenic ^{36}Cl (Liu et al., 1994).

The major difficulty in using carbonate crusts derives from uncertainties surrounding a vital assumption that the sampled carbonate deposit is 'closed' to postdepositional modification. Virtually all carbonate crust chronometric methods require that nothing happens to the carbonate minerals once they are deposited. Yet, because carbonate is extremely mobile in the terres-

Table 7.3 Different methods that have been used to assess rock varnish chronometry

Method	Synopsis of method
Accumulation of Mn and Fe	As more varnish accumulates, the mass of manganese and iron gradually increases. Occasionally this old idea is resurrected (Lytle et al., 2002), but it has long ago been demonstrated to yield inaccurate results in tests against independent control (Bard, 1979; Dorn, 2001).
Appearance	The appearance of a surface darkens over time as varnish thickens and increases in coverage. However, much of this darkening has to do with exposure of inherited coatings, and with the nature of the underlying weathering rinds, that do not permit accurate or precise assignment of ages based on visual appearance. There is no known method that yields reliable results.
Cation-ratio dating	Rock varnish contains elements that are leached (washed out) rapidly (Dorn and Krinsley, 1991; Krinsley, 1998). Over time, a ratio of leached to immobile elements decline over time (Dorn, 2001). If the correct type of varnish is used, the method performs well in blind tests (Loendorf, 1991). This method has also seen use in such places as China (Zhang et al., 1990), Israel (Patyk-Kara et al., 1997), and South Africa (Whitley and Annegarn, 1994), Yemen (Harrington, 1986) and elsewhere.
Foreign Material Analysis	Rock carvings made historically may have used steel. The presence of steel remains embedded in a carving would invalidate claims of antiquity, whereas presence of such material as quartz would be consistent with prehistoric antiquity (Whitley et al., 1999).
Lead Profiles	20th century lead and other metal pollution is recorded in rock varnish, because the iron and manganese in varnish scavenges lead and other metals. This leads to a 'spike' in the very surface micron from 20th century pollution. Confidence is reasonably high, because the method (Dorn, 1998: 139) has been replicated (Fleisher et al., 1999; Thiagarajan and Lee, 2004; Hodge et al., 2005; Wayne et al., 2006) with no publications yet critical of the technique that can discriminate 20th century from pre-20th century surfaces.
Organic Carbon Ratio	Organic carbon exists in an open system in the rock varnish that covers petroglyphs. This method compares the more mobile carbon and the more stable carbon. The method is best used in soil settings (Harrison and Frink, 2000), but it has been applied experimentally to rock varnish in desert pavements (Dorn et al., 2001).
^{14}C carbonate	Calcium carbonate sometimes forms over varnish, and can be radiocarbon dated, providing a minimum age for such features as rock art. The method has been used in Australia (Dragovich, 1986a) and eastern California (Smith and Turner, 1975; Cerveny et al., 2006).
^{14}C organic	The hope is that carbon trapped by coating provides minimum age for the petroglyph. First developed in 1986, two independent investigators working in a blind test (Dorn, 1997; Watchman, 1997) both found organic carbon that pre-dates and post-dates the exposure of the rock surface. The only person who still uses organic carbon of unknown residues in radiocarbon dating (Watchman, 2000; Huyge et al., 2001), Watchman now admits that he has not tested results against independent controls (Watchman, 2002; Whitley and Simon, 2002a; Whitley and Simon, 2002b).
^{14}C oxalate	The inorganic mineral oxalate (e.g., whewellite: $CaC_2O_4 \cdot H_2O$) sometimes deposits on top of or underneath rock varnish (Watchman et al., 2000). Because this mineral contains datable carbon, the radiocarbon age can provide a minimum age for the underlying or overlying varnish. The most reliable research on radiocarbon dating of oxalates in rock surface contexts has been conducted in west Texas (Rowe, 2001; Spades and Russ, 2005) and in a rock art shelter (Watchman et al., 2005).
Uranium-series dating	Since radionuclides are enhanced in varnish (Marshall, 1962), uranium-series isotopes show potential (Knauss and Ku, 1980). Complications surround acquiring the necessary amount of material from the basal layers and concerns over accounting for the abundant thorium that derives from clay detritus instead of radioactive decay.
Laminations (VML)	Climate fluctuations change the pattern of varnish microlaminations (VML). The confidence level is high, because the method (Liu, 2003; Liu, 2008; Liu and Broecker, 2007) has been replicated in a rigorous blind test (Marston, 2003), and the method is based on analyses of over ten thousand rock microbasins.

trial weathering environment, closed systems are rare (Stadelman, 1994). Some of the most stable components of carbonate crust are not actually carbonate, but silica associated with and sealing carbonate (Ludwig and Paces, 2002) and playing a role in generating laminations that are better for dating (Wang et al., 1996).

The great potential of carbonate crusts to serve as a geomorphic tool, depends on the identification of those microsettings where silica helps generate a less open system. This is illustrated where carbonate crusts undergo a change in environment such that they actually interlayer with rock varnish. Radiocarbon dating carbonate superimposed over rock varnish (Smith and Turner, 1975; Dragovich, 1986a) was originally thought to be a very rare possibility. However, both anthropogenic and natural processes can flip boulders, generating dating potential when the subaerial position of rock varnish and pedogenic position of carbonate crusts are inverted (Cerveny et al., 2006). An example comes from rock cairns (cf. Fig. 7.2B) where carbonate crusts are exposed to the atmosphere and also where formerly varnished surfaces are thrust deep enough into the soil to form a carbonate crust (Fig. 7.11C). Dating the carbonate formed over the varnish provides a minimum age for the rotation, and the most reliable minimum ages for this flipping process derives from the silica-rich laminated carbonate crust (Fig. 7.11C2) (Cerveny et al., 2006).

Oxalate

Oxalate crusts (Fig. 7.11B) offer tremendous potential for geomorphic research, because oxalate is a carbon-bearing mineral. Both radiocarbon dating (Russ et al., 1990; Watchman et al., 2005) and stable carbon isotope palaeoclimatic analyses (Russ et al., 2000; Beazley et al., 2002) have yet to reveal the problems of an open system associated with carbonate. Although oxalate dating has yet to be used as a geomorphic tool, having been tried mostly on oxalate crusts formed over rock paintings, ^{14}C dating provides opportunity to study such topics as mass wasting when an oxalate-streaked face topples and is buried by cliff retreat (cf. Fig. 7.1).

Lithobionts

Measuring the progressive growth of lichens or lichenometry has been used extensively in Arctic and alpine settings (Lock et al., 1979; Matthews and Shakesby, 1983; Worsley, 1990). Most investigators focus on the largest lichen, assuming that its size indicates the age of colonization — and ideally substrate exposure. Comparing size against a dating curve yields a calibrated age. Although many lichenometry researchers interpret these dating curves in terms of an initial rapid juvenile growth, biologists believe that the 'great growth' is explained by a high mortality rate of early colonists (Loso and Doak, 2006). This difference in the interpretation of empirical curves does not deny the proven utility of this method in cool-wet regions where epilithic lichens thrive (Fig. 7.7).

Deserts lack moisture, placing a severe limitation on the activity of lichens (Fig. 7.7). Thus, lichenometry in drier regions has only seen speculative use, most often focused in archaeology (Joubert et al., 1983). Some work has been completed in semi-arid areas, such as basalt terraces in Lesotho, where methods were modified to study lichen cover on scarp faces (Grab et al., 2005). Lichen growth on rock falls in semi-arid eastern California has been used to infer past tectonic events (Bull, 1995). However, lichenometry has not seen substantive use in desert geomorphology.

Since moisture and rock hardness influence the weathering efficiency of lithobionts (Viles, 1995), it is a logical next step to infer that climatically distinct patterns of weathering might relate to particular climates — assuming that rock type can be controlled. Extensive research in Israel, controlling carbonate lithology, reveals distinctive weathering patterns generated by the different lithobionts of endolithic lichens, epilithic lichens, and cyanobacteria (Fig. 7.12) (Danin et al., 1982; Danin, 1983; Danin and Garty, 1983; Danin et al., 1983). Contemporary lithobiontic weathering in the driest parts of the Negev, for example, only creates small-scale pitting from microcolonial fungi and cyanobacteria, but there are also 'puzzle pattern' weathering features on these hyperarid carbonate rocks (Fig. 7.12). These puzzle patterns match forms produced by lichens that only grow in wetter parts of Israel. The conclusion reached was that a former wetter climates fostered the growth of and weathering efficiency of these more mesic lithobionts in the heart of the Negev (Danin, 1985; Danin, 1986). Thus, there exists substantive potential of weathering patterns to map out palaeoclimatic boundaries of desert weathering efficiencies.

Future Research Directions

Academic roots have greatly influenced methods, conclusions and the overall agenda of previous desert rock coating research. For example, the geological focus of researchers in the late 19th and early 20th century on rocks led them to favour the incorrect hypothesis that coatings were 'sweated' out of the underlying rock and 'baked' on rock surfaces. It was not until electron microscopes showed incredibly distinct contacts

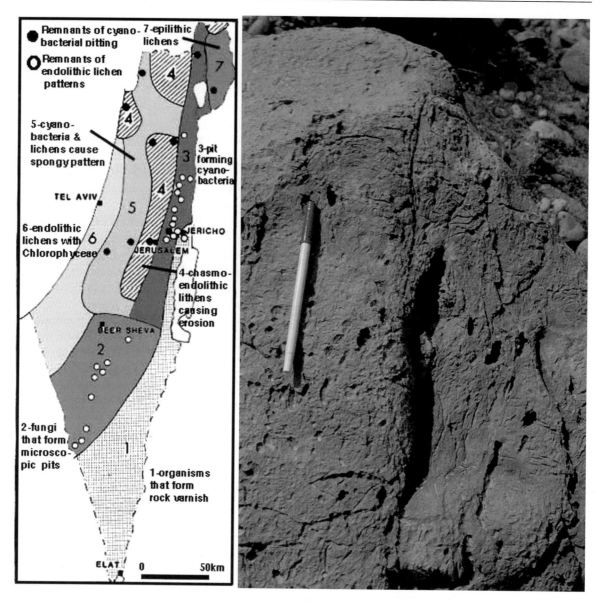

Fig. 7.12 Generalized map of the distribution of dominant lithbiontic weathering in Israel (modified from Danin 1986: 245). Rock varnish forming bacteria dominate in the driest parts of Israel, but as moisture increases so does the pattern of carbonate weathering — all the way to far northern Israel where epilithic lichens dominate. The photograph a southern Negev Desert carbonate boulder shows centimetre-scale pitting and 'puzzle pattern' weathering forms generated by long-dead and absent lithobionts made under a wetter climate. Sub-millimetre pitting by microcolonial fungi and cyanobacteria is the contemporary lithobiont weathering process slowly erasing the palaeoforms

between coating and rock (Potter and Rossman, 1977) that the almost universal accretionary nature of rock coatings came to be fully recognized (Dorn, 1998). Microbial ecologists, in contrast, have concentrated on culturing organisms and studying organic remains in association with rock coatings, with a concomitant natural tendency to downplay mineralogically driven processes.

Funding has also influenced rock coating research. In an example of the tail of cash wagging the dog of research, NASA's agenda to search for life Mars has driven an explosion of interest in rock varnish

(DiGregorio, 2002; Gorbushina et al., 2002; Johnson et al., 2002; Mancinelli et al., 2002; Allen et al., 2004; Kuhlman et al., 2005; Spilde et al., 2005; Perry et al., 2006; Perry and Lynne, 2006). This is nothing new, since funding and sociological concerns have often driven 'normal science' (Fuller, 2000) between paradigm shifts.

It is most unlikely that the most significant rock coating research in the future will rest with targeted agency agendas or within a single disciplinary perspective. Thus, I list the top four interdisciplinary research agendas that I think would have the greatest potential impact on rock coatings research in desert.

(1) *Use varnish microlaminations (VML) to answer the difficult desert geomorphology questions.* Tanzhuo Liu has just finished a decade of technique development that allows desert geomorphology researchers to date Holocene (Liu, 2008; Liu and Broecker, 2007) and late Quaternary (Liu and Broecker, 2008a, 2008b) landforms. In contrast to cosmogenic nuclides that suffer from high inherent costs and concerns over the 'inheritance' of nuclide build-up in any transported sediment (Robinson, 2002; Cockburn and Summerfield, 2004), VML can be used to tackle virtually any landform that hosts subaerial varnish. There are a host of classic and unsolved desert research questions ripe for answering through VML.

(2) *The study of rock coatings in the context of priceless rock art.* There can be little doubt that anthropogenic factors and natural erosion continue to result in the destruction of countless numbers of rock art engraved or painted on desert rock surfaces (ICOMOS, 2000; Bertilsson, 2002; J.PaulGettyTrust, 2003; Varner, 2003; Keyser et al., 2005). Many laboratory scientists favour an interventionist strategy to preserve art by treatments such as organosilicone-polyurethanes (Puterman et al., 1998), acrylic copolymers (Brugnara et al., 2004), polymeric membranes (Drioli et al., 1995), *in situ* polymerization (Vicini et al., 2004), and intrapore precipitation of calcite (Tiano, 2004). Field scientists, in contrast, do not generally advocate active intervention by subtracting lithobionts or adding stabilizing agents (Dolanski, 1978; Pope, 2000; Zhang et al., 2001; Pope et al., 2002; DeAngelis et al., 2003; Tratebas

et al., 2004). There is also a major perceptual difference between those studying building stones, who start with the premise that the host rock is unweathered stone from a quarry, and those studying natural rock art, who start with the premise that the host rock is already in a state of decay. Unfortunately, there is very little basic research on the role of rock coatings in the stabilization of or weathering of this priceless art. Much more research is needed, for example, on processes by which coatings stabilize stone surfaces by oxalate (Del Monte et al., 1987; Zhang et al., 2001), rock varnish (Gordon and Dorn, 2005a), heavy metals (Tratebas et al., 2004), silica glaze (Gordon and Dorn, 2005b), or simply understanding the spatial context of weathering and rock coatings (Barnett et al., 2005; Wasklewicz et al., 2005).

(3) *Map the geography of rock coatings.* The normal strategy in rock coating research has been to utilize a microanalytical technique at a few sites, or even to bring to bear a suite of expensive tools at just a single site. Oberlander (1994: 118) emphasized that 'researchers should be warned against generalizing too confidently from studies of single localities.' Even though a spatial perspective on geochemistry helped prevent dead-ends in geochemical research (Perel'man, 1961, 1966, 1967; Fortescue, 1980; Perel'man, 1980), very few mapping studies of rock coatings have yet to be conducted (Danin, 1986; White, 1990; Christensen and Harrison, 1993; White, 1993a; Dorn, 1998; Palmer, 2002). Successful models for rock coating research must be able to explain simple geographical questions such as why, for example, rock varnish grows with iron films, silica glaze, phosphate skins, and oxalate crusts in the Khumbu of Nepal (Dorn, 1998: 360–361) and with dust films, carbonate crust, phosphate film, silica glaze, and oxalate crusts in Tibet (Dorn, 1998: 367–369). The recent paper entitled 'baking black opal in the desert sun: the importance of silica in desert varnish' postulated an untenable model for varnish formation simply by not considering the geography of varnish in locales with little or no light and heat (Anderson and Sollid, 1971; Douglas, 1987; Dorn and Dragovich, 1990; Whalley et al., 1990; Dorn et al., 1992; Douglas et al., 1994; Villa et al., 1995; Dorn, 1998).

(4) Falsify abiotic hypotheses, if possible. A general uncertainty envelopes research on the genesis of almost all rock coatings. There are often both abiotic and biotic processes capable of creating a geochemical barrier to fix manganese, iron, phosphate, carbonate and other coating constituents on rock surfaces. Even though the preponderance of evidence might favour a biotic mechanism, as is the case in rock varnish, the role of abiotic fixation simply cannot be ruled out at the present time. This uncertainty comes home as a giant problem if a rock coating is to be used as an indicator of ancient life on Earth (Crerar et al., 1980; Dorn and Dickinson, 1989), or on Mars (DiGregorio, 2002; Gorbushina et al., 2002; Mancinelli et al., 2002; Allen et al., 2004; Kuhlman et al., 2005; Spilde et al., 2005; Perry et al., 2006; Perry and Lynne, 2006). Just as the use of varnish as a bioindicator of ancient life on Mars is untenable until abiotic origins are falsified for Martian conditions, a clever strategy to falsify (or confirm) abiotic origins on Earth would similarly aid terrestrial research.

Acknowledgments Thanks to Arizona State University for providing sabbatical support and to the late James Clark for his brilliance on the microprobe.

References

Adams, J. B., Palmer, F., and Staley, J. T., 1992, Rock weathering in deserts: Mobilization and concentration of ferric iron by microorganisms, *Geomicrobiology Journal* **10**: 99–114.

Allen, C., Probst, L. W., Flood, B. E., Longazo, T. G., Scheble, R. T., and Westall, F., 2004, Meridiani Planum hematite deposit and the search for evidence of life on Mars — iron mineralization of microorganisms in rock varnish, *Icarus* **171**: 20–30.

Amundson, R., Wang, Y., Chadwick, O., Trumbore, S., McFadden, L., McDonald, E., Wells, S., and DeNiro, M., 1994, Factors and processes governing the C-14 content of carbonate in desert soils, *Earth and Planetary Sciences Letters* **125**: 385–305.

Anderson, J. L., and Sollid, J. L., 1971, Glacial chronology and glacial geomorphology in the marginal zones of the glaciers, Midtadlsbreen and Nigardsbreen, South Norway, *Norsk Geografisk Tidsskrift* **25**: 1–35.

Arocena, J. M., and Hall, K., 2003, Calcium phosphate coatings on the Yalour Islands, Antarctica: Formation and geomorphic implication, *Arctic Antarctic and Alpine Research* **35**: 233–241.

Banfield, J. F., Barker, W. W., Welch, S. A., and Taunton, A., 1999, Biological impact on mineral dissolution: Application of the lichen model to understanding mineral weathering in the rhizosphere, *Proceedings National Academy of Sciences* **96**: 3404–3411.

Bard, J. C., 1979, The development of a patination dating technique for Great Basin petroglyphs utilizing neutron activation and X-ray fluorescence analyses. Ph.D. Dissertation, Dissertation thesis, 409 pp., University of California, Berkeley.

Bargar, K. E., 1978, Geology and thermal history of Mammoth Hot Springs, Yellowstone National Park, Wyoming, *U.S. Geologicay Survey Bulletin* **1444**: 1–55.

Barnett, T., Chalmers, A., Díaz-Andreu, M., Longhurst, P., Ellis, G., Sharpe, K., and Trinks, I., 2005, 3D laser scanning for recording and monitoring rock art erosion, *INORA* **41**: 25–29.

Beazley, M. J., Rickman, R. D., Ingram, D. K., Boutton, T. W., and Russ, J., 2002, Natural abundances of carbon isotopes (^{14}C, ^{13}C) in lichens and calcium oxalate pruina: Implications for archaeological and paleoenvironmental studies, *Radiocarbon* **44**: 675–683.

Bell, J. W., Peterson, F. F., Dorn, R. I., Ramelli, A. R., and Ku, T. L., 1991, Late Quaternary surficial geology in Crater Flat, Yucca Mountain, southern Nevada, *Geological Society America Abstracts with Program* **23(2)**: 6.

Benson, L., 1994, Carbonate deposition, Pyramid Lake Subbasin, Nevada: 1. Sequence of formation and elevational distribution of carbonate deposits (tufas), *Palaeogeography, Palaeoecology, Palaeoclimatology* **109**: 55–87.

Benson, L., Kashgarian, M., and Rubin, M., 1995, Carbonate deposition, Pyramid Lake subbasin, Nevada: 2. Lake levels and polar jet stream positions reconstructed from radiocarbon ages and elevations of carbonates (tufas) deposited in the Lahontan basin, *Palaeogeography, Palaeoclimatology, Palaeoecology* **117**: 1–30.

Benzerara, K., Chapon, V., Moreira, D., Lopez-Garcia, P., Guyot, F., and Heulin, T., 2006, Microbial diversity on the Tatahouine meteorite, *Meteoritics & Planetary Science* **41**: 1259–1265.

Bertilsson, R., 2002, Rock art at Risk, *http://www.international. icomos.org/risk/2002/rockart2002.htm* **(accessed 3/8/05)**.

Bjelland, T., Saebo, L., and Thorseth, I. H., 2002, The occurrence of biomineralization products in four lichen species growing on sandstone in western Norway, *Lichenologist* **34**: 429–440.

Bjelland, T., and Thorseth, I. H., 2002, Comparative studies of the lichen-rock interface of four lichens in Vingen, western Norway, *Chemical Geology* **192**: 81–98.

Bourman, R. P., and Milnes, A. R., 1985, Gibber plains, *Australian Geographer* **16**: 229–232.

Broecker, W. S., and Liu, T., 2001, Rock varnish: recorder of desert wetness?, *GSA Today* **11 (8)**: 4–10.

Browne, B. A., and Driscoll, C. T., 1992, Soluble aluminum silicates: stoichiometry, stability, and implications for environmental geochemistry, *Science* **256**: 1667–1669.

Brugnara, M., Degasperi, E., Volpe, C. D., Maniglio, D., Penati, A., Siboni, S., Toniolo, L., Poli, T., Invernizzi, S., and Castelvetro, V., 2004, The application of the contact angle in monument protection: new materials and methods, *Colloids and Surfaces A: Physicochemical and Engineering Aspects* **241**: 299–312.

Bull, W. B., 1991, *Geomorphic Responses to Climatic Change*, Oxford University Press, Oxford, 326pp.

Bull, W. B., 1995, Dating San Andreas fault earthquakes with lichenometry, *Geology* **24**: 111–114.

Butzer, K. W., Fock, G. J., Scott, L., and Stuckenrath, R., 1979, Dating and context of rock engravings in South Africa, *Science* **203**: 1201–1214.

Cariati, F., Rampazzi, L., Toniolo, L., and Pozzi, A., 2000, Calcium oxalate films on stone surfaces: Experimental assessment of the chemical formation, *Studies in Conservation* **45**: 180–188.

Carter, C. L., Dethier, D. P., and Newton, R. L., 2003, Subglacial environment inferred from bedrock-coating siltskins, Mendenhall Glacier, Alaska, USA, *Journal of Glaciology* **49**: 568–576.

Casey, W. H., Westrich, H. R., Banfield, J. F., Ferruzzi, G., and Arnold, G. W., 1993, Leaching and reconstruction at the surfaces of dissolving chain-silicate minerals, *Nature* **366**: 253–256.

Cerveny, N. V., Kaldenberg, R., Reed, J., Whitley, D. S., Simon, J., and Dorn, R. I., 2006, A new strategy for analyzing the chronometry of constructed rock features in deserts, *Geoarchaeology* **21**: 181–203.

Chen, Y., and Polach, J., 1986, Validity of C-14 ages of carbonates in sediments, *Radiocarbon* **28 (2A)**: 464–472.

Christensen, P. R., and Harrison, S. T., 1993, Thermal infrared emission spectroscopy of natural surfaces: application to desert varnish coating on rocks, *Journal of Geophysical Research* **98 (B11)**: 19,819–19,834.

Chukhrov, F. V., Zvyagin, B. B., Ermilova, L. P., and Gorshkov, A. I., 1973, *New Data on Iron Oxides in the Weathering Zone*, Division de Ciencias C.S.I.C., Madrid, 333–341p.

Cockburn, H. A. P., and Summerfield, M. A., 2004, Geomorphological applications of cosmogenic isotope analysis, *Progress in Physical Geography* **28**: 1–42.

Collins, J. F., and Buol, S. W., 1970, Effects of fluctuations in the Eh-pH environment on iron and/or manganese equilibria, *Soil Science* **110**: 111–118.

Conca, J. L., 1985, Differential weathering effects and mechanisms, Dissertation thesis, 251pp., California Institute of Technology, Pasadena.

Conca, J. L., and Rossman, G. R., 1982, Case hardening of sandstone, *Geology* **10**: 520–525.

Cooks, J., and Fourie, Y., 1990, The influence of two epilithic lichens on the weathering of quartzite and gabbro, *South African Geographer* **17 (1/2)**: 24–33.

Coudé-Gaussen, G., Rognon, P., and Federoff, N., 1984, Piegeage de poussières éoliennes dans des fissures de granitoides due Sinai oriental, *Compte Rendus de l'Academie des Sciences de Paris* **II**: 369–374.

Cremaschi, M., 1996, The desert varnish in the Messak Sattafet (Fezzan, Libryan Sahara), age, archaeological context and paleo-environmental implication, *Geoarchaeology* **11**: 393–421.

Crerar, D. A., Fischer, A. G., and Plaza, C. L., 1980, Metallogenium and biogenic deposition of manganese from Precambrian to recent time, in: *Geology and Geochemistry of Manganese. Volume III. Manganese on the bottom of Recent Basins*, I. M. Varentsov and G. Grasselly, ed., E. Schweizerbart'sche Verlagsbuchhandlung, Stuttgart, pp. 285–303.

Curtiss, B., Adams, J. B., and Ghiorso, M. S., 1985, Origin, development and chemistry of silica-alumina rock coatings from the semiarid regions of the island of Hawaii, *Geochemica et Cosmochimica Acta* **49**: 49–56.

Dandurand, J., Gout, R., Hoefs, J., Menschel, G., Schott, J., and Usdowski, E., 1982, Kinetically controlled variations of major components and carbon isotopes in a calcite-precipitating stream, *Chemical Geology* **36**: 299–315.

Danin, A., 1983, Weathering of limestone in Jerusalem by cyanobacteria, *Zeitschrift für Geomorphologie* **27**: 413–421.

Danin, A., 1985, Palaeoclimates in Israel: evidence form weathering patterns of stones in and near archaeological sites, *Bulletin of the American Schools of Oriental Research* **259**: 33–43.

Danin, A., 1986, Patterns of biogenic weathering as indicators of palaeoclimates in Israel, *Proceedings of the Royal Society of Edinburgh* **89B**: 243–253.

Danin, A., and Garty, J., 1983, Distribution of cyanobacteria and lichens on hillsides of the Negev highlands and their impact on biogenic weathering, *Zeitschrift für Geomorphologie* **27**: 423–444.

Danin, A., Gerson, R., and Garty, J., 1983, Weathering Patterns on Hard Limestone and Dolomite by Endolithic Lichens and Cyanobacteria: Supporting Evidence for Eolian Contribution to Terra Rossa Soil, *Soil Science* **136**: 213–217.

Danin, A., Gerson, R., Marton, K., and Garty, J., 1982, Patterns of limestone and dolomite weathering by lichens and blue-green algae and their palaeoclimatic significance, *Palaeogeography, Palaeoclimatology, Palaeoecology* **37**: 221–223.

DeAngelis, F., Ceci, R., Quaresima, R., Reale, S., and DiTullio, A., 2003, Investigation by solid-phase microextraction and gas chromatography/mass spectrometry of secondary metabolites in lichens deposited on stone, *Rapid Communications in Mass Spectrometry* **17**: 526–531.

Del Monte, M., Sabbioni, C., and Zappa, G., 1987, The origin of calcium oxalates on historical buildings, monuments and natural outcrops, *The Science of the Total Environment* **67**: 17–39.

Diaz, T. A., Bailley, T. L., and Orndorff, R. L., 2002, SEM analysis of vertical and lateral variations in desert varnish chemistry from the Lahontan Mountains, Nevada, *Geological Society of America Abstracts with Programs* **May 7–9 Meeting**: <///gsa.confex.com/gsa/2002RM/finalprogram/abstract_33974.htm>.

DiGregorio, B. E., 2002, Rock varnish as a habitat for extant life on Mars, in: *Instruments, Methods, and Missions for Astrobiology IV*, R. B. Hoover, G. V. Levin, R. R. Paepe and A. Y. Rozanov, eds., NASA, SPIE Vol. 4495, pp. 120–130.

Dixon, J. C., Darmody, R. G., Schlyter, P., and Thorn, C. E., 1995, Preliminary investigation of geochemical process responses to potential environmental change in Kärkevagge, Northern Scandinavia, *Geografiska Annaler* **77A**: 259–267.

Dolanski, J., 1978, Silcrete skins—their significance in rock art weathering, in: *Conservation of Rock Art*, C. Pearson, ed., ICCM, Sydney, pp. 32–36.

Dorn, R. I., 1990, Quaternary alkalinity fluctuations recorded in rock varnish microlaminations on western U.S.A. volcanics, *Palaeogeography, Palaeoclimatology, Palaeoecology* **76**: 291–310.

Dorn, R. I., 1997, Constraining the age of the Côa valley (Portugal) engravings with radiocarbon dating, *Antiquity* **71**: 105–115.

Dorn, R. I., 1998, *Rock coatings*, Elsevier, Amsterdam, 429p.

Dorn, R. I., 2001, Chronometric techniques: Engravings, in: *Handbook of Rock Art Research*, D. S. Whitley, ed., Altamira Press, Walnut Creek, pp. 167–189.

Dorn, R. I., 2009, Rock varnish and its use to study climatic change in geomorphic settings, in: *Geomorphology of Desert Environments. 2nd Edition.*, A. J. Parsons and A. Abrahams, ed., Springer, in press.

Dorn, R. I., and Dickinson, W. R., 1989, First paleoenvironmental interpretation of a pre-Quaternary rock varnish site, Davidson Canyon, south Arizona, *Geology* **17**: 1029–1031.

Dorn, R. I., and Dragovich, D., 1990, Interpretation of rock varnish in Australia: Case studies from the Arid Zone, *Australian Geographer* **21**: 18–32.

Dorn, R. I., and Krinsley, D. H., 1991, Cation-leaching sites in rock varnish, *Geology* **19**: 1077–1080.

Dorn, R. I., Krinsley, D. H., Liu, T., Anderson, S., Clark, J., Cahill, T. A., and Gill, T. E., 1992, Manganese-rich rock varnish does occur in Antarctica, *Chemical Geology* **99**: 289–298.

Dorn, R. I., and Meek, N., 1995, Rapid formation of rock varnish and other rock coatings on slag deposits near Fontana, *Earth Surface Processes and Landforms* **20**: 547–560.

Dorn, R. I., and Oberlander, T. M., 1981, Microbial origin of desert varnish, *Science* **213**: 1245–1247.

Dorn, R. I., and Oberlander, T. M., 1982, Rock varnish, *Progress in Physical Geography* **6**: 317–367.

Dorn, R. I., Stasack, E., Stasack, D., and Clarkson, P., 2001, Through the looking glass: Analyzing petroglyphs and geoglyphs with different perspectives, *American Indian Rock Art* **27**: 77–96.

Douglas, G. R., 1987, Manganese-rich rock coatings from Iceland, *Earth Surface Processes and Landforms* **12**: 301–310.

Douglas, G. R., McGreevy, J. P., and Whalley, W. B., 1994, Mineralogical aspects of crack development and freeface activity in some basalt cliffs, County Antrim, Northern Ireland, in: *Rock weathering and landform evolution*, D. A. Robinson and R. B. G. Williams, ed., Wiley, Chichester, pp. 71–88.

Dragovich, D., 1986a, Minimum age of some desert varnish near Broken Hill, New South Wales, *Search* **17**: 149–151.

Dragovich, D., 1986b, Weathering of desert varnish by lichens, *21st I.A.G. Proceedings*: 407–412.

Drake, N. A., Heydeman, M. T., and White, K. H., 1993, Distribution and formation of rock varnish in southern Tunisia, *Earth Surface Processes and Landforms* **18**: 31–41.

Drioli, E., Gagliardi, R., Donato, L., and Checcetti, A., 1995, CoPVDF membranes for protection of cultural heritages, *Journal of Membrane Science* **102**: 131–138.

Dunkerley, D. L., 1987, Deposition of tufa on Ryans and Stockyard Creeks, Chillagoe Karst, North Queensland: the role of evaporation, *Helictite* **25**: 30–35.

Engel, C. G., and Sharp, R. S., 1958, Chemical data on desert varnish, *Geological Society of America Bulletin* **69**: 487–518.

Ericksen, G. E., 1981, Geology and origin of the Chilean nitrate deposits, *U.S. Geological Survey Professional Paper* **1188**.

Fajardo-Cavazos, P., and Nicholson, W., 2006, Bacillus endospores isolated from granite: Close molecular relationships to globally distributed Bacillus spp. from endolithic and extreme environments, *Applied and Environmental Microbiology* **72**: 2856–2863.

Farr, T., and Adams, J. B., 1984, Rock coatings in Hawaii, *Geological Society of America Bulletin* **95**: 1077–1083.

Fisk, E. P., 1971, Desert glaze, *Journal Sedimentary Petrology* **41**: 1136–1137.

Fleisher, M., Liu, T., Broecker, W., and Moore, W., 1999, A clue regarding the origin of rock varnish, *Geophysical Research Letters* **26** (1): 103–106.

Flood, B. E., Allen, C., and Longazo, T., 2003, Microbial fossils detected in desert varnish, *Astrobiology* **2(4)**.

Folk, R. L., 1978, Angularity and silica coatings of Simpson Desert sand grains, Northern Territory, Australia, *Journal of Sedimentary Petrology* **48**: 611–624.

Folk, R. L., 1993, SEM imaging of bacteria and nanobacteria in carbonate sediments and rocks, *Journal of Sedimentary Petrology* **63**: 990–999.

Forbes, E. A., Posner, A. M., and Quirk, J. P., 1976, The specific adsorption of divalent Cd, Co, Cu, Pb, and Zn on goethite, *Journal of Soil Science* **27**: 154–166.

Fortescue, J. A. C., 1980, *Environmental Geochemistry. A Holistic Approach*, Springer-Verlag, New York, 347p.

Fortin, D., and Langley, S., 2005, Formation and occurrence of biogenic iron-rich minerals, *Earth-Science Reviews* **72**: 1–19.

Frazier, C. S., and Graham, R. C., 2000, Pedogenic transformation of fractured granitic bedrock, southern California, *Soil Science Society of America Journal* **64**: 2057–2069.

Fullagar, R. L. K., 1991, The role of silica in polish formation, *Journal of Archaeological Science* **18**: 1–24.

Fuller, S., 2000, *Thomas Kuhn: A Philosophical History for Our Times*, University Chicago Press, Chicago, 496 p.

Fyfe, F., Schultze, S., Witten, T., Fyfe, W., and Beveridge, T., 1989, Metal interactions with microbial biofilms in acidic and neutral pH environments, *Applied and Environmental Microbiology* **55**: 1249–1257.

Fyfe, W. S., 1996, The biosphere is going deep, *Nature* **273**: 448.

Gehrmann, C., Krumbein, W. E., and Petersen, K., 1988, Lichen weathering activities on mineral and rock surfaces, *Studia Geobotanica* **8**: 33–45.

Gilbert, G. K., 1877, *Geology of the Henry Mountains*, U.S. Geological and Geographical Survey, Washington D.C., 160p.

Gile, L. H., Peterson, F. F., and Grossman, R. G., 1966, Morphological and genetic sequences of carbonate accumulation in desert soils, *Soil Science* **101**: 347–360.

Glazovskaya, M. A., 1968, Geochemical landscape and types of geochemical soil sequences, *Transactions of 9th International Congress Soil Science, Adelaide* **IV**: 303–311.

Glazovskaya, M. A., 1973, Technogiogeomes – the intial physical geographic objects of landscape geochemical prediction, *Soviet Geography Review and Translation* **14(4)**: 215–228.

Golubic, S., Friedmann, E., and Schneider, J., 1981, The lithobiontic ecological niche, with special reference to microorganisms, *Journal Sedimentary Petrology* **51**: 475–478.

Gorbushina, A. A., Krumbein, W. E., and Volkmann, M., 2002, Rock surfaces as life indicators: New ways to demonstrate life and traces of former life, *Astrobiology* **2(2)**: 203–213.

Gordon, S. J., and Dorn, R. I., 2005a, In situ weathering rind erosion, *Geomorphology* **67**: 97–113.

Gordon, S. J., and Dorn, R. I., 2005b, Localized weathering: Implications for theoretical and applied studies, *Professional Geographer* **57**: 28–43.

Goudie, A., 1972, Climate, weathering, crust formation, dunes, and fluvial features of the Central Namib Desert, near Gobabeb, South West Africa, *Madoqua Series II* **1**: 15–28.

Goudie, A., 1983, Calcrete, in: *Chemical sediments and geomorphology: precipitates and residua in the near surface environment*, A. Goudie and K. Pye, ed., Academic Press, New York, pp. 93–131.

Goudie, A. S., and Cooke, R. U., 1984, Salt efflorescences and saline lakes: a distributional analysis, *Geoforum* **15**: 563–582.

Grab, S., van Zyl, C., and Mulder, N., 2005, Controls on basalt terrace formation in the eastern Lesotho highlands, *Geomorphology* **67**: 473–485.

Graham, R. C., and Franco-Vìzcaino, E., 1992, Soils on igneous and metavolcanic rocks in the Sonoran Desert of Baja California, Mexico, *Geoderma* **54**: 1–21.

Ha-mung, T., 1968, The biological nature of iron-manganese crusts of soil-forming rocks in Sakhalin mountain soils, *Microbiology* **36**: 621–624.

Haberland, W., 1975, Untersuchungen an Krusten, Wustenlacken und Polituren auf Gesteinsoberflachen der nordlichen und mittlerent Saharan (Libyen und Tchad), **21**: 1–77.

Harrington, C. D., 1986, Investigation of desert rock varnish on a cobble from a stone burial mound at al-Farrah, al-Jubah Quadrangle, Yemen Arab Republic, in: *Geological and Archaeological Reconnaissance in the Yemen Arab Republic, 1985, the Wadi Al-Jubah archaeological project. Volume 4*, W. C. a. o. Overstreet, ed., American Foundation for the Study of Man, Washington D.C., pp. 41–46.

Harrington, C. D., and Raymond, R., Jr., 1989, *SEM Analysis of Rock Varnish Chemistry: A Geomorphic Age Indicator*, San Francisco Press, San Francisco, 567–570 p.

Harrison, R., and Frink, D. S., 2000, The OCR carbon dating procedure in Australia: New dates from Wilinyjibari Rockshelter, southeaster Kimberley, Western Australia, *Australian Archaeology* **51**: 6–15.

Harry, K. G., Bierman, P. R., and Fratt, L., 1992, *Lithic Procurement and Rock Varnish Dating: Investigations at CA-KER-140, a Small Quarry in the Western Mojave Desert*, Statistical Research, Tucson, 129p.

Hayden, J., 1976, Pre-altithermal archaeology in the Sierra Pinacate, Sonora, Mexico, *American Antiquity* **41**: 274–289.

Hazel, F., Schock, R. V., and Gordon, M., 1949, Interaction of ferric ions with silicic acid, *Journal of the American Chemical Society* **71**: 2256–2257.

Helms, J. G., McGill, S. F., and Rockwell, T. K., 2003, Calibrated, late Quaternary age indices using clast rubification and soil development on alluvial surfaces in Pilot Knob Valley, Mojave Desert, southeastern California, *Quaternary Research* **60**: 377–393.

Hetu, B., Vansteijn, H., and Vandelac, P., 1994, Flows of frost-coated clasts – A recently discovered scree slope grain-flow type, *Geographie Physique et Quaternaire* **48**: 3–22.

Hobbs, W. H., 1917, The erosional and degradational processes of deserts, with especial reference to the origin of desert depressions, *Annals of the Association of American Geographers* **7**: 25–60.

Hodge, V. F., Farmer, D. E., Diaz, T. A., and Orndorff, R. L., 2005, Prompt detection of alpha particles from Po-210: another clue to the origin of rock varnish?, *Journal of Environmental Radioactivity* **78**: 331–342.

Holland, H. D., 1978, *The Chemistry of the Atmosphere and Oceans*, Wiley, New York, 351p.

Hooke, R. L., Yang, H., and Weiblen, P. W., 1969, Desert varnish: an electron probe study, *Journal Geology* **77**: 275–288.

Huguen, C., Benkhelil, J., Giresse, P., Mascle, J., Muller, C., Woodside, J., and Zitter, T., 2001, Clasts from 'mud volcanoes' from the eastern Mediterranean, *Oceanologica Acta* **24**: 349–360.

Hungate, B., Danin, A., Pellerin, N. B., Stemmler, J., Kjellander, P., Adams, J. B., and Staley, J. T., 1987, Characterization of manganese-oxidizing (MnII–>MnIV) bacteria from Negev Desert rock varnish: implications in desert varnish formation, *Canadian Journal Microbiology* **33**: 939–943.

Hunt, A. G., and Wu, J. Q., 2004, Climatic influences on Holocene variations in soil erosion rates on a small hill in the Mojave Desert, *Geomorphology* **58**: 263–289.

Huyge, D., Watchman, A., De Dapper, M., and Marchi, E., 2001, Dating Egypt's oldest 'art': AMS ^{14}C age determinations of rock varnishes covering petroglyphs at El-Hosh (Upper Egypt), *Antiquity* **75**: 68–72.

ICOMOS, 2000, International Scientific Committee on Rock Art, *http://www.international.icomos.org/risk/isc-rockart_2000.htm* (accessed 3/5/05).

Israel, E. J., Arvidson, R. E., Wang, A., Pasteris, J. D., and Jolliff, B. L., 1997, Laser Raman spectroscopy of varnished basalt and implications for in situ measurements of Martian rocks, *Journal of Geophysical Research-Planets* **102 (E12)**: 28705–28716.

J.PaulGettyTrust, 2003, Conservation of rock art, *http://www.getty.edu/conservation/education/rockart/* accessed July 9, 2005.

Jessup, R. W., 1951, The soils, geology, and vegetation of northwestern South Australia, *Royal Society of South Australia Transactions* **74**: 189–273.

Johnson, J. R., Christensen, P. R., and Lucey, P. G., 2002, Dust coatings on basaltic rocks and implications for thermal infrared spectroscopy of Mars, *Journal of Geophysical Research-Planets* **107 (E6)**: Art No. 5035.

Jones, C. E., 1991, Characteristics and origin of rock varnish from the hyperarid coastal deserts of northern Peru, *Quaternary Research* **35**: 116–129.

Jones, D., and Wilson, M. J., 1985, Chemical activity of lichens on mineral surfaces – a review, *International Biodeterioration* **21(2)**: 99–104.

Joubert, J. J., Kriel, W. C., and Wessels, D. C. J., 1983, Lichenometry: its potential application to archaeology in southern Africa, *The South African Archaeological Society Newsletter* **6(1)**: 1–2.

Keyser, J. D., Greer, M., and Greer, J., 2005, Arminto petroglyphs: Rock art damage assessment and management considerations in Central Wyoming, *Plains Anthropologist* **50**: 23–30.

Kim, J. G., Lee, G. H., Lee, J., Chon, C., Kim, T. H., and Ha, K., 2006, Infiltration pattern in a regolith-fractured bedrock profile: field observations of a dye stain pattern, *Hydrological Processes* **20**: 241–250.

Klute, F., and Krasser, L. M., 1940, Über wüstenlackbildung im Hochgebirge, *Petermanns geographische Mitteilungen* **86**: 21–22.

Knauss, K. G., and Ku, T. L., 1980, Desert varnish: potential for age dating via uranium-series isotopes, *Journal of Geology* **88**: 95–100.

Konhauser, K. O., Fyfe, W. S., Ferris, F. G., and Beveridge, T. J., 1993, Metal sorption and mineral precipitation by bacteria in two Amazonian river systems: Rio Solimoes and Rio Negro, Brazeil, *Geology* **21**: 1103–1106.

Konhauser, K. O., Fyfe, W. S., Schultze-Lam, S., Ferris, F. G., and Beveridge, T. J., 1994, Iron phosphate precipitation by epilithic microbial biofilms in Arctic Canada, *Canadian Journal Earth Science* **31**: 1320–1324.

Krauskopf, K. B., 1956, Dissolution and precipitation of silica at low temperatures, *Geochimica et Cosmochimica Acta* **10**: 1–26.

Krauskopf, K. B., 1957, Separation of manganese from iron in sedimentary processes, *Geochimica et Cosmochimica Acta* **12**: 61–84.

Krinsley, D., 1998, Models of rock varnish formation constrained by high resolution transmission electron microscopy, *Sedimentology* **45**: 711–725.

Krinsley, D., Dorn, R. I., and Anderson, S., 1990, Factors that may interfere with the dating of rock varnish, *Physical Geography* **11**: 97–119.

Krinsley, D. H., Dorn, R. I., and Tovey, N. K., 1995, Nanometer-scale layering in rock varnish: implications for genesis and paleoenvironmental interpretation, *Journal of Geology* **103**: 106–113.

Krumbein, W. E., 1969, Über den Einfluss der Mikroflora auf die Exogene Dynamik (Verwitterung und Krustenbildung), *Geologische Rundschau* **58**: 333–363.

Krumbein, W. E., 1971, Biologische Entstehung von wüstenlack, *Umschau* **71**: 210–211.

Krumbein, W. E., 1979, Photolithotrophic and chemoorganotrophic activity of bacteria and algae as related to beachrock formation and degradation (Gulf of Aqaba, Sinai), *Geomicrobiology Journal* **1**: 139–203.

Krumbein, W. E., and Jens, K., 1981, Biogenic rock varnishes of the Negev Desert (Israel): An ecological study of iron and manganese transformation by cyanobacteria and fungi, *Oecologia* **50**: 25–38.

Kuhlman, K. R., Allenbach, L. B., Ball, C. L., Fusco, W. G., La_Duc, M. T., Kuhlman, G. M., Anderson, R. C., Stuecker, T., Erickson, I. K., Benardini, J., and Crawford, R. L., 2005, Enumeration, isolation, and characterization of ultraviolet (UV-C) resistant bacteria from rock varnish in the Whipple Mountains, California, *Icarus* **174**: 585–595.

Kuhlman, K. R., Fusco, W. G., Duc, M. T. L., Allenbach, L. B., Ball, C. L., Kuhlman, G. M., Anderson, R. C., Erickson, K., Stuecker, T., Benardini, J., Strap, J. L., and Crawford, R. L., 2006, Diversity of microorganisms within rock varnish in the Whipple Mountains, California, *Applied and Environmental Microbiology* **72**: 1708–1715.

Lapeyrie, F., 1988, Oxalate synthesis from soil biocarbonate on the mycorhizzal fungus *Paxillus involutes*, *Plant and Soil* **110**: 3–8.

Laudermilk, J. D., 1931, On the origin of desert varnish, *American Journal of Science* **21**: 51–66.

Lee, D. H., Tien, K. G., and Juang, C. H., 1996, Full-scale field experimentation of a new technique for protecting mudstone slopes, Taiwan, *Engineering Geology* **42**: 51–63.

Lee, M. R., and Bland, P. A., 2003, Dating climatic change in hot deserts using desert varnish on meteorite finds, *Earth and Planetary Science Letters* **206**: 187–198.

Lee, M. R., and Parsons, I., 1999, Biomechanical and biochemical weathering of lichen-encrusted granite: Textural controls on organic-mineral interactions and deposition of silica-rich layers, *Chemical Geology* **161**: 385–397.

Leeder, M. R., Harris, T., and Kirkby, M. J., 1998, Sediment supply and climate change: implications for basin stratigraphy, *Basin Research* **10**: 7–18.

Li, F. X., Margetts, S., and Fowler, I., 2001, Use of 'chalk' in rock climbing: sine quanon or myth?, *Journal of Sports Sciences* **19**: 427–432.

Linck, G., 1901, Über die dunkelen Rinden der Gesteine der Wüste, *Jenaische Zeitschrift für Naturwissenschaft* **35**: 329–336.

Linck, G., 1928, Über Schutzrinden, *Chemie die Erde* **4**: 67–79.

Liu, B., Phillips, F. M., Elmore, D., and Sharma, P., 1994, Depth dependence of soil carbonate accumulation based on cosmogenic ^{36}Cl dating, *Geology* **22**: 1071–1074.

Liu, T., 1994. Visual microlaminations in rock varnish: a new paleoenvironmental and geomorphic tool in drylands. Ph.D. Dissertation, Visual microlaminations in rock varnish: a new paleoenvironmental and geomorphic tool in drylands, Ph.D. thesis, 173pp., Arizona State University, Tempe.

Liu, T., 2003, Blind testing of rock varnish microstratigraphy as a chronometric indicator: results on late Quaternary lava flows in the Mojave Desert, California, *Geomorphology* **53**: 209–234.

Liu, T., 2008, VML Dating Lab, *http://www.vmldatinglab.com/* <**accessed November 14, 2008**>.

Liu, T., and Broecker, W. S., 2000, How fast does rock varnish grow?, *Geology* **28**: 183–186.

Liu, T., and Broecker, W., 2007, Holocene rock varnish microstratigraphy and its chronometric application in drylands of western USA, *Geomorphology* **84**: 1–21.

Liu, T., and Broecker, W.S., 2008a, Rock varnish microlamination dating of late Quateary geomorphic features in the drylands of western USA. *Geomorphology* **93**: 501–523.

Liu, T., and Broecker, W.S., 2008b, Rock Varnish evidence for latest Pleistocene millennial-scale wet events in the drylands of western United States. *Geology* **36**: 403–406.

Liu, T., Broecker, W. S., Bell, J. W., and Mandeville, C., 2000, Terminal Pleistocene wet event recorded in rock varnish from the Las Vegas Valley, southern Nevada, *Palaeogeography, Palaeoclimatology, Palaeoecology* **161**: 423–433.

Liu, T., and Dorn, R. I., 1996, Understanding spatial variability in environmental changes in drylands with rock varnish microlaminations, *Annals of the Association of American Geographers* **86**: 187–212.

Lock, W. W., Andrews, J. T., and Webber, P. J., 1979, A manual for lichenometry, *British Geomorphological Research Group Technical Bulletin* **26**: 1–47.

Loendorf, L. L., 1991, Cation-ratio varnish dating and petroglyph chronology in southeastern Colorado, *Antiquity* **65**: 246–255.

Loso, M. G., and Doak, D. F., 2006, The biology behind lichenometric dating curves, *Oecologia* **147**: 233–229.

Lou, G., and Huang, P. M., 1988, Hydroxyl-aluminosilicate interlayers in montmorillonite: implications for acidic environments, *Nature* **335**: 625–627.

Lowenstam, H. A., 1981, Minerals formed by organisms, *Science* **211**: 1126–1131.

Lucas, A., 1905, *The blackened rocks of the Nile cataracts and of the Egyptian deserts*, National Printing Department, Cairo, 58p.

Ludwig, K. R., and Paces, J. B., 2002, Uranium-series dating of pedogenic silica and carbonate, Crater Flat, Nevada, *Geochimica et Cosmochimica Acta* **66**: 487–506.

Lytle, F. W., Pingitore, N. E., Lytle, N. W., Ferris-Rowley, D., and Reheis, M. C., 2002, The possibility of dating petroglyphs from the growth rate of desert varnish, *Nevada Archaeological Association Meetings Abstracts.*

Machette, M. N., 1985, Calcic soils of the southwestern United States, *Geological Society of America Special Paper* **203**: 1–21.

Mancinelli, R. L., Bishop, J. L., and De, S., 2002, Magnetite in desert varnish and applications to rock varnish on Mars, *Lunar and Planetary Science* **33**: 1046.pdf.

Mansfield, G. R., and Boardman, L., 1932, Nitrate deposits of the United States, *U.S. Geological Survey Bulletin* **838**.

Marshall, C. E., 1977, *The physical chemistry and mineralogy of soils*, Wiley, New York p.

Marshall, R. R., 1962, Natural radioactivity and the origin of desert varnish, *Transactions of the American Geophysical Union* **43**: 446–447.

Marston, R. A., 2003, Editorial note, *Geomorphology* **53**: 197.

Matsukura, Y., Kimata, M., and Yokoyama, S., 1994, Formation of weathering rinds on andesite blocks under the influence of volcanic glases around the active crater Aso Volcano, Japan, in: *Rock Weathering and Landform Evolution*, D. A. Robinson and R. B. G. Williams, ed., Wiley, London, pp. 89–98.

Matthews, J. A., and Shakesby, R. A., 1983, The status of the Little Ice Age in Southern Norway. Relative-age dating of Neoglacial moraines with Schmidt hammer and lichenometry, *Boreas* **13**: 333–346.

McFadden, L. D., Ritter, J. B., and Wells, S. G., 1989, Use of multiparameter relative-age methods for age estimation and correlation of alluvial fan surfaces on a desert piedmont, eastern Mojave Desert, *Quaternary Research* **32**: 276–290.

McKeown, D. A., and Post, J. E., 2001, Characterization of manganese oxide mineralogy in rock varnish and dendrites using X-ray absorption spectroscopy, *American Mineralogist* **86**: 701–713.

McKnight, D. M., Kimball, B. A., and Bencala, K. E., 1988, Iron photoreduction and oxidation in an acidic mountain stream, *Science* **240**: 637–640.

Meek, N., and Dorn, R. I., 2000, Is mushroom rock a ventifact?, *Califonia Geology* **November/December**: 18–20.

Merrill, G. P., 1906, *A treatise on rocks, rock-weathering, and soils*, Macmillan, New York, 400 p.

Monger, H. C., and Buck, B. J., 1999, Stable isotopes and soil-geomorphology as indicators of Holocene climate change, northern Chihuahuan Desert, *Journal of Arid Environments* **43**: 357–373.

Moore, C. B., and Elvidge, C. D., 1982, Desert varnish, in: *Reference handbook on the deserts of North America*, G. L. Bender, ed., Greenwood Press, Westport, pp. 527–536.

Moreiras, S. M., 2006, Chronology of a probable neotectonic Pleistocene rock avalanche, Cordon del Plata (Central Andes),Mendoza,Argentina, *Quaternary International* **148**: 138–148.

Mottershead, D. N., and Pye, K., 1994, Tafoni on coastal slopes, South Devon, U.K., *Earth Surface Processes and Landforms* **19**: 543–563.

Mustoe, G. E., 1981, Bacterial oxidation of manganese and iron in a modern cold spring, *Geological Society of America Bulletin* **92**: 147–153.

Nagy, B., Nagy, L. A., Rigali, M. J., Jones, W. D., Krinsley, D. H., and Sinclair, N., 1991, Rock varnish in the Sonoran Desert: microbiologically mediated accumulation of manganiferous sediments, *Sedimentology* **38**: 1153–1171.

Nanson, G. C., Jones, B. G., Price, D. M., and Pietsch, T. J., 2005, Rivers turned to rock: Late Quaternary alluvial induration influencing the behaviour and morphology of an anabranching river in the Australian monsoon tropics, *Geomorphology* **70**: 398–420.

Nishiizumi, K., Kohl, C., Arnold, J., Dorn, R., Klein, J., Fink, D., Middleton, R., and Lal, D., 1993, Role of in situ cosmogenic nuclides ^{10}Be and ^{26}Al in the study of diverse geomorphic processes, *Earth Surface Processes and Landforms* **18**: 407–425.

Oberlander, T. M., 1988, Salt crust preservation of pre-late Pleistocene slopes in the Atacama Desert, and relevance to the age of surface artifacts., *Program Abstracts, Association of American Geographers, Phoenix Meeting*: 143.

Oberlander, T. M., 1989, Slope and pediment systems, in: *Arid Zone Geomorphology*, D. S. G. Thomas, ed., Belhaven Press, London, pp. 56–84.

Oberlander, T. M., 1994, Rock varnish in deserts, in: *Geomorphology of Desert Environments*, A. Abrahams and A. Parsons, ed., Chapman and Hall, London, pp. 106–119.

Palmer, E., 2002. Feasibility and implications of a rock coating catena: analysis of a desert hillslope. M.A. Thesis, Masters thesis, Arizona State University, Tempe.

Palmer, F. E., Staley, J. T., Murray, R. G. E., Counsell, T., and Adams, J. B., 1985, Identification of manganese-oxidizing bacteria from desert varnish, *Geomicrobiology Journal* **4**: 343–360.

Paradise, T. R., 2005, Petra revisited: An examination of sandstone weathering research in Petra, Jordan, *Geological Society of America Special Paper* **390**: 39–49.

Paraguassu, A. B., 1972, Experimental silicification of sandstone, *Geological Society of America Bulletin* **83**: 2853–2858.

Patyk-Kara, N. G., Gorelikova, N. V., Plakht, J., Nechelyustov, G. N., and Chizhova, I. A., 1997, Desert varnish as an indicator of the age of Quaternary formations (Makhtesh Ramon Depression, Central Negev), *Transactions (Doklady) of the Russian Academy of Sciences/Earth Science Sections* **353A**: 348–351.

Perel'man, A. I., 1961, Geochemical principles of landscape classification, *Soviet Geography Review and Translation* **11** (3): 63–73.

Perel'man, A. I., 1966, *Landscape geochemistry. (Translation No. 676, Geological Survey of Canada, 1972)*, Vysshaya Shkola, Moscow, 388p.

Perel'man, A. I., 1967, *Geochemistry of epigenesis*, Plenum Press, New York, 266 p.

Perel'man, A. I., 1980, Paleogeochemical landscape maps, *Geochemistry International* **17**(1): 39–50.

Perry, R. S., and Adams, J., 1978, Desert varnish: evidence of cyclic deposition of manganese, *Nature* **276**: 489–491.

Perry, R. S., Dodsworth, J., Staley, J. T., and Engel, M. H., 2004, Bacterial diversity in desert varnish, in: *Third European Workshop on Exo/Astrobiology, Mars: The Search for life*, vol. SP-545, ed., European Space Agency Publications, Netherlands, pp. 259–260.

Perry, R. S., and Kolb, V. M., 2003, Biological and organic constituents of desert varnish: Review and new hypotheses, in: *Instruments, methods, and missions for Astrobiology VII*, vol. 5163, R. B. Hoover and A. Y. Rozanov, ed., SPIE, Bellingham, pp. 202–217.

Perry, R. S., Kolb, V. N., Lynne, B. Y., Sephton, M. A., McLoughlin, N., Engel, M. H., Olendzenski, L., Brasier, M., and Staley, J. T., 2005, How desert varnish forms?, in: *Astrobiology and Planetary Missions*, R. B. Hoover, G. V. Levin, A. Y. Rozanov and G. R. Gladstone, ed., SPIE, SPIE Vol. 2906, pp. 276–287.

Perry, R. S., Lynne, B. Y., Sephton, M. A., Kolb, V. M., Perry, C. C., and Staley, J. T., 2006, Baking black opal in the desert sun: The importance of silica in desert varnish, *Geology* **34**: 737–540.

Perry, R. S., and Lynne, Y. B., 2006, New insights into natural records of planetary surface environments: The role of silica in the formation and diagenesis of desert varnish and siliceous sinter, *Lunar and Planetary Science* **37**: 1292.

Polynov, B. B., 1937, *The Cycle of Weathering. [Translated from the Russian by A. Muir]*, Nordemann Publishing, New York, pp. 220.

Pope, G. A., 2000, Weathering of petroglyphs: Direct assessment and implications for dating methods, *Antiquity* **74**: 833–843.

Pope, G. A., Meierding, T. C., and Paradise, T. R., 2002, Geomorphology's role in the study of weathering of cultural stone, *Geomorphology* **47**: 211–225.

Potter, R. M., 1979. The tetravalent manganese oxides: clarification of their structural variations and relationships and characterization of their occurrence in the terrestrial weathering environment as desert varnish and other manganese oxides. Ph.D. Dissertation, Ph.D. Dissertation thesis, 245 pp., California Institute of Technology, Pasadena.

Potter, R. M., and Rossman, G. R., 1977, Desert varnish: The importance of clay minerals, *Science* **196**: 1446–1448.

Potter, R. M., and Rossman, G. R., 1979a, The manganese- and iron-oxide mineralogy of desert varnish, *Chemical Geology* **25**: 79–94.

Potter, R. M., and Rossman, G. R., 1979b, Mineralogy of manganese dendrites and coatings, *American Mineralogist* **64**: 1219–1226.

Potter, R. M., and Rossman, G. R., 1979c, The tetravalent manganese oxides: identification, hydration, and structural relationships by infrared spectroscopy, *American Mineralogist* **64**: 1199–1218.

Probst, L., Thomas-Keprta, K., and Allen, C., 2001, Desert varnish: preservation of microfossils and biofabric (Earth and Mars?), *GSA Abstracts With Programs* http://gsa.confex.com/gsa/2001AM/finalprogram/abstract_27826.htm.

Probst, L. W., Allen, C. C., Thomas-Keprta, K. L., Clemett, S. J., Longazo, T. G., Nelman-Gonzalez, M. A., and Sams, C., 2002, Desert varnish – preservation of biofab-

rics and implications for Mars, *Lunar and Planetary Science* **33**: 1764.pdf.

Puterman, M., Jansen, B., and Kober, H., 1998, Development of organosilicone-polyurethanes as stone preservation and consolidation materials, *Journal of Applied Polymer Science* **59**: 1237–1242.

Reneau, S. L., 1993, Manganese accumulation in rock varnish on a desert piedmont, Mojave Desert, Califonria, and application to evaluating varnish development, *Quaternary Research* **40**: 309–317.

Robbins, E. I., D'Agostino, J. P., Ostwald, J., Fanning, D. S., Carter, V., and Van Hoven, R. L., 1992, Manganese nodules and microbial oxidation of manganese in the huntley Meadows wetland, Virginia, USA, *Catena Supplement* **21**: 179–202.

Robbins, L. L., and Blackwelder, P. L., 1992, Biochemical and ultrastructural evidence for the origin of whitings: A biologically induced calcium carbonate precipitation mechanism, *Geology* **20**: 464–468.

Robert, M., and Tessier, D., 1992, Incipient weathering: some new concepts on weathering, clay formation and organization, in: *Weathering, Soils & Paleosols*, I. P. Martini and W. Chesworth, ed., Amsterdam, Elsevier, pp. 71–105.

Robinson, D. A., and Williams, R. B. G., 1992, Sandstone weathering in the High Atlas, Morocco, *Zeitschrift fur Geomorphologie* **36**: 413–429.

Robinson, S. E., 2002, Cosmogenic nuclides, remote sensing, and field studies applied to desert piedmonts, Dissertation thesis, 387pp., Arizona State University, Tempe.

Rodriguez-Navarro, C., Rodriguez-Gallego, M., BenChekroun, K., and Gonzalez-Munoz, M. T., 2003, Conservation of ornamental stone by Myxococcus xanthus-induced carbonate biomineralization, *Applied and Environmental Microbiology* **69**: 2182–2193.

Rowe, M. W., 2001, Dating by AMS radiocarbon analysis, in: *Handbook of Rock Art Research*, D. S. Whitley, ed., Altamira Press, Walnut Creek, pp. 139–166.

Russ, J., Hyman, M., Shafer, H. J., and Rowe, M. W., 1990, Radiocarbon dating of prehistoric rock paintings by selective oxidation of organic carbon, *Nature* **348**: 710–711.

Russ, J., Loyd, D. H., and Boutton, T. W., 2000, A paleoclimate reconstruction for southwestern Texas using oxalate residue from lichen as a paleoclimate proxy, *Quaternary International* **67**: 29–36.

Russ, J., Palma, R. L., Lloyd, D. H., Boutton, T. W., and Coy, M. A., 1996, Origin of the whewellite-rich rock crust in the Lower Pecos region of Southwest Texas and its significance to paleoclimate reconstructions, *Quaternary Research* **46**: 27–36.

Saiz-Jimenez, C., and Hermosin, B., 1999, Thermally assisted hydrolysis and methylation of the black deposit coating the ceiling and walls of Cueva del Encajero, Quesada, Spain, *Journal of Analytic and Applied Pyrolysis* **49**: 349–357.

Scheffer, F., Meyer, B., and Kalk, E., 1963, Biologische ursachen der wüstenlackbildung, *Zeitschrift für Geomorphologie* **7**: 112–119.

Scheidegger, A., Borkovec, M., and Sticher, H., 1993, Coating of silica sand with goethite: preparation and analytical identification, *Geoderma* **58**: 43–65.

Schelble, R., McDonald, G., Hall, J., and Nealson, K., 2005, Community structure comparison using FAME analysis of desert varnish and soil, Mojave Desert, California, *Geomicrobiology Journal* **22**: 353–360.

Schiavon, N., 1993, Microfabrics of weathered granite in urban monuments, in: *Conservation of stone and other materials*, vol. 1, M.-J. Thiel, ed., E & FN Spon, London, pp. 271–278.

Smith, B. J., 1994, Weathering processes and forms, in: *Geomorphology of Desert Environments*, A. D. Abrahams and A. J. Parsons, ed., Chapman & Hall, London, pp. 39–63.

Smith, B. J., and Warke, P. A., *Processes of Urban Stone Decay*, Donhead Publishing, London, 1996, p. 274.

Smith, B. J., and Whalley, W. B., 1988, A note on the characteristics and possible origins of desert varnishes from southeast Morocco, *Earth Surface Processes and Landforms* **13**: 251–258.

Smith, G. A., and Turner, W. G., 1975, *Indian Rock art of Southern California with Selected Petroglyph Catalog*, San Bernardino County Museum Association, Redlands, pp. 150.

Smoot, N. C., 1995, Mass wasting and subaerial weathering in guyot formation: the Hawaiian and Canary Ridges as examples, *Geomorphology* **14**: 29–41.

Souza-Egipsy, V., Wierzchos, J., Sancho, C., Belmonte, A., and Ascaso, C., 2004, Role of biological soil crust cover in bioweathering and protection of sandstones in a semi-arid landscape (Torrollones de Gabarda, Huesca, Spain), *Earth Surface Processes and Landforms* **29**: 1651–166.

Spades, S., and Russ, J., 2005, GC-MS analysis of lipids in prehistoric rock paints and associated oxalate coatings from the lower Pecos region, Texas, *Archaeometry* **45**: 115–126.

Spilde, M. N., Boston, P. J., and Northrup, D. E., 2002, Subterranean manganese deposits in caves: Analogies to rock varnish?, *Geological Society of American Abstracts with Programs http://gsa.confex.com/gsa/2002AM/finalprogram/abstract_46060.htm.*

Spilde, M. N., Boston, P. J., Northrup, D., and Dichosa, A., 2005, Surface and subsurface manganese microbial environments, *SPIE Annual Salt Lake City Meeting.*

Stadelman, S., 1994, Genesis and post-formational systematics of carbonate accumulations in Quaternary soils of the Southwestern United States, Ph.D. Dissertation thesis, 124 pp., Texas Tech University, Lubbock.

Staiger, J. W., Marchant, D. R., Schaefer, J. M., Oberholzer, P., Johnson, J. V., Lewis, A. R., and Swanger, K. M., 2006, Plio-Pleistocene history of Ferrar Glacier, Antarctica: Implications for climate and ice sheet stability, *Earth and Planetary Science Letters* **243**: 489–503.

Staley, J. T., Adams, J. B., Palmer, F., Long, A., Donahue, D. J., and Jull, A. J. T., 1991, Young 14Carbon ages of rock varnish coatings from the Sonoran Desert, *Unpublished Manuscript.*

Staley, J. T., Palmer, F., and Adams, J. B., 1982, Microcolonial fungi: common inhabitants on desert rocks?, *Science* **215**: 1093–1095.

Sterflinger, K., Krumbein, W. E., Lallau, T., and Rullkötter, J., 1999, Microbially mediated orange patination of rock surfaces, *Ancient Biomolecules* **3**: 51–65.

Stevenson, J. J., 1881, *Report on the U.S. Geographical Surveys West of the One Hundredth Meridian (Wheeler Survey), V. III. Supplement-Geology*, U.S. Army Engineer Department, Washington, 420p.

Stretch, R. C., and Viles, H. A., 2002, The nature and rate of weathering by lichens on laval flows on Lanzarote, *Geomorphology* **47**: 87–94.

Thiagarajan, N., and Lee, C. A., 2004, Trace-element evidence for the origin of desert varnish by direct aqueous atmospheric deposition, *Earth and Planetary Science Letters* **224**: 131–141.

Thoma, S. G., Gallegos, D. P., and Smith, D. M., 1992, Impact of fracture coatings on fracture/matrix flow interactions in unsaturated, porous media, *Water Resources Research* **28**: 1357–1367.

Tiano, P., 2004, Innovative treatments for stone conservation, *Corrosion Reviews* **22**: 365–280.

Tratebas, A. M., Cerveny, N., and Dorn, R. I., 2004, The effects of fire on rock art: Microscopic evidence reveals the importance of weathering rinds, *Physical Geography* **25**: 313–333.

Trueman, N., 1965, The phosphate, volcanic and carbonate rocks of Christmas Island, *Journal Geological Society Australia* **12**: 261–283.

Urrutia, M. M., and Beveridge, T. J., 1994, Formation of fine-grained metal and silicate precipitates on a bacterial surface (*Bacillus subtilis*), *Chemical Geology* **116**: 261–280.

Urzì, C., Krumbein, W. E., and Warscheid, T., 1992, On the question of biogenic color changes of Mediterranean monuments (coatings – crust – microstromatolite – patina – scialbatura – skin – rock varnish), in: *II. International Symposium for the Conservation of Monuments in the Mediterranean Basin*, D. Decrouez, J. Chamay and F. Zezza, ed., Musée d'Histoire Naturelle, Geneva, pp. 397–420.

Vardenoe, W. W., 1965, A hypothesis for the formation of rimstone dams and gours, *National Speleological Society Bulletin* **27**: 151–152.

Varner, G. R., 2003, The destruction of America's Cultural Resources, *http://www.authorsden.com/visit/viewarticle.asp?AuthorID=1215&id=10400* (**accessed 3/8/05**).

Vicini, S., Princi, E., Pedemonte, E., Lazzari, M., and Chiantore, O., 2004, In situ polymerization of unfluorinated and fluorinated acrylic copolymers for the conservation of stone, *Journal of Applied Polymer Science* **91**: 3202–3213.

Viles, H., 1995, Ecological perspectives on rock surface weathering: towards a conceptual model, *Geomorphology* **13**: 21–35.

Viles, H. A., 2001, Scale issues in weathering studies, *Geomorphology* **41**: 61–72.

Viles, H. A., and Goudie, A. S., 1990, Tufas, travertines and allied carbonate deposits, *Progress in Physical Geography* **14**: 19–41.

Viles, H. A., and Goudie, A. S., 2004, Biofilms and case hardening on sandstones from Al-Quawayra, Jordan, *Earth Surface Processes and Landforms* **29**: 1473–1485.

Villa, N., Dorn, R. I., and Clark, J., 1995, Fine material in rock fractures: aeolian dust or weathering?, in: *Desert aeolian processes*, V. Tchakerian, ed., Chapman & Hall, London, pp. 219–231.

von Humboldt, A., 1812, *Personal Narrative of Travels to the Equinoctial Regions of America During the Years 1799-1804 V. II (Translated and Edited by T. Ross in 1907)*, George Bell & Sons, London, 521 p.

Wang, Y., McDonald, E., Amundson, R., McFadden, L., and Chadwick, O., 1996, An isotopic study of soils in chronological sequences of alluvial deposits, Providence Mountains, California, *Geological Society of America Bulletin* **108**: 379–391.

Wang, Y. X., and Hua, P. D., 1998, The environment, composition, and protection of Dazu rock inscriptions, *Environmental Geology* **33**: 295–300.

Wasklewicz, T., Staley, D., Volker, H., and Whitley, D. S., 2005, Terrestrial 3D laser scanning: A new method for recording rock art, *INORA* **41**: 16–25.

Watchman, A., 1985, *Mineralogical analysis of silica skins covering rock art*, Australian National Parks and Wildlife Service, Canberra, 281–289 p.

Watchman, A., 1990, A summary of occurrences of oxalate-rich crusts in Australia, *Rock Art Research* **7**: 44–50.

Watchman, A., 1992, Composition, formation and age of some Australian silica skins, *Australian Aboriginal Studies* **1992** (**1**): 61–66.

Watchman, A., 1997, Differences of Interpretation for Foz Côa Dating Results, *National Pictographic Society Newsletter* **8** (**1**): 7.

Watchman, A., 2000, A review of the history of dating rock varnishes, *Earth-Science Reviews* **49**: 261–277.

Watchman, A., 2002, A reply to Whitley and Simon, *INORA* **34**: 11–12.

Watchman, A., David, B., and McNiven, I., Flood, J., 2000, Microarchaeology of engraved and painted rock surface crusts at Yiwaralarlay (The Lightning Brothers site), Northern Territory, Australia, *Journal of Archaeological Science* **27**: 315–325.

Watchman, A., O'Connor, S., and Jones, R., 2005, Dating oxalate minerals 20–45 ka, *Journal of Archaeological Science* **32**: 369–374.

Watson, A., 1988, Desert gypsum crusts as palaeoenvironmental indicators: A micropetrographic study of crusts from southern Tunisia and the central Namib Desert, *Journal of Arid Environments* **15**: 19–42.

Wayne, D. M., Diaz, T. A., Fairhurst, R. J., Orndorff, R. L., and Pete, D. V., 2006, Direct major-and trace-element analyses of rock varnish by high resolution laser ablation inductively-coupled plasma mass spectrometry (LA-ICPMS), *Applied Geochemistry* **21**: 1410–1431.

Weaver, C. E., 1978, Mn-Fe Coatings on saprolite fracture surfaces, *Journal of Sedimentary Petrology* **48**: 595–610.

Weed, R., and Norton, S. A., 1991, Siliceous crusts, quartz rinds and biotic weathering of sandstones in the cold desert of Antarctica, in: *Diversity of environmental biogeochemistry (Developments in Geochemistry, Vol. 6)*, J. Berthelin, ed., Elsevier, Amsterdam, pp. 327–339.

Whalley, W. B., 1978, Scanning electron microscope examination of a laboratory-simulated silcrete, in: *Scanning electron microscopy in the study of sediments*, W. B. Whalley, ed., Geo-Abstracts, Norwich, pp. 399–405.

Whalley, W. B., Gellatly, A. F., Gordon, J. E., and Hansom, J. D., 1990, Ferromanganese rock varnish in North Norway: a subglacial origin, *Earth Surface Processes and Landforms* **15**: 265–275.

White, C. H., 1924, Desert varnish, *American Journal of Science* **7**: 413–420.

White, K., 1990, Spectral reflectance characteristics of rock varnish in arid areas, *School of Geography University of Oxford Research Paper* **46**: 1–38.

White, K., 1993a, Image processing of Thematic Mapper data for discriminating piedmont surficial materials in the Tunisian Southern Atlas, *International Journal of Remote Sensing* **14**: 961–977.

White, K., 1993b, Mapping the distribution and abundance of gypsum in South-Central Tunisia from Landsat Thematic Mapper data, *Zeitschrift für Geomorphologie* **37**: 309–325.

Whitley, D. S., 2001, Rock art and rock art research in worldwide perspective: an introduction, in: *Handbook of Rock art Research*, D. S. Whitley, ed., Altamira Press, Walnut Creek, pp. 7–51.

Whitley, D. S., 2005, *Introduction to Rock Art Research*, Left Coast Press, Walnut Creek, 215 p.

Whitley, D. S., and Annegarn, H. J., 1994, Cation-ratio dating of rock engravings from Klipfontein, Northern Cape Province, South Africa, in: *Contested Images: Diversity in Southern African Rock art Research*, T. A. Dowson and J. D. Lewis-Williams, eds., University of Witwatersrand Press, Johannesburg, pp. 189–197.

Whitley, D. S., and Simon, J. M., 2002a, Recent AMS radiocarbon rock engraving dates, *INORA* **32**: 11–16.

Whitley, D. S., and Simon, J. M., 2002b, Reply to Huyge and Watchman, *INORA* **34**: 12–21.

Whitley, D. S., Simon, J. M., and Dorn, R. I., 1999, Vision quest in the Coso Range, *American Indian Rock Art* **25**: 1–31.

Whitney, K. D., and Arnott, H. J., 1987, Calcium oxalate crystal morphology and development in *Agaricus bisporus*, *Mycologia* **79**: 180–187.

Williams, R., and Robinson, D., 1989, Origin and distribution of polygonal cracking of rock surfaces, *Geografiska Annaler* **71A**: 145–159.

Worsley, P., 1990, Lichenometry, in: *Geomorphological Techniques, Second Edition*, A. Goudie, ed., Unwin Hyman, London, pp. 422–428.

Zák, K., and Skála, R., 1993, Carbon isotopic composition of whewellite ($CaC_2O_4 \cdot H_2O$) from different geological environments and its significance, *Chemical Geology (Isotope Geosciences Section)* **106**: 123–131.

Zanin, Y. N., 1989, Phosphate-bearing weathering crusts and their related deposits, in: *Weathering: Its products and deposits. Volume II. Products – Deposits – Geotechnics*, A. Barto-Kryiakidis, ed., Theophrastus Publications, Athens, pp. 321–367.

Zhang, B. J., Yin, H. Y., Chen, D. Y., Shen, Z. Y., and Lu, H. M., 2001, The crude ca-oxalate conservation film on historic stone, *Journal of Inorganic Materials* **16**: 752–756.

Zhang, Y., Liu, T., and Li, S., 1990, Establishment of a cation-leaching curve of rock varnish and its application to the boundary region of Gansu and Xinjiang, western China, *Seismology and Geology (Beijing)* **12**: 251–261.

Zhou, B. G., Liu, T., and Zhang, Y. M., 2000, Rock varnish microlaminations from northern Tianshan, Xinjiang and their paleoclimatic implications, *Chinese Science Bulletin* **45**: 372–376.

Zhu, X., Wang, J., and Chen, J., 1985, Preliminary studies on the 'litholac' soaking on the Gobi rock surface in Gobi Desert, in: *Quaternary Geology and Environment of China*, Q. R. A. o. China, ed., China Ocean Press, Bejing, p. 129.

Zorin, Z. M., Churaev, N., Esipova, N., Sergeeva, I., Sobolev, V., and Gasanov, E., 1992, Influence of cationic surfactant on the surface charge of silica and on the stability of aqueous wetting films, *Journal of Colloid and Interface Science* **152**: 170–182.

Chapter 8

Rock Slopes

Alan D. Howard and Michael J. Selby

Introduction

The very slow chemical and physical weathering rates in desert areas coupled with a relatively high efficiency of wash processes, due to the general sparseness of vegetation, results in more widespread occurrence of slopes with little or no regolith than in areas with humid climates. This chapter outlines the processes and landforms occurring on desert slopes that are either massive bedrock or are scarps and cuestas in layered rocks dominated by outcropping resistant rock layers.

Slopes in Massive Rocks

Hillslopes formed on bodies of massive rock – that is, with few joints and high intact strength – are found in most of the world's deserts. These deserts vary in age, past climatic regimes, structure, and geological history. It seems inherently improbable, therefore, that all hillslopes will have one set of controls on their form, or even that the same controls have prevailed throughout the evolution of the hillslope. In this section the following influences on hillslope form will be considered: rock mass strength and slope forms; structural influences; sheeting; bornhardts; karst in siliceous rocks; etch processes and inherited forms.

A.D. Howard (✉)
Department of Environmental Sciences, University of Virginia, Charlottesville, VA 22904-4123, USA
e-mail: ah6p@cms.mail.virginia.edu

Rock Mass Strength and Slope Forms

There has been a general recognition by geomorphologists that rock resistance to processes of erosion has played a part in controlling the form of hillslopes, particularly in arid regions, but attempts to study such relationships in detail were dependent upon the development of rock mechanics as a discipline. The work of tunnelling and mining engineers has been important in the establishment of methods of investigation (see, for example, Terzaghi 1962, Brown 1981, Hoek and Bray 1981, Brady and Brown 1985, Bieniawski 1989). The application of rock mechanics approaches to the understanding of hillslopes has been particularly fruitful in arid and semi-arid areas where excellent exposure of rock outcrops is available (Selby 1980, 1982a,b,c,d, Moon and Selby 1983, Moon 1984a, 1986, Allison and Goudie 1990a,b).

The underpinning of the geomorphological studies listed above was the formulation of a semiquantitative method of evaluating the strength of rock masses (Selby 1980). The method involves a five-level ordinal rating of seven characteristics of exposed rock related to mass strength. These are intact rock strength (typically measured by Schmidt hammer), degree of weathering, spacing of joints, orientation of joints relative to the slope surface, joint width, joint continuity, and degree of groundwater outflow. The individual ratings of these characteristics are incorporated into a weighted sum characterizing rock mass strength. Details of this method have been referred to in accessible sources, such as Selby (1982d) and Gardiner and Dackombe (1983). The original method is fundamentally unchanged but Moon (1984b) has suggested some refinements to measurements and

Abrahams and Parsons (1987) have improved the definition of the envelope defining strength equilibrium slopes.

The contention that strength equilibrium slopes are widespread is confirmed by independent work and has a number of implications. (a) Rock slopes in strength equilibrium retreat to angles which preserve that equilibrium: if the process of retreat brings to the slope surface rocks of lower mass strength the slope angle will become lower; if higher strength rock masses are exposed, slope angles will steepen; only if rock mass strength is constant will slopes retreat parallel to themselves. (b) Only if controls other than those of mass strength supervene will the above generalization be invalid. (c) Rock slopes evolve to equilibrium angles relatively quickly. If this were not so, slope angles too high or too low for equilibrium would be more common than they are. (d) Gentle rock slopes remain essentially uneroded until either weakened by weathering or steepened by undercutting to the strength equilibrium gradient. Application of the technique is useful for distinguishing the effects of controls other than rock mass strength on slope inclination.

The assessment of strength equilibrium for hillslopes forming escarpments and inselbergs in South Africa, the Namib Desert, and Australia and on a great variety of lithologies has demonstrated that there is no simplistic model for slope evolution of the kind implied in such terms as parallel retreat or downwearing. If such ideas have any validity, it can only be in application to a particular variation in mass strength into a rock mass as the slopes on it evolve.

In all of the early work on the form of rock slopes the units for assessment were chosen as having a uniform slope and angle or an obvious uniformity of rock properties. Such a selection process has certain advantages, but it limits the application of statistical assessment of parameters. Allison and Goudie (1990a) have introduced the use of a 'kennedy parameter' for measuring slope shape and a fixed distance of 5 m over which slope angle changes are recorded. This method permits assessment of slope curvature. The kennedy parameter has a value of 0.5 for straight slopes, increasingly concave slopes have values tending towards 0.0 and increasingly convex slopes towards 1.0. Symmetrical concavo-convex or convexo-concave slopes also have a value of 0.5. Profile records must therefore include profile plots if errors are to be avoided. The same workers also advocate the use of the sine of the slope angle if frequency distributions and statistical analyses are to be applied to slope profile data.

Structural Influences

The rock mass strength rating system is insensitive of the condition in which critical joints or weak bedding planes dip steeply out of a slope – that is, where the stability of the overall slope is controlled by deep-seated structural influences rather than processes and resistance operating at the scale of individual joint blocks. If joints are infilled with weak materials, such as clays, the critical angle for stability could be as low as 10° (see Selby 1982d, pp. 72 and 158). A careful assessment of the shear strength along critical joints is then required. Basic methods are described by Brown (1981) and their application to geomorphic studies by Selby (1987).

Moon (1984a) has studied steep slopes in the Cape Mountains and recognized both unbuttressed and buttressed slopes with units which are as steep as 90°. Buttressed slopes are supported by rock units lower on the hillslope whenever the dominant joints fail to outcrop. This happens most obviously where bedding joints plunge at the same angle as the hillslope angle. Buttressed slopes have slope angles less than their apparent strength based upon the rock mass strength parameters due to lack of exposed dominant jointing. Such slopes demonstrate the importance of a lack of weathering along the joints and the lack of dilation which would otherwise allow sliding to take place. The effective friction angle along critical joints is assumed to exceed 55° in some cases, as there are few signs of mass failure of scarp faces. The elimination of buttressing by weathering, buckling, or other processes is an essential prerequisite to the development of strength equilibrium.

Rock slopes out of equilibrium with the rock mass strength rating can also result from rapid undercutting (by stream erosion of mass wasting lower on the slope), form dominance by non-mass-wasting processes (such as solution of limestones), rocks with a regolith cover, exhumation (for example, structural planes on resistant caprocks exposed by erosion of overlying weak rocks, former Richter slopes denuded of their debris cover, and some bornhardts), and slopes dominated by sheeting fracturing.

Sheeting

The formation of planar or gently curved joints conformable with the face of a cliff or valley floor is called sheeting. To merit this name there usually is evidence that the joints form more than one layer of separating rock. Such evidence is obtained from exposures which reveal parallel shells of rock separating from the parent rock mass, which is massive. Sheeting joints have been described from several rock types which can form massive bodies; granite and hard, dense sandstones (Bradley 1963) are the most common. In deserts domed inselbergs, called bornhardts, and high cliffs are the most common features which develop sheet structures (Figs. 8.1 and 8.2).

At least four major hypotheses have been proposed to account for sheeting: (a) sheeting results from stress release; (b) sheet structures are developed in granitic rocks by the formation of stretching planes during intrusion into the crust; (c) sheeting is the result of faulting with the development of secondary shears; and (d) lateral compression within the crust creates dome-like forms with concentric jointing being developed during the compression. The expansion of rock during weathering produces thin slabs of rock which, if confined laterally, may arch and create small-scale features which are similar to the larger-scale forms created by crustal stresses.

The unloading or stress release hypothesis was expressed in its most persuasive form by Gilbert (1904). He, like many geologists, was impressed by the

Fig. 8.2 Sheeting on the surface of a granite dome, showing various degrees of separation. Gross Spitzkoppe, Namibia (photo M.J. Selby)

evidence of rock bursts in mines and deep tunnels, of the springing up of rock slabs after retreat of glaciers and by the common occurrence of sheeting on the walls of deglaciated valleys and in quarries. Work in quarries (Dale 1923, Jahns 1943) showed clearly that the thickness of sheets increases with depth into the rock mass. Sheet thicknesses range from less than one metre to more than ten metres and transect rock structures, or even dykes in rock bodies, which have few or no other joint sets. Sheet structures terminate laterally where they meet other joints or weak rock units.

Stresses in the upper crust are usually described as being derived from four major sources: gravitational, tectonic, residual, and thermal (see Selby et al. 1988 for a review). Gravitational stresses at a point within a rock body are induced by the weight of the column of rock above that point. Rocks under load tend to expand transversely to the direction of the applied load, with the resulting transverse stress having a magnitude which is approximately one-third that of the vertical stress. Even if the overburden is removed the tendency for transverse expansion to occur is locked into the rock mass as a residual stress (Haxby and Turcotte 1976). Tectonic stresses result from convergence, and their presence is indicated by thrust faulting and conjugate joint sets. Stress fields in areas of convergence may yield horizontal stresses which exceed the local vertical stresses (McGarr and Gay, 1978). Thermal stresses result from the prevention of expansion or contraction of a solid during heating or cooling. The magnitude of the stresses in a confined rock body which cools through 100°C may exceed the tensile strength of the rock (Voight and St Pierre 1974). The stresses from

Fig. 8.1 A sheet of arkosic sandstone has separated from the dome of Ayers Rock, central Australia. The bedding of the rock lies in a vertical plane and has no influence on the sheeting. Note the tafoni on the dome face, which have been interpreted as having an origin within a weathering mantle (photo M.J. Selby)

gravitational, tectonic, and thermal effects may all be locked in the rock body as residual stresses if they cannot be relieved by expansion of the rock, internal deformation, or the development of joints.

In weak rocks and soils, continuous joint systems do not develop freely because the high void space and the presence of many microcracks and plastic clays within their structure prevent fracture propagation. In dense, strong rocks, by contrast, joints can propagate with few impediments (for a discussion of fracture mechanics and crack propagation see Einstein and Dershowitz 1990).

The tendency of rock bodies to expand as a result of gravitational loading and thermal stress from cooling has been analysed in some detail by finite-element analyses. Such analyses have now been applied to a number of geomorphic phenomena at scales ranging from small cliffs and individual slopes to mountain peaks and mountain massifs (see, for example, Yu and Coates 1970, Sturgul et al. 1976, Lee 1978, Augustinus and Selby 1990). It is evident from such analyses that all of the four forms of stress described above can create a tendency for rock masses to expand laterally and for stresses to be concentrated at particular sites along cliff faces, usually at the bases of cliffs (Yu and Coates 1970). Furthermore, the magnitude of the stresses commonly exceeds the tensile strength of the rock. As fractures are propagated in directions which are normal to the direction of the principal stress (Einstein and Dershowitz 1990), it is evident that stress relief can be a major cause of sheeting which can operate when the confining pressure of surrounding rock masses is removed by erosion.

Yatsu (1988) has offered a dissenting view of the role of residual or 'locked-in' stresses on development of sheeting and rock bursts, maintaining that steady-state creep can erase such stresses over the long timescales required for exposure of formerly deeply buried rock. Yatsu emphasized the role of neotectonic stress fields and gravitational stress related to present-day topography.

The evidence for steady-state creep is likely to be found in thin sections taken from rock at shallow depth. Such sections may be expected to show alignment of minerals such as micas and development of shears in crystals and silica overgrowths in orientations which are unrelated to original rock microstructure and are aligned downslope. No such evidence has been reported.

Neotectonic stress fields would be expected to create preferred alignments of sheeting joints, but sheeting occurs parallel to valley walls and dome surfaces with many orientations in a small area. Furthermore, many areas of sheeting occur in cratons far from areas of Cenozoic tectonism.

Gravitational stress is caused by existing overburdens and is relevant to flanks of steep-sided peaks and ridges and to walls of deep valleys. It is not relevant to areas of low relative relief where many examples of stress relief are found. It should be noted also that, although a few studies of stresses in individual boulders have been carried out, residual stresses have been measured in joint-bounded columns of basaltic lava (Bock 1979). Much further work is needed, but Bock's work has the implication that residual stresses, of thermal or gravitational origin, may be locked into small rock units.

The alternative mechanisms listed in the second paragraph of this section are not necessarily excluded from being contributing factors to sheeting but are, in essence, special cases which can apply to only a few cases in specific environments.

Mass Wasting Influences

Many steep bedrock slopes in desert environments exhibit chutes – linear depressions oriented downslope. Although most of these are developed along zones of structural weakness, their deepening and downslope integration may be enhanced by the motion of rockfalls and avalanches through the chutes. Such features are most elegantly developed as 'spur and gully' terrain in arctic and alpine environments where competition from runoff processes is negligible (Matthes 1938, Blackwelder 1942, Rapp 1960a,b, Luckman 1977, 1978, Akerman 1984, Rudberg 1986) and on the sides of steep scarps on Mars (Sharp and Malin 1975, Blasius et al. 1977, Lucchitta 1978). In arctic and alpine terrain dry rockfalls, debris avalanches, and snow avalanches appear to be capable of rock erosion (Matthes 1938, Blackwelder 1942, Rapp 1960a,b, Peev 1966, Gardiner 1970, 1983, Hewett 1972, Luckman 1977, 1978, Corner 1980, O'Loughlin and Pearce 1982, Ackroyd 1987). Lucchitta (1978) and Patton (1981) cited examples from desert areas, but competition from fluvial erosion makes the evidence for rockfall and avalanche erosion equivocal.

Howard (1990) has modelled gully erosion by avalanching using avalanche rheology and an approach similar to that used from modelling of snow avalanches (Perla et al. 1980, Dent and Lang 1980, 1983, Martinelli et al. 1980) and some debris avalanches (Cannon and Savage 1988, McEwen and Malin 1990). These models generally assume that the avalanche moves as a unit with the net downslope surface shear τ being expressed by a relationship

$$\tau = \rho_b gh(\cos\theta \tan\Phi - \mathrm{Sin}\,\theta) - \rho_b C_f V^n \quad (8.1)$$

where h is the avalanche thickness (often assumed to be constant during the avalanche motion), g is the gravitational acceleration, θ is the slope angle, Φ is a friction angle, V is the avalanche velocity, ρ_b is the flow bulk density, n is an exponent, and C_f is a friction coefficient. A theoretical basis for C_f is not firmly established and may represent air drag, internal frictional dissipation, and 'ploughing' of surface material (Perla et al. 1980). Some models utilize a 'turbulent' friction with $n = 2$, and others assume 'laminar' friction with $n = 1$ (Buser and Frutiger 1980, Dent and Lang 1980, 1983, Lang and Dent 1982, Perla et al. 1980, Martinelli et al. 1980, McClung and Schaerer 1983, Schiewiller and Hunter 1983, McEwen and Malin 1990, Cannon and Savage 1988). Rather than finding analytical solutions for travel distance, in these models avalanche motion is generally routed downslope over the existing terrain, with deposition where V drops to zero. These models also often account for lack of flow contact in overhangs and momentum loss at sudden decreases of slope angle. Howard (1990) showed that chutes can be created either through the action of debris motion triggering failure of partially weathered bedrock or regolith under conditions where they would otherwise be stable or through direct scour of the substrate.

Bornhardts

Bornhardts are steep-sided, domical hills with substantial surfaces of exposed rock (Willis 1936; Fig. 8.3). In detail they vary considerably in form from being nearly perfect hemispheres, through cylinders with domed tops, to elongated ovoids. These different forms have been given a variety of local and general descriptive names (see Twidale 1981, 1982a,b).

Fig. 8.3 Gross Spitzkoppe, Namibia, is a granite bornhardt which rises 600 m above the surrounding desert plain. Sheeting is occurring on the main face and at depth in the central section which is in deep shadow. On the *right*, sheets are being subdivided by cross-joints (photo M.J. Selby)

The name bornhardt honours the German geologist Wilhelm Bornhardt who gave evocative accounts of the landscapes of East Africa where granite domes stand above extensive plains (Bornhardt 1900, Willis 1936). The association of granite with isolated domical hills standing above plains is clearly implicit in the original designation, but hills with similar forms occur on other rock types of which Ayres Rock and the Olgas of central Australia are among the best known. Ayres Rock is formed of a coarse-grained arkose – that is, a sandstone rich in feldspars. The Olgas, on the other hand, consist of a massive conglomerate, the matrix of which is as resistant as the boulder- and gravel-sized clasts (Twidale 1978) (Figs. 8.3 and 8.4).

Fig. 8.4 Domes of the Olgas, central Australia, showing the effect of major joints in subdividing the massif into elongated domes (photo M.J. Selby)

The outstanding feature of bornhardts is their massive form with few major or continuous joints passing through the body of rock, but the margins of bornhardts can often be seen to be determined by bounding joints. Large rock masses may be partitioned by widely spaced joints which then separate the mass into a series of domes. Such partitioning can be seen in the Pondok Mountains of the Namib Desert (Selby 1982a) and in the Olgas (Fig. 8.4). Sheeting, with separation of curved plates of rock conformable with the dome surface, is a feature of many bornhardts. It is easily recognized that such jointing will permit survival of the dome form even after successive sheets are broken by cross joints, and weathering has caused disintegration of the resulting slabs. Sheet jointing, however, is not always concentric with the dome surface, and where it has a curvature with a shorter radius or longer radius than that of the dome, the result will be a steepening or flattening of the ultimate dome form.

Bornhardts have been reported from virtually all climatic zones, but they are especially abundant in the humid, subhumid, and arid landscapes of the Gondwana cratons where large exposed intrusions of granitic rocks are common. In areas of younger granitic and gneissic intrusions, granitic domes form the cores of mountain massifs and may be revealed by valley erosion along their flanks, as in the Yosemite Valley, California, or as growing intrusions in areas of active uplift, as in Papua New Guinea (Ollier and Pain 1981). In other continental settings diapiric intrusions form domed hills, as at Mount Duval, Australia (Korsch 1982). Such features, and the recognition that granite domes are created as part of the intrusion process and revealed as cover beds are removed, as in the inland Namib escarpment (Selby 1977), indicate clearly that some dome forms are entirely structural in origin and have forms maintained by sheeting processes (Fig. 8.5).

There is no evidence, or process known, for a structural origin for domes of sedimentary rock; for them, and possibly for some domes of igneous rock, other mechanisms have to be considered. It is widely recognized that rounded core stones develop from cuboid joint blocks within weathering profiles. On much larger scales, it is often assumed that similar rounding of corners will form domical rock masses within weathering profiles. Such processes may be aided, or even made possible, by stress release joints opening along the edges of large blocks and thus promoting the formation of rounded forms.

Fig. 8.5 Domes of granite being stripped of their cover rocks which are unweathered schists. This exposure is one of the best pieces of evidence for the creation of domes by intrusive emplacement and the inheritance of that form in the landscape as cover rocks are removed by erosion (photo M.J. Selby)

Construction of railways, roads, and mine tunnels (Boyé and Fritsch 1973) has revealed that many domes occur as compartments of solid relatively unjointed rock within a deeply weathered regolith. The widely accepted hypothesis that many bornhardts are resistant bodies of rock which have survived attack by subsurface weathering and then been revealed by erosional stripping of the regolith (Falconer 1911) is clearly demonstrable where domes stand above erosion surfaces below which deep regoliths survive. For example, in central Australia the exposure of Ayres Rock and the Olgas can be put at the end of the Mesozoic, yet these inselbergs stand above plains that have silcretes of early to middle Tertiary age at the surface (Twidale 1978).

Much of the controversy about bornhardts has been concerned with why bodies of rock should be relatively free of major joints while the rock around them is more closely jointed and therefore subject to penetration by water and chemical alteration. The focus of the debate has been on the lack of joints which has commonly been ascribed to the massive bodies being in a state of compression due to horizontal stresses in the crust (e.g. Twidale 1981, 1982a,b). The arguments, however, will be convincing only when stress levels in the rock are measured. Standard methods exist (Brady and Brown 1985) and have been used by mining engineers for some time. The difficulty is that the material around the bornhardt usually has either been removed by erosion or been deeply weathered, so comparable data cannot be obtained from the dome and its surroundings. In the Namib Desert, however, emerging domes and their surrounds are available for study.

It is evident from the above discussion that domed hills are not unique to any climatic environment. They are found in many of the world's deserts, but primarily because these lie on the surfaces of cratons which have been deeply weathered and stripped. The distinct feature of domes in deserts is their excellent exposure, with limited weathering of their surfaces by modern processes because the domes shed to their margins such rain as falls on them. Even the tafoni, pits, and other superficial features are usually attributed to weathering within the regolith. The process of stripping, whether it be by downwearing or backwearing of the surrounding material is irrelevant to understanding the origins of the domes. Similarly, the mineralogy and petrology of the rock is relevant only in so far as any bornhardt that survives has had to resist the processes acting on it.

Karst in Siliceous Rocks

Karst forms, possibly inherited from past, more humid environments, may be important parts of some present desert environments. Such an area is the Bungle Bungle of the south-eastern Kimberley region of north-western Australia (Young 1986, 1987b, 1988) and similar areas have been reported from the Sahara (Mainguet 1972). In both of these areas the rocks are quartzose sandstones and quartzites.

The siliceous rocks of the south-eastern Kimberley region are nearly horizontally bedded and of Devonian age. Their karstic features include complex fields of towers and sharp ridges, cliff-lined gullies, steep escarpments with cave and tube systems and conical hills standing above flat-floored valleys which debouch on to pediments which may have near right-angle junctions with the scarps. The local relief between the pediment surface and the ridge crests is seldom greater than 300 m (Fig. 8.6). The present climate is semi-arid with a dry season lasting from April to November and a wetter season occurring between December and March. The best estimate for the annual rainfall is 600 mm. On pediment surfaces, to the south, are widely spaced sand dunes now fixed by stable vegetation but indicating a formerly dry climate; the last arid phase ended 10,000–15,000 years BP (Wyrwoll 1979).

Studies by Young (1988) have indicated the processes which have contributed to, or perhaps controlled, the development of the present landscape. The primary evidence comes from scanning electron

Fig. 8.6 Conical hills and sinuous ridges of the Bungle Bungle, northern Australia, rising above a pediment (photo M.J. Selby)

microscope analyses of the rocks. The sandstones have quartz grains cemented by quartz overgrowths which did not eliminate all primary porosity. As a result water could circulate through the rock and etch the grains and dissolve much of the overgrowth silica. The sandstones have, in some formations, been left as granular interlocking grains with few cementing bonds.

The rock bodies now have relatively high compressive strength (35–55 MPa), as applied loads are transmitted by point-to-point contacts between grains, but very low tensile or shear strengths because of weak cementation. The sandstones are now friable and are readily broken down into small blocks or single grains by sediment-laden water. On cliff faces, bedding forms stand out clearly where small-scale fretting and granular exfoliation has undercut more coherent rock units (Fig. 8.7). Stream channels have obviously followed major joints and, in many canyons, streams have undercut valley walls. Below cliffs and escarpments the boundary between the cliff and the pediment is extremely sharp and has angles as high as 90°. Sheet flows crossing the pediments can readily transport the dominantly sand-size grains carried by wet-season flows.

Stream flow and sheet wash are clearly active and the presence of streams in the month of May suggests substantial storage of water in the rocks and the possibility of sapping as a significant factor in slope development. Case hardening is common on most of the slopes to depths of 1–10 mm, but whether the deposition of silica and clays to form the casing is still active or is a relic from former dry periods is unclear. There are a few sites in the Bungle Bungle, and in Hidden Valley near Kununurra to the north (see

Fig. 8.7 Kings Canyon, central Australia, with bedding forms of the sandstone exposed (photo M.J. Selby)

Young 1987b), where thin slab exfoliation has occurred in the relatively recent past and left bare unfretted faces which do not show evidence of case hardening; this may indicate either inactivity of case hardening or just very slow hardening.

The evidence for solution and weakening of the rocks is very clear, but the period over which this has occurred is much more difficult to establish. Caves and dolines in sandstone have usually been attributed to removal of silica in acidic solutions under tropical humid conditions (Pouyllau and Seurin 1985) where there is an abundance of organic molecules. Palaeoclimatic data from Australia indicate that much of the continent had a rainforest cover under which deep weathering profiles and lateritized surfaces developed until at least the middle Miocene when grasslands became common (Kemp 1981). However, laboratory studies of the solubility of silica show that this is greatest when pH is high and that it is enhanced by high chloride concentrations (Young 1987a). Potassium chloride crystals have been found in close association with etched quartz grains in the Kimberley sandstones. Whether much, or all, of the solution took place in humid or in arid conditions is not clear, but it seems probable that it may have occurred in both of these extreme tropical environments.

Some of the landforms described in this section have very similar forms to those found closer to the heart of the Australian arid zone near Alice Springs, where annual rainfall is in the range of 200–300 mm (Slatyer 1962). The horizontally bedded sandstones of Kings Canyon area (Fig. 8.7), for example, exemplify a joint-controlled incised landscape in which the canyon walls have a generally convex form and stand above flat-floored valleys and pediments.

Local controls by hard bands in the rock create a waterfall and plunge-pool long profile to channel beds. Local rockfalls are of greater significance here than they are in the Bungle Bungle. Whether or not solution of the silicate rocks has played any part in the evolution of these landforms is unknown.

The case for calling landforms developed under dominant solution processes 'karstic' has been made by Jennings (1983). They are relatively common in the tropical zone of the Gondwana continents and are of significance in several desert areas. The extent to which they are wholly or only partly inherited from humid environments is unclear.

Etch Processes and Inherited Forms

Discussions of the development of hillslopes in terms of rock resistance to modern processes and of the effect of modern processes are inevitably limited in their relevance to slopes which may have a history of 100 m.y. or more. The idea of very long periods for landscape development in the core zones of the continents is not new and has a history which can be traced back at least to Suess (German edition, 1885–1909; English edition, 1904–1924) who introduced the word 'shield' for describing the exposed nuclear core of each major land mass. The addition of the word 'craton', by Stille (1924) to distinguish the long-stabilized foundations of Precambrian age of every continent, added to the recognition of great geological age for extensive parts of all continents. Only the platforms of relatively undeformed sedimentary rocks deposited predominantly during marine transgressions, the marginal mobile belts, and the hotspot-generated zones of volcanism and rifting disturb the pattern of continental stability. The relevance of these comments to desert hillslopes is that many of the world's most extensive deserts are on cratons and may therefore have had upland masses with a long evolutionary history for their hillslopes.

Remote sensing has given a clear indication of both the extent of bedrock exposures and of features veneered by thin cover beds on desert cratonic surfaces (for example, Brown et al. 1989; Burke and Wells 1989). Broad swells and undulations of cratonic surfaces, and continental rifting with elevation of rift margins may be attributable to mantle convection. It has been suggested by Fairbridge (1988) that tectono-eustasy related to seafloor spreading has given rise to a thalossostatic condition linked to a biostatic regime (Erhart 1956) marked by a worldwide moist and warm climate with low relief, continuous vegetation cover, abundant non-seasonal rainfall, and strong biochemical weathering. This thalossostatic condition alternated with a disturbed or rhexistatic state in which high-standing continental surfaces (epeirostatic conditions) with low base levels, rejuvenation of fluvial systems, land erosion, seasonal rainfalls, zonally contrasting climates, formation of deserts, and monsoonal weather patterns were major features. Cratonic regions, according to this concept, tend to be characterized throughout geological history by alternation of chemical leaching, weathering, and duricrust formation with periods of erosion, desiccation, and offshore deposition.

Whether or not this grand but simple scheme is valid, it does add a possible explanation for the clear evidence of periodic stripping of weathering mantles from cratonic surfaces with the consequence that many hillforms, and especially those of inselbergs and related forms, have features that have been attributed to weathering beneath deep regoliths, followed by erosional stripping of the mantles and exposure of bedrock.

The idea of nearly world-wide cratonic weathering mantles was most vigorously expressed by Budel (1957) and has subsequently been vigorously espoused as the basis of many geomorphological explanations by Ollier (1988), Thomas (1989a,b), and Twidale (1990). The essence of the theory is that deep weathering mantles are developed through the action of meteoric waters penetrating along joints and fissures and progressively forming a mantle of saprolite which may extend to depths of tens or even hundreds of metres. The processes of alteration are fundamentally chemical, with the production of a residual soil that is leached of soluble material and depleted of the colloids removed by drainage waters. That the saprolite is formed without loss of volume is indicated by the survival in it of relict joints, veins of quartz and other structural forms derived from the original bedrock. The loss of mobile elements leaves a low-density residuum.

Weathering mantles commonly develop two major zones: an upper unsaturated zone which has a red coloration and is oxidized, and a lower saturated zone

which has white to pale green coloration and a reducing environment. The boundary between the two zones is the water table which fluctuates in level seasonally or as a result of storms and droughts. At depth, a zone is reached where joints are closed, water cannot penetrate, and therefore weathering cannot operate; this is the basal weathering front. All that is required for deep weathering is deep penetration of ground water in the liquid state (i.e. not frozen) and time. There is no necessary relationship between climate at the ground surface and chemical weathering at depth. At depths ranging from about one to ten metres, depending upon climatic zones and penetration of fresh water, soil temperatures are constant at the mean annual surface temperature. At greater depths the geothermal heat flow controls soil temperatures. The availability of groundwater depends mostly on fresh inputs from percolation but is commonly available as 'old' water in arid and semi-arid zones.

It has been pointed out by Habermehl (1980) and Ollier (1988) that the Great Artesian Basin of Australia has groundwater extending to depths of 3,000 m, so weathering by hydrolysis can also extend to this depth. Stable isotope studies show that the water is of meteoric origin (not connate) and flow rates indicate that it takes two million years for the groundwater to flow from recharge areas to discharge zones in artesian springs. The present land surface is a desert, but weathering of the basal rock is occurring under the control of geothermal heat and Pliocene, or early Pleistocene, meteoric water. The Great Artesian Basin is, no doubt, an extreme example but it illustrates the general principle.

Deep weathering profiles have often been associated with current humid tropical climates, perhaps because the surfaces of cratons which are now within the tropics have many land surfaces formed upon deeply weathered mantles. There is increasing evidence that some weathering profiles are of great age; in Australia, for example, Mesozoic and early Cenozoic ages are recognized (Idnurm and Senior 1978). More widely it has been recognized that relics of deep weathering profiles occur in areas which are far from the tropics. Hack (1980) noted saprolite in the Piedmont and Blue Ridge areas of the Appalachians with an average depth of 18 m and a maximum depth which may exceed 90 m; Hall (1986) identified preglacial saprolites in Buchan, Scotland, up to 50 m deep, and in the Gaick area of the Grampian Mountains up to 17 m (Hall 1988); Bouchard (1985) has found saprolites up to 15 m thick at protected sites covered by Quaternary ice sheets in Canada.

The relevance of all of these observations for an understanding of hillslope development in the modern desert zones is that many features now visible may have developed at the basal weathering front and later been exposed at the ground surface as regolith was stripped in periods of rhexistasy.

Etch Forms of Deserts

The recognition that boulders, and other minor and major forms, have developed within weathering mantles and at the weathering front has a long history which goes back to the beginning of modern geology (Twidale 1990). These forms when exposed by mantle stripping are collectively known as etch forms (Wayland 1933). Weathering fronts are particularly sharp in granitic and other crystalline rocks of low permeability. The front is more diffuse, and essentially a zone, in the weaker sedimentary rocks. Crystalline rocks commonly form the bedrock of cratons, and the various associations of boulders recognized as boulder-strewn platforms (Oberlander 1972), tors or koppies (Fig. 8.8) are readily identified as remnants of corestones because corestones are common in exposures through weathering mantles. Perhaps more significant was the proposal by Falconer (1911) that the inselberg landscapes of northern Nigeria were shaped not by epigene processes but by chemical action of waters acting at the weathering front. Furthermore, he recognized that the

variations in depth of penetration of the front were controlled by the spacing of joints in the bedrock. The result of mantle stripping was therefore an irregular land surface with inselbergs standing above plains with residual mantles of varying depth. Such ideas were elaborated into a system of geomorphological evolution of the cratonic surfaces by Budel (1957, 1982).

The evidence in favour of the concept of etchforms developing on cratonic surfaces is now very strong. It indicates the development of currently exposed land surfaces over periods of 100 m.y. or more. This timespan far exceeds the period of existence of the world's major deserts which are mid to late Tertiary in origin (see Selby 1985 for a review). The obvious conclusion is that many of the major hill mass and plain landforms had an origin beneath deep weathering mantles of ancient origin. Subsequent mantle stripping has exposed them to subaerial processes of weathering and erosion. In the time since exposure, denudation of bare rock has been sufficient for many hillslopes to achieve slope angles which are in equilibrium with the mass strength of their rocks, but on hillslopes formed of massive rocks with little pre-existing jointing the possible controls on form include survival of etch forms; survival of the original form through stress release joints developing nearly parallel to the original rock surface; the development of cross fractures which permit an incomplete adjustment to the slope angle controlled by rock mass strength; survival of structural influences and controls; dominance by solution processes; control by talus covers and formation of Richter-denudation slope units; undercutting of upper slope units by sapping, stream action, or recession of an underlying weak rock unit.

Slopes in Layered Rocks

Fig. 8.8 Devils Marbles, central Australia, which have been interpreted as residual etchforms developed within a weathering mantle which has now been stripped (photo M.J. Selby)

Many desert areas are underlain by generally flat-lying sedimentary rocks of varied composition, sometimes intermingled with tabular intrusive or extrusive volcanic rocks. Examples include the Colorado Plateau in the south-western United States, North Africa, the Arabian peninsula, and portions of the other major deserts. The erosion of such sedimentary sequences creates a landscape of scarps or cuestas capped by the more resistant rock units. By contrast, a few desert areas, such as the Zagros Mountains region of Iran, have complex patterns of cuestas, hogbacks, strike valleys, etc. developed in strongly folded or faulted layered sedimentary rocks. The discussion here will focus on the simple cuesta landforms of flat or inclined beds, although the general principles are applicable in areas of more complicated structure.

The classification of rock slope types introduced by Selby (1980, 1987) and Moon (1984b) can also be applied to cuesta landforms. Figure 8.9 shows a classification of slope elements on an eroded anticline in layered rocks. Strength equilibrium slopes (shown by =) occur primarily on cliff faces eroded by small rockfalls, but where such cliffs are actively eroded by stream undercutting (+U+) they may be steeper than strength equilibrium. Scarp faces dominantly eroded by landslides may be less steep than strength equilibrium and those dominantly wasted by rockfalls involving the entire cliff face may be steeper than strength

equilibrium. Structurally controlled slopes (–S–) occur where erosion of weaker layers exposes the top surface of resistant rocks. Normal slopes in weak rock (typically badlands with shallow regolith) are shown by –E–. Rampart slopes in weak rocks below cliffs of resistant rock will generally be talus (–D–) or Richter slopes (not shown).

In a general sense cuestas are an automatic adjustment of the landscape to permit rocks of varied erosional resistance to be eroded at roughly equivalent rates (Hack 1966). When a resistant rock layer is first exposed the erosion rate diminishes on the exposed top of the resistant layer as the overlying weaker rocks are removed, often creating lithologically controlled near-planar upland surfaces called stripped plains. As the resistant layer becomes more elevated relative to surrounding areas of weaker rocks, the caprock is eventually breached, exposing underlying weaker rock units,

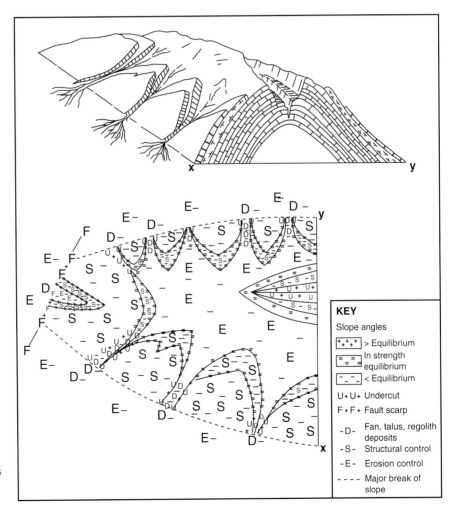

Fig. 8.9 Identification of landform elements on cuestas in folded rocks (from Selby 1987, Fig. 15.13)

KEY

Slope angles

$\boxed{^+\!+_+^+}$ > Equilibrium

$\boxed{=\ =\ =}$ In strength equilibrium

$\boxed{^-\,_-^-}$ < Equilibrium

U+U+ Undercut

F+F+ Fault scarp

–D– Fan, talus, regolith deposits

–S– Structural control

–E– Erosion control

- - - - Major break of slope

whose rapid erosion creates a steep scarp and accelerated caprock erosion by virtue of the steep gradients relative to the superjacent stripped plain. This discussion of such scarps is organized into two general headings: evolution of scarps in profile and evolution of scarp planforms. Examples are taken primarily from the Colorado Plateau, south-western United States. Additional illustrations of the features described here and a road log of sapping and geologic features are given in Howard and Kochel (1988).

Evolution of Scarps in Profile

Resistant sandstones, limestones, and volcanic flows and sills in desert areas are generally exposed as bare rock slopes except where mantled with aeolian sands or alluvium. However, areas of very low relief, such as stripped plains, may be mantled with sandy, cobbly, poorly horizonated soils and scrubby vegetation. Two morphological end members characterize the resistant rock exposures, low to high relief slopes in massive rock (in particular, the rounded sandstone exposures termed 'slickrock') and cliff or scarp slopes developed where the caprocks are being undermined. Emphasis in this section is placed on scarp-front processes and morphology. An intermediate landform type, termed

segmented cliffs by Oberlander (1977), will also be discussed.

The more readily weathered and eroded rock units, generally shales or poorly cemented sandstones or alluvium, are commonly eroded into badlands where they are thick (Chapter 10). But in areas with thick interbedded caprock-forming units the easily eroded strata are usually exposed primarily on the subcaprock slopes or ramparts. Terms used here to describe the characteristic parts of an escarpment are shown in Fig. 8.10. Several alternative terms have been used to denote the rampart, including subtalus slope (Koons 1955), debris-covered slope (Cooke and Warren 1973), footslope (Ahnert 1960, Oberlander 1977), lower slope (Schmidt 1987), and substrate ramp (Oberlander 1989).

The exposure of weaker strata (generally shales or poorly cemented and/or highly fractured sandstones) beneath massive sandstones causes undermining of the sandstone, leading to cliff development and rapid scarp backwasting. A far greater volume of rock is initially broken up by scarp retreat than by erosion on backslopes when considering average rates over large areas. Because of the rapid retreat of scarp slopes, the cliffs generally eat back into pre-existing backslopes. Figure 8.11 shows an example where updip exposure of shale near the stream level (right side of picture) has caused development of cliffs and their backwasting into sandstone slickrock slopes; such undermining was

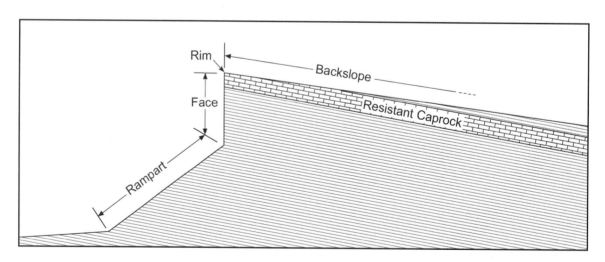

Fig. 8.10 Escarpment components. The face is the vertical to near-vertical cliff developed at the *top* of the escarpment mostly in the caprock but occasionally extending into the non-resistant layer beneath. The rampart is a slope of lesser inclination

extending from the bottom of the face to the base of the escarpment which is commonly partly mantled with debris from backwasting of the face (from Howard 1970, Fig. 55)

Fig. 8.11 Landforms of De Chelly Sandstone north of Kayenta, Arizona along US Highway 163. Slickrock slopes to left give way to vertical scarps at right where undermining is active due to exposure of shaly Organ Rock Tongue at the base of the scarps (photo A. Howard)

discussed by Ahnert (1960) and Oberlander (1977). The relief developed on scarp backslopes depends upon the erodibility of the caprock unit compared with overlying units, the thickness of the caprock, the types of weathering and erosional processes acting on the exposed rock, and rates of basel level lowering.

Processes of Scarp Erosion

Wasting of caprocks occurs primarily by rockfall, block-by-block undermining, and slumping. Some sandstone caprocks (e.g. the Morrison Formation of the American South-west) are undermined block-by-block by weathering and erosion of the underlying shale without rapid fall of the undermined blocks (Fig. 8.12). The blocks may be repeatedly lowered with little downslope sliding or rolling, but typically the blocks slide and occasionally roll a short distance upon being undermined. Block-by-block undermining requires a relatively thin caprock, well-developed jointing, and shale that weathers easily by addition of water (e.g. the smectitic Morrison Shale). Slumping is prevalent on relatively few escarpments, where it may dominate as the mechanism of scarp retreat (Fig. 8.13). The Toreva block slumps are a classic example (Reiche 1937), and other examples have been discussed by Strahler (1940) and Watson and Wright (1963). Conditions leading to slumping failure have not been firmly established, but a low shear strength of the unweathered subcaprock unit is probably the major factor. Low shear strength can result from low bulk strength or a high degree of fracturing and/or abundant bedding plane partings. Other factors may be deep

Fig. 8.12 Block-by-block undermining of channel sandstones in shales of Morrison Formation, near Hanksville, Utah (**a** and **b**) (photo A. Howard)

Fig. 8.13 Slump failure of Morrison–Summerville escarpment near Hanksville, Utah (photo A. Howard)

Fig. 8.14 Portion of escarpment of Emery Sandstone over Mancos Shale on North Caineville Mesa near Hanksville, Utah, along Utah Highway 24. Note extensive deposits of rockfall debris that are dissected by continuing erosion of shale on rampart. Cliff in alcove extends far into the underlying Mancos Shale. Seepage lines are present at the base of the Emery Sandstone in the alcove. High debris blankets the *right* and *left* of the alcove that extend upward to the base of the sandstone are strongly dissected at their base. Presence of the debris blanket inhibits further scarp backwasting until the debris is weathered and eroded. These high debris blankets are probably a Bull Lake equivalent (early Wisconsinan?) pluvial deposit. The picture is taken from the top of a sandstone debris-capped Bull Lake pediment that formerly merged with the dissected talus deposits. The Bull Lake pediment formerly extended behind the photographer to merge with Fremont River gravels (now terraces about 60 m above present river level)

weathering of the subcaprock unit by groundwater flow and high pore water pressures. Landslide and rockfall processes were summarized in Anderson and Richards (1987) and Brunsden and Prior (1984).

Rockfall is the most common form of scarp retreat (Fig. 8.14), involving events ranging from calving of individual blocks to the failure and fall of a wide segment of the face, resulting in a rock avalanche on the scarp rampart. Debris produced by rockfalls with high potential energy may result in powdering of a large percentage of the original rock (Schumm and Chorley 1966), but on most scarps the coarse debris produced by the rockfalls must be weathered and eroded before further scarp retreat can occur (Fig. 8.15). Weathering processes acting on the debris are similar to those occurring on slickrock slopes, including splitting or shattering, granular disintegration, and solution of cement (or the rock *en masse* in the case of limestones). The necessity for weathering of scarp-front debris before further erosion of the subcaprock unit leads to a natural episodic nature of rockfalls and scarp morphology, as outlined by Koons (1955), Schipull (1980), and Schmidt (1987) (Fig. 8.14). Where caprocks are eroded primarily by large rockfalls continued erosion of the subcaprock unit at the margins or base of the rockfall eventually raise the debris blanket into relief, sometimes forming subsidiary small escarpments where the debris blanket is subject to further mass wasting. Thus old rockfalls stand well above surrounding slopes of both exposed subcaprock unit and younger rockfalls. When these old rockfalls are contiguous with the scarp face, they prevent the development of high relief at the cliff face, and thereby inhibit further rockfalls until the talus

is weathered and eroded. In some cases the talus at the foot of the rampart becomes isolated from the scarp face as erosion of the subcaprock unit continues, forming a talus flatiron (Koons 1955, Schipull 1980, Gerson 1982, Schmidt 1987). Talus flatirons thus formed have been termed non-cyclic flatirons as contrasted with similar features resulting from climatic fluctuations (Schmidt 1989a) (Chapter 22). Schmidt (1987, 1989a) suggested that flatirons are best developed in scarps with heterogeneous subcaprock strata, including beds of variable resistance and slope inclination. However, flatirons are also well developed on the scarps of the Colorado Plateau composed of massive sandstones over homogeneous marine shales. Gerson and Grossman (1987) noted that flatirons are absent on desert scarps lacking a strong caprock over a weaker subcaprock unit.

Most prominent scarps on the Colorado Plateau are formed of massive sandstone underlain by shale or other easily weathered rock, so that backwasting is caused by a combination of weathering of exposed

Fig. 8.15 Recent rockfall in Navajo Sandstone in the Inscription House area of the Navajo indian reservation showing abundant rockfall debris (photo A. Howard). Cliff above rockfall is approximately 30 m tall

caprocks (e.g. off-loading fracturing, freeze–thaw, and groundwater sapping) and loss of bulk strength of the underlying layer accompanied by erosion of the scarp rampart. However, some of the incompetent layers producing scarps are strong in bulk but are eroded primarily because of denser fracturing relative to more massive (but not necessarily stronger) overlying sandstones (Oberlander 1977, Nicholas and Dixon 1986). Creep of subcaprock shales has been implicated in breakup of caprocks (block gliding) in humid environments (e.g. Zaruba and Mencl 1982) but has not been noted on desert scarps. It is possible that slight creep or off-loading expansion in shales may locally be a factor in development of off-loading fractures in overlying sandstone caprocks. Gravity-induced creep of evaporite beds has been suggested as the mechanism responsible for creating the miniature horst and graben structure of the Needles District in Canyonlands National Park, Utah (McGill and Stromquist 1975).

Although rockfall and slumping are the major transport processes in scarp backwasting, weathering and erosion by groundwater commonly can be as, or more, important in weakening the scarp face than is undermining by erosion of the rampart. The role of groundwater sapping is discussed separately below.

The amount of caprock talus exposed on scarp ramparts depends in part on the planform curvature of scarps, being greater in re-entrants where caprock debris converges on the lower rampart and lesser in front of headlands or projections (or around small buttes)

where debris is spread radially. The amount of debris is also controlled by spatial variations in rates of scarp retreat (generally higher at the head of re-entrants) and by the volume of caprock eroded per unit amount of backwasting, which is higher in re-entrants and lower at headlands.

Scarp ramparts are eroded by a variety of processes, including normal slope and rill erosion where the subcaprock unit is exposed and weathering and erosional processes act on caprock detritus. Weathering processes on talus include frost and/or hydration splitting or shattering, spalling due to salt crystallization (primarily on unexposed surfaces), granular disintegration, and solution. The relative mix of these weathering processes depends upon the size and composition of the talus. Calcareous-cemented sandstones, common on the Colorado Plateau, are primarily shattered in large blocks, but succumb to granular disintegration in small blocks, yielding easily eroded sand-sized detritus. On the other hand, siliceous-cemented sandstones do not weather by granular disintegration, and large blocks of caprock commonly remain behind as the scarp retreats (Fig. 8.12b).

Where caprock debris is copious or very resistant, the talus material may be reworked several times before its final removal due to continuing erosion of the subcaprock unit, forming elevated blankets eroded at its margins as subsidiary scarps (e.g. the flatirons discussed above), development of individual rocks on subcaprock unit pedestals (damoiselles), or less dramatic undermining and rolling of individual boulders. Slopewash and gullying are important in removing sand- to gravel-sized weathering products, and some talus blankets are extensively modified by, or even emplaced by, wet debris flows (Gerson and Grossman 1987).

Absolute and Relative Rates of Scarp Erosion

A variety of methods has been used to date regional rates of scarp retreat in desert areas, including archaeological dating (Sancho et al. 1988), beheading of consequent valleys (Schmidt 1980, 1989b), stratigraphic relationships with dated volcanic or sedimentary deposits (Lucchitta 1975), and age of faulting initiating scarp retreat (Yair and Gerson 1974). Inferred retreat rates vary over two decades from about 0.1 to 10 m

1000 y^{-1} (Schmidt 1988, 1989b, Oberlander 1989). The primary long-term controls over erosion rates are rate of base level lowering and rock dip, as discussed below. However, short-term erosion rates are strongly influenced by climate variations and climate-related changes in local base level.

The relative rates of erosion on different parts of cuestas can be illustrated by considering, as a first approximation, that the form elements maintain a constant gradient and a constant position relative to the stratigraphic layers through time. These assumptions require a constancy of both stream erosion and slope processes through time, which probably approximates the long-term average behaviour of scarp erosion but not the short-term changes due to climatic fluctuations. The assumption of a constant position of slope elements relative to the stratigraphy (Fig. 8.16b) is clearly a closer approximation to scarp evolution than the assumption that slope elements retain a constant position through time (i.e. a constant rate of vertical erosion on all elements of the scarp (Fig. 8.16a)).

In horizontal stratified rock these assumptions predict a rate of vertical erosion proportional to the slope tangent, whereas the horizontal rate of erosion (lateral backwasting) is identical on all slope elements (Fig. 8.16b). This implies an infinite rate of downwasting for a vertical cliff, which is an artifact of considering cliff retreat as continuous erosion rather than as discrete events such as rockfalls. Therefore, as mentioned above, on a typical escarpment the downwasting of the

slickrock slope on top of the caprock is very slow compared with both cliff retreat and vertical erosion below the rim.

The relative rates of erosion on various slope elements are also affected by the structural dip. If all form elements erode at an equal rate parallel to the structure (that is, in a downdip direction) with constant gradient, then the instantaneous rate of vertical downwasting V_t is given by the structural dip d, the slope angle s, and the rate of downdip backwasting of the escarpment D_t. Where the slope is inclined with the dip (Fig. 8.17a),

$$V_t = D_t(\sin d - \cos d \tan s), \text{ for } 90° > d > s > 0°$$
$$(8.2)$$

and (Fig. 8.17b),

$$V_t = D_t(\tan s \cos d - \sin d), \text{ for } 90° > s > d > 0°$$
$$(8.3)$$

Where the slope opposes the structural dip (Fig. 8.17a):

$$V_t = D_t(\sin d + \cos d \tan s), \text{ for } 90° > s > 0°. \quad (8.4)$$

During continued downcutting by streams draining the escarpment, the relief should adjust until downdip exposure of new caprock and updip removal by backwasting are roughly balanced. Therefore the rate of vertical reduction of the rim should be independent of the dip (maintaining a constant relief through time), whereas horizontal retreat of the escarpment would be inversely proportional to the tangent of the dip, and

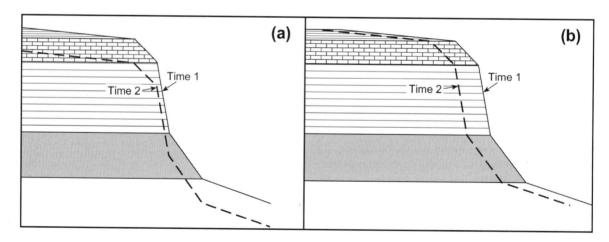

Fig. 8.16 Erosion of a compound hillslope under the assumptions of (**a**), uniform rate of downwasting on all slope elements, and (**b**), maintenance of correlations between slope elements and stratigraphic units. Both assume that gradients of the respective elements remain constant through time. Under the first assumption (**a**) the volume of material removed increases with decrease of gradient, whereas the volume decreases in (**b**) (from Howard 1970, Fig. 56)

(a)

(b)

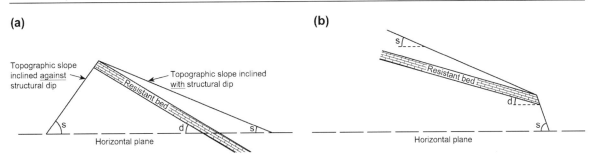

Fig. 8.17 Definition of angles and slope elements on escarpments in dipping strata (from Howard 1970, Fig. 58)

the volume of caprock eroded per unit time would be inversely proportional to the sine of the dip. The very rapid rate of horizontal retreat predicted for low dips does not occur because the escarpment becomes segmented by erosion along drainage lines into isolated mesas and buttes whose local relief, distance from the main escarpment, and rate of backwasting increase through time. Nevertheless, these considerations imply that, in general, a greater volume of rock must be eroded per unit time from gently dipping scarps than from steeper ones. The gradients and total relief on a given scarp should increase where the structural dip decreases to maintain relatively constant rates of vertical reduction of the rim. Figure 8.18 compares relief of escarpments on two sandstones in the Henry Mountains area, Utah, as a function of the reciprocal of the sine

of the dip, showing that there is a relationship of the type predicted for the Emery Sandstone. The relationship is poor for the low escarpment of Ferron Sandstone. Areal variations in fluvial incision at the scarp base since the Bull Lake glacial maximum (varying from about 60 m along Fremont River to a few metres in remote locations) may be responsible for the large scatter for this low scarp.

The overall height and steepness of a given scarp should be controlled by one of three factors: (a) development of sufficient relief to trigger caprock mass wasting (which will be related to caprock resistance); (b) the rate of weathering and erosion of caprock debris on scarp ramparts; and (c) the rate of erosion of the subcaprock unit.

Caprock resistance has been suggested as a controlling factor by Schumm and Chorley (1966), Nicholas and Dixon (1986) (who related scarp backwasting rates to degree of fracturing of the lower caprock unit), and Schmidt (1989b) (who found a strong relationship between regional backwasting rates and the product of caprock thickness and a measure of caprock resistance). To the degree that scarps adjust over long time intervals to balance rate of caprock removal with long-term rates of base level lowering (i.e. V_t in Equations 8.2, 8.3, 8.4 is areally uniform), relationships between backwasting *rate* and caprock resistance of the type found by Nicholas and Dixon (1986) and Schmidt (1989b) cannot be universally valid (although *local* variations in back-wasting rates on a given scarp may be related to caprock resistance as found by Nicholas and Dixon (1986) due to exposure of zones of caprock of differing resistance to erosion). Rather, the back-wasting rate D_t will adjust to be a function only of rock dip and regional erosion rate V_t. As suggested above, overall scarp height and steepness adjust to equalize erosion rates, and these should be related to caprock resistance and thickness (the scarps

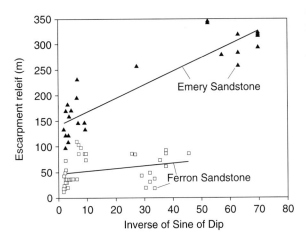

Fig. 8.18 Relief of escarpments in dipping strata as a function of the angle of dip for two sandstones sandwiched between over- and underlying shales. Relief is measured from crest of escarpment to base of scarp rampart (see Fig. 8.10). Each data point is a measurement from a separate location along the western edge of the Henry Mountains, Utah. A least-squares regression line has been fitted to observations from each sandstone scarp (based on Howard 1970, Fig. 59)

in the thin Ferron Sandstone are lower than those of the massive, thick Emery Sandstone (Fig. 8.18)). It seems reasonable that a threshold scarp steepness and/or height would be required to trigger the more energetic types of mass wasting processes, such as rock avalanche and landsliding. Block-by-block undermining small rockfalls and caprock weakening by groundwater sapping are less clearly related to overall scarp relief.

Koons (1955) emphasized that accumulated caprock talus protects the caprock from further large rockfalls until it is removed. Koons (1955) and Howard (1970) suggested that the length of the rampart self-adjusts over the long run to provide a surface area sufficient to weather talus at the rate that it is supplied. Schumm and Chorley (1966) introduced the talus weathering ratio, which they defined as the ratio of the rate of talus production from the cliff to the rate of talus destruction on the rampart. They noted that a ratio greater than unity leads to a moribund scarp choked in its own detritus, similar to the models of Lehmann (1933) and others (summary in Scheidegger 1991, pp. 130–4). Schumm and Chorley (1966) suggested that ratios less than unity characterize certain nearly debris-free scarps encountered on the Colorado Plateau. However, over the long run, continuing scarp retreat implies the talus weathering ratio equals unity, since only as much talus can be weathered as is produced (Gerson and Grossman 1987, Howard and Kochel 1988). This concept can be illustrated by letting P equal the volumetric rate of supply of caprock talus per unit time per unit width of scarp, K equal the potential volumetric erosion rate (either weathering- or erosion-limited) of talus per unit area of scarp rampart, and L the required rampart length. Then for balance of addition and removal

$$L = P/K \qquad (8.5)$$

However, as discussed below, this ratio may vary considerably as a result of climatic fluctuations as well as local short-term imbalances of P and K, as discussed by Koons (1955). Since talus on ramparts is composed of caprock debris, scarp relief controlled by talus weathering will be indirectly controlled by caprock resistance.

The rate of erosion of the subcaprock bedrock (commonly shales) has been cited as the controlling factor for scarp erosion rates by Gerson and Grossman (1987) and Schipull (1980). Strictly speaking, if the rate of subcaprock bedrock erosion were the dominant factor in scarp retreat, scarps should not be higher or different in form than other slopes in the subcaprock unit where no caprock is present. None the less, it is true that base level control is transmitted to the scarp via the channels and slopes developed in the subcaprock unit.

It is likely that these three factors vary in relative importance in controlling scarp form from scarp to scarp, from place to place on the same scarp, and through time as climate and/or base level control varies. Scarps that backwaste largely by rockfall are most likely to have planforms controlled over the long run by requirements for weathering of caprock debris, leading to the observation that 50–80% of the scarp front is covered by talus at equilibrium (Gerson and Grossman 1987). As will be discussed further below, many scarps on the Colorado Plateau are quite stable under the present climatic regime, so that the ramparts are nearly bare of talus, and tall cliffs have developed in both the caprock and in the subcaprock shales. Thus both base level control and talus weathering at present have little influence on scarp form, and the backwasting that does occur is largely due to caprock weathering and small rock-falls. On the other hand, for escarpments characterized by block-by-block undermining (Fig. 8.12) caprock resistance is less important than erosion of the subcaprock shales. In fact, where the caprocks consist of hard-to-weather silica-cemented sandstones, caprock boulders may be gradually let down by undermining and rolling while the escarpment continues to retreat, leaving piles of large caprock fragments (Fig. 8.12b).

Schmidt (1989b) noted that very resistant caprocks often include less resistant beds in the subcaprock strata that, in the absence of the more resistant overlying bed, would independently form scarps. In sections of scarp that are linear or indented, debris shed from the overlying caprock largely prevents development of subsidiary lower scarps, but in front of headlands, where rapid retreat of the main caprock and radial dispersal of its detritus leave the subcaprock units largely free of talus, the lower resistant units commonly form low scarps.

Backslopes on Caprock

In the classic case of a scarp composed of a thin resistant layer sandwiched between thick, easily eroded strata (Fig. 8.10) the caprock is exposed as a low-relief stripped plain on the scarp backslope, and caprock erosion occurs primarily at the scarp face. However, when the caprock is thick, erosional sculpting of the backslope may be as, or more, important than the lateral attack at scarp faces. Landform development on thick, homogeneous rocks is discussed in the Slopes in Massive Rocks section, but processes and landforms on slickrock slopes are outlined here because of their importance in development of segmented scarps.

The striking and unusual slickrock slopes occur on desert sandstone exposures as low, generally rolling relief on bare rock slopes (Figs. 8.11 and 8.19). Hill forms are generally convex to convexoconcave and rather irregular due to the prevalence of small-scale structural and lithologic controls exerted by the exposed rock upon weathering and erosional processes. Slickrock slopes occur most commonly on the backslopes of cuestas, but high relief forms occur in thick, massive sandstones such as the Navajo Sandstone. Where strong structural control by jointing or faulting occurs, the fractures tend to be eroded into furrows or valleys, and the sandstone landscape takes on a reticulated or maze-like appearance as at Arches National Park, Utah. Doelling (1985) noted that sandy colluvium collecting along depressions developed on joints accelerates weathering of the sandstone by providing a moist environment; thus the influence of fractures on the topography is enhanced by a positive feedback on weathering rates. Small-scale horst-and-graben development associated with extension caused by flow of underlying evaporites has created the 'needles' section of Canyonlands National Park, Utah (McGill and Stromquist 1975).

Slickrock slopes are weathering limited (Carson and Kirkby 1972, pp. 104–6) in that transport processes are potentially more rapid than weathering processes. That is, loose debris is removed from the slopes as fast as it is produced by weathering so that little or no loose residuum covers the bedrock. On slickrock slopes the bedding is emphasized by the grain-by-grain loosening or disintegration of thin surface crusts or whole layers of the sandstone exposed on these weathering-limited slopes, particularly on exposures of the massively cross-bedded Navajo Sandstone (Fig. 8.20). Coarser sand layers with fewer grain-to-grain contacts weather and loosen most readily, aiding differential surface expression of minor lithologic variations (Hamilton 1984). Despite these microscale lithologic controls, the slickrock slopes generally show only minor form control by bedding and the fairly planar to rounded slopes cut across bedding planes (Figs. 8.11 and 8.19).

The major reason for development of smooth, generally convex slopes is the development of exfoliation or sheeting fractures in massive, poorly jointed (referring here to pre-existing regional or systematic jointing) sandstones such as the Navajo Sandstone (Bradley 1963). The fracturing may be due to stress relief (Bradley 1963), expansive stresses due to weathering of near-surface beds (Hamilton 1984), and possibly freeze–thaw. These mechanisms produce lenticular sheets of thickness ranging from a few centimetres to a metre or so, with more widely spaced

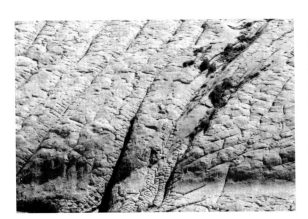

Fig. 8.19 Polygonal superficial fracturing of Navajo Sandstone along US 160 west of Kayenta, Arizona superimposed upon weathered joints. Individual polygons range from 1 to several meters in size (photo A. Howard)

Fig. 8.20 Expression of bedding in slickrock slopes in Navajo Sandstone near Utah Highway 12 west of Boulder, Utah (photo A. Howard)

fractures below these, fading out within 10–20 m from the surface. Thus the exfoliation and sheeting joints form 'a crude, somewhat subdued replica of the surface form' (Bradley 1963, p. 521). Although grain-by-grain removal of sand grains loosened by solution of the calcite cement or peeling of thin (<1 cm) weathered rinds seems to be the dominant erosional process, deeper weathering is indicated locally by the presence of shallow jointing perpendicular to the sandstone surface. In massive sandstones these cracks create a network pattern with a scale of 1–5 m that clearly wrap around existing topography, creating an 'elephant hide' pattern (Fig. 8.19). Where strong layering is exposed in cross-section, the cracks follow bedding planes and also create fractures cutting across the bedding, forming a 'checkerboard' or 'waffle' pattern as at Checkerboard Mesa, Zion National Park, Utah. The effective depth of these fractures is probably about 1/5–1/2 their lateral spacing. Hamilton (1984, pp. 32–4) suggested the fractures result from cyclic near-surface volume changes resulting from thermal cycling, wetting and drying, or freeze-thaw.

In addition to exfoliation, any weathering process that acts through some depth from the surface will tend to erode away projecting masses due to the greater surface area relative to volume and lead to a 'grading' of surface slopes, with a characteristic scale of action of the same order of magnitude as the depth of weathering (presumably a few centimetres to a few metres). Such processes may include solution of cement, weathering of feldspars and clays, disruption of the rock along microfractures and between grains due to differential volume changes produced by temperature changes, freeze-thaw, or shrink–swell of clays.

Segmented Scarps

Many areas of moderate relief on sandstones on the Colorado Plateau exhibit a complex topography embodying both elements of slickrock morphology and of scarps. Such landscapes developed in the Slick Rock member of the Entrada Sandstone at Arches National Park, Utah, were the object of a comprehensive study by Oberlander (1977). In this area slickrock slopes are interrupted by nearly vertical cliffs which Oberlander termed slab walls due to their erosion by failure along sheeting (off-loading) fractures parallel to the scarp face. The slab walls terminate at their base at inden-

tations developed in thin weak zones (partings) whose weathering and erosion cause the slab wall backwasting (Fig. 8.21). Partings that readily weather (effective partings) are either closely spaced bedding planes with highly fractured sandstone sandwiched in between, or are one or more thin (2–5 cm) layers of fissile ferriferous shale. The partings commonly are of limited horizontal extent, so that slab walls die out laterally (Fig. 8.21). Some slopes may have more than one slab wall where partings occur at two or more levels.

Oberlander presented convincing evidence that slope erosion occurs both by erosion of slickrock slopes and slab-wall backwasting. This, coupled with intersection of new partings and lateral dying out of other partings during slope retreat, leads to progressive changes in slope profile form (Fig. 8.22). In Oberlander's model the gradient of slickrock slopes below effective partings largely depends upon the relative rates of scarp backwasting by parting erosion and the rate of weathering and erosion on the slickrock slopes, with gentler slickrock slopes associated with rapid parting erosion. Sometimes backwasting at a parting may cease, either due to playing out of the parting or local conditions less conducive to parting erosion. In such cases continued erosion of the slickrock slope below the parting leads to development of a near-vertical slope below the parting; such slopes were called secondary walls by Oberlander. Such inactive slab walls also commonly develop alveolar

Fig. 8.21 Segmented scape in Entrada Sandstone at Arches National Park, Utah, showing slickrock slopes, slab walls, and effective partings at base of slab walls. Note that the prominent slab wall in the middle of the scarp at left dies out in the right side of the photo due to pinching out of the effective parting (photo A. Howard)

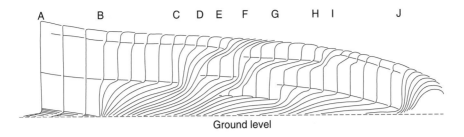

Ground level

Fig. 8.22 Development of scarp forms in massive sandstone through time and space. For simplicity a constant ground level (*dashed line*) is assumed during scarp retreat, along with equal thicknesses of removal from major slab walls in each unit of time. At A a through-going cliff is present due to sapping above a thin-bedded substrate. At B the thin-bedded substrate passes below ground level, scarp retreat slows, and effective intraformational partings assume control of scarp form. Effective partings close at C, E, F, G, H, and J, leading to local slab wall stagnation and rounding into slickrock. Partings that open at D, E, F, G, and I, initiate growth of new slab walls. Note that the effect of former partings in rocks that have been removed continues to be expressed in the form of slickrock ramps and concave slope breaks. Lowering of ground level during backwearing would cause cliff extension upward from major contact (caption and illustration from Oberlander 1977, Fig. 8)

weathering, discussed further below. An important conclusion of Oberlander's study is that thin partings in otherwise massive bedrock cause a complicated slope form (in particular the slab wall), so that slope breaks are not necessarily an indication of lithologic contrast above and below the break, but may imply only a thin discontinuity.

Similar slope forms occur in other sandstone units on the Colorado Plateau, especially the Navajo Sandstone and the Cedar Mesa Sandstone at Natural Bridges National Monument, Utah. In these formations the slab wall is commonly strongly overhung into a thin two-dimensional arch or alcove presumably backwasted along sheeting fractures. One puzzling aspect of these prominent indentations is a general paucity of mass wasting debris on the lower floor. Schumm and Chorley (1966) cited the ready breakup of the wasted debris as an explanation, but slab wall failures from relatively short cliffs yield abundant debris (Fig. 8.15), and the alcoves are a relatively protected environment. Another possible explanation is present-day inactivity of parting erosion and resulting slab failure due to aridity. The effects of climatic change on scarp morphology are discussed further below.

Hoodoos and Demoiselles

A few scarps are sculpted into highly intricate forms, such as occur at Bryce Canyon, Utah, and smaller forests of boulders on shale pedestals called hoodoos (Fig. 8.23). These forms occur where the caprock unit is discontinuous but massive and the underlying shales

Fig. 8.23 Hoodoos in Entrada Sandstone near Hanksville, Utah. Hoodoos are capped by resistant concretions, whereas nearby poorly cemented sandy shales exhibit regolith-mantled badland slopes. Hoodoos are often supported by thin pedestals of shale (photo A. Howard)

are readily eroded where exposed to rain but easily protected from weathering by slight overhangs. Many scarps in the south-western United States are composed of cliffs that extend well below the caprock unit into the underlying shales (Fig. 8.24). These cliffs are remarkably stable, having persisted and grown vertically (downward) throughout the Holocene (and late Pleistocene?), showing the efficacy of overhangs as small as a few tens of centimetres in restricting surface weathering of shales in a desert environment. Typically the caprock units are concretions or discontinuous interbeds of cemented sandstones or limestones. They are embedded in shales or shaly, poorly cemented sandstones that typically erode into badlands where relief is sufficient and caprocks are absent. Exposure of the concretion or resistant interbed greatly reduces

Fig. 8.24 Typical view of Morrison–Summerville escarpment at Sandslide Point off Utah Highway 95 near Hanksville, Utah, showing the cliff face extending into the weakly resistant Summerville Formation, the badland rampart in the same formation abutting the escarpment face, and the absence of debris on the rampart. Note that the rampart is shallowly underlain by bedrock and is not a colluvial apron (photo A. Howard)

erosion rates while the surrounding shale badlands continue to erode, producing a dichotomy in the shales between the vertical slopes protected by the caprock and the subjacent badlands. Development of stucco-like coatings of clay and lime by vertical drainage on the shale cliffs may be a contributing factor in protecting the shale from erosion (Lindquist 1979). Rates of vertical erosion in the shale badlands may be 40 times the rate of lateral retreat of the vertical shale cliffs (Lindquist 1979.

Role of Sapping Processes in Scarp Erosion and Morphology

Various geomorphologists have suggested that rock weathering and erosion at zones of groundwater discharge have contributed to the backwasting of scarps and valleys in sandstones exposed on the

Colorado Plateau (Gregory 1917, Bryan 1928, Ahnert 1960, Campbell 1973, Laity and Malin 1985). General discussions of groundwater sapping and its landforms have been provided in Higgins (1984), Howard et al. (1988), Baker et al. (1990), and Higgins and Osterkamp (1990). Laity and Malin (1985, p. 203) defined sapping as 'the process leading to the undermining and collapse of valley head and side walls by weakening or removal of basal support as a result of enhanced weathering and erosion by concentrated fluid flow at a site of seepage'. Higgins (1984) distinguished between 'spring sapping' caused by concentrated water discharge and 'seepage erosion' resulting from diffuse discharge at lithologic contacts or other lithologic boundaries. This discussion addresses both the role of seepage groundwater in scarp erosion and the development of deeply incised valleys in sandstone by spring sapping.

Such definitions of sapping and seepage erosion are complicated by marginal and transitional situations. Scarp erosion processes that are clearly not sapping erosion include plunge-pool undermining and rock weathering by moisture delivered to the scarp face by precipitation, condensation, or absorption of water vapour. However, other circumstances are not as clear-cut. For example, water penetrating into tensional and exfoliation joints close to cliff faces may cause rockfalls as a result of freezing or water pressure but would probably not be classified as sapping by most geomorphologists. Similarly, corrasional erosion of shale beneath sandstone by water penetrating along wide fractures (termed subterranean wash by Ahnert 1960) is similar to piping, but probably should not be included as a process of groundwater sapping. On the other hand, rockfall caused by weathering of shales beneath a sandstone scarp in which water is delivered by flow along joints within the sandstone is more likely to be considered to be sapping, even in the absence of obvious water discharge along the scarp face. Weathering processes resulting from intergranular flow within sandstone would generally be considered sapping.

Groundwater flow plays an uncertain role in weathering of the shales and weakly cemented layers whose erosion causes scarp retreat in overlying sandstones. Oberlander (1977, 1989) mentioned spring sapping of shale partings as a possible process of scarp retreat in segmented scarps. Schumm and Chorley (1966), while providing experiments and observations on surface weathering of caprock units, essentially avoided the issue of processes of caprock undermining. Koons (1955) was similarly vague. Ahnert (1960) clearly felt that sapping processes are of general importance in scarp retreat in sandstone-shale scarps of the Southwest USA, but he provided little evidence. Laity and Malin (1985) suggested that disruption of surface exposures by salt crystal growth where seepage emerges and sloughing of thin sheets of the bedrock are the major processes of sapping erosion in massive sandstones, and that sapping is usually concentrated in thin zones above less permeable boundaries within or below the sandstone. This backwasting and undermining of the overlying sandstone then occasions development of slab failure and, locally, alcove development associated with development of exfoliation jointing as outlined by Bradley (1963). Laity and Malin primarily discussed spring sapping processes occurring at canyon headwalls, and it is uncertain the degree to which they felt sapping or seepage erosion occurs more generally on sandstone scarps. Observations reported below suggest an important role of shallow groundwater circulation in scarp retreat in sandstone–shale sequences of the Southwest USA under present climates.

Many scarps are indented by V-shaped re-entrants or canyons excavated by erosion along streams passing over the escarpment. Such erosion is generally considered to result from corrasion of the bed (e.g. Howard and Kerby 1983) or from plunge-pool action. However, examination of canyon heads in sandstone suggests that sapping processes may play a role in channel erosion, at least for washes with drainage areas less than several square kilometres. Washes passing over scarps in sandstone commonly occupy only a fraction of the total scarp width. In addition, the scarps are commonly overhung when developed in massive sandstone and plunge pools are rare and small below the waterfalls. Even steep streams on thin sandstone beds sandwiched between shale layers exhibit overhangs considerably wider than the stream bed and show little development of plunge pools. Much of the erosion of the sandstone beds in such cases may occur due to sapping, with a very localized water source from the overlying stream. Additionally, small washes developed on slickrock slopes above scarps are commonly interrupted by solution pits and the washes exhibit flutes and furrows that suggest that solutional removal of calcite cement is more important than mechanical corrasion in bed erosion.

Weathering and erosion of sandstone by the effects of crystal growth occur at a variety of scales in sheltered locations. Steep slopes and scarps in sandstone are frequently interrupted by rounded depressions, often overlapping, which intersect sharply with the general slope (Fig. 8.25). Such alveolar weathering, or tafoni, occurs not only in sandstone, but also in granites, tuff, and other massive rocks (Mustoe 1982, 1983). Both accelerated erosion in the hollows and case hardening of the exposed portions of the slope (Conca and Rossman 1982) may contribute to development of tafoni. Salt accumulations are often quite apparent in the cavernous hollows, and the backwasting results in spalling of sheets of weathered rock up to a few centimetres in thickness. Mustoe (1983) noted high soluble cation contents in the spall detritus in tafoni and the presence of the mineral gypsum. Laity (1983) and Laity and Malin (1985) found calcite deposition on spalling walls. The mechanisms by which such mineral deposition may contribute to the spalling include pressures exerted by crystal growth, thermal expansion and contraction of the crystal filled rock, and expansion and contraction due to hydration of deposited minerals (Cooke and Smalley 1968). Freeze–thaw disruption on the moist seepage faces may also contribute, and spalling may be aided by the weight of accumulated winter ice (Laity and Malin 1985).

A surface protected from surface runoff is a necessary condition for tafoni and alcove development. On steep sandstone scarps surface runoff commonly flows as sheets down the scarp, held by surface tension on slightly overhung slopes. Such runoff paths are commonly accentuated by rock varnish. Where

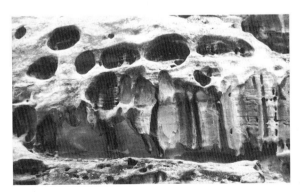

Fig. 8.25 Alveolar weathering, or tafoni, developed in Navajo Sandstone at Capital Reef National Monument, along Utah Highway 24. Note uneroded ribs where surface wash occurs. Image height about 10 m (photo A. Howard)

such runoff paths cross zones of tafoni development, backwasting is inhibited, and the tafoni are separated by columns that often resemble flowstone columns in caves (Fig. 8.25). Surface runoff might inhibit salt fretting simply by solution and removal of salts brought to the surface by evaporating groundwater or more actively by case hardening of the exposed surface by deposition of clays or calcite (Conca and Rossman 1982).

Two intergrading types of sapping landforms develop on massive sandstones. The more exotic form is the development of tafoni on steep scarps and on large talus blocks. Such tafoni may literally riddle certain steep slopes (Fig. 8.25), with the tafoni concentrated along certain beds that are either more susceptible to the salt fretting or receive greater ground water discharge. Talus blocks generally develop tafoni on their lower, overhung portions. The concentration of tafoni development at the base of such blocks may be due both to the protection from surface wash as well as upward wicking of salts from underlying soils or shales, a process that contributes to weathering of the bases of tombstones (Mustoe 1983, Hamilton 1984). Oberlander (1977) pointed out that tafoni develop most strongly on scarps initially steepened by basal undermining but presently no longer backwasting because alluviation or aeolian deposition covers and protects the basal backwasting face. Thus generalized tafoni development on a scarp indicates relative inactivity of backwasting by surface attack or basal undercutting.

On the other hand, large alcoves are common in massive sandstones and are often actively retreating as a result of sapping erosion (Fig. 8.26). However, direct sapping usually is localized to zones less than 2 m thick along permeability discontinuities where the discharge of groundwater is concentrated (Fig. 8.27), although at major valley heads the seepage zone may be 20–25 m thick. These sapping zones generally backwaste by processes similar to those of tafoni, and locally tafoni are superimposed upon the sapping face. In addition to salt fretting, backwasting by groundwater discharge can also occur by cement dissolution and by weathering of shale beneath or interbedded in the sandstone. The retreat of the active zone of sapping undermines the sandstone above, with the result that occasional rockfalls occur (Fig. 8.15). In massive sandstone the undermining occasions the development of exfoliation sheeting fractures, resulting in large arches or alcoves, with the deepest parts of the alcoves presumably corresponding to the most rapid sapping at-

Fig. 8.26 Theatre-headed valley in Navajo Sandstone near Utah Highway 95 along North Wash (southern Henry Mountains area). Note cottonwood trees and *dark* figure in wash at bottom centre of photo for scale. The active seep is the *dark* band at the base of the alcove. Note the evidence of offloading fracturing. The stream passing over the *top* of the alcove is clearly inadequate in size to have created the alcove as a plunge pool. Water draining down the face of the alcove from the lip of the falls has created the dark streaks. However, delivery of water along the alcove walls occurs only locally, and is insufficient to account for the backwasting of the alcove (photo A. Howard)

tack (Fig. 8.27). In well-jointed sandstones, such as the Wingate Sandstone, arches and overhanging cliffs are less common, and the role of scarp retreat by sapping processes is not as obvious but may be just as important.

Fig. 8.27 Seepage face at the theater head alcove shown in Fig. 9.26. Zone of seepage is 1–2 m thick. Floor of alcove is highly weathered sandstone debris supporting a vegetative cover. The seepage face is encrusted with ferns, mosses, algae, and other phreatophytes. Salt efflorescences occur above moist zone. Note person for scale at arrow (photo A. Howard)

The major aquiclude for the Navajo Sandstone is the underlying Kayenta Formation, and the major seeps develop at this discontinuity (Figs. 8.26 and 8.27). However, thin shales and limestone interbeds (interdunal deposits) create minor aquicludes within the Navajo (Laity 1988, Kochel and Riley 1988), leading to frequent development of multiple levels of seeps and associated alcoves at valley headwalls.

A distinction may be made between wet sapping with a damp rock face and an effluent discharge and dry sapping, where the rock face is generally incrusted with mineral salts (Laity and Malin 1985). In general, tafoni are associated with dry sapping because of their localized development, whereas large alcoves are generally associated with a more regional groundwater flow and exhibit faces that are at least seasonally wet.

Neither active seepage nor deposition of mineral crusts on protected sandstone walls are necessarily correlated with rapid weathering and backcutting of the sandstone walls. For example, the Weeping Wall at Zion National Park, Utah, is an impressive seep emerging from the Navajo Sandstone, but the associated alcove and canyon are small. Many other examples of fairly high discharge rates but only minor, or non-existent, alcoves can be found throughout the Colorado Plateau. Too rapid a seepage may in fact discourage deposition of salts. Although rapid seepage can also cause backwasting by dissolution of calcite or gypsum cement, this would only occur if the groundwater were under-saturated. Similarly, many examples of thick mineral incrustations at seeps lacking evidence of backwasting can be found on sandstones throughout the Southwest USA. Several factors control whether minerals deposited by evaporating seepage are deposited intergranularly within the rock (encouraging exfoliation and granular disintegration) or at the rock surface (with little resulting sapping), including the type and concentration of salts, the average and variance of water discharge to the surface, the distribution of pore sizes and their interconnectivity, the presence and size of fractures, the temperature regime at the rock face, local humidity and winds, and the frequency of occurrence of wetting of the rock face by rain or submergence (if along a stream or river). The interaction of these factors is poorly understood. Discrepancies between size of alcove or sapping valley and the magnitude of the seep can also result from differences in the length of time that sapping has been active. Despite these cautions, it is reasonable to expect within a specific physiographic, structural, and stratigraphic setting that the degree of alcove development and degree of headward erosion of sapping valleys would correlate with the discharge of the seeps involved (Laity and Malin 1985).

In summary, most evidence suggests that sapping processes are common in sandstones of the Colorado Plateau and the processes result in the widespread development of tafoni and alcoves. The major caution is the evidence cited by Oberlander suggesting that groundwater sapping is unimportant in erosion of effective partings at thin shale layers in the Entrada and other sandstones. One of several possibilities may account for this difference in interpretation. The effective shale partings may, in fact, be so readily weathered by atmospheric moisture due to high salt content or other factors that seepage is not required for such partings as it is for cavernous weathering and alcove development elsewhere. The Entrada Sandstone exhibits considerable variation in lithology and cementation both vertically and areally, so that in the Arches National Park the Slick Rock Member may be relatively impermeable. However, well-developed arches and tafoni with evidence of active salt fretting are found a few tens of kilometres south in this same member. Another possibility is that backwasting along effective partings is not very active under present climatic conditions, but was more active during the late Pleistocene. The evidence for relatively inactive scarp retreat (and, by extension, seepage erosion) under present climatic conditions is discussed below.

Climatic Change and Scarp Morphology

Most recent discussions of scarp morphology and associated weathering and erosional processes have implied that the morphological elements can be explained by presently active processes (e.g. Koons 1955, Schumm and Chorley 1966, Oberlander 1977). However, other authors have maintained that escarpments in the Colorado Plateau region are undergoing considerably less caprock erosion under the present climate than during the late Pleistocene (Reiche 1937, Ahnert 1960, Howard 1970).

Given our poor understanding of present scarp processes and little historical control, it is not surprising that different conclusions have been drawn about present activity and the presence or the lack of a direct relationship between present processes and present scarp form. For example, Schumm and Chorley (1964, 1966) cited numerous examples of historical rockfalls as evidence for present activity of scarps. However, a simple analysis suggests just the opposite. Inferred long-term rates of scarp retreat on the Colorado Plateau average about 3 m per 1000 years (Schmidt 1989b). If rockfalls occur randomly through time and characteristically involve a block of caprock 1 m thick and 10 m in horizontal dimension, then one should expect about three rockfalls per year on 10 km of scarp front. If the characteristic rockfall is larger, say 3 m thick and 100 m long, then there should be a rockfall every ten years. Since there are tens to hundreds of thousands of kilometres of scarp front on the Colorado Plateau, occurrences of large rockfalls should be much more frequent than appears to be the case, possibly by as much as an order of magnitude.

The escarpment of Emery Sandstone near Caineville, Utah, is a case in point. Extensive old rockfall deposits on North and South Caineville Mesas, dissected to a depth of as much as 50 m at their lower and lateral margins, were interpreted by Howard (1970) to be Bull Lake equivalent in age (Illinoian?) because some of the debris blankets interfinger with pediment deposits of Bull Lake age, whereas the remainder extend no further downslope than the level of the former pediment surface (Fig. 8.28). The old debris blankets have been dissected by a similar amount and project concordantly laterally along the

Fig. 8.28 Pleistocene-age (Early Wisconsinan?) rockfall on escarpment of Emery Sandstone over Mancos Shale on South Caineville Mesa near Hanksville, Utah. Note heavy dissection of lower end of rockfall blanket. Note remnant of pediment of same age in front of escarpment on left side of photo (photo A. Howard)

escarpment. Much of the scarp rampart is now devoid of extensive debris, so that extensive badlands have developed on the underlying Mancos Shale. In places the cliff capped by the Emery Sandstone extends downward by as much as 70 m into the underlying shale (Fig. 8.14). Koons (1955) demonstrated that development of cliff faces in the non-resistant subcaprock unit is a normal occurrence in the cycle of rockfall, debris blanket erosion, and erosion of the subcaprock unit which triggers the next rockfall. However, for reasons outlined above, extensive cliff development in the subcaprock unit may also indicate stagnation of caprock mass wasting while erosion of the subcaprock unit continues on the scarp rampart.

Schmidt (1988, 1989a) and Gerson and Grossman (1987) noted similar remnant debris blankets related to climatic change and, in particular, talus flatirons and associated pediment flatirons (Fig. 8.29). Schmidt (1988, 1989a) termed climatically controlled flatiron development cyclical flatirons as contrasted with the non-cyclic flatirons discussed previously, and also cited the presence of associated dissected pediments (pediment flatirons) as a distinguishing characteristic of cyclic flatirons.

An outstanding example of a presently stagnant scarp morphology is the escarpment capped by the lower units of the Morrison Formation overlying the thin bedded sand and shale beds of the Summerville Formation near Hanksville, Utah. Much of the escarpment is capped by a thin gypsum layer which is stable enough to support vertical cliffs up to 30 m high in the Summerville Formation (Fig. 8.24). The gypsum cap has been eroded into small crenellations mimicked in the underlying Summerville, indicating that the caprock protects the vertical face from weathering and erosional processes. The rampart generally consists of debris-free badlands likewise carved in the Summerville Formation, which meet the vertical face at an abrupt angle. The rampart is bordered downslope by dissected pluvial (Bull Lake?) pediments paved with agate derived from the caprock, giving evidence that abundant debris was transported from the escarpment during that time. Where a caprock of gypsum is present, even a thick sequence of overlying sandstone sheds little debris on to the rampart. But where the gypsum unit is discontinuous or missing (Fig. 8.30), the overlying sandstone wastes by undermining and rockfall. However, during the Bull Lake pluvial climate the gypsum caprock evidently succumbed more

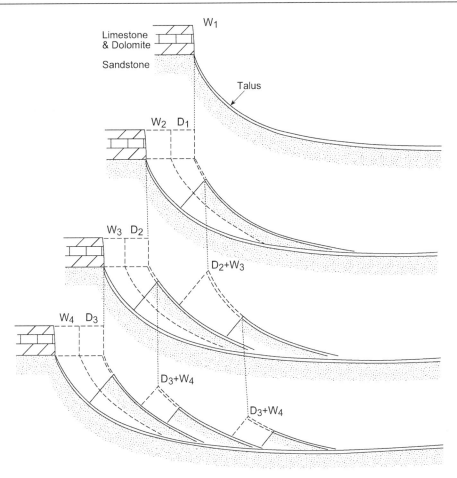

Fig. 8.29 Talus flatirons developed by alternating wet (W) and dry (D) climates. Talus aprons and associated pediments develop during enhanced weathering and transport of wet periods and are dissected during dry epochs (from Gerson and Grossman, 1987, Fig. 17–8)

Fig. 8.30 View of northwest face of Goatwater Point, along Utah Highway 95 near Hanksville, Utah, showing rapid backwasting of the sandstones capping the Morrison–Summerville escarpment in the absence of the gypsum caprock. Vertical cliffs in the Summerville Formation occur only where the gypsum unit is present, such as on the left edge and centre of the picture (photo A. Howard)

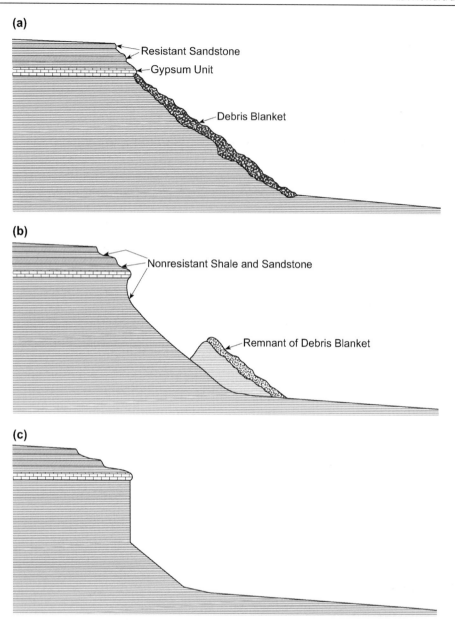

Fig. 8.31 Cross-sections through the Morrison–Summerville escarpment showing presumed Pleistocene conditions compared with the present-day profile. During the Pleistocene pluvials, backwasting of the caprock units by rockfall and block-by-block undermining was probably rapid, and the debris-covered rampart extended up to the caprock without tall vertical cliffs in the underlying Summerville Formation (cross-section **a**). Between rockfalls, weathering and erosion eventually removed the sandstone and gypsum debris, and a cliff would begin to form in the Summerville Formation (**b**). Because of the greater supply of water and lower temperatures, vertical cliffs higher than those in profile (**b**) were probably unstable, for another rockfall would return the escarpment to the conditions at (**a**). However, under the drier and warmer Holocene conditions, the gypsum caprock has become very resistant, and backwasting of the escarpment has essentially halted, with the result that the escarpment rampart is rapidly eroding away, leaving behind high cliffs in the Summerville Formation (**c**) (from Howard 1970, Fig. 81)

readily to rockfall and occasional slumping (Fig. 8.31). The scarps illustrated in photos 3 and 22 of Schumm and Chorley (1966) are probably further examples of such presently inactive scarps.

In canyon areas, a cliff in a massive sandstone often is underlain by a nearly bare rock slope with a gradient of about 30–40°. No obvious lithologic break separates these slope elements. In some cases, such as at Canyon de Chelly, Arizona, examples can be found where such bare rock slopes yield laterally to rock slopes of similar gradient but mantled with rockfall debris (Fig. 8.32). One explanation for such features is that they are Richter slopes, forming at the characteristic friction angle of rockfall debris under conditions where removal of rockfall debris (by weathering or erosion at the scarp base) just balances production at the superjacent cliff face (Richter 1901, Bakker and Le Heux 1952, Cotton and Wilson 1971, Selby 1982d, pp. 204–8, Scheidegger 1991, p. 132; also equivalent to the case of Schumm and Chorley's (1966) unity weathering ratio). However, at least in the Colorado Plateau examples, it is difficult to imagine how such a scattered debris cover protects the underlying rock from weathering and erosion. One possible explanation is that the moving talus has erosive capability which planes the rock down to the friction angle of the moving talus. Erosion by rock and ice avalanches has been used to explain spur-and-gully development on steep arctic and alpine mountain slopes (Rapp 1960a,b, Howard 1989, 1990). However, in this case erosion tends to become localized in chutes.

Fig. 8.32 Portion of a canyon wall in a portion of Canyon De Chelly, Arizona, showing prominent vertical cliffs in upper parts of the De Chelly Sandstone with slickrock slope of 30–40° inclination below the cliffs and also development in the De Chelly Sandstone. Note partial mantling of lower slopes with talus and the suggestion that other slickrock slopes were formerly mantled (photo A. Howard)

A more direct explanation is that the bare rock slopes were formerly mantled with rock debris produced by cliff retreat and that the debris mantle protected the underlying rock from weathering and erosion, forming threshold slopes (Carson 1971, Carson and Petley 1970). The cover of debris on such slopes is now obviously inadequate to protect the rock from weathering and erosion with the implication that scarp backwasting is presently largely inactive, permitting the weathering and stripping of the former debris mantle. Somewhat similar forms can be found in the areas of segmented scarps in Entrada Sandstone at Arches National Monument (the area studied by Oberlander 1977) (Fig. 8.21). Oberlander interpreted the low-gradient slickrock slopes to be a normal feature of erosion of segmented scarps (Fig. 8.22), but it is possible that some of the slickrock slopes below vertical cliffs may have formerly been mantled by a debris blanket under a different climate (a scattering of debris can be seen on many of these segmented slopes).

Ahnert's (1960) historical interpretation of scarp morphology contrasts sharply with the 'dynamic equilibrium' interpretations of Schumm and Chorley (1966) and Oberlander (1977). Ahnert suggested that scarp retreat occurs by erosion of shales and other weakly resistant layers by sapping-related processes that have only been active during pluvial episodes characterized by abundant, gentle precipitation. By contrast, he felt that slickrock slopes are eroded largely by sheetwash, which has been active during the sparse but intense rainfall of interpluvials. Ahnert felt that the generally sharp breaks between slickrock slopes and subjacent scarps were evidence of the non-contemporaneous origin of the two features, whereas Oberlander (1977) and Howard and Kochel (1988) felt that such sharp breaks can result from contemporaneous slickrock erosion and scarp retreat. Ahnert's assertion that slickrock slopes are eroded primarily by sheetwash misses the importance of the initial weathering that must occur on these weathering-limited slopes. This weathering, which makes debris capable of slopewash transport, is probably not optimal under present climatic conditions of brief, intense rainstorms, but is favoured by winter precipitation and freeze–thaw.

Despite these objections to Ahnert's interpretations, the examples presented previously suggest that there have been changes throughout the Pleistocene and

Holocene in the relative rates of the major processes producing cuesta landforms: scarp undermining by sapping, undermining, or surface-directed weathering; weathering and erosion of rockfall debris; weathering and erosion of slickrock slopes; erosion of interbedded shale layers; and, locally, slumping and landsliding. Changes in the relative importance of these processes have produced landforms that in many cases can only be understood by knowledge of the temporal process changes (i.e. relict landforms). These examples of remnant effects of past climates add a complicating element in unravelling the geomorphic history of cuesta landforms and interpretation of scarp morphologies which would greatly benefit from application of modern techniques of geomorphic surface dating.

The greater rainfall during the Bull Lake pluvial maximum would suggest greater sapping activity then relative to now, but that same moisture supply would also contribute to surface-directed weathering, particularly freeze–thaw weathering. Also, more abundant groundwater does not necessarily produce more sapping in sandstones because if seepage is sufficient and humidity high enough to permit runoff of the seepage, little salt accumulation will occur. However, other groundwater-related weathering processes would be enhanced by greater available moisture, such as hydration and leaching of shale interbeds. Thus during the pluvial epochs we might expect more debris production by freeze–thaw, less alveolar weathering, but greater back-wasting at the contact between sandstones and underlying shales and along shale interbeds. The present paucity of rockfall debris, rather than indicating nearly complete breakup and rapid weathering of the debris, as suggested by Schumm and Chorley (1966), may simply indicate relative inactivity of many scarps in present climates.

Another prominent feature of the theatre headwalls at seeps is the paucity of rockfall debris. This commonly contrasts with abundant rockfall debris along the sides of the valley downstream from the seep (Fig. 8.26). Laity (1988) suggested that both initial pulverization of rockfall debris and rapid weathering in the moist environment of the headwall account for the paucity of debris. However, recent rockfalls in alcoves generate abundant debris, and the rockfall debris along valley walls also suggests that the backwasting of scarp walls produces a considerable volume of coarse debris that must be further weathered before it is transported from the rampart and backwasting can

continue. The paucity of rockfall debris at headwater scarps is therefore interpreted as further evidence of greater sapping activity during past pluvial periods and relative inactivity during the present climatic regime.

In the western desert of Egypt, particularly the Gilf Kebir, scarp retreat via groundwater sapping during wet episodes of the Pleistocene and present erosional inactivity has been inferred by Peel (1941), Maxwell (1982), and Higgins and Osterkamp (1990).

Quantitative Models of Scarp Profile Evolution

A variety of quantitative models of scarp profiles has been developed over the years. Most notable are the models of scarp backwasting and talus accumulation that predict the evolutionary slope history and the sub-talus bedrock profile (Lehmann 1933, Bakker and Le Heux 1952, Gerber and Scheidegger 1973, and others: see summary in Scheidegger 1991, pp. 130–4). These have been developed for the case of a horizontal, fixed base level with no provision for weathering and removal of talus except through a constant ratio of rock eroded to talus produced.

Other slope evolution models permit resistant layers, but some express resistance solely in terms of resistance to weathering and do not allow for the effects of mass movement and surface wash (e.g. Scheidegger 1991, pp. 144–7; Aronsson and Linde 1982), or they include creep-like mass movement but not the effects of surface wash, rockfall, and talus weathering (Ahnert 1976, Pollack 1969).

Developing of realistic scarp profile models will prove challenging owing to the range of processes and materials that must be considered. The first component must be the accounting of weathering and failure of the caprock, addressing rates of weathering, effects of slope steepness and profile, and predicting size and frequency of rockfall, undermining, or slumps, possibly including the role of groundwater sapping. A second component should follow the distribution, comminution during emplacement, weathering, and erosion of caprock debris, as well as the protective influence of the debris on underlying bedrock. The third component should address erosion of the subcaprock unit where exposed. Finally, initial and boundary conditions must be addressed, including stratigraphy, dip, base level changes, and possible process variations through time. A final problem with profile evolution on ramparts with rockfall talus is that production and erosion of the talus is inherently three-dimensional (e.g. the development of flatirons).

Evolution of Scarp Planform

The planimetric form of canyons and escarpments is the most obvious signature of the erosional processes involved in scarp retreat in layered rocks. The processes of scarp erosion create characteristic shapes of planform features such as re-entrants, projections, and inset canyons. For example, some scarps have rounded projections, or headlands, whereas others are sharply terminated. Similarly, some scarps are inset with deep, narrow canyons, while others have shallow, broadly rounded reentrants. The scarp form is determined by the spatial distribution of the processes discussed earlier as well as the lithologic and structural influences such as rock thickness and dip. Several generalities about scarp planform have been noted for many years.

(a) In areas of strong structural dip (more than a few degrees) scarps tend to be oriented updip with a linear to gently curving planform that closely follows the structural strike, locally interrupted by through-flowing streams. At least two factors contribute to the linearity of scarps with steep dip. Drainage divides tend to be at or very close to the scarp crest, so that little fluvial or sapping erosion occurs. In addition, escarpment height is a strong function of crest location, so that portions of the scarp that lag in erosion rapidly become higher than surrounding locations, tending to enhance average rates of retreat. Thus the remarkable linearity of scarps in steeply dipping rocks is *prima facie* evidence for the sensitive dependency of scarp backwasting rates on scarp relief.

(b) Where the structural dip is slight, scarp planforms are highly textured, with deep embayments and canyons as well as headlands and detached mesas. Scarps may face either updip or downdip, with downdip portions being more deeply embayed because drainage areas on top of such scarps are generally larger.

(c) Resistant rock units generally become first breached by stream erosion, forming scarps,

where structural elevations are high (e.g. at the top of anticlines) and/or where gradients are steepest (e.g. on dip-slope stripped plains on the resistant unit). Both factors are combined in the common scenario of breaching of caprocks just below the crest of monoclines.

Since caprock erosion occurs dominantly by scarp retreat, the scarp planform reflects areal variations in rate of scarp retreat. One factor controlling scarp retreat rates is the erosional resistance of the caprock and/or subcaprock units. Nicholas and Dixon (1986) examined the role of variations in inherent caprock resistance (e.g. compressive strength, slake durability, cementation) and caprock fabric (fracture orientation and spacing) in controlling relative rates of caprock erosion. They found that embayments in the Organ Rock Formation escarpment, Utah, have a higher density of fractures at the base of the caprock than occurs beneath headlands. However, caprock resistance is nearly equal on headlands and escarpments. Accordingly, they suggest that areal variations in caprock fabric are a dominant control on scarp planform (Fig. 8.33).

The other factor controlling scarp form is areal variations in process rates. Erosion of the scarp face by rockfall, slumping, and undermining of the caprock, weathering of the caprock debris, and erosion of subcaprock units will be termed scarp backwasting. Deep re-entrant canyons are clearly created by rapid erosion along a linear zone, generally by fluvial erosion or groundwater sapping. Structure often plays an indirect

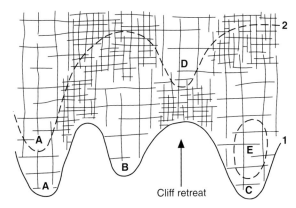

Fig. 8.33 Plan view of scarp retreat, showing variable rates of backwasting governed by differences in fracture density. Particularly unfractured masses may persist as isolated buttes, as at A, B, D and E (from Nicholas and Dixon 1986, Fig. 8)

role by channelling surface or subsurface water along fractures. The role of areal variations in intensity of these three classes of processes and in rock resistance in creating scarp planforms is discussed below based upon both field evidence and theoretical modelling. The discussion will concentrate on the development of Schumm and Chorley's (1964) compound scarps consisting of one scarp-forming unit sandwiched between easily eroded rocks.

Scarp Backwasting

Natural scarps commonly exhibit a planform characterized by sharply pointed or cuspate projections (headlands, spurs) and broadly concave re-entrants (embay-

ments). The role of essentially uniform rates of scarp retreat in creating pointed headlands and broad embayments was first discussed by Dutton (1882, pp. 258–9) and Davis (1901, pp. 178–80), who noted the nearly uniform spacing between successively lower scarps on the walls of the Grand Canyon. Attack of an escarpment by areally uniform backwasting gradually makes embayments more shallow and the planform of the scarp face close to linear (Fig. 8.34). Similar observations were made by Howard (1970) and Schipull (1980). Lange (1959) systematically studied the consequences of uniform erosional attack (uniform decrescence) on two- and three-dimensional surfaces, showing that uniform erosion acting on any arbitrary, rough surface produces rounded re-entrants and cuspate spurs. He also showed that scarp retreat in the

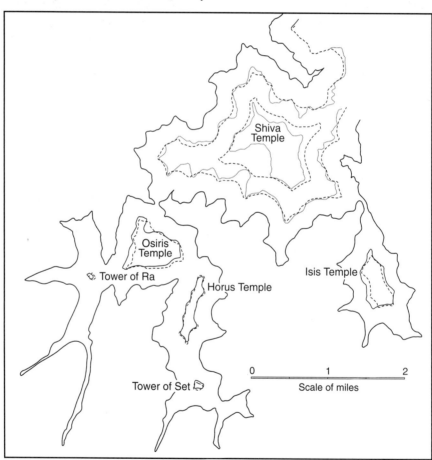

Fig. 8.34 Process of uniform erosional attack illustrated on the Redwall, Coconino, and Kaibab escarpments within the Grand Canyon. The rim outline of both the Coconino and Kaibab escarpments (*dotted lines*) may be approximately duplicated by uniform erosional attack of the Redwall escarpment rim (*dashed lines*). In a few cases the hollows, or re-entrants, are deeper than would be produced by uniform backwasting of the next lower escarpment; sapping backwasting may be involved (from Lange 1959, Fig. 7.17)

Grand Canyon is uniform to a first approximation, with some tendency towards more rapid erosion of embayments (Fig. 8.34). Furthermore, uniform decrescence and uniform addition of material (accrescence) are not reversible (Lange 1959), so that the position of an escarpment rim during the past cannot be uniquely inferred from its present configuration, even if subject to uniform retreat in the past, especially in the case of former spurs that first eroded into outliers and then wasted away.

A simple numerical model of scarp backwasting (Howard 1988, 1995) lumps local variations in process rates and rock resistance into a single 'erodibility', which is assumed to vary randomly from location to location. The erodibility is assumed to be a self-similar fractal with variations at both large and small scales. The spatial variability of erodibility is characterized by three parameters, an average value, a variance, and a rate of change of variance with scale. The rate of lateral backwasting is assumed to be linearly dependent upon the erodibility. Figure 8.35 shows a plan view of successive scarp positions starting from a square 'mesa'. All simulations discussed here assume a scarp such as is shown in Fig. 8.10, with a single caprock over shale and a superjacent stripped plain. As expected, scarps produced by this type of

backwasting are characterized by sharply terminating projections and broad re-entrants (as in the Grand Canyon, Fig. 8.34), with re-entrants produced by erosion through more erodible rocks.

The projections of some natural scarps are rounded rather than sharply pointed. This suggests that the lateral erosion rates on the projections are enhanced relative to straight or concave portions of the scarp. Several mechanisms may account for this. The rate of erosion of many scarps is limited by the rate that debris shed from the cliff face can be weathered and removed from the underlying rampart on the less resistant rocks. Debris from convex portions of scarps (projections) is spread over a larger area of rampart, so that weathering rates may be enhanced. Additionally, headlands commonly stand in higher relief than re-entrants, leading to longer and/or steeper ramparts and enhanced erosion rates. Finally, some scarp caprock may be eroded by undermining due to outward creep or weathering of the rocks beneath the caprock. Both processes are enhanced at convex scarp projections. Figure 8.36 shows a model scarp in which backwasting rates are enhanced by scarp convexity and restricted by concavity, and Fig. 8.37 shows a similar natural scarp.

By contrast, some natural scarps exhibit narrow, talon-like headlands that are unlikely to have resulted

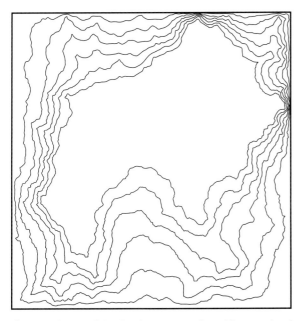

Fig. 8.35 Stages of simulated scarp retreat by uniform erosional attack of caprock with spatially variable resistance. The simulation starts from an assumed square mesa and the lines show successive stages of retreat of the edge of the scarp. Note that the lines are not contours

Fig. 8.36 Plan view of simulated scarp evolution by uniform erosional attack with an assumed dependency of retreat rate on planform convexity. Note rounding of scarp headlands

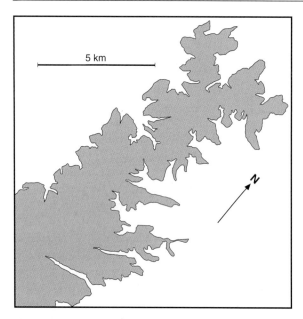

Fig. 8.37 Plan view of a natural scarp with rounded headlands in sandstone underlain by shale in north-eastern New Mexico (1:100000 Capulin Mountain quadrangle). Note that fluvial incision has created numerous sharply terminated embayments. The uneroded portion of the scarp is patterned

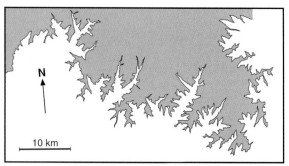

Fig. 8.38 Natural scarp with sharply pointed projections and sharply pointed, gradually narrowing box canyon embayments due to stream incision (scarp of the Redwall limestone in Grand Canyon National Park: Grand Canyon 1:100000 quadrangle). The uneroded portion of the scarp is patterned

from pure uniform retreat (Figs. 8.34 and 8.38). Occasional thin headlands may result from narrowing of divides between canyons by uniform back-wasting, but where such sharp cusps are frequent it is likely that they result from slowed erosion of narrow projections. Impeding of erosion on such headlands may be due to a lack of uplands contributing surface or subsurface drainage to the scarp face, thus inhibiting chemical and physical weathering of the cliff face or its undermining. Scarps that have been broken up into numerous small buttes, such as occurs at Monument Valley, Arizona, may be further examples of inhibition of scarp retreat where upland areas are small. An alternative explanation may be that these headlands occur in locations with minimal caprock jointing, in the manner suggested by Nicholas and Dixon (1986).

Almost all escarpments in gently dipping rocks exhibit deep re-entrants and, as erosion progresses, break up into isolated mesas and buttes. This indicates erosion concentrated along generally linear zones of either structural weakness, fluvial erosion, or sapping (often acting in combination). One indicator of the role of either or both fluvial erosion and groundwater sapping is the asymmetry of scarps in gently tilted rocks. The planform of segments where the scarp faces up-

dip is generally similar to that expected by uniform decrescence, since little drainage passes over the scarp, but segments facing downdip ('back scarps' of Ahnert 1960) are deeply indented as a result of fluvial or sapping erosion (Fig. 8.39) (Ahnert 1960, Laity and Malin 1985).

Nicholas and Dixon (1986) emphasized the role of variable density of fracturing of incompetent and caprock layers in controlling the rate of scarp retreat and the development of re-entrants, projections, and isolated buttes. Their evidence suggests that variable density of fracturing is important on certain scarps, but it probably controls primarily the small-scale planform features. Variable density of fracturing cannot account for the remarkable asymmetry of gently dipping scarps (Fig. 8.39) and the general association of embayments with the drainage basins on the overlying stripped surface of the caprock.

Fluvial Downcutting

One process creating deep re-entrants is downcutting by streams which originate on the top of the escarpment and pass over its front. Such caprock erosion is localized to the width of the stream, which is generally very small compared with overall scarp dimensions, so that the heads of fluvially eroded canyons are quite narrow, but increase gradually in width downstream by scarp backwasting, as noted by Ahnert (1960) and Laity and Malin (1985). Examples of such gradually

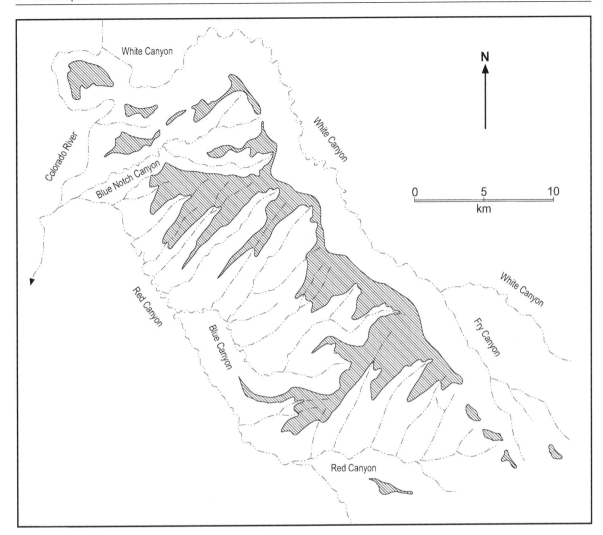

Fig. 8.39 Plan view of an escarpment of gently dipping Wingate Sandstone showing smooth, nearly straight planform on updip side and deeply embayed downdip side. Strata dip about 1.5° due west. Mesa is located to south-west of Utah Highway 95 between the Colorado River and Natural Bridges National Monument

narrowing, deep re-entrants are common on the Colorado Plateau (Figs. 8.37 and 8.38).

Scarp erosion by combined fluvial erosion and scarp backwasting can be numerically modelled by superimposing a stream system on to a simulated scarp and allowing it to downcut through time. If a resistant layer is present, the stream develops a profile characterized by a low gradient section on top of the resistant layer and a very steep section (rapids or falls in natural streams) which rapidly cuts headward. In the model shown here, erosion is assumed to be proportional to the product of the bed erodibility and the average channel bed shear stress (Howard

and Kerby 1983). Between streams, scarps are assumed to erode by the backwasting process described above. Figure 8.40 shows a simulated scarp eroded by both fluvial downcutting and scarp backwasting, and Fig. 8.41 is similar but with convexity-enhanced scarp backwasting. Figures 8.37 and 8.38 are natural scarps similar to Figs. 8.41 and 8.40, respectively. Scarp morphology is similar to that produced by scarp backwasting, with the addition of several deep re-entrants characterized by a gradually narrowing width and projections that are sharply pointed in the linear case and rounded in the convexity-enhanced case.

Fig. 8.40 Plan view of successive stages of simulated erosion of an initially square mesa by combined action of fluvial downcutting and uniform scarp retreat

All of the fluvial models combine scarp backwasting with downcutting and scarp retreat due to fluvial erosion, with valleys gradually enlarging downstream due to longer duration since fluvial backwasting passed

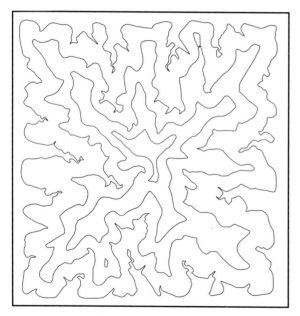

Fig. 8.41 Plan view of successive stages of simulated erosion of an initially square mesa by combined action of fluvial downcutting and uniform erosion with a curvature rate dependence

through that location. The angle subtended by the valley walls ψ is related to the ratio of the scarp backwasting rate R_b to the stream headcutting rate R_s by the expression

$$R_b/R_s = \sin \psi/2. \qquad (8.6)$$

The rate of downstream increase of valley width has been used by Schmidt (1980, 1989b) to estimate scarp recession rates.

Groundwater Sapping

Groundwater sapping is an important process in scarp retreat throughout the Colorado Plateau (Ahnert 1960) and probably contributes to the scarp backwasting process discussed above. However, the deep, narrow, bluntly terminated canyon networks of the type discussed by Laity and Malin (1985) and Howard and Kochel (1988) are the scarp form most clearly dominated by sapping processes.

Ahnert (1960) and Laity and Malin (1985) suggested that while fluvial erosion produces V-shaped canyon heads that widen consistently downstream, sapping produces U-shaped, theatre-headed canyons of relatively constant width downstream (the terms U- and V-shaped refer here to valley planform, not valley cross-section). Other planform features diagnostic of sapping-dominated canyon extension and widening include theatre-shaped heads of first-order tributaries with active seeps (Figs. 8.26 and 8.42), relatively constant valley width from source to outlet (Figs. 8.42 and 8.43), high and steep valley sidewalls, pervasive structure control, and frequent hanging valleys. The most direct evidence of sapping processes are the numerous alcoves, both in valley heads and along sidewalls, and the many springs on the valley walls and bottoms. Although many valley headwalls occur as deeply undercut alcoves, some terminate in V or half-U shapes with obvious extension along major fractures (Fig. 8.43). Although sapping action is fairly evident where alcoves occur at valley heads, sapping may also occur in fracture-controlled headwalls. Evidence for erosion by streams passing over the valley headwalls is slight, and plunge pools are not obvious. The valley extension along fracture traces suggests control by groundwater flowing along the fractures, and a preponderance of the major valley heads are

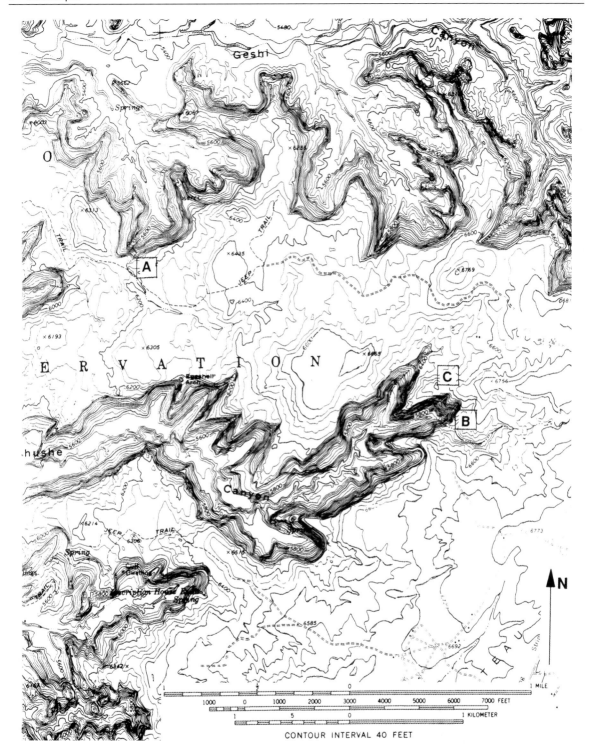

Fig. 8.42 Topographic map of box canyons eroded into Navajo Sandstone in the Inscription House area of the Navajo indian reservation, Arizona (part of the Inscription House Ruin 1:24000 quadrangle). Note that the backslope drainage at B does not enter the canyon at its head and that the canyon headwalls at C extend nearly to the backslope divide with little contributing drainage area. The extensive upslope drainage enters the large alcove at A at a minor niche. Also note the rounded, theatre-like canyon headwalls, most of which have active seeps

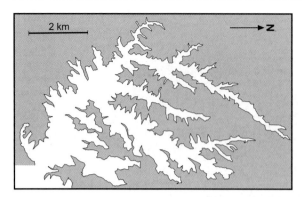

Fig. 8.43 Planform of cuesta in Navajo Sandstone at Betatakin National Monument, Arizona, showing rounded canyon headwalls, weak branching, and nearly uniform canyon width. Reentrants locally show strong control of orientation by fractures (Kayenta 1:100000 quadrangle). The uneroded portion of the scarp is patterned

oriented updip, which is consistent with sapping by a regional downdip groundwater flow. One striking feature of these canyon networks is frequent discrepancy between the extent of headward canyon growth and the relative upland area contributing drainage to the canyon head, including situations where the upland drainage enters the side rather than the end of the canyon (Fig. 8.42). Such circumstances suggest that goundwater flow rather than surface runoff controls headward extension of the canyons. Another indication of sapping control is the extension of canyons right up to major topographic divides (Fig. 8.42), sometimes causing the divides to be displaced. In contrast to surface runoff, groundwater flow can locally cross topographic divides. Divide migration can also occur in surface runoff drainage basins (see discussion of badlands), but is associated with the presence of slopes that are well graded from stream to divide. In the case of headward migration of canyons, streams above the scarp are perched on the upper slickrock slopes, so that base-level control is very indirect.

The pattern of valley development and the relative contribution of sapping processes versus fluvial erosion is influenced by many structural, stratigraphic, and physiographic features. Infiltration may be restricted where overlying aquicludes were not stripped from the sandstone (Laity and Malin 1985, Laity 1988). The permeability of the sandstone is influenced not only by primary minerals but by diagenetic cements and overgrowths. These secondary minerals vary considerably from layer to layer and location to

location (Laity 1983, 1988). Kochel and Riley (1988) and Laity (1988) discussed the important role that thin aquicludes (generally interdune deposits) and bedding strikes and dips have in controlling groundwater flow through the sandstone. Primary and off-loading (sheeting) fractures are important avenues of groundwater migration (Laity and Malin 1985, Laity 1988, Kochel and Riley 1988). On the other hand, dense primary fracturing, such as occurs in the Wingate Sandstone, restricts development of alcoves and probably diminishes weathering by cement solution and salt fretting (Laity 1988), although sapping processes may still be important but lack the spectacular alcoves commonly developed in the Navajo Sandstone. Relatively flat uplands on sandstones should encourage infiltration, whereas steep slickrock slopes probably lose much precipitation to runoff. Weathering pits and thin covers of windblown sand may also encourage infiltration.

A simulation model of scarp backwasting by groundwater sapping assumes a planar caprock aquifer (that may be tilted) with uniform areal recharge and a self-similar areal variation in permeability. The rate of sapping backwasting E is assumed to depend upon the amount by which the linear discharge rate at the scarp face q exceeds a critical discharge q_c: $E = k(q - q_c)^a$, where k and a are coefficients. Scarp erosion is simulated by solving for the groundwater flow, eroding the scarp by a small amount, with repeated iterations. Figure 8.44 shows a simulated scarp with $q_c = 0$ and $a = 1$. Figure 8.45 shows a case with $q_c = 0$ and $a = 2$. Figure 8.46 shows a case with $q_c = 0$ and $a = 2$ and superimposed scarp backwasting. Figure 8.43 is a natural scarp in Navajo Sandstone formed largely by sapping. Valleys developed by groundwater sapping tend to be linear or crudely dendritic, with nearly constant width and rounded rather than sharp terminations, in accord with the predictions of Ahnert (1960) and Laity and Malin (1985).

Deep re-entrants that are essentially constant in width with rounded ends imply either that almost all valley enlargement occurs at the headward end of the valley (see, for example, the simulations in Figs. 8.44 and 8.45) or that a period of rapid headward canyon growth (either by sapping or fluvial incision) is succeeded by a period of uniform scarp retreat. This latter case is exemplified by the simulations in Fig. 8.46 combining groundwater sapping and uniform backwasting in which rapid valley erosion by sapping

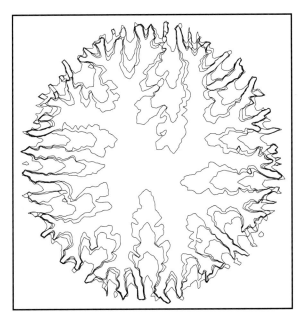

Fig. 8.44 Plan view of successive stages of simulated erosion of an initially round mesa by groundwater sapping. Caprock aquifer is assumed to undergo uniform recharge from above and water flows along the base of the caprock at the contact with underlying shale until it emerges at the edges of the mesa. Erosion rate is proportional to seepage discharge rate per unit width of scarp face

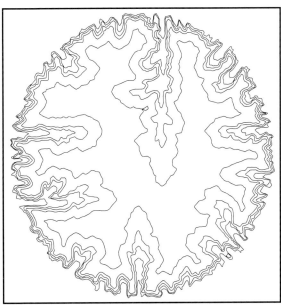

Fig. 8.46 Simulated sapping erosion of an initially round mesa as in Fig. 8.43, but with superimposed uniform scarp retreat

early in the simulation is succeeded by a dominance of uniform backwasting in later stages of erosion as upland areas contributing to groundwater sapping diminish in size.

Summary of Scarp Modelling

These simulations illustrate a common evolutionary sequence in scarp erosion. Early stages are characterized by a planform determined largely by the agency first exposing the scarp, generally fluvial downcutting. Thus scarp walls parallel the stream course with small re-entrants along tributaries draining over the scarp. Continuing tributary erosion by fluvial or sapping processes with modest rates of scarp backwasting between streams eventually creates a maximally crenulate scarp planform. As erosion continues and upland areas diminish, rates of stream or sapping backcutting diminish relative to more uniform scarp backwasting (which may accelerate as relief increases) so that the scarp planform becomes characterized by broad re-entrants, sharp cusps, and often isolated small buttes during the last stages before complete caprock removal. This is illustrated in Figs. 8.40 and 8.46. This sequence can occur sequentially on a single large

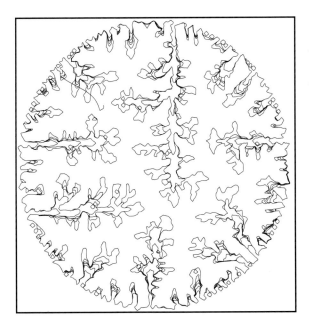

Fig. 8.45 Simulated sapping erosion of an initially round mesa as in Fig. 8.42, but with erosion rate proportional to the square of seepage discharge rate

Fig. 8.47 Simulated sapping erosion of a slightly tilted square mesa by groundwater sapping. The caprock tilts to the bottom. Note the strongly asymmetrical development of re-entrant canyons

mesa, as takes place in the simulations, but it is also illustrated by spatial sequences during long-continued erosion of a given caprock from newly exposed scarps in structurally low areas to the last caprock remnants in structurally high areas.

Even slight structural dips create a strong asymmetry of rates of fluvial and sapping headcutting on down- and updip facing portions of the scarp, since drainage divides tend to be located close to the updip crest. General scarp backwasting is less affected by dip, so that updip scarp segments are nearly linear, whereas downdip sections are highly crenulate. This is illustrated in the natural scarp in Fig. 8.39 and the simulated sapping erosion of a tilted scarp in Fig. 8.47.

The planform of escarpments gradually develops through scarp retreat of hundreds to thousands of metres. Thus it is unlikely that planforms reflect to any great extent Quaternary climatic oscillations. Longer-term climatic changes may have influenced scarp planforms, but such influences would probably be hard to detect.

Conclusions

Weathering and erosional processes on exposed bedrock are generally either very slow-acting or occur as large and often rapid mass wasting events. As a result, process studies of the sort that have been conducted on badland and rock-mantled slopes are rare. Furthermore, the slow development of bedrock landforms often implies a strong influence of climate changes on slope form. Among the issues that need further study are (a) the relative roles of stress-release fracturing and contemporary gravitational, neotectonic, or weathering-induced stresses in generating broadly rounded bedrock exposures, (b) weathering processes involved in groundwater sapping, (c) controls on rates of scarp backwasting and scarp height and steepness, and (d) processes of talus weathering and removal on scarp ramparts. We anticipate that the study of landform development on bedrock will benefit greatly from application of modern techniques for dating of geomorphic surfaces, including use of rock varnish (Chapters 7 and 21) and thermoluminescence, fission-track, and radiogenic or cosmogenic isotopes.

References

Abrahams, A.D. and A.J. Parsons 1987. Identification of strength equilibrium rock slopes: further statistical considerations. *Earth Surface Processes and Landforms* **12**, 631–5.

Ackroyd, P. 1987. Erosion by snow avalanche and implications for geomorphic stability, Torlesse Range, New Zealand. *Arctic and Alpine Research* **19**, 65–80.

Ahnert, F. 1960. The influence of Pleistocene climates upon the morphology of cuesta scarps on the Colorado Plateau. *Annals of the American Association of Geographers* **50**, 139–56.

Ahnert, F. 1976. Darstellung des Structureinflusses auf de Oberflachenformen in theoretischen Modell. *Zeitschrift für Geomorphologie Supplement Band* **24**, 11–22.

Akerman, H.J. 1984. Notes on talus morphology and processes in Spitsbergen. *Geografiska Annaler* **66A**, 267–84.

Allison, R.J. and A.S. Goudie 1990a. The form of rock slopes in tropical limestone and their associations with rock mass strength. *Zeitschrift für Geomorphologie* **34**, 129–48.

Allison, R.J. and A.S. Goudie 1990b. Rock control and slope profiles in a tropical limestone environment: the Napier Range of Western Australia. *Geographical Journal* **156**, 200–11.

Anderson, M.G. and K.S. Richards (eds) 1987. *Slope stability – geotechnical engineering and geology*. Chichester: Wiley.

Aronsson, G.K. and K. Linde 1982. Grand Canyon – a quantitative approach to the erosion and weathering of a stratified bedrock. *Earth Surface Processes and Landforms* **7**, 589–99.

Augustinus, P.C. and M.J. Selby 1990. Rock slope development in McMurdo Oasis, Antarctica, and implications for interpretation of glacial history. *Geografiska Annaler* **72A**, 55–62.

Baker, V.R., R.C. Kochel, J.E. Laity and A.D. Howard 1990. Spring sapping and valley network development. In *Groundwater geomorphology*, C.G. Higgins and D.R. Coates (eds), 235–65. Geological Society of America Special Paper **252**.

Bakker, J.P. and J.W.N. Le Heux 1952. A remarkable new geomorphological law. *Koninklijke Nederlandsche Academic van Wetenschappen* **B55**, 399–410, 554–71.

Bieniawski, Z.T. 1989. *Engineering rock mass classifications*. New York: Wiley.

Blackwelder, E. 1942. The process of mountain sculpture by rolling debris. *Journal of Geomorphology* **4**, 324–8.

Blasius, K.R., J.A. Cutts, J.E. Guest and H. Masursky 1977. Geology of the Valles Marineris: first analysis of imaging from Viking 1 orbiter primary mission. *Journal of Geophysical Research* **82**, 4067–91.

Bock, H. 1979. Experimental determination of the residual stress field in a basaltic column. *Proceedings of the International Congress on Rock Mechanics Montreux* **1**, 45–9; **3**, 136–7.

Bornhardt, W. 1900. *Zur Oberflachengestaltung und Geologie Deutsch-Ostafrikas*. Berlin: Reimer.

Bouchard, M. 1985. Weathering and weathering residuals on the Canadian Shield. *Fennia* **163**, 327–32.

Boyé, M. and P. Fritsch 1973. Degagement artificiel d'un dome cristallin au Sud Cameroun. In *Cinq etudes de geomorphologie et de palynologie travaux et documents de geographic tropicale*. CEGET-CNRS **8**, 31–63.

Bradley, W.C. 1963. Large-scale exfoliation in massive sandstones of the Colorado Plateau. *Bulletin of the Geological Society of America* **74**, 519–28.

Brady, B.H.G. and E.T. Brown 1985. *Rock mechanics for underground mining*. London: George Allen & Unwin.

Brown, E.T. 1981. *Rock characterization testing and monitoring*: *ISRM suggested methods*. Oxford: Pergamon.

Brown, G.F., D.L. Schmidt and A.C. Huffman 1989. Geology of the Arabian Peninsula: shield area of western Saudi Arabia. *U.S. Geological Survey Professional Paper* 560-A.

Brunsden, D. and D.B. Prior (eds) 1984. *Slope instability*. Chichester: Wiley.

Bryan, K. 1928. Niches and other cavities in sandstones at Chaco Canyon, N. Mexico. *Zeitschrift für Geomorphologie* **3**, 125–40.

Budel, J. 1957. Die 'doppelten Einebnungsflachen' in den feuchten Tropen. *Zeitschrift für Geomorphologie* **1**, 201–88.

Budel, J. 1982. *Climatic geomorphologie* (translated by L. Fischer and D. Busche). Princeton: Princeton University Press.

Burke, K. and G.L. Wells 1989. Trans-African drainage system of the Sahara: was it the Nile? *Geology* **17**, 743–7.

Buser, O. and H. Frutiger 1980. Observed maximum run-out distance of snow avalanches and the determination of the friction coefficients mu and epsilon. *Journal of Glaciology* **94**, 121–30.

Campbell, I.A. 1973. Controls of canyon and meander forms by jointing. *Area* **5**, 291–6.

Cannon, S.H. and W.Z. Savage 1988. A mass-change model for the estimation of debris-flow runout. *Journal of Geology* **96**, 221–7.

Carson, M.A. 1971. An application of the concept of threshold slopes to the Laramie Mountains, Wyoming. *Institute of British Geographers Special Publication* **3**, 31–47.

Carson, M.A. and M.J. Kirkby 1972. *Hillslope form and process*. Cambridge: Cambridge University Press.

Carson, M.A. and D.J. Petley 1970. The existence of threshold slopes in the denudation of the landscape. *Transactions of the Institute of British Geographers* **49**, 71–95.

Conca, J.L. and G.R. Rossman 1982. Case hardening of sandstone. *Geology* **10**, 520–33.

Cooke, R.U. and I.J. Smalley 1968. Salt weathering in deserts. *Nature* **220**, 1226–7.

Cooke, R.U. and A. Warren 1973. *Geomorphology in deserts*. Berkeley: University of California Press.

Corner, C.D. 1980. Avalanche impact landforms in Troms, north Norway. *Geografiska Annaler* **62A**, 1–10.

Cotton, C.A. and A.T. Wilson 1971. Ramp forms that result from weathering and retreat of precipitous slopes. *Zeitschrift für Geomorphologie* **15**, 199–211.

Dale, T.N. 1923. The commercial granites of New England. *U.S. Geological Survey Bulletin* **738**, 488p.

Davis, W.M. 1901. An excursion into the Grand Canyon of the Colorado. *Harvard Museum of Comparative Zoology and Geology* **5**, 105–201.

Dent, J.D. and T.E. Lang 1980. Modeling of snow flow. *Journal of Glaciology* **26**, 131–40.

Dent, J.D. and T.E. Lang 1983. A biviscous modified Bingham model of snow avalanche motion. *Annals of Glaciology* **4**, 42–6.

Doelling, H.H. 1985. *Geology of Arches National Park*. Utah Geological and Mineral Survey, Map 74.

Dutton, C.E. 1882. Tertiary history of the Grand Canyon district. *U.S. Geological Survey Monograph* 2.

Einstein, H.H. and W.S. Dershowitz 1990. Tensile and shear fracturing in predominantly compressive stress fields – a review. *Engineering Geology* **29**, 149–72.

Erhart, H. 1956. *La genese des sols. Esquise d'une theorie geologique*. Paris: Masson.

Fairbridge, R.W. 1988. Cyclical patterns of exposure, weathering and burial of cratonic surfaces, with some examples from North America and Australia. *Geografiska Annaler* **70A**, 277–83.

Falconer, J.D. 1911. *The geology and geography of northern Nigeria*. London: Macmillan.

Gardiner, J. 1970. Geomorphic significance of avalanches in the Lake Louise area, Alberta, Canada. *Arctic and Alpine Research* **2**, 135–44.

Gardiner, J. 1983. Observations on erosion by wet snow avalanches, Mount Rae area, Alberta, Canada. *Arctic and Alpine Research* **15**, 271–4.

Gardiner, V. and R. Dackombe 1983. *Geomorphological field manual*. London: George Allen and Unwin.

Gerber, E.K. and A.E. Scheidegger 1973. Erosional and stress-induced features on steep slopes. *Zeitschrift für Geomorphologie Supplement Band* **18**, 38–49.

Gerson, R. 1982. Talus relics in deserts: a key to major climatic fluctuations. *Israel Journal of Earth Sciences* **31**, 123–32.

Gerson, R. and S. Grossman 1987. Geomorphic activity on escarpments and associated fluvial systems in hot deserts. In *Climate history, periodicity, predictability*, M.R. Rampino, J.S. Sanders, R.E. Newman and L.K. Konigsson (eds), 300–22. New York: Van Nostrand Reinhold.

Gilbert, G.K. 1904. Domes and dome structures of the High Sierra. *Bulletin of the Geological Society of America* **15**, 29–36.

Gregory, H.E. 1917. Geology of the Navajo country. *U.S. Geological Survey Professional Paper* 93.

Habermehl, M.A. 1980. The Great Artesian Basin, Australia. *BMR Journal of Geology and Geophysics* **5**, 9–38.

Hack, J.T. 1966. Interpretation of Cumberland escarpment and Highland rim, South-central Tennessee and northeast Alabama. *U.S. Geological Survey Professional Paper* 524-C.

Hack, J.T. 1980. Rock control and tectonism; their importance in shaping the Appalachian Highlands. *U.S. Geological Survey Professional Paper* 1126-A–J, pp. B1-B17.

Hall, A.M. 1986. Deep weathering patterns in north-east Scotland and their geomorphological significance. *Zeitschrift für Geomorphologie* 30, 407–22.

Hall, A.M. 1988. The characteristics and significance of deep weathering in the Gaick area, Grampian Highlands, Scotland. *Geografiska Annaler* 70A, 309–14.

Hamilton, W.L. 1984. *The sculpturing of Zion*. Zion National Park: Zion Natural History Association.

Haxby, W.F. and D.L. Turcotte 1976. Stresses induced by the addition or removal of overburden and associated thermal effects. *Geology* 4, 181–4.

Hewett, K. 1972. The mountain environment and geomorphic processes. In *Mountain geomorphology*, H.O. Slaymaker and H.J. McPherson (eds), 17–34. Vancouver: Tantalus Press.

Higgins, C.G. 1984. Piping and sapping: development of landforms by groundwater flow. In *Groundwater as a geomorphic agent*, R.G. LaFleur (ed), 18–58. Boston: Allen & Unwin.

Higgins, C.G. and W.R. Osterkamp 1990. Seepage-induced cliff recession and regional denudation. In *Groundwater geomorphology*, C.G. Higgins and D.R. Coates (eds), 291–317, Geological Society of America Special Paper 252.

Hoek, E. and J.W. Bray 1981. *Rock slope engineering*. London: Institution of Mining and Metallurgy.

Howard, A.D. 1970. A study of process and history in desert landforms near the Henry Mountains, Utah. Unpublished Ph.D. dissertation. Baltimore: Johns Hopkins University.

Howard, A.D. 1988. Groundwater sapping experiments and modeling. In *Sapping features of the Colorado Plateau*, A.D. Howard, R.C. Kochel and H.E. Holt (eds), 71–83, Washington: National Aeronautics and Space Administration SP-491.

Howard, A.D. 1989. Miniature analog of spur-and-gully landforms in Valles Marineris scarps. *Reports of the planetary geology and geophysics program – 1988.* Washington: National Aeronautics and Space Administration TM 4130, 355–7.

Howard, A.D. 1990. Preliminary model of processes forming spur-and-gully terrain. *Reports of the planetary geology and geophysics program – 1989.* Washington: National Aeronautical and Space Administration TM 4210, 345–7.

Howard, A.D. 1995. Simulation modeling and statistical classification of escarpment planforms, *Geomorphology* 12, 187–214.

Howard, A.D. and G.R. Kerby 1983. Channel changes in badlands. *Bulletin of the Geological Society of America* 94, 739–52.

Howard, A.D. and R.C. Kochel 1988. Introduction to cuesta landforms and sapping processes on the Colorado Plateau. In *Sapping features of the Colorado Plateau*. A.D. Howard, R.C. Kochel and H.E. Holt (eds), 6–56. Washington: National Aeronautics and Space Administration SP-491.

Howard, A.D., R.C. Kochel and H.E. Holt (eds) 1988. *Sapping features of the Colorado Plateau*. Washington: National Aeronautics and Space Administration SP-491.

Idnurm, M. and B.R. Senior 1978. Palaeomagnetic ages of Late Cretaceous and Tertiary weathered profiles in the Eromanga Basin, Queensland. *Palaeogeography, Palaeoclimatology, Palaeoecology* 24, 263–77.

Jahns, R.H. 1943. Sheet structure in granites, its origin and use as a measure of glacial erosion in New England. *Journal of Geology* 51, 71–98.

Jennings, J.N. 1983. Sandstone pseudokarst or karst. In *Aspects of Australian sandstone landscapes*, 21–30. Australian and New Zealand Geomorphology Group, Special Publication 1.

Kemp, E.M. 1981. Tertiary palaeogeography and the evolution of Australian climate. In *Ecological biogeography of Australia*, A. Keast (ed), 33–49. The Hague: W. Junk.

Kochel, R.C. and G.W. Riley 1988. Sedimentologic and stratigraphic variations in sandstones of the Colorado Plateau and their implications for groundwater sapping. In *Sapping features of the Colorado Plateau*, A.D. Howard, R.C. Kochel and H.E. Holt (eds), 57–62. Washington: National Aeronautics and Space Administration SP-491.

Koons, D. 1955. Cliff retreat in the southwestern United States. *American Journal of Science* 253, 44–52.

Korsch, R.J. 1982. Mount Duval: geomorphology of a near-surface granite diapir. *Zeitschrift für Geomorphologie* 26, 151–62.

Laity, J.E. 1983. Diagenetic controls on groundwater sapping and the valley formation, Colorado Plateau, as revealed by optical and electron microscopy. *Physical Geography* 4, 103–25.

Laity, J. 1988. The role of groundwater sapping in valley evolution on the Colorado Plateau. In *Sapping features of the Colorado Plateau*. A.D. Howard, R.C. Kochel and H.E. Holt (eds), 63–70. Washington: National Aeronautics and Space Administration SP-491.

Laity, J.E. and M.C. Malin 1985. Sapping processes and the development of theater-headed valley networks in the Colorado Plateau. *Bulletin of the Geological Society of America* 96, 203–17.

Lang, T.E. and J.D. Dent 1982. Review of surface friction, surface resistance, and flow of snow. *Reviews of Geophysics and Space Physics* 20, 21–37.

Lange, A.L. 1959. Introductory notes on the changing geometry of cave structures. *Cave Studies* 11, 66–90.

Lee, C.F. 1978. Stress relief and cliff stability at a power station near Niagara Falls. *Engineering Geology* 12, 193–204.

Lehmann, O. 1933. Morphologishe Theorie der Vervitterung von Steinschlag Wandern. *Vierteljahrschrift der Naturforschende Gesellschaft in Zürich* 87, 83–126.

Lindquist, R.C. 1979. Genesis of the erosional forms of Bryce Canyon National Park. *Proceedings of the 1st conference on scientific research in the National Parks* 2, 827–34. Washington, DC: National Park Service.

Lucchitta, B.K. 1978. Morphology of chasma walls, Mars. *Journal of Research, U.S. Geological Survey* 6, 651–62.

Lucchitta, I. 1975. Application of ERTS images and image processing to regional geologic problems and geologic mapping in northern Arizona – Part IV B, the Shivwits Plateau. *National Aeronautics and Space Administration Technical Report* 32-1597, 41–72.

Luckman, B.H. 1977. The geomorphic activity of snow avalanches. *Geografiska Annaler* 59A, 31–48.

Luckman, B.H. 1978. Geomorphic work of snow avalanches in the Canadian Rocky Mountains. *Arctic and Alpine Research* 10, 261–76.

Mainguet, M. 1972. *Le modelegdes gres*. Paris: Institut Geographique National.

Martinelli, M. Jr, T.E. Lang and A.I. Mears 1980. Calculations of avalanche friction coefficients from field data. *Journal of Glaciology* **26**, 109–19.

Matthes, F.E. 1938. Avalanche sculpture in the Sierra Nevada of California. *International Association of Scientific Hydrology Bulletin* **23**, 631–7.

Maxwell, T.A. 1982. Erosional patterns of the Gilf Kebir Plateau and implications of the origin of Martian canyonlands. In *Desert landforms of southwest Egypt: a basis for comparison with Mars*, F. El-Baz and T.A. Maxwell (eds), 207–39. Washington; National Aeronautics and Space Administration CR-3611.

McClung, D.M. and P.A. Schaerer 1983. Determination of avalanche dynamics friction coefficients from measured speeds. *Annals of Glaciology* **4**, 170–3.

McEwen, A.S. and M.C. Malin 1990. Dynamics of Mount St. Helens' 1980 pyroclastic flows, rockslide-avalanche, lahars, and blast. *Journal of Volcanology and Geothermal Research* **37**, 205–31.

McGarr, A. and N.C. Gay 1978. State of stress in the earth's crust. *Annual Review of Earth and Planetary Sciences* **6**, 405–36.

McGill, G.E. and A.W. Stromquist 1975. Origin of graben in the Needles District, Canyonlands National Park, Utah. *Four Corners Geological Society Guidebook, 8th Field Conference*, 235–43.

Moon, B.P. 1984a. The form of rock slopes in the Cape Fold Mountains. *The South African Geographical Journal* **66**, 16–31.

Moon, B.P. 1984b. Refinement of a technique for measuring rock mass strength for geomorphological purposes. *Earth Surface Processes and Landforms* **9**, 189–93.

Moon, B.P. 1986. Controls on the form and development of rock slopes in fold terrane. In *Hillslope processes*, A.D. Abrahams (ed), 225–43. Boston: Allen & Unwin.

Moon, B.P. and M.J. Selby 1983. Rock mass strength and scarp forms in southern Africa. *Geografiska Annaler* **65A**, 135–45.

Mustoe, G.E. 1982. Origin of honeycomb weathering. *Bulletin of the Geological Society of America* **93**, 108–15.

Mustoe, G.E. 1983. Cavernous weathering in the Capitol Reef Desert, Utah. *Earth Surface Processes and Landforms* **8**, 517–26.

Nicholas, R.M. and J.C. Dixon 1986. Sandstone scarp form and retreat in the Land of Standing Rocks, Canyonlands National Park, Utah. *Zeitschrift für Geomorphologie* **30**, 167–87.

Oberlander, T.M. 1972. Morphogenesis of granitic boulder slopes in the Mojave Desert California. *Journal of Geology* **80**, 1–20.

Oberlander, T.M. 1977. Origin of segmented cliffs in massive sandstones of southeastern Utah. In *Geomorphology in arid regions*, D.O. Doehring (ed), 79–114. Binghamton, NY: Publications in Geomorphology.

Oberlander, T.M. 1989. Slope and pediment systems. In *Arid zone geomorphology*, D.S.G. Thomas (ed), 56–84. New York: Halsted Press.

Ollier, C.D. 1988. Deep weathering, groundwater and climate. *Geografiska Annaler* **70A**, 285–90.

Ollier, C.D. and C.F. Pain 1981. Active gneiss domes in Papua New Guinea, new tectonic landforms. *Zeitschrift für Geomorphologie* **25**, 133–45.

O'Loughlin, C.L. and A.J. Pearce 1982. Erosional processes in the mountains. In *Landforms of New Zealand*, J.M. Soons and M.J. Selby (eds), 67–79. Auckland: Longman Paul.

Patton, P.C. 1981. Evolution of the spur and gully topography on the Valles Marineris wall scarps. *Reports of the planetary geology program 1981*. National Aeronautics and Space Administration Technical Memorandum 84211, 324–5.

Peel, R.F. 1941. Denudational landforms of the central Libyan Desert. *Journal of Geomorphology* **4**, 3–23.

Peev, C.D. 1966. Geomorphic activity of snow avalanches. *International Association of Scientific Hydrology Publication 69*, 357–68.

Perla, R., T.T. Cheng and D.M. McClung 1980. A two-parameter model of snow-avalanche motion. *Journal of Glaciology* **26**, 197–207.

Pollack, H.N. 1969. A numerical model of Grand Canyon. *Four Corners Geological Society, Geology and Natural History of the Grand Canyon Region*, 61–2.

Pouyllau, M. and M. Seurin 1985. Pseudo-karst dans des roches greso-quartzitiques de la formation Roraima. *Karstologia* **5**, 45–52.

Rapp, A. 1960a. Recent development of mountain slopes in Karkevagge and surroundings, northern Scandinavia. *Geografiska Annaler* **42**, 73–200.

Rapp, A. 1960b. Talus slopes and mountain walls at Templefjorden, Spitzbergen. *Norsk Polarinstitutt Scrifter* **119**, 96 pp.

Reiche, P. 1937. The Toreva Block, a distinctive landform type. *Journal of Geology* **45**, 538–48.

Richter, E. 1901. Geomorphologische Untersuchungen in den Hochalpen. *Dr. A. Petermann's Mitteilungen aus Justus Perthes' geographischer Ansalt, Ergänzungsheft* 132.

Rudberg, S. 1986. Present-day geomorphological processes in Prins Oscars Land, Svalvard. *Geografiska Annaler* **68A**, 41–63.

Sancho, C., M. Gutierrez, J.L. Pena and F. Burillo 1988. A quantitative approach to scarp retreat starting from triangular slope facets, central Ebro Basin, Spain. *Catena Supplement* **13**, 139–46.

Scheidegger, A.E. 1991. *Theoretical geomorphology* (3rd edn). Berlin: Springer.

Schiewiller, T. and K. Hunter 1983. Avalanche dynamics. Review of experiments and theoretical models of flow and powder-snow avalanches. *Journal of Glaciology* **29**, 283–5.

Schipull, K. 1980. Die Cedar Mesa – Schichtstufe aud dem Colorado Plateau – ein Beispiel für die Morphodynamik arider Schichtstufen. *Zeitschrift für Geomorphologie* **24**, 318–31.

Schmidt, K.-H. 1980. Eine neue Metode zur Ermittlung von Stufenruckwanderungsraten, dargestellt am Beispiel der Black Mesa Schichtstufen, Colorado Plateau, USA. *Zeitschrift für Geomorphologie* **24**, 180–91.

Schmidt, K.-H. 1987. Factors influencing structural land-form dynamics on the Colorado Plateau – about the necessity of calibrating theoretical models by empirical data. *Catena Supplement* **10**, 51–66.

Schmidt, K.-H. 1988. Rates of scarp retreat: a means of dating neotectonic activity. In *The Atlas system of Morocco – studies on its geodynamic evolution*, V.H. Jacobshagen (ed), 445–62. Berlin: Lecture Notes in Earth Science **15**.

Schmidt, K.-H. 1989a. Talus and pediment flatirons – erosional and depositional features of dryland cuesta scarps. *Catena Supplement* **14**, 107–18.

Schmidt, K.-H. 1989b. The significance of scarp retreat for Cenozoic landform evolution on the Colorado Plateau, USA. *Earth Surface Processes and Landforms* **14**, 93–105.

Schumm, S.A. and R.J. Chorley 1964. The fall of threatening rock. *American Journal of Science* **262**, 1041–54.

Schumm, S.A. and R.J. Chorley 1966. Talus weathering and scarp recession in the Colorado Plateau. *Zeitschrift für Geomorphologie* **10**, 11–36.

Selby, M.J. 1977. Bornhardts of the Namib Desert. *Zeitschrift für Geomorphologie* **21**, 1–13.

Selby, M.J. 1980. A rock mass strength classification for geomorphic purposes: with tests from Antarctica and New Zealand. *Zeitschrift für Geomorphologie* **24**, 31–51.

Selby, M.J. 1982a. Form and origin of some bornhardts of the Namib Desert. *Zeitschrift für Geomorphologie* **26**, 1–15.

Selby, M.J. 1982b. Rock mass strength and the form of some inselbergs in the Central Namib Desert. *Earth Surface Processes and Landforms* **7**, 489–97.

Selby, M.J. 1982c. Controls on the stability and inclinations of hillslopes formed on hard rock. *Earth Surface Processes and Landforms* **7**, 449–67.

Selby, M.J. 1982d. *Hillslope materials and processes*. Oxford: Oxford University Press.

Selby, M.J. 1985. *Earth's changing surface*. Oxford: Clarendon Press.

Selby, M.J. 1987. Rock slopes. In *Slope stability: geotechnical engineering and geomorphology*, M.G. Anderson and K.S. Richards (eds), 475–504. Chichester: Wiley.

Selby, M.J., P. Augustinus, V.G. Moon and R.J. Stevenson 1988. Slopes on strong rock masses: modelling and influences of stress distributions and geomechanical properties. In *Modelling geomorphological systems*, M.G. Anderson (ed), 341–74. Chichester: Wiley.

Sharp, R.P. and M.C. Malin 1975. Channels on Mars. *Bulletin of the Geological Society of America* **86**, 593–609.

Slatyer, R.O. 1962. Climate of the Alice Springs area. *Land Research Series* **6**, CSIRO, Melbourne, 109–28.

Stille, H. 1924. *Grundfragen der vergleichenden Tektonic*. Berlin: Gebruder Borntraeger.

Strahler, A.N. 1940. Landslides of the Vermillion and Echo Cliffs, northern Arizona. *Journal of Geomorphology* **3**, 285–300.

Sturgul, J.R., A.E. Scheidegger and Z. Grinshpan 1976. Finite-element model of a mountain massif. *Geology* **4**, 439–42.

Suess, E. 1904–1924. *The face of the Earth* (translated by H.B.C. Sollas), 5 vols. Oxford: Clarendon Press. (German edition 1885–1909, Vienna: Tempsky.)

Terzaghi, K. 1962. Stability of steep slopes in hard unweathered rock. *Geotechnique* **12**, 251–70.

Thomas, M.F. 1989a. The role of etch processes in land-form development: I. etching concepts and their applications. *Zeitschrift für Geomorphologie* **33**, 129–42.

Thomas, M.F. 1989b. The role of etch processes in land-form development: II. etching and the formation of relief. *Zeitschrift für Geomorphologie* **33**, 257–74.

Twidale, C.R. 1978. On the origin of Ayres Rock, central Australia. *Zeitschrift für Geomorphologie Supplement Band* **31**, 177–206.

Twidale, C.R. 1981. Granite Inselbergs. *Geographical Journal* **147**, 54–71.

Twidale, C.R. 1982a. The evolution of bornhardts. *American Scientist* **70**, 268–76.

Twidale, C.R. 1982b. *Granite landforms*. Amsterdam: Elsevier.

Twidale, C.R. 1990. The origin and implications of some erosional landforms. *Journal of Geology* **98**, 343–64.

Voight, B. and B.H.P. St Pierre 1974. Stress history and rock stress. *Proceedings of the Third Congress of the International Society for Rock Mechanics, Denver* **2**, 580–2.

Watson, R.A. and H.E. Wright, Jr, 1963. Landslides on the east flank of the Chuska Mountains, northwestern New Mexico. *American Journal of Science* **261**, 525–48.

Wayland, E.J. 1933. Peneplains and some other erosional platforms. *Annual Report and Bulletin of the Protectorate of Uganda Geological Survey and Department of Mines, Note* **1**, 77–9.

Willis, B. 1936. *East African plateaus and rift valleys. Studies in comparative seismology*. Washington, DC: Carnegie Institution.

Wyrwoll, K.H. 1979. Late Quaternary climates of Western Australia: evidence and mechanisms. *Journal of the Royal Society of Western Australia* **62**, 129–42.

Yair, A. and R. Gerson 1974. Mode and rate of escarpment retreat in an extremely arid environment (Sharm el Sheikh, southern Sinai Peninsula). *Zeitschrift für Geomorphologie* **21**, 106–21.

Young, R.W. 1986. Tower karst in sandstone Bungle Bungle massif, northwestern Australia. *Zeitschrift für Geomorphologie* **30**, 189–202.

Yatsu, E. 1988. *The nature of weathering: an introduction*. Tokyo: Sozosha.

Young, A.R.M. 1987a. Salt as an agent in cavernous weathering. *Geology* **15**, 962–6.

Young, R.W. 1987b. Sandstone landforms of the tropical East Kimberley region, northwestern Australia. *Journal of Geology* **95**, 205–18.

Young, R.W. 1988. Quartz etching and sandstone karst: examples from the East Kimberleys, northwestern Australia. *Zeitschrift für Geomorphologie* **32**, 409–23.

Yu, Y.S. and D.F. Coates 1970. Analysis of rock slopes using the finite element method. *Department of Energy, Mines and Resources, Mines Branch, Mining Research Centre, Research Report* R229, Ottawa.

Zaruba, Q. and V. Mencl 1982. *Landslides and their control* (2nd edn). Amsterdam: Elsevier.

Chapter 9

Rock-Mantled Slopes

Anthony J. Parsons, Athol D. Abrahams, and Alan D. Howard

Introduction

Desert hillslopes below the angle of repose are dominated by the weathering characteristics of the underlying lithology, and specifically by the rate of production of fine material compared to the rate of removal. The previous chapter considered hillslopes underlain by massive rocks, or those in layered rocks dominated by outcropping resistant layers. These lithologies are weathering limited, and give rise to hillslopes where a surficial layer of weathered material is thin or absent. On more readily weathered lithologies a more-or-less continuous layer of debris is found. This layer of debris is subject to pedogenic processes. This chapter deals with such hillslopes.

The size distribution of particles forming this layer is a function of the composition and weathering characteristics of the underlying lithology. Consequently, rock-mantled slopes lie along a continuum. At one end of the continuum they grade into rock slopes. At the other end of the continuum, where the underlying lithology is, itself, dominated by fine-grained materials, they may grade into badlands (Chapter 10). On both massive rocks and on badlands vegetation in sparse or absent. In contrast, the layer of weathered material on rock-mantled slopes provides a substrate for vegetation to grow. Consequently, these hillslopes often have a vegetation cover and the processes acting upon them are affected by this vegetation (see Chapter 3).

A.J. Parsons (✉)
Sheffield Centre for International Drylands Research,
Department of Geography, University of Sheffield, Sheffield
S10 2TN, UK
e-mail: a.j.parsons@sheffield.ac.uk

Slope Form and Adjustment

Although rock-mantled slopes may exist at any gradient below the angle of repose, a distinction is often drawn between those with gradients in excess of 10° and those below this gradient. This distinction arises because of the sharp boundary between upland areas and piedmonts – the piedmont junction – that is characteristic of desert environments and which usually occurs at about this angle. Above this gradient, rock-mantled slopes have been termed boulder-controlled slopes (Bryan 1922), debris-covered slopes (Melton 1965) or debris slopes (Abrahams et al. 1985). However, many characteristics of these slopes persist through the piedmont junction, so here we retain the more general term rock-mantled slopes and consider hillslopes both steeper and gentler than 10°.

Typically the profiles of rock-mantled slopes are convex-rectilinear–concave, though either the rectilinear or concave elements may be missing. Generally the upper convexity is narrow, and the profile is dominated by either the rectilinear or concave element. The rectilinear element tends to dominate on slopes affected by stream undercutting (Strahler 1950), and the concave element on slopes unaffected by this process. However, even in the latter circumstances, a rectilinear element may be present and occupy a significant proportion of the profile, especially where the slope is long or steep.

The coarse debris mantling these slopes is often embedded within and/or resting upon a matrix of fines, particularly toward the base of slopes or where gradients are gentle. These fines are produced by the chemical and physical breakdown of the coarse debris. As the weathering particles become finer, they are preferentially transported downslope by hydraulic processes,

A.J. Parsons, A.D. Abrahams (eds.), *Geomorphology of Desert Environments*, 2nd ed.,
DOI 10.1007/978-1-4020-5719-9_9, © Springer Science+Business Media B.V. 2009

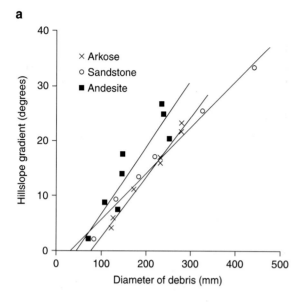

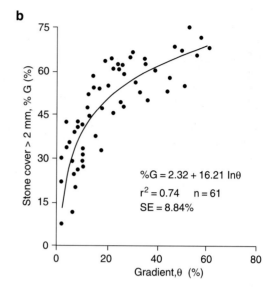

Fig. 9.1 Graphs of hillslope gradient against various measures of particle size. (**a**) Graph of hillslope gradient against the mean size of the ten largest debris particles for three debris slopes on three rock types in southern Arizona (after Akagi 1980). (**b**) Graph of stone cover against hillslope gradient for 12 debris slopes in Walnut Gulch Experimental Watershed, Arizona

so that usually the proportion of fines increases and the proportion of coarse debris decreases in this direction. Inasmuch as gradient also tends to decrease down-slope, positive correlations between gradient and various measures of particle size often obtain, particularly on weak to moderately resistant rocks (e.g., Cooke and Reeves 1972, Kirkby and Kirkby 1974, Ak-

agi 1980, Abrahams et al. 1985, Simanton et al. 1994) (Fig. 9.1).

A more detailed picture of the variation in mean particle (fines plus debris) size with gradient down a slope profile is presented in Fig. 9.2. This profile is located in the Mojave Desert, California, and is underlain by closely jointed latitic porphyry (Fig. 9.3). Beginning

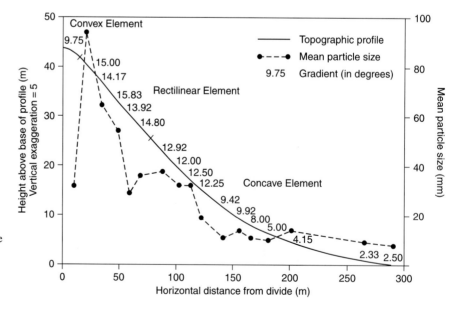

Fig. 9.2 Debris slope profile showing the downslope variation in mean particle size (sample size 100) and gradient (measured length 5 m). The debris slope is depicted in Fig. 9.3

Fig. 9.3 Photograph of well-adjusted debris slope underlain by latitic porphyry in Turtle Valley, Mojave Desert, California

at the divide, mean particle size increases with gradient down the upper convexity. Note, however, that the particles are much larger than at comparable gradients on the basal concavity. This is because the weathering mantle is thinner and bedrock outcrops are more common on the convexity. Downslope from the convexity is a substantial rectilinear element. Mean particle size is at a maximum at the top of this element and decreases down the element. The decrease continues down the long concave element. This downslope pattern of change in particle size is representative of many, if not most, rock-mantled slopes without basal streams, including those that are both much steeper and much gentler than this example, and it appears to be primarily due to the selective transport of fine sediment by hydraulic processes.

That hydraulic processes play a dominant role in removing sediment from and fashioning many rock-mantled slopes is suggested by a study of the relation between gradient S and mean particle size $\overline{D}$ for slopes underlain by weak to moderately resistant rocks in the Mojave Desert, California. In this study, Abrahams et al. (1985) found that plan-planar slopes on different rocks have S–$\overline{D}$ relations with similar slope coefficients but different intercepts (Fig. 9.4). However, on a given rock the slope coefficient varies with slope planform, being greater for plan-concave slopes than for plan-convex ones (Fig. 9.5).

To explain their findings Abrahams et al. assumed that sediment transport by hydraulic processes can be characterized by an equation of the form

$$G \propto X^m S^n / \overline{D}^p \qquad (9.1)$$

where G is sediment transport rate, X is horizontal distance from the divide, and m, n, and p are positive

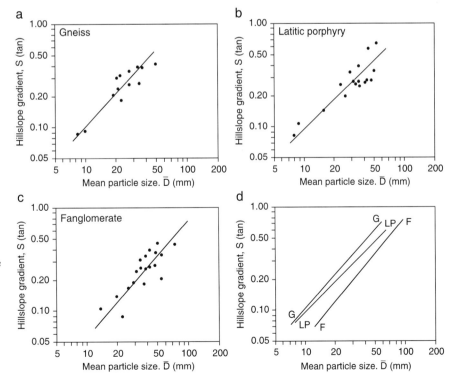

Fig. 9.4 Graphs of hillslope gradient against mean particle size for plan-planar debris slopes underlain by (**a**) gneiss, (**b**) latitic porphyry, and (**c**) fanglomerate. The fitted lines in (**a**), (**b**), and (**c**) are reproduced in (**d**) for comparative purposes (after Abrahams et al. 1985)

Fig. 9.5 Graphs of hillslope gradient against mean particle size for two (**a, b**) plan-convex debris slopes and their basal pediments and two (**c, d**) plan-concave slopes and their basal alluvial fans (after Abrahams 1987). Note that the S–$\overline{D}$ relations are much steeper for the plan-concave slopes than for the plan-convex ones

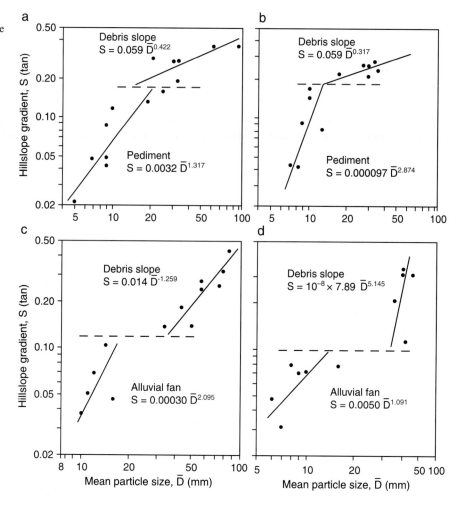

coefficients. Now if slopes are formed by and adjusted to hydraulic processes, Equation (9.1) may be manipulated to ascertain how the $S–\overline{D}$ relation varies with slope planform. Because $\overline{D} \propto X^q$, where $q < 0$,

$$G \propto \overline{D}^{[(m/n)-p]} S^n \qquad (9.2)$$

Rearranging Equation (9.2), one obtains

$$S \propto G^{1/n} / \overline{D}^{[(m/q)-p]/n} \qquad (9.3)$$

From Equation (9.1) it can be seen that m is larger for plan-concave slopes, where overland flow converges, than for plan-convex slopes, where overland flow diverges. The larger the value of m, the larger is the exponent (i.e. slope coefficient) of $\overline{D}$ in Equation (9.3), and the steeper is the $S–\overline{D}$ relation. Thus the analysis predicts that plan-concave slopes have steeper $S–\overline{D}$ relations than plan-convex ones. The agreement

between the analysis and observed variation in the $S – \overline{D}$ relation with planform implies that the slopes are formed by and adjusted to hydraulic processes.

In the above sediment transport equation (Equation 9.1), the arithmetic mean particle size $\overline{D}$ was used as the measure of particle size because, of the several measures of particle size tested, it correlated most highly with gradient. Variable $\overline{D}$ worked best in this instance possibly because it represented resistance to flow, and the contribution by each piece of rock to flow resistance was additive rather than multiplicative. However, there are other measures of particle size, and in different circumstances they may be better predictors of G than is $\overline{D}$. For example, $\overline{D}$ is very insensitive to size of fines. Therefore where this property has a significant effect on G, a more sensitive sediment size variable should be used.

Laboratory experiments by Poesen and Lavee (1991) showed that the proportion of the surface covered with coarse debris (i.e. percentage stone cover) and the size of debris (stones) have an important influence on G (Fig. 9.6). Usually G decreases as stone cover increases due to increased resistance to flow and increased protection of the underlying fines. However, where stones are larger than about 50 mm and cover less than 70% of the surface, the opposite is true because the stones tend to concentrate the flow. Most interesting is the fact that for a given stone cover, G consistently increases with stone size, again because the stones tend to concentrate the flow. These findings by Poesen and Lavee suggest that although Equation (9.1) may be a useful start to the modelling of slope form, the situation on actual slopes is probably far more complex, and that a great deal more work is required to elucidate the effect of rock-fragment size and cover on sediment transport rate.

Two studies have provided direct evidence of the movement of stones on rock-mantled hillslopes in the American South-west. Kirkby and Kirkby (1974) painted lines across 12 hillslopes with gradient up to 20° in the Sonoran Desert of southern Arizona. During a two-month period they measured after each rainstorm the movement of all particles with diameters ≥ 1 mm. Field observations confirmed that the processes moving these particles were rainsplash and unconcentrated overland flow, and statistical analyses indicated that the distance moved was directly related to hillslope gradient and inversely related to grain size.

Abrahams et al. (1984) analysed 16 years of stone movement on two hillslopes with gradients up to 24° in the Mojave Desert, California. They found that the distance each particle moved was directly related to both length of overland flow (a surrogate for overland flow discharge) and hillslope gradient and inversely related to particle size. These results were interpreted as indicating that the stones, which ranged in size up to 65 mm, were moved mainly by hydraulic action. Citing Kirkby and Kirkby's findings as well as their own, Abrahams et al. (1984, p. 369) concluded 'that hydraulic action is probably the dominant process transporting coarse debris down hillslopes with gradients up to at least 24° over most of the Mojave and Sonoran Deserts'.

In a further study of the character and likely process of movement of debris on rock-mantled hillslopes, Abrahams et al. (1990) investigated the fabric of coarse particles mantling a debris slope on Bell Mountain in the Mojave Desert. The slope is typical of debris slopes in the Mojave Desert underlain by closely jointed or mechanically weak rocks. Samples of rod- and disc-shaped particles from five sites ranging in gradient from 11.7° to 33.17° were found to display essentially the same fabric: particles tend to be aligned down-slope and to lie flat on the ground surface. There is no evidence of imbrication signifying sliding or creep nor of transverse modes indicating rolling. Abrahams et al. concluded that the fabric is probably produced by hydraulic action, and that this process is mainly

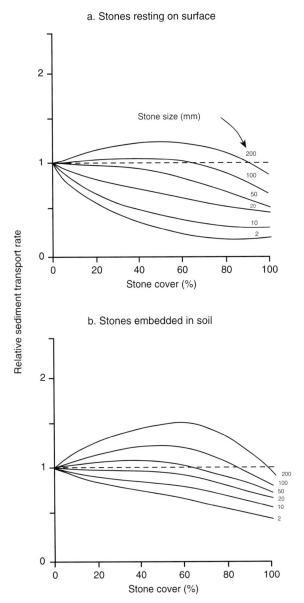

Fig. 9.6 Graphs showing relations between sediment transport rate and stone cover for different stone sizes: (**a**) stones resting on the soil surface, and (**b**) stones partially embedded in the soil. The graphs are generalizations of the experimental results of Poesen and Lavee (1991)

responsible for moving coarse particles on gradients up to 33° on these debris slopes. Cumulative size distributions of the particles sampled at the two sites with gradients greater than 28° reveal that about 25% of the particles are larger than 64 mm and that the largest particle in each sample has a diameter in excess of 300 mm. It is difficult to imagine particles of this size being entrained by overland flow a few millimetres deep and transported as bed load. Abrahams et al. (1990) suggested that such particles may be moved downslope by a process termed runoff creep. De Ploey and Moeyersons (1975) observed this process on steep hillslopes in Nigeria and then replicated it in a laboratory flume. Their flume experiments disclosed that under the influence of overland flow (a) blocks shifted and tilted downslope when smaller gravel particles on which they were resting became wet and collapsed; (b) pebbles moved forward and tilted downslope during liquefaction of the underlying soil layer; (c) erosion of underlying finer material caused pebbles to settle downslope; and (d) scour on the upslope side of pebbles resulted in their being drawn into the holes and tilted upslope.

Piedmont Junctions

The sharp transition zone between upland areas and the piedmont characteristic of deserts has been variously referred to as the transition slope (Fair 1948), the nickpoint (Rahn 1966), the break in slope (Kirkby and Kirkby 1974), the piedmont angle (Twidale 1967, Young 1972, pp. 204–8, Cooke and Warren 1973,

p. 199), and the piedmont junction (Mabbutt 1977, p. 82, Parsons and Abrahams 1984). In this chapter we use the term piedmont junction. At many locations the piedmont junction marks the boundary between the operation of different processes: for example, where an alluvial fan abuts against a hillslope. At other locations, the morphology of the piedmont junction is manifestly influenced by geological structure (e.g. Twidale 1967) or subsurface weathering (e.g. Twidale 1962, Mabbutt 1966). We are concerned with none of these situations here. Rather we focus on piedmont junctions that are simply concavities in slope profiles. These piedmont junctions occur at the transition between a pediment and its backing hillslope and may be defined as extending from 15° on the lower part of the backing hillslope to 5° on the upper part of the pediment (Kirkby and Kirkby 1974).

Piedmont junctions vary greatly in concavity. At one extreme are features that are so concave that they take the form of a true break in slope and can be identified only as a point on the hillslope profile (Fig. 9.7a). At the other extreme are features whose concavity is so slight that they can reach lengths of 750 m (Fig. 9.7b) (Kirkby and Kirkby 1974). Numerous workers have noted that piedmont junctions in many locations conform to two general tendencies. First, under a given climate they tend to vary in concavity from one rock type to another (e.g. Kirkby and Kirkby 1974, Mabbutt 1977, pp. 85–7). Second, on a given rock type they tend to decrease in concavity as precipitation increases (e.g. Bryan 1940, Fair 1947, 1948, Young 1972, p. 208, Mabbutt 1977, p. 85). The latter

Fig. 9.7 Photographs of piedmont junctions, showing (**a**) a narrow, highly concave one formed on widely jointed quartz monzonite, Mojave Desert, California, and (**b**) a broad, gently concave one developed on gneiss, Mojave Desert, California

tendency is of course implicit in the fact that piedmont junctions (i.e. pronounced concavities) are generally associated with desert landscapes and not humid ones.

Both these tendencies derive from the fact that on slopes adjusted to present-day processes of sediment transport, gradient varies directly with particle size, which in turn varies inversely with distance downslope at different rates on different rock types and in different climates. Early workers claimed that gradient was related to particle size (e.g. Lawson 1915, Bryan 1922, Gilluly 1937), and in the latter part of the twentieth century this relation was verified quantitatively (e.g. Kirkby and Kirkby 1974, Abrahams et al. 1985). M.J. Kirkby (Carson and Kirkby 1972, pp. 346–7, Kirkby and Kirkby 1974) was perhaps the first to point out that different rock types have different comminution sequences and that, because hydraulic processes selectively transport finer particles further downslope than coarser ones, the comminution sequence is reflected in the downslope rate of change in particle size and, hence, gradient. At one extreme are rocks, such as basalts and schists, that have fairly continuous comminution sequences from boulder- to silt-sized particles. The slopes that form in these rocks exhibit a progressive decrease in particle size accompanied by a steady decline in gradient downslope, forming a broad and gently curving piedmont junction. At the other extreme are rocks, such as widely jointed granites, that are characterized by markedly discontinuous comminution sequences in which boulders disintegrate directly into granules and sands. On these rocks, steep backing slopes mantled with boulders give way abruptly downslope to gentle pediments covered with granules and sands, and an extremely narrow, almost angular piedmont junction is produced. It is therefore evident that given the relationship between hillslope gradient and particle size, the concavity of piedmont junctions in a desert climate depends on the comminution sequence of the underlying rock.

The same line of reasoning may be applied to explaining the variation in piedmont junction concavity with climate. In desert climates, particle size decreases across piedmont junctions in accordance with the comminution sequence of the underlying rock, as explained above. In humid climates, on the other hand, soils with fine-grained A horizons are developed on both the backing hillslopes and the footslopes, and the size of surface particles decreases very little, if at all, downslope (e.g. Furley 1968, Birkeland 1974, p. 186). Because the decrease in particle size across the pied-

Fig. 9.8 Photograph of a steep, poorly adjusted debris slope developed on widely jointed quartz monzonite in the Mojave Desert, California

mont junction is more pronounced in desert climates than in humid ones, the piedmont junctions are typically narrower and more concave.

The foregoing discussion applies to slopes that are adjusted to present-day processes, at least in the vicinity of the piedmont junction. However, not all slopes are so adjusted. Where they are not, particle size may be unrelated to gradient, and the preceding analysis is irrelevant. The situation most commonly encountered is where a pediment covered with fines and presumably adjusted to contemporary processes is backed by a weathering-limited slope that is clearly not adjusted to current transport processes (Fig. 9.8). The form of the backing slope might be controlled by rock mass strength (Selby 1980, 1982a, b, pp. 199–203) or rock structure (Oberlander 1972) or inherited from a previous climate. In such circumstances, the concavity of the piedmont junction cannot be understood in terms of contemporary hydraulic processes. About all that can be said about piedmont junctions of this type is that they tend to be more concave than most. The reason for this is that in a given (desert) climate, steep backing slopes are more likely to become weathering-limited than are gentle ones, and piedmont junctions with steep backing slopes are likely to be more concave than those with gentle backing slopes.

Processes on Rock-Mantled Slopes

Hydraulic Processes

Virtually all runoff from desert hillslopes occurs in the form of overland flow that is generated when the rainfall intensity exceeds the surface infiltration rate. Such rainfall-excess overland flow is widely termed Hortonian overland flow (Horton 1933). Because hydraulic processes are, therefore dependent on infiltration rates, understanding infiltration is central to understanding runoff and erosion on rock-mantled slopes.

Infiltration on Rock-Mantled Slopes

As in all environments, infiltration through the surface layer of rock-mantled hillslopes is controlled by its physical and chemical properties (see, for example, Mills et al., 2006) However, what is particularly important for hydraulic processes on desert hillslopes is the great variation in infiltration that they exhibit, both spatial and temporal. This variation may reflect variation in surface or subsurface properties. Although much less is known about the role of the latter than the former, Perrolf and Sandstrom (1995), in a study undertaken in Botswana and Tanzania, showed that variations in subsoil conditions were responsible for only a fivefold variability in infiltration, compared to twentyfold differences due to variations in surface conditions. Among the properties controlling infiltration and runoff are the ratio of bedrock to soil, surface and subsurface stone size, stone cover, vegetation and surface sealing.

Given the widespread occurrence of bedrock outcrops on many desert hillslopes, an important control of infiltration and runoff is the ratio of bedrock to soil. Figure 9.9 shows the infiltration curves for rocky and soil-covered surfaces at Sede Boqer and the Hovav Plateau in the northern Negev, Israel (Yair 1987). The infiltration capacities are lower for the bedrock than for the soil-covered surfaces at both sites. The difference is especially pronounced for the Sede Boqer site because the rock is a smooth, massive crystalline limestone, whereas at the Hovav Plateau it is densely jointed and chalky. Data from natural rainfall events at Sede Boqer (Yair 1983) indicate that the threshold level of daily rainfall necessary to generate runoff in the rocky areas is 1–3 mm, whereas it is 3–5 mm for the colluvial soils. As rain showers of less than 3 mm represent 60% of the rain events, the frequency and magnitude of runoff events are both much greater on the rocky than on the soil-covered areas.

Even where bedrock outcrops are absent, desert soils are typically stony. The effect of surface stones

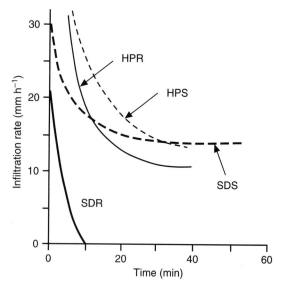

Fig. 9.9 Infiltration curves for rocky and soil-covered surfaces in the northern Negev, Israel: Sede Boqer (rainfall intensity 36 mm h^{-1}) massive limestone (SDR), Sede Boqer stony colluvium soil (SDS), Hovav Plateau (rain intensity 33 mm h^{-1}) densely jointed and chalky limestone (HPR), and Hovav Plateau stoneless colluvial soil (HPS) (after Yair 1987)

on runoff is quite complex and has been the subject of numerous field and laboratory studies (e.g. Jung 1960, Seginer et al. 1962, Epstein et al. 1966, Yair and Klein 1973, Yair and Lavee 1976, Box 1981, Poesen et al. 1990, Abrahams and Parsons 1991a, Lavee and Poesen 1991, Poesen and Lavee 1991). Figure 9.10, which is based on laboratory experiments by Poesen and Lavee (1991, Fig. 3), summarizes the state of knowledge for surfaces devoid of vegetation. Basically, surface stones affect runoff by two groups of mechanisms. First, increasing stone size and stone cover increasingly protect the soil surface from raindrop impact and thereby inhibit surface sealing and reduce runoff. Increasing stone size and stone cover also increase depression storage which promotes infiltration. Second, increasing stone size and stone cover result in greater quantities of water being shed by the stones (stone flow) and concentrated in the interstone areas, where the water overwhelms the ability of the underlying soil to absorb it and runs off in increasing amounts. Both groups of mechanisms operate simultaneously. In general, it appears that as stone size increases the second group dominates. As a result, runoff increases with stone size irrespective of stone cover. The relation between runoff and stone cover is less straightforward. Where stone sizes and stone covers are small, the

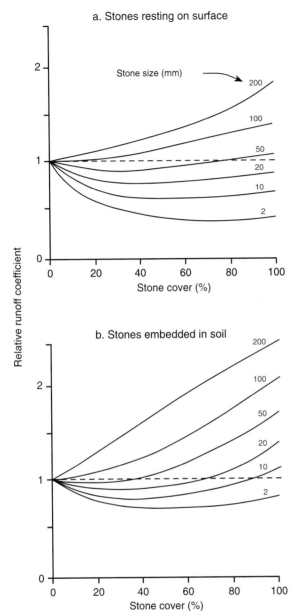

Fig. 9.10 Graphs showing relations between runoff coefficient and stone cover for different stone sizes: (**a**) stones resting on the soil surface and (**b**) stones partially embedded in the soil. The graphs are generalizations of the experimental results of Poesen and Lavee (1991)

are embedded in the soil runoff rates are higher than where they are resting on the surface. Interestingly, for intermediate (mean) stone sizes (i.e. 20–50 mm), the sign of the relation between runoff and stone cover may actually change from negative for stones resting on the surface to positive for stones embedded in the soil. (Poesen 1990, Poesen et al. 1990).

Figure 9.10 applies to areas devoid of vegetation. Where there is a significant vegetation cover, particularly of shrubs, the controls of infiltration and runoff are quite different. This is reflected in the correlation between infiltration and stone cover. Abrahams and Parsons (1991b) noted that both positive and negative correlations between infiltration and stone cover have been reported for semiarid hillslopes in the American South-west. They observed that positive correlations (Tromble 1976, Abrahams and Parsons 1991a) have been obtained when infiltration measurements were confined to (bare) stone-covered areas between shrubs (lower curves in Fig. 9.10), and they attributed these correlations to increasing stone cover progressively impeding surface sealing. In contrast, negative correlations have been found when infiltration was measured in shrub as well as intershrub areas (e.g. Tromble et al. 1974, Simanton and Renard 1982, Wilcox et al. 1988, Abrahams and Parsons 1991b). Abrahams and Parsons ascribed these correlations to infiltration rates under shrubs being greater than those between shrubs (Lyford and Qashu 1969), and percentage stone cover being negatively correlated to percentage shrub canopy (Wilcox et al. 1988). The mechanisms giving rise to higher infiltration rates under shrubs than between them are summarized in Fig. 9.11. As might be expected, positive correlations have been recorded between infiltration rate and percentage plant canopy (Kincaid et al. 1964, Simanton et al. 1973, Tromble et al. 1974).

In recent years, the role of surface crusts in controlling infiltration has achieved greater recognition. These crusts are of two types: mechanical and biological. Mechanical crusts are formed by one or more of raindrop impact, trapped gas bubbles (forming a vesicular crust) and evaporation (forming chemical crusts, e.g. Romao and Escudero 2005). Biological crusts are created by an association between soil particles and cyanobacteria, lichens and mosses. Airborne silts and clays are trapped by sticky cyanobacterial sheaths, resulting in a thin surface layer of silts and clays that are often lacking where biological crusts are absent. In

first group of mechanisms dominates, and runoff is negatively related to stone cover. However, for other combinations of stone size and stone cover the second group dominates, and runoff is positively related to stone cover. Stone position also affects runoff. A comparison of Figs. 9.10a, b reveals that where stones

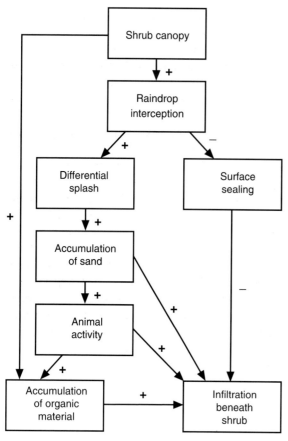

Fig. 9.11 Causal diagram showing the mechanisms whereby a shrub's canopy promotes infiltration under the shrub at Walnut Gulch, Arizona (after Abrahams and Parsons 1991a)

general, it is believed that the existence of crusts decreases surface permeability and infiltration (e.g. Wang et al. 2007), but debate still exists on the effects of biological crusts (see Belnap 2006, for a review). All crusts are fragile and may be disturbed by faunal activity, vehicles and penetration by large raindrops. Quantification the effects of crusts on infiltration relative to other factors remains unresolved (Belnap 2006).

In addition to the effects of crusts, many dryland soils are hydrophobic. This hydrophobicity may arise from the effects of fires, where spatial variability in fire intensity results in spatial variability in hydrophobicity (e.g. Ferreira et al. 2005), or from chemicals exuded by desert plants and the effects of organic debris (Cammeraat et al. 2002).

Finally, spatial variability in infiltration may result from faunal activity. Animals affect infiltration in many ways from grazing and consequent reduction in vegetation cover to disturbance of soil crusts, but probably their greatest influence on infiltration is in their cre-

ation of macropores through burrowing. In deserts, termites are probably the most significant producers of macropores. Mando and Miedema (1997), in an experimental study in Burkina Faso to assess the effects of termites on degraded soils, found 60% of macropores to be due to termite activity, accounting for 38.6% of total soil porosity in the top 7 cm of the soil. Similarly, working in Senegal, Sarr et al. (2001) showed that infiltration rates were about 80% lower in plots from which termites had been excluded. In contrast, however, Debruyn and Conacher (1994) found that micropores produced by ants only affected infiltration rates when the soil is saturated and water is ponded on the surface.

Not only do desert hillslopes exhibit spatial variability in infiltration, but they also exhibit temporal variability. At a seasonal scale variations in antecedent soil mositure may affect infiltration (Fitzjohn et al. 1998), and this effect may vary with rainfall intensity (Castillo et al. 2003). Simanton and Renard (1982) identified differences between spring and autumn infiltration which they attributed to the effects of frost and wetting and drying in the winter and rainfall in the summer. Processes in the winter loosened the soil surface, whereas those in the summer compacted it. Over longer timespans fire-induced hydrophobicity, for example, changes. Although it is generally found to decline, Cerda and Doerr (2005) found topsoil hydrophobicity actually increased with time under *Pinus halepensis*.

The spatial variability in the factors controlling point infiltration rates means that overland flow is not generated uniformly over a desert hillslope but preferentially from those parts of the hillslope where the infiltration rates are lowest. Overland flow generated then travels downslope where it may encounter other areas whose infiltration capacity remains higher than the rainfall intensity. Some or all of the flow may infiltrate into these areas (Smith and Hebbert 1979, Hawkins and Cundy 1987). As a result of this runon infiltration, runoff per unit area may decrease with the size of the area being investigated. This phenomenon is illustrated in Fig. 9.12 for a small piedmont watershed at Walnut Gulch, Arizona (Kincaid et al. 1966).

Not only may spatial variability in point infiltration cause a decrease in runoff per unit area as scale increases so, too, may temporal variability in rainfall input. Most studies of infiltration have used ring infiltrometers or constant-intensity simulated rainfall and have thus neglected the impact of temporal variability

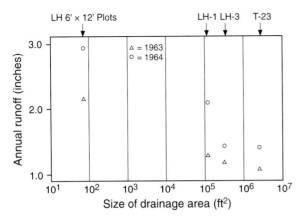

Fig. 9.12 Graph of annual runoff against size of drainage area for runoff plots and very small watersheds ($>10^5$ ft^2) at Walnut Gulch, Arizona (after Kincaid et al. 1966)

of rainfall input on runon infiltration. Wainwright and Parsons (2002) showed in a modelling study that temporal variability in rainfall input is likely to be a larger contributor to runon infiltration than is spatial variability in point infiltration. Similarly, De Lima and Singh (2002) and Reaney et al. (2007) demonstrated that storms delivering the same rainfall amount could yield different amounts of runoff depending on the temporal distribution of the rainfall.

Unconcentrated Runoff

Runoff on desert hillslopes generally first appears as an unconcentrated sheet of water with threads of deeper, faster flow diverging and converging around surface protuberances, rocks, and vegetation. As a result of these diverging and converging threads, flow depth and velocity may vary markedly over short distances, giving rise to changes in the state of flow. Thus over a small area the flow may be wholly laminar, wholly turbulent, wholly transitional, or consist of patches of any of these three flow states. Much research on the hydraulics of unconcentrated overland flow has focused on the controls on resistance offered to the flow by the rough surface that characterises rock-mantled hillslopes. Resistance to overland flow may be quantified by the dimensionless Darcy–Weisbach friction factor

$$f = 8ghS/V^2 \qquad (9.4)$$

where g is the acceleration of gravity, h the mean depth of flow, S the energy slope, and V the mean flow velocity. Flow resistance f consists of grain resistance,

form resistance, rain resistance and sediment-transport resistance. Grain resistance f_g is imparted by soil particles and microaggregates that protrude into the flow less than about ten times the thickness of the viscous sublayer (Yen 1965). Form resistance f_f is exerted by microtopographic protuberances, stones, and vegetation that protrude further into the flow and control the shape of the flow cross-sections (Sadeghian and Mitchell 1990). Rain resistance f_r is due to velocity retardation as flow momentum is transferred to accelerate the raindrop mass from zero velocity to the velocity of the flow (Yoon and Wenzel 1971). Finally, sediment-transport resistance f_t is due to velocity retardation as flow momentum is transferred to accelerate sediment mass from zero velocity to its transport velocity (Abrahams and Li 1998). For laminar flow on gentle slopes f_r may attain 20% of f (Savat 1977). However, generally it is a much smaller proportion, and the proportion becomes still smaller as the state of flow changes from transitional to turbulent (Yoon and Wenzel 1971, Shen and Li 1973). Because f_r is typically several orders of magnitude less than f on desert hillslopes (Dunne and Dietrich 1980), and similarly f_t is likely to be small on rough hillslopes (Abrahams and Li 1998), the following discussion will focus on f_g and f_f.

Resistance to flow generally varies with the intensity of flow, which is represented by the dimensionless Reynolds Number

$$R_e = 4Vh/\nu \qquad (9.5)$$

where ν is the kinematic fluid viscosity. Laboratory experiments and theoretical analyses since the 1930s have established that where f is due entirely to grain resistance the power relation between f and R_e for shallow flow over a plane bed is a function of the state of flow. The relation has a slope of -1.0 where the flow is laminar and a slope close to -0.25 where it is turbulent. This relation between f and R_e (or surrogates thereof) for plane beds has been widely used in models of hillslope runoff. However, the surfaces of desert hillslopes are rarely, if ever, planar, and the anastomosing pattern of overland flow around microtopographic protuberances, rocks, and vegetation attests to the importance of form resistance. If form resistance is important, its influence might be expected to be reflected in the shape of the f–R_e relation.

This was first recognized in a set of field experiments conducted by Abrahams et al. (1986) on small runoff plots located in intershrub areas on piedmont hillslopes at Walnut Gulch, southern Arizona. Although the plot surfaces were mantled with gravel, clipped plant stems occupied as much as 10% of their area, and the steeper plots had quite irregular surfaces. Analyses of 14 cross-sections yielded f–R_e relations that were positively sloping, negatively sloping, and convex-apward (Fig. 9.13). These shapes were attributed to the progressive inundation of the roughness elements (i.e. gravel, plant stems, and microtopographic protuberances) that impart form resistance. So long as these elements are emergent from the flow, f increases with R_e as the upstream wetted projected area of the elements increases. However, once the elements become submerged, f decreases as

R_e increases and the ability of the elements to retard the flow progressively decreases.

In a second set of field experiments on small plots sited in gravel-covered intershrub areas at Walnut Gulch, Abrahams and Parsons (1991c) obtained the regression equation

$$\log f = -5.960 - 0.306 \log R_e + 3.481 \log \%G + 0.998 \log D_g \tag{9.6}$$

where $\%G$ is the percentage of the surface covered with gravel, D_g is the mean size of the gravel (mm), and $R^2 = 0.61$. Of the independent variables in Equation (8.4), $\%G$ was by far the single best predictor of f, explaining 50.1% of the variance. The dominance of $\%G$ implies that $f_f >> f_g$ on these gravel-covered hillslopes. This was confirmed using a procedure developed by Govers and Rauws (1986) for calculating the relative magnitudes of f_g and f_f in overland flow. For the 73 experiments performed on the small plots the modal and median values of $\% f_g$, which denotes grain resistance expressed as a percentage of total resistance, were 4.55% and 4.53% (Fig. 9.14). Thus on these gravel-covered hillslopes, f_g is typically about one-twentieth of f_f. This conclusion has important implications for sediment transport which will be explored below.

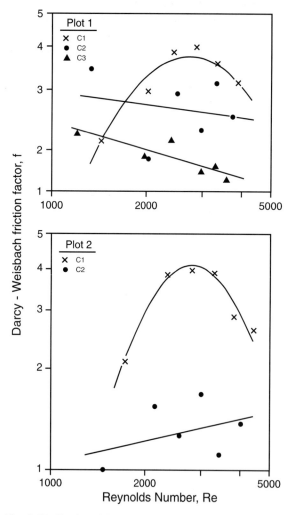

Fig. 9.13 Graphs of Darcy–Weisbach friction factor against Reynolds Number for five cross-sections on two runoff plots at Walnut Gulch, Arizona. The cross-sections are denoted by C1, C2, etc. (after Abrahams et al. 1986)

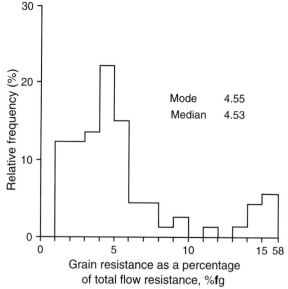

Fig. 9.14 Relative frequency distribution of grain resistance expressed as a percentage of total flow resistance for 73 experiments on 8 runoff plots at Walnut Gulch, Arizona

The findings of Abrahams et al. (1986) and Abrahams and Parsons (1991c) are supported by laboratory experiments by Gilley et al. (1992) in which varying rates of flow were introduced into a flume covered with different concentrations and sizes of gravel. Gilley et al. also recorded positively sloping, negatively sloping, and convex-upward f–R_e relations which they attributed to the progressive inundation of the gravel. In addition, they obtained regression equations of the form

$$\log f = \log a_1 - a_2 \log R_e + a_3 \log \%G \qquad (9.7)$$

for each gravel size class. The R^2 value for each regression exceeded 0.94. These flume experiments confirm the important role of gravel cover in controlling resistance to overland flow through its influence on form resistance. The importance of rock-fragment size was shown in a study by Bunte and Poesen (1994). These authors found significant differences between pebbles (with mean b-axis of 15 mm) and cobbles (mean b-axis 86 mm) in the changes to flow hydraulics with percentage cover, particularly at low cover percentages (Fig. 9.15). They attributed these differences to the more reticular flow around the smaller pebbles.

Because rock-mantled hillslopes are also typically vegetated, rock fragments are not the only contributor to form roughness. In a laboratory experiment, Dunkerley et al. (2001) showed that for a given percentage cover, plant litter increases resistance to flow more than do rock fragments. Dunkerley (2003) used a similar argument to that used by Bunte and Poesen (1994) to explain this effect, and it is notewothy that the rock-fragment cover percentages used in the experiments are relatively low, so the results are consistent with the differences found by Bunte and Poesen. Furthermore, Dunkerley et al. (2001) showed that for flows wholly within the laminar range f consistently declined with Re, in contrast to observations of more complex relationships, and argued that flow regime is a significant control on the form of the relationship. Away from the laboratory, flow on natural hillslopes is almost always a mixture of wholly laminar, wholly turbulent and wholly transitional, termed by Abrahams et al. (1986) *composite flow*. In more recent work, Dunkerley (2004) has argued that deriving a simple, unweighted average value for f for such composite flow biases the value towards

the high-resistance shallow laminar flow, and that estimates of average f should be weighted according to discharge in the different elements of composite flow.

The hydraulics of unconcentrated overland flow over entire hillslopes were investigated by Parsons et al. (1990, 1996) using simulated rainfall on plots 18 m wide and approximately 30 m long located on shrub-covered and grass-covered piedmont hillslopes at Walnut Gulch. The h, V, indunated width w and discharge Q values were computed for two measured sections situated 12.5 and 21 m from the top of the plot on the shrub-covered plot, and 6, 12 and 20.5 m from the top of the plot on the grassland plot. (Figures 9.16 and 9.17, respectively) At-a-section h–Q w-Q and V–Q relations show that increases in Q are accommodated by increases in h and w. On both plots, increases in discharge result in minimal change to mean velocity. Downslope hydraulic relations differ strikingly from at-a-section relations and between the two vegetation types. Under equilibrium (steady state) runoff conditions on the shrubland, f decreases rapidly as Q increases, permitting increases in Q to be accommodated almost entirely by increases in V. The decrease in f is due to the progressive downslope concentration of flow into fewer, larger threads. Under non-equilibrium conditions, downslope hydraulic relations are different from those at equilibrium, but f always decreases downslope. This is the result of low flows following pathways formed by higher flows that concentrate downslope. On the grassland downslope increases in Q are accommodated more-or-less equally by increases in V and h because flow does not concentrate downslope to the degree that it does on the shrubland.

Concentrated Runoff

The tendency for threads within unconcentrated overland flow to increase in depth and velocity downslope coupled with the convergence (and divergence) of these threads around obstructions may lead to the formation of small channels (rills). Such features are very common on desert hillslopes, particularly where the underlying material is easily eroded. However, in contrast to their study on agricultural land, there are few studies of rills on desert hillslopes. Abrahams et al. (1996) undertook a study of rills on a shrub-covered hillslope in southern Arizona. Although they

Fig. 9.15 Differences in flow
hydraulics for surfaces
partially covered with
pebbles and cobbles (after
Bunte and Poesen 1994)

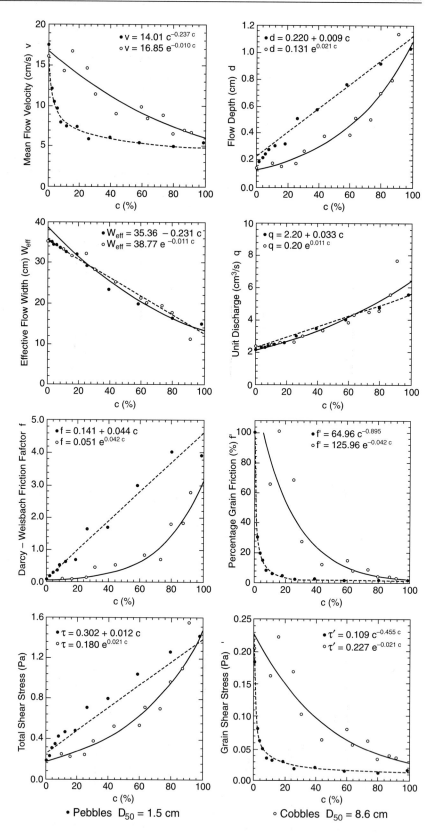

• Pebbles $D_{50} = 1.5$ cm   ○ Cobbles $D_{50} = 8.6$ cm

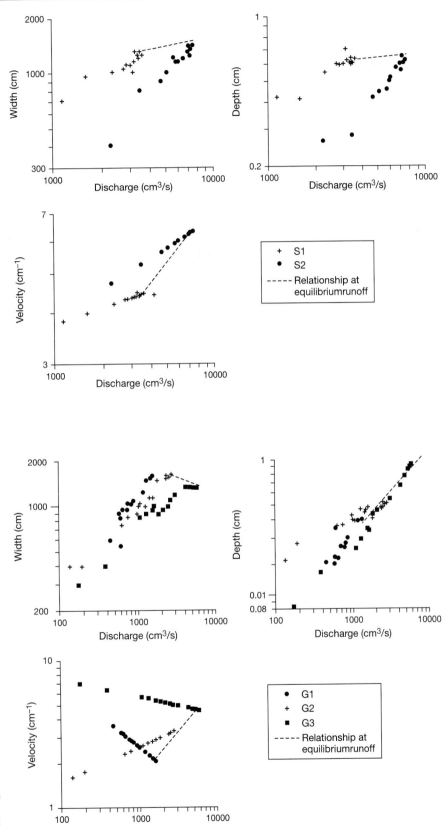

Fig. 9.16 Flow hydraulics at two cross sections on a large shrubland plot at Walnut Gulch, Arizona (after Parsons et al. 1996)

Fig. 9.17 Flow hydraulics at three cross sections on a large grassland plot at Walnut Gulch, Arizona (after Parsons et al. 1996)

found only a small difference in at-a-station hydraulic geometry between these rills and their agricultural counterparts that was not statistically significant, they argued that the difference was real and reflected the tendency for these rills to be wider and shallower than those on agricultural land. Furthermore, whereas Govers (1992) argued that a general relationship existed for rills in which flow velocity u depended on discharge Q and was unaffected by slope S or soil materials, the rills studied by Abrahams et al. showed a consistently lower velocity for a given discharge than the data presented by Govers (Fig. 9.18), and, in a multiple regression equation to predict flow velocity, Abrahams et al. found that both slope and percentage of gravel-sized particles (≥ 2 mm) $\%G$ were significant independent variables in an equation to predict rill flow velocity

$$\log u = 0.672 + 0.330 \log Q - 0.00415 \%G \\ + 0.0664 \log S, \qquad (9.8)$$

which had an R^2 of 0.859. In a more recent laboratory experiment, Rieke-Zapp et al. (2007) also found increasing resistance to flow with increasing percentage of gravel cover. However this effect was less evident at higher discharges and at the higher of the two slopes at which their experiments were conducted.

Erosion by Hydraulic Processes

Rates

There are very few data on rates of erosion by hydraulic processes on desert hillslopes. A survey by Saunders and Young (1983) indicated that rates exceed 1 mm y^{-1} on normal rocks in semi-arid climates but are less than 0.01 mm y^{-1} in arid climates. These rates of erosion for semi-arid climates are amongst the highest in the world. Although debris flows may be an important agent of erosion on slopes steeper than 30°, Young and Saunders (1986) concluded that hydraulic action is the predominant denudational process in semi-arid climates, and probably in arid ones as well. Within a given climate, however, there is considerable variability in rates of hydraulic erosion, even over a single hillslope. This variability is largely the result of differences in surface properties affecting runoff generation and sediment supply. Among these properties are stone size, stone cover, vegetation cover, and biotic activity.

Erosion by Unconcentrated Runoff

Controlling Factors

Abrahams and Parsons (1991a) investigated the relation between hydraulic erosion and gradient at Walnut

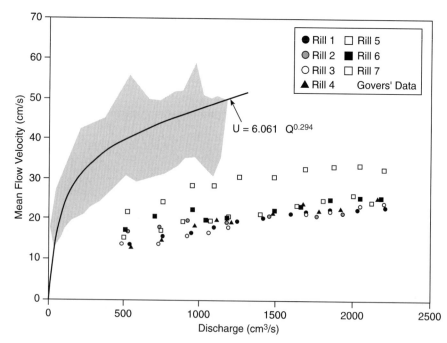

Fig. 9.18 Graph of mean flow velocity against discharge for seven rills in shrubland at Walnut Gulch, Arizona compared to Govers' (1992) data (shaded) and best-fit equation (after Abrahams et al. 1996)

Fig. 9.19 Curves fitted to graphs of (**a**) sediment yield and (**b**) runoff coefficient against gradient for three sets of experiments denoted by E1, E2, and E3 at Walnut Gulch, Arizona. Experiments E1 and E2 were conducted on plots underlain by Quaternary alluvium, with the ground vegetation being clipped for E1 but not for E2. Experiment E3 was performed on plots underlain by the Bisbee Formation

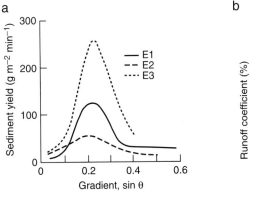

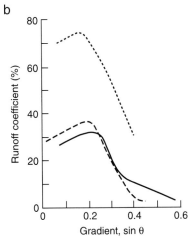

Gulch by conducting three sets of field experiments on small runoff plots under simulated rainfall on two different substrates. Each set of experiments yielded a convex-upward sediment-yield–gradient relation with a vertex at about 12° (Fig. 9.19). The key to understanding this relation is the relation between runoff and gradient. On slopes less than 12° runoff increases very slowly with gradient, so sediment yield increases with gradient mainly in response to the increase in the downslope component of gravity. On slopes steeper than 12° runoff decreases rapidly as gradient increases. This decrease in runoff outweighs the increase in the downslope component of gravity and causes sediment yield to decrease.

Although sediment yield is curvilinearly related to gradient, it is actually controlled in a complex way by a combination of stone size, surface roughness, and gradient. The nature of this control is outlined in Fig. 9.20 (Abrahams et al. 1988). Where gradients exceed 12° sediment yield is positively correlated with runoff which, in turn, is negatively correlated with gradient, stone size, and surface roughness (Yair and Klein 1973). Where gradients are less than 12° runoff is almost constant, and sediment yield is positively correlated with these variables. Thus the controls of sediment yield depend on the range of gradient being considered. Where gradients exceed 12° stone size and surface roughness have a strong influence on runoff and, through runoff, affect sediment yield. On the other hand, where gradients are less than 12°, stone size and surface roughness have little effect on runoff. However, they are correlated with gradient, and gradient determines sediment yield. The interesting question raised

by these results for slopes steeper than 12° is if stoniness increases with gradient causing runoff and erosion to decrease, how does one explain the increase of stoniness with gradient? The most likely explanation is that the small plot experiments that produced the above results do not take into account overland flow from upslope which would presumably be highly effective in eroding the steeper portions of desert hillslopes.

The relation between stone cover and sediment yield on semi-arid hillslopes has been investigated by Iverson (1980) and Simanton et al. (1984) using simulated rainfall. For 21 plots in the Mojave Desert, California, Iverson obtained a correlation of −0.56 between sediment yield and percentage stones (>2 mm) in the surface soil. However, these plots ranged in gradient from

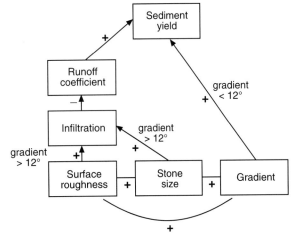

Fig. 9.20 Causal diagram showing the factors controlling the runoff coefficient and sediment yield on desert hillslopes

4° to 25°, which contributed greatly to the scatter. In a better controlled study in which all the plots had similar gradients (5.1–6.8°), Simanton et al. obtained a correlation of −0.98 between sediment yield and percentage stone cover (>5 mm) for eight plots at Walnut Gulch, Arizona. These negative correlations can be attributed to several factors: the stones protect the soil structure against aggregate breakdown and surface sealing by raindrop impact; enhance infiltration and diminish runoff; increase surface roughness which decreases overland flow velocities; and reduce soil detachment and, hence, interrill erosion rates (Poesen 1990).

Laboratory experiments by Poesen and Lavee (1991), however, suggest that the correlation between sediment yield and stone cover is not always negative. Figure 9.6, which is a generalization of Poesen and Lavee's results, indicates that the correlation becomes positive where the stones are embedded in the soil and are larger than 50 mm. In these circumstances, the increasing stone-flow effect outweighs the increasing protection-from-raindrop-impact and flow-retardation effects as stone cover increases, and the increasing concentration of water between the stones results in greater flow detachment and transport of soil particles. However, once stone cover increases above about 70%, sediment yield begins to decline toward a minimum at 100%, when the stone cover affords complete protection of the soil beneath. Poesen and Lavee's experiments also show that for a given stone cover, sediment yield consistently increases with stone size due to increasing stone flow.

Simanton and Renard (1982) used simulated rainfall to examine seasonal variations in the erosion of three soils at Walnut Gulch. In the spring the soil surface is loose due mainly to freeze–thaw during the preceding winter, whereas in the autumn it is compacted as a result of summer thunderstorms. Nevertheless, sediment yields in the spring are not always greater than those in the autumn. Figure 9.21a shows that the change in sediment yield is closely related to the change in runoff, which is inversely related to the percentage of the surface covered with stones (>2 mm). This relation can be attributed at least in part to an increase in stone cover inhibiting surface sealing. However, between spring and autumn there is also an increase in vegetation cover in response to the summer rains. This increase is negatively correlated with the change in runoff and sediment yield (Fig. 9.21b), suggesting that the change in

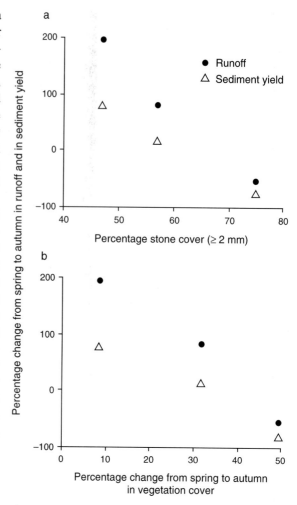

Fig. 9.21 Graphs of percentage change from spring to autumn in runoff and sediment yield against (**a**) percentage stone cover and (**b**) percentage change from spring to autumn in vegetation cover for three soils at Walnut Gulch, Arizona

sediment yield is also a function of differences in summer vegetation growth.

The bulk of desert flora consists of ephemerals and annuals that germinate in response to rainfall events (Thomas 1988). This is very significant geomorphologically, as ephemerals typically appear 2–3 days after a rainfall event and annuals a few days later. Thus major rainfalls that trigger growth at the end of dry periods are erosionally very effective. Conversely, erosion rates at the end of wet periods, during which the plant canopy has thickened, are generally much lower than at other times. The plant canopy impedes soil erosion in a variety of ways, including protecting the ground surface from raindrop impact, which promotes infiltration and reduces soil detachment, and slowing overland flow. However, given that interrill erosion is governed by soil detachment rates (even if it is not detachment-limited) and that detachment on most desert hillslopes is accomplished mainly by raindrop impact, the principal mechanism whereby the plant canopy reduces erosion is probably through its influence on soil detachment.

Semi-arid ecosystems are dominated by either shrubs or grasses. Although semi-arid grasses are often clumped, they are more effective than shrubs as interceptors of rainfall (Thomas 1988). As a consequence, erosion rates at Walnut Gulch are two to three times greater for watersheds with predominantly shrub cover than for those with predominantly grass cover, even though runoff rates are similar (Kincaid et al. 1966). In general, erosion rates in semi-arid environments are inversely related to plant canopy or biomass. Kincaid et al. (1966) provide an example of such a relation for grass-covered watersheds at Walnut Gulch, whereas Johnson and Blackburn (1989) offer one for sagebrush-dominated sites in Idaho.

On some desert hillslopes, biological activity, in the way of digging and burrowing by animals or insects, plays a significant part in determining spatial and temporal variations in erosion rates. In a study conducted at the Sede Boqer experimental site, northern Negev, Israel, Yair and Lavee (1981) recorded intense digging and burrowing by porcupines and isopods (woodlice). Porcupines seeking bulbs for food break the soil crust which otherwise, due to its mechanical properties and cover of soil lichens and algae, inhibits soil erosion. Thus fine soil particles and loose aggregates are made available for transport by overland flow. Similarly, burrowing isopods produce small faeces which disintegrate easily under the impact of raindrops. Measurements of sediment produced by this biological activity on different plots revealed amounts that were of the same order of magnitude as eroded from the plots during a single rainy season (Fig. 9.22). Erosion rates were greater on the Shivta than on the Drorim and Netzer Formations (a) because of the proximity of biotic sediment to the measuring stations at the slope base, and (b) because of the higher magnitude and frequency of overland flow on the massive Shivta Formation. Yair and Lavee (1981) investigated the availability of biotic sediment across the northern Negev and found that it increased from 3 to 70 g m^{-1} as mean annual precipitation increased from 65 to 310 mm. These authors also

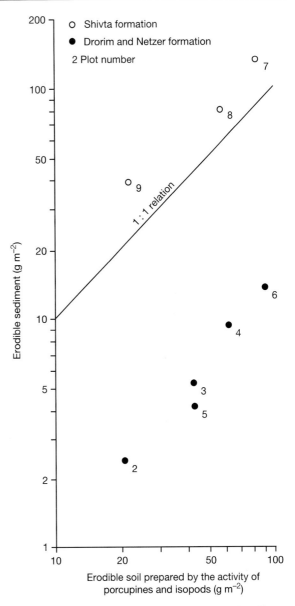

Fig. 9.22 Graph of eroded sediment against erodible soil prepared by porcupines and isopods for the Sede Boqer experimental site, northern Negev, Israel. Data from Yair and Shachak (1987, Table 10.4)

noted that biotic sediment may be produced in desert environments by a variety of animals and insects other than porcupines and isopods, including moles, prairie dogs, and ants (Table 9.1). In a study of the effects of small mammels on sediment yield in the Chihuahuan Desert, Neave and Abrahams (2001) showed that, under simulated rainfall, sediment concentration in runoff from small plots established on degraded grassland and in intershrub spaces correlated with the mean diameter

Table 9.1 Biological activity and sediment production in northern Negev, Israel (data from Yair and Lavee 1981)

Plot name	Annual rainfall (mm)	Sediment production (g m^{-2})			
		Isopods	Porcupines	Moles	Total
Yattir	310	41.3	30.2	100	171.6
Dimona	110	4.5	3.5	0	8.0
Shivta	100	5.3	3.3	0	8.6
Sede Boqer	93	11.9	8.7	0	20.6
Mount Nafha	85	2.1	6.6	0	8.7
Tamar-Zafit	65	0	2.9	0	2.9

of animal disturbances. They argued that small mammels disturb surface crusts on soils and scatter sediment that is then available for entrainment into runoff.

Raindrop Detachment and Erosion

On desert hillslopes where the vegetation cover is generally sparse (Thomas 1988), the impact of rain drops is an important mechanism in the erosion process. Raindrop impact gives rise to rainsplash and rain dislodgement. Each of these processes will be discussed in turn.

Rainsplash occurs when raindrops strike the ground surface or a thin layer of water covering the ground and rebound carrying small particles of soil in the splash droplets. On a horizontal surface the mass of material splashed decreases exponentially with distance from the point of impact (Savat and Poesen 1981, Torri et al. 1987). The presence of a thin film of water appears to promote splash. Although Palmer (1963) reported that maximum splash occurs when the ratio of water depth to drop diameter is approximately 1, other workers have found that the maximum occurs at much smaller ratios (Ellison 1944, Mutchler and Larson 1971). Mass of soil splashed then decreases as the ratio increases (Poesen and Savat 1981, Park et al. 1982, Torri et al. 1987).

The most complete model currently available for predicting the net downslope splash transport rate for vertical rainfall has been proposed by Poesen (1985):

$$Q_{rs} = \frac{E \cos\theta}{R\gamma_b}[0.301\sin\theta + 0.019D_{50}^{-0.220}$$
$$(1 - \exp^{-2.42\sin\theta})] \qquad (9.9)$$

where Q_{rs} is net downslope splash transport rate (m^3 m^{-1} y^{-1}), E is kinetic rainfall energy (J m^{-2} y^{-1}),

R is resistance of the soil to splash (J kg^{-1}), γ_b, is bulk density of the soil (kg m^{-3}), θ is slope gradient (degrees of arc), and D_{50} is median grain size (mm).

This model indicates that Q_{rs} is positively related to rainfall kinetic energy corrected for surface gradient and negatively related to the bulk density of the soil and its resistance to splash. Resistance is, in turn, a function of D_{50} with a minimum at about $100\,\mu$m (Fig. 9.23). Coarser particles are more difficult to splash by virtue of their greater mass, while finer particles are more susceptible to compaction, are more cohesive, and promote the formation of a water layer that impedes splash

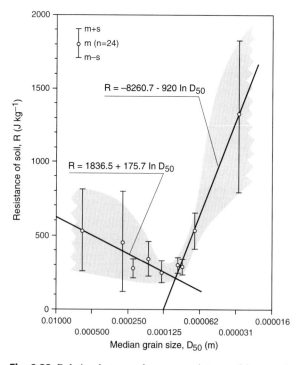

Fig. 9.23 Relation between the mean resistance of loose sediments to rainsplash and their median grain size. The standard deviation (s) of experimental values is shown with each mean (m)

(Poesen and Savat 1981). The first two terms inside the brackets respectively represent the effects of gradient and particle size on the mean splash distance, whereas the expression inside the parentheses reflects the influence of gradient on the difference between the volumes of soil splashed upslope and downslope. The model is based on laboratory experiments. However, Poesen (1986) assembled field data from a number of sources suggesting that it produces order-of-magnitude estimates of splash transport on bare slopes.

There have been few field studies of rainsplash in desert environments. Kirkby and Kirkby (1974) monitored painted stone lines near Tucson, Arizona, over a two-month period during the summer thunderstorm season. They found that mean travel distance due to rainsplash and unconcentrated overland flow increases with gradient and decreases with particle size (Fig. 9.24). By multiplying the travel distances in Fig. 9.25 by the grain diameters, they obtained the mass transport for each grain size per unit area. Then combining these data with data on storm frequency and assuming that rainsplash is completely suppressed under vegetation, Kirkby (1969) produced

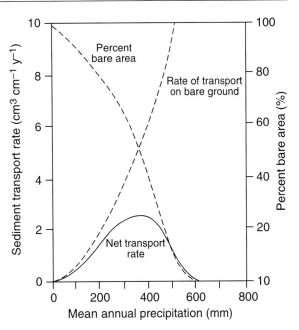

Fig. 9.25 Generalized relations of sediment transport rate, percentage bare area, and net transport rate (calculated as the product of the former two variables) to mean annual precipitation for the southern United States (after Kirkby 1969)

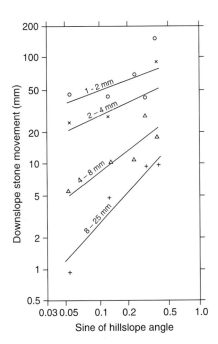

Fig. 9.24 Graph of downslope stone movement by rain-splash and unconcentrated overland flow against hillslope gradient for different stone sizes. Data were collected by Kirkby and Kirkby (1974) from painted stone lines on hillslopes near Tucson, Arizona

Fig. 9.25, which shows that erosion by rainsplash and unconcentrated overland flow reaches a maximum at annual precipitations of 300–400 mm. In other parts of the world, the maximum may occur at somewhat different precipitations, reflecting differences in the distribution of intense storms, seasonality of rainfall, and vegetation characteristics.

Kotarba (1980) monitored splash transport on two plots with gradients of 12° and 15° on the Mongolian steppe. During a single summer he found that length of transportation was closely related to rainfall intensity and grain diameter for particles coarser than 2.5 mm but weakly related to these variables for particles 2.0–2.5 mm in size. He attributed these weak correlations to the finer particles being transported by wind as well as splash. Kotarba pointed out that although the region has an annual precipitation of about 250 mm, 80% of which occurs as rainfall during June, July, and August, overland flow is confined to very limited areas, and rainsplash is the dominant erosional agent. Stones as large as 12 mm were moved by this process. The average displacement of stones during a single summer was 2–4 cm, with some stones travelling as far as 50 cm.

Martinez et al. (1979) measured rainsplash under simulated rainfall at six sites in southern Arizona. They found that mass of splashed material decreases as the proportion and size of stones in the surface pavement increase. Moreover, the presence of an undisturbed stone pavement seems to dampen the effect of increasing rainfall intensity, causing mass of splashed material to increase with rainfall intensity to the 0.48 power, whereas for bare agricultural fields the exponent is usually in the range of 1.5–2.5 (Meyer 1981, Watson and Laflen 1986). Finally, these authors noted that rainsplash is greatest for particles with diameters between 100 and 300 μm (i.e. fine sand), consistent with the laboratory findings of Poesen and Savat (1981).

Parsons et al. (1991b) pointed out that on many semi-arid hillslopes, shrubs are located atop small mounds of fine material, whereas the intervening intershrub areas are swales with a desert pavement surface. Applying simulated rainfall to seven shrubs at Walnut Gulch, Arizona, they showed that these mounds were formed largely by differential rainsplash – that is, to more sediment being splashed into the areas beneath shrubs than is splashed outward. Parsons et al. also demonstrated that both the splashed sediment and the sediment forming the mounds were richer in sand than the matrix soil in the intershrub areas, reflecting the tendency of rainsplash to preferentially transport sand-sized particles.

Rain dislodgement refers to the movement of soil particles by raindrops where the particles are not transported in splash droplets. Ghadiri and Payne (1988) showed that a large proportion of the splash corona (and hence detached sediment) fails to separate into droplets and falls back into the impact area. This proportion increases as the layer of water covering the surface becomes deeper. Moeyersons (1975) coined the term splash creep for the lateral movement of gravel by raindrop impact. He observed that stones as large as 20 mm could be moved in this manner, and using simulated rainfall he demonstrated that splash creep rate increases with gradient and rainfall intensity (Fig. 9.26).

Raindrop-detached sediment includes that which is dislodged by raindrops as well as that carried in splash droplets. In laboratory experiments, Schultz et al. (1985) found that the total weight of detached sediment was 14–20 times greater than that transported by splash. This finding underscores the fact that the most important role of raindrop impact is not in directly transporting sediment but in detaching soil particles from the surface prior to their removal by overland flow.

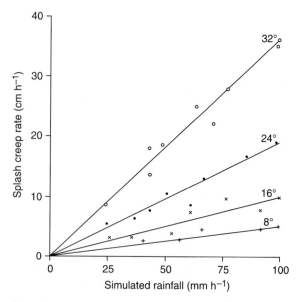

Fig. 9.26 Graph of splash creep rate measured in the laboratory against rainfall intensity for different slope gradients (after Moeyersons 1975)

Detachment of soil particles from the soil mass may be due to raindrop impact or flowing water. Numerous studies have established that on agricultural lands under unconcentrated runoff detachment occurs chiefly by raindrop impact (Borst and Woodburn 1942, Ellison 1945, Woodruff 1947, Young and Wiersma 1973, Lattanzi et al. 1974, Quansah 1985). To ascertain whether the same is true on undisturbed semi-arid hillslopes, five paired runoff plots, one covered with an insect screen to absorb the kinetic energy of the falling raindrops and the other uncovered, were established at Walnut Gulch, Arizona. Assuming that there is no raindrop detachment on the covered plots, the proportion of the sediment load detached by flowing water may be estimated by dividing the sediment load from the covered plot by that from the uncovered plot. As can be seen in Table 9.2, the proportions for all the plots are less than or equal to 0.25. These results support the proposition that raindrop impact is the dominant mode of detachment.

Work by Parsons and Abrahams (1992), however, has indicated that on a semi-arid hillslope at Walnut Gulch, the erosion rate in areas of unconcentrated runoff is less than the detachment capacity. Two types of evidence support this contention. First, particle size analyses reveal that splashed sediment is coarser than sediment being transported by overland flow, signify-

Table 9.2 Sediment yields for covered and uncovered runoff plots, Walnut Gulch, Arizona

Plot number	Status	Gradient degrees	Percentage vegetation	Percentage stones	Sediment yield, G (gm^{-2}min^{-1})	$\dfrac{G \text{ for covered plot}}{G \text{ for uncovered plot}}$
1	Covered	7.7	38.1	43.8	6.0	0.19
1	Uncovered	7.5	44.8	33.3	31.9	
2	Covered	11.7	15.2	41.9	6.4	0.10
2	Uncovered	11.5	19.1	44.8	61.0	
3	Covered	16.0	27.6	48.6	1.2	0.086
3	Uncovered	17.5	18.1	55.2	13.4	
4	Covered	14.0	25.7	54.3	2.3	0.087
4	Uncovered	14.0	35.2	38.1	26.2	
5	Covered	17.0	18.1	48.6	8.9	0.25
5	Uncovered	17.7	34.3	44.8	36.3	

ing that the coarser detached particles are not eroded from the hillslope (Parsons et al. 1991a). Second, although raindrop detachment occurs over the entire hillslope except where overland flow is too deep, the detached sediment is transported downslope only where it is splashed into or dislodged within overland flow competent to transport it. Using a simulation model and detailed measurements of the cross-slope variation in overland flow depth and velocity, Abrahams et al. (1991) showed that there are significant proportions of the hillslope where soil is detached but there is no flow competent to transport it downslope. Inasmuch as overland flow on all desert hillslopes displays across-slope variations in depth and velocity, soil detachment rates probably always exceed actual sediment transport rates. Thus the notion that interrill erosion is detachment-limited appears to be an oversimplification. In reality the erosion rate will always be smaller than the detachment capacity. However, the magnitude of the disparity is difficult to estimate and is probably highly variable over both time and space.

Detachment and Erosion Under Concentrated Runoff

Although soil detachment in areas of unconcentrated runoff on desert hillslopes appears to occur mainly by raindrop impact, as flow paths become longer and threads of flow deeper, flow detachment may come to dominate in these threads both because critical flow shear stresses are exceeded and because the deeper water protects the soil surface from raindrop impact. This line of reasoning is supported by Roels' (1984a, b) findings on a rangeland hillslope in the Ardeche drainage basin, France. Roels observed that natural ir-

regularities in the ground surface cause runoff to concentrate into interrill flow paths, which range in length up to 20 m, and he termed the longest of these flow paths prerills. He derived separate regression models for soil loss from the prerill and non-prerill interrill areas. In prerill sites 87% of the variation in soil loss was accounted for by a runoff erosivity factor $REF = Q \times Q_p^{0.33}/A$, where Q is the runoff volume, Q_p the peak discharge, and A the drainage area. In contrast, in non-prerill interrill areas, 85% of the variation in soil loss was explained by a rainfall erosivity factor $EA \times IM$, where EA is the excess rainfall amount and IM is the maximum 5-minute rainfall intensity. These results imply that flow detachment dominates in the prerill portions of these areas, whereas raindrop detachment dominates elsewhere. Unless the prerills migrate laterally or are periodically infilled by other processes, flow detachment will inevitably cause them to evolve into rills.

Understanding how and when flow detachment becomes effective has, however, proved less than straightforward. In simple terms, it can be defined as when shear stress exerted by the flowing water exceeds the shear strength of the soil. Several authors have attempted to identify critical values for threshold mean shear stresses or shear velocities (e.g., Rauws and Govers, 1988; Slattery and Bryan, 1992), but Nearing (1994) has pointed out that mean shear stress exerted by shallow flow is of the order of a few pascals, whereas mean shear strength of soils is typically measured in the order of kilopascals. Nearing proposed that the solution to this apparent conundrum lay in the overlapping distributions of the two values (Nearing 1991). Based on this approach, Parsons and Wainwright (2006) undertook an analysis of the probability of incision on shrubland and grassland hillslopes at Walnut Gulch, southern Arizona (Fig. 9.27). They demonstrated that probabilities for incision on both shrubland and grassland were similar for similar discharges and similar degrees of soil moisture, and that the explanation for incision on shrubland hillslopes but not on grassland that is apparent at Walnut Gulch lies in the more frequent higher discharges on the shrubland compared to the grassland. These more frequent higher discharges on the shrubland compared to the grassland are a result of soil and microtopographic differences between the two vegetation communities (Parsons et al., 1996).

Gravitational Proceses

The movement of weathered detritus under the influence of gravity encompasses a very wide range of phenomena differing in depth and mass of material being mobilized, rates of motion, transport mechanisms, and relative volumes of debris, water, ice and air. Only a limited number of these phenomena are common on rock-mantled hilllslopes in arid environments. The most important of these are debris flows, but a limited amount of movement of dry debris also occurs.

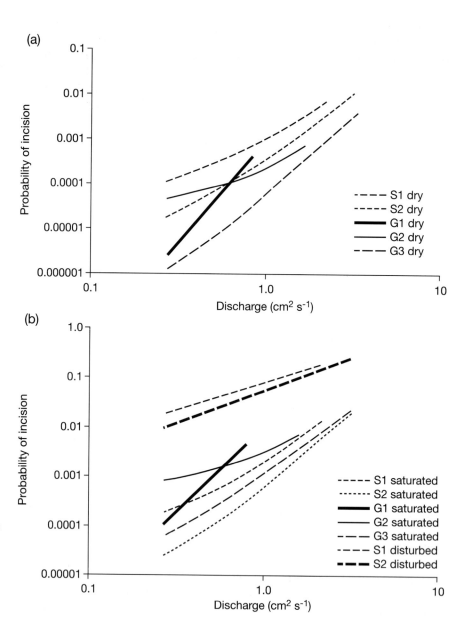

Fig. 9.27 Probabilities of incision at cross section on grassland and shrubland plots at Walnut Gulch, Arizona under dry, saturated and disturbed soil conditions (after Parsons and Wainwright 2006). See also Figs. 9.16 and 9.17

Movement of Dry Debris

The movement of dry debris as particle-by-particle sliding under gravity is known as dry ravel and has been claimed to be a dominant mechanism of sediment transport on steep hillslopes in arid and semi-arid environments (Gabet 2003). Inasmuch as most coarse particles on debris slopes are weathered in-situ from the underlying bedrock the question of the triggering mechanism to make previously stable particles become unstable naturally arises. The most commonly cited cause for the initiation of dry ravel on debris slopes is fire (e.g. Florsheim et al. 1991, Cannon et al. 1998, Roering and Gerber 2005). The destruction of vegetation removes the support for fine material that often accumulates behind plants and it is argued that this initiates instability in the coarse fragments.

Debris Flows

The commonest phenomena due to gravitational processes on rock-mantled hillslopes, certainly in terms of amount of sediment moved, are debris flows. A full discussion of the mobilization of debris flows is beyond the scope of this chapter, which will focus on conditions on desert rock-manted hillslopes that influence this mobilization. For further discussion of debris-flow processes, the reader is referred to the review by Iverson et al. (1997). Debris flows generally exhibit a consistent range of behaviour (Blackwelder 1928, Johnson 1970, Fisher 1971, Costa 1984, Johnson and Rodine 1984). The flows occur as a series of blunt-nosed pulses with the first pulse commonly being the largest. The maximum depth of each pulse occurs near the nose, with a long tailing flow that is commonly more fluid than the nose, often changing to hyper-concentrated or water flood flows during the waning stages. What characterizes debris flows from either of these types of flow is the synergistic transfer of momentum by both solids and fluids (Iverson et al. 1997). Debris flows are noted for the wide range of grain sizes transported and the tendency for large boulders to be concentrated near the flow surface and at the leading edge of the flow. Deposits from debris flows often show inverse grading. Characteristics of debris flow deposits are considered in Chapter 14. The general rule for debris flows is that failure will occur in the layer of weathered material on rock-mantled hillslopes if stresses obey Coulomb's (1773) rule

$$\tau > \sigma' \tan \varphi + c \qquad (9.10)$$

where τ is the mean shear stress acting on the failure surface, σ' is the mean effective normal stress acting on the failure surface, φ is the angle of internal friction of the weathered layer, and c is the cohesion of the weathered layer. Cohesion mainly depends upon electrostatic forces between clay particles and secondary mineralisation in the weathered layer. The latter can be significant, especially where petrocalcic horizons have developed, but, in the absence of such horizons, many rock mantles are effectively almost cohesionless because of their low clay content. The angle φ is a function of the weathered layer, and hence the weathering characteristics of the underlying lithology. Consequently, the frequency of debris flows on these hillslopes is strongly dependent on the underlying lithology. Effective normal stress accounts for pore-fluid pressure p such that

$$\sigma' = \sigma + p \qquad (9.11)$$

where σ is the total normal stress.

Although it cannot be assumed that ϕ and c are fixed quantities for a given weathered mantle (see Iverson et al. 1997), the key requirement for mobilization of debris flows is sufficient water to result in high pore pressures. On desert hillslopes this requirement is almost certain to be met as a result of rainfall; groundwater inflow is unlikely. However, as Iverson et al. (1997, p.101) point out, mobilization of debris flows by rainfall on steep hillslopes 'presents a mechanical difficulty'. How do the slopes remain stable for long enough to become nearly saturated? The solution proposed by Iverson et al. is to suggest that prolonged rainfall at an intensity greater than the saturated hydraulic conductivity of the mantle layer will create a saturated zone at the ground surface that will propagate downward. This layer may remain tension-saturated after rainfall ceases so that a subsequent burst of high-intensity rainfall may cause positive pore pressure to develop almost instantaneously. Although heavy rainfall is widely associated with reports of debris flows on desert hillslopes (e.g. Coe et al. 1997, Cannon et al. 2001) data to support this hypothesized mechanism are lacking. An alternative supply of large amounts of water may be from upslope where percentages of bedrock may be higher and infiltration rates lower. Such differences in infiltration between upslope and downslope portions of desert hillslopes are not uncommon (Yair 1987). An alternative triggering mechanism is proposed by Blijenberg et al. (1996) who postulate that microscale mass movements that occur in response to rainfall events could play a significant role in triggering mass movements.

Three other factors may contribute to the occurrence of debris flows on rock-mantled hillslopes. First, the sparse vegetation of desert environments limits the cohesive strength that is imparted to soil layers by plant roots. However, although the sparseness of vegetation makes rock-manted hillslopes inherently unstable, it also removes one of the triggers to debris-flow mobilization, namely the sudden loss of strength as cohesion is suddenly and dramatically lost when roots are broken. Secondly, marked differences in infiltration capacity often exist on debris slopes, particularly where petrocalcic horizons exist. The existence of such hoizons may provide temporary perched water tables during storm events. Finally, debris flows in deserts commonly occur shortly after fire has destroyed the natural vegetation cover (Sidle et al. 1985, Wohl and Pearthree 1991, Cannon et al. 1998). Fire may contribute to failure by reducing evapotranspiration while creating a hydrophobic fire-sealed soil layer that promotes surface soil saturation and runoff (Wells 1981, 1987, Laird and Harvey 1986, Campbell et al. 1987, Wells et al. 1987). However, Florsheim et al. (1991) noted that wildfire often is not followed by large debris flows even though sediment yield is increased, and they suggest that rainfall intensity and duration is much more important in triggering debris flow than is wildfire.

The mathematical analysis of debris-flow mobilization is typically undertaken using the infinite-slope model (see Iverson et al. 1997, Fig. 9). The insights that the model provide are important because they indicate the relationships that exist among the forces controlling the mobilization of debris flows. However, this model, as Iverson et al. point out, is not amenable to testing because it makes predictions that cannot be applied to any naturally occurring hillslope. Consequently, many gaps still remain in our understanding of the conditions under which debris flows are triggered on rock-mantled slopes.

Climate Change

A discussion of rock-mantled slopes would be incomplete if it did not consider the effects of climatic change. Major climatic fluctuations have probably occurred in every desert during the Cenozoic (Chapter 28) and have strongly influenced the form of many debris slopes (Chapter 22). The imprint of these former climates appears to be most pronounced where rock resistance is greatest. This is well illustrated by Oberlander's (1972) classic study of boulder-covered slopes on resistant quartz monzonite in the Mojave Desert, California. These slopes consist of a 'jumble of subangular to spheroidal boulders of a variety of shapes and sizes clearly derived from plane-faced blocks bounded by intersecting joints' (Oberlander 1972, p. 4) (Fig. 9.6). Oberlander argued that the boulders formed as corestones within a deep weathering profile under a wetter climate, and that these corestones became stranded on bedrock slopes as the supporting matrices of fines were removed under the more arid climate of the late Tertiary and the Quaternary. Not surprisingly, Oberlander could find no correlation between hillslope gradient and boulder size. Other investigators too have reported an absence of any relation between gradient and debris size on slopes underlain by resistant rocks (e.g. Melton 1965, Cooke and Reeves 1972, Kesel 1977), suggesting that these slopes also owe much of their form and sedimentology to climatic change.

Legacies from past climates are probably more prevalent on rock-manted hillslopes than is generally realized. Certainly, Oberlander's description of boulder-clad slopes in the Mojave Desert applies to similar hillslopes in most granitic terranes. Erosion on such slopes is often characterized as weathering-limited (e.g. Young 1972, p. 206, Mabbutt 1977, p. 41). However, if this were wholly the case, such slopes would be more or less devoid of fine material because, by definition, such material should be removed as rapidly as it is produced. Instead, what we often find are boulders or bedrock outcrops protruding from a matrix of fines that becomes progressively more extensive downslope. Parsons and Abrahams (1987) investigated this phenomenon in the Mojave Desert and concluded that the presence of the fines indicates adjustment by the hillslope to extant hydraulic processes, and that the degree of

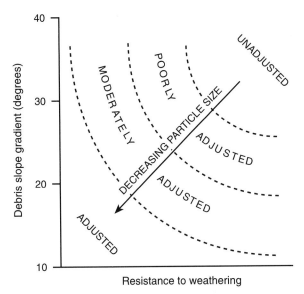

Fig. 9.28 Graph of debris slope gradient against resistance to weathering, showing how the degree of debris slope adjustment varies with these variables (after Parsons and Abrahams 1987)

adjustment is inversely related to slope gradient and rock resistance (Fig. 9.28). It is interesting to note that inasmuch as particle size decreases as degree of adjustment increases, the hillslopes studied by Parsons and Abrahams display strong correlations between gradient and particle size, even though they are far from being adjusted to contemporary hydraulic processes, especially in their steeper parts.

Conclusion

As with most of the landforms and processes in deserts, our knowledge of rock-mantled slopes is deficient and there are many opportunities for further field observations and experimentation, laboratory investigations, and theoretical modelling. Our understanding of weathering processes of regolith generation is poor, and the relative roles of present and past climates is uncertain. Mechanisms of debris mobilization, deposition, and further weathering are reasonably well understood in a qualitative sense, but the long-term interaction of processes and materials to create specific types of desert slopes is poorly characterized.

The deficiency in our understanding of rock-mantled hillslopes is important because the majority of desert hillslopes fall into this category. Understanding the form and processes of these hillslopes is, therefore, important for understanding a significant component of desert landscapes. Furthermore, they give rise to a variety of hydrologic and geomorphic phenomena such as flash flooding, extreme soil erosion, and hazardous debris flows. More generally, they exert a major control over the flux of water and sediment that passes through desert river systems, across active piedmonts, and into closed lake basins. Thus an understanding of many, if not most, desert geomorphic systems must begin with a comprehension of the processes operating on these desert hillslopes.

Acknowledgments The experiments reported in Table 9.2 were funded by NSF grant SES-8812587.

References

Abrahams, A.D. and G. Li 1998. Effect of saltating sediment on flow resistance and bed roughness in overland flow. *Earth Surface Processes and Landforms* **23**, 953–960.

Abrahams, A.D., Li, G. and A.J. Parsons 1996. Rill hydraulics on a semiarid hill slope, southern Arizona. *Earth Surface Processes and Landforms* **21**, 35–47.

Abrahams, A.D. and A.J. Parsons 1991a. Relation between sediment yield and gradient on debris-covered hillslopes, Walnut Gulch, Arizona. *Bulletin of the Geological Society of America* **103**, 1109–13.

Abrahams, A.D. and A.J. Parsons 1991b. Relation between infiltration and stone cover on a semiarid hillslope, southern Arizona. *Journal of Hydrology* **122**, 49–59.

Abrahams, A.D. and A.J. Parsons 1991c. Resistance to overland flow on desert pavement and its implications for sediment transport modeling. *Water Resources Research* **27**, 1827–36.

Abrahams, A.D., A.J. Parsons, R.U. Cooke and R.W. Reeves 1984. Stone movement on hillslopes in the Mojave Desert, California: a 16-year record. *Earth Surface Processes and Landforms* **9**, 365–70.

Abrahams, A.D., A.J. Parsons and P.J. Hirsch 1985. Hillslope gradient-particle size relations: evidence for the formation of debris slopes by hydraulic processes in the Mojave Desert. *Journal of Geology* **93**, 347–57.

Abrahams, A.D., A.J. Parsons and S.-H. Luk 1986. Resistance to overland flow on desert hillslopes. *Journal of Hydrology* **88**, 343–63.

Abrahams, A.D., A.J. Parsons and S.-H. Luk 1988. Hydrologic and sediment responses to simulated rainfall on desert hillslopes in southern Arizona. *Catena* **15**, 103–17.

Abrahams, A.D., A.J. Parsons and S.-H. Luk 1991. The effect of spatial variability in overland flow on the downslope pattern of soil loss on a semiarid hillslope, southern Arizona. *Catena* **18**, 255–70.

Abrahams, A.D., N. Soltyka, A.J. Parsons and P.J. Hirsch 1990. Fabric analysis of a desert debris slope: Bell Mountain, California. *Journal of Geology* **98**, 264–72.

Akagi, Y. 1980. Relations between rock-type and the slope form in the Sonoran Desert, Arizona. *Zeitschrift für Geomorphologie* **24**, 129–40.

Belnap, J. 2006. The potential roles of biological soil crusts in dryland hydrologic cycles. *Hydrological Processes* **20**, 3159–3178.

Birkeland, P.W. 1974. *Pedology, Weathering, and Geomorphological Research*. Oxford: Oxford University Press.

Blackwelder, E. 1928. Mudflow as a geologic agent in semi-arid regions. *Bulletin of the Geological Society of America* **39**, 465–84.

Blijenberg, H.M., P.J. DeGraaf, M.R. Hendriks, J.F. DeRuiter and A.A.A.Van Tetering 1996. Investigation of infiltration characteristics and debris flow initiation conditions in debris flow source areas using a rainfall simulator. *Hydrological Processes* **10**, 1527–1543.

Borst, H.L. and R. Woodburn 1942. The effect of mulching and methods of cultivation on run-off and erosion from Muskingum silt loam. *Agricultural Engineering* **23**, 19–22.

Box, J.E. 1981. The effects of surface slaty fragments on soil erosion by water. *Soil Science Society of America Journal* **45**, 111–16.

Bryan, K. 1922. Erosion and sedimentation in the Papago Country, Arizona. *U.S. Geological Survey Bulletin* 730-B, 19–90.

Bryan, K. 1940. The retreat of slopes. *Annals of the Association of American Geographers* **30**, 254–68.

Bunte, K. and J. Poesen 1994. Effects of rock-fragment size and cover on overland flow hydraulics, local turbulence and sediment yield on an erodible soil surface. *Earth Surface Processes and Landforms* **19**, 115–135.

Cammeraat, L.H., S.J. Willott, S.G. Compton and L.D. Incoll 2002. The effects of ants' nests on the physical, chemical and hydrological propoerties of a rangeland soil in semi-arid Spain. *Geoderma* **105**, 1–20.

Campbell, A.G., P.M. Wohlgemuth and W.W. Wells 1987, Post-fire response of a boulder bed channel. *International Association of Scientific Hydrology Publication* **165**, 277–8.

Cannon, S.H., P.S. Powers and W.Z. Savage 1998. Fire-related hyperconcentrated and debris flows on Storm King Mountain, Glenwood Springs, Colorado, USA. *Environmental Geology* **35**, 210–218.

Cannon, S.H., E.R. Bigio and E. Mine 2001. A process for fire-related bedris flow initiation, Cerro Grande fire, New Mexico. *Hydrological Processes* **15**, 3011–3023.

Carson, M.A. and M.J. Kirkby 1972. *Hillslope form and Process*. Cambridge: Cambridge University Press.

Castillo, V.M., A. Gomez-Plaza and M. Martinez-Mena 2003. The role of antecedent soil water content in the runoff response of semiarid catchments: a simulation approach. *Journal of Hydrology* **284**, 114–130.

Cerda, A. and S.H. Doerr 2005. Influence of vegetation recovery on soil hydrology and erodibility following fire: an 11-year investigation. *International Journal of Wildland Fire* **14**, 423–437.

Coe, J.A., P.A. Glancy and J.W. Whitney 1997. Volumetric analysis and hydrologic characterization of a modern debris flow near Yucca Mountain, Nevada. *Geomorphology* **20**, 11–28.

Cooke, R.U. and R.W. Reeves 1972. Relations between debris size and the slope of mountain fronts and pediments in the Mojave Desert, California. *Zeitschrift für Geomorphologie* **16**, 76–82.

Cooke, R.U. and A. Warren 1973. *Geomorphology in Deserts*. Berkeley: University of California Press.

Costa, J.E. 1984. Physical geomorphology of debris flows. In *Developments and Applications of Geomorphology*, J.E. Costa and P.J. Fleisher (eds), 268–317. Berlin: Springer.

Debruyn, L.A.L. and A.J. Conacher 1994. The effect of ant biopores on water infiltration in soils in undisturbed bushland and on farmland in a semiarid environment. *Pedobiologia* **38**, 193–207.

De Lima, J.P. and V.P. Singh 2002. The influence of moving rainstorms on overland flow. *Advances in Water Resources* **25**, 817–828.

De Ploey, J. and J. Moeyersons 1975. Runoff creep of coarse debris: experimental data and some field observations. *Catena* **2**, 275–88.

Dunkerley, D.L. 2003. Organic litter: dominance over stones as a source of interrill flow roughness on low-gradient desert slopes at Fowlers Gap, arid Western NSW, Australia. *Earth Surface Processes and Landforms* **28**, 15–29.

Dunkerley, D. 2004. Flow threads in surface run-off: implications for the assessment of flow properties and friction coefficients in soil erosion and hydraulic investigations. *Earth Surface Processes and Landforms* **29**, 1011–1026.

Dunkerley, D., P. Domelow and D. Tooth 2001. Frictional retardation of laminar flow by plant litter and surface stones on dryland surfaces: a laboratory study. *Water Resources Research* **37**, 1417–1423.

Dunne, T. and W.E. Dietrich 1980. Experimental study of Horton overland flow on tropical hillslopes. 2. Hydraulic characteristics and hillslope hydrographs. *Zeitschrift für Geomorphologie Supplement Band* **35**, 60–80.

Ellison, W.D. 1944. Studies of raindrop erosion. *Agricultural Engineering* **25**, 131–6, 181–2.

Ellison, W.D. 1945. Some effects of raindrops and surface flow on soil erosion and infiltration. *Transactions of the American Geophysical Union* **26**, 415–29.

Epstein, E., W.J. Grant and R.A. Struchtemeyer 1966. Effects of stone on runoff, erosion, and soil moisture. *Soil Science Society of America Proceedings* **30**, 638–40.

Fair, T.J.D. 1947. Slope form and development in the interior of Natal, South Africa. *Transactions of the Geological Society of South Africa* **50**, 105–19.

Fair, T.J.D. 1948. Hill-slopes and pediments of the semi-arid Karoo. *South African Geography Journal* **30**, 71–9.

Ferreira, A.J.D., C.O.A. Coelho, A.K. Boulet, G. Leighton-Boyce, J.J. Keizer and C.J. Ritsema 2005. Influence of burning intensity on water repellency and hydrological processes at forest and shrub sites in Portugal. *Australian Journal of Soil Research* **43**, 327–336

Fisher, R.V. 1971. Features of coarse-grained, high-concentration fluids and their deposits. *Journal of Sedimentary Petrology* **41**, 916–27.

Fitzjohn, C., J.L. Ternan and A.G. Williams 1998. Soil moisture variability in a semi-arid gully catchment: implications for runoff and erosion control. *Catena* **32**, 55–70.

Florsheim, J.L., E.A. Keller and D.W. Best 1991. Fluvial sediment transport in response to moderate storm flows following chaparral wildfire, Ventura County, Southern California. *Bulletin of the Geological Society of America* **103**, 504–11.

Furley, P.A. 1968. Soil formation and slope development 2. The relationship between soil formation and gradient angle in the Oxford area. *Zeitschrift für Geomorphologie* **12**, 25–42.

Gabet, E.J. 2003. Sediment transport by dry ravel. *Journal of Geophysical Research – Solid Earth* **B108(B1)**, Art. No. 2049.

Ghadiri, H. and D. Payne 1988. The formation and characteristics of splash following raindrop impact on soil. *Journal of Soil Science* **39**, 563–75.

Gilley, J.E., E.R. Kottwitz and G.A. Weiman 1992. Darcy-Weisbach roughness coefficients for gravel and cobble surfaces. *Journal of Irrigation and Drainage Engineering* **118**, 104–12.

Gilluly, J. 1937. Physiography of the Ajo region, Arizona. *Bulletin of the Geological Society of America* **48**, 323–48.

Govers, G. 1992. Relationship between discharge, velocity, and flow area for rills eroding in loose, non-layered materials. *Earth Surface Processes and Landforms* **17**, 515–528.

Govers, G. and G. Rauws 1986. Transporting capacity of overland flow on plane and on irregular beds. *Earth Surface Processes and Landforms* **11**, 515–24.

Hawkins, R.H. and T.W. Cundy 1987. Steady-state analysis of infiltration and overland flow for spatially-varied hillslopes. *Water Resources Bulletin* **23**, 215–16.

Horton, R.E. 1933. The role of infiltration in the hydrologic cycle. *Transactions of the American Geophysical Union* **14**, 446–60.

Iverson, R.M. 1980. Processes of accelerated pluvial erosion on desert hillslopes modified by vehicular traffic. *Earth Surface Processes* **5**, 369–88.

Iverson, R.M., M.E. Reid and R.G. LaHusen 1997. Debris-flow mobilization from landslides. *Annual Review of Earth and Planetary Sciences*, **25**, 85–138.

Johnson, A.M. 1970. *Physical Processes in Geology*. San Francisco: Freeman & Cooper.

Johnson, A.M. and J.R. Rodine 1984. Debris flow. In *Slope Instability*, D. Brunsden and D.B. Prior (eds), 257–361. New York: Wiley.

Johnson, C.W. and W.H. Blackburn 1989. Factors contributing to sagebrush rangeland soil loss. *Transactions of the American Society of Agricultural Engineers* **32**, 155–60.

Jung, L. 1960. The influence of the stone cover on runoff and erosion on slate soils. *International Association of Scientific Hydrology Publication* **53**, 143–53.

Kesel, R.H. 1977. Some aspects of the geomorphology of inselbergs in central Arizona, USA. *Zeitschrift für Geomorphologie* **21**, 120–46.

Kincaid, D.R., J.L. Gardner and H.A. Schreiber 1964. Soil and vegetation parameters affecting infiltration under semiarid conditions. *International Association of Scientific Hydrology Publication* 65, 440–53.

Kincaid, D.R., H.B. Osborn and J.L. Gardner 1966. Use of unit-source watersheds for hydrologic investigations in the semiarid Southwest. *Water Resources Research* **2**, 381–92.

Kirkby, A. and M.J. Kirkby 1974. Surface wash at the semi-arid break in slope. *Zeitschrift für Geomorphologie Supplement Band* **21**, 151–76.

Kirkby, M.J. 1969. Erosion by water on hillslopes. In *Water, Earth and Man*, R.J. Chorley (ed.) 229–38. London: Methuen.

Kotarba, A. 1980. Splash transport in the steppe zone of Mongolia. *Zeitschrift für Geomorphologie Supplement Band* **35**, 92–102.

Laird, J.R. and M.D. Harvey 1986. Complex-response of a chaparral drainage basin to fire. *International Association of Scientific Hydrology Publication* **159**, 165–83.

Lattanzi, A.R., L.D. Meyer and M.F. Baumgardner 1974. Influences of mulch rate and slope steepness on interrill erosion. *Soil Science Society of America Proceedings* **38**, 946–50.

Lavee, H. and J. Poesen 1991. Overland flow generation and continuity on stone-covered soil surfaces. *Hydrological Processes* **5**, 345–60.

Lawson, A.C. 1915. The epigene profiles of the desert. *University of California Publications, Bulletin of the Department of Geology* **9**(3), 23–48.

Lyford, F.P. and H.K. Qashu 1969. Infiltration rates as affected by desert vegetation. *Water Resources Research* **5**, 1373–6.

Mabbutt, J.A. 1966. Mantle-controlled planation of pediments. *American Journal of Science* **264**, 78–91.

Mabbutt, J.A. 1977. *Desert Landforms*. Canberra: Australian National University Press.

Major, J.J. and T.C. Pierson 1992. Debris flow rheology: experimental analysis of fine-grained slurries. *Water Resources Research* **28**, 841–57.

Mando A. and R. Miedema 1997. Termite-induced changes in soil structure after mulching degraded (crusted) soil in the Sahel. *Applied Soil Ecology* **6**, 241–249.

Martinez, M., L.J. Lane and M.M. Fogel 1979. Experimental investigation of soil detachment by raindrop impacts. *Proceedings of the Rainfall Simulation Workshop, Tucson, Arizona, March 7–9, 1979*. USDA Agricultural Reviews and Manuals ARM-W-10, 153–5. Oakland, CA: U.S. Department of Agriculture.

Melton, M.A. 1965. Debris-covered hillslopes of the southern Arizona desert – consideration of their stability and sediment contribution. *Journal of Geology* **73**, 715–29.

Meyer, L.D. 1981. How rainfall intensity affects interrill erosion. *Transactions of the American Society of Agricultural Engineers* **24**, 1472–5.

Mills, A.J., M.V. Fey, A. Grongtoft, A. Petersen and T.V. Medinski 2006. Unravelling the effects of soil properties on water infiltration: segmented quartile regression on a large data set from arid south-west Africa. *Australian Journal of Soil Research* **44**, 783–797.

Moeyersons, J. 1975. An experimental study of pluvial processes on granite gruss. *Catena* **2**, 289–308.

Mutchler, C.K. and C.L. Larson 1971. Splash amounts from waterdrop impact on a smooth surface. *Water Resources Research* **7**, 195–200.

Oberlander, T.M. 1972. Morphogenesis of granite boulder slopes in the Mojave Desert, California. *Journal of Geology* **80**, 1–20.

Nearing, M.A. 1991. A probabilistic model of soil detachment by shallow turbulent flow. *Transactions of the American Society of Agricultural Engineers* **34**: 81–85.

Nearing, M.A. 1994. Detachment of soil by flowing water under turbulent and laminar conditions. *Soil Science Society of America Journal* **58**: 1612–1614.

Neave M. and A.D. Abrahams 2001. Impact of small mammal disturbances on sediment yield from grassland and shrubland ecosystems in the Chihuahuan Desert. *Catena* **44**, 285–303.

Palmer, R.S. 1963. The influence of a thin water layer on waterdrop impact forces. *International Association of Scientific Hydrology Publication* **64**, 141–8.

Park, S.W., J.K. Mitchell and G.D. Bubenzer 1982. Splash erosion modeling: physical analyses. *Transactions of the American Society of Agricultural Engineers* **25**, 357–61.

Parsons, A.J. and A.D. Abrahams 1984. Mountain mass denudation and piedmont formation in the Mojave and Sonoran Deserts. *American Journal of Science* **284**, 255–71.

Parsons, A.J. and A.D. Abrahams 1987. Gradient–particle size relations on quartz monzonite debris slopes in the Mojave Desert. *Journal of Geology* **95**, 423–32.

Parsons, A.J. and A.D. Abrahams 1992. Controls on sediment removal by interrill overland flow on semi-arid hillslopes. *Israel Journal of Earth Sciences*, **41**, 177–88.

Parsons A.J. and J. Wainwright 2006. Depth distribution of interrill overland flow and the formation of rills. *Hydrological Processes* **20**, 1511–1523.

Parsons, A.J., A.D. Abrahams and S.-H. Luk 1990. Hydraulics of interrill overland flow on a semi-arid hillslope, southern Arizona. *Journal of Hydrology* **117**, 255–73.

Parsons, A.J., A.D. Abrahams and S.-H. Luk 1991a. Size characteristics of sediment in interrill overland flow on a semi-arid hillslope, southern Arizona. *Earth Surface Processes and Landforms* **16**, 143–52.

Parsons, A.J., A.D. Abrahams and J.R. Simanton 1991b. Microtopography and soil-surface materials on semi-arid piedmont hillslopes, southern Arizona, *Journal of Arid Environments* **22**, 107–15.

Parsons A.J., J. Wainwright and A.D. Abrahams 1996. Runoff and erosion on semi-arid hillslopes. In: M.G. Anderson and S.M. Brooks (eds.) *Advances in Hillslope Processes*, Wiley, Chichester, pp.1061–1078.

Perrolf K. and K. Sandstrom 1995. Correlating landscape characteristics and infiltration – a study of surface sealing and subsoil conditions in semi-arid Botswana and Tanzania. *Geografiska Annaler Series A – Physical Geography* **77A**, 119–133.

Poesen, J. 1985. An improved splash transport model. *Zeitschrift für Geomorphologie* **29**, 193–211.

Poesen, J. 1986. Field measurements of splash erosion to validate a splash transport model. *Zeitschrift für Geomorphologic Supplement Band* **58**, 81–91.

Poesen, J. 1990. Erosion process research in relation to soil erodibility and some implications for improving soil quality. In *Soil degradation and Rehabilitation in Mediterranean Environmental Conditions*, J. Albaladejo, M.A. Stocking and E. Diaz (eds), 159–70. Madrid: Consejo Superior de Investigaciones Cientificas.

Poesen, J. and H. Lavee 1991. Effects of size and incorporation of synthetic mulch on runoff and sediment yield from interrills in a laboratory study with simulated rainfall. *Soil and Tillage Research* **21**, 209–23.

Poesen, J. and J. Savat 1981. Detachment and transportation of loose sediments by raindrop splash. Part II. Detachability and transportability measurements. *Catena* **8**, 19–41.

Poesen, J., F. Ingelmo-Sanchez and H. Mucher 1990. The hydrological response of soil surfaces to rainfall as affected by cover and position of rock fragments in the top layer. *Earth Surface Processes and Landforms* **15**, 653–71.

Quansah, C. 1985. Rate of soil detachment by overland flow, with and without rain, and its relationship with discharge, slope steepness, and soil type. In *Soil Erosion and Conservation*, S.A. El-Swaify, W.C. Moldenhauer and A. Lo (eds), 406–23. Ankeny, IA: Soil Conservation Society of America.

Rahn, P.H. 1966. Inselbergs and nickpoints in southwestern Arizona. *Zeitschrift für Geomorphologie* **10**, 215–24.

Rauws G. and G. Govers G 1988. Hydraulic and soil mechanical aspects of rill generation on agricultural soils. *Journal of Soil Science* **39**, 111–124.

Reaney, S.M., L.J. Bracken and M.J. Kirkby, 2007. Use of the connectivity runoff model (CRUM) to investigate the influence of storm characteristics on runoff generation and connectivity in semi-arid areas. *Hydrological Processes* **21**, 894–906.

Rieke-Zapp, D., J. Poesen and M.A. Nearing 2007. Effects of rock fragments incorporated in the soil matrix on concentrated flow hydraulics and erosion. *Earth Surface Processes and Landforms* **32**, 1063–1076.

Roels, J.M. 1984a. Surface runoff and sediment yield in the Ardeche rangelands. *Earth Surface Processes and Landforms* **9**, 371–81.

Roels, J.M. 1984b. Modelling soil losses from the Ardeche rangelands. *Catena* **11**, 377–89.

Roering, J.J. and M. Gerber 2005. Fire and the evolution of steep, soil-mantled landscapes. *Geology*, **33**, 349–352.

Romao, R.L. and A.Escudero 2005. Gypsum physical crusts and the existence of gypsophytes in semi-arid central Spain. *Plant Ecology* **181**, 127–137.

Sadeghian, M.R. and J.K. Mitchell 1990. Hydraulics of microbraided channels: resistance to flow on tilled soils. *Transactions of the American Society of Agricultural Engineers* **33**, 458–68.

Sarr, M., C. Agbogbam, A. Russell-Smith and D. Masse 2001. Effects of soil and faunal activity on water infiltration rates in a semi-arid fallow of Senegal. *Applied Soil Ecology* **16**, 283–290.

Saunders, I. and A. Young 1983. Rate of surface processes on slopes, slope retreat and denudation. *Earth Surface Processes and Landforms* **8**, 473–501.

Savat, J. 1977. The hydraulics of sheet flow on a smooth surface and the effect of simulated rainfall. *Earth Surface Processes* **2**, 125–40.

Savat, J. and J. Poesen 1981. Detachment and transportation of loose sediments by raindrop splash. Part I. The calculation of absolute data on detachability and transportability. *Catena* **8**, 1–17.

Schultz, J.P., A.R. Jarrett and J.R. Hoover 1985. Detachment and splash of a cohesive soil by rainfall. *Transactions of the American Society of Agricultural Engineers* **28**, 1878–84.

Seginer, I., J. Morin and A. Shachori 1962. Runoff and erosion studies in a mountaneous Terra Rosa region in Israel. *Bulletin of the International Association of Scientific Hydrology* **7**(4), 79–92.

Selby, M.J. 1980. A rock mass strength classification for geomorphic purposes: with tests from Antarctica and New Zealand. *Zeitschrift für Geomorphologie* **24**, 31–51.

Selby, M.J. 1982a. Controls on the stability and inclinations of hillslopes formed on hard rock. *Earth Surface Processes and Landforms* **7**, 449–67.

Selby, M.J. 1982b. *Hillslope Materials and Processes*. Oxford: Oxford University Press.

Shen, H.W. and R.-M. Li 1973. Rainfall effect on sheet flow over smooth surface. *Journal of the Hydraulics Division, Proceedings of the American Society of Civil Engineers* **99**, 771–92.

Sidle, R.C., A.J. Pearce and C. O'Loughlin 1985. *Hillslope Stability and Land Use*. American Geophysical Union Water Resources Monograph 11.

Simanton, J.R. and K.G. Renard 1982. Seasonal change in infiltration and erosion from USLE plots in southeastern Arizona. *Hydrology and Water Resources in Arizona and the Southwest* **12**, 37–46.

Simanton, J.R., K.G. Renard and N.G. Sutter 1973. *Procedure for Identifying Parameters Affecting Storm Runoff Volumes in a Semiarid Environment*. U.S. Agricultural Research Service Report ARS-W-1.

Simanton, J.R., E. Rawitz and E. Shirley 1984. Effects of rock fragments on erosion of semiarid rangeland soils. *Soil Science Society of America Special Publication* **13**, 65–72.

Simanton, J.R., K.G. Renard, L.J. Lane and C.E. Christiansen 1994. Spatial distribution of surface rock fragments along catenas in a semiarid area. *Catena*. **23**, 29–42.

Slattery M.C. and R.B. Bryan 1992. Hydraulic conditions for rill initiation under simulated rainfall: a laboratory experiment. *Earth Surface Processes and Landforms* **17**, 127–146.

Smith, R.E. and R.H.B. Hebbert 1979. A Monte Carlo analysis of the hydrologic effects of spatial variability of infiltration. *Water Resources Research* **15**, 419–29.

Strahler, A.N. 1950. Equilibrium theory of erosional slopes, approached by frequency distribution analysis. *American Journal of Science* **248**, 673–96, 800–14.

Thomas, D.S.G. 1988. The biogeomorphology of arid and semiarid environments. In *Biogeomorphology*, H.A. Viles (ed.), 193–221. Oxford: Blackwell.

Torri, D., M. Sfalanga and M. Del Sette 1987. Splash detachment: runoff depth and soil cohesion. *Catena* **14**, 149–55.

Tromble, J.M. 1976. Semiarid rangeland treatment and surface runoff. *Journal of Range Management* **29**, 215–5.

Tromble, J.M., K.G. Renard and A.P. Thatcher 1974. Infiltration for three rangeland soil-vegetation complexes. *Journal of Range Management* **27**, 318–21.

Twidale, C.R. 1962. Steepened margins of inselbergs from northwestern Eyre Peninsula, South Australia. *Zeitschrift für Geomorphologie* **6**, 51–9.

Twidale, C.R. 1967. Origin of the piedmont angle as evidenced in South Australia. *Journal of Geology* **75**, 393–411.

Wainwright, J. and A.J. Parsons 2002. The effect of temporal variations in rainfall on scale dependency in runoff coefficients. *Water Resources Research* **38**, 10.1029/2000WR000188.

Wang, X-P., X-R. Li, H-L. Xiao, R. Berndtsson and Y-X. Pan 2007. Effects of surface characteristics on infiltration patterns in an arid shrub area. *Hydrological Processes* **21**, 72–79.

Watson, D.A. and J.M. Laflen 1986. Soil strength, slope, and rainfall intensity effects on interrill erosion. *Transactions of the American Society of Agricultural Engineers* **29**, 98–102.

Wells, W.G. 1981. Some effects of brush fires on erosion processes in southern California. *International Association of Scientific Hydrology Publication* 165, 305–42.

Wells, W.G. 1987. The effects of fire on the generation of debris flows in southern California. In *Debris Flows/Avalanches: Processes, Recognition and Mitigation*, J.E. Costa and G.F. Weiczorek (eds), 105–14. Reviews in Engineering Geology 7. Boulder: Geological Society of America.

Wells, W.G., P.M. Wohlgemuth and A.G. Campbell 1987. Post-fire sediment movement by debris flows in the Santa Ynez Mountains, California. *International Association of Scientific Hydrology Publication* **165**, 275–6.

Wilcox, B.P., M.K. Wood and J.M. Tromble 1988. Factors influencing infiltrability of semiarid mountain slopes. *Journal of Range Management* **41**, 197–206.

Wohl, E.E. and P.P. Pearthree 1991. Debris flow as geomorphic agents in the Huachuca Mountains of southeastern Arizona. *Geomorphology* **4**, 273–92.

Woodruff, C.M. 1947. Erosion in relation to rainfall, crop cover, and slope on a greenhouse plot. *Soil Science Society of America Proceedings* **12**, 475–8.

Yair, A. 1983. Hillslope hydrology water harvesting and areal distribution of some ancient agricultural systems in the northern Negev Desert. *Journal of Arid Environments* **6**, 283–301.

Yair, A. 1987. Environmental effects of loess penetration into the northern Negev Desert. *Journal of Arid Environments* **13**, 9–24.

Yair, A. and M. Klein 1973. The influence of surface properties on flow and erosion processes on debris covered slopes in an arid area. *Catena* **1**, 1–18.

Yair, A. and H. Lavee 1976. Runoff generative process and runoff yield from arid talus mantled slopes. *Earth Surface Processes* **1**, 235–47.

Yair, A. and H. Lavee 1981. An investigation of source areas of sediment and sediment transport by overland flow along arid hillslopes. *International Association of Hydrological Sciences Publication* 133, 433–46.

Yair, A. and J. Rutin 1981. Some aspects of the regional variation in the amount of available sediment produced by isopods and porcupines, northern Negev, Israel. *Earth Surface Processes and Landforms* **6**, 221–34.

Yair, A. and M. Shachak 1987. Studies in watershed ecology of an arid area. In *Progress in desert research*, L. Berkofsky and M.G. Wurtele (eds), 145–93. Totowa, NJ: Rowman and Littlefield.

Yen, B.-C. 1965. Discussion of 'Large-scale roughness in open channel flow'. *Journal of the Hydraulics Division, Proceedings of the American Society of Civil Engineers* **91**(5), 257–62.

Yoon, Y.N. and H.G. Wenzel 1971. Mechanics of sheet flow under simulated rainfall. *Journal of the Hydraulics Division, Proceedings of the American Society of Civil Engineers* **97**, 1367–86.

Young, A. 1972. *Slopes*. London: Longman.

Young, A. and I. Saunders 1986. Rates of surface processes and denudation. In *Hillslope Processes*, A.D. Abrahams (ed.), 3–27. Boston: Allen & Unwin.

Young, R.A. and J.L. Wiersma 1973. The role of rainfall impact in soil detachment and transport. *Water Resources Research* **9**, 1629–36.

Chapter 10

Badlands and Gullying

Alan D. Howard

Introduction

Badlands have fascinated geomorphologists for the same reasons that they inhibit agricultural use: lack of vegetation, steep slopes, high drainage density, shallow to non-existent regolith, and rapid erosion rates. Badlands appear to offer in a miniature spatial scale and a shortened temporal scale many of the processes and landforms exhibited by more normal fluvial landscapes, including a variety of slope forms, bedrock or alluvium-floored rills and washes, and flat alluvial expanses similar to large-scale pediments. Although badlands evoke an arid image, they can develop in nearly any climate in soft sediments where vegetation is absent or disturbed. For example, short-lived badlands have been documented in the temperate climate of the eastern United States in sediment waste piles (Schumm 1956a) and borrow pits (Howard and Kerby 1983).

Gullies are closely related to badlands in that they record accelerated erosion into regolith or soft sediment by processes that are generally identical to those occurring in badlands. Large gully systems can evolve into badlands. Gullying occurs primarily in response to local disturbance of a vegetation cover by climate change, fire, or land-use changes and some badlands, such as the classic ones in South Dakota, have evolved through incision into a vegetated uplands.

This chapter will emphasize the relationship between process and form in badlands and gullied land-scapes and their long-term evolution through both descriptive and quantitative modelling.

Badlands

General reviews of badlands and badland processes are provided by Campbell (1989) and Bryan and Yair (1982a), including discussion of the climatic, geologic, and geographic setting of badlands, sediment yields, host rock and regolith variations among badlands, field measurements of processes.

Badland Regolith and Processes

The often (but not universally) rapid landform evolution in badlands provides the prospect of direct observational coupling of process and landform evolution in both natural and man-induced badlands. However, Campbell (1989) and Campbell and Honsaker (1982) caution about problems of scaling between processes on badland slopes and channels to larger landforms. Furthermore, although badland slopes commonly exhibit smooth profiles and are-ally uniform surface texture, (Figs. 10.1 and 10.2), recent studies have demonstrated that weathering, mass-wasting, and water erosional processes on badland slopes exhibit complex spatial and temporal variability and that weathering and erosional processes are largely hidden from direct observation in cracks and micropipes (Bryan et al. 1978, Yair et al. 1980, Bouma and Imeson 2000, Kuhn and Yair 2004, Kasanin-Grubin and Bryan 2007).

A.D. Howard (✉)
Department of Environmental Sciences, University of Virginia, Charlottesville, VA 22904-4123, USA
e-mail: ah6p@cms.mail.virginia.edu

A.J. Parsons, A.D. Abrahams (eds.), *Geomorphology of Desert Environments*, 2nd ed., DOI 10.1007/978-1-4020-5719-9_10, © Springer Science+Business Media B.V. 2009

Fig. 10.1 Badlands in Mancos Shale near Caineville, Utah. *Top* of North Caineville Mesa in background is 360 m above alluvial surface in middle distance, and is capped by the 60 m thick Emery Sandstone. Note the sharp-crested, straight-sloped badlands in Mancos Shale in the middle distance which rise abruptly 30–50 m above the Holocene alluvial surface. Level ridge crests marked by '*' are remnants of the Early Wisconsin (Bull Lake) pediment (photo by A. Howard)

Fig. 10.2 Rounded badland slopes in Morrison Formation near Hanksville, Utah (photo by A. Howard)

An Example: Badlands in the Henry Mountains Area, Utah

Shales of Jurassic and Cretaceous age exposed in the vicinity of the Henry Mountains, Utah, have in places been sculpted into dramatic badlands. The climate in the desert areas at about 1500 m above M.S.L. is arid, with about 125 mm of annual precipitation, most of which occurs as summer thunderstorms. These badlands provide premier examples of the erosional conditions favouring badlands, badland slope form and process, relationship between fluvial and slope processes, and variations of badland form with rock type. In particular, the great thickness and uniform lithology of the

Upper Cretaceous Mancos Shale (Hunt 1953) results in an unusual spatial uniformity of slope form and process (Figs. 10.1 and 10.2), first described by Gilbert (1880). Badlands developed in this area on the Mancos Shale, the Morrison Formation, and the Summerville Formation will be used here as examples. These badlands are discussed more fully by Howard (1970, 1997).

Badland Regolith

Badlands generally have very thin regoliths in arid regions, ranging from about 30 cm to essentially unweathered bedrock. Many badland areas share a similar regolith profile. The top 1–5 cm is a surface layer exhibiting dessication cracks when dry. This surface layer is a compact crust with narrow polygonal cracking in the case of shales with modest shrink-swell (<25%) behavior, such as the Mancos Shale badlands in Utah (Fig. 10.3), the Brule Formation badlands in South Dakota (Schumm 1956a,b), and portions of the Dinosaur Badlands, Canada (Bryan et al. 1978, 1984, Hodges and Bryan 1982). With higher shrink-swell behaviour the surface layer is broken into irregular, loose 'popcorn' fragments with large intervening voids, as exemplified in badlands on the Chadron Formation in South Dakota, the Morrison (Fig. 10.4) and Chinle Formations in Utah, and portions of the Dinosaur Badlands, Canada. The surface layer may be thicker (10 or more centimeters) in such cases. Although the surface layer may contain a few partially weathered fragments of shale, vein fillings, nodules, etc., it is primarily composed of disaggregated and remolded shale, silt and sand weathered from the shale which is leached of highly soluble components (particularly when derived from marine shales such as the Mancos Shale (Laronne 1982).

Beneath the surface layer is a sublayer (crust) averaging about 5–10 cm in thickness which may range from a dense, amorphous crust (Hodges and Bryan 1982, Gerits et al. 1987) to a loose, granular layer (Howard 1970, Schumm and Lusby 1963). Part of the apparent variability of surface layer and crust may be due to differences among researchers in locating the division between these units. Below the crust there occurs a transitional 'shard layer' ranging from 10 to 40 cm in thickness consisting of partially disaggregated and weathered shale chips grading to firm, unweathered shale.

Fig. 10.3 Regolith profile in Mancos Shale badlands. Note smooth, mud-cracked surface crust, shard layer, and essentially unweathered shale at depth of about 15 cm. This is a marine shale dominated by Illite clay mineralogy. Note watch for scale (photo by A. Howard)

However, there is considerable variability in badland regoliths. Well-cemented sandstone layers outcrop as bare rock, often creating ledges or caprocks. Slightly cemented sand layers (often with $CaCO_3$ cement) generally form steep slopes with regolith 1–5 mm thick or less (Gerits et al. 1987, Hodges and Bryan 1982, Bowyer-Bower and Bryan 1986). These sandstones may develop permanent rill networks (Kasanin-Grubin and Bryan 2007). Carman (1958) described a regolith in fluted badlands in a clay matrix-supported conglomerate composed of a hard veneer of sandy clay 10–15 cm thick over a softer interior layer and a hard core. Badlands on shaly sandstones in the Summerville Formation have a non-cohesive surface layer of sand grains and shale chips which grades fairly uniformly to unweathered bedrock within 10–20 cm. Some badland surfaces are cemented with a biological crust of algae or lichens (Yair et al. 1980, Finlayson et al. 1987). In badlands where the shale has non-dispersive clays or physical weathering processes dominate, surface crusts may be absent, with a thin shard layer grading to solid shale within a few centimetres (Figs. 10.5 and 10.6). Badlands in marine shales, such as the Mancos Shale, may have salt recrystallization in the sublayers and local areas of salt surface crusts (Laronne 1982). Due to the frequent occurrence of rapid mass wasting on steep badland slopes, recently denuded areas have atypically thin regoliths with poor development of the surface layer.

Systematic variations in regolith properties may be associated with slope aspect. Churchill (1981), in

Fig. 10.4 Popcorn surface crust on Morrison Formation shale. Clays in these badlands are dominated by Smectite (Montmorillonite) clays with high shrink-swell potential (photo by A. Howard)

Fig. 10.5 Badlands near Lumbier, Navarra, Spain. Note rounded divides and linear lower slope profiles. (Photo by A. Howard)

Fig. 10.6 Surface texture of badlands shown in Fig. 10.5. Regolith consists primary of shards up to 1 cm in size with unaltered bedrock within a few centimeters (photo by A. Howard)

a comparison of north- and south-facing slopes in Brule Formation Badlands (South Dakota), found the north-facing slopes to be more gently sloping, more densely rilled, with deeper regoliths, and characterized by infrequent mass-wasting involving much of the regolith (20–30 cm). By contrast, south-facing slopes have frequent shallow (3–10 cm thick) sloughing failures. Churchill suggested that slower evaporation promotes deeper infiltration on north-facing slopes. Yair et al. (1980) and Kuhn and Yair (2004) found that north-facing slopes in the Zin badlands of the Negev desert to have a rough, lichen-covered surface, a deeper regolith, few rills, and relatively high runoff and erosion. South-facing slopes are smoother, with greater runoff, frequent mudflows, and pipe development, but they experience less runoff erosion overall.

Weathering processes of badland regolith development involve relatively modest changes in mineralogy, since the source rocks are poorly lithified sedimentary rocks. Simple wetting may be sufficient to disaggregate the shale and disperse the clay minerals. For example, unweathered blocks of Morrison Formation smectitic shales will completely disaggregate and disperse within a few minutes to a few hours in just enough water to completely saturate the sample (this may be a considerable relative volume of water because of the pronounced swelling). Rapid slaking accompanies this disaggregation. The shallow regolith on these slopes testifies to the absorptive capacity and impermeability of the surface layers. On the other hand, samples of unweathered Mancos Shale, a marine shale in which illite is dominant, decompose

more slowly, with less swelling and incomplete dispersion. A yellowish liquor of dissolved salts (mostly sodium and calcium sulfates: Laronne 1981, 1982) is produced, the accumulation of which apparently inhibits further disintegration. Leaching of these solutes is apparently required for complete weathering, and salt concentrations decrease upwards within the regolith (Laronne 1982). Some clayey sandstones may have a calcite cementing which requires dissolution before erosion, and the regolith is only a few millimeters thick (Bowyer-Bower and Bryan 1986). On the other hand, marls composed of more than 50% calcite and gypsum may be essentially free from cementation. Consequently clay mineral dispersion is all that is required to form a disaggregated regolith (Finlayson et al. 1987, Imeson and Verstraten 1985). Some unconsolidated sediments require no weathering prior to entrainment by rainsplash and runoff, so that regolith is absent or only seasonally present due to frost heaving (for example, the artificial Perth Amboy badlands of Schumm 1956a).

The structure and appearance of badland regolith may change appreciably in response to the sequence of precipitation events. Kasanin-Grubin and Bryan (2007) document a change in surface texture over two years on badlands in Alberta, Canada from an open, popcorn surface (similar to Fig. 10.4) to a dense, cracked crust (similar to Fig. 10.3) over a two year period in response to snowfalls and low intensity, long-duration rainfall. They also conducted experiments demonstrating that low duration rainstorms produce popcorn textures through shrinkage and swelling while long-duration rainstorms produce coherent crusts through clay dispersion. Greater runoff and deeper rill erosion accompany the formation of coherent crusts.

The physical and chemical properties of clay minerals, grain size distribution, density, and cementation affect the weathering by wetting and leaching, runoff characteristics, sediment and solute yields, and badland form. The properties of shales relevant to their erosional behaviour can be measured by a number of simple physico-chemical tests of cation-exchange capacity, solute extracts (concentration and chemistry), Atterberg (consistency) limits, vane shear strength, dispersion index, aggregate stability, shrink-swell behavior (Imeson et al. 1982, Imeson and Verstraten 1988, Hodges and Bryan 1982, Bryan et al. 1984, Gerits et al. 1987, Bouma and Imeson, 2000, Faulkner et al., 2000, 2003).

Erosional Processes on Badland Slopes

Discussion of slope erosional processes is conventionally divided into mass-wasting (creep, sliding, and flowage) and wash processes (runoff and rainsplash erosion). This discussion follows that division, although these processes are so interrelated and intergrading that such divisions are primarily for convenience.

Mass-Wasting

The rounded slopes of the Morrison (Fig. 10.2), Chinle, and Chadron formations have been attributed to the dominance of creep as a slope-forming process (Davis 1892, Schumm 1956b, Howard 1970). These formations contain clays with high dispersive and shrink-swell potential. The volume changes occurring during wetting together with the loss of bulk strength encourage downslope sag and creep of surface layers. Schumm (1963) found surface creep rates to be proportional to the sine of the slope gradient for slope gradients less than 40° on Mancos Shale, and such a relationship is likely to be a good approximation to mass-wasting rates for low slope gradients.

Many badland sideslopes are close to their maximum angle of stability, as long, narrow, shallow slips are common occurrences. Because these slopes are generally at angles of 40° or more, and failures occur during rainstorms, slope stability can be analyzed by the well known infinite slope model for saturated regolith and flow parallel to the slope (e.g., Lambe and Whitman 1969, pp. 352–356, Carson 1969, p. 87, Carson and Kirkby 1972, pp. 152–159). Figure 10.7 shows the definitions for an analysis similar to Lambe and Whitman's except that uniform surface flow is also allowed. Conditions for failure at a depth d_r below the surface are assumed to be approximated by a linear relationship:

$$\tau_f = c + \sigma_e \tan \varphi, \tag{10.1}$$

where τ_f is the shear stress on the failure plane, σ_e is the effective normal stress on the failure plane, c is the saturated cohesion, and ϕ is the saturated friction angle for the regolith material.

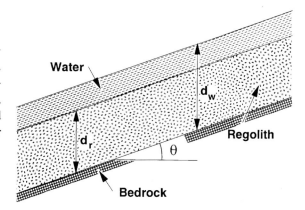

Fig. 10.7 Definition of terms for infinite slope failure with water flowing parallel to the slope surface

Based upon Fig. 10.3 for the case when $d_w \geq d_r$ the shear and effective normal stresses on the potential failure plane are:

$$\sigma_e = \rho_b\, d_r\, g \cos^2 \theta \tag{10.2}$$

and

$$\tau = \left[\rho_t\, d_r + \rho_w(d_w - d_r)\right]g \sin \theta \cos \theta \tag{10.3}$$

where ρ_b and ρ_t are the buoyant and total densities of the regolith, respectively, ρ_w is the water density, g is the gravitational constant, θ is the slope angle, and d_w is the depth of water flow above the regolith surface. For the case when $d_w < d_r$

$$\sigma_e = \left[\rho_u(d_r - d_w) + \rho_b\, d_w\right]g \cos^2 \theta \tag{10.4}$$

and

$$\tau = \left[\rho_u(d_r - d_w) + \rho_t\, d_w\right]g \sin \theta \cos \theta \tag{10.5}$$

where ρ_u is the unsaturated unit weight of the regolith (which is likely to be very close to ρ_t).

Failure occurs when $\tau \geq \tau_f$. For the shallow badland regoliths only slight cohesion is necessary to permit slopes greater than 40° if water flow depths are small. For typical values of regolith density complete saturation with flow parallel to the surface reduces maximum slope angles by a factor of about two (Lambe and Whitman, 1969). Under such conditions badland slopes may fail as mudflows (Yair et al. 1980, Battaglia et al. 2002, Kuhn and Yair 2004, Desir and Marín 2006, Nadal-Romero et al., 2007).

Wash Processes

Wash processes on badlands occur only during and immediately after rainstorms. If a rainstorm starts with a dry regolith (the usual case) much of the initial rainfall is absorbed by the regolith and initially penetrates deeply along cracks and between popcorn aggregates, if present. Runoff from badland slopes is delayed for several minutes after rainfall initiates; for example, Bryan et al. (1984) note a delay of 10–15 min after initiation of sprinkling at a rate of 10–20 mm h^{-1}. The depth of wetting is limited either by absorption along crack and fissure edges or by ponding on top of the dense crust layer, if present (Hodges and Bryan 1982). Swelling of the regolith clays rapidly begins to close the cracks and gaps, so that flow is increasingly restricted to lateral flow in larger cracks and, in many cases to micropipes developed near the crust-shard layer boundary (Hodges and Bryan 1982) (Fig. 10.8). The flow in the larger cracks which receive appreciable flow from upslope may be able to erode their walls and keep pace with regolith swelling (Engelen 1973). Most of these will be ephemeral rills, which are partly or wholly eradicated by continued swelling

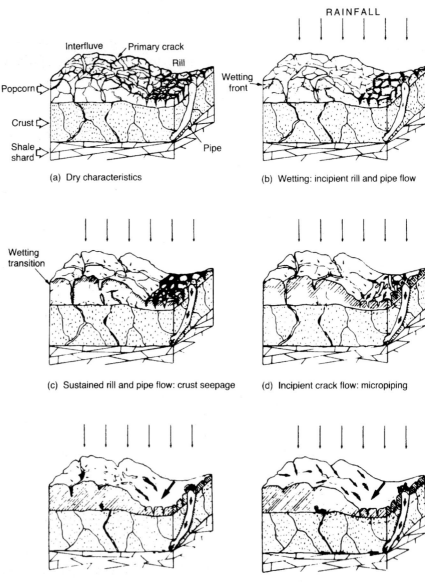

Fig. 10.8 Regolith structure and runoff behavior on shale badland slopes. Diagrams (**a–f**) portray successive stages in wetting and runoff development. (from Hodges and Bryan 1982, Fig. 2.7)

Fig. 10.9 Development of a perched water table on top of shard layer in badland regolith and lateral drainage to rills. Note development of a saturated, soft layer between the rills which may fail, forming a pipe and eventually a new rill. (from Imeson and Verstraten 1988, Fig. 10.1)

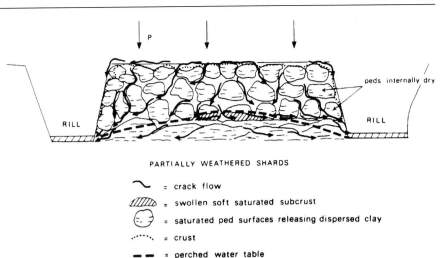

PARTIALLY WEATHERED SHARDS

= crack flow

= swollen soft saturated subcrust

= saturated ped surfaces releasing dispersed clay

= crust

= perched water table

and subsequent drying and cracking of the surface layer. Sediment is added to the flow both by shear and dispersion of clay minerals, and the sediment load may range up to mudflow concentrations. Flow in the cracks and micropipes eventually feeds into rills or onto alluvial surfaces. Badland interflow has been compared to intergranular flow to drains (the rills) (Gerits et al. 1987, Imeson and Verstraten 1988) (Fig. 10.9). Surface rills may initiate through liquefaction of saturated, dispersed crust above the shard layer with accompanying runout, pipe formation and surface collapse in the zones of greatest depth of saturation between surface rills (Fig. 10.9).

Hodges and Bryan (1982) pointed out that in Dinosaur Badlands, Canada, runoff occurs more rapidly and more completely on silty surfaces which locally encrust badland rills than on the cracked surface layers of badland slopes. Such silty layers in rills are uncommon on badlands in the Henry Mountains region, but alluvial surfaces are generally covered by such sediment and have rapid runoff response. Cerdà and García Fayos (1997) also note more complete runoff from alluvial surfaces than badland slopes. A few very gentle badland slopes in Morrison Shale exhibit a banding alternating between typical popcorn-textured badland slopes and short alluvial surfaces (Fig. 10.10). The absence of shrink-swell cracking of the alluvial surfaces suggests that the silt layers are very effective at preventing infiltration. The banding may be an expression of layering in the shale or it may indicate a natural bifurcation at low slope gradient between normal badland slopes with swelling enhanced by water shed from upslope alluvial patches and allu-

Fig. 10.10 Low, terraced hillslope on shale in Morrison Formation. White bands are small segments of alluvial surface (photo by A. Howard)

vial surfaces graded downstream to the swollen shale bands.

In some badlands much of the slope drainage is routed through deep, corrasional pipe networks (see reviews in Campbell 1989, Harvey 1982, Jones 1990, Parker et al. 1990, several papers in Bryan and Yair 1982b, Faulkner et al. 2007). Piping in susceptible lithologies is encouraged by prominent jointing and steeper hydraulic gradients via underground rather than surface paths caused by layering in the bedrock or though dissection of an originally flat upland (Campbell 1989). Faulkner et al. (2007) discuss a badlands landscape in which the initial response to downcutting by the master drainage is development of a deep piping network which then, through collapse, forms an integrated surface network of channels. Piping is rare in the Henry Mountains badlands discussed here.

Rainsplash

Rainsplash erosion is important on badlands both as a direct agent of detachment and splash transport and indirectly as detachment contributes to runoff erosion and affects the flow hydraulics of shallow flows. A considerable experimental literature exists on rainsplash erosion mechanics (see Chapter 9); the major controlling variables are raindrop size and velocity, rain intensity, and regolith characteristics. Few direct measurements exist of the contribution of rainsplash to badland slope erosion. Howard (1970) and Moseley (1973) attributed narrow, rounded divides on some badlands to the action of rainsplash rather than creep, and the rounded slopes with very thin regolith on the badlands in Fig. 10.5 may be due to diffusive rainsplash erosion. Carson and Kirkby (1972, p. 221) also suggest divide convexity may be caused by rainsplash on some arid slopes. The influence of rainsplash may be limited at the beginning of rainstorms by the cohesiveness of the dry clay, and biotic crusts, where present, protect shale regolith from rainsplash entrainment (Yair et al. 1980, Finlayson et al. 1987). In contrast to rock-mantled slopes, rainsplash has modest direct influence on overland flow hydraulics due to the concentration of flow into cracks and micropipes.

Runoff Erosion

Runoff on badlands seldom exemplifies the classic characteristics of overland flow on agricultural land because of the concentration of flow into cracks, micropipes, and ephemeral rills. The rate of erosion in such channelled overland- and inter-flow, as well as in larger rills and washes functionally depends upon flow conditions and resistance of the bedrock or regolith to weathering or detachment. The processes of erosion are poorly understood but may involve direct detachment from the bedrock, scour by sediment, and weathering processes such as leaching and wetting with dispersion.

Several approaches have been used to quantify runoff erosion in both rills and interrill areas. The most common approach on agricultural slopes is to estimate sediment transport rates using bedload or total load transport formulas assuming that the flow is loaded to capacity in the sand size ranges (transport-limited conditions). These are often empirically corrected for rainsplash effects based upon results of plot experiments in non-cohesive sediments or weakly cohesive soils (e.g. Meyer and Monke 1965, Komura 1976, Gilley et al. 1985, Julian and Simons 1985, Everaert 1991, Kinnell 1990, 1991). Detachment by rainsplash is assumed to assure capacity transport even where soils are moderately cohesive. Theoretical and empirical runoff erosion models commonly assume capacity transport (e.g. Kirkby 1971, 1980a, Carson and Kirkby 1972 pp. 207–219, Smith and Bretherton 1972, Hirano 1975), but such assumptions often overestimate actual transport rates severalfold (Dunne and Aubrey 1986). Due to the cohesion and the small sand-sized component of badland regoliths, coupled with the steep slope gradients, runoff and interflow are likely to carry bed sediment loads well below capacity.

A few researchers have recognized that flow on steep slopes is commonly detachment limited and suggest that the detachment (or deposition) by the flow D_f is related to an intrinsic detachment capacity D_c (for zero sediment load), the actual sediment load G and the flow transport capacity T_c (Foster and Meyer 1972, Meyer 1986, Lane et al. 1988, Foster 1990)

$$D_f = D_c(1 - G/T_c) \qquad (10.6)$$

This relationship implies an interaction between deposition and entrainment on the bed. However, rills and badland slopes seldom show evidence of sediment redeposition or partial surface mantling until flow reaches alluvial washes or alluvial surfaces (miniature pediments). The downstream transition from bare regolith to sand- and silt-mantled alluvial surfaces is generally abrupt (Schumm 1956a, b, Smith 1958, Howard 1970). The high roughness and possibly the greater grain rebound on badland regolith may make transport capacity greater than for an alluvial surface at the same gradient (Howard 1980). This suggests that on steep badland slopes and rills actual detachment D_f can reasonably be assumed to equal the intrinsic detachment D_c. This approach is used in the following models.

Howard and Kerby (1983) successfully related areal variations in observed rates of erosion of bedrock channels on shales to the pattern that would be expected if erosion rates were determined by dominant shear stress in the channel. Following Howard (1970) they suggested that erosion rate $\partial y / \partial t$ (detachment rate) is determined by shear stress τ:

$$\frac{\partial y}{\partial t} = K_c(\tau - \tau_c)^\beta \qquad (10.7)$$

where τ_c is a critical shear stress, β is an exponent, and the constant of proportionality K_c depends upon both flow durations and bedrock erodibility. Foster (1982) and Foster and Lane (1983) assume a similar relationship (with $\beta = 1$) for rill detachment capacity. Numerous experiments on fluvial erosion of cohesive deposits indicate scour rates that correlate with the applied shear stress (Parthenaides 1965, Parthenaides and Paaswell 1970, Akky and Shen 1973, Parchure and Mehta 1985, Ariathurai and Arulandan 1986, Kuijper et al. 1989, Knapen et al. 2007). Assuming certain hydraulic geometry and resistance equations, and further assuming that the dominant values of shear stress greatly exceed the critical value, then Equation (10.7) can be re-expressed as a function of local gradient S and contributing drainage area A_d (Howard 1994b, 1997, 2007):

$$\frac{\partial y}{\partial t} = K_c(K_e\,A_d^m\,S^n - \tau_c)^\beta \qquad (10.8)$$

where the constant of proportionality, K_e, likewise depends upon constants in the hydraulic geometry equations. Theoretical values for m and n for the assumed hydraulic geometry relationships are $\sim$0.3 and $\sim$0.7, respectively, with observed values being 0.45 and 0.7.

In some badland washes and in much of the throughflow in badland regolith the limiting factor may be the rate of decrease of shear strength of surface rinds either due to weathering of the bedrock or to wetting and dispersion of regolith crusts and shards. Even though the weathering rate is a limiting factor, flow conditions can still affect erosion rates, as illustrated in a simple model by (Howard 1994a, 1998) that assumes that flocs can be removed when weathering, progressing from the surface inward, reduces cohesion to below the applied shear stress. The resulting erosion rate increases as a joint function of shear stress and intrinsic weathering rate.

Both of these models suggest an erosion rate that increases with some measure of the strength of the ef-

fective flow. However, data on erosion rates in badland channels is limited, and that on the accompanying flow is essentially nonexistent.

In some cases flow in bedrock rills may become so sediment laden that they exhibit Bingham flow properties with development of levees on rills and depositional mudflow fans. The common occurrence of localized slumps and draping flows on badland slopes might be appropriately characterized by Bingham flow as well.

Requisite conditions for rill development and the overall control of drainage density in badlands has been a continuing theme in badlands geomorphology, starting with Schumm's (1956a) introduction of the 'constant of channel maintenance', a characteristic length from the divide to the head of rills. Rills have been discussed in two contexts. The first is the critical hydraulic conditions required for the transition from dispersed overland flow to channelized rill flow (see general discussions in Bryan 1987, Gerits et al. 1987, Rauws and Govers 1988, and Torri et al. 1987). Several criteria have been proposed, including critical slope gradients (Savat and De Ploey 1982), a critical Froude number (Savat 1976, 1980, Savat and De Ploey 1982, Hodges 1982, Karcz and Kersey 1980), and a critical shear stress or shear velocity (Moss and Walker 1978, Moss et al. 1979, 1982, Savat 1982, Chisci et al. 1985, Govers 1985, Govers and Rauws 1986, Rauws 1987). On the steep badland slopes with strongly channelized flow in cracks, microrills and pipes much of the flow on badland slope exceeds critical conditions for rill initiation by any of these criteria. Rill initiation on badland slopes has also been related to unroofing of tunnels and micropipes (Hodges and Bryan 1982) which may be related as much to saturation, swelling and softening of badland regolith as to the hydraulic factors mentioned above (see Fig. 10.5, Gerits et al. 1987, Imeson and Verstraten 1988).

Runoff thus seems to be capable of rill initiation over most of badland slope surfaces, with the possible exception of well-cemented sandstone layers with thin regoliths. The development and maintenance of semipermanent rills requires a balance between the tendency of runoff to incise and other processes that tend to destroy small rills (Schumm 1956a, Kirkby 1980b). Such processes include shrink-swell of the surface layers, which destroys microrills (Engelen 1973), needle-ice growth (Howard and Kerby 1983), mass wasting by creep and shallow slides (Schumm 1956a, b, Howard and Kerby 1983) and rainsplash (Howard 1970, Moseley 1973, Dunne and Aubrey 1986). On badlands in humid climates a well-defined seasonal cycle of rill creation and obliteration occurs (Schumm 1956a, Howard and Kerby 1983). In arid regions advance and retreat of rill systems on slopes may occur over much longer time scales, but may also be dramatically altered by a single heavy summer rainfall or a winter mass-wasting event. In the discussion below the emphasis is on 'permanent' rills and gullies that have persisted long enough to have created well-defined valleys. In the simulation modelling discussed below the location of permanent rills is determined by the balance between runoff erosion using Equation 10.8 and diffusive processes (rainsplash or mass-wasting).

Badland Architecture and Evolution

This section emphasizes the spatial process variations that determine the overall architecture of both slope and channel features of badlands. Furthermore, the temporal evolution of badlands will be examined both through reference to the specific case of the Henry Mountains area and through the use of simulation models.

Evolution and Areal Distribution of Badlands

The rapid erosion of badland slopes means that they occur only where high relief has been created in shaly rocks. In the Henry Mountains area this has occurred through erosional removal of a protective caprock or through rapid master stream downcutting. The ramparts of sandstone cuestas feature local badlands in sub-caprock shales, and more extensive badland areas occur where buttes have been recently denuded of their caprock. However, the badlands on cuesta ramparts are well developed only during relatively arid epochs (such as the present) when mass-wasting of caprocks is relatively quiescent (see discussion in Chapter 8).

The master stream is the Dirty Devil-Fremont River system which, during the Quaternary, has had a history of alternating stability or slight aggradation during pluvial epochs with rapid downcutting, followed by stability at the close of non-pluvial epochs (Howard 1970, 1997). During the pluvial epochs the stable base level coupled with physical weathering and mass-wasting of the sandstone cuestas resulted in development of extensive talus slopes on the ramparts of the escarpments coupled with gravel-veneered alluvial surfaces (pediments) mantling the shales. Thus badlands were rare during pluvial epochs, probably occurring only locally on scarp ramparts or caprock-stripped buttes. The post-pluvial (Bull Lake) dissection of river terraces and alluvial surfaces underlain by Mancos Shale has created the spectacular badland landscape near Caineville, Utah (Fig. 10.1). The river apparently downcut about 65 m shortly after the close of the Bull Lake pluvial (of Early Wisconsin age, Anderson et al. 1996), followed by stability at about its present level (Howard 1970, 1997). As a result, a wave of dissection has moved headwards towards the sandstone cuestas to which the alluvial surfaces were graded, producing a sequence of landforms from scattered Bull Lake alluvial surface remnants near the scarps (remaining primarily where the capping gravels were thickest) through a zone of high-relief (50 to 60-m) badlands to modern alluvial surfaces near the master drainage where the badlands have been completely eroded (Fig. 10.11). Shale areas that are either remote from the master drainage or have been protected from stream downcutting by downstream sandstone exposures are either undissected Pleistocene alluvial surfaces or very-low-relief badlands.

Similar post-glacial downcutting has been implicated in forming the shale badland landscapes of the Great Plains of the United States and Canada (Bryan et al. 1987, Campbell 1989, Wells and Gutierrez 1982, Slaymaker 1982), with a similar progression from undissected uplands (often capped by a protective grassland cover) through high-relief badlands to modern alluvial surfaces. In semi-arid and humid regions, badlands occur primarily where a former vegetation has been removed from shales or easily eroded regolith. Thus such occurrences are similar to relief generation through a protective caprock and its subsequent removal.

Relationship Between form and Processes

Badlands exhibit a surprisingly wide range of slope form. A contrast between steep, straight-sloped badlands with very narrow divides and a convex form with generally gentler slopes was noted quite early in the

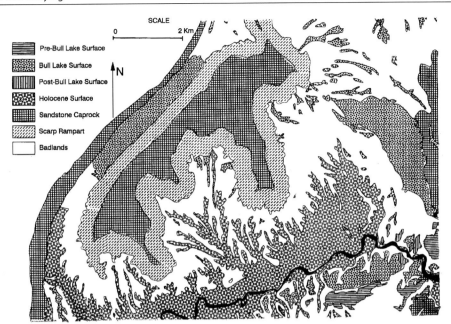

Fig. 10.11 Geomorphic map of part of the Caineville Area, near the Henry Mountains, Utah. Bull Lake surfaces are Early Wisconsin in age. Bull Lake surface includes terraces of the Fremont River and alluvial surfaces graded to these terraces. The Fremont River is the black sinuous line near the bottom of the illustration. The large area of Bull Lake surface at the right edge is preserved because it drains across a sandstone caprock that acts as a local baselevel. Pre- and Post-Bull Lake surfaces are Fremont River terraces. Holocene surface is Fremont River floodplain and alluvial surface graded to modern river level. Scarp rampart includes both rockfall-mantled shale and high-relief badlands (based on Howard 1970 Plate 6)

literature on badlands and is exemplified in the classic South Dakota badlands by slopes in the Brule and Chadron formations, respectively (Schumm 1956b). In the Henry Mountains area a similar contrast occurs between badlands on the Mancos Shale and those in the Morrison Formation (compare Figs. 10.1 and 10.2). Badlands in the Summerville Formation are similar to the Mancos Shale badlands in having straight slopes and narrow divides, but they have significantly lower maximum slope gradients (Table 10.1). Many badlands have quite complicated slope forms due to contrasting lithology and interbedded resistant layers (e.g., the Dinosaur badlands of Canada). Occasionally, pinnacle forms of badlands are found that are characterized by extremely high drainage density, knife-edge divides, and generally concave slope form (Fig. 10.12). The following discussion of badland erosional processes links these differences in slope form to variations in lithology and climate.

The broadly rounded upper slopes of the Morrison, Chinle, and Chadron formations are probably due to creep processes (see above). The popcorn texture of surface aggregates allows relatively large relative

Table 10.1 Comparison of slope angles (in degrees) on badlands in Mancos Shale, the Summerville Formation, and the Morrison Formation

Formation	Angle-of-repose slopes[1]	Unstable slopes[2]	Average slope angle[3]
Summerville	31.5	31.7	27
Mancos	32	34.8	40
Morrison	–	34.8	–

[1] Slopes constructed of unweathered shale shards.
[2] Constructional talus slopes of mass-wasted regolith below vertical cliffs in shale.
[3] Average gradient of long natural slopes in areas of moderate to high relief.

movements of the aggregates as their edges become wet and slippery. However, lower sideslopes sometimes exceed 40°, locally resulting in rapid flowage of the popcorn surface layer off the slope. This results in long, narrow tracks of exposed subsoil on the lower slopes which rapidly develop a new surface layer of popcorn aggregates. Convex slopes develop through creep in any circumstance where the creep rate increases monotonically with slope gradient and where

Fig. 10.12 Pinnacle badland slopes in Cathedral Valley, near Caineville, Utah. The steep slopes were formerly protected by a caprock. Note the low-gradient slopes in the same formation at the base of the slope (Entrada Sandstone) (photo by A. Howard)

creep is the dominant erosional process; this was first elaborated by Davis (1892) and Gilbert (1909).

Even on these rounded slopes runoff erosion becomes increasingly important downslope, becoming dominant in rills and locally within shallow pipes and cracks in the thin regolith.

The Mancos badlands have a nearly linear profile with narrow, rounded divides, which range in width from less than 0.5 m in high-relief badlands to 1–2 m in low-relief areas (Fig. 10.1). Because of the very thin regolith on these narrow divides and a tendency for development of a shale-chip surface armoring, Howard (1970) and Moseley (1973) attributed the divide rounding to rainsplash. This process is effective on narrow divides even at low gradient because the maximum splash distance is greater than the divide width.

Virtually all high-relief badland slopes in the Henry Mountains region have nearly constant gradient on their lower portions, even on the broadly convex slopes in the Morrison Formation. These maximum gradients are usually within a few degrees of the angle of repose of dry detritus weathered from the formations. Maximum gradients on the Summerville Formation, with its loose weathered layer, are 3–5° less than the angle of repose, probably due to seepage flow decreasing the maximum stable slope angle (implied

in Equations (10.1, 10.2, 10.3, 10.4 and 10.5); see Lambe and Whitman 1969, p. 354). However, in the Morrison and Mancos badlands the slopes are 3–10° higher than the repose angle for loose weathered shale due to cohesion. Consequently, many lower slopes are on the verge of failing by flowage and slipping. Many such slopes do occasionally fail, involving only the thin surface layer (5–10 cm) and leaving long, narrow tracks of exposed subsoil on the lower slopes. Such flows are numerous on steep slopes on the Morrison and Summerville Formations, but are rare on the Mancos badlands. However, on the Mancos badlands, whole sections of hillside appear to slip or slump short distances downhill during rainstorms, producing tension cracks arranged in waves suggesting differential movement and, rarely, extensive shallow slumping (Fig. 10.13). Tension cracks are the more numerous and wider the steeper the gradient, particularly on slopes undercut by meandering washes.

On low-gradient portions of slopes, creep-like movement of the surface layer predominates, which is generally modeled as a linear function of slope gradient (Culling 1963, Carson and Kirkby 1972, Howard 1994b, Fernandes and Dietrich 1997). However, for the gradients approaching the limiting slope angle (which in actuality is temporally and spatially

Fig. 10.13 Steep Mancos Shale slopes showing evidence of accelerated rate of mass wasting. The regolith mass at "@" derived from the scar immediately above. Elsewhere in this picture are numerous scars and regolith masses suggestive of episodic slumping or flow. New regolith rapidly forms whenever it becomes exposed. Contrast the lack of rills on this slope with the Mancos Shale slopes in Fig. 10.1 (photo by A. Howard)

process, producing divide convexity. As the total volume of mass wasting debris increases downslope, equivalent rates of erosion may require gradients approaching the limiting slope angle where slippage or flowage becomes important. In these lower slope regions the incremental addition of weathered material along the slope can be accommodated by a very slight increase in gradient, thereby creating a nearly straight profile. Such lower slopes are essentially equivalent to the threshold slopes of Carson (1971) and Carson and Petley, (1970) except that they are probably best modelled by a rapid but continuous increase in mass wasting rate as the limiting angle is approached rather than by an abrupt threshold (Howard, 1994b, Roering et al. 2001a).

Indirect evidence for threshold slopes in steep badlands comes from areal variations in badland form. Average hillslope gradients in badlands of the Summerville and Mancos Formations show little variation with slope length except for very short slopes (Fig. 10.14). However, the drainage density exhibits a complicated relationship to relief ratio, generally increasing with relief ratio up to a value of about 0.5, but decreasing in very high relief badlands (Fig. 10.15). The relatively low drainage density of areas with very high relief ratio may be explained by the onset of sliding and slumping on steep slopes. In areas of high relief large increases of the rate of erosion of a slope at its base should be accompanied by only a slight change in slope gradient (Fig. 10.14). But the small increase

variable), mass wasting rates can be functionally represented by a rapid rate increase (Kirkby (1984, 1985b), Howard (1994b), Roering et al. 1999, 2001a, b).

Nearly linear lower slopes on regolith-mantled, high-relief badlands would be expected if mass-wasting determines slope form. Close to the divide, where creep rates are low (due to modest amounts of regolith supplied from upslope) and corresponding gradients are low, mass wasting rate follows the sine relationship. Consequently, gradients increase rapidly downslope. Rainsplash erosion is also a diffusive

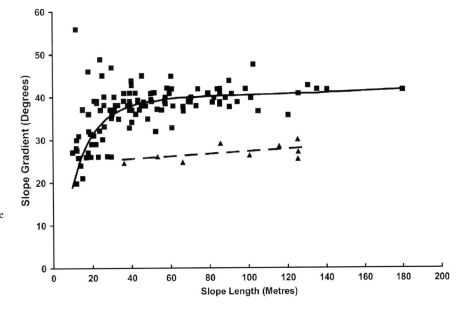

Fig. 10.14 Plot of hillslope gradient versus slope length in badlands in the Summerville (*triangles*) Formation and Mancos Shale (*squares*) (based on Howard 1970, Fig. 43). The short slopes with very high gradients occur on undercut banks along meandering washes

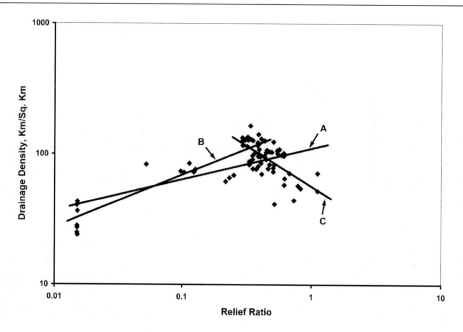

Fig. 10.15 Drainage density versus relief ratio for badland areas on Mancos Shale. Measurements made on areal photographs with a scale of approximately 1:12,000. Relief ratio determined by inscribing a 150 m diameter circle around a badland divide and measuring the maximum relief photogrammetrically and dividing by 75 m. Drainage density is defined as total length of all hollows and valleys visible on the photograph in the circle divided by the area of the circle. Ephemeral rills lacking defining ridge lines are not included. Regression lines are A for all points, B for values of relief ratio less than 0.35 and C for values greater for 0.2. All are statistically significant

in slope angle increases the efficiency of erosion on the slope relative to the channel, so that the critical drainage area necessary to support a channel increases, with a resulting decrease in drainage density. Average slope length L and drainage density D are related to the average slope angle θ by (Schumm 1956a, p. 99)

$$L \, \text{Cos} \, \theta = 2/D. \qquad (10.9)$$

Because slope angles vary little with relief ratio in steep badlands, the decrease in drainage density accompanying increase in relief ratio is accompanied by a proportional increase in slope length. This decrease in drainage density with increasing erosion rates for landscapes with threshold slopes has been theoretically justified and demonstrated in simulation modeling by Howard (1997) and Tucker and Bras, (1998). By contrast, in areas of low overall gradient, increase in relief ratio is accompanied by increases in drainage density. Therefore, slope length decreases nearly in proportion since the cosine term is near unity. Modelling by Howard, (1997) suggests that this occurs when slopes are eroded by classic

linear diffusion and channel erosion occurs with shear stresses not much above the threshold of detachment (τ_c in Equations 10.7 and 10.8).

At least four types of surfaces occur on badlands: (1) slopes with exposed bedrock; (2) regolith-mantled slopes; (3) Rills and washes with exposed bedrock; and (4) alluvium-mantled surfaces. These surfaces generally sharply abut against each other, with the transitions corresponding to thresholds in the relative importance of processes. Most badland slopes are regolith-mantled, albeit shallowly. However, on steep slopes or where bedrock is resistant to weathering the surface is irregular, expressing variations in weathering characteristics of the rock, jointing, and stratigraphy (Fig. 10.16). In contrast, minor differences in lithology of the rocks underlying regolith-mantled badlands (non-undercut areas in Fig. 10.13) do not affect slope form, for the geometry is determined by processes of downslope transport of the weathered debris. The threshold between bare and regolith-mantled slopes is commonly equated with weathering- versus transport-limited conditions, respectively (Culling 1963, Carson and Kirkby 1972 pp. 104–106). However, there are

Fig. 10.16 Near-vertical slopes in shale of the Morrison Formation formed by past undercutting by the Fremont River. Differences in bedding are strongly expressed in vertical slopes in contrast to the smoother, gentler slopes mantled by 20+ cm of regolith (photo by A. Howard)

four factors that may limit overall slope erosion rate, the potential weathering rate PW, the potential mass wasting rate PM, the potential detachment rate by runoff (combined splash and runoff detachment) PD, and the potential fluvial transport rate PT. On regolith-mantled slopes $PW > (PM + PD)$ and $PD < PT$. Either mass-wasting or runoff detachment may be quantitatively dominant on such slopes; measurements by Schumm (1963) suggest runoff dominance. The use of the term transport-limiting for these mantled slopes is inaccurate, because the runoff component of erosion is detachment- rather than transport-limited. Bare bedrock slopes have $PW < (PM + PD)$ and $PD < PT$, so that the term weathering-limiting is appropriate. Alluvial surfaces have $PD > PT$, and generally $PM \approx 0$ because of the low gradients. Bedrock-floored rills and washes have similar conditions to bare rock slopes except that $PM \approx 0$. The rock-mantled slopes considered in Chapter 9 would appear to be similar to badland regolith-mantled slopes. However, the surface layer is often a lag pavement that greatly restricts runoff erosion, so that the overall rate of erosion is often determined either by (1) the rate of breakdown of the pavement by weathering or (2) the rate of upward migration of fines due to freeze-thaw or bioturbation, or (3) mass-wasting rates.

The supply of moisture is the primary factor determining the weathering rate of the soft rocks forming badlands. For vertically falling precipitation the interception per unit surface area of slope diminishes with the cosine of the slope angle (in actuality some water attacks vertical or overhanging slopes, for escarpment caprocks often project beyond underlying shales). This suggests that a critical gradient separates mantled slopes on which rates of mass wasting increase with slope gradient from steeper bedrock slopes which erode less rapidly as gradients increase (at least until slope relief is great enough to cause bulk failure in the shale bedrock).

Badlands with steep slopes and exposed bedrock commonly develop pinnacle forms (alternatively termed needle-like, serrate, or fluted) (Fig. 10.12). Examples include the Brice Canyon pinnacles, the badlands described by Carman (1958), the spires of Cathedral Valley, Utah, and portions of the badlands of South Dakota. Pinnacle badlands commonly are initiated as a result of near-vertical slopes developing in non-resistant shales lying beneath a resistant caprock which erodes very slowly as compared to downwearing in the surrounding badlands. Eventually the caprock is weathered away or fails, and the underlying shales rapidly erode, developing the fluted form due to rapid rill incision because of the steep relief. Drainage densities are exceedingly high on these slopes and owing to the absence of a mantling regolith and mass wasting processes divides are knife-edged. Slope profiles are generally concave, indicating the dominant role of runoff (both unconcentrated and in rills) in erosion. Although the steepness of the fluted slopes is partially due to the high initial relief, the slopes at divides generally steepen during development of fluted badlands because rill erosion rapidly erodes slope bases and weathering and erosion rates on sideslopes decrease with increasing slope gradient due to less interception of moisture. As the fluted slopes downwaste, they are generally replaced by mantled badland slopes with lower drainage density at a very sharp transition, as a Bryce Canyon.

Fig. 10.17 Runoff on alluvial surfaces on Morrison Formation Shales. Photo was taken during waning flow stages (photo by A. Howard)

Fluvial Processes and Landforms of Badland Landscapes

Howard (1980) distinguished three types of fluvial channels, bedrock, fine-bed alluvial, and coarse-bed alluvial. In most cases these types of channel are separated by clear thresholds in form and process. In badlands coarse-bed alluvial channels usually occur only where gravel interbeds are present.

Alluvial Surfaces

Few contrasts in landscape are as distinct as that between badland and alluvial surfaces on shaly rocks. Low-relief, alluvium-mantled surfaces in badland areas have been referred to by a variety of names, including miniature pediment (Bradley 1940, Schumm 1956b, 1962, Smith 1958), pseudo- and peri-pediment (Hodges 1982), and alluvial surface (Howard 1970, Howard and Kerby 1983). The last term is used here because of its neutral connotation regarding the numerous and conflicting definitions of 'pediment'.

Flow on alluvial surfaces is either unconfined or concentrated in wide, braided washes inset very slightly below the general level of the alluvial surface (Fig. 10.17). Alluvial surfaces may be of any width compared to that of the surrounding badlands, contrasting with the confinement of flow in bedrock rills and washes. The alluvium underlying most alluvial surfaces is bedload carried during runoff events and redeposited as the flow wanes, but adjacent to major washes overbank flood deposits may predominate. Although the surface of alluvial surfaces and washes is very smooth when dry, during runoff events flow is characterized by ephemeral roll waves and shallow chute rilling which heals during recessional flows (Hodges 1982).

An alluvial surface and badlands commonly meet at a sharp, angular discordance (Figs. 10.18 and 10.19). At such junctures slopes, rills and small washes with inclinations up to 45° abut alluvial surfaces with gradients of a few degrees. In most instances the alluvial surfaces receive their drainage from badlands upstream and are lower in gradient than the slopes or washes. The layer of active alluvium beneath an alluvial surface may be as thin as 2 mm, but increases to 10 cm or

Fig. 10.18 Sharp junction between steep badland slopes and alluvial surface in Mancos Shale. Arrows are at the level of Bull Lake terrace. Fremont River flows just in front of the terrace shown by the arrows (photo by A. Howard)

more below larger braided washes. Below the active alluvium there occurs either a thin weathered layer grading to bedrock or, in cases where the alluvial surface has been aggrading, more alluvium.

Generally alluvial surfaces and their alluvial washes obey the same pattern of smaller gradients for larger contributing areas. Howard (1970, 1980) and Howard and Kerby (1983) showed that the gradient of alluvial surfaces is systematically related to the contributing drainage area per unit width (or equivalent length) L in badlands both in arid and humid climates:

$$S = CL^z, \qquad (10.10)$$

where the exponent z may range from -0.25 to -0.3. The equivalent length for washes on alluvial surfaces is simply the drainage area divided by the channel width, but on unconfined flow on smaller alluvial surfaces it is the drainage area contributing to an arbitrary unit width perpendicular to the gradient. The proportionality constant C can be functionally related to areal variations in sediment yield, runoff, and alluvium grain size. Howard (1980) and Howard and Kerby (1983) showed that the value of the exponent z is consistent with the assumption that alluvial surface gradients are adjusted to transport sand-sized bedload at high transport rates.

Badland alluvial surfaces are therefore comparable to sand-bedded alluvial river systems in general in that gradients decrease with increasing contributing area (downstream). The major contrast is the presence of unchannelled areas in the headward portion of alluvial surfaces. Their presence is probably related to the absence of vegetation and flashy flow, which discourages formation of banks and floodplains. The alluvial surfaces contrast with alluvial fans in that the former are through-flowing systems with only minor losses of water downstream and contributions of water and sediment from the entire drainage area. Furthermore, the alluvial surfaces are generally slowly lowering. If they are aggrading, they do so at a very slow rate. Thus the downstream spreading of discharge characteristic of alluvial fans (and deltas) generally does not occur on alluvial surfaces.

Alluvial surfaces are surfaces of transportation, with a gradient determined by the long-term balance between supplied load and discharge (Smith 1958). Thus they are graded surfaces in the sense of Mackin (1948). Howard (1982) and Howard and Kerby (1983) discussed the applicability of this concept in the context of seasonal and long-term changes in the balance of sediment load and discharge, and the limits to the concept of grade. Interestingly, most badland alluvial

Fig. 10.19 Downstream transition between bedrock wash and alluvial surface in Mancos Shale badlands. Slope gradients are about the same whether grading to bedrock wash or to alluvial surface (photo by A. Howard)

surfaces slowly lower through time, as indicated both by direct measurements (Schumm 1962, Howard and Kerby 1983) and by the presence of rocks on pedestals of shale surrounded by an alluvial surface that has lowered around it. Thus the alluvial surfaces are not destroyed by slow lowering of base level, but they become readily dissected by bedrock washes if base level drops too rapidly (Fig. 10.20). Howard (1970) and Howard and Kerby (1983) present evidence that larger alluvial channels are capable of more rapid erosion than smaller ones without being converted to bedrock channels. Howard and Kerby (1983) suggest that the maximum erosion rate is proportional to about the 0.2 power of drainage area. It is uncertain whether the erosion (or aggradation) rate on alluvial surfaces systematically influences their gradients.

Bedrock Channels

Badland washes and rills are usually floored by slightly weathered bedrock. Beneath the smallest rills the weathered zone is about as thick as that on adjacent unrilled slopes, although this thickness varies with the seasonal rill cycle (Schumm 1956a, 1964, Schumm and Lusby 1963). In larger badland washes, weathering products are rapidly removed, exposing bedrock. Deposits of alluvium occur only locally in scour depressions. Larger bedrock washes often display marked meandering with the wavelength increasing systematically with the 0.4 power of drainage area (Howard 1970), which is consistent with meander wavelength in alluvial streams. In contrast to the alluvial surfaces, the gradient of bedrock washes is not uniquely related to the size of contributing drainage area and their gradient is steeper than alluvial washes of equivalent area (Howard 1970, Howard and Kerby 1983).

Slope-Channel Interactions in Badlands

Erosion in large bedrock streams may be nearly independent of the sediment load supplied by slope erosion (e.g., Equations 10.7 and 10.8), so that the nature of the surrounding slopes has minor influence on stream erosion rates. Thus, in general, evolution of hillslopes follows rather than leads that of the stream network. However, the low-order tributaries, including rills, interact with adjacent slopes and their gradients are largely determined by hillslope gradients. The smallest rills are ephemeral (Schumm 1956a, Schumm and Lusby 1963, Howard and Kerby 1983) and have gradients essentially equal to hillslope gradients.

The nature of hillslope-channel interactions is poorly understood. Permanent channels occur only where runoff is sufficient such that streams erode as rapidly as the surrounding slopes and with a lesser gradient. Diffusive processes (creep and rainsplash) are more efficient than channel erosion for small contributing areas. This is suggested by models of bedrock channel erosion (e.g., Equation 10.8) where, for a given erosion rate, gradients would have to approach infinity as contributing area approaches zero. In contrast, mass wasting slope processes and rainsplash function even at divides. Ephemeral rills are generally not inset into the slope because seasonal (or

Fig. 10.20 Badlands and dissected alluvial surface in Mancos Shale. The former boundary between badland slopes and the alluvial surface occurred at the light-dark transition (photo by A. Howard)

year-to-year) mass wasting by needle ice or shallow slips and direct frost heaving episodically destroys them. Permanent rills and washes are distinguished by (and can be defined by) being inset within a definable drainage basin, so that they have a lower gradient than surrounding slopes. Even so, such permanent rills may occasionally be partially infilled by mass-wasting debris.

The sharpness of the junction between badland slopes and adjacent alluvial surface (e.g., Figs. 10.1 and 10.18) as well as its spatio-temporal persistence has fascinated generations of geomorphologists, and a variety of hypotheses have been offered for its origin and maintenance. Badlands rising from alluvial surfaces have gradients nearly equal to nearby slopes terminating at bedrock channels (Fig. 10.19), implying nearly equal rates of erosion, even though the alluvial surfaces may be stable or very slowly degrading. Moreover the slope form (straight-sided or broadly convex) remains the same. In Fig. 10.19 successive profiles across the valley in a downstream direction may be similar to changes in profiles through time at one location, assuming that the alluvial surface remains at a stable level. The alluvial surfaces expand at the expense of the adjacent slope, as measured by Schumm (1962).

Schumm (1962), Smith (1958), Emmett (1970), and Hodges (1982) suggested that the abrupt contact is created and maintained by erosion concentrated at the foot of the slope, due perhaps to spreading of dis-

charge from the rills at the base of the slope (a type of lateral planation), re-emergence of interflow, and changes of flow regime (subcritical to supercritical) at the slope-alluvial surface junction. Engelen (1973) suggested that spreading of water emerging onto alluvial surfaces from ephemeral micro-rills may be as important as spreading flow from more permanent rills. Hodges (1982) provided experimental information and observations on flow on badland slopes, rills, and alluvial surfaces. These data indicate the complicated nature of flow on these surfaces, including ephemeral rilling of the alluvial surface.

Howard (1970) suggested that the abrupt change of gradient at the head of an alluvial surface can be maintained without recourse to special processes acting at this location. Simulation modelling of coupled bedrock and alluvial channel evolution (using an equation similar to Equation (10.8) for bedrock erosion) indicated that bedrock channels maintained a nearly uniform gradient as they downcut until they were abruptly replaced by alluvial surface when elevations dropped to the level that the gradients were just sufficient to transport supplied alluvium (Fig. 10.21, discussed below). Although these simulations were targeted to the bedrock channel system on badlands, a similar situation may pertain to unrilled badland slopes. Water erosion is the dominant erosional process at the base of badland slopes (Schumm and Lusby 1963) and may be governed by similar rate laws as permanent rills and washes. As noted above, most seemingly unrilled

Fig. 10.21 Simulated badland landscape replicating the evolution of the Mancos Shale badlands near Caineville, Utah, based upon Howard (1997). Top boundary is a drainage divide, lateral boundaries are periodic, and bottom boundary is the base level control, conceived to be the Fremont River. The simulation started with initial conditions of a sloping alluvial surface at the level of the Bull Lake surface. Rapid incision then occurred to the modern level of the Fremont River followed by base level stability. The narrow divides near "#" are close to the level of the initial alluvial surface. Note that a new alluvial surface is extending headward from the fixed baselevel as badland slopes erode and retreat. As in Figs. 10.17 and 10.18, steep slopes are maintained as the badland slopes undergo nearly parallel retreat

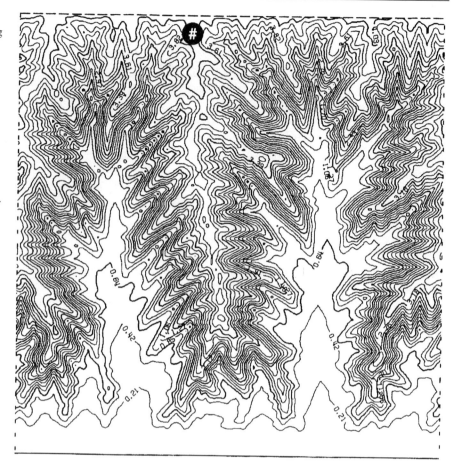

badland slopes are primarily drained by concentrated flow either through an ephemeral surface network of crack flows (which are obliterated during the swelling and reshrinking of the surface layer) or as interflow in deeper cracks and micropipes.

Theoretical Models of Badland Evolution

There has been a rich history of quantitative modelling of slope and channel processes. These models have not primarily been directed towards specific issues related to badland slopes, but they have a general formulation that could be adapted to specific processes and materials in badlands. Early models were primarily applied to evolution of slopes in profile only, either for ease of analytical solutions or for numerical solutions with modest computational demands. A variety of approaches has been used, but two end members can be identified. Some investigators, as exemplified by the approaches of Kirkby (Kirkby 1971, 1976a,b, 1985a,b, Carson and

Kirkby 1972, Kirkby and Bull 2000b) have attempted to quantify almost all of the processes thought to be acting on slopes and their distribution both on the surface and throughout the regolith. This approach offers the promise of detailed understanding of the spatio-temporal evolution of regolith and soils as well as the potential for addressing site-specific issues of erosion processes, slope hydrology, surface water geochemistry, and soil and water pollution. Some attempts have been made to apply such models to cover areal as well as profile characteristics of slopes, including issues of initiation of drainage (Kirkby 1986, 1990). However, the generality of such models restricts their use over large spatial or temporal domains due to present limitations of computer resources and the large number of parameters that must be specified.

Another approach has been to deliberately simplify models in order to address issues of drainage basin morphology and landscape evolution. Ahnert (1976, 1977) has been a pioneer in developing 3-D models of landscape evolution, although most of his efforts have been directed to cross-sectional evolution of slope profiles (Ahnert 1987a,b, 1988). Recent advances in computer capabilities have permitted spatially explicit modeling of drainage basin evolution, and general reviews of these models are presented in Coulthard (2001) and Willgoose (2005). Typically these models include components for representing weathering, diffusive slope evolution by rainsplash and mass wasting, fluvial detachment, and sediment transport and deposition. Landform evolution models have been utilized, for example, to explore controls on drainage density (Howard 1997, Tucker and Bras 1998), effects of different processes of slope and channel coupling (Tucker and Bras 1998), effects of climate change and tectonic forcing (Tucker and Slingerland 1997, Coulthard et al. 2000, Tucker 2004, Gasparini et al. 2007), continental shelf incision during sea level lowstands (Fagherazzi et al. 2004), formation of alluvial fans (Coulthard et al. 2002, Garcia-Castellanos 2002), and incision of gullies (Howard 1999, Kirkby and Bull 2000a, Istanbulluoglu et al. 2005).

Howard (1997) applied a spatially explicit model to explore the evolution of the badland landscape in Mancos Shale discussed above. The model (Howard 1994b, 1997, 2007) incorporates bedrock weathering, diffusive mass wasting including enhanced transport rates as a critical slope steepness is approached (Roering et al. 1999, 2001a,b), shear stress dependent fluvial detachment (Equations 10.7 and 10.8), and sediment transport and deposition in alluvial channels using a bedload transport model. For the badland evolution simulations, weathering was assumed to be able to keep pace with erosion so that a regolith cover is assumed to be present on all slopes and regolith thickness is assumed not to be a limiting factor in determining the erosion rate. In accord with the observation that high relief Mancos Shale slopes are nearly linear in profile and exhibit instability suggestive of rapid mass wasting near the limiting angle of regolith stability (e.g. Fig. 10.13), the badland slopes were modelled such that they were eroding in the region of strong non-linear dependence of erosion rate on gradient. See Howard (1994b, 1997) for model details.

The specific issue examined by Howard (1997) is which history of river incision by the Fremont River best explains the pattern of slopes and alluvial surfaces in the Mancos Shale badlands in the area shown in Fig. 10.11. The existence of broad modern alluvial surfaces grading to the Fremont River coexisting with high-relief badlands created by dissection of an Early Wisconsin alluvial surface was postulated by Howard (1970) to be best explained by fairly rapid downcutting from the Early Wisconsin level of the Fremont river about 60 m above the present level to the level of the modern river followed by subsequent relative stability of base level. This scenario was explored in simulation modelling and compared with an alternative hypothesis that the river level lowered gradually from its Early Wisconsin level. Figure 10.21 shows the result of a simulation with the preferred hypothesis of rapid downcutting followed by stability. In early stages during the rapid downcutting, incision occurs first near the base level control (the lower boundary, corresponding to the Fremont River), with high relief slopes and steep channels gradually extending further inland. Extensive areas of nearly undissected original alluvial surface remained in headwater areas. Following cessation of downcutting, the downstream channel gradients decline until they reach a transport-limited condition. At this point an alluvial surface develops and extends headward, gradually replacing the badland slopes which undergo nearly parallel retreat (Fig. 10.20). The simulation duplicates the expected pattern of remote divides remaining near the level of the Early Wisconsin alluvial surface (e.g., at "#" in Fig. 10.20), high relief badlands occupying most of the simulation domain away from the main drainage (the lower boundary of the simulation domain), and an advancing modern alluvial surface near the main drainage. Simulations with gradual lowering of the river level failed to produce broad modern alluvial surfaces coexisting with high-relief badlands.

Landform evolution modeling of badlands thus has the potential to investigate both the landscape morphology resulting from different sets of process assumptions as well as different evolutionary scenarios.

Gullies

Localized accelerated erosion, including gully erosion, is a recurrent problem in most landscapes with deep regolith or soft bedrock. Gullying is most commonly associated with landscapes with climates near the semi-arid to arid transition or in regions with strongly seasonal precipitation where rapid erosion is triggered by disruption of the vegetation cover by agricultural mismanagement, construction activities, fire, local landslides, or climatic change. However, rilling and gullying can occur even in humid climates with an erodible substrate lacking a coherent vegetation cover. Badlands can be viewed as the extreme end of gullying in which large areas become intensely dissected with sparse vegetation.

At the small end of the size spectrum agricultural fields are commonly plagued by rilling and ephemeral gullies (gullies small enough to be filled in by yearly ploughing, e.g., Poesen et al. (2006)). At the large end of the spectrum are incised river channels (e.g. Schumm et al. 1984, Wang et al. 1997, Darby and Simon 1999). The emphasis here will be in the middle scale, where localized deep incision attacks valley slopes and low-order valleys. General reviews of gullying are provided by Bull and Kirkby (1997), Poesen et al. (2003, 2006), and Valentin et al. (2005), including issues of mitigation and societal impacts not emphasized by this review that emphasizes gullying processes and the role of gullies in landform evolution.

Role of Vegetation

The discussion here will concentrate on the most common circumstance where channelized incision occurs through a vegetated substrate into weak subsurface regolith or sediment. Gully erosion generally initiates when applied fluvial shear stresses exceed the strength of the vegetation cover, and the vegetation is unable to recover between flow events. Thus erosion can occur when the shear resistance of the vegetation is diminished or as a result of increased stress from overland flow due to change in infiltration capacity or topographic changes concentrating flow, and less commonly, to extreme storm events. Once accelerated erosion has been initiated, return to a normal rate of erosion on all parts of the landscape may not occur until specific restorative actions are taken, such as seeding, fertilization, surface stabilization, grade control structures, and land recontouring. Incised valley bottoms are eventually stabilized due to lowered erosion rates and enhanced moisture availability (e.g., Ireland et al. 1939). In some landscapes natural revegetation on gully walls and headcuts occurs when

slope gradients drop below critical values for rapid failure or when a year or two of frequent, low intensity rainstorms encourages vegetation establishment (e.g., Harvey 1992, Alexander et al. 1994). Gully systems considered here develop in low-order channels, headwater hollows, and the lower portions of slopes (e.g., Fig. 10.22).

In most landscapes in humid climates the natural vegetation significantly restricts erosion by runoff due to root cohesion, an open soil structure that encourages infiltration, and a high surface roughness that diminishes runoff velocity (Thornes 1985). The importance of this vegetative cover is made apparent when it is removed during construction, when erosion rates may increase by a factor of several hundred to ten thousand (Guy 1965, 1972, Wolman 1967, Vice et al. 1968, Howard and Kerby 1983). The deep regolith present throughout much of the southeastern United States could only have developed beneath a protective vegetation cover persisting during much of the Cenozoic (Pavich 1989). Accelerated erosion in this region due to poor agricultural practices from colonial times through the beginning of this century transferred immense quantities of sediment from hillslopes to valley bottoms, often resulting in deep gullying and valley sedimentation of a metre or more (Ireland et al. 1939, Happ et al. 1940, Trimble 1974, Jacobson and Coleman 1986).

Vegetation cover is often undervalued in terms of its control over landscape evolution. Its resistance to erosion, however, may be of the same order of magnitude as the underlying bedrock. In the Piedmont region of

Fig. 10.22 A gully system advancing through a broad hollow in the coast ranges of Northern California east of Point Reyes, USA. Some revegetation of the channel bottom is occurring even though the gully headwalls are still advancing (photo by A. Howard)

the southeastern United States, bedrock is commonly exposed in low-order stream channels, whereas deep regolith occurs beneath slopes and divides. If these landscapes approximate steady state denudation, then erosion of the hillslopes beneath a vegetative cover by creep and water erosion occurs only as rapidly as bedrock scour in nearby channels.

A vegetated landscape can thus be viewed as a three-layer structure: a thin vegetation cover, a more erodible regolith, and underlying bedrock. In some cases either the outer or inner layer effectively may be absent as with unvegetated badlands in the case of the former or deep unconsolidated sediments in the case of the latter. Where bedrock is exposed both outer layers are absent. Such a three-layer structure is incorporated in the present model. In some cases it may be necessary to incorporate a more complicated structure. For example, the B-horizon in soils in the southeastern U.S. often has greater erosional resistance than the overlying A-horizon, producing appreciable scarps in the clay-rich layer during gully incision (Ireland et al. 1939). Hardpans, fragipan, calcrete, silcrete, and ferricrete layers may play similar roles in other soils (Poesen and Govers 1990).

Reduction of erosional resistance through disturbance of the vegetation cover, either directly by fire, cultivation, overgrazing, etc., or indirectly by landslides, is a common cause of accelerated erosion (Thornes 1985, Foster 1990). In some cases, the enhanced erosion may instead arise from high stresses imposed by storm runoff, in part due to higher specific runoff yield and reduced hydraulic resistance where vegetation density is diminished (Graf 1979, Bull 1997). In order for erosion to continue to the point of producing a gullied landscape, reestablishment of vegetation must be inhibited; this inhibition is the second requirement for a accelerated erosional regime. Several factors related to the rapid erosion following vegetation disturbance are responsible for this inhibition: physical undermining of vegetation by erosion; removal of the seed and nutrient reservoir of the upper soil layers, rapid drying of unvegetated soil, and high temperatures on sunlit soils, among others. The inhibition of vegetation recovery in areas of high erosion rates is incorporated in some ecological models (Williams et al. 1984, Biot 1990, Collins et al. 2004, Istanbulluoglu and Bras 2005).

Accelerated erosion may also be triggered by rapid entrenchment of master drainage channels, creating migrating knickpoints on tributaries that encroach into headwater hollows and slopes. The entrenchment may be due either to human interference, such as channelization (e.g., Daniels 1960, Daniels and Jordan 1966) or overgrazing, or to climatic change (e.g., Brice 1966, Faulkner et al. 2007); a general review of the controversy surrounding climatic versus anthropogenic control over gullying in the US Southwest is provide d by Cooke and Reeves (1976). Master-channel incision can be caused by vegetation degradation within the channel, for example, in the grass-covered cienega channels of the Southwest US (Melton 1965, Schumm and Hadley 1957, Bull 1997) and the Australian dambo channels (Boast 1990, Prosser and Slade 1994, Prosser et al. 1994, 1995, Rutherford et al. 1997). In this instance gully extension is just a larger example of the simulations presented here. Alternatively, incision may be due to unrelated causes such as climate change or base level change.

Erosional Processes

A variety of processes has been observed to operate at gully headwalls, including fluvial incision, seepage erosion (Howard and McLane 1988), piping (Jones 1971, 1981, Bryan and Yair 1982a, Prosser and Abernethy 1996, Fartifteh and Soeters 1999, Faulkner et al. 2000, 2007, Díaz et al. 2007), plunge-pool erosion (Stein and Julien 1993; Moore et al. 1994; Robinson and Hanson 1994, 1995, Bennett et al. 1997; Hanson et al. 1997, Stein et al. 1997: Bennett 1999, Bennett et al. 2000, Alonso et al. 2002, Bennett and Alonso 2006), and mass-wasting (Bradford et al. 1973, 1978, Blong et al. 1982; Poesen and Govers 1990: Istanbulluoglu et al. 2005; Tucker et al. 2006). General reviews are provided by Gardner (1983), Poesen and Govers (1990), Bocco (1991), Dietrich and Dunne (1993), and Bull and Kirkby (1997). The relative importance of each process varies amongst different geologic, vegetation, and climatic environments, and each process depends somewhat differently upon flow and material properties. Ephemeral gullies commonly erode by knickpoint recession of a plunge pool whereas larger gullies erode by a variety of process including plunge-pool incision, seepage erosion, piping, and mass wasting.

Predicting the initiation and temporal evolution of gully systems is difficult because of the multiplicity of processes and the threshold behaviour of gully incision. The most common approach is diagnostic rather than prognostic, assessing the spatial pattern of established gully systems. The most common approach is to examine the contributing area – slope gradient relationship at either the site of initial gullying or at its maximum extent. This approach was pioneered by Brice (1966), Patton and Schumm (1975) and Begin and Schumm (1979), put on a mechanistic foundation for a range of processes by Dietrich et al., (1992, 1993) and Montgomery and Dietrich (1994), and has been used in a number of studies of gullies (Moore et al. 1988, Vandaele et al. 1996, Vandekerckhove et al. 1998, Desmet et al. 1999, Nachtergaele et al. 2001, Morgan and Mngomezulu 2003, Poesen et al. 2003, Beechie et al. 2007, Gabet and Bookter 2007). The observed relationships between gradient and contributing area at gully heads are generally consistent with control by a shear stress threshold (e.g. Eqn. 10.8).

Gully Models

A variety of models has been used to predict the rate of gully erosion utilizing a wide range of approaches. A representative model is the Ephemeral Gully Erosion Model (EGEM) (Woodward 1999) which uses a detachment capacity proportional to applied shear stress (Equation 10.7) and a net detachment rate decreasing as sediment transport rate approaches capacity (Equation 10.6). A hydrological flow routing model is also utilized. Other theoretical approaches to gully erosion are presented by Kirkby and Bull (2000a), Casalí et al. (2003), Kirkby et al. (2003), Torri and Borselli (2003) and Sidorchuk (2005). Nachtergaele et al. (2001) and Capra et al. (2005) applied the EGEM model to a field example of ephemeral gully erosion, but found that the model had limited predictive capability, presumably due to difficulties in adequately characterizing the values of input parameters required by the model. Poesen et al. (2003, 2006) provides a comprehensive review of gully modeling approaches ranging between empirical and theoretical, concluding that extant theoretical models lack validation and calibration. Empirical models, however, while performing better for specific settings, cannot be extrapolated.

Simulating Gully Evolution

Most of the theoretical models discussed above target the evolution of individual gullies and lack the capability of predicting their areal distribution. Only a few models have attempted to explicitly model the spatial development of gullies, including Howard (1999), Kirkby and Bull (2000a), and Istanbulluoglu et al. (2005). At their present stage of development, such models are exploratory rather than useful predictive tools, but they are useful in understanding the important controls on gully development. A brief summary of the Howard (1999) model is presented here.

The Howard (1999) model assumes that erosion rates due to the dominant processes correlate with shear stress (Equation 10.7) or alternatively with drainage area and gradient (Equation 10.8). Often seepage, piping, mass wasting and plunge-pool erosion at headwalls depend upon relief generated by channel incision occurring downstream from the headwall, and therefore indirectly upon the shear stress or stream power. Where seepage erosion dominates, however, drainage area may not be a good measure of subsurface discharge (Coelho Netto and Fernandes 1990).

When the vegetation is undisturbed the critical shear stress, τ_c (Eqn. 10.7), is assumed to be areally and temporally constant. However, the critical shear stress is assumed to also depend upon the rate of erosion in a threshold manner. This critical shear stress is assumed to have a high value, τ_{cu}, so long as the local erosion rate, E (or $-\frac{\partial y}{\partial t}$), is lower than a critical value, E_u. However, if erosion locally exceeds this rate, the critical shear stress drops to a lower value, τ_{cd}, until the erosion rate drops to a low value E_d, when it is assumed that the vegetation is able to become reestablished and the critical shear stress is reestablished at the undisturbed value, τ_{cu}. In addition, the intrinsic erodibility, K_c, may increase for bare (K_{cd}) versus vegetated land surface (K_{cu}). When the erosion rate is governed by the disturbed critical shear stress, the specific runoff yield is assumed to be greater by a factor R than that of the undisturbed landscape, due to the reduction in infiltration capacity that accompanies vegetation disturbance; enhanced runoff rates from sites of vegetation disturbance or removal has been noted in several studies (Graf 1979; Prosser and Slade 1994; Parsons et al. 1996; Bull 1997). Thus, two erosional states are possible at each location in the landscape, normal and accelerated (Fig. 10.23). As is illustrated in this figure, there may be a range of steady-state erosion rates, E_n, with ($E_d < E_n < E_u$), such that the whole landscape might be in either the normal or the accelerated state depending upon the initial conditions.

Direct experimental measurements on undisturbed grasslands suggests values of critical shear stress of about 100–240 Pascals (Newtons/m^2) for undisturbed grasslands, dropping to about 70 Pascals for heavily disturbed vegetation (Reid 1989; Prosser and Slade 1994; Prosser and Dietrich 1995; Prosser et al. 1995). Bare soils have a critical shear stress of only about 0–40 Pascals (Prosser et al. 1995, Reid 1989, Knapen et al. 2007), suggesting that the ratio of τ_{cu}/τ_{cd} ranges upwards from 3 to 10 or more. Additionally, increase in vegetation density appears to decrease the erodibility coefficient, K_c, in Equation (10.7) (De Baets et al. 2006).

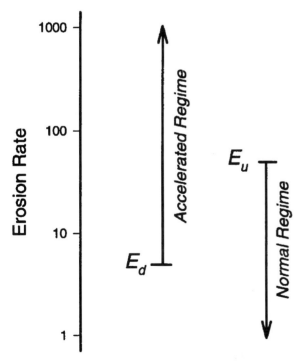

Fig. 10.23 Conceptual diagram of the limits of stability of the normal and accelerated erosional regimes in gullied landscapes. Units for erosion rates are arbitrary and the transition erosion rates E_u and E_d are discussed in the text

The initial conditions for the simulations assume that erosion has been operating in the undisturbed state and that the landscape is in steady state adjustment to an erosion rate E_n ($E_n < E_d < E_u$) with a critical shear stress τ_{cu} (e.g., broadly rounded slopes and shallow hollows in Fig. 10.24). A short-lived disturbance of the vegetation cover is assumed to occur due, perhaps, to fire, cultivation, or overgrazing. The duration of this disturbance, T_d, is assumed to be very short compared with the timescale for development of the steady state landscape, T_e. During the disturbance interval the critical shear stress is assumed to drop to a value, τ_{cx} that is less than τ_{cu} but may be greater or lesser than τ_{cd}. Following this period of disturbance the vegetation is assumed to recover, so that the critical shear stress returns to its undisturbed value, τ_{cu}, except for those portions of the landscape where $E > E_u$. In the disturbed areas fluvial erosion is governed by the parameters values τ_{cd} and K_{td}, until the erosion rate should drop to a value $E < E_d$, whereupon the undisturbed values τ_{cu} and K_{tu} are reestablished.

A short-duration lowering of the critical shear stress for fluvial erosion may be sufficient to trigger accelerated erosion in landscapes with highly erodible regolith sandwiched between a resistant vegetation cover and bedrock. The accelerated erosion may continue over periods many times the duration of the initial disturbance. Incision rates are greatest on steep hillslopes and in hollows and low-order valleys (Fig. 10.24a), whereas divides and upper slopes may revert to normal erosion rates after the short duration of vegetational stress. Erosion in high-order valleys is limited by base level control. The locus of maximum erosion gradually works headward as a shock front and eventually may consume the divide areas, even though the disturbance initiating accelerated erosion might have lasted only a short while. If the vegetation undergoes no additional disturbance, erosion will eventually reduce gradients to the point that erosion rates will revert to the normal state. A different fate may occur if the regolith is thin, because bedrock will be exposed, and reversion to the previous state of thick regolith protected by vegetation would be difficult.

The proportion of the landscape which is triggered into accelerated erosion depends upon: (1) how much the erosion resistance offered by the vegetation cover is temporarily reduced (the ratio τ_{cx}/τ_{cu} in the model); (2) the relative shearing resistance and intrinsic erodibility of the regolith compared to the vegetated surface

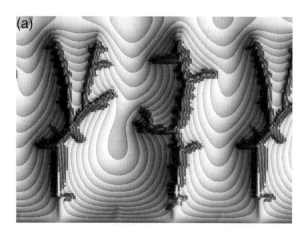

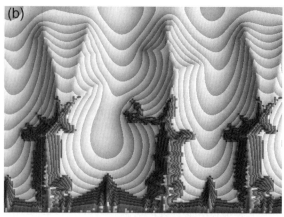

Fig. 10.24 Simulated gullying of vegetated landscape (based on Howard 1999). Boundary conditions as in Fig. 10.21. Smooth, rounded slopes and channelways are vegetated, darker areas with closely spaced contours are gullies. (**a**) Gullies developed as a result of incision through vegetation into underlying easily-eroded regolith or shale. (**b**) Gullies developed as a result of rapid downcutting of the master drainage along the lower boundary

(τ_{cd}/τ_{cu} and K_{td}/K_{tu}); (3) the rate of induced erosion which assures that the vegetation cannot become reestablished by recovery of the normal vegetation (E_u); (4) the erosion rate at which vegetation recovery is assured following the cessation of disturbance (E_d); and 5) the degree to which specific runoff is enhanced in areas with disturbed vegetation (R).

The model also illustrates the two major approaches to controlling accelerated erosion: (1) enhancement of the vegetation cover growth potential to reduce E_d, reduce runoff (R), and increase E_u; and (2) protection of the surface to increase erosional resistance (increase τ_c). Because steep, rapidly-eroding erosional fronts are created by the initial disturbance, uplands are progressively consumed by the advancing headwall even though the vegetation cover may be fully recovered. Thus structural treatments in the gullies (grade control structures, bank stabilization, etc.) are commonly required to halt their advance.

The model simulations clearly suggest that, once initiated, gullying in thick, erodible regolith or weak rocks may continue until all slopes are consumed by the wave of dissection if there is no intervention. In some environments, such unconstrained gullying and badland development may gradually consume the undisturbed areas. One well-known example is the badlands of South Dakota, which originated due to downcutting by the White River through Cenozoic shale and mudstone. These badlands have gradually advanced over thousands of years as a steep erosional front consuming a rolling, grassy upland (Fig. 10.25). Dissection of the loess terrain of China has continued over a similar timespan (Zhu Xianmo 1986). Gradual encroachment of gullies and badlands into vegetated slopes has continued over hundreds of years in Spain and Italy. In light of such examples, Bocco (1991) and Hudson (1985) suggest that gullying is not self-healing. On the other hand, inactive, naturally revegetated gullies can be found in many environments, such as the coastal ranges of California (Fig. 10.26). Graf (1977) suggests that the rate of gully extension follows a negative exponential function, decreasing to zero as the gully approaches an equilibrium length, and Graf (1977), Prosser and Abernethy (1996), and Rutherford et al. (1997) present supporting data. This equilibrium length occurs when contributing area and slope drop below threshold values (e.g., Equation 10.8).

Gullying can also be initiated by a rapid incision of the master drainage, resulting in headward knickpoint migration (Fig. 10.24b), resulting in a pattern of incision radiating from the master drainage channels.

Gullying clearly ceases in many landscapes before the entire landscape is consumed. An example is shown in Fig. 10.26. The steep, grassy landscape in the northern California Coast Ranges shown in Fig. 10.26a has been dissected by two gullies occupying colluvial hollows. The gullying may have been initiated by vege-

Fig. 10.25 An advancing erosional front in the Badlands of South Dakota, USA. Remnants of the gently-rolling grassy upland surface are visible. The gullying was triggered by late Pleistocene downcutting of the master drainage, possibly aided by a drier Holocene climate that has reduced vegetation cover. As the upland is consumed by the narrow zone of advancing bad- land slopes, an alluvial surface develops and extends headward. The alluvial surface in turn becomes vegetated and stabilized. In some locations additional master stream downcutting has caused dissection of the alluvial surfaces as a second erosional front. (photo by A. Howard)

tation disturbance or small landslides in the colluvial fill. The left gully in Fig. 10.26a is undergoing active extension (Fig. 10.26b), whereas the right gully (which is probably older) has become stabilized by vegetation (Fig. 10.26c). The factors leading to the restabilization are uncertain, but may include diminished incision rates when incision of the main channel reached bedrock, the greater moisture supply in the gully floor encouraging vegetation regrowth, and wet seasons with abundant moisture but no strong storms.

Several circumstances can encourage gully stabilization. Headward erosion of gullies in the landscape studied by Prosser and Abernethy (1996) cease when the gullies incised into alluvium reach the base of the hillslopes, which are shallowly underlain by bedrock. Exposure of bedrock may prevent plunge-pool erosion and enhance revegetation by reducing erosion rates in valley bottoms and encouraging seepage discharge along gully walls.

The simple Howard (1999) model assumes spatially and temporally invariant erosion rate thresholds and spatially invariant critical shear stress. Spatial variation may occur in both erosion thresholds and critical shear stress. In semi-arid landscapes vegetation density and growth rates may be greatest in hollows and valley bottoms. As a result, the critical shear stress, (τ_c), may be higher in valleys as might be the erosion rate, E_u required to initiate gully development. In some circumstances, gullies might develop on convex hillslopes while hollows and valleys remain stable despite their greater source drainage area. Gully incision also may increase the delivery of moisture to headwalls by groundwater seepage or baseflow, thus encouraging revegetation due to a higher erosion rate reversion to normal condition, E_d, not accounted for in the present model. This erosion rate transition, E_d, may also be time-dependent, increasing during wet periods. Revegetation can also be encouraged during periods of low erosion rates in intervals lacking major storm runoff. Finally, supply of subsurface water to headwalls extended by groundwater seepage may drop to zero as the channels become less incised upon reaching the upper portions of hollows and sideslopes.

A more elaborate treatment of precipitation forcing and vegetative growth is presented by Istanbulluoglu et al. (2005) and Tucker et al. (2006) involving simulation of individual precipitation events and vegetative growth and disturbance dynamics. High intensity rains encourage incision by increasing the applied shear stress (high values of K_c in Equation 10.7). Frequent low intensity events encourage vegetation growth by increasing τ_c, E_u, and E_d, whereas droughts would have the opposite effect.

Some landscapes may experience natural epicycles of gully erosion without external forcing, possibly analogous to the cycles of colluvial infilling and landslide excavation in hollows in steep landscapes of the Pacific Rim (Dietrich and Dunne 1978, Reneau and Dietrich 1991, Benda and Dunne 1997). In fact, the partial excavation of hollows by landslides can trigger further incision by gully extension. Such a mechanism might have triggered the gullies shown in Figs. 10.22 and 10.26. Landsliding may not be a necessary trigger, however, because colluvial infilling of hollows may provide a sufficient imbalance. Colluvial filling decreases the concavity of hollows, which increases gradients at the base of the hollow where contributing

drainage area is the greatest, so that shear stresses are increased. In addition, the decreased concavity may reduce available moisture and thus decrease the vegetation density.

A similar approach can be utilized to predict entrenchment in larger vegetated streams or unchanneled valleys, such as in cienegas (Leopold and Miller 1956, Melton 1965, Bull 1997) and dambos (Boast 1990, Prosser and Slade 1994, Prosser et al. 1994). If channels have modest maximum flow intensities and a high frequency of low flows, vegetation (often grasses and sedges) may cover the channel bed. This reduces shear stresses on the bed and encourages sedimentation. However, as with the gullying discussed above, rapid incision into the vegetation cover can occur due to either high flow intensities or reduced vegetation density following droughts. This leads to a tendency for valley

Fig. 10.26 Stable and active gullies in the coastal hills north of San Francisco, California, USA. (**a**) The left gully is actively extending headwards, whereas the right gully has become stablilized. Both gully systems occupy colluvial hollows. (**b**) View looking across the head of the left gully in (**a**). Note the abrupt headwall, the steep slopes undergoing rapid mass wasting, and the high drainage density on the headwall. Headwall retreat occurs both by runoff and shallow slumping as well as deeper rotational slumps. (**c**) View across the right gully in (**a**), showing the high vegetation density in the hollow, which has inhibited downcutting and promoted revegetation of the headwalls. Some localized headwall retreat is still occurring, as below and to the left of the figures. The causes for the contrast between the two gully systems is discussed in the text. (photos by A. Howard)

bottoms to alternate between periods of aggradation when flows are broad and shallow through abundant vegetation versus cycles of entrenchment in narrow, unvegetated incised channels. Autogenic cycles of incision and aggradation influenced by vegetation may alternate not only temporally, but spatially as well in the guise of discontinuous gullies (Schumm and Hadley 1957, Leopold and Miller 1956, Blong 1970, Bull 1997).

Conclusions

Despite the apparent morphologic simplicity of badland and gully systems and the similarity in features over a wide range of environments and lithology, the formative processes are surprisingly diverse and remain difficult to quantify. For example, runoff and erosion of badland slopes involves a complex combination of runoff, infiltration through cracks, flow in macropores, clay swelling and dispersion, chemical leaching, and a strong dependence of these processes on precipitation history. Similarly, erosion of gully heads may occur through vegetation disturbance, plunge pool scour, mass wasting, seepage erosion, and piping. Further understanding of badlands will require studies conducted at a range of temporal and spatial scales. Approaches include unraveling the erosional history, runoff plot experiments conducted either in the field regolith or laboratory, study of weathering processes and regolith structure, and development of mathematical models. Badlands and gully systems offer a unique opportunity for development and testing of quantitative landform models, because we not only have the landform morphology to compare with theoretical models, but processes are rapid enough that rates of landform change can be measured with reasonable accuracy over periods of just a few years.

References

Ahnert, F. 1976. Brief description of a comprehensive three-dimensional process-response model of landform development. *Zeitschrift für Geomorphologie Supplement Band* **25**, 29–49.

Ahnert, F. 1977. Some comments on the quantitative formulation of geomorphological processes in a theoretical model. *Earth Surface Processes* **2**, 191–202.

Ahnert, F. 1987a. Approaches to dynamic equilibrium in theoretical simulations of slope development. *Earth Surface Processes and Landforms* **12**, 3–15.

Ahnert, F. 1987b. Process-response models of denudation at different spatial scales. *Catena Supplement* **10**, 31–50.

Ahnert, F. 1988. Modelling landform change. In *Modelling Geomorphological Processes*, M.G. Anderson (Ed.), 375–400. Chichester: John Wiley & Sons.

Akky, M.R. and Shen, C. K. 1973. Erodibility of a cement-stabilized sandy soil. In *Soil erosion: causes and mechanisms*. U.S. Highway Research Board Special Report 135, 30–41.

Alexander, R., Harvey, A. M., Calvo, A., James, P. A. and Cerda, A. 1994, Natural stabilization mechanisms on badland slopes, Tibernas, Almeria, Spain. In *Environmental Change in Drylands: Biogeographical and Geomorphological Perspectives*, Millington, A.C. and Pye, K. (Eds), 85–111, Wiley, Chichester.

Alonso, C., Bennett, S.J. and Stein, O.R. 2002. Predicting head cut erosion and migration in concentrated flows typical of upland areas. *Water Resources Research* **38**, doi:10.1029/2001WR001173.

Anderson, R. S., Repka, J. L. and Dick, G. S., 1996. Explicit treatment of inheritance in dating depositional surfaces using in situ ^{10}Be and ^{26}Al, *Geology* **24**, 47–51.

Ariathurai, R. and Arulandan, K. 1986. Erosion rates of cohesive soils. *Journal of the Hydraulics Division, American Society of Civil Engineers* **104**, 279–298.

Battaglia, S., Leoni, L. and Sartori, F. 2002. Mineralogical and grain size composition of clays developing calanchi and biancane erosional landforms. *Geomorphology* **49**, 153–170.

Beechie, T.J., Pollock, M.M. and Baker, S. 2007. Channel incision, evolution and potential recovery in the Walla Walla and Tucannon River basins, northwestern USA. *Earth Surface Processes and Landforms*, doi: 10.1002/esp.1578.

Begin, Z.B. and Schumm, S.A. 1979. Instability of alluvial valley floors: A method for its assessment, *Transactions of the American Society of Agricultural Engineers* **22**, 347–350.

Benda, L. and Dunne, T. 1997. Stochastic forcing of sediment supply to channel networks from landsliding and debris flow, *Water Resources Research* **33**, 2849–2863.

Bennett, S.J. 1999. Effect of slope on the growth and migration of headcuts in rills. *Geomorphology* **30**, 273–290.

Bennett, S.J., Alonso, C., Prasad, S.N. and Römkens, M.J.M. 2000. A morphological study of headcut growth and migration in upland concentrated flows. *Water Resources Research* **36**, 1911–1922.

Bennett, S.J. and Alonso, C.V. 2006. Turbulent flow and bed pressure within headcut scour holes due to plane reattached jets. *Journal of Hydraulic Research* **44**, 510–521.

Bennett, S.J., Alonso, C.V., Prasad, S.N. and Römkens, M.J.M. 1997. Dynamics of head-cuts in upland concentrated flows. In *Management of landscapes disturbed by channel incision*, Wang, S.Y., Langendoen, E.J. and Shields, F.D. Jr. (Eds.), 510–515, University of Mississippi, Oxford, Miss.

Biot, Y. 1990. THEPROM – an erosion productivity model. In *Soil erosion on agricultural land*, Boardman, J., Foster, I.D.L. and Dearing, J.A. (Eds), 465–479, John Wiley & Sons, Chichester.

Blong, R.J. 1970. The development of discontinuous gullies in a pumice catchment. *American Journal of Science* **268**, 369–384.

Blong, R.J., Graham, O.P. and Veness, J.A. 1982. The role of sidewall processes in gully development; some N.S.W. examples. *Earth Surface Processes and Landforms* **7**, 381–385.

Boast, R. 1990. Dambos: A review. *Progress in Physical Geography* **14**, 153–177.

Bocco, G. 1991. Gully erosion: processes and models. *Progress in Physical Geography* **15**, 392–406.

Bouma, N.A. and Imeson, A.C. 2000. Investigation of relationships between measured field indicators and erosion processes on badland surfaces at Petrer, Spain. *Catena* **40**, 141–171.

Bowyer-Bower, T.A.S. and Bryan, R. B. 1986. Rill initiation: concepts and evaluation on badland slopes. *Zeitschrift für Geomorphologie Supplement Band* **59**, 161–175.

Bradford, J.M., Farrell, D.A. and Larson, W.W. 1973. Mathematical evaluation of factors affecting gully stability. *Soil Science Society American Proc.*, **37**, 103–107.

Bradford, J.M., Piest, R.F. and Spomer, R.G. 1978. Failure sequence of gully headwalls in western Iowa. *Soil Science Society of America Journal* **42**, 323–238.

Bradley, W.H. 1940. Pediments and pedestals in miniature. *Journal of Geomorphology* **3**, 244–554.

Brice, J.C. 1966. Erosion and deposition in the loess-mantled Great Plains, Medicine Creek drainage basin, Nebraska. *U. S. Geological Survey Professional Paper 352H*, 255–335.

Bryan, R.B. 1987. Processes and significance of rill development. *Catena Supplement* **8**, 1–15.

Bryan, R.B., Campbell, I.A. and Yair, A. 1987. Postglacial geomorphic development of the Dinosaur Provincial Park badlands, Alberta. *Canadian Journal of Earth Sciences* **24**, 135–146.

Bryan, R.B., Imeson A.C. and Campbell, I.A. 1984. Solute release and sediment entrainment on microcatchments in the Dinosaur Park badlands, Alberta Canada. *Journal of Hydrology* **71**, 79–106.

Bryan, R. and Yair, A. 1982a. Perspectives on studies of badland geomorphology. In *Badland Geomorphology and Piping*, Bryan, R. B. and Yair, A. (eds), 1–13. Norwich: Geo Books, Norwich, England.

Bryan, R. and Yair, A. (eds). 1982b. *Badland geomorphology and piping*. Norwich: Geo Books.

Bryan, R.B., Yair, A. and Hodges, W. K. 1978. Factors controlling the initiation of runoff and piping in Dinosaur Provincial Park badlands, Alberta, Canada. *Zeitschrift für Geomorphologie Supplement Band* **34**, 48–62.

Bull, L.J. and Kirkby, M.J. 1997. Gully processes and modelling. *Progress in Physical Geography* **21**, 354–374.

Bull, W.B. 1997. Discontinuous ephemeral streams. *Geomorphology* **19**, 227–276.

Campbell, I.A. 1989. Badlands and badland gullies. In *Arid Zone Geomorphology*, Thomas, D.S.G. (ed), 159–193. New York: Halstead Press.

Campbell, I.A. and Honsaker, J. L. 1982. Variability in badlands erosion; problems of scale and threshold identification. In *Space and Time in Geomorphology*, Thorn, C. E. (ed), 59–79. London: George Allen & Unwin.

Capra, A., Mazzara, L.M. and Scicolone, B. 2005. Application of the EGEM model to predict ephemeral gully erosion in Sicily, Italy. *Catena* **59**, 133–146.

Carman, M.F., Jr. 1958. Formation of badland topography. *Geological Society of America Bulletin* **69**, 789–790.

Carson, M.A. 1969. Models of hillslope development under mass failure. *Geographical Analysis* **1**, 76–100.

Carson, M.A. 1971. An application of the concept of threshold slopes to the Laramie Mountains, Wyoming. *Institute of British Geographers Special Publication* **3**, 31–47.

Carson, M.A. and Kirkby, M. J. 1972. *Hillslope form and Process*, 475pp. Cambridge: Cambridge University Press.

Carson, M.A. and Petley, D. J. 1970. The existence of threshold slopes in the denudation of the landscape. *Transactions of the Institute of British Geographers* **49**, 71–95.

Casalí, J., López, J.J. and Giráldez, J.V. 2003. A process-based model for channel degradation: application to ephemeral gully erosion. *Catena* **50**, 435–447.

Cerdà, A. and García Fayos, P. 1997. The influence of slope angle on sediment, water and seed losses on badland landscapes. *Geomorphology* **18**. 77–90.

Chisci, G., Sfalanga, M. and Torri, D. 1985. An experimental model for evaluating soil erosion on a single-rainstorm basis. In *Soil Erosion and Conservation*, Swaify, S. A., Moldenhauer, W. C. and Lo, A. (eds), 558–565. Ankeny, Iowa: Soil Conservation Society of America.

Churchill, R.R. 1981. Aspect-related differences in badlands slope morphology. *Annals of the Association of American Geographers* **71**, 374–388.

Coelho Netto, A.L. and Fernandes, N.F. 1990. Hillslope erosion, sedimentation, and relief inversion in SW Brazil: Bananal, SP. *International Association of Scientific Hydrology Publication* **192**, 174–182.

Collins, D.B.G., Bras, R.L. and Tucker, G.E. 2004. Modeling the effects of vegetation-erosion coupling on landscape evolution. *Journal of Geophysical Research* **109**, F03004, doi:10.1029/2003JF000028.

Cooke, R.U. and Reeves, W.R.1976. *Arroyos and Environmental Change in the American South-West*, Clarendon Press, Oxford.

Coulthard, T.J. 2001. Landscape evolution models: a software review. *Hydrological Processes* **15**, 165–173.

Coulthard, T.J., Kirkby, M.J. and Macklin, M.G. 2000. Modelling geomorphic response to environmental change in an upland catchment. *Hydrological Processes* **14**, 2031–2045.

Coulthard, T.J., Macklin, M.G. and Kirkby, M.J. 2002. A cellular model of Holocene upland river basin and alluvial fan evolution. *Earth Surface Processes and Landforms* **27**, 269–288.

Culling, W.E.H. 1963. Soil creep and the development of hillside slopes. *Journal of Geology* **71**, 127–161.

Daniels, R.B. 1960. Entrenchment of the Willow Drainage Ditch, Harrison County, Iowa. *American Journal of Science* **258**, 161–176.

Daniels, R.B. and Jordan, R.H. 1966. Physiographic history and the soils, entrenched stream systems and gullies, Harrison County, Iowa. *U.S. Department of Agriculture Technical Bulletin* **1348.**

Darby, S.E. and Simon, A. (eds). 1999. *Incised River Channels: Processes, forms, Engineering and Management*. Wiley, Chichester, 442pp.

Davis, W.M. 1892. The convex profile of bad-land divides. *Science* **20**, 245.

De Baets, S., Poesen, J., Gyssels, G. and Knapen, A. 2006. Effects of grass roots on the erodibility of topsoils during concentrated flow. *Geomorphology* **76**, 54–67.

Desir, G. and Marín, C. 2006. Factors controlling the erosion rates in a semi-arid zone (Bardenas Reales, NE Spain). *Catena*, doi:10.1016/j.catena.2006.10.004.

Desmet, P.J.J., Poesen, J., Govers, G. and Vandaele, K. 1999. Importance of slope gradient and contributing area for optimal prediction of the initation and trajectory of ephemeral gullies. *Catena* 37, 377–392.

Díaz, A.R., Sanleandro, P.M., Soriano, A.S., Serrato, F.B. and Faulkner, H. 2007. The causes of piping in a set of abandoned agricultural terraces in southeast Spain. *Catena* 69, 282–293.

Dietrich, W.E. and Dunne, T. 1978. Sediment budget for a small catchment in mountainous terrain. *Zeitschrift für Geomorphology Supplement* 29, 191–206.

Dietrich, W.F. and Dunne, T. 1993. The channel head, In *Channel network hydrology*, Beven, K. and Kirkby, M.J. (eds), 175–219, Wiley, Chichester.

Dietrich, W.F., Wilson, C.J, Montgomery, D.R., McKean, J. and Bauer, R. 1992. Erosion threshold and land surface morphology. *Geology* 20, 675–679.

Dietrich, W.E., Wilson, C.J., Montgomery, D.R., McKean, J. 1993. Analysis of erosion thresholds, channel networks, and landscape morphology using a digital terrain model, *Journal of Geology* 101. 259–278.

Dunne, T. and Aubrey, B.F. 1986. Evaluation of Horton's theory of sheetwash and rill erosion on the basis of field experiments. In *Hillslope Processes*, Abrahams, A. D. (ed), 31–53. Boston: Allen & Unwin.

Emmett, W.W. 1970. The hydraulics of overland flow on hillslopes. *U.S. Geological Survey Professional Paper* 662-A.

Engelen, G.B. 1973. Runoff processes and slope development in Badlands National Monument, South Dakota. *Journal of Hydrology* 18, 55–79.

Everaert, W. 1991. Empirical relations for the sediment transport capacity of interrill flow. *Earth Surface Processes and Landforms* 16, 513–532.

Fagherazzi, S., Howard, A.D. and Wiberg, P.L. 2004. Modeling fluvial erosion and depositon on continental shelves during sea level cycles. *Journal of Geophysical Research* 109, F03010, doi:10.1029/2003JF000091.

Fartifteh, J. and Soeters, R. 1999. Factors underlying piping in the Basilicata region, southern Italy. *Geomorphology* 26, 239–251.

Faulkner, H., Alexander, R. and Wilson, B.R. 2003. Changes to the dispersive characteristics of soils along an evolutionary slope sequence in the Vera badlands, southeast Spain: implications for site stabilisation. *Catena* 50, 243–254.

Faulkner, H., Alexander, R. and Zukowskyj, P. 2007. Slope-channel coupling between pipes, gullies and tributary channels in the Mocatán catchment badlands, Southeast Spain. *Earth Surface Processes and Landforms*, doi: 10.1002/esp. 1610.

Faulkner, H., Sípivey, D. and Alexander, R. 2000. The role of some site geochemical processes in the development and stabilisation of three badland sites in Almería, Southern Spain. *Geomorphology* 35, 87–99.

Fernandes, N.F. and Dietrich, W.E. 1997. Hillslope evolution by diffusive processes: the timescale for equilibrium adjustments, *Water Resources Research* 33, 1307–1318.

Finlayson, B.L., Gerits, J. and van Wesemael, B. 1987. Crusted microtopography on badland slopes in southeast Spain. *Catena* 14, 131–144.

Foster, G.R. 1982. Modeling the erosion process. In *Hydrologic Modeling of Small Watersheds*, C.T. Hahn, H.P. Johnson and Brakensiek, D. L. (eds), 297–382. St. Joseph, Michigan: American Society of Agricultural Engineers.

Foster, G.R. 1990. Process-based modelling of soil erosion by water on agricultural land. In *Soil Erosion on Agricultural Land*, Boardman, J., Foster, I. D. L. and Dearing, J. A. (eds), 429–445. Chichester: John Wiley & Sons.

Foster, G.R. and Lane, L.J. 1983. Erosion by concentrated flow in farm fields. In *Proceedings of the D.B. Simons Symposium on Erosion and Sedimentation, 9.65–9.82*. Fort Collins, Colorado: Colorado State University.

Foster, G.R. and Meyer, L.D. 1972. A closed-form soil erosion equation for upland areas. In *Sedimentation (Einstein)*, Shen, H.W. (ed), 12,1–12,19. Fort Collins, Colorado: Colorado State University.

Gabet, E.J. and Bookter, A. 2007. A morphometric analysis of gullies scoured by post-fire progressively bulked debris flows in southwest Montana, USA. *Geomorphology*, doi:10.1016/j.geomorph.2007.03.016.

Garcia-Castellanos, D. 2002. Interplay between lithospheric flixure and river transport in foreland basins. *Basin Research* 14, 89–104.

Gardner, T.W. 1983. Experimental study of knickpoint migration and longitudinal profile evolution in cohesive, homogeneous material. *Geological Society of America Bulletin* 94, 664–672.

Gasparini, N.M., Whipple, K.X. and Bras, R.L. 2007. Predictions of steady state and transient landscape morphology using sediment-flux-dependent river incision models. *Journal of Geophysical Research* 112, F03S09, doi:10.1029/2006JF000567.

Gerits, J., Imeson, A.C., Verstraten, J.M. and Bryan, R. B. 1987. Rill development and badland regolith properties. *Catena Supplement* 8, 141–160.

Gilbert, G.K. 1880. *Report on the Geology of the Henry Mountains*. Washington: U.S. Geographical and Geological Survey of the Rocky Mountain Region.

Gilbert, G.K. 1909. The convexity of hilltops. *Journal of Geology* 17, 344–351.

Gilley, J.E., Woolhiser, D.A. and McWhorter, D. B. 1985. Interrill soil erosion – Part I: Development of model equations. *Transactions of the American Society of Agricultural Engineers* 28, 147–153, 159.

Govers, G. 1985. Selectivity and transport capacity of thin flows in relation to rill erosion. *Catena* 12, 35–49.

Govers, G. and Rauws, G. 1986. Transporting capacity of overland flow on plane and on irregular beds. *Earth Surface Processes and Landforms* 11, 515–524.

Graf, W.L. 1977. The rate law in fluvial geomorphology. *American Journal of Science* 277, 178–191.

Graf, W.L. 1979. The development of montane arroyos and gullies. *Earth Surface Processes* 4, 1–14.

Guy, H.P. 1965. Residential construction and sedimentation at Kensington, Maryland. In *Federal Inter-Agency Sedimentation Conference Proceedings 1963*, U.S. Department of Agriculture Miscellaneous Publication 970, 30–37.

Guy, H.P. 1972. Urban sedimentation – in perspective. *Proceedings of the American Society of Civil Engineers, Journal of the Hydraulics Division* 98, 2009–2016.

Hanson, G.J., Robinson, K.M. and Cook, K.R.1997. Experimental flume study of headcut migration. In *Management of Landscapes Disturbed by Channel Incision*, Wang, S.Y., Langendoen, E.J. and Shields, F.D. Jr. (eds.), 503–509, Oxford, Miss: University of Mississippi.

Happ, S.C. Rittenhouse, G. and Dobson, G.C. 1940. Some principles of accelerated stream and valley sedimentation. *U.S. Department of Agriculture Technical Bulletin* **695**, 133pp.

Harvey, A. 1982. The role of piping in the development of badlands and gully systems in south-east Spain. In *Badland Geomorphology and Piping*, Bryan, R. and Yair, A. (eds), 317–335. Norwich: Geo Books.

Harvey, A.M. 1992. Process interactions, temporal scales and the development of hillslope gully systems: Howgill Fells, northwest England, *Geomorphology*, **5**, 323–344.

Hirano, M. 1975. Simulation of developmental process of interfluvial slopes with reference to graded form. *Journal of Geology* **83**, 113–123.

Hodges, W.K. 1982. Hydraulic characteristics of a badland psuedo-pediment slope system during simulated rainstorm experiments. In *Badland Geomorphology and Piping*, Bryan, R. and Yair, A. (eds), 127–151. Norwich: Geo Books.

Hodges, W.K. and Bryan, R.B. 1982. The influence of material behavior on runoff initiation in the Dinosaur Badlands, Canada. In *Badland Geomorphology and Piping*, Bryan, R. and Yair, A. (eds), 13–46. Norwich: Geo Books.

Howard, A. D. 1970. A study of process and history in desert landforms near the Henry Mountains, Utah. Unpublished PhD dissertation, 198pp. Baltimore: Johns Hopkins University.

Howard, A.D. 1980. Thresholds in river regime. In *Thresholds in Geomorphology*, Coates, D. R. and Vitek, J. D. (eds), 227–258. London: George Allen & Unwin.

Howard, A.D. 1982. Equilibrium and time scales in geomorphology: application to sand-bed alluvial channels. *Earth Surface Processes and Landforms* **7**, 303–325.

Howard, A. D. 1994a. Badlands. In *Geomorphology of Desert Environments*, Abrahams, A. D. and Parsons, A. J. (eds), 213–242. London: Chapman & Hall.

Howard, A. D. 1994b. A detachment-limited model of drainage basin evolution. *Water Resources Research* **30**, 2261–2285.

Howard, A.D. 1997. Badland morphology and evolution: Interpretation using a simulation model. *Earth Surface Processes and Landforms* **22**, 211–227.

Howard, A.D. 1998. Long profile development of bedrock channels: Interaction of weathering, mass wasting, bed erosion and sediment transport. *American Geophysical Union Geophysical Monograph* **107**, 297–319.

Howard, A.D. 1999. Simulation of gully erosion and bistable landforms. In *Incised Channels*, Darby, S. and A. Simon, A. (eds), *Incised Channels*, 277–300. New York: Wiley.

Howard, A.D. 2007. Simulating the development of martian highland landscapes through the interaction of impact cratering, fluvial erosion, and variable hydrologic forcing. *Geomorphology* **91**, 332–363.

Howard, A.D. and Kerby, G. 1983. Channel changes in badlands. *Geological Society of America Bulletin* **94**, 739–752.

Howard, A.D. and McLane, C.F. 1988. Erosion of cohesionless sediment by groundwater seepage. *Water Resources Research*, **24**, 1659–1674.

Hudson, N.W. 1985. *Soil Conservation*, London: Batsford.

Hunt, C.B. 1953. Geology and Geography of the Henry Mountains Region. *U. S. Geological Survey Professional Paper* 228.

Imeson, A.C. and Verstraten, J. M. 1985. The erodibility of highly calcareous soil material from southern Spain. *Catena* **12**, 291–306.

Imeson, A.C. and Verstraten, J. M. 1988. Rills on badland slopes: a physico-chemically controlled phenomenon. *Catena Supplement* **12**, 139–150.

Imeson, A.C., Kwaad, F.J.P.M. and Verstraten, J.M. 1982. The relationship of soil physical and chemical properties to the development of badlands in Morocco. In *Badland Geomorphology and Piping*, Bryan, R. and Yair, A. (eds), 47–69. Norwich: Geo Books, Norwich.

Ireland, H.A., Sharpe, C.F.S. and Eargle, D.H. 1939. Principles of gully erosion in the Piedmont of South Carolina. *U.S. Department of Agriculture Technical Bulletin*, **633**, 143pp.

Istanbulluoglu, E. and Bras, R.L. 2005. Vegetation-modulated landscape evolution: Effects of vegetation on landscape processes, drainage density, and topography. *Journal of Geophysical Research* **110**, F02012, doi:10.1029/2004JF000249).

Istanbulluoglu, E., Bras, R.L. and Flores-Cervantes, H. 2005. Implications of bank failures and fluvial erosion for gully development: Field observations and modeling. *Journal of Geophysical Research* **110**, F01014, doi:10.1029/2004JF000145.

Jacobson, R.B. and Coleman, D.J. 1986. Stratigraphy and recent evolution of Maryland Piedmont flood plains *American Journal of Science*, **268**, 613–637.

Jones, J.A.A. 1971. Soil piping and stream channel initiation. *Water Resources Research* **7**, 602–610.

Jones, J.A.A. 1981. *The Nature of Soil Piping – A Review of Research*, British Geomorphology Research Group Monograph **3**, Geobooks, Norwich.

Jones, J.A.A. 1990. Piping effects in drylands. *Geological Society of America Special Paper* **252**, 111–138.

Julian, P.Y. and Simons, D. B. 1985. Sediment transport capacity of overland flow. *Transactions of the American Society of Agricultural Engineers* **28**, 755–762.

Karcz, I. and Kersey, D. 1980. Experimental study of free-surface flow instability and bedforms in shallow flows. *Sedimentary Geology* **27**, 263–300.

Kasanin-Grubin, M. and Bryan, R. 2007. Lithologic properties and weathering response on badland hillslopes. *Catena* **70**, 68–78.

Kinnell, P.I.A. 1990. Modelling erosion by rain-impacted flow. *Catena Supplement* **17**, 55–66.

Kinnell, P.I.A. 1991. The effect of flow depth on sediment transport induced by raindrops impacting shallow flows. *American Society of Agricultural Engineers Transactions* **34**, 161–168.

Kirkby, M.J. 1971. Hillslope process-response models based on the continuity equation. *Institute of British Geographers Special Publication* **3**, 15–30.

Kirkby, M.J. 1976a. Soil development models as a component of slope models. *Earth Surface Processes* **2**, 203–230.

Kirkby, M.J. 1976b. Deterministic continuous slope models. *Zeitschrift für Geomorphology Supplementband* **25**, 1–19.

Kirkby, M.J. 1980a. Modelling water erosion processes. In *Soil Erosion*, Kirkby, M.J. and Morgan, R.P.C. (eds), 183–216. Chichester: John Wiley & Sons.

Kirkby, M. J. 1980b. The stream head as a significant geomorphic threshold. In *Thresholds in Geomorphology*, Coates, D.R. and Vitek, J.D. (eds), 53–73. London: George Allen & Unwin.

Kirkby, M.J. 1984. Modelling cliff development in South Wales: Savigear re-viewed. *Zeitschrift für Geomorphologie* **28**, 405–426.

Kirkby, M.J. 1985a. The basis for soil profile modelling in a geomorphic context. *Journal of Soil Science* **36**, 97–122.

Kirkby, M.J. 1985b. A model for the evolution of regolith-mantled slopes. In *Models in Geomorphology*, Woldenberg, M. J. (ed), 213–237. Boston: Allen & Unwin.

Kirkby, M.J. 1986. A two-dimensional simulation model for slope and stream evolution. In *Hillslope Processes*, Abrahams, A. D. (ed), 203–222. Winchester: Allen & Unwin.

Kirkby, M.J. 1990. A one-dimensional model for rill inter-rill interactions. *Catena Supplement* **17**, 133–146.

Kirkby, M.J., Bull, L.J., Poesen, J., Nachtergaele, J. and Vandekerckhove, L. 2003. Observed and modelled distributions of channel and gully heads – with examples from SE Spain and Belgium. *Catena* **50**, 415–434.

Kirkby, M.J. and Bull, L.J. 2000a. Some factors controlling gully growth in fine-grained sediments: a model applied in southeast Spain. *Catena* **40**, 127–146.

Kirkby, M.J. and Bull, L.J. 2000b. Some factors controlling gully growth in fine-grained sediments; a model applied in Southeast Spain. In *Badlands in Changing Environments*, Torri, D., Poesen, J., Calzolari, C. and Rodolfi, G. (eds), 127–146. Catena-Verlag Rohdenburg, Cremlingen-Destedt.

Knapen, A., Poesen, J., Govers, G., Gyssels, G. and Nachtergaele, J. 2007. Resistance of soils to concentrated flow erosion: A review. *Earth-Science Reviews* **80**, 75–109.

Komura, S. 1976. Hydraulics of slope erosion by overland flow. *Journal of the Hydraulics Division, American Society of Civil Engineers* **102**, 1573–1586.

Kuhn, N.J. and Yair, A. 2004. Spatial distribution of surface conditions and runoff generation in small arid watersheds, Zin Valley Badlands, Israel. *Geomorphology* **57**, 183–200.

Kuijper, C., Cornelisse, J.M. and Winterwerp, J. C. 1989. Research on erosive properties of cohesive sediments. *Journal of Geophysical Research* **94**, 14341–14350.

Lambe, T.W. and Whitman, R.V. 1969. *Soil mechanics*, 553pp. New York: Wiley.

Lane, L.J., Shirley, E.D. and Singh, V.P. 1988. Modelling erosion on hillslopes. In *Modelling Geomorphological Systems*, M.G. Anderson (ed), 287–308. Chichester: John Wiley & Sons.

Laronne, J.B. 1981. Dissolution kinetics of Mancos Shale – associated alluvium. *Earth Surface Processes and Landforms* **6**, 541–552.

Laronne, J.B. 1982. Sediment and solute yield from Mancos Shale hillslopes, Colorado and Utah. In *Badland Geomorphology and Piping*, Bryan, R. and A. Yair, A. (eds), 181–192. Norwich: Geo Books.

Leopold, L.B. and Miller, J.P. 1956. Ephemeral streams. Hydraulic factors and their relation to the drainage net. *U.S. Geological Survey Professional Paper* **282–A**.

Mackin, J.H. 1948. Concept of the graded river. *Geological Society of America Bulletin* **59**, 463–512.

Melton, M.A. 1965. The geomorphic and paleoclimatic significance of alluvial deposits in southern Arizona, *Journal of Geology* **73**, 1–38.

Meyer, L.D. 1986. Erosion processes and sediment properties for agricultural cropland. In *Hillslope Processes*, Abrahams, A.D. (ed), 55–76. Winchester: Allen & Unwin.

Meyer L.D. and Monke, E.J. 1965. Mechanics of soil erosion by rainfall and overland flow. *Transactions of the American Society of Agricultural Engineers* **8**, 572–577, 580.

Montgomery, D.R. and Dietrich, W.E. 1994. landscape dissection and drainage area-slope thresholds. In *Process Models and Theoretical Geomorphology*, Kirkby, M.J. (ed.), 221–245, John Wiley & Sons, Chichester.

Moore, I.D., Burch, G.J. and MacKenzie, D.H. 1988. Topographic effects on the distribution of surface soil water and the location of ephemeral gullies. *Transactions American Society of Agricultural Engineers* **31**, 1098–1107.

Moore, J.S., Temple, D.M. and Kirsten, H.A.D. 1994. Headcut advance threshold in earth spillways. *Bulletin of the Association of Engineering Geology* **31**, 277–280.

Morgan, R.P.C. and Mngomezulu, D. 2003. Threshold conditions for initiation of valley-side gullies in the Middle Veld of Swaziland. *Catena* **50**, 401–414.

Moseley, M.P. 1973. Rainsplash and the convexity of badland divides. *Zeitschrift für Geomorphologie Supplementband* **18**, 10–25.

Moss, A.J., Green, P. and Hutka, J. 1982. Small channels: their experimental formation, nature and significance. *Earth Surface Processes and Landforms* **7**, 401–416.

Moss, A.J. and Walker, P.H. 1978. Particle transport by continental water flow in relation to erosion, deposition, soils and human activities. *Sedimentary Geology* **20**, 81–139.

Moss, A.J., Walker, P.H. and Hutka, J. 1979. Raindrop-stimulated transportation in shallow water flows: an experimental study. *Sedimentary Geology* **22**, 165–184.

Nachtergaele, J., Poesen, J., Steegen, A., Takken, I., Beuselinck, L., Vandekerckhove, L. and Govers, G. 2001. The value of a physically based model versus an empirical approach in the prediction of ephemeral gully erosion for loess-derived soils. *Geomorphology*, **40** , 237–252.

Nadal-Romero, E., Regüés, D., Martí-Bono, C. and Serrano-Muela, P. 2007. Badland dynamics in the Central Pyrenees: temporal and spatial patterns of weathering processes. *Earth Surface Processes and Landforms*, **32**, 888–904.

Parchure, T.M. and Mehta, A.J. 1985. Erosion of soft cohesive sediment deposits. *Journal of the Hydraulics Division, American Society of Civil Engineers* **111**, 1308–1326.

Parker, G.G., Sr., Higgins, C.G. and Wood, W.W. 1990, Piping and pseudokarst in drylands. *Geological Society of America Special Paper* **252**, 77–110.

Parsons, A.J., Wainwright, J. and Abrahams, A.D. 1996. Runoff and erosion on semi-arid hillslopes, In *Advances in Hillslope Processes*, Anderson, M. G. and Brooks, S.M. (eds), 1061–1078. Chichester: John Wiley & Sons.

Parthenaides, E. 1965. Erosion and deposition of cohesive soils. *Journal of the Hydraulics Division, American Society of Civil Engineers* **91**, 105–139.

Parthenaides, E. and Paaswell, R.R. 1970. Erodibility of channels with cohesive banks. Journal of the Hydraulics Division, American Society of Civil Engineers **96**, 755–771.

Patton, P.C. and Schumm, S.A., 1975. Gully erosion, Northwestern Colorado; A threshold phenomenon. *Geology* **3**, 88–89.

Pavich, M.J., 1989. Regolith residence time and the concept of surface age of the Piedmont "peneplain". *Geomorphology* **2**, 181–196.

Poesen, J. and Govers, G. 1990. Gully erosion in the loam belt of Belgium: typology and control measures. In *Soil Erosion on Agricultural Land*, Boardman, J., Foster, I.D.L. and Dearing, J.A., (eds), 513–530. Chichester: John Wiley & Sons.

Poesen, J., Nachtergaele, J., Verstraeten, G. and Valentin, C. 2003. Gully erosion and environmental change: importance and research needs. *Catena* **50**, 91–133.

Poesen, J., Vanwalleghem, T., de Venta, J., Verstraeten, G. and Martínez-Casasnovak, J.A. 2006. Gully erosion in Europe. In *Soil Erosion in Europe*, Boardman, J. and Poesen, J. (eds), 515–536. Chichester: Wiley.

Prosser, I.P. and Abernethy, B. 1996. Predicting the topographic limits to a gully network using a digital terrain model and process thresholds. *Water Resources Research* **32**, 2289–2298.

Prosser, I.P., Chappell, J. and Gillespie, R. 1994. Holocene valley aggradation and gully erosion in headwater catchments, southeastern highlands of Australia. *Earth Surface Processes and Landforms* **19**, 465–480.

Prosser, I.P. and Dietrich, W.E. 1995. Field experiments on erosion by overland flow and their implication for a digital terrain model of channel initiation. *Water Resources Research* **31**, 2867–2876.

Prosser, I.P., Dietrich, W.E. and Stevenson, J. 1995. Flow resistance and sediment transport by concentrated overland flow in a grassland valley. *Geomorphology* **13**, 71–86.

Prosser, I.P. and Slade, C.J. 1994. Gully formation and the role of valley-floor vegetation, southeastern Australia. *Geology* **22**, 1127–1130.

Rauws, G. 1987. The initiation of rills on plane beds of non-cohesive sediments. *Catena Supplement* **8**, 107–118.

Rauws, G. and Govers, G. 1988. Hydraulic and soil mechanical aspects of rill generation on agricultural soils. *Journal of Soil Science* **39**, 111–124.

Reid, L.M. 1989. Channel formation by surface runoff in grassland catchments, unpublished Ph.D. thesis, University of Washington, Seattle, 139pp.

Reneau, S.L. and Dietrich, W.E. 1991. Erosion rates in the southern Oregon Coast range: evidence for an equilibrium between hillslope erosion and sediment yield, *Earth Surface Processes And Landforms* **16**, 307–322.

Robinson, K.M. and Hanson, G.J. 1994. A deterministic headcut advance model. *Transactions American Society of Agriculture Engineers* **37**, 1437–1443.

Robinson, K.M. and Hanson, G.J. 1995. Large-scale headcut erosion testing. *Transactions American Society of Agricultural Engineers* **38**, 429–434.

Roering, J.J., Kirchner, J.W. and Dietrich, W.E. 1999. Evidence for nonlinear, diffusive sediment transport on hillslopes and implications for landscape morphology. *Water Resources Research* **35**, 853–870.

Roering, J.J., Kirchner, J.W. and Dietrich, W.E. 2001a. Hillslope evolution by nonlinear, slope-dependent transport; steady state morphology and equilibrium adjustment timescales. *Journal of Geophysical Research* **106**, 16,499–16,513.

Roering, J.J., Kirchner, J.W., Sklar, L.S. and Dietrich, W.E. 2001b. Hillslope evolution by nonlinear creep and landsliding: An experimental study. *Geology* **29**, 143–146.

Rutherford, I.D., Prosser, I.P. and Davis, J. 1997. Simple approaches to predicting rates of gully development. In *Management of Landscapes Disturbed by Channel Incision*,

Wang, S.Y., Langendoen, E.J. and Shields, F.D. Jr. (eds), 1125–1130. Oxford, Miss: University of Mississippi.

Savat, J. 1976. Discharge velocities and total erosion of a calcareous loess: a comparison between pluvial and terminal runoff. *Revue Geomorphologie Dynamique* **24**, 113–122.

Savat, J. 1980. Resistance to flow in rough supercritical sheetflow. *Earth Surface Processes* **5**, 103–122.

Savat, J. 1982. Common and uncommon selectivity in the process of fluid transportation: field observations and laboratory experiments on bare surfaces. *Catena Supplement* **1**, 139–160.

Savat, J. and De Ploey, J. 1982. Sheetwash and rill development by surface flow. In *Badland Geomorphology and Piping*, Bryan, R. and Yair, A. (eds), 113–125. Norwich: Geo Books.

Schumm, S.A. 1956a. Evolution of drainage systems and slopes in badlands at Perth Amboy, New Jersey. *Geological Society of America Bulletin* **67**, 597–646.

Schumm, S.A. 1956b. The role of creep and rainwash on the retreat of badland slopes. *American Journal of Science* **254**, 693–706.

Schumm, S.A. 1962. Erosion on miniature pediments in Badlands National Monument, South Dakota. *Geological Society of America Bulletin* **73**, 719–724.

Schumm, S.A. 1963. Rates of surficial rock creep on hillslopes in western Colorado. *Science* **155**, 560–561.

Schumm, S.A. 1964. Seasonal variations of erosion rates and processes on hillslopes in western Colorado. *Zeitschrift für Geomorphologie Supplementband* **5**, 215–238.

Schumm, S.A. and Hadley, R.F. 1957. Arroyos and the semiarid cycle of erosion. *American Journal of Science* **255**, 164–174.

Schumm, S.A., Harvey, M.D. and Watson, C.C. 1984. *Incised Channels; Morphology, Dynamics and Control*. Littleton, Co: Water Resources Publications, 200pp.

Schumm, S.A. and Lusby, G.C. 1963. Seasonal variation of infiltration capacity and runoff on hillslopes in western Colorado. *Journal of Geophysical Research* **68**, 3655–3666.

Sidorchuk, A. 2005. Stochastic components in the gully erosion modelling. *Catena* **63**, 299–317.

Slaymaker, O. 1982. The occurrence of piping and gullying in the Penticton glacio-lacustrine silts, Okanagan Valley, B.C. In *Badland Geomorphology and Piping*, Bryan, R.B. and Yair, A. (eds), 305–316. Norwich: GeoBooks.

Smith, K.G. 1958. Erosional processes and landforms in Badlands National Monument, South Dakota. *Bulletin of the Geological Society of America* **69**, 975–1008.

Smith, T.R. and Bretherton, F.P. 1972. Stability and the conservation of mass in drainage basin evolution. *Water Resources Research* **8**, 1506–1529.

Stein, O.R. and Julien, P.Y. 1993. A criterion delineating the mode of headcut migration. *American Society Civil Engineers, Journal of the Hydraulics Division* **119**, 37–50.

Stein, O.R., Julien, P.J. and Alonso, C.V. 1997. Headward advancement of incised channels. In *Management of Landscapes Disturbed by Channel Incision*, Wang, S.Y. Langendoen, E.J. and Shields, F.D. Jr., 497–503. Oxford, Mississippi: University of Mississippi.

Thornes, J.B. 1985. The ecology of erosion. *Geography* **70**, 222–236.

Torri, D. and Borselli, L. 2003. Equation for high-rate gully erosion. *Catena* **50**, 449–467.

Torri, D., Sfalanga, M. and Chisci, G. 1987. Threshold conditions for incipient rilling. *Catena Supplement* **8**, 97–115.

Trimble, S.W. 1974. *Man-Induced Soil Erosion on the Southern Piedmont, 1700–1970*, Soil Conservation Society, Ankemy, Iowa, 180pp.

Tucker, G.E. 2004. Drainage basin sensitivity to tectonic and climatic forcing; implications of a stochastic model for the role of entrainment and erosion thresholds. *Earth Surface Processes and Landforms* **29**, 185–205.

Tucker, G.E. and Bras, R.L. 1998. Hillslope processes, drainage density, and landscape morphology. *Water Resources Research* **34**, 2751–2764.

Tucker, G.E. Arnold, J.R., Bras, R.L., Flores, H., Istanbulloglu, E. and Sólyom, P. 2006. Headwater channel dynamics in semiarid rangelands, Colorado high plains, USA. *Bulletin of the Geological Society of America* **118**, 959–974.

Tucker, G.E. and Slingerland, R. 1997. Drainage basin responses to climate change. *Water Resources Research* **33**, 2031–2047.

Valentin, C., Poesen, J. and Li, Y. 2005. Gully erosion: Impacts, factors and control. *Catena* **63**, 132–153.

Vandaele, K., Poesen, J., Govers, G. and van Wesemael, B. 1996. Geomorphic threshold conditions for ephemeral gully incision. *Geomorphology* **16**, 161–173.

Vandekerckhove, L., Poesen, J., Wijdenes, D.O. and de Figueiredo, T. 1998. Topographical thresholds for ephemeral gully initiation in intensively cultivated areas of the Mediterranean. *Catena* **33**, 271–292.

Vice, R.B., Guy, H.P. and Ferguson, G.E. 1968. Sediment movement in an area of suburban highway construction – Scott Run Basin, Virginia – 1961–64. *U.S. Geology Survey Water-Supply Paper* **1591-E**, 41pp.

Wang, S.S.Y., Langendoen, E.J. and Shields, F.D.J. (eds.). 1997. *Management of Landscapes Disturbed by Channel Incision: Stabilization, Rehabilitation, Restoration*. Center for Computational Hydroscience and Engineering, University of Mississippi, University, Mississippi, 1134pp.

Wells, S.G. and Gutierrez, A.A. 1982. Quaternary evolution of badlands in the southeastern Colorado Plateau, U.S.A. In *Badland Geomorphology and Piping*, R. Bryan and A. Yair (eds), 239–257. Norwich: Geo Books.

Willgoose, G. 2005. Mathematical modeling of whole landscape evolution. *Annual Review of Earth and Planetary Sciences* **33**, 443–459.

Williams, J.R., Jones, C.A. and Dyke, P.T. 1984. A modeling approach to determining the relationship between erosion and soil productivity. *Transactions of the American Society of Civil Engineers* **27**, 129–144.

Wolman, M.G. 1967. A cycle of sedimentation and erosion in urban river channels. *Geografiska Annaler* **49A**, 385–395.

Woodward, D.E. 1999. Method to predict cropland ephemeral gully erosion. *Catena* **37**, 393–399.

Yair, A., Lavee, H., Bryan, R.B. and Adar, E. 1980. Runoff and erosion processes and rates in the Zin Valley badlands, northern Negev, Israel. *Earth Surface Processes* **5**, 205–225.

Zhu Xianmo (ed.), 1986. *Land Resources on the Loess Plateau of China*, Northwest Inst. Soil and Water Conservation., Academia Sinica, Shaanxi Science and Technique Press.

Part IV
Rivers

Chapter 11

Catchment and Channel Hydrology

John B. Thornes[†]

Introduction

Solar radiation, wind, and water are the driving agents of desert landscapes. Water has four major roles to play. First, in sustaining any life forms that exist; secondly, as a chemical substance which interacts with other chemical substances, notably salts; third, as a medium of transport of mass; and, fourth, as a direct source of energy. The last role, though small by comparison with the roles of solar and wind energy, may none the less be critical in determining the threshold of operation of runoff, through its impact on infiltration. Since infiltration is one of the major thresholds in dryland morphological development, the factors which control it have a role out of all proportion to the energy involved.

We have come to take for granted the well-known hydrological cycle of arid lands (as illustrated in Fig. 11.1). The hydrological cycle of desert environments has the same inputs and outputs, and there is the same requirement for the conservation of mass and energy.

It is the relative importance of the different components which is critical. In temperate environments percolation, throughflow, saturated flow, and groundwater play a most significant role. In dry environments infiltration normally only occurs to shallow depths, soil moisture is consistently low or very low, and overland flow is important, with deep percolation and groundwater being relatively unimportant except in suballuvial or externally recharged aquifers. This relative importance also varies spatially and through time more than in temperate environments. For example, under even sparse vegetation infiltration becomes relatively much more important than between vegetation. Under seasonal control, infiltration may be much more important in the wetter than in the drier season, and so on. As this balance changes so does the relative role of different geomorphic processes. The differences between desert and temperate hydrology from a geomorphological point of view are therefore more complex than simply a shift in the magnitude and frequency of events.

The major differences, which are illustrated and developed in the remainder of this chapter, can be summarized as follows.

(a) Rainfall occurs at high intensities, with low overall amounts, at irregular intervals, often with a strong seasonal bias and usually with a very large inter-annual variability.

(b) The rain falls on ground with a sparse or non-existent vegetation cover, which is irregular in its distribution and especially adapted to collect rainfall. Interception rates are low and highly variable and rapid direct evaporation of excess surface water is characteristic.

(c) Infiltration is largely controlled by the bare surface characteristics which range from sands to organic crusts and from stones to chemical precipitates.

(d) Losses due to evapotranspiration are dominated by soil-water availability and controlled by profile characteristics as well as by atmospheric stress; subsurface water movement may be significantly affected by regolith chemistry.

(e) Overland flow is relatively more likely when storms occur and the terrain over which it occurs may be exceptionally rough.

J.B. Thornes (✉)
Department of Geography, King's College London, Strand, London WC2R 2LS, UK
†deceased

A.J. Parsons, A.D. Abrahams (eds.), *Geomorphology of Desert Environments*, 2nd ed., DOI 10.1007/978-1-4020-5719-9_11, © Springer Science+Business Media B.V. 2009

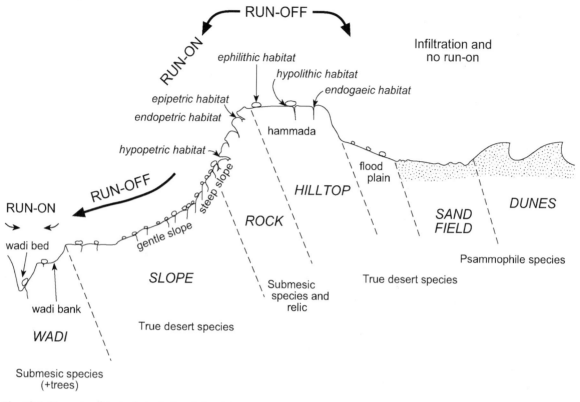

Fig. 11.1 Elements of the hydrological cycle in arid lands (after Shmida et al. 1986)

(f) Channel flow is ephemeral and, hence, significantly influenced by boundary conditions, especially transmission losses to alluvium.

(g) Groundwater obeys the same rules as in temperate environments, but suballuvial aquifers assume a more significant hydrological role with ephemeral channel flow.

Precipitation

Magnitudes at Different Timescales

From a geomorphological point of view, annual rainfalls tend to determine the overall character of the environment through the vegetation condition because, to a first approximation, the gross vegetation biomass and productivity are determined by rainfall for areas with less than about 600 mm (Walter 1971, Leith and Whittaker 1975). Above this figure nutrients become more significant in limiting plant growth. Below it

the physiological response of plants in biomass terms to gross precipitation may actually be quite variable because ultimately it is the available moisture for plant growth that is the controlling factor. This availability reflects evapo-transpiration amounts and the water use efficiency of particular species (Woodward 1987). Shrubs and dwarf shrubs are the most conspicuous life forms in desert regions. At the wetter margin shrubs such as *Artemisia*, *Atriplex*, *Cassia*, *Ephedra*, *Larrea*, and *Retama* prevail, whereas in extreme deserts dwarf shrub communities are present, such as the Chenopodiacae, Zygophillaceae, and the Asteraceae (such as *Artemisia*). Typically the semi-deserts have 150–350 mm of rain, whereas the extreme deserts have less than 70 mm, according to Shmida (1985).

Generally here we follow Le Houerou (1979) in restricting deserts to environments with less than 400 mm y^{-1}, which in North Africa and in southern Europe corresponds to the northern limit of steppe vegetation, such as *Stipa tenacissima* and *Artemisia herba-alba*. Le Houerou differentiates extreme desert lands as those with less than 100 mm, corresponding to the northern border of typical desert plants, such as *Calligonium comosum* and *Cornulaca monacantha*.

Annual rainfall magnitudes are also reflected in altitudinal variations in plant cover. Whittaker and Niering (1964), for example, show the clear gradation from desert to forest developed in the Santa Catalina mountains of Arizona.

There have been several models which relate geomorphic processes or responses to annual rainfall amounts and temperature (Peltier 1950, Langbein and Schumm 1958, Carson and Kirkby 1972), and these consistently show low total magnitudes of rainfall having low rates of geomorphic activity but with sharply increasing rates up to a maximum at about 300 mm.

There is an enormous interannual variability in total amounts in dry areas. Figure 11.2 shows interannual variations of rainfall totals for Murcia, southeast Spain, for 124 years, expressed as rainfall anomalies (mean annual rainfall/standard deviation). Here the mean annual rainfall is 300 mm, and the interannual coefficient of variation 35%, a figure typical of semi-arid environments. Bell (1979) has shown that, in general, annual precipitation variability is highest in zones of extreme aridity and that the lowest variations are at the poleward margins of the deserts, where frontal rainfall forms a greater proportion of the total annual rainfall.

As a general rule the standardized rainfall anomaly decreases as the mean annual rainfall increases.

The geomorphic significance of seasonal rainfall distribution varies with evapotranspiration because these together determine soil moisture conditions and infiltration rates, decomposition rates for organic matter, soil erodibility, evaporation rates, solute movement, and chemical precipitates.

Figure 11.3 shows the annual pattern of soil moisture for the Murcia site shown in Fig. 11.2. The relative importance of seasonal rainfall totals depends on how much the plant cover varies. Typically in arid and semi-arid conditions the seasonal impact is principally on annual herbs. In semi-arid environments these may account for as little as 5% of the total cover, the amount increasing as the conditions become drier. The seasonality of rainfall is generally expressed in some ratio of monthly to annual rainfall. For example the Fournier index (Fournier 1960) expressed seasonality by the index p^2/P, where p is the rainfall of the wettest month and P the annual rainfall. By regression analysis he showed that this index is related to total sediment yield. The yield increases with seasonality and relief, and the rate of increase is greater in desert areas. More recently Kirkby and Neale (1987) indicated that it is seasonality, expressed as the degree of rainfall concentration, which leads to the characteristic Langbein and Schumm (1958) curve.

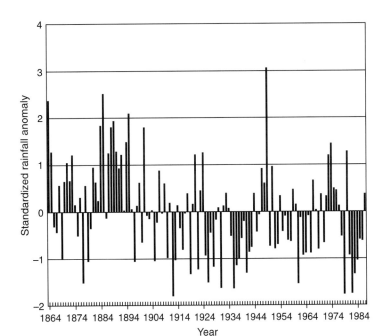

Fig. 11.2 Annual rainfall anomalies for Murcia, south-east Spain (after Thornes 1991)

Fig. 11.3 Seasonal
fluctuations in soil moisture
for the El Ardal site, Murcia,
1989-1990 under different
ground conditions based on
data for the MEDALUS
project

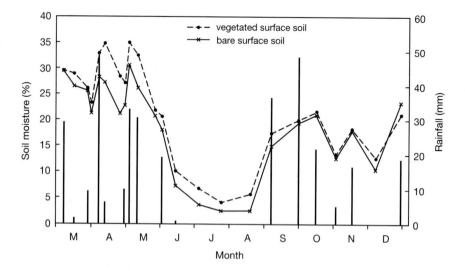

In areas of extreme aridity the rainfall is not strongly seasonal in amount except in the Gobi–Tarim basin, where cyclonic storms are blocked by the persistent Siberian High in winter. Towards the desert margins rainfall seasonality increases and most significant is the distribution of rainfall in relation to temperature. Around the central Australian desert, for example, long periods without rain are usual in the northern winter and the southern summer. In the Mojave Desert rain occurs in winter, in the Chihuahuan Desert in the summer, and the Sonoran Desert has both winter and summer rainfalls.

Event rainfall magnitudes are important in determining soil moisture characteristics, but generally in desert areas it is the high intensity that is geomorphologically important, as discussed below.

The idea of coupling magnitude and frequency was first developed comprehensively in geomorphology by Wolman and Miller (1960). Put simply it draws attention to the fact that it is not only the magnitude of the force applied by a process but also how frequently the force is applied which determines the work done by the process. In temperate environments their results suggest that it is the medium scale events of medium frequency that perform most geomorphic work. Ahnert (1987) has used the same concept to express the magnitude and frequency of climatic events. Plotting the event magnitude on the vertical axis and the log (base 10) of the recurrence interval of the event on the horizontal axis provides an alternative representation of the relationship. By

fitting a regression model to the data he obtained the relationship

$$P_{24} = Y + A \log_{10}(\text{RI}) \qquad (11.1)$$

where P_{24} is the daily precipitation of an event with recurrence interval of RI (in years), and the recurrence interval is obtained from $\text{RI} = (N + 1)/\text{rank}$ by magnitude in a list of n daily rainfall totals. Ahnert showed that Y and A are characteristics of the rainfall, and he used them as indices of magnitude and frequency. The map of California (Fig. 11.4a) indicates that both A and Y are generally low in the Mojave Desert. The indices can be obtained for other time intervals, for partial series (such as P-threshold), and for other variables such as frost frequency. Moreover, it is possible to derive the magnitude and frequency product and plot this against return period. For example, Ahnert calculated the curve for $(P - T_r)F$, where P is the daily rainfall, T_r a threshold value, and F the frequency of the event of magnitude $(P - T_r)$, and plotted this against recurrence interval. The family of curves (Fig. 11.4b) shows that in desert areas the overall product is low and that the maximum of the product occurs in tens of years rather than several times a year. This implies that most work is done by rather less frequent events, but not by very rare events. De Ploey et al. (1991) developed this idea further to provide an index of cumulative erosion potential.

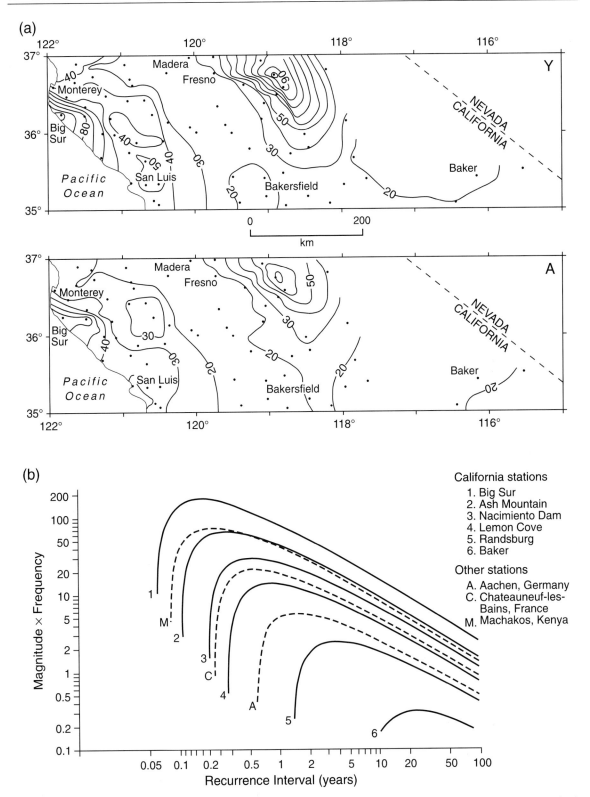

Fig. 11.4 Frequency and magnitude distribution of rainfall (after Ahnert 1987). (**a**) Regional distribution of values of *A* and *Y* in California. (**b**) The product of magnitude and frequency for a variety of stations

In temperate environments there are a large number of small events and a much smaller number of large events, so the distribution tends to negative exponential and gives a straight line on the semi-log plot. Ahnert (1987) used this relationship to explore the characteristics of storm events, and this provides a useful comparison of the models for desert and temperate environments (Fig. 11.4). This pattern of magnitude and frequency can also be expressed for individual seasons.

Rainfall Intensity

The intensity of rainfalls is especially important in determining the canopy storage and gross interception losses as well as the production of overland flow. Intensity also determines the compaction of soils and thus indirectly affects the infiltration capacity (Roo and Riezebos 1992). One of the simplest measures of intensity is the mean rainfall per rain day. It provides the key parameter for providing the exponential distribution of rainfall magnitudes and deriving the excess above a critical threshold value (Thornes 1976). It is sensitive to the definition of rain day, but generally a figure of 0.2 mm is taken as the threshold. Average figures for desert areas are between 5 and 10 mm, and these tend to increase to the margins. These figures do not differ greatly from temperate areas and are generally less in the extreme deserts.

While mean rainfall per rain day is a useful index when only daily rainfall amounts are available (which is the case for large parts of the Earth and especially desert regions), the absolute intensity $(mm\,h^{-1})$ generally increases as shorter periods are considered. Although record intensities occur in humid tropical areas, intensities for arid areas are generally higher than those of temperate regions. For example, Dhar and Rakhecha (1979), working in the Thar Desert of India where average annual rainfall is 310 mm, found that one-day point rainfalls of the order of 50–120 mm have a recurrence interval of 2 years and that rainfalls associated with monsoon depressions can be from 250 to 500 mm in a single day. The peak daily rainfall can be several times the annual average. Berndtsson (1987), for example, observed that on 25 September 1969, 400 mm of rain fell at Gabes, central Tunisia, which is about five times the mean annual

rainfall. In semi-arid south-eastern Spain the storms of 18–19 October 1973 produced rainfalls equal to the mean annual amount (300 mm) in a period of about 10 h. In this semi-arid environment rainfall intensities of 70 mm h^{-1} typically recur about every 5 years.

Using the 1-h duration and 2-year return period, Bell (1979) obtained figures typically of the order of 10–20 mm for central desert areas and 20–50 mm along desert margins. Again, this suggests that it is the desert margins where runoff effects are most likely to be geomorphologically important. He estimated the 'maximum probable' 1-h rainfalls for the Australian arid region to be of the order of 180 mm at the poleward edge to 280 at the equatorwards margin. Similar figures have been estimated for the desert areas of the United States by Herschfield (1962), and Bell (1979) believed that they are probably typical of the arid zone in other parts of the world. Typically the 2-year 1-h value has an intensity about ten times that of the 2-year 24-h rainfall.

Temporal Variability

For geomorphic purposes most data are of relatively short duration. Typically 40–50 years of daily rainfall records are available for the arid areas of the world. Autographic data, from which storm profiles can be constructed, are even rarer. For developing models of the impact of rainfall series it is therefore sometimes necessary to simulate series which have the same characteristics as the actual data. The problem is that small sample sizes preclude normal forcasting techniques and the variability of total rainfall in an event is quite high.

Within individual storm events there are often significant variations in intensity. Figure 11.5 illustrates a storm profile for the semi-arid Cuenca area of central Spain. The total rainfall for the storm was 24 mm, with an average intensity of 6 mm h^{-1} while the high burst in minutes 35–39 produced an intensity of 36 mm h^{-1}. Within-storm rainfall has been modelled by Jacobs et al. (1988) and a description of their approach is found below.

The distribution of events is reflected in the seasonal distribution of cumulative rainfall totals and these have a strong effect on plant growth in dry areas. For

Fig. 11.5 Rainfall storm profile for an autumn storm in Cuenca, Spain

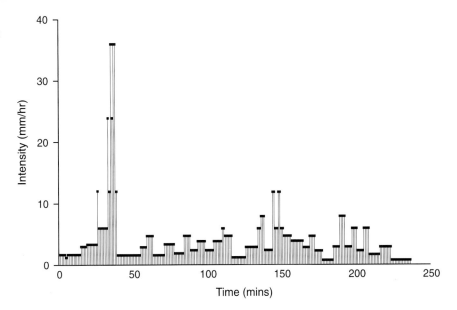

example Fig. 11.6 shows the contrasts in rainfall patterns within the year for normal (1981), exceptionally wet (1961) and exceptionally dry years (1979, 1982) in south-east Spain. Lack of autumn rains inhibits the germination of seedlings, whereas lack of soil moisture in spring limits plant growth.

The seasonal distribution of event magnitudes and durations is also important from a runoff and infiltration point of view. Generally event magnitudes and durations reflect prevailing weather types, with cyclonic storms often giving long-duration, low-intensity rain-

falls and convective (summer) storms giving short-duration, high-intensity rainfalls.

A simple seasonal model is provided by the Markov chain, in which sequences of rain and no-rain days and the rainfall magnitudes for these days are generated. Thornes and Brandt (1994) have provided a synthetic generator for the semiarid regions of south-east Spain using this principle. The sequence of rain or non-rain days is obtained from a Markov probability matrix of transitions from rain-rain, rain–dry, dry–rain, and dry–dry days derived from the historical records on a sea-

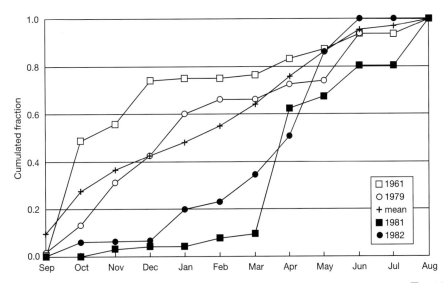

Fig. 11.6 Seasonal variations in the annual distribution of rainfall for the province of Murcia, Spain. Key: □ = 1961; ○ = 1979; + = mean; ■ = 1981; ● = 1982

sonal basis. The magnitudes are then obtained from a two-parameter gamma distribution, again with seasonal parameters. Both the transition matrices and the gamma parameters were found to be stable over time in this environment.

A more sophisticated model was used by Duckstein et al. (1979) to investigate and simulate the structure of daily series. They define an event as a day having a precipitation greater than a constant amount, say 0.2 mm. The statistics of interest are then the number of events per unit time, the time between events, the depth of precipitation, the duration of precipitation, and the maximum intensity of rainfall in an event. A sequence is a number of rainfall events separated by three or fewer days. Then the dry spell duration is a run of days without rain, and the average seasonal maximum of the run of dry days defines extreme droughts. Finally, interarrival time is the time between the beginning of one sequence and the beginning of the next; in statistical terminology it is the renewal time.

For convective storms occurring in summer the events are independent and short (cf. Smith and Schreiber 1973). The probability of n events of j days duration is then given by the Poisson mass function

$$f_n(j) = \exp(-m)m_j/j! \qquad (11.2)$$

where m is the mean number of events per season.

The probability that the interarrival time T is t days is given by

$$f_T(t) = u \exp(-ut) \qquad (11.3)$$

where u is estimated from $1/\overline{T}$, with $\overline{T}$ the mean inerarrival time between events. For Tucson, Arizona, u is estimated to be 0.27. Under some conditions, especially where the storms are more frequent (e.g. where a cyclonic element prevails), interarrival times may be better described by a gamma rather than an exponential distribution. Finally, the event magnitudes are also independent. For example in Granada province, southern Spain, Scoging (1989) found that 90% of summer storm rainfall lasted less than an hour and that the magnitude of the rainfall in an event is independent of its duration. Under these circumstances, which are most common for convective thunderstorms, the gamma distribution again provides a suitable model for rainfall magnitudes.

The amounts, intensities, and durations of storms are also found to vary significantly in the long term in dry as they do in temperate areas, and interest in these fluctuations has been heightened in recent years as the result of the potential implications of global warming for already dry areas.

Berndtsson (1987), for example, found that in Tunis from 1890 to 1930 there was a massive fall in mean annual rainfall and that after 1925 there was an oscillation of durations of 10–15 years. Similar phenomena have been recorded in other statistical analyses. Conte et al. (1989) demonstrated that oscillations in rainfall in the Western Mediterranean in the period after 1950 could be related to the Mediterranean Oscillation in barometric pressure. Thornes (1990) also observed both the steep decline in rainfall in the period 1890–1934 and oscillations in rainfall anomalies in the post-1950 period in the southern Iberian Peninsula. The rainfall decline, at 3 mm y^{-1}, is of the order of magnitude currently predicted for climatic changes over the next 60 years by climatologists from general circulation models. The rainfall anomaly oscillations are in phase with those recorded by Conte et al. and almost exactly out of phase with those described for the Sahel.

Spatial Variability

If arid zone rainfall is difficult to predict in time, it is almost as difficult to predict in space. The spatial distribution of storms tends to be random in subtropical deserts. There is usually a poor correlation of rainfall amounts at stations only 5 km apart. It is rare for a single station to experience more than one storm in a day, and the spacing between concurrent storms is typically 50–60 km.

The development of rainfall in convective storms is quite complex and accounts for much of the locally highly concentrated rainfall amounts and intensities that accompany desert storms. Well-defined storm cells are born and then decay, and the cellular structure can give rise to well-defined rates of attenuation of rainfall with distance from the cell centres. Jacobs et al. (1988) examined summer data from a network of 102 gauges distributed over a 154 km^2 area in Walnut Gulch Experimental Watershed, Arizona for the period 1970–1977, and they attempted to model the intensity and cumulative depth over temporal realizations of the process or for a particular event over space. The procedure assumes that a storm is born at maximum intensity and then decays exponentially through time, with the initial intensity being an independent exponentially distributed random variable. The spatial distribution of storm centres is modelled as a Poisson process, and the time of birth of each cell relative to storm outset is taken to be exponential. From these assumptions the variance and covariance structures of rainfall intensity and depth are derived analytically and appear to provide a reasonable representation of the structure of the storm patterns observed.

In comparison with the Walnut Gulch data, one of the best sets of rainfall data in a desert area in the world, the storm depths modelled by Jacobs et al. were found to be more homogeneous and isotropic over time. Zawadeski (1973) had already shown that storms are typically isotropic under about 10 km but increasingly anisotropic over larger distances.

Interception

Our understanding of surface hydrological processes is very largely in terms of specific surfaces, such as bare soils, plants of various species and architecture, and more recently in terms of stone cover. In fact, desert surfaces usually comprise complex mixtures of these different types, just as satellite images represent complex mosaics of land use (Fig. 11.7). This means that both the understanding and modelling of such surfaces tend to be difficult. Moreover the very nature of desert soils with their chemical crusts, concentrated aggregates, complex surface roughness, and particular plant types means that hillslope hydrology (as opposed to hydraulics) of such regions is still in its relative infancy. Earlier treatises on desert hydrology, for example, have tended to simply reiterate the received wisdom from temperate soils, which is, in reality, completely inappropriate.

Desert plants are structurally rather different from temperate and humid tropical plants in their life forms. They often have low leaf areas, especially adapted leaves and stems, and wide spacing (see below). The morphologies of several rangeland species of shrubs and grasses (e.g. *Cassia* sp., *Stipa tenacissima*) are designed to intercept rainfall and channel the water down as stemflow. This water infiltrates adjacent to the tree bole and produces enhanced water storage at depth. Slatyer (1965) showed that with mulga scrub once the interception store of 2–3 mm has been filled, stemflow accounts for up to 40% of the rain falling on the canopy, and falls of 15–20 mm of rain can cause about 100 l of stemflow from adult trees. Stemflow equal to

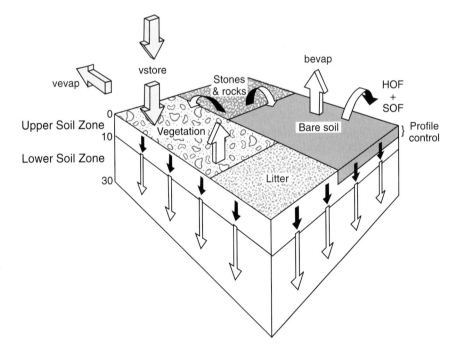

Fig. 11.7 Compound character of surfaces in semi-arid and arid environments, showing major fluxes of evaporation (vevap, bevap), transpiration (trans), interception (vstore) and Hortonian and saturated overland flow (HOF, SOF)

30–40% of intercepted rain has also been found in semi-arid environments in the Zaragoza area of Spain by Gonzalez-Hidalgo (1991).

Evans et al. (1976), on the other hand, reported that about 90% of rainfall was converted into throughfall for creosote bush and about 65% for bursage. However, the percentage for different storms ranged from 67% to greater than 100% and from 26% to greater than 100% for creosote bush and bursage, respectively. Generally, in rangeland, losses from the canopy through the direct evaporation of intercepted water are thought to be unimportant (Johns et al. 1984). However, though the total loss to gross interception may be small, its role in redirecting water to the concentrated root zone below the stem may be very significant. The concentration of nutrients in the root zone is partly a testimony to this concentration of wash from the canopy itself. De Ploey (1982) has attempted to provide a simple model to estimate the interception properties of plant cover on the basis of stem density and stem angle and has simulated the behaviour of stems in the laboratory with a reasonable degree of success.

Stone cover is also important in determining the interception of water, but the picture is somewhat confused for two reasons. First, losses to infiltration are estimated as the residual between water rate application by sprinkler and runoff from experimental micro-plots. Second, the loss rate is influenced by whether the stones are lying on the surface or embedded within it (Poesen et al. 1990). Our field experiments indicate that stones may intercept and store considerable volumes of water depending on their surface characteristics. Therefore, they probably account for much of the losses attributed to infiltration in sprinkler experiments. This question is further discussed below. The same is true of crusting. As yet we do not know the relative proportions attributable to interception, infiltration, and evaporation in determining the residual sum available for runoff.

Unsaturated Zone Processes

Soils in deserts are as variable as, if not more variable than, their temperate counterparts. The effects of soil texture are emphasized, as profile control is dominant in the overall water balance. For example, Hillel and Tadmoor (1962), in a comparison of four desert soils in the Negev, showed that sandy soils wetted deeply, have the largest storage, the highest infiltration rates, and the lowest evaporation when compared with rocky slope soils and loessic and clay soils. The latter had very low storage as a result of surface crusting, poor absorption, and high vaporation rates. As a result, the plant growth capabilities were markedly different. Others too have shown that in sandy soils the upward movement of water vapour, facilitated by the open pore structure, may lead to condensation of water near the surface in winter-cold desert soils in amounts of the order of 15–25 mm.

A major driving force in desert soils is the high rate of near-surface evaporation. Water moves up through the soil as a result of the thermal gradient mainly through vapour transfer (Jury et al. 1981). With available moisture the near-surface potential rates may reach 1500–2000 mm y^{-1}, at 2 m the rates are about 100 mm y^{-1}, and at 3–4 m they are practically negligible. As a result of this steep gradient, upward movement of water and, if available, salts is the dominant process in summer months, as in the *takyr* soils of Turkmenistan.

The stone content of desert soils generally increases with depth, though such soils may also have surface armouring induced by a variety of processes Chapter 5). Field bulk density may be quite high for stoney soils, typically of the order of 1.8 compared with values of 1.0–1.4 for non-stoney soils. Mehuys et al. (1975) found little difference between stoney and non-stoney soils in the relationship between conductivity and soil tension, but strong differences in the relationship between conductivity and soil moisture. In two out of three cases the conductivity is higher in stoney than in non-stoney soils with milar soil moistures.

Infiltration

Infiltration rates on bare weathered soils in dry environments are often relatively high when surfaces are not compacted, especially where vegetation or a heavy surface litter occurs. Runoff is often critically determined by the controls of infiltration, especially vegetation at plot scales (Francis and Thornes 1990). In experiments using a simple sprinkler system at rates of 55 mm h^{-1}, with an expected recurrence interval of about 10 years, infiltration rates may be as high as 50 mm h^{-1} with runoff accounting for as little as 2–3% of the rainfall. However, under all circumstances the infiltration rates are highly variable.

In general, infiltration rates increase with litter and organic matter content of soils. One of the earliest investigations, by Lyford and Qashu (1969), showed that infiltration rates decreased with radial distance from shrub stems due to lower bulk density and higher organic matter content under the shrubs. Later investigations by Blackburn (1975) on semi-arid rangeland in Nevada examined the effects of coppice dunes (i.e. vegetation bumps) on infiltration, runoff, and sediment production. Using a sprinkler simulator at 75 mm h^{-1} on small plots, he analysed the average infiltration rates at the end of 30 min and found that the coppice dune rate was three to four times that for the dune interspace areas. The soil characteristics of the interspaces were most important in determining runoff rates, and infiltration rates were strongly negatively correlated with percentage bare ground. Similarly, Swanson and Buckhouse (1986), in a comparison of three subspecies of big sagebrush (*Artemisia tridentata, A. wyomingensis*, and *A. vaseyana*) occupying three different biotopes in Oregon, found no significant differences in final infiltrability between habitat type and climax understorey species. However, infiltration in shrub canopy zones had generally higher rates than shrub interspaces, confirming the results obtained by Lyford and Qashu (1969) and Blackburn (1975).

There is often an interaction between infiltration capacity, vegetation cover and aspect in dry environments. For example, Faulkner (1990), examining infiltration rates on sodic soils in Colorado, has found that aspect has a strong influence on infiltration rates through vegetation cover, though some caution needs to be exercised here because of the role of piping in infiltration rate and runoff control in the bare badland areas.

Gifford et al. (1986) examined the infiltrability of soils under grazed and ungrazed (20 years) crested wheatgrass (*Agropyron desertorum*) in Utah using both sprinkling and ring infiltrometers. They found that infiltration rates measured by the ring infiltrometer were 2.3–3.2 times higher than those measured by the sprinkling infiltrometer (cf. Scoging and Thornes 1979) and that season and grazing/nongrazing were the main sources of variation. The spring infiltrability values were double the summer values, and the ungrazed were three times the grazed values.

Infiltration rates are generally reduced by crusting, which may result from algal growth, rainfall compaction, and precipitate development. Thin algal crusts have an important role in stabilizing the soil surface, though there seems to be some disagreement as to its hydrological significance. Some workers claim that the crust, by limiting the impact of raindrops on a bare soil, actually prevents the development of a true rainbeat crust and thus the algae increase infiltration. A contrary view, however, is that the biological crusts have a strong water repellency and so increase runoff. Alexander and Calvo (1990) and Yair (1990) have investigated the organic crusting and attest to its significance in controlling surface runoff. Alexander and Calvo found that lichens provide crusting on north-facing but not on south-facing slopes and that lichens produce more rapid ponding and runoff but not more erosion. Yair reported that removal of the crust from sandy dune soils produced a dramatic increase in the time to runoff and that the runoff coefficient was an order of magnitude lower, largely due to reinfiltration.

On the other hand, compaction by raindrop impact leading to sealing significantly reduces infiltration (Morin and Benyami 1977). Roo and Riezebos (1992) have attempted to provide a model for the impact of crust development on infiltration amounts.

Finally, stones control the infiltration rates into soils to some extent. The general conclusion of the many field and laboratory studies on infiltration rates under a rock cover indicates a generally positive correlation between infiltration and stone cover, but Poesen et al. (1990) argued that this depends on whether the rocks are partially buried or resting on the surface. In the former case there tends to be a negative correlation because the stones intercept the rainfall and lead to higher evaporation and runoff rates, whereas in the latter case they tend to inhibit crusting and so encourage infiltration. Working on shrub lands, both Tromble et al. (1974) and Wilcox et al. (1988) obtained negative correlations between infiltration and stone cover, but the latter accounted for this in terms of the reduced opportunity for plant cover development, which enhances infiltration as noted above. In a field experiment Abrahams and Parsons (1991) also obtained a negative correlation between stone cover and infiltration rate and again explained this in terms of the higher infiltration rates under bushes where stones were generally absent (see Fig. 9.11). They concluded that, at least for shrub covered pediments, inverse relations between infiltration and stone cover could be used in modelling.

Evaporation from Surfaces and Transpiration from Plants

Most literature for dry environments suggests that there is a high correlation between evapotranspiration and rainfall (Scholl 1974) and that in general the total annual evapotranspiration for desert environments is equal to the cumulative annual infiltration into the soil for the year. This is more or less equal to the annual precipitation minus the annual runoff. Moreover, at least once per year the soil moisture becomes more or less negligible and deep drainage is more or less negligible relative to total annual evapotranspiration. Therefore, the principal interest is in the temporal distribution of evapotranspiration during the year.

Actual estimates of the relative balance of the various components give a clear idea of the importance of evapotranspiration in the overall water budget. Renard (1969) suggested that evaporation and evapotranspiration from Walnut Gulch is about 85–90% of the incoming rainfall, and in the study by Floret et al. (1978) in Tunisia evaporation from bare soil alone was calculated to be between 40 and 70% of the total depending on the year. Evans et al.'s (1981) comparative analysis of bare soil and different plant species (creosote, mesquite, and sagebrush) revealed no significant differences between the bare soil and the vegetated sites for any extended period of time. Nor were there any differences between plant species, and

most of the evapotranspiration took place in periods immediately following rainfall. Then the rates varied between 2.3 and 10 mm d^{-1}, with rates of 1 mm d^{-1} for extended periods of time between rainfall events and rates of less than 0.1 mm d^{-1} for the driest parts of the year.

These results emphasize the ability of desert plants to exist at high water stresses. Final water contents correspond to suctions of −10 to −50 bars. Typically the plant stress is uniformly distributed throughout the plant at dawn and in equilibrium with the soil suction. As the day develops the plant stresses increase so that for *Larrea tridentata stresses* of 40–65 bars are reached, and in *Ceratoides lanata*, for example, 120 bars potential is not uncommon.

The evapotranspiration process is clearly crucially important for dry environments, even when vegetation cover is small, and exercises a significant control on the total water budget. Since this budget is controlled by the plant cover, which also plays a major role in the hydrology and hydraulics of overland flow, we return to the vegetation cover next.

A general classification of xerophytic plants is given in Fig. 11.8 (Levitt 1972). Xerophytic plants can escape drought seasons by an ephemeral habit or resist drought either by avoiding it by reducing water consumption or by using the maximum possible when it is available. The latter strategy is a characteristic especially of phreatophytes, such as mesquite and tamarisk. Other plants are able to tolerate drought by mechanical means. Mosses and lichens, for example, can expand again after dehydration. Such poikilo-hydrous plants have the capacity to take up water

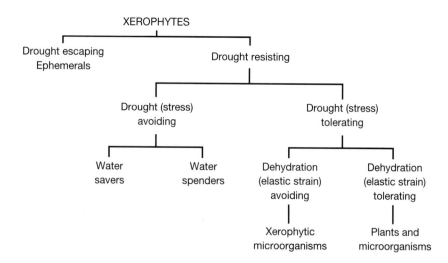

Fig. 11.8 Structure of xerophytic plant types (after Levitt 1972)

instantaneously and survive extreme and prolonged desertification.

Specific observations on desert vegetation have been made by Nobel (1981) in the Sonoran Desert. He investigated the desert grass *Hilaria rigida*, which is a clump grass of the C4 type and has amongst the highest photosynthetic rates so far reported. In this species, clumps of increasing size tend to be further from a random point and further from their nearest neighbour of the same species. This larger spacing suggests pre-emption of groundwater. This conclusion is partly borne out by the fact that leaves on small clumps (less than 10 culms) have a threefold higher rate of transpiration per unit leaf area than do leaves on larger clumps (over 200 culms), indicating that the water transpired does not vary much with clump size. Given that the amount of CO_2 fixed per unit ground area does not differ much with clump size, the pattern of large and small clumps can be quite stable. This may also be indicated by the low interdigitation of root systems for different clumps. This high level of adjustment suggests that clump life could be quite long. Somewhat similar conclusions were reached by Phillips and McMahon (1981).

Some species are evergreen, while others are facultatively drought deciduous. Several studies have indicated that the leaf area index (LAI) declines as water potential declines, and Woodward (1987) has developed a model which predicts LAI in terms of a cumulated water deficit, arguing that when the water deficit falls below -80 mm abscission begins (Fig. 11.9).

In addition to changes of leaf shape and plant structure, other special plant physiological adaptations occur which give desert plants more efficient water use. For example, succulent plants possess pathways for the fixation of CO_2 that enable assimilation to proceed with the stomata open only at night, when potential for evaporation from cladodes and leaves is minimal (Fuchs 1979).

The plant cover and biomass is a function of the production and distribution of the photosynthetic production, and this is usually controlled by the evapotranspiration rate. In modelling the impact of climatic change on erosion in which the plant cover is a most critical variable, the key issue in arid and semi-arid areas is the plant water balance and its impact on net production (i.e. the difference between the plant material produced by photosynthesis and that consumed for maintenance). It is generally assumed that the plant biomass

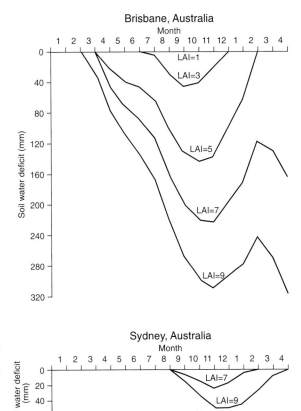

Fig. 11.9 Predicted monthly soil-water deficit for soils at two sites in Australia in accordance with Woodward's (1987) model. LAI = leaf area index (after Woodward 1987)

comes into equilibrium with the available water (cf. Eagleson 1979, Thornes 1990, and see below).

Lane et al. (1984) demonstrated for the Mojave Desert that the best predictors of annual plant productivity were linear relations with mean annual precipitation ($r^2 = 0.51$), seasonal precipitation (January–May, $r^2 = 0.74$), annual transpiration ($r^2 = 0.84$), and seasonal transpiration ($r^2 = 0.90$).

Net plant productivity in the south Tunisian steppe is approximately 1000 kg ha^{-1} y^{-1} of dry matter or approximately 5.5 kg ha^{-1} mm^{-1} of rainfall. Wet years produced 1550 kg of dry matter per hectare and dry years about 1060. The difference is due to the distribution of rainfall. This distribution is especially important when the annuals and perennials occur at different times of the year. Floret et al. (1978) found that annuals peak in May, while perennials peak in June and July. For maximum production of annuals, rainfall should be regularly distributed between autumn and the end of spring, while bushes do better if the deep soil layers are still wet in mid-May.

Modelling vegetation growth and water consumption for sparse vegetation can be illustrated by the ARFEQ model of Floret et al. (1978). These authors modelled primary production and water use in a south Tunisian steppe, comprising dominantly *Rhatherium suaveolus*. This is a zone of typically 170–180 mm y^{-1} rainfall, falling mainly in September and March. The soil is a gypsous sandy clay with an organic matter content of the order of 0.4–0.7%. The cover of perennial species is about 35% *Stipa* sp., *Aristida* sp., and *Plantago albicans*. *Rantherium* accounts for a further 30% with about 26 500 tussocks ha^{-1}. The surfaces typically have 1600 kg ha^{-1} of above-ground phytomass and 50–80 kg ha^{-1} of litter on the soil.

In the ARFEQ model bare soil drainage is an inverse function of the storage above field capacity; potential evapotranspiration is calculated as a function of the measured Piche evaporation; and actual evapotranspiration is related to potential through the dimensionless water content of the water content of the upper horizon. For the plant cover, total transpiration is obtained from a comparison of the atmospheric demand and the soil-water stress. The atmospheric demand is expressed in terms of LAI, stomatal conductance, and Piche evaporation. The soil-water resistance is obtained from the difference between soil-water stress and leaf stress mediated by a root resistance, which is inversely proportional to the amount of root mass in a given soil layer. Leaf-water potential is then assumed to meet the requirement that there is an equilibrium between the atmospheric demand and the root extraction. This is a common assumption for water-limited plant covers (i.e. that the plant water use is in equilibrium with the applied stress). In practice this assumption may often be doubtful given the high interannual and longer-term variability. Below we examine how changes in water availability and redistribution lead to vegetation cover extinction and, hence, to changes in the hydrological regime over the longer term.

From these hydrological considerations, gross photosynthesis is assumed to be a function of water status, LAI, and incoming solar radiation (Feddes 1971), and the photosynthetic rate less respiration and senescence is used to determine dry matter production. The results of the model suggest that approximately 0.26–0.54 g of dry matter are produced per kilogram of H_2O, rising in spring to 1.26 for annuals and 1.39 for perennials.

While still in the domain of unsaturated control, it is worth noting that there have been significant new developments in estimating the impact of large areas of bare surface on the now widely accepted Penman–Monteith equation (Monteith 1981), though the developments are still largely restricted to agricultural crops, such as millet. The Penman–Monteith model is based on the estimation of potential evapotranspiration controlled by atmospheric demand through the resistances of plant stomata (stomatal resistance), the canopy overall, and the atmosphere above the canopy to water vapour transport. However, it has been observed recently that a substantial amount of water is lost from the surface by evaporation from dry soils. For example, Gregory (1991) reported that as much as 75%, and commonly of the order of 35%, is lost by evaporation from bare soil surfaces between row crops. There seem to be two possible approaches to dealing with this problem: the first is to have separate estimators for crops and bare areas; and the second it to have a combined model which takes both into account. With regard to the first approach, Ritchie (1972) developed a model which assumed that after rain, soil evaporation is determined by net radiation at the soil surface. However, following this initial phase, evaporation from the bare soil becomes increasingly limited by the ability of water to diffuse through the drying soil surface. In this second phase, Ritchie set cumulative soil evaporation to be inversely proportional to the square root of time. Many observations of desert soils have empirically supported the use of this proportionality even though this model allows no interaction between the soil and the canopy. Two particular models are important in the second approach: ENWATBAL developed by Lascano et al. (1987) and the sparse crop model of Shuttleworth and Wallace (1985). In the latter the near-surface layer (Fig. 11.10) comprises an upper layer, which loses moisture directly from the canopy, and an interactive soil substrate, which has its own surface resistance and an aerodynamic resistance between the soil and the canopy. The combined evapotranspiration is then set up as:

$$LE_c + LE_s = C_cPM_c + C_sPM_s \qquad (11.4)$$

in which L is the latent heat of vaporization, E_c the canopy transpiration, E_s the bare soil evaporation, and

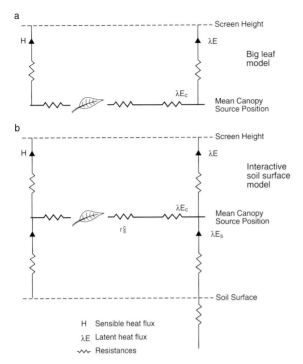

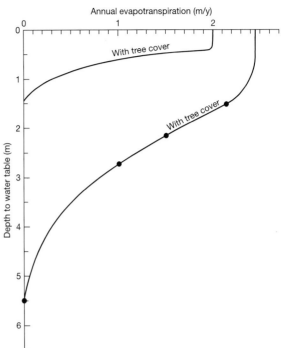

Fig. 11.10 The Shuttleworth and Wallace (1985) model of canopy and bare soil: (**a**) classical 'big leaf' model, (**b**) interactive soil substrate model. E is total latent heat flux, with subscripts c and s representing canopy and bare ground, respectively, and r is the canopy resistance

Fig. 11.11 Gila valley phreatophytes water loss by evapotranspiration over depth, with (*solid lines* with *circles*) and (hypothetically) without salt cedar cover

PM_c and PM_s the respective Penman–Monteith estimates. The coefficients C_c and C_s are then empirical coefficients which are functions of the aerodynamic resistance and bulk stomatal resistance of the crop and the soil (Wallace 1991).

Phreatophytes

It was argued in the 1960s that phreatophytes in the American South-west were major consumers of water and that the elimination of phreatophyte vegetation might be considered as a suitable method of water harvesting. Bouwer (1975) considered that potentially 6.4 million hectares of phreatophytes could be consuming $30 \times 10^9 \, m^3$ of water. The uptake of groundwater by plant roots varies diurnally and seasonally, but fundamentally the evapotranspiration rate appears to be related to the depth of water table. Bouwer (1975) has modelled it as such on the assumption that the system is in steady state. The deeper the roots the greater the water table can sink and still maintain essentially potential evapotranspiration rates. Because roughness and total leaf area tend to increase with increasing plant height, and trees and shrubs are generally deeper rooted than grasses, potential evapotranspiration for deep-rooting vegetation is higher than that for shallow-rooting plants, and higher still than that for bare soil.

Van Hylckama (1975) has estimated the water loss with and without salt cedar cover in the lower Gila valley, Arizona (Fig. 11.11). His analysis suggests that with salt cedar the surface groundwater rate would yield about $2.5 \, m \, y^{-1}$ of water loss with trees and $2.0 \, m \, y^{-1}$ without. This difference increases sharply as the water table falls. Bouwer, however, noted that the reduction in channel seepage due to phreatophyte control increases and the actual amount of water which can be saved decreases as distance from the channel decreases.

Hillslope Runoff

In drylands overland flow is patchy in occurrence, reticular in character, and often terminates before reaching the channel by reinfiltrating on the hillslope

(Yair and Lavee 1985). Its characteristics are largely determined by the disposition of stones, vegetation, and macroscale roughness. Conversely, it plays an important role in the location of plant activity at the drier margins, and there may be a complex interaction between overland flow, reinfiltration, plant cover, and erosion. This reticular flow has been the subject of much recent research, since the important observations of Emmett (1970). Abrahams et al. (1986), Thornes et al. (1990), Dunne et al. (1991), and Baird et al. (1992) have all drawn attention to the impact of surface topographic stones and vegetation in concentrating the flow into rills and eventually gullies. Reticular flow contrasts significantly with sheet flow. In the past overland flow has often been modelled as sheet flow by means of the kinematic cascade model, which assumes that an overland flow of uniform depth is propagated down a hillslope comprising a series of smooth planes. The significance of reticular flow lies in the restriction of flow-induced infiltration and wash erosion to the actual flow-wetted area at the surface, which in

turn has important implications for a wide range of processes. In Chapter 9 the generation and propagation of reticular runoff in hydraulic terms is discussed in detail. Here we restrict ourselves to the main controls on the quantity and timing of slope runoff in terms of the water reaching the channel system.

On slopes mantled with coarse debris, the stone content exerts a substantial control on hillslope runoff process. Yair and Lavee (1974) examined runoff from coarse scree slopes using rainfall simulators at three intensities: 60 mm h^{-1} for 10 min, 30 mm h^{-1} for 10 min, and 15 mm h^{-1} for 20 min. In these experiments the runoff was generated mainly in the lower areas (i.e. within the gullies). The explanation given was that the larger blocks (15–20 cm) in these areas yield more runoff and concentrate the water on small underlying patches of fine-grained materials, thus exceeding the infiltration rates in these fine-grained receiving areas.

Stones, like vegetation (Fig. 11.12), concentrate the flow so that velocities are higher, infiltration localized, and scour restricted in space. Baird et al. (1992) have

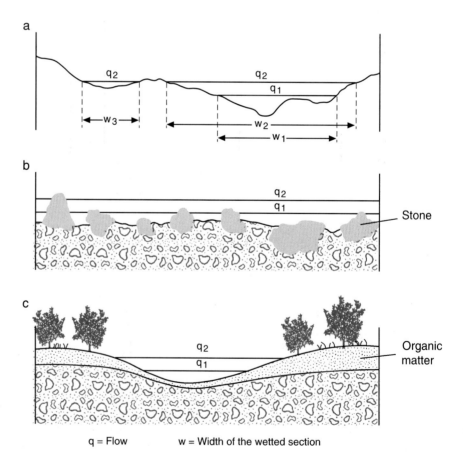

Fig. 11.12 Diagrams illustrating roughness elements. In (**a**) rills provide different flow depths and widths. In (**b**) the roughness elements are formed by surface rocks. In (**c**) the roughness elements are created by bushes and generate different infiltration amounts according to the thickness and disposition of organic matter

q = Flow   w = Width of the wetted section

modelled this process following the free volume concept of Thornes et al. (1990). In their model the rainfall is assumed to be distributed across the entire surface but water entering a slope section is allowed to fill the hollows to a depth equivalent to its volume. Assuming a normally distributed ground roughness, they avoid describing the detailed topography in each cell and generate a distribution of flow velocities on the basis of depth and slope distributions for the wetted areas. As the flow increases in depth, it drowns surfaces of different infiltration properties and infiltration rates evolve dynamically for different flow conditions. Results of the operation of this model show significant changes in the hydrograph and in sediment yields when compared with unrilled slopes (Fig. 11.13).

At the hillslope scale, Yair and Lavee (1985) demonstrated a complex interaction between runoff generation, slope properties, and vegetation cover on slopes in the northern Negev at Sede Boqer. They observed that flow from bare upper-slope rock surfaces passes on to lower slopes with increasing amounts of colluvium. There is a progressive diminution of water availability to plants further downslope, resulting in a change in species composition as well as plant cover density. In the upper part of the colluvium, perennial species are dominated by Mediterranean species, whereas lower down Saharo-Arabian species with a lower water demand dominate.

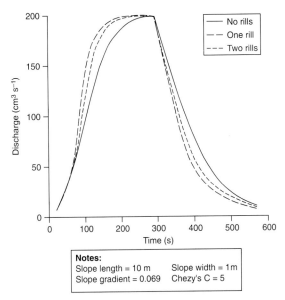

Fig. 11.13 Results from the RETIC model showing that the existence of rilling tends to steepen-up both the rising and receding limbs of the hydrograph

The distribution of species according to water availability and erosion competition was modelled by Thornes (1990) on the assumption of a wedgeshaped colluvium of increasing thickness downslope. This model yields an accumulation of water with increasing distance downslope, common in more humid Mediterranean environments, so that plants with a lower water use efficiency are located at the base of the slope (Fig. 11.14). Kirkby (1990), on the other hand, modelled the actual runon and reinfiltration conditions at Sede Boqer using a digital simulation model. Here the microtopography is generated by an exponential distribution, infiltration and evaporation follow storage-controlled laws, and water is routed kinematically. Some typical results are shown in Fig. 11.15a, giving the simulated overland flow for the bedrock surface and soil-covered sections and illustrating the pronounced effects of reinfiltration in the latter. The effects of redeposition are also felt so that the peak of erosion is in mid-slope, the net effect being to sweep the sediment in a downslope direction through time, leading to a redistribution of the whole complex of evaporation, recharge, and plant growth in that direction, which is offset only by the deposition of loess on the slope. The effects of the progressive movement of the system downslope are illustrated by the changes in the estimated relative plant densities through time shown in Fig. 11.15b. After 100 years the peak densities are predicted at about 28 m and by 500 years at about 42 m. At a given time the rise in plant cover continues downslope to a peak until no more water is available for evapotranspiration, after which it falls again dramatically in a downslope direction (Fig. 11.15b). The effect of this sweeping of sediment is to concentrate the vegetation in a narrow band in the zone of maximum infiltration losses. This peak vegetation cover actually moves slightly upslope through time. This complex interaction between vegetation, infiltration, and erosion at the hillslope scale is thought to replicate the actual conditions described for Sede Boqer quite well.

Much of the observation and modelling of dryland processes has been at the point, plot, or hillslope scale, and the problem of scaling up these observations and models to a regional scale has hardly been mentioned. An exception to this is the work of Pickup (1988) in Australia. Although this work is essentially directed at the problem of grazing-induced erosion, through the idea of scour–transportfill (STF) sequences, the

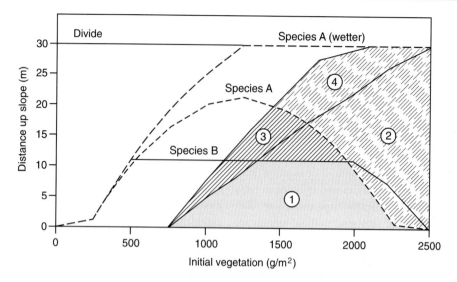

Fig. 11.14 The distribution of two species competing for water on a slope as a function of the initial vegetation cover. Species B is restricted to the lower part of the slope (below 12 m) almost regardless of the initial conditions. Species A can extend further upslope and under wet conditions can reach the divide, provided that the initial vegetation exceeds about 1200 g m^{-2}. In zone 1 both species are unchanging in biomass. In zone 3 (with diagonal hatching) species A is stable under dry conditions but species B will decrease. In zones 2 and 4 species A is stable and of unchanging biomass under wet conditions, but unstable and decreasing under dry conditions (after Thornes 1990)

approach is essentially hydrological in character. The spatial organization of runoff controls was also examined in a semi-arid mountain catchment by Faulkner (1987). This work also serves to remind us of the importance of snowmelt in semi-arid mountains as well as cold deserts. The hypothetical model of domains is shown in Fig. 11.16a, where the relative dominance of different processes is dependent on aspect and vegetation response. Faulkner proceeded to simulate the runoff production from the different surfaces in terms of snowmelt and rainfall runoff generation using established techniques. Figure 11.16b shows not only the importance of snowmelt on the hydrograph but also the different spatial impact its contribution has on the pattern of channel runoff.

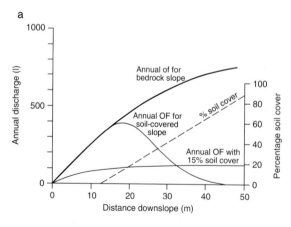
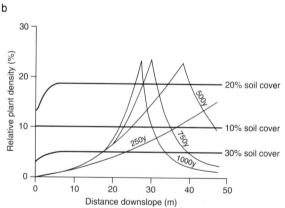

Fig. 11.15 Kirkby prediction of overland flow (OF) and vegetation density: (**a**) annual runoff on bedrock and on a colluvial slope with increasing colluvium in a downslope direction; (**b**) estimated plant cover and its evolution as erosion and deposition take place through runon from bedrock to a colluvial surface (after Kirkby 1990)

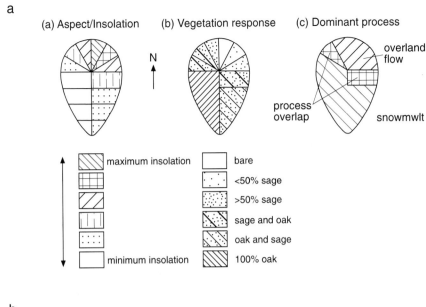

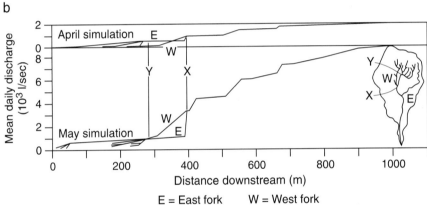

Fig. 11.16 (**a**) Hypothetical process domains in a mountain semi-arid snowmelt-dominated mountain catchment. (**b**) Simulated downstream changes in mean daily discharges in the pre-melt and melt seasons in Alkali Creek, Colorado (after Faulkner 1987)

At the regional scale the measured runoff redistribution of rainfall represents a complex range of factors. The various sources of losses to the surface affect the nature and timing of runoff in semi-arid environments on the hillslopes as well as in the channels. Pilgrim et al. (1979) estimated for the Fowler's Gap Arid Zone Research Station site in New South Wales initial losses prior to runoff of 5.35 mm with a continuing loss of typically 2 mm h^{-1}. However, as Wheater and Bell (1983) point out, catchment losses vary according to the scale of the runoff processes. In this regard, Shanen and Schick (1980) showed that in the Middle East initial loss estimates are 2.5 mm for small plots, 5.5 mm for 1–7 ha catchments, and 7–8 mm for a

3.45 km^2 catchment. This is a reflection of the general law that percentage runoff decreases with size of catchment, but in dry lands the losses are generally more acute. There is a mixture here too of direct losses from the catchment surface and channel transmission losses. The latter become particularly important with the development of an infiltrating channel fill (see below).

In recent years it has become generally more recognised following the work of Le Bissonais on runoff in relation to erosion in N.W. France that the surface characteristics of soils notably infiltration and crusting are not only important in runoff at several different scales but that they can be characterised and mapped using remotely sensed imagery. Coupled with GIS this is an important innovation in hillslope hydrology for large catchments. A simple example was Bull's (1991) distinction between loose sediment and rocky slopes or plateaux. In a direct application to arid zone hydrology on a very large scale Lange (1999) used six terrain types to represent hydrologically relevant surface characteristics in modelling runoff for the 1400 km^2 Nahel (Wadi) Zin in southern Israel. Each terrain type was studied in the field by obtaining infiltration parameters on small plots (Yair 1990) and detailed analyses of natural rainstorms on instrumented slopes. For the remaining terrain types, infiltration parameters were assessed by investigations of the stability of top-soil crusts and stony pavements (Yair 1992).

It was recognised long-ago (Thornes and Gilman 1983) that different Tertiary lithologies played a key role in runoff generation in Bronze-Age times in South-East Spain. It is the formal coupling of G.I.S. field studies and remote sensing in desert catchments that underlies the importance of recent work, especially recognising the intrinsic difficulties of working in and obtaining parameters for large desert areas. I raises the issues of what parameters are relevant and at what scales for defining hydrological similar surfaces.

Dalen et al. (2005) addressed precisely this question in their study of simulated and actual runoff in a dryland catchment in S.E. Spain. This proceeded through the development of a new runoff model base on the Green-Ampt algorithm coupled to the SCS Curve Number approach. This was used with a simulation model to examine the effects of rainfall intensity and geomorphological catchment characteristics on runoff at different scales, ranging from hill-slopes to moderate size catchments (up to 100 km^2). Vegetation was not taken into account. Nor was the aerial distribution of infiltration or crusting, but it is recognised as 'of great importance'.

It was found that spatial variations in rainfall intensity during the storm were most important in controlling run-off because of re-infiltration of run-off on long slopes. The 'effective intensity' decreases as (slope length)$^{-0.5}$ for major storms.

The second most important effect is produced by catchment geomorphic characteristics. Shape effects the distribution of travel times and channel cut-and-fill sequences affect transmission losses (Thornes 1977). Fresh incisions, such as badlands quickly convey water to the outlet.

Although crusting and infiltration properties were not specifically modelled, evidence from nearby areas demonstrate unequivocally that they are key controls on runoff (Li et al. 2005).

The HYSS approach seems destined to revolutionise hill-slope hydrology and thereby attempts to estimate runoff for large desert catchments.

Overall Water Balance

Accepting that these complexities arise in runoff regimes and that they have strong distributed patterns at the local scale, there is nevertheless an important role for a general model of the overall water balance, especially in the light of attempts to connect general circulation models to models of surface change. One of the most significant developments in this respect has been Eagleson's (1979) attempt to provide a hydrological model for the principal hydrological balance components in the overall annual hydrological cycle. This complex model rests on a statistical dynamic formulation of the vertical (point) water budget through equations expressing the infiltration, exfiltration, transpiration, percolation, and capillary rise from the water table both during and between storms. By asserting that the vegetation growth is in equilibrium and never in a stressed condition, Eagleson derives the expectation of the annual evapotranspiration and the optimal vegetation cover density for any location for which the required parameters are available. Typically the overall partition of the water budget and its division into climatic regimes are shown in Fig. 11.17a as a function of rainfall m_{PA}. The E outside the brackets indicates that the values are the statistically expected values. The solid lines indicate the major controls. Actual evapotranspiration $E^*_{T_A}$ increases with rainfall to a maximum when it equals potential evapotranspiration $E^*_{P_A}$. Infiltration I_A increases with rainfall until it is limited by saturated percolation to groundwater, which is a function of the mean duration of the rainy season m_τ and the saturated hydraulic conductivity $K(1)$. The uppermost, straight, solid line is the line through which mean annual rainfall m_{P_A} is equal to mean annual potential evapotranspiration $E^*_{P_A}$, so to the right of this line there is always excess water.

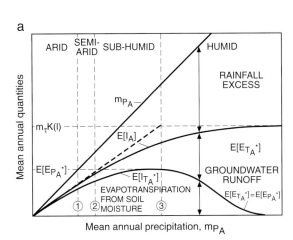

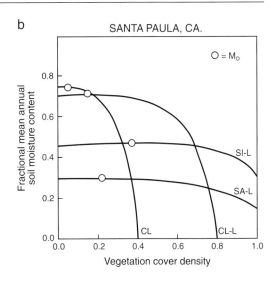

Fig. 11.17 (**a**) The hydrological components and classification of hydrological regimes according to Eagleson's model. (**b**) The relationship between plant cover density, soil type, and soil moisture in the same model

The vertical dashed lines associated with numbers just above the horizontal axis define climatic zones. The lower bound of the semi-arid zone (1) occurs when the mean annual rainfall is greater than the mean actual annual evapotranspiration, recalling that the latter includes the effects of seasonality, vegetation cover, soil type, and so on. The upper bound of this zone (2) separates the semi-arid from the subhumid regimes. In the former, infiltration and evapotranspiration are profile- (soil-) controlled, whereas in the latter evapotranspiration is climate-controlled. Above this (above rainfall at 3) infiltration is greater than actual evapotranspiration and the soil will be fully humid. Figure 11.17(b) shows the response of vegetation cover to soil moisture for different soil types (silt-loam, sandy loam, clay-loam, and clay). The point marked M_o is the optimum vegetation cover for the prevailing soil characteristics and soil moisture in an unstressed condition in Santa Paula, California (Eagleson 1979).

Groundwater

Although deep percolation from rainfalls is generally small and, therefore, groundwater is generally relatively unimportant from a geomorphic point of view, this is not true of suballuvial aquifers. Here the aquifer is recharged from transmission losses from channel flow and, therefore, it has important implications for runoff and transporting capacity. Moreover, the suballuvial rockhead configuration may lead to groundwater resurgence and availability of water to phreatophytes. Figure 11.18a shows the configuration of the alluvial fill of the Mojave River, California. It can be seen from this diagram that the suballuvial aquifer thins and shallows in the vicinity of Victorville, to the south of Bell Mountain, and again near Barstow. These subsurface 'narrows' force the water near to the surface, causing a significant growth of phreatophytes, which have a significant effect on channel behaviour and morphology. Figure 11.18b shows the pattern of precipitation at Lake Arrowhead and the resulting flow at Victorville and Barstow for the spring of 1969. By the time the flow had reached Barstow, the transmission losses to the bed had eliminated the lower flows. Finally, Fig. 11.18c shows the evolution of ground water in the suballuvial aquifer. This graph indicates the remarkable recovery of levels by recharge in the 1969 and 1978 floods. Buono and Lang (1980) reported that by the time the flow had reached Victorville 43% of it had been lost, and that 50% had been lost by the time it reached Barstow.

In another study of a much smaller channel system, Butcher and Thornes (1978) attempted to model the impact of transmission losses on channel flow and the survival of flow to the main channel. They showed that

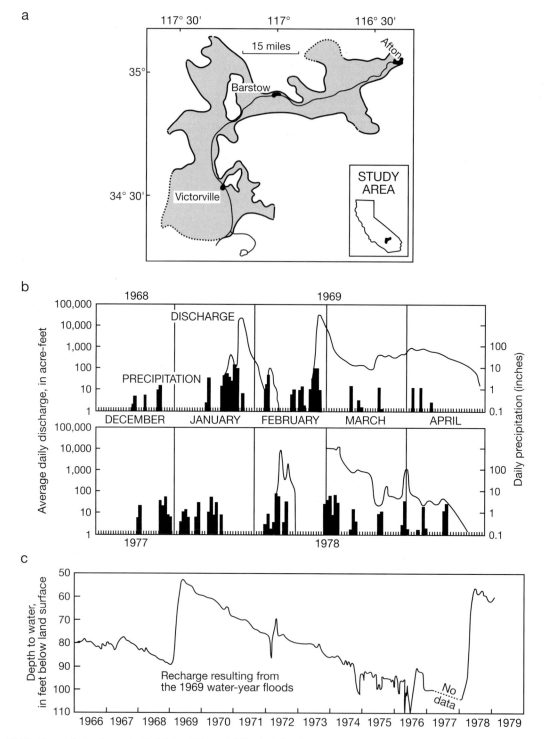

Fig. 11.18 Transmission losses in the Mojave River, California (after Buono and Lang 1980). (**a**) General location map showing alluvial fill. (**b**) Precipitation and flow for the floods of 1969 and 1978. (**c**) Changes in the water level below the surface at Baker

coupled with the transmission losses, the propensity for flows to survive in smaller channels and reach the main channel flow system also depended on the contributions of yet smaller channels to the system in question. The pattern of contributions of smaller subcatchments, called the area increment function, determined the likelihood that a flow would survive as far as the next tributary input. Butcher and Thornes modelled flows using a kinematic cascade routing model (KIN-GEN) developed by Woolhiser (1976) coupled with a transmission loss model for reaches between major tributaries based on a simple Kostiakov-type equation and the opportunity time for flow permitted by the passage of the flood wave. Smaller infiltration rates were found to produce disproportionately greater survival lengths. This application to a specific channel was generalized by Thornes (1977), who combined three principles – a differential equation for channel losses developed from Burkham's (1970) model, the hydraulic geometry of ephemeral channels (Leopold and Miller, 1956), and an area increment function based on Shreve (1974) – to solve the equations for the probability of survival downstream given various parameters derived from field studies. The general spatial scheme is shown in Fig. 11.19, together with typical theoretically derived survival curves. The implications of this model for geomorphic phenomena are illustrated in the next chapter. An excellent summary of the modelling of transmission losses was given by Lane (1980), and Lane (1982) developed a semi-empirical approach to transmission losses in the context of distributed modelling for small semi-arid watersheds.

Channel Flow

McMahon (1979) summarized the characteristics of runoff in desert environments on the basis of an examination of 70 annual flow records and peak discharge series from six desert zones. He found that runoff in desert channels was more varied than in humid channels, a fact confirmed by McMahon et al. (1987). Examining global runoff variations, these authors found that as runoff decreases, variability (defined as the standard deviation/mean annual runoff) increases. Consequently, extrapolation from humid zones is not acceptable. In testing the carryover effect, it was found that in desert channels the first-order serial correlation between annual flows (for 50 records) is 0.01. McMahon summarized the situation as follows:

(a) flood events are irregular and of short duration;
(b) data from gauging records are extremely poor;
(c) high stream velocities ($4 \, \text{m s}^{-1}$) are common;
(d) high debris loads occur during the passage of flood waves;
(e) gauging station controls are sandy and unstable;
(f) drainage boundaries are indefinite;
(g) overbank flows occur frequently;
(h) underflow is often a large part of flood events.

These conclusions are borne out by a few examples. Costa (1987) itemized the 12 largest floods ever measured in the United States. All occurred in semi-arid to arid areas, with mean annual rainfalls ranging from 114 to 676 mm; 10 had mean annual rainfalls less than 400 mm. This finding conflicts with the results of Wolman and Gerson (1978) who concluded that differences in maximum possible runoff from a single severe storm in different physiographic regions and climatic regimes appear to be insignificant. However, the United States experience does not suggest that these floods were due simply to high rainfall intensity. Rather it indicates that for a given intensity, rains in semi-arid and arid environments produce more runoff per unit area than they do in temperate environments. This finding reflects the whole panoply of surface infiltration and runoff controls dealt with earlier in the chapter. Mean velocity in these floods ranged from 3.47 to $9.97 \, \text{m s}^{-1}$, and shear stresses and unit stream powers were several hundred times greater than those produced in larger rivers.

The analysis of more individual extreme events is even more perplexing. Wheater and Bell (1983) found that problems with data are severe and that flood hydrology in arid zones was (and still is) largely unquantified. They exemplify the difficulties involved with reference to Wadi Adai in Oman where their reconstruction of a specific major flood depended heavily on a mixture of modelling and parameter estimation from sparse data. The wadi has a catchment area of $370 \, \text{km}^2$, channel slopes of 0.13–0.50, and a channel width of 50–150 m. On the basis of intermittent flood records stretching back to 1873, the flood of 3 May 1981 had a recurrence interval of 100–300 years. The main flood

a

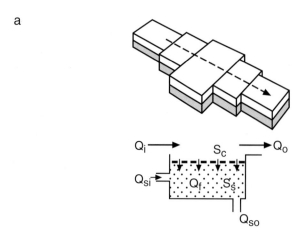

b

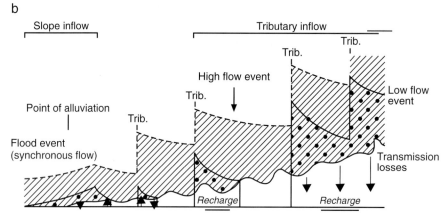

c

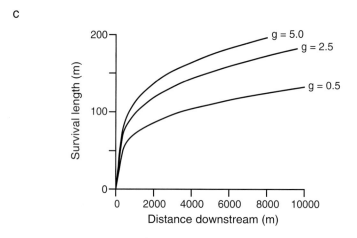

Fig. 11.19 (**a**) The effects of transmission losses on flow (after Thornes 1977). Q is surface discharge, Q_s is subsurface discharge, Q_f is transmission losses, S_c is channel sediment discharge and S_s is subsurface sediment discharge. The subscripts i and o denote inflow and outflow, respectively. (**b**) Hypothetical pattern of flows for two flood events. (**c**) Distribution of survival lengths with variation in the parameter g, which is the rate of change of discharge with area

peak arrived at 0300 h, the recession started at 0530 h, and there was hardly any flow by 1700 h. Flow velocities were estimated to be 3.6–4.5 m s^{-1}.

Descriptions abound of the characteristic features of flow in ephemeral channels, though the number of well-documented sites at which they have been properly investigated is still quite small. Among these the best are the Walnut Gulch Experimental Watershed, Arizona (Renard 1970) and the Nahal Yael Watershed, Israel (Schick 1986). The events in ephemeral channels, separated by zero flows, are often asynchronous across the channel network and are sometimes accompanied by well-developed bores and periodic surges in flow. Moreover, the recession limbs are usually very steep. All these features have been modelled, sometimes in combination, and Lane (1982) has considered their application to Walnut Gulch. We now consider the modelling of some of the components. The geomorphic impact of these phenomena is considered in the following chapter.

The development of a steep bore at the front of ephemeral channel flow is caused by a combination of intense storms, rapid translation of the water under conditions of high drainage density, and the interaction of wave movement and transmission losses. The second of these is addressed in Chapter 12. The development of a bore is analogous to the release of water in an irrigation ditch or in a dam burst. At the front edge of the wave the water is initially shallow and the bed friction high. This is compounded by the exhaust of air from the bed as infiltration forces it out, sometimes trapping it between the advancing subsurface wetting front and the groundwater.

Smith (1972) modelled these effects using the kinematic wave approach, which assumes that consideration of the momentum of the flow can be neglected. He predicted the advance rate, surface profiles, and modifications with time to kinematic wave flow over an initially dry infiltrating plane. Although the kinematic wave approach has been the subject of much recent criticism, it appears to be appropriate for small catchments, where it is possible to resolve the physical detail without compromising the deterministic nature of the model (Ponce 1991). Smith's approach was to set up the continuity equation to take account of the transmission losses – that is

$$\mathrm{d}h/\mathrm{d}t + \mathrm{d}(uh)/\mathrm{d}x = q(x,t) \qquad (11.5)$$

where h is flow depth, u the local velocity, x the distance along the channel, t is time, and $q(x, t)$ the local inflow (e.g. from tributaries or valley-side walls) or outflow (i.e. transmission losses). Figure 11.20a shows the basic setup of his model. A sharp fronted flood wave is advancing from left to right across an inclined channel bed. Beneath the bed the graphs show that infiltration rate is greatest near the front of the advancing wave and that the cumulative infiltration increases away from the advancing front. Laboratory experiments in a glass-sided flume by the author (Thornes 1979) clearly reveal a sharp wetting front that is quite even in sand-sized sediments, but has a fractal appearance in more variable sediments. The kinematic approximation is given by the stage-discharge equation

$$Q = \alpha \cdot h^{m+1} \qquad (11.6)$$

in which α and m are constants. The wave front itself is assumed to move as a kinematic shock. For infiltration, Smith set

$$-q(x) = K(T)^{-a} + f_0 \qquad (11.7)$$

in which $q(x)$ is the infiltration rate at x, T is the time since surface ponding, K and a are coefficients related to the characteristics of the channel bed materials, and f_0 is the long-term transmission loss rate. This is essentially the Kostiakov equation. Typical advance rates of the front are shown for the model (which replicated field data of Criddle (1956) very well) in Figure 11.20b. The curve suggests that the velocity of the wave front (the gradient of the line) increases with time, the average rate over this period being about 1 m min^{-1} in this particular case. Typical figures in semi-arid channels seem to be of the order of 1–3 km h^{-1} (as illustrated for Walnut Gulch in Fig. 11.20c), and the model replicates well the steepening of the wave front downstream. Behind the peak flow of the hydrograph, which comes almost immediately after the wave front, the deeper water provides a relatively lower roughness and a higher velocity, and so the wave front grows.

Infiltration also takes its toll on the recession limbs of ephemeral channel flood waves, which are also usually steeper than in temperate channels. The recession limb can be imagined as a wedge-shaped reservoir moving down channel, with the tail of the flow catching up with the front as transmission losses draw in

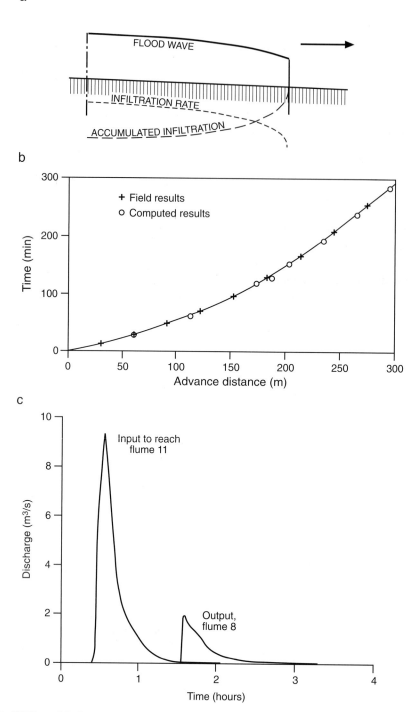

Fig. 11.20 Smith's (1972) model of wave front propagation in an infiltrating channel: (**a**) general set up of model, (**b**) downstream progress of flood wave compared with field data, and (**c**) actual propagation of a flood wave in Walnut Gulch between two stations

the shallower water of the recession limb, as has been modelled by Peebles (1975).

Not all rising limbs are associated with a single wave of water, and Foley (1978) described the onset of flow to be a series of translatory waves of small amplitude building up to flood depth. Periodic surges are also a characteristic of steady ephemeral channel flow, especially in larger channels. These appear as sudden increases in stage and may move as waves faster than the actual flow of the water itself, usually, but not always in a downstream direction. They can be responsible for extensive damage and also for some special sediment transport and depositional effects exclusive to this kind of flow. Leopold and Miller (1956) provided an excellent description of such a flow in which they observed a series of bores each 0.15–0.30 m high and between 0.5 and 1 min apart. One possible explanation is that the waves represent asynchronous contributions from different parts of the channel network. However, Leopold and Miller (1956) ascribed them to a type of momentum wave associated with the hydraulics of the channel itself. Such surges have also been simulated in the laboratory by Brock (1969), and R.J. Heggen (unpublished data 1986) described changes in width and velocity as well as height in these phenomena. Generally, under Green's Law, the elevation of a sinusoidal wave decreases as channel width increases, so the surges tend generally to attenuate downstream.

Conclusion

From this chapter it is evident that while the essential controls of arid and semi-arid hydrology are much the same as those of temperate environments, it is the time–space intensity of the hydrological processes which leads to significant differences in the corresponding geomorphic processes. Above all it is the time and space distribution of precipitation which leads to a much less ordered patterning of processes and the resulting forms. Even in arid environments, and certainly in semi-arid environments, the spatio-temporal diversity of rainfall leads to a distinctive pattern of plant water use, plant growth, and the resulting biological activity, both on hillslopes and, to a lesser extent, in channels. These two interrelated elements, water and biological growth, are the key to understanding past, present, and future deserts.

References

Abrahams, A.D. and A.J. Parsons 1991. Relation between infiltration and stone cover on a semiarid hillslope, southern Arizona. *Journal of Hydrology* **122**, 49–59.

Abrahams, A.D., A.J. Parsons and S.H. Luk 1986. Resistance to overland flow on desert slopes. *Journal of Hydrology* **88**, 343–63.

Ahnert, F. 1987. An approach to the identification of morpho-climates. In *International Geomorphology 1986*, Volume II, V. Gardiner (ed.), 159–88. Chichester: Wiley.

Alexander, R.W. and A. Calvo 1990. The influence of lichens on slope processes in some Spanish badlands. In *Vegetation and Erosion*, J.B. Thornes (ed.), 385–98. Chichester: Wiley.

Baird, A., J.B. Thornes and G. Watts 1992. Extending overland flow models to problems of slope evolution and the representation of complex slope surface topographies. In *Overland Flow: Hydraulics and Erosion Mechanics*, A.J. Parsons and A.D. Abrahams (eds), 199–211. London: University College Press.

Bell, F.C. 1979. Precipitation. In *Arid Land Ecosystems*, Volume I, D.W. Goodall and R.A. Perry (eds), 373–93. Cambridge: Cambridge University Press.

Berndtsson, R. 1987. Spatial and temporal variability of rainfall and potential evapotranspiration in Tunisia. *International Association of Scientific Hydrology Publication* 168, 91–100.

Blackburn, W.H. 1975. Factors influencing infiltration and sediment production of semi-arid rangeland in Nevada. *Water Resources Research* **11**, 929–37.

Bouwer, H. 1975. Predicting reduction in water losses from open channels by phreatophyte control. *Water Resources Research* **11**, 96–101.

Brock, R.R. 1969. Development of roll-wave trains in open channels. *Journal Hydraulics Division, Proceedings of the American Society of Civil Engineers* **95**, 1401–27.

Bull, W.B. 1991. *Geomorphic Response to Climate Change*, New York: Oxford University Press.

Buono, A. and D.J. Lang 1980. Aquifer recharge from the 1969 and 1978 floods in the Mojave River basin, California. In *Water Resources Investigations Open File Report 80–207*. Washington, DC: U.S. Government Printing Office, U.S. Geological Survey.

Burkham, D.E. 1970. A method for relating infiltration rates to stream flow rates in perched streams. *U.S. Geological Survey Professional Paper* 700-D, 226–71.

Butcher, G.C. and J.B. Thornes 1978. Spatial variability in runoff processes in an ephemeral channel. *Zeitschrift für Geomorphologie Supplement Band* **29**, 83–92.

Carson, M.A. and M.J. Kirkby 1971. *Hillslope form and process*. Cambridge: Cambridge University Press.

Conte, M., A. Giuffrida and S. Tedesco 1989. The Mediterranean oscillation. In *Conference on Climate and Water*, L. Huttunen (ed.), 121–38. Helsinki: Government of Finland Printing House.

Costa, 1987. Hydraulics and basin morphology of the largest flash floods in the conterminous United States. *Journal of Hydrology*, **93**, 313–38.

Criddle, C.L. 1956. Methods for evaluating irrigation systems. In *Agricultural Handbook*, 82, Washington, DC: U.S. Department of Agriculture.

Dalen, E.N., M.J. Kirkby, P.J. Chapman and L.J. Bracken 2005. Runoff generation in S.E. Spain. Poster presentation to the European Geophysical Union Meeting, Vienna.

De Ploey, J. 1982. A stemflow equation for grasses and similar vegetation. *Catena* **9**, 139–52.

De Ploey, J., M.J. Kirkby and F. Ahnert 1991. Hillslope erosion by rainstorms – a magnitude and frequency analysis. *Earth Surface Processes and Landforms* **16**, 399–410.

Dhar, O.N. and P.R. Rakhecha 1979. Incidence of heavy rainfall in the Indian desert region. *International Association for Scientific Hydrology Publication* **128**, 13–8.

Duckstein, L., M. Fogel and I. Bogardi 1979. Event based models of precipitation. *International Association of Scientific Hydrology Publication* **128**, 51–64.

Dunne, T., W. Zhang and B.F. Aubry 1991. Effects of rainfall, vegetation and microtopography on infiltration and runoff. *Water Resources Research* **27**, 2271–87.

Eagleson, P.S. 1979. Climate, soil and vegetation. *Water Resources Research* **14**, 705–77.

Emmett, W.W. 1970. The hydraulics of overland flow on hillslopes. *U.S. Geological Survey Professional* Paper 662–A.

Evans, D.D., T.W. Sammnis and B. Asher 1976. Plant growth and water transfer interactive processes under desert conditions. *US/IBP Desert Biome Research Memoir* 75–44. Logan, UT: Utah State University.

Evans, D.D., T.W. Sammnis and D.R. Cable 1981. Actual evapotranspiration under desert conditions. In *Water in Desert Ecosystems*, D.D. Evans and J.L. Thames (eds), 195–219. Stroudsburg, PA: Dowden, Hutchinson and Ross.

Faulkner, H. 1987. Gully evolution in response to flash flood erosion, western Colorado. In *International Geomorphology 1986*, Volume I, V. Gardiner (ed.), 947–73. Chichester: Wiley.

Faulkner, H. 1990. Vegetation cover density variations and infiltration patterns on piped alkali sodic soils. Implications for modelling overland flow in semi-arid areas. In *Vegetation and Erosion*, J.B. Thornes (ed.), 317–46. Chichester: Wiley.

Feddes, R.A. 1971. Water, heat and crop growth. Thesis, Agricultural University of Wageningen.

Floret, C., R. Pontanier and S. Rambal 1978. Measurement and modelling of primary production and water use in a south Tunisian steppe. *Journal of Arid Environments*, **82**, 77–90.

Foley, M.G. 1978. Scour and fill in steep sand-bed ephemeral streams. *Bulletin of the Geological Society of America* **89**, 559–70.

Fournier, F. 1960 *Climat et erosion*. Paris: Presses Universitaire.

Francis, C.F. and J.B. Thornes 1990. Runoff hydrographs from three Mediterranean vegetation cover types. In *Vegetation and erosion*, J.B. Thornes (ed.) 363–84. Chichester: Wiley.

Fuchs, M. 1979. Atmospheric transport processes above arid land vegetation. In *Arid Land Ecosystems*, Volume 1, D.W. Goodall and R.A. Perry (eds), 393–433. Cambridge: Cambridge University Press.

Gifford, G.F., M. Merzougi and M. Achouri 1986. Spatial variability characteristics of infiltration rates on a seeded rangeland site in Utah, USA. In *Rangeland: A Resource Under Siege*, P.J. Joss, P.W. Lynch and O.B. Williams (eds), 46–7. Canberra: Australian Academy of Sciences.

Gonzalez-Hidalgo, C. 1991. Aspect, vegetation and erosion on slopes in the Violada area, Zaragosa. Unpublished Ph.D. thesis, University of Zaragosa.

Gregory, P.J. 1991. Soil and plant factors affecting the estimation of water extraction by crops. *International Association for Scientific Hydrology Publication* **199**, 261–73.

Herschfield, D.M. 1962. Extreme rainfall relationships. *Journal of Hydraulics Division, Proceedings of the American Society of Civil Engineers* **88**, 73–92.

Hillel, D. and N. Tadmoor 1962. Water regime and vegetation in the central Negev Highlands of Israel. *Ecology* **43**, 33–41.

Jacobs, B.L., I. Rodriguez-Iturbe and P. Eagleson 1988. Evaluation of homogenous point process description of Arizona thunderstorm rainfall. *Water Resources Research* **24**, 1174–86.

Johns, G.G., D.J. Tongway and G. Pickup 1984. Land and water processes. In *Management of Australia's Rangelands*, G.N. Harrington, A.D. Wilson and M.D. Young (eds), 25–40. Canberra: Commonwealth Scientific and Industrial Research Organization.

Jury, W.A., J. Letey and L.H. Stolzy 1981. Flow of water and energy under desert conditions. In *Water in Desert Ecosystems*, D.D. Evans and J.L. Thames (eds), 92–113. Stroudsburg, PA: Dowden, Hutchinson and Ross.

Kirkby, M.J. and R.H. Neale 1987. A soil erosion model incorporating seasonal factors. In *International Geomorphology 1986*, Volume II, V. Gardiner (ed.), 189–210. Chichester: Wiley.

Kirkby, M.J. 1990. A simulation model for desert runoff and erosion. *International Association for Scientific Hydrology Publication* 129, 87–104.

Lange, J., C. Leibundgut, N. Greenbaum and A.P. Schick 1999. A non-calibrated rainfall-runoff model for large, arid catchments. *Water Resources Research* **35(7)**, 2161–2172

Lane, L.J. 1980. Transmission losses. *National Engineering Handbook*, Section 4, Chapter 19. Washington, DC: U.S. Printing Office, U.S. Department of Agriculture, Soil Conservation Service.

Lane, L.J. 1982. Distributed model for small semi-arid watersheds. *Journal Hydraulics Division. Proceedings of the American Society of Civil Engineers* **108**, 1114–31.

Lane, L.J., E.M. Romney and T.E. Hakenson 1984. Water balance calculations and net production of perennial vegetation in the northern Mojave desert. *Journal Range Management* **37**, 12–18.

Langbein, W.B. and S.A. Schumm 1958. Yield of sediment in relation to mean annual precipitation. *Transactions of the American Geophysical Union* **39**, 1076–84.

Lascano, R.J., C.H.M. van Bavel, J.L. Hatfield and D.R. Upchurch 1987. Energy and water balance of a sparse crop: simulated and measured soil evaporation. *Soil Science Society of America Journal* **51**, 1113–21.

Le Houerou, N. 1979. North Africa. In *Arid Land Ecosystems*, Volume 1, D.W. Goodall and R.A. Perry (eds), 83–107. Cambridge: Cambridge University Press.

Leith, H. and R.H. Whittaker 1975. *The Primary Productivity of the Biosphere*, Ecological Studies 14. New York: Springer.

Leopold, L.B. and J.P. Miller, 1956. Ephemeral streams – hydraulic factors and their relationship to the drainage net. *U.S. Geological Survey Professional Paper* 282-A.

Levitt, J. 1972. *Responses of Plants to Environmental Stress*. New York: Academic Press.

Li, X.-Y., A. González and A. Solé Benet 2005. Laboratory methods for the estimation of infiltration rate of soil crusts in the Tabernas Desert badlands. *Catena* **60**, 255–266.

Lyford, F.P. and H.K. Qashu 1969. Infiltration rates as affected by desert vegetation. *Water Resources Research* **5**, 1373–6.

McMahon, T.A. 1979. Hydrological characteristics of arid zones. *International Association of Scientific Hydrology Publication* **128**, 105–23.

McMahon, T.A., B.L. Finlayson and R. Srikanthan 1987. Runoff variability: a global perspective. *International Association of Scientific Hydrology Publication* **168**, 3–12.

Mehuys, G.R., L.H. Stolzy, J. Letey and L.V. Weeks 1975. Effect of stones on the hydraulic conductivity of relatively dry desert soils. *Soil Science Society of America Proceedings* **39**, 37–42.

Monteith, J. 1981. Evaporation and surface temperature. *Quarterly Journal of the Royal Meteorological Society* **107**, 1–27.

Morin, J. and Y. Benyami 1977. Rainfall infiltration into bare soils. *Water Resources Research* **13**, 813–7.

Nobel, P.S. 1981. Spacing and transpiration of various sized clumps of a desert grass *Hilaria rigida*. *Journal of Ecology* **69**, 735–42.

Peebles, R.W. 1975. Flow recession in the ephemeral stream. Unpublished Ph.D. thesis, University of Arizona, Tucson.

Peltier, L.C. 1950. The geographical cycle in periglacial regions. *Annals of the Association of American Geographers* **50**, 214–36.

Phillips, D.L. and J.A. McMahon, 1981. Competition and spacing patterns in desert shrubs. *Journal of Ecology* **69**, 97–115.

Pickup, G. 1988. Modelling arid zone soil erosion at the regional scale. In *Essays in Australian Fluvial Geomorphology*, R.F. Warner (ed.), 1–18. Canberra: Academic Press.

Pilgrim, D.H., I. Cordery and D.G. Doran 1979. Assessment of runoff characteristics in arid western New South Wales. *International Association for Scientific Hydrology Publication* **128**, 141–50.

Poesen, J., F. Ingelmo-Sanchez and H. Mucher 1990. The hydrological response of soil surfaces to rainfall as affected by cover and position of rock fragments in the top layer. *Earth Surface Processes and Landforms*, **15**, 653–72.

Ponce, V.M. 1991. The kinematic wave controversy. *Journal of Hydraulic Engineering* **117**, 511–25.

Renard, K.J. 1969. Evaporation from an ephemeral stream bed: discussion. *Journal of the Hydraulics Division, Proceedings of the American Society of Civil Engineers* **95**, 2200–4.

Renard, K.J. 1970. The hydrology of semi-arid rangeland watersheds. *U.S. Department of Agriculture, Agricultural Research Service*, 41–162. Washington, DC: U.S. Government Printing Office.

Ritchie, J.T. 1972. Model for predicting evaporation from a row crop with incomplete cover. *Water Resources Research* **8**, 1204–13.

Roo, A.P.J. and H. Th. Riezebos 1992. Infiltration experiments on loess soils and their implications for modelling surface runoff and soil erosion. *Catena* **19**, 221–41.

Schick, A.P. 1986. Hydrologic aspects of floods in extreme arid climates. In *Flood geomorphology*, V.R. Baker, R.C. Kochel, R.C. and P.C. Patton (eds), 189–203. Wiley: New York.

Scholl, D.G. 1974. Soil moisture flux and evapotranspiration determined from soil hydraulic properties in a chaparral stand. *Soil Science Society America Journal* **40**, 14–18.

Scoging, H. 1989. Run-off generation and sediment mobilisation by water. In *Arid Zone Geomorphology*, D.S.G. Thomas (ed.) 87–116. London: Belhaven.

Scoging, H. and J.B. Thornes 1979. Infiltration characteristics in a semi-arid environment. *International Association Scientific Hydrology Publication* **128**, 159–68.

Shanen, L. and A.P. Schick 1980. A hydrological model for the Negev Desert Highlands – effects of infiltration, runoff and ancient agriculture. *Hydrological Sciences Bulletin* **25**, 269–82.

Shmida, A. 1985. Biogeography of desert flora. In *Ecosystems of the World*, Volume 12A, *Hot Desert and Arid Shrublands*, Evenari, M., I. Noy-Meir and D.W. Goodall (eds) 23–77. Amsterdam: Elsevier.

Shmida, A., M. Evenari and I. Noy-Meir 1986. Hot desert ecosystems an integrated view. In *Ecosystems of the World*, Volume 12B, *Hot Deserts and Arid Shrublands*, M. Evenari, I. Noy-Meir and D.W. Goodall (eds), 379–88. Amsterdam: Elsevier.

Shreve, R.L. 1974 Variations of main stream length with basin area in river networks. *Water Resources Research* **10**, 1167–77.

Shuttleworth, W.J. and J.S. Wallace 1985. Evaporation from sparse crops – an energy combination theory. *Quarterly Journal Royal Meteorological Society* **111**, 839–55.

Slatyer, R.O. 1965. Measurements of precipitation interception by an arid zone plant community (*Acacia aneura*). *UNESCO Arid Zone Research* **25**, 181–92.

Smith, R.E. 1972. Border irrigation advance and ephemeral flood waves. *Proceedings American Society of Civil Engineers* **98(IR2)**, 289–307.

Smith R.E. and H.A. Schreiber 1973. Point processes of seasonal thunderstorm rainfall 1. Distribution of rainfall events. *Water Resources Research* **9**, 871–84.

Swanson, S.R. and J.C. Buckhouse 1986. Infiltration on Oregon lands occupied by three subspecies of big sagebrush *Artemisia*. *U.S. Department of Agriculture, Agricultural Research Service* INT-200, 286–91. Washington, DC: U.S. Government Printing Office.

Thornes, J.B. 1976. *Semi-Arid Erosional Systems*, Geography Research Papers No. 7. London: London School of Economics.

Thornes, J.B. 1977. Channel changes in ephemeral streams, observations, problems and models. In *River Channel Changes*, K.J. Gregory (ed.), 317–55. Chichester: Wiley.

Thornes, J.B. 1979. Fluvial Processes. In *Process in Geomorphology*, C.E. Embleton and J.B. Thornes (eds), 213–72. London: Arnold.

Thornes, J.B. 1990. The interaction of erosional and vegetational dynamics in land degradation: spatial outcomes. In *Vegetation and Erosion*, Thornes, J.B. (ed.) 41–53. Chichester: Wiley.

Thornes, J.B. 1991. Environmental change and hydrology. In *El Agua en Andalucia III*, V. Giraldez (ed.), 555–70. Cordoba: University of Cordoba.

Thornes, J.B. and J.C. Brandt. 1994. Erosion-vegetation competition in a stochastic environment undergoing climatic change. In *Environmental Change in Drylands*, A.C. Millington and K. Pye (eds). 305–20. Chichester: Wiley.

Thornes, J.B. and Gilman A. (1983) Potential and actual erosion around archaeological sites in south-east Spain' In J. de Ploey (ed.), *Rainfall Simulation, Runoff and Soil Erosion*, Catena Supplement No.4, Catena, Cremelingen.

Thornes, J.B., C.F. Francis, F. Lopez-Bermudez and A. Romero-Diaz 1990. Reticular overland flow with coarse particles and vegetation roughness under Mediterranean conditions. In *Strategies to Control Desertification in Mediterranean Europe*, Rubio, J.L. and J. Rickson (eds), 228–43. Brussels: European Community.

Tromble, J.M., K.G. Renard and A.P. Thatcher 1974. Infiltration for three rangeland soil-vegetation complexes. *Journal Range Management* **41**, 197–206.

Van Hylckama, T.E.A. 1975. Water use by salt cedar in the lower Gila River valley (Arizona) *U.S. Geological Survey Professional Paper* 491-E.

Wallace, J.S. 1991. Measurement and modelling of evaporation from a semi-arid system. *International Association Scientific Hydrology Publication* **199**, 131–48.

Walter, H. 1971. *Ecology of Tropical and Subtropical Vegetation*. Edinburgh: Oliver and Boyd.

Wheater, H.S. and N.C. Bell 1983. Northern Oman flood study. *Proceedings of the Institute of Civil Engineers* **75**, 453–73.

Whittaker, R.H. and W.A. Niering 1964. The vegetation of the Santa Catalina Mountains, Arizona. 1. Ecological classification and distribution of species. *Journal of the Arizonian Academy of Science* **3**, 9–34.

Wilcox, B.P., M.K. Wood and J.M. Tromble 1988. Factors influencing infiltrability of semiarid mountain slopes. *Journal of Range Management* **41**, 197–206.

Wolman, M.G. and R. Gerson 1978. Relative scales of time and effectiveness of climate in watershed geomorphology. *Earth Surface Processes* **3**, 189–208.

Wolman, M.G. and J.P. Miller 1960. Magnitude and frequency of forces in geomorphic processes. *Journal Geology* **68**, 54–74.

Woodward, F.I. 1987. *Climate and Plant Distribution*. Cambridge: Cambridge University Press.

Woolhiser, D.A. 1976. Overland flow. In *Unsteady Open Channel Flow*. K. Mahmood and V. Yevjevich (eds), 485–508. Fort Collins, CO: Water Resources Publications.

Yair, A. 1990. Runoff generation in a sandy area – the Nizzana Sands, Western Negev, Israel. *Earth Surface Processes and Landforms* **15**, 597–609.

Yair, A. 1992. The control of headwater area on channel runoff in a small arid watershed in Parsons, A.J. and Abrahams, A. (Ed) *Overland Flow*, London, University College Press, 53–68.

Yair, A. and H. Lavee 1974. Areal contribution to runoff on scree slopes in an extreme arid environment – a simulated rainfall experiment. *Zeitschrift für Geomorphologie Supplement Band* **21**, 106–21.

Yair, A. and H. Lavee 1985. Runoff generation in arid and semi-arid environments. In *Hydrological Forecasting*, M.G. Anderson and T. Burt (eds), 183–220. Chichester: Wiley.

Zawadeski, I.I. 1973. Statistical properties of precipitation patterns. *Journal of Applied Meteorology* **12**, 459–71.

Chapter 12

Dryland Rivers: Processes and Forms

D. Mark Powell

Introduction

Dryland alluvial rivers vary considerably in character. In terms of processes, high energy, sediment-laden flash floods in upland rivers contrast dramatically with the low sediment loads and languid flows of their lowland counterparts while from a form perspective, the unstable wide, shallow and sandy braid plains of piedmont rivers are quite different from the relatively stable, narrow, deep and muddy channels of anastomosing systems (Nanson et al. 2002). It is also apparent that few, if any, morphological features are unique to dryland rivers. The variety of dryland river forms and the absence of a set of defining dryland river characteristics makes it difficult to generalise about dryland rivers and raises questions about whether it is necessary (or even desirable) to consider dryland river systems separately from those in other climatic zones. Indeed, as noted in the introduction to this volume, the recent shift away from the study of morphogenesis within specific climatic regimes (e.g. Tricart and Cailleux 1972) towards the study of geomorphological processes *per se* (e.g. Bates et al. 2005) has largely undermined the distinctiveness of desert geomorphology. This is not to say rivers draining different climatic regions do not differ in aspects of their behaviour. They clearly do, as exemplified in several reviews of tropical (Gupta 1995), periglacial (McEwen and Matthews 1998) and dryland (Graf 1988; Knighton and Nanson 1997; Reid and Frostick 1997; Tooth 2000a) fluvial geomorphology. However, given the diversity of dryland river morphologies, and the fact that many of the forms are shared by rivers that drain other climatic zones, it is far from clear how far dryland rivers can be categorised as a distinctive group and whether such a categorisation provides a suitable basis for developing an understanding of them. On this basis, rather than attempt to understand dryland rivers as a distinctive and definable group of rivers, this chapter seeks explanations for the character (the diversity, distinctiveness and, in some cases, the uniqueness) of dryland rivers in terms of the operation of geomorphological processes as they are mediated by the climatic regime. Because the multivariate and indeterminate nature of river channel adjustment makes it difficult to describe directly the three-dimensional subtleties of channel form, the chapter follows the approach of Ferguson (1981) and concentrates on three separate two-dimensional views in turn: the channel cross-section (size and shape), planform geometry and longitudinal profile. Adjustments to the configuration of channel bed sediments are also considered. Since the form of alluvial rivers evolves in response to the movement of bed material, the chapter starts by considering the dynamics of solute/sediment transport in dryland rivers. A discussion of dryland river hydrology can be found in the preceding chapter.

Solute and Sediment Transport

Although the geomorphological effectiveness of fluvial activity in dryland environments is widely recognised (e.g. Graf 1988; Bull and Kirkby 2002), our understanding of key processes is far from complete. Relatively little is known about the transport of solutes in dryland streams. The paucity of information

D. Mark Powell (✉)
Department of Geography, University of Leicester, University Road, Leicester LE1 7RH, UK
e-mail: dmp6@le.ac.uk

A.J. Parsons, A.D. Abrahams (eds.), *Geomorphology of Desert Environments*, 2nd ed., DOI 10.1007/978-1-4020-5719-9_12, © Springer Science+Business Media B.V. 2009

on water chemistry reflects the limited importance of solution to dryland denudational processes (Meybeck 1976) and the fact that solute transport has little effect on channel form and stability. Recent work, however, has highlighted a growing awareness of the importance of floodwater chemistry for aquatic ecology (Grimm et al. 1981; Davies et al. 1994; Costelloe et al. 2005; Smolders et al. 2004), the cycling of nutrients (Sheldon and Thoms 2006) and the properties of alluvial soils (Jacobson et al. 2000a) and geochemical sediments (McCarthy and Ellery 1995; Khadkikar et al. 2000; Nash and McLaren 2003) in dryland environments. Much more is known about the transport of sediments. In terms of the movement of bed material, a basic distinction can be made between sand- and gravel-bed rivers (Parker 2008, p. 178–182). In general, sand-bed rivers are dominated by high excess shear stresses and suspended sediment transport while gravel-bed rivers are dominated by low excess shear stresses and bedload transport. Most work in dryland streams has focussed on understanding the dynamics of suspended sediment transport. Direct measurements of bedload are notoriously difficult to make and the logistical and practical difficulties are enhanced considerably by the uncertainty, unpredictability and infrequency of rainfall and runoff in dryland environments. Consequently there are few data sets documenting the dynamics of bedload transport in dryland rivers.

Dissolved Load

Since dissolved materials mix readily in turbulent flow, solute concentrations in streams and rivers are conventionally determined from a single, mid-stream sample. Solute monitoring programmes usually adopt discrete sampling methods (either manual or automatic) although an increasing number of water-quality parameters can be measured continuously *in situ*. Concentrations of total dissolved solids, for example, are routinely derived from measurements of electrical conductivity (specific conductance SC; $\mu S\,cm^{-1}$) which are converted to ionic concentrations with the aid of ion-specific calibration curves. Since the relationship between electrical conductivity and ionic concentration is temperature dependent, conductance values are usually adjusted to a standard temperature

of 25°C. Concentrations are expressed in units of mass per volume ($mg\,l^{-1}$) or mass per mass (parts per million; ppm). The units are usually used interchangeably even though a density correction should be applied to account for variations in fluid density due to temperature and solute concentration (USGS 1993). The unit of micro moles per litre ($\mu mol\,l^{-1}$) is used for chemical mass-balance calculations.

The relationship between solute concentration (Cc; $mg\,l^{-1}$) and discharge (Q; $m^3\,s^{-1}$) is usually modelled as the power function

$$Cc = aQ^b \qquad (12.1)$$

in which the empirical coefficients a and b are fitted by ordinary least squares regression. Typically, b < 0 indicating that solute concentrations decrease with discharge (Walling and Webb 1983) reflecting the dilution effect by stormflow of low ionic status. Classic dilution effects have been observed in dryland settings. Hem (1985) for example attributed a decline in electrical conductance in the San Francisco River, Arizona to stormflow dilution of heavily mineralised perennial spring waters (Fig. 12.1a). A mixing model utilising mass balance equations for three sources of runoff (spring water, baseflow and storm runoff) provides a good fit to the observed data. Dilution concepts may, however, be less useful in ephemeral streams due to the absence of base flow and the high velocities of overland flow which limit the length of time runoff has to react to near-surface rock and soil minerals. In the ephemerally flowing Nahal Eshtemoa in Southern Israel, for example, marked variations in solute concentrations are only observed during the rising stages of the flood pulse and are attributed to the flushing of solutes from the watershed at the onset of the event (Fig. 12.1b). Thereafter, the ionic concentrations and composition of rainfall and runoff are broadly comparable. Similar flushing effects have been observed in the Gila River near Fort Thomas in Arizona (Hem 1948) and may explain the chemical changes observed during flash floods in Sycamore Creek (Fig. 12.1c) and KR Wash, also in Arizona (Fisher and Minckley 1978; Fisher and Grimm 1985). The data from Sycamore Creek also highlights the contrasting behaviour of different solutes during flood events.

In a spatial context, solute concentrations in dryland rivers have been found to increase downstream due to flow attenuation by transmission losses (Jacobson

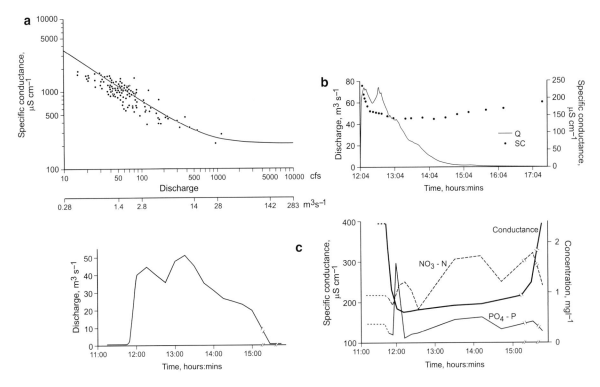

Fig. 12.1 Chemical characteristics of flood waters in (**a**) San Fransisco River, Arizona (after Hem 1985), (**b**) Nahal Eshtemoa, Israel (after Alexandrov 2005) and (**c**) Sycamore Creek, Arizona (after Fisher and Minckley 1978). The curve in (**a**) represents the fit of a three-source mixing model to the data

et al. 2000b). In terms of organics, several authors have noted that dryland rivers transport high concentrations of dissolved and particulate organic matter (Jones et al. 1997; Jacobson et al. 2000b). The comparatively high organic matter loadings of dryland streams have been attributed to lower mineralisation rates, limited sorption of dissolved organic matter in sandy soils and the rapid concentration of runoff into channels (Mulholland 1997).

Suspended Sediment

The properties of suspended sediment are usually measured using extracts of water-sediment mixtures (Edwards and Glysson 1999) but they can also be measured *in situ* using optical sensors (Gippel 1995). There are significant spatial and temporal heterogeneities in suspended sediment concentrations so measurements must be made within an appropriate sampling frame-

work to prevent sampling bias (Meade et al. 1990; Hicks and Gomez 2003).

Although suspended sediment loads reflect a wide range of climatic, topographic, lithological and anthropogenic controls (Lvovich et al. 1991), rivers draining areas of low precipitation are frequently distinctive in terms of high suspended sediment concentrations (Walling and Kleo 1979; Alexandrov et al. 2003). One of the most sediment-laden rivers on Earth is Rio Puerco in semi-arid New Mexico. Concentrations in excess of 600,000 ppm are routinely measured at the USGS gauging station near Bernardo and the 50-year average annual suspended sediment concentration of 113,000 ppm ranks fourth highest in a global comparison of sediment load data for selected world rivers (Gellis et al. 2004). In fact, the transport of hyperconcentrations of suspended sediment (defined as those in excess of 400,000 ppm) is a frequent occurrence in many dryland rivers (Lane 1940; Beverage and Culbertson 1964; Gerson 1977; Stoddart 1978; Walling 1981; Lekach and Schick 1982; Xu 1999).

As for solutes, the relation between concurrent measurements of suspended sediment concentration (C_s; mg l^{-1} or ppm) and discharge is conventionally modelled by a power function (Equation 12.1). Unlike the solute case, however, the exponent (b) is typically greater than zero indicating that suspended sediment concentrations increase with discharge (Fig. 12.2).

Frostick et al. (1983) note that suspended sediment rating curves for dryland streams are associated with higher coefficients (a) and lower exponents (b) than those derived for humid-temperate streams. The difference in coefficients is indicative of the transport of larger suspended sediment loads at low discharges. The difference in exponents indicates that dryland suspended sediment concentrations are less sensitive to changes in discharge. In fact, sediment concentrations in dryland environments increase at a rate less than a proportionate increase in discharge (b < 1) which is in contrast to humid-temperate streams for which b normally lies in the range 2–3 (Leopold and Maddock 1953).

Measurements of suspended sediment incorporate both wash load and suspended bed material load. The former is fine-grained sediment, typically fine sands, silts and clays, delivered to the channel with hillslope runoff. The latter comprises coarser material sourced from the instream sediments. Although the two components cannot be separated unequivocally, an arbitrary distinction can often be made on the basis of sediment size by comparing bed material and suspended sediment grain size distributions (Fig. 12.3).

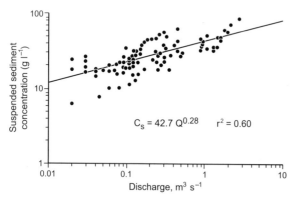

Fig. 12.2 Suspended sediment rating curve for the Katiorin catchment, central Kenya (after Sutherland and Bryan 1990). r^2 is the coefficient of determination of the fitted rating relation

Suspended Bed Material Load

It has long been appreciated that bed material is suspended into the flow by the action of coherent turbulent flow structures or eddies (e.g. Sutherland 1967; Jackson 1976). More recent work has illuminated the hydrodynamics of particle suspension in considerable detail. Over smooth boundaries (e.g. planar beds of sand-sized sediment), eddies originate as hairpin vortices from alternate zones of high and low speed within the viscous sublayer (see review by Smith 1996). Similar structures are observed in flows over gravel-sized sediment and over bedforms due to the shedding of wakes from individual clasts (Kirkbride 1993) and from the shear layer that forms due to flow separation downstream of bedform crests (McLean et al. 1996). Of particular importance to the suspension of bed material is the violent ejection of low momentum fluid from the bed during turbulent motions (Lapointe 1992; Garcia et al. 1996). These flow ejection events, or 'bursts', lift particles into the flow and oppose the tendency for the uplifted grains to settle under the influence of gravity.

According to this model, a particle will remain suspended in the flow providing the vertical turbulent velocity fluctuations exceed the particle's fall velocity. Consequently, the competence of a turbulent flow to transport sediment in suspension is commonly defined by the criterion

$$v'/\omega_0 > 1 \tag{12.2}$$

in which v' is the maximum root-mean-square vertical turbulent velocity fluctuation (m s^{-1}) and ω_0 is the mean settling velocity of the suspended sediment (m s^{-1}). For shear turbulence, there is abundant experimental evidence that the upward components of vertical velocity fluctuations (v_{up}'; m s^{-1}) are, on average, greater than the downward components (v_{dn}'; m s^{-1}) and that v' is proportional to the shear velocity, u_* (m s^{-1}; McQuivey and Richardson 1969; Kreplin and Eckelmann 1979). Using the assumptions $v_{up}' = 1.6v'$ and $v' = 0.8u_*$, Bagnold (1966) expressed Equation 12.2 in terms of Shields' dimensionless bed shear stress, τ^*,

$$\tau^* = \frac{\tau}{(\rho_s - \rho_f)gD} \tag{12.3}$$

where τ = bed shear stress (N m^{-2}), ρ_s = density of sediment (kg m^{-3}), ρ_f = density of flow (kg m^{-3}),

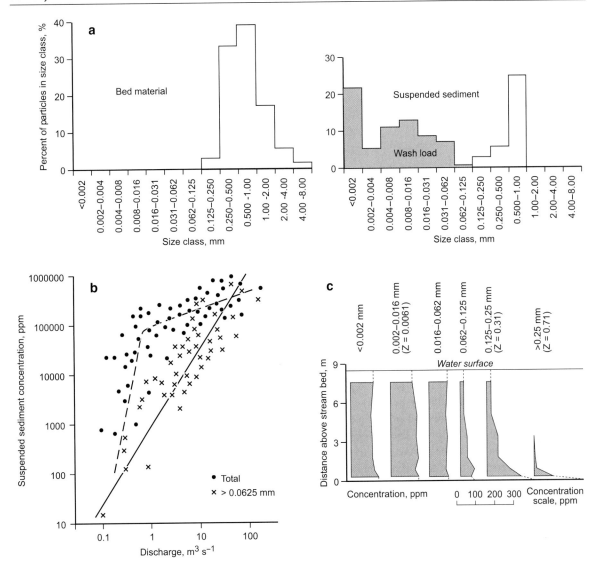

Fig. 12.3 Using grain size to distinguish between wash material and bed material in suspended sediments. (**a**) Bed material and suspended sediment size distributions in the Rio Grande at Otowi Bridge, New Mexico (after Nordin and Beverage 1965). Suspended sediment finer than 0.125 is not represented in the bed and is assumed to be wash load. (**b**) Wash load and suspended sediment concentrations in the Paria River at Lees Ferry, Arizona 1954–1965 (after Gregory and Walling 1973). Wash load concentrations frequently depart from simple bivariate rating relations (Equation 12.1) because of catchment controlled variations in sediment supply. (**c**) Vertical distribution of different sediment sizes in the Mississippi River at St Louis, Montana (after Colby 1963). Vertical concentration gradients are commonly uniform for wash load and steeply decreasing away from the bed for suspended bed material load. Values of the Rouse number (Z) from Allen (1997 p. 197)

g = acceleration due to gravity (m s^{-2}) and D = particle size (m) to give a suspension threshold, τ_s^*:

$$\tau_s^* = \frac{\omega_o^2}{1.56\,(\rho_s/\rho_f - 1)gD} \qquad (12.4)$$

For grain and flow densities of 2750 and 1000 kg m^{-3} respectively, the expression simplifies to

$$\tau_s^* = \frac{0.4\,\omega_o^2}{gD} \qquad (12.5)$$

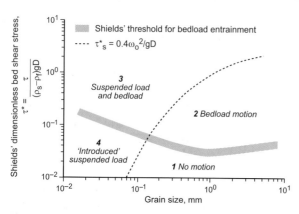

Fig. 12.4 Plot of the Bagnold (1966) suspension criterion along with the Shields bedload entrainment function (after Leeder et al. 2005). Alternative suspension criteria are reviewed by Garcia and Parker (1991)

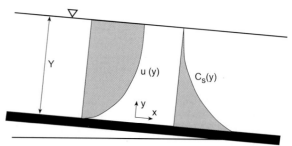

Fig. 12.5 Sketch of velocity (u; m s^{-1}) and sediment concentration profiles in steady, uniform open-channel flows (after Wright and Parker 2004a). The vertical profiles of velocity and sediment concentration can be modelled by the law of the wall (Equation 12.7) and the Rouse equation (Equation 12.8) respectively. Y and y are flow depth (m) and height above bed (m) respectively and x is distance downstream (m)

The Bagnold suspension criterion (Equation 12.5) is plotted in Fig. 12.4 along with the Shields curve for the initiation of bedload movement as defined by Miller et al. (1977). Taken together, the two competence criteria define four fields of sediment transport. Fields 1–3 relate to the transport of bed material. Field 4 relates to the transport of material already held in suspension (i.e. the wash load). Since the transport of wash load is dependent on rates of sediment supply from catchment hillslopes rather than the competence of the flow, a functional understanding of the suspended bed material load in relation to channel hydraulics is restricted to field 3 (Vetter 1937; Einstein and Chien 1953).

Methods for predicting the transport rate of suspended bed material from flow and sediment characteristics are based on models for the vertical concentration and velocity profile in steady uniform flow (e.g. Einstein 1950; van Rijn 1984; Fig. 12.5).

The suspended sediment transport rate per unit width (q$_s$) is given by

$$q_s = \int_{y'}^{Y} uCs\, \delta y \qquad (12.6)$$

where u = flow velocity (m s^{-1}); y' is a near-bed reference height (m), Y = flow depth (m) and y = height above the bed (m). The vertical velocity profile for a steady uniform flow is given by

$$\frac{u_y}{u_*} = \frac{1}{\kappa} \log e \left(\frac{y}{y_o} \right) \qquad (12.7)$$

where u_y = velocity at height y (m s^{-1}), u_* = shear velocity (m s^{-1}), κ is Von Karman's constant (≈ 0.4) and y_o is the roughness length (m). Notwithstanding recent advances in the dynamics of sediment suspensions outlined above, models for the vertical suspended sediment concentration profile rely on classical diffusion theory. The Rouse equation balances the downward settling of grains under gravity with their upward diffusion due to turbulence to yield (Rouse 1937):

$$\frac{Cs_y}{Cs_{y'}} = \left[\frac{(Y - y)}{y} \frac{(Y - y')}{y'} \right]^Z \qquad (12.8)$$

where Cs_y is the concentration at height y (mg l^{-1}), $Cs_{y'}$ is the concentration at height y' (mg l^{-1}) and Z is a dimensionless suspension parameter known as the Rouse number. Since

$$Z = \omega_0 / \beta \kappa u_* \qquad (12.9)$$

where β is the sediment diffusion coefficient (commonly assumed ≈ 1), the Rouse number models the concentration gradient by expressing the interaction between the upward-acting turbulent forces and the downward-acting gravitational forces. As shown in Fig. 12.3c, low values of Z model the near-uniform concentrations that result from fine particles (low ω_0) and high flow intensities (high u_*). Conversely, higher values of Z model the stronger concentration gradients generated by larger particles and lower flow intensities.

Application of Equation 12.8 requires definition of $Cs_{y'}$, a reference concentration at height y' above the bed. In the formulation of Einstein (1950), $Cs_{y'}$ is defined at a distance $y' = 2D$ using the Einstein bedload equation. This, however, has been shown to overpredict near-bed concentrations (Samaga et al. 1986). A number of alternative entrainment functions have subsequently been developed. Of these, Garcia and Parker (1991) conclude that the functions of Smith and McLean (1977) and van Rijn (1984), together with their newly developed relation, performed best when tested against a standard set of data.

Significant improvements to the models for u_y and Cs_y (Equations 12.7 and 12.8) have resulted from a consideration of density stratification and bedform effects (e.g. McLean 1991; 1992; Wright and Parker 2004a,b). Sediment-induced density gradients dampen turbulence and reduce the flux of mass and momentum within the water column. The result is an increase in mean flow velocity and a decrease in mean sediment concentration. Since the concentration effect dominates, the net effect is a reduction in transport rate. Stratification also results in finer distributions of suspended sediment because the largest sizes have the strongest concentration gradients and are affected the most by the reduction in vertical mixing. Bedform effects reflect the hydraulic consequences of form drag. For a given mean flow velocity, form drag increases

the total drag (and hence the carrying capacity of the flow) while decreasing skin friction (the ability of the flow to entrain sediment). Because the former usually dominates, the net effect is also a reduction in transport rate. In the suspended load equation of Wright and Parker (2004b), density stratification effects in both velocity and concentration profiles and the effects of bedforms on flow resistance are addressed using relations based on Wright and Parker (2004a) and Engelund and Hansen (1967) respectively. Estimates of near-bed sediment concentrations are made using a modified version of the entrainment function presented in Garcia and Parker (1991). The relation was tested using the data of Toffaleti (1968). Although the model yields reasonably good predictions of suspended sediment concentrations and size distributions, the test is restricted to relatively low concentrations ($Cs < 600 \, \mathrm{mg \, l^{-1}}$). It remains to be seen whether the equation can successfully predict the higher suspended sediment concentrations commonly found in dryland fluvial systems. One model that has been tested in a dryland environment is that due to Laursen (1958). The model was tested by Frostick et al. (1983) in the Il Kimere, a sand-bed stream in the semi-arid province of northern Kenya. The semi-empirical relation makes good predictions of suspended sediment concentrations for those size classes that make up the bed material (Fig. 12.6a). Applications of the

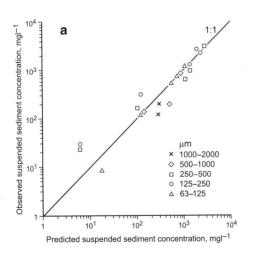

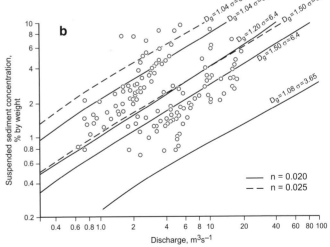

Fig. 12.6 Relations between observed and predicated concentrations of suspended bed material in (**a**) Il Kimere Kenya (after Frostick et al. 1983) and (**b**) Walnut Gulch, Arizona (after Renard and Laursen 1975). Predictions due to Laursen (1958). Figure 12.6b shows the sensitivity of the predicted concentrations to variations in bed material size (as modelled by the mean (D_g) and standard deviation (σ) of the size distribution) and channel roughness (as modelled by Manning's roughness coefficient; n)

model in Walnut Gulch in SE Arizona were also reasonable, (Fig. 12.6b) though the model is very sensitive to poorly constrained input parameters such as bed material size distribution and channel roughness.

It is worth noting that several sediment- and hydraulic-related phenomena of particular relevance to dryland fluvial systems have yet to be incorporated into models of suspended sediment transport. For example, dryland flow events are often unsteady and commence as a flood bore travelling over a dry bed. Recent research into the turbulence characteristics of unsteady flows has demonstrated that turbulence is higher on the rising limb of a hydrograph than it is on the falling limb (Song and Graf 1996; Nezu et al. 1997) with potential consequences for differential suspension of bed material during rising and falling flood stages. Other workers have highlighted the potential for turbulence-induced scouring at the front of advancing bores (Capart and Young 1998). This may explain the finding that peak suspended sediment concentrations in floods that propagate over a dry bed are often associated with the bore rather than the peak discharge (e.g. Frostick et al. 1983; Dunkerley and Brown 1999; Jacobson et al. 2000b). If so, it suggests that the increase in turbulence at the bore is more than sufficient to counteract any reduction in transport capacity due to the entrainment of air into the bore and the consequent reduction in relative sediment density (Chanson 2004). Finally, Dunkerley and Brown (1999) speculate that the infiltration of sediment suspensions into unsaturated porous bed material may be an important mechanism controlling suspended sediment concentrations in dryland rivers. Confirmation of this phenomenon and elucidation of the controls requires careful study of the infiltration of sediment suspensions into unsaturated porous bed materials.

Wash Load

Since a considerable proportion of the sediment carried in suspension by dryland rivers is fine-grained wash load, the suspended sediment dynamics of many dryland streams are complicated by issues pertaining to the availability of sediment on hillslopes. Figure 12.7, for example, shows the variation in

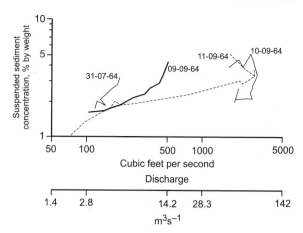

Fig. 12.7 Variations in suspended sediment concentrations during four storms in Walnut Gulch, Arizona (after Renard and Laursen 1975). Compare with Fig. 12.6b

suspended sediment concentrations with discharge during four storm events in Walnut Gulch in south eastern Arizona. It is apparent that concentrations vary by almost an order of magnitude at any specified discharge. Since suspended sediment concentrations for individual events are higher during rising stages than they are at similar discharges during falling stages, much of the scatter can be attributed to clockwise hysteresis in flood-period suspended sediment transport.

Similar storm-period variations in dryland suspended sediment concentrations have been observed in Upper Los Alamos Canyon, New Mexico (Malmon et al. 2004), several central Kenyan rivers (Syrén 1990; Sutherland and Bryan 1990; Ondieki 1995), the Nahal Eshtemoa, Israel (Alexandrov et al. 2003; 2006), Sycamore Creek, Arizona (Fisher and Minckley 1978) and the Burdekin River, Queensland (Amos et al. 2004) and have been attributed to the flushing and subsequent depletion of the most readily mobilised sediment following the generation of runoff on hillslopes and in channels. Sediment supply issues are also important at longer-time scales. Khan (1993) attributed the seasonal decline in suspended sediment concentrations in the Sukri and Guhiya Rivers in western Rajasthan, India to the progressive exhaustion of fine grained sediment deposited on catchment hillslopes by aeolian processes during the preceeding dry season (see also Amos et al. 2004). Time-conditioned processes of sediment accumulation and subsequent depletion have also been shown to control suspended sediment concentrations in the hyper-arid Nahal Yael in southern Israel (Lekach and Schick 1982). In this case, much of the sediment load is sourced from the products of hillslope weathering rather than aeolian deposition.

In many of these studies, the inter- and intra-event time-dependencies in suspended sediment concentrations are strongest for, or even exclusive to, the fine fractions sourced from outside the channel (i.e. the wash load). In other dryland streams, however, hydrologic control has been shown to extend across the full range of grain sizes so that the behaviour of wash and suspended bed material is not so different. In Il Kimere, for example, suspended sediment concentrations for individual size classes, including wash material, show good correlations when rated against discharge (Frostick et al. 1983; Fig. 12.8a). Since coarser fractions are associated with progressively steeper rating relations, the suspended sediment size distribution changes systematically with the flow (Reid and Frostick 1987; Fig. 12.8b).

The hydraulic control of overall suspended sediment concentrations and grain size in Il Kimere is attributed to the combined influence of abundant and readily transportable sediment of all sizes on sparsely vegetated hillslopes and in unarmoured sandy channel fills and the efficiency and effectiveness by which overland flow routes sediment into the channel network. Similar factors were invoked by Belperio (1979) to explain the high correlation observed between wash load concentration and discharge in the Burdekin River, Australia.

Notwithstanding these studies, the general implication of the work discussed above is that catchment-controlled sediment supply issues are significant controls on suspended sediment behaviour in dryland environments. As illustrated by Alexandrov et al. (2003; 2006) an improved understanding of suspended sediment dynamics in dryland streams requires the development of supply-based models that account for the distribution of sediment sources and the spatio-temporal complexity of rainfall-runoff patterns within

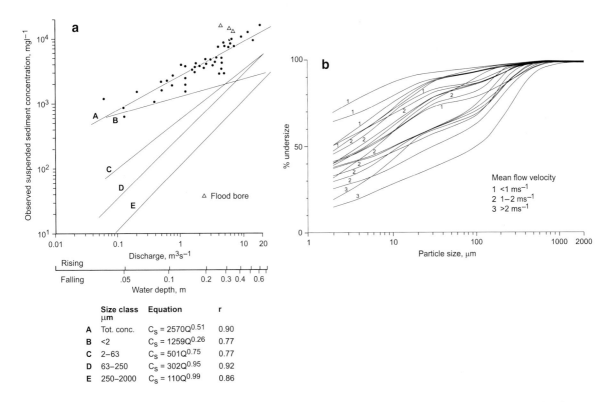

Size class μm		Equation	r
A	Tot. conc.	$C_S = 2570Q^{0.51}$	0.90
B	<2	$C_S = 1259Q^{0.26}$	0.77
C	2–63	$C_S = 501Q^{0.75}$	0.77
D	63–250	$C_S = 302Q^{0.95}$	0.92
E	250–2000	$C_S = 110Q^{0.99}$	0.86

Fig. 12.8 Relation between (**a**) flow and suspended sediment concentration by size class (after Frostick et al. 1983) and (**b**) flow and suspended sediment size distribution (after Reid and Frostick 1987) for the Il Kimere Kenya. The curves in b are labelled according to contemporary mean flow velocities. The progressive shift and change in shape with decreasing flow velocity reflects the dropping-out of coarse bed-material entrained by turbulent suspension at peak flows and the increasing dominance of finer sediment generated by wash processes on catchment hillslopes

dryland catchments. Such models should also account for drainage net influences on water and suspended sediment delivery which have been shown to control the sedimentological character of channel fills (Frostick and Reid 1977) and the type of hysteretic pattern exhibited by suspended sediment rating curves (Heidel 1956).

Bedload Transport

A wide variety of indirect and direct approaches have been used to study bedload transport processes in dryland rivers. Indirect approaches, including reservoir sedimentation studies and particle tracing programmes, are useful in that they do not require personnel to be onsite during flow events. This is of considerable advantage given the ephemeral discharge regime of many dryland fluvial systems. They also provide data that integrate hydrologic, hydraulic and sedimentolgical responses over a wider range of spatial and temporal scales than is usually possible using direct methods. As a consequence, however, much detail relating to the hydrodynamics of bedload transport processes is lost which can compromise understanding (Schick and Lekach 1993). Such information can be gained from direct and contemporaneous measurements of bedload transport rates and hydraulic parameters during flow events. Although this is an onerous and difficult undertaking in environments where floods are infrequent and unpredictable and where access may be restricted, many of the practical and logistical constraints can be overcome by using automated sampling technologies.

Indirect Measurement Methods

Reservoirs are effective sediment traps and conventional terrestrial and/or bathymetric surveys of reservoir sedimentation provide a well-tested methodology for assessing sediment delivery processes and yields in dryland catchments (Laronne 2000; Haregeweyn et al. 2005; Griffiths et al. 2006). Although most studies do not distinguish between sediment delivered as bedload and as suspended load, such a distinction can often be made since the coarser bedload is generally deposited in prograding

deltaic lobes at the reservoir entrance whilst the finer suspended sediment disperses and settles throughout the reservoir. A reservoir survey was used to quantify the bedload yield of Nahal Yael in the hyper arid southern Negev Desert (Schick and Lekach 1993). The volume of sediment stored within the reservoir delta over a 10-year period was equivalent to a bedload yield of $116 \text{ t km}^{-2} \text{ yr}^{-1}$ which represented two-thirds of the total sediment yield for the 0.5 km^2 catchment. Although bedload is commonly believed to be more significant in dryland environments than it is in humid-temperate environments (Schumm 1968), the ratio of bedload to suspended load is generally less than 0.5 (Graf 1988 p. 139; Powell et al. 1996). The dominance of the bedload contribution to Nahal Yael's sediment yield can be attributed to high magnitude events, steep hillslopes and channels, an abundant supply of coarse-grained sediment on debris-mantled hillslopes and in channel bars and strong hillslope-channel coupling.

Particle tracing techniques (Hassan and Erginzinger 2003) can be used with relative ease in ephemeral rivers because the nature of the discharge regime facilitates tracer relocation and recovery after flood events. Most work has focused on the movement of gravel-sized sediment because of the technical difficulties associated with tagging and tracing finer particles. Detailed tracer-studies of bedload movement in gravel-bed dryland streams have been undertaken in the Negev and Judean Deserts of Israel (Hassan 1990, 1993; Hassan et al. 1991). The results indicate that the travel distances of individual particles during individual events are not correlated with particle size (Fig. 12.9a). Although this conclusion is consistent with field studies in humid temperate environments (e.g. Stelczer 1981) and reflects the stochastic nature of sediment transport (Einstein 1937), it should be noted that the narrow tracer distributions rather precludes an examination of the relative mobility of different sizes. In terms of mean travel distances, the data conform to models of size selective bedload transport in which mean travel distances of particles in the ith size class ($\overline{L}_i$, m) decrease with increasing mean particle size of that class (D_{gi}, m), though significant departures are observed from the simple $\overline{L}_i \propto 1/D_{gi}$ relation that arises from traditional force balance analyses (Fig. 12.9b). In particular, particle travel distances for the finer sizes are relatively insensitive to particle size. This result has been confirmed by

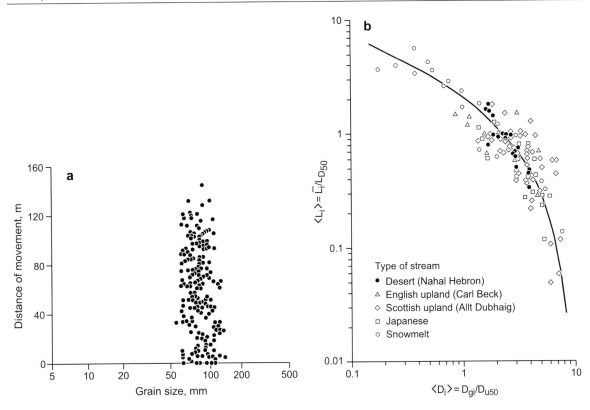

Fig. 12.9 Distance of travel versus particle size (after Hassan and Church 1992). (**a**) Data for individual clasts, in Nahal Hebron, Israel. (**b**) Relation between scaled travel distance ($\langle L_i \rangle = \overline{L}_i/\overline{L}_{D_{50}}$) and scaled particle size ($\langle D_i \rangle = D_{gi}/D_{u50}$) for gravel-bed rivers from a range of hydroclimatic regimes. $\overline{L}_{D_{50}}$ is the mean distance of movement for the particle size group that includes the median size of the surface bed material (D_{50}; m) and D_{u50} is the median grain size of the subsurface material (m)

Wilcock (1997) and Ferguson and Wathen (1998) and is attributed to the trapping-action of the bed-surface pocket geometry which principally affects the finer sizes (Einstein 1950).

Distributions of particle travel distances were found to conform to the Poisson-based model of Einstein (1937) and Sayre and Hubbell (1965; EHS) and to the two parameter gamma function (Fig. 12.10). The former yielded skew-peaked distributions, whilst the latter gave monotonic (Fig. 12.10a–c) and skew-peaked distributions (Fig. 12.10d–h). The monotonic distributions were associated with relatively small events in which a large number of particles moved only a short distance. The skew-peaked distributions were generated by the larger events in which particle movements were more significant. The skewed models did not fit the data as well as the monotonic models. It is suggested, therefore, that the distributions are only suitable for modelling the local dispersion of sediment. More complex models are required to model

the longer travel distances of large events because of complex bedload-bedform interactions such as the movement of sediment into storage within bars (see also Leopold et al. 1966; Hassan et al. 1999). Particle travel distances are also affected by the sedimentological environment: particles locked within the surface layer, or buried within the subsurface material, travel, in general, shorter distances than unconstrained particles (Hassan 1993).

Tracers are often buried (e.g. Hassan 1990; Hassan and Church 1994) as a result of scour and fill of the stream-bed. Although scour and fill are characteristic of all alluvial rivers, they are of particular significance in many dryland environments where there is often an unlimited supply of sand and fine gravel that is readily entrained by infrequent, but intense flooding (Leopold and Maddock 1953; Colby 1964; Foley 1978). Perhaps the most extensive study of scour and fill in a dryland channel is due to Leopold et al. (1966) who measured stream-bed scour and fill at 51 cross-sections

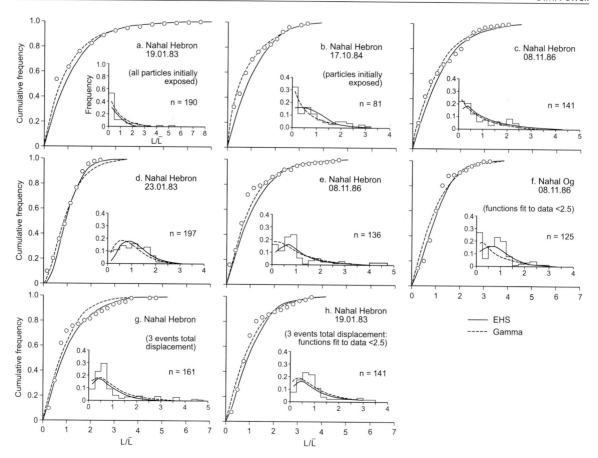

Fig. 12.10 Distributions of particle travel distances in two gravel-bed streams in Israel (after Hassan et al. 1991). Travel distances (L; m) have been normalised by the mean distance of movement ($\overline{L}$; m). n is the number of data. Only those particles that moved are considered. Similar results are obtained for the full data set

within a 10 mile section of a predominantly sand-bedded arroyo in New Mexico. The results suggest that the bed was scoured extensively during flood events (mean scour depths varied with the square root of discharge per unit channel width) but that compensating fill maintained the channel in approximate balance. More intensive investigations into the variability and pattern of stream-bed scour and fill at channel-reach scales were conducted by Powell et al. (2005, 2006, 2007). These studies deployed dense arrays of scour chains in three low-order channels of the Walnut Gulch catchment in SE Arizona. Detailed statistical analyses demonstrated that mean depths of scour increased with event magnitude and that many populations of scour depths were exponentially distributed (Fig. 12.11a). Exponential model parameters (α; cm^{-1}) collapse onto a general trend when rated against shear stress in ex-

cess of a threshold shear stress for entrainment (τ_c), thereby providing a means to estimate depths of scour in comparable streams (Fig. 12.11b). In terms of spatial patterns, active bed reworking at particular locations within the reaches resulted in downstream patterns of alternate shallower and deeper area of scour (Fig. 12.11c). During each event, compensating fill returned the streams to preflow elevations indicating that the streams were in approximate steady state over the period of the study (Fig. 12.11d). The results support the suggestion of Butcher and Thornes (1978) that sediment storage does not exert a significant control on sediment transfers through steep headwaters of dryland channels.

Because of the ephemeral discharge regime, the beds of dryland streams are readily accessible and post-event measurements of particle travel distances and

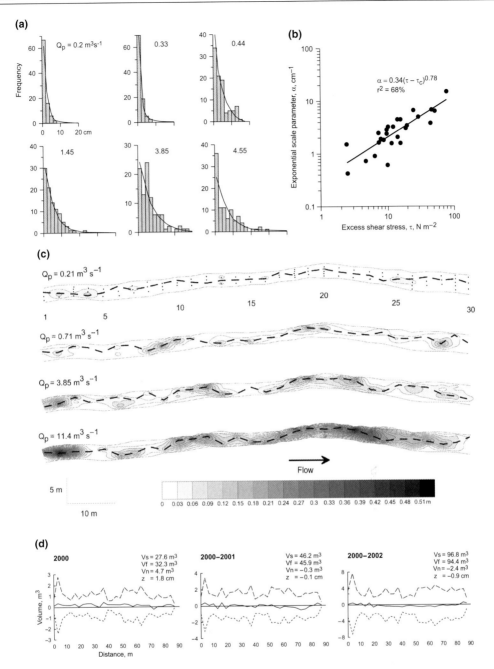

Fig. 12.11 Changes in stream bed elevations during individual events in a low order tributary of Walnut Gulch, Arizona (the main channel of Powell et al. 2005). (**a**) Distributions of scour depths for six events (after Powell et al. 2005). The events are ordered by peak discharge (Q_p; m^3 s^{-1}). Distributions are modelled using the one parameter exponential model. (**b**) Least squares relationship between exponential model parameter and excess shear stress (after Powell et al. 2005). The relationship incorporates data from two additional channels and provides a means to estimate depths of scour in similar streams. (**c**) Spatial patterns of stream-bed scour and fill for four flow events (after Powell et al. 2006). Cross-section and scour chain locations are shown in the top illustration. The *dashed line* shows the locus of the maximum depth of scour. (**d**) Cumulative patterns of volumetric scour (Vs; m^3), fill (Vf; m^3), net change (Vn=Vs−Vf; m^3) and average change in stream-bed elevation (z; m) at the end of three flood seasons (after Powell et al. 2007). Aggradation and degradation fluctuated with no persistent temporal trend so that sediment transfers did not lead to significant and progressive change to the volume of sediment stored within the reach

depths of stream-bed scour provide an attractive and relatively inexpensive means to quantify rates of bed material movement. The method utilises the relation

$$q_{bm} = u_b\, z_s(1 - p)\rho_s \qquad (12.10)$$

where q_{bm} is the mass transport rate of bed material (kg m^{-1} s^{-1}), u_b is the virtual rate of particle travel (m s^{-1}), z_s and w_s are the active depth and width of the stream bed respectively (m), and p is the porosity of the sediment (Hassan et al. 1992; Haschenburger and Church 1998). Data from the Nahal Yatir in southern Israel (see below) have been used to evaluate the method in a dryland environment. Post-event estimates of bedload yield based on the displacement of gravel-tracers and the depth of scour and fill obtained by scour chains are found to be very similar to that derived from a bedload rating relation derived by direct monitoring (Laronne pers comm.).

Direct Measurement Methods

Direct monitoring during flood events provides numerous opportunities to develop further insights into the dynamics of bedload transport in dryland environments. Of particular significance are the studies undertaken in Nahal Yatir and Nahal Eshtemoa, two neighbouring upland gravel-bed rivers in the Northern Negev Desert, Israel. In these streams, contemporaneous measurements of bedload discharge (q_b; kg m^{-1} s^{-1}) and shear stress during flash floods have been obtained using automatic sediment transport monitoring stations comprising a number of Birkbeck-type slot samplers (Reid et al. 1980; Laronne et al. 1992) and stage recorders (Fig. 12.12).

The studies show that the two streams are subject to intense bedload activity (Fig. 12.13a). Maximum recorded channel average transport rates are about 7 kg m^{-1} s^{-1} (Reid et al. 1995, 1998) while rates as high as 12.6 kg m^{-1} s^{-1} are recorded at individual samplers in Nahal Eshtemoa (Powell et al. 1999). The high transport rates reflect the high transport stages ($= \tau/\tau_c$) generated by the flash floods. In Nahal Eshtemoa, for example, all but three of the 19 flow events monitored over a four-year study period generated transport stages of three or more and over 50% generated transport stages greater than five (Powell et al. 2003). As shown in Fig. 12.13a, the relationships between channel-average shear stress and contemporary channel average shear stress for nine events in Nahal Eshtemoa and four events in Nahal Yatir are unusually well defined. Moreover, the predictions of several engineering formulae correspond closely to the observed data suggesting that the measured transport rates approximate the transport capacity of the flow (Reid et al. 1996; Powell et al. 1999). The transport of capacity loads and the simplicity and consistency of the bedload response recorded in these two dryland streams is in marked contrast to that observed in many humid-temperate perennial streams. The differences may be explained by the fact that the beds of Nahal Eshtemoa and Nahal Yatir are not armoured (see

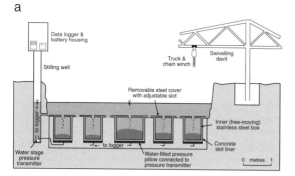

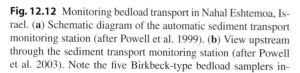

Fig. 12.12 Monitoring bedload transport in Nahal Eshtemoa, Israel. (**a**) Schematic diagram of the automatic sediment transport monitoring station (after Powell et al. 1999). (**b**) View upstream through the sediment transport monitoring station (after Powell et al. 2003). Note the five Birkbeck-type bedload samplers installed across the width of the channel beneath the bridge and the stage recorders extending up the approach reach. (**c**) Flood bore advancing over the bedload samplers in Nahal Eshtemoa (after Powell et al. 2003)

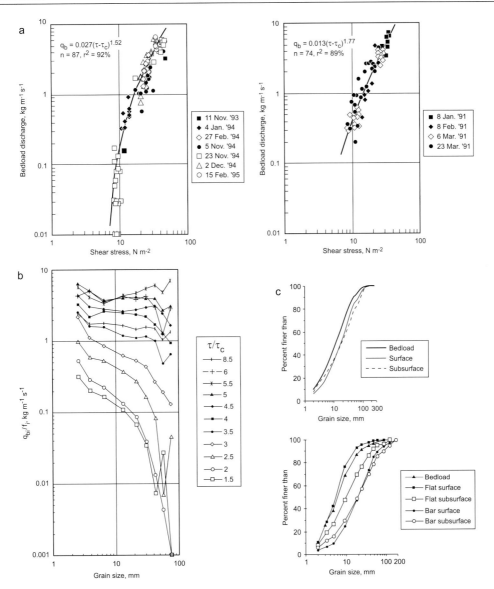

Fig. 12.13 Bedload transport rates and grain size distributions in Nahal Eshtemoa and Nahal Yatir, Israel. (**a**) Channel average bedload transport rates as a function of channel average shear stress in Nahal Eshtemoa (*left*; after Reid et al. 1998) and Nahal Yatir (*right*; after Reid et al. 1995). The curves passing through the data are the ordinary least squares rating relations (the coefficients have been adjusted to eliminate bias due to the log-log transformations). Zero transport rates in Nahal Eshtemoa plotted as 0.01 kg m^{-1} s^{-1}. (**b**) Ratio of mean transport rates for the ith size class (q_{bi}) with the frequency of occurrence in the bed (f_i) as a function of the geometric mean grain size (D_{gi}) of each size fraction in Nahal Eshtemoa (after Powell et al. 2001). The near-absence of the coarsest size fractions in the bedload at $\tau/\tau_c < 2$, the over-representation of the finer fractions and the under-representation of coarse fractions at c. $\tau/\tau_c = 3$ and the equivalence of bedload and bed material grain size distributions at $\tau/\tau_c > 4$ is indicative of partial transport (Wilcock and McArdell 1993), size selective transport (Ashworth et al. 1992) and equal mobility (Parker and Toro-Escobar 2002) respectively. (**c**) Bed material and bedload grain size distributions in Nahal Eshtemoa (*top*; after Powell et al. 2003) and Nahal Yatir (*bottom*; after Reid et al. 1995). The bedload size distribution in Nahal Eshtemoa represents the calibre of the material transported out of the catchment over a four year period estimated using the transport relation of Powell et al. (2001). The bedload size distribution in Nahal Yatir represents the sediment that accumulated in the centre sampler during four events as reported in Laronne et al. (1994). The terms 'bar' and 'flat' refer to contrasting sedimentary units within the reach (see Fig. 12.22a)

below) which reduces well known sedimentological constraints on sediment mobility and availability (Laronne et al. 1994; Reid and Laronne 1995).

It is worth noting that other dryland streams demonstrate more complex bedload responses to changes in flow strength. In Nahal Yael in southern Israel, for example, coarse grained sediment waves were found to migrate through the measuring section every 40–50 min (Fig. 12.14). The origin of the waves is not known, but may be related to catchment and network controls on sediment delivery to the channels. Other workers have highlighted the effect that unsteady flows have on bedload transport rates due to the inability of the bed to adjust as quickly as the flow (e.g. Plate 1994; Lee et al. 2004). The implications of this and related work for sediment transport in flashy dryland streams awaits evaluation.

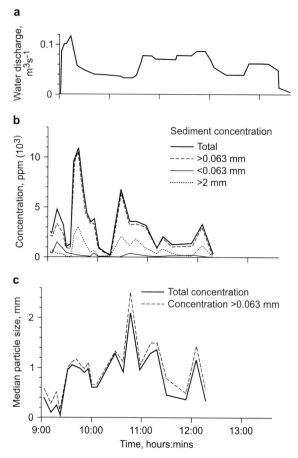

Fig. 12.14 Variation in (**a**) water discharge, (**b**) concentration of sediment load and (**c**) median particle size of sediment load during the event of 20 February 1970 in Nahal Yael, Israel (after Lekach and Schick 1983). Bedload transport occurred as a series of waves that formed independently of the pulses in discharge

In Nahal Eshtemoa, the bedload is fine grained at low flow but coarsens with increasing shear stress, converging with the grain size distribution of the bed at high flows (Fig. 12.13b). The shift in bedload grain size distribution with increasing flow strength accords with the widely held view that transport is partial and size selective at low excess shear stresses but approaches a condition of equal mobility at high levels of excess shear stress (see review by Gomez 1995). Since flow duration increases with decreasing flow magnitude, Wilcock and McArdell (1997) suggest that partial transport is the dominant transport regime in gravel-bed rivers and results in sediment loads that are considerably finer than the bed material (see also Leopold 1992; Lisle 1995). In Nahal Eshtemoa, partial and size selective transport occurs for 73% of the time the channel is competent to transport bedload. The size distribution of the bedload modelled over a four year period, however, is only slightly finer than that of the bed material (Powell et al. 2001, 2003; Fig. 12.13c). Even though partial and size selective transport conditions dominate and produce bedload size distributions that are finer than the size distribution of the bed material for the majority of the time the stream is geomorphologically active, the rate of transport of the coarser fractions that occurs at high transport stages almost serves to compensate, rendering the size distribution of the annual bedload not that much finer than the bed material. A similar evolution in bedload grain size is observed in Nahal Yatir though the finer bed material ensures that the partial transport domain is largely absent and that the bed is fully mobilised for a greater proportion of time. As a consequence, bed material and bedload grain size distributions also show a close correspondence (Laronne et al. 1994; Fig. 12.13c). A similar dynamic is observed in Goodwin Creek, a seasonal stream in north-central Mississippi (Kuhnle and Willis 1992).

Several authors have questioned whether there are differences in the dynamics of bedload transport between ephemeral and perennial rivers (e.g. Almedeij and Diplas 2003, 2005). Reid et al. (1995) compared bedload transport rates recorded in a number of perennial and ephemeral/seasonal rivers (Fig. 12.15a). They noted that Oak Creek (Oregan, USA; perennial), Turkey Brook (England, UK; perennial) and Nahal Yatir (Israel; ephemeral) define a relatively consistent relation, but that data from East Fork River (Wyoming, USA; perennial), Torlesse Stream (New

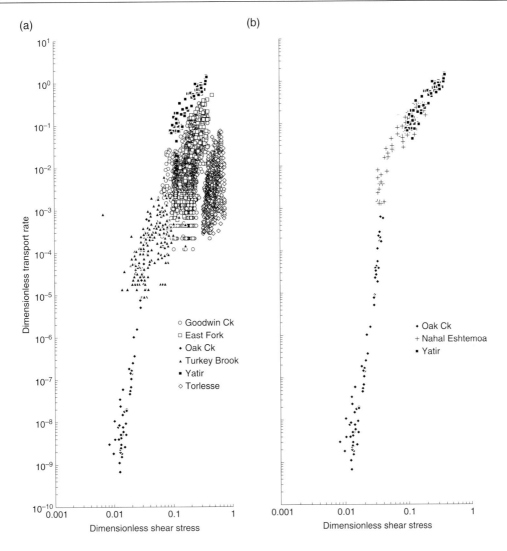

Fig. 12.15 Dimensionless bedload transport rates as a function of dimensionless shear stress in (**a**) one ephemeral (Yatir), one seasonal (Goodwin Creek) and four perennial streams (after Reid et al. 1995) and (**b**) Nahal Yatir, Nahal Eshtemoa and Oak Creek

Zealand; perennial) and Goodwin Creek (Mississippi, USA; seasonal) are shifted to the right, suggesting a different dynamic. However, Goodwin Creek and East Fork River contain significant amounts of sand which can be expected to augment transport rates in a non-linear manner (Wilcock et al. 2001; Wilcock and Crowe 2003). After accounting for the effect of sand on gravel transport rates, Wilcock and Kenworthy (2002) demonstrate that Oak Creek, Goodwin Creek and the East Fork River collapse onto a single curve. The implication of these comparisons is that bedload transport rates measured in perennial and ephemeral rivers fall on different parts of a single continuum that

represents that the bedload-shear stress response of gravel-bed rivers, a conclusion further supported by the fact that the data from Nahal Eshtemoa (Israel; ephemeral) dovetails with the data from Oak Creek, Turkey Brook and Nahal Yatir (Fig. 12.15b). This issue is considered further in the context of stream-bed armours (see below).

Channel Morphology

The morphology of alluvial channels develops through spatially and temporally variable patterns of erosion, transport and deposition. Much of the research on alluvial channel forms has been conducted in humid-temperate rivers and is based on the identification and analysis of equilibrium channel forms. Four aspects of channel morphology are usually considered: (i) the shape and size of the channel cross-section; (ii) the configuration of the channel bed; (iii) the river longitudinal profile and slope and (iv) the channel pattern. This conceptual framework is adopted here though it is recognised that our ability to make rational generalisations about dryland river forms is hampered by that fact that many dryland rivers fail to exhibit equilibrium behaviour.

Channel Equilibrium and Formative Events

Equilibrium concepts are relevant to medium timescales over which, it is reasoned, rivers develop a relatively stable and characteristic morphology that allows them to transmit the imposed water and sediment discharges (Mackin 1948; Leopold and Bull 1979). Explanations for the form of channels in equilibrium are usually sought in terms of a single 'dominant' or 'formative' discharge, a statistically- or morphologically-based construct that replaces the frequency distribution of flows. Wolman and Miller (1960) defined the dominant discharge as the flow which cumulatively transports the most sediment. They argued that the geomorphological effectiveness of a particular discharge magnitude is the product of

the sediment transported by an event of that magnitude and its frequency of occurrence. Using a sediment transport law and flood frequency distribution parameterised for humid-temperate conditions, they demonstrated that the most effective flood is defined by an event of moderate magnitude and frequency (Fig. 12.16a). Other workers have defined dominant discharge in terms of the flow that determines particular channel parameters such as the cross-sectional capacity of the flow (Wolman and Leopold 1957) or the wavelength of meander bends (Ackers and Charlton 1970).

The extensive debate that surrounds the concepts of dominant discharge and equilibrium adjustment is beyond the scope of this review (see Phillips (1992) and other papers from the 23rd Binghampton Symposium; Thorn and Welford 1994; Bracken and Wainwright 2006). It is worth noting, however, that the explanatory power of the two concepts in dryland environments is often questioned. The hydrological regime of many dryland rivers generates large differences between high and low flows and pronounced spatial and temporal discontinuities in process operation which makes the definition of formative discharges and the recognition of equilibrium forms difficult (Thornes 1980; Schick et al. 1987; Bourne and Pickup 1999; Hooke and Mant 2000; Coppus and Imeson 2002). Moreover, many dryland rivers appear not to exhibit equilibrium behaviour. As explained below, this contrast between dryland and humid-temperate river behaviour is due to fundamental differences in magnitude-frequency relationships (Baker 1977) and relaxation times (Wolman and Gerson 1978; Brunsden and Thornes 1979).

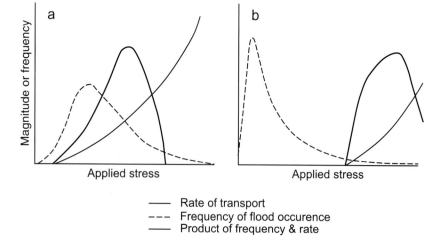

Fig. 12.16 (a) Generalised magnitude-frequency relationships of Wolman and Miller (1960). (b) Modifications to the magnitude-frequency relationships of Wolman and Miller (1960). After Baker (1977)

—— Rate of transport
--- Frequency of flood occurence
—— Product of frequency & rate

Baker (1977) re-evaluated Wolman and Miller's (1960; Fig. 12.16a) magnitude-frequency analysis in a dryland context by arguing that in dryland environments, the mode of the flood frequency distribution shifts left (because the ratio of high to small flow events is larger) and that the sediment transport law shifts to the right (because the sediments are generally coarser and have higher entrainment thresholds). The resultant increase in the magnitude and decrease in the frequency of the flow that transports the most sediment implies that the rare catastrophic event is more important in shaping dryland streams (Fig. 12.16b). Moreover, Wolman and Gerson (1978) recognised that the geomorphological effectiveness of high magnitude flows is further enhanced in dryland rivers by the limited occurrence of high frequency/low magnitude flows and the absence of sediment-trapping vegetation that facilitates channel recovery in humid-temperate environments (Fig. 12.17a–c). Although active channels may show some short-term adjustment to the prevailing hydrological regime, these equilibrium channels are often superimposed on a palimpsest disequilibrium morphology produced by more infrequent, higher magnitude events (Rhoads 1990) while in hyper-arid environments, channel recovery may be virtually non-existent such that successively larger floods leave permanent imprints on the landscape (Fig. 12.17d).

These climatically controlled contrasts in river behaviour can be characterised by the transient form ratio (TF) defined as the ratio of the mean relaxation time to the mean recurrence interval of significant channel disturbing events (Brunsden and Thornes 1979). Fluvial systems for which TF < 1 can develop equilibrium channel forms because they adjust to new conditions or recover from flood-induced channel change before the next major disturbance occurs (Fig. 12.17a,b). Many dryland rivers, however, may be characterised by TF > 1 with the result that channel forms display either dis-equilibrium or non-equilibrium behaviour (Stevens et al. 1975; Rhoads 1990; Bourne and Pickup; 1999; Beyer 2006). As defined by Renwick, (1992), the former represents instances when the development of an equilibrium state is precluded by long relaxation times (Fig. 12.17c) while the latter occurs when systems display no net tendency toward an equilibrium state (Fig. 12.17d).

Tooth and Nanson (2000a) argue that non- or dis-equilibrium channel behaviour is not a character-

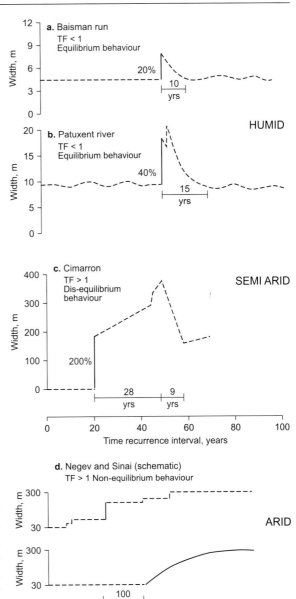

Fig. 12.17 Temporal changes in channel width in rivers from different hydro-climatic regimes showing relaxation (recovery) times following storm disturbances (after Wolman and Gerson 1978). Channel recovery following major flood disturbances occurs in response to sedimentation during frequent low-magnitude restorative flows and occurs most rapidly and effectively in humid-temperate environments. TF is the transient form ratio of Brunsden and Thornes (1979). Equilibrium terminology after Renwick (1992)

istic of all dryland rivers. They suggest that that the potential for dryland rivers to develop an equilibrium channel form is a function of local hydrological, geomorphological and sedimentological conditions.

High energy environments with low erosion thresholds (steep, low-order rivers subject to short-lived high magnitude flash floods carrying large amounts of coarse bedload) favour the development of non- (dis-) equilibrium channels whilst lower energy environments with higher erosion thresholds (medium – large low gradient rivers with resistant, confining banks subject to long duration floods) favour the development of equilibrium forms. The latter conditions typify the medium-large sized catchments of the Northern Plains and Channel Country of central Australia where channels meet several criteria said to characterise equilibrium conditions (stability despite the occurrence of large floods, sediment transport continuity, strong correlations between channel form and process; an adjustment to maximise sediment transport efficiency).

Cross-Sectional Form

The dominant control on the cross-sectional dimensions of a river is discharge. This is perhaps best illustrated by Ferguson's (1986) observation that channel width and depth increase systematically with increasing bankfull discharge as it varies over nine orders of magnitude from small laboratory channels to the world's largest rivers. Empirical geomorphological investigations of the relationships between channel geometry and stream discharge have traditionally followed the downstream hydraulic geometry approach of Leopold and Maddock, (1953) in which downstream changes in width (w; m), depth (y; m) and velocity (u; m s^{-1}) are expressed as power functions of an assumed dominant discharge (a discharge at a specified frequency of occurrence (Q_x; m^3 s^{-1}):

$$w = aQ_x^b \qquad (12.11)$$

$$y = cQ_x^f \qquad (12.12)$$

$$u = kQ_x^m \qquad (12.13)$$

The exponents $b = 0.5$, $f = 0.4$ and $m = 0.1$ defined for streams in the American Midwest using the mean annual flood are often used to characterise the downstream adjustment of humid temperate perennial streams to increasing discharge. The exponent set indicates that width increases faster than depth (gener-

ating downstream changes in channel shape as indexed by the width:depth ratio) and that velocity increases downstream (contradicting traditional Davisian assumptions). Although comparative data from dryland environments are sparse, compilations of hydraulic geometry exponents suggest some regional variation according to climatic regime (Park 1977; ASCE 1982). A study of the downstream adjustment of ephemeral channels in New Mexico, USA shows that although the increase in width is about the same as that observed in humid-temperate perennial rivers, the increase in velocity is more rapid and the increase in depth is less rapid (Leopold and Miller 1956; Fig. 12.18). The different response of the ephemeral channels is attributed to a downstream increase in suspended sediment concentrations that decreases turbulence and bed erosion.

The erodibility of channel banks exerts important secondary controls on cross-sectional adjustment. Since dryland weathering processes do not produce significant amounts of cohesive silts and clays, bank materials often lack the strength to resist processes of bank erosion. As a result, channels tend to respond to floods by widening, rather than deepening, their cross-section. Schumm (1960), for example, showed that width:depth ratios are negatively correlated with the silt-clay content of perimeter sediments (an index of bank shear strength and erodibility; Fig. 12.19).

Merritt and Wohl (2003) examined the downstream adjustment of Yuma Wash in SW Arizona to an event with a discharge estimated at c. 20% of the maximum probable flood. They found that increases in width were substantial ($b = 0.78$) whereas the increases in depth and velocity were modest ($f = 0.15$ and $m = 0.14$). They attributed the rapid increase in channel width to the low cohesion of the bank material which comprised less than 3% silt and clay. The behaviour of Yuma wash contrasts with channels in the northern Negev Desert which are able to maintain relatively deep and narrow cross-sections despite the high transport stages generated by flash floods (Laronne and Reid 1993; Reid et al. 1998; Fig. 12.12b). The absence of significant bank retreat in these channels has been attributed to the cohesive properties of the loess-rich soils (Powell et al. 2003). In non-cohesive sediments, flood-induced increases of channel width can be dramatic. Schumm and Lichty (1963) describe how a major flood in 1914

Fig. 12.18 Downstream hydraulic geometry relations of ephemeral channels in New Mexico (after Leopold and Miller 1956). Data were collected at discharges of unknown frequency

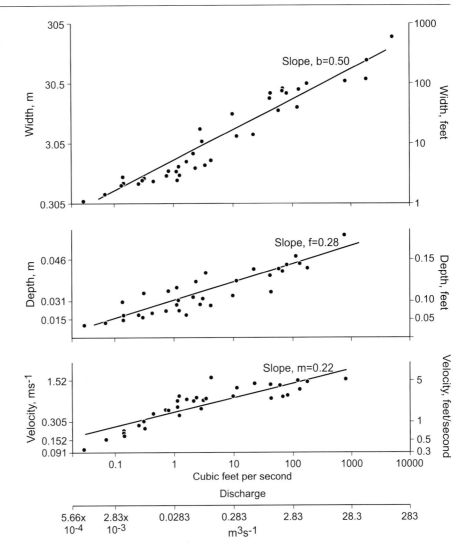

in SW Kansas began a process of channel widening that increased the average width of the Cimarron River from 15 m in 1874 to 366 m by 1939. Similar transformations of channel width have been observed in other sand-bed rivers of the SW USA including Plum Creek, a tributary of the South Platte River, Colorado (Osterkamp and Costa 1987), the Santa Cruz River in Arizona (Parker 1993) and the Gila River in SE Arizona (Burkham 1972; Huckleberry 1994; Hooke 1996; Fig. 12.20).

Bank stability is also controlled by vegetation. Although vegetation is generally sparse in semi-arid environments, many dryland rivers support dense stands of riparian vegetation which influence bank stability through mechanical and hydrological effects (Simon and Collinson 2002; Simon et al. 2004). The latter are likely to be particularly important in semi-arid environments where banks are typically unsaturated and susceptible to changes in soil moisture levels (Katra et al. 2007). Stabilising effects include root-binding of sediment which increases the tensile strength and elasticity of soils and helps to distribute shear stresses rather like the bars in reinforced concrete or the fibres in a carbon fibre material (Tal et al. 2003) and enhanced canopy interception and evapotranspiration which results in better drained bank materials with reduced bulk weight and lower positive porewater pressures. Riparian vegetation also increases flow resistance, thereby decreasing flow velocities and the shear stress available for erosion (Thornes 1990; Wilson et al. 2005). Destabilising effects of vegetation include bank loading by the weight of trees and higher near-surface moisture

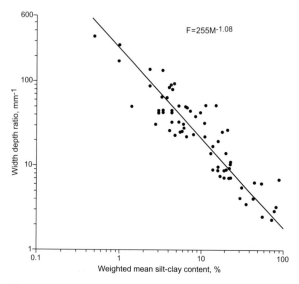

Fig. 12.19 Relation between width:depth ratio (F) and weighted mean percent silt-clay content in the channel boundary (M; after Schumm 1960)

contents after rainfall due to increased soil infiltration capacities. Vegetation plays an important role in narrowing channels after they have been widening by major flood events. The effects of vegetation on channel recovery are discussed below in the context of channel planform adjustment.

In a spatial context, Wolman and Gerson (1978) demonstrate that the rate of change of channel width with increasing drainage is more rapid in dryland rivers than it is in humid-temperate rivers, at least in catchments up to about $100\,km^2$ (Fig. 12.21a). Interestingly, dryland channels draining larger catchments maintain near constant widths. This has been attributed to various factors including an imposition of an upper limit on stream discharge caused by the limited areal extent of storm events (Sharon 1972, 1981; Renard et al. 1993; Goodrich et al. 1995) and/or transmission losses (see Chapter 11) and in hyper-arid environments, the lack of channel recovery between events (Fig. 12.17d). Where transmission losses exceed tributary inflows, the resultant downstream decrease in discharge downstream can lead to concomitant reductions in channel width and depth (e.g. Dunkerley 1992; Fig. 12.21b) leading to the termination of channelised flow and bedload transport in broad low gradient surfaces known as floodouts (Tooth 1999; Fig. 12.21c). It is not known whether the hydraulic geometry exponents that model the downstream

increase in channel dimensions under conditions of increasing discharge (Fig. 12.18) also describes the downstream decrease in channel dimensions observed under conditions of decreasing discharge.

The complexity of channel width adjustment in large dryland rivers is demonstrated by Tooth (2000b) who documented changes in channel character along the length of the Sandover, Bundey (Sandover-Bundey) and Woodforde Rivers in central Australia (Fig. 12.20c). In all three rivers, distinct form-process associations define four contrasting fluvial environments: confined upland, piedmont, lowland zones and unconfined floodout zones. Channel widths tend to increase throughout the upland and piedmont zones where integrating channel networks cause discharge to increase downstream. In the lowland zone, transmission losses exceed tributary recharge and widths and depths decrease downstream until the flows and sediments dissipate in the floodout zone. Although the rivers exhibit variable patterns of downstream channel change and several unusual channel characteristics (e.g. anabranching and aggrading floodout zones), Tooth (2000b) concludes from a qualitative review of channel pattern parameters that many aspects of channel form (including channel width) are 'strongly correlated to and sensitively adjusted to tributary inputs of water and sediment' (Tooth 2000b p. 200).

Where systematic relationships between cross-section channel geometry and discharge exist, they suggest a functional adjustment of channel form to the imposed discharge, the nature of which should be amenable to rational explanation using hydraulic and sediment transport principles. The development of deterministic solutions for the geometry of river cross-sections is hampered by the fact that the degrees of freedom for alluvial channel adjustment exceed the number of available equations. The traditional approach is to assume that width, depth, velocity and either slope or sediment load adjust to the other of these two variables plus discharge and grain size (Ferguson 1986) such that a solution is provided by solving the flow continuity relationship, a flow resistance law, a sediment transport equation and assuming either (i) a threshold channel, (ii) maximum efficiency criterion in conjunction with a bank stability criterion or (iii) by fitting an empirical relation to one variable (see Ferguson 1986 for a review of approaches). The relative merits of the different approaches are subject to some debate in part, because they all fail to account

(a)

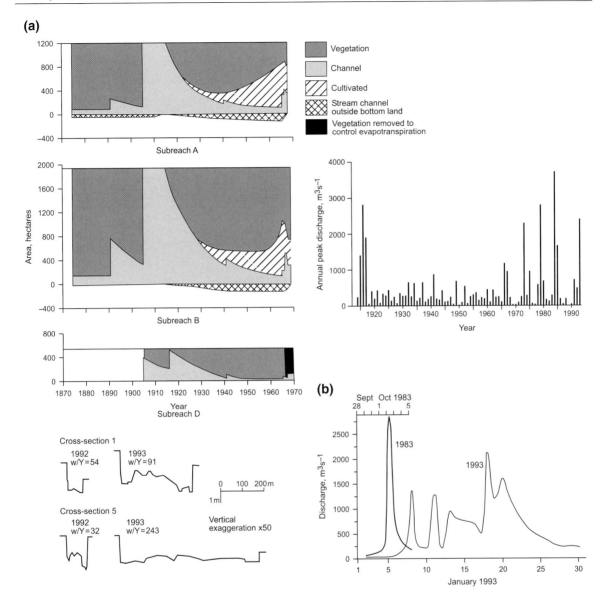

Fig. 12.20 Discharge records and changes in channel width in the Gila River, Arizona, USA (**a**) 1875–1968 near Safford (after Burkham 1972) and (**b**) 1993–1993 near Florence (after Huckleberry 1994). Note that the channel did not widen appreciably during the 1983 flood

for a variety of real-world complications (Eaton and Millar 2004; Millar 2005). However, they all have the same qualitative outcome in which the steady-state morphology is associated with a characteristic value of dimensionless shear stress (Ferguson 1986). The explanatory power of the approach has yet to be tested in a dryland river showing regularity in cross-section adjustment.

There remains considerable uncertainty as to how channels adjust their cross-sections. Much is known about the geotechnical and hydraulic forces that control bank stability and retreat, and attempts have been made to couple models of specific bank erosion processes (fluvial entrainment and mass wasting of bank materials and the downstream transport of failed bank materials) to predict cross-section adjustment in alluvial channels (e.g. Simon et al. 2000). Further work, however, is needed better to understand how hydraulic and gravitational processes interact to control rates of bank retreat and channel widening and to incorporate

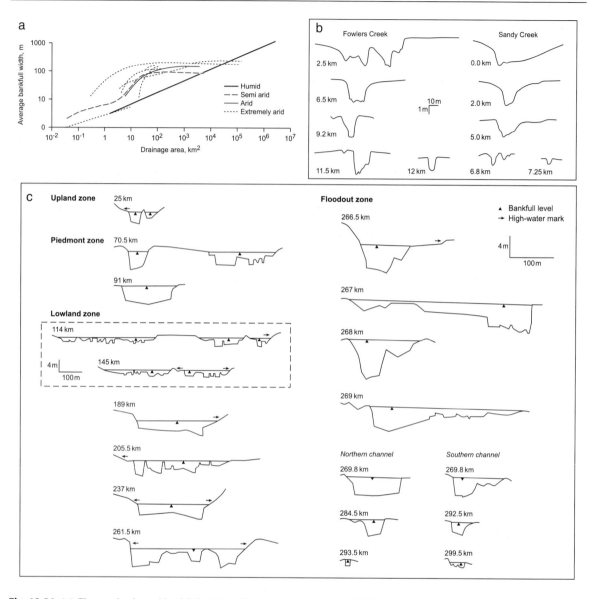

Fig. 12.21 (**a**) Changes in channel bankfull width with increasing drainage area in different climatic settings (after Wolman and Gerson 1978). (**b**) Downstream changes in channel width in Fowler and Sandy Creeks, western NSW, Australia (after Dunkerley 1992). (**c**) Downstream changes in channel width in the Bundey (Sandover-Bundey) River in the northern plains of central Australia (after Tooth 2000b)

this understanding into existing models of flow, sediment transport and morphological change (e.g. Darby et al. 2002).

Bed Configuration and Texture

Bed configuration and texture represent two of the most adjustable components of channel form with potential for regulating the short-term and mutual adjustment of water flow, sediment supply and grain size at a range of spatial and temporal scales. The complex relationships between flow, sediment transport and bedform geometry that facilitate this adjustment are beyond the scope of this review. Comment is therefore restricted to some important aspects of the bed morphology and sedimentology of dryland streams.

Bed Configuration

In humid-temperate environments, single-thread channels with coarse heterogeneous sediments on low–moderate slopes ($< 2\%$) often develop an alternating pattern of coarse-grained topographic highs (riffles) and finer-grained topographic lows (pools) with a wavelength of about 5–7 channel widths (Keller and Melhorn 1978). On steeper slopes, the bedform evolves to a step-pool sequence with a wavelength of about 2 channel widths (Chin 2002). Riffles and steps are significant sources of flow resistance that concentrate energy losses at particular locations along the course of a river (Church and Jones 1982; Abrahams et al. 1995; Chin and Phillips 2007). Since the development of riffles-pools and steps-pools reflects a significant aspect of channel adjustment, they are widely regarded as equilibrium channel forms.

The undulating topography of the pool-riffle sequence is conspicuously absent from many single-thread dryland rivers though their sediments still appear to be distributed in patterns associated with a typical sequence. Reid et al. (1995) for example, describe how the bed material of Nahal Yatir, a gravel-bed river in the Northern Negev Desert, is characterised by an alternating pattern of comparatively coarse 'bars' ($D_{50} = 20$ mm) and longer, planar, finer 'flats' ($D_{50} = 6$ mm; Fig. 12.22a). These bedforms have little or no topographic expression and their positions are stable over time, despite the passage of competent floods. The neighbouring Nahal Hebron has a comparable sedimentology comprising a 'barely discernible alternation of gravel bars and granular-sandy pools' (Hassan 1993, p. 109). Similar patterns of sediment sorting appear in mixtures of sand and gravel. Local concentrations of gravel on otherwise planar, sandy beds have, for example, been described in the arroyos of northern New Mexico (Leopold et al. 1966, Fig. 151) and in the channels of the East Rudolf sedimentary basin in Northern Kenya (Frostick and Reid 1977, p. 2). Comparable alternating sequences of coarser and finer sediments have been identified in steeper channels that might have been expected to form steps and pools (Bowman 1977, Fig. 12.22b).

Intriguingly, the gravel accumulations of Leopold et al. (1966) have a spacing of five-seven times the channel width and are likened to riffles that formed as a kinematic wave. Little was known about steps and pools at the time of Bowman's (1977) work, but his descriptions indicate that although the channels lack the stair-case morphology of a conventional step-pool system, the coarser segments share many other characteristics including a wavelength of about twice the channel width, steep gradients, high roughness, gravelly-bouldery composition and an association with infrequent discharges and near- or super-critical flow (Montgomery and Buffington 1997). These observations suggest that single-thread ephemeral stream channels develop distinct patterns of longitudinal sediment sorting that are analogous to the pool-riffle

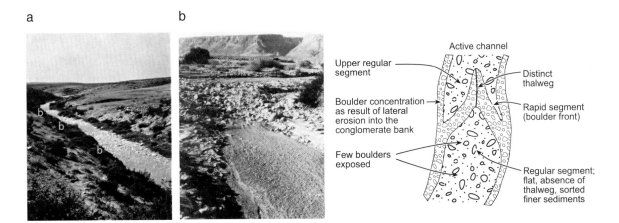

Fig. 12.22 Sedimentary units in dryland streams in southern Israel. (**a**) Longitudinal alternation of coarse channel bars ('b' placed on the adjacent channel bank) and finer 'flats' in Nahal Yatir (after Reid et al. 1995). (**b**) Stepped bed morhology in Nahal Zeelim (after Bowman 1977)

and step-pool sequences of their perennial counterparts. However, the sedimentary characteristics and
dynamics of semi-arid fluvial systems have not been
widely studied and the extent to which dryland rivers
develop symmetrically repeating bed configurations,
the origin, form and function of which can be likened
to, or distinguished from, the continuum of channel
morphologies identified in humid-temperate streams
(e.g. Montgomery and Buffington 1997) is not known.

Small-Scale Bedforms in Sand- and Gravel-Bed Channels

Alluvial rivers develop a wide range of smaller scale
bedforms that provide additional sources of flow
resistance and also reduce particle mobility. They
include ripples, dunes, and antidunes in sand-sized
sediments and particle clusters, transverse ribs and
bedload sheets in gravels and in mixtures of sands
and gravels (Allen 1982; Best 1996). The occurrence
of different bedforms is usually determined using
bedform phase diagrams. These are largely based
on laboratory experiments, heavily biased towards
sand-sized sediments and define equilibrium bedform
regimes in terms of sediment mobility (grain size or
fall velocity) and flow intensity (velocity, shear stress,
stream power). Such diagrams need to be applied
with caution in dryland streams for two reasons. First,
research has shown that bedforms in sand-bed rivers
have a minimum relaxation time in which they are
able to respond to changes in flow conditions (Simons
and Richardson 1963; Allen 1973). The implication of
this work is that equilibrium bedforms are unlikely to
develop in dryland rivers with flashy discharge regimes
(Jones 1977). Second, most research conducted on
gravel bedforms is either conducted under, or guided
by, field conditions observed in humid-temperate
rivers in which the majority of sediment transport
occurs under the regime of partial transport during
perennial flows (Hassan and Reid 1990; Hassan
and Church 2000; Wittenberg and Newson 2005;
Oldmeadow and Church 2006). Compared to our
understanding of sediment sorting and the autogenetic
modification of bed surface grain size (see below) very
little is known about the structural sedimentology of
gravelly beds, and the extent to which the ephemeral
discharge regime and the high rates of sediment

supply and transport restricts the development of
bed structures in dryland streams remains to be
assessed. In this context, it is interesting to note that
Hassan (2005) found little evidence of imbrication or
other surface structures such as stone cells or particle
clusters on the surfaces of channel bars in Nahal Zin
in the Negev Desert, Israel. Marked contrasts in the
structural sedimentology of dryland ephemeral and
humid-temperate perennial rivers were also recorded
by Wittenberg (2002).

Bed Texture Adjustment in Gravel-Bed Rivers

Most gravel-bed rivers develop a coarse surface
armour layer that overlies finer subsurface material
(Fig. 12.23a). Several workers, however, have noted
that the surface and subsurface sediments of many
dryland, gravel-bed rivers are not markedly different
(Schick et al. 1987; Laronne et al. 1994; Hassan 2005;
Laronne and Shlomi 2007; Hassan et al. 2006,
Fig. 12.23b). The weakly- or un-armoured nature of
alluvial gravels in dryland environments has been
attributed to high rates of sediment supply and bedload
discharge, the limited duration of flash flood recession
limbs and, of course, the absence of baseflow (Laronne
et al. 1994). Some of these issues are explored below.

Since the surface of a gravel-bed stream becomes
finer with increasing transport stage, eventually approaching the grain size of the substrate in the limit
of large τ/τ_c (Andrews and Parker 1987), the weak armouring of many dryland channels is consistent with
the high transport stages they sustain. Parker (2008) illustrates the dynamic with reference to the unarmoured
Nahal Yatir and Sagehen Creek, a perennial stream in
the Sierra Nevada of California with a well armoured
bed. In the following discussion, f_{ui}, D_{ug}, D_{b50}, D_{bg},
τ_{50}^* and τ_{c50}^* denote the fraction of subsurface material
in the ith grain size range, the subsurface mean grain
size (m), the median bedload grain size (m), the mean
bedload grain size (m), the dimensionless shear stress
for D_{50} and the critical dimensionless shear stress for
D_{50} respectively. The analysis uses ACRONYM2 of
Parker (1990a,b) to predict the surface size distribution and shear velocity required to transport a given
bedload size distribution and transport rate. The input bedload size distribution for both streams was ap-

a b

Fig. 12.23 Contrasts in surface armouring between (**a**) River Wharfe (humid-temperate perennial river, England, UK) and (**b**) Nahal Yatir (dryland ephemeral river, Israel)

proximated by the subsurface size distribution of Sage-hen Creek since in the normalised form of f_{ui} versus D_{gi}/D_{ug}, the grain size distribution also approximates that of Nahal Yatir. The simulation was conducted using a range of transport rates. The predicted values of the ratio D_{50}/D_{b50} and the predicted grain size distributions of the static and mobile armours at different values of τ_{50}^* are shown in Fig. 12.24, together with estimates of τ^* at bankfull flows in Sagehan Creek and at low and high flows in Nahal Yatir. D_{50}/D_{b50} ratios in Sagehan Creek show little change with increasing dimensionless shear stress and the mobile armour is considerably coarser than the bedload at bankfull dimensionless shear stresses ($\tau^* = 0.059$). In contrast, the mobile armour in the Yatir has become much closer to the bedload than the static armor at $\tau_{50}^* = 0.1$ and the armouring has vanished relative to the bedload at $\tau_{50}^* = 0.3$. The evolution and convergence of the surface grain size distribution from a static armour at low

transport rates (when $\tau_{50}^* \approx \tau_{c50}^*$) to that of the bedload at very high transport rates (when $\tau_{50}^* >> \tau_{c50}^*$) is clearly shown in Fig. 12.24b. It is worth highlighting that the model assumes that the bedload size distribution is constant and approximates the subsurface size distribution. Parker (2008) notes that this is generally not the case (e.g. Fig. 12.13b) and that the bedload grain size dependence on shear stress will reduce the convergence of bedload and bed surface size distributions. Nevertheless, the clear implication is that the lack of armour in dryland streams represents a dynamic sedimentological response to high transport rates generated by high dimensionless shear stresses.

This explanation for the lack of a coarse surface layer dryland river gravels is based on a view of sediment transport dynamics that regards armouring to be a natural consequence of the transport of sediment mixtures at values of shear stresses that prevail in most gravel-bed rivers. Dietrich et al. (1989), however, view

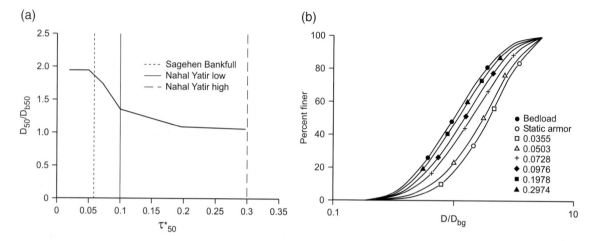

Fig. 12.24 (a) Modelled variation in ratio of median surface grain size (D$_{50}$) to median bedload grain size (D$_{b50}$) with τ^*_{50}. (b) Normalised bedload grain size distribution (assumed) and pre-dicted size distributions for static and mobile armours at values of τ^*_{50} shown in the legend. After Parker (2008)

surface coarsening to be related to the balance between rates of sediment supply and transport. They argue that transport rates in excess of sediment supply induce selective erosion and surface coarsening (e.g. as seen downstream of dams in regulated rivers) while a surplus of sediment forces the deposition of the finer fractions and a consequent reduction in surface grain size (Lisle and Hilton 1992). As a result, surface grain size can be used as an indicator of sediment supply (Buffington and Montgomery 1999). From this perspective, the near equivalence of surface, subsurface and bedload size distributions in dryland channels reflects the high rates of sediment supply from the sparsely vegetated hillslopes. Data collected by Hassan et al. (2006) from arid and humid-temperate streams with a range of sediment supply regimes confirms that many dryland streams are weakly armoured and that sediment supply is a first order control on surface texture and the development of a coarse surface layer. They also suggest that hydrograph shape plays a secondary role. Flume experiments conducted to investigate the influence of hydrograph characteristics on the development of channel armours demonstrate the importance of flow duration and hydrograph symmetry for the development of armoured surfaces.

Channel Pattern

Although rivers exhibit spatial and temporal transitions in channel pattern in response to variations in the magnitude, frequency and sequencing of flood events, they are often classified by their planform geometry into single-thread (straight and meandering) or multi-thread (braided and anabranching) forms. Straight and anabranching channels are relatively uncommon which suggests that they develop under a restricted set of environmental conditions. The former, for example, tend only to occur in locations where channel alignment is forced by geological controls as illustrated by the drainage of Walnut Gulch in SE Arizona (Murphy et al. 1972). Walnut Gulch and its tributaries drain alluvial fills of Tertiary and Quaternary age and are characterised by wide, shallow, sinuous single- and multi-thread courses. However, the channels are essentially straight where they are bounded by outcrops of caliche and traverse resistant conglomerates with marked and abrupt changes of direction signalling entrenched, fault-controlled drainage. Meandering is the most frequently occurring channel planform at the global scale (Knighton 1998, p. 231). Within dryland environments, however, braided channels are more common than meandering channels (Graf 1988, p. 201). Braided rivers are characterised by frequent shifts in channel position and those in drylands are no exception (Graf 1981, 1983). As a result, the braided channel form has often been regarded as disequilibrium aggradational response to high sediment loads. However, even though individual channels may be transient, the fact that braiding appears to be favoured by particular environmental conditions (high and variable discharges, steep slopes, dominant

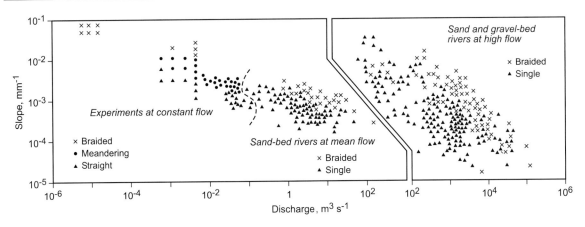

Fig. 12.25 Compilation of field and laboratory data showing the dependency of channel pattern on slope and discharge (after Ferguson 1987). Note that thresholds for meandering in various laboratory models match well and a parallel, higher threshold for braiding is consistent with the experimental results and the data from sand-bed rivers at mean discharge

bedload transport and erodible banks) suggests that it represents a valid equilibrium form.

Field and laboratory data suggest that channel pattern is controlled by discharge and slope (Leopold and Wolman 1957; Schumm and Khan 1972). The compilation of data shown in Fig. 12.25 indicates firstly, that a channel with a given discharge and bed material has one threshold slope (S; m m^{-1}) above which it will meander and a higher threshold slope above which it will braid and secondly, that the threshold slopes decrease as discharge increases. These inverse slope-discharge thresholds have been widely interpreted as thresholds of specific stream power (ω; W m^{-2}) defined as the time rate of potential energy expenditure per unit bed area:

$$\omega = \rho_f g Q S / w \qquad (12.14)$$

Ferguson (1981) and Carson (1984) substituted for w using the downstream hydraulic relation w = aQ$^{0.5}$ (Equation 12.11) to show that the meandering-braided threshold corresponds to a constant stream power of 30–50 W m^{-2} (the exact value depends on the value of a). Begin (1981) similarly demonstrated that the slope-discharge threshold also represents a constant shear stress.

The dependency of channel pattern on stream power provides an explanation for the common occurrence of braided rivers in dryland environments. Although flood events in dryland environments are relatively infrequent, they often generate high stream powers as a result of steep slopes and high discharges. In terms of slope, many dryland streams have high gradients because they flow down pediments and alluvial fans which are steep in comparison to the gradients of valley- and basin-floors. In terms of discharge, high magnitudes are favoured by the effectiveness and efficiency by which rainfall is converted to runoff and concentrated into channels in dryland environments (Baker 1977; Osterkamp and Friedman 2000). In fact, the magnitudes of infrequent flood events in dryland rivers are often much greater than those found in rivers draining humid-temperate catchments of similar size (Costa 1987). Beard (1975), for example, used a flood potential index to demonstrate that the dryland regions of the southwest USA are more susceptible to high magnitude flood events than more humid central and eastern regions. A similar conclusion was reached by Crippen and Bue (1977) who compiled envelope curves for potential maximum flood flows for 17 flood regions of the coterminous USA though their generalisation does not hold for medium–large basins ($> 2,600$ km^2) in which the larger rainfall amounts of more humid regions generate larger floods (Graf 1988, p.90). The tendency for small-medium sized basins in arid and semi-arid environments to have larger floods than similar sized basins in more humid environments is evident in a comparison of flood frequency curves from different climatic regions (Baker 1977; Farquharson et al. 1992; Fig. 12.26).

Channel pattern also reflects sedimentary controls. For example, Schumm (1963) found that channel sinu-

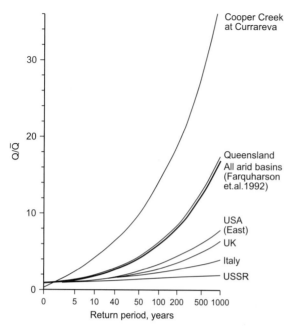

Fig. 12.26 Compilation of flood-frequency relations for rivers from diverse climates showing the more variable flood regime of arid and semi-arid rivers compared to their humid counterparts (after Knighton and Nanson 1997). Note that the magnitude of the mean annual flood ($\overline{Q}$) event in semi-arid environments is only a few percent of the rare event whereas mean annual events in humid temperate environments are much closer to the rare event. (The mean annual event has a recurrence interval of 2.33 years if the distribution conforms to the EV1 Gumbel distribution)

osity of sand-bed rivers in the Great Plains of the USA increased with the percentage slit clay in the bed and banks and Kellerhalls (1982) and Carson (1984) found that sand-bed rivers braid at lower slopes than gravel-bed rivers with similar discharges. In these studies, grain size is being used as a surrogate for bank strength (as in studies of cross-sectional adjustment discussed earlier) with the implication that channel pattern is dependent on the erodibility of the channel banks as well as the erosivity of the flow (Fig. 12.27).

As already noted, dryland sediments are not generally rich in cohesive silts and clays and unless riparian vegetation is sufficient to stabilise the channel banks and/or discourage the formation of new channels, channel widening will promote the development of a braided channel pattern through the instability of sediment transport in wide channels (Parker 1976). Indeed, Murray and Paola (1994) regard the braided channel form as the inevitable consequence of unconstrained flow over a non-cohesive bed. As such,

vegetation is a primary determinant of channel pattern in dryland environments. The importance of vegetation in controlling dryland channel patterns can be illustrated by reference to the introduction and spread of Tamarisk throughout the SW USA. Tamarisk (commonly known as Saltcedar) was introduced into the SW USA from the Mediterranean basin in the mid 1800s. Because of its competitive advantages over native riparian species (Brotherson and Field 1987; Howe and Knopf 1991), Tamarisk colonised and spread rapidly along riparian corridors and by the late 1990s, had became established in nearly every semi-arid drainage basin within the SW USA (Randall and Marinelli 1996). The result was a marked decrease in channel geometry throughout the SW USA typified by a 27% reduction in the average width of major streams of the Colorado plateau (Graf 1978). Moreover, in rivers widened and destabilised by large flood events, vegetation regrowth and encroachment (including the invasion of tamarisk) resulted in channel narrowing, floodplain reconstruction and the conversion of multi-thread rivers to single-thread forms (Schumm and Lichty 1963; Burkham 1972; Eschner et al. 1983; Martin and Johnson 1987; Friedman et al. 1996; VanLooy and Martin 2005). Although the mechanisms underlying these channel adjustments have not been demonstrated directly in the field, it is generally reasoned that vegetation first colonises and stabilises the surfaces of flood deposits such as channel bars and dunes and which then grow by vertical and lateral accretion of sediment during low magnitude events. Rates of accretion are enhanced by the increased roughness of the vegetated surfaces which reduce flow velocities and encourage sedimentation. In the absence of destructive high flows, these areas of incipient flood plain grow and coalesce, leading to the aggradation and abandonment of the surrounding channels. Over time, a new floodplain develops that is composed of a mosaic of coalesced islands, abandoned channels and areas of floodplain that build up adjacent to the low water channel.

This model of channel evolution is supported by laboratory (Tal and Paola 2007) and cellular (Tal et al. 2003) models of braided rivers which demonstrate how channels choked by vegetation and/or vegetation-induced sedimentation cause reductions in channel width, braiding index and channel mobility (Fig. 12.28). Ultimately, vegetation eliminates weak flow paths, thereby concentrating water into a single

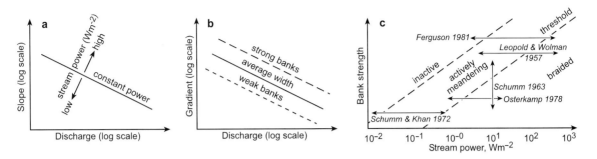

Fig. 12.27 Influence on bank strength on the thresholds of channel pattern (after Ferguson 1987). (**a**) Threshold of constant stream power (traditional slope-discharge threshold). (**b**) Anticipated shift in threshold according to bank strength. (**c**) Empirical relationships between channel pattern, stream power and bank strength. Schumm and Khan (1972) = unvegetated sandy laboratory channels; Osterkamp (1978) = sand bed channels with some silt and clay and riparian vegetation; Schumm (1963) channels with variable silt and clay; Leopold and Wolman (1957) mixture of sand- and gravel-bed channels; Ferguson (1981) = well vegetated banks of gravel and/or cohesive sediments

dominant channel. Progressive reductions in total channel width leads to the establishment of a new self-organised steady state in which the flow removes vegetated areas as fast as they are produced. The results also suggest that colonisation by vegetation is not easily reversible so that the morphological effects are likely to be long-lived. This concurs with the findings of several field studies conducted in dryland environments that suggest that vegetation encroachment raises the threshold for channel adjustment with the result that subsequent floods are not able to widen the channel as they might have previously (Eschner et al. 1983; Hooke 1996). The decreased ability of the channels to adjust to large flood events may lead to increases in the magnitude and frequency of overbank flooding and floodplain sedimentation resulting in further increases in the stability of the channel form through positive feedback (Graf 1978).

Braiding can also be understood as a response to maintain transport competence in relation to the imposed grain size (e.g. Henderson 1961; Carson 1984) and/or transport capacity in relation to the imposed sediment load (e.g. Kirkby 1977; Bettess and White 1983; Chang 1985). From these perspectives, the common occurrence of braided rivers in dryland environments reflects the availability of large amounts of coarse sediment and its movement as bedload. Such approaches provide important links with the underlying causes of braiding, namely local aggradation (often linked to the stalling of bedload sheets, channel bars or loss of competence in flow expansions), bar growth (by vertical and lateral accretion) and subsequent dissection (Ashmore 1991).

Anabranching channels differ from braided channels in that the system of multiple channels is separated by vegetated or otherwise stable islands which are emergent at stages up to the bankfull discharge (Nanson and Knighton 1996). They form a diverse range of channel forms that are associated with flood-dominated channel regimes, resistant banks and mechanisms that induce channel avulsion (Makaske 2001). Low energy, fine grained anabranching systems known as anastomosing channels are found in the Lake Eyre basin (also known as the Channel Country) of east-central Australia (Nanson et al. 1988; Gibling et al. 1998) and in the Red Desert of Wyoming (Schumann 1989). The type is well-represented by Cooper Creek in the Lake Eyre Basin which maintains an active belt of anastomosed channels up to 10 km wide for distances of several hundred kilometres. The floodplain is made up of a well integrated primary system of one-four channels supplemented by subsidiary channels that become active at progressively higher discharges. Anastomosis serves to concentrate stream flow and maximise the transport of bed sediments in regions where there is little opportunity to increase channel gradients (Nanson and Huang 1999; Jansen and Nanson 2004). In other parts of central and northern Australia, channel forms are dominated by ridge-forming anabranching rivers (Tooth and Nanson 2000b). These channels are characterised by a low sinuosity belt of subparallel channels separated by narrow, flow-aligned, sandy ridges vegetated by teatrees (*Melaleuca glomerata*). The ridges develop as a result of spatial patterns of erosion and sedimentation induced by the growth of teatrees within the channel.

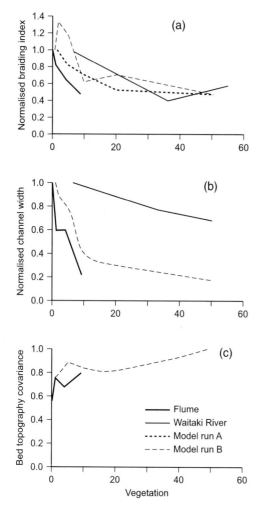

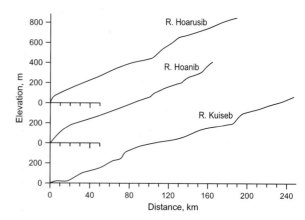

Fig. 12.29 Longitudinal profiles of three rivers in the Namib Desert (after Vogel 1989)

Fig. 12.28 Changes in braiding index (**a**), channel width (**b**) and bed topography correlation coefficient (**c**) with vegetation in the braided Waitaki River, New Zealand, a laboratory flume and a cellular automata model (after Tal et al. 2003). Vegetation in the field, flume and cellular model is represented by fractional vegetation cover of the braid plain, the density of alfalfa stems and a vegetation strength parameter respectively. The bed topography covariance is measure of the channel mobility rate; higher values of covariance indicate lower channel mobility rates

Channel Gradient and the Longitudinal Profile

The longitudinal profile represents the final aspect of channel adjustment to be considered. Longitudinal profiles are typically upwardly concave. It is generally recognised, however, that the long profiles of many dryland rivers are less concave than their humid-temperate counterparts (e.g. Langbein 1964) and may even be linear or upwardly convex (Schumm 1961; Vogel 1989; Fig. 12.29). These differences can be

explained by considering the effect of tributary inputs of water and sediment on profile concavity.

In perennial rivers, sediment concentrations usually decrease downstream because tributary inputs increase the supply of water more than they increase the supply of sediment. As a result, stream slopes reduce by incision and an upwardly concave profile tends to develop (Gilbert 1877; Wheeler 1979; Hey and Thorne 1986). In dryland channels, however, the ratio of sediment to streamflow often increases downstream due to transmission losses (e.g. Leopold and Miller 1956). Since sediment is carried by progressively less flow, there is a tendency for aggradation and the development of a convex profile. Profile concavity due to tributary inputs has been modelled by Sinha and Parker (1996). Although Sinha and Parker's model cannot be tested due to a lack of data concerning rates of tributary water and sediment input along river profiles, it does confirm that under conditions of high transport rates, a downstream declining concentration of bed material load is a necessary condition for profile concavity and that convex profiles develop if sediment concentrations increase downstream. The form of the long profile is also a function of sediment calibre and profile concavity has long been associated with the streamwise fining of sediment (Hack 1957; Ikeda 1970; Cherkauer 1972; Parker 1991a,b). However, since downstream fining is the result of particle abrasion and sorting, grain size can control (e.g. Shulits 1941; Yatsu 1955; Sinha and Parker 1996), and be controlled by (e.g. Hoey and Ferguson 1994) profile concavity. The downstream trends of particle size in dryland streams have not been widely studied (notable exceptions include the work of Frostick and Reid 1980; Rhoads 1989). Research in humid-temperate streams has, however, demonstrated that sediment inputs from tributary inputs and non-alluvial sources often preclude the development of systematic downstream fining trends (e.g. Ferguson et al. 2006). Since high drainage densities and easily erodible banks are characteristic of many dryland rivers, it is quite possible that they may be typified by random grain size variations. Such conditions can be expected to inhibit the development of profile concavity as illustrated by the work of Rice and Church (2001) who studied the profile form of sedimentary links, reaches of river unaffected by inputs of water and sediment. Their results show that under conditions of constant downstream discharge, links with a strong downstream fining trend exhibit concave profile forms (the greater the rate of grain size diminution, the greater the concavity) whilst links exhibiting a weak or no downstream fining trend exhibit convex profile forms.

Summary

As shown by this review, our understanding of dryland rivers has been transformed in recent years through detailed field, laboratory and modelling studies of contemporary runoff and sediment transport processes. However, fundamental questions remain as to how channel morphology and change is related to the time distribution of flow events, variations in sediment supply and their interactions with channel boundary conditions (topography, sedimentology and vegetation) over longer time scales. These questions are not new (e.g. Douglas 1982; Lane and Richards 1997) nor are they restricted to dryland environments (e.g. Kirkby 1999) but they lie at the heart of developing an integrated theory of dryland river behaviour that combines explanations for processes and forms as well as morphological change (Graf 1988). Much work remains to be done to reconcile a kinematic understanding of channel evolution with a dynamic understanding of process mechanics at scales larger than the channel reach and longer than the duration of individual flow events.

References

Abrahams, A.D., G. Li and J.F. Atkinson 1995. Step-pool streams: Adjustment to maximum flow resistance. *Water Resources Research*, **31**, 2593–2602.

Ackers, P. and F.H. Charlton 1970. Meander geometry arising from varying flows. *Journal of Hydrology*, **11**, 230–252.

Alexandrov, Y. 2005. *Suspended Sediment Dynamics in Semiarid Ephemeral Channels, Nahal Eshtemoa, Israel*. Unpublished PhD thesis, Ben Gurion University of the Negev, Israel.

Alexandrov, Y., J.B. Laronne and I. Reid 2003. Suspended sediment concentration and its variation with water discharge in a dryland ephemeral channel, northern Negev, Israel. *Journal of Arid Environments*, **53**, 73–84.

Alexandrov, Y., J.B. Laronne and I. Reid 2006. Intra-event and inter-seasonal behaviour of suspended sediment in flash floods of the semi-arid northern Negev, Israel. *Geomorphology*, **85**, 85–97.

Allen, J.R.L. 1973. Phase differences between bed configuration and flow in natural environments and their geological relevance. *Sedimentology*, **20**, 323–329.

Allen, J.R.L. 1982. *Sedimentary Structures: Their Character and Physical Basis*. Amsterdam: Elsevier.

Allen, P.A. 1997. *Earth Surface Processes*. London: Blackwell Science.

Almedeij, J.H. and P. Diplas 2003. Bedload transport in gravel-bed streams with unimodal sediment. *Journal of Hydraulic Engineering*, **129**, 896–904.

Almedeij, J.H. and P. Diplas 2005. Bed load sediment transport in ephemeral and perennial gravel bed streams. *EOS, Transactions American Geophysical Union*, **86**, 429–434.

Amos, K.J., J. Alexander, A. Horn, G.D. Pocock and C.R. Fielding 2004. Supply-limited sediment transport in a high discharge event of the tropical Burdekin River, north Queensland, Australia. *Sedimentology*, **51**, 145–162.

Andrews, E.D. and G. Parker 1987. Formation of a coarse surface layer as the response to gravel mobility. In *Sediment Transport in Gravel-bed Rivers*, C.R. Thorne, J.C. Barthurst and R.D. Hey (eds) 269–300. Chichester: Wiley.

ASCE 1982. Relationships between morphology of small streams and sediment yield. Task Committee on Relations Between Morphology of Small Streams and Sediment Yield of the Committee on Sedimentation of the Hydraulics Division. *Proceedings of the American Society of Civil Engineers, Journal Hydraulics Division*, **108**, 1328–1365.

Ashmore, P.E. 1991. How do gravel bed rivers braid? *Canadian Journal of Earth Sciences*, **28**, 326–341.

Ashworth, P.J., R.I. Ferguson, P.E. Ashmore, C. Paola, D.M. Powell and K.L. Prestegaard 1992. Measurements in a braided river chute and lobe 2. Sorting of bedload during entrainment, transport and deposition. *Water Resources Research*, **28**, 1887–1896.

Bagnold, R.A. 1966. An approach to the sediment transport problem from general physics. *United States Geological Survey Professional Paper*, **421I**.

Baker, V. 1977. Stream channel response to floods, with examples from central Texas. *Bulletin Geological Society of America*, **88**, 1057–1071.

Bates, P.D., S.N. Lane and R. Ferguson 2005. *Computational Fluid Dynamics: Applications in Environmental Hydraulics*. Chichester: Wiley.

Beard, L. 1975. *Generalized Evaluation of Flash Flood Potential*. Center for Research in Water Resources Report CRWR-124. Austin: University of Texas.

Begin, Z.B. 1981. The relationship between flow-shear stress and stream pattern. *Journal of Hydrology*, **52**, 307–319.

Belperio, A.P. 1979. The combined use of wash load and bed material load rating curves for the calculation of total load: An example from the Burdekin River, Australia. *Catena*, **6**, 317–329.

Best, J.L. 1996. The fluid dynamics of small-scale alluvial bedforms. In *Advances in Fluvial Dynamics and Stratigraphy*, P.A. Carling and M.R. Dawson (eds) 67–125. Chichester: Wiley.

Bettess, R. and W.R. White 1983. Meandering and braiding of alluvial channels. *Proceedings Institute Civil Engineers, Part 2*, **75**, 525–538.

Beverage, K.P. and J.K. Culbertson 1964. Hyperconcentrations of suspended sediment. *Proceedings of the American Society of Civil Engineers, Journal Hydraulics Division*, **90**, 117–128.

Beyer, P.J 2006. Variability in channel form in a free-flowing dryland river. *River Research and Applications*, **22**, 203–217.

Bourne, M. and G. Pickup. 1999. Fluvial form variability in arid central Australia. In *Varieties of Fluvial Form*, A.J. Miller and A. Gupta (eds) 249–272. Chichester: Wiley.

Bowman, D. 1977. Stepped-bed morphology in arid gravelly channels. *Bulletin Geological Society of America*, **88**, 291–298

Bracken, L.J. and J. Wainwright 2006. Geomorphological equilibrium: Myth and metaphor? *Transactions Institute British Geographers*, **31**, 167–178.

Brotherson, J. and D. Field 1987. Tamarix: Impacts of a successful weed. *Rangelands*, **9**, 110–112.

Brunsden, D. and J.B. Thornes 1979. Landscape sensitivity and change. *Transactions, Institute of British Geographers*, **4**, 463–484.

Buffington, J.M. and D.R. Montgomery 1999. Effects of sediment supply on surface textures of gravel-bed rivers. *Water Resources Research*, **35**, 3523–3530.

Bull, L.J. and M.J. Kirkby 2002. *Dryland Rivers: Hydrology and Geomorphology of Semi-Arid Channels*. Chichester: Wiley.

Burkham, D.E. 1972. Channel changes of the Gila River in Safford Valley, Arizona 1946–70. *United States Geological Survey Professional Paper*, **655G**.

Butcher, G.C. and J.B. Thornes, 1978. Spatial variability in runoff processes in an ephemeral channel. *Zeitschrift Fur Geomorphologie Supplementband*, **29**, 83–92.

Capart, H. and D.L. Young 1998. Formation of a jump by the dam-break wave over a granular bed. *Journal of Fluid Mechanics*, **372**, 165–187.

Carson, M.A. 1984. The meandering-braided river threshold; a reappraisal. *Journal of Hydrology*, **73**, 315–334.

Chang, H.H. 1985. River morphology and thresholds. *Journal of Hydraulic Engineering*, **111**, 503–519.

Chanson, H. 2004. Experimental study of flash flood surges down a rough sloping channel. *Water Resources Research*, **40**, W03301, doi:10.1029/2003WR002662.

Cherkauer, D.S. 1972. Longitudinal profiles of ephemeral streams in southeastern Arizona. *Bulletin Geological Society America*, **83**, 353–365.

Chin A. 2002. The periodic nature of step-pool mountain streams. *American Journal of Science*, **302**, 144–167.

Chin, A. and J.D. Phillips 2007. The self-organization of step-pools in mountain streams. *Geomorphology*, **83**, 346–358.

Church, M. and D. Jones 1982. Channel bars in gravel-bed rivers. In *Gravel-bed Rivers*, R.D. Hey, J.C. Bathurst and C.R. Thorne (eds) 291–324. Chichester: Wiley.

Colby, B.R. 1963. Fluvial sediments – a summary of source, transportation, deposition and measurement of sediment discharge. *Bulletin United States Geological Survey*, **1181A**.

Colby, B.R. 1964. Scour and fill in sand-bed streams. *United States Geological Survey Professional Paper*, **462D**.

Coppus, R. and A.C. Imeson. 2002. Extreme events controlling erosion and sediment transport in a semi-arid sub-Andean valley. *Earth Surface Processes and Landforms*, **27**, 1365–1375.

Costa, J.E. 1987, A comparison of the largest rainfall-runoff floods in the United States, with those of the Peoples Republic of China. *Journal of Hydrology*, **96**, 101–115.

Costelloe, J.F., J. Powling, J.R.W. Reid, R.J. Shiel and P. Hudson 2005. Algal diversity and assemblages in arid zone rivers of the Lake Eyre Basin, Australia. *River Research and Applications*, **21**, 337–349.

Crippen, J. and C. Bue 1977. Maximum flood flows in the conterminous United States. *United States Geological Survey Water Supply Paper*, **1887**.

Darby, S.E., A.M. Alabyan and M.J. Van de Wiel 2002. Numerical simulation of bank erosion and channel migration in meandering rivers, *Water Resources Research*, **38**, 1163, doi:10.1029/2001WR000602.

Davies, B.R., M.C. Thoms, K.F. Walker, J.H. O'Keefe and J.A. Gore 1994. Dryland rivers: Their ecology, conservation and management. In *The Rivers Handbook: Hydrological and Ecological Principles, Volume 2*, P. Calow and G.E. Petts (eds) 484–512. Oxford: Blackwell.

Dietrich, W.E., J.W. Kirchner, H. Ikeda and F. Iseya 1989. Sediment supply and the development of the coarse surface layer in gravel-bedded rivers. *Nature*, **340**, 215–217.

Douglas, I. 1982. The unfulfilled promise: Earth surface processes as the key to landform evolution. *Earth Surface Processes and Landforms*, **7**, 101.

Dunkerley, D.L. 1992. Channel geometry, bed material, and inferred flow conditions in ephemeral stream systems, Barrier Range, western NSW Australia. *Hydrological Processes*, **6**, 417–433.

Dunkerley, D.L. and K. Brown 1999. Flow behaviour, suspended sediment transport and transmission losses in a small (sub-bank-full) flow event in an Australian desert stream. *Hydrological Processes*, **13**, 1577–1588.

Eaton, B.C. and R.G. Millar 2004. Optimal alluvial channel width under a bank stability constraint. *Geomorphology*, **62**, 35–45.

Edwards, T.K. and G.D. Glysson 1999. *Field methods for measurement of fluvial sediment*. U.S. Geological Survey, Techniques of Water-Resources Investigations, Book 3, Chapter C2.

Einstein H.A. 1937. The bed load transport as probability problem. Mitteilung der Versuchsanstalt fuer Wasserbau an der Eidgenössischen Technischen Hochschule. Zürich; 110. (English translation by W.W. Sayre (1972). In *Sedimentation*, H.W. Shen (ed) Appendix C, Fort Collins, Colorado: H.W. Shen.

Einstein, H.A. 1950. The bed-load function for sediment transportation in open channel flows. *United States Department Agriculture, Technical Bulletin*, **1026**.

Einstein, H.A. and N. Chien 1953. Can the rate of wash load be predicted from the bed-load function? *Transactions of the American Geophysical Union*, **34**, 876–882.

Engelund, F. and E. Hansen 1967. *A Monograph on Sediment Transport in Alluvial Streams*. Copenhagen: Teknisk Forlag.

Eschner, T.R. 1983. Hydraulic geometry of the Platte River near Overton, south-central Nebraska, *United States Geological Survey Professional Paper*, **1277C**.

Farquharson, F.A.K., J.R. Meigh and J.V. Sutcliffe 1992. Regional flood frequency analysis in arid and semi-arid areas. *Journal of Hydrology*, **138**, 487–501.

Ferguson, R.I. 1981. Channel forms and channel changes. In *British Rivers*, J. Lewin (ed) 90–125. London: Allen and Unwin.

Ferguson, R.I. 1986. Hydraulics and hydraulic geometry. *Progress in Physical Geography*, **10**, 1–31.

Ferguson, R.I. 1987. Hydraulic and sedimentary controls of channel pattern. In *River Channels: Environment and Process*, K.S. Richards (ed) 129–158. Oxford: Blackwell.

Ferguson, R.I. and S.J. Wathen 1998. Tracer-pebble movement along a concave river profile: Virtual velocity in relation to grain size and shear stress. *Water Resources Research*, **34**, 2031–2038.

Ferguson R.I., J.R. Cudden, T.B. Hoey and S.P. Rice 2006. River system discontinuities due to lateral inputs: Generic styles and controls. *Earth Surface Processes and Landforms*, **31**, 1149–1166.

Fisher, S.G. and N.B. Grimm. 1985. Hydrologic and material budgets for a small Sonoran Desert watershed during three consecutive cloudburst floods. *Journal of Arid Environments*, **9**, 105–118.

Fisher, S.G. and W.L. Minckley 1978. Chemical characteristics of a desert stream in flash flood. *Journal of Arid Environments*, **1**, 25–33.

Foley, M.G. 1978. Scour and fill in steep, sand bed ephemeral streams. *Bulletin Geological Society of America*, **89**, 559–570.

Friedman J.M., W.R. Osterkamp and W.M. Lewis Jr. 1996. The role of vegetation and bed-level fluctuations in the process of channel narrowing. *Geomorphology*, **14**: 341–351.

Frostick, L.E. and I. Reid 1977. The origin of horizontal laminae in ephemeral stream channel-fill. *Sedimentology*, **24**, 1–9.

Frostick, L.E. and I. Reid 1980. Sorting mechanisms in coarse-grained alluvial sediments: Fresh evidence from a basalt plateau gravel, Kenya. *Journal Geological Society of London*, **137**, 431–441.

Frostick, L.E., I. Reid and J.T. Layman 1983. Changing size distribution of suspended sediment in arid-zone flash floods. *Special Publication International Association Sedimentologists*, **6**, 97–106.

Garcia, M. and G. Parker 1991. Entrainment of bed sediment into suspension. *Proceedings of the American Society of Civil Engineers, Journal Hydraulics Division*, **117**, 414–435.

Garcia, M., Y. Niño and F. López 1996. Laboratory observations of particle entrainment into suspension by turbulent bursting. In *Coherent Flow Structures in Open Channels*, P.J. Ashworth, S.J. Bennett, J.L. Best and S.J. McLelland (eds) 63–86. Chichester: Wiley.

Gellis, A.C., M.J. Pavich, P.R. Bierman, E.M. Clapp, A. Ellevein and S. Aby 2004. Modern sediment yield compared to geologic rates of sediment production in a semi-arid basin, New Mexico: Assessing the human impact. *Earth Surface Processes and Landforms*, **29**, 1359–1372.

Gerson, R. 1977. Sediment transport for desert watersheds in erodible materials. *Earth Surface Processes*, **2**, 343–361.

Gibling, M.R., G.C. Nanson and J.C. Maroulis 1998. Anastomosing river sedimentation in the Channel Country of central Australia. *Sedimentology*, **45**, 595–619.

Gilbert, G.K. 1877. *Report on the Geology of the Henry Mountains, Washington DC*. United States Geological and Geographical Survey, Rocky Mountain Region, Washington D.C.: General Printing Office.

Gippel, C.J. 1995. Potential of turbidity monitoring for measuring the transport of suspended solids in streams. *Hydrological Processes*, **9**, 83–97.

Gomez, B. 1995. Bedload transport and changing grainsize distribution. In *Changing River Channels*, A.M. Gurnell and G.E. Petts (eds) 177–199. Chichester: Wiley.

Goodrich, D.C., J.M. Faures, D.A. Woolhiser, L.J. Lane and S. Sorooshian 1995. Measurement and analysis of small-scale convective storm rainfall variability. *Journal of Hydrology*, **173**, 283–308.

Graf, W.F. 1978. Fluvial adjustment to the spread of tamarisk in the Colorado Plateau region. *Bulletin Geological Society of America*, **89**, 1491–1501.

Graf, W.L. 1981 Channel instability in a braided, sand bed river. *Water Resources Research*, **17**, 1087–1094.

Graf, W.L. 1983. Flood-related channel change in an arid-region river. *Earth Surface Processes and Landforms*, **8**, 125–139.

Graf, W.L. 1988. *Fluvial Processes in Dryland Rivers*. Berlin: Springer-Verlag.

Gregory, K.J. and D.E. Walling 1973. *Drainage Basin Form and Process: A Geomorphological Approach*, London: Arnold.

Griffiths, P.G., R. Hereford and R.H. Webb, 2006. Sediment yield and runoff frequency of small drainage basins in the Mojave Desert, U.S.A. *Geomorphology*, **74**, 232–244.

Gupta, A. 1995. Magnitude, frequency, and special factors affecting channel form and processes in the seasonal tropics. In *Natural and Anthropogenic Influences in Fluvial Geomorphology*, J.E. Costa, A.J. Miller, K.W. Potter and P.R. Wilcock (eds) 125–136. Washington D.C.: American Geophysical Union.

Grimm, N.B., S.G. Fisher and W.L. Minckley 1981. Nitrogen and phosphorus dynamics in hot desert streams of southwestern USA. *Hydrobiologia*, **83**, 303–312.

Hack, J.T. 1957. Studies of longitudinal stream profiles in Virginia and Maryland. *United States Geological Survey Professional Paper*, **294B**, 59–63.

Haregeweyn, N., J. Poesen, J. Nyssen, G. Verstraeten, J. de Vente, G. Govers, S. Deckers and J. Moeyersons 2005. Specific sediment yield in Tigray-Northern Ethiopia: Assessment and semi-quantitative modelling, *Geomorphology* **69**, 315–331.

Haschenburger, J. and M. Church, 1998. Bed material transport estimated from the virtual velocity of sediment. *Earth Surface Processes and Landforms*, **23**, 791–808.

Hassan, M.A. 1990. Scour, fill and burial depths of coarse bed material in gravel bed streams. *Earth Surface Processes and Landforms*, **15**, 341–356.

Hassan, M.A., 1993. Structural controls on the mobility of coarse material. *Israel Journal of Earth Science*, **41**, 105–122.

Hassan, M.A., 2005. Gravel bar characteristics in arid ephemeral streams. *Journal of Sedimentary Research*, **75**, 29–42.

Hassan, M.A. and P. Erginzinger 2003. Use of tracers in fluvial geomorphology. In *Tools in Fluvial Geomorphology*, G.M. Kondolf and H. Piégay (eds) 397–424. Chichester: Wiley.

Hassan, M.A. and M. Church 1992. The movement of individual grains on the streambed. In *Dynamics of Gravel-Bed Rivers*, P. Billi, R.D. Hey, C.R. Thorne and P. Tacconi (eds) 159–175. Chichester: Wiley.

Hassan, M.A. and M. Church, 1994. Vertical mixing of coarse particles in gravel bed rivers: A kinematic model. *Water Resources Research*, **30**, 1173–1185.

Hassan, M.A. and M. Church, 2000. Experiments on surface structure and partial sediment transport. *Water Resources Research*, **36**, 1885–1895.

Hassan, M.A. and I. Reid 1990. The influence of microform bed roughness elements on flow and sediment transport in gravel bed rivers. *Earth Surface Processes and Landforms*, **15**, 739–750.

Hassan, M.A., M. Church and P.J Ashworth, 1992. Virtual rate and mean distance of travel of individual clasts in gravel bed channels. *Earth Surface Processes and Landforms*, **17**, 617–627.

Hassan, M.A., M. Church and A.P Schick 1991. Distance of movement of coarse particles in a gravel bed stream. *Water Resources Research*, **27**, 503–511.

Hassan, M.A., R. Egozi and G. Parker 2006. Experiments on the effect of hydrograph characteristics on vertical grain sorting in gravel bed rivers. *Water Resources Research*, **42**, WO9408, doi:10.1029/2005WR004707.

Hassan, M.A., A.P. Schick and P.A. Shaw 1999. The transport of gravel in an ephemeral sandbed river. *Earth Surface Processes and Landforms*, **24**, 623–640.

Heidel, S.G. 1956. The progressive lag of sediment concentration with flood waves. *Transactions of the American Geophysical Union*, **37**, 56–66.

Hem, J.D. 1948. Fluctuations in concentrations of dissolved solids in some southwestern streams. *Transactions of the American Geophysical Union*, **29**, 80–84.

Hem, J.D. 1985. Study and interpretation of the chemical characteristics of natural waters. *United States Geological Survey Water Supply Paper*, **2254**.

Henderson, F.M. 1961. Stability of alluvial channels. *Journal Hydraulics Division American Society Civil Engineers*, **87**, 109–138.

Hey, R.D. and C.R. Thorne, 1986. Stable channels with mobile gravel beds. *Journal of Hydraulic Engineering*, **112**, 671–689.

Hicks, D.M. and B. Gomez. 2003. Sediment transport. In *Tools in Fluvial Geomorphology*, G.M. Kondolf and H. Piégay (eds) 425–461. Chichester: Wiley.

Hoey, T. and R.I. Ferguson 1994. Numerical simulation of downstream fining by selective transport in gravel bed rivers: Model development and illustration. *Water Resources Research*, **30**, 2251–2260.

Hooke, J.M. 1996. River responses to decadal-scale changes in discharge regime: The Gila River, SE Arizona. In *Global Continental Changes: The Context of Palaeohydrology*, J.

Branson, A.G. Brown and K.J. Gregory (eds) 191–204. Geological Society Special Publication, **115**.

Hooke, J.M. and J.M. Mant 2000. Geomorphological impacts of a flood event on ephemeral channels in SE Spain. *Geomorphology*, **34**, 163–180.

Howe, W. and F. Knopf 1991. On the imminent decline of Rio Grande cottonwoods in central New Mexico. *The Southwestern Naturalist*, **36**, 218–224.

Huckleberry, G. 1994. Contrasting channel response to floods on the middle Gila River, Arizona. *Geology*, **22**, 1083–86.

Ikeda, H. 1970. On the longitudinal profiles of the Asake, Mitaki and Utsube Rivers, Mie Prefecture. *Geographical Review of Japan*, **43**, 148–159.

Jackson, R.G. 1976. Sedimentological and fluid-dynamic implications of the turbulent bursting phenomenon in geophysical flows. *Journal of Fluid Mechanics*, **77**, 531–561.

Jacobson, P.J., K.M. Jacobson, P.L. Angermeier and D.S. Cherry 2000a. Hydrologic influences on soil properties along ephemeral rivers in the Namib Desert. *Journal of Arid Environments*, **45**, 21–34.

Jacobson, P.J., K.M. Jacobson, P.L. Angermeier and D.S. Cherry 2000b. Variation in material transport and water chemistry along a large ephemeral river in the Namib Desert. *Freshwater Biology*, **44**, 481–491.

Jansen, J. and G. Nanson, Anabranching and maximum flow efficiency in Magela Creek, northern Australia. *Water Resources Research*, **40**, 1–12.

Jones, C.M. 1977. Effects of varying discharge regimes on bedform sedimentary structures in modern rivers. *Geology*, **5**, 567–570.

Jones, J.B. Jr, J.D. Schade, S.G. Fisher and N.B. Grimm 1997. Organic matter dynamics in Sycamore Creek, a desert stream in Arizona, USA. *Journal of the North American Benthological Association*, **16**, 78–82.

Katra, I., D.G. Blumberg, L. Hanoch and S. Pariente 2007. Spatial distribution dynamics of topsoil moisture in shrub microenvironment after rain events in arid and semi-arid areas by means of high-resolution maps. *Geomorphology*, **86**, 455–464.

Keller E.A. and W.N. Melhorn 1978. Rhythmic spacing and origin of pools and riffles. *Bulletin Geological Society of America*, **89**, 723–730.

Kellerhalls, R. 1982. Effect of river regulation on channel stability. In *Gravel-bed Rivers*, R.D. Hey, J.C. Bathurst, and C.R. Thorne (eds) 685–715. Chichester: Wiley.

Khadkikar, A.S., L.S. Chamyal and R. Ramesh. 2000. The character and genesis of calcrete in late Quaternary alluvial deposits, Gujarat, western India, and its bearing on the interpretation of ancient climates. *Palaeogeography, Palaeoclimatology, Palaeoecology*, **162**, 239–261.

Khan, M.A. 1993. Suspended sediment and solute characteristics of two desert rivers of India. *Annals of Arid Zone*, **32**, 151–156.

Kirkbride, A.D. 1993. Observations of the influence of bed roughness on turbulence structure in depth limited flows over gravel beds. In *Turbulence: Perspectives on Flow and Sediment Transport*, N.J. Clifford, J.R. French, and J Hardisty (eds) 185–196. Wiley: Chichester.

Kirkby, M.J. 1977. Maximum sediment efficiency as a criterion for alluvial channels. In *River Channel Changes*, K.J. Gregory (ed) 429–442. Chichester: Wiley.

Kirkby, M.J. 1999. Towards an understanding of varieties of fluvial form. In *Varieties of Fluvial Form*, A.J. Miller and A. Gupta (eds) 507–516. Chichester: Wiley.

Knighton, A.D. 1998. *Fluvial Forms and Processes: A New Perspective*. London: Arnold.

Knighton, A.D. and G. Nanson 1997. Distinctiveness, diversity and uniqueness in arid zone river systems. In *Arid Zone Geomorphology: Process, Form and Change in Drylands*, D.S.G. Thomas (ed) 185–204. Chichester: Wiley.

Kreplin, H. and H. Eckelmann, H. 1979. Behaviour of the three fluctuating velocity components in the wall region of a turbulent channel flow. *Physics of Fluids*, **22**, 1233–1239.

Kuhnle, R.A. and J.C. Willis 1992. Mean size distribution of bed load on Goodwin Creek. *Journal of Hydraulic Engineering*, **118**, 1443–1446.

Lane, E.W. 1940. Notes on limit of sediment concentration. *Journal of Sedimentary Petrology*, **10**, 95–96.

Lane, S.N. and K.S. Richards 1997. Linking river channel form and process: Time, space and causality revisited. *Earth Surface Processes and Landforms*, **22**, 249–260.

Langbein, W.B. 1964. Profiles of rivers of uniform discharge. *United States Geological Survey Professional Paper*, **501B**, 119–122.

Lapointe, M. 1992. Burst-like sediment suspension events in a sand bed river. *Earth Surface Processes and Landforms*, **17**, 253–270.

Laronne, J.B. 2000. Event-based deposition in the ever-emptying Yatir Reservoir, Israel. *International Association of Hydrological Sciences Publication*, **261**, 285–302.

Laronne, J. and I. Reid 1993. Very high rates of bedload sediment transport by ephemeral desert rivers. *Nature*, **366**, 148–150.

Laronne, J.B. and Y. Shlomi 2007. Depositional character and preservation potential of coarse-grained sediments deposited by flood events in hyper-arid braided channels in the Rift Valley, Arava, Israel. *Sedimentary Geology*, **195**, 21–37.

Laronne J.B, I. Reid, Y. Yitshack and L.E. Frostick 1992. Recording bedload discharge in a semiarid channel, Nahal Yatir, Israel. In *Erosion and Sediment Transport Monitoring Programmes in River Basins*, J. Bogen, D.E. Walling and T.J. Day (eds) 79–86. Oxfordshire: International Association of Hydrological Sciences Press.

Laronne J.B., I. Reid, Y. Yitshak and L.E. Frostick 1994. The non-layering of gravel streambeds under ephemeral flood regimes. *Journal of Hydrology*, **159**, 353–363.

Laursen, E.M. 1958. The total sediment load of streams. *Proceedings of the American Society of Civil Engineers, Journal Hydraulics Division*, **84**, 1530–1536.

Lee, K.T., Y-L., Liu and K-H. Cheng 2004. Experimental investigation of bedload transport processes under unsteady flow conditions. *Hydrological Processes*, **18**, 2439–2454.

Leeder, M.R., T.E. Gray and J. Alexander 2005. Sediment suspension dynamics and a new criterion for the maintenance of turbulent suspensions. *Sedimentology*, **52**, 683–691.

Lekach, J. and A.P. Schick 1982. Suspended sediment in desert floods in small catchments. *Israel Journal of Earth Science*, **31**, 144–156.

Lekach, J. and A.P. Schick 1983. Evidence for transport of bedload in waves: Analysis of fluvial sediment samples in a small upland desert stream channel. *Catena*, **10**, 267–280.

Leopold, L.B. 1992. The sediment size that determines channel morphology. In *Dynamics of Gravel-Bed Rivers*, P. Billi, R.D.

Hey, C.R. Thorne and P. Tacconi (eds) 297–311. Chichester: Wiley.

Leopold, L.B. and W.B. Bull 1979. Base level, aggradation and grade. *Proceedings American Philosophical Society*, **123**, 168–202.

Leopold, L.B. and T. Maddock 1953. The hydraulic geometry of stream channels and some physiographic implications. *United States Geological Survey Professional Paper*, **252**, 1–57.

Leopold, L.B. and J.P. Miller 1956. Ephemeral streams: Hydraulic factors and their relation to drainage net. *United States Geological Survey Professional Paper*, **282A**.

Leopold, L.B. and M.G. Wolman 1957. River channel patterns – braided, meandering and straight. *United States Geological Survey Professional Paper*, **282B**.

Leopold, L.B., W.W. Emmett and R.M. Myrick 1966. Channel and hillslope processes in a semi-arid area, New Mexico, *United States Geological Survey Professional Paper*, **352G**, 193–253.

Lisle, T.E. 1995. Particle size variations between bed load and bed material in natural gravel bed channels. *Water Resources Research*, **31**, 1107–1118.

Lisle, T.E. and S. Hilton 1992. The volume of fine sediment in pools: An index of sediment supply in gravel-bed streams. *Water Resources Bulletin*, **28**, 371–383.

Lvovich, M.I., G.Ya. Karasik, N.L. Bratseva, G.P. Medvedeva and A.V. Meleshko 1991. *Contemporary Intensity of the World Land Intracontinental Erosion*. USSR Academy of Sciences, Moscow.

Mackin, J.H. 1948. Concept of the graded river. *Bulletin Geological Society America*, **59**, 463–512.

Makaske, B. 2001. Anastomosing rivers: A review of their classification, origin and sedimentary products. *Earth Science Reviews*, **53**, 149–196.

Malmon, D.V., S.L. Reneau and T. Dunne 2004. Sediment sorting and transport by flash floods. *Journal of Geophysical Research*, **109**, F02005, doi:10.1029/2003JF000067.

Martin, C.W. and W.C. Johnson 1987. Historical channel narrowing and riparian vegetation expansion in the Medicien Lodge River Basin, Kansas, 1871–1983. *Annals of the American Society of Geographers*, **77**, 436–449.

McCarthy, T.C. and W.N. Ellery 1995. Sedimentation on the distal reaches of the Okavango Fan, Botswana, and its bearing on calcrete and silcrete (Ganister) formation. *Journal of Sedimentary Research*, **A65**, 77–90.

NcEwen, L.J. and J.A. Mathews 1998. Channel form, bed material and sediment sources of the Sprongdøla, southern Norway: Evidence for a distinct periglacio-fluvial system. *Geografiska Annaler*, **80A**, 16–36.

McLean, S.R. 1991. Depth-integrated suspended-load calculations. *Proceedings of the American Society of Civil Engineers, Journal Hydraulics Division*, **117**, 1440–1458.

McLean, S.R. 1992. On the calculation of suspended load for non-cohesive sediments. *Journal of Geophysical Research (Oceans)*, **97(C4)**, 5759–5770.

McLean, S.R., J.M. Nelson and R.L. Shreve 1996. Flow-sediment interactions in separating flows over bedforms. In *Coherent Flow Structures in Open Channels*, P.J. Ashworth, S.J. Bennett, J.L. Best and S.J. McLelland (eds) 203–226. Chichester: Wiley.

McQuivey, R.S. and E.V. Richardson 1969. Some turbulence measurement in open-channel flow. *Proceedings of the American Society of Civil Engineers, Journal Hydraulics Division*, **95**, 209–223.

Meade, R.H., T.R. Yuzyk and T.J. Day 1990. Movement and storage of sediment in rivers of the United States and Canada. In *The Geology of North America, Vol 0-1 Surface Water Hydrology*, M.G. Wolman and H.C. Riggs (eds) 255–280. Boulder, Colorado: The Geological Society of America.

Merritt, D.M. and E.E. Wohl 2003. Downstream hydraulic geometry and channel adjustment during a flood along an ephemeral, arid-region drainage. *Geomorphology*, **52**, 165–180.

Meybeck, M. 1976. Total dissolved transport by world's major rivers. *Hydrological Sciences Bulletin*, **21**, 265–284.

Miller, M.C., I.N. McCave and P.D. Komar 1977. Threshold of sediment motion under unidirectional currents. *Sedimentology*, **24**, 303–314.

Millar, R.G. 2005. Theoretical regime equations for mobile gravel-bed rivers with stable banks. *Geomorphology*, **64**, 207–220.

Montgomery, D.R. and J.M. Buffington 1997. Channel-reach morphology in mountain drainage basins. *Bulletin Geological Society of America*, **109**, 596–611.

Mulholland, P.J. 1997. Dissolved oxygen matter concentration and flux in streams. *Journal of the North American Benthological Association*, **16**, 131–141.

Murphy, J.B., L.J. Lane and M.H. Diskin 1972. Bed material characteristics and transmission losses in an ephemeral stream. In *Hydrology and Water Resources in Arizona and the Southwest*. Proceedings Arizona Section American Water Resources Association and the Hydrology Section, Arizona Academy of Science, May, Prescott, Arizona.

Murray, A.B. and C. Paola 1994. A cellular model of braided streams. *Nature*, **371**, 54–57.

Nanson, G.C. and D.A. Knighton 1996. Anabranching rivers: Their cause, character and classification. *Earth Surface Process and Landforms*, **21**, 217–239.

Nanson, G.C. and H.Q. Huang 1999. Anabranching rivers: Divided efficiency leading to fluvial diversity. In *Varieties of Fluvial Form*, A.J. Miller and A. Gupta (eds) 477–494. Chichester: Wiley.

Nanson, G.C., R.W. Young, D.M. Price and B.R. Rust 1988. Stratigraphy, sedimentology and late Quaternary chronology of the Channel Country of western Queensland. In *Fluvial Geomorphology of Australia*, R.F. Warner (ed) 151–175. Sydney: Academic Press.

Nanson, G.C., S. Tooth and A.D. Knighton 2002. A global perspective on dryland rivers: Perceptions, misconceptions and distinctions. In *Dryland Rivers: Hydrology and Geomorphology of Semi-Arid Channels*, L.J. Bull and M.J. Kirkby (eds) 17–54. Chichester: Wiley.

Nash, D.J. and S.J. McLaren 2003. Kalahari valley calcretes: Their nature, origins, and environmental significance. *Quaternary International*, **111**, 3–22.

Nezu, I., A. Kadota and H. Nakagawa 1997. Turbulent structure in unsteady depth-varying open-channel flows. *Journal Hydraulic Engineering*, **123**, 752–763.

Nordin C.F. and J.P. Beverage 1965. Sediment transport in the Rio Grande, New Mexico. *United States Geological Survey Professional Paper*, **462F**.

Oldmeadow, D.F. and M. Church 2006. A field experiment on streambed stabilization by gravel structures. *Geomorphology*, **78**, 335–350.

Ondieki, C.M. 1995. Field assessment of flood event suspended sediment transport from ephemeral streams in the tropical sub-arid catchments. *Environmental Monitoring and Assessment*, **35**, 43–55.

Osterkamp, W.R. 1978. Gradient, discharge, and particle-size relations of alluvial channels of Kansas, with observations on braiding. *American Journal of Science*, **278**, 1253–1268.

Osterkamp, W.R. and J.E. Costa 1987. Changes accompanying an extraordinary flood on a sandbed stream. In *Catastrophic Flooding*, L. Mayer and D. Nash, (eds) 201–224. Boston: Allen and Unwin.

Osterkamp, W.R and J.M. Friedman 2000. The disparity between extreme rainfall events and rare floods – with emphasis on the semi-arid American West. *Hydrological Processes*, **14**, 2819–2829.

Park, C.C. 1977. World-wide variations in hydraulic geometry exponents of stream channels: An analysis and some observations. *Journal of Hydrology*, **33**, 133–146.

Parker, G. 1976. On the cause and characteristic scales of meandering and braiding in rivers. *Journal of Fluid Mechanics*, **76**, 459–480.

Parker, G. 1990a. Surface-based bedload transport relation for gravel rivers. *Journal Hydraulic Research*, **28**, 417–436.

Parker, G. 1990b. *The ACRONYM Series of PASCAL Programs for Computing Bedload Transport in Gravel-bed Rivers*. External Memorandum M-200, St Anthony Falls Laboratory, University of Minnesota, Minneapolis, Minnesota, USA.

Parker, G. 1991a. Selective sorting and abrasion of river gravel. I: Theory. *Journal of Hydraulic Engineering*, **117**, 131–149.

Parker, G. 1991b. Selective sorting and abrasion of river gravel. II: Applications. *Journal of Hydraulic Engineering*, **117**, 150–171.

Parker, G. 2008. Transport of gravel and sediment mixtures. In *Sedimentation Engineering: Processes, Measurements, Modeling, and Practice*, M. Garcia, (ed.) 165–252. ASCE Manuals and Reports on Engineering Practice No. 110.

Parker, G. and C.M. Toro-Escobar 2002. Equal mobility of gravel in streams: The remains of the day. *Water Resources Research*, **38**, 1264, doi:10.1029/2001WR000669.

Parker, J.T.C. 1993. Channel Change on the Santa Cruz River, Pima County, Arizona, 1936–1986. *United States Geological Survey Open-File Report*, **93–41**.

Phillips, J.D. 1992. Nonlinear dynamical systems in geomorphology: Revolution or evolution? *Geomorphology*, **5**, 219–29.

Plate, E.J. 1994. The need to consider non-stationary sediment transport. *International Journal of Sedimentary Research*, **9**, 117–123.

Powell, D.M., I. Reid and J.B. Laronne, 1999. Hydraulic interpretation of cross-stream variations in bedload transport rates recorded in two straight alluvial channels. *Journal of Hydraulic Engineering*, **125**, 1243–1252.

Powell, D.M., I. Reid and J.B. Laronne 2001. Evolution of bedload grain size distribution with increasing flow strength and the effect of flow duration on the calibre of bedload sediment yield in ephemeral gravel-bed rivers. *Water Resources Research*, **37**, 1463–1474.

Powell, D.M., I. Reid and J.B. Laronne 2003. The dynamics of bedload sediment transport in Nahal Eshtemoa, an ephemeral gravel-bed river in the Northern Negev Desert, Israel. *Advances in Environmental Monitoring and Modelling*, **1**, 1–27.

Powell, D.M., I. Reid, J.B. Laronne and L.E. Frostick 1996. Bed load as a component of sediment yield from a semi-arid watershed of the northern Negev'. In *Erosion and Sediment Yield: Global and Regional Perspectives*, D.E. Walling and B.W. Webb (eds) 389–387 Oxfordshire: International Association of Hydrological Sciences Press.

Powell, D.M., R. Brazier, J.W. Wainwright, A.J. Parsons and J. Kaduk 2005. Streambed scour and fill in low-order dryland streams. *Water Resources Research*, **41**, W05019: doi:10.1029/2004WR003662.

Powell, D.M., R. Brazier, J.W. Wainwright, A.J. Parsons and M. Nichols 2006. Spatial patterns of stream-bed scour and fill in dryland sand-bed rivers. *Water Resources Research*, **42**, W08412, doi:10.1029/2005WR004516.

Powell, D.M., R. Brazier, J.W. Wainwright, A.J. Parsons and M. Nichols 2007. Sediment transport and storage in dryland headwater streams. *Geomorphology*, **88**, 152–166.

Randall, J. and J. Marinelli 1996. *Invasive Plants: Weeds of the Global Garden*. New York: Brooklyn Botanitacal Garden.

Reid I. and L.E. Frostick 1987. Flow dynamics and suspended sediment properties in arid zone flash floods. *Hydrological Processes*, **1**, 239–253.

Reid, I. and L.E. Frostick 1997. Channel forms, flows and sediments in deserts. In *Arid Zone Geomorphology: Process, Form and Change in Drylands*, D.S.G. Thomas (ed) 205–230. Chichester: Wiley.

Reid I and J.B. Laronne 1995. Bedload sediment transport in an ephemeral stream and a comparison with seasonal and perennial counterparts. *Water Resources Research*, **31**, 773–781.

Reid, I., L.E. Frostick and J.T. Layman. 1980. The continuous measurement of bedload discharge. *Journal of Hydraulic Research*, **18**, 243–249.

Reid, I., J.B. Laronne and D.M. Powell 1995. The Nahal Yatir bedload database: Sediment dynamics in a gravel-bed ephemeral stream. *Earth Surface Processes and Landforms*, **20**, 845–857.

Reid, I., J.B. Laronne and D.M. Powell 1998. Flash flood and bedload dynamics of desert gravel-bed streams. *Hydrological Processes*, **12**, 543–557.

Reid, I., D.M. Powell and J.B. Laronne 1996. Prediction of bedload transport by desert flash floods. *Journal of Hydraulic Engineering*, **122**, 170–173.

Renard, K.G. and E.M. Laursen 1975. Dynamic behaviour model of ephemeral stream. *Proceedings of the American Society of Civil Engineers, Journal Hydraulics Division*, **101**, 511–528.

Renard, K.G., L.J. Lane, J.R. Simanton, W.E. Emmerich, J.J. Stone, M.A. Weltz, D.C. Goodrich and D.S. Yakowitz 1993. Agricultural impacts in an arid environment: Walnut Gulch Studies. *Hydrological Science and Technology*, **9**, 145–190.

Renwick, W.H 1992. Equilibrium, disequilibrium and non equilibrium landforms in the landscape. *Geomorphology*, **5**, 265–76.

Rhoads, B.L., 1989. Longitudinal variations in the size and sorting of bed material along six arid-region mountain streams. In *Arid and Semi-Arid Environments – Geomorphological and Pedological Aspects*, A. Yair and S.M. Berkowicz (eds), *Catena Supplement*, **14**, 87–105.

Rhoads, B.L. 1990. Hydrological characteristics of a small desert mountain stream: Implications for short-term magnitude and frequency of bedload transport. *Journal of Arid Environments*, **18**, 151–163.

Rice, S.P. and M. Church 2001. Longitudinal profiles in simple alluvial systems. *Water Resources Research*, **37**, 417–426.

Rouse, H. 1937. Modern conceptions of the mechanics of turbulence. *Transactions of the American Society of Civil Engineers*, **102**, 436–505.

Samaga, B.R., K.G. Ranga Raju and R.J. Garde 1986. Suspended load transport of sediment mixtures. *Proceedings of the American Society of Civil Engineers, Journal Hydraulics Division*, **112**, 1019–1035.

Sayre, W.W. and D.W. Hubbell, 1965. Transport and dispersion of labeled bed material: North Loupe River, Nebraska. *United States Geological Survey Professional Paper*, 433C.

Schick, A.P. and J. Lekach, 1993. Storage versus input-output accuracy: An evaluation of two ten-year sediment budgets, Nahal Yael, Israel. *Physical Geography*, **14**, 225–236.

Schick, A.P., J. Lekach and M.A Hassan 1987. Bedload transport in desert floods – observations in the Negev. In *Sediment Transport in Gravel-Bed Rivers*, C.R Thorne, I.C. Bathurst, R.D. Hey (eds) 617–642. Chichester: Wiley.

Schumann, R.R. 1989. Morphology of Red Creek Wyoming, an arid region anastomosing channel system. *Earth Surface Processes and Landforms*, **14**, 277–288.

Schumm, S.A., 1960. The shape of alluvial channels in relation to sediment type. *United States Geological Survey Professional Paper*, **352B**.

Schumm, S.A. 1961. Effect of sediment characteristics on erosion and deposition in ephemeral stream channels. *United States Geological Survey Professional Paper*, **352C**, 31–68.

Schumm, S.A. 1963. Sinuosity of alluvial rivers on the Great Plains. *Bulletin Geological Society America*, **74**, 1089–1100.

Schumm, S.A. 1968. River adjustment to altered hydrologic regimen – Murrumbidgee River and palaeochannels, Australia. *United States Geological Survey Professional Paper*, **598**.

Schumm, S.A. and H.R. Khan 1972. Experimental study of channel patterns. *Bulletin Geological Society America*, **83**, 1755–1770.

Schumm, S.A. and R.W. Lichty 1963. Channel widening and floodplain construction along the Cimarron river in southwestern Kansas. *United States Geological Survey Professional Paper*, **352D**.

Sharon, D. 1972. The spottiness of rainfall in a desert area. *Journal of Hydrology*, **17**, 161–175.

Sharon, D. 1981. The distribution in space of local rainfall in the Namib Desert. *Journal of Climatology*, **1**, 69–75.

Sheldon, F. and M.C. Thoms 2006. In-channel complexity: The key to the dynamics of organic matter in large dryland rivers. *Geomorphology*, **77**, 270–285.

Shulits, S. 1941. Rational equation of riverbed profile. *Transactions American Geophysical Union*, **22**, 622–631.

Simon, A. and A.J. Collinson, 2002. Quantifying the mechanical and hydrological effects of riparian vegetation on streambank stability. *Earth Surface Processes and Landforms*, **27**, 527–546.

Simon, A., S.J. Bennett and V.S. Neary 2004. Riparian vegetation and fluvial geomorphology: Problems and opportunities. In *Riparian Vegetation and Fluvial Geomorphology: Hydraulic, Hydrologic, and Geotechnical Interactions*, S.J. Bennett, and A. Simon (eds) part 1. Water Science and Application Series Volume 8, Washington D.C.: American Geophysical Union.

Simon, A., A. Curini, S.E. Darby and E.J. Langendoen, 2000. Bank and near-bank processes in an incised channel. *Geomorphology*, **35**, 193–217.

Simons, D.B. and E.V. Richardson 1963. Forms of bed roughness in alluvial channels. *Transactions American Society Civil Engineering*, **128**, 284–302.

Sinha, S.K. and G. Parker 1996. Causes of concavity in longitudinal profiles of rivers. *Water Resources Research*, **32**, 1417–1428.

Smith, C.R. 1996. Coherent flow structures in smooth-wall turbulent boundary layers: Facts, mechanisms and speculation. In *Coherent Flow Structures in Open Channels*, P.J. Ashworth, S.J. Bennett, J.L. Best and S.J. McLelland (eds) 1–40. Chichester: Wiley.

Smith, J.D. and S.R. McLean 1977. Spatially averaged flow over a wavy surface. *Journal of Geophysical Research*, **82**, 1735–1746.

Smolders, A.J.P., K.A. Hudson-Edwards, G. Van der Velde and J.G.M. Roelofs 2004. Controls on water chemistry of the Pilcomayo river (Bolivia, South America). *Applied Geochemistry*, **19**, 1745–1758.

Song, T. and W.H. Graf 1996. Velocity and turbulence distribution in unsteady open-channel flows. *Journal of Hydraulic Engineering*, **122**, 141–154.

Stelczer K. 1981. *Bed Load Transport: Theory and Practice*. Water Resources Publications. Littleton: Colorado.

Stevens, M.A., D.B. Simons and E.V. Richardson 1975. Nonequilibrium channel form. *Proceedings of the American Society of Civil Engineers, Journal Hydraulics Division*, **101**, 557–566.

Stoddart, D.R. 1978. Geomorphology in China. *Progress in Physical Geography*, **2**, 187–236.

Sutherland, A.J. 1967. Proposed mechanism for sediment entrainment by turbulent flows. *Journal of Geophysical Research*, **72**, 191–198.

Sutherland, R.A. and R.B. Bryan 1990. Runoff and erosion from a small semi-arid catchment, Baringo District, Kenya. *Applied Geography*, **10**, 91–109.

Syrén, P. 1990. Estimated transport of the suspended load of the River Malewa, Kenya, between 1931–1959. *Geografiska Annaler*, **72A**, 285–299.

Tal, M. and C. Paola 2007. Dynamic single-thread channels maintained by the interaction of flow and vegetation. *Geology*, **35**, 347–350.

Tal, M., K. Gran, A.B. Murray, C. Paola and D.M. Hicks 2003. Riparian vegetation as a primary control on channel characteristics in multi-thread rivers. In *Riparian Vegetation and Fluvial Geomorphology: Hydraulic, Hydrologic, and Geotechnical Interactions*, S.J. Bennett, and A. Simon (eds) 45–58. Water Science and Application Series Volume 8, Washington D.C.: American Geophysical Union.

Thorn, C.E. and M.R. Welford 1994. The equilibrium concept in geomorphology, *Annals of the Association of American Geographers*, **84**, 666–696.

Thornes, J.B. 1980. Structural instability and ephemeral stream channel behaviour. *Zeitschrift für Geomorphologie Supplementband*, **36**, 233–244.

Thornes, J.B. 1990. Effects of vegetation on river bank erosion and stability. In *Vegetation and Erosion*, J.B. Thornes (ed) 125–143. Chichester: Wiley.

Toffaleti, F.B. 1968. A procedure for the computation of the total river sand discharge and detailed distribution, bed to surface. *Technical Report No. 5*, Committee on Channel Stabilisation, Corps of Engineers.

Tooth, S. 1999. Floodouts in Central Australia. In *Varieties of Fluvial Form*, A.J. Miller and A. Gupta (eds) 219–247. Wiley: Chichester.

Tooth, S. 2000a. Process, form and change in dryland rivers: A review of recent research. *Earth-Science Reviews*, **51**, 67–107.

Tooth, S. 2000b. Downstream changes in dryland river channels: The Northern Plains of arid central Australia. *Geomorphology*, **34**, 33–54.

Tooth, S. and G.C. Nanson 2000a. Equilibrium and nonequilibrium conditions in dryland rivers. *Physical Geography*, **21**, 183–211.

Tooth, S. and G.C. Nanson 2000b. The role of vegetation in the formation of anabranching channels in an ephemeral river, Northern plains, arid central Australia. *Hydrological Processes*, **14**, 3099–3117.

Tricart, J. and A. Cailleux 1972. *Introduction to Climatic Geomorphology*. (Translated by C.J.K. de Jonge), London, Longman.

USGS 1993. Policy and technical guidance for conversion of sediment concentration from parts per million (ppm) to mulligrams per liter (mg/L). *United Stages Geological Survey Office of Surface Water Technical Memorandum* No **93.21**.

van Rijn, L.C. 1984. Sediment transport, Part II: Suspended load transport. *Proceedings of the American Society of Civil Engineers, Journal Hydraulics Division*, **110**, 1613–1641.

VanLooy J.A. and C.W. Martin 2005. Channel and vegetation change on the Cimarron River, Southwestern Kansas, 1953–2001. *Annals of the Association of American Geographers*, **95**, 727–739.

Vetter, C.P. 1937. Why desilting works for the all-American canal. *Engineering News Record*, **118**, 321.

Vogel, J.C. 1989. Evidence of past climatic change in the Namib Desert. *Palaeogeography, Palaeoclimatology, Palaeoecology*, **70**, 355–366.

Walling, D.E. 1981. Yellow River which never runs clear. *Geographical Magazine*, **53**, 568–575.

Walling, D.E. and A.H.A. Kleo 1979. Sediment yields of rivers in areas of low precipitation: A global review. In *Hydrology of Areas of Low Precipitation*. Proceedings of the Canberra Symposium, International Association of Hydrological Sciences Publication, **128**, 479–493.

Walling, D.E. and B. Webb. 1983. The dissolved loads of rivers: A global overview. In B.W. Webb (ed) *Dissolved Loads of Rivers and Surface Water Quantity/Quality Relationships*. International Association of Hydrological Sciences Publication No. **141**, 3–20.

Wheeler, D.A. 1979. The overall shape of longitudinal profiles of streams. In *Geographical Approaches to Fluvial Processes*, A.F. Pitty (ed) 241–260. Norwich: Geobooks.

Wilcock, P.R. 1997. The components of fractional transport rate. *Water Resources Research*, **33**, 247–258.

Wilcock, P.R. and J.C. Crowe 2003. A surface-based transport model for sand and gravel. *Journal of Hydraulic Engineering*, **129**, 120–128.

Wilcock, P.R. and S.T. Kenworthy 2002. A two fraction model for the transport of sand/gravel mixtures. *Water Resources Research*, **38**, 1194–2003.

Wilcock, P.R. and B.W. McArdell 1993. Surface-based fractional transport rates: Mobilization thresholds and partial transport of a sand-gravel sediment. *Water Resources Research*, **29**, 1297–1312.

Wilcock, P.R. and B.W. McArdell 1997. Partial transport of a sand/gravel sediment. *Water Resources Research*, **33**, 235–245.

Wilcock, P.R., S.T. Kenworthy and J.C. Crowe 2001. Experimental study of the transport of mixed sand and gravel. *Water Resources Research*, **37**, 3349–3358.

Wilson C.A.M.E., T. Stoesser and P.D. Bates 2005. Open channel flow through vegetation. In *Computational Fluid Dynamics: Applications in Environmental Hydraulics*, P.D. Bates, S. Lane and R. Ferguson (eds) 395–428. Chichester: Wiley.

Wittenberg, L. 2002. Structural patterns in coarse gravel river beds: Typology, survey and assessment of the role of grain size and river regime. *Geografiska Annaler*, **84A**, 25–37.

Wittenberg, L. and M.D. Newson 2005. Particle clusters in gravel-bed rivers – an experimental morphological approach to bed material transport and stability concepts. *Earth Surface Processes and Landforms*, **30**, 1351–1368.

Wolman, M.G. and Gerson, R. 1978. Relative scales of time and effectiveness in watershed geomorphology, *Earth Surface Processes*, **3**, 189–208.

Wolman, M.G. and L.B. Leopold 1957. River flood plains: Some observations on their formation. *United States Geological Survey Professional Paper*, **282C**.

Wolman, M.G. and Miller, J.P. 1960. Magnitude and frequency of forces in geomorphic processes. *Journal of Geology*, **68**, 54–74.

Wright, S. and G. Parker 2004a. Density stratification effects in sand bed rivers. *Proceedings of the American Society of Civil Engineers, Journal Hydraulics Division*, **130**, 783–795.

Wright, S and G. Parker 2004b. Flow resistance and suspended load in sand-bed rivers: Simplified stratification model. *Proceedings of the American Society of Civil Engineers, Journal Hydraulics Division*, **130**, 796–805.

Xu, J. 1999. Grain-size characteristics of suspended sediment in the Yellow River, China. *Catena*, **38**, 243–263.

Yatsu, E. 1955. On the longitudinal profile of the graded river. *Transactions American Geophysical Union*, **36**, 655–663.

Part V
Piedmonts

Chapter 13

Pediments in Arid Environments

John C. Dohrenwend and Anthony J. Parsons

Introduction

Pediments, gently sloping erosional surfaces of low relief developed on bedrock, occur in a wide variety of lithologic, neotectonic, and climatic settings. Reported on six continents, their distribution spans the range of subpolar latitudes from the Arctic to the Antarctic and the range of climate from hyperarid to humid tropical (Whitaker 1979). On the Cape York Peninsula in tropical north-east Queensland, a tectonically quiescent region of Precambrian granitic and metamorphic rocks overlain by gently dipping Mesozoic and Cenozoic sediments, pediments are a dominant landscape component occurring as fringing piedmonts and strath valleys of the crystalline rocks and as broad erosional plains on the sedimentary rocks (Smart et al. 1980). In the Gran Sabana of south-east Venezuela, a broadly upwarped region of gently dipping early Proterozoic strata, pediments comprise the floors of broad subsequent valleys developed within less resistant parts of the section (Dohrenwend et al. 1995). In the late Tertiary and Quaternary landscape of the Basin and Range geomorphic province of the south-west United States, pediments (developed within a variety of neotectonic settings and on a broad range of igneous, metamorphic, and sedimentary lithologies) occupy piedmont slopes, mountain passes, and broad topographic domes (Hadley 1967, Oberlander 1972, Cooke and Mason 1973, Moss 1977, Dohrenwend 1987a). Clearly, pediments are azonal, worldwide phenomena that tend to form under condi-

tions of relative geomorphic stability where processes of erosion and deposition are locally balanced over the long term such that mass transport and general surface regrading dominate landscape evolution. However, pediments are most conspicuous in arid and semi-arid environments where vegetation densities are low and deep weathering is limited. Hence, they have been studied most intensively in these regions and are generally perceived as an arid land phenomenon.

Pediments have long been the subject of geomorphological scrutiny. Furthermore, 'the origin of these landforms has been controversial since Gilbert (1877) first recognized and described them' (Tator 1952, p. 294). Indeed, 'pediments have attracted more study and controversy, and have sparked the imagination of more geomorphologists, than most other landforms in deserts' (Cooke and Mason 1973, p. 188). In consequence, 'conflicting views concerning the origin of rock pediments in deserts ... probably comprise the largest corpus in the literature on arid landforms' (Oberlander 1997, p.136). This attention and controversy stem from a variety of factors which are inherent in the very nature of these landforms.

(a) Pediments are counter-intuitive landforms. To many, it is intuitive that uplands should be erosional features underlain by resistant bedrock, whereas the adjacent piedmont plains should be depositional features underlain by sedimentary deposits derived from the uplands. To others, it is intuitive that most landscapes should be adjusted such that variations in lithology/structure and form are highly correlated. In reality, most pediments are, at least in part, surfaces of erosion and transportation where the underlying lithology/

J. C. Dohrenwend (✉)
Southwest Satellite Imaging, PO Box 1467, Moab, UT 84532
USA
e-mail: dohrenwend@scinternet.net

A.J. Parsons, A.D. Abrahams (eds.), *Geomorphology of Desert Environments*, 2nd ed.,
DOI 10.1007/978-1-4020-5719-9_13, © Springer Science+Business Media B.V. 2009

structure may be the same as or very similar to that of the adjacent uplands.

(b) As defined in geomorphology, a pediment is a nearly planar surface of mass transport and/or laterally uniform erosion which functions as a zone of transition between a degrading upland and a stable base level or slowly aggrading lowland. As such, pediments comprise a fairly general class of landforms that occur within many different climatic regimes and geomorphic settings. They are the product of a variety of processes, and the relative significance of these processes varies from one climatic-geomorphic setting to another.

(c) Within any particular climatic-geomorphic setting, the processes which act to form and maintain the pediment operate both non-linearly and discontinuously. It follows, therefore, that both the distribution of form and the operation of process will probably be chaotic (in the true mathematical sense) within the zone of transition represented by the pediment. Such systems cannot be clearly and comprehensively described by linear models of cause and effect.

(d) As transitional landforms, pediments are partly active, partly inactive, partly dissected, and partly buried. Their boundaries are irregular, gradational, and poorly defined. In most cases, they are at least partly obscured by discontinuous veneers of alluvial deposits and/or deeply weathered bedrock; indeed, many actively forming pediments undoubtedly lie unrecognized beneath continuous mantles of such materials. These characteristics commonly frustrate attempts of measurement, description, and analysis.

Unfortunately, the net result of this long history of study is not altogether clear or cogent, and has not produced a clear understanding of the processes responsible for pediment development. In recent years, perhaps in response to the apparent intractability of the development of a clear understanding of pediment-forming processes, attention given to this landform has waned.

Definitions

The confusion and controversy which pervade the study of pediments encompass even the basic definition of the term. Originally used as an architectural term, pediment refers to the vertically oriented triangular termination (or gable) of a gently pitched ridge roof (Dinsmoor 1975, p. 394, Janson 1969, p. 90). When borrowed from architecture and used in the field of geomorphology, however, the term quickly acquired an entirely different meaning (i.e. a gently sloping erosional surface of low relief). Morphologically, the gently sloping erosion surface of geomorphology corresponds much more closely to a gently pitched roof than to the vertical termination of that roof.

Of more fundamental significance, however, is the fact that pediments are morphologically complex landforms that form within a wide variety of geomorphic environments. It is hardly surprising, therefore, that the variability in pediment definition is nearly as great as the variability in pediments themselves (Tator 1953, Cooke and Mason 1973, p. 188, Whitaker 1979, Oberlander 1989).

Only the most general definitions can accommodate the broad range of form and wide variety of geomorphic environments associated with pediments. The definition proposed by Whitaker (1979, p. 432) is a case in point. 'A pediment is a terrestrial erosional footslope surface inclined at a low angle and lacking significant relief. ... It usually meets the hillslope at an angular nickline, and may be covered by transported material'. Unfortunately, such definitions are so broad that they limit the utility of the term as a basis for meaningful geomorphic inquiry and comparison. Qualifying modifiers such as those suggested by Tator (1953, p. 53) partly overcome this problem. 'The term pediment should be retained for the broad (but individually distinct) degradational surface produced by subaerial processes (including running water) in dry regions. Qualifications as to physiographic location may be expressed by the use of mountain, piedmont, or flat-land ... Additional terminology should include the words suballuvial, for alluviated erosion levels, and subaerial, for non-alluviated levels'. However, the problem of describing pediments in a consistent and scientifically productive manner remains.

Other definitions increase precision but at the expense of general applicability. Consider for example, two definitions that apply specifically to piedmont pediments in the south-western United States:

The term pediment may be restricted to that portion of the surface of degradation at the foot of a receding slope, which (1) is underlain by rocks of the upland and which is either bare or mantled by a layer of alluvium not exceeding in thickness the depth of stream scour during flood, (2) is essentially a surface of transportation, experiencing neither marked vertical downcutting nor excessive deposition, and (3) displays a longitudinal profile normally concave, but which may be convex at its head in later stages of development (Howard 1942, p. 8).

    ... pediments are composed of surfaces eroded across bedrock or alluvium, are usually discordant to structure, have longitudinal profiles usually either concave upward or rectilinear, slope at less than 11°, and are thinly and discontinuously veneered with rock debris. The upper limits of pediments are usually mountain/piedmont junctions, although pediments may meet along watersheds; pediments are generally masked downslope by alluvium, and their lower boundaries are the lines at which the alluvial covers become continuous (Cooke 1970, p. 28).

These definitions are useful general descriptions of piedmonts; however, they are not entirely satisfactory when applied to pediment domes and terrace pediments in the same region. Moreover, although generally similar, they are not in complete agreement. Cooke placed the downslope boundary of the pediment where alluvial cover becomes continuous, whereas Howard defined it as the point where the thickness of cover equals the depth of effective stream scour. Tator (1953) concurred that the thickness of alluvial cover commonly approximates the depth of scour. However, Bull (1977) suggested that the downslope limit be defined where the thickness of alluvial cover exceeds one percent of the piedmont length. Such wide definitional disparities obstruct comparative analysis and synthesis of the pediment literature.

For the purposes of the present discussion, a pediment is descriptively defined as a gently sloping erosional surface developed on bedrock or older unconsolidated deposits. This erosion surface may be subaerially exposed or covered by a discontinuous to continuous veneer of alluvial deposits. Its downslope limit may be defined (following the suggestion of Bull, 1977) as that point where the deposit thickness exceeds a small fraction of the pediment length (e.g. 0.5–1.0%) or some arbitrarily defined maximum thickness (e.g. 25–30 m), whichever is less. The bedrock may include essentially any lithologic type with any structural attitude. An erosional surface developed on piedmont or basin fill deposits is no different genetically from an erosional surface developed on plutonic or metamorphic rocks. Indeed, at least some pediment surfaces extend without interruption across high-angle contacts between crystalline bedrock and

partly indurated alluvium (Dohrenwend et al. 1986). (Of course, where the underlying sediments are neither indurated nor deformed, discrimination between pediment and alluviated piedmont may be difficult.)

Classification

That pediments are highly variable landforms is further illustrated by the variety of schemes that have been proposed to classify them. Cooke and Mason (1973) offered a classification based on general geomorphic environment: (a) an apron pediment is located between watershed and base level, usually between an upland and a depositional plain; (b) a pediment dome occurs on upland slopes (and crests) that are not surmounted by a mountain mass; and (c) a terrace pediment is developed adjacent to a relatively stable base level such as a through-flowing stream. Twidale (1983) and Bourne and Twidale (1998) presented a morphogenic classification based on relations between surface material and underlying lithology which are used to infer pediment forming processes: (a) a mantled pediment occurs where crystalline bedrock is veneered by a residual weathering mantle and which is inferred to have been formed by subsurface weathering of the crystalline bedrock and wash removal of the resulting debris; (b) a rock pediment forms where crystalline bedrock is exposed at the surface and which is inferred to respresent the exposed weathering front of a formerly mantled pediment; and (c) a covered pediment is characterized by a veneer of coarse debris covering an erosional surface that cuts discordantly across sedimentary strata. Oberlander (1989) advocated a two-part classification based on the relations between the pediment, its adjacent upland, and the underlying lithology and structure: (a) a glacis pediment (*glacis d'erosion en roches tendres*; Dresch 1957, Tricart 1972) is an erosional surface that bevels less resistant materials but does not extend into adjacent uplands underlain by materials of greater resistance, whereas (b) a rock pediment occurs where there is no change in lithology between the erosional surface and the adjacent upland. Applegarth (2004) distinguished between bedrock pediments and alluvial slopes which exhibit a bedrock platform but where this platform is buried 'under alluvial debris at a depth measured in meters' (p. 225). Applegarth's distinction draws into focus the fact that landforms exist on a continuum and that strict definitions are impossible. Pediments form part of the low-angle piedmont, often surrounding mountains. The supply of sediment to the piedmont and that rate at which it is moved across it will determine the depth of alluvial cover and the degree to which the gradient of the bedrock surface accords with the topographic gradient. At one end of the continuum lies a bare bedrock surface; at the other lies a thick accumulation of sediment in which the gradient of the underlying bedrock surface may or may not accord with the topographic gradient. Somewhere along this continuum lies the distinction geomorphologists make between pediments and alluvial fans. The various classifications provide general conceptual frameworks for the study and analysis of pediments; however, they tend to be somewhat parochial in perspective and are not entirely consistent or compatible with one another.

The Pediment Association

Geomorphically, the pediment is only part of an open erosional-depositional system termed the pediment association by Cooke (1970). This system includes the pediment, the upland area tributary to it, and the alluvial plain to which it is tributary (Fig. 13.1). Johnson (1932b, p. 399) conceptualized the system as 'three concentric zones in each of which the dominant action of streams differs from that in the other two: (1) an inner zone, the zone of degradation, corresponding closely to the mountainous highland, in which vertical down-cutting of the streams reaches its maximum importance; (2) An intermediate zone, the zone of lateral

corrasion, surrounding the mountain base, in which the lateral cutting by streams attains its maximum relative importance. This is the zone of pediment formation. (3) An outer zone, the zone of aggradation, where upbuilding by deposition of alluvium has its maximum relative importance'.

Morphologically, the pediment is the most stable component of this open system. As surfaces of fluvial transport (in arid and semi-arid environments, at least) which operate over long periods of time at or very close to the threshold for critical power in streams, pediments preserve little evidence of their own evolutionary history. Consequently, relatively little can be learned regarding pediment formation from the pediment surface itself. With the exception of those few places where relict pediment surfaces are preserved beneath lava flows, or other similar caprock materials that can be precisely dated (Oberlander 1972, Dohrenwend et al. 1985, 1986), geologic evidence of pediment formation is best preserved within other components of the pediment association: the geomorphic record preserved in relic upland landforms, the stratigraphic record contained within deposits of the alluvial plain, and the ongoing process/landform transition of the piedmont junction. Thus, analysis of pediment formation is best approached, in most cases, through study of the entire pediment association.

Pediment Morphology

Certainly one of the most remarkable physical attributes of any pediment is the generally planar and

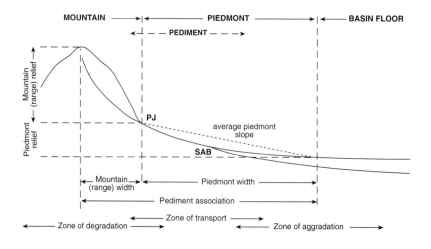

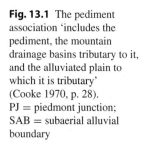

Fig. 13.1 The pediment association 'includes the pediment, the mountain drainage basins tributary to it, and the alluviated plain to which it is tributary' (Cooke 1970, p. 28). PJ = piedmont junction; SAB = subaerial alluvial boundary

Fig. 13.2 The broad, gently sloping, largely undissected surface of Cima Dome. One of the largest pediments in the eastern Mojave Desert, Cima Dome is approximately 16 km across and 600 m high. Aerial view is south-east

featureless character of most (or at least part of) its surface (Figs. 13.2 and 13.3). Consider, for example, this description of the pediments fringing the Sacaton Mountains of southern Arizona: 'The surface which encircles the mountains is remarkably smooth, being scarred only by faint channels rarely more than a foot deep. Near the mountains it has a slope of 200–250 feet per mile but at a distance of a few miles it is so flat that films of water cover broad areas after heavy rains' (Howard 1942, p. 16). Despite this general simplicity, detailed examination of pediment morphology commonly reveals substantial complexity; '... a pediment which is a clean, smooth bedrock surface is rare indeed' (Cooke and Mason 1973, p. 196). In many instances, the proximal pediment surface is characterized by an irregular mosaic of exposed bedrock and veneers of thin alluvial cover and (or) residual regolith. On active pediments, exposed bedrock areas may be shallowly dissected and gently undulating with several metres of local relief, whereas mantled areas are typically only slightly dissected to undissected with generally less than 1–2 m of local relief. Either type of surface may be interrupted by isolated knobs, hills, and ridges of bedrock that stand above the general level of the surrounding erosional plain.

Significant aspects of pediment morphology include: (a) the character and form of pediment boundaries including the upslope limit (piedmont junction) and downslope limit (upslope limit of alluvial cover);

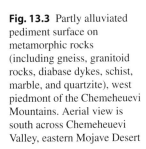

Fig. 13.3 Partly alluviated pediment surface on metamorphic rocks (including gneiss, granitoid rocks, diabase dykes, schist, marble, and quartzite), west piedmont of the Chemeheuevi Mountains. Aerial view is south across Chemeheuevi Valley, eastern Mojave Desert

(b) characteristic surface profiles, both longitudinal and transverse; (c) patterns of surface drainage and drainage dissection; (d) the form and distribution of major surface irregularities including inselbergs and tors; and (e) the character and distribution of the associated regolith. These morphologic characteristics are determined, at least in part, by a variety of influences including lithology and structure, relative size and shape of the piedmont and its associated upland, tectonic history, and climate. Consequently, they may vary widely even within the limits of an individual pediment.

Form and Character of the Mountain Front

The boundary between an upland area, mountain or mountain range, and its associated piedmont is commonly termed the mountain front. However, 'mountain front is an unfortunate term inasmuch as it suggests an almost linear and continuously outfacing boundary between the mountain mass and the pediment, whereas virtually all mountain fronts are indented to some degree by embayments' (Parsons and Abrahams 1984, p. 256). These embayments are, for the most part, widened valley floors that tend to form along the mountain front whenever the rate of erosional retreat of the valley sides exceeds the rate of incision of the valley floor. Such conditions are common where the mountain mass is no longer undergoing rapid uplift and local base level is either stable or rising. Consequently, the sinuosity (degree of embayment) of a tectonically inactive mountain front tends to increase with time (Bull and McFadden 1977).

The embayed nature of inactive mountain fronts has been recognized by many workers, and its significance relative to the formation and expansion of pediments is well established (e.g. Bryan 1922, Johnson 1932a, Sharp 1940, Howard 1942, Parsons and Abrahams 1984). Indeed, Lawson (1915, p. 42) was among the first to describe the general progression of embayment development. 'The contour of the subaerial [mountain] front for the greater part of the time of its recession is not in reality a straight, or even a gently sinuous line, but is actually indentate. At the indentations are cañons or gullies and from these emerge the greater part of the detritus which forms the embankment [alluvial plain] and which is distributed radially from an apex in, or at the mouth of, every cañon or gully. ... The identations may be slight in the early stages of front recession, is most pronounced in the middle stages and becomes less intricate in the later stages.' Parsons and Abrahams (1984, p. 258) further define the process as follows, 'For a particular mountain mass during a particular period of time, the relative effectiveness of divide removal [between embayments] and mountain front retreat in extending a piedmont will depend on the relative rates of migration of piedmont junctions in embayments and along mountain fronts and the relative lengths of embayments and mountain fronts'.

Form and Character of the Piedmont Junction

The junction of the mountain slope and piedmont is commonly termed the piedmont junction. In many areas, this boundary is marked by an abrupt and pronounced break in slope. It may be particularly abrupt and well-defined where (a) an active fault bounds the mountain front, (b) marginal streams flow along the base of the upland slope, (c) slopes are capped with resistant caprock, (d) slopes are coincident with bedrock structure (dykes, joints, fault line, steep-dipping to vertical beds, etc.), or (e) a pronounced contrast exists between debris sizes on the hillslope and on the piedmont surface. In other areas, particularly along the dip-slope flanks of tilted range blocks, piedmont surfaces may merge smoothly, almost imperceptibly, with the range slope and, in the case of gentle asymmetric tilting, may even extend up to and be truncated by the range crest. In yet other areas, particularly where a proximal pediment is pervasively dissected, the piedmont junction may be represented by a diffuse transition zone as much as several hundred metres wide and highly irregular both in plan and in profile. In such locations, the junction may be so irregular that it cannot be mapped as a discrete boundary. Indeed, even in those areas where the junction appears to be relatively well-defined when viewed from a distance, closer examination commonly reveals considerable complexity in both plan and profile.

The local character of the piedmont junction is determined in part by the nature of the transition from relatively diffuse erosional processes on the hillslope to somewhat more concentrated fluvial processes on the piedmont surface. As Bryan (1922, pp. 54–5) observed '... the angle of slope of the mountain is controlled by the resistance of the rock to the dislodgement of joint blocks and the rate at which these blocks disintegrate. ... Fine rock debris is moved down the mountain slope by rainwash and carried away from the foot of the slope by rivulets and streams that form through the concentration of the rainwash. ... The angle of slope of the pediment, however, is due to corrasion by the streams, and this corrasion is controlled by the ability of the water to transport debris. ... The sharpness of the angle between the mountain slope and the pediment is one of the most remarkable results of the division of labor between rainwash on the mountain slopes and streams on the pediment'. Where this transition is profound, the piedmont junction is abrupt; where the transition is more gradual, the junction is typically less well defined. In those areas where the junction separates hillslope from piedmont interfluve the abruptness of the transition appears to be related to hillslope lithology and structure. 'In quartz monzonite areas there is often a marked contrast between debris sizes on mountain fronts and on pediments, which may account for the distinct break of slope between the two landforms in these areas; on other rock types, where the particle-size contrast is less marked, the change in slope is less abrupt' (Cooke and Mason 1973, p. 195).

In the extreme case where no transition in process occurs, there is no piedmont junction. The "sharp break in slope" between the pediment surface and the mountain front ... exists only in interfluvial areas. ... The course of any master stream channel from a given drainage basin in the mountains onto the pediment surface and thence to the basin floor below has no sharp break in slope. In the absence of constraints, such as recent structural disturbances, any stream channel will exhibit a relatively smooth, concave upward, longitudinal profile that accords with the local hydraulic geometry. ... The interfluvial areas, however, generally do exhibit a marked change in slope. ... The reason for the existence of such a zone is precisely that it is an interfluvial area; the dominant process that operates on the mountain front is not fluvial' (Lustig 1969, p. D67).

The Alluvial Boundary and the Suballuvial Surface

The transition between exposed bedrock and alluvial cover on pediments is not well documented. The very nature of this boundary impedes detailed mapping, measurement or quantitative description. The few general descriptions available show that the subaerial alluvial boundary is gradational, discontinuous, and highly irregular (e.g. Sharp 1957, Tuan 1959, Dohrenwend et al. 1986). Several factors contribute to this complexity. (a) The suballuvial bedrock surface (termed the suballuvial bench by Lawson, 1915) subparallels the surface of the alluvial mantle; its longitudinal profile is nearly rectilinear and its average slope is only very slightly steeper than the average slope of the subaerial surface (Cooke and Mason 1973). (b) On many pediments in the Mojave and Sonoran deserts, alluvial thicknesses of a few metres or less are typical for distances of at least 1 km downslope from exposed bedrock surfaces (Howard 1942, Tuan 1959, Cooke and Mason 1973, Dohrenwend unpublished data). (c) The subaerial alluvial boundary appears to be highly transient, migrating up and downslope across the piedmont in response to long-term changes in stream power and sediment load. Upslope migration is documented by buried palaeochannels of similar dimensions to active surface channels (Cooke and Mason 1973, Dohrenwend unpublished data), whereas downslope migration and proximal exhumation are suggested by the pervasive dissection of exposed bedrock surfaces (Cooke and Mason 1973).

Longitudinal Slope

The longitudinal slope of pediments typically ranges between 0.5° and 11° (Tator 1952, Cooke and Mason 1973, p. 193, Thomas 1974, p. 217). Slopes in excess of 6° are uncommon except in proximal areas. In the south-west United States, average longitudinal slopes typically range between 2° and 4° and rarely exceed 6° (Table 13.1).

There is general agreement that the characteristic form of the longitudinal profile of most pediments is concave-upward (e.g. Bryan 1922, Johnson 1932b,

Table 13.1 Representative pediment slope data (degrees), south-western United States

Area	Range of longitudinal slopes	Mean longitudinal slope	Number of pediments	Reference
Western Mojave Desert (granitic)	0.5–5.37	2.5	37	Cooke (1970)
Western Mojave Desert (non-granitic)	0.75–5.15	2.8	16	Cooke (1970)
Eastern Mojave Desert (granitic)	0.8–5.3	2.0	5	Dohrenwend (unpublished data)
Eastern Mojave Desert (granitic domes)	1.0–6.0	3.4	9	Dohrenwend (unpublished data)
Eastern Mojave Desert (non-granitic)	0.7–4.4	1.9	5	Dohrenwend (unpublished data)
Mojave Desert (granitic)	1.6–8.6*	3.4	17	Mammerickx (1964)
Sonoran Desert (granitic)	1.95–5.0*	3.5	4	Mammerickx (1964)
Mojave and Sonoran deserts (non-granitic)	1.25–4.4*	2.8	5	Mammerickx (1964)
South-eastern Arizona	0.5–2.2	–	–	Bryan (1922)
	0.5–3.3	–	–	Gilluly (1937)
Ruby–East Humboldt Mountains, Nevada (west flank)	0.8–3.8	–	–	Sharp (1940)
Ruby–East Humboldt Mountains, Nevada (east flank)	4.3–5.4	–	–	Sharp (1940)

* Range of mean values.

Lustig 1969, p. D67, Thomas 1974, p. 217). Although Davis (1933), Gilluly (1937) and Howard (1942) described convex-upward profiles for the crestal areas of pediment passes and pediment domes; rectilinear to concave longitudinal profiles typify most slopes on these landforms except in the immediate vicinity of the crest (Sharp 1957, Dohrenwend unpublished data; Fig. 13.2). The pediments of arid and semi-arid regions are generally considered to be graded surfaces of fluvial transport where the surface slope is adjusted to the discharge and load of the upland/piedmont drainage, and the characteristic concave longitudinal profile is the result of this adjustment to fluvial processes. Systematic relations between pediment slope and stream size clearly demonstrate this adjustment. In proximal areas, for example, pediment slope is steeper immediately downslope from intercanyon areas than downstream from valley mouths (Bryan 1922). Also, as Gilluly (1937, p. 332) observed in the Little Ajo Mountains of south-west Arizona, 'the pediment is, for the most part, concave upward in profile, generally more steeply along the smaller stream courses [3.25°], less steeply along the larger streams [0.75°]'. Local variations in pediment slope that are closely related to lithologic variations also testify to this adjustment. 'The longitudinal profile (concave) should not be assumed to be a smooth curve in all cases. It is rather, similar to the average stream profile, segmented, the smoother

portions being developed across areas of homogeneous rock. Across heterogeneous rocks the profile is segmental, steeper on the more resistant and gentler on the less resistant types' (Tator 1952, p. 302).

The influences of lithology and climate on pediment slope are not well defined; however, they appear to be relatively insignificant. On the basis of a morphometric analysis, Mammerickx (1964) concluded that bedrock is not a significant factor in determining average pediment slopes in the Sonoran and Mojave Deserts. She supported this conclusion by documenting the apparently random asymmetry of a number of pediment domes in the region. 'On any dome it is difficult to imagine that the amount of rain or debris supplied to the different slopes is different, yet they are asymmetrical. The asymmetry further is not systematic in any particular direction' (Mammerickx 1964, p. 423). Comparison of slope means and ranges for granitic versus non-granitic pediments in the south-west United States (Table 13.1) also suggests the validity of this conclusion. The influence of climate on pediment slope also appears to be relatively insignificant. The typical range of longitudinal slopes appears to be essentially the same in wet–dry tropical environments as it is in arid and semi-arid environments (Thomas 1974, p. 223–4).

The influence of tectonism on pediment slope is similarly ill-defined but may be considerably more significant. Cooke (1970) documented a significant positive correlation between the relief:length ratio of the pediment association and pediment slope for 53 pediment associations in the western Mojave Desert. Because the relief:length ratio of the pediment association is very likely at least partly determined within this region by the local history and character of tectonic movement, the possibility of an indirect influence of tectonism on the slope of these pediments is indicated by this correlation. This tentative conclusion is supported by a comparative analysis of the general morphometric characteristics of several diverse neotectonic domains within the south-west Basin and Range province (Dohrenwend 1987b). Average values of piedmont slope for each of these domains show a strong positive correlation with average values of range relief, which in turn show a strong positive correlation with a morphometric index of relative vertical tectonic activity (Table 13.2). This suggests that average piedmont slope is largely determined by the general morphometric character of each domain (e.g. initial range relief, range width and spacing, etc.) which in turn is largely determined by the style, character, and timing of local tectonism.

Transverse Surface Profile

Although the longitudinal pediment profile is, for the most part, slightly concave-upward, the form of the pediment surface transverse to slope may assume any one of several general forms; convex-upward or fan-shaped (Johnson 1932b, Rich 1935, Howard 1942), concave-upward or valley-shaped (Bryan 1936, Gilluly 1937, Howard 1942), or essentially rectilinear (Tator 1952, Dohrenwend unpublished data). These general transverse forms are usually better developed in proximal areas of the pediment (Tator 1952). However, undulating surfaces displaying irregular combinations of convex, concave, and rectilinear slope elements superimposed on one or more of these general forms are also characteristic where proximal areas are pervasively dissected. Undissected distal areas approach a gently undulating to nearly level transverse form (Tator 1952).

Arguments based on the presence or absence of a particular transverse surface form have been used to advocate various conceptual models of pediment development. For example, Johnson (1932b) consid-

ered rock fans (a proximal pediment with a general fan shape in plan and a convex-up transverse profile) to be compelling evidence of lateral planation by streams; whereas, Rich (1935) argued that such forms may also be produced by repeated stream capture induced by unequal incision of drainage with upland sources as compared to drainage without upland sources. Howard (1942) suggested that the transverse profiles of pediment embayments may be either convex, concave, or a combination of these forms depending on the relative activity of trunk and tributary streams within the embayment. As these and other models suggest, the tendency for water flow to channelize and then to concentrate via the intersection of channels acts to produce a variety and complexity of transverse surface forms.

Patterns of Drainage and Drainage Dissection

According to Cooke and Mason (1973, p. 197), the general pattern of drainage on pediments is essentially identical to the characteristic drainage patterns of alluvial fans and bajadas. Three interrelated types of drainage commonly develop: (a) distributary networks which radiate from proximal areas to gradually dissipate in medial and/or distal parts of the piedmont; (b) frequently changing, complexly anastomosing networks of shallow washes and rills in medial piedmont areas (Fig. 13.4); and (c) in areas of falling base level, integrated subparallel networks which are generally most deeply incised in distal areas of the alluviated piedmont plain. The transitions between these various drainage types are typically gradational and diffuse.

Detailed examination of pediment drainage, however, reveals a somewhat more complex situation than that portrayed by this generalized model (Fig. 13.5). Although many large pediments are generally smooth and regular with less than a few metres of local relief, a more complex morphology occurs where shallow drainageways locally incise the pediment surface into irregular patchworks of dissected and undissected topography. The undissected areas are mostly flat with shallow, ill-defined, discontinuous to anastomosing drainageways and low indistinct interfluves, whereas dissected areas are typically scored by well-defined

Table 13.2 Morphometric summary of the west-central Basin and Range province average range and piedmont dimensions

Tectonic domain	Average range relief, Rr* (km)	Average piedmont relief, Pr (km)	Average range width, Rw (km)	Average piedmont width, Pw (km)	Average slope of piedmont association (Rr + Pr)/(Rw + Pw)	Averagae piedmont slope[†] (Pr/Pw)
Central Great Basin	0.73	0.21	6.20	5.90	0.077	0.034
South-east Great Basin	0.63	0.24	3.15	6.00	0.095	0.040
South-west Great Basin	1.38	0.29	6.90	4.90	0.142	0.059
Northern Mojave Desert	0.49	0.30	2.85	8.20	0.072	0.037
Walker Lane Belt						
North-west Goldfield Block	0.64	0.27	3.30	6.10	0.096	0.044
North-east Goldfield Block	0.49	0.24	3.25	7.70	0.066	0.031
Spring Mountains Block	1.19	0.51	9.25	11.9	0.080	0.043

* Average range relief $Rr = 0.15Ti + 0.16$, where the index of relative vertical tectonic activity $Ti = (Rr/Pr) + (Rw/Pw)$; $r = 0.767$, $p = 0.044$.
† Average piedmont slope $(Pr/Pw) = 0.022Rr + 0.024$; $r = 0.829$, $p = 0.021$.

Fig. 13.4 The west flank of the Granite Springs pediment dome (eastern Mojave Desert) has been dissected into a washboard of low rounded interfluves by a fine-textured network of shallow, subparallel to complexly interconnecting washes. Aerial view is north

subparallel drainageways that form shallow regularly spaced valleys separated by rounded interfluves (Dohrenwend et al. 1986, p. 1051). 'If the dissection is not deep (less than 3 feet), the ground surface of the pediment is undulatory and the stream pattern is braided. If the pediment is deeply dissected (more than 3 feet) the stream pattern is essentially parallel, weakly integrated, and not braided' (Mammerickx 1964, p. 423). In the specific case of Cima Dome in the eastern Mojave Desert, 'dissection has been greatest on the east and southeast slopes. ... Other parts of the dome, particularly the upper slopes and lower south flank, have washes between residual knobs and ridges. These are discontinuous and seemingly due to channelization of runoff between the residuals. ...

Some of the lowermost smooth alluvial slopes have small channels as much as 1 foot deep, 2–3 feet wide, and 100–200 feet long that start abruptly and end in a lobate tongue of loose grus ...' (Sharp 1957, p. 277).

One of the more noteworthy features of pediment drainage, at least within the desert regions of western North America, is the characteristic tendency for dissection of proximal areas (Fig. 13.6). On some pediments, proximal dissection is largely restricted to small isolated areas along the mountain front; on others, particularly those surrounding small residual mountain masses, it forms broad continuous zones of intricately scored topography that extend as much as 2 or 3 km downslope from the mountain front. This dissection likely results from several diverse influences

Fig. 13.5 'Chaotic' patchwork of dissected and undissected bedrock surfaces on Cretaceous monzogranite, west flank of the Granite Springs pediment dome, eastern Mojave Desert. Pliocene lava flows of the Cima volcanic field cap the deeply embayed erosional escarpment in the background

Fig. 13.6 Localized
dissection of a proximal
pediment surface cut on
Mesozoic monzogranite.
'Sediment-starvation' has
lowered the critical power
threshold of this
piedmont-sourced drainage
system causing intense
dissection (of as much as
20 m) of the pediment
surface. Adjacent undissected
areas are traversed by
sediment-satiated streams fed
by large drainage basins in
the adjacent uplands. Aerial
view is south-east across the
south piedmont of the Granite
Mountains, eastern Mojave
Desert

including short-term climate change (Bryan 1922) and the tendency for differential incision of drainage with upland sources as compared with drainage where sources are limited to the piedmont (Rich 1935, Denny 1967). However, regrading of proximal piedmont areas in response to evolutionary reduction of the associated mountain mass may be the dominant influence, particularly where a broad continuous pediment surrounds a small upland mass (Johnson 1932a,b, Howard 1942, Cooke and Mason 1973). 'As the mountains dwindle in size, precipitation and the volume of sediment available for streams decrease. The decrease in load more than compensates for the loss of water, so that the dwindling waters are gradually rejuvenated. They thus lower their gradients, but the lowering is greatest in their headwater regions because farther down the slope the load is increased by pediment debris' (Howard 1942, p. 135). The occurrence of proximal dissection on many pediments indicates that pediment regrading does not necessarily proceed via uniform downwearing of the proximal surface but rather may proceed via shallow incision by a diffuse network of small stream channels.

Inselbergs

Inselbergs are isolated hills that stand above a surrounding erosional plain (Fig. 13.7). They vary widely

Fig. 13.7 Numerous widely
scattered tors and small
inselbergs (developed on
Mesozoic monzogranite)
interrupt the otherwise gently
sloping pediment plain which
forms the east piedmont of
the Granite Mountains in the
eastern Mojave Desert

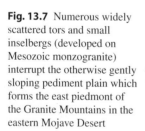

in abundance and size from one pediment to other. Inselbergs as tall as 180 m have been described in the Mojave and Sonoran Deserts (e.g. Sharp 1957, Moss 1977); however, heights of 10 to 50 m appear to be much more common in this region (e.g. Tuan 1959). Kesel (1977) reported that inselbergs of the Sonoran Desert have convex-concave slopes where the basal concavity comprises from 50 to 70% and the crestal convexity from 15 to 50% of the profile. One of the more significant characteristics of inselbergs relative to pediment formation is their distribution relative to the pediment surface. They commonly occur as extensions of major range front interfluves, and their size and number tend to decrease with increasing distance from the mountain front (Cooke and Mason 1973, p. 196).

Regolith

The regolith associated with pediment surfaces is, of course, widely variable. Transported regolith, predominantly wash and channelized flow deposits, also includes colluvial deposits within and adjacent to interfluvial piedmont junctions and thin deposits of desert loess on some inactive surface remnants. Residual regolith is mostly *in situ* weathered bedrock. The maximum thickness of transported regolith on pediments is, as discussed previously, a matter of definition. However, average thicknesses in many proximal to medial areas are limited to a few metres (Gilluly 1937, Tator 1952, Mabbutt 1966, Cooke and Mason 1973). As discussed above, these deposits commonly form thin discontinuous to continuous mantles across the bedrock surface, and surface relief is typically very similar in both form and magnitude to that of the underlying bedrock. Channels as much as 1–2 m deep and 10 m wide are common on both surfaces; however, both are mostly planar and relatively featureless. The thickness of residual regolith of most pediments is difficult to measure but appears to be quite variable (Oberlander 1974, Moss 1977). At least several tens of metres of saprolitic bedrock typically underlie the surfaces of many of the granitic piedmonts in the south-west Basin and Range province (Oberlander 1972, 1974, Moss 1977, Dohrenwend et al. 1986).

Influences on Pediment Development

Lithologic and Structural Influences

It has long been recognized that local lithologic and structural variations strongly influence pediment development (Gilbert 1877, p. 127–8, Davis 1933, Bryan and McCann 1936, Tator 1952, Warnke 1969). On the piedmonts flanking the Henry Mountains of southern Utah, for example, Gilbert (1877, p. 128) observed that in 'sandstones flat-bottomed cañons are excavated, but in the great shale beds broad valleys are opened, and the flood plains of adjacent streams coalesce to form continuous plains. The broadest plains are as a rule carved from the thickest beds of shale'. Gilluly (1937, pp. 341–2) reported a similar relation in the Little Ajo Mountains of southwestern Arizona. 'The development of pediments is apparently conditioned by the lithology of the terrane, the softer and more readily disintegrated rocks forming more extensive pediments than do the harder and more resistant rocks. .... They are commonly better developed on [weathered] granitic rocks and soft sediments than on other rocks.'

Among the most conspicuous expressions of local lithologic/structural influence are the inselbergs, tors and other positive surface irregularities that so commonly interrupt the pediment plain (Tuan 1959, Kesel 1977). These erosional remnants are characteristically associated with rocks that, by virtue of their lithology, texture, and/or structure, are more resistant to weathering and erosion than those underlying the adjacent pediment. Such relations are particularly well expressed on Cima Dome in the eastern Mojave Desert where '. . . major knobs and hillocks rising 50–450 feet above the slopes of [the dome] are underlain by rocks other than the pervasive quartz monzonite. . . . At scattered places, dikes and silicified zones in the quartz monzonite form linear ridges or abrupt risers a few to 25 feet high. . . . On upper parts of the dome are numerous knobs and hillocks as much as 35 feet high composed of exceptionally massive quartz monzonite' (Sharp 1957, p. 277).

On a more regional scale, numerous reports from the literature suggest that although pediments occur on many different rock types in arid regions of the south-west United States, the more extensive pediments may be preferentially developed on less resistant lithologies. In the Mojave and Sonoran Deserts, extensive

Fig. 13.8 Dissected *glacis d'erosion* cut across late Tertiary alluvial and lacustrine strata, Stewart Valley, Nevada. Aerial view is south

bedrock pediments bevel deeply weathered, coarse-crystalline granitic rocks in many areas (Davis 1933, Gilluly 1937, Tuan 1959, Mammerickx 1964, Warnke 1969, Cooke 1970, Oberlander 1972, 1974, Moss 1977, Dohrenwend et al. 1986) (Figs. 13.2, 13.4 and 13.7). In the west-central Great Basin (Gilbert and Reynolds 1973, Dohrenwend 1982a,b), pediments occur primarily on middle to upper Cenozoic terrigenous sedimentary rocks (predominantly partly indurated fluvial and lacustrine deposits, Fig. 13.8) and on less resistant volcanic rocks (predominantly lahars and non-welded ash flow tuffs, Fig. 13.9). Pediments also bevel upper Cenozoic terrigenous sediments along the Ruby and East Humboldt ranges of north-east Nevada (Sharp 1940), in the Furnace Creek area of

Death Valley (Denny 1967), in the San Pedro, Sonoita, and Canada del Oro basins of south-east Arizona (Melton 1965, Menges and McFadden 1981), along the lower valley of the Colorado River (Wilshire and Reneau 1992), and along the valley of the Rio Grande (Denny 1967). Applegarth (2004) examined mountain slope morphology to discriminate between pediments and bedrock-covered alluvial slopes, and showed that bedrock pediments were characterised by mountain slopes that had fewer joints, larger clasts and steeper gradients than those backing alluvial slopes. Coupled with his finding that catchments upslope of pediments were smaller than those upslope of alluvial slopes, Applegarth's findings point to sediment supply as a key control on the distribution of pediments.

Fig. 13.9 Dissected remnants of a rock pediment and *glacis d'erosion* cut across Mesozoic granitic rocks, Tertiary volcanic rocks, and Tertiary terrigenous sediments on the back-tilted flank of the Wassuk Range, west-central Nevada. View is north-east towards Lucky Boy Pass (elevation c. 8000 ft) and Corey Peak (elevation 10 520 ft)

In many other areas, however, extensive pediments cut across a variety of more resistant sedimentary, volcanic and metamorphic rocks (Tuan 1959, Mammerickx 1964, Cooke and Reeves 1972, Dohrenwend 1987a). Moreover, regional mapping and analysis of the general distribution of pediments within the tectonically active central and western Great Basin (Dohrenwend et al. 1996a) indicates that, within this region at least, pediment development is not clearly related to the distribution of more easily erodible rock types. Rather, the relative extent of rock types exposed on bedrock pediments accords well with the relative extent of those same rock types within the adjacent ranges (Fig. 13.10). This suggests that local tectonic stability may be the dominant control

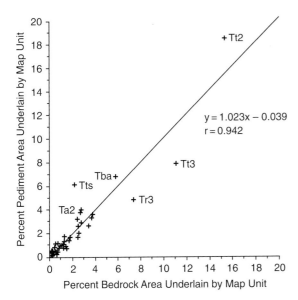

Fig. 13.10 Graph comparing the relative extent of exposed bedrock pediments versus the relative abundance of bedrock in upland areas for each bedrock unit (with total surface exposures greater than $100\,km^2$) on the geologic map of Nevada (Stewart and Carlson 1977). In essentially all cases, the relative abundance of exposed bedrock pediments underlain by a specific rock type accords well with the relative abundance of that same rock type within the upland areas of the region. Thus, it would appear that pediment development in Nevada has not been strongly influenced by lithologic distribution (from Dohrenwend et al. 1996a). **Ta2** = Tertiary andesite flows and breccias and related rocks of intermediate composition (17–34 Ma); **Tba** = Tertiary basalt and andesite flows (6–17 Ma); **Tr3** = Tertiary rhyolite flows and shallow intrusive rocks (6–17 Ma); **Tts** = Tertiary ash-flow tuffs and tuffaceous sedimentary rocks (6–17 Ma); **Tt2** = Tertiary welded and non-welded silicic ash-flow tuffs (17–34 Ma); **Tt3** = Tertiary welded and non-welded silicic ash-flow tuffs (6–17 Ma)

of pediment development in this tectonically active region.

Tectonic Influences

The bedrock pediments of the south-west United States are located, for the most part, within stable or quasi-stable geomorphic environments where erosional and depositional processes have been approximately balanced for relatively long periods of time. Although pediments have not been systematically mapped or correlated with tectonic environment across the entire region, a clear correspondence between tectonic stability and pediment development is apparent.

General geomorphic analyses of the Basin and Range province document a variety of regional morphometric and geomorphic variations (including the distribution of pediments versus alluvial fans) that are clearly related to variations in the distribution and style of Quaternary faulting (Lustig 1969, p. D68, Bull 1977, Bull and McFadden 1977). For example, Lustig's (1969) morphometric and statistical analysis of Basin and Range morphology concludes that pediments are more abundant in areas characterized by low average values of range relief, height, width, length, and volume, whereas alluvial fans are more abundant where these values are large. This conclusion is generally supported by a regional neotectonic evaluation of the south-west Basin and Range province based on a geomorphic classification of the relative tectonic activity of mountain-fronts (Bull 1977, Bull and McFadden 1977). This study documents several distinct regions of contrasting neotoctonic activity that may be distinguished, in part, by a relative lack or abundance of pediments. Active and moderately active range fronts (with large alluvial fans, few if any pediments, steep hillslopes on all rock types, elongate drainage basins, and low mountain front sinuosities) define a region of pronounced dip-slip faulting between the Sierra Nevada and Death Valley in the south-west Great Basin; and moderately active to slightly active range fronts (with both fans and pediments, relatively equant drainage basins, broader valleys, and moderate mountain front sinuosities) characterize a region of strike-slip faulting in the south-central Mojave Desert. In contrast, inactive range fronts (with broad mountain front pediments, large embayments, and high mountain front sinuosities) distinguish regions of post-middle Miocene tectonic stability in the western and eastern parts of the Mojave Desert (Bull, 1977, Dohrenwend et al. 1991). Thus, the pediments of the Basin and Range province are generally larger and more continuous in areas of greater vertical tectonic stability.

This strong relation between tectonic stability and pediment development is also reflected throughout the south-west Basin and Range province by a general spatial association between extensive pediments and many areas of shallow Miocene detachment faulting (Dohrenwend 1987a). Morphometric comparisons of basin and range morphology (Dohrenwend 1987b), range front morphology (Bull 1977, Bull and Mc-Fadden 1977), and the regional distribution of young faults (Dohrenwend et al. 1996b) indicate significantly less post-detachment vertical tectonism in areas with extensive pediments than in other parts of the province. Moreover, a generally low local topographic relief (c. 100–250 m?) has been inferred for areas of shallow deformation along closely spaced listric normal faults in this region (Zoback et al. 1981). This combination of relative tectonic stability and relatively low initial tectonic relief has permitted the formation of numerous pediments that typically encircle their associated uplands and extend downslope as much as 5–10 km from the mountain front.

Regional mapping of pediments throughout the state of Nevada (which includes many of the more tectonically active regions of the Basin and Range province) indicates that pediment development there also has been controlled primarily by spatial and temporal variations in late Cenozoic tectonic activity (Dohrenwend et al. 1996a). Pediment areas with exposed bedrock or thin alluvial cover occupy approximately 15% of the intermontane piedmonts and basins of the region. The more extensive and continuous of these pediments are located in the south-central and western parts of the state in areas of relatively shallow Miocene detachment faulting that apparently have undergone relatively little post-Miocene vertical tectonic movement (see above). Typically, these areas are characterized by a relatively subdued topography consisting of low narrow ranges, broad piedmonts, and small shallow basins. Conversely, pediments are smaller and less abundant within the more tectonically active areas of the state. Included within these tectonically active areas are the longest, most continuous, and most active dip-slip fault zones; the longest and widest

ranges; and the largest, deepest, and most continuous Cenozoic basins in the Great Basin.

Within tectonically active areas, pediments are largely confined to local settings of relative geomorphic stability. Exposed bedrock pediments typically occupy proximal piedmont areas immediately adjacent to the range front; however, in some cases pediments may extend from range front to basin axis. Particularly favourable settings include range embayments and both low broad passes within and narrow gaps between ranges. Such settings are especially well suited for pediment development if they are also situated on the backtilted flanks of large asymmetric range blocks or gently upwarped structural highs. Smaller pediments occur on the tilted flanks of small blocks along major strike-slip faults, on upfaulted piedmont segments, and within small embayments adjacent to active fault-bounded range fronts (Dohrenwend 1982c).

Although all of these diverse geomorphic settings are related by a common condition of relative long-term stability, they are quite different in many other respects. Hence, the pediments formed within each of these local geomorphic environments are morphologically distinct. Comparison of the pediments on the east and west flanks of the Ruby–East Humboldt Range in north-eastern Nevada illustrates this relation between pediment morphology and local neotectonic–geomorphic environment (Sharp 1940). The east flank of this asymmetrically west-tilted range is bounded by a major range front fault system. Thus the drainage basins and piedmonts of the east flank are smaller and steeper than those of the west flank, and Pleistocene glaciation was less extensive on the east flank than on the west flank. Consequently, the large pediments of the west flank extend across the backtilted flank of the range, whereas the smaller pediments of the east flank are confined to embayments upslope from the range-bounding fault. Seven surfaces (pediments and partial pediments) occur on the west flank of the range, and these pediments are cut mostly on soft basin deposits. In contrast, the pediments of the east flank are, with one exception, cut entirely on the hard rocks of the mountain block. 'Remnants of pediments on the west flank extend as far as 5 miles from the foot of the mountain slope, and the undissected surfaces were probably even more extensive. The east flank pediments are narrower, at least as exposed, and seldom extend more than 1.5 miles from the mountains. The west flank pediments have gentler gradients and are more nearly smooth than those of the east side. The pediments of the west side are mantled by a comparatively uniform cover of stream gravel; those of the east side have only local patches of stream gravel...' (Sharp 1940, pp. 362–3).

Climatic Influences

The influence of climate on pediment development has not been systematically studied and, therefore, is not very well documented or understood. Pediments occur within the broad range of climates from subpolar to humid tropical. They are particularly abundant and well developed, however, in arid/semi-arid and tropical wet–dry regimes. Indeed pediments are so characteristic of these two rather disparate regimes that they represent somewhat of an embarrassment to advocates of climatic geomorphology (Chorley et al. 1984, p. 486). Descriptive reports in the literature indicate that the pediments of most climatic realms are generally similar in form and landscape position. For example, 'broad, gently sloping [erosional] surfaces which extend from the base of hillslopes, and which in some cases pass beneath alluvial accumulations and in others terminate at a break in slope leading down to the river channel or floodplain, undeniably exist in the tropics' (Thomas 1974, p. 218). These tropical pediments are characterized by generally smooth, concave-up longitudinal profiles with declivities ranging between $1°$ and $9°$. Also they are most commonly interpreted to function as surfaces of transport between hillslope and riverine plain. These similarities of form, position, and function notwithstanding, it is not at all clear that the processes of mass transport which maintain these surfaces are at all similar from one climatic regime to another. However, it is apparent that these surfaces are for the most part typically associated with either local or regional settings of long-term geomorphic stability.

Because long-term geomorphic stability appears to be a general requirement for pediment formation, it is very likely that the development of many pediments, particularly larger ones, may transgress periods of unidirectional climate change and that such changes may profoundly affect pediment development. For example, several lines of evidence indicate that the granitic pediments of the Mojave and Sonoran Deserts were fully developed by late Miocene time and since

that time have been modified primarily by partial stripping of a thick saprolitic regolith (Oberlander 1972, 1974, Moss 1977). According to Oberlander, the late Miocene Mojave–Sonoran region was probably an open woodland interspersed with grassy plains that extended across extensive cut and fill surfaces of low relief surmounted by steep-sided hills. Pediment formation within this landscape appears to have involved an approximate balance between regolith erosion (probably by slope wash, rill wash and channelized flow processes) and regolith renewal (by chemical breakdown of granitic rock along a subsurface weathering front). This balance was apparently upset by increasing aridity resulting in a loss of vegetative cover which triggered regolith stripping and concomitant formation of the relatively smooth pediment surfaces that characterize the modern landscape. Evidence supporting this general scenario includes saprolite remnants preserved beneath late Tertiary lava flows, the elevated positions of these lava-capped palaeosurfaces above present erosion surfaces, continuity between the relic saprolite and boulder mantles on hillslopes, and general grusification as much as 40 m below present pediment surfaces. Interestingly enough, it is generally agreed that the pediments of many tropical and subtropical areas have also developed 'across pre-weathered materials or have been significantly modified, at least, by erosional stripping of deeply weathered materials (Mabbutt 1966, Twidale 1967, Thomas 1974, p. 218–20).

Pediment Development and Time

It is generally perceived that extensive pediments are a characteristic feature of old landscapes (Lawson 1915, Bryan 1922, Davis 1933, Howard 1942) and that their formation requires a condition of general landscapes stability wherein erosional and depositional processes are approximately balanced over long periods of time (cyclic time of Schumm and Lichty 1965). There is little doubt that the more extensive pediments of the Basin and Range province are, at least in part, relics of considerable age. Throughout this region, local burial of piedmonts and pediments by late Cenozoic lava flows provides convincing evidence of their long-term morphologic stability, even in regions of neotectonic activity (Figs. 13.11 and 13.12, Table 13.3). Moreover,

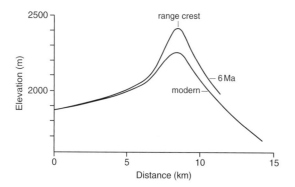

Fig. 13.11 Longitudinal profiles comparing modern and late Miocene erosional surfaces on the west flank of the Reveille Range. Since latest Miocene time, rates of incision and downwearing have ranged between 20 and 35 m per million years in crestal and upper flank areas and between 5 and 20 m per million years in lower flank and proximal piedmont areas. Medial and distal piedmont areas have undergone less than 5 m of net downwearing

ubiquitous deep weathering beneath extensive pediments in the Mojave and Sonoran Deserts indicates at least pre-Quaternary ages for the original surfaces of these pediments (Oberlander 1972, 1974, Moss 1977).

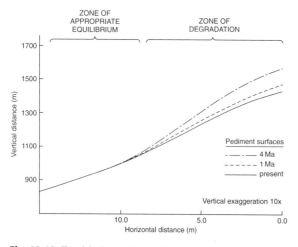

Fig. 13.12 Empirical model of pediment dome evolution in the eastern Mojave Desert based on a compilation of elevation differences between active pediment surfaces and relict surfaces capped by dated lava flows. The 1 Ma and 4 Ma surfaces were reconstructed using a smoothed plot of average downwearing rates versus distance from dome summits. Upper flank areas have been eroded, midflank areas have remained in a state of approximate equilibrium, and lower flank areas (not shown) have probably aggraded. Horizontal distance is measured from the dome summit (from Dohrenwend et al. 1986)

Average Rates of Downwearing and Backwearing

One reliable approach for estimating average erosion rates in the Basin and Range province involves the documentation and analysis of erosion surfaces capped by Tertiary and Quaternary basaltic lava flows. Average downwearing rates can be estimated by measuring the average vertical distance between an erosional palaeosurface capped by a lava flow and the adjacent modern erosion surface, then dividing this height difference by the K/Ar age of the lava flow. Using this approach, average downwearing rates have been estimated for several widely separated upland and piedmont areas within the Great Basin and Mojave Desert. For upland areas, these estimates range between 8 and 47 m per million years for periods of 0.85–10.8 m.y. (Table 13.3a); whereas for proximal piedmont areas, the estimates range from less than 2–13 m per million years for periods of 1.08–8.9 m.y. (Table 13.3b). During similar time periods, medial piedmont areas have remained largely unchanged (Table 13.3c). These data document a general evolutionary scenario of upland downwearing and proximal piedmont regrading which has been regulated in part by the long-term stability of the medial piedmont which serves as a local base level for the upper part of the pediment association (Figs. 13.11 and 13.12).

More recently, the use of ^{10}Be and ^{26}Al cosmogenic nuclides has provided another method for estimating rates of downwearing over periods of 10^3–10^5 years. Nichols et al. (2002, 2005), working in the Mojave Desert, have obtained rates of 38 (^{10}Be) and 40 (^{26}Al) m per million years for the Chemehuevi Mountain, and 31 and 33 m per million years, respectively, for Granite Mountain. Proximal pediment downwearing of the Chemehuevi Mountain pediment was estimated at 10 and 21 m per million years. These rates are in accord with those obtained from the dating of lava flows.

Average rates of backwearing or slope retreat can be estimated in a similar manner by measuring the average horizontal distance between the maximum possible original extent of a lava flow which caps an erosional palaeosurface and the present eroded margin of that flow, then dividing this horizontal distance by the K/Ar age of the lava flow. Estimated rates of slope retreat determined by this method range between 37 and 365 m per million years (Tables 13.3a,b). Of course, these estimates of average backwearing should be used

Table 13.3 (a) Erosion rates in upland areas of the south-west Basin and Range province inferred from comparisons between active and basalt-capped relic erosion surfaces

Location	Minimum age of relic surface (Ma)	Maximum average downwearing rate (m per m.y.)	Maximum average rate of slope retreat (m per m.y.)	Reference
Fry Mountains, southern Mojave Desert	8.90 ± 0.9	8	–	Oberlander (1972)
Cima volcanic field, eastern Mojave Desert	4.48 ± 0.15	11	290	Dohrenwend et al. (1986) Turrin et al. (1985)
Cima volcanic field, eastern Mojave Desert	3.85 ± 0.12	29	365	Dohrenwend et al. (1986) Turrin et al. (1985)
Cima volcanic field, eastern Mojave Desert	0.85 ± 0.05	30	320	Dohrenwend et al. (1986) Turrin et al. (1985)
White Mountains, south-west Great Basin	10.8	24	–	Marchand (1971)
Buckboard Mesa, southern Great Basin	2.82	47	–	Carr (1984)
Columbus Salt Marsh, central Great Basin	3.04 ± 0.2	21	37	Dohrenwend (unpublished data)
Reveille Range, central Great Basin	5.70 ± 0.20	34	230	Ekren et al. (1973)
Reveille Range, central Great Basin	3.79 ± 0.34	20	200	Dohrenwend et al. (1985, unpublished data)

Table 13.3 (b) Erosion rates in proximal piedmont areas of the south-west Basin and Range province inferred from comparisons between active and basalt-capped relic erosion surfaces

Location	Minimum age of relict surface (Ma)	Maximum average downwearing rate (m per m.y.)	Maximum average rate of slope retreat (m per m.y.)	Reference
Fry Mountains, southern Mojave Desert	8.90 ± 0.90	<2	–	Oberlander (1972)
Cima volcanic field, eastern Mojave Desert	3.64 ± 0.16	11	85	Dohrenwend et al. (1986)
Reveille Range, central Great Basin	5.76 ± 0.32	8	230	Dohrenwend et al. (1985, unpublished data)
Reveille Range, central Great Basin	5.58 ± 0.30	2.5	190	Dohrenwend et al. (1985, unpublished data)
Lunar Crater volcanic field, central Great Basin	2.93 ± 0.30	9	–	Turrin and Dohrenwend (1984, unpublished data)
Lunar Crater volcanic field, central Great Basin	1.08 ± 0.14	13	–	Turrin and Dohrenwend (1984, unpublished data)
Quinn Canyon Range, central Great Basin	8.5 ± 0.7	15	–	Dohrenwend (unpublished data)
Columbus Salt Marsh, central Great Basin	3.04 ± 0.2	13	50	Dohrenwend (unpublished data)

Table 13.3 (c) Late Cenozoic basaltic volcanic fields on piedmonts in the south-west Basin and Range province where net vertical erosion/deposition has been low (generally less than 5 m in most medial and/or distal piedmont areas) since lava flow emplacement

Area	Latitude	Longitude	Age range (Ma)	Reference
Buffalo Valley, northern Great Basin	40.35°N	117.3°W	0.9–3.0	Dohrenwend (1990a)
Reveille Range, central Great Basin	38.1°N	116.2°W	3.8–5.8	Dohrenwend et al. (1985) Dohrenwend (1990b)
Lunar Crater volcanic field, central Great Basin	38.25°N	116.05°W	c. 0.1–2.9	Turrin and Dohrenwend (1984) Dohrenwend (1990c)
Quinn Canyon Range, central Great Basin	37.9°N	115.95°W	7.8–9.2	Dohrenwend (unpublished data)
Clayton Valley, western Great Basin	37.8°N	117.65°W	0.2–0.5	Dohrenwend (1990d)
Columbus Salt Marsh, central Great Basin	38.1°N	118.05°W	2.8–3.2	Dohrenwend (unpublished data)
Crater Flat, southern Great Basin	37.1°N	116.5°W	c. 0.1–3.7	Carr (1984), Turrin et al. (1991)
Big Pine, south-west Great Basin	37.05°N	118.25°W	<0.1–1.2	Gillespie (1990)
Saline Valley, south-west Great Basin	36.85°N	117.7°W	0.5–0.7	Dohrenwend (unpublished data)
Cima volcanic field, eastern Mojave Desert	35.25°N	115.75°W	c. 0.1–1.1	Dohrenwend et al. (1986) Dohrenwend (1990e)

only as general approximations. The armouring effect of blocky basalt talus inhibits the backwearing of basalt capped hillslopes; moreover, substantial uncertainties are inherent in determining the original extent of most lava flows. However, these average backwearing rates are comparable with the 0.1 to 1-km-permillion-year rates of range front retreat and pediment formation that have been previously estimated or assumed for diverse areas of the Basin and Range province (Wallace 1978, Mayer et al. 1981, Menges and McFadden 1981, Saunders and Young 1983).

If one assumes that the estimated rates of downwearing and backwearing compiled in Table 13.3 are, in fact, representative of erosion rates in arid regions of the south-western United States, then a typical upland in this region has probably undergone from as little as 50 m to as much as 250 m of downwearing and from 0.2 to 2 km of backwearing since late-Miocene time. During this period, the medial areas of the piedmont fringing this typical upland would have experienced little morphologic change. Hence, it is very likely that many of the larger pediments of the Mojave and Sonoran Deserts (which typically are at least several kilometres wide) have been forming since at least late Miocene time. This pattern of landscape evolution is particularly well documented on the slopes and piedmonts of the Reveille Range in south-central

Nevada (Fig. 13.11). Since latest Miocene time, denudation rates along the crest and upper flanks of the Reveille Range have ranged between 20 and 35 m per million years. Downwearing of lower flank and upper-piedmont areas has ranged between 5 and 20 m per million years; and most middle and lower piedmont areas have remained in a state of approximate erosional equilibrium (Dohrenwend et al. 1985, 1986).

Direct Evidence of Pediment Age

The local burial of Tertiary pediments remnants by late Miocene and Pliocene basalt flows provides convincing evidence for a long history of pediment evolution (Oberlander 1972, 1974, Dohrenwend et al. 1985, 1986). North of the San Bernardino Mountains in the south-western Mojave Desert, late Miocene lava flows cover pediment remnants cut across saprolitic weathering profiles. These remnants now stand as much as 60 m above adjacent pediment surfaces (Oberlander 1972). In the vicinity of Cima dome in the north-eastern Mojave Desert, Pliocene lava flows bury remnants of ancient pediments cut across less intensely weathered materials (Fig. 13.13); and flows younger than 1.0 Ma bury pediment remnants locally capped by soils similar to those developed

Fig. 13.13 Early Pliocene basaltic lava flows of the Cima volcanic field preserve remnants of a large Miocene pediment dome cut across deeply weathered monzogranite and Tertiary terrigenous sedimentary rocks. Aerial view north-east towards the 80 to 120-m-high erosional scarp which bounds the western edge of the Pliocene flows. A continually evolving pediment surface extends downslope from the base of this deeply embayed escarpment. This modern pediment is pervasively dissected by a finely textured, subparallel network of shallow (<3 m deep) washes

Fig. 13.14 Quaternary basaltic lava flows (c. 0.1–1.0 Ma) of the Cima volcanic field veneer remnants of a continually evolving pediment surface formed on Cretaceous monzogranite and middle Tertiary terrigenous sediments. Deep incision of a c. 1.0 Ma flow complex (*left* foreground) reflects the delicate erosional–depositional balance of this active pediment surface. Aerial view is north-west

on adjacent Quaternary surfaces (Fig. 13.14). In this area, differences in height between buried surface remnants and adjacent active pediment surfaces are systematically related to the ages of the overlying lava flows (i.e. at equal distances downslope from the crest of the pediment dome, the older the buried remnant the greater its height above the adjacent active surface). Moreover, the ubiquitous presence of deep weathering profiles underlying other extensive granitic pediments in the Mojave and Sonoran Deserts suggests at least pre-Quaternary ages for the original pediment surfaces (Oberlander 1972, 1974, Moss 1977). These relations indicate that, since late Miocene time, pediment surfaces cut across deeply weathered granitic rocks have evolved more or less continuously by progressive stripping of a thick late Tertiary weathering mantle (Oberlander 1972, 1974, Moss 1977, Dohrenwend et al. 1986). Hence, these pediments are, at least in part, ancient forms of late Miocene age.

long-lived landforms (such as pediments) that have evolved through complex histories of tectonic and climate change. The various processes operating on such landforms and the relative effectiveness of these processes undoubtedly has changed dramatically through time. Even when a cause and effect relation between form and process does exist, such a relation may well be unidirectional (i.e. the form determines the process, but the process did not produce the form). For example, sheetflood erosion cannot produce a planar pediment surface because a planar surface is necessary for sheetflooding to occur (Paige 1912, Cooke and Warren 1973, p. 203). Similarly, weathering and subsequent removal of the weathered debris are unlikely to produce a pediment surface but will act to maintain a pre-existing surface if these processes are uniformly applied (Lustig 1969, p. D67). Thus, the assumption that the processes presently operating to maintain or modify pediment form are the same as those responsible for pediment formation is not justified.

Pediment Processes

As Cooke and Mason (1973, p. 203) aptly pointed out, attempts to deduce the formation of an existing landform from observation of the processes presently operating on that landform are often unsuccessful because this approach commonly confuses cause and effect. Mere spatial association does not establish a cause and effect relation between form and process. Indeed, such relations are particularly unlikely in the case of

Processes on Pediment Surfaces

In the desert regions of western North America, pediments serve as integral parts of the piedmont, and the processes operating on pediment surfaces are much the same as those operating on other parts of the piedmont plain (Dohrenwend 1987a). Processes which act to maintain or modify pediment surfaces include surface and subsurface weathering, unchannelized flow (sheet flow, sheetflooding, rill wash), and channelized flow (gully wash, debris flow, and stream flow). In most cases, the relatively diffuse processes of weathering and unchannelized flow will tend to be more uniformly distributed in space than the more concentrated processes of channelized flow. Where the combined effect of these processes acts uniformly over the entire pediment surface, the surface will be maintained as an active pediment. Where the combined effect is not uniformly applied, however, incision will occur most likely in response to decreases in stream sediment load (Fig. 13.6) and/or increases in stream discharge. Surfaces abandoned by fluvial action as a result of incision form isolated islands of relative stability where the processes of subsurface and surface weathering (particularly salt weathering), soil formation, desert loess accumulation, and stone pavement development are locally dominant.

Under conditions of tectonic and base level stability, the piedmont surface is divisible into three general process zones: an upper zone of erosion, an intermediate zone of transportation, and a lower zone of aggradation (Johnson 1932a,b, Cooke and Warren 1973, p. 197, Dohrenwend et al. 1986). These zones shift upslope and downslope across the piedmont in response to changes in the overall upland/piedmont drainage system, and the boundaries between zones are both gradational and complex. Their precise position at any given time is determined by the relative rates of debris supply and removal across the piedmont plain. If the rate of debris supply to the piedmont increases or the rate of removal decreases, these zones tend to migrate upslope; conversely, if the rate of supply decreases or the rate of removal increases, they tend to migrate downslope (Cooke and Mason 1973). On most piedmonts in arid and semi-arid environments, the rates of debris supply and removal are largely determined by channelized flow processes; thus the relative extent and position of each general process zone reflects the average location, over time, of the threshold of critical power for each component of the piedmont drainage system. On average, stream power exceeds critical power in the zone of erosion, approximately equals critical power in the zone of transportation, and falls short of critical power in the zone of aggradation (Bull 1979). This general scenario is supported by (a) regional patterns of proximal piedmont dissection in the Mojave and Sonoran Deserts; (b) long-term trends of piedmont and pediment erosion documented by spatial relations be-

tween volcanic landforms and active piedmont surfaces (Dohrenwend et al. 1985, 1986, Dohrenwend 1990a,c); and (c) morpho-stratigraphic relations in the Apple Valley area of the western Mojave Desert (Cooke and Mason 1973).

These conceptual divisions of pediment surfaces are supported by rates of denudation obtained from [10]Be and [26]Al measurements. On the Chemehuevi Mountain pediment in the Mojave Desert Nichols et al. (2005) obtained denudation rates of 22 m per million years at a distance of 1 km from the mountain front compared to 8 m per million years 3 km from the mountain front. Nuclide data indicate that the lower part of the pediment is an area of sediment transport and deposition, with the interfluve areas effectively stable.

In the zone of erosion, channelized flow processes typically dominate and commonly result in shallow to moderate dissection of the proximal piedmont. However, surface regrading may also proceed without significant incision via the operation of unchannelized flow and lateral shifting of channelized flow. In the zone of transportation, stream power and critical power are essentially balanced over the long term such that net surface erosion is nearly undetectable. Surface processes in this zone are probably dominated by lateral shifting of channelized flow in shallow anastomosing gullies and washes (Rahn 1967, Cooke and Mason 1973, Dohrenwend et al. 1986). Rates of sediment transport obtained from [10]Be and [26]Al data are found to be highest in confined bedrock channels (10s of metres per year), intermedaite (metres per year) in incised alluvium, but only decimetres per year in highly permeable active alluvial surfaces (Nichols et al., 2005).

Subsurface Weathering

Subsurface weathering processes have strongly influenced pediment development in many regions, particularly in humid tropical, tropical wet–dry, and semi-arid environments (Ruxton 1958, Mabbutt 1966, Thomas 1974, p. 218–26, Oberlander 1989). Such processes are critical to the formation or modification of many pediments, particularly those developed on crystalline bedrock (Oberlander 1974, 1989, Moss 1977). Clearly, weathered and disaggregated bedrock can be much more readily loosened, entrained, and eroded by wash and stream action than unweathered bedrock.

Indeed, the apparent predilection of pediments for areas underlain by granitic rocks is due, in large measure, to the susceptibility of these rocks to subsurface weathering and to the particular mechanical characteristics of the residual grus. This well-sorted, sand-sized, non-cohesive material forms non-resistant channel banks that are highly susceptible to erosion by laterally shifting channelized flow. Hence, granitic pediments possess a highly effective mechanism of self-regulation which tends to suppress fluvial incision. Wherever the threshold of critical power is exceeded by stream flow, erosion of the channel banks rapidly increases sediment load and raises the critical power threshold to the level of available stream power. The profound contrast in surface morphology between granitic and metamorphic piedmonts in the Mojave Desert serves as a graphic illustration of this phenomenon. Granitic piedmonts are, in most cases, essentially undissected and relatively featureless plains of active transport, whereas metamorphic piedmonts are more commonly dissected into intricately nested surfaces of diverse age.

Controlled largely by the availability of moisture, the rates and distribution of subsurface weathering processes are closely related to local geomorphic setting and to shallow subsurface form and structure. Conditions favouring moisture accumulation and retention will typically intensify weathering processes. On arid and semi-arid piedmonts, moisture accumulation and retention commonly occurs (a) beneath footslopes within the piedmont junction, (b) along the alluvium–colluvium and bedrock interface, (c) along buried bedrock channels, and (d) in areas of intersecting bedrock fractures. Hence, the surface described by the subsurface weathering front is typically highly irregular, and stripping of the residual regolith in areas of proximal dissection commonly reveals complex and intricate etching of the bedrock surface.

Piedmont Modification by Subsurface Weathering and Regrading

Subsurface weathering and the subsequent erosional modification of piedmonts or other surfaces of low relief has played a central role in the expansion and/or regrading of many pediments (Mabbutt 1966, Oberlander 1972, 1974, 1989, Moss 1977, Dohrenwend

et al. 1986). In the Flinders Ranges of South Australia, for example, 'the abrupt change in slope in the piedmont zone (junction) is caused principally . . . by differential weathering at the foot of the scarp. . . . The bedrock 1.5 km from the scarp is weathered only to a moderate degree. . . . Close under the scarp, however, similar strata are much more intensely altered. Kaolin is abundant, bedding only vaguely discernable, and the rock has lost much of its strength. . . . Runoff from the often bare hillslopes percolates into the rock strata where it reaches the plain, causing [this] intense weathering' (Twidale 1967, p. 113). Also in central Australia, '. . . grading on . . . granitic and schist pediments has largely acted through the mantles, whereby the smoothness of the depositional profiles has been transmitted to suballuvial and part-subaerial pediments. . . . On granitic pediments . . . suballuvial notching and levelling proceed through weathering in the moist subsurface of the mantle. On schist pediments control of levelling by the mantle is less direct in that its upper surface is the plane of activity of ground-level sapping and of erosion by rainwash and sheetflow. . . . [These processes] are most active near the hill foot, where mantling and stripping alternate more frequently; the lower parts of both schist and granitic pediments . . . appear to be largely and more permanently suballuvial . . .' (Mabbutt 1966, pp. 90–1). Although subsurface weathering processes have strongly influenced pediment development in many areas and profoundly modified pediment surfaces in many others, it would appear unlikely that these processes actually 'control' pediment development, at least in arid and semi-arid environments. The preservation of tens of metres of intensely weathered bedrock beneath active granitic pediments demonstrates that these landforms are transport-limited (not weathering- or detachment-limited). The present form of the active surface is the product of fluvial erosion and transport; it is not closely related to either the position or the form of the subsurface weathering front. Moreover, even though stripping of the proximal areas of many pediments has exposed the intricately etched (weathering-limited) surfaces of formerly buried weathering fronts, deeply weathered regolith remains beneath adjacent mantled surfaces of fluvial transport. Thus even this pediment regrading is primarily controlled by fluvial processes; only the detailed form of the proximal pediment has been significantly influenced by the position of the weathering front.

Models of Pediment Formation

Within the general context of the pediment association (i.e. degrading upland, aggrading alluvial plain, and intervening zone of transition and transport), pediment development can be usefully perceived as the necessary result of upland degradation (Lustig 1969, p. D67). As the pediment association evolves and mass is transferred from the upland to the alluvial plain, the diminishing upland is replaced by an expanding surface of transportation, the pediment. Hence, most models of pediment development implicitly link upland degradation and pediment development; and the major differences between these models centre primarily on the issues of mountain front (or piedmont junction) retreat and upland degradation. Accordingly, the various models that have been presented in the literature can be grouped according to the dominant style of upland degradation proposed. These styles are

(a) range front retreat where channelized fluvial processes (mainly lateral planation) predominate (Gilbert 1877, p. 127–9, Paige 1912, Blackwelder 1931, Johnson 1931, 1932a,b, Field 1935, Howard 1942);

(b) range front retreat where diffuse hillslope and piedmont processes predominate (Lawson 1915, Rich 1935, Kesel 1977);

(c) range front retreat assisted by valley development where the relative dominance of diffuse hillslope and piedmont processes versus channelized flow varies according to general geomorphic environment (Bryan 1922, 1936, Gilluly 1937, Sharp 1940);

(d) range degradation dominated by drainage basin development (valley deepening and widening, embayment formation and enlargement) (Lustig 1969, p. D67, Wallace 1978, Bull 1979, 1984, Parsons and Abrahams 1984).

This trend in dominant styles of upland degradation clearly shows a progressive maturation of Gilbert's original (1877) model. Thus, as Parsons and Abrahams (1984) point out, these models are not mutually exclusive; indeed, they are more complementary than contradictory.

Range Front Retreat by Lateral Planation

Models of lateral planation emphasize the importance of channelized fluvial processes. Erosion of the mountain mass and retreat of the mountain front is considered to be effected primarily by the lateral shifting of debris-laden streams issuing from within the mountains. The diffuse action of weathering, slopewash, and rill wash on hillslopes is acknowledged but is not considered essential to pediment formation. The resulting pediment surface is continually regraded by the lateral shifting of these streams across the piedmont.

> Whenever the load [of a stream] reduces the downward corrasion to little or nothing, lateral corrasion becomes relatively and actually of importance. . . . The process of carving away the rock so as to produce an even surface, and at the same time covering it with an alluvial deposit, is the process of planation. . . . The streams . . . accomplish their work by a continual shifting of their channels; and where the plains are best developed they employ another method of shifting. . . . The supply of detritus derived from erosion . . . is not entirely constant. . . . It results from this irregularity that the channels are sometimes choked by debris and . . . turned aside to seek new courses upon the general plain. . . . Where a series of streams emerge from the adjacent mountain gorges upon a common plain, their shiftings bring about frequent unions and separations, and produce a variety of combinations (Gilbert 1877, pp. 127–9).
>
> . . . the essence of the theory lies in the broader conception that rock planes of arid regions are the product of stream erosion rather than in any particular belief as to the relative proportions of lateral and vertical corrasion. . . . Every stream is, in all its parts, engaged in the three processes of (a) vertical downcutting or degrading, (b) upbuilding or aggrading, and (c) lateral cutting or planation. . . . The gathering ground of streams in the mountains, where greater precipitation occurs, will normally be the region where vertical cutting is at its maximum. . . . Far out from the mountain mass conditions are reversed. Each stream must distribute its water and its load over an everwidening sector of country. The water disappears, whether by evaporation or by sinking into the accumulating alluvium. Aggradation is at its maximum. . . . Between . . . [these regions] . . . there must be a belt or zone where the streams are essentially at grade. . . . Thus from the center outward are (1) the zone of degradation, (2) the zone of lateral corrasion, and (3) the zone of aggradation. Heavily laden streams issuing from the mountainous zone of degradation are from time to time deflected against the mountain front. This action combined with the removal of peripheral portions of the interstream divides by lateral corrasion just within the valley mouths, insures a gradual recession of the face or faces of the range. Such recession will be aided by weathering as well as by rain and rill

wash; but it must occur even where these processes are of negligible importance (Johnson 1932a, pp. 657–8).

> ... lateral planation may be carried on by streams of all sizes, by distributaries as well as by tributaries, by individual channels of a braided stream, and by sheetfloods (Howard 1942, p. 107).

Range Front Retreat by Diffuse Hillslope and Piedmont Processes

A second group of models identifies diffuse hillslope processes (mainly weathering, slopewash, and rill wash) coupled with removal of the resultant debris from the piedmont junction by sheetwash and rill wash as the dominant mechanism of mountain front retreat and pediment extension. Lateral planation by mountain streams is acknowledged as locally significant but is not considered to be an essential process. Sheetflooding is considered to be the dominant process of pediment surface regrading.

> If we examine a typical region in an old age stage of the arid cycle, where scattered residuals of former mountain masses stand on broad rock pediments. ... Obviously the mountains are wasting away by weathering, and just as obviously the weathered products must be carried away over a gradient steep enough to permit them to be moved. The rock, though weathered, cannot be removed below the line of this gradient. Consequently as the mountains waste away, a sloping rock plain, representing the lower limit of wasting, must encroach upon them from all sides. ... The gradient of this rock plain must be that necessary for the removal of the waste material – no more and no less – therefore, the rock will be covered by only a thin and discontinuous veneer of waste in transit (Rich 1935, p. 1020).

> The prevailing conditions of stream load force the desert streams to corrade laterally, as Johnson has pointed out, and in many places such corrasion contributes actively to the formation of pediments, especially along the sides of canyons debouching from the desert mountains. Nevertheless, lateral corrasion need be only a contributing, and not an essential factor in the formation of pediments, and a very minor factor in the retreat of mountain fronts (Rich 1935, p. 1021).

Range Front Retreat Assisted by Valley Development

A third group of models builds on the complementary nature of the previous models and accommodates their differences in emphasis by recognizing that the relative importance of weathering, unchannelized flow, or channelized flow processes varies according to specific geomorphic setting.

> The conclusions reached as to the formation of pediments under the various geological, topographic, and climatic conditions of the Ruby–East Humboldt region are as follows:
>
> (1) Pediments are formed by lateral planation, weathering, rill wash, and rain wash. The relative efficacy of these various processes is different under geologic, topographic, and climatic conditions.
>
> (2) Lateral planation is most effective along large permanent streams and in areas of soft rocks.
>
> (3) Weathering, rill wash, and rain wash are most effective in areas of ephemeral streams, hard rocks, and a low mountain mass.
>
> (4) All variations from pediments cut entirely by lateral planation to those formed entirely by the other processes are theoretically possible, although in this area the end members of the series were not observed and perhaps do not actually exist in nature (Sharp 1940, p. 368).

Range Degradation by Drainage Basin Development

The first three groups of models focus on the evolutionary retreat of the mountain front. In contrast, the models of this group address the broader issue of the overall degradation of the mountain mass. Their emphasis is on the evolutionary development of upland valleys through non-uniform erosion of the mountain mass by the concentrated action of channelized streamflow. It is significant to note that these models do not require parallel retreat of the mountain front as the primary mode of pediment expansion. These more comprehensive models emphasize the dominant role of fluvial erosion in the degradational evolution of the entire pediment association, and they illustrate the increasing sophistication of geomorphic theory as applied to the analysis of landscape evolution.

> The many discussions of pediment surfaces have focused on the wrong landform. ... There is no question that the processes of subaerial and suballuvial weathering occur on pediment today, nor that fluvial erosion also occurs. A pediment must exist prior to the onset of these processes, however, and in this sense the origin of the pediments resides in the adjacent mountain mass and its reduction through time. ... Given stability for a sufficient period of time, the consequences of mountain re-

duction must inevitably include the production of a pediment, whether in arid or nonarid regions. The nature of the surface produced may vary, and it may be mantled by, or free of, alluvium. However, it simply represents an area that was formerly occupied by a mountain. ... The only real "pediment problem" is how the reduction or elimination of the mountain mass occurs (Lustig 1969, p. D67).

... it is obvious that the pediment grows most rapidly along the major streams. In every indentation in the mountain front and in places where streams emerge from the canyons onto the plains the rate of formation of the pediment is rapid, and consequently extensions of the pediment into the mountains are common. ... The erosion of the mountains at the headwaters of many streams is much faster than in the lower portions of the same streams. ... Consequently the headwater slopes may recede more rapidly than the side walls of the valleys (Bryan 1922, pp. 57–8).

The rates of mountain front retreat are basically unknown, but by any reasonable assessment they are slow in relation to rates of processes that are operative in drainage basins. This is clearly true because the headwater region of any given drainage basin also consists of steep walls that are virtually identical to those of the mountain front in interfluvial areas. In these headwater regions the same processes of weathering and removal of debris occur. Hence, the rates of retreat of the bounding walls in the headwaters of drainage basins must be at least as great as the rate of retreat of the mountain front in interfluvial areas. Also, however, the drainage basins represent the only parts of any mountain range that are subjected to concentration of flow and to its erosional effects, and these basins must therefore be the principal loci of mountain mass reduction (Lustig 1969, p. D67).

During the valley downcutting that occurs after mountain-front uplift, the width of the valley floor will approximate stream width at high discharges. Valley-floor width decreases upstream from the front because of the decrease in the size of the contributing watershed. ... With the passage of geologic time, the stream will widen its valley as it approaches the threshold of critical power. As lateral cutting becomes progressively more important, the stream will not spread over the entire valley floor during high discharges. The approximation of a threshold condition migrates gradually upstream as the upstream reaches downcut. ... The configurations of the ... pediment embayments ... are functions of the rates of lateral cutting and/or hillslope retreat along the stream and the time elapsed since lateral erosion became predominant at various points along the valley. More than a million years may be needed to form pediment embayments (Bull 1979, pp. 459–60). ... it seems likely that, where a significant proportion of the piedmont junction occurs within embayments, the actual rate of piedmont formation will be much greater than that achieved by mountain front retreat alone. In embayments the rate of piedmont formation will depend upon the length of the embayments as well as upon the rate of piedmont junction migration (Parsons and Abrahams 1984, p. 258).

General Model of Pediment Formation

Fundamental Concepts

A number of fundamental geomorphological concepts provide insights useful for developing a comprehensive understanding of pediments and their formation.

(a) Analysis of a geomorphic system is in large part determined by the temporal and spatial limits that are used to define the system (Schumm and Lichty 1965). For example, the concept of dynamic equilibrium provides useful insights concerning the tendency for adjustment of slope declivity to lithology and/or structure within an upland area: however, the concept is less useful when applied to the problem of upland degradation over 'cyclic time' (unless mass continues to be added to the system through uplift). Depending on the time and space perspective of the analysis, processes may appear to be steady or fluctuating, continuous or discontinuous, and forms may appear to be stable, unstable, or quasi-stable. When considering the evolution of a landscape or the interrelations among its components, it is essential to define the system within those scales of space and time that are appropriate to the problem at hand. Regarding the specific problem of pediment formation, the appropriate spatial scale is very likely the pediment association (Cooke 1970) and the appropriate temporal scale is cyclic time (Schumm and Lichty 1965). Viewed from this perspective, emphasis is placed on interdependent changes between landscape components and processes in a changing open system (Bull 1975).

(b) Most geomorphic systems are multivariate and operate both non-linearly and discontinuously; therefore, these systems may respond complexly to mass and energy inputs (whether these inputs are non-linear or linear, continuous or discontinuous). 'When the influence of external variables such as isostatic uplift is combined with the effects of complex response and geomorphic thresholds, it is clear that denudation, at least during the early stage of the geomorphic cycle, cannot be a progressive process. Rather, it should be comprised of episodes of erosion separated by periods of relative stability, a complicated sequence of events' (Schumm 1975, p. 76).

(c) Process transitions and thresholds further complicate the operation of most geomorphic systems; and the behaviour of such systems may be particularly complex at or near these transitions. In the specific case of pediments, the locus of pediment formation is generally considered to be the piedmont junction, a zone of abrupt transition for both morphology and process. Moreover, the pediment itself serves as the transition between range and alluvial plain, and over the long term, pediment drainage operates at or very close to the threshold of critical power in streams. 'In the western Mojave Desert, evidence for the movement of the upper limits of alluvium across piedmont plains is provided by such features as channels buried beneath alluvium downslope of the limits, and upslope of the limits by the truncation of soil and weathering profiles and the presence of alluvial outliers. Movement of the boundary need not be accompanied by dissection of the plain, but this is frequently the case' (Cooke 1970, p. 37). Like many boundary zones in non-linear systems, this continuously shifting transition is both complex and chaotic. Hence, the timing and duration of periods of erosion, transport, or aggradation at any point in the system may very well be indeterminant, particularly where fluvial components of the system are operating at or are approaching the threshold of critical power.

(d) Form and process are closely interrelated within the pediment association. At any point in time, form is an initial constraint on process. As the distribution of form changes through time in response to process, the distribution of process changes accordingly. For example, '... the relative importance of divide removal [between embayments] and mountain front retreat ... is strongly dependent on the size and relief of the associated mountain mass, and hence ... it is likely to change in a systematic way as the mountain mass diminishes in size through time. Specifically ... as a mountain mass diminished in size there is a tendency for ... divide removal to become progressively less important relative to mountain front retreat' (Parsons and Abrahams 1984, p. 265). Similarly, the relative importance of lateral planation very likely also

diminishes with time. 'In an arid or desert region, flow over a long time and detrital load are most likely to exist where an ephemeral stream emerges from a canyon having a considerable length and a fairly large drainage basin within a mountain area. Such conditions prevail in the cycle of the erosion from youth to postmaturity. In old age, however, the detrital load is small, the dissection of the original mountain mass is far advanced, and the streams are easily diverted out of their channels. All of these factors tend to minimize lateral planation' (Bryan 1936, p. 775). Thus within the pediment association, different processes operate at different rates and times in different parts of the system and the distribution of these processes changes in space through time.

(e) As an open geomorphic system, the pediment association is 'a complex multivariate phenomenon which responds at different rates to different external variables such that it need not be in equilibrium with all variables at once' (Palmquist 1975, p. 155). Thus, the various components of the pediment association are not necessarily adjusted to one another or to the system as a whole. Rather, as interdependent components of an evolving open system, the progressively degrading uplands and aggrading alluvial plains of the pediment association are examples of landforms that are not necessarily attracted toward a steady state. 'For many landforms, height, volume, or other dimensions change progressively with time instead of tending toward a time-independent size or configuration' (Bull 1975, p. 112).

(f) Pediments may be formed by many different combinations of processes operating within and constrained by a variety of tectonic, climatic, and lithologic/structural environments. Moreover, as slowly developing long-lived landforms, pediments may be subjected to substantial changes in these environmental factors as they develop. Thus many pediments may be, in part at least, forms inherited from former conditions. This possibility severely limits the profitable application of morphometric analysis to the question of pediment development. Unless a high degree of similarity of initial size and shape and of subsequent morphogenetic history can be demonstrated, morphological differences between pediments or their associated uplands cannot be used to develop general inferences about range degradation or pediment formation.

Factors Affecting Pediment Formation and Expansion

A model of pediment development which applies equally well to all possible geomorphic situations is likely to be either oversimplified with respect to any specific situation or unmanageably complex if all possible situations are analysed in detail; consequently, the following summary focuses on piedmont pediments in arid and semi-arid environments.

(a) As piedmont surfaces, pediments serve as zones of transition (and transport) between uplands and alluvial plains. In arid and semi-arid regions, the redistribution of mass between the degrading upland (where stream power generally exceeds critical power) and the aggrading alluvial plain (where stream power is generally less than critical power) proceeds mainly by physical processes. Erosion within the upland is driven primarily by the action of concentrated water flow, development of integrated drainage networks from rills to rivers and the concomitant formation and expansion of valleys and embayments (Lustig 1969, p. D67, Wallace 1978, Bull 1979). The more diffuse and uniformly distributed processes of hillslope weathering and erosion operate within this general geomorphic framework of fluvially carved valleys and embayments. Moreover, the great bulk of material removed from the mountain mass is transported via streamflow, and debris transport across the pediment surface as well as deposition on the alluvial plain are likewise dominated by fluvial action.

(b) Long-term geomorphic stability appears to be a fundamental requirement for the formation and maintenance of extensive bedrock pediments. Within the south-western United States, pediments are larger and more abundant in areas of low vertical tectonic activity where base level is either stable or slowly rising. Such conditions facilitate the approximate long-term balance between erosional and depositional processes on piedmonts that is essential for pediment

development. 'The pediment, instead of being abnormal and restricted to certain localities, is widespread throughout ... [the desert regions of western North America] and is the type of plain normally developed during quiescent periods. Alluvial fans on the other hand, probably cannot be made under static conditions. They are built where normal gradients have been changed by faulting, warping, lateral erosion, or other special causes' (Blackwelder 1929, p. 168).

(c) Pediment formation is unidirectional through time. 'When tectonic stability exists, a geomorphic system is partially closed to materials. ... Thus the [general] behavior of the system during denudation becomes predictable; a continued decrease in relief must occur. Therefore, a dynamic equilibrium within a drainage basin or a hillslope as a whole cannot exist for material flux except when the rate of uplift equals the rate of denudation. ... The imbalance in material flux means that changes in the size, elevation or form of the system must occur' (Palmquist 1975, pp. 155–6). Although individual components of the pediment association (e.g. hillslopes, valley bottoms, stream channels, piedmont junctions, etc.) may be generally adjusted to lithology and process over the short term; each component and its relations to the other components change systematically and asymmetrically over the long term. As a fluvially dominated open system, the pediment association is attracted towards a stationary state; it is 'indeed directed through time' (Montgomery 1989, p. 51).

(d) Pediment formation (expansion, maintenance, and modification) varies in time and space, both non-linearly and discontinuously. At any point in time, processes of pediment formation vary in space (Sharp 1940). For example, as a pediment develops and expands along a mountain front, it is also maintained as a surface of transportation in more distal areas. Expansion involves fluvial processes (in channels, valleys, and embayments) and hillslope processes (on valley and embayment sideslopes and on mountain front interfluves). Maintenance involves weathering, sheet flow, and streamflow processes on the piedmont. Moreover, the relative significance of these processes also changes over time (Bryan 1936). Initially when the mountains are large and the piedmonts narrow, upland degradation predominates; eventually as the mountains become smaller and the piedmonts broader, pediment maintenance and regrading dominate. The rates at which these processes operate also vary (both in time and space). Degradation in response to uplift is initially rapid and gradually slows (Morisawa 1975). Other factors being equal, larger streams generally erode more rapidly and, therefore, occupy larger deeper valleys with lower gradients than do smaller streams (Bull 1979). Consequently, the effects of fluvial action are seldom distributed uniformly along the mountain front. The mountain front is distinctly non-linear in plan (Lawson 1915) and, in most instances, does not retreat parallel to itself (Parsons and Abrahams 1984).

(e) The retreat of mountain front spurs appears to be less important with regard to the evolution of mountain fronts (and pediment development) than fluvial downwearing and embayment development (Lustig 1969, p. D67, Bull 1984). In the absence of significant fluvial erosion, tectonically inactive mountain fronts may remain relatively unchanged over periods of as much as several million years (Dohrenwend et al. 1985, Mayer 1986, Harrington and Whitney 1991). Strongly asymmetric ranges clearly illustrate the dominance of fluvial erosion on mountain front degradation. The small subparallel drainages along the scarp flank erode relatively slowly inducing little change; whereas the large integrated drainages of the backtilted flank erode rapidly inducing drainage basin development, valley widening, embayment enlargement, valley sideslope convergence and spur decline. The presence of deeply dissected, partly stripped remnants of older, higher, and more steeply sloping relic surfaces along many mountain fronts also argues against general mountain front retreat as a primary mode of mountain mass reduction.

(f) Medial piedmonts are very stable and often persist with little dissection or regrading over periods of several million years (Dohrenwend et al. 1985, 1986). Although the threshold of critical power in streams may be crossed abruptly, a generally long-term balance between stream power and critical power appears to be maintained across a considerable breath of many piedmont surfaces. Where a drainage system is able to adjust channel form and roughness in response to limited variations in discharge and sediment load, it may accommodate such changes while at the same time continuing to operate at the critical power threshold. Also where the piedmont is broad, changes in discharge or load can be accommodated, at least in part, by upslope or downslope migration of the erosion-transport and transport–deposition limits of the critical power threshold (Cooke and Mason 1970). The stability of the medial piedmont implies that it is the local base level for mountain degradation, proximal piedmont dissection, and pediment formation.

(g) The critical requirement for pediment formation (and maintenance) is that all piedmont drainage must operate, on average, at or very close to the threshold of critical power across a broad zone of transition between upland and lowland. Within this zone, episodes of dissection and aggradation must be limited in both time and space. Because pediment formation is a slow process that is largely concentrated along the junction between upland and piedmont, pediment formation and maintenance must proceed simultaneously in different parts of the system. Upland slopes may retreat and/or decline by a variety of processes, but conditions must be maintained such that an expanding pediment replaces the shrinking mountain mass. Embayment formation and expansion, inselberg development, pediment formation and maintenance all result from the erosional–depositional balance of the medial piedmont.

(h) Although long-term change is relatively uniformly distributed across the pediment; short-term changes may be non-uniformly and 'chaotically' distributed. The precise character of these short-term changes is determined in part by the system's ability to 'self regulate' responses to short-term variations in discharge or sediment load. This self regulation is significantly influenced by the availability of readily erodible and transportable materials. If such materials are abundant and uniformly distributed across the piedmont, the tendency for lateral erosion usually exceeds the tendency for vertical incision and the system will regrade uniformly in space without significant dissection. However, if these materials are not abundant and/or are not uniformly distributed, the tendency for vertical incision may equal or exceed the tendency for lateral erosion and the system will very likely regrade locally and discontinuously.

(i) Proximal dissection is a common characteristic of both pediments and alluviated piedmonts. This regrading occurs in response to a sustained increase in stream power (duration and/or intensity of flow) or a sustained decrease in critical power (amount and/or calibre of available load). Possible causes include short-term climate changes (Bryan 1922), long-term undirectional climate change, tectonic tilting of the piedmont, complex response of the upland–proximal-piedmont drainage system (Rich 1935, Denny 1967), and isostatic adjustment of the piedmont association (Howard 1942, Cooke and Mason 1973). Over the long term, however, the most likely cause would appear to be the redistribution of mass within the pediment association. Regrading is constrained by the stability of the medial piedmont; if the medial piedmont is undissected and morphologically stable then proximal dissection must be the result of change within the adjacent upland. As uplands diminish in size, both stream discharge and sediment load decrease but the effect of the decrease in load appears to predominate.

It is a general misconception that most alluviated piedmonts are primarily if not entirely depositional landforms. In fact after initial formation, even bajadas may be as much zones of regrading and transport as they are loci of deposition. Proximal dissection is common on both pediments and bajadas, and the tendency for dissection by piedmont drainage underscores the delicate erosional–depositional balance of many proximal and medial piedmont areas. Field measurement of badlands dissection (Schumm 1956, 1962) and comparison of the relative positions of adjacent modern and relict piedmont surfaces capped by late Tertiary lava flows (Dohrenwend et al. 1985, 1986) also document the erosional predilection of proximal piedmont areas.

(j) Subsurface weathering undoubtedly facilitates pediment formation via conversion of resistant bedrock into non-resistant regolith that can be disaggregated into readily loosened and easily transported detritus. However, it would appear unlikely that the operation of sursurface weathering processes actually controls pediment development

in those situations where the overall mode of landscape evolution proceeds via upland retreat and replacement by expanding piedmonts (e.g. Bryan 1922, Johnson 1932b, Lustig 1969, p. D67, Parsons and Abrahams 1984). As the preservation of tens of metres of intensely weathered bedrock beneath granitic pediments in the south-western United States clearly demonstrates, many pediment surfaces are not closely related to either the position or the form of the subsurface weathering front. Rather, the existence of an alluviated piedmont, pediment, or similar surface of low relief would seem to be a required precondition for development of a quasi-planar weathering front and for subsequent planation of its deeply weathered regolith mantle. Hence, exhumation of the weathering front, mantle-controlled planation, and other similar processes are probably most effective where the overall mode of landscape change is one of stepwise weathering and stripping in a region of low relief (e.g. Mabbutt 1966).

Conclusion

There is a large literature on pediments. This literature has provided a wealth of information on the form of pediments and the settings in which they are found. We know much about the character and variability of pediments. In contrast, information about pediment processes is notably less. Even setting aside the difficult problem of the relationship of contemporary processes to pediment development, knowledge of these processes, *per se*, and their significance for contemporary sediment budgets in deserts is weak, and much needs to be done. In large measure, the process paradigm of geomorphology in the second half of the twentieth century appears to have passed pediments by. The origin of pediments and their long-term evolution has been the subject of much speculation, but this speculation has remained rooted in the approach that characterised geomorphology of the early part of the twentieth century. The problem is an immense one. Such evidence as we have indicates that pediment evolution is slow. Consequently, these landforms must have developed through times of significantly varying climate. In the study of badlands, Howard (1997) uses a model of drainage-basin evolution (Howard, 1994) to simulate landscape

evolution. There is concordance between the output of
the model and the real landscape, but he notes that this
concordance should be regarded as preliminary. 'The
simulation model process rate laws and model parameters have not been directly validated and calibrated
by field observation. Long term process observations
would be valuable. In particular, rates of erosion in
bedrock rills and channels could be measured and related to drainage area, channel gradient and rainfall history to calibrate an erosion law' (p. 224). Although the
problem is a simpler one, inasmuch as badlands evolve
rapidly and the vicissitudes of climate can reasonably
be ignored, the approach would seem to provide a way
out of the current *impasse* in advancing understanding
of pediment evolution (e.g. Strudley et al., 2006). Suitably constrained by data on rates of sediment movement and downwearing (e.g. Nichols et al., 2005) numerical modelling of this type could provide a way of
testing the various models that have been proposed for
pediment development.

References

Applegarth, M.T. 2004. Assessing the influence of mountain
slope morphology on pediment form, south-central Arizona.
Physical Geography **25**, 225–236.

Blackwelder, E. 1929. Origin of the piedmont plains of the Great
Basin, *Bulletin of the Geological Society of America* **40**,
168–9.

Blackwelder, E. 1931. Desert plains. *Journal of Geology* **39**,
133–40.

Bourne, J.A. and C.R. Twidale 1998. Pediments and alluvial
fans: genesis and relationships in the western piedmot of
the Flinders Ranges, South Australia. *Australian Journal of
Earth Sciences* **45**, 123–135.

Bryan, K. 1922. Erosion and sedimentation in the Papago country, Arizona. *U.S Geological Survey Bulletin* **730**, 19–90.

Bryan, K. 1936. The formation of pediments. *16th International
Geological Congress Report*, Washington, DC (1933), 765–
75.

Bryan, K. and F.T. McCann 1936. Successive pediments and terraces of the upper Rio Puerco in New Mexico. *Journal of
Geology* **44**, 145–72.

Bull, W.B. 1975. Landforms that do not tend toward a steady
state. In *Theories of landform development*, W.N. Melhorn
and R.C. Flemal (eds.), 111–28. Binghamton, NY: Publications in Geomorphology.

Bull, W.B. 1977. The alluvial fan environment. *Progress in Physical Geography* **1**, 222–70.

Bull, W.B. 1979. Threshold of critical power in streams. *Bulletin
of the Geological Society of America Part I* **90**, 453–64.

Bull, W.B. 1984. Tectonic geomorphology. *Journal of Geological Education* **32**, 310–24.

Bull, W.B. and L.D. McFadden 1977. Tectonic geomorphology
north and south of the Garlock fault, California. In *Geomorphology in arid regions*, D.C. Doehring (ed.), 115–37. Binghamton, NY: Publications in Geomorphology.

Carr, W.J. 1984. Regional structural setting of Yucca Mountain,
southwestern Nevada, and late Cenozoic rates of tectonic activity in part of the southwestern Great Basin, Nevada and
California. *US Geological Survey Open-File Report*, OFR-
84–854.

Chorley, R.J., S.A. Schumm and D.E. Sugden 1984. *Geomorphology*. London: Methuen.

Cooke, R.U. 1970. Morphometric analysis of pediments and associated landforms in the western Mojave Desert, California.
American Journal of Science **269**, 26–38.

Cooke, R.U. and P. Mason 1973. Desert Knolls pediment and
associated landforms in the Mojave Desert, California. *Revue
de Geomorphologie Dynamique* **20**, 71–8.

Cooke, R.U. and R.W. Reeves 1972. Relations between debris
size and the slope of mountain fronts and pediments in the
Mojave Desert, California. *Zeitschrift für Geomorphologie*
16, 76–82.

Cooke, R.U. and A. Warren 1973. *Geomorphology in deserts*.
Berkeley, CA: University of California Press.

Davis, W.M. 1933. Granitic domes in the Mohave Desert, California. *San Diego Society of Natural History Transactions* **7**,
211–58.

Denny, C.S. 1967. Fans and pediments. *American Journal of Science* **265**, 81–105.

Dinsmoor, W.B. 1975. *The architecture of Ancient Greece,
an account of its historic development*. New York:
Norton.

Dohrenwend, J.C. 1982a. Surficial geology, Walker Lake 1°
by 2° quadrangle, Nevada–California. *US Geological Survey Miscellaneous Field Studies Map*, MF-1382-C, scale
1:250,000.

Dohrenwend, J.C. 1982b. Late Cenozoic faults, Walker Lake 1°
by 2° quadrangle, Nevada–California. *US Geological Survey Miscellaneous Field Studies Map*, MF-1382-D, scale
1:250,000.

Dohrenwend, J.C. 1982c. Tectonic control of pediment distribution in western Great Basin. *Geological Society of America
Abstracts with Programs* **14**, 161.

Dohrenwend, J.C. 1987a. The Basin and Range. In *Geomorphic
systems of North America*, W. Graf (ed.), 303–42. Boulder,
CO.: Geological Society of America, The Geology of North
America, Centennial Special Volume 2.

Dohrenwend, J.C. 1987b. Morphometric comparison of tectonically defined areas within the west-central Basin and Range.
US Geological Survey Open-File Report 87–83.

Dohrenwend, J.C. 1990a. Buffalo Valley volcanic field, Nevada.
In *Volcanoes of North America*, C.A. Wood and J. Kienle
(eds.), 256–7. Cambridge: Cambridge University Press.

Dohrenwend, J.C. 1990b. Cima, California. In *Volcanoes of
North America*, C.A. Wood and J. Kienle (eds.), 240–1. Cambridge: Cambridge University Press.

Dohrenwend, J.C. 1990c. Clayton Valley, Nevada. In *Volcanoes
of North America*, C.A. Wood and J. Kienle (eds.), 261. Cambridge: Cambridge University Press.

Dohrenwend, J.C. 1990d. Lunar Crater, Nevada. In *Volcanoes
of North America*, C.A. Wood and J. Kienle (eds.), 258–9.
Cambridge: Cambridge University Press.

Dohrenwend, J.C. 1990e. Reveille Range, Nevada. In *Volcanoes of North America*, C.A. Wood and J. Kienle (eds.), 260–1. Cambridge: Cambridge University Press.

Dohrenwend, J.C., B.D. Turrin and M.F. Diggles 1985. Topographic distribution of dated basaltic lava flows in the Reveille Range, Nye County, Nevada: implications for late Cenozoic erosion of upland areas in the Great Basin. *Geological Society of America Abstracts with Programs* **17**, 351.

Dohrenwend, J.C., S.G. Wells, L.D. McFadden and B.D. Turrin 1986. Pediment dome evolution in the eastern Mojave Desert, California. In *International geomorphology*, V. Gardiner (ed.), 1047–62. Chichester: Wiley.

Dohrenwend, J.C., W.B. Bull, L.D. McFadden, G.I. Smith 1991. Quaternary Geology of the Basin and Range province in California. In *Quaternary nonglacial geology: conterminous United States*, Morrison, R.B. (ed.), 321–52. Boulder, CO.: Geological Society of America, The Geology of North America, K-2.

Dohrenwend, J.C., G. Yanez and G. Lowry 1995. Cenozoic landscape evolution of the southern part of the Gran Sabana Sur, Southeastern Venezuela – Implications for the occurrence of gold and diamond placers: U.S. Geological Survey Bulletin 2124-K, K1–K17.

Dohrenwend, J.C., R.C. Jachens, B.C. Moring and P.C. Schruben 1996a. An analysis of Nevada's metal-bearing mineral resources, Chapter 8. Indicators of subsurface basin geometry. In *An analysis of Nevada's metal-bearing mineral resources: Nevada Bureau of Mines and Geology Open File, Report 96-2, p. 8–1 to 8–8 and Plate 8.1*, Singer, D.A. (ed.), (scale 1:1,000,000).

Dohrenwend, J.C., B.A. Schell, C.M. Menges, B.C. Moring and M.A. McKittrick 1996b. An analysis of Nevada's metal-bearing mineral resources, Chapter 9. Reconnaissance photogeologic map of young (Quaternary and late Tertiary) faults. In *An analysis of Nevada's metal-bearing mineral resources: Nevada Bureau of Mines and Geology Open File, Report 96-2, p. 9–1 to 9–12 and Plate 9.1*, Singer, D.A. (ed.), (scale 1:1,000,000).

Dresch, J. 1957. Pediments et glacis d'erosion, pediplains et inselbergs. *Information Géographique* **21**, 183–96.

Ekren, E.B., C.L. Rogers and G.L. Dixon 1973. Geologic and Bouguer gravity map of the Reveille Quadrangle, Nye County, Nevada. *US Geological Survey Miscellaneous Geologic Investigations Map* I-806, scale 1:48,000.

Field, R. 1935. Stream carved slopes and plains in desert mountains. *American Journal of Science* **29**, 313–22.

Gilbert, C.M. and M.W. Reynolds 1973. Character and chronology of basin development, western margin of the Basin and Range province. *Bulletin of the Geological Society of America* **84**, 2489–510.

Gilbert, G.K. 1877. *Report on the geology of the Henry Mountains*. US Geographical and Geological Survey of the Rocky Mountain Region. Washington, DC: U.S. Department of the Interior.

Gillespie, A.R. 1990. Big Pine, California. In *Volcanoes of North America*, C.A. Wood and J. Kienle (eds.), 236–7. Cambridge: Cambridge University Press.

Gilluly, J. 1937. Physiography of the Ajo region, Arizona. *Bulletin of the Geological Society of America* **48**, 323–48.

Hadley, R.F. 1967. Pediments and pediment-forming processes. *Journal of Geological Education* **15**, 83–9.

Harrington, C.D. and J.W. Whitney 1991. Quaternary erosion rates on hillslopes in the Yucca Mountain region, Nevada. *Geological Society of America Abstracts with Programs* **23**, A118.

Howard, A.D. 1942. Pediment passes and the pediment problem. *Journal of Geomorphology* **5**, 3–32, 95–136.

Howard, A.D. 1994. A detachment-limited model of drainage basin evolution. *Water Resources Research* **30**, 2261–2285.

Howard, A.D. 1997. Badland morphology and evolution: interpretation using a simulation model. *Earth Surface Processes and Lanforms* **22**, 211–227.

Janson, H.W. 1969. *History of art: a survey of the major visual arts from the dawn of history to the present day*. Englewood Cliffs, NJ: Prentice-Hall.

Johnson, D.W. 1931. Planes of lateral corrasion. *Science* **73**, 174–7.

Johnson, D.W. 1932a. Rock planes of arid regions. *Geographical Review* **22**, 656–65.

Johnson, D.W. 1932b. Rock fans of arid regions. *American Journal of Science* **23**, 389–416.

Kesel, R.H. 1977. Some aspects of the geomorphology of inselbergs in central Arizona, USA. *Zeitschrift für Geomorphologie* **21**, 119–46.

Lawson, A.C. 1915. The epigene profiles of the desert. *University of California Department of Geology Bulletin* **9**, 23–48.

Lustig, L.K. 1969. Trend surface analysis of the Basin and Range province, and some geomorphic implications. *US Geological Survey Professional Paper* 500-D.

Mabbutt, J.A. 1966. Mantle-controlled planation of pediments. *American Journal of Science* **264**, 78–91.

Mammerickx, J. 1964. Quantitative observations on pediments in the Mojave and Sonoran deserts (Southwestern United States). *American Journal of Science* **262**, 417–35.

Marchand, D.E. 1971. Rates and modes of denudation, White Mountains, eastern California. *American Journal of Science* **270**, 109–35.

Mayer, L. 1986. Tectonic geomorphology of escarpments and mountain fronts. In *Active tectonics*, R.E. Wallace (ed.), 125–35. Washington, DC: National Academy Press.

Mayer, L., M. Mergner-Keefer and C.M. Wentworth 1981. Probability models and computer simulation of landscape evolution. *US Geological Survey Open-File Report* 81–656.

Melton, M.A. 1965. The geomorphic and paleoclimatic significance of alluvial deposits in southern Arizona. *Journal of Geology* **73**, 1–38.

Menges, C.M. and L.D. McFadden 1981. Evidence for a latest Miocene to Pliocene transition from Basin-Range tectonic to post-tectonic landscape evolution in southeastern Arizona. *Arizona Geological Society Digest* **13**, 151–60.

Montgomery, K. 1989. Concepts of equilibrium and evolution in geomorphology: the model of branch systems. *Progress in Physical Geography* **13**, 47–66.

Morisawa, M. 1975. Tectonics and geomorphic models. In *Theories of landform development*, W.N. Melhorn and R.C. Flemal (eds.), 199–216. Binghamton, NY: Publications in Geomorphology.

Moss, J.H. 1977. Formation of pediments: scarp backwearing of surface downwasting? In *Geomorphology in arid regions*, D.O. Doehring (ed.), 51–78. Binghamton, NY: Publications in Geomorphology.

Nichols, K.K., P.R. Bierman, R. LeB Hooke, E.M. Clapp and M. Caffee 2002. Quantifying sediment transport on desert piedmonts using ^{10}Be and ^{26}Al. *Geomorphology* **45**, 105–125.

Nichols, K.K., P.R. Bierman, M.C. Eppes, M. Caffee, R. Finkel and J. Larsen 2005. Late Quaternary history of the Chemehuevi Mountain piedmont, Mojave Desert, deciphered using ^{10}Be and ^{26}Al. *American Journal of Science* **305**, 345–368.

Nichols, K.K., P.R. Bierman, M. Caffee, R. Finkel and J. Larsen 2005. Cosmogenically enabled sediment budgeting. *Geology* **33**, 133–136.

Oberlander, T.M. 1972. Morphogenesis of granitic boulder slopes in the Mojave Desert, California. *Journal of Geology* **80**, 1–20.

Oberlander, T.M. 1974. Landscape inheritance and the pediment problem in the Mojave Desert of southern California. *American Journal of Science* **274**, 849–75.

Oberlander, T.M. 1989. Slope and pediment systems. In *Arid zone geomorphology*, D.S.G. Thomas (ed.), 56–84. London: Belhaven.

Oberlander, T.M. 1997. Slope and pediment systems. In *Arid Zone Geomorphology*, D.S.G. Thomas (ed.), 135–163. Chichester: John Wiley & Sons.

Paige, S. 1912. Rock-cut surfaces in the desert regions. *Journal of Geology* **20**, 442–50.

Palmquist, R.C. 1975. The compatibility of structure, lithology and geomorphic models. In *Theories of landform development*, W.N. Melhorn and R.C. Flemal (eds.), 145–68. Binghamton, NY: Publications in Geomorphology.

Parsons, A.J. and A.D. Abrahams 1984. Mountain mass denudation and piedmont formation in the Mojave and Sonoran Deserts. *American Journal of Science* **284**, 255–71.

Rahn, P.H. 1967. Sheetfloods, streamfloods, and the formation of pediments. *Annals of the Association of American Geographers* **57**, 593–604.

Rich, J.L. 1935. Origin and evolution of rock fans and pediments. *Bulletin of the Geological Society of America* **46**, 999–1024.

Ruxton, B.P. 1958. Weathering and sub-surface erosion in granites at the piedmont angle, Balos, Sudan. *Geological Magazine* **95**, 353–77.

Saunders, I. and A. Young 1983. Rates of surface processes on slopes, slope retreat and denudation. *Earth Surface Processes and Landforms* **8**, 473–501.

Schumm, S.A. 1956. The role of creep and rainwash on the retreat of badland slopes. *American Journal of Science* **254**, 693–706.

Schumm, S.A. 1962. Erosion on miniature pediments in Badlands National Monument, South Dakota. *Bulletin of the Geological Society of America* **73**, 719–24.

Schumm, S.A. 1975. Episodic erosion: a modification of the geomorphic cycle. In *Theories of landform development*, W.N. Melhorn and R.C. Flemal (eds.), 69–86. Binghamton, NY: Publications in Geomorphology.

Schumm, S.A. and R.W. Lichty 1965. Time, space and causality in geomorphology. *American Journal of Science* **263**, 110–9.

Sharp, R.P. 1940. Geomorphology of the Ruby–East Humboldt Range, Nevada. *Bulletin of the Geological Society of America* **51**, 337–72.

Sharp, R.P. 1957. Geomorphology of Cima Dome, Mojave Desert, California. *Bulletin of the Geological Society of America* **68**, 273–90.

Smart, J., K.G. Grimes, H.F. Doutch and J. Pinchin 1980. The Carpentaria and Karumba Basins, north Queensland. *Australian Bureau of Mineral Resources, Geology and Geophysics Bulletin* 202.

Stewart, J.H. and J.E. Carlson 1977. Geologic map of Nevada. *Nevada Bureau of Mines and Geology Map 57*, scale 1:500,000.

Strudley, M.W., A.B. Murray and P.K. Haff 2006. Emergence of pediments, tors, and piedmont junctions from a bedrock weathering-regolith thickness feedback. *Geology* **34**, 805-808.

Tator, B.A. 1952. Pediment characteristics and terminology (part I). *Annals of the Association of American Geographers* **42**, 295–317.

Tator, B.A. 1953. Pediment characteristics and terminology (part II). *Annals of the Association of American Geographers* **43**, 47–53.

Thomas, M.F. 1974. *Tropical geomorphology*. New York: Wiley.

Tricart, J. 1972. *Landforms of the humid tropics, forests and savannas*. London: Longman.

Tuan, Yi-Fu 1959. Pediments in southeastern Arizona. *University of California Publications in Geography* **13**.

Turrin, B.D. and J.C. Dohrenwend 1984. K-Ar ages of basaltic volcanism in the Lunar Crater volcanic field, northern Nye County, Nevada: implications for Quaternary tectonism in the central Great Basin. *Geological Society of America Abstracts with Programs* **16**, 679.

Turrin, B.D., J.C. Dohrenwend, R.E. Drake and G.H. Curtis 1985. Potassium–argon ages from the Cima volcanic field, eastern Mojave Desert, California. *Isochron West* **44**, 9–16.

Turrin, B.D., D. Champion and R.J. Fleck 1991. ^{40}Ar/^{39}Ar age of Lathrop Wells volcanic center, Yucca Mountain, Nevada. *Science* **253**, 654–7.

Twidale, C.R. 1967. Hillslopes and pediments in the Flinders Ranges, South Australia. In *Landform studies from Australia and New Guinea*, J.N. Jennings and J.A. Mabbutt (eds.), 95–117. Cambridge: Cambridge University Press.

Twidale, C.R. 1983. Pediments, peniplains, and ultiplains. *Revue de Geomorphologie Dynamique* **32**, 1–35.

Wallace, R.E. 1978. Geometry and rates of change of fault-generated range fronts, north-central Nevada. *US Geological Survey Journal of Research* **6**, 637–50.

Warnke, D.A. 1969. Pediment evolution in the Halloran Hills, central Mojave Desert, California. *Zeitschrift für Geomorphologie* **13**, 357–89.

Whitaker, C.R. 1979. The use of the term 'pediment' and related terminology. *Zeitschrift für Geomorphologie* **23**, 427–39.

Wilshire, H.G. and S.L. Reneau 1992. Geomorphic surfaces and underlying deposits of the Mohave Mountains piedmont, lower Colorado River, Arizona. *Zeitschrift für Geomorphologie* **36**, 207–26.

Zoback, M.L., R.E. Anderson and G.A. Thompson 1981. Cenozoic evolution of the state of stress and style of tectonism of the Basin and Range province of western United States. *Philosophical Transactions of the Royal Society of London, Series A* **300**, 407–34.

Chapter 14

Processes and Forms of Alluvial Fans

Terence C. Blair and John G. McPherson

Introduction

Alluvial fans are a conspicuous conical landform commonly developed where a channel emerges from a mountainous catchment to an adjoining valley (Figs. 14.1 and 14.2). Although present in perhaps all global climates, fans in deserts have been the most studied due to their excellent exposure and ease of access. Drew (1873), working in the upper reaches of the Indus River valley in the western Himalaya of India, provided the earliest illustrations and scientific description of desert alluvial fans (pp. 445–447):

> The accumulations to which I give this name [alluvial fans] are of great prevalence in Ladākh, and are among the most conspicuous forms of superficial deposits. They are found at the mouths of side-ravines, where they debouch into the plain of a wider valley... The radii of the fans are about a mile long; the slope of the ground along these radii (which are each in the direction of greatest slope) is five or six degrees. The fan is properly a flat cone, having its apex at the mouth of the ravine... The mode of formation is not difficult to trace. Granting the stream of the side-ravine to be carrying down such an amount of detritus as to cause it to be accumulating, rather than a denuding stream, and there being such a relation between the carrying power of the water and the size of the material as to allow of this remaining at a marked slope, we have before us all of the conditions necessary.

Scientific publications on desert alluvial fans pre-dating 1960 are few, but have increased exponentially in number since then. The spectacular fans of west-ern North America have been the dominant focus of research, but notable work on fans has also increased in other arid and semi-arid regions, including in Peru, Argentina, Chile, southern Europe, east-central Africa, the Middle East, Iran, Pakistan, India, China, and Mongolia. The growth of fan research has been fuelled by multi-disciplinary needs, including for environmental and geological hazards mitigation (neotectonics, slope stability, flood control, urban planning, groundwater cleanup, hazardous waste disposal), civil engineering (construction of highways, dams, and other infrastructure), delineating groundwater resources, understanding climate change, archeological studies, petroleum geology and rock-record interpretation, mining of aggregate and ores, and continued basic research in geomorphology, sedimentology, hydrogeology, and engineering geology. More recently, desert-like alluvial fans imaged on Mars have created a challenge to understand the fan-forming surficial processes of other planets (e.g. Moore and Howard 2005).

This chapter provides: (a) an up-to-date synthesis of the literature on desert alluvial fans, (b) a framework for understanding fan processes, form, and evolution, and (c) a discussion of the issues that plague fan research. The latter point is especially critical because 'alluvial fan' is used unscientifically by a faction of authors for virtually any subaerial environment on an arbitrary basis. This synthesis thus emphasizes fundamental concepts on the processes and forms of fans, many of which are exemplified by our case studies in the southwestern USA. Though illustrated by fans in deserts, the concepts provided herein also are applicable to fans forming under other climates (Blair and McPherson 1994b).

T.C. Blair (✉)
Blair and Associates LLC, 1949 Hardscrabble Place, Boulder, CO 80305, USA
e-mail: tcblair@aol.com

A.J. Parsons, A.D. Abrahams (eds.), *Geomorphology of Desert Environments*, 2nd ed., DOI 10.1007/978-1-4020-5719-9_14, © Springer Science+Business Media B.V. 2009

Fig. 14.1 Aerial photographs of selected alluvial fans from California, including: (**a**) Trail Canyon fan of southwestern Death Valley, (**b**) Tuttle Canyon fan of southern Owens Valley, (**c**) Grotto Canyon fan of northern Death Valley, and (**d**) South Badwater fan in southeastern Death Valley

Fig. 14.2 Overview photographs of alluvial fans, including: (**a**) Trail Canyon fan of southwestern Death Valley, (**b**) Tuttle and Lone Pine Canyon fans of southern Owens Valley, (**c**) Grotto Canyon fan of northern Death Valley, (**d**) South Badwater fan of southeastern Death Valley, (**e**) small fan north of Badwa-ter, Death Valley, and (**f**) the Rifle Range fan near Hawthorne, Nevada. The Trail Canyon, South Badwater, North Badwater, and Rifle Range fans are bordered distally by playas, the Tut-tle and Lone Pine fans by an axial river, and the Grotto Canyon fan by an aeolian erg

General Features

Alluvial fans are aggradational sedimentary deposits shaped overall like a segment of a cone radiating downslope from a point where a channel emerges from a mountainous catchment (Drew 1873, Bull 1977) (Figs. 14.1, 14.2, and 14.3). Thus, fans constitute a sedimentary environment with a conspicuous mor-phology. Alluvial fans are arcuate in plan view, either spanning outward from the range front in a 180° semicircle, or having a more restricted pie-piece plan resulting from lateral coalescence with other fans to form a bajada. Cross-fan profiles display a plano-convex geometry, and radial profiles either exhibit a

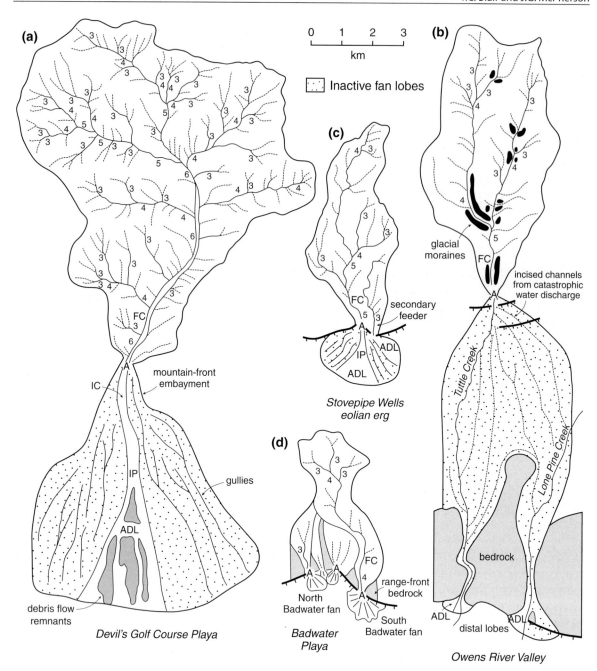

Fig. 14.3 Plan-view diagram of fans illustrated in Fig. 14.1 and their catchment drainage nets based on 1:24,000 topographic maps, aerial photographs, and fieldwork, including: (**a**) Trail Canyon fan, (**b**) Tuttle Canyon fan, (**c**) Grotto Canyon fan, and (**d**) Badwater fans. First-order channels are not depicted in the catchments due to scale. Second-order channels are dashed, and higher-order channels are solid lines labelled by order. The feeder channel (*FC*) of the catchment leads to the fan apex (*A*). Other labelled features are the fan incised channel (*IC*), intersection point (*IP*), and the active depositional lobe (*ADL*)

constant slope like a cone segment, or have half of a plano-concave-upwards geometry (Fig. 14.4). Fans typically extend 0.5–10.0 km from the mountain front (Anstey 1965, 1966), with larger fans reaching nearly 20 km (Blair, 2003). Desert vegetation on alluvial fans varies from sparse cacti and other xerophytes in hyperarid settings to more dense grass, shrubs, or trees in semi-arid regions.

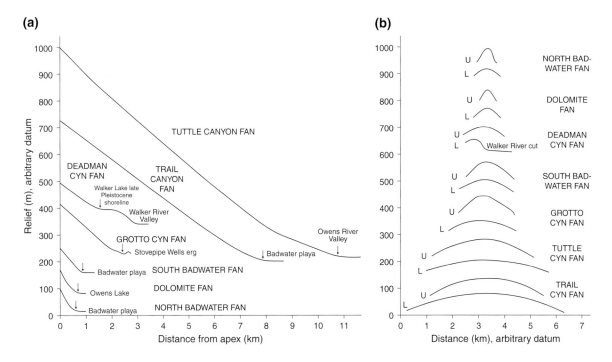

Fig. 14.4 Radial profiles (**a**) and cross profiles (**b**) of fans illustrated in Fig. 14.1 plus of the Dolomite fan of Owens Valley, California, and the Deadman Canyon fan of Walker Lake, Nevada.

Vertical exaggeration is 10X. Arrows above the radial profiles denote boundaries between the fans and neighbouring environments. Cross profiles illustrate upper (*U*) and lower (*L*) transects

Alluvial fans develop by the accumulation of sediment where a channel exits an upland drainage area (Drew 1873). Fans deposits typically are coarse-grained and poorly sorted due to the: (a) relatively short sediment-transport distance, (b) involvement of mass-wasting and flash-flood processes instigated by high relief, and (c) rapid loss of flow capacity in the piedmont (foot of mountain) zone. Fan deposits mainly consist of virgin (first-cycle) sediment that is produced through weathering and erosion of uplifted bedrock. Although fans can locally be dominated by fine sediment, they overall comprise some of the coarsest deposits on Earth. Their description, as used herein, thus requires grain-size terminology that details gravel and megagravel, with classes such as pebbles, cobbles, boulders, blocks, and megablocks defined by the length of the intermediate axis (d_I) of the clast (Fig. 14.5).

Fans comprise the most common element of a spectrum of high-sloping deposits found in the piedmont. Steep mountain-flanking deposits lacking fan morphology are called alluvial slopes (Hawley and Wilson 1965), and high-sloping deposits rimming volcanoes are called volcanic aprons. Wind-blown

deposits that mantle mountain fronts are termed aeolian ramps. Glacial drift and moraines may also occur in desert piedmont settings (e.g. Fig. 14.6a). Fans are bordered distally by volcanic, aeolian, fluvial, lacustrine, or marine environments (Fig. 14.3). They are easily differentiated from neighbouring river environments by their textures and characteristic fan form, including high radial slope values that contrast with the $< 0.5°$ slope of rivers (McPherson et al. 1987, Blair and McPherson 1994b).

Alluvial fans are directly linked with an upland catchment, also called a drainage basin, that constitutes the area from which water and sediment are discharged to a specific fan (Fig. 14.6b, c, and d). The area, relief, bedrock type, and other features may notably differ between even adjoining catchments. Catchments are separated from each other by drainage divides. Rather than being fixed, these divides, and thus catchment area, change with time due to dynamic factors such as headward erosion, stream piracy, and slope failures. Catchments of desert fans vary in elevation from ~sea level to over 6000 m, and, therefore, may be sparsely vegetated or thickly forested, and can contain valley glaciers and lakes. Thus, because of relief,

PARTICLE LENGTH (dl)				GRADE	CLASS	FRACTION	
km	m	mm	φ			Unlithified	Lithified
1075			−30	very coarse	Megalith	Megagravel	Mega-conglomerate
538			−29				
269			−28	coarse			
134			−27	medium			
67.2			−26	fine			
33.6			−25	very fine			
16.8			−24	very coarse	Monolith		
8.4			−23	coarse			
4.2			−22	medium			
2.1			−21	fine			
1.0	1048.6		−20	very fine			
0.5	524.3		−19	very coarse	Megablock		
0.26	262.1		−18	coarse			
	131.1		−17	medium			
	65.5		−16	fine			
	32.8		−15	very coarse	Block		
	16.4		−14	coarse			
	8.2		−13	medium			
	4.1	4096	−12	fine			
	2.0	2048	−11	very coarse	Boulder	Gravel	Conglomerate
	1.0	1024	−10	coarse			
	0.5	512	−9	medium			
	0.25	256	−8	fine			
		128	−7	coarse	Cobble		
		64	−6	fine			
		32	−5	very coarse	Pebble		
		16	−4	coarse			
		8	−3	medium			
		4	−2	fine			
		2	−1	very fine			
		1	0	very coarse	Sand	Sand	Sandstone
		0.50	1	coarse			
		0.25	2	medium			
		0.125	3	fine			
		0.063	4	very fine			
		0.031	5	coarse	Silt	Mud	Mudstone or Shale
		0.015	6	medium			
		0.008	7	fine			
		0.004	8	very fine			
		0.002	9		Clay		
		0.001	10				
		0.0005	11				
		0.0002	12				
		0.0001	13	?			

Fig. 14.5 Index of grain size terminology used in chapter (after Blair and McPherson 1999)

desert fans may have catchments with other microclimates such as alpine tundra.

All but incipient catchments have a converging drainage net with typically short channel segments. The catchment drainage net can be characterized by stream order (Horton 1945, Strahler 1957), with first-order channels (those lacking tributaries) and the area they drain located at divides between catchments, or between sectors of a single catchment. Second-order channels form farther downslope where two first-order channels join, third-order channels form where two second-order channels join, etc. (Fig. 14.3). Channel

Fig. 14.6 Photographs of various catchment slope materials. (**a**) View of terminal moraine (*M*) ∼500 m high deposited in the lower part of the Bishop Creek fan catchment and upper piedmont prior to deglaciation, Owens Valley, California. Glaciation widely stripped colluvium from the granodiorite bedrock comprising the catchment slopes. (**b**) View of Cartago fan of southern Owens Valley, with exfoliated granodiorite cliffs forming slopes in the upper catchment, and with colluvium mantling the lower slopes. (**c**) Interstratified limestone and shale bedrock extensively mantled by colluvium comprise the catchment slopes of the Lead Canyon fan, south-central New Mexico. (**d**) View of Titus Canyon fan, northern Death Valley, with a large catchment of both stratified carbonate bedrock and colluvial slopes. The fan possess a proximal incised channel (*arrow*) feeding a distal active depositional lobe (light coloured). (**e**) Exposure of colluvial deposits in the lower catchment of the Copper Canyon fan, Walker Lake, Nevada (geologist for scale). (**f**) Downslope view of the lower catchment of the Tuttle Canyon fan, Owens Valley. Bouldery glacial moraines (*M*) lead to the fan apex (*A*). A trail (*arrow*) is marked for scale

ordering requires topographic maps with sufficient detail to delineate the drainage net. Our studies show that maps can be no smaller than 1:24,000 scale and with 12.2 m (40 ft) contours to achieve an accurate depiction of all of the channels. The converging pattern of the catchment drainage net causes all channels to funnel to the highest-order channel, called the feeder channel, that leads to the fan. The feeder channel commonly is oriented at a high angle to the mountain front. Usually only one feeder channel is present, although some fans may have secondary feeders (e.g. Fig. 14.3c). Fan catchments in deserts typically have ephemeral first, second, or up to about eighth-order feeders. The feeder channel is perennial for some desert fans, such as the Lone Pine fan in Owens Valley, California (Blair 2002), where snowmelt or icemelt from high in the catchment sustain base flow.

The key elements of an alluvial fan are the apex, incised channel, intersection point, active depositional lobe, older surfaces, and headward-eroding gullies. The apex of a fan is the point at the mountain front where the feeder channel emerges from the catchment (Drew 1873). This point represents the most proximal and usually the highest part of the fan. The apex is obvious where the mountain front is sharp, but is less distinct where the feeder channel has carved an embayment. The incised channel, also called the fanhead trench (Eckis 1928), is a downslope extension of the catchment feeder channel onto the fan (Fig. 14.3). It usually is a single trunk stream that merges downslope with the fan surface. Incised channels are not always present or well developed, occurring most commonly on fans with longer radii, or those that have intra-fan fault scarps. They usually terminate in the proximal or medial part of the fan, but can extend to the distal margin in cases where distal-fan fault scarps are present, or where base level in adjoining environments has notably dropped. The down-fan position where an incised channel ends is called the intersection point (Hooke 1967). Flows through the incised channel laterally expand onto the fan surface at this point. The fan segment downslope from the intersection point is the site of sediment aggradation in an area termed the active depositional lobe (Fig. 14.3). The arc length of this lobe is a function of its radius and angle of expansion. Lobe expansion angles may be 180° on small fans, but more typically are between 15° and 90°. Old fan surfaces with varnished pavements typically are present lateral to the active depositional lobe or incised channel. Headward-eroding gullies are common features on the distal fan, particularly in inactive areas away from the active lobe. Headward erosion by these gullies, either as single channels or a downslope-converging network, may eventually progress sufficiently upslope to intersect the incised channel, possibly causing autocyclic switching of the active depositional lobe to another part of the fan (Denny 1967).

Conditions for Alluvial Fan Development

Three conditions necessary for optimal alluvial fan development are: (a) a topographic setting where an upland catchment drains to a valley, (b) sufficient sediment production in the catchment to construct the fan, and (c) a triggering mechanism, usually sporadic high water discharge and less commonly earthquakes, to incite the transfer of catchment sediment to the fan. The most common topographic setting for fans is marginal to uplifted structural blocks bounded by faults with significant dip-slip. An example of this setting is in the extensional Basin and Range province of the western USA, where an upland with adjoining lowland configuration is tectonically developed and maintained. Other topographic settings conducive to the development of alluvial fans are where tributary channels enter a canyon or valley (e.g. Drew 1873, Blair 1987a, Webb et al. 1987, Florsheim 2004), or where bedrock exposures possessing topographic relief form by differential erosion (Sorriso-Valvo 1988, Harvey 1990).

Sediment production in a catchment, the second condition for fan development, typically is met given time because of the presence of relief, and because of incessant weathering of rocks at the Earth's surface. Sediment yield from a catchment increases exponentially with relief due to the effect of gravity on slope erosion (Schumm 1963, 1977, Ahnert 1970). The types of rock weathering in desert catchments that produce sediment are: (a) physical disintegration, including fracturing, exfoliation, ice- or salt-crystal growth in voids, and root wedging, and (b) chemical alteration, encompassing reactions such as hydrolysis, dissolution, and oxidation (e.g. Ritter 1978). Weathering is greatly promoted along structurally controlled mountain fronts because of tectonic fracturing, which exposes significantly more rock surface area to alteration than in unfractured rocks. Thus, the catchments of alluvial fans along faulted mountain fronts overall are ideal sediment producers due to their location on a structural block where relief is maintained, and where tectonic fracturing is common. In contrast, catchments in non-tectonic settings, such as along paraglacial valleys, may have previously deposited sediment available for building a fan, although it may become depleted (e.g. Ryder 1971). Catchments developed on bedrock spurs created by variable erosion, or from tectonism that has ceased, may also generate fans for which sediment supply is limited. Fans in these latter settings may have processes similar to those formed in tectonic settings, but their evolutionary scenarios may differ due to the lack of maintained relief and sediment supply.

The third necessity for fan development is a mechanism to move catchment sediment to the fan. The key processes achieving this transport are related to water input and mass wasting. These processes are promoted by flood conditions resulting from heavy or prolonged precipitation, rapid icemelt or snowmelt, or the rapid release of impounded water due to failure of a natural dam (e.g. McGee 1897, Beaty 1963, 1990, Leggett et al. 1966, Caine 1980, Cannon and Ellen 1985, Wieczorek 1987, Costa 1988, Blair 2001, 2002). The topography and shape of fan catchments make them prone to generating catastrophic floods. Mountains induce precipitation by causing vertical airflow that triggers condensation (Houghton et al. 1975, Hayden 1988). Precipitation that falls in these catchments is quickly funnelled through the short segments of the converging drainage net to the feeder channel, giving rise to flows with the potential to move extremely coarse sediment (e.g. French 1987, Patton 1988, Blair and McPherson 1994b, 1999). Flash-flood potential is greatest in catchments with high relief, multiple high-order chan-

nels, and a rotund shape (Strahler 1957). Flash floods also are important for inducing mass-wasting events that rapidly increase sediment discharge and create new first-order channels (Patton 1988).

Net fan aggradation requires that discharge from the catchment loses competency and capacity upon reaching the fan. This loss results from: (a) a lessening of slope at the fan site, or (b) decreases in both flow depth and velocity due to lateral expansion caused by the loss of confining channel walls either at the apex, or at the intersection point. Deposition is instigated on many fans by a pronounced slope decrease (e.g. Trowbridge 1911, Beaty 1963), although such a change does not always exist (Bull 1977). This relationship is illustrated by comparing the slope of the 1-km-long segments of the lower feeder channel and upper fan (thus 1 km on either side of the apex) for 132 fans in Death Valley, California, using 1:24000-scale topographic maps. The slope of the lower feeder channel segment is significantly greater (> 1°) than the upper fan in 40% of these cases, is within 1° in 56% of the cases, and in 4% the upper fan is steeper than the lower feeder channel (Fig. 14.7). Thus, in 60% of these cases, fan aggradation must result from flow expansion rather than decreasing slope on the upper fan.

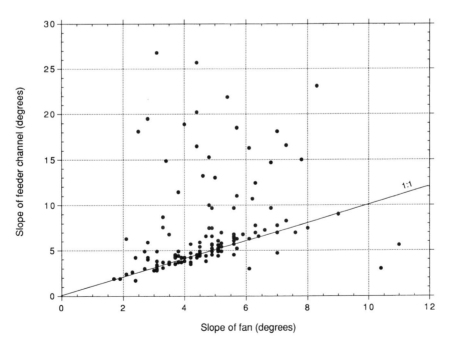

Fig. 14.7 Cross-plot of the slope of the 1-km-long segment of the feeder channel upslope from the fan apex versus the slope of the 1-km-long segment of the upper fan immediately downslope from the apex for 132 Death Valley fans based on 1:24,000 scale topographic maps

Not all channels issuing from mountain fronts construct alluvial fans. Perennial rivers with highly integrated catchments, such as the Kern, San Joaquin, and Feather Rivers of California, and the Arkansas and Platte Rivers of Colorado, maintain channel banks and flow competency. Instead of building fans at the piedmont, these rivers instead either maintain bedload-dominated fluvial tracts, or bypass the piedmont by incision. Alternatively, the piedmont linked to glaciated or recently deglaciated catchments may feature moraines or other till deposits instead of alluvial fans (Fig. 14.6a).

Given the necessary conditions for development, it follows that alluvial fans are most prevalent in active tectonic belts, where relief is maintained, and catastrophic discharge and sediment production are promoted. Of the tectonic belts, those characterized by normal and strike-slip faulting are most conducive to fan development, such as the Basin and Range Province of the western USA, the East African-Red Sea-Dead Sea rift system, piggyback basins of compressional settings such as in northern Chile, and escape zones of the western Himalaya. Relief also is tectonically developed in fold and thrust belts, though fans therein likely are more ephemeral due to the lateral instability of the range fronts.

Primary Processes on Alluvial Fans

Inasmuch as alluvial fans are aggradational deposits, their understanding requires a knowledge of the processes that transport sediment to and within this environment. Sedimentary processes active on alluvial fans are of two types called primary and secondary (Blair and McPherson 1994a, b). Primary processes are those that transport sediment from the catchment or range front to the fan. They include rockfalls, rock slides, rock avalanches, earth flows, colluvial slips, debris flows, incised-channel floods, and sheetfloods. Primary processes overall cause fan construction, and enlarge the catchment by sediment removal. They mainly are triggered by intense precipitation or earthquakes, and thus are mostly infrequent and of short duration, but have high impact with respect to fan aggradation. Secondary processes, in contrast, modify sediment previously deposited on a fan by any of the primary processes. They include overland flow, wind

erosion or deposition, bioturbation, soil development, weathering, faulting, and toe erosion. Secondary processes typically result in fan degradation, and, except for faulting and some overland flows, are associated with normal or noncatastrophic conditions. Although they are of limited importance to fan construction, secondary processes dominate the fan surface except in areas recently affected by a primary depositional event due to the infrequency of the latter (Blair 1987a).

All primary fan processes are instigated by failure of catchment slopes, and the downslope transfer of the destabilized material. Catchment slopes consist of two unique material types, bedrock and colluvium. Bedrock is usually present as cliffs or flanks that form steep (50–90°) slopes in the catchment (Figs. 14.2 and 14.6b, c, d). In contrast, colluvium constitutes the commonly gravel-dominated, unsorted to poorly sorted, fine to coarse sediment loosened from the bedrock cliffs through weathering, and deposited along the lower cliffs or in the catchment channels (Fig. 14.6c and e) (e.g. Drew 1873, Sharpe 1938, Rapp and Fairbridge 1968, Rahn 1986, Turner 1996). Slopes formed of colluvium commonly are near the angle of repose (30–40°), but range from 15° to 56° (Campbell 1975). Colluvium can be repositioned in a catchment by flows through the drainage net or, less commonly but more drastically, by glaciers to form moraines (Fig. 14.6a and f). Fan catchments may have slopes underlain principally by bedrock, principally by colluvium, or, most typically, by some of each.

Different primary processes are activated in a fan catchment depending on whether the failed slope material is of bedrock or colluvium, and on the transport mechanism that is instigated. Failed slope material can be transported from the catchment to the fan by: (a) fluid-gravity processes (i.e. water flows), in which colluvial particles are moved by the force of water, (b) sediment-gravity processes, in which colluvial particles and any contained fluids are transported by the force of gravity acting directly on the sediment, or (c) rock-gravity processes, in which commonly disintegrating bedrock is transported by the force of gravity acting directly on the bedrock (e.g. Middleton and Hampton 1976, Blair and McPherson 1999). Sediment may also be moved to the fan by ice-gravity processes, in which colluvium is pushed or carried by the force of gravity acting on glacial ice, and deposited as moraines or till (Derbyshire and Owen 1990, Blair and McPherson 1999, Blair 2001, 2002). Ice-gravity processes, however, are only noteworthy to piedmont sedimentation in highly glaciated settings, and commonly produce landforms other than fans (e.g. Fig. 14.6a). Primary fan processes thus can be grouped into three key types: (a) rock gravity flows generated by failure of bedrock slopes in the catchment, (b) sediment-gravity flows generated by failure of colluvial slopes in the catchment, and (c) fluid-gravity flows generated by failure of colluvial slopes in the catchment (Blair and McPherson 1994a, b).

Rock-Gravity Processes from Bedrock Slope Failures

Four types of rock-gravity processes instigated by destabilized bedrock comprising either the range front or fan catchment slopes are rockfalls, rock slides, rock avalanches, and earth flows. All four are initiated by failure, under the force of gravity, of bedrock usually exposed in the upper catchment. Rockfalls, rock slides, and rock avalanches represent a gradational spectrum of processes related to the brittle failure of bedrock, whereas earth flows constitute a more ductile style of failure by certain rock types.

Rockfalls, Rock Slides, and Rock Avalanches

The processes of rockfall, rock slide, and rock avalanche all derive from the progressive lowering of the internal friction and shear strength of brittle bedrock exposures due to: (a) fracturing and weathering, (b) slope steepening due to downcutting or undercutting, or (c) ground motion from earthquakes (Hadley 1964, Morton 1971, Keefer 1984, 1999, Plafker and Ericksen 1984, Statham and Francis 1986, Cotecchia 1987, Hencher 1987, Beaty and DePolo 1989, Harp and Keefer 1989, Sidle and Ochiai 2006). Unlike other primary processes, water is not relevant to the transport of rockfalls, rock slides, or rock avalanches. Conditions that promote these processes are where: (a) range-front faults are splayed, and bedrock blocks are tectonically sized, (b) syntectonic or inherited fracture patterns in an uplifted mass are oriented nearly parallel to the mountain front or feeder channel, (c) the range front is composed of sedimentary strata dipping at a high angle, with failure

likely along bedding planes, (d) other discontinuities, such as metamorphic foliation planes, dip at a high angle near the range front, (e) slopes are oversteepened through increasing relief, or (f) large exfoliation blocks are developed.

Rockfall, the simplest of the processes initiated by failure of bedrock slopes, encompasses the downward rolling or skipping under the force of gravity of individual particles, especially gravel (Drew 1873). The clasts usually are angular in outline due to their liberation from fractures or exfoliation planes. Rockfall deposits commonly mantle the base of cliffs in the catchment or at the range front, forming colluvial talus slopes or, if funnelled, talus cones or incipient alluvial fans (Fig. 14.8a, b, and c) (e.g. Turner 1996, Turner and Makhlouf 2002). Rockfall clasts may also roll or bounce directly from the range front to the fan (Beaty 1989, Beaty and DePolo 1989), where they constitute outsized clasts (Fig. 14.8d). Besides rolling or skipping as individual particles, transport on talus cones and fans may also occur more collectively as grain flows in tongues or chutes (Fig. 14.8b and c) (e.g. Bertran and Texier 1999).

Rock slides, in contrast, constitute larger blocks or megablocks that break away from bedrock cliffs along faults or geological discontinuities, and move either rapidly or slowly downslope as a coherent mass above a basal glide plane (Varnes 1978). As implied by their name, rock slides are differentiated by their sliding transport mechanism, as opposed to the rolling or skipping motion of rockfalls (Sharpe 1938, Mudge 1965). The commonly curved excavation of the bedrock formed by detachment of a rock slide is called the slide scar. Slide scars typically become new fan catchments because they constitute a concavity in the bedrock wherein subsequent precipitation will converge. Rockfall deposits commonly are present along the base of a slide scar due to brecciation from shear along the detachment. Rock-slide deposits accumulate either near the base of the bedrock slopes, or as megaclasts on the fan (Fig. 14.8e).

The failure of a large, fractured bedrock cliff more typically occurs as a rapid and catastrophic downward fall accompanied by partial to widespread disintegration and pulverization to produce a brecciated mass called a rock avalanche (Figs. 14.8f and 14.9a, b) (Harrison and Falcon 1937, Mudge 1965, Shreve 1968, Browning 1973, Hsü 1975, Hunt 1975, Porter and Orombelli 1980, Nicoletti and Sorriso-Valvo 1991,

Fig. 14.8 Photographs of rock-gravity flow deposits. (**a**) Colluvial cone in Death Valley with prevalent chutes. Older chutes are darker from rock varnish. (**b**) Rockfall talus cone with active grain-flow chutes located along the Fraser River near Lillooet, British Columbia. (**c**) Incipient fan in Eureka Valley, California, with toe cuts that reveal a scallop-shaped stratigraphy resulting from the accumulation of sediment in rockfall chutes (e.g. *arrow*). (**d**) View of loose boulders transported by falling and rolling to the piedmont from a fault-bounded range front, Owens Valley, California. The particles are shaped by fractures in the granitic bedrock. (**e**) Two megablocks of carbonate rock (*arrows*) with d_l of 255 and 380 m transported to the piedmont as rock slides from the range front; South Titanothere Canyon fan, Death Valley. These megaclasts are being buried by subsequent gravelly rockfall talus (*T*). (**f**) View of 90 m high mounded rock-avalanche deposits (*centre*) on the distal Panamint Canyon fan, California

Blair 1999a, c). Rather than rolling or sliding, avalanches mainly achieve motion by transforming into a dry granular flow. This transformation occurs as the rock mass disintegrates during the fall stage, or during impact with the piedmont. Brecciation may progress further during transport, especially along the base where shear is greatest due to friction and to the weight of the overlying mass. Rock avalanches are known to move at speeds of about 25–100 m/s (Erismann and Abele 2001). They typically undergo deposition in the piedmont zone due to a lessening of slope, lateral expansion, and basal and lateral friction. Total runout distance also is related to the kinetic energy of the fall and the potential energy acquired during downslope movement, both of which increase with the height of the fall (Hsü 1975, Melosh 1987; Fauque and Strecker 1988; Hart 1991). Failure of a part of a range-front facet between two fan catchments can create a new catchment and fan (Blair 1999a).

Rock-avalanche deposits on alluvial fans have variable but diagnostic features and forms. They typically comprise massive units 10 to > 100 m thick present as: (a) irregular or conical forms, (b) arcuate to U-shaped levée-snout forms with either low or high length-to-width ratios, or (c) continuous lobes (Figs. 14.8f and 14.9a, b) (Hadley 1964, Burchfiel 1966, Gates 1987, Fauque and Strecker 1988, Evans et al. 1989, 1994, Blair 1999a, c, Hermanns and Strecker 1999, Philip and Ritz 1999, Hewitt 2002). The upslope ends of these deposits either extend directly from the range front, or are detached from it by small to great distances. Where exposed, the basal avalanche and underlying fan deposits are deformed, including by thrusting and injection. Large rock avalanches may produce a fan and catchment in a single event, as exemplified by the 1925 Gros Ventre avalanche in Wyoming, the 1970 Huasaracan avalanche in Peru, and the 1987 Val Pola avalanche in Italy (Blackwelder 1912, Voight 1978, Plafker and Ericksen 1984, Costa 1991, Erismann and Abele 2001, Govi et al., 2002, Schuster et al. 2002). Other large cases, such as at Hebgen Lake, Montana in 1959 (Hadley 1964), did not produce a fan or catchment because of the breadth of the collapsed range-front bedrock, and the lack of funneling of the moved material.

Irrespective of depositional form, rock-avalanche deposits diagnostically consist of pervasively shattered, angular clasts of gravel to megagravel separated by variable amounts of cataclastic matrix (Fig. 14.9c), leading to the term 'megabreccia' (e.g. Longwell 1951, Burchfiel 1966, Shreve 1968). Clasts of all sizes are internally shattered, but remain intact or are only slightly expanded. Finer clasts derive from the disaggregation of shattered clasts. Detailed textures such as jigsaw breccia and crackle breccia have been differentiated (Yarnold and Lombard 1989). Avalanche clasts have a composition that directly matches the bedrock source. Particles are commonly of a single lithology due to the homogeneous bedrock from which avalanches are derived, giving rise to the term 'monolithologic breccia.'

Earth Flows

Earth flows constitute the glacier-like, slow or episodic downslope movement of a small to large volume of bedrock as a partially ductile mass. Earth flows are generated where fine-grained bedrock, especially bedrock bearing water-sensitive expandable clay minerals, comprise upland slopes (Varnes 1978, Keefer and Johnson 1983). Earth-flow movement most commonly occurs during sustained wet-season rainfall following a dry period. When wetted, the sensitive bedrock loses shear strength and gains plasticity. Earth flows are thus activated by failure, under the force of gravity, of fine-grained bedrock on steep slopes through water infiltration, and aided by loading (Fleming et al. 1999). Places and rock types where earth flows are documented to be common are upland slopes of: (a) Neogene mudstone and Mesozoic melange in the San Francisco Bay area, (b) Cretaceous to Eocene shale in the dry interior of British Columbia, Canada, and (c) Cenozoic volcanic tuff sequences in the American Rocky Mountains (Fig. 14.9d, e) (Krauskopf et al. 1939, Crandell and Varnes 1961, Keefer and Johnson 1983, Bovis 1985, 1986, Shaller 1991, Bovis and Jones 1992, Varnes and Savage 1996, Fleming et al. 1999).

Earth-flow movement is related to increased pore-water pressure, and is accommodated by sliding along a basal glide plane developed along an internal discontinuity, such as between weak and strong layers, or zones of saturation (e.g. Keefer and Johnson 1983). Lateral expansion and contraction of the mass during flow shows that it undergoes internal ductile deformation, whereas the presence of fissures, lateral ridges, and thrusts attest to brittle behaviour (Fig. 14.9d, e). The cavity at the upslope end of an earth flow, from which the mass has detached, is called the head scar. Once generated, earth flows may be reactivated in whole or in part by lubrication from subsequent water input that commonly is concentrated in the head scar. Lateral expansion of the distal end of high-volume earth flows reaching the piedmont

Fig. 14.9 Photographs of fans with various types of rock-gravity and sediment-gravity flows. (**a**) View on North Long John fan, Owens Valley, of a prominent rock avalanche with levées leading to a 108-m-high frontal snout (*S*). The avalanche originated by failure of bedrock previously present in the range-front scar (*C*). (**b**) Downslope view of the avalanche mounds (*M*) and levées (*L*) that are partly dissected by gullies; Rose Creek fan delta, Walker Lake, Nevada. (**c**) Exposure of the North Long John rock avalanche showing an unstratified and unsorted, very angular texture of muddy, cobbly, fine to coarse pebble gravel.

Scale bar is 15 cm long. (**d**) Cross-valley view of the Slumgullion earth flow, southwestern Colorado, derived from a head scar (*H*) in volcanic bedrock. The earth flow has expanded at its distal end (*D*) to form a fan ∼2 km wide. Photograph provided courtesy of the U.S. Geological Survey. (**e**) Overview of the alluvial fan formed by the Pavilion earth flow, south-central British Columbia. (**f**) Fan formed from slips of colluvium mantling the steep catchment, Conundrum Creek valley near Aspen, Colorado. The lower catchment and fan were later incised by water flow

can produce an alluvial fan, such as for the Carlson earth flow in Idaho (Shaller 1991), the Pavilion in British Columbia (Fig. 14.9e), and the Slumgullion in Colorado (Fig. 14.9d). The presence of the thickest

part of an earth flow at its distal end, including on a fan, is due to the continued movement of the upper part of the mass to the zone of accumulation, where it stabilizes.

Sediment-Gravity Processes from Colluvial Slope Failures

Colluvial Slips

Sediment-gravity processes on alluvial fans generated from colluvial slope failures include colluvial slips, debris flows, and noncohesive sediment-gravity flows. Colluvial slips (also called colluvial slides or slumps) consist of intact to partially disaggregated masses of destabilized cohesive colluvium that move either slowly or rapidly downslope above a detachment horizon (Varnes 1978, Cronin 1992). The detachment may develop at a zone of weakness within the colluvium, but more commonly it constitutes the contact between colluvium and the underlying bedrock, where infiltrated water is perched. Colluvial slips can be triggered in a dry or unsaturated state, such as in response to earthquakes (Keefer 1984), but most often are initiated by the addition of rainfall or snowmelt in volumes that saturate the sediment (Caine 1980, Ellen and Fleming 1987, Reneau et al. 1990). Colluvial slopes fail in this state because of the lowered resisting forces related to increased hydrostatic pore pressure and decreased shear strength (Campbell 1975, Hollingsworth and Kovacs 1981, Mathewson et al. 1990). The presence of clay in the colluvium promotes failure by lowering permeability and providing strength to the interstitial fluid phase.

Colluvial slips commonly accumulate in the catchment at the base of the failed slope or within the drainage net. They also can build an alluvial fan in cases where they are funnelled through a low-order catchment directly to the piedmont. Fans built mainly by colluvial slips are common in the high elevation of Colorado, where they develop at the base of bedrock shoots (Fig. 14.9f). These fans are active mostly during spring snowmelt, especially below shoots that had their tree cover damaged by snow avalanches or fire. The deposits are of thick to massive beds of muddy gravel texturally like the colluvium.

Debris Flows

Debris flow is the most important sediment-gravity process type with respect to the volume of material delivered to alluvial fans. Such flows consist of a mixture of sedimentary particles spanning from clay to gravel, along with entrained water and air, that move downslope in a viscous state under the force of gravity (Blackwelder 1928, Sharp and Nobles 1953, Johnson 1970, 1984, Johnson and Rahn 1970, Fisher 1971, Hooke 1987, Coussot and Meunier 1996, Iverson 1997). Debris flows are instigated as failed colluvium in the catchment in response to the addition of a notable amount of water that undergoes rapid infiltration and runoff to form a flash flood. Water input is from: (a) rapid precipitation, such as from a thunderstorm, (b) heavy rainfall following previous sustained rainfall that saturated the colluvium, or (c) rapid snowmelt or icemelt from warming air temperature (Costa 1984, 1988). As with colluvial slips, the generation of a debris flow is promoted where colluvium contains mud. The presence of mud induces failure and debris-flow initiation by lowering the permeability of the colluvium, allowing hydrostatic pore pressure to increase and overcome shear strength, leading to the rapid downslope movement of a mud-bearing mass. Slopes that promote debris-flow initiation typically are at ~27–56°, with slopes > 56° too steep to maintain a colluvial mantle, and slopes < 27° having a lower propensity for failure (Campbell 1975). Because flash floods are infrequent, and the accumulation of colluvial sediment requires time, the recurrence interval of debris flows is relatively long, varying from about 300 to 10000 years (Costa 1988, Hubert and Filipov 1989). Debris flows locally are more frequent where conditions are presently ideal for their generation (e.g. Jian and Defu 1981, Jian and Jingrung 1986, Cerling et al. 1999).

More specifically, debris flows are initiated from colluvium by one of two mechanisms, both of which cause disaggregation and dilatancy as the failed mass transforms into a cohesive flow resembling wet concrete (Johnson and Rahn 1970, Campbell 1974, 1975, Costa 1988). One way that transformation is achieved is through the disaggregation of a wet colluvial slip as it moves downslope. The change from a slip to a flow occurs as shear expands from the base to throughout the mass, and clasts begin to move independently. The second triggering method is by rapid erosion where flashy runoff intersects colluvium in the drainage net (e.g. Cannon et al. 2001). Such erosion can cause the undercutting and sloughing of sediment along the channel sides, leading to quick sediment bulking of the flow. Water in this case dissipates its energy by dispersing clasts through churning, tossing, and mixing to produce a debris flow (Johnson 1970, 1984). The presence of mud in the colluvium provides cohesive strength to the ensuing flow. The generation of debris flows either by slip or erosion is promoted in fan catchments by the presence of steep and poorly sorted colluvial slopes, combined with a converging drainage net that concentrates colluvial sediment and overland flows in the same area (e.g. Reneau et al. 1990).

Debris flows move as a viscous mass in a non-Newtonian laminar manner, causing them to be nonerosive even though they can transport clasts weighing several tons (Johnson 1970, Rodine and Johnson 1976). Debris flows have been observed to move at speeds of 1–13 m/s (Sharp and Nobles 1953, Curry 1966, Morton and Campbell 1974, Li and Luo 1981, Li and Wang 1986). Individual flows typically are 1–10 m thick within catchment channels, and thin by lateral expansion upon reaching the fan. They improve flow efficiency in the catchment by shearing off their sides or base in rough zones, outer bends, or where the channel rapidly widens. A debris flow event can entail a single pulse, but more commonly consists of several surges caused by the episodic addition of sediment via multiple slips or sloughs, or from the repeated generation and breaching of jams between boulders, logs, and pathway elements (Blackwelder 1928, Fryxell and Horberg 1943, Sharp and Nobles 1953, Johnson 1984).

Particles in debris flows are supported by the high density and strength of the mass related to cohesive, dispersive, and buoyant forces (Middleton and Hampton 1976, Costa 1984, 1988). The differential response of boulders to buoyant and dispersive forces generated by small differences in density between them and the rest of the material, along with kinetic sieving, cause boulders in a moving debris flow to become concentrated at the top (Johnson 1970). Friction at the base of a debris flow makes it move more slowly than the top, resulting in the progressive conveyance of the boulder-rich upper tread to the front of the flow. Frontal boulder accumulations typically are interlocked and lack pore-fluid pressure, forcing the flow either to cease, or to push aside the frontal boulders (Johnson 1984, Iverson 1997, Major and Iverson 1999).

Once initiated, a debris flow moves until gravity forces decrease to the point where they no longer can overcome the shear strength. Debris-flow mobility is aided by the lubricating effect of the muddy pore fluid, which facilitates motions of grains past one another, and mediates grain contacts (Iverson 1997). The cessation of a debris flow primarily results from thinning to the point where the plastic yield strength equals the shear strength (Johnson 1970), a process promoted by expansion, and aided by dewatering and a lessening of slope. These conditions, and thus deposition, are most common where debris flows depart from the catchment onto a fan. Debris flows may bypass the upper fan if an incised channel is present because the channel walls prevent flow expansion. Deposition in this case commences when the flow reaches the active depositional lobe (Fig. 14.10a). As in the catchment, debris flows may also be halted on the fan prior to critical thinning due to the damming of coarse clasts at the margins, or the jamming of clasts and logs with flow-path obstacles such as boulders or trees. The extent of the deposits on a fan from a single debris-flow event, including radial run-out distance, ultimately is a function of the debris-flow volume given that the continued addition of sediment prevents critical thinning (Blair 2003).

Debris-flow deposits on alluvial fans consist of unsorted to extremely poorly sorted, muddy to gravelly sediment present mainly as levées and lobes (Blackwelder 1928, Sharp and Nobles 1953, Beaty 1963, 1974, Johnson 1970, Costa 1984, Blair and McPherson 1998, 2008). Levées constitute sharply bounded, radially oriented ridges typically 1–2 m wide and 1–4 m tall present in parallel pairs separated by a 2–10 m gap (Fig. 14.11a, b). They represent the boulder-rich fraction of a debris flow that was selectively conveyed to the flow front during motion, and then pushed aside and sheared from the lateral margins as the flow continued downslope. Paired levées may extend for tens to hundreds of metres down the fan, and be joined at their ends by a snout. Levées can be slightly sinuous and widen along their outer curves where the debris flow followed a previous drainage. They are characteristically boulder-rich, with the a-b plane of elongated boulders aligned about parallel to the trend of the levée. This fabric is produced as the boulders are pushed aside from the front of the flow. Boulders and other clasts usually are interlocked (clast-supported) on the outer levée margins (away from the interlevée area), and are supported by abundant matrix on their inner sides (Fig. 14.11a, b). Mangled tree logs and limbs also may be present.

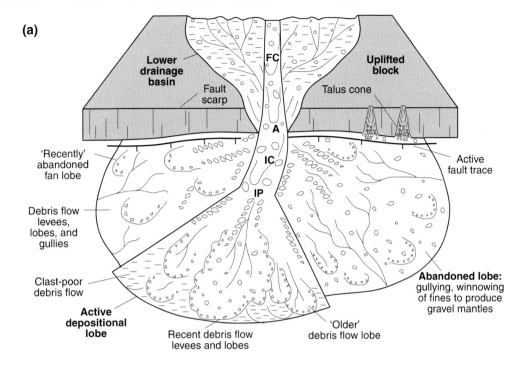

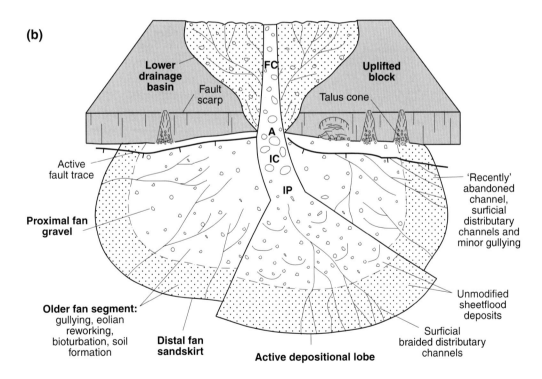

Fig. 14.10 Schematic diagrams of the common processes on alluvial fans, including for those dominated by debris flows (**a**) or sheetfloods (**b**). Labels denote the catchment feeder channel (*FC*), fan apex (*A*), incised channel (*IC*), and intersection point (*IP*)

Fig. 14.11 Photographs of debris-flow deposits. (**a**) Downfan view from apex of Dolomite fan, Owens Valley, of two sets of paired levées 150–250 cm tall deposited during a 1984 debris-flow event. Note the clast-rich outer levées on either side of an older ridge (*O*). (**b**) View of inside of 1984 debris-flow levée on the proximal Dolomite fan; shovel for scale. (**c**) View of 50 cm thick clast-rich and matrix-rich debris-flow lobe deposited in 1984 on the Dolomite fan. (**d**) View of 1984 debris-flow lobe on the medial Dolomite fan consisting of clast-rich (*R*) and clast-poor (*P*) phases that were partly winnowed (*W*) during falling

flood stage. (**e**) Overview of 1984 debris-flow deposits (light) on the Dolomite fan consisting of proximal levées (*upper arrow*) and distal lobes (*lower arrow*). The latter terminate at the fan toe, where a depositional slope gap is apparent. Older levées are visible in pre-1984 deposits of the proximal fan. (**f**) View of 1990 clast-poor debris-flow lobe on the Copper Canyon fan, Walker Lake, Nevada. Note that the flow delicately divided around the desert plants despite the concentration gravel clasts at the lobe margin

Debris-flow deposits most abundantly are present on alluvial fans as radially elongated lobes 1–100 m wide and ~0.05–2 m thick (Fig. 14.11c, d). Lobes are commonly continuous for hundreds of metres

downfan beginning at the apex of the fan or active depositional lobe, or at the distal ends of paired levées (Fig. 14.11e). Lobes in the latter case represent material that passed between the levées, and continued to flow farther downslope. The margins of the lobes usually are sharp and steep (> 45°), but delicately wrap around pre-flood features such as plants (Fig. 14.11f). Texturally, debris-flow lobes are mud-rich and matrix-supported, and are either clast-rich or clast-poor. Unlike levées, boulders are uncommon in lobe deposits. Clast-rich lobes consist of muddy, pebbly cobble gravel with sharp lateral and distal margins outlined by a commonly clast-supported fringe of coarse to very coarse pebbles, cobbles, and possibly wood fragments (Fig. 14.11c, d). Clast-poor debris-flow lobe deposits, also called mudflows, are similar to clast-rich lobes except that they lack abundant coarse pebbles and cobbles, consisting instead of muddy very fine to medium pebble gravel, or pebbly mud (Figs. 14.11f and 14.12a). Like clast-rich lobes, they have sharp margins typically fringed by wood and pebbles, although clast-poor lobes tend to be thinner (< 50 cm), and become mud-cracked during desiccation.

Debris-flow deposits on a given alluvial fan may be dominantly of (a) levées, (b) clast-rich lobes, (c) clast-poor lobes, or (d) levées in the proximal zone that give way downfan to lobes (Fig. 14.10a). Case studies reveal that the presence or absence of the various forms is a function of the typical texture of the catchment colluvium (Blair and McPherson 1994b, 1998, 2008, Blair 1999f, 2003). For example, surface forms and facies reveal that the Dolomite alluvial fan of Owens Valley, California, consists mostly of levées in the proximal fan and clast-rich lobes in the distal fan (Blair and McPherson 1998). A similar pattern was produced on this fan during the afternoon of 15 August 1984 by a multiphase debris flow triggered from a strong thunderstorm. A total of ~50,000 m³ of sediment was deposited during this event as multiple paired levées extending for 200–300 m on the proximal fan, and giving way to continuous 200–400 m long lobes on the distal fan (Fig. 14.11e). The lobes extend downslope from the ends of the conjugate levées, and consist mostly of a clast-rich phase that is overlain and offlapped by clast-poor phase. Boulders are abundant in the levées but are essentially absent in the lobes. Seven major levée-lobe tracts, each representing a single surge, were deposited during this debris flow event.

The transition from levées to lobes for each of the 1984 debris flow surges on the Dolomite fan took place in response to the selective depletion of boulders (Blair and McPherson 1998). This depletion was caused by

the preferred movement during flow of boulders to the upper part of the debris flow due to buoyancy and dispersion, and then to the front due to higher velocity of the upper versus lower treads of the flow in response to basal drag. Boulders concentrated at the front of this debris flow were pushed aside by the main flow mass, and then were sheared off and left behind as levées (e.g. Sharp 1942). This process in the Dolomite debris flow ensued until the boulder content of a given surge was depleted. A loss of boulders, typically at a distance of ~250 m from the fan apex, caused the boulder-deficient central part of the debris flow to continue downslope beyond the ends of the levées, where it expanded and then accumulated as a clast-rich lobe. A clast-poor lobe stage, representing the more dilute debris flow tail, capped and offlapped the downslope ends of the clast-rich lobes. Deposition of proximal levées and linked distal lobes by each of the seven major debris-flow surges was caused by a generally consistent texture per surge of muddy pebble cobble gravel with some boulders. A greater content of boulders in the colluvium would have caused a greater dominance of levées on the fan, whereas a lower boulder content would have caused a dominance of lobes. The generally consistent texture of the seven tracts implies that each surge was instigated by freshly sloughed colluvium of roughly similar content.

Another attribute of debris flow fans exemplified by the Dolomite case is the relationship between process and surface slope. The levée-dominated proximal part of the Dolomite fan has a surface slope of 9–12°, whereas the zone with clast-rich lobes has a slope of 3–5°, and the clast-poor lobes 2–3° (Blair and McPherson 1998). The proximal to distal inflection of slope takes place over a short distance to create two pronounced radial segments. The coincidence of this slope inflection with the depletion of boulders in the deposits, and with a change from levées to lobes, shows that the slope is a product of the process. More specifically, the 9–12° slope of the levéed segment reflects the highest magnitude of slope for which transport is not possible given the shear strength of the typical debris-flow mix yielded from the catchment. Similarly, flows producing the clast-rich lobes ceased motion at slopes of 3–5°, and clast-poor lobes at 2–3°. These relationships show that the fan slope results from the dominant aggradational process, not vice-versa, and that the process in the case of the Dolomite fan is a function of the sedimentary texture, including boulder content, yielded during a given colluvial failure in the catchment.

Alluvial fans dominated by clast-poor debris flows (mudflows) are a special case representing situations where coarse gravel is not shed in abundance from the catchment despite relief. This scenario occurs where fan catchments are underlain by easily disaggregated shale, siltstone, sandstone, or fine-grained volcanic rocks, or where intense tectonic shear has pulverized more coherent bedrock (Blair and McPherson 1994b, Blair 2003). Examples of this fan type, such as the Cucomungo Canyon fan in Eureka Valley, California (Fig. 14.12a), have slopes of 2–4°, indicating that this is the common slope range at which the flow shear strength is no longer exceeded by gravity forces.

Noncohesive Sediment-Gravity Flows

Noncohesive sediment-gravity flows (NCSGFs) form where catastrophic water discharge in a catchment intersects sandy and gravelly colluvium containing little mud. Such material is loose because mud is the main agent providing cohesive strength to colluvium. Water erosion of noncohesive colluvium normally produces a fluid-gravity flow, but catastrophic discharge intersecting abundant loose colluvium can cause erosion and bulking so extreme that the water-sediment mixture rapidly transforms into a NCSGF. NCSGFs are similar to debris flows in many respects, but low mud content precludes cohesive strength. Clasts, including boulders and blocks, are instead supported during transport by dispersive, buoyant, and structural-grain forces. Like debris flows, transportation is laminar, and thus NCSGF are nonerosive.

NCSGFs are not well understood, but can be characterized by two examples. A documented NCSGF occurred 15 July 1982 on the Roaring River alluvial fan in Rocky Mountain National Park, Colorado (Jarrett and Costa 1986, Blair 1987a). This event was triggered by failure of a man-made dam containing a reservoir in the upper catchment. The impounded water was rapidly released when the dam failed, and within a few hours the ensuing flash flood moved about 280,000 m³ of sediment to the fan. Much of the sediment was eroded from mud-poor glacial moraines present along the feeder channel that consist of sand, pebbles, cobbles, boulders, and blocks derived from the colluvium mantling gneissic bedrock. Extreme erosion where the turbulent

flash flood intersected the moraine produced a NCSGF that deposited two fan lobes 220 and 600 m in length, and 11–13 m thick. The lobe margins are partly delineated by jams between boulders, logs, and upright trees (Fig. 14.12b). Paired but non-parallel boulder-rich levées 1.5 m tall were sheared from the margins of the first lobe, and extend for 70 m from the fan apex. Significant sedimentation during this event also occurred from sheetflooding (next section) that followed deposition of the NCSGF lobes.

Major NCSGF events have been reconstructed for near-surface deposits of the Tuttle, Lone Pine, and several other fans of the Sierra Nevada piedmont along Owens Valley, California (Figs. 10.1b and 10.2b) (Blair 2001, 2002). Glaciers, moraines, and cirques are present in the upper catchments of these fans, and moraines built of repositioned colluvium are present in the lower catchment or on the upper fan as a result of more expansive glaciation during latest Pleistocene time (Figs. 14.3b and 14.6f). The moraines consist of mud-poor sand, gravel, and blocks derived from granitic bedrock. Massive matrix-supported lobes of sandy cobbly boulder gravel 3 to > 8 m thick span across most of these ~10 km long fans (Fig. 14.12c and d). Fine to medium blocks (d_I of 4–15 m) also are present. The lobes formed in response to rapid erosion of moraines in the catchment due to the failure of moraine or ice dams that naturally impounded water. The rapid release of the water caused extreme erosion of the moraine and sediment down stream, transforming the material into NCSGFs that were catastrophically deposited across the fans. The surficial NCSGF deposit of the Tuttle fan has a volume of ~140 million m³. Following NCSGF deposition, continued catastrophic flood discharge eroded channels through these deposits (Fig. 14.12d), and moved this sediment to the distal fan where it built sheetflood lobes.

Fluid-Gravity Processes from Colluvial Slope Failures

Fluid-gravity flows, or water flows, are Newtonian fluids characterized by a lack of shear strength, and by the maintenance of sediment and water in separate phases during transport (Costa 1988). Turbulence causes sediment to move either as suspended or quasi-suspended

Fig. 14.12 Photographs of various fan flow types. (**a**) View of 1997 mudflow deposits 35 cm thick and > 20 m wide on the medial Cucomungo fan, Eureka Valley, California; arrows point to steep margin. These deposits extend outward from a central channel (*C*), enveloping plants and aggrading a planar bed upon the relatively smooth surface of 1984 mudflow deposits. (**b**) Up-fan view of the distal margin of NCSGF on the Roaring River fan, Colorado. The margin consists of a boulder-log jam ~ 2 m tall. (**c**) NCSGF exposure 4 m tall on the Tuttle Canyon fan, Owens Valley; fieldbook for scale. Note the matrix support, lack of sorting, and presence of calcium carbonate coatings on clast undersides (*C*). (**d**) Downslope view of sand-bedded channel (*C*) 10 m wide and 3 m deep cut by catastrophic discharge into NCSGF deposits of the Tuttle Canyon fan. (**e**) Incised channel of a fan derived from the Smith Mountain pluton, Death Valley. The channel bed consists of boulders, lateral to which sandy gravel was deposited. (**f**) Downfan view of incised-channel wall of the proximal Warm Spring Canyon fan, Death Valley. The wall contains a side relict of older incised-channel deposits (*IC*) inset within debris-flow sequences (*D*)

load, or by the rolling or saltating of particles as bed-load along the flow base. Sediment concentration in fluid-gravity flows typically is ≤ 20% by volume, but may reach ~ 47% (Costa 1988). Flows in the latter case, called hyperconcentrated, achieve low shear strength and have reduced fall velocity of particles, but retain water flow properties, including transport from turbulence, and the maintenance of sediment and water in separate phases (Nordin 1963, Beverage and Culbertson 1964, Costa 1988, Coussot and Meunier 1996). More recently, others have erroneously applied 'hyperconcentrated flow' or 'hyperconcentrated flood flow' to debris flows, or to a mechanically uncertain flow proposed to be transitional between water flow and debris flow.

Fluid-gravity flows are generated in a fan catchment as a result of the concentration of water as overland flow in the drainage net due to input from precipitation, snowmelt, or natural dam failures. Overland flow is achieved when the infiltration capacity of the colluvium is exceeded, and runoff begins. Colluvium mantling catchment slopes is eroded by runoff via undercutting and entrainment. Should the eroded colluvium contain mud, the water-sediment mixture then typically transforms into a sediment-gravity flow, such as a debris flow. The generation of a fluid-gravity flow from catchment runoff under these conditions thus requires the presence of generally mud-free colluvium. Two types of fluid-gravity flows, incised-channel floods and sheetfloods, are important primary fan processes generated this way.

Incised-Channel Floods

Incised-channel floods constitute the continued confined transfer of a flash flood from the feeder channel across the fan. If present, the walls of an incised channel prevent the flood from expanding, causing flow depth and competency achieved in the feeder channel to be maintained. Deposits from an incised-channel flood thus are mainly of the coarsest clasts (commonly cobbles and boulders) derived from the catchment, or eroded from the walls or floor of the channel. Incised channels typically have a floor of coarse clasts 0.5–3 m thick, and may contain drapes of finer sediment deposited during falling flood stage (Fig. 14.12e) (Blair 1987a, 1999d, 2000). The lateral repositioning of the incised channel coincident with incision can produced terraces or stranded margins (Fig. 14.12f) (e.g., Blair 1999f). Incised channels primarily serve as conduits for discharge across the upper fan. Their deposits thus mostly comprise an armoured channel bed set within other primary deposits such as debris flows, NCSGFs, or sheetfloods.

Sheetfloods

A sheetflood is a short-duration, catastrophic expanse of unconfined water (McGee 1897, Bull 1972, Hogg 1982). Sheetfloods are instigated by torrential precipitation such as from a thunderstorm, or from the release of impounded water due to the failure of a natural dam (Blair 1987a, Gutiérrez et al. 1998, Meyer et al. 2001). Sheetfloods readily develop on alluvial fans where flash flood discharge from the catchment is able to expand. Expansion is promoted by the conical surface of a fan, and begins either at the fan apex or on an active depositional lobe located downslope from an incised channel (Fig. 14.10b).

The characteristics of alluvial-fan sheetflooding are illustrated by the catastrophic event of 15 July 1982 on the Roaring River fan in Rocky Mountain National Park, Colorado (Blair 1987a). An aerial photograph taken while the sheetflood was underway shows water and sediment discharge over a 320-m-long active depositional lobe with an expansion angle of 120°, and fed by an incised channel 160 m long present between two previously deposited NCSGF lobes. The length of this lobe was restricted by the opposing valley margin. At the time of the photograph, transverse upslope-breaking waves typical of supercritical flow (Froude number > 1) were present on the water surface in 43 radially oriented trains 3–6 m wide, 10–250 m long, and with wavelengths of 5–25 m. The trains covered ~20% of the lobe area, and were developed in the deepest part of the sheetflood. Hydraulic calculations from the waves show that the sheetflood had an average depth of 0.5 m, a velocity of 3–6 m/s, maximum water discharge of 45.6 m^3/s, and a Froude number of 1.4–2.8. Post-flood evaluation of this lobe showed a 2° to 5° sloping surface of cobbly pebble beds 3–6 m wide and 10–20 cm thick separated by more widespread pebbly sand 5–20 cm thick (Fig. 14.13a). Trenches revealed a sheetflood sequence up to 5 m thick consisting mostly of rhythmically alternating beds with the same textures and thickness as the surface units (Fig. 14.13b). Clasts within the couplet beds are clast-supported and imbricated, and the pebbly sand is laminated. The surface was partly incised during falling flood stage, forming channels within which coarse clasts were concentrated. Much of the sheetflood surface was remolded during subsequent noncatastrophic discharge into shallow channels

Fig. 14.13 Views of alluvial fan sheetflood deposits. (**a**) Photograph taken soon after sheetflood deposition in 1982 on the Roaring River fan in Colorado. The surface is smooth, slopes 4°, and has not been modified by falling-stage channel incision. Gravel beds ~2 m wide and more widespread sand beds are apparent. (**b**) Vertical trench in 1982 sheetflood deposits of the Roaring River fan near the site of previous photograph showing alternating couplets of cobble–pebble gravel and laminated granular sand with a slope similar to the fan surface. The dark silty horizon near the top of the trench was deposited ten months after the flood by secondary spring snowmelt discharge. (**c**) View of a 12-m high exposure of the proximal Anvil fan, Death Valley, dominated by planar-bedded couplets 5–25 cm thick of sandy very fine to medium pebble gravel and sandy cobbly coarse to very coarse pebble gravel. Beds slope 5° downfan, to the right. (**d**) Close-up view of Anvil fan sequence 2 m thick of sheetflood couplets containing three wedge-planar backset beds ~20 cm thick; arrows at base. Backsets dip 7–15° upfan, (leftward). (**e**) Exposure 7 m thick on Anvil fan of sheetflood couplets and backset beds separated by recessively weathered sand and bouldery cobbly pebble gravel; dark, e.g. arrows. The recessive beds are from secondary reworking. They divide the section into 7 sets 50–200 cm thick, each representing the deposits of a single sheetflood. (**f**) Oblique view of fan in western Arizona displaying transverse ribs of sediment-deficient sheetflood origin; road for scale. Photograph provided courtesy of S.G. Wells

wherein a sediment lag accumulated upon the couplet sequences, armouring them from further erosion.

Alternating cobbly pebble gravel and pebbly sand couplets also dominate exposures of other waterlaid fans, such as the Anvil and Hell's Gate fans in Death Valley, and the distal Tuttle and Lone Pine fans in Owens Valley (Fig. 14.13c) (Blair 1999b, d, 2000, 2001, 2002). These beds are not horizontal, but instead are oriented at a 2–5° slope. Backset (upslope-dipping) cross beds dipping 5–28° formed by supercritical flow also are found within couplet sequences in some of the fans, and couplet sequences commonly are divided by cobble lags that rest upon erosional surfaces (Fig. 14.13d and e). Sheetflood couplets have been documented in other fan sequences (Van de Kamp 1973, Harvey 1984b, Gomez-Pujol 1999, Meyer et al. 2001), and in the rock record (Blair 1987b, Blair and Raynolds 1999).

The sheetflood process is known from the alluvial fan case studies, and is supported by hydraulic studies in flumes (Gilbert 1914, Fahnestock and Haushild 1962, Kennedy 1963, Jopling and Richardson 1966, Simons and Richardson 1966, Shaw and Kellerhals 1977, Blair 1987a, Langford and Bracken 1987, Blair 1999b, d, 2000, 2001). The deposition of multiple (5–20) couplets during a single sheetflood is related to the autocyclic nature of trains of water and sediment waves, called standing waves, that form in supercritical flow. Supercritical standing waves rhythmically develop and terminate numerous times in a single flood. More specifically, they: (a) initiate, (b) lengthen in extent and heighten in magnitude, (c) migrate upslope, (d) become unstable and oversteepen, and then (e) terminate either by gently rejoining the flood, or more commonly by violently breaking and shooting downslope. Backset-bed (antidune) units accumulate during the first three of these stages as bedforms that are in phase with the surface waves. These bedforms are preserved if the standing wave terminates by gently remingling with the flow. The couplets are produced from the more common violent termination, called breaking, of the standing wave train. Such termination results in a rapid and turbulent washout that erodes the antidune bedforms, and temporarily suspends fine pebbles and sand. The coarse component of the couplet beds is deposited in a ~3–6 m wide tract as the washout bore of the breaking wave shoots downslope. The laminated fine couplet member accumulates more widely thereafter

by fallout of the quasi-suspended load. Irregularities in bedding are locally caused by flow-path obstacles, such as boulders, that incite flow separation and scour. Because standing waves develop in the relatively deepest part of the sheetflood, new wave trains initiate in a position lateral to recent depositional tracts. Repetitive generation and washout of standing waves during a single sheetflood causes autocyclic lateral and vertical amalgamation across the fan or active lobe of numerous couplets ± backset beds. Coarser gravel lags bounding sheetflood sequences form by surficial incision of channels during falling flood stage, or during later non-catastrophic discharge. These lags thus delineate the deposits of individual catastrophic sheetfloods (Fig. 14.13e), and account for nearly all of the time but little of the stratigraphy of the stacked sheetflood sequences (Blair 1999d).

The recurrence of a sheetflood at a given fan position is unknown but is probably long, as such floods are rare. Notable sheetflood events occurred on the same ~25° sector of the Furnace Creek fan of Death Valley in 1984 and 2004, probably exemplifying a more active case given the large catchment of this fan and the presence of a flood-control berm. Variable tectonic tilting of sheetflood sequences of the nearby Hell's Gate fan developed along the Northern Death Valley fault shows that sheetflood events there were 2–8 times more frequent than the seismic events that caused their tilting (Blair 2000).

A variant of alluvial-fan sheetflood deposition called transverse ribs also has been described. These deposits constitute beds with crests that are sinuous and laterally discontinuous, and with troughs lacking sediment, wherein older fan deposits are exposed (Fig. 14.13f) (Koster 1978, McDonald and Day 1978, Rust and Gostlin 1981, Wells and Dohrenwend 1985). These features likely represent deposition from supercritical flows that carried a lower volume of sediment, akin to a starved ripple. The transverse ribs described by Wells and Dohrenwend have wavelengths of 2–6 m, and indicate multiple, sediment-deficient sheetflood events with estimated flow velocities of 0.3–0.6 m/s.

So why are sheetfloods on alluvial fans supercritical? By solving the Froude and Manning equations for slope, the critical slope (S_c) at which water flows change from subcritical to supercritical (Froude number of ~1.0) for a given bed texture is approximated by the equation $S_c = n^2 g / D^{1/3}$, where n = the Manning roughness coefficient, g = the gravitational acceleration constant of $9.8\,\mathrm{m/s^2}$, and D = average flow depth in metres (Blair and McPherson 1994b). A plot of S_c for the Manning coefficients corresponding to sandy pebble to slightly bouldery cobble textures typical of alluvial fan sediment (0.024–0.040) shows that supercritical flow conditions prevail given the common range of slopes (2–5°) found on sheetflood-dominated alluvial fans (Fig. 14.14). The S_c curves further show that in the range of slopes typical of sandy and gravelly braided rivers in sedimentary basins (0.1–0.4°), flows only are supercritical with relatively great depth ($\geq 2.2\,\mathrm{m}$ for n of 0.032 and higher), consistent with observations of floods in rivers. The S_c curves thus identify two fundamentally unique water flow environments, with a nearly vertical limb corresponding to alluvial fan slopes, and a nearly horizontal limb corresponding to the river slopes in basins (Fig. 14.14) (Blair and McPherson 1994b). These curves inflect in the range of slopes (0.5–1.5°) uncommon to either alluvial fans or rivers in sedimentary basins.

Are sheetfloods on alluvial fans supercritical because they respond to the slope conditions of the fan, or, as the prominent process, do they create the fan slope? The 2–5° slope values of the surface of alluvial fans built by sheetfloods, and the 2–5° orientation of the sheetflood couplet units within the deposits, indicate that they are directly related. Antidune and plane-bed deposition are known from flume studies to be promoted by high water and sediment discharge, the latter caused by the proportional increase of flow capacity with increasing water discharge (Gilbert 1914, Simons and Richardson 1966). Supercritical flow conditions are promoted during high water and sediment discharge events because it is the flow state in which the overall resistance to flow is minimal, and in which the transport of large volumes of sediment is most efficiently accomplished (Simons and Richardson 1966). Fan catchments are ideal for generating high water and sediment discharge events due to the rapid concentration of high-volume precipitation across steep colluvium-mantled slopes. Thus, to achieve efficiency, sediment-bearing sheetfloods are supercritical and deposit sediment at slopes of 2–5°, and this depositional slope is reflected as

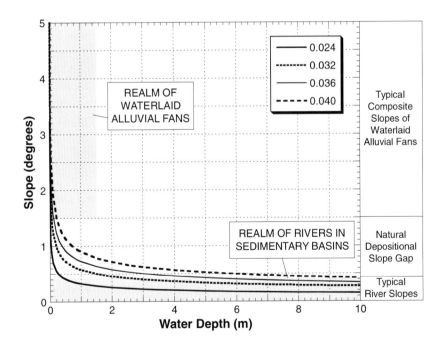

Fig. 14.14 Critical slope (S_c) curves versus water depth for Manning roughness values ranging from 0.024 to 0.040 (after Blair and McPherson 1994b). Curves are derived from the critical slope equation (see text). Turbulent supercritical hydraulic conditions exist for depth-slope scenarios above and right of the S_c curves, and subcritical conditions characterize the field below and left of curves. The typical depth-slope scenario for alluvial fans and gravelly rivers relate to opposite limbs of the S_c curves, with inflections occurring in the natural depositional slope gap. Also note that hydraulic conditions are supercritical for alluvial fans, and are commonly subcritical for rivers

the fan surface slope (Blair 1987a, 1999d). Like fans built of debris flows, sheetflood fans thus have a slope reflecting the mechanics of the main primary process. In contrast, deposition of transverse ribs may be controlled by the slope traversed by less common sediment-deficient sheetfloods.

Secondary Processes on Alluvial Fans

The long period between recurring primary depositional episodes on alluvial fans makes surficial sediment susceptible to modification by secondary processes, including surface or ground water, wind, bioturbation, neotectonics, particle weathering, and pedogenesis.

Surficial Reworking by Water

Discharge from rainfall or snowmelt in a fan catchment only infrequently produces a primary depositional event. On most occasions, water discharge to the fan is non-catastrophic and may carry limited sediment. Such discharge can slowly infiltrate through the permeable fan sediment, or it may pass through the incised channel and across the fan. These water discharge events, called sheetflows (Jutson 1919) or overland flows (Horton 1945), are commonly capable of winnowing fine sediment from the surface of previous primary deposits (Fig. 14.15a, b). Overland flows also can be triggered by precipitation directly on the fan, where winnowing further results from the impact of raindrops. The winnowed sediment typically is of sand, silt, or clay size, but can range to pebbles and cobbles (e.g. Beaumont and Oberlander 1971). This sediment is moved either farther downslope, or off the fan. A large part of mud sequences in playas, such as Badwater in Death Valley, is derived from winnowing of the adjoining fans.

Surficial reworking of primary deposits by overland flows is the most common secondary process on alluvial fans, producing rills, gullies, or coarse-grained mantles. Rills and gullies form by overland flow related to the falling stage of a wet primary process, less catastrophic catchment discharge, or rainfall directly on the fan. Rills are initiated by the convergence of overland flow as a result of slight topographic or textu-

ral irregularities. Such features, typically < 0.5 m deep and ~1 m wide, span downslope as distributaries from the fan apex or head of an active depositional lobe (Fig. 14.15c). Overland flow also erodes gullies that, in contrast to rills, initiate on the distal fan or at a fault scarp, and lengthen upslope through headward erosion. These features also contrast with rills by having either a single thread or contributary pattern, and by greater depth (Fig. 14.15d, e) (Denny 1965, 1967). Rills and gullies are floored by thin lags of coarse clasts left behind from winnowing, or by finer sediment that is in transit. These beds comprise minor lenses within troughs eroded into deposits from the primary processes. Another product of surficial winnowing by water is the concentration of gravel a few clasts thick called a gravel or boulder mantle (Figs. 14.15f and 14.16a). These surfaces can develop on a fan lacking recent primary events, where winnowing is extensive. They may cap an inactive lobe, or dominate the fan.

Calcite encrustation of the base or sides of channels or gullies on an alluvial fan may also result from the passage of water saturated with calcium and carbonate. Chemical precipitation from this discharge can produce a calcite crust termed case hardening. Lattman and Simonberg (1971) concluded from studies near Las Vegas, Nevada, that case hardening best develops on fans with catchments underlain by carbonate or basic igneous bedrock, the weathering of which supplies the necessary ions. This process can tightly cement freshly exposed deposits in as little as 1–2 years (Lattman 1973).

Wind Reworking and Deposition

Clay, silt, sand, and very fine pebbles on the fan surface are susceptible to wind erosion, as shown by dust plumes in the atmosphere over deserts on windy days (Fig. 14.15d). One effect of wind is to winnow the fine fraction (Tolman 1909, Denny 1965, 1967, Hunt and Mabey 1966, Al-Farraj and Harvey 2000). Protruding clasts may be carved into ventifacts by abrasion from the passage of wind-carried sand. Unlike other secondary processes, wind transport can also add sand or silt to the fan that is derived from non-catchment sources, such as from adjoining dunes, lakes, rivers, or deltas. This sediment can accumulate as: (a) aeolian

Fig. 14.15 Photographs of the products of secondary fan processes. (**a**) Side view of a gully on the Rose Creek fan delta, Hawthorne, Nevada, showing debris-flow deposits that have been winnowed at the surface by overland flow and wind to form a desert pavement. Scale bar is 15 cm long. (**b**) Surficial fine-fraction winnowing of sheetflood deposits (exposures), leaving a widespread lag of varnished cobbles; Anvil fan, Death Valley. (**c**) Aeolian sand (light-coloured) transported by wind has accumulated in rills and upon undissected gravel of the Furnace Creek fan surface, Death Valley. (**d**) Gullies are prominent on the distal part of the Hell's Gate fan, Death Valley. Note the dust plume in the valley centre generated by strong northerly winds achieving gusts of 80 km/h. (**e**) View of Carroll Creek fan, Owens Valley, containing a distal fault scarp *(F)*. Gullies are prominently eroding headward from the scarp into the older part of the fan. (**f**) Extensive winnowing of bouldery debris-flow deposits of the Shadow Rock fan, Deep Springs Valley, California, has produced a varnished, surficial boulder mantle

drift on the irregular fan surface (Fig. 14.15c), (b) nebka around plants, (c) sandsheet deposits initiated by irregular topography, or (d) sand dune complexes that either initiate or migrate onto a fan (e.g. Anderson and Anderson 1990, Blair et al. 1990, Blair and McPherson 1992). Sandsheets can form relatively thick and laterally continuous blankets that disrupt or even overwhelm fan sedimentation.

Bioturbation and Groundwater Activity

Plants and burrowing insects, arthropods, or rodents are present in surficial fan deposits of even the most arid deserts. Plant life can be sustained by rare precipitation, dew, or shallow groundwater. Plant roots may extend for a metre or more into the fan sediment, disrupting the original primary stratification and potentially homogenizing the deposits (Fig. 14.16b). Desert plants also provide a habitat for animals. Colonies of rodents amid desert plants on fans may alter primary deposits by disrupting stratification and dispersing sediment (Fig. 14.16c). Burrowers may also disturb desert pavement, exposing previously protected sediment to wind erosion.

Alluvial fans serve as important conduits of groundwater from the mountains to the valley floor, where aquifers can be recharged (e.g. Listengarten 1984, Houston 2002). Shallow groundwater flow may create conditions conducive to plant growth. The slow movement of groundwater rich in dissolved solids can also cause the precipitation of cements such as calcite in the fan sediment (e.g. Bogoch and Cook 1974, Jacka 1974, Alexander and Coppola 1989), whereas travertine precipitation may occur where groundwater issues to the surface at springs (Hunt and Mabey 1966). Distal fan sediment near playas or marine embayments may be cemented or disrupted by evaporite crystal growth in pores due to evaporative draw. Groundwater flow may also destabilize slopes, instigating slumping. Subsidence along faults, cracks, or fissures may also occur on a fan due to depression of the water table.

Neotectonics

Tectonic offset of alluvial fans is common where they are developed along seismically active mountain fronts. Faults can cause nearly vertical offset of fan sediment as scarps 0.5–10.0 m high. Fault scarps usually are present near and trend parallel to the mountain front, but they also can be oriented obliquely, and cut across the medial or distal fan (Figs. 14.15e and 14.16d) (Longwell 1930, Wallace 1978, 1984a, 1984b, Beehner 1990, Reheis and McKee 1991). Scarps create unstable, high-angle slopes that will degrade by freefall or slumping of sediment to produce fault-slope colluvial wedges (e.g.

Wallace 1978, Nash 1986, Berry 1990, Nelson 1992). Faulting may disrupt groundwater flow through the fan sediment, possibly initiating springs (Alexander and Coppola 1989). Scarps also instigate the development of headward-eroding gullies (Fig. 14.15e). Fan deposits can also develop fissures during earthquake motion. Mountain-front or intra-fan faults with a strike-slip component, such as along the San Andreas or North Death Valley faults of California, alternately can cause tilting or folding of fan sediment (e.g. Butler et al. 1988, Rockwell 1988, Blair 2000), or significant lateral offset of the fan from its catchment in a process called beheading (e.g. Harden and Matti 1989). Rotational tilting of fan deposits also is a common feature caused by dip-slip along listric mountain front faults (e.g. Hooke 1972, Rockwell 1988).

Weathering and Soil Development

Many types of physical and chemical weathering modify fan sediment, including salt crystal growth in voids, exfoliation, oxidation, hydrolysis, and dissolution (Hunt and Mabey 1966, Ritter 1978, Goudie and Day 1980). These reactions take place both on the surface of clasts, and within them along fractures, foliation planes, or bedding planes. The net effect of surface weathering is the reduction of clast size, as exemplified by clasts near the evaporitic playa in Death Valley (Fig. 14.16e). Given time, even boulders can be reduced to fine sediment by weathering (Fig. 14.16a). Alteration of fan deposits by oxidation and hydrolysis also takes place immediately below the surface, where unstable grains such as feldspar or ferromagnesian minerals alter to clay minerals or haematite (Walker 1967, Walker and Honea 1969).

Another common clast modification involves the precipitation on exposed surfaces of thin hydrous ferric manganese oxide coatings called desert varnish (Figs. 14.15f and 14.16a) (Hunt et al. 1966). Variable darkness of the varnish and its microstratigraphy allow for the differentiation of relative ages of fan lobes, with the oldest lobes the darkest (e.g. Hooke 1967, Dorn and Oberlander 1981, Dorn 1988, Liu and Broecker 2007). Radiocarbon dating of varnish provides potential for obtaining absolute exposure ages of surfaces, although problems with this method remain, such as isolating enough carbon (Dorn et al. 1989, Bierman and Gille-

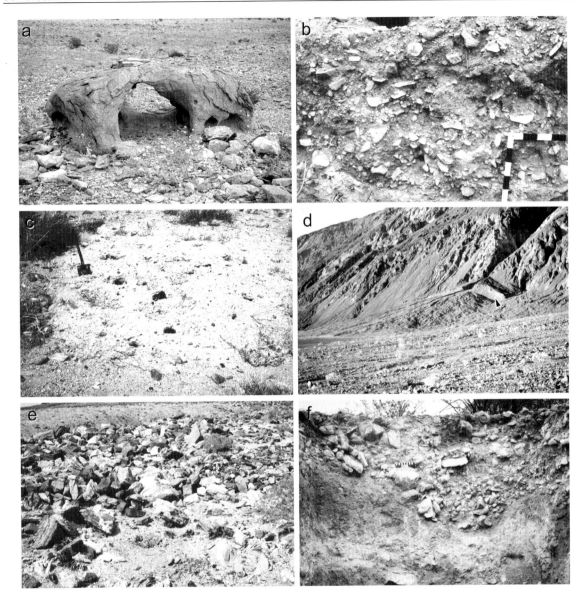

Fig. 14.16 Photographs of the products of secondary fan processes. (**a**) View of exhumed debris-flow units on the Hanaupah fan, Death Valley, where upon a varnished desert pavement has developed, and clasts have become extensively weathered, including the hollowing and disintegration of boulders 1 m long. (**b**) Gully cut of a proximal fan along the western Jarilla Mountains, south-central New Mexico. Gravelly fan and sandy aeolian deposits are intermixed by plant rooting, resulting in a loss of primary stratification. Darkened Bt (argillic) and white Btk (argillic and calcic) soil horizons are developed, and pedogenic carbonate (white) extensively coats gravel clasts. (**c**) Overview of a rodent colony comprising numerous burrows in near-surface fan sediment, Walker Lake, Nevada; shovel for scale. (**d**) Fan in Death Valley that has been offset near the apex by a fault scarp (*arrow*). Colluvium is accumulating at the base of the scarp. (**e**) View of metamorphic cobbles of the distal North Badwater fan, Death Valley, that are disintegrating due to salt crystal growth along foliation planes. (**f**) Vertical trench revealing well-developed carbonate soil horizons in a fan along the western Jarilla Mountains, New Mexico. Cobbles and pebbles of a filled gully are visible in the centre of the photograph

spie 1991, Bierman et al. 1991, Reneau et al. 1991). Optical luminescence and cosmogenic-isotope dating are other methods employed to determine the age of fan surfaces, though with questionable success due to inheritance issues related to sediment exposure prior to fan deposition (e.g. Nishiizumi et al. 1993, White et al. 1996, Matmon et al. 2005, Robinson et al. 2005, Le et al. 2007).

Fan sediment also serves as a parent material for developing soils, especially on the inactive lobes where the surface is stable. Soils develop by the translocation of clay or solutes from the surface to shallow depth by infiltrating water, usually from precipitation directly on the fan, that then dries in the vadose zone to produce horizons enriched in clay, carbonate, silica, or gypsum (Fig. 14.16b, f). Although organic-enriched (*Ao*) or aeolian-related vesicular (*Av*) horizons may form on fans, B soil horizon are the most prevalent. These horizons form where infiltrating surface water desiccates, at which depth translocated clay attaches to grains, and solutes precipitate. B horizons enriched in clay (argillic, *Bt*), calcium carbonate (calcic, *Bk*), or mixed clay and carbonate (*Btk*) are the most common types present in fans of southwestern USA and Spain (Gile and Hawley 1966, Walker et al. 1978, Gile et al. 1981, Christenson and Purcell 1985, Machette 1985, Harvey 1987, Wells et al. 1987, Mayer et al. 1988, Reheis et al. 1989, Berry 1990, Blair et al. 1990, Harden et al. 1991, Slate 1991, Ritter et al. 2000). Gypsiferous and siliceous horizons are less commonly documented in fan sediment (Reheis 1986, Al-Sarawi 1988, Harden et al. 1991). Soil-profile development in fan sediment is a function of the time that a fan surface has been stable, the local or aeolian flux of the materials from which the horizon is composed, and the typical amount and wetting depth of precipitation (Machette 1985, Reheis 1986, Mayer et al. 1988). The presence of plant roots in sediment promotes soil development by providing pathways for infiltrating water. The extent of soil-horizon development in a given area is useful for determining the relative age and correlation of fan deposits (e.g. Wells et al. 1987, Slate 1991).

Extensively developed soil horizons can produce lithified zones, including calcrete (also called caliche or petrocalcic horizon), silcrete, and gypcrete. Subsequent exhumation of tightly cemented soil horizons armours the fan from further secondary erosion, and expedites the downslope movement of overland flow

(Lattman 1973, Gile et al. 1981, Van Arsdale 1982, Wells et al. 1987, Harvey 1990).

Significance of Distinguishing Primary and Secondary Processes

Primary versus secondary processes on alluvial fans have long been understood (e.g. Beaty 1963, Denny 1967), but much confusion remains because of the failure of many to appreciate their differences. The greatest problem has resulted from assuming that secondary processes, which usually dominate the fan surface because of their frequency, are the principal processes constructing the fan, rather than realizing that they mostly surficially remould and mask primary deposits. Outgrowths of this erroneous view include the ideas of sieve lobes and braided streams on fans, and equating fan activity to climatic influences.

Sieve Lobes on Alluvial Fans

The concept of sieve-lobe deposition on alluvial fans originated from laboratory studies of small-scale (radii ≤ 1-m-long) features constructed of granules and sand that morphologically resembled fans (Hooke 1967). In these experiments, a lobate deposit was identified as forming by the rapid infiltration of sediment-laden water into the permeable sand substrate. This feature, termed a sieve lobe, formed in the sand box where water was unable to achieve further transport due to infiltration. Although acknowledging that these studies may lack significance to real fans, it was concluded that sieve lobes like those of the laboratory comprise extensive deposits on seven natural fans in California (Hooke 1967). These fans were reported to be constructed of material derived from catchments underlain by bedrock that did not weather to produce permeability-reducing fine sediment. Based on this work, the sieve lobe mechanism has become established as one of the major processes operative on alluvial fans (e.g. Bull 1972, 1977, Spearing 1974, Dohrenwend 1987). An evaluation of the exposed stratigraphy of the type fans identified as possessing sieve lobes, however, revealed that

Fig. 14.17 Photographs of misidentified fan processes. (**a**) Oblique view of one of the type 'sieve lobes' (photograph centre), North Badwater fan, Death Valley. (**b**) View of 1.5-m-high channel wall cut in the vicinity of the previous photograph revealing matrix-rich debris-flow deposits below a zone containing the proposed sieve lobe. The fine sediment has been removed from the surficial part of the debris-flow deposits by overland flow, leaving an arcuate mass of winnowed gravel incorrectly called a sieve lobe. (**c**) Overview of the Roaring River fan, Colorado. Secondary post-flood braided distributary channels (*D*) were carved into the fan surface underlain solely by sheetflood deposits. (**d**) View of light-coloured, clast-poor debris flows on the Trail Canyon fan, Death Valley. These flows give a misleading braided distributary appearance to the fan when viewed from a distance

they are constructed by debris flows (Fig. 14.17a, b). The catchment bedrock in these cases has weathered to produce abundant mud, at odds with the sieve lobe concept. Instead of developing through sieving, these features represent the surficial part of clast-rich debris-flows lobes from which the matrix has been removed by secondary overland flows (Blair and McPherson 1992). Matrix is abundant at depth in the deposits in all of these cases.

Braided Distributary Channels on Alluvial Fans

Perhaps the greatest misconception concerning alluvial fans is the widely held belief that they are constructed from braided distributary channels by the same processes that are operative in braided streams. This view results from the apparent presence of channels with a braid-like pattern on many fan surfaces. The idea was popularized by Bull (1972) who wrote (p. 66):

> Most of the water-laid sediments [of alluvial fans] consist of sheets of sand, silt, and gravel deposited by a network of braided distributary channels. ... The shallow distributary channels rapidly fill with sediment and then shift a short distance to another location. The resulting deposit commonly is a sheetlike deposit of sand, or gravel, that is traversed by shallow channels that repeatedly divide and rejoin. ... In general, they [the deposits] may be cross-bedded, laminated, or massive. The characteristics of sediments deposited by braided streams are described in detail by Doeglas (1962).

Our examination of the stratigraphic sequence of numerous fans displaying shallow braid-like channels reveals that, in every case, these features are surficial and formed either by slight rilling during waning flood stage, or through later secondary surficial winnowing and remoulding of primary deposits, including of debris flows and sheetfloods (e.g. Fig. 14.17c). Additionally, the stratotypes of Doeglas (1962) for braided streams that were concluded by Bull (1972) to form on fans, such as lower-flow-regime planar and trough crossbedding, are not found in exposures of waterlaid fans. The calculation of critical slopes shows that such structures cannot form on fans (Fig. 14.14). Other issues are illustrated by two fans reported to be built by braided distributary channels, including the Trail and Hanaupah Canyon fans of Death Valley (Fig. 14.1a and 14.2a) (e.g. Nummedal and Boothroyd 1976, Richards 1982). Such channels on the Hanaupah fan actually have a contributary, rather than distributary pattern (see Richards 1982, p. 249) that formed by headward erosion of gullies into debris-flow deposits of the inactive part of the fan. The light-coloured 'channels' on the Trail Canyon fan also are shallow secondary features cut into debris flows, and the most vivid of these features comprise clast-poor debris-flow lobes (Fig. 14.17d).

The failure to understand the origin of the apparent 'braided distributary channels' has caused the inability of many to recognize the major constructive processes, such as sheetfloods and debris flows. The surficial reworking of primary deposits, combined with their long recurrence intervals, is probably why it was not until Blackwelder (1928) that the importance of debris flows in fan construction was recognized. Similarly, the masking of sheetfloods by surficial remoulding was concluded by Blair (1987a) to be the reason that it was previously not appreciated as principal fan-building process.

Implications for Climate Change-Process Model for Alluvial Fans

The downcutting of incised channels on alluvial fans (i.e. fanhead trenching) and the dissection of the fan surface by rilling and gullying have been attributed by many authors to be a consequence of climatic change. Fan sequences undergoing dissection are believed to have accumulated during more moist periods in the past, and thus represent

fossil fans (e.g. Blissenbach 1954, Lustig 1965, Melton 1965, Williams 1973, Nilsen 1985, Harvey 1984a, b, 1987, 1988, Dohrenwend 1987, Dorn et al. 1987, Dorn 1988). Although the validity of this climate-response hypothesis has been questioned (e.g. Rachocki 1981), and the difficulty of establishing time stratigraphy and climate-sensitivity parameters remains, this hypothesis has been widely reiterated. It is based on the idea that moister past conditions caused greater sediment production in the catchment, and that aggradation concurrently took place as a result of the expedient transfer of this sediment to the fan. As a corollary, this model claims that sediment production in the catchment is retarded during periods of greater aridity, causing water flows to depart the catchment without sediment, thereby eroding the fan. At odds with this view is the fact that the fans used to support it, such as those of the Panamint piedmont in Death Valley, have been the sites of historical aggradation (e.g. Fig. 14.17d), and that the catchment of these and many other desert fans remain well-stocked with colluvium. Primary aggradational events are documented on fans around the globe, casting further doubt on the climate-caused fossil-fan theory. Additionally, the moister late Pleistocene conditions of fans in settings such as Death Valley were still more arid than the present setting of many other areas with alluvial fans. Finally, as shown by case studies, rills and gullies are intrinsic secondary processes active, irrespective of climate, during periods between infrequent primary events.

Controls on Fan Processes

At least five factors influence fan processes, including catchment bedrock lithology, catchment shape, neighbouring environments, climate, and tectonism. Although complicated due to interactions, the impact of each of these variables can be examined.

Catchment Bedrock Lithology

The type of bedrock underlying the catchment is the main control on the primary processes of alluvial fans (Blair and McPherson 1994a, b, 1998, Blair 1999b, d, f). Rocks of differing lithology yield contrasting sediment suites and volumes due to their variable response to weathering. Bedrock in desert settings optimal for fan development, especially tectonically maintained mountain fronts, yields sediment in varying size and volume depending on: (a) the style of fracturing in proximity to faults, (b) the presence or absence of internal discontinuities such as bedding planes or foliation planes, and (c) the reaction to chemical weathering and non-tectonic types of physical weathering. These effects can be exemplified by a survey of sediment found in fan catchments underlain by various bedrock in the southwestern USA.

Granitic to dioritic plutons and gneissic bedrock split into particles ranging from sand to very coarse boulders due to jointing, fracturing, exfoliation, and granular disintegration. The coarse sediment size of fans derived from plutons results from a commonly uniform joint pattern developed due to a homogeneous, coarsely crystalline fabric. Gneissic rocks typically yield more bladed, platy, or oblate boulders due to anisotropy from the metamorphic foliation. Boulders from both of these lithotypes are either angular, reflecting the joint pattern, or are more rounded depending upon the degree of weathering along the clast edges. Very fine pebbles and sand (grus) also are commonly produced from granitic or gneissic rocks related to physical disaggregation of crystals typically of this size. Clay-sized sediment is only a minor product in arid settings because it forms either through tight tectonic shearing (Blair 1999c, 2003), or more importantly by hydrolysis of feldspar and accessory minerals. Such reactions are slow especially in hyper-arid deserts, and thus mud yield requires long residence time in the colluvium.

Bedrock consisting of tightly cemented dense sedimentary rocks, such as quartzite, undergoes significant brittle fracture in proximity to mountain-front faults, producing angular pebbles, cobbles, and boulders. Little sand, silt, or clay is generated in this situation due to effective cementation of the matrix grains. Dense carbonate rocks also respond to tectonism in a brittle fashion, producing bladed, platy, or oblate clasts. If present, interstratified soft sedimentary rocks such as shale add a clay fraction to the colluvium, and cause the intervening brittle rocks to fracture and weather to produce tabular clasts.

Finer-grained catchment bedrock, such as pelitic metamorphic rocks, shale, mudstone, or volcanic tuff, commonly weather to yield sediment varying in size from boulders to clay, with an abundance of the finer

sizes but a deficiency of sand. This size suite results from the formation of gravel due to fracturing, and of silt and clay, but not sand, from disaggregation. Thus, thickly mantled colluvial slopes comprising cobbles, pebbles, and clay are commonly developed on bedrock of this type.

The various weathering styles of bedrock, and the colluvial textures they produce, promote different modes of erosion and sediment transport. Fractured brittle bedrock slopes yield rockfalls, rock slides, and rock avalanches, whereas water-sensitive fine-grained bedrock such as shale may yield earth flows. Bedrock that weathers to produce abundant gravel and sand but little mud sheds colluvium with low cohesion. Failure of these slopes in response to water input incites incised-channel and sheetfloods, and more rarely, NCSGFs, on the coupled alluvial fan. Finer-grained rocks, such as shale, pelitic metamorphic rocks, and volcanic rocks, more commonly yield mud as well as gravel, forming cohesive colluvium. The failure of such colluvium with the addition of water typically produces debris flows.

The contrasting processes resulting from the cohesive versus noncohesive catchment colluvium are illustrated by two examples. The adjoining Anvil and Warm Springs Canyon fans and their catchments in Death Valley are forming under identical tectonic, climatic, and topographic conditions, but the Anvil fan is built mainly by sheetfloods, and the Warm Spring fan by debris flows (Blair 1999b, d, f). This difference is the result of contrasting bedrock types underlying their catchments, which leads to cohesive colluvium in the Warm Spring fan catchment and noncohesive colluvium in the Anvil catchment. Differing bedrock lithotypes can also affect the rate of colluvium erosion, as exemplified by the Nahal Yael catchment in Israel. Bull and Schick (1979) concluded that the grusy colluvium from granitic rocks has been mostly stripped, whereas areas underlain by amphibolite have been only partly stripped due to greater cohesion.

Catchment Shape and Pre-Existing Geology

The overall shape and evolution of a catchment can impact the operative sedimentary processes on an alluvial fan. Catchment shape affects side slopes, feeder-channel profile, relief, propensity for flash flood promotion, and sediment-storage capacity. Slope angles, along with bedrock type, may determine whether rockfalls, rock avalanches, colluvial slips, debris flows, or flash floods are promoted. The presence or absence of storage capacity in the catchment also affect sediment delivery to the fan site. Relief and area may determine the volume of sediment that can be generated and transported in a single flow, whereas the elevation of the catchment affects the chances of receiving significant precipitation from either rainfall or snowfall. The orientation of the catchment with respect to sunlight, or the track of major storms, can also influence weathering, erosion, and transport activity, and thereby fan aggradation.

The ability of catchments to rapidly transmit or store sediment varies with their area, which can range from < 1 to > 100 km². The smallest catchments may consist only of a single valley carved along a fracture in bedrock, with dislodged clasts rapidly transferred to the fan. Feeder-channel incision and widening can proceed with time, allowing storage of colluvium or of primary deposits such as debris flows. Progressively greater storage occurs with catchment enlargement because sediment can be maintained either as side-sloping colluvium or as deposits on the floor of the drainage net. The volume of sediment stored in a feeder channel depends on the long profile and width. The long profiles of feeder channels can range from consistently steep to step-like (Fig. 14.18). The reaches with reduced slope in stepped feeder-channel profiles may induce deposition from passing flows, the volume of which increases as a function of channel width. Two

examples of feeder channels that have stepped profiles containing zones of sediment storage are the Coffin and Copper Canyon fans of Death Valley. Feeder channel erosion is dominant in the reaches of these feeder channels with slopes of > 7°, whereas sediment deposition has occurred in reaches sloping ≤ 7° (Fig. 14.18). Storage of sediment may lessen aggradational rates on the fan, but alternatively may allow for the generation of high-volume primary flows capable of constructing large alluvial fans (e.g. Blair 2003).

The ability of the feeder channel to store sediment appears to increase with catchment size, probably reflecting structural complexities in the underlying bedrock. This relationship is illustrated by the 20 adjoining fans in the vicinity of Copper Canyon in Death Valley. The catchments of these fans vary in area from < 0.5 to 60 km² (Fig. 14.19). The largest fans, numbers 2, 5, 8, 12, and 14, have relatively large catchments with fourth- to sixth-order feeder channels, whereas the smaller catchments have second- or third-order feeders. Most stored sediment in this sector of the Black Mountains occurs in the two largest catchments (Coffin and Copper), and is concentrated along their highest-order channels.

The initial shape of a fan catchment and its subsequent evolution are largely a product of: (a) inherited (pre-existing) local and regional structures and discontinuities such as faults, joints, and geological contacts, (b) newly imposed (neotectonic) structural discontinuities, and (c) bedrock lithology. In general, fractures or other discontinuities, whether inherited or neotectonic, become the locus of catchment development because these zones erode more quickly relative to adjoining

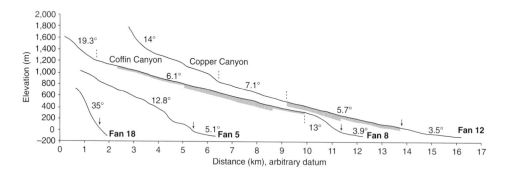

Fig. 14.18 Long profiles of the feeder channels for four fans in the vicinity of Copper Canyon, Death Valley (see Fig. 14.19 for plan-view locations). Average slopes of the feeder channel segments are labelled. The shaded pattern denotes segments with stored sediment, with pattern thickness depicting greatest sediment storage. Vertical exaggeration is 2.5X; vertical datum is mean sea level

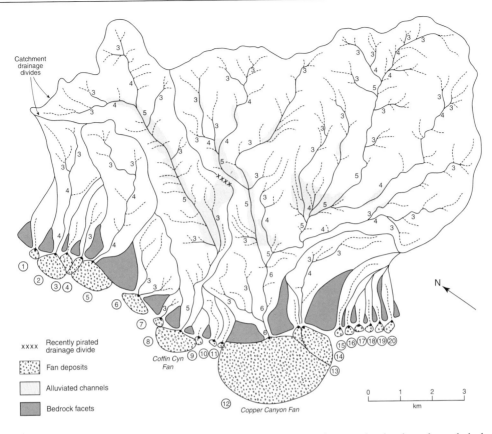

Fig. 14.19 Twenty fans (circled numbers) and their catchment drainage nets in the vicinity of Copper Canyon, Death Valley, based on 1:24,000 scale topographic maps with 12.2 m (40 ft) contours. First-order channels in the catchments are not shown due to map scale; second-order channels are dashed; and higher-order channels are solid lines labelled by order. Alluviated channels are after Drewes (1963)

ones, and thus focus overland flow. This factor is exemplified by comparing the catchment drainage net of the 20 fans in the Copper Canyon vicinity of Death Valley (Fig. 14.19) with their underlying geology (Fig. 14.20). These maps show that the position and orientation of feeder channels of the largest fans coincide with faults trending at a high angle to the mountain front. This relationship is due to enhanced weathering and erosion along the pre-existing structures. Another example of how inherited geology affects catchment and fan development is provided by the Copper Canyon fan. This catchment is centred on a down-dropped block, the surface of which is composed of relatively soft Miocene-Pliocene sedimentary rocks (fan 12, Figs. 14.19 and 14.20). The resultant high sediment yield has produced the largest fan in this sector of the Black Mountains piedmont. By contrast, nearby small fans have catchments developed upon relatively unfractured anticlinal flanks of Precambrian metasedimentary rock.

Effects of Adjoining Environments

Aeolian, fluvial, volcanic, lacustrine, or marine environments that border alluvial fans can impact fan processes by modifying the conditions of deposition. Aeolian sandsheet or dunes on fans (Fig. 14.21a) limit the runout distance of water or debris flows, causing aggradation in more proximal settings. The presence of dunes also can cause unconfined water flows to channelize by creating topographic barriers. For example, along the western Jarilla Mountain piedmont of south-central New Mexico, aeolian sandsheet deposition has been so high relative to fan activity that fan sedimentation has been reduced to flows in isolated arroyos cut into the aeolian deposits (Fig. 14.16f) (Blair et al. 1990).

Fluvial environments, usually in the form of longitudinally oriented rivers, may affect fans by eroding

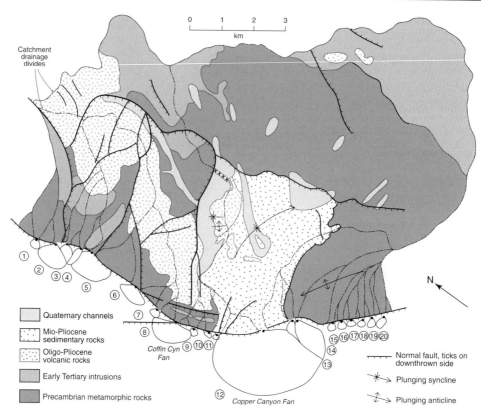

Fig. 14.20 Bedrock and structural geology underlying the catchments of 20 alluvial fans in the vicinity of Copper Canyon, Death Valley, that are depicted in Fig. 14.19 (geology after Drewes 1963)

their distal margins. An example is where Death Valley Wash has eroded the toes of converging fans in northern Death Valley (Fig. 14.21b). Headward-eroding gullies extend up-fan from these toe cuts. Another example is from near Schurz, Nevada, where the Walker River is eroding the distal Deadman Canyon fan, and distributing this sediment into its floodplain or delta (Fig. 14.21c) (Blair and McPherson 1994c). Progressive downcutting of an axial channel can also cause the fan to build at progressively lower steps to produce a telescoped pattern (Drew 1873, Bowman 1978, Colombo 2005).

The presence of lacustrine or marine water bodies marginal to fans can affect fan development in several ways. Subaerial processes may transform into other flow types, such as subaqueous debris flows or underflows, upon reaching a shoreline (Sneh 1979, McPherson et al. 1987). A fan may be significantly eroded by wave or longshore currents where it adjoins a water body, and become the site of beach, longshore drift, or shoreface accumulations (e.g. Link

et al. 1985, Beckvar and Kidwell 1988, Newton and Grossman 1988, Blair and McPherson 1994b, 2008, Blair 1999e, Ibbeken and Warnke 2000). Pronounced changes in the depth of lakes, such as for Walker Lake, Nevada, can cause significant fan erosion over a large elevation range, and, upon lake-level fall, leave oversteepened slopes (Fig. 14.21d). Falling lake level also can induce the development of headward-eroding gullies, or a telescoped progradational pattern (e.g. Harvey et al. 1999, Harvey 2005). Flat-lying evaporitic peritidal or playa settings may develop in the distal fan area (Hayward 1985, Purser 1987), promoting particle weathering, and inciting primary fan flows to undergo deposition due to a pronounced slope reduction (Fig. 14.21e).

Volcanism can strongly affect fan processes by emplacing ash either in the catchments or directly on the fan, causing an interference of flows at the fan site, and potentially instigating debris flows on steep catchment slopes. Volcanic flows emanating from mountain-front faults may also cause barriers to sediment transport on

Fig. 14.21 Effects of neighbouring environments on alluvial fans. (**a**) Aeolian sandsheet deposits flanking the distal Shadow Rock and Trollheim fans, Deep Springs Valley, California. (**b**) View of eroded toes (*dark vertical walls*, photograph centre) of converging fans in northern Death Valley. Erosion has been caused by the concentration of discharge in the longitudinal Death Valley Wash. (**c**) Erosion into distal Deadman Canyon fan (*vertical wall*) by Walker River near Schurz, Nevada. (**d**) Extensive erosion by waves and longshore currents of fans adjoining Walker Lake, Nevada; vehicle (*lower arrow*) for scale. The now-exposed lower fan segment (downslope of *upper arrow*) has a greater slope than the upper segment due to erosion. (**e**) Aerial view of the Coffin Canyon fan prograding into Badwater playa in Death Valley; road for scale. Debris flow tongues reaching this playa undergo deposition due to the rapid lowering of slope. (**f**) Aerial view of fan of the Sierra Nevada piedmont in central Owens Valley widely capped by rugged basalt flows (*F*), and possessing a cinder cone (*C*) near the apex. The fan toe is bordered by Owens River

the fan surface, in extreme cases armouring the entire fan with almost unerodible basalt, such as on fans in the vicinity of the Poverty Hills in Owens Valley, California (Fig. 14.21f).

Climatic Effects

Climate is widely believed to have a major control on alluvial fans because water availability impacts factors such as weathering, sediment generation, and vegetation. However, although many conclusions have been speculated, the effect of climate on fans remains unestablished. Two directions have emerged, one evaluating climatic variables and the second associating specific fan processes to climate.

Effects of Climate Variables on Fans

Three interrelated climatic variables discussed with regard to alluvial fans are precipitation, temperature, and vegetation. These variables may be relevant to fans inasmuch as they affect bedrock weathering rates, sediment yield, and the recurrence interval of primary events. The most basic aspect of precipitation is the mean annual amount because, without rainfall, weathering and vegetation would be limited, and sediment transport would be restricted to dry mechanisms such as rockfall, rock slides, and rock avalanches. However, weathering and fan aggradation can occur even with minimal precipitation.

Two other perhaps more significant aspects of precipitation are the intensity of individual events and their frequency (e.g. Leopold 1951, Caine 1980, David-Novak et al. 2004). Both of these variables affect infiltration capacity in the catchments, which must be exceeded before overland flow and potential sediment transport can occur. Infiltration capacity is exceeded, and discharge events are generated, either by intense rainfall, or by less intense rainfall following antecedent precipitation (Ritter 1978, Cannon and Ellen 1985, Wieczorek 1987). Discharge arises in the latter case if the infiltration capacity is unable to return to its original value through percolation or evaporation. Rainfall intensity and frequency strongly affect fans by inciting primary processes. High-intensity thunderstorms probably are the most important mechanism for instigating such events in deserts, followed by extended periods of rainfall.

Another aspect of rainfall intensity and frequency that may impact alluvial fans is short-term global variations in ocean circulation patterns, such as the El Niño Southern Oscillation that causes greater precipitation along the Pacific coast of the Americas. Slope failures are documented to have been more frequent in California during the 1982–1984 and 1997–1998 El Niño episodes (Ellen and Wieczorek 1988, Coe et al. 1998, Jayko et al. 1999, Gabet and Dunne 2002).

Fan aggradation was also more frequent during historical El Niño periods in Peru, and such activity there during the last 38000 years has been proposed to be mostly related to such periods (Keefer et al. 2003). A similar speculation has been made for late Quaternary deposits in the southwestern USA (Harvey et al. 1999).

The effect temperature has on fans is more poorly understood. It likely is significant, however, given that chemical weathering rates increase exponentially with temperature. Temperature gradients caused by the orographic conditions in a catchment may result in an initial decrease in weathering rates upslope due to decreasing temperature, followed by an increase in weathering at higher altitudes where freeze-thaw or heating-cooling fatigue processes become important. This trend may be complicated by the tendency for weathering rates to increase as precipitation increases with altitude.

Vegetation has long been rated an important factor concerning sediment yield from a catchment. Inasmuch as vegetation reflects climate, it is a climate variable. One factor commonly attributed to catchment plant cover is an increase in clay production due to the enhanced chemical weathering caused by organic acids near roots, and due to the greater preservation of soil moisture (e.g. Lustig 1965). Plant roots also affect sediment slope stability by strengthening its resistance to gravity as a result of increased shear strength (Greenway 1987). Differing plant types will have a variable effect on slopes due to the multitude of styles and depth of root penetration, and to the density of the ground cover (Terwilliger and Waldron 1991). Contrarily, plants may serve to produce long-term instability by causing the slopes to become steeper than they would be if the plant cover did not exist. This effect is illustrated by the common failure of slopes after vegetation is disturbed, such as after forest or brush fires (e.g. Wells 1987, Meyer and Wells 1996, Cannon and Reneau 2000, Cannon 2001).

The Question of 'Wet' and 'Dry' Alluvial Fans

A second approach with respect to climate and fans has been the attempt to relate the prevalence of certain fan processes to broad desert (dry) or non-desert (wet) categories. The main claim of this hypothesis is that debris flows prevail on fans in arid and semi-arid climates, and water flows prevail in wetter climates (e.g. Schumm 1977, McGowen 1979, Miall 1981). This hypothesis continues to be widely held even though, for nearly a century, data from fans have disproved it. For example, a catalogue of the global occurrence of historical debris flows on fans shows that they have been long known to develop under all climatic conditions (Costa 1984, Blair and McPherson 1994b). Additionally, fans dominated by sheetflooding have been shown to be present under some of the driest (Death Valley) and wettest (South Island, New Zealand) climates on earth (Blair and McPherson 1994b, Blair 1999d).

This climate hypothesis has been taken further to purport that river plains or deltas, such as the Copper River delta of Alaska, the Kosi River of India, the swampy Okavango River delta of Botswana, and even the Mississippi River delta, represent the 'humid-climate type' of alluvial fans (Boothroyd 1972, Boothroyd and Nummedal 1978, McGowen 1979, Nilsen 1982, Fraser and Suttner 1986, McCarthy et al. 2002). This hypothesis: (a) fails to acknowledge that fans as described in this chapter exist in humid as well as desert climates, (b) arbitrarily designates some river systems as fans and others as rivers, and (c) fails to recognize the scientific uniqueness of fans versus rivers in terms of hydraulic conditions, processes, forms, and facies (see Blair and McPherson 1994b for discussion).

Tectonic Effects

The most common and favourable conditions for the development and long-term preservation (including into the rock record) of alluvial fans exist in tectonically active zones that juxtapose mountains and lowland valleys. Sediment yield exponentially increases with relief (e.g. Ahnert 1970), and tectonism can both create and maintain relief. Without continued tectonism, fans may be minor or short-lived features characterized by secondary reworking. A possible example of fans of this style are those in southeast Spain developed adjacent to compressional structures formed prior to middle Miocene time (Harvey 1984a, 1988). This setting contrasts with that of an extensional basin, where relief and mountain-to-valley topographic configuration can be maintained for tens of millions of years, and where individual fans may be sites of net aggradation for 1–7 million years or more (e.g.

Blair and Bilodeau 1988). Extensional and translational tectonic settings are best for fan development and preservation, whereas the lateral movement and recycling of the mountain front makes compressional regimes less ideal.

More detailed characteristics of tectonism, including rates and occurrence of uplift, down-throw, and lateral displacement, can influence the overall form and development of a fan (e.g. Fig. 14.15a). Examples of fans reflecting small-scale variations in tectonism are those along the range-front fault in southeastern Death Valley (Figs. 14.19 and 14.20). Secondary and inherited structures have impacted sediment yield and catchment development there, as demonstrated by the variations in size of the fans and their catchments. Another common relationship is that fans along the tectonically active side of half grabens overall are smaller than those of the opposing inactive side due to greater tectonic subsidence (e.g. Hunt and Mabey 1966). A larger-scale example of tectonic variation in the Basin and Range province affecting fan development is provided by maps of fault activity (Thenhaus and Wentworth 1982, Wallace 1984b). Such maps delineate sub-provinces defined by whether the age of the most recent faulting was historical, pre-historical Holocene, late Quaternary, or pre-late Quaternary. The best developed fans, such as those of Death Valley, are present in sub-provinces with the most recent faulting.

Tectonism also affects fans by influencing climate and catchment vegetation. Adjustments to elevation directly impact these variables, possibly affecting weathering or erodibility of the catchment bedrock. As a result, factors such as sediment supply rate, calibre, and flash-flood frequency may be altered, and these alterations may affect the primary sediment transport mechanisms.

Alluvial Fan Forms

Two classes of alluvial fan forms, constituent and composite, have been differentiated on the basis of origin and scale (Blair and McPherson 1994a). Constituent forms are produced directly from the primary and secondary processes building and modifying the fan, or from external influences such as faulting or interactions with neighbouring environments. Individually, they comprise a small part of the fan. Constituent forms contrast with the overall fan morphology, or the composite form, representing the consequential or resultant shape of all of the constituent forms.

Constituent Morphology

If present, the incised channel usually is the most prominent constituent form of an alluvial fan. It ranges from 5 to 150 m in width, has nearly vertical walls 1–20 m high, and extends 10's to 1000's of metres down-fan from the apex (Figs. 14.2, 14.3, and 14.6d). Outburst floods may incise additional channels of similar size on the fan (Fig. 14.12d). Smaller forms may be present within incised channels, including terraces, boulders bars, debris-flow plugs, and gullies (Fig. 14.12e and f). Alternately, a smooth channel floor may result from abundant sand or clast-poor debris flows. Other large constituent forms that may exist on fans are rock-slide and rock-avalanche deposits. These features can rise ∼10–100 m above the surface of the fan, and span 10's to 1000's of metres in plan view (Figs. 14.8e, f and 14.9a, b).

Many constituent forms have 0.5–5 m of relief and a lateral extent of 10's of metres. Rockfall clasts, rock-avalanche tongues, boulder-log jams, and debris-flow levées cause irregularities, particularly in the proximal fan, with relief of 1–4 m (Figs. 14.11 and 14.22). Individual debris-flow lobes may be 2–10 m across, and extend radially for 10's to 100's of metres with 1 m of relief. Erosive secondary forms, including gullies and winnowed mantles (Fig. 14.15e, f), may also produce features of this scale. Fault scarps and toe cutting can create walls ∼1–5 m high (Fig. 14.16d). Volcanic flows may protrude 0.5 m or more above the fan surface, and extend for 100's of metres. Lower-relief forms (≤ 0.5 m high) that extend radially for 10's to 100's of metres also are common on fans, especially rills produced by secondary erosion (Fig. 14.15c). Windblown sand and rodent colonies may form mounds of this scale (e.g. Fig. 14.16c). Sediment-deficient sheetfloods can produce widely distributed transverse ribs 20 cm high, and sediment-laden sheetflood and clast-poor debris flows may produce expansive nearly smooth areas. Smooth fan surfaces also can be formed by overland flow and aeolian winnowing (Fig. 14.16a).

Constituent forms generally are not shown in cross-fan or radial-fan profiles because their relief is less than the typical resolution of topographic maps. Constituent forms such as incised channels, levees, and lobe bound-

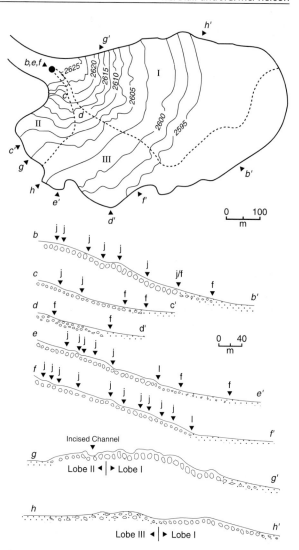

Fig. 14.22 Topographic map (*top*) and radial and cross-profiles of the Roaring River fan, Colorado, with some constituent forms labelled (after Blair 1987a). Arrows point to segment boundaries from boulder-log jams (*j*), facies changes (*f*), or lobe boundaries (*l*). Vertical exaggeration of profiles is 2.5X

aries, may be discernible from maps with ∼ 1 m contour intervals (Fig. 14.22), or from databases generated from LiDAR.

Composite Morphology

The overall fan form, or composite morphology, is characterized by plan-view shape, presence or absence of incised channel, and radial and cross-fan profiles. Composite forms are more studied than the constituent forms because they can more easily be evaluated using maps and aerial imagery.

Plan-View Shape and Incised Channels

Alluvial fans have a semi-circular plan view shape where they aggrade without lateral constriction, and a pie-pieced shape where they laterally coalesce (Fig. 14.1). Lateral constriction from coalescence causes the fan radii to become elongated perpendicular to the range front. Alternately, the presence of lakes, oceans, aeolian sandsheets, rivers, or fans from the opposing range front can limit the radial length of a specific fan in a trend perpendicular to the range front (e.g. Fig. 14.21d).

Although a constituent form, the presence or absence of a prominent incised channel on a fan is an important aspect of the composite form. Incised channels promote fan progradation by transferring flows from the catchment to active depositional lobes positioned progressively farther from the range front. Incised channels, therefore, are more likely to exist and have greater length on fans with longer radii. Such fans also are more likely to possess large abandoned surfaces in the bypassed proximal zone.

Cross-Fan Profiles

Cross-fan profiles are a poorly studied element of the composite fan form. Cross-fan profiles overall have a plano-convex geometry, although variations exist. Such profiles from the upper fan, for example, have greater amplitude than those from the lower fan (Fig. 14.4b). The height and relative smoothness of the cross profiles may vary between fans due to differences in relief caused by variations in the primary processes or the degree of lateral coalescence. On the Roaring River fan, for example, cross profiles are asymmetric due to the distribution of lobes (Fig. 14.22). Cross-profile width may also vary between fans due to catchment evolution, overlap by adjoining fans, erosion from adjoining environments, or differing constituent forms.

Radial Profiles

The radial profile is a significant element of the composite fan form because it is determined by and influences primary processes. Radial profiles are characterized by slope pattern and magnitude. The radial pattern may exhibit: (a) a constant slope like that of a cone segment, (b) a distally decreasing slope depicting a half of a plano-concave-upward form, or (c) a segmented slope, where the surface inflects distally to one or more less-steep segments (Fig. 14.4a). Each of these profiles may either be smooth, or have notable irregularities resulting from the presence of rock avalanches, rock slides, or steps from boulder-log jams, fault scarps, or telescopic boundaries. The radial profile may alternately increase distally from erosion (Fig. 14.21d). Segmented profiles have been attributed to either a change in slope caused by tectonism, base-level change, or climate change (Bull 1964a). In the case of fans in Fresno County, California, Bull (1962) favoured a slope change due to catchment uplift. The Roaring River and Dolomite fans show that segmentation can result from intrinsic factors such as a change from debris-flow levées to lobes, or crossing boundaries between lobes or constituent forms (Figs. 14.11d and 14.22).

The average slope of alluvial fans ranges from 2° to 35°, with most between 2° and 20°. Except for faulting and aggradational or erosional effects of neighbouring environments, the fan radial profile is a product of the dominant sedimentary processes, and is thus called the depositional profile or the depositional slope (Blair and McPherson 1994b). Fans built entirely of rockfall talus have depositional slopes of 25–35°, whereas the addition of debris-flow or colluvial slips can reduce the average to 15–30°. Fans dominated by colluvial slips, debris-flow levées, or NCSGFs with boulder-log jams commonly have a depositional slope of 9–18°; fans dominated by unconfined NCSGFs or clast-rich debris-flow lobes typically have slopes of 4–9°; and those dominated by clast-poor lobes commonly average 2–4°. A change from proximal levées to distal lobes, such as in the Dolomite fan case, produces a fan with a 9–15° proximal segment that inflects to a distal 4–7° segment. Sheetflood fans commonly have a radial slope that decreases progressively from about 5° proximally to 2–3° distally as a result of textural fining (Lawson 1913, Blissenbach 1952, Blair 1987a). Gravel-poor sheetflood fans and the distal part of mudflow fans have average slopes of 1.5–3°. Thus, besides processes, facies, settings, and scale, the depositional slope of a fan is another parameter that separates them from rivers in sedimentary basins, which typically slope $\leq 0.4°$ (Blair and McPherson 1994b).

A faction of authors claims that fans with slopes lower than those discussed above are common (Harvey et al. 2005). Fans for which this claim is made are undocumented or misclassified, with recent articles exemplifying the latter. The first example is the swampy terminus of the Okavango River in the Kalahari Desert of Botswana. This feature has long been identified as the Okavango delta, but some now claim that it is an alluvial fan (McCarthy et al. 2002). The nearly flat (0.0015°) slope, lack of conical form, total dominance by a swamp with channels, and numerous other features provide sound evidence that it is not an alluvial fan. The second example is a 'low-angle fan' designated to exist in the Hungarian plain. A location map of the nearly flat valley with the outlined 'fan' shows that it is crossed in the 'upper fan' segment by the high-sinuosity Tisza River (Gábris and Nagy 2005), indicating that this feature is not an alluvial fan. The third example is a 'fan' in Iran that is claimed to have a slope of 0.5° (Arzani 2005). Photographs provided in the article show that the 'fan' mostly comprises lake beds. Adding a wide interval of flat lake beds to an alluvial segment greatly reduces the average slope of the transect, but does not demonstrate a low-sloping fan. A fourth example is from the western Sierra Nevada of California, where rivers are claimed to have built low-sloping fans (e.g. Weissmann et al. 2005). However, features indicative of fans in this area are not evident in facies along channel walls, on topographic maps, or when examining these flat forms on the surface or in the subsurface via ground-penetrating radar (e.g. see Figs. 3, 5, and 6 in Bennett et al. 2006). The problem in all of these examples is error in basic landform recognition and map reading that is then compounded by argument for expanding the definition of 'alluvial fans' to include the misidentified features despite their contrasting characterisitics. If the definition of alluvial fan were expanded to include these and other misidentified features, than basically all sub-aerial environments and deposits would be unscientifically classified as alluvial fans.

Morphometric Relationships Between Fans and Their Catchments

Much attention has been given to relating morphometric attributes of alluvial fans and their catchments.

To date, such analysis has been based on topographic maps with scales providing insufficient accuracy. In the western USA, for example, the most detailed maps available before the 1980s, when most of the cited fan-catchment datasets were generated, had a scale of 1:62,500 and a contour interval of 40 or 80 ft. Such maps do not allow for an accurate delineation of the fan toe, which is why fans were erroneously concluded to have slopes that are gradational with neighbouring environments despite having clear slope-break boundaries on aerial photographs (e.g. Figs. 14.1 and 14.2). The reported results in these datasets thus are problematic because they may assign far too much or too little area to the fan, and because extending the fan to include flatter neighbouring environments gives inaccurately long radial lengths that lead to low slope determinations. Interpolation between contours to determine the fan apex elevation also adds error to fan height potentially equal to ± the value of the contour interval. Datasets from elsewhere have been generated from similar or poorer scale maps, or by using maps of unknown scale. Two relationships have nonetheless emerged from these studies, including area-area and area-slope.

Fan Area Versus Catchment Area

The most widely compared features of a fan and its catchment are their respective plan-view areas, which have a broad positive correlation (Fig. 14.23) (e.g. Bull 1962, 1964a, 1977, Denny 1965, Hawley and Wilson 1965, Hooke 1968, Beaumont 1972, Hooke and Rohrer 1977, French 1987, Lecce 1988, 1991, Mather et al. 2000). This relationship usually is given quantitatively as $A_f = cA_d^n$, where A_f is the fan area, A_d is the catchment (drainage) area, and c and n are empirically determined 'constants' that are not constant. The exponent n in widely referenced datasets varies from 0.7 to 1.1, and c ranges from 0.1 to 2.1 (Harvey 1989). The results produced by variations in the 'constants' alone differ so enormously as to not be relevant. For example, a catchment with an area of $50\,km^2$ is predicted by the area-area equation to equate to a fan with an area ranging from $1.6\,km^2$ to $155.3\,km^2$. This variation is why cross plots display a wide band of data (Fig. 14.23). Attempts have been made to isolate the effect of the variables on this relationship, but with unsatisfactory results. For example, Bull (1964a, b) concluded in his study of Fresno County, California fans that catchments underlain by erodible bedrock, such as shale, produce larger fans per unit of catchment area than catchments underlain by more resistant sandstone. In contrast, Lecce (1988, 1991) found that fans along the nearby western White Mountain front have greatest area per unit of catchment area where derived from very resistant bedrock such as quartzite.

The association of small fans with small catchments and large fans with large catchments is intuitively obvious because the movement of sediment from the catchment to the fan serves to increase the size of each. But wide data scatter and the need for changing 'constants' prove the invalidity of the commonly used fan area-catchment area relationship. As noted long ago (Lustig 1965), the problem with this equation is that it only compares the plan-view areas of three-dimensional objects, which is only valid mathematically if the vertical dimension of fans and catchments are constant, a condition that is never met. Catchment relief clearly is not constant, and in most cases fan thickness is unknown, but is highly unlikely to be constant. Further, due to dilatancy as rock transforms to sediment, even catchment volume

versus fan volume are not directly comparable. Besides elementary mathematical and map-scale problems, the fan area-catchment area relationship is further complicated by a wide array of fan and catchment features that are dynamic or have a transient nature, including relief, altitude, stream piracy, tectonic beheading, sedimentary processes, variable inherited geological discontinuities, bedrock lithology, aerial constriction or overlap of fans, and the interplay with environments that border the fans.

Fan Slope Versus Catchment Area

Drew (1873) observed that fans with relatively large catchments have lower average slopes than those with smaller catchments, and this relationship has been studied by others (e.g. Bull 1962, 1964a, Denny 1965, Hawley and Wilson 1965, Melton 1965, Hooke 1968, Beaumont 1972, French 1987, Harvey 1987, Lecce 1988, 1991). Although a weak correlation is obtained (e.g. Fig. 14.23 in Blair and McPherson 1994a), wide scatter shows that issues exist. For example, compiled data show that fans with

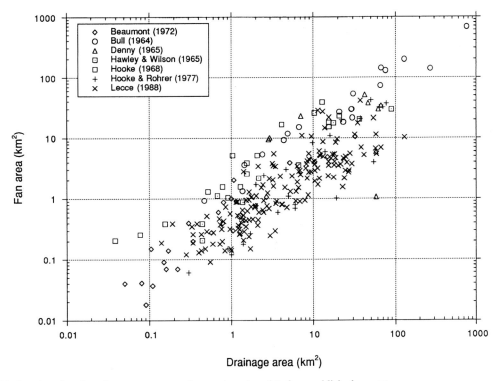

Fig. 14.23 Log–log plot of catchment area versus fan area based on data from published sources

a 2° slope relate to catchments with an area of between 0.5 and 110 km², and fans with a 10° slope relate to catchments with an area of 0.1 and 80 km², providing little discrimination. Major problems, including all those discussed for the area-area relationship, also apply to the catchment area-fan slope relationship, and additional pitfalls result from the inability to obtain accurate slope determinations from small-scale topographic maps.

Types of Alluvial Fans

Based on our research, literature synthesis, and reconnaissance of many fans around the world, we identify 13 fan types defined by their dominant primary processes and textures. The dominant processes also are reflected in their constituent and composite morphology. The 13 fan types are grouped into three classes based on whether the catchment slope material, from which the dominant primary processes are triggered, consists of bedrock (BR), cohesive colluvium (CC), or noncohesive colluvium (NC).

Type BR Alluvial Fans

BR alluvial fans are dominated by deposits resulting from the failure of bedrock slopes present in the catchment or at the range front. Four types are differentiated, including those dominated by rockfall (BR-1), rock slides (BR-2), rock avalanches (BR-3), or earth flows (BR-4) (Table 14.1). Rockfall fans require the funnelling of talus through a notch in the bedrock, and then progradation by movement in chutes. These fans typically are steep (20–35°) with a straight radial pattern or one that lessens near the toe, and have a restricted (< 0.5 km) radial length (Fig. 14.8a, b, and c). Fans dominated by rock slides probably are rare given that slides usually do not accumulate in a way that produces a fan form. They thus dominate fans only where rock slides volumetrically constitute an important part of small fan, such as one otherwise built by rockfall (Fig. 14.8e). Their profiles are irregular because of the presence of rock mounds. Fans dominated by rock avalanches (BR-3) are more common, with highly disintegrated masses producing an overall smooth composite form but with rubbly

Table 14.1 Types of Alluvial Fans

Type BR Alluvial Fans (from Bedrock)

BR-1: Fans dominated by rockfalls
BR-2: Fans dominated by rock slides
BR-3: Fans dominated by rock avalanches
BR-4: Fans dominated by earth flows

Type CC Alluvial Fans (from Cohesive Colluvium)

CC-1: Fans dominated by debris-flow levées
CC-2: Fans dominated proximally by debris-flow levées and distally by lobes
CC-3: Fans dominated by clast-rich debris-flow lobes
CC-4: Fans dominated by clast-poor debris-flow lobes
CC-5: Fans dominated by colluvial slips or slides

Type NC Alluvial Fans (from Noncohesive Colluvium)

NC-1: Fans dominated by sandy clast-poor sheetflood deposits
NC-2: Fans dominated by sand-poor but clast-poor sheetflood deposits
NC-3: Fans dominated by gravel-poor but sandy sheetflood deposits
NC-4: Fans dominated by noncohesive sediment-gravity flows

constituent forms, as exemplified by the Gros Ventre avalanche fan in Wyoming. Avalanches that remain more intact can create a fan with profile irregularities, such as the North Long John and Rose Creek fans (Fig. 14.9a, b). These fans have constituent forms such as large-scale levées, lobes, and mounds. Earth-flow fans (BR-4), derived from fine-grained bedrock, build fans with greater radial length and lower slope (3–10°) than other BR types, such as the Slumgullion fan in Colorado and the Pavilion fan in British Columbia (Fig. 14.9d, e). Constituent forms include radial and concentric crevasse-like troughs and ridges.

Type CC Alluvial Fans

Type CC alluvial fans are those constructed principally by debris flows. They may or may not have an incised channel in their upper segment (Fig. 14.10a). CC fans all derive from the failure, in response to the addition of water, of colluvial slopes in the catchment containing sufficient mud volume to be cohesive. Five CC fans types are differentiated, including those with primary processes and deposits dominated by debris-flow levées (CC-1), proximal debris-flow levées and distal debris-flow lobes (CC-2), clast-rich debris-flow lobes

(CC-3), clast-poor debris-flow lobes (CC-4), and colluvial slips (CC-5) (Table 14.1; Figs. 14.9f and 14.11). Colluvial-slip fans are the steepest of these types, with an average radial slope of 15–20°. CC-1 fans also are steep at 9–18°, and CC-2 fans have a steep (9–18°) proximal segment and a lower-sloping (3–6°) distal segment. CC-3 fan slopes typically average 3–8°, and CC-4 fans 2–4°. The slope pattern of CC-2 fans is segmented, whereas the others have a straight pattern or one that slightly decreases distally. Colluvial slip (CC-5) fans are relatively smooth except for rills or gullies, and the other CC fans have constituent levées or lobes in addition to rills and gullies. Secondary processes dominate the inactive parts of the CC fans, forming desert pavement or boulder mantles. Rockfalls, rock slides, and rock avalanches may comprise a small part of these fans.

Type NC Alluvial Fans

Type NC alluvial fans are constructed principally by sheetflooding (Figs. 14.10b and 14.13). They may or may not have an incised channel in their upper segment. NC fans derive from the failure, through water erosion, of colluvial slopes in the catchment that lack sufficient mud content to be cohesive. Four types of NC fans are identified. Three are dominated by sheetflood deposits and are differentiated by texture, including those composed of gravel and sand (NC-1), sand-poor gravel (NC-2), or gravel-poor sand (NC-3) (Table 14.1). These textural divisions are designated because they provide information on the nature of the catchment bedrock, including composition and the prevalence of tectonic fracturing versus granular disaggregation. NC-1 and NC-2 are the most common of the three sheetflood types. NC-1 fans typically are derived from coarsely crystalline plutonic or gneissic bedrock, and NC-2 fans from tightly cemented brittle rock such as quartzite or carbonate. NC-3 fans arise from bedrock composed mostly of poorly cemented sandstone, where disaggregation dominates over tectonic fracturing. NC-1 and NC-2 fans typically have slopes that decrease distally from about 5° to 2.5° due to decreasing clast size. NC-3 fans have average slopes of 2–3°. Such fans may display distributary rills and headward-eroding gullies, and have a desert pavement, coarse mantle, or minor rockfall, rock-slide, or rock-avalanche deposits.

Type NC-4 fans are a special case generated from noncohesive colluvium. They are dominated by NC-SGFs, which presently are only known to form by the rapid erosion of clay-poor colluvium during catastrophic discharge triggered from dam failures, such as for the Roaring River and Tuttle fan cases (Fig. 14.12b, c, and d). These fans have depositional slopes of 4° to 8°. Typical constituent forms are one or more incised channels, levées, boulder shadow zones, boulder-log jams, rills, and gullies.

Alluvial Fan Evolutionary Scenarios

Given the concepts provided in this chapter and a freshly formed topographic setting, an idealized three-stage scenario for fan development with time can be envisaged. This scenario reflects the progressive enlargement of the fan and its catchment. Incipient or Stage 1 fans are a product of initial catchment development within uplifted bedrock. They typically are related to either inherited or neotectonic weaknesses in the bedrock. Rockfall cones and fans (BR-1) commonly form at this stage by the funnelling of loose clasts through a notch developed on a geological discontinuity. The enlargement of a talus cone to an incipient fan requires conditions in the bedrock favourable for funnelling. Incipient BR-1 fans achieve slopes lower than the ~35° slope of talus cones by progradation via chutes, and possibly by contributions from rock slides, small rock avalanches, colluvial slips, or debris flows (Fig. 14.2e). Alternatively, a catchment may be initiated by the catastrophic brittle failure of a weathered bedrock mass that moves to the fan site as a rock slide or rock avalanche to form a BR-2 or BR-3 fan. Such events produce a range-front cavity mantled by sediment, wherein water is subsequently concentrated. A third variant is for fine-grained bedrock that deforms more ductilely to create an earth-flow fan (BR-4) with a head scarp catchment. Another option is that cohesive colluvium accumulating along local depressions in the bedrock may fail when saturated by water to generate a colluvial-slip (CC-5) fan (Fig. 14.9f). All of these incipient fan scenarios involve catchment initiation through bedrock weathering, and the formation of a drainage net of first- to second-order. The incipient fan can take the form of a steep conical feature with a short radius, or it may comprise a more irregular platform for subsequent fan development.

Stage 2 encompasses the generation of the more common composite fan morphology. Fans of this stage are dominated either by debris flow or sheetflood processes, and more rarely by rock avalanches or NCS-GFs. Rockfall, rock-slide, and colluvial slips may contribute small amounts of sediment. Stage 2 fans commonly have radial lengths of < 4 km, although some NCSGF fans reach ~10 km in length. Stage 2 fan slopes vary from ~10° to 18° where dominated by debris-flow levées, but more commonly have a 4–10° average slope. The change from Stage 1 to Stage 2 is related to: (a) the enlargement of the catchment as the drainage net develops from water erosion to include second- to perhaps fifth-order channels, and (b) the ability of the catchment to accumulate and store colluvium along slopes and within the drainage net that can then be episodically moved to the fan. The catchment during this stage still has significant bedrock exposures, but also builds noteworthy colluvial slopes.

Stage 3 is characterized by the development of an incised channel ~1 km or more long through which flows are transferred across the upper fan to an active depositional lobe. Stage 3 fans typically are dominated either by debris flow lobes or sheetfloods, depending upon the cohesiveness of the catchment colluvium. The slope of such fans is 2–8° due to the dominant primary processes and to the effect of the incised channel on fan progradation. Progradation ensues by progressive lengthening of the incised channel. The development of a Stage 3 fan results from the transfer of a significant volume of sediment from the catchment to the fan. Thus, Stage 3 fans usually have relatively large catchments with a well-developed drainage net characterized by fifth- to perhaps eighth-order feeder channels. The catchment slopes typically are of bedrock in the upper reaches, and of widespread colluvium lower down.

The rate of progression through the three idealized stages varies from one fan to another, and may contrast even between adjoining fans due to differences in the catchment, such as the location of old faults or more erodible bedrock. An example from the vicinity of Titus Canyon in northern Death Valley has adjacent fans displaying all three stages of evolution. The largest and most advanced (Stage 3) is the Titus Canyon fan (Fig. 14.6d). This fan has a catchment built around a prominent structure oriented at a high angle to the range front (Reynolds 1974). Many other geological discontinuities in this catchment also induce weathering and the generation of flows that transport material to the fan, causing it to enlarge rapidly. In contrast, neighbouring fans display developmental Stages 1 or 2 depending on the position of their catchments with respect to minor discontinuities in the uplifted block.

Conclusions

Alluvial fans are a prominent conical landform developed through sediment aggradation where a channel draining an upland catchment emerges to an adjoining valley. Analysis of fans thus requires an understanding of sedimentary processes and products in this environment, and how they create the diagnostic fan form. Fans optimally develop in tectonic settings where fault offset creates relief, enhances weathering to yield sediment, and promotes flash floods that move sediment to the fan. Two types of processes, primary and secondary, are active on alluvial fans. Primary processes are those that transport sediment from the catchment to the fan, whereas secondary processes mostly modify sediment previously deposited on the fan by any of the primary processes.

Primary processes that construct fans include those triggered by the failure of bedrock (BR) slopes in the catchment or range front, and those related to the failure of either cohesive (CC) or noncohesive (NC) colluvial slopes in the catchment. Fans typically are dominated by the deposits of a single primary process, including rockfalls (BR-1), rock slides (BR-2), rock avalanches (BR-3), earth flows (BR-4), debris-flow levées (CC-1), debris-flow levées and lobes (CC-2), clast-rich debris-flow lobes (CC-3), clast-poor debris-flow lobes (CC-4), colluvial slips (CC-5), sandy gravelly sheetfloods (NC-1), sand-deficient gravelly sheetfloods (NC-2), gravel-deficient sandy sheetfloods (NC-3), or noncohesive sediment-gravity flows (NC-4). CC and NC fans may also have an incised channel. Primary processes are infrequent and of short duration, but have high impact with respect to fan aggradation. They also impart the prevailing composite fan form, including radial and cross-fan profiles. Primary processes related to bedrock failure and colluvial slip are important during the early stage of fan and catchment evolution, whereas debris flow and sheetflood fans, and more rarely NCSGF-dominated fans, develop after the incipient stage when the catchment has evolved to shed and store colluvium. Colluvium is noncohesive unless mud (especially clay) is present. Therefore, cohesive colluvium, and thus CC fans, are related to catchment bedrock that weathers to yield clay, and noncohesive colluvium and the linked NC fans are derived from catchments underlain by bedrock that yields comparatively little clay under the specific weathering conditions. The impact of cohesion is determined with the passage of a flash flood, wherein cohesive colluvium can transform into a debris flow, and noncohesive colluvium develops into a sediment-laden water flood.

Except for local aeolian deposition or volcanism, secondary processes are not important to fan aggradation, but commonly dominate the fan surface due to their prevalence during the long intervals between successive primary events. They include: (a) rilling and gullying from overland water flow, (b) desert pavement and sand-drift or nebkha deposits from wind transport, (c) desert varnish and particle weathering from hydrolysis, oxidation, salt crystal growth, and exfoliation, (d) bioturbation from plants and animals, (d) groundwater modifications, (e) soil development, (f) tectonic deformation, and (g) modification by adjoining environments, such as toe erosion or volcanism. Secondary processes, especially overland flow, can mask the features of, and thus be mistaken for, the primary processes.

References

Ahnert F (1970) Functional relationships between denudation, relief, and uplift in large mid-latitude drainage basins. American Journal of Science 268: 243–263

Al-Farraj A, Harvey AM (2000) Desert pavement characterisitics on wadi terrace and alluvial fan surfaces: Wadi Al-Bih, U.A.E. and Oman. Geomorphology 35: 279–297

Al-Sarawi M (1988) Morphology and facies of alluvial fans in Kadhmah Bay, Kuwait. Journal of Sedimentary Petrology 58: 902–907

Alexander D, Coppola L (1989) Structural geology and dissection of alluvial fan sediments by mass movements: an example from the southern Italian Apennines. Geomorphology 2: 341–361

Anderson SP, Anderson RS (1990) Debris-flow benches: dune-contact deposits record paleo-sand dune positions in north Panamint Valley, Inyo County, California. Geology 18: 524–527

Anstey RL (1965) Physical characteristics of alluvial fans. Natick, MA: Army Natick Laboratory, Technical Report ES-20.

Anstey RL (1966) A comparison of alluvial fans in west Pakistan and the United States. Pakistan Geographical Review 21: 14–20

Arzani N (2005) The fluvial megafan of Abarkoh basin (central Iran): an example of flash-flood sedimentation in arid lands. In: Harvey A, Mather AE, Stokes M (eds.) Alluvial fans: geomorphology, sedimentology, dynamics. Geological Society Special Publication 251: 41–60

Beaty CB (1963) Origin of alluvial fans, White Mountains, California and Nevada. Association of American Geographers Annals 53: 516–535

Beaty CB (1974) Debris flows, alluvial fans, and revitalized catastrophism. Zeitschrift für Geomorphologie 21: 39–51

Beaty CB (1989) Great boulders I have known. Geology 17: 349–352

Beaty CB (1990) Anatomy of a White Mountain debris flow – the making of an alluvial fan. In: Rachocki AH, Church M (eds.) Alluvial fans – a field approach. Wiley, New York, pp. 69–90

Beaty CB, DePolo CM (1989) Energetic earthquakes and boulders on alluvial fans: Is there a connection? Seismological Society of America Bulletin 79: 219–224

Beaumont P (1972) Alluvial fans along the foothills of the Elburz Mountains, Iran. Palaeogeography, Palaeoclimatology, Palaeoecology 12: 251–273

Beaumont P, Oberlander TM (1971) Observations on stream discharge and competence at Mosaic Canyon, Death Valley, California. Geological Society of America Bulletin 82: 1695–1698

Beckvar N, Kidwell SM (1988) Hiatal shell concentrations, sequence analysis, and sealevel history of a Pleistocene alluvial fan, Punta Chueca, Sonora. Lethaia 21: 257–270

Beehner TS (1990) Burial of fault scarps along the Organ Mountains fault, south-central New Mexico. Association of Engineering Geologists Bulletin 27: 1–9

Bennett GL, Weissmann GS, Baker GS, Hyndman DW (2006) Regional-scale assessment of sequence-bounding paleosol on fluvial fans using ground-penetrating radar, eastern San Joaquin Valley, California. Geological Society of America Bulletin 118: 724–732

Berry ME (1990) Soil catena development on fault scarps of different ages, eastern escarpment of the Sierra Nevada, California. Geomorphology 3: 333–350

Bertran P, Texier JP (1999) Sedimentation processes and facies on a semi-vegetated talus, Lousteau, southwestern France. Earth Surface Processes and Landforms 24: 177–187

Beverage JP, Culbertson JK (1964) Hyperconcentrations of suspended sediment. Proceedings of the American Society of Civil Engineers, Journal of the Hydraulics Division, 90(HY6): 117–128

Bierman P, Gillespie A (1991) Range fires: Accuracy of rock-varnish chemical analyses: implications for cation-ratio dating. Geology 19: 196–199

Bierman P, Gillespie A, Huehner S (1991) Precision of rock-varnish chemical analyses and cation-ratio ages. Geology 19: 135–138

Blackwelder E (1912) The Gros Ventre slide, an active earth-flow. Geological Society of America Bulletin 23: 487–492

Blackwelder E (1928) Mudflow as a geologic agent in semi-arid mountains. Geological Society of America Bulletin 39: 465–484

Blair TC (1987a) Sedimentary processes, vertical stratification sequences, and geomorphology of the Roaring River alluvial fan, Rocky Mountain National Park, Colorado. Journal of Sedimentary Petrology 57: 1–18

Blair TC (1987b) Tectonic and hydrologic controls on cyclic alluvial fan, fluvial, and lacustrine rift-basin sedimentation, Jurassic-lowermost Cretaceous Todos Santos Formation, Chiapas, Mexico. Journal of Sedimentary Petrology 57: 845–862

Blair TC (1999a) Alluvial-fan and catchment initiation by rock avalanching, Owens Valley, California. Geomorphology 28: 201–221

Blair TC (1999b) Cause of dominance by sheetflood versus debris-flow processes on two adjoining alluvial fans, Death Valley, California. Sedimentology 46: 1015–1028

Blair TC (1999c) Form, facies, and depositional history of the North Long John rock avalanche, Owens Valley, California. Canadian Journal of Earth Sciences 36: 855–870

Blair TC (1999d) Sedimentary processes and facies of the water-laid Anvil Spring Canyon alluvial fan, Death Valley, California. Sedimentology 46: 913–940

Blair TC (1999e) Sedimentology of gravelly Lake Lahontan highstand shoreline deposits, Churchill Butte, Nevada. Sedimentary Geology, 123: 199–218

Blair TC (1999f) Sedimentology of the debris-flow-dominated Warm Spring Canyon alluvial fan, Death Valley, California. Sedimentology 46: 941–965

Blair TC (2000) Sedimentology and progressive tectonic unconformities of the sheetflood-dominated Hell's Gate alluvial fan, Death Valley, California. Sedimentary Geology 132: 233–262

Blair TC (2001) Outburst-flood sedimentation on the proglacial Tuttle Canyon alluvial fan, Owens Valley, California, U.S.A. Journal of Sedimentary Research 71: 657–679

Blair TC (2002) Alluvial-fan sedimentation from a glacial outburst flood, Lone Pine, California, and contrasts with meteorological flood fans. In: Martini IP, Baker VR, Garzon G (eds.) Flood and megaflood processes and deposits, Recent and ancient examples. International Association of Sedimentologists Special Publication 32: 113–140

Blair TC (2003) Features and origin of the giant Cucomungo Canyon alluvial fan, Eureka Valley, California. In: Chan MA, Archer AW (eds.) Extreme depositional environments: Mega end members in geologic time. Geological Society of America Special Paper 370: 105–126

Blair TC, Bilodeau WL (1988) Development of tectonic cyclothems in rift, pull-apart, and foreland basins: sedimentary response to episodic tectonism. Geology 16: 517–520

Blair TC, McPherson JG (1992) The Trollheim alluvial fan and facies model revisited. Geological Society of America Bulletin 104: 762–769

Blair TC, McPherson JG (1994a) Alluvial fan processes and forms. In: Abrahams AD, Parsons A (eds.) Geomorphology of desert environments. Chapman Hall, London, pp. 354–402

Blair TC, McPherson JG (1994b) Alluvial fans and their natural distinction from rivers based on morphology, hydraulic processes, sedimentary processes, and facies. Journal of Sedimentary Research A64: 451–490

Blair TC, McPherson JG (1994c) Historical adjustments by Walker River to lake-level fall over a tectonically tilted half-graben floor, Walker Lake Basin, Nevada. Sedimentary Geology 92: 7–16

Blair, TC, McPherson JG (1998) Recent debris-flow processes and resultant form and facies of the Dolomite alluvial fan, Owens Valley, California. Journal of Sedimentary Research 68: 800–818

Blair TC, McPherson JG (1999) Grain-size and textural classification of coarse sedimentary particles. Journal of Sedimentary Research 69: 6–19

Blair TC, McPherson JG (2008) Quaternary sedimentology of the Rose Creek fan delta along Walker Lake, Nevada, U.S.A., and relevance to fan-delta facies models. Sedimentology 55: 579–615

Blair TC, Raynolds RG (1999) Sedimentology and tectonic implications of the Neogene synrift Hole in the Wall and Wall Front members, Furnace Creek Basin, Death Valley, California. In: Wright L, Troxel B (eds.) Cenozoic basins of the Death Valley region. Geological Society of America Special Paper 333: 127–168

Blair TC, Clark JC, Wells SG (1990) Quaternary stratigraphy, landscape evolution, and application to archeology; Jarilla piedmont and basin floor, White Sands Missile Range, New Mexico. Geological Society of America Bulletin 102: 749–759

Blissenbach E (1952) Relation of surface angle distribution to particle size distribution on alluvial fans. Journal of Sedimentary Petrology 22: 25–28

Blissenbach E (1954) Geology of alluvial fans in semi-arid regions. Geological Society of America Bulletin 65: 175–190

Bogoch R, Cook P (1974) Calcite cementation of a Quaternary conglomerate in southern Sinai. Journal of Sedimentary Petrology 44: 917–920

Boothroyd JC (1972) Coarse-grained sedimentation on a braided outwash fan, northeast Gulf of Alaska. University of South Carolina Coastal Research Division Technical Report 6-CRD.

Boothroyd JC, Nummedal D (1978) Proglacial braided outwash: a model for humid alluvial fan deposits. In: Miall AD (ed.) Fluvial sedimentology. Canadian Association of Petroleum Geologists Memoir 5, Calgary, pp. 641–668

Bovis MJ (1985) Earthflows in the Interior Plateau, southwest British Columbia. Canadian Geotechnical Journal 22: 313–334

Bovis MJ (1986) The morphology and mechanics of large-scale slope movement, with particular reference to southwest British Columbia. In: Abrahams AD (ed.) Hillslope processes. Allen & Unwin, Boston, pp. 319–341

Bovis MJ, Jones P (1992) Holocene history of earthflow mass movements in south-central British Columbia: the influence of hydroclimatic changes. Canadian Journal of Earth Science 29: 1746–1755

Bowman D (1978) Determinations of intersection points within a telescopic alluvial fan complex. Earth Surface Processes 3: 265–276

Browning JM (1973) Catastrophic rock slide, Mount Huasaracan, north-central Peru, May 31, 1970. American Association of Petroleum Geologists Bulletin 57: 1335–1341

Bull WB (1962) Relations of alluvial fan size and slope to drainage basin size and lithology in western Fresno County, California. U.S. Geological Survey Professional Paper 450-B

Bull WB (1964a) Geomorphology of segmented alluvial fans in western Fresno County, California. U.S. Geological Survey Professional Paper 352-E

Bull WB (1964b) History and causes of channel trenching in western Fresno County, California. American Journal of Science 262: 249–258

Bull WB (1972) Recognition of alluvial fan deposits in the stratigraphic record. In: Rigby JK, Hamblin WK (eds.) Recognition of ancient sedimentary environments. Society of Economic Paleontologists and Mineralogists Special Publication 16: 63–83

Bull WB (1977) The alluvial fan environment. Progress in Physical Geography 1: 222–270

Bull WB, Schick AP (1979) Impact of climatic change on an arid watershed: Nahal Yael, southern Israel. Quaternary Research 11: 153–171

Burchfiel BC (1966) Tin Mountain landslide, south-eastern California, and the origin of megabreccia. Geological Society of America Bulletin 77: 95–100

Butler PR, Troxel BW, Verosub KL (1988) Late Cenozoic history and styles of deformation along the southern Death Valley fault zone, California. Geological Society of America Bulletin 100: 402–410

Caine N (1980) The rainfall intensity–duration control of shallow landslides and debris flows. Geografiska Annaler 62A: 23–27

Campbell RH (1974) Debris flows originating from soil slips during rainstorms in southern California. Quarterly Journal of Engineering Geologists 7: 339–349

Campbell RH (1975) Soil slips, debris flows, and rainstorms in the Santa Monica Mountains and vicinity, southern California. U.S. Geological Survey Professional Paper 851

Cannon SH (2001) Debris-flow generation from recently burned watersheds. Environmental and Engineering Geoscience 7: 301–320

Cannon SH, Ellen S (1985) Abundant debris avalanches, San Francisco Bay region, California. California Geology, December: 267–272

Cannon SH, Reneau SL (2000) Conditions for generation of fire-related debris flows, Capulin Canyon, New Mexico. Earth Surface Processes and Landforms 25: 1103–1121

Cannon SH, Kirkham RM, Parise M (2001) Wildfire-related debris-flow initiation processes, Storm King Mountain, Colorado. Geomorphology 39: 171–188

Cerling TE, Webb RH, Poreda RJ, Rigby AD, Melis TS (1999) Cosmogenic ^{3}He ages and frequency of late Holocene debris flows Prospect Canyon, Grand Canyon, USA. Geomorphology 27: 93–111

Christenson GE, Purcell C (1985) Correlation and age of Quaternary alluvial-fan sequences, Basin and Range province, southwestern United States. In: Weide DL (ed.) Soils and Quaternary geology of the southwestern United States. Geological Society of America Special Paper 203: 115–122

Coe JA, Godt JW, Wilson RC (1998) Distribution of debris flows in Alameda County, California triggered by 1998 El Niño rainstorms: a repeat of January 1982? EOS 79: 266

Colombo F (2005) Quaternary telescopic-like alluvial fans, Andean Ranges, Argentina. In: Harvey A, Mather AE, Stokes M (eds.) Alluvial fans: geomorphology, sedimentology, dynamics. Geological Society Special Publication 251: 69–84

Costa JE (1984) Physical geomorphology of debris flows. In: Costa JE, Fleisher PJ (eds.) Developments and applications of geomorphology. Springer, Berlin, pp. 268–317

Costa JE (1988) Rheologic, geomorphic, and sedimentologic differentiation of water floods, hyperconcentrated flows, and debris flows. In: Baker VR, Kochel RC, Patton PC (eds.) Flood geomorphology. Wiley, New York, pp. 113–122

Costa JE (1991) Nature, mechanics, and mitigation of the Val Pola landslide, Valtellina, Italy, 1987–1988. Zeitschrift für Geomorphologie 35: 15–38

Cotecchia V (1987) Earthquake-prone environments. In: Anderson MG, Richards KS (eds.) Slope stability. Wiley, Chichester, pp. 287–330

Coussot P, Meunier M (1996) Recognition, classification and mechanical description of debris flows. Earth-Science Reviews 40: 209–227

Crandell DR, Varnes DJ (1961) Movement of the Slumgullion earth flow near Lake City, Colorado. U.S. Geological Survey Professional Paper 424, pp. 136–139

Cronin VS (1992) Compound landslides: Nature and hazard potential of secondary landslides within host landslides. In: Slossen JE, Keene AG, Johnson JA (eds.) Landslides/landslide mitigation. Geological Society of America Reviews in Engineering Geology 9: 1–9

Curry RR (1966) Observation of alpine mudflows in the Tenmile Range, central Colorado. Geological Society of America Bulletin 77: 771–776

David-Novak HB, Morin E, Enzel Y (2004) Modern extreme storms and the rainfall thresholds for initiating debris flows on the hyperarid western escarpment of the Dead Sea, Israel. Geological Society of America Bulletin 116: 718–728

Denny CS (1965) Alluvial fans in the Death Valley region, California and Nevada. U.S. Geological Survey Professional Paper 466

Denny CS (1967) Fans and pediment. American Journal of Science 265: 81–105

Derbyshire E, Owen LA (1990) Quaternary alluvial fans in the Karakoram Mountains. In: Rachocki AH, Church M (eds.) Alluvial fans – a field approach. Wiley, New York, pp. 27–54

Doeglas DJ (1962) The structure of sedimentary deposits of braided rivers. Sedimentology 1: 167–193

Dohrenwend JC (1987) Basin and Range. In: Graf WL (ed.) Geomorphic systems of North America. Geological Society of America Centennial Special 2: 303–342

Dorn RI (1988) A rock-varnish interpretation of alluvial-fan development in Death Valley, California. National Geographic Research 4: 56–73

Dorn RI, Oberlander TM (1981) Rock varnish origin, characteristics, and usage. Zeitschrift für Geomorphologie 25: 420–436

Dorn RI, DeNiro MJ, Ajie HO (1987) Isotopic evidence for climatic influence of alluvial-fan development in Death Valley, California. Geology 15: 108–10

Dorn RI, Jull AJT, Donahue DJ, Linick TW, et al. (1989) Accelerator mass spectrometry radiocarbon dating of rock varnish. Geological Society of America Bulletin 101: 1363–1372

Drew F (1873) Alluvial and lacustrine deposits and glacial records of the Upper Indus Basin. Geological Society of London Quarterly Journal 29: 441–471

Drewes H (1963) Geology of the Funeral Peak Quadrangle, California, on the east flank of Death Valley. U.S. Geological Survey Professional Paper 413

Eckis R (1928) Alluvial fans in the Cucamonga district, southern California. Journal of Geology 36: 111–141

Ellen SD, Fleming RW (1987) Mobilization of debris flows from soil slips, San Francisco Bay region, California. In: Costa JE, Wieczorek GF (eds.) Debris flows/avalanches: process, recognition, and mitigation. Geological Society of America Reviews in Engineering Geology 7: 31–40

Ellen SD, Wieczorek GF (1988) Landslides, floods, and marine effects of the storm of January 3–5 in the San Francisco Bay region, California. U.S. Geological Survey Professional Paper 1434

Erismann TH, Abele G (2001) Dynamics of rockslides and rockfalls. Springer, Berlin

Evans SG, Clague JJ, Woodsworth GJ, Hungr O (1989) The Pandemonium Creek rock avalanche, British Columbia. Canadian Geotechnical Journal 26: 427–446

Evans SG, Hungr O, Enegren EG (1994) The Avalanche Lake rock avalanche, Mackenzie Mountains, Northwest Territories, Canada: description, dating, and dynamics. Canadian Geotechnical Journal 31: 749–768

Fahnestock RK, Haushild WL (1962) Flume studies of the transport of pebbles and cobbles on a sand bed. Geological Society of America Bulletin 73: 1431–1436

Fauque L, Strecker MR (1988) Large rock avalanche deposits (Strurzstrome, sturzstroms) at Sierra Aconquija, northern Sierras Pampeanas, Argentina. Ecologae Geologae Helvetae 81: 579–592

Fisher RV (1971) Features of coarse-grained, high-concentration fluids and their deposits. Journal of Sedimentary Petrology 41: 916–927

Fleming RW, Baum RL, Giardino M (1999) Map and description of the active part of the Slumgullion landslide, Hinsdale County, Colorado. U.S. Geological Survey pamphlet to accompany Geologic Investigations Series Map I-2672

Florsheim JL (2004) Side-valley tributary fans in high-energy river floodplain environments: sediment sources and depositional processes, Navarro River basin, California. Geological Society of America Bulletin 116: 923–937

Fraser GS, Suttner L (1986) Alluvial fans and fan deltas. International Human Resources Development Corporation, Boston

French RH (1987) Hydraulic processes on alluvial fans. Elsevier, Amsterdam

Fryxell FM, Horberg L (1943) Alpine mudflows in Grand Teton National Park, Wyoming. Geological Society of America Bulletin 54: 457–472

Gabet EJ, Dunne T (2002) Landslides on coastal sage-scrub and grassland hillslopes in a severe El Niño winter: The effects of vegetation conversion and sediment delivery. Geological Society of America Bulletin 114: 983–990

Gábris G, Nagy B (2005) Climate and tectonically controlled river style changes on the Sajó-Hernád alluvial fan (Hungary). In: Harvey A, Mather AE, Stokes M (eds.) Alluvial fans: geomorphology, sedimentology, dynamics. Geological Society Special Publication 251: 61–67

Gates WB (1987) The fabric of rockslide avalanche deposits. Association of Engineering Geologists Bulletin 24: 389–402

Gilbert GK (1914) The transportation of debris by running water. U.S. Geological Survey Professional Paper 86

Gile LH, Hawley JW (1966) Periodic sedimentation and soil formation on an alluvial-fan piedmont in southern New Mexico. Soil Science Society of America Proceedings 30: 261–268

Gile LH, Hawley JW, Grossman RB (1981) Soils and geomorphology in the Basin and Range area of southern New Mexico – guidebook to the Desert Project. New Mexico Bureau of Mines and Mineral Resources Memoir 39

Gomez-Pujol L (1999) Sedimentologia i evolució geomorfológica quaternária del ventall alluvial des Caló (Betlem, Artá, Mallorca). Bolletin Societat d'História Natural de les Balears 42: 107–124

Goudie AS, Day MJ (1980) Disintegration of fan sediments in Death Valley, California by salt weathering. Physical Geography 1: 126–137

Govi M, Gulla G, Nicoletti PG (2002) Val Pola rock avalanche of July 28, 1987, in Valtellina (central Italian Alps). In: Evans SG, DeGraff JV (eds.) Catastrophic landslides: effects, occurrences, mechanisms. Geological Society of America Reviews in Engineering Geology XV: 71–89

Greenway DR (1987) Vegetation and slope stability. In: Anderson MG, Richards KS (eds.) Slope stability. Wiley, Chichester, pp. 187–230

Gutiérrez F, Gutiérrez M, Sancho C (1998) Geomorphological and sedimentological analysis of a catastrophic flash flood in the Arás drainage basin (central Pyrenees, Spain). Geomorphology 22: 265–283

Hadley JB (1964) Landslides and related phenomena accompanying Hebgen Lake earthquake of August 17, 1959. U.S. Geological Survey Professional Paper 435

Harden JW, Matti JC (1989) Holocene and late Pleistocene slip rates on the San Andreas fault in Yucaipa, California, using displaced alluvial-fan deposits and soil chronology. Geological Society of America Bulletin 101: 1107–1117

Harden JW, Slate JL, Lamothe P, Chadwick OA, et al. (1991) Soil formation on the Trail Canyon alluvial fan. U.S. Geological Survey Open-File Report 91–296

Harp EL, Keefer DK (1989) Earthquake-induced landslides, Mammoth Lakes area, California. In: Brown WM (ed.) Landslides in central California. 28th International Geological Congress Field Trip Guidebook T381: 49–53

Harrison JV, Falcon NL (1937) The Saidmarrah landslip, southwest Iran. Geographical Journal 89: 42–47

Hart MW (1991) Landslides in the Peninsular Ranges, southern California. In: Walawender MJ, Hanan BB (eds.) Geological excursions in southern California and Mexico. Geological Society of America annual meeting field guidebook, pp. 349–371

Harvey AM (1984a) Aggradation and dissection sequences on Spanish alluvial fans: influence on morphological development. Catena 11: 289–304

Harvey AM (1984b) Debris flow and fluvial deposits in Spanish Quaternary alluvial fans: implications for fan morphology. In: Koster EH, Steel RJ (eds.) Sedimentology of gravels and conglomerate. Canadian Society of Petroleum Geologists Memoir 10: 123–132

Harvey AM (1987) Alluvial fan dissection: relationship between morphology and sedimentation. In: Frostick L, Reid I (eds.) Desert sediments: ancient and modern. Geological Society of London Special Publication 35: 87–103

Harvey AM (1988) Controls of alluvial fan development: the alluvial fans of the Sierra de Carrascoy, Murcia, Spain. Catena Supplement 13: 123–137

Harvey AM (1989) The occurrence and role of arid zone alluvial fans. In: Thomas DSG (ed.) Arid zone geomorphology. Belhaven, London, pp. 136–158

Harvey AM (1990) Factors influencing Quaternary fan development in southeast Spain. In: Rachocki AH, Church M (eds.) Alluvial fans – a field approach. Wiley, New York, pp. 247–70

Harvey AM (2005) Differential effects of base-level, tectonic setting and climatic change on Quaternary alluvial fans in the northern Great Basin, Nevada, USA. In: Harvey A, Mather AE, Stokes M (eds.) Alluvial fans: geomorphology, sedimentology, dynamics. Geological Society Special Publication 251: 117–131

Harvey AM, Wigand PE, Wells SG (1999) Response of alluvial fan systems to the late Pleistocene to Holocene climatic transition: contrasts between the margins of pluvial Lakes Lahontan and Mojave, Nevada and California, USA. Catena 36: 255–281

Harvey A, Mather AE, Stokes M (2005) Alluvial fans: geomorphology, sedimentology, dynamics- introduction, a review of alluvial fan research. In: Harvey A, Mather AE, Stokes M (eds.) Alluvial fans: geomorphology, sedimentology, dynamics. Geological Society Special Publication 251: 1–8

Hawley JW, Wilson WE (1965) Quaternary geology of the Winnemucca area, Nevada. University of Nevada Desert Research Institute Technical Report 5

Hayden BP (1988) Flood climates. In: Baker VR, Kochel RC, Patton PC (eds.) Flood geomorphology. Wiley, New York, pp. 13–26

Hayward AB (1985) Coastal alluvial fans (fan deltas) of the Gulf of Aqaba (Gulf of Eilat), Red Sea. Sedimentary Geology 43: 241–260

Hencher SR (1987) The implications of joints and structures for slope stability. In: Anderson MG, Richards KS (eds.) Slope stability. Wiley, Chichester, pp. 145–186

Hermanns RL, Strecker MR (1999) Structural and lithological controls on large Quaternary rock avalanches (sturzstroms) in arid northwestern Argentina. Geological Society of America Bulletin 111: 934–948

Hewitt K (2002) Styles of rock-avalanche depositional complexes conditioned by very rugged terrain, Karakoram Himalaya, Pakistan. In: Evans SG, DeGraff JV (eds.) Catastrophic landslides: effects, occurrences, mechanisms. Geological Society of America Reviews in Engineering Geology XV: 345–377

Hogg SE (1982) Sheetfloods, sheetwash, sheetflow, or . . .?. Earth Science Reviews 18: 59–76

Hooke RL (1967) Processes on arid-region alluvial fans. Journal of Geology 75: 438–460

Hooke RL (1968) Steady-state relationships of arid-region alluvial fans in closed basins. American Journal of Science 266: 609–629

Hooke RL (1972) Geomorphic evidence for late Wisconsin and Holocene tectonic deformation, Death Valley,

California. Geological Society of America Bulletin 83: 2073–2098

Hooke RL (1987) Mass movement in semi-arid environments and the morphology of alluvial fans. In: Anderson MG, Richards KS (eds.) Slope stability. Wiley, Chichester, pp. 505–529

Hooke RL, Rohrer WL (1977) Relative erodibility of source-area rock types, as determined from second-order variations in alluvial-fan size. Geological Society of America Bulletin 88: 117–182

Hollingsworth R, Kovacs GS (1981) Soil slumps and debris flows: Prediction and protection. Association of Engineering Geologists Bulletin 18: 17–28

Horton RE (1945) Erosional development of streams and their drainage basins; hydrophysical approach to quantitative morphology. Geological Society of America Bulletin 56: 275–370

Houghton JG, Sakamoto CM, Gifford RO (1975) Nevada's weather and climate. Nevada Bureau of Mines and Geology Special Publication 2

Houston J (2002) Groundwater recharge through an alluvial fan in the Atacama Desert, northern Chile: mechanisms, magnitudes and causes. Hydrological Processes 16: 3019–3035

Hsü KJ (1975) Catastrophic debris streams (sturzstroms) generated by rockfalls. Geological Society of America Bulletin 86: 129–140

Hubert JF, Filipov AJ (1989) Debris-flow deposits in alluvial fans on the west flank of the White Mountains, Owens Valley, California. Sedimentary Geology 61: 177–205

Hunt CB (1975) Death Valley: geology, ecology, and archeology: University of California Press, Berkeley

Hunt CB, Mabey DR (1966) Stratigraphy and structure, Death Valley, California. U.S. Geological Survey Professional Paper 494-A

Hunt CB, Robinson TW, Bowles WA, Washburn AL (1966) Hydrologic basin, Death Valley, California. U.S. Geological Survey Professional Paper 494–B

Ibbeken H, Warnke DA (2000) The Hanaupah fan shoreline deposit at Tule Spring, a gravelly shoreline deposit of Pleistocene Lake Manly, Death Valley, California, USA. Journal of Paleolimnology 23: 439–447

Iverson RM (1997) The physics of debris flows. Reviews of Geophysics 35: 245–296

Jacka AD (1974) Differential cementation of a Pleistocene carbonate fanglomerate, Guadalupe Mountains. Journal of Sedimentary Petrology 44: 85–92

Jarrett RD, Costa JE (1986) Hydrology, geomorphology, and dam-break modeling of the July 15, 1982 Lawn Lake and Cascade Lake dam failures, Larimer County, Colorado. U.S. Geological Survey Professional Paper 1369

Jayko AS, de Mouthe J, Lajoie KR, Ramsey DW, Godt JW (1999) Map showing locations of damaging landslides in San Mateo County, California, resulting from 1997-98 El Niño rainstorms. U.S. Geological Survey Miscellaneous Field Studies Map MF-2325-H

Jian L, Defu L (1981) The formation and characteristics of mudflow and flood in the mountain area of the Dachao River and its prevention. Zeitschrift für Geomorphologie 25: 470–484

Jian L, Jingrung W (1986) The mudflows in Xiaojiang Basin. Zeitschrift für Geomorphologie Supplementband 58: 155–164

Johnson AM (1970) Physical processes in geology. Freeman-Cooper, San Francisco

Johnson AM (1984) Debris flow. In: Brunsden D, Prior DB (eds.) Slope instability. Wiley, New York, pp. 257–361

Johnson AM, Rahn PH (1970) Mobilization of debris flows. Zeitschrift für Geomorphologie Supplementband 9: 168–186

Jopling AV, Richardson EV (1966) Backset bedding developed in shooting flow in laboratory experiments. Journal of Sedimentary Petrology 36: 821–824

Jutson JT (1919) Sheet-flows or sheetfloods and their association in the Niagara District of sub-arid south-central western Australia. American Journal of Science 198: 435–439

Keefer DK (1984) Landslides caused by earthquakes. Geological Society of America Bulletin 95: 406–421

Keefer DK (1999) Earthquake-induced landslides and their effects on alluvial fans. Journal of Sedimentary Research 69: 84–104

Keefer DK, Johnson AM (1983) Earth flows: morphology, mobilization, and movement. U.S. Geological Survey Professional Paper 1264

Keefer DK, Mosely ME, deFrance SD (2003) A 38000-year record of floods and debris flows in the Ilo region of southern Peru and its relation to El Niño events and great earthquakes. Palaeogeography, Palaeoclimatology, Palaeoecology 194: 41–77

Kennedy JF (1963) The mechanics of dunes and antidunes in erodible bed channels. Journal of Fluid Mechanics 16: 521–544

Koster EH (1978) Transverse ribs: Their characteristics, origin, and hydraulic significance. In: Miall AD (ed.) Fluvial sedimentology. Canadian Society of Petroleum Geologists Memoir 5: 161–186

Krauskopf KB, Feitler S, Griggs AB (1939) Structural features of a landslide near Gilroy, California. Journal of Geology 47: 630–648

Langford R, Bracken B (1987) Medano Creek, Colorado, a model for upper-flow-regime fluvial deposition. Journal of Sedimentary Petrology 57: 863–870

Lattman LH (1973) Calcium carbonate cementation of alluvial fans in southern Nevada. Geological Society of America Bulletin 84: 3013–3028

Lattman LH, Simonberg EM (1971) Case-hardening of carbonate alluvium and colluvium, Spring Mountains, Nevada. Journal of Sedimentary Petrology 41: 274–281

Lawson AC (1913) The petrographic designation of alluvial fan formations. University of California Publications in Geology 7: 325–334

Le K, Lee J, Owen LA, Finkel R (2007) Late Quaternary slip rates along the Sierra Nevada frontal fault zone, California: slip portioning across the western margin of the eastern California shear zone-Basin and Range Province. Geological Society of America Bulletin 119: 240–256

Lecce SA (1988) Influence of lithology on alluvial fan morphometry, White and Inyo Mountains, California and Nevada. M.A. thesis, Arizona State University, Tempe

Lecce SA (1991) Influence of lithologic erodibility on alluvial fan area, western White Mountains, California and Nevada. Earth Surface Processes and Landforms 16: 11–18

Leggett RF, Brown RJE, Johnson GH (1966) Alluvial fan formation near Aklavik, Northwest Territories, Canada. Geological Society of America Bulletin 77: 15–30

Leopold LB (1951) Rainfall frequency: an aspect of climate variation. Transactions of the American Geophysical Union 32: 347–357

Li J, Luo D (1981) The formation and characteristics of mudflow in the mountain area of the Dachao River and its prevention. Zeitschrift für Geomorphologie 25: 470–484

Li J, Wang J (1986) The mudflows of Xiaojiang Basin. Zeitschrift für Geomorphologie Supplementband 58: 155–164

Link MH, Roberts MT, Newton MS (1985) Walker Lake Basin, Nevada: an example of late Tertiary(?) to recent sedimentation in a basin adjacent to an active strike-slip fault. In: Biddle KT, Christie-Blick N (eds.) Strike-slip deformation, basin formation, and sedimentation. Society of Economic Paleontologists and Mineralogists Special Publication 37: 105–125

Listengarten VA (1984) Alluvial cones as deposits of groundwater. International Geology Review 26: 168–177

Liu T, Broecker W (2007) Holocene rock varnish microstratigraphy and its chronometric application in the drylands of western USA. Geomorphology 84: 1–21

Longwell CR (1930) Faulted fans of the Sheep Range, southern Nevada. American Journal of Science 20: 1–13

Longwell CR (1951) Megabreccia developed downslope from large faults. American Journal of Science 249: 343–355

Lustig LK (1965) Clastic sedimentation in Deep Springs Valley, California. U.S. Geological Survey Professional Paper 352–F

Machette MN (1985) Calcic soils of the southwestern United States. In: Weide DL (ed.) Soils and Quaternary geology of the southwestern United States. Geological Society of America Special Paper 203: 115–122

Major JJ, Iverson RM (1999) Debris-flow deposition: effects of pore-fluid pressure and friction concentrated at flow margins. Geological Society of America Bulletin 111: 1424–1434

Mather AE, Harvey AM, Stokes M (2000) Quantifying long-term catchment changes of alluvial fan systems. Geological Society of America Bulletin 112: 1825–1833

Mathewson CC, Keaton JR, Santi PM (1990) Role of bedrock ground water in the initiation of debris flows and sustained post-flow stream discharge. Association of Engineering Geologists Bulletin 27: 73–83

Matmon A, Schwartz DP, Finkel R, Clemmens S, Hanks T (2005) Dating offset fans along the Mojave section of the San Andreas fault using cosmogenic ^{26}Al and ^{10}Be. Geological Society of America Bulletin 117: 795–807

Mayer L, McFadden LD, Harden JW (1988) Distribution of calcium carbonate in desert soils: a model. Geology 16: 303–306

McCarthy TS, Smith ND, Ellery WN, Gumbricht T (2002) The Okavango Delta- semi-arid alluvial-fan sedimentation related to incipient rifting. In: Renault RW, Ashley GM (eds.) Sedimentation in Continental Rifts. Society for Sedimentary Geology (SEPM) Special Publication 73: 179–193

McDonald BC, Day TJ (1978) An experimental flume study on the formation of transverse ribs. Geological Survey of Canada Paper 78-1A: 441–451

McGee WJ (1897) Sheetflood erosion. Geological Society of America Bulletin 8: 87–112

McGowen JH (1979) Alluvial fan systems. In: Galloway WE, Kreitler CW, McGowen JH (eds.) Depositional and groundwater flow systems in the exploration for uranium. Texas Bureau of Economic Geology Research Colloquium, Austin, pp. 43–79

McPherson JG, Shanmugam G, Moiola RJ (1987) Fan deltas and braid deltas: Varieties of coarse-grained deltas. Geological Society of America Bulletin 99: 331–40

Melosh HJ (1987) The mechanics of large avalanches. In: Costa JE, Wieczorek GF (eds.) Debris flows/avalanches: process, recognition, and mitigation. Geological Society of America Reviews in Engineering Geology 7: 41–50

Melton (1965) The geomorphic and paleoclimatic significance of alluvial deposits in southern Arizona. Journal of Geology 73: 1–38

Meyer GA, Wells SG (1996) Fire-related sedimentation events on alluvial fans, Yellowstone National Park, U.S.A. Journal of Sedimentary Research 67: 776–791

Meyer GA, Pierce JL, Wood SH, Jull AJT (2001) Fire, storms, and erosional events on the Idaho batholith. Hydrological Processes 15: 3025–3038

Miall AD (1981) Analysis of fluvial depositional systems. American Association of Petroleum Geologists Educational Course Note Series 20

Middleton GV, Hampton MA (1976) Subaqueous sediment transport and deposition by sediment gravity flows. In: Stanley DJ, Swift DJP (eds.) Marine sediment transport and environmental management. Wiley, New York, pp. 197–218

Moore JM, Howard AD (2005) Large alluvial fans on Mars. Journal of Geophysical Research 110: 24 p.

Morton DM (1971) Seismically triggered landslides in the area above San Fernando Valley. U.S. Geological Survey Professional Paper 733

Morton DM, Campbell RH (1974) Spring mudflows at Wrightwood, southern California. Quartery Journal of Engineering Geology 7: 377–384

Mudge MR (1965) Rockfall-avalanche and rockslide-avalanche deposits at Sawtooth Ridge, Montana. Geological Society of America Bulletin 76: 1003–1014

Nash DB (1986) Morphologic dating and modeling degradation of fault scarps. In: Active tectonics: studies in geophysics. National Academy Press, Washington DC, pp. 181–194

Nelson AR (1992) Lithofacies analysis of colluvial sediments-an aid in interpreting the recent history of Quaternary normal faults in the Basin and Range Province, western United States. Journal of Sedimentary Petrology 62: 607–621

Newton MS, Grossman EL (1988) Late Quaternary chronology of tufa deposits, Walker Lake, Nevada. Journal of Geology 96: 417–433

Nicoletti PG, Sorriso-Valvo M (1991) Geomorphic controls of the shape and mobility of rock avalanches. Geological Society of America Bulletin 103: 1365–1373

Nilsen TH (1982) Alluvial fan deposits. In: Scholle P, Spearing D (eds.) Sandstone depositional environments. American Association of Petroleum Geologists Memoir 31: 49–86

Nilsen TH (1985) Introduction and Editor's comments. In: Nilsen TH (ed.) Modern and ancient alluvial fan deposits. Van Nostrand Reinhold, New York, pp. 1–29

Nishiizumi K, Kohl CP, Arnold JR, Dorn R, Klein J, Fink D, Middleton R, Lal D (1993) Role of in situ cosmogenic nuclides ^{10}Be and ^{26}Al in the study of diverse geomorphic processes. Earth Surface Processes 18: 407–425

Nordin CF (1963) A preliminary study of sediment transport parameters, Rio Puerco near Bernardo, New Mexico. U.S. Geological Survey Professional Paper 462-C

Nummedal D, Boothroyd JC (1976) Morphology and hydrodynamic characteristics of terrestrial fan environments. University of South Carolina Coastal Research Division Technical Report 10-CRD

Patton PC (1988) Drainage basin morphometry and floods. In: Baker VR, Kochel RC, Patton PC (eds.) Flood geomorphology. Wiley, New York, pp. 51–64

Philip H, Ritz JF (1999) Gigantic paleolandslide associated with active faulting along the Bogd fault (Gobi-Altay, Mongolia). Geology 27: 211–214

Plafker G, Ericksen GE (1984) Nevados Huascaran avalanches, Peru. In: Brunsden D, Prior DB (eds.) Slope instability. John Wiley and Sons, New York, pp. 277–314

Porter SC, Orombelli G (1980) Catastrophic rockfall of September 12, 1717 on the Italian flank of the Mont Blanc massif. Zeitschrift für Geomorphologie 24: 200–218

Purser BH (1987) Carbonate, evaporite, and siliciclastic transitions in Quaternary rift sediments of the northwestern Red Sea. Sedimentary Geology 53: 247–267

Rachocki A (1981) Alluvial fans. Wliey, Chichester

Rahn PH (1986) Engineering geology. Elsevier, New York

Rapp A, Fairbridge RW (1968) Talus fan or cone; scree and cliff debris. In: Encyclopedia of geomorphology. Reinhold, New York, pp. 1106–1109

Reheis MC (1986) Gypsic soils on the Kane alluvial fans, Big Horn County, Wyoming. U.S. Geological Survey Professional Paper 1590–C

Reheis MC, Harden JW, McFadden LD, Shroba RR (1989) Development rates of late Quaternary soils, Silver Lake playa, California. Soil Science Society of America Proceedings 53: 1127–1140

Reheis MC, McKee EH (1991) Late Cenozoic history of slip on the Fish Lake Valley fault zone, Nevada and California. U.S. Geological Survey Open-File Report 91–290

Reneau SL, Dietrich WE, Donahue DJ, Jull AJT, et al. (1990) Late Quaternary history of colluvial deposition and erosion in hollows, central California Coastal Ranges. Geological Society of America Bulletin 102: 969–982

Reneau SL, Oberlander TM, Harrington CD (1991) Accelerator mass spectrometry radiometric dating of rock varnish: discussion. Geological Society of America Bulletin 103: 310–311

Reynolds MW (1974) Geology of the Grapevine Mountains, Death Valley, California. In: Guidebook to the Death Valley region, California and Nevada, Geological Society of America Cordilleran Section Field Trip 1: 92–99

Richards K (1982) Rivers- form and process in alluvial channels. Methuen, Inc., London

Ritter DF (1978) Process geomorphology. William C. Brown, Dubuque

Ritter JB, Miller JR, Husek-Wulforst J (2000) Environmental controls on the evolution of alluvial fans in Buena Vista Valley, north central Nevada, during late Quaternary time. Geomorphology 36: 63–87

Robinson RAJ, Spencer JQG, Strecker MR, Richter A, Alonso RN (2005) Luminesence dating of alluvial fans in intramontane basins of NW Argentina. In: Harvey A, Mather AE, Stokes M (eds.) Alluvial fans: geomorphology, sedimentology, dynamics. Geological Society Special Publication 251: 153–168

Rockwell T (1988) Neotectonics of the San Cayetano fault, Transverse Ranges, California. Geological Society of America Bulletin 100: 500–513

Rodine JD, Johnson AM (1976) The ability of debris heavily freighted with coarse clastic materials to flow on gentle slopes. Sedimentology 23: 213–234

Rust BR, Gostlin VA (1981) Fossil transverse ribs in Holocene alluvial fan deposits, Depot Creek, South Australia. Journal of Sedimentary Petrology 51: 441–444

Ryder JM (1971) Some aspects of the morphometry of paraglacial alluvial fans in south-central British Columbia. Canadian Journal of Earth Sciences 8: 1252–1264

Schumm SA (1963) The disparity between the present rates of denudation and orogeny. U.S. Geological Survey Professional Paper 454–H

Schumm SA (1977) The fluvial system. Wiley, New York

Schuster RL, Salcedo DA, Valenzuela L (2002) Overview of catastrophic landslides of South America in the twentieth century. In: Evans SG, DeGraff JV (eds.) Catastrophic landslides: effects, occurrences, mechanisms. Geological Society of America Reviews in Engineering Geology XV: 1–34

Shaller PJ (1991) Analysis of a large, moist landslide, Lost River Range, Idaho, U.S.A. Canadian Geotechnical Journal 28: 584–600.

Sharp RP (1942) Mudflow levees. Journal of Geomorphology 5: 222–227

Sharp RP, Nobles LH (1953) Mudflow of 1941 at Wrightwood, southern California. Geological Society of America Bulletin 64: 547–560

Sharpe CFS (1938) Landslides and related phenomena: a study of mass movements of soil and rock. Columbia University Press, New York

Shaw J, Kellerhals R (1977) Paleohydraulic interpretation of antidune bedforms with applications to antidunes in gravel. Journal of Sedimentary Petrology 47: 257–266

Shreve RL (1968) The Blackhawk landslide. Geological Society of America Special Paper 108

Sidle RC, Ochiai H (2006) Landslides processes, prediction, and land use. Washington DC, American Geophysical Union Water Resources Monograph 18

Simons DB, Richardson EV (1966) Resistance to flow in alluvial channels. U.S. Geological Survey Professional Paper 422-J

Slate JL (1991) Quaternary stratigraphy, geomorphology, and ages of alluvial fans in Fish Lake Valley. In: Pacific Cell Friends of the Pleistocene Guidebook to Fish Lake Valley, Nevada and California, pp. 94–113

Sneh A (1979) Late Pleistocene fan-deltas along the Dead Sea rift. Journal of Sedimentary Petrology 49: 541–552

Sorriso-Valvo R (1988) Landslide-related fans in Calabria. Catena Supplement 13: 109–121

Spearing DA (1974) Alluvial fan deposits. Geological Society of America Summary Sheets of Sedimentary Deposits, sheet 1

Statham I, Francis SC (1986) Influence of scree accumulation and weathering on the development of steep mountain slopes. In: Abrahams, AD (ed.) Hillslope processes. Allen & Unwin, Boston, pp. 245–268

Strahler AN (1957) Quantitative analysis of watershed geomorphology. Transactions of the American Geophysical Union 38: 913–920

Terwilliger VJ, Waldron LJ (1991) Effects of root reinforcement on soil-slip patterns in the Transverse Ranges of southern California. Geological Society of America Bulletin 103: 775–785

Thenhaus PC, Wentworth CM (1982) Map showing zones of similar ages of surface faulting and estimated maximum earthquake size in the Basin and Range province and selected adjacent areas. U.S. Geological Survey Open-File Report 82–742

Tolman CF (1909) Erosion and deposition in the southern Arizona bolson region. Journal of Geology 17: 136–163

Trowbridge AC (1911) The terrestrial deposits of Owens Valley, California. Journal of Geology 19: 706–747

Turner AK (1996) Colluvium and talus. In: Turner AK, Schuster RL (eds.) Landslides. National Academy of Sciences Press, Washington DC, pp. 525–554

Turner BR, Makhlouf I (2002) Recent colluvial sedimentation in Jordan: fans evolving into sand ramps. Sedimentology 49: 1283–1298

Van Arsdale R (1982) Influence of calcrete on the geometry of arroyos near Buckeye, Arizona. Geological Society of America Bulletin 93: 20–26

Van de Kamp PC (1973) Holocene continental sedimentation in the Salton Basin, California: A reconnaissance. Geological Society of America Bulletin 84: 827–848

Varnes DJ (1978) Slope movement types and processes. In: Schuster RL, Krizek RJ (eds.) Landslides, analysis and control. Transportation Research Board, National Academy of Sciences, Washington D.C., Special Report 176: 11–33

Varnes DJ, Savage WZ (1996) The Slumgullion earth flow: a large-scale natural laboratory. U.S. Geological Survey Professional Paper 2130

Voight B (1978) Lower Gros Ventre slide, Wyoming, U.S.A., In: Voight B (ed.) Rockslides and avalanches, 1. Elsevier, Amsterdam, pp. 112–166

Walker TR (1967) Formation of red beds in modern and ancient deserts. Geological Society of America Bulletin 67: 353–368

Walker TR, Honea RM (1969) Iron content of modern deposits in the Sonoran Desert: a contribution to the origin of red beds. Geological Society of America Bulletin 80: 535–544

Walker TR, Waugh B, Crone AJ (1978) Diagenesis in first-cycle desert alluvium of Cenozoic age, southwestern United States and northwestern Mexico. Geological Society of America Bulletin 89: 19–32

Wallace RE (1978) Geometry and rates of change of fault-generated range-fronts, north-central Nevada. U.S. Geological Survey Research Journal 6: 637–650

Wallace RE (1984a) Fault scarps formed during the earthquakes of October 2, 1915, in Pleasant Valley, Nevada, and some tectonic implications. U.S. Geological Survey Professional Paper 1274-A

Wallace RE (1984b) Patterns and timing of late Quaternary faulting in the Great Basin Province and relation to some regional tectonic features. Journal of Geophysical Research 89: 5763–5769

Webb RH, Pringle PT, Rink GR (1987) Debris flows from tributaries of the Colorado River, Grand Canyon National Park, Arizona. U.S. Geological Survey Open-File Report 87–118

Weissmann GS, Bennett GL, Lansdale AL (2005) Factors controlling sequence development on Quaternary alluvial fans,

San Joaquin Valley, California, USA. In: Harvey A, Mather AE, Stokes M (eds.) Alluvial fans: geomorphology, sedimentology, dynamics. Geological Society Special Publication 251: 169–186

Wells SG, Dohrenwend JC (1985) Relict sheetflood bedforms on late Quaternary alluvial fan surfaces in the southwestern United States. Geology 13: 512–516

Wells SG, McFadden LD, Dohrenwend JC (1987) Influence of late Quaternary climatic changes on geomorphic and pedogenic processes on a desert piedmont, eastern Mohave Desert, California. Quaternary Research 27: 130–146

Wells WG (1987) The effects of fire on the generation of debris flows in southern California. In: Costa JE, Wieczorek GF (eds.) Debris flows/avalanches: process, recognition, and mitigation. Geological Society of America Reviews in Engineering Geology 7: 105–114

White K, Drake N, Millington A, Stokes S (1996) Constraining the timing of alluvial fan response to late Quaternary climatic changes, southern Tunisia. Geomorphology 17: 295–304

Wieczorek GF (1987) Effect of rainfall intensity and duration on debris flows in central Santa Cruz Mountains, California. In: Costa JE, Wieczorek GF (eds.) Debris flows/avalanches: process, recognition, and mitigation. Geological Society of America Reviews in Engineering Geology 7: 93–104

Williams GE (1973) Late Quaternary piedmont sedimentation, soil formation and paleoclimates in arid South Australia. Zeitschrift für Geomorphologie 17: 102–125

Yarnold JC, Lombard JP (1989) Facies model for large block avalanche deposits formed in dry climates. In: Colburn IP, Abbott PL, Minch J (eds.) Conglomerates in basin analysis. Pacific Section Society of Economic Paleontologists and Mineralogists Symposium Book 62: 9–32

Part VI
Lake Basins

Chapter 15

Hemiarid Lake Basins: Hydrographic Patterns

Donald R. Currey[†] and Dorothy Sack

> *'The Great Basin [of western North America]: . . . contents almost unknown, but believed to be filled with rivers and lakes which have no communication with the sea . . .'*
> Brevet Capt. J.C. Frémont, Corps of Topographical Engineers (1845).

Introduction

Lakes and other mappable bodies of standing water exist at the atmosphere-lithosphere interface. Over shorter time intervals, water body configurations (hydrography) respond directly to atmospheric (hydroclimatic) forcing. Over longer intervals, hydrography also reflects tectonic and volcanic forcing from the lithosphere. Hydrographic patterns in lake basins, in turn, strongly influence and even control many geomorphic and stratigraphic patterns (e.g. Mabbutt 1977). These linked patterns (hydroclimatic + tectonic → hydrographic → geomorphic + stratigraphic) make lakes and kindred water bodies superb instruments for gauging environmental change and recording palaeoenvironmental history.

Large quantities of lacustrine hydrographic and geomorphic information occur in spatial and temporal patterns, several of which are summarized in this and the following chapter. Although it is often convenient to view water bodies and associated features two-dimensionally (e.g. in plan or cross section), lacustrine spatial patterns are inherently three-dimensional (Fig. 15.1). Similarly, lacustrine temporal patterns that derive from changing spatial patterns are inherently four-dimensional. Therefore, X (easting), Y (northing), and Z (elevation) coordinates obtained by georeferencing technologies, such as a global positioning system (GPS), together with ages measured by chronoreferencing technologies, such as accelerator mass spectrometer (AMS) analysis of cosmogenic isotopes and optically stimulated luminescence (OSL) analysis of clastic sediments, are essential elements of hydrographic and geomorphic patterns.

Hemiarid Lake Basins

Hydrographic and topographic closure are fundamental, and quite distinct, aspects of surface water distribution in all drainage basins. By definition, every lake basin has topographic closure, or containment, but only those that have 'no communication with the sea' (Frémont 1845) also possess hydrographic closure. Most humid areas have hydrographically open drainage basins. Humid areas without topographic closure typically convey surface water in stream networks, while those with topographic closure contain externally drained lakes as part of the stream system (Sack 2001). Basins with hydrographic closure are described as internally drained, or basins of interior drainage. Areas that are hydrographically closed yet topographically open have little surface water and exist, for example, in hyperarid coastal deserts. Drainage basins that are closed both topographically and hydrographically and also contain or have contained bodies of standing water are hemiarid lake basins (Fig. 15.2), the subject of this and the following chapter.

D. Sack (✉)
Department of Geography, Ohio University, Athens, OH 45701, USA
e-mail: sack@ohio.edu

[†] deceased

A.J. Parsons, A.D. Abrahams (eds.), *Geomorphology of Desert Environments*, 2nd ed., DOI 10.1007/978-1-4020-5719-9_15, © Springer Science+Business Media B.V. 2009

Fig. 15.1 Block diagram of a narrow sector in a hypothetical lake basin showing (**a**) water and sediment transfer from (1) the zone of net runoff and net denudation, through (2) the zone of maximum water and sediment transfer, to (3) the shoreline, and into (4) the zone of water body equilibration and net sedimentation; and (**b**) four-tiered environmental/palaeoenvironmental column in the basin

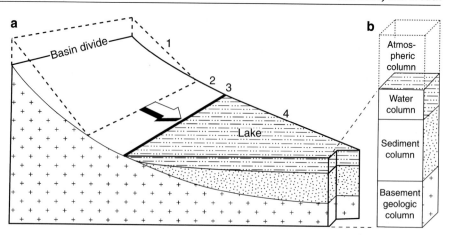

Hemiarid lake basins have dual hydroclimates. Highlands are nonarid with cumulative water balance surpluses, that is, annual precipitation is greater than annual evapotranspiration. Lowlands are arid, semiarid, or hyperarid with cumulative water balance deficits because annual precipitation is less than annual potential evapotranspiration. Highlands in hemiarid basins, therefore, are runoff producers; they provide water for stream flow and recharge the groundwater, both of which are conducted toward the lowlands. The lowlands mainly collect and evaporate (consume) surface and subsurface water, most of which is directed to them from the highlands. Water bodies in hemiarid lake basins tend to be classic examples of self-regulating, climate-driven systems (Thorn 1988). By undergoing sometimes dramatic size fluctuations in response to climate changes, water losses and gains maintain dynamic equilibrium (e.g. Mifflin and Wheat 1979, Street-Perrott and Harrison 1985). Gains, mainly from highland runoff and direct precipitation onto the lowland water body, increase lake surface area which promotes greater water loss by surface evaporation. The negative feedback role of surface area in water-body self-regulation is illustrated in Fig. 15.3.

Hydroclimatic zonation within hemiarid lake basins is mainly a function of elevation and slope aspect, but the areal extent of each zone is a planimetric variable (Fig. 15.4). The water-surplus higher part of a hemiarid lake basin has an area that can be expressed dimensionlessly by the surplus area ratio (SAR), where SAR is water-surplus area as a fraction of total basin area. The lower limit of the water-surplus zone, and upper limit of the water-deficit zone, is the hydroclimate equilibrium line altitude (HELA), which is the average elevation above sea level at which annual precipitation and evapotranspiration are equal. The water-deficit lower part of a hemiarid lake basin has an area that can be expressed by the deficit area ratio (DAR), where DAR + SAR = 1. Water-deficit subzones (Fig. 15.4) include dry lowlands, wet lowlands (except lakes and ponds), and open water (lakes and ponds). The area of open water can be expressed dimensionlessly by the water area ratio (WAR), where WAR is the area of open water as a fraction of total basin area. Completely nonarid basins (SAR = 1.0) with topographic closure invariably lack hydrographic closure, that is, they contain externally drained bodies of fresh water. Completely arid basins (DAR = 1.0) seldom contain water bodies larger than groundwater-fed brine pools.

Hydrographic features of several types, and represented by numerous names, generally connote hemiarid lake basins. These include alkali lakes, badwaters, bitter lakes, bolson lakes, brackish lakes, brine lakes, closed-basin lakes, dead seas, desert lakes, dry lakes, endoreic lakes, ephemeral lakes, hypersaline lakes, impermanent lakes, inland salt lakes, inland seas, intermittent lakes, landlocked

	Hydrographically	
	open basin	closed basin
Topographically open basin	Uncontained and externally drained (river system)	Uncontained and internally drained (influent system)
Topographically closed basin	Contained and externally drained (lake)	Contained and internally drained (hemiarid)

Fig. 15.2 Hemiarid lake basins (*lower right*) are closed both topographically and hydrographically, that is, they have both containment and internal drainage

Fig. 15.3 Water balance transfers in hemiarid lake basins, where runoff originates in nonarid highlands (*upper left*), passes through water-deficit dry uplands and lowlands (*middle*), and terminates in hydroclimatically arid lowlands with standing water (*lower right*). Transfers are dimensionally equal to length (depth of water) per unit time. Negative feedback regulates runoff per unit area of lowland standing water, keeping inputs (mainly runoff) to the standing water and outputs (evaporation) from the standing water in dynamic equilibrium

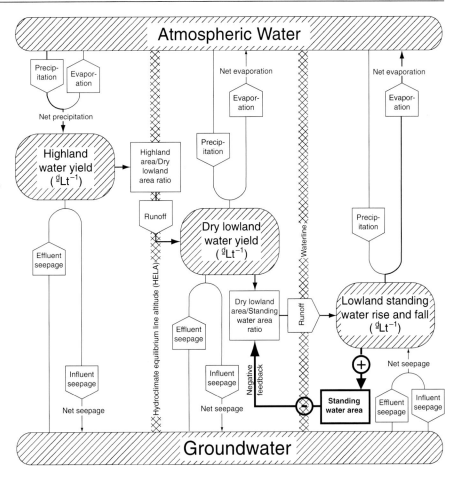

lakes, lost lakes, mud lakes, oasis lakes, pan lakes, playa lakes, saline lakes, salt marsh lakes, salton seas, sink lakes, soda lakes, temporary lakes, terminal lakes, undrained lakes, and many others. Terrestrial water bodies below sea level indicate hemiarid lake basins unless the water derives mainly from seawater or groundwater. In addition, many so-called pluvial lakes of Pleistocene and Holocene age (e.g. Morrison 1968, Reeves 1968, Smith and Street-Perrott 1983) occurred in what were and still are hemiarid lake basins, although the sites of some of these palaeolakes have been breached and dissected by through-flowing drainage (e.g. Meek 1989).

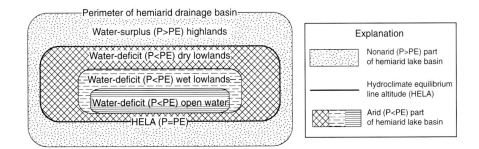

Fig. 15.4 Schematic plan view of a hypothetical hemiarid lake basin showing water-surplus (nonarid) and water-deficit (arid) hydroclimatic zones; P = annual precipitation and PE = annual potential evapotranspiration (Mather 1978). This basin has a surplus area ratio (SAR) of 0.5 and a water area ratio (WAR) of 0.1

Table 15.1 Structural (tectonic and volcanic) origins of lake containment basins,* where underlying causes of topographic closure are structural blockage (B) of basin outlets and structural lowering (L) of basin floors. Very shallow (1), shallow (2), moderately deep (3), and deep (4) topographic closure depths are typical of occurrences in North American deserts and semideserts

Structural causes of topographic closure	Typical closure depths			
	(1)	(2)	(3)	(4)
Tectonic lake containment basins:				
major half-grabens (L)			•	•
major grabens (L)			•	•
transcurrent faulting sag ponds (B)	•	•		
transcurrent-faulting-blocked valleys (B)	•	•		
horst-dammed valleys (B)			•	
anticline-dammed valleys (B)			•	
diapir-dammed valleys (B)		•	•	
doubly plunging syncline basins (L)			•	
ring-fracture-bounded basins (L)		•	•	
volcano-tectonic collapse basins (L)			•	•
Volcanic lake containment basins:				
calderas (L)			•	•
craters (L)		•	•	
maars (L)		•	•	
volcano-dammed valleys (B)			•	
lahar-dammed valleys (B)		•	•	
ash-flow-dammed valleys (B)			•	
lava-flow-dammed valleys (B)		•	•	
collapsed lava tubes (L)	•	•		

* Compiled from Hutchinson (1957, p. 156–163).

Hydroclimatic zones and hydrographic features in hemiarid lake basins are always subject to change. Figure 15.5 shows five annual hydroclimatic outcomes – much drier, drier, little net change, wetter, much wetter – that can result from nine possible combinations of deviation in annual precipitation, annual evaporation, or both. Annual precipitation and evaporation values, in turn, derive from 81 possible combinations of seasonal changes in those variables, which are represented in the figure by the outer tier of nine lettered cells (seasonal precipitation) and nine numbered cells (seasonal evaporation). A G3 seasonal scenario, for example, would be expected to produce a much wetter annual outcome, as would any combination of a G, H, or I seasonal precipitation trend with a 1, 2, or 3 seasonal evaporation trend. If the location is a subtropical desert, a G3 scenario would probably involve significant intensification of monsoonal circulation (Magee et al. 1995, Schuster et al. 2003, DeVogel et al. 2004).

Basin Origins

Hemiarid lake basins exist in regions where aridity and tectonic activity coincide (Fig. 15.6). Tectonic activity, with or without syntectonic volcanism, is generally the underlying cause of topographic closure (e.g. Thornbury 1965, Eaton 1982, May et al. 1999). Exogenic geomorphic processes, which can enhance or degrade topographic closure, are commonly important secondary causes.

Tectonism and volcanism cause topographic closure in two ways: (a) by blocking basin outlets, and (b) by lowering basin floors (Table 15.1). The greatest depths of topographic closure typically result from block faulting in regions of extensional tectonics and from subsidence due to partial emptying of magma chambers beneath volcanic terranes (e.g. Le Turdu et al. 1999).

Exogenic geomorphic processes also cause topographic closure by blocking basin outlets and lowering basin floors (e.g. Trauth and Strecker 1999) (Table 15.2). These geomorphic contributions to topographic closure are usually limited to low-relief embellishments of pre-existing structural basins. Some geomorphic processes cause structural basins and systems of interconnecting structural basins to become segmented into shallow, topographically closed subbasins (e.g. Hunt et al. 1966, Peterson 1981). Large alluvial fans and compound barrier beaches are particularly important as sills between subbasins (e.g. Russell 1885, Benson and Thompson 1987, Sack 2002).

Fig. 15.5 Hydroclimatic change in hemiarid lake basins, with five annual outcomes appearing in nine large cells at *lower right*. These are produced by 81 seasonal scenarios portrayed by groups of smaller cells at *upper right* and *lower left*. Symbols: $-$ = decrease, 0 = no change, and $+$ = increase annually or seasonally; see text for further explanation

Summer precipitation change

Winter precipitation change		$-$	0	$+$	$-$	0	$+$	$-$	0	$+$
	$-$	A	B				D			
	0	C				E				G
	$+$				F				H	I

Summer evaporation change

Winter evaporation change		$-$	0	$+$
	$-$	1	2	
	0	3		
	$+$			
	$-$			4
	0		5	
	$+$	6		
	$-$			
	0			7
	$+$		8	9

Annual evaporation change / Annual precipitation change

Annual evaporation change	Annual precipitation change $-$	0	$+$
$-$	Little net change	Wetter	Much wetter
0	Drier	Little net change	Wetter
$+$	Much drier	Drier	Little net change

Basin Changes

A hemiarid lake basin can contain lakes up to a maximum size (basin capacity) that is controlled by the height of the basin threshold, and it can contain shallow water bodies that are identifiable as lakes down to a minimum size (basin capability) that is controlled by the shape of the basin floor. Many geomorphic and structural factors affect basin capacity and basin capability. Geomorphic processes may degrade or aggrade basin thresholds and basin floors, but threshold degradation and aggradational widening and flattening of basin floors predominate (Fig. 15.7). As a result,

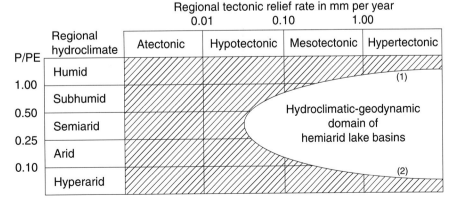

Fig. 15.6 Hemiarid lake basins occur in regions where aridity and tectonic activity coincide; P and PE as in Fig. 15.4. Hypertectonism extends the domain of hemiarid lake basins by causing (1) intrazonal rainshadows in humid zones, and (2) intrazonal orographic precipitation in hyperarid zones

Regional tectonic relief rate in mm per year

	P/PE	Regional hydroclimate	Atectonic 0.01	Hypotectonic 0.10	Mesotectonic 1.00	Hypertectonic
	1.00	Humid				(1)
	0.50	Subhumid				
	0.25	Semiarid			Hydroclimatic-geodynamic domain of hemiarid lake basins	
	0.10	Arid				
		Hyperarid				(2)

Table 15.2 Geomorphic origins of lake containment basins,* where immediate causes of topographic closure are geomorphic blockage (B) of basin outlets or geomorphic lowering (L) of basin floors. Very shallow (1), shallow (2), moderately deep (3), and deep (4) topographic closure depths are typical of occurrences in North American deserts and semideserts

Geomorphic causes of topographic closure†	Typical closure depths			
	(1)	(2)	(3)	(4)
Mass wasting lake containment basins:				
landslide-dammed basins (B)	•	•	•	
slump rotation basins (L)	•	•		
Glacial lake containment basins:				
ice-dammed basins (B)				
moraine-dammed basins (B)	•	•	•	
glacially scoured basins (L)				
kettles (L)	•	•		
Periglacial lake containment basins:				
cryofluction-dammed basins (B)				
thermokarst basins (L)				
nivation hollows (L)	•			
Solution lake containment basins:				
travertine- and sinter-rimmed pools (B)	•	•		
sinkholes (L)	•	•		
cavern pools (L)	•	•		
Fluvial lake containment basins:				
natural-levee-dammed basins (B)	•	•		
alluvial-fan-dammed basins (B)	•	•	•	•
delta-dammed basins (B)	•	•	•	
abandoned waterfall plunge pools (L)	•	•		
abandoned channels (L)	•	•		
flood-scoured pools (L)	•	•		
Aeolian lake containment basins:				
interdune swales (B)	•	•		
dune-dammed basins (B)	•	•		
blowouts (L)	•	•		
deflation basins (L)	•	•	•	
Seashore lake containment basins:				
barrier-beach-enclosed lagoons (B)				
barrier-reef-enclosed lagoons (B)				
shore platform tide pools (L)				
Lakeshore lake containment basins:				
barrier-beach-enclosed lagoons (B)	•	•		
barrier-complex-dammed subbasins (B)	•	•	•	•
Phytogenic lake containment basins:				
bog-rimmed pools (B)	•			
Zoogenic lake containment basins:				
beaver ponds (B)	•	•		
animal wallows (L)	•			
Cosmogenic lake containment basins:				
meteorite impact craters (L)	•	•	•	

* Compiled from Hutchinson (1957, p. 156–163).
† The absence of any indication of closure depth adjacent to a particular lake type signifies that lake type is not found in deserts.

some basins lose part or all of their capacity to contain large lakes and become incapable of containing small lakes. As basin capacity and capability diminish with time, the range of lake sizes that a basin can accommodate becomes narrower and there is a decrease in the range of hydroclimatic variation that the lake basin can document in its geomorphic and stratigraphic record.

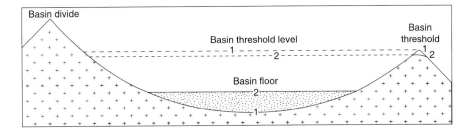

Fig. 15.7 Schematic cross section of a hypothetical lake basin showing basin threshold (bt) and basin floor (bf) changes produced by geomorphic and structural activity (Tables 15.3 and 15.4) through time (1 = earlier; 2 = later). Basin fill sediments overlie bf$_1$ and underlie bf$_2$. Topographic closure decreased due to both the threshold lowering from bt$_1$ to bt$_2$ and the basin floor aggrading from bf$_1$ to bf$_2$

Basin threshold height, which ultimately controls basin capacity, is subject to change from a variety of geomorphic and structural causes (Table 15.3). The greatest change in threshold height is likely to occur the first time a lake transgresses to its basin threshold. In that situation, observations from several hemiarid lake basins in western North America suggest the following sequence of geomorphic events, listed in order of occurrence: (1) piping and sapping by pre-flood subsurface overflow beneath and downstream from the threshold, (2) channel incision by pre-flood surface overflow downstream from the threshold, (3) headward erosion of the pre-flood overflow channel, (4) hydraulic failure when headward erosion reaches the threshold, (5) scouring of the flood path into a deep flood channel upstream and downstream from the threshold, (6) scouring an overdeepened kolk in the flood channel floor near the site of the former threshold, (7) landsliding along the flanks of the flood channel, (8) alluviation of the flood channel by waning-flood and postflood overflow, (9) segmentation of the flood palaeochannel by small alluvial fans that build from either side, and (10) colluviation of the flood palaeochannel by slopewash that includes reworked loess.

Basin floor elevation and breadth, which are also subject to change from a variety of geomorphic and structural causes (Table 15.4), tend to increase as basin fill sediments aggrade through time. As a result, the capability of many hemiarid lake basins to accommodate small lakes gradually decreases. In theory, decreasing basin capability should be offset by processes, such as faulting, volcanism, and to a lesser extent aeolian deflation and abrasion (Magee et al. 1995, Hoelzmann et al. 2001), that create new basin floor relief. In practice, however, aggrading sediments are readily distributed over very low gradients on basin floors that alternate between subaqueous and subaerial conditions and that are subject to recurring hydroaeolian planation. This repeatedly restores and enlarges basin floor horizontality in many hemiarid lake basins, and repeatedly flattens many sites that might otherwise contain small lakes.

Table 15.3 Basin threshold morphodynamics*

Causes of basin threshold change	Threshold height change†
Geomorphic causes:	
fan toe accretion	+
loess accretion	+
aeolian sand accretion	+
aeolian deflation and abrasion	−
cryptoluvial piping and sapping	−
surface overflow incision	−
surface overflow headward erosion	−
hydraulic failure of threshold	−
flood channel scouring	−
flood channel kolk overdeepening	−
flood-triggered landsliding	+
palaeochannel alluviation	+
palaeochannel colluviation	+
Structural (tectonic and volcanic) causes:	
far-field isostatic deflection	+ or −
near-field hydroisostatic deflection	+ or −
near-field lithoisostatic deflection	−
near-field glacioisostatic deflection	0
extensional seismotectonic displacement	+ or −
transcurrent seismotectonic displacement	+ or −
compressional seismotectonic displacement	+ or −
volcanic flow emplacement	+ or −
cinder cone construction	+
tephra deposition	+

* Adapted from Currey (1990).
† Symbols: + = threshold raised; 0 = insignificant in most hemiarid lake basins; and − = threshold lowered.

Table 15.4 Basin floor morphodynamics*

Causes of basin floor change	Height change†	Breadth change‡
Geomorphic causes:		
proluvial (fan toe sandflat) deposition	+	−
deltoid (Mabbutt 1977) deposition	+	0 or −
deltaic deposition	+	0 or −
offshore deposition	+	0
lakeshore erosion	+	0 or +
lakeshore deposition	+	0 or −
evaporite deposition	+	0 or +
evaporite dissolution	−	0 or −
spring bog deposition	+	0
spring mound deposition	+	0
hydroaeolian planation	+	+
aeolian erosion	−	0 or +
aeolian deposition	+	0 or −
Structural (tectonic and volcanic) causes:		
far-field isostatic deflection	+ or −	−
near-field isostatic deflection	+ or −	−
seismotectonic displacement	+ or −	−
salt diapir doming	+	−
magma chamber inflation or deflation	+ or −	−
volcanic flow emplacement	+	−
volcano accretion	+	−
volcano tectonic subsidence	−	−
tephra deposition	+	0

* Adapted from Currey (1990).
† Symbols: + = basin floor raised; 0 = level of basin floor unchanged; and − = basin floor lowered.
‡ Symbols: + = basin floor enlarged; 0 = area of basin floor unchanged; and − = basin floor reduced in size.

Basin Hydrographic Connections

Intrabasin hydrographic connections consist of waterways that move water, solutes, and sediments from one area to another within a hemiarid lake basin. Interbasin hydrographic connections are cascades that transport water, solutes, and sediments from one hemiarid lake basin to another. It is common for two or more basins to be connected by subsurface cascades (e.g. Harrill et al. 1988); it is less common for two or more hemiarid lake basins to be connected by surface cascades (e.g. Hubbs et al. 1974, Mifflin and Wheat 1979, Williams and Bedinger 1984). Indeed, surface cascades occur only (a) infrequently, when threshold-controlled 'minipluvial' highstands happen in basins with shallow topographic closure, (b) rarely, when threshold-controlled 'plenipluvial' highstands occur in basins with deep topographic closure, or (c) extrinsically, when runoff acquired through drainage diversion or stream capture leads to threshold control (e.g. Bouchard et al. 1998).

There are two kinds of intrabasin waterways: (a) input waterways act as conduits directing runoff into lakes from the surrounding drainage basin, and (b) throughput waterways form links between distinct morphological subdivisions within lakes. Lakes that lack major perennial or intermittent input waterways are simple lakes, whereas complex lakes receive inflow from one or more major input waterway (Fig. 15.8).

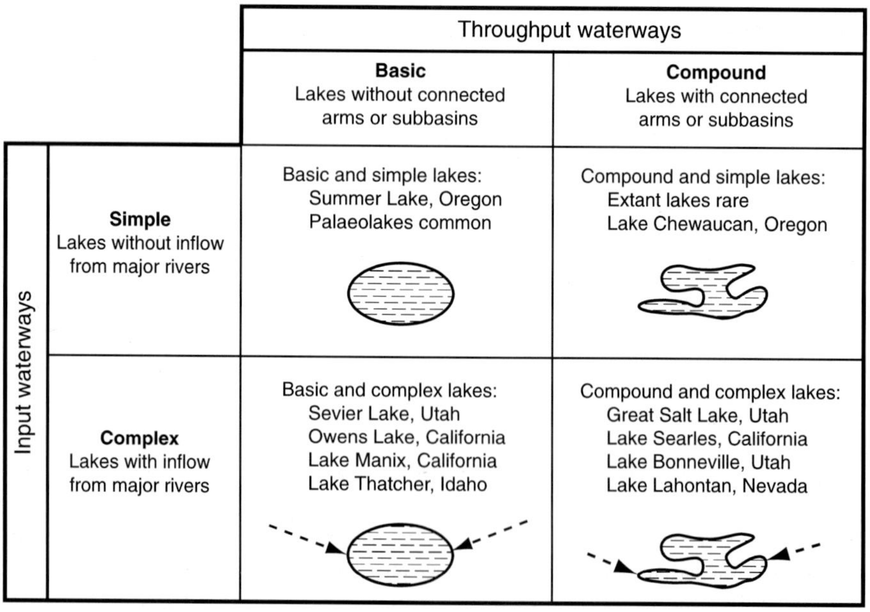

Fig. 15.8 Hemiarid lakes are described as complex or simple according to the presence or absence of input waterways, and as compound or basic depending on the presence or absence of strait-connected arms and/or sill-connected subbasins (modified from Currey 1990). These concepts are illustrated with plan-view sketches of primary attributes. Listed example lakes include extant lakes and palaeolakes

		Throughput waterways	
		Basic Lakes without connected arms or subbasins	**Compound** Lakes with connected arms or subbasins
Input waterways	**Simple** Lakes without inflow from major rivers	Basic and simple lakes: Summer Lake, Oregon Palaeolakes common	Compound and simple lakes: Extant lakes rare Lake Chewaucan, Oregon
	Complex Lakes with inflow from major rivers	Basic and complex lakes: Sevier Lake, Utah Owens Lake, California Lake Manix, California Lake Thatcher, Idaho	Compound and complex lakes: Great Salt Lake, Utah Lake Searles, California Lake Bonneville, Utah Lake Lahontan, Nevada

The presence or absence of throughput waterways distinguishes compound from basic lakes. Compound lakes consist of strait-connected arms or sill-connected subbasins, and tend to be large lakes. Intrabasin waterways strongly influence lacustrine geomorphic patterns in lakes that are compound and complex, like former Lake Lahontan (Benson and Thompson 1987, Benson and Paillet 1989).

At a very general level, all hemiarid lake basins occupy one of four possible positions in subsurface and surface cascades (Fig. 15.9) and many basins have attributes that reflect their cascade position. For example, cascade-head basins are typically low-salinity environments, whereas cascade-terminus basins often contain brines and saline sediments. In surface cascades, water, solutes, and usually sediments move from one basin to another at comparatively high, even torrential, rates for relatively brief periods. In subsurface cascades, water and solutes, but not sediments, move from basin to basin at low rates, sometimes moistening and saliniz-ing basin floors from beneath, for extended periods. As they enter, cross, and leave hemiarid lake basins, sur-face and subsurface cascades interact with basin mor-phology in distinctive patterns.

Certain cascade patterns are repeated spatially and temporally in many hemiarid lake basins. Spa-tially, five patterns prevail in hydrologically and topographically closed terminal and near-terminal basins (Fig. 15.10, right-hand column). Such basins seldom have completely dry surfaces and commonly store solute and sediment yields from large upstream regions. Temporally, five stages of cascade evolution are common in upstream basins (Fig. 15.10, middle column): (1) tectonically young basins commonly store sediments and spill mainly water and solutes; (2) with geomorphically reduced topographic closure, basins increasingly pass sediments downstream as

well; (3) with deepening threshold incision, stored materials are released at rates that range from gradual to catastrophic; (4) some basins are eventually opened completely by through-flowing surface drainage; and (5) when stream discharge wanes, through-flowing drainage can be limited to groundwater and solutes in the subsurface. Regional cascades of water, solutes, and sediments that are typical of hemiarid lake basins are illustrated in Fig. 15.11.

Hydrographic Continuum

Despite water's long term regional scarcity, it exists in a remarkable variety of forms on the floors of hemiarid lake basins (e.g. Langbein 1961, Houghton 1986). These forms belong to a hydrographic continuum that can be subdivided into several definable, but intergradational, hydrographic realms and subrealms (Table 15.5). In essence megalakes, lakes, and ponds are runoff-collecting and runoff-evaporating water bodies. Aquatic wetlands, saturated wetlands, and unsaturated wetlands are groundwater-discharging and groundwater-evaporating basin floors. Playas and microplayas are stormwater-wetted, stormwater-infiltrating, and stormwater-evaporating basin floors.

Ecologically, lacustrine habitats are limited mainly to water bodies with depths greater than 2 m; shallower water bodies are the realm of palustrine habitats (Cow-ardin et al. 1979). Geomorphically, well developed ev-idence of lacustrine processes is limited mainly to wa-ter bodies with depths greater than 4 m. In shallower water bodies (a) aerodynamic turbulence causes fre-quent mixing of the water column and vigorous hy-drodynamic stirring of the offshore sediment record, and (b) gently shelving topography suppresses onshore propagation of waves in other than low energy bands of the wave spectrum, which effectively limits onshore and longshore transport of beach-forming sediments.

In low-lying areas that contain little or no standing water much of the time, as well as in low-lying areas that are adjacent to lakes, mixed wetlands form complex mosaics on the floors of many hemiarid lake basins. In order of increasing salinity, mixed wetlands commonly range from spring-fed aquatic marshes, to seep-fed muddy marshes, to alkali meadows, to saline mudflats, and ultimately to phreatic saltflats (Table 15.5).

| | Relative to cascade head ||
	Most headward basin H	Basin downstream from head h
Relative to cascade terminus		
Basin upstream from terminus t	Cascade-head basin Ht	Mid-cascade basin ht
Terminal basin T	Cascade with one basin HT	Cascade-terminus basin hT

Fig. 15.9 Basin positions in regional cascades of water, solutes, and sediments

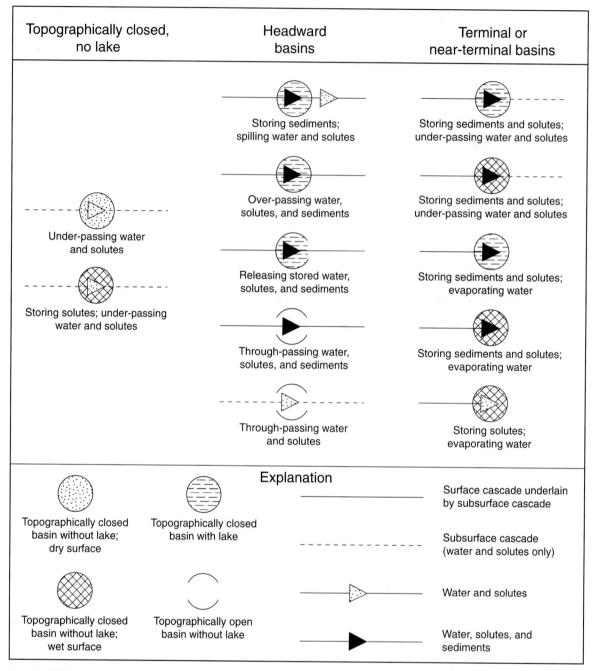

Fig. 15.10 Plan view symbols depicting cascades of water, solutes, and sediments in hemiarid lake basins. Direction of flow in each case (indicated by triangles) is *left to right*

Hydrographic Change

Hydrographic change is a hallmark of hemiarid lake basins (e.g. Street and Grove 1979). It takes many forms and has many causes. Transgression, stillstand, or regression each can result from several causes (Fig. 15.12). Spatial, temporal, and kinematic analysis of net hydrographic change and its components is fundamental to lacustrine geomorphology, thus hydrographic change is viewed here from basinwide spatial, temporal, and spatiotemporal perspectives.

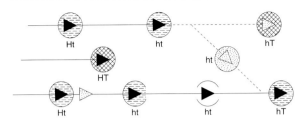

Fig. 15.11 Schematic plan of two hypothetical regional cascades, a one-basin cascade (HT) and an eight-basin cascade that has two terminal basins (hT). For explanation of symbols see Figs. 15.9 and 15.10

Hydrographic change occurs within spatial limits that are set by basin morphometry. As described above in the section on basin changes, the upper limit of potential hydrographic change, that is, basin capacity, is threshold controlled and the lower limit, basin capability, is controlled by the basin floor. A large range of uninterrupted hydrographic change is possible only in hemiarid lake basins with deep topographic closure and small basin floors (Fig. 15.13).

Hydrographic responsiveness expresses the vertical (surface elevation) change that a water body undergoes in response to a given hydrologic (surface area) change. Steep-sided basins (e.g. Fig. 15.13) are responsive to hydroclimatic change, while basins with very low sloping sides are relatively unresponsive – that is, hydrographic responsiveness is a direct function of basin slope (Street-Perrott and Harrison 1985, Bartov et al. 2002).

Hydrographic fitness expresses how well late Pleistocene and Holocene highstands and lowstands in hemiarid lake basins fit their containment basins (Fig. 15.14). Full geomorphic and stratigraphic records of hydrographic change are found in continuously fit basins, such as the Lake Lahontan basin in Nevada and California, in which the water body is contained below the threshold at highstands and above the minimum level for lacustrine processes at lowstands. Less than complete records of hydrographic change are found in overfit basins, including the Lake

Table 15.5 Hydrographic realms and subrealms on the floors of hemiarid lake basins have characteristic surface water and groundwater hydrology, and characteristic geomorphic and geodynamic features

Hydrographic realms and subrealms	Surface water and groundwater hydrology	Geomorphic and geodynamic features
Lakes: megalakes	Permanent* bodies of open standing water; static water surfaces >100 m above basin floors over large areas (>1000 km^2); total areas >10 000 km^2; far-field tributary networks, usually with surface runoff from more than one climatic region and several highland life zones	Well developed evidence of shore (backshore, foreshore, nearshore) and offshore sedimentation; tributary networks usually derive terrigenous sediments from several geologic terranes; measurable evidence of near-field hydroisostatic deflection; possible evidence of lithoisostatic deflection near depocentres
lakes	Persistent* or permanent bodies of open standing water; static water surfaces >4 m above basin floors; usually with surface runoff from more than one life zone	Well developed evidence of shore sedimentation; commonly evidence of offshore sedimentation; terrigenous sediments commonly from more than one geologic terrane; little evidence of near-field isostatic deflection, but possible evidence of far-field deflection
Quasi-lakes: ponds	Transient* or persistent bodies of open standing water; static water surfaces 1 to 4 m above basin floors; standing water mostly from surface runoff	Too shallow for well developed evidence of shore sedimentation; too shallow for unreworked evidence of offshore sedimentation; hydroaeolian planation and, in shallow basins, brim-full sedimentation tend to transform ponds into wetlands and playas; evaporite ponds floored with soluble salts occur in subbasins where evaporating brines are replaced by saline surface waters, sometimes from adjacent water bodies by restricted inflow across low barriers
Wetlands: aquatic	Ground surfaces covered by transient layers of local standing water <1 m deep; standing water mostly from groundwater discharge	Range from biotically productive aquatic marshes dominated by emergent hydrophytes to clear evaporite pools floored with soluble salts and saline pools floored with saline muds

Table 15.5 (continued)

Hydrographic realms and subrealms	Surface water and groundwater hydrology	Geomorphic and geodynamic features
saturated	Surficial materials water saturated; groundwater tables that coincide with basin floors discharge water and solutes	Range from muddy marshes and spring bogs dominated by hydrophytes to nonvegetated phreatic saltflats (= saltpans, etc.) where perennial evaporite pavements are renewed or enlarged by recurring precipitation of soluble salts from saline groundwaters that saturate basin floors; unless renewed as phreatic saltflats, limnogenous saltflats (evaporite pavements formed by desiccation of antecedent lakes) undergo dissolution into saline mudflats or playas, or burial by younger muds
unsaturated	Surficial materials damp, but usually drier than field capacity; groundwater tables that underlie basin floors discharge capillary water and solutes through evaporative pumping, also termed the wick effect	Range from alkali meadows dominated by salt-tolerant grasses and shrubs to sparsely vegetated saline mudflats; undergo precipitation of efflorescent salts during seasons of strong evaporative pumping, followed by dissolution of efflorescent salts during seasons of surface wetting
Drylands: playas	Surficial materials usually drier than wilting point; water tables much below basin floors; capillary fringes do not intercept basin floors; wetted briefly by rain or snowmelt, and flooded occasionally by stormwater runoff from surrounding piedmont and upland source areas	Relatively salt-free, fine-grained surfaces (= nonsaline mudflats, claypans, etc.) where stormwater runoff events spread suspended sediments widely and dump coarser sediments locally; in absence of effective effluent seepage or evaporative pumping, occasional influent seepage effectively leaches soluble salts from playa surfaces
microplayas	Very small playas, usually with areas $<1\,\mathrm{km}^2$	Playas that develop in local depressions such as blowouts and former lagoons, rather than on basin floors of more general extent

* See Fig. 15.15.

Bonneville-Great Salt Lake basin in Utah and the Lake Russell-Mono Lake basin in California. Water bodies in overfit basins lie above the minimum level for lacustrine processes at lowstands but become threshold-controlled at highstands. Far from complete records of hydrographic change are found in misfit basins, like Lake Searles, California, which display spilling lakes at highstands and also experience lowstand levels below the minimum for lacustrine processes. Underfit basins, such as Death Valley, California, likewise encode incomplete records. Although they are well contained at highstands, lowstands are too shallow to leave significant lacustrine evidence. In addition to hydrographic fitness, lakeshore configuration and sediment supply are important factors in the geomorphic recording of hydrographic change. These topics are reviewed fully in the next chapter.

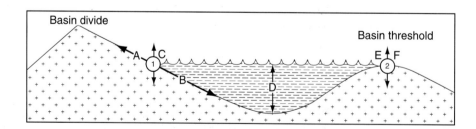

Fig. 15.12 Schematic profile through a lake basin, showing possible net hydrographic change (A = transgression and B = regression) at a shoreline locality (1) due to possible combinations of upward or downward neotectonic deformation (C), hydrologic change of the lake (D), and geomorphic change (E) or neotectonic deformation (F) of the lake threshold (2)

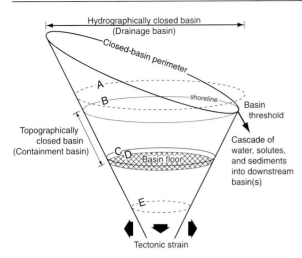

Fig. 15.13 'Dixie cup' (conical graben) model of a hypothetical hemiarid lake basin showing (A) off-scale highstand inferred by extrapolating from geomorphically recorded hydrographic history, (B) threshold-controlled shoreline, (B–C) hydrographically fit lake stages, (C) low-water limit of lacustrine processes, (C–D) ponds and aquatic wetlands, (D) saturated and unsaturated wetlands on basin floor, and (E) off-scale lowstand inferred by extrapolating from geomorphically recorded hydrographic history (after Currey 1990)

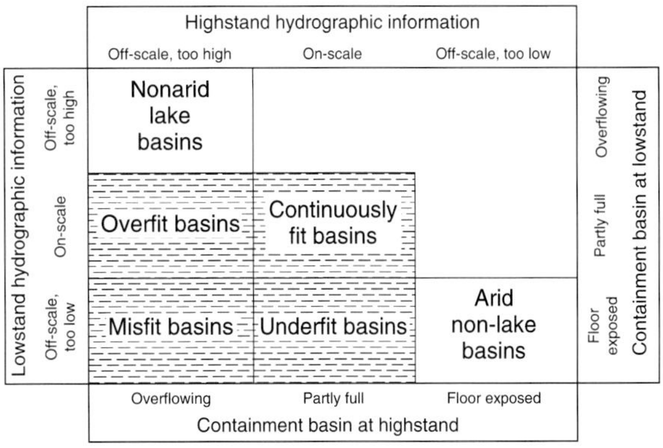

Fig. 15.14 Hydrographic fitness defined in terms of late Quaternary highstands and lowstands (after Currey 1990); hemiarid lake basin cells (shaded) may be continuously fit, overfit, misfit, or underfit

Temporal measures of hydrographic change in hemiarid lake basins include persistence and recurrence of open standing water (Fig. 15.15). Persistence is the average length in years of periods of continuous inundation expressed as a proportion of recurrence. Recurrence is the length of the average continuous inundation interval plus the average length in years of continuous subaerial periods. Water bodies vary greatly in their persistence and recurrence, which are particularly important as factors in the geomorphic development of basin floors. Transient lake-playa regimes with relatively short inundation and subaerial periods (high- to medium-frequency cycles of lake-playa alternation, Fig. 15.15) strongly favour geomorphic processes that tend to flatten and widen basin floors, and thus reduce the capability of such basins to contain small lakes.

Temporal domains of hydrographic change in hemiarid lake basins are historic, protohistoric, and prehistoric and consist of written records of change, archaeological evidence related to change, and change prior to human habitation, respectively. In all three temporal domains the basic tool for storing, correlating, and displaying basinwide spatiotemporal information about hydrographic change is the hydrograph (Fig. 15.16).

Hydrographs have spatiotemporal coordinate systems, typically with y-axis water levels, areas, or volumes plotted as functions of x-axis time. Figure 15.16 spans historic, protohistoric, and latest prehistoric time. In accordance with graphical norms, time is plotted flowing from left to right.

Hypsographs have spatial coordinate systems but can also display selected spatiotemporal information (Fig. 15.17). Dimensionless spatial coordinates derived from indices of comparative water body morphometry facilitate direct size comparisons of present and past water bodies in one or more basins (Benson and Paillet 1989, Bengtsson and Malm 1997). Two such indices are the palaeolake height index (PHI), where ϕ is a water body's depth at a particular time as a fraction of its greatest late Quaternary depth, and the palaeolake surface index (PSI), where ψ is a water body's area at a particular time as a fraction of its greatest late Quaternary area (Currey 1988). Figure 15.17 spans a greater spatiotemporal range than, and provides a basinwide frame of reference for, Fig. 15.16.

Published graphical representations of hydrographic change in hemiarid lake basins are not documents of absolute certainty and authority. Clearly, the reliability of such representations can be no better than the quality of the geomorphic information (see following chapter) on which they are based. They are most appropriately viewed as thought-provoking working hypotheses of explanation that will continue to be refined as research proceeds.

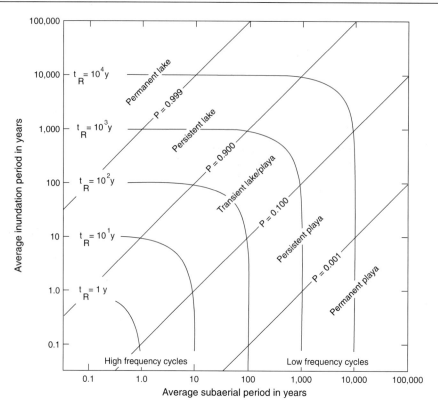

Fig. 15.15 Persistence (P) of standing water on floors of lake basins is a dimensionless number from 0.0 to 1.0 that expresses average continuous inundation period in years as a fraction of recurrence. Recurrence (t_R) is time in years of average continuous inundation period plus average continuous subaerial period (after Currey 1990). Frequencies of episodic inundation range from ultra-high to ultra-low

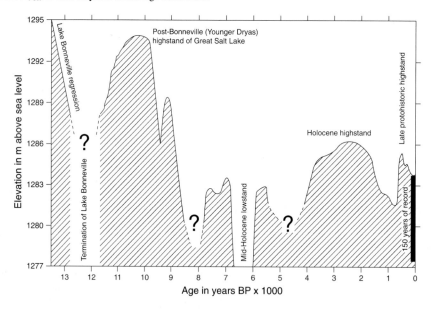

Fig. 15.16 Example hydrograph showing upper envelope of lake levels (upper limit of static water at high stages) in the basin of Great Salt Lake, Utah, during the last 13,000 years (after Murchison 1989). Lower envelope, delimiting low stages during that time, is poorly known. Earliest known human habitation in the basin occurred between 11 and 10 ka at Danger Cave (1314 m; Jennings 1957). Historic hydrologic record has been summarized by Arnow and Stephens (1990)

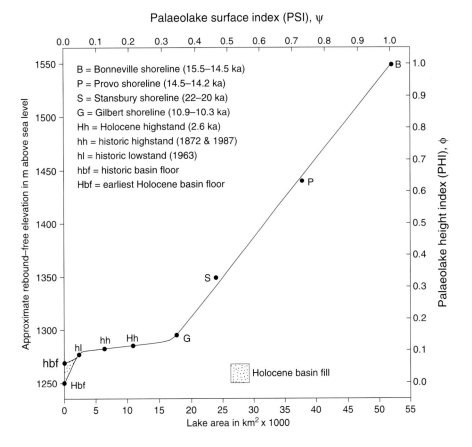

Fig. 15.17 Example hypsograph showing containment basin topography (*sloping line*) and selected late Pleistocene and Holocene lake levels (*plotted points*) in the Lake Bonneville basin, western USA (after Currey 1990); chronology from Currey and Burr (1988), Oviatt (1988), Oviatt et al. (1990), and Benson et al. (1992). Estimated rebound-free elevations (Currey and Oviatt 1985) simplify basinwide studies. In this example, PHI is scaled from 0.0 on the terminal Pleistocene basin floor to 1.0 at the highest late Quaternary (threshold-controlled Bonneville) shoreline; PSI is scaled from 0.0 at no standing water, which is unlikely in this basin, to 1.0 at the Bonneville shoreline

Acknowledgments Much of the research on which this chapter is based was supported by NSF grant EAR-8721114. Several of the ideas expressed here have taken shape during many years of fruitful association with University of Utah graduate students and faculty, and with esteemed colleagues in archaeology, geodynamics, geological science, geotechnical engineering, soils, the Utah Geological Survey, and the U.S. Geological Survey. In addition, the second author deeply appreciates having had the opportunity to learn from and work with the late D.R. Currey.

References

Arnow T, Stephens D (1990) Hydrologic characteristics of the Great Salt Lake, Utah: 1847–1986. US Geol Surv Water-Supply Pap 2332

Bartov Y, Stein M, Enzel Y, Agnon A, Reches Z (2002) Lake levels and sequence stratigraphy of Lake Lisan, the late Pleistocene precursor of the Dead Sea. Quat Res 57:9–21

Bengtsson L, Malm J (1997) Using rainfall-runoff modeling to interpret lake-level data. J Paleolimn 18:235–248

Benson LV, Paillet FL (1989) The use of total lake-surface area as an indicator of climatic change: Examples from the Lahontan basin. Quat Res 32:262–275

Benson LV, Thompson RS (1987) Lake-level variation in the Lahontan basin for the past 50,000 years. Quat Res 28:69–85

Benson L, Currey D, Lao Y, Hostetler S (1992) Lake-size variations in the Lahontan and Bonneville basins between 13,000 and 9000 ^{14}C yr B.P. Palaeogeogr Palaeoclim Palaeoecol 95:19–32

Bouchard DP, Kaufman DS, Hochberg A, Quade J (1998) Quaternary history of the Thatcher basin, Idaho, reconstructed from the [87]Sr/[86]Sr and amino acid composition of lacustrine fossils: Implications for the diversion of the Bear River into the Bonneville basin. Palaeogeogr Palaeoclim Palaeoecol 141:95–114

Cowardin LM, Carter V, Golet FC, LaRoe ET (1979) Classification of wetlands and deepwater habitats of the United States. US Fish and Wildl Serv FWS/OBS-79/31

Currey DR (1988) Isochronism of final Pleistocene shallow lakes in the Great Salt Lake and Carson Desert regions of the Great Basin. Am Quat Assoc Program Abstr 10:117

Currey DR (1990) Quaternary palaeolakes in the evolution of semidesert basins, with special emphasis on Lake Bonneville and the Great Basin, U.S.A. Palaeogeogr Palaeoclim Palaeoecol 76:189–214

Currey DR, Burr TN (1988) Linear model of threshold-controlled shorelines of Lake Bonneville. Utah Geol Miner Surv Misc Publ 88–1:104–110

Currey DR, Oviatt CG (1985) Durations, average rates, and probable causes of Lake Bonneville expansions, stillstands, and contractions during the last deep-lake cycle, 32,000 to 10,000 years ago. In: Kay PA, Diaz HF (eds) Problems of and prospects for predicting Great Salt Lake levels. University of Utah, Salt Lake City

DeVogel SB, Magee JW, Manley WF, Miller GH (2004) A GIS-based reconstruction of late Quaternary paleohydrology: Lake Eyre, arid central Australia. Palaeogeogr Palaeoclim Palaeoecol 204:1–13

Eaton GP (1982) The Basin and Range province: Origin and tectonic significance. Annu Rev Earth Planet Sci 10:409–440

Frémont, JC (1845) Map of an exploring expedition to the Rocky Mountains in the year 1842 and to Oregon and North California in the years 1843–44. In: Jackson D, Spence ML (eds) The expeditions of John Charles Frémont – map portfolio 1970 map 3. University of Illinois Press, Urbana

Harrill JR, Gates JS, Thomas JM (1988) Major ground-water flow systems in the Great Basin region of Nevada, Utah, and adjacent states. US Geol Surv Hydrol Investig Atlas HA-694-C

Hoelzmann P, Keding B, Berke H, Kröpelin S, Kruse H-J (2001) Environmental change and archaeology: Lake evolution and human occupation in the eastern Sahara during the Holocene. Palaeogeogr Palaeoclim Palaeoecol 169:193–217

Houghton SG (1986) A trace of desert waters, the Great Basin story. Howe, Salt Lake City

Hubbs CL, Miller RR, Hubbs LC (1974) Hydrographic history and relict fishes of the north-central Great Basin. Calif Acad Sci Mem 7

Hunt CB, Robinson TW, Bowles WA, Washburn AL (1966) Hydrologic basin, Death Valley, California. US Geol Surv Prof Pap 494-B

Hutchinson GE (1957) A treatise on limnology, vol 1, geography, physics, and chemistry. Wiley, New York

Jennings JD (1957) Danger Cave. University Utah Anthropol Pap 27

Langbein WB (1961) Salinity and hydrology of closed lakes. US Geol Surv Circ 52

Le Turdu C, Tiercelin J-J, Gibert E, Travi Y, Lezzar K-E, Richert J-P, Massault M, Gasse F, Bonnefille R, Decobert M, Gensous B, Jeudy V, Tamrat E, Mohammed MU, Martens K, Atnafu B, Chernet T, Williamson D, Taieb M (1999) The Ziway-Shala lake basin system, main Ethiopian rift:

Influence of volcanism, tectonics, and climatic forcing on basin formation and sedimentation. Palaeogeogr Palaeoclim Palaeoecol 150:135–177

Mabbutt JA (1977) Desert landforms. MIT Press, Cambridge MA

Magee JW, Bowler JM, Miller GH, Williams DLG (1995) Stratigraphy, sedimentology, chronology and palaeohydrology of Quaternary lacustrine deposits at Madigan Gulf, Lake Eyre, South Australia. Palaeogeogr Palaeoclim Palaeoecol 113:3–42

Mather JR (1978) The climatic water budget in environmental analysis. Lexington Books, Lexington KY

May G, Hartley AJ, Stuart FM, Chong G (1999) Tectonic signatures in arid continental basins: An example from the Upper Miocene-Pleistocene, Calama basin, Andean fore-arc, northern Chile. Palaeogeogr Palaeoclim Palaeoecol 151:55–77

Meek N (1989) Geomorphic and hydrologic implications of the rapid incision of Afton Canyon, Mojave Desert, California. Geol 17:7–10

Mifflin MD, Wheat MM (1979) Pluvial lakes and estimated pluvial climates of Nevada. Nev Bur Mines Geol Bull 94

Morrison RB (1968) Pluvial lakes. In: Fairbridge RW (ed.) The encyclopedia of geomorphology. Reinhold, New York

Murchison SB (1989) Fluctuation history of Great Salt Lake, Utah, during the last 13,000 years. Ph.D. dissertation, University of Utah, Salt Lake City

Oviatt CG (1988) Late Pleistocene and Holocene lake fluctuations in the Sevier Lake basin, Utah, U.S.A. J Paleolimn 1:9–21

Oviatt CG, Currey DR, Miller DM (1990) Age and paleoclimatic significance of the Stansbury shoreline of Lake Bonneville, northeastern Great Basin. Quat Res 33:291–305

Peterson FF (1981) Landforms of the Basin and Range province defined for soil survey. Nev Agric Exp Stn Tech Bull 28

Reeves CC (1968) Introduction to paleolimnology. Elsevier, Amsterdam

Russell IC (1885) Geologic history of Lake Lahontan. US Geol Surv Monogr 11

Sack D (2001) Shoreline and basin configuration techniques in paleolimnology. In: Last WM, Smol JP (eds) Tracking environmental change using lake sediments, volume 1, basin analysis, coring, and chronological techniques. Kluwer, Dordrecht

Sack D (2002) Fluvial linkages in Lake Bonneville subbasin integration. Smithson Inst Contrib Earth Sci 33:129–144

Schuster M, Duringer P, Ghienne J-F, Vignaud P, Beauvilain A, Mackaye HT, Brunet M (2003) Coastal conglomerates around the Hadjer El Khamis inselbergs (western Chad, central Africa): New evidence for Lake Mega-Chad episodes. Earth Surf Process Landf 28:1059–1069

Smith GI, Street-Perrott FA (1983) Pluvial lakes of the western United States. In: Porter SC (ed) Late-Quaternary environments of the United States, volume 1, the late Pleistocene. University of Minnesota Press, Minneapolis

Street FA, Grove AT (1979) Global maps of lake-level fluctuations since 30,000 yr B.P. Quat Res 12:83–118

Street-Perrott FA, Harrison SP (1985) Lake levels and climate reconstruction. In: Hecht AD (ed) Paleoclimate analysis and modeling. Wiley, New York

Thorn CE (1988) Introduction to theoretical geomorphology. Unwin Hyman, London

Thornbury WD (1965) Regional geomorphology of the United States. Wiley, New York

Trauth MH, Strecker MR (1999) Formation of landslide-dammed lakes during a wet period between 40,000 and 25,000 yr B.P. in northwestern Argentina. Palaeogeogr Palaeoclim Palaeoecol 153:277–287

Williams TR, Bedinger MS (1984) Selected geologic and hydrologic characteristics of the Basin and Range province, western United States. US Geol Surv Misc Investig Map I-1522D

Chapter 16

Hemiarid Lake Basins: Geomorphic Patterns

Donald R. Currey[†] and Dorothy Sack

When the lakes of arid regions become extinct, either by reason of evaporation or sedimentation, evidence of their former existence remains inscribed on the inner slopes of their basins or concealed in the strata deposited over their bottoms. These records as a rule are much more lasting than those left by lakes in humid lands. . . .

Israel C. Russell (1895, pp. 94–95).

Introduction

Hemiarid ('half arid') lake basins are drainage basins that have arid lowlands, nonarid highlands, and topographic and hydrologic closure. As a result of these characteristics, hemiarid lake basins contain or have contained nonoutlet lakes in their lowest reaches. A rich body of lacustrine and palaeolake evidence is stored in many hemiarid lake basins, providing the basis for reconstructing basin hydrography and the underlying hydrology, hydroclimate, and tectonics. Much of this lacustrine evidence occurs in regional and local patterns of geomorphology, sedimentology, and stratigraphy, which in lacustrine geoscience, with an emphasis on the depositional record, differ more in etymology than substance. Regional and local geomorphic patterns are equally important in effecting the hydrographic and related reconstructions from hemiarid lake basins. Regional geomorphic patterns constitute the predictive basis for making local geomorphic observations, and local patterns are the observational basis for building and testing regional models.

Figure 16.1 illustrates the importance of regional and local geomorphic patterns, and the underlying tectonic, hydrographic, and hydrologic factors, in two substantially different hypothetical basins. The hypertectonic hemiarid lake basin (Fig. 16.1, A-A′) has highlands that yield significant runoff, lowlands that are typically segmented into closed subbasins by active alluvial fans, and subbasin floors that are likely to be locally depressed by active faulting. In contrast, the hypotectonic arid basin (Fig. 16.1, B-B′) has inselberg uplands that are unlikely to yield significant runoff, lowlands in which subbasins have long been aggraded and amalgamated, and a basin floor that in all likelihood is large and essentially flat. The occurrence of a lake or the geomorphic evidence of one or more Quaternary palaeolakes, large or small, has a substantially higher probability in the hypertectonic basin than in the hypotectonic basin. This chapter on hemiarid lake basins, therefore, considers lacustrine geomorphic, and related sedimentologic and stratigraphic, patterns in hypertectonic basins.

Lakebeds

From their greatest depths to their highest surges, lakes are in continuous to occasional contact with sublacustrine materials. Portions of lake-substrate interfaces have been referred to as beds, bottoms, floors, shores, mudlines, and other terms. In this

D. Sack (✉)
Department of Geography, Ohio University, Athens, OH 45701, USA
e-mail: sack@ohio.edu

[†] deceased

A.J. Parsons, A.D. Abrahams (eds.), *Geomorphology of Desert Environments*, 2nd ed., DOI 10.1007/978-1-4020-5719-9_16, © Springer Science+Business Media B.V. 2009

Fig. 16.1 Schematic profiles showing regional (A-A′ and B-B′) and local (1–12) geomorphic patterns that are typical of hypertectonic hemiarid lake basins and hypotectonic arid basins

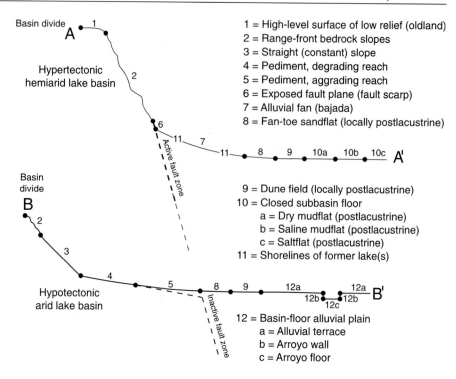

1 = High-level surface of low relief (oldland)
2 = Range-front bedrock slopes
3 = Straight (constant) slope
4 = Pediment, degrading reach
5 = Pediment, aggrading reach
6 = Exposed fault plane (fault scarp)
7 = Alluvial fan (bajada)
8 = Fan-toe sandflat (locally postlacustrine)

9 = Dune field (locally postlacustrine)
10 = Closed subbasin floor
       a = Dry mudflat (postlacustrine)
       b = Saline mudflat (postlacustrine)
       c = Saltflat (postlacustrine)
11 = Shorelines of former lake(s)

12 = Basin-floor alluvial plain
       a = Alluvial terrace
       b = Arroyo wall
       c = Arroyo floor

account, all lake-substrate interfaces in hemiarid basins are termed lakebeds. Conceptually, lakebeds are the desktops on which lakes encode and inscribe information about their environments, including their morphometry (Håkanson 1981), energy fluxes (Bergonzini et al. 1997, Menking et al. 2004), material fluxes, and sedimentology (Sly 1978, Håkanson and Jansson 1983). Information regarding lakebed depth (bathymetry), net balance of deposition and erosion (Håkanson and Jansson 1983), sediment size, mineralogy (Jones and Bowser 1978), and geochemistry (Ku et al. 1998, Parker et al. 2006) is of prime importance geomorphically.

A regional geomorphic model applicable to lakebeds in many hemiarid basins consists of topologically concentric, bathymetric environment zones (Table 16.1). Each isochronic lakebed that marks a successive stage in a lacustrine sequence is a special case of the general model. In each case the nearshore, foreshore, and backshore zones form a geomorphically dynamic lakeshore (Fig. 16.2). Lakebed sediment sizes, which are characteristically well sorted in accordance with turbulent energies, typically increase from the deepest part of the offshore zone to the middle of the foreshore zone.

Several spatial and temporal factors are represented in lakebed geomorphic patterns and their component sediment sources, pathways, and sinks. Given adequately georeferenced and chronoreferenced databases, individual isochronic lakebeds can be resolved into sets of linear patterns (e.g. mappable source-to-sink sediment pathways) and areal patterns (mappable polygons); successions of lakebeds can be resolved into volumetric patterns (mappable polyhedrons). Although minimal spatiotemporal resolution can be sufficient to distinguish among pre-lacustrine, co-lacustrine, and postlacustrine geomorphic features, high spatiotemporal resolution is required to quantify the kinematics of geomorphic and hydrographic change (e.g. Currey and Burr 1988).

Lakebed Sediment Sources

Lakebeds are surfaces underlain by lacustrine sediments that range in thickness from greater than 1 m in many offshore and lakeshore areas of net deposition, to shallow veneers on erosional platforms, to virtually nothing on steeply sloping bedrock headlands. Lacustrine sediments are of three types: (a) those that originate outside the water column, (b) those that originate in the water column, and (c) those that originate in fluid-filled pores under the water column. The

Table 16.1 Offshore, nearshore, foreshore, backshore, and rearshore bathymetric environment zones in and adjacent to lakeshores in hemiarid lake basins*

Zones and zone transitions	Typical environments
Offshore deep	Deepest places in lakes; lakebeds commonly anoxic and fetid, with sediments commonly >1% plankton residues
Offshore zone	Depths typically >6 m; gently sloping lakebeds blanketed by marls (carbonate micrite muds), diatomaceous marls, terrigenous silts and clays, or volcanic ashes; soluble salts can precipitate offshore during major regressions
Offshore perimeter	Landward transition from low to moderate turbulence and from muddy to sandy lakebeds; basinward limit of lakeshore
Nearshore zone	Depths typically 2–6 m; low-gradient, basinward-sloping aprons of fine clastic sands, charaliths, or mats and heads of algaloid tufa; commonly ooid-producing; soluble salts can precipitate nearshore from cold, high-salinity waters
Nearshore perimeter	Landward transition from moderate-energy sandy lakebeds to high-energy gravelly lakebeds
Lower foreshore zone	Depths typically <2 m; moderately steep, basinward-sloping, surf-constructed subaqueous beach faces of medium to coarse sands and gravels; commonly bouldery; lithoid and algaloid tufas locally abundant as beach matrix and as coatings on rocky lakeshores; soluble salts can precipitate on lower foreshore from cold, high-salinity waters
Lower foreshore perimeter	Waterline; landward limit of open water under static water conditions, i.e. basinward edge of land
Upper foreshore zone	Heights typically <2 m above static water level; swash-constructed subaerial beach faces are landward extensions of subaqueous beach faces; locally bouldery
Upper foreshore perimeter	Crests of basinward-sloping, backset-stratified beach faces; landward limit of swash
Backshore zone	Inland-sloping foreset-stratified washover beach slopes, lagoons, primary dunes (foredunes), deltaic lowlands, mudflats, saltflats, and marshes directly impacted by overwash processes and episodic onshore surges
Backshore perimeter	Inland transition from direct to indirect lacustrine geomorphic impacts, i.e. inland limit of foredune accretion and flooding by onshore surges; inland limit of lakeshore
Rearshore zone	Secondary dunes derived from foredunes; floodplains and groundwater tables affected by fluctuations of lacustrine base levels
Rearshore perimeter	Inland limit of indirect lacustrine geomorphic impacts, i.e. inland limit of secondary dunes and base level effects

*See Fig. 16.2.

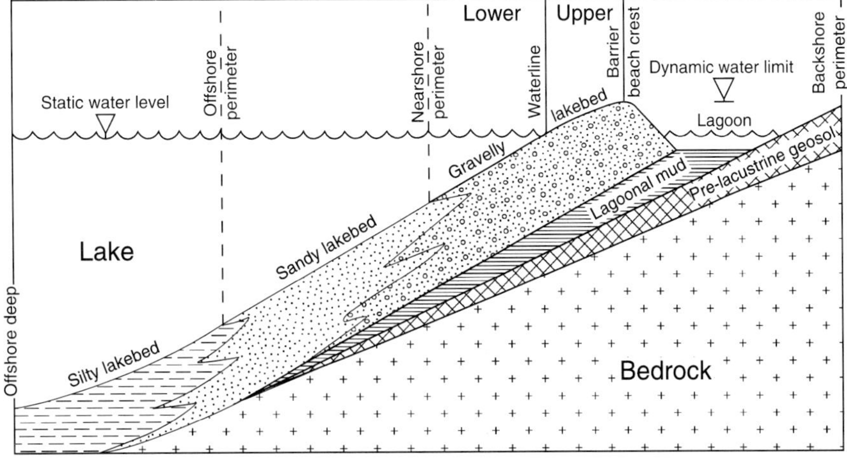

Fig. 16.2 Schematic cross section of a lakebed in a hemiarid lake basin, showing bathymetric environment zones in and near an idealized barrier beach lakeshore. See Table 16.1 for details

three types are commonly referred to as allogenic, endogenic, and authigenic, respectively, although the latter two are sometimes grouped together because of their common links to lake biochemistry (Jones and Bowser 1978). Here, sediments that have their origins on land are termed terrigenous, following usage in marine geology, and those that have their origins in and under lakes are termed limnogenous.

Terrigenous lacustrine sediments (Table 16.2) comprise all clastic materials that are transported into lakes by geomorphic agents, as well as those that lakes acquire by lakeshore erosion. Fluviolacustrine materials are commonly the dominant terrigenous sediments on the lakebeds of hemiarid basins with major streams. Away from the deltas of major streams and in hemiarid lake basins without major streams, alluviolacustrine materials are commonly the dominant terrigenous sediments, particularly in lakeshore deposits.

Limnogenous sediments (Table 16.3) comprise all materials that originate physicochemically or biochemically at the top of, within, at the bottom of, or under the water column in hemiarid lake basins, regardless of water depth. Of the lacustrine chemical sediments (Eugster and Kelts 1983), carbonates are the most ubiquitous in hemiarid lake basins, far exceeding organic matter and typically equalling or exceeding soluble salts. Micritic marls, with amounts and sizes of admixed terrigenous fines that reflect proximity to land-based sediment sources, are the

most common offshore carbonate sediments (Galat and Jacobsen 1985, Oviatt et al. 1994). Various forms of tufa (Morrison 1964, Benson et al. 1995, Ku et al. 1998) and ooidal sands (Eardley 1966) are common carbonate sediments in lakeshore zones.

In many hemiarid lake basins algaloid tufas (Table 16.3, Fig. 16.3) are among the more intriguing and varied surficial materials. Algaloid tufas are laminated calcite, aragonite, and calcite-aragonite mixtures that form on algae-colonized substrates in sunlit, wave-agitated water saturated with calcium bicarbonate. Large algaloid tufa forms include mounds, domes, and pinnacles, as well as extensive mats (hardgrounds) on soft sediments. Smaller forms include cauliform discs and heads, radiating (dendritic) discs and heads, polyp clusters, smooth rinds and rippled rinds on bedrock and detached rocks, rinds enveloping metre-size pods of gravel, bun-shaped oncoids, and pearl-like pisolites.

Lakeshore Sediment Pathways

Most limnogenous sediments are deposited beneath or at their point of origin in the formative water column. Alternatively, most terrigenous sediments, as well as coarse limnogenous sediments such as ooidal sands, are transported significant distances before undergoing long-term deposition. The finer terrigenous sediments,

Table 16.2 Terrigenous lacustrine sediments of importance in hemiarid basins are supplied by lakeshore erosion (E) or by short- to long-distance transportation (T)

Sediment types	Sediment origins
Litholacustrine	(E) Clastic sediments that lakes acquire by eroding cliffs (free faces) in resistant bedrock and bluffs (straight slopes) in nonresistant bedrock; includes clasts acquired by rockfalls and rockslides from cliffs and bluffs
Colluviolacustrine	(E) Clastic sediments that lakes acquire by stripping unconsolidated colluvial materials off slopes that are underlain by bedrock and consolidated sediments
Alluviolacustrine	(E) Clastic sediments that lakes acquire by eroding bluffs in alluvial fans and bajadas; includes clasts acquired by mass wasting of bluffs
Retrolacustrine	(E) Clastic sediments (intraclasts) that lakes acquire by reworking and redepositing terrigenous and limnogenous sediments of earlier lakebeds
Fluviolacustrine	(T) Clastic sediments, including glacial outwash, that are introduced into lakes by drainage networks of many sizes; partly dispersed offshore and partly localized inshore (in low-gradient, suspended-load deltas, bedload fan deltas, sandy underflow fans, and aggraded-prograded estuaries)
Aerolacustrine	(T) Airborne mineral dusts (silts and clays), biodusts (palynomorphs and humic matter), organic ashes, volcanic ashes, and salts that settle out of the atmospheric column onto lake surfaces; insoluble dusts settle out of the water column onto lakebeds
Cryolacustrine	(T) Ice-rafted dropstones (commonly pebbles and cobbles) that are released during spring breakup of shore-fast lake and river ice; ice-shoved beach deposits
Anthropolacustrine	(T) Materials (structures and refuse) that are placed in lakes by people
Glaciolacustrine	Sediments that are carried into lakes by glacier ice are rare or nonexistent in hemiarid lake basins

Table 16.3 Limnogenous sediments of importance in hemiarid lake basins

Sediment types	Sediment origins
Micrites	Silt-size carbonate crystals that form in calcium-bicarbonate-saturated surface waters, and settle out as marl in the summer, when CO_2 is depleted by phytoplankton (mainly algal) photosynthesis and thermal degassing; calcite micrites (and tufas) tend to precipitate in low-salinity hard waters, and aragonite micrites (and tufas) tend to precipitate at higher salinities
Plankton residues	Organic matter (sapropel or proto-kerogen) that results from anaerobic decay of settled-out plankton (mainly algae); common in fine-grained sediments on poorly oxygenated lakebeds
Microfossils*	Opaline frustules of diatoms can be particularly abundant, with sediments ranging from diatomaceous muds and marls to pure diatomites; pollen grains are commonly preserved in water-saturated sediments
Macrofossils*	Mollusca shells are common in the deposits of fresh-to-brackish shallow waters; charaliths (diminutive straw-like calcified filaments of macrophyte algae, particularly of the genus *Chara*) are commonly washed into nearshore deposits from foreshore substrates; ostracod shells are common in fine-grained sediments on well-oxygenated lakebeds; fish bones and scales are less common, but occur widely in the deposits of fresh-to-brackish waters
Ooids	Spheroidal ooids (aragonite-encapsulated sand grains) and cylindrical ooids (aragonite-encapsulated fecal pellets, e.g. of the brine shrimp *Artemia*) that form in gently shoaling calcium-bicarbonate-saturated nearshore waters; commonly wash onshore to form beaches and dunes
Algaloid† tufas	Phycolites (general name for all algaloid structures) comprise (a) spongiostromes (laminated structures formed by algae growing in smooth mats), including stromatolites (attached to substrate) and oncolites (unattached, free to roll), and (b) dendrolites (arborescent or near-arborescent structures formed by algae growing in tufted mats), including dendritic, cellular, and 'coralline' tufa
Tufacretes	Conglomerate- or concrete-like, calcite- and aragonite-cemented gravels and coarse sands that form in shallow, calcium-bicarbonate-saturated waters where moderate (not copious) supplies of coarse clasts are available and where breaking waves accelerate warm-season degassing of CO_2; tufacrete landforms include platform pavements and ledges, slope-draping slabs, and foreshore beachrock
Salt beds	Offshore and nearshore beds of soluble salts that settle out after crystallizing in seasonally supersaturated surface waters, e.g. halite (NaCl) when evaporation is high and mirabilite ($Na_2SO_4 \cdot 10H_2O$) when temperatures are low; foreshore salt beds include mirabilite that washes onshore and is preserved in the interstices of more stable beach materials; backshore salt beds include halite that precipitates on lake-fringing saltflats

*Including subfossils of late Quaternary age.
†Richard Rezak's classification (Morrison 1964, p. 48).

mainly fluviolacustrine clays, silts, and fine sands, are commonly transported by runoff that discharges directly into offshore waters by hypopycnal flow as near-surface suspensions of clays and silts, and by hyperpycnal underflow in near-bottom density currents laden with mixtures of clay, silt, and fine sand. The coarser terrigenous sediments, consisting mainly of alluviolacustrine and litholacustrine gravels and coarse sands (Table 16.2), and coarse limnogenous sediments occur mostly in lakeshore (nearshore, foreshore, and backshore) zones. There they are reworked and redeposited many times as they move from sediment sources to sediment traps along lakeshore pathways.

The four basic lakeshore sediment pathways (Fig. 16.4) in hemiarid lake basins trend (a) landward from the foreshore zone to the backshore zone and sometimes beyond, (b) landward from the nearshore zone to the foreshore zone, (c) basinward from the foreshore zone to the nearshore or offshore zone, and (d) in the direction of net littoral drift within the foreshore and nearshore zones. These pathways are known as inland transport, onshore transport, offshore transport, and longshore transport, respectively.

It is common for lakeshore sediment pathways in hemiarid lake basins to be well defined geomorphically. In inland transport, foreshore medium sands are carried into lagoons by overwash and onto accreting foredunes by saltation. In onshore transport, nearshore ooids and fragments of tufa and beachrock commonly are carried onto foreshore beach faces by the onshore component of breaking waves (swash). In offshore transport, foreshore fines are winnowed into deeper water by breaking waves, and foreshore sands and gravels pass into deeper water on the updrift sides of groin-like headlands and at the ends of spits and cuspate barriers. In longshore transport, foreshore sands and gravels can be carried several kilometres by the longshore component of breaking waves (littoral drift),

Fig. 16.3 Dense, coral-like tufa in the Lake Bonneville basin deposited on what was a steep, subaqueous slope of lacustrine gravel

without benefit of topographic gradient. Longshore transport has moved large (128–256 mm) cobbles at least 2 km on high-energy beaches in the Great Basin.

It is also common for geomorphic conditions to vary greatly from one lakeshore sector to another. In all but the smallest hemiarid lake basins, this produces distinctive basinwide and local patterns of lakeshore segmentation (Fig. 16.5). Lakeshore segments are distinguished by two criteria, their longshore transport directions and longshore sediment budgets. Lakeshore segments join at segment boundaries, called nodes, which are points where (a) longshore transport directions reverse, or (b) longshore sediment budgets change algebraic sign (from net erosion to net deposition or vice versa). Erosional lakeshore segments have deficit longshore sediment budgets in which total longshore outputs exceed total longshore inputs, with the difference equal to sediments entrained by lakeshore erosion. Depositional lakeshore segments have surplus longshore sediment budgets in which total sediment inputs exceed total longshore outputs, with

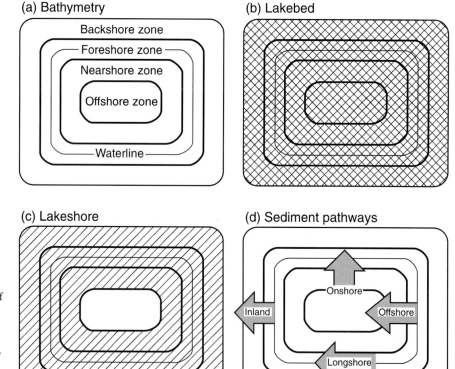

Fig. 16.4 Schematic plans of (**a**) bathymetric zones (Table 16.1 and Fig. 16.2), showing (**b**) lakebed and (**c**) lakeshore areas of (**d**) inland, onshore, offshore, and longshore sediment movement

Fig. 16.5 Map of hypothetical lakeshore showing longshore pathways from sediment sources to depositional sites. L = lagoon and B = barrier beach. Longshore vector polygon and resultant longshore vector are plotted within lake area. N = number of lakeshore segments (= number of longshore vectors), ΣL = total length of longshore vectors, L_R = length of resultant longshore vector, S_R = strength of resultant longshore vector (L_R as fraction of ΣL), and Θ_R = azimuth of resultant longshore vector. Symbols indicating erosional segments project landward; those marking depositional segments project lakeward

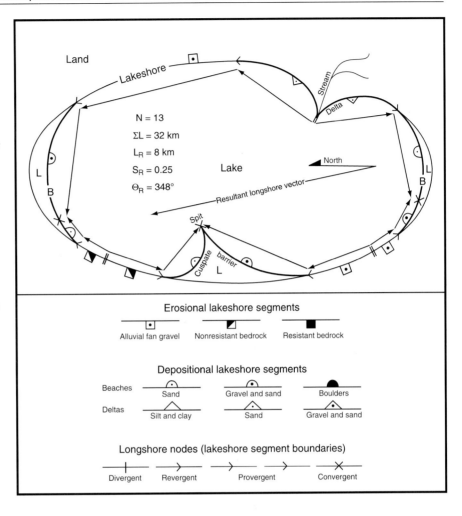

the difference equal to sediments stored in beaches and deltas.

Lakeshore segment boundaries (Fig. 16.5, longshore nodes) occur where longshore transport directions diverge or converge (divergent and convergent nodes), where longshore transport budgets change from deposition to erosion (revergent nodes), and where longshore transport budgets change from erosion or steady state to deposition (provergent nodes). In addition to the longshore criteria that are used to define lakeshore segmentation, features such as lithology and geomorphic expression (original geomorphic development together with subsequent geomorphic preservation) can be noteworthy features of lakeshore segments and subsegments (Wilkins and Currey 1997, Adams and Wesnousky 1999, Carter et al. 2006).

Lakebed Sediment Sinks

The marine geomorphic literature sometimes distinguishes among coastal sediments that reside in (a) stores, such as temporary sediment storage in active beaches, (b) traps, which constitute long-term sediment storage as in beaches at reentrants, and (c) sinks, representing permanent sediment losses from beaches onto land and into deep water (Davies 1980). Similar distinctions can sometimes be made in hemiarid lake basins, although relatively steep stream and slopewash geomorphic gradients tend to limit postlacustrine residence times of lake sediments in beaches and, most notably, in blanket deposits above basin floors. However, the corollary, that the residence times of lake sediments, including the ubiquitous reworked sediments, in basin-floor sinks are essentially unlimited, does not necessarily follow. This stems from the fact that the floors of many hemiarid lake basins (a) are prone to vigorous subaqueous and aeolian scour during lowstands (Magee et al. 1995, Nanson et al. 1998), and (b) eventually undergo tectonic rejuvenation by tilting and/or faulting (Amit et al. 1999).

Sediment sinks in hemiarid lake basins can be viewed from several perspectives. For example, physiography may be used as a basis for identifying and differentiating depositional subenvironments. Alluvial fans, fan-toe sandflats, dry mudflats, saline mudflats, salt pans, perennial saline lakes, aeolian dunes, perennial streams, ephemeral streams, springs, spring-fed ponds, and shorelines represent major depositional environments in saline lake basins (Hardie et al. 1978, Eugster and Kelts 1983). Distributary-channel deltoids (Mabbutt 1977) are important in some desert lowlands, such as where the terminal reach of the Amargosa River intermittently aggrades Badwater Basin on the floor of Death Valley, California (Hunt et al. 1966). Two spatial perspectives that focus on lakebed sediment sinks are outlined here, the zone-sector-polygon model and the bubble model.

The zone-sector-polygon subenvironment model (Table 16.4 and Fig. 16.6) describes the continuum of lakebed sediment sinks in terms of two orthogonal spatial patterns: (a) topologically concentric bathymetric zones, and (b) topologically radial sediment budget sectors. In this model, the four bathymetric environment zones outlined previously (Fig. 16.2 and Table 16.1) intersect 12 (or more) longshore sediment budget sectors to define 48 (or more) depositional subenvironment polygons. The structure of this model lends itself to GIS applications, including geomorphic mapping, in hemiarid lake basins.

'Bubble models' of depocentres represent another approach for highlighting major patterns of lakebed sediment sinks. Models (e) through (k) in Fig. 16.7 derive specifically from observations in North American hemiarid lake basins. Underflow depocentres (Fig. 16.7e) occur offshore from river mouths with sediment-laden hyperpycnal flow (e.g. Oviatt 1987, Oviatt et al. 2003). Fringing-beach depocentres (Fig. 16.7f) are found mainly in small basins with easily eroded piedmont gravels. Depocentres fed by wind-driven longshore drift (Fig. 16.7g,j) occur in almost every hemiarid lake basin, often as major beaches at gravel-dominated sites and beach-fed dunes at sandy sites. Bidirectional longshore drifting (Fig. 16.7h) has formed major beaches at the north and south ends of many Great Basin lakes due to orographically channelled southerly and northerly winds (e.g. Sack 1990, plate 1). A major dune field was constructed of sand from the front of a windward-projecting late Pleistocene delta (Fig. 16.7i) inland from at least one Great Basin palaeolake (Sack 1987). Ooidal beaches and beach-fed dunes of Holocene age are widespread in gently shelving, mainly west-facing sectors of Great Salt Lake (Fig. 16.7k) (Dean 1978, Currey 1980).

In many hemiarid basins, lakebed landforms and sediments are regionally mantled and locally buried by subsequent aeolian, colluvial, and pedogenic surficial materials. Aeolian deposits, including loess (ubiquitous in at least minor quantities), mud-pellet dunettes (often anchored aerodynamically to shrubs), and lunettes (around downwind margins of playas), and secondary sand dunes and sand sheets reworked from beach-fed primary dunes (Davies 1980), commonly date from lake regressions and lowstands. Colluvial deposits, including coverhead that accumulates on erosional shorelines (Sharp 1978) and coverbeds that accrete around the basin during waning-ice-age intervals (Kleber 1990), are poorly sorted, loess-enriched surficial materials emplaced by creep and slopewash. Colluvial surficial deposits commonly contain pedogenic carbonates and translocated clays (e.g. Birkeland et al. 1991), particularly where steppe vegetation has restricted geomorphic activity and contributed to pedogenesis.

Lakebed Landforms

Sediment sources, pathways, and sinks in hemiarid lake basins are embodied in landforms that record pre-lacustrine, co-lacustrine, and postlacustrine geomorphic history. Lakebed landforms can be viewed in many ways; here they are presented as modules in planimetric and stratigraphic patterns, and as indicators of postlacustrine change.

Lacustrine geomorphic information is recorded mainly in lakeshore landforms, particularly depositional landforms. Geomorphic resolution of lake history tends to increase with increasing supplies of lakeshore terrigenous sediments (and ooidal sands) and increasing depth of lakeshore embayments (Fig. 16.8).

Table 16.4 Zone-sector-polygon model of lacustrine depositional subenvironments, in which 48 subenvironment polygons
(Fig. 16.6) are defined by the intersections of four bathymetric
environment zones (Fig. 16.2 and Table 16.1) and 12 longshore
environment (sediment dynamics) sectors. Paired upper case letters denote typically strong linkages between depositional subenvironments and longshore sediment dynamics; lower case letters
denote typically weak linkages

Longshore environment (sediment dynamics) sectors	Bathymetric environment zones			
	Offshore (O)	Nearshore (N)	Foreshore (F)	Backshore (B)
	Depositional subenvironment polygons			
Sediment source (net output) sectors:				
nonresistant bedrock (N)	on	NN	FN	BN
earlier lakebeds (E)	oe	NE	FE	BE
alluvial fan gravels (A)	oa	NA	FA	BA
stream mouth gravels (S)	OS	NS	FS	BS
river mouth sands and silts (R)	OR	NR	FR	BR
ooid-supplied shore (O)	oo	NO	FO	BO
Sediment transfer (steady state) sector:				
longshore transport (L)	ol	NL	FL	BL
Sediment sink (net input) sectors:				
beach accretion sink (B)	ob	NB	FB	BB
inland aeolian sink (I)	oi	NI	FI	BI
deep water sink (D)	OD	ND	FD	bd
Energy- and sediment-starved sectors:				
wave-starved shallows (W)	ow	nw	FW	BW
clast-starved resistant bedrock (C)	oc	nc	FC	BC

Fig. 16.6 Zone-sector-
polygon model of lacustrine
depositional subenvironments
arrayed in conceptual plan,
with 48 subenvironment
polygons defined by
intersections of four
concentric, bathymetric
environment zones (offshore,
nearshore, foreshore,
backshore; Table 16.1 and
Fig. 16.2) and 12 radial
sectors (longshore sediment
dynamics sectors). See
Table 16.4 for explanation of
symbols

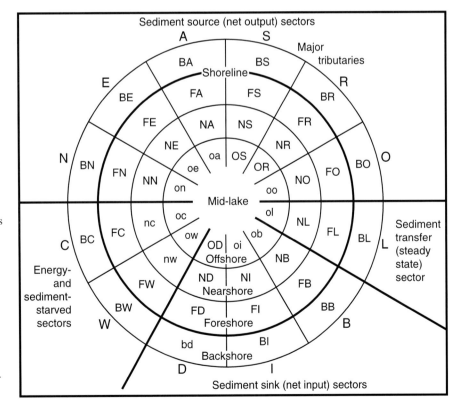

Fig. 16.7 'Bubble' models showing areal patterns of sediment redistribution in lacustrine depocentres; (**a–d**) are after Longmore (1986); (**e–k**) are additional patterns observed in western North America

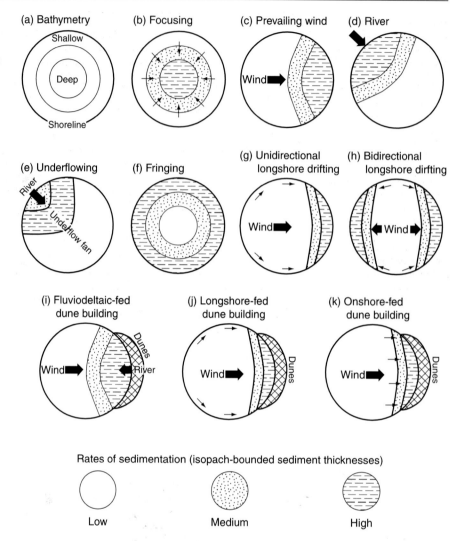

Straits are significant sites of geomorphic information (e.g. Burr and Currey 1992) in large hemiarid lake basins; major embayments, including the concave ends of elongate lakes, are important in most basins; corner bays, coves, and pockets with useful information occur in most basins and on islands (Fig. 16.9). Two exceptions to the geomorphic importance of lakeshore embayments and terrigenous sediment supply should be noted. It has long been recognized that (a) cuspate progradation (e.g. Gilbert 1885, Zenkovich 1967, Carter 1988) commonly produces detailed geomorphic records in lakeshore sectors that are straight or even convex lakeward, and (b) tufa accretion (e.g. Russell 1895, Morrison 1964, Benson et al. 1992, Benson et al. 1995, Ku et al. 1998) commonly produces detailed lakeshore geomorphic records during

intervals favourable to biogeochemical production of limnogenous sediments.

Stratigraphic sequences in hemiarid lake basins can be viewed as dominated by two geomorphic end members (Fig. 16.10) that alternate in basins with large lake-size variation. In the shallow water geomorphic pattern (Fig. 16.10a), basin floors undergo progressive enlargement and overall flattening by hydroaeolian planation, in which six phases of geomorphic activity occur repeatedly as conditions alternate between shallow (typically less than 2 m) inundation and subaerial exposure (Merola et al. 1989, Currey 1990): (1) wind-driven bodies of shallow standing water entrain suspended sediment by subaqueous and circumaqueous (lateral) scour; (2) deposition of suspended sediment occurs through the processes of settling in still water

Fig. 16.8 Forty-five common lakeshore geomorphic environments, arrayed *left to right* in order of increasing terrigenous sediment supply, and *bottom to top* in order of increasing embayment depth; uncommon or null sets are stippled. Geomorphic resolution of lakeshore history is largely a function of sediment supply and embayment depth. Lakeshore geomorphic environments merge laterally in many combinations. Headland environments, for example, are commonly flanked by hemi-embayment environments (see Fig. 16.9)

Lakeshore planimetric configuration (Embayment depth increases generally upward)	Lakeshore terrigenous sediment source (Sediment supply increases to right)				
	Litholacustrine or alluviolacustrine source			Fluviolacustrine source	
	Resistant bedrock	Nonresistant bedrock	Alluvial fan gravels	Stream-mouth gravels	Stream-mouth sands
Strait	fringing beaches	set of spits or cuspate barrier	set of spits or cuspate barrier	*(stippled)*	*(stippled)*
Estuary	*(stippled)*	*(stippled)*	*(stippled)*	estuarine fan delta	estuarine delta and underflow fan
Major embayment (Bay)	bayhead beach	bayhead or baymouth beach ridges	bayhead or baymouth beach ridges	bayhead fan delta	bayhead delta and underflow fan
Hemi-embayment (Corner bay)	inner-corner beach	outer-corner spit, inner-corner barrier	outer-corner spit, inner-corner barrier	inner-corner fan delta	inner-corner delta
Minor embayment (Cove)	cove-head beach	cove-mouth barrier	cove-mouth barrier	cove-head fan delta	cove-head delta
Incipient embayment (Pocket)	pocket beach	pocket beach	pocket beach	pocket beach	*(stippled)*
Broadly concave lakeshore	fringing beach	fringing beach ridge	fringing beach ridge	fringing beach ridges	beach-ridge plain
Straight lakeshore	fringing beach	cliff-base beach	bluff-base beach	fringing beach ridges	beach-ridge plain
Broadly convex lakeshore	wave-cut cliff and platform	wave-cut bluff and platform	wave-cut bluff and platform	arcuate fan delta or foreland	arcuate delta or foreland
Cuspate depositional lakeshore	*(stippled)*	cuspate barrier or double tombolo	cuspate barrier or double tombolo	cuspate fan delta or foreland	cuspate delta or foreland
Bold rocky headland	wave-cut cliff and platform	wave-cut bluff and platform	Geomorphic Record Resolution High Low Erosional Depositional { Beach / Delta		

and stranding in evaporating water; (3) desiccation reduces newly deposited suspended sediment to subaerial clasts of curled and cracked dry mud; (4) aeolian erosion deflates mud clasts and abrades desiccated surfaces; (5) aeolian deposition of mud clasts constructs foredunes as upwind-opening lunettes that partially encircle deflated areas, and as downwind-opening antilunettes that are largely surrounded by extensive deflated areas; and (6) post-aeolian diagenesis sinters and cements aeolian mud clasts. Materials in (5) and (6) are reworked by repetitions of (1)–(4). Draping lacustrine strata, especially lakeshore strata, are common in the deep water geomorphic pattern (Fig. 16.10b), where basin-floor planation is eclipsed by lakebed differentiation into offshore, nearshore, foreshore, and backshore bathymetric zones (Table 16.1 and Fig. 16.2).

After a lake cycle, portions of the lakebed geomorphic record in hemiarid lake basins may be (a) stratigraphically preserved but geomorphically obliterated through burial by nonlacustrine sediments, or (b) partially to wholly obliterated from geomorphic and stratigraphic preservation by erosion and by sediment reworking. Aeolian and alluvial fan processes are predominant obliterators in this

Fig. 16.9 Map of a
hypothetical island with one
hemi-embayment (corner
bay), two minor embayments
(coves), and three incipient
embayments with pocket
beaches

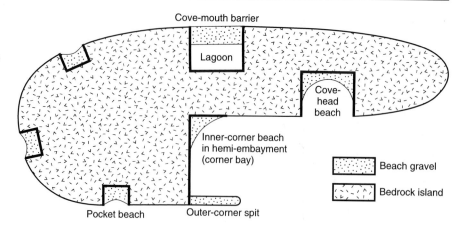

setting (Sack 1995). For example, lakeshore sand is commonly entrained and reworked into aeolian dunes (Hoelzmann et al. 2001). Mud pellets that accumulate as dunettes or lunettes and gypsum pellets that form dunes and sand sheets which bury lakebeds in one location are typically deflated from subaerially exposed fine-grained, basin-floor lakebeds upwind in the same lake basin (Fig. 16.11) (Sack 1994, Magee et al. 1995). In some basins, expanses of small yardangs mark substantial aeolian erosion of basin-floor lakebeds (Hoelzmann et al. 2001). When a lake cycle ends, ephemeral and intermittent streams as well as debris flows return to the highland-bordering alluvial fans that during the lake cycle were scored with shoreline bluffs and mined by waves and currents for downdrift coastal depositional landforms. Postlake alluvial fan processes may bury part of the shoreline record from the piedmont zone, but frequently breach and erode

lakeshore landforms, reworking the sediment into stream and alluvial fan deposits lower on the piedmont (Fig. 16.12).

This postlacustrine geomorphic change in co-lacustrine, as well as pre-lacustrine, lakeshore landforms is perceptible across a wide range of spatial scales. At the local end of the scale spectrum the change is revealed in surveyed profiles and cross sections, such as the erosional shoreline examples of Fig. 16.13. There, geomorphic modification is time-dependent change in which postlacustrine colluvial and alluvial wedges grow at the expense of buried mid-slopes and waning upper slopes. Regionally, in maps and imagery, landform decay is time-dependent plani-metric change in which postlacustrine geomorphic activity, commonly fluvial and mass wasting processes related to alluvial fan activity, gradually obliterates lateral segments of lakeshore, as in the generalized

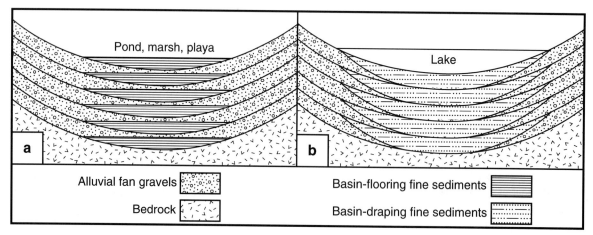

Fig. 16.10 Schematic cross sections of hemiarid lake basins with depositional sequences produced by recurring (**a**) shallow-water and (**b**) deep-lake cycles (after Currey 1990)

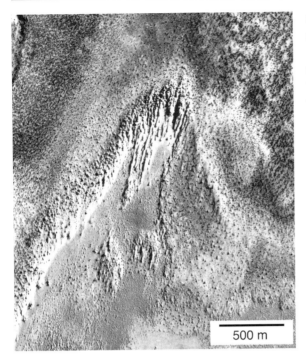

Fig. 16.11 Wind-blown clasts of sand-sized, carbonate-rich, mud pellets from a Lake Bonneville subbasin floor are deposited downwind in a source-bordering lunette

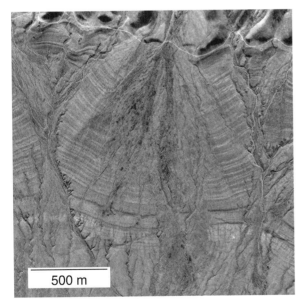

Fig. 16.12 On the highland-bordering piedmont zone of this hypertectonic hemiarid lake basin, postlake (Holocene) alluvial fan processes have obliterated segments of the Lake Bonneville shoreline geomorphic record

shoreline examples of Fig. 16.14. A useful gauge of geomorphic decay is given by the expression $1 - \pi_S$, where π_S is the shoreline preservation index, a dimensionless number from 0 to 1 that expresses visible shoreline length as a fraction of reconstructed total shoreline length (Sack 1995).

Palaeolake Studies

Despite postlacustrine subaerial processes, hemiarid lake basins commonly display considerable evidence of having contained standing water bodies in the late Pleistocene and Holocene. Studies of these palaeolakes draw on paradigms and techniques of the geosciences and biological sciences (e.g. Spencer et al. 1984, Benson et al. 1990, Oviatt et al. 1999, Last and Smol 2001, Balch et al. 2005) and have applications that relate to many science questions and public policy issues. The goals of palaeolake studies are to reconstruct palaeolake histories and to constrain reconstructions of regional and global environmental change. Selected aspects of palaeolake studies that are relevant to geomorphology are outlined here.

Palaeolake histories, and the histories of hemiarid lake basins, are reconstructed from lacustrine and non-lacustrine material evidence (bold rectangular boxes in Fig. 16.15). Much of this evidence is contained in depositional landforms, that is, in morphostratigraphic records. Morphostratigraphy, in which the methods of geomorphology and stratigraphy are employed interactively, is fundamental to successful research design in geomorphic studies of palaeolakes and their basins (Fig. 16.16). Morphostratigraphy is particularly useful in palaeolake studies that stress shoreline history as a basis for reconstructing hydrographic and tectonic history.

Spatial analysis of basinwide palaeolimnology, as illustrated by the idealized Great Basin palaeolake in Fig. 16.17, provides a unifying framework for reconstructing individual stages of palaeolake history. In qualitative terms, the proximal, medial, and distal reaches of compound palaeolakes tend to have distinctive patterns of hydrology, circulation, sedimentation, and lithofacies (Table 16.5) that help to explain geomorphic and stratigraphic patterns. In more quantitative terms, the proximal, medial, and distal reaches of many palaeolakes can be viewed as loci in ternary fields that depict relationships among

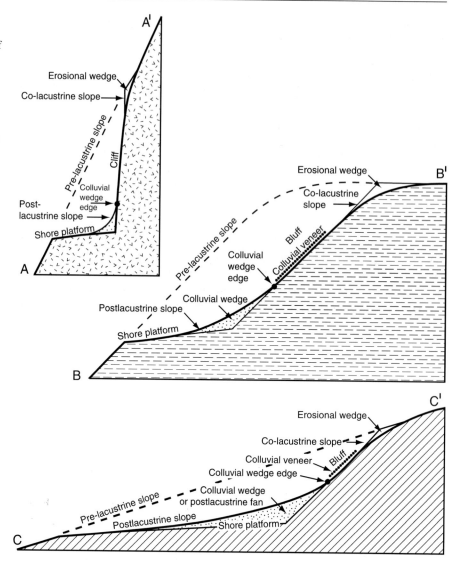

Fig. 16.13 Schematic cross sections showing postlacustrine modification of co-lacustrine slopes, with patterns that are typical of erosional shorelines in resistant bedrock (A-A′), nonresistant bedrock (B-B′), and alluvial fan gravels (C-C′). Colluvial wedge edge (large dot = upper limit of positive colluvium mass balance) migrates upslope toward middle of co-lacustrine cliff (i.e. in resistant bedrock) or bluff (i.e. in nonresistant bedrock and sediments) as a function of postlacustrine time

palaeolimnologic variables, including inputs to the lacustrine water balance (Fig. 16.18a), origins of net horizontal motion in the epilimnion (Fig. 16.18b), and origins of sediment at the bottom of the water column (Fig. 16.18c).

In contrast to basinwide palaeolimnology, basinwide (or subbasin-wide) stratigraphy, as illustrated by the idealized Great Basin graben in Fig. 16.19, provides the framework for reconstructing long intervals of palaeolake history. The large variations in water body size that are so characteristic of hemiarid lake basins result in depositional sequences with complex lateral and vertical changes of lithofacies, biofacies, and chemofacies. This complexity tends to be cyclical, which lends itself to what the *North American Strati-*

graphic Code (NACSN 2005) terms allostratigraphic classification. In Fig. 16.19, four allostratigraphic units of alloformation rank, each including two or three lithofacies, can be differentiated on the basis of laterally traceable discontinuities, in this case buried soils and disconformities (NACSN 2005). Stratigraphic markers, including regional discontinuities that generally date from palaeolake lowstands, assume great importance in hemiarid lake basins (e.g. Table 16.6) for two reasons: (a) in some cases they are the lower and upper boundaries of alloformations, and (b) in many other cases they are indispensable tools for correlating intra-alloformation events within individual basins and from one basin to another.

Fig. 16.14 Schematic plans showing postlacustrine decay of successively younger concentric shorelines. Patterned after late Quaternary shorelines in Nevada (e.g. Mifflin and Wheat 1979). Shoreline (**a**) probably formed about 140 ka, (**b**) about 15 ka, and (**c**) about 2.6 ka

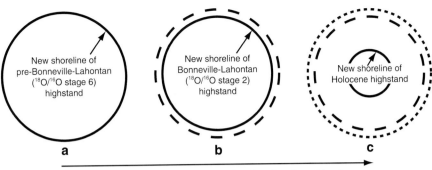

Geomorphic decay of shorelines as a function of postlacustrine time

Advances in palaeolake studies have closely paralleled refinements in geochronology (e.g. Easterbrook 1988, Forman 1989, Sack 1989, Birkeland et al. 1991, Noller et al. 2000, Björck and Wohlfarth 2001). In the last two decades, increasing spatiotemporal resolution has made hemiarid lake basins prime venues for studies of geomorphic kinematics (time rates of past landform changes) (Sack 1995), hydrographic kinematics (time rates of past waterbody changes), and crustal-upper mantle geodynamics (Bills and May 1987, Bills et al. 1994, 2002). Because radiocarbon dating has been such an important tool in late Quaternary palaeolake studies (e.g. Table 16.7), refinements and improved understanding of that technology are noteworthy (Benson 1993, Benson et al. 1995, Geyh et al. 1999, Fornari et al. 2001, Pigati et al. 2004). The shift from β-decay counting to accelerator mass spectrometry (AMS) measurement of ^{14}C has allowed ages of good precision to be obtained from samples containing milligram quantities of carbon. Applications include AMS ^{14}C dating of organic carbon traces in lake sediments (e.g. Thompson et al. 1990, Peck et al. 2002) and in lakeshore rock varnish (e.g. Dorn et al. 1990, Liu and Broecker 2007). As calendaric calibration of ^{14}C ages has improved (e.g. Bard et al. 1990, Stuiver and Reimer 1993, Stuiver and van der Plicht 1998), so has precision in studies of palaeolake kinematics and geodynamics.

Other geochronology advances that have proven useful for palaeolake studies include various types of luminescence dating (e.g. Smith et al. 1990, Kuzucuoglu et al. 1998, Oviatt et al. 2005, Stone 2006) dating, and geochronometry using cosmogenic isotopes other than ^{14}C. For example, cosmogenic helium (^{3}He) and chlorine (^{36}Cl) have seen successful experimental use in dating basalt erosion and halite deposition, respectively, in the Lake Bonneville region, but ^{3}He was recently unsuccessful in the study of another Great Basin palaeolake (Carter et al. 2006).

Within-basin and between-basin comparisons of isotopic ages of possibly synchronic (coeval) features are commonly of interest in palaeolake studies. A dimensionless basis for comparing paired ages, that is, the ages before present of two samples from one palaeolake basin or of one sample from each of two palaeolake basins, is the index of temporal accordance ι. This index ranges from 0 where paired ages are completely dissimilar (discordant) to 1 where paired ages are identical (accordant). In Table 16.8, $\iota = 1 - [(\text{older age} - \text{younger age})/(\text{older age} + \text{younger age})]$.

Beginning with the work of Russell (1885) and Gilbert (1890), late Quaternary hydrographic change in hemiarid lake basins has traditionally been portrayed by hydrographs in which surface elevation, maximum depth, average depth, area, volume, or mass of standing water is plotted as a function of relative or absolute time. With increasing spatiotemporal resolution of lakeshore morphostratigraphy have come increasingly detailed palaeolake histories, often with several temporal wavelengths and spatial amplitudes of hydrographic change appearing as cycles within cycles (Fig. 16.20). First- and second-order temporal wavelengths, sometimes termed episodes and phases (NACSN 2005), are measured in millennia. Third- and fourth-order temporal wavelengths, sometimes termed oscillations (e.g. Currey 1990) and stands, are mostly measured in centuries and decades (e.g. Currey and Burr 1988). 'Instantaneous' singularities in palaeolake history are commonly referred to as events, with some of the more notable being seismic, pyroclastic, and catastrophic flood (e.g. Jarrett and Malde 1987) events.

Fig. 16.15 Flowchart showing elements of palaeolake history (rounded boxes), which are reconstructed from the geologic and geomorphic records (bold rectangular boxes) in lake basins and surrounding regions by response-process inference that is essentially counter to process → response causality (after Currey 1990)

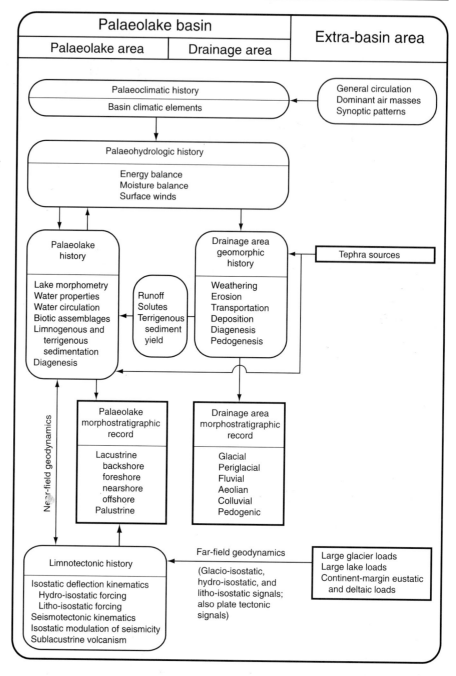

Geomorphic and palaeoenvironmental change that may not be evident in traditional hydrographs can often be represented explicitly in geomorphically annotated hydrographs. In Fig. 16.20, annotations highlight several hypothetical but plausible features. Ooid beaches indicate saline stages, possibly with brine shrimp. Prominent gravel beaches mark transgressive stages, when pre-lacustrine alluvial fans and colluvium-mantled slopes were initially worked into beaches. Tufa-cemented stonelines denote regressive stages, when water hardness increased and older beach gravels were locally reworked and cemented. Intrenched and prograded deltas indicate a falling local base level, that is, a regression. A salinity crisis near the close of the lacustral episode is represented by fetid, organic-rich sediments that accumulated offshore

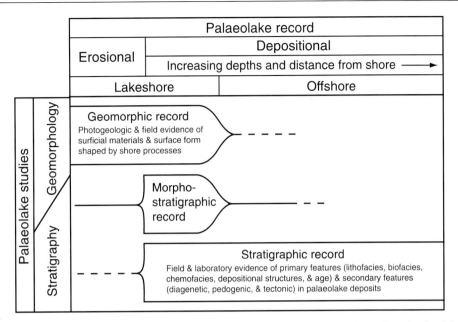

Fig. 16.16 Morphostratigraphy combines the methods of geomorphology and stratigraphy in studies of palaeolake depositional landforms (after Currey and Burr 1988)

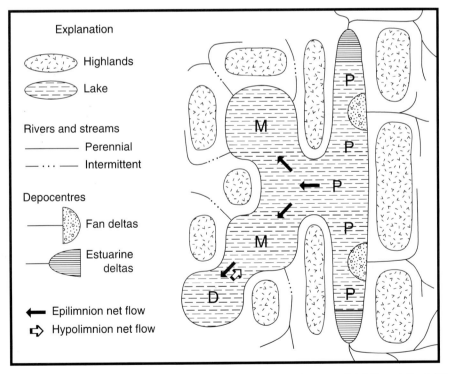

Fig. 16.17 Map of hypothetical palaeolake in hemiarid basin showing proximal (P), medial (M), and distal (D) reaches (after Currey 1990). Arrows show general directions of net flow (see Table 16.5 and Fig. 16.18)

Table 16.5 Conditions typical of proximal, medial, and distal reaches in closed-basin lakes*

	Proximal reach	Medial reach	Distal reach
Hydrology	Local runoff dominant source of water-balance input; epilimnion salinity relatively low	Transbasin flow important source of water-balance input	Transbasin flow dominant source of water-balance input; epilimnion salinity relatively high
Circulation	Water-balance-driven net outflow in epilimnion	Water-balance-driven net throughflow in epilimnion	Water-balance-driven net inflow in epilimnion
Sedimentation	High rates of terrigenous and very low rates of limnogenous sedimentation	Low rates of terrigenous and limnogenous sedimentation	Low rates of limnogenous and very low rates of terrigenous sedimentation
Lithofacies	Fluviodeltaic clastics prevalent	Offshore micrite (typically calcite) and lakeshore clastics	Offshore micrite (typically aragonite) and lakeshore carbonates

*Adapted from Currey (1990). See Figs. 16.17 and 16.18.

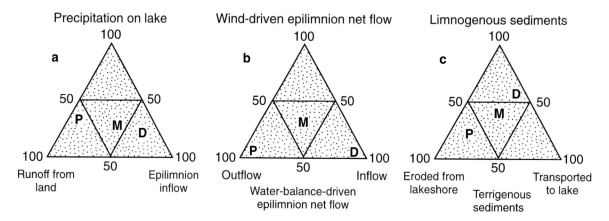

Fig. 16.18 Loci of proximal (P), medial (M), and distal (D) reaches of hypothetical palaeolake (Fig. 16.17 and Table 16.5) in ternary fields depicting lake dynamics: (**a**) water balance inputs as a percentage of long-term receipts, (**b**) sources of net horizontal motion in epilimnion as a percentage of long-term net epilimnion flow, and (**c**) sources of sediments at bottom of water column as a percentage of long-term sedimentation (after Currey 1990)

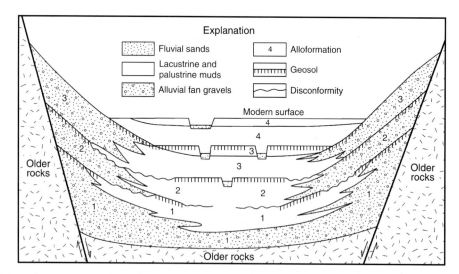

Fig. 16.19 Schematic cross section illustrating allostratigraphic classification of lacustrine muds and other sediments in a graben (after NACSN 2005). Patterned after Cache Valley, Utah, in the Lake Bonneville basin (Williams 1962)

Table 16.6 Stratigraphic markers of regional importance in Lake Bonneville studies: B = marker beds (distinctive sediment layers), H = marker horizons (distinctive surfaces between sedi- ment layers), and P = marker profiles (sediment layers imprinted by subaerial palaeoenvironments or palaeoseismicity)

Stratigraphic markers	B	H	P	Examples
Offshore micrites:				Oviatt et al. 1994
white marl	•			
with abundant dropstones	•			
Coquinas:				Oviatt 1987
gastropod	•			
ostracod	•			
Tephra:				Oviatt and Nash 1989
silicic	•			
basaltic	•			
Evaporites:				Eugster and Hardie 1978
mirabilite	•			Eardley 1962
halite	•			
Stonelines:				
foreshore	•			
fluviolacustrine	•			
buried desert pavement	•	•		Currey 1990
Tufa:				
tufacrete ledges	•	•	•	
heads and mounds	•	•		Carozzi 1962
beachrock and hardground	•			Spencer et al. 1984
Transgressive basal layers:				
gravels on truncated geosols	•	•	•	
basal organics (muck and wood)	•			Scott et al. 1983
Oolitic sand layers	•			Eardley et al. 1973
Buried aeolian deposits:				
loessal colluvium	•	•	•	
aeolian sand	•	•		NACSN 2005
Discontinuities:				
erosional unconformities		•		
nondepositional unconformities		•		
Red beds:				
in situ		•	•	
transported	•			Currey 1990
Soils:				Birkeland et al. 1991
buried soils (geosols)		•	•	Eardley et al. 1973
relict (residual) soils		•	•	
K horizons (caliche)		•	•	Machette 1985
natric horizons		•	•	
Root tubules (rhizoliths)		•	•	Eardley et al. 1973
Desiccation cracks		•	•	Currey 1990
Seismically convoluted bedding			•	

when freshwater plankton populations were annihilated by (and pickled in) increasingly concentrated brines. Red beds mark emergence of the basin floor at the close of the lacustral episode, when previously anoxic hypolimnion sediments were reddened by oxidation of Fe^{+2} sulphides to Fe^{+3} oxides. Lastly, salt beds are former dissolved solids forced from solution by volumetric reduction at the close of the lacustral episode.

Applied Palaeolake Studies

Palaeolake studies have applications in many fields, including climatology, hydrogeology, wetlands and shorelands management, mineral industries, and limneotectonics. Many of these applications are based on the principle of converse uniformitarianism (conformitarianism), which states that the past is the key to the

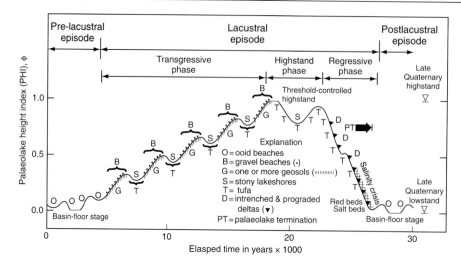

Fig. 16.20 Geomorphically annotated hydrograph (*heavy line*) of a hypothetical palaeolake cycle in a hydrographically closed basin. Patterned in part after Lake Bonneville (e.g. Currey et al. 1984). Hydrograph depicts decamillennium-scale change (lacustral episode and phases) and millennium-scale oscillations (subphases); century-scale fluctuations are not shown except as implied by beach dots; decade-scale variation lies within the hydrograph line width

present and future. The importance of palaeolakes as sources of proxy climate data in atmospheric research, including global change studies and general circulation models, is well documented (e.g. Street-Perrott and Harrison 1985, Benson et al. 1998, Broecker et al. 1998, Lin et al. 1998, Stager et al. 2002). In the western United States, studies of Quaternary

lakes provide crucial information in hydrogeologic evaluations of potential sites for long-term storage of high-level nuclear waste and other sensitive projects (e.g. Williams and Bedinger 1984, Sargent and Bedinger 1985).

Wetlands management in hemiarid lake basins seeks to minimize the adverse effects that hydrographic

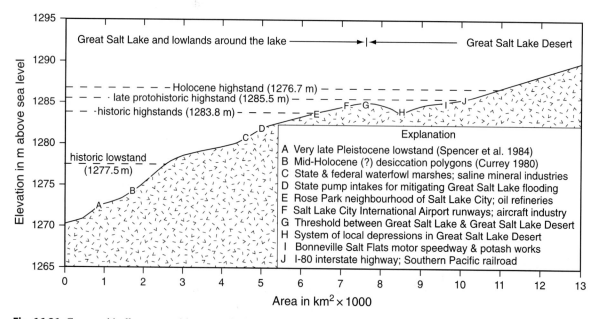

Fig. 16.21 Geomorphically annotated hypsograph showing singular levels in the lowest part of the Great Salt Lake basin, Utah (after Currey 1990). Average gradient across area depicted is only 0.0002

Table 16.7 Organic carbon (O) and carbonate carbon (A = aragonite, C = calcite, D = dolomite) materials yielding radiocarbon ages of relevance to Lake Bonneville studies

Material	O	A	C	D
Wood fragments	•			
Charcoal flecks	•			
Spring bog peat	•			
Marsh muck	•			
Seeds	•			
Geosol humates	•			
Biodust in rock varnish	•			
Bone collagen	•			
Neotoma excreta	•			
Mollusca, shell proteins	•			
Mollusca, snail shells		•		
Mollusca, clam shells		•		
Ostracods, shells		•		
Ostracods, shell proteins	•			
Ooids		•		
Artemia faeces in ooids	•			
Marl, *Chara* phytoliths		•	•	
Marl, micrite		•	•	
Plankton residues in mud	•			
Tufa, algaloid		•	•	
Algal residues in tufa	•			
Tufacrete matrix		•	•	
Travertine, cave			•	
Travertine, spring			•	
Caliche			•	
Root tubule rhizoliths			•	
Saline mudflat carbonate				•

Table 16.8 Index of temporal accordance (IOTA)

Temporal accordance	IOTA (ι)
High (paired ages plausibly synchronous)	$1.00 \geq \iota > 0.99$
Problematic (paired ages possibly synchronous)	$0.99 > \iota > 0.90$
Low (paired ages doubtfully synchronous)	$0.90 > \iota \geq 0$

fluctuations can have on vital aquatic environments, including those that support waterfowl, recreation and tourism, and saline industries (e.g. Fig. 16.21). Shorelands management in hemiarid lake basins seeks to minimize, by measures such as planning, zoning, dyking, and pumping, the potential impacts of rising lake levels on transportation and communication networks, public utilities, industrial development, and urban sprawl. Increasingly, wetlands and shorelands management strategies are founded on baseline information from palaeolake studies.

Many hemiarid lake basins contain significant mineral resources, chiefly potash, soda ash, sodium chloride, borates, nitrates, magnesium, lithium, uranium, sand and gravel, volcanics, zeolites, diatomite, and fossil fuels (e.g. Reeves 1978, Gwynn 2002). Palaeolake studies and mineral industries tend to be mutually supportive, with the former providing information that often is of value in assessing mineral reserves and development options, and the latter contributing to the corpus of palaeolake knowledge by making available subsurface information that frequently has been obtained at great expense (e.g. Smith 1979, Smith et al. 1983).

Limneotectonics (lacustrine neotectonics) is the use of palaeolake levels as long baseline tiltmeters and palaeolake beds as local slipmeters to measure co-lacustrine and postlacustrine rates of crustal motion, particularly vertical motion (Currey 1988, Bills et al. 1994, 2002). A palaeolake level is the originally horizontal upper bounding surface (atmosphere-lake interface) of a palaeolake stage of known age, as reconstructed from lakeshore morphostratigraphic evidence that is analogous to the raised shorelines (e.g. Rose 1981) of many sea coasts. A palaeolake bed is the lower bounding surface (lake-substrate interface) of a palaeolake stage of known age, as observed in the lakebed stratigraphic record.

In many hemiarid lake basins, total neotectonic vertical deformation is the algebraic sum of (a) seismotectonic vertical displacement caused by faulting, and (b) one or more (near-field, far-field, hydro-, glacio-, litho-) isostatic vertical deflection signals generated by crustal loading and/or unloading. An important goal of limneotectonics is to assess the contribution made by each of these sources of crustal motion (e.g. Fig. 16.22).

Palaeolakebeds that have been displaced by faulting are common in hemiarid lake basins that owe their existence to hypertectonism (Fig. 16.1), as does the basin of Lake Bonneville-Great Salt Lake. Beneath Great Salt Lake, seismic reflection surveys (Mikulich and Smith 1974) and many boreholes have helped to define the geometry of subbottom palaeolake beds and, thereby, the late Cenozoic kinematics of subbottom tectonic blocks (e.g. Pechmann et al. 1987). In backhoe trenches across postlacustrine fault scarps at many localities in the Lake Bonneville-Great Salt Lake region, detailed studies of offset palaeolake beds (and of scarp-burying colluvial wedges) have provided insights into the magnitudes and recurrence intervals of late Quaternary seismic events (e.g. Machette et al. 1991).

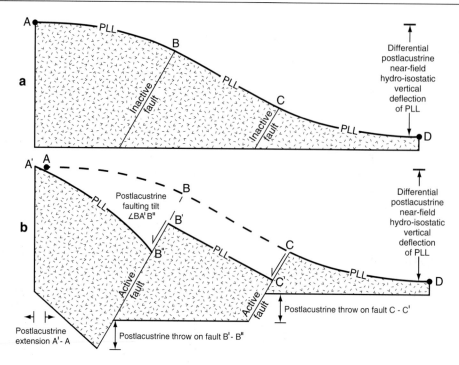

Fig. 16.22 Schematic profiles (not to scale) of regional and subregional deformation of an originally horizontal palaeolake level (PLL) of known age (after Currey 1990). A–D is regional trend of a hydro-isostatically deflected PLL. A and D are PLL localities of maximum and minimum observed isostatic rebound, that is, A is the isostatic deflection centroid and D is a distal point on an isostatic deflection radius. (**a**) Total postlacustrine neotectonic deformation comprises only near-field hydro-isostatic deflection. (**b**) Total postlacustrine neotectonic deformation comprises regional isostatic deflection (dashed) overprinted with subregional seismotectonic displacement, including rotational displacement of tectonic block A′-B″ and translational displacement of tectonic block B′-C′

At several sites in the Lake Bonneville-Great Salt region, maximum late Quaternary seismicity and maximum late Quaternary isostatic rebound seem to have coincided in time, particularly where the orientation of seismotectonic extension corresponded with the orientation of isostatic deflection radii (Fig. 16.22b). This suggests a pattern in which the tempo of seismotectonic displacement is modulated by the vertical direction and tempo of hydro-isostatic deflection. Specifically, where the traces of active extensional (normal) faults are tangent to isolines (isobases) of near-field hydro-isostatic deflection, seismicity may be (a) suppressed during times of sustained isostatic subsidence, and (b) enhanced during subsequent times of maximum isostatic rebound.

Palaeolake levels that have undergone near-field hydro-isostatic deflection occur in hemiarid lake basins with histories of large water load changes, as in the Lake Lahontan (Mifflin and Wheat 1971) and Lake Bonneville (Crittenden 1963, Currey 1982, Bills et al. 2002) regions of the Great Basin. Subsequent to Lake Bonneville's highstand about 15 ka, when the lake had a depth of over 370 m and a volume of almost 10,000 km^3, about 74 m of differential uplift occurred within the near-field area (52,000 km^2) enclosed by the highest shoreline (Currey 1990). About 16 m of additional differential uplift can be detected as far-field effects in small lake basins up to 120 km beyond the highest Bonneville shoreline (e.g. Bills and Currey 1998, May et al. 1991). Post-Bonneville rebound has totalled less than 1 m since 2.6 ka, as gauged by the essentially horizontal shoreline that marks the Holocene highstand of Great Salt Lake.

The maximum rate of hydro-isostatic vertical deflection in the Bonneville basin is estimated to have been -7.3 cm y^{-1} at about 15 ka (Currey and Burr 1988). This is somewhat less than maximum rates of glacio-isostatic rebound in Fennoscandia and Canada, where regional crustal uplift exceeded 10 cm y^{-1} in very late Pleistocene and early Holocene time (Lajoie 1986). Where (a) known water loading and unloading histories have acted on (b) unknown Earth rheologies to produce (c) known hydro-isostatic deflection histories, limneotectonics can be a powerful tool in Earth rheology modelling (Bills and May 1987), that is, limneotectonics can be a rich source of insight into the thickness and flexural rigidity of the Earth's crust and the viscosity layering of the upper mantle.

Geomorphic patterns of palaeolake evidence in hemiarid lake basins direct researchers to many important applications. As the embodiment of sediment sources, pathways, and sinks, depositional lacustrine landforms are especially noteworthy sources of information on the palaeohydrography, and underlying palaeohydrology, palaeoclimatology, and tectonics, of hypertectonic hemiarid terrain. As the appreciation of the relevance to the present and future of the record of past environments continues to grow, so should the research opportunities for scientists studying hemiarid lake basins.

Acknowledgments Much of the research on which this chapter is based was supported by NSF grant EAR-8721114. A large debt is owed to supportive colleagues, including G. Atwood, M. Berry, B. Bills, T. Burr, J. Czarnomski, B. Everitt, B. Haslam, M. Isgreen, J. Keaton, G. King, M. Lee, K. Lillquist, D. Madsen, S. Murchison, G. Nadon, J. Oviatt, J. Petersen, A. Reesman, G. Tackman, and G. Williams. In addition, the second author deeply appreciates having had the opportunity to learn from and work with the late D.R. Currey.

References

Adams KD, Wesnousky SG (1999) The Lake Lahontan highstand: Age, surficial characteristics, soil development, and regional shoreline correlation. Geomorphology 30:357–392

Amit R, Zilberman E, Porat N (1999) Relief inversion in the Avrona playa as evidence of large-magnitude historical earthquakes, southern Arava Valley, Dead Sea Rift. Quat Res 52:76–91

Balch DP, Cohen AS, Schnurrenberger DW, Haskell BJ, Valero Garces V, Beck JW, Cheng H, Edwards RL (2005) Ecosystem and paleohydrological response to Quaternary climate change in the Bonneville basin, Utah. Palaeogeogr Palaeoclim Palaeoecol 221:99–122

Bard E, Hamelin B, Fairbanks RG, Zindler A (1990) Calibration of the ^{14}C timescale over the past 30,000 years using mass spectrometric U-Th ages from Barbados corals. Nature 345:405–410

Benson LV (1993) Factors affecting ^{14}C ages of lacustrine carbonates: Timing and duration of the last highstand lake in the Lahontan basin. Quat Res 39:163–174

Benson LV, Currey DR, Dorn RI, Lajoie KR, Oviatt CG, Robinson SW, Smith GI, Stine S (1990) Chronology of expansion and contraction of four Great Basin lake systems during the past 35,000 years. Palaeogeogr Palaeoclim Palaeoecol 78:241–286

Benson L, Currey D, Lao Y, Hostetler S (1992) Lake-size variations in the Lahontan and Bonneville basins between 13,000 and 9000 ^{14}C yr B.P. Palaeogeogr Palaeoclim Palaeoecol 95:19–32

Benson LV, Kashgarian M, Rubin M (1995) Carbonate deposition, Pyramid Lake subbasin, Nevada: 2. Lake levels and polar jet stream positions reconstructed from radiocarbon ages and elevations of carbonates (tufas) deposited in the Lahontan basin. Palaeogeogr Palaeoclim Palaeoecol 117:1–30

Benson LV, Lund SP, Burdett JW, Kashgarian M, Rose TP, Smoot JP, Schwartz M (1998) Correlation of Late-Pleistocene lake-level oscillations in Mono Lake, California, with North Atlantic climatic events. Quat Res 49:1–10

Bergonzini L, Chalié F, Gasse F (1997) Paleoevaporation and paleoprecipitation in the Tanganyika basin at 18,000 years B.P. inferred from hydrologic and vegetation proxies. Quat Res 47:295–305

Bills BG, Currey DR (1988) Lake Bonneville: Earth rheology inferences from shoreline deflections. EOS 69:1454

Bills BG, Currey DR, Marshall GA (1994) Viscosity estimates for the crust and upper mantle from patterns of lacustrine shoreline deformation in the eastern Great Basin. J Geophys Res 99:22059–22086

Bills BG, May GM (1987) Lake Bonneville: Constraints on lithospheric thickness and upper mantle viscosity from isostatic warping of Bonneville, Provo, and Gilbert stage shorelines. J Geophys Res 92:11493–11508

Bills BG, Wambeam TJ, Currey DR (2002) Geodynamics of Lake Bonneville. In: Gwynn JW (ed) Great Salt Lake: An overview of change. Utah Department of Natural Resources Special Publication, p.7–32

Birkeland PW, Machette MN, Haller KM (1991) Soils as a tool for applied Quaternary geology. Utah Geol Mineral Surv Misc Publ 91–3

Björck S, Wohlfarth B (2001) ^{14}C chronostratigraphic techniques in paleolimnology. In: Last WM, Smol JP (eds) Tracking environmental change using lake sediments. Kluwer, Dordrecht

Broecker WS, Peteet D, Hajdas I, Lin J, Clark E (1998) Antiphasing between rainfall in Africa's Rift Valley and North America's Great Basin. Quat Res 50:12–20

Burr T, Currey DR (1992) Hydrographic modelling at the Stockton Bar. Utah Geol Surv Misc Publ 92–3:207–219

Carozzi AV (1962) Observations of algal biostromes in the Great Salt Lake, Utah. J Geol 70:246–252

Carter DT, Ely LL, O'Connor JE, Fenton CR (2006) Late Pleistocene outburst flooding from pluvial Lake Alvord into the Owyhee River, Oregon. Geomorphology 75:346–367

Carter RWG (1988) Coastal environments. Academic Press, London

Crittenden MD (1963) New data on the isostatic deformation of Lake Bonneville. US Geol Surv Prof Pap 454-E

Currey DR (1980) Coastal geomorphology of Great Salt Lake and vicinity. Utah Geol Mineral Surv Bull 116:69–82

Currey DR (1982) Lake Bonneville: Selected features of relevance to neotectonic analysis. US Geol Surv Open-File Rep 82–1070

Currey DR (1988) Seismotectonic kinematics inferred from Quaternary paleolake datums, Salt Lake City seismopolitan region, Utah. US Geol Surv Open-File Rep 88–673: 457–461

Currey DR (1990) Quaternary palaeolakes in the evolution of semidesert basins, with special emphasis on Lake Bonneville and the Great Basin, U.S.A. Palaeogeogr Palaeoclim Palaeoecol 76:189–214

Currey DR, Burr TN (1988) Linear model of threshold-controlled shorelines of Lake Bonneville. Utah Geol Mineral Surv Misc Publ 88–1:104–110

Currey DR, Atwood G, Mabey DR (1984) Major levels of Great Salt Lake and Lake Bonneville. Utah Geol and Mineral Surv Map 73

Davies JL (1980) Geographical variation in coastal development, 2nd edn. Longman, London

Dean LE (1978) Eolian sand dunes of the Great Salt Lake basin. Utah Geol 5:103–111

Dorn RI, Jull AJT, Donahue DJ, Linick TW, Toolin LJ (1990) Latest Pleistocene lake shorelines and glacial chronology in the western Basin and Range Province, U.S.A.: Insights from AMS radiocarbon dating of rock varnish and paleoclimatic implications. Palaeogeogr Palaeoclim Palaeoecol 78:315–331

Eardley AJ (1962) Glauber's salt bed west of Promontory Point, Great Salt Lake. Utah Geol Mineral Surv Spec Stud 1, p. 12

Eardley AJ (1966) Sediments of Great Salt Lake. Utah Geol Soc Guideb 20:105–120

Eardley AJ, Shuey RT, Gvosdetsky V, Nash WP, Picard MD, Grey DC, Kukla GJ (1973) Lake cycles in the Bonneville basin, Utah. Geol Soc Am Bull 84:211–216

Easterbrook DJ (ed) (1988) Dating Quaternary sediments. Geol Soc Am Spec Pap 227, p. 165

Eugster HP, Hardie LA (1978) Saline lakes. In Lerman A (ed) Lakes: Chemistry, geology, physics. Springer, New York

Eugster HP, Kelts K (1983) Lacustrine chemical sediments. In: Goudie AS, Pye K (eds) Chemical sediments and geomorphology. Academic Press, London

Forman SL (ed) (1989) Dating methods applicable to Quaternary geologic studies in the western United States. Utah Geol Mineral Surv Misc Publ 89–7

Fornari M, Risacher F, Féraud G (2001) Dating of paleolakes in the central Altiplano of Bolivia. Palaeogeogr Palaeoclim Palaeoecol 172:269–282

Galat DL, Jacobsen RL (1985) Recurrent aragonite precipitation in saline-alkaline Pyramid Lake, Nevada. Arch Hydrobiol 105:137–159

Geyh MA, Grosjean M, Núñez L, Schotterer U (1999) Radiocarbon reservoir effect and the timing of the late-glacial/early Holocene humid phase in the Atacama Desert (northern Chile). Quat Res 52:143–153

Gilbert GK (1885) The topographic features of lake shores. US Geol Surv Annu Rep 5:69–123

Gilbert GK (1890) Lake Bonneville. US Geol Surv Monogr 1

Gwynn JW (2002) Great Salt Lake, an overview of change. Utah Department of Natural Resources, Salt Lake City

Håkanson L (1981) A manual of lake morphometry. Springer, Berlin

Håkanson L, Jansson M (1983) Principles of lake sedimentology. Springer, Berlin

Hardie LA, Smoot JP, Eugster HP (1978) Saline lakes and their deposits: A sedimentological approach. In: Matter A, Tucker ME (eds) Modern and ancient lake sediment. Blackwell Scientific, Oxford

Hoelzmann P, Keding B, Berke H, Kröpelin S, Kruse H-J (2001) Environmental change and archaeology: Lake evolution and human occupation in the eastern Sahara during the Holocene. Palaeogeogr Palaeoclim Palaeoecol 169:193–217

Hunt CB, Robinson TW, Bowles WA, Washburn AL (1966) Hydrologic basin, Death Valley, California. US Geol Surv Prof Pap 494-B

Jarrett RD, Malde HE (1987) Paleodischarge of the late Pleistocene Bonneville flood, Snake River, Idaho, computed from new evidence. Geol Soc Am Bull 99:127–134

Jones BF, Bowser CJ (1978) The mineralogy and related chemistry of lake sediments. In: Lerman A (ed) Lakes: Chemistry, geology, physics. Springer, New York

Kleber A (1990) Upper Quaternary sediments and soils in the Great Salt Lake area, U.S.A. Z Geomorphol 34:271–281

Ku T-L, Luo S, Lowenstein TK, Li J, Spencer RJ (1998) U-series chronology of lacustrine deposits in Death Valley, California. Quat Res 50:261–275

Kuzucuoglu C, Parish R, Karabiyikoglu M (1998) The dune systems of the Konya Plain (Turkey): Their relation to environmental changes in central Anatolia during the late Pleistocene and Holocene. Geomorphology 23:257–271

Lajoie KR (1986) Coastal tectonics. In: Wallace RE, Chairman, Active tectonics, Studies in Geophysics, 95–124. National Academy Press, Washington

Last WM, Smol JP, (eds) (2001) Tracking environmental change using lake sediments. Kluwer, Dordrecht

Lin JC, Broecker WS, Hemming SR (1998) A reassessment of U-Th and ^{14}C ages for late-glacial high-frequency hydrological events at Searles Lake, California. Quat Res 49:11–23

Liu T, Broecker WS (2007) Holocene rock varnish microstratigraphy and its chronometric application in the drylands of western USA. Geomorphology 84:1–21

Longmore ME (1986) Modern and ancient sediments, data base for management of aquatic ecosystems and their catchments. In: De Deckker P, Williams WD (eds) Limnology in Australia. W. Junk, Dordrecht

Mabbutt JA (1977) Desert landforms. MIT Press, Cambridge MA

Machette MN (1985) Calcic soils of the southwestern United States. Geol Soc Am Spec Pap 203:1–21

Machette MN, Personius SF, Nelson AR, Schwartz DP, Lund WR (1991) The Wasatch fault zone, Utah, segmentation and history of Holocene earthquakes. J Struct Geol 13:137–149

Magee JW, Bowler JM, Miller GH, Williams DLG (1995) Stratigraphy, sedimentology, chronology and palaeohydrology of Quaternary lacustrine deposits at Madigan Gulf, Lake Eyre, South Australia. Palaeogeogr Palaeoclim Palaeoecol 113:3–42

May GM, Bills BG, Hodge DS (1991) Far-field flexural response of Lake Bonneville from paleopluvial lake elevations. Phys Earth Planet Inter 68:274–284

Menking KM, Anderson RY, Shafike NG, Syed KH, Allen BD (2004) Wetter or colder during the last glacial maximum?

Revisiting the pluvial lake question in southwestern North America. Quat Res 62:280–288

Merola JA, Currey DR, Ridd MK (1989) Thematic mapper-laser profile resolution of Holocene lake limit, Great Salt Lake Desert, Utah. Remote Sens Environ 27:229–240

Mifflin MD, Wheat MM (1971) Isostatic rebound in the Lahontan basin, northwestern Great Basin. Geol Soc Am Abstr with Program 3:647

Mifflin MD, Wheat MM (1979) Pluvial lakes and estimated pluvial climates of Nevada. Nev Bur Mines Geol Bull 94, p. 57

Mikulich MJ, Smith RB (1974) Seismic reflection and aeromagnetic surveys of the Great Salt Lake, Utah. Geol Soc Am Bull 85:991–1002

Morrison RB (1964) Lake Lahontan: Geology of the southern Carson Desert, Nevada. US Geol Surv Prof Pap 401

(NACSN) North American Commission on Stratigraphic Nomenclature (2005) North American stratigraphic code. Am Assoc Pet Geol Bull 89:1547–1591

Nanson GC, Callen RA, Price DM (1998) Hydroclimatic interpretation of Quaternary shorelines on South Australian playas. Palaeogeogr Palaeoclim Palaeoecol 144:281–305

Noller JS, Sowers JM, Lettis WR, (eds) (2000) Quaternary geochronology, methods and applications. American Geophysical Union, Washington

Oviatt CG (1987) Lake Bonneville stratigraphy at the Old River Bed, Utah. Am J Sci 287:383–398

Oviatt CG, Nash WP (1989) Late Pleistocene basaltic ash and volcanic eruptions in the Bonneville basin, Utah. Geol Soc Am Bull 101:292–303

Oviatt CG, Habiger GD, Hay JE (1994) Vertical and lateral variability in the composition of Lake Bonneville marl: A potential key to the history of lake-level fluctuations and paleoclimate. J Paleolimn 11:19–30

Oviatt CG, Thompson RS, Kaufman DS, Bright J, Forester RM (1999) Reinterpretation of the Burmester core, Bonneville basin, Utah. Quat Res 52:180–184

Oviatt CG, Madsen DB, Schmitt, DN (2003) Late Pleistocene and early Holocene rivers and wetlands in the Bonneville basin of western North America. Quat Res 60:200–210

Oviatt CG, Miller DM, McGeehin JP, Zachary C, Mahan S (2005) The Younger Dryas phase of Great Salt Lake, Utah, USA. Palaeogeogr Palaeoclim Palaeoecol 219:263–284

Parker AG, Goudie AS, Stokes S, White K, Hodson MJ, Manning M, Kennet D (2006) A record of Holocene climate change from lake geochemical analyses in southeastern Arabia. Quat Res 66:465–476

Pechmann JC, Nash WP, Viveiros JJ, Smith RB (1987) Slip rate and earthquake potential of the East Great Salt Lake fault, Utah. EOS 68:1369

Peck JA, Khosbayar P, Fowell SJ, Pearce RB, Ariunbileg A, Hansen BCS, Soninkhishig N (2002) Mid to late Holocene climate change in north central Mongolia as recorded in the sediments of Lake Telmen. Palaeogeogr Palaeoclim Palaeoecol 183:135–153

Pigati JS, Quade J, Shahanan TM, Haynes CV Jr (2004) Radiocarbon dating of minute gastropods and new constraints on the timing of late Quaternary spring-discharge deposits in southern Arizona, USA. Palaeogeogr Palaeoclim Palaeoecol 204:33–45

Reeves CC (1978) Economic significance of playa lake deposits. In: Matter A, Tucker ME (eds) Modern and ancient lake sediments. Blackwell Scientific, Oxford

Rose J (1981) Raised shorelines. In: Goudie A (ed) Geomorphological techniques. Allen & Unwin, London

Russell IC (1885) Geologic history of Lake Lahontan. US Geol Surv Monogr 11

Russell IC (1895) Lakes of North America. Ginn, Boston

Sack D (1987) Geomorphology of the Lynndyl Dunes, west-central Utah. Utah Geol Assoc Publ 16:291–299

Sack D (1989) Reconstructing the chronology of Lake Bonneville. In: Tinkler KJ (ed) History of geomorphology. Unwin Hyman, London

Sack D (1990) Quaternary geologic map of Tule Valley. Utah Geol Mineral Surv Map 124

Sack D (1994) Geologic map of the Coyote Knolls quadrangle, Millard County, Utah. Utah Geol Surv Map 162

Sack D (1995) The shoreline preservation index as a relative-age dating tool for late Pleistocene shorelines: An example from the Bonneville basin, U.S.A. Earth Surf Process Landf 20:363–377

Sargent KA, Bedinger MS (1985) Geologic and hydrologic characterization and evaluation of the Basin and Range province relative to the disposal of high-level radioactive waste – Part II, Geologic and hydrologic characterization. US Geol Surv Circ 904-B

Scott WE, McCoy WD, Shroba RR, Rubin M (1983) Reinterpretation of the exposed record of the last two cycles of Lake Bonneville, western United States. Quat Res 20:261–285

Sharp RP (1978) Coastal southern California. Kendall/Hunt, Dubuque IA

Sly PG (1978) Sedimentary processes in lakes. In: Lerman A (ed) Lakes: Chemistry, geology, physics. Springer, New York

Smith BW, Rhodes EJ, Stokes S, Spooner NA, Aitken MJ (1990) Optical dating of sediments: Initial quartz results from Oxford. Archaeometry 32:19–31

Smith GI (1979) Subsurface stratigraphy and geochemistry of late Quaternary evaporites, Searles Lake, California. US Geol Surv Prof Pap 1043

Smith GI, Barczak VJ, Moulton GF, Liddicoat JC (1983) Core KM-3, a surface-to-bedrock record of late Cenozoic sedimentation in Searles Valley, California. US Geol Surv Prof Pap 1256

Spencer RJ, Baedecker MJ, Eugster HP, Forester RM, Goldhaber MB, Jones BF, Kelts K, McKenzie J, Madsen DB, Rettig SL, Rubin M, Bowser SJ (1984) Great Salt Lake and precursors, Utah: The last 30,000 years. Contrib Mineral Petrolog 86:321–334

Stager JC, Mayewski PA, Meeker LD (2002) Cooling cycles, Heinrich event 1, and the desiccation of Lake Victoria. Palaeogeogr Palaeoclim Palaeoecol 183:169–178

Stone T (2006) Last glacial cycle hydrological change at Lake Tyrrell, southeast Autstralia. Quat Res 66:176–181

Street-Perrott FA, Harrison SP (1985) Lake levels and climate reconstruction. In: Hecht AD (ed) Paleoclimate analysis and modeling. Wiley, New York

Stuiver M, Reimer P (1993) Extended ^{14}C data base and revised CALIB 3.0 ^{14}C age calibration program. Radiocarb 35: 215–230

Stuiver M, van der Plicht J, (eds) (1998) INTCAL98: Calibration issue. Radiocarb 40:1041–1164

Thompson RS, Toolin LJ, Forester RM, Spencer RJ (1990) Accelerator-mass spectrometer (AMS) radiocarbon dating of Pleistocene lake sediments in the Great Basin. Palaeogeogr Palaeoclim Palaeoecol 78:301–313

Wilkins DE, Currey DR (1997) Timing and extent of late Quaternary paleolakes in the Trans-Pecos closed basin, west Texas and south-central New Mexico. Quat Res 47:306–315

Williams JS (1962) Lake Bonneville, geology of southern Cache Valley, Utah. US Geol Surv Prof Pap 257-C

Williams TR, Bedinger MS (1984) Selected geologic and hydrologic characteristics of the Basin and Range province, western United States. US Geol Surv Misc Investig Map I-1522D

Zenkovich VP (1967) Processes of coastal development. Wiley, New York

Part VII
Aeolian Surfaces

Chapter 17

Aeolian Sediment Transport

William G. Nickling and Cheryl McKenna Neuman

Introduction

Aeolian processes, involving the entrainment, transport, and deposition of sediment by the wind, are important geomorphic processes operating in arid regions. This chapter, in association with Chapters 18, 19, and 20 form an integrated unit that discusses the fundamentals of aeolian sediment entrainment and transport, dune morphology and dynamics, wind erosion processes and aeolian landforms, and the significance of dust transport.

Aeolian processes involve complex interactions between the wind and the ground surface. Understanding these processes requires knowledge of relevant surface characteristics (e.g. texture, degree of cohesion, crusting, and vegetation cover), as well as the dynamics of airflow over the surface. The aeolian sediment transport system can be subdivided into two classes based on grain size characteristics. That is: (a) material of sand size ($>50\,\mu m$); and (b) silt- and clay-sized particles ($<50\,\mu m$) or dust. There are major differences in the processes by which these two types of aeolian sediment are entrained, transported, and deposited.

In recent years, important advances have been made in our understanding of the processes of aeolian sediment transport. The field, however, remains grounded in the classical work of R.A. Bagnold, especially with regard to the transport of sand. The recognition that aeolian processes occur on other planets, and in particular Mars, has provided a stimulus for studies into the fundamental processes that govern the entrainment and transport of sediment by wind. Increasing concern over desertification of arid and semi-arid lands has also resulted in extensive research into those factors responsible for wind erosion on susceptible soils and in particular, the processes controlling the entrainment and transport of dust.

In the past, the study of aeolian processes tended to be carried out in academia primarily by earth scientists, but has changed dramatically over the past two decades with these processes being investigated by researchers from a wide range of disciplines including, earth, atmospheric, biological and planetary sciences, and engineering. Recent advances have come by a combination of careful wind tunnel and field experiments in association with the numerical modelling of processes.

Despite our improved understanding of wind transport mechanics, there has been a tendency for field and wind tunnel researchers and modellers to view the aeolian transport system from a *transport limited* perspective, particularly with regard to sand transport. That is, given a certain wind speed or bed shear stress, the sediment flux can be estimated assuming that the surface can supply an unlimited amount of sediment. However, recent research has shown that most natural eroding surfaces tend to be *supply limited*. In these cases the total sediment transport rate is controlled by the ability of the surface to supply grains to the air stream, often resulting in lower total transport rates than would be predicted by most theoretical or empirical models for a given wind speed. In general, most natural surfaces tend to have higher threshold (shear) velocities and lower sediment supply rates because of textural and surficial factors (e.g. surface crusts, vegetation, surface moisture), which tend to hold individual grains in place. Moreover, there can be a great spatial and temporal variability in these surficial and textural

W.G. Nickling (✉)
Wind Erosion Laboratory, Department of Geography,
University of Guelph, Guelph, Ontario N1G 2W1, Canada
e-mail: nickling@uoguelph.ca

A.J. Parsons, A.D. Abrahams (eds.), *Geomorphology of Desert Environments*, 2nd ed.,
DOI 10.1007/978-1-4020-5719-9_17, © Springer Science+Business Media B.V. 2009

controls that is not taken into account in most predictive models that generally assume a steady state condition for atmospheric, surficial, and textural conditions. These complexities are common in drylands and tend to control aeolian transport systems in these and other environments.

Transport Modes

There are varied modes of aeolian transport, each depending primarily on the size of the particles entrained. From his observations in the field and in wind tunnel simulations, Bagnold (1941) was first to classify particle motion into three main categories: suspension, saltation and creep, which are shown schematically in Fig. 17.1. Suspension refers to the motion of particles that are light and therefore easily kept aloft for relatively long distances. By definition, the terminal velocity u_f of particles suspended in an airflow is small in relation to the velocity at which the air parcels are dispersed vertically by turbulence, which can be represented by the shear (or friction velocity) u_*. In order for very small particles to remain in suspension for long periods and be carried great distances, the ratio of u_f to u_* is typically less than 0.1 ($u_f/u_* < 0.1$, Fig. 17.2) (Hunt and Nalpanis 1985). Although the size range of particles transported is dependent on the u_f/u_* ratio, suspended dust particles are typically less than 70 μm in diameter. Suspension can be further classified according to the residence time in the atmosphere (Tsoar and Pye 1987). Long term suspension refers to events lasting several days, during which suspended particles may travel thousands of kilometres from the source area. Such particles usually have an upper size limit of about 20 μm. Coarser dust particles

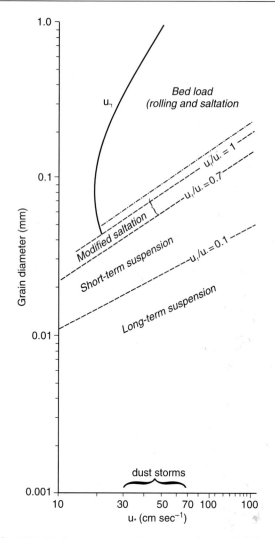

Fig. 17.2 Modes of transport for quartz spheres of different diameters at different wind shear velocities (After Tsoar and Pye 1987)

(20 μm < d < 70 μm) remain in short term suspension for minutes to hours, and may travel a few metres to several kilometres at most (0.1 < u_f/u_* < 0.7). Larger particles (approximately 70–500 μm) bounce or saltate across the surface, following long, parabolic trajectories. For an idealized case, assuming that the initial take-off speed is in the same order as the shear velocity, the characteristic trajectory height h may be approximated as $0.81u_*^2/g$ and the length l as $10.3u_*^2/g$ (Owen 1964). Such particles have a high probability of rebounding from the surface to a height sufficient for acceleration by the wind once again. This mode of transport, described as saltation, underlies the initiation and growth of most aeolian bedforms. It may comprise

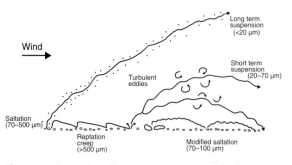

Fig. 17.1 Modes of aeolian sediment transport (After Pye 1987)

as much as 95% of the total mass transport in an aeolian system. Pure saltation, referring to the smooth trajectories in Fig. 17.1, occurs when the turbulent vertical component of velocity has no significant effect on the particle trajectory. A sharp distinction does not exist between saltation and true suspension. Instead, there is a transitional state (termed modified saltation) that is characterized by semi-random particle trajectories influenced by both inertia and settling velocity (Hunt and Nalpanis 1985; Nalpanis 1985).

The impact of saltating particles with the surface may cause short-distance, low energy movement of adjacent grains in reptation. Surface creep refers to particles that are pushed or rolled along without losing contact with the surface. Usually larger than 500 μm in diameter, such particles are too heavy to be lifted by the wind.

The Surface Wind

Over most natural surfaces, airflow is turbulent and consists of eddies of different sizes that move with different speeds and directions. Turbulent eddies transfer momentum from one 'layer' to another, such that each 'layer' has a different average velocity and direction. This process is known as 'turbulent mixing' and the definition of wind profiles in this way is known as the mixing length model (Prandtl 1935).

At a given point above the surface in a turbulent flow, the wind can be described in terms of a standard coordinate system representing the streamwise (u), crosswise (w) and vertical (v) components of velocity (Fig. 17.3). Over flat, relatively uniform terrain, u is usually aligned horizontally and parallel to the surface.

Using this coordinate system, the time series representing the instantaneous velocities of u, v and w can be described in terms of the associated time averaged means and fluctuating (turbulent) components (Fig. 17.3):

$$
\begin{aligned}
u &= \bar{u} + u' \\
v &= \bar{v} + v' \\
w &= \bar{w} + w'
\end{aligned}
\qquad (17.1)
$$

where u, w and v are the instantaneous velocities measured in the positive x (streamwise or horizontal), y (vertical) and z (crosswise) directions, $\bar{u}$, $\bar{v}$ and $\bar{w}$ are the corresponding time averaged velocities, and u', v' and w' are the fluctuating velocities representing turbulence.

As a result of viscous frictional effects, the horizontal wind speed near the bed is retarded giving rise to a distinctive log-linear change in wind speed with height (Fig. 17.4a,b). This portion of the wind profile defines the lower atmospheric boundary layer. If the surface is composed of small particles ($< 80\,\mu$m), it is aerodynamically smooth (particle friction Reynolds number $Re_p \leq 5$). In this case a very thin laminar sublayer, usually <1 mm thick, develops adjacent to the bed even for flows in which most of the boundary layer is turbulent. By contrast, when surface particles or other roughness elements are relatively large, the surface is aerodynamically rough ($Re_p \geq 70$). The laminar sublayer ceases to exist and is replaced by a viscous sublayer for which the velocity profile is not well defined (Middleton and Southard 1984). Under conditions of neutral atmospheric stability, the velocity profile above the viscous sublayer for aerodynamically rough surfaces is characterized by the Prandtl–von Karman equation:

$$
u/u_* = 1/\kappa\, ln(z/z_0) \qquad (17.2)
$$

where u is the velocity at height z, z_0 is the aerodynamic roughness length of the surface, u_* is the shear velocity and κ is von Karman's constant (≈ 0.4). In these conditions, the wind profile plots as a straight line on semi-logarithmic axes, with the intercept on the ordinate axis representing z_0 (Fig. 17.4b). No matter how strong the wind blows, all the velocity profiles tend to converge to approximately the same intercept or z_0 value (Figs. 17.4b and 17.5). Thus u increases

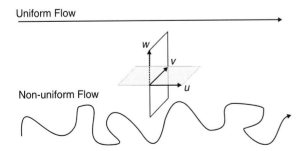

Fig. 17.3 Nature of fluid motion relative to the standard coordinate system used in many studies of fluid transport (air and water)

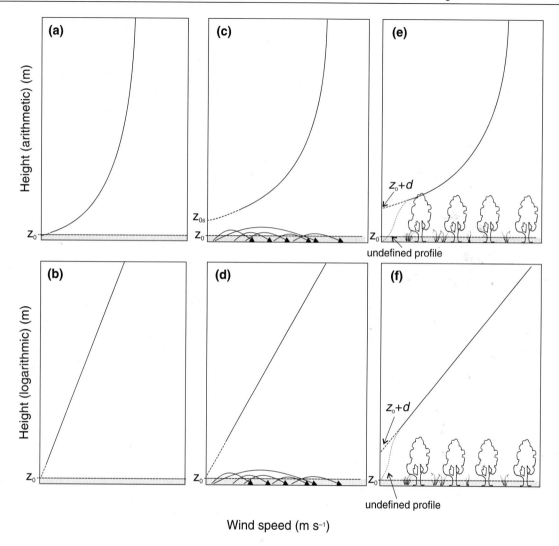

Fig. 17.4 Wind profile over an aerodynamically rough surface, showing the aerodynamic roughness length (z_0), plotted on an arithmetic (**a**) and logarithmic (**b**) height scales. Representative wind profiles in the presence of saltation showing the added roughness (z_{os}) due to saltation (**b** and **c**). Representative wind profiles over a surface with large roughness elements showing the displacement (**d**) of the wind profile (**e** and **f**)

with increasing wind speed at a given height (z) above the surface.

The shear velocity (u_*) is proportional to the slope of the wind velocity profile when plotted on a logarithmic height scale (Fig. 17.4b) and is related to the shear stress (τ) at the bed and the air density (ρ) by

$$u_* = (\tau/\rho)^{1/2} \qquad (17.3)$$

or

$$\tau = \rho u_*^2 \qquad (17.4)$$

For relatively smooth sand surfaces, z_0 is approximately 1/30 the mean particle diameter, but varies with the shape and distance between individual particles or other roughness elements (Fig. 17.6). The value of z_0 increases to a maximum of about 1/8 particle diameter when the roughness elements are spaced about two times their diameter (Greeley and Iversen 1985). Typical values of z_0 for desert surfaces are shown in Table 17.1.

Where the surface is covered by tall vegetation or high densities of other large roughness elements (e.g. large boulders), the wind velocity profile is displaced upwards from the surface to a new reference plane

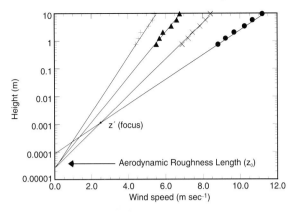

Fig. 17.5 Change in boundary-layer wind profiles over the surface of an alluvial fan in Death Valley with different wind speeds (measured at the highest anemometer). Note that the roughness length remains the same although wind speed at a given height increases

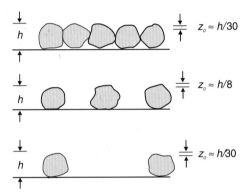

Fig. 17.6 Relationship between aerodynamic roughness length z_0 and non-erodible roughness element spacing (After Greeley and Iversen 1985)

that is a function of the height, density, porosity, and flexibility of the roughness elements (Oke 1978) (Fig. 17.4e,f). The upward displacement is termed the zero plane displacement height (d). The wind profile equation then becomes

$$u/u_* = 1/\kappa \, ln[(z-d)/z_0] \qquad (17.5)$$

Table 17.1 Typical values of aerodynamic roughness (z_o) for desert surfaces (from Lancaster et al. 1991)

Surface type	Aerodynamic roughness (m)
Playas	0.00013–0.00015
Alluvial fans	0.00084–0.00167
Desert pavements	0.00035
Lava flows	0.0148–0.02860

Although the exact nature of the relationship among z_0, d_0, surface roughness, and particle size is poorly understood (Lancaster et al. 1991), in the presence of large roughness elements (vegetation, boulders, etc.), the ratio of the mean element height h to z_0 is a function of the roughness density (λ) (Raupach et al. 1993):

$$\frac{z_0}{h} = f\lambda \qquad (17.6)$$

Roughness density is defined as:

$$\lambda = \frac{nbh}{S} \qquad (17.7)$$

where b and h are the breadth and height of the roughness elements, n is the number of elements and S is the ground area that the elements occupy.

As roughness density increases, there is an initial sharp increase in z_0/h, reaching a maximum value at some critical λ (Fig. 17.7). As λ continues to increase, the elements become increasingly closer together making the surface both physically and aerodynamically smoother, with a resultant decrease and eventual relatively constant value of z_0/h associated with the newly defined (rougher) surface (Fig. 17.7).

The increase in roughness density is also associated with a change in the flow over and around the roughness elements. This change is characterized by the three distinct flow regimes: isolated roughness flow, wake interference flow, and skimming flow (Lee and Soliman 1977). As the wind flows around an object it creates an area of flow separation and deceleration

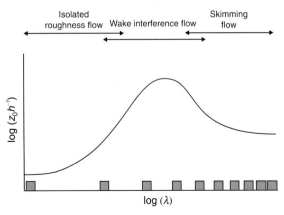

Fig. 17.7 Schematic representation of the change in z_0/h with element density (*grey blocks*) for the three flow regimes (After Raupach 1992)

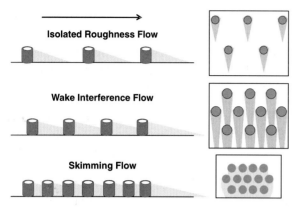

Fig. 17.8 Schematic diagram showing the development of wakes shed from roughness elements (*grey area*) and their interaction with downwind elements for three different flow regimes (modified from Wolfe and Nickling 1993)

directly downwind of the element termed the wake region (Fig. 17.8). Characteristic spacings and surface covers associated with these flow regimes are shown in Table 17.2. When elements are widely spaced (low values of λ), individual wakes to not impinge on adjacent elements and is termed *isolated roughness flow* (Lee and Soliman 1977; Wolfe and Nickling 1993) As the element spacing decreases, the wake regions of the elements interfere with each other (*wake interference flow*), until the spacing is small enough that the wake regions completely overlap over the surface, creating stable vortices between the elements such that the wind appears to 'skim' over the elements (*skimming flow*) (Lee and Soliman 1977).

When the wind shear velocity is great enough to move sand particles, the near-surface wind velocity profile is altered because momentum is extracted from the wind near the surface by the saltating sand grains.

Table 17.2 Flow regimes and associated roughness element concentrations from flow experiments and staggered configuration (after Lee and Soliman 1977)

Flow regimes	Roughness element description		
	Spacing to height ratio (S_P/h)	Percent cover (C) %	Roughness concentration (λ)
Isolated Roughness Flow	>3.5	<16	<0.082
Wake Interference Flow	3.5–2.25	16–40	0.083–0.198
Skimming Flow	<2.25	>40	>0.198

The velocity distributions with height remain a straight line but tend to converge to a focus (z') 0.2–0.4 cm above the surface (Fig. 17.5). Below this point, the wind profile is non-logarithmic, and wind speed in this zone may actually decrease as u_* increases (Bagnold 1941). The focus z' may represent the mean saltation height of uniformly sized grains (Bagnold 1941) and is approximately ten times the mean grain diameter of the bed (Zingg 1953). The apparent bed roughness z_0' is dependent on u_* during sediment transport (Owen 1964) (Fig. 17.4c,d). Thus

$$z_0' = a(u_*^2/2g) \tag{17.8}$$

where $a = 0.02$ and g is the acceleration of gravity.

Owen argued that the effective bed roughness scales with the height of the saltation layer. However, wind tunnel studies by White and Schultz (1977) and Gerety and Slingerland (1983) show that there is a wide range of saltation trajectories and that the $u_*/2g$ parameter is not a good predictor of saltation layer thickness, although it may be useful as an index of effective roughness height (Gerety 1985).

Reynolds Stress

Turbulent flow can be described as a collection of vortices of varying size, generally increasing from the surface up through the boundary layer, that are superimposed on the mean flow. There is a transfer of momentum between eddies, with larger eddies generating smaller eddies that carry momentum towards the surface from the upper boundary layer (van Boxel et al. 2004). The momentum exchange across a plane parallel to the streamwise flow can be characterized by the Reynolds stress (RS):

$$RS = -\overline{\rho u'w'} \tag{17.9}$$

In that RS is a measure of shear stress, shear velocity (u$_*$) can be defined by

$$u_{*RS} = \sqrt{\left|\overline{u'w'}\right|} \tag{17.10}$$

(Kaimal and Finnigan 1994).

In theory, under uniform flow conditions, the wind profile derived shear stress (τ) (Equations 17.3 and 17.4) should be equal to the time averaged Reynolds stress (RS) at the same point in the flow field (Tennekes and Lumley 1972; Walker 2005). However, in practice, profile derived shear stresses rarely match near surface estimates of Reynolds stress based on the measurement of the horizontal and vertical components of wind velocity using fast response instrumentation described below (Wiggs et al. 1996; Butterfield 1999; Walker 2000; Wiggs 2001). This results from the fact that τ and RS are only comparable if the flow vectors for each of the parameters are exactly parallel (see Walker [2005] for a detailed discussion).

Measurement of Wind Speed

Traditionally, wind speed in aeolian transport studies has been measured using 3-cup anemometers mounted on towers (Fig. 17.9a). A wide range of anemometer types is available commercially that vary in size, quality, and sensitivity. In general smaller anemometers with light cups and quality bearings are used for detailed measurements close to the surface but have the disadvantage that they may be damaged by the abrasion of saltating sand or dust particles entering into the bearings. Wind profiles are typically measured with 3–10 anemometers mounted with logarithmic spacing on towers from a few centimeters above the surface up to 5 or 10 m (Fig. 17.9a). Wind direction

is normally measured using sensitive wind vanes mounted at one or more heights at the same location as the anemometers. Output from the anemometers and vanes are most often collected on various types of data loggers.

Although cup anemometry is a commonly used and reliable method, it only allows for the measurement of the time-averaged horizontal component of the wind flow. Recent research (e.g. Wiggs et al. 1996; Butterfield 1999; Walker 2000, 2005; Wiggs 2001; van Boxel et al. 2004) has begun to focus on the role of turbulence in the entrainment and transport of sediment, following the long standing approach taken in the fluvial literature (e.g. Clifford et al. 1993). One of the oldest, most reliable and frequently used methods for the measurement of turbulence in clear airflows is the hot-wire anemometer (constant temperature or constant current anemometer). These instruments typically consist of a very thin platinum or tungsten wire 4–10 μm diameter and 1 mm in length held tightly between the two arms of a small mounting post. The wire can be either heated by applying constant current or maintained at a constant temperature by varying the current. In either case, when the probe is placed into a moving fluid, heat is lost to the fluid by convection. The speed and direction of the fluid motion are directly related to the heat loss and the orientation of the wire to the fluid motion. A single wire probe can only be used to measure one directional component of velocity at a time. However, using a cross wire probe, having two wires mounted in an X configuration, two velocity

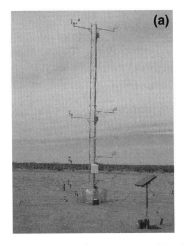

Fig. 17.9 (**a**) A typical 10 m meterological mast with sensitive 3-cup anemometers and wind vanes mounted logarithmically (**b**) three dimensional (*upper*) and two dimensional (*lower*) sonic anemometers mount on an 'H' frame

vectors can be measured simultaneously (most commonly the streamwise and vertical components). These types of probes have been used frequently in wind tunnel experiments to investigate wind flow patterns over dunes and other landforms (e.g. Wiggs et al. 2002; Walker and Nickling 2003).

Using hot-wire anemometry, sampling frequencies up to 10 kHz or more are easily achieved for each flow component, allowing for the measurement of small rapidly moving turbulent structures. The limitation of these probes; however, is their fragile nature that precludes their use in sediment laden flows, and particularly, in the presence of saltating sand that can easily break the wires in the probes, which are very expensive. In an attempt to overcome this limitation, several suppliers now manufacture ruggedized or hot film probes. In this case, a layer of conducting platinum is placed over a very small solid ceramic or glass core, which can be cylindrical, conical or wedge shaped. The more rigid core helps maintain the integrity of the conducting film. These ruggedized instruments have been used successfully in field (e.g. Bauer and Namikas 1998; Rasmussen and Sørensen 1999) and wind tunnel experiments (e.g. Bauer et al. 2004) examining the role of turbulence in saltation transport.

In recent years, several investigators have begun to use sonic anemometers in aeolian transport studies (e.g. Leenders et al. 2005; Anderson and Walker 2006). These instruments, which are commonly used in boundary layer meteorological studies, can provide the three components of velocity by measuring the transit time of high frequency sound pulses (Fig. 17.9b). They come in several different configurations and sizes, are very accurate, quite rugged, have no moving parts, and measure the components of wind velocity at very high frequencies (see van Boxel et al. 2004 and Walker 2005 for detailed reviews on the operation and use of sonic anemometers).

Entrainment of Sediment by the Wind

The Threshold of Motion

Grains will be moved by the wind when the fluid forces exceed the effects of the weight of the particle and cohesion between adjacent particles (Fig. 17.10). The

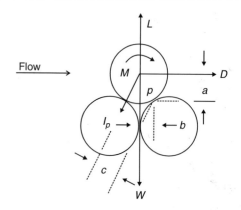

Fig. 17.10 Schematic view of forces on a spherical particle at rest: D = aerodynamic drag, L = lift, M = moment, I_p = interparticle force, and W = weight. Moment arms about a point p given by a, b, c (After Greeley and Iversen 1985)

drag and lift forces, as well as the resultant moment, are caused by the fluid flow around and over the exposed particles. The lift force results from the decreased fluid static pressure at the top of the grain as well as the steep velocity gradient near the grain surface (Bernoulli effect). The weight and cohesive forces are related to physical properties of the surface particles including their size, density, mineralogy, shape, packing, moisture content, and the presence or absence of bonding agents such as soluble salts.

As drag and lift on the particle increases, there is a critical value of u_* when grain movement is initiated. This is the fluid threshold shear velocity or u_{*t} (Bagnold 1941):

$$u_{*t} = A\{[\rho_P - \rho/\rho]gd\}^{1/2} \qquad (17.11)$$

where A is an empirical coefficient dependent on grain characteristics, ρ_P is the grain density, and d is the grain diameter. The value of A is approximately 0.1 for particle friction Reynolds numbers $Re_P > 3.5$.

During the downwind saltation of grains, their velocity and momentum increase before they fall back to the surface. On striking the surface, the moving particles may bounce off other grains and become re-entrained into the airstream or embedded in the surface. In both cases, momentum is transferred to the surface in the disturbance of one or more stationary grains. As a result of the impact of saltating grains, particles are ejected into the airstream at shear velocities lower than that required to move a stationary grain by direct fluid pressure. This new lower threshold

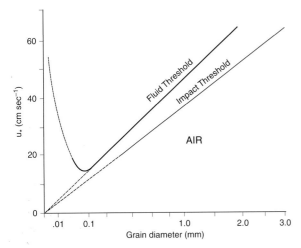

Fig. 17.11 Relationship between fluid and impact threshold shear velocity and particle size (After Bagnold 1941)

required to move stationary grains after the initial movement of a few particles is the dynamic or impact threshold (Bagnold 1941). The dynamic or impact threshold for a given sediment follows the same relationship as the fluid threshold (Equation 17.11) but with a lower coefficient A of 0.08 (Fig. 17.11).

Processes of Sediment Entrainment

Bagnold (1941) suggested that once the critical threshold shear velocity is reached, stationary surface grains begin to roll or slide along the surface by the direct pressure of the wind. As these particles begin to gain speed, they start to bounce off the surface into the air stream and initiate saltation. More recent work (e.g. Chepil 1959; Iversen and White 1982) indicates that particles moving into saltation do not usually roll or slide along the surface prior to upward movement into the air stream. Initial movement into saltation is caused by instantaneous air pressure differences near the surface which act as lift forces. As wind velocity is increased, particles begin to vibrate with increasing intensity and at some critical point leave the surface instantaneously. Lyles and Krauss (1971) observed that the vibrations were seldom steady and occurred in flurries of three to five vibrations (1.8 ± 0.3 Hz for sand grains of 0.59–0.84 mm) before the particles instantaneously leave the bed. It appears therefore that microscale turbulence may play a role in threshold

processes. Williams et al. (1990) suggest that fluid threshold is dependent on the turbulent structure of the boundary layer, and that initial disturbance is related spatially and temporally to semi-organized flurries of activity. The flurries appear to be associated with sweep and burst sequences that are well known in boundary layers. Grains dislodged by sweeps of movement are likely to become the agents for further dislodgements by collision.

Nickling (1988), using a sensitive laser system to detect initial grain motions and count individual grain movements during wind tunnel experiments, found that when velocity is slowly increased over the sediment surface, the smaller or more exposed grains are first entrained by fluid drag and lift forces in either surface creep or saltation. As the velocity continues to rise, larger or less exposed grains are also moved by fluid forces. Saltating grains rebound from the surface and may eject one or more stationary grains at shear velocities lower than that required to entrain them by direct fluid pressures and lift forces. The newly ejected particles move downwind and impact the surface, displacing an even larger number of stationary grains. As a result, there is a cascade effect with a few grains of varying sizes and shapes being entrained primarily by drag and lift forces. These grains set in motion a rapidly increasing number of grains by saltation impact. This progression from fluid to dynamic threshold, based on number of grain movements, is characterized by the relationship between the number of grain movements and u_* measured in the wind tunnel tests (Fig. 17.12). The shear velocity associated with the minimum radius of curvature on the grain count plots closely approximates the fluid threshold determined from Bagnold's threshold equation. It is clear that the threshold process is complex and involves localized near-bed turbulent flow structures that dislodge grains which act as 'seeds' for downwind ejections of other grains. Initiation of widespread grain motion involves a cascading effect, such that the number of grains in motion increases exponentially. Interactions between the developing saltation cloud and the wind result in extraction of momentum from the near-surface wind and a reduction in the nearbed wind speed so that the saltating grains are accelerated less and impact the bed with less energy. As a result, the number of grains dislodged at the bed decreases, and the transport rate approaches a dynamic equilibrium value termed 'steady state saltation' by Anderson and Haff (1988)

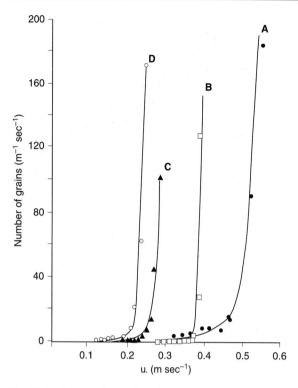

Fig. 17.12 Increase in number of grain entrainments with increasing shear velocity for four sands measured with a laser particle counting system. Sands A–D have different mean sizes (190–770 μm) and sorting characteristics that are reflected in the shape and relative positions of the grain count curves (After Nickling 1988)

or 'equilibrium saltation' by Owen (1964). Equilibrium appears to be reached very rapidly, with a characteristic time period of 1–2 s.

Entrainment Threshold and Saltation Intermittency

Although saltation is often viewed as a steady state process, particularly with regard to modelling, recent field and laboratory research suggests that saltation is a highly unsteady process (Stockton and Gillette 1990; Butterfield 1991 1993; Stout and Zobeck 1997; McKenna-Neuman et al. 2000; Wiggs et al. 2004; Davidson-Arnott et al. 2005). The intermittent nature of sand transport over short time periods (seconds to minutes) results from the temporal fluctuations of instantaneous wind speed and direction (gustiness) as well as surface conditions (e.g surface moisture, crust

development) that limits the supply of sand grains to the air stream (Nickling 1989; Wiggs et al. 2004; Davidson-Arnott et al. 2005).

Recent developments in instrumentation for the rapid and simultaneous measurement of wind speed and saltation, as discussed above, have made it possible to study the effects of unsteady flow on saltation transport. The nature of the temporal fluctuations in wind and sediment transport has been analyzed using concepts of intermittency, in which the proportion of time that a system is active can be described by the intermittency function γ (Stout and Zobeck 1997). These authors define a relative wind strength index (s)

$$s = \bar{u} - u_t/\sigma \qquad (17.12)$$

where $\bar{u}$ is the mean wind speed, u_t is the threshold velocity for sand transport and σ is the standard deviation of the wind speed distribution.

Stout and Zobeck (1997) define two transport conditions with regard to wind gustiness, which can be expressed in terms of the relative wind strength index (s). If $s < 0$ then the mean wind speed is less than the threshold wind speed (i.e. $\bar{u} < u_t$). In this case, however, occasional gusts may exceed u_t and sand transport may occur for a short period of time. In contrast, where $s > 0$ there may be short periods when no sand transport occurs because the wind speed drops below the threshold wind speed despite the fact that $\bar{u} > u_t$.

As a result of the fluctuating wind speeds and associated intermittent nature of saltation, it is difficult to define a finite threshold value for a given surface as suggested by Bagnold (1941) and others who assume steady winds and a uniform, unchanging surface. Using the concept of intermittency, Stout and Zobeck (1997) and Stout (1998) provide a method for the determination of transport threshold (u_t) in the presence of unsteady winds (Time Fractional Equivalence Method, TFEM) where the frequency of mass transport is matched to the proportion of time the wind speed exceeds threshold.

Wiggs et al. (2004) carried out a field study to evaluate the intermittency concept and the TFEM. As part of this study they presented a new method to calculate the threshold value that overcomes the rather laborious iterative procedure presented by Stout and Zobeck (1997), while maintaining the underlying concepts of the TFEM. Wiggs et al. (2004) found that although the TFEM is an objective technique for threshold determination under unsteady winds, this method explained only a small proportion of the recorded sand transport events. Use of different sampling periods, time average shear velocity (u_*) and height of wind measurement failed to improve the prediction of threshold.

Numerical simulations (Fig. 17.13) and direct field measurements suggest that that saltation typically lags behind wind velocity fluctuations by 1–2 s (Anderson and Haff 1988; Sterk et al. 1998; McEwan and Willetts 1991, 1993; Butterfield 1991, 1993). These observations suggest that an above threshold wind speed may be reached before entrainment and that saltation transport may continue for some period of time because of inertial effects even if wind speed drops below threshold (Jackson 1996; Jackson and Nordstrom 1997). Wiggs et al. (2004) found that sand transport events are somewhat better correlated to wind speed fluctuations if a 1 s lag is considered. They also found that the use of time average wind speed and sand transport data consistently explained a much higher proportion of sand transport events as a result of the smoothing out of the internal variability of the wind and transport data. In addition to the temporal variability of the wind, Wiggs et al. (2004) and Davidson-Arnott et al. (2005) found that surficial factors such as surface moisture content, that vary both temporally and spatially, can have a profound effect on threshold and saltation intermittency. As a result of the temporal and spatial variability of atmospheric and surface conditions, as well as the range in size of surface grains, it is very unlikely that a given surface can be defined by a finite threshold value. Rather most natural surfaces are likely better represent by a range of threshold values that may vary from point to point and from one instant to the next (Nickling 1984; Wiggs et al. 2004).

Grain Size Effects

Although fluid threshold can be closely defined for a uniform sediment size greater than approximately 0.1 mm, it cannot be defined for most natural sediments because they usually contain a range of grain sizes and shapes which vary in grain density and packing. As a result, fluid and dynamic thresholds should be viewed as threshold ranges that are a function of the size, shape, sorting, and packing of the surface sediments. The relationship between threshold shear velocity and particle diameter based on Equation (17.11) shows that for a particle friction Reynolds number $Re_p > 3.5$ (grain diameter $>80\,\mu$m) u_{*t} increases with the square root of grain diameter (Fig. 17.11). In this case, the grains protrude into the airflow and are therefore aerodynamically 'rough' so that the drag acts directly on the grains. For $Re_P < 3.5$, the grains lie within the laminar sublayer. In this case, the drag is distributed more evenly over the surface which is aerodynamically 'smooth', and the value of the threshold parameter A rises rapidly. As a result u_{*t} is no longer proportional to the square root of grain diameter but is dependent on the value of A. Wind tunnel studies (e.g. Bagnold 1941; Chepil 1945a) confirm that very high wind speeds are required to entrain fine-grained materials.

Iversen et al. (1976), Greeley et al. (1976), and Iversen and White (1982) have questioned the assumption that the coefficient A in the Bagnold threshold model is a unique function of particle friction Reynolds

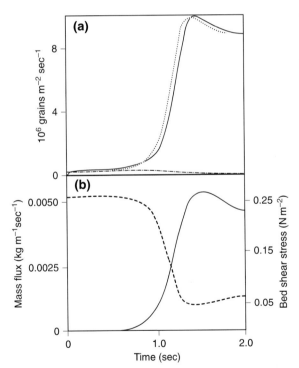

Fig. 17.13 Computer simulations of saltation for 250-μm grains. (**a**) Numbers of grains impacting (*dotted line*), being ejected (rebound and splash, *solid line*), and being entrained (*dot dashed*) as a function of time. Note that equilibrium is reached in a very short time (about 2 s). (**b**) Corresponding history of total mass flux (*solid*) and shear stress at the bed (*dashed*) (After Anderson and Haff 1988)

number. Through a series of detailed theoretical and wind tunnel investigations, with a wide range of particle sizes and densities, Greeley and his coworkers confirmed Bagnold's earlier assumption that A is nearly constant for larger particles, but increases rapidly for small particles (Fig. 17.11). This suggests that A is not solely a function of particle diameter for grains less than about 80 µm in diameter. For very small grains interparticle forces such as electrostatic charges and moisture films appear to be important. Based on regression analyses of their extensive wind tunnel data, Iversen and White (1982) derived a series of threshold equations that account for Reynolds number effects due to particle size, as well the increasing role of interparticle forces as grain size decreases.

Particles with grain diameters less than approximately 80 µm are inherently resistant to entrainment by wind, yet major dust storms are common occurrences in many parts of the world. This is because dust particles (silt and clay size) are frequently ejected by the impact of saltating sand grains (Chepil and Woodruff 1963), and not by direct fluid pressures and lift forces (Chepil 1945b; Gillette 1974; Nickling and Gillies 1989). Abrasion by saltating particles breaks up surface crusts and aggregates, releasing fine particulates into the air stream (Chepil 1945c; Gillette et al. 1974; Nickling 1978; Hagen 1984; Shao et al. 1993; Shao and Lu 2000). The vertical emission of small particulates, expressed as a ratio of the amount of soil moved in saltation and creep, is proportional to approximately u_*^4 (Gillette 1977; Nickling and Gillies 1989). Based on theoretical reasoning and wind tunnel testing, Shao et al. (1993) argue that since dust emission is primarily a function of saltation bombardment, which is a function of the cube of shear velocity ($q \, \alpha \, u_*^3$), it follows that dust emission should also be a function of u_*^3. Sandsized aggregates of silt and clay may also be transported in saltation but tend to break up or abrade during transport, providing fines which are carried away in suspension (Gillette and Goodwin 1974; Gillette 1977; Nickling 1978; Hagen 1984).

Surface Roughness

The presence of vegetation, gravel lag deposits, and other non-erodible roughness elements is important to both the entrainment and transport of sediment by wind. Roughness elements act to attenuate wind erosion by physically covering a portion of a vulnerable surface and by extracting a portion of the wind's momentum (Wolfe and Nickling 1993). However, low spatial densities of roughness elements may lead to an increase in entrainment and transport around the roughness elements because of the development and shedding of eddies (Logie 1982). However, as the roughness density increases beyond some critical level, erosion tends to decrease.

Several researchers have attempted to develop empirical and physically based models that can that account for the effect that the increased surface roughness has on reducing wind erosion (Lettau 1969; Arya 1975; Raupach 1992; Marticorena et al. 1995). Several of these models are directly or indirectly based on the shear stress partitioning theory presented by Schlichting (1936). Schlichting (1936) argued that the total drag force imparted upon a surface with roughness elements (F) can be separated into the force exerted on the surface (F_S) and the force on the roughness elements (F_R):

$$F = F_R + F_S \qquad (17.13)$$

Dividing by the surface area affected, the components of the total shear stress can be resolved:

$$\tau = \tau_R + \tau_S \qquad (17.14)$$

where τ is the total stress, τ_R is the shear stress on the roughness elements and τ_S is the shear stress on the underlying surface in the absence of roughness elements.

Marshall (1971) undertook extensive wind tunnel experiments examining the partition of shear stresses between roughness elements and the surface in the context of wind erosion. He measured τ and τ_R, for arrays of solid roughness elements with densities and spatial arrangements representative of vegetation in arid regions, leading to a relationship between τ and the stress partition. Marshall's (1971) results provided the first comprehensive data set on the drag partitioning, and set the framework for investigating the effect of non-erodible roughness elements on wind erosion (Gillette and Stockton 1989; Raupach 1992).

Stockton and Gillette (1990) expressed the partitioning of shear stress in terms of the threshold friction velocity (u_{*t}), which is the u_* at which sediments become entrained for a given surface. Higher densities of roughness tend to increase u_{*t} (Lancaster and Baas 1998) from the value of u_{*t} that exists for a similar bare surface. Therefore, the amount of force dissipated by the roughness elements can be specified as:

$$R_t = \frac{u_{*t S}}{u_{*t R}} \qquad (17.15)$$

where $u_{*t S}$ represents the u_{*t} for a bare, erodible surface, and $u_{*t R}$ is the value for the same surface with non-erodible roughness elements present. Field and laboratory studies have verified the decrease in values of R_t with increasing λ (Musick and Gillette 1990; Musick et al. 1996; Lancaster and Baas 1998). Raupach (1992) modelled the shear stress partition by relating the reduction in τ_S to the characteristics of wakes generated in the lee of roughness elements based on the geometric and drag properties of cylinders in turbulent flow, along with λ. Raupach et al. (1993) then linked the Raupach (1992) model to aeolian sediment entrainment by associating it with the friction velocity ratio, such that:

$$R = \frac{u_{*t S}}{u_{*t R}} = \left[\frac{\tau'_s}{\tau}\right]^{1/2} = \left[\frac{1}{(1 - \sigma\lambda)(1 + \beta\lambda)}\right]^{1/2} \qquad (17.16)$$

with τ'_S defined as the shear stress acting on the exposed intervening surface, β defined as the ratio of the drag coefficient for an individual roughness element to the drag coefficient of the unobstructed surface (C_R/C_S). The ratio of roughness element basal area to frontal area (σ) is incorporated into the model such that $\sigma\lambda$ is the basal area per unit ground area. The drag partition prediction agrees well with R values from wind measurements of individual drag forces in wind tunnels (Marshall 1971; Crawley and Nickling 2003), field studies measuring drag forces (Wolfe and Nickling 1996; Wyatt and Nickling 1997; Gillies et al. 2006a), wind tunnel and field studies where $u_{*t S}/u_{*t R}$ was measured (Lyles and Allison 1976; Musick and Gillette 1990; Lancaster and Baas 1998), and computer simulations (Li and Shao 2003).

Because of the spatially and temporally variable distribution of shear stress over rough surfaces, Equation (17.16) was modified by Raupach et al. (1993) to account for the fact that the threshold of particle movement is determined not by the spatially averaged stress on the exposed surface, but the maximum stress acting on any point of the surface (τ''_S). Consequently, an additional parameter (m) was included to account for the spatial averaging of shear stress:

$$R'' = \left[\frac{r''_S}{\tau}\right]^{1/2} = \left[\frac{1}{(1 - m\sigma\lambda)(1 + m\beta\lambda)}\right]^{1/2} \qquad (17.17)$$

By definition Raupach et al. (1993), the m parameter ($0 < m < 1$) scales the maximum shear stress to the shear stress present on a surface with a lower roughness density, such that:

$$\tau''_S(\lambda) = \tau'_S(m\lambda) \qquad (17.18)$$

Raupach et al. (1993) suggested $m \approx 0.5$ for flat, erodible, homogeneous surfaces. Although the shear stress partitioning concept is relatively straightforward, the direct measurement of the shear stress applied to the elements or the surface is difficult, especially in field situations where the size and distribution of roughness elements can vary greatly. Bradley (1969a,b) designed a type of drag plate to measure the force of the wind on the surface in field situations and used this instrument to measure the change in shear stress as wind passed from a smooth to a rough surface. Similarly, Luttmer (2002) and Namikas (2002) developed field drag plates to be used in aeolian transport studies. Despite this work, few direct measurements of shear stress partitioning have been made in areas covered by sparse vegetation or other largescale roughness elements in natural environments.

Parameterization of the Raupach model by Wyatt and Nickling (1997) in a field study with shrubs as roughness elements resulted in a lower m value than expected ($m = 0.16$). They argue that the derived low value of m might have been a consequence of inadequate spatial measurements of τ''_S.

Crawley and Nickling (2003), in a detailed wind tunnel experiment using a sensitive tiered force balance, evaluated the m parameter using Equation (17.17) and found that τ''_S was a multiple of τ'_S; however the resulting m values resulted in a gross overestimation of R'' when compared to the measured ratios. As a result, Crawley and Nickling (2003) suggested that the definition of m given by Raupach et al. (1993) was likely physically incorrect and needed reevaluation. Importantly, the lack of a precise physical definition did not allow for the independent evaluation of the m parameter by Crawley and Nickling (2003).

King et al. (2008) examining the drag partition for staggered arrays with different sized elements but with a constant λ found that τ_S' increased with the width of the roughness elements rather than with element height (h).

Overall, the Raupach et al. (1993) model provides good drag partition predictions for staggered arrays; however, natural surfaces rarely display this configuration. It is uncertain if the drag partitioning of non-staggered roughness configurations will behave in the same manner as staggered arrays. Sparsely distributed roughness elements under isolated roughness or wake interference flow will dissipate a portion of the wind's momentum; however, accelerated flow around the roughness may act to increase the shear stress on the surface. Even at higher roughness densities, some distributions of roughness may act to increase local wind shear on the intervening surface, whereas more uniform distributions may affect the wind more evenly over an area.

Bryant (2004) and Gillies et al. (2006a,b) carried out one of the most detailed field studies to investigate

the role of surface roughness on shear stress partitioning and sediment transport. The study was carried out on a flat bare surface that was covered with artificial roughness elements (5 gallon plastic buckets, $0.26 \times 0.30 \times 0.36$ m) with six roughness densities (λ) ranging from 0.016 to 0.095 (Fig. 17.14). The drag force τ_S' on individual elements was measured with a specially designed force balance to which individual buckets were attached at six locations along the dominant wind direction within the array. Shear stress on the surface (τ_R) within an array was measured with Irwin sensors (Fig. 17.15). The Irwin sensor is a simple, omnidirectional skin friction meter that measures the near surface vertical pressure gradient (Irwin 1980). The differential in dynamic pressure is measured between two ports, one at the surface and the other at a height of 0.00175 m above the surface (Fig. 17.15). Once calibrated, the Irwin sensor can be used to measure surface shear stresses at frequencies greater than 10 Hz (Irwin 1980; Wu and Stathopoulos 1994) and has been used successfully in a variety of flow conditions and surface roughness configurations (Irwin 1980; Wu

Fig. 17.14 Roughness arrays used in a field experiment investigating shear stress partitioning and sediment at the USDA Experimental Range, New Mexico. Five gallon plastic buckets were used as roughness elements. The arrays covered approximately 1600 m^2 using from 400 to 1900 buckets for the different roughness densities (After Gillies et al. 2006b)

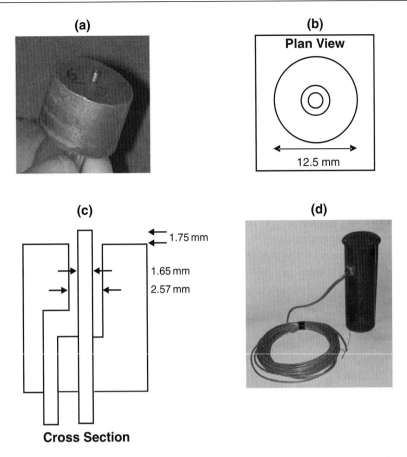

Fig. 17.15 Schematic diagrams and photos of Irwin Sensors used to measure near surface shear stress (τ_0). The brass sensor (**a**) is mounted flush with the wind tunnel floor or the lid of a plastic tube (**d**) used in field applications. The stainless steel pressure ports on the sensor (**c**) are connected to a sensitive differential pressure transducer with 1.0 mm diameter plastic tubing. The pressure transducer, which is housed in the plastic tube is read by a data logger or A–D board in a laptop computer (After Brown et al. 2008)

and Stathopoulos 1994; Monteiro and Viegas 1996; Crawley and Nickling 2003).

Results from this study, where the size (*b* and *h*) of the artificial elements was held constant for increasing values of λ, indicated that the spacing between the elements has a significant role in how stress is partitioned between roughness elements and the surface. Sediment transport through the arrays was shown to be strongly controlled by the size and distribution of the roughness elements, which affect the force available on the intervening surface to transport grains. Gillies et al. (2006b) found that sediment transport decreased in a predictable exponential manner through the array, with the exponent of the relationship being a function of λ. Element size and aspect ratio however, appears

to control the magnitude of the sediment transport reduction more than can be attributed solely to λ. That is for a given λ, large elements and particularly those which are wider, tend to cause a more rapid reduction in saltation flux within the array. Typically, most saltating particles are transported below 0.3 m with a large percentage moving between 0 and 0.1 m. The interaction between larger roughness elements for similar λ values in the presence of saltating grains with this vertical distribution of mass transport creates a condition that is more restrictive to the downwind movement of grains for the smaller elements.

Recent work by Gillette and Pitchford (2004), Okin and Gillette (2001, 2004) and Okin (2005) indicates that the spatial arrangement of roughness elements also has an important effect on the shear stress partitioning and sediment flux that goes beyond that which can be attributed solely to the λ. Okin and Gillette (2001, 2004) found that the preferential growth of mesquite in the Chihuahuan Desert leads to the formation of 'streets' of bare soil oriented in the direction of the prevailing winds, which leads to a higher sediment and dust flux than vegetation oriented in a regular pattern. The streets that tend to be aligned with the prevailing wind produced greater sediment fluxes in these bare corridors than those openings in the canopy that were not aligned with the wind. Gillette and Pitchford (2004) also found that sediment transport increased with downwind distance along the streets. Once a 'street' forms, a positive feedback mechanism prevents vegetation from growing in this area, thereby promoting the further development of the street.

Recently Brown and Nickling (2007) carried out a controlled wind tunnel experiment to evaluate shear stress partitioning in arrays having differing element distributions but similar λ's. A sensitive tiered drag balance was used to independently and simultaneously measure the drag on the arrays of roughness elements and the drag on the intervening surface (Fig. 17.16). Irwin sensors (Fig. 17.15) recorded point measurements of surface shear stress within the arrays. Roughness arrays consisted of small cylinders in four different spatial arrangements at four roughness densities (Fig. 17.17). Results indicate that the surface protection increased similarly with roughness density regardless of the roughness arrangement supporting recent theoretical arguments of Raupach et al. (2006). Point measurements of shear stress revealed that the roughness configuration had a small impact on the distribution of shear stress at the surface, and that a consistent relationship between the maximum and average shear stress existed. The wind tunnel measurements compared favourably to those predicted by the Raupach et al. (1993) model as well as related field and wind tunnel experiments.

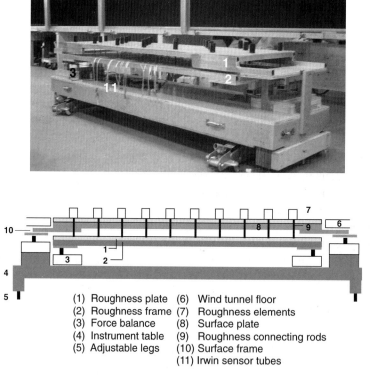

(1) Roughness plate   (6) Wind tunnel floor
(2) Roughness frame   (7) Roughness elements
(3) Force balance   (8) Surface plate
(4) Instrument table (9) Roughness connecting rods
(5) Adjustable legs   (10) Surface frame
                      (11) Irwin sensor tubes

Fig. 17.16 Tiered force balance used in wind tunnel experiments to measure, independently and simultaneously, the drag on the roughness elements and the intervening surface (After Brown et al. 2008)

Fig. 17.17 Roughness arrays used in wind tunnel experiments to investigate shear stress partitioning over various surfaces: (**a**) regular staggered array (**b**) arrays with elongated 'streets' (**c**) clumped and (**d**) random (After Brown et al. 2008)

Sand Transport by Wind

As noted above, there are three distinct modes of aeolian transport (Figs. 17.1 and 17.2) that depend primarily on the grain size of the available sediment. Very small particles (<60–$70\,\mu$m) are transported in suspension and kept aloft for relatively long distances by turbulent eddies in the wind, where as larger particles (approximately 70–$500\,\mu$m) move downwind by saltation. The impact of saltating particles with the surface may cause short-distance movement of adjacent grains in reptation. Larger ($>500\,\mu$m) or less exposed particles may be pushed or rolled along the surface by the impact of saltating grains in surface creep.

Particle Saltation and Splash

The collision of saltators with a loose bed results in the transfer of a portion of the energy gained from the wind during flight to particles on the surface that surround the point of impact (Fig. 17.18). For some saltators, the portion of energy lost represents only a small fraction of their total energy so that there is a high probability of continuing downwind in a series of undiminished rebounds. This transport regime is referred to as steady-state or successive saltation (Rumpel 1985). At the other end of the saltation spectrum, particles that are ejected from the surface at very low velocities may only travel a few millimetres before coming to rest, causing no further particle ejections. Such motion was first described as reptation by Ungar and Haff (1987). Intermediate states of motion are associated with particles that either gain momentum after a series movements of increasing trajectory length, that eventually may result in steady state saltation, or conversely with grains that lose momentum with each impact causing the motion to decline into creep and then cease. Both the particle lift-off velocity and angle determine the fate of subsequent particle motion initiated by impact, herein referred to as splash. The number and velocity distribution of particles ejected is described statistically by the 'splash function' (Ungar and Haff 1987; Werner 1988). Splash is recognized as a major contributor to saltation transport.

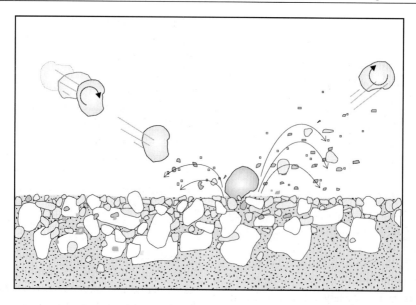

Fig. 17.18 Schematic diagram showing the impact of a relative-ley large saltating grain splashing up numerous low energy ejecta of varying size moving short distances in repatation and a few high energy ejecta moving in saltation or modified saltation. Ripple wave lengths are believed to scale with the movement of the reptating particles

Numerous direct measurements of particle lift-off, acceleration, impact, and rebound have been obtained in wind tunnel studies (e.g. Bagnold 1941; Chepil 1945a; White and Schultz 1977; Mitha et al. 1985; Willetts and Rice 1985, 1986, 1989; Rice et al. 1995, 1996a; Dong et al. 2002; Wang et al. 2006) with increasing accuracy and detail tied directly to the evolution of Particle Image Velocimetry (PIV) technology. Such measurements have the purpose of not only enhancing knowledge of the processes that underlie aeolian transport processes, but also improving parameterization within numerical simulation studies. For example, White and Schultz (1977) observed that particle trajectories were higher than those predicted by theoretical equations of motion at the time of their study. When they incorporated lift forces generated by the spinning of the particle as it moves through the air (Magnus effect), the theoretical trajectories obtained were in much better agreement with those observed in wind tunnel experiments.

Observations based on conventional high-speed photography by Rice et al. (1995) suggest that the impact angle for saltators is low and remarkably consistent, on average between $10°$ and $15°$, as compared to that for rebounding particles with mean of $23–40°$ and standard deviation of $15–30°$. The velocity ratio (rebound/impact) was not found to depend upon par-

ticle size with values between 0.5 and 0.6 for particle impact velocities between 3 and $4\,\mathrm{ms}^{-1}$. Such impacts ejected between 2 and 6 particles out of the surface at lift-off velocities below 10% of that for the saltating particles. In general, Rice et al. (1995) determined that the probability of creating a splash (P_S) increased with the diameter of the impacting saltator. In comparison, phase Doppler anemometer measurements obtained within 1 mm of the surface by Dong et al. (2002) suggest that both the angle and speed of impacting and splashed particles vary more widely than previously reported, and are well represented by a Weibull distribution function. These authors attribute high observed mean impact angles (40–78) and lift-off angles (39–94) to mid-air collisions at very low height, an effect not detected by conventional high-speed photography and ignored in theoretical simulation models. Such mid-air collisions appear to result in a large proportion of backward-impacting particles.

In parallel with this empirical work, Anderson and Haff (1991), and Haff and Anderson (1993) carried out computational simulations of splash using discrete element methods. Although limited to two-dimensional systems consisting of spheres, these simulations were able to derive splash functions, which emulated contemporary wind tunnel measurements by Rice and her coworkers.

The Saltation Cloud

Wind tunnel and field studies, as well as numerical models of the cloud of saltating and reptating grains, show that most sediment is transported close to the ground with an exponential decline in sediment and mass flux concentration with height (e.g. Bagnold 1941; Sharp 1964; Williams 1964; Nickling 1978; Nickling 1983; Anderson and Hallett 1986; Nalpanis et al. 1993; Zou et al. 2001; Wang et al. 2006). Over sand surfaces in wind tunnels, most particles travel within the lower 1–2 cm. Sharp (1964) found that 50% by weight of sediment transported across an alluvial surface in the Coachella Valley travelled within 13 cm of the ground, and 90% below 60 cm. There is no clear upper limit to the saltation layer, but most studies suggest that it is approximately ten times the mean saltation height. As observed by Bagnold (1941) and Sharp (1964), saltating grains are found at much greater heights over hard gravel or pebble surfaces. Williams (1964) and Gerety (1985) also observed a decline in average grain size with height. However, the size of sediment transported at a given height tends to increase with u_*.

Numerical Simulation

The dynamics of saltation, described qualitatively in the preceding sections, are represented in two separate numerical models. The first full simulation was designed by Ungar and Haff (1987) and refined by Anderson and Haff (1991), followed by the large-eddy scheme of Shao and Li (1999). Both models are complex and computationally demanding. The framework for each encompasses three subsystems and their linkages. The first subsystem addresses the atmospheric boundary-layer flow that provides the kinetic energy required to drive the mass transport phenomena. The second addresses the physics of the particle trajectories, and the third, simulates the partitioning of energy arising from particle impacts on the surface. Both numerical models construct feedback loops between all three subsystems so that the self-limiting nature of the aeolian transport phenomenon is correctly represented. That is, saltating particles extract momentum from the airflow as they rise away from the surface and partition this acquired momentum between rebound, ejection and creep in their return to the surface. The loss of fluid momentum limits the capacity of the flow to entrain and accelerate additional particles. The principal difference between the Anderson and Haff (1991) and Shao and Li (1999) schemes lies in the treatment of the boundary-layer flow. The earlier model is designed for mean flow conditions and describes steady state saltation. The model of Shao and Li (1999) is expanded to address the effects of turbulent flows, and produces a convincing simulation of the growth, overshoot and eventual equilibrium of the mass transport rate observed in empirical wind tunnel studies by Shao and Raupach (1992). Because of the complexity of both numerical simulations, their principal use to date has been targeted toward theory development and the development of new avenues for empirical investigation. Shao (2000) provides a comprehensive review of the dynamics and modelling of saltation, and so, further details are not repeated herein.

Wind Profile Adjustments to Saltation

Owen (1964) hypothesized that within the saltation layer the total shear stress (τ) is partitioned between the fluid (τ_a) and the moving particles (τ_p) as shown in Fig. 17.19. Above this layer, τ is constant and reflects the 'added' roughness associated with the saltation cloud. This partitioning of the vertical stress results in an upward curvature in the wind profile when shown on a semi-logarithmic plot (Fig. 17.20), as now confirmed by numerous wind tunnel experiments (e.g. McKenna-Neuman and Nickling 1994;

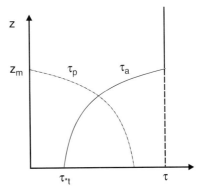

Fig. 17.19 Toward the bed surface, the total shear stress τ is partitioned between the grain borne stress τ_g and the air τ_a, as first hypothesized by Owen and represented in all subsequent models of saltation. Hypothetical vertical profiles of these stresses are illustrated here

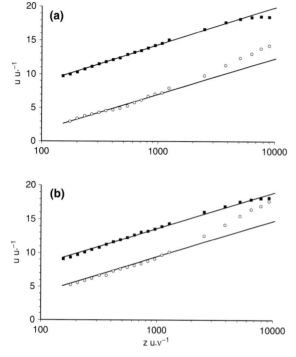

Fig. 17.20 Selected velocity profiles illustrate the variation in u/u∗ with dimensionless distance zu∗/v for stable surfaces (*solid squares*) as compared to mobile beds (*open circles*). Plots A and B correspond to sand particles of diameter 200 μm and 420 μm respectively for identical freestream velocities of 13 ms⁻¹. Solid lines indicate the log model least squares approximation to velocity data within the lower 15% of the boundary-layer depth. (After McKenna Neuman and Maljaars 1997)

Spies et al. 1995; McKenna Neuman and Maljaars Scott 1997; Bauer et al. 2004). Owen (1964) further suggested that for steady state saltation, the fluid drag on the surface (τ_{ao}) is reduced to a value just slightly exceeding the threshold for particle motion. In the field, the boundary layer can be very thick (e.g. 100s of meters) relative to the saltation cloud so that measurement of the total stress is relatively straightforward. In wind tunnel experiments, the depth of flow affected by vertical variation in τ_p can correspond to most, if not the whole, of the thin boundary layer. The numerical simulation scheme of Shao and Li (1999) does suggest that while wind profiles are modified by saltating particles, they are adjusted within seconds to a new equilibrium with near constant τ_a within a few centimetres of the bed surface. Experimentalists similarly have observed, in wind tunnel studies of saltation, that very near the bed surface the wind profile does remain logarithmic

so that τ_a attains a constant value near the threshold for particle motion (e.g. Spies et al. 1995; McKenna Neuman and Maljaars Scott 1997; Bauer et al. 2004). Unfortunately, to date, there is no consensus on a standard protocol for measurement of τ_a from wind profiles affected by saltation. No doubt, this has contributed to a lack of agreement between wind tunnel experiments carried out at varied facilities, and field versus simulation data. Although informal debate has taken place concerning the utility of u_*, no other variable has been identified that is more suitable and can be as conveniently measured.

Sand Transport Prediction

In comparison to the full numerical simulations, there exist a number of analytical sand transport equations for steady state saltation that are widely applied owing to their computational simplicity. A detailed summary of available transport equations has been compiled by Greeley and Iversen (1985) and a subset is presented in Table 17.3. Shao (2000) provides a thorough review of the physics and derivation of these models. To facilitate meaningful comparison, note that the mass transport rate is expressed in dimensionless form as $\frac{qg}{\rho u_*^3}$ in the table. All models share a common structure such that the magnitude of the horizontal mass transport rate q is determined primarily by the cube of the friction velocity (u_*), as represented in the left side of the dimensionless equation. The early models developed by Bagnold (1941) and Zingg (1953) suggest that this basic relation is modified by the particle diameter relative to a standard diameter for dune sand (D = 250 μm), as represented in the right side of the equation listed in Table 17.3. That is, for a constant shear velocity, higher mass transport rates are associated with larger particle sizes. Later models all express the effect of surface texture through inclusion of the threshold friction velocity (u_{*t}) required for particle entrainment, relative to u_* (e.g. Kawamura 1964; Owen 1964; Lettau and Lettau 1978; White 1979; Sørensen 1991, 2004). For time averaged flow conditions, q must equal 0 when $u_* < u_{*t}$, while u_{*t} is largely determined by the particle diameter (Equation 17.11). Each sand transport equation contains at least one parameter that must be determined empirically and encompasses numerous unspecified factors (e.g. particle size distribution

Table 17.3 Common mass transport rate expressions

Source	Dimensionless rate: $\frac{qg}{\rho u_*^3} =$
Bagnold (1941)	$C\sqrt{\dfrac{d}{D}}$ where C = 1.5 for uniform sand 1.8 naturally graded sand 2.8 poorly sorted sand 3.5 pebbly surface
Zingg (1953)	$C\sqrt[3]{\dfrac{d}{D}}, C = 0.83$
Owen (1964)	$\left(0.25 + 0.33\dfrac{u_{*_t}}{u_*} \cdot \dfrac{U_F}{u_{*_t}}\right)\left(1 - \left\{\dfrac{u_{*_t}}{u_*}\right\}^2\right)$
Lettau and Lettau (1978)	$C\left(1 - \left\{\dfrac{u_{*_t}}{u_*}\right\}\right), C = 4.2$
Kawamura (1951); White (1979) Independently derived with varied values for C	$C\left(1 + \dfrac{u_{*_t}}{u_*}\right)^2\left(1 - \dfrac{u_{*_t}}{u_*}\right)$ where C = 2.78 (Kawamura) 2.61 (White)
Sørensen (2004)	$\left(1 - \left\{\dfrac{u_*}{u_{*_t}}\right\}^{-2}\right)\left(\alpha + \beta\left\{\dfrac{u_*}{u_{*_t}}\right\}^{-2} + \gamma\left\{\dfrac{u_*}{u_{*_t}}\right\}^{-1}\right)$ where α, β and γ depend upon distributions of particle size and shape

and packing, turbulent fluctuations, inter-particle cohesion). Suggested values from the original model calibrations also are provided in Table 17.3.

It is widely recognized that the performance of these sand transport equations is highly variable from setting to setting (e.g. Sarre 1987). Even when compared for identical conditions, differences between the model outcomes can be large (Fig. 17.21). The lack of agreement arises in part because saltation is a stochastic process. The simplistic transport equations summarized in Table 17.3 describe the mean condition for which a steady, uniform flow of air continuously drives a horizontally homogenous cloud of sand. Researchers measuring saltation transport in the field and in wind tunnels well recognize that the transport phenomenon is far more complex both temporally and spatially.

For example, true sand drift is locally organized into streamers that mirror turbulent eddy structures formed in the lower boundary layer. Baas (2006), and Baas and Sherman (2005, 2006) have recently carried out detailed real-time measurements of these phenomena on beaches using a one-dimensional array of piezoelectric impact sensors oriented normal to the mean wind direction. Similarly, the majority of surfaces within arid regions are not smooth and flat, as assumed in idealized models and wind tunnel simulations, but rather contain non-erodible roughness elements varying in scale from pebbles to large shrubs.

These elements modify the partitioning of the total fluid stress, deflect and trap saltating particles, as well as initiate and alter the development of coherent flow structures in the atmospheric inertial sub-layer. A limited number of studies have addressed the empirical simulation and modelling of aeolian transport over such complex, partially sheltered surfaces (e.g. Lyles et al. 1974; Greeley et al. 1995; McKenna-Neuman and Nickling 1995; Nickling and McKenna Neuman 1995; McKenna Neuman 1998; Lancaster and Baas 1998 and Al-Awadhi and Willetts 1999). These include a recent field experiment carried out by Gillies et al. (2006b) in which the most sophisticated shear stress and particle transport measurements were carried out to this date for large arrays of buckets (60 m × 60 m × 60 m) set out on a desert surface in New Mexico, USA (Fig. 17.14).

As noted above, a stochastic model developed by Stout and Zobeck (1997) for practical determination of u_{*_t} gives further recognition to the fact that rarely is the mass transport rate continuous through time, but much more commonly, the phenomena is intermittent. During selected time intervals, transport will occur as expected for $u_* > u_{*_t}$, but also when $u_* < u_{*_t}$, and infrequently may not occur even though $u_* > u_{*_t}$. The degree of intermittency is inversely related to the wind strength. The transport phenomena becomes less intermittent for mean wind speeds well exceeding

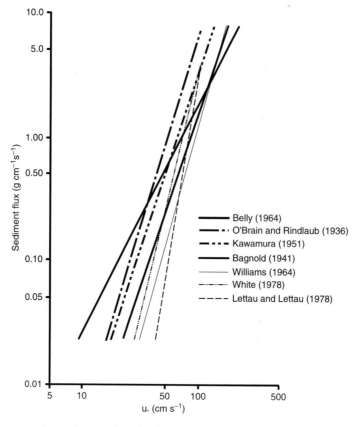

Fig. 17.21 Comparison among sediment flux equations for similar sand sizes (0.25 mm)

threshold, so that turbulent fluctuations in the momentum flux are unlikely to drop below that required to sustain particle motion.

In summary, while each of the analytical sand transport equations summarized in Table 17.3 is published with recommended values for the parameters it contains, these values are rarely useful in practice. In most applications, the model will need to be recalibrated for the setting in which it is to be used, particularly for circumstances where the underlying assumptions are not met. As a case in point, it is unlikely that steady-state transport is achieved in most wind tunnels. In selected facilities, an external particle feed is often used to initiate saltation at the entrance to the working section in order to hasten the attainment of steady-state saltation. Similarly, for beds of particles that are non-uniform in size, shape, and packing, it is well recognized in empirical studies that u_{*t} is effectively a range and not a discrete value (Nickling 1988; Wiggs et al. 2004). A great challenge exists

for the development of sand transport equations that retain structural simplicity while relaxing fundamental assumptions that constrain their relevance.

Sand Transport Measurement

The validation and calibration of analytical and numerical models of aeolian transport is made difficult by the large uncertainty associated with direct measurement of the mass transport rate. In general, aeolian sediment traps can be sub-divided into two broad groups depending on the orientation of the intake orifice, either horizontal or vertical (inclusive of point samplers and vertically integrating traps). Many trap designs exist (Fig. 17.22), as described and compared in a number of review papers (e.g. Shao et al. 1993; Nickling and McKenna-Neuman 1997; Al-Awadhi and Cermak 1998; Rasmussen and Mikkelsen 1998; Goossens et al. 2000).

Several key design criteria determine the overall accuracy and versatility. Ideally, the sampler should be isokinetic such that the instantaneous wind speed through the sampling orifice is equal to the ambient wind speed along the flow streamline immediately upwind. Distortion of the flow must be minimal such that there is no discrimination between small particles that tend to follow curved streamlines and larger particles that tend to cross streamlines because of their high inertia. The collected sample should have the same particle size distribution as the incident flow. Small bedforms, inclusive of ripples, must migrate unimpeded into the sampling orifice without evidence of scouring of the surface. The trap efficiency should not be affected by the shifting orientation of the wind flow. Finally, any given trap design ought to be relatively inexpensive, compact and portable. This allows the placement of a large number on a given desert surface so that adequate spatial resolution can be obtained without either flow interference or alteration of the aerodynamic roughness of the surface. In practice, no such ideal trap exists. Trap efficiencies are usually unknown for a given field application and can vary both temporally and spatially.

Recently electronic instrumentation has become available for the field measurement of saltation. Two of these instruments, the Sensit^{TM} (Fig. 17.22d) and the Safire (Sabatech Inc., Fig. 17.22e), use a piezoelectric crystal to measure grain impacts. When a sand grain moving in saltation strikes the crystal it generates an electrical pulse, the size of which is proportional to the grain size and particle velocity which is read by a data logger (typically 1–10 Hz), either as a pulse or analogue voltage signal. The saltiphone (Fig. 17.22f) is a similar instrument that uses a small sensitive microphone to measure grain impacts, which can also be read as pulse or analogue voltage output. All three of the instruments are omni-directional. However, the saltiphone sampling head is rotated into the wind by two vanes, whereas the Sensit^{TM} and Safire are fixed instruments that measure saltation impacts on a circular piezoelectric crystal mounted on the instrument body. The instruments have great utility but do require careful calibration and maintenance because of deterioration of the piezoelectric crystals or microphone with time, and the observed directional dependence of the output signal in the case of the Sensit^{TM} and Safire sensors (Baas 2004; Van Pelt et al. 2006).

Bonding Agents

As described in earlier sections of this chapter, fundamental physical and numerical models of wind erosion are founded on the concept of transport limitation; that is, the mass moved in unit time is self-regulated by the associated partitioning of a portion of the total fluid stress to that borne by the particles. Such conditions can be attained in wind tunnel studies, but are rarely met in nature. The transport limited case represents an upper limit for q while assuming an unrestricted supply of particles for the airflow to carry. A great deal of work has been carried out over the last decade concerning surface constraint on particle emissions associated with inter-particle forces. The following sections address the most important bonding agents associated with particle supply limitation: soil moisture, silt and clay, organic matter, and precipitated soluble salts. High friction velocities are required for the entrainment and transport of either very damp or aggregated particles, though in the extreme case of continuous surface crusts, entrainment usually only arises through rupture from the impact of loose particles. An emphasis is placed upon reviewing the physical processes, experimental work, modelling approaches, and future research needs. Many authors have provided extensive reviews of the regional to global distribution and significance of these factors (e.g. Rice and McEwan 2001; Livingstone and Warren 1996).

Moist Particles

Field observations and wind tunnel studies have shown that surface moisture content is an extremely important variable controlling both the entrainment and flux of sediment by the wind. Experiments by Belly (1964) show that gravimetric moisture contents (w_c) of approximately 0.6% can more than double the threshold velocity of medium-sized sands. Above approximately 5% gravimetric moisture content, sand-sized material is inherently resistant to entrainment by most natural winds. Wind tunnel experiments on agricultural soils showed an exponential relationship between the increase in moisture content and threshold velocity (Azizov 1977).

The physical mechanisms underlying the effect of moisture on inter-particle cohesion have received a

Fig. 17.22 Various sediment traps and electronic sensors for the measurement of saltation (**a**) Bagnold type trap modified to orient into the wind (Nickling 1978), (**b**) wedge type trap that automatically faces into the wind and continuously weighs sediment (After Gillies et al. 2006b) (**c**) self aligning BSNE type traps (Fryrear 1986), (**d**) Sensit^{TM} (**e**) Safire and (**f**) Saltiphone (Cornelis et al. 2004b)

great deal of attention in the soil physics literature and now are reasonably well understood. The total potential energy of soil water is described as the matric potential (Ψ_m) and results from both capillary and adsorptive forces. Under arid conditions, the gravitational potential can be ignored, and the pressure potential is negative relative to atmospheric. Thus, Ψ_m represents the work per unit volume that must be carried out by externally applied forces to transfer water vapour *reversibly* and *isothermally* within the soil voids to the liquid phase. The relation of the matric potential to the humidity of the air within the soil voids is modelled by the Kelvin equation:

$$\psi_m = \frac{R}{V_w} T \ln\left(\frac{e}{e_s}\right) \qquad (17.19)$$

where R is the ideal gas constant ($8.314 \, \text{mol}^{-1} \, \text{K}^{-1}$), T is temperature (K), Vw is the molar volume of water ($1.8 \times 10^{-5} \, \text{m}^3 \, \text{mol}^{-1}$), and e/e_s represents the relative humidity (RH), giving Ψ_m in Pa. The matric potential can be reduced by either increasing relative humidity or reducing air temperature. The Kelvin equation is the basis of the soil psychrometer method for the determination of the soil matric potential, as well as the method of equilibration over salt solutions for sorption curve determination at large values of Ψ_m. The highly structured water adjacent to particle surfaces has a depressed freezing point temperature, so that this model may be applied at $T < 0°C$, though the degree of supercooling is difficult to determine.

When a completely dry soil is exposed to an atmosphere containing water vapour at $35\% < RH < 40\%$, a single layer of water is adsorbed onto the charged faces of mineral surfaces by hydrogen bonding. This water is held in place by very strong forces and has a thickness of approximately 0.3 nm. As the RH increases, additional water is adsorbed in an ice-like structure, with the second layer completed at about 60% RH. These estimates are based upon data from clean, smooth quartz plates, and may well vary with the surface roughness of natural soil particles. Considering only van der Waals forces, the simplest expression relating the adsorbed water film thickness (δ) to the matric potential is given as follows

$$\delta = A_h/6\pi\Psi_m \qquad (17.20)$$

where A_h is the Hamaker constant approximated as -1.9×10^{-19} J (Iwamatsu and Horii 1996; Tuller and Or 2005). For any two sedimentary particles that share a thin film of adsorbed water where they rest against one another, it is reasonable to assume as a first approximation, that the contact area (A_c) increases with the film thickness (δ), and that the inter-particle cohesion developed will vary as $\Psi_m A_c$.

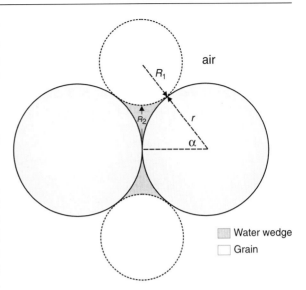

Fig. 17.23 Free body diagram illustrating the geometry of a water lens at the point of contact between two spheres. The radii of the contact area (R_2) and the meniscus (R_1) determine the magnitude of the capillary force that draws the particles together

For larger amounts of pore water that accumulate in the form of stable lenses at the interparticle contacts, the magnitude of cohesion can be estimated directly from an explicit model for an ideal soil consisting of spheres of equal diameter (Fig. 17.23), as developed by Haines (1925) and modified by Fisher (1926):

$$F_c = 2\pi R_2 \gamma_s + \pi R_2^2 \gamma_s \left(\frac{1}{R_1} - \frac{1}{R_2}\right) \qquad (17.21)$$

where R_1 represents the radius of curvature of the air-water interface, and R_2 is the radius of the contact area between the water lens and the particle. The surface tension of water is represented by the constant γ_s, while the pressure deficiency in the water wedge relative to the atmosphere is given as $\gamma_s (1/R_1 - 1/R_2)$. As the gravimetric water content in medium sand rises above approximately 1% (McKenna Neuman and Nickling 1989), the contact area (πR_2^2) expands much more rapidly than the pressure deficit drops (i.e. R_1 and R_2 become large). As a result, the inter-particle cohesion becomes so large that entrainment by most natural winds, exclusive of extreme phenomena, is impossible. The range of application of the Haines-Fisher model for spheres varies from water contents just below the point at which the discrete lenses begin to coalesce, down to very low values at which the lenses become unstable (Allberry 1950).

Application of the Haines-Fisher model is impractical because of the complex contact geometry found within natural soils. A number of researchers have attempted to simplify parameterization of the model so that it may be applied in field and wind tunnel settings (McKenna Neuman and Nickling 1989; McKenna-Neuman and Maljaars Scott 1998; Cornelis et al. 2004a,b; Ravi et al. 2006). For example, McKenna Neuman and Nickling (1989) assume a conic geometry for the inter-particle contacts giving

$$F_c = G \frac{\pi \gamma_s^2}{\psi_m} \qquad (17.22)$$

where G is a dimensionless geometric coefficient describing the shape of the contacts between grains with empirically determined values between 0.2–0.4. Introduction of a capillary force moment into Bagnold's (1941) entrainment model provides the following general expression for the threshold velocity of a moist surface:

$$u_{*tw} = A[(\rho_p - \rho)/\rho]^{1/2}\{(6 \sin 2\alpha)/$$
$$[\pi d^3 (\rho_p - \rho)g \sin \alpha]F_c + 1\}^{1/2} \qquad (17.23)$$

where α is the angle of internal friction and d is the grain diameter. Results from wind tunnel tests compare well with those predicted by Equation 17.23 (Fig. 17.24); however, the model underestimates u_{*t} at high matric potentials approaching those associated with adsorbed water.

In parallel with this theoretical work, many researchers have carried out direct empirical investigations of the role of capillary water in the deflation of sedimentary surfaces in both field and laboratory settings (e.g. Chepil 1956; Bisal and Hsieh 1966; Smalley 1970; Azizov 1977; Logie 1982; Hotta et al. 1984; Brazel et al. 1986; Jackson and Nordstrom 1997; McKenna Neuman and Maljaars Scott 1998; Fécan et al. 1999; Yang and Davidson-Arnott 2005; Wiggs et al. 2004). From this work, it is evident that water content is one of the most formidable particle emission controls to measure accurately, and to represent in models that are appropriately distributed in space and time. This is especially true for beach settings, as concluded in numerous field studies (e.g. Arens 1996; Hotta et al. 1984; Sherman et al. 1998; Svasek and Terwindt 1974; Wiggs et al. 2004; Yang and Davidson-Arnott 2003.). At the interface between the mineral particles and the airflow in the boundary layer above, water is drawn up through the capillary fringe in response to evaporation, and specifically, the increased tension at which this water is held at the surface (Durar et al. 1995). The amount of moisture under consideration in the topmost few millimeters of sand is frequently small, less than about 5% by weight. It can change rapidly through time, over the course of a few minutes to hours, and through distance, over a few centimetres to metres.

There are considerable methodological challenges associated with measurement of the total force that binds particles together at the very instant that they are entrained by the wind. From slabs of sediment extracted from the surface that well exceed $10^3 \, \mu$m in thickness, it is likely that the gravimetric water content of the topmost particles entrained (diameter $\sim 10^2 \, \mu$m) is grossly overestimated, especially where the vertical vapour pressure gradient is large. Depending upon the sampling procedures, the error in measurement can be large in comparison to the mean determined. Because of this, model validation and calibration has not been resolved from empirical work, and there remains a great deal of uncertainty regarding the relative importance of adsorbed and pendular water as controls on particle emission. For example, theoretical models of entrainment and wind tunnel simulations both suggest that particles cannot be entrained by fluid drag alone at gravimetric water contents (w_c) exceeding 1–2%. Whereas in the field, direct measurements of w_c at threshold have been observed to reach 5% and more (e.g. Wiggs et al. 2004).

Empirical studies in a very wide range of settings all demonstrate that for a given wind speed, the mass transport rate (q) for damp surfaces can vary by several orders of magnitude. Sherman et al. (1998) compared measurements of q on a beach at Kerry, Ireland with predictions from five transport models adjusted for slope and mean gravimetric water content. None performed satisfactorily. The authors report that 'The influence of sediment moisture content appeared to be the critical factor in degrading model viability' (p. 113). Despite a high degree of control in wind tunnel simulation, McKenna Neuman and Langston (2006) similarly observed order of magnitude variation in the mass transport rate over damp surfaces at constant flow velocity. Incomplete characterization of the spatial distribution of moisture in the surface, organization of the saltation cloud into streamers, and fetch constraint all contributed to the uncertainty in q observed for this work. To

Fig. 17.24 Comparison of theoretical and observed threshold velocities with increases in soil moisture tension (decrease in gravimetric moisture content). Theoretical curves for open and close packing arrangements computed using Equation (17.13) (After McKenna-Neuman and Nickling 1989)

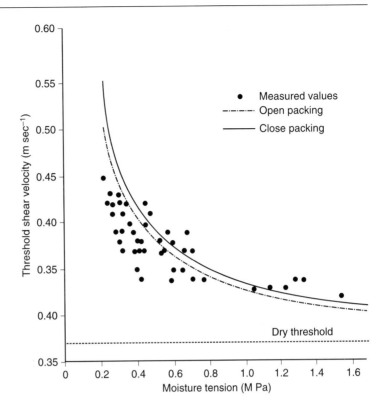

complicate matters further, McKenna Neuman and Maljaars Scott (1998) suggest that the number and speed of ejected grains is not necessarily reduced by the presence of moisture. In wind tunnel experiments involving coarse grains and high wind velocities, the authors observed that q increased on damp surfaces, presumably because the particle collisions were more elastic than on a dry, cohesionless surface.

Before further progress can be made in the understanding and modelling of this control on the particle supply to the airflow, it is imperative that researchers continue to make significant improvements in its measurement. Many researchers have commented on the drawbacks of existing approaches. Though highly accurate, manual measurement of w_c is labour only intensive and provides information only for a point that is fixed in both space and time. The neutron probe provides continuous measurements, but has insufficient accuracy ($+/-12\%$) and averages water content over an unrealistically large volume for application to aeolian transport. Most recently, Atherton et al. (2001) and Yang and Davidson-Arnott (2003) have obtained very good temporal resolution using ThetaProbe technology. These instruments are based

on the impedance of a sensing rod array that varies with the moisture content of the sediment in which it is inserted. The measurements are rapid (5–10 s) and the sensing volume is small at 35 cm³. Still, these devices require careful calibration and they are subject to error as the surface either deflates or accretes. McKenna Neuman and Langston (2003, 2006) report on a simple remote sensing technique in which the surface brightness is sampled as a surrogate for the distribution of pore water content. The technique is based on the principal that light penetrates deeper into wet surfaces, which then appear dark, as a result of the change in the refractive index at the particle/water interface as compared to the solid/air interface. A considerable amount of parallel work has been undertaken recently to develop a physical model of this phenomena in porous materials (Banninger et al. 2005) to measure light penetration as affected by texture and wetness (Ciani et al. 2005), and to develop fibre optic technology suitable for such applications (Garrido et al. 1999). The remote sensing approach provides information about the spatial distribution of the moisture phenomena, and when monitoring is continuous, indicates temporal variation when ambient light conditions are suitable.

Aggregated Particles and Surface Crusts

In many arid regions, discrete sedimentary particles aggregate into erosion-resistant structural units that persist over many timescales, and ultimately, may form a continuous sheet known as a surface crust. Crusting is a key factor in reducing soil erosion in arid and semi-arid regions, and is currently being investigated as a means of stabilizing fallow fields (Rajot et al. 2003). Even weak crusts (modulus of rupture <0.07 mPa) significantly increase the threshold velocity for wind erosion, as reported by Gillette et al. (1980, 1982) for surfaces in the Mojave Desert. Unfortunately, disturbance of crusted surfaces by off-road vehicles and cattle trampling now is recognized as a growing problem in many arid regions of the world that contributes widely to dust emission (Nickling and Gillies 1989).

The soil constituents that contribute to particle aggregation and crusting are highly variable in nature, found under a wide range of environmental conditions, and can include any or all of the following: silts and clay minerals (physical), organic materials (biological), and precipitated soluble salts (chemical).

The effectiveness of physical bonding agents depends on their relative proportion in relation to the quantity of sand-sized material. For example, soils with 20–30% clay, 40–50% silt, and 20–40% sand are least affected by abrasion (Chepil and Woodruff 1963). Physical crusts can be formed by raindrop impact on bare soils (Chen et al. 1980) so that the fine clay particles cement the larger particles together as the soil dries (Rajot et al. 2003). In dryland soils, the clay minerals often originate from dust precipitation as compared to rock weathering in more temperate settings.

Even low concentrations of soluble salts can significantly increase threshold velocity by the formation of cement-like bonds between individual particles (Nickling 1978, 1984; Nickling and Ecclestone 1981). Sodium chloride is more effective than magnesium and calcium chloride in reducing soil movement because sodium tends to produce a surface crust that protects the underlying soils (Lyles and Schrandt 1972). Salt crusts form when dissolved salts infiltrate the soil surface and re-crystallize as the moisture is evaporated (Nickling 1978; Pye 1980; Gillette et al. 1980, 1982). This crust type is particularly prevalent on playas underlain by shallow reserves of groundwater.

The presence of organic matter increases the ability of particles to form aggregates that are less susceptible to entrainment by wind than the individual grains (Chepil 1951), but very high organic contents in dry soils increase erosion potential because of the very loose soil structure.

Of the three crust types, biological crusts are perhaps the most complex. Although biotic crusts are identified in all parts of the world inclusive of semi-arid and arid regions (Isichei 1990; West 1990; Johansen 1993), they are especially prevalent in humid coastal settings (van den Anker et al. 1985; Pluis and de Winder 1989; Pluis and van Boxel 1993). These crusts owe their existence to two binding mechanisms: (i) extracellular excretions, such as mucilage and polysaccharides, that act as cementing agents, and (ii) filamentous growths that blanket and entangle loose particles. Growth of the crust is strongly tied to moisture availability in the environment. In sandy soils which have little fine material to bind the larger particles together, biotic crusts are particularly important in diminishing the potential for soil erosion (Leys and Eldridge 1998). From field experiments with a portable wind tunnel, Belnap and Gillette (1997, 1998) have shown that following a past disturbance, the wind velocity required for the onset of erosion is entirely dependent upon the stage of biotic crust development.

A growing body of work has addressed the physical characteristics of crusts in relation to the threshold for entrainment, their response to particle impact, and the mechanisms by which they disintegrate. Crust strength varies with the composition and distribution of the binding media, and therefore, is more often than not spatially heterogeneous. This heterogeneity is important since wind-borne particles that strike the surface will rupture weaker areas of the crust, provided the impact force exceeds the binding force. This abrasion, in turn, liberates greater numbers of saltating particles and creates further ruptures downwind (McKenna Neuman et al. 1996, 2005; McKenna Neuman and Maxwell 2002). The emission of dust from crusted playa surfaces also has been shown to be dependent upon the abrasive action of saltating sand grains that set loose the fine fraction (Cahill et al. 1996; Houser and Nickling 2001).

Rice (1996, 1997) propose two analytical models describing crust rupture by particle impact. The first model addresses the probability distributions of (i) the kinetic energy associated with grain impact $P[E_i]$, and (ii) the local energy required to break the surface crust $P[E_s]$ (Fig. 17.25). Assuming that the kinetic energy

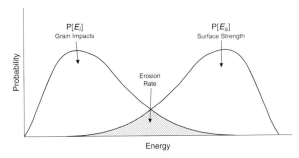

Fig. 17.25 Model of crust rupture associated with the overlap of the distribution of kinetic energy of impacting particles with the distribution of binding energies for the particles bound together in the crust. (After Rice et al. 1999)

of any given particle impact must exceed the total penetration energy required for crust rupture, the rate of erosion is determined by the degree of overlap of the tails of these two distributions. In theory, rupture occurs primarily at the weakest points on the surface under the most energetic impacts. In the model developed by Rice and her co-workers, the impacting particle must penetrate the surface to the depth at which the maximum penetration force (F_{max}) is measured. In the example of a 300 μm diameter particle impacting a tension wetted soil (Rice et al. 1997), this depth is in the order of 2–6 grain diameters.

The second model is based on a comparison between (i) F_{max} and (ii) the force (F_i) delivered to the crust by an impacting grain of mass (m) with a component velocity (v). Local surface failure and therefore erosion will occur where $F_i > F_{max}$. Rice et al. (1997) show that the impact force may be estimated from:

$$F_i = v\sqrt{\frac{m M_e}{2}} \qquad (17.24)$$

where M_e is the modulus of deformation. In practice, M_e can be determined from pin penetrometer data as $F_{max}/\delta z$, given δz is the amount of surface deformation (i.e. depth of penetration) at F_{max}. Equation 17.24 implies that a more rigid surface will experience a greater impact force. Therefore, a strong but stiff surface may be no less susceptible to erosion than one which is weak but pliable.

Clearly, both models depend upon detailed knowledge of the stress-strain relations for a given crust, and particularly the determination of F_{max}. For the earliest measurements of crust strength, a load was applied incrementally to a thin crust sample causing flexure,

similar to well established tests developed for concrete beams (Fig. 17.26a). Such tests were first carried out by Richards (1953), then later by Gillette et al. (1982) for salt crusts and by McKenna Neuman and Maxwell (1999) for biotic crusts. This method best represents the prying up and tearing away of large flakes of crust only a few millimetres in thickness. It emulates a very advanced stage of destruction whereupon localized areas of crust are fully penetrated by impacting grains so that undercutting of the remaining, intact crust occurs.

Coinciding with the development of an idealized model of crust breakdown under particle impact, pin penetrometry (Fig. 17.26b) became the standard method for measuring F_{max} (e.g. Rice et al. 1996b, 1997, 1999; Rice and McEwan 2001). The technique entails measuring the load applied by a pin as it penetrates a given distance into a crust. It is considered more representative of grain-scale ruptures than bending a crust sample. The probability density function of crust strength is based on the maximum load measured at numerous locations on the surface. The modulus of elasticity is obtained from the slope of the linear stress-strain relation prior to rupture. E_s is computed as the sum of the cross product of the penetration depth increment and each penetration pressure up to the maximum value (Rice et al. 1997).

In a test of their conceptual model of crust degradation under particle impact (Fig. 17.25) Rice and McEwan (2001) investigated the strength properties of four aggregate crusts containing 12%, 24%, 36%, and 48% loam soil mixed with sand. In each case, the distribution of penetration energy overlapped with that of the estimated kinetic energy of an impacting particle, suggesting that crust rupture should occur. Comparison of the cumulative mass loss for each crust in wind tunnel studies also demonstrated that weak crusts deteriorate more easily than strong crusts. Variations of the penetrometer method have since been adapted in several studies including a recently designed field pin penetrometer (Fig. 17.26c,d) (e.g. Houser and Nickling 2001; McKenna Neuman and Maxwell 2002).

Unfortunately, wind tunnel experiments by McKenna Neuman and Maxwell (2002) and McKenna Neuman et al. (2005) demonstrate that deflation of biotic crusts also occurs over an extended period of time when the probability distributions do not overlap significantly. While the penetrometer method of strength testing usually establishes a well-defined maximum,

Fig. 17.26 Instrumentation used for the measurement of soil strength (**a**) laboratory beam balance. The sample is placed on two sharp edges separated by a known distance. Force is applied by a thin blade to the centre of the sample briquette by a screw feed. The load is measured by a balance on which the apparatus sits (**b**) laboratory pin penetrometer. A pin (approximately 1 mm in diameter) is driven into the soil sample at a known rate. The load is recorded by a balance on which the sample rests (**c**) portable field penetrometer for in situ field testing and (**d**) drive mechanism and load cell that records (10 Hz) the force on the pin as it enters the soil

herein termed the ultimate crust strength, each penetration typically contains localized maxima and minima in the stress-strain curve that make interpretation difficult (Fig. 17.27). Work by granular physicists (e.g. Geng et al. 2001) shows that beneath the point of load application, stress is not distributed uniformly among supporting particles, but is concentrated into chains that branch out in tree-like formations that are dependent upon the packing arrangement (Fig. 17.28). The stress magnitude decreases both horizontally and vertically away from the point of load application (Geng et al. 2001). Micro-fissures in the binding media between particles are most likely to develop along these chaotic stress chains. As the penetrometer pin pushes deeper into the crust, experiments from granular physics with photoelastic particles (Fig. 17.28) suggest that the stress chains will continually change in pattern and density. Localized areas within the bed that are subjected to a high density of intersecting stress chains, as integrated over time, should eventually become weak as compared to others area of low density. In this regard, the ultimate strength measured is not strictly independent of the method of observation. The vertical penetration record (Fig. 17.27) might then be viewed as a suitable analogue of the effect of accumulated instantaneous stresses applied to a bed surface via saltating grains, especially in the context of simultaneous impacts. This interpretation falls in line with earlier suggestions that repeated impacts on a crust surface weakens the inter-particle bonds, analogous to the fatigue that occurs in metals (McKenna Neuman and Maxwell 2002; McKenna Neuman et al. 2005; Langston and McKenna Neuman 2005), pavement, and concrete under repeated strain.

Several other potential deficiencies of the penetrometer method are identified by McKenna Neuman and Rice (2002). While saltating particles generally impact at angles between 10 and 30°, application of the load in penetrometry is normal to the surface. The slow rate of penetration also is a poor simulation of a particle impact. Still, the method does provide a useful measure of relative crust strength that usually correlates well with rankings of crust stability obtained from wind tunnel studies. McKenna Neuman and Rice further report that the speed of penetration does not appear to be as important as perceived. In a series of laboratory experiments, separate measurements were made at a large number of puncture sites of the total energy required to penetrate a relatively thick (8 mm) aggregate

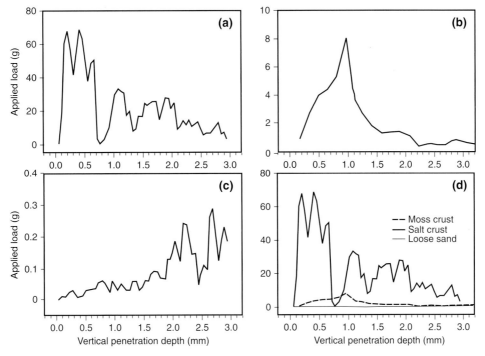

Fig. 17.27 Comparison of sample stress-strain curves for a (**a**) salt crust, (**b**) biotic crust and (**c**) loose sand as measured by a pin penetrometer (**d**) compares all three. (After Langston and McKenna Neuman 2005)

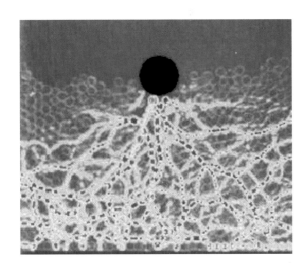

Fig. 17.28 In granular materials, force is transmitted along a network forming 'stress chains'. These chains can exist immediate to regions where there is little or no force. The particles in this image are made of a photoelastic material. When viewed through polarizers, high forces show up as bright regions as compared to regions of no force that remain dark. (Image provided by Behringer's Nonlinear Flow Group, Duke University, Department of Physics.)

crust formed from bentonite (12%) mixed with sand (88%). The rate of advancement of the pin was varied by as much as one order of magnitude from one puncture site to another. No systematic difference was found between the distributions of penetration energy for each of five, pre-selected penetration rates (37, 167, 299, and 437 μm s^{-1}). Figure 17.29 shows the overlap in the distributions for the slowest and fastest rates of penetration.

Further work is needed to compare the strength properties of various crust types (physical, biological and chemical) and to provide insight into the importance of crust strength versus elasticity. Langston and McKenna Neuman (2005) suggest that the elasticity of the crust appears to be at least as important as the ultimate strength in determining relative resistance to abrasion. Although much stronger, the salt crusts were found to break down and erode sooner than biotic crusts in wind tunnel studies. The notable elasticity of biotic crusts appears to afford them some protection against abrasion. The particle capture efficiency of rough crust surfaces has not been explored as yet in rigorous experiments.

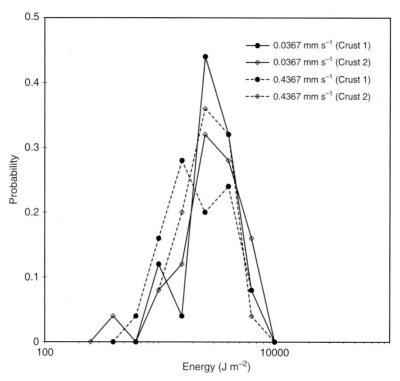

Fig. 17.29 An order of magnitude difference in the penetration rate appears to have no substantive effect upon the distribution of energy (E_s) required for crust rupture. (After Rice and McKenna Neuman 2002)

Conclusions and Future Research Needs

Although the past decade has brought significant advances in our understanding of the complex processes of sediment transport by the wind, there remain many challenging problems for the future. In order to reduce the conceptual and experimental complexities to manageable proportions, most theoretical and empirical research to date has focused on very simple systems in which sediments are assumed to be composed of a single grain size, and the wind is assumed to be blowing with a steady velocity. Important questions, such as the formation of small-scale stratification types (e.g. wind ripple laminae), and the ejection of dust from desert surfaces, require knowledge about grain behaviour in mixed grain size beds. This will include studies of saltation with a wide range of grain trajectories and consideration of the development of the bed in space and time (e.g. armouring and the development of wind ripples). It is clear that the surface wind is not steady in time and space and that saltation responds rapidly to changes in wind velocity. When coupled with the strongly non-linear relationship between sediment flux

and wind shear velocity, calculation of long-term transport rates requires much more detailed information on the high frequency nature of wind velocity fluctuations. These problems present important challenges to the aeolian research community. Whereas increased computing power will enable more realistic simulations of aeolian processes, there are extreme difficulties to be overcome in the experimental and field techniques required to verify and parameterize these models. As Anderson et al. (1991) note, theory has caught up with wind tunnel and field experimental techniques in many areas. More cooperation between theoretical and experimental approaches is clearly indicated for the next generation of aeolian transport studies.

Notation

A	threshold coefficient
A_c	surface contact area
A_h	Hamaker constant
a	empirical constant

b	roughness element breadth	w	crosswise component of wind velocity
C	coefficient (mass transport equations)	w'	instantaneous fluctuation from mean crosswise wind velocity
D	reference grain diameter (250 μm)	w_C	gravimetric water content
d	grain diameter	z	height above the surface
d	zero plane displacement height	z_0	aerodynamic roughness length
F_C	capillary force	α	angle of internal friction
F_i	impact force	κ	von Karman's constant (≈ 0.4)
F_{max}	maximum penetration force	μ	dynamic viscosity of air
F_R	wind force on roughness element(s)	ρ	air density
F_S	wind force on ground surface	ρ_p	particle density
G	vertical particle flux	σ	standard deviation, or ratio of roughness element basal area to frontal area
g	gravitational acceleration		
h	roughness element height		
l	saltation path length	τ	shear stress
M_e	Modulus of deformation	τ_R	shear stress on the roughness elements
n	number of roughness elements	τ_S	shear stress on the underlying surface in the absence of roughness elements
$P[E_i]$	kinetic energy of grain impact		
$P[E_s]$	local energy required to rupture the surface crust	τ'_S	shear stress acting on the exposed intervening surface
P_s	probability of creating a splash	τ''_S	maximum stress acting on any point on the surface
q	sand transport rate (mass flux)		
R	soil ridge roughness factor	τ_p	grain borne shear stress
R_1	radius of curvature of the air-water interface	λ	roughness density
		Ψ_m	matric potential
R_2	radius of the contact area between the water lens and the particle	γ_s	surface tension of water
Re_p	particle friction Reynolds number	β	ratio of drag coefficient for an individual roughness element to the drag coefficient of an unobstructed surface (CR/CS)
	RH Relative Humidity		
	RS Reynolds stress		
R_t	threshold ratio (unsheltered/sheltered)	δ	adsorbed water film thickness
S	ground area	δ_z	distance of surface deformation
s	relative wind speed index		
T	temperature		
U_f	particle settling velocity		
u_t	threshold velocity for sand transport		
u	horizontal component of wind velocity		
u'	instantaneous fluctuation from mean horizontal wind velocity		
u_*	wind shear velocity or friction speed		
u_{*t}	threshold shear velocity		
u_{*ts}	for a bare, erodible surface		
u_{*tr}	for a sheltered surface		
u_{*tw}	threshold shear velocity for moist surface		
V	velocity		
v	vertical component of wind velocity		
v'	instantaneous fluctuation from vertical wind velocity		
V_w	molar volume of water		

References

Aaby, B. 1997. Mineral dust and pollen as tracers of agricultural activities. *PACT (Journal of the European Study Group on Physical, Chemical and Mathematical Techniques Applied to Archaeology)* **52**, 115–122.

Aagaard, T., R. Davidson-Arnott, B. Greenwood and J. Nielsen. 2004. Sediment supply from shoreface to dunes: linking sediment transport measurements and long-term morphological evolution. *Geomorphology* **60**(1–2), 205–224.

Abdal, M.S. and M.K. Suleiman. 2002. Soil conservation as a concept to improve Kuwait environment. *Archives of Nature Conservation and Landscape Research (Archiv für Naturschutz und Landschaftsforschung)* **41**(3–4), 125–131.

Al-Awadhi, J.M. and J.E. Cermak. 1998. Sand traps for field measurement of aeolian sand drift rate in the Kuwaiti desert.

In: S.A. Omer, R. Misak and D. Al-Ajmi (eds.), *Sustainable Development in Arid Zones*. Rotterdam, Balkema, **1**, pp. 177–188.

Al-Awadhi, J.M. and B.B. Willetts. 1999. Sand transport and deposition within arrays of nonerodible cylindrical elements. *Earth Surface Processes and Landforms* **24**, 423–435.

Allberry, E.C. 1950. On the capillary forces in an idealized soil. *Journal of Agricultural Science* **40**, 134–142.

Anderson, R.S. and B. Hallett. 1986. Sediment transport by wind: toward a general model. *Bulletin of the Geological Society of America* **97**, 523–535.

Anderson, R.S. and P.K. Haff. 1991. Wind modification and bed response during saltation of sand in air. *Acta Mechanica Supplementum* **1**, 21–52.

Anderson, R.S., M. Sørensen and B.B. Willetts. 1991. A review of recent progress in our understanding of aeolian sediment transport. *Acta Mechanica Supplementum* **1**, 1–19.

Anderson, J.L. and I.J. Walker. 2006. Airflow and sand transport variations within a backshoreparabolic dune plain complex: NE Graham Island, British Columbia, Canada. *Geomorphology* **77**(1–2), 17–34.

Arens, S.M. 1996. Rates of aeolian transport on a beach in a temperate humid climate. *Geomorphology* **17**, pp. 3–18.

Atherton, R.J., A.J. Baird and G.F.S. Wiggs. 2001. Inter-tidal dynamics of surface moisture content on a meso-tidal beach. *Journal of Coastal Research* **17**, 482–489.

Azizov, A. 1977. Influence of soil moisture on the resistance of soil to wind erosion. *Soviet Soil Science* **9**, 105–108.

Baas, A.C.W. 2004. Evaluation of saltation flux impact responders (Safires) for measuring instantaneous aeolian sand transport intensity. *Geomorphology* **17**, 482–489.

Baas, A.C.W. 2006. Challenges in aeolian geomorphology: investigating aeolian streamers. Accepted for publication in Geomorphology (special issue).

Baas, A.C.W. and D.J. Sherman. 2005. The formation and behavior of aeolian streamers. *Journal of Geophysical Research* **110**, F03011, doi:10.1029/2004JF000270.

Baas, A.C.W. and D.J. Sherman. 2006. Spatio-temporal variability of aeolian sand transport in a coastal environment. *Journal of Coastal Research* **22**, 1198–1205.

Bryant, J. 2004. Wind flow characteristics over rough surfaces, M.Sc. Thesis, University of Guelph, Guelph, Ontario, Canada.

Bagnold, R.A. 1941. *The physics of blown sand and desert dunes*. London: Chapman & Hall.

Banninger, D., P. Lehmann, H. Fluhler and J. Tolke. 2005. Effect of water saturation on radiative transfer. *Vadose Zone* **4**, 1152–1160.

Bauer, B.O. and S.L. Namikas. 1998. Design and field test of a continuously weighing, tippingbucket assembly for aeolian sand traps. *Earth Surface Processes and Landforms* **23**(13), 1171–1183.

Bauer, B.O., C.A. Houser and W.G. Nickling. 2004. Analysis of velocity profile measurements from wind-tunnel experiments with saltation. *Geomorphology* **59**(1–4), 81–98.

Belly, Y. 1964. *Sand movement by wind*. U.S. Army Corps of Engineers, Coastal Engineering Research Center, Technical Memo 1, Addendum III, 24pp.

Belnap, J. and D.A. Gillette. 1997. Disturbance of biological soil crusts: Impacts on potential wind erodibility of sandy desert

soils in southeastern Utah. *Land Degradation and Development* **8**(4), 355–362

Belnap, J. and D.A. Gillette. 1998. Vulnerability of desert biological crusts to wind erosion: the influences of crust development, soil texture, and disturbance. *Journal of Arid Environments* **39**, 133–142.

Bisal, F. and J. Hsieh. 1966. Influence of moisture on erodibility of soil by wind. *Soil Science* **3**, 143–146.

Bradley, E.F. 1969a. A micrometeorological study of velocity profiles and surface drag in the region modified by a change in surface roughness. *Quarterly journal of the Royal Meteorological Society* **96**, 361–369.

Bradley, E.F. 1969b. A shearing stress meter for micrometeorological studies. *Quarterly Journal of the royal Meteorological Society* **94**, 380–387.

Brazel, A.J., W.G. Nickling and J. Lee. 1986. Effect of antecedent moisture conditions on dust storm generation in Arizona. In W.G. Nickling (ed.), *Aeolian Geomorphology*. Proceedings of the 17th Annual Binghamton Symposium, Allen & Unwin, pp. 261–271.

Brown, S., W.G. Nickling, and J.A. Gillies. 2008. A wind tunnel examination of shear stress partitioning for an assortment of surface roughness distributions, *Journal of Geophyscal Research* **113**, F02S06, doi:10.1029/2007JF000790.

Butterfield, G.R. 1991. Grain transport rates in steady and unsteady turbulent airflows. *Acta Mechanica Supplementum* **1**, 97–122.

Butterfield, G.R. 1993. Sand transport response to fluctuating wind velocity. In: Clifford, N.J., French, J.R. and Hardisty, J. (eds.), *Turbulence: Perspectives on Flow and Sediment Transport*. New York, John Wiley and Sons, pp. 305–335.

Butterfield, G.R. 1999. Near-bed mass flux profiles in aeolian sand transport: high-resolution measurements in a wind tunnel. *Earth Surface Processes and Landforms* **24**(5), 393–412.

Cahill, T.A., T.E. Gill, J.S. Reid, E.A. Gearhart and D.A. Gillette. 1996. Saltating particles, playa crusts, and dust aerosols at Owens (Dry) Lake, California. *Earth Surface Processes and Landforms* **21**(7), 621–640.

Chen, Y., J. Tarchitzky, J. Brouwer, J. Morin and A. Banin. 1980. Scanning electron microscope observations on soil crusts and their formation. *Soil Science* **130**, 49–55.

Chepil, W.S. 1945a. Dynamics of wind erosion: I. Nature of movement of soil by wind. *Soil Science* **60**, 305–320.

Chepil, W.S. 1945b. Dynamics of wind erosion: II. Initiation of soil movement. *Soil Science* **60**, 397–411.

Chepil, W.S. 1945c. Dynamics of wind erosion: IV. The translocating and abrasive action of the wind. *Soil Science* **61**, 169–177.

Chepil, W.S. 1951. Properties of soil which influence wind erosion: I. The governing principle of surface roughness. *Soil Science* **69**, 149–162.

Chepil, W.S. 1956. Influence of moisture on erodibility of soil by wind. *Soil Science Society Proceedings* **20**, 288–292.

Chepil, W.S. 1959. Equilibrium of soil grains at the threshold of movement by wind. *Proceedings of the Soil Science Society of America* **23**, 422–428.

Chepil, W.S. and N.P. Woodruff. 1963. The physics of wind erosion and its control. *Advances in Agronomy* **15**, 211–302.

Ciani, A., K.-U. Goss and R.P. Schwarzenbach. 2005. Light penetration in soil and particulate minerals. *European Journal of Soil Science* **56**, 561–574.

Clifford, N.J., J.R. French and J. Hardisty (eds). 1993. *Turbulence: Perspectives on Flow and Sediment Transport*. Chichester: John Wiley and Sons, 360pp.

Cornelis, W.M., D. Gabriels and R. Hartmann. 2004a. A conceptual model to predict the deflation threshold shear velocity as affected by near-surface soil water: I. Theory. *Soil Science Society of America Journal* **68**, 1154–1161.

Cornelis, W.M., D. Gabriels and R. Hartmann. 2004b. A conceptual model to predict the deflation threshold shear velocity as affected by near-surface soil water: II. Calibration and Verification. *Soil Science Society of America Journal* **68**, 1162–1168.

Crawley, D.M. and W.G. Nickling. 2003. Drag partition for regularly-arrayed rough surfaces. *Boundary-Layer Meteorology* **107**(2), 445–468.

Davidson-Arnott, RG.D., K. MacQuarrie and T. Aagaard. 2005. The effect of wind gusts, moisture content and fetch length on sand transport on a beach. *Geomorphology* **68**(1–2), 115–129.

Dong, Z., X. Liu, F. Li, H. Wang, and A. Zhao. 2002. Impact/entrainment relationship in a saltating cloud. *Earth Surface Processes and Landforms* **27**(6), 641–658.

Durar, A.A., J.L. Steiner, S.R. Evett and E.L. Skidmore. 1995. Measured and simulated surface soil drying. *Agronomy Journal* **87**(2), 235–244.

Fécan, F., B. Marticorena and G. Bergametti. 1999. Parameterization of the increase of the aeolian erosion threshold wind friction velocity due to soil moisture for arid and semiarid areas. *Annales Geophysicae* **17**(1), 149–157.

Fisher, R.A. 1926. On the capillary forces in an ideal soil. Correction of formulae given by W.B. Haines. *Journal of Agricultural Science* **16**, 492–505.

Fryrear, D.W. 1986. A field dust sampler. *Journal of Soil and Water Conservation* **41**(22), 117–120.

Garrido, F., M. Ghodrati and M. Chendorain. 1999. Small-scale measurement of soil water content using a fiber optic sensor. *Soil Science Society of America Journal* **63**, 1505–1512.

Geng, J., D. Howell, E. Longhi and R.P. Behringer. 2001. Footprints in sand: The response of a granular material to local perturbations. *Physical Review Letters* **83**(3), 035506-1-035506-4.

Gerety, K.M. and R. Slingerland. 1983. Nature of the saltating population in wind tunnel experiments with heterogeneous size-density sands. In: M.E. Brookfield and T.S. Ahlbrandt (eds.), *Eolian sediments and processes*. Amsterdam, Elsevier, pp. 115–131.

Gerety, K.M. 1985. Problems with determination of $u*$ from wind-velocity profiles measured in experiments with saltation. In: O.E. Barndorff-Nielsen, J.T. Møller, K.R. Rasmussen and B.B. Willetts (eds.), *Proceedings of International Workshop on the Physics of Blown Sand*. Aarhus, University of Aarhus, pp. 271–300.

Gillette, D.A. 1974. On the production of soil wind erosion aerosols having the potential for long-term transport. *Journal of Atmospheric Research* **8**, 735–744.

Gillette, D.A. and P.A. Goodwin. 1974. Microscale transport of sand-sized soil aggregates eroded by wind. *Journal of Geophysical Research* **79**, 4080–4084.

Gillette, D.A. 1977. Fine particulate emissions due to wind erosion. *Transactions of the American Society of Agricultural Engineers* **20**, 890–897.

Gillette, D.A., J. Adams, A. Endo and D. Smith. 1980. Threshold velocities for the input of soil particles into the air by desert soils. *Journal of Geophysical Research* **85**, 5621–5630.

Gillette, D.A., J. Adams, D. Muhs and R. Kihl. 1982. Threshold friction velocities and rupture moduli for crusted desert soils for the input of soil particles in the air. *Journal of Geophysical Research* **87**, 9003–9015.

Gillette, D.A. and P.H. Stockton. 1989. The effect of nonerodible particles on the wind erosion of erodible surfaces. *Journal of Geophysical Research* **94**(12), 885–893.

Gillette, D.A. and A.M. Pitchford. 2004. Sand flux in the northern Chihuahuan desert, New Mexico, USA, and the influence of mesquite-dominated landscapes. *Journal of Geophysical Research* **109**, F04003.

Gillies, J.A., W.G. Nickling and J. King. 2006a. Shear stress partitioning in large patches of roughness in the atmospheric inertial sublayer, *Boundary-Layer Meteorology* **122**, doi:10.1007/s10546-006-9101-5.

Gillies, J.A., W.G. Nickling and J. King. 2006b. Aeolian sediment transport through large patches of roughness in the atmospheric inertial sublayer. *Journal of Geophysical Research* **111**, F02006, doi:10.1029/2005JF000434.

Goossens, D., Z. Offer and G. London. 2000. Wind tunnel and field calibration of five Aeolian sand traps. *Geomorphology* **35**(3–4), 233–252.

Greeley, R., B.R. White, R.N. Leach, J.D. Iversen and J. Pollack. 1976. Mars: wind friction speeds for particle movement. *Geophysical Research Letters* **3**(8), 417–420.

Greeley, R. and J.D. Iversen. 1985. *Wind as a geological process*. Cambridge: Cambridge University Press.

Greeley, R., D.G. Blumberg, A.R. Dobrovolskis, L.R. Gaddis, J.D. Iversen, N. Lancaster, K.R. Rasmussen, R.S. Saunders, S.D. Wells and B.R. White. 1995. Potential transport of windblown sand: influence of surface roughness and assessment with radar data. In: *Desert aeolian processes, V.P.E.* Tchakerian (ed.). London, Chapman and Hall, pp. 75–100.

Haff, P.K. and R.S. Anderson. 1993. Grain scale simulations of loose sedimentary beds: the example of grain-bed impacts in aeolian saltation. *Sedimentology* **40**, 175–198.

Hagen, L. 1984. Soil aggregate abrasion by impacting sand and soil particles. *Transactions of the American Society of Agricultural Engineers* **27**, 805–808.

Haines, W.B. 1925. Studies in the physical properties of soils. II. A note on the cohesion developed by capillary forces in an ideal soil. *Journal of Agricultural Science* **15**, 525–535.

Hotta, S., S. Kubota, S. Katori and K. Horikawa. 1984. Sand transport by wind on a wet sand surface. *Proceedings of the 19th International Conference on Coastal Engineering*, US Army Corps of Engineers.

Houser, C.A. and W.G. Nickling. 2001. The factors influencing the abrasion efficiency of saltating grains on a clay-crusted playa. *Earth Surface Processes and Landforms* **26**, 491–505.

Hunt, J.C.R. and P. Nalpanis. 1985. Saltating and suspended particles over flat and sloping surfaces, I. Modelling concepts. In: O.E. Barndorff-Nielson, J.T. Møller, K.R. Rasmussen and B.B. Willetts (eds), *Proceedings of international workshop on the physics of blown sand*. Aarhus, University of Aarhus, pp. 9–36.

Irwin, H.P.A.H. 1980. A simple omnidirectional sensor for wind tunnel studies of pedestrian level winds, *Jour-*

nal of Wind Engineering and Industrial Aerodynamics, **7**, 219–239.

Isichei, A.O. 1990. The role of algae and cyanobacteria in arid lands. A review. *Arid Soil Research and Rehabilitation* **4**, 1–17.

Iversen, J.D., J.B. Pollack, R. Greeley and B.R. White. 1976. Saltation threshold on Mars: the effect of interparticle force, surface roughness, and low atmospheric density. *Icarus* **29**, 383–393.

Iversen, J.D. and B.R. White. 1982. Saltation threshold on Earth, Mars and Venus. *Sedimentoloy* **29**, 111–119.

Iwamatsu, M. and K. Horii. 1996. Capillary condensation and adhesion of two wetter surfaces. *Journal of Colloid Interface Science* **182**, 400–406.

Jackson, D.W.T. 1996. A new, instantaneous aeolian sand trap design for field use. *Sedimentology* **43**(5), 791–796.

Jackson, N.L. and K.F. Nordstrom 1997. Effects of time-dependent moisture content of surface sediments on aeolian transport rates across a beach, Wildwood, New Jersey, U.S.A. *Earth Surface Processes and Landforms* **22**, 611–621.

Jackson, D.W.T. and J. McCloskey. 1997. Preliminary results from a field investigation of aeolian sand transport using high resolution wind and transport measurements. *Geophysical Research Letters* **24**(2), 163–166.

Johansen J.R. 1993. Cryptogamic crusts of semiarid and arid lands of North America. *Journal of Phycology* **29**, 140–147.

Kaimal, J.C. and J.J. Finnigan. 1994. *Atmospheric boundary layer flows, their structure and measurement*. Oxford, Oxford University Press, p. 289.

Kawamura, R. 1951. *Study of sand movement by wind. Institute of Science and Technology*, Tokyo, Report 5(3–4), Tokyo, Japan, pp. 95–112.

Kawamura, R. 1964. Study of sand movement by wind. In: *Hydraulic Eng. Lab. Tech. Rep.* Number HEL-2-8, Berkeley, University of California, pp. 99–108.

King, J., W.G. Nickling and J.A. Gillies. 2008. Investigations of the law-of-the-wall over sparse roughness elements. *Journal of Geophysical Research* **113**, F02S07, doi:10.1029/2007JF000804.

Lancaster, N., R. Greeley and K.R. Rasmussen. 1991. Interaction between unvegetated desert surfaces and the atmospheric boundary layer: a preliminary assessment. *Acta Mechanica Supplement* **2**, 89–102.

Lancaster, N. and A. Baas. 1998. Influence of vegetation cover on sand transport by wind: field studies at Owens Lake, California. *Earth Surface Processes and Landforms* **23**(1), 69–82.

Langston, G. and C. McKenna Neuman. 2005. An experimental study on the susceptibility of crusted surfaces to wind erosion: A comparison of the strength properties of biotic and salt crusts. *Geomorphology* **72**, 40–53.

Lee, J.A. 1991. The role of desert shrub size and spacing on wind profile parameters. *Physical Geography* **12**(1), 72–89.

Lee, B.E. and B.F. Soliman. 1977. An investigation of the forces on three-dimensional bluff bodies in rough wall turbulent boundary layers. *Transactions of the ASME, Journal of Fluids Engineering* **99**, 503–510.

Leenders, J.K., J.H. van Boxel and G. Sterk. 2005. Wind forces and related saltation transport. *Geomorphology* **71**(3–4), 357–372.

Lettau, K. and H.H. Lettau. 1978. Experimental and micrometeorological field studies on dune migration. In: K. Lettau and H.H. Lettau (eds.), *Exploring the World's Driest Climate*. Madison, WI, University of Wisconsin, Institute for Environmental Studies, pp. 110–147.

Leys, J.F. and D.J. Eldridge. 1998. Influence of cryptogamic crust disturbance on wind erosion of sand and loam rangeland soils. *Earth Surface Processes and Landforms* **23**, 963–974.

Li, A. and Y. Shao. 2003. Numerical simulation of drag partition over rough surfaces. *Boundary-Layer Meteorology* **108**(3), 317–342. doi:10.1023/A:1024179025508.

Livingstone, I. and A. Warren. 1996. *Aeolian geomorphology: an introduction*. London, Longman, 211pp.

Logie, M. 1982. Influence of roughness elements and soil moisture of sand to wind erosion. *Catena* **1**, 161–173.

Luttmer, C. 2002. The partition of drag in salt grass communities, M.S. thesis, Guelph, Guelph, ON, Canada, 2002.

Lyles, L. and R.K. Krauss. 1971. Threshold velocities and initial particle motion as influenced by air turbulence. *Transactions of the American Society of Agricultural Engineers* **14**, 563–566.

Lyles, L. and R.L. Schrandt. 1972. Wind erobility as influenced by rainfall and soil salinity. *Soil Science* **114**, 367–372.

Lyles, L., R.L. Schrandt and N.F. Schmeidler. 1974. How aerodynamic roughness elements control sand movement. *Transactions of the American Society of Agricultural Engineers* **17**(1), 134–139.

Lyles, L. and B.E. Allison. 1976. Wind erosion: the protective role of simulated standing stubble. *Transactions of the American Society of Agricultural Engineers* **19**(1), 61–64.

Monteiro, J.P. and D.X. Viegas. 1996. On the use of Irwin and Preston wall shear stress probes in turbulent incompressible flows with pressure gradients. *Journal of Wind Engineering and Industrial Aerodynamics* **64**, 15–29, 1996.

Marshall, J.K. 1971. Drag measurements in roughness arrays of varying densities and distribution. *Agricultural Meteorology* **8**, 269–292.

McEwan, I.K. and B.B. Willetts. 1991. Numerical model of the saltation cloud. *Acta Mechanica Supplementum* **1**, 53–66.

McEwan, I.K. and B.B. Willetts. 1993. Adaptation of the near-surface wind to the development of sand transport. *Journal of Fluid Mechanics* **252**, 99–115.

McKenna Neuman, C. and W.G. Nickling. 1989. A theoretical and wind tunnel investigation of the effect of capillary water on the entrainment of sediment by wind. *Canadian Journal of Soil Science* **69**, 79–96.

McKenna-Neuman, C. and W.G. Nickling. 1994. Momentum extraction with saltation: Implications for experimental evaluation of wind profile parameters. *Boundary-Layer Meteorology* **68**(1–2), 35–50.

McKenna-Neuman, C. and W.G. Nickling. 1995. Aeolian sediment flux decay: non-linear behavior on developing deflation lag surfaces. *Earth Surface Processes and Landforms* **20**(5), 423–435.

McKenna Neuman, C., C. Maxwell and J.W. Boulton. 1996. Wind transport of sand surfaces crusted with photoautotrophic microorganisms. *Catena* **27**(3–4), 229–247.

McKenna Neuman, C. and M. Maljaars Scott. 1997. Wind tunnel measurement of boundarylayer response to sediment transport. *Boundary Layer Meteorology* **84**, 67–83.

McKenna Neuman, C. 1998. Particle transport and adjustments of the boundary layer over rough surfaces with an unrestricted, upwind supply of sediment. *Geomorphology* **25**, 1–17.

McKenna-Neuman, C. and, M. Maljaars Scott 1998. A wind tunnel study of the influence of pore water on aeolian sediment transport. *Journal of Arid Environments* **39**(3), 403–419.

McKenna Neuman, C. and C. Maxwell. 1999. A wind tunnel study of the resilience of three fungal crusts to particle abrasion during aeolian sediment transport. *Catena* **38**(2), 151–173.

McKenna-Neuman, C., N. Lancaster and W.G. Nickling. 2000. The effect of unsteady winds on sediment transport on the stoss slope of a transverse dune, Silver Peak, NV, USA. *Sedimentology* **47**(1), 211–226.

McKenna Neuman, C. and A. Rice. 2002. *Mechanics of crust rupture and erosion*. Fifth International Conference on Aeolian Research, Lubbock Texas.

McKenna Neuman, C. and C. Maxwell. 2002. Temporal aspects of the abrasion of microphytic crusts under impact. *Earth Surface Processes and Landforms* **27**, 891–908.

McKenna Neuman, C.L. and G. Langston. 2003. *Spatial analysis of surface moisture content on beaches subject to aeolian transport*. Canadian Coastal Conference.

McKenna Neuman, C., C. Maxwell and C. Rutledge. 2005. Spatial analysis of crust deterioration under particle impact. *Journal of Arid Environments* **60**(2), 321–342.

McKenna Neuman, C. and G. Langston. 2006. Measurement of water content as a control of particle entrainment by wind. *Earth Surface Processes and Landforms* **31**, 303–317.

Middleton, G.V. and J.B. Southard. 1984. *Mechanics of sediment movement*. Tulsa, OK: Society of Economic Paleontologists and Mineralogists.

Mitha, S., M.Q. Tran, B.T. Werner and P.K. Haff. 1985. *The grain-bed impact process in aeolian saltation*. Brown bag preprint series in basic and applied science BB-36. Pasadena, CA: Department of Physics, California Institute of Technology.

Musick, H.B. and D.A. Gillette. 1990. Field evaluation of relationships between a vegetation structural parameter and sheltering against wind erosion. *Land degradation and rehabilitation* **2**, 87–94.

Musick, H.B., S.M. Trujillo and C.R. Truman. 1996. Wind-tunnel modelling of the influence of vegetation structure on saltation threshold. *Earth Surface Processes and Landforms* **21**(7), 589–606.

Nalpanis, P. 1985. Saltating and suspended particles over flat and sloping surfaces, II. Experiments and numerical simulations. In: O.E. Barndorff-Nielsen, J.T. Møller, K.R. Rasmussen and B.B. Willetts (eds), *Proceedings of International Workshop on the Physics of Blown Sand*. Aarhus, University of Aarhus, pp. 37–66.

Nalpanis, P., J.C.R. Hunt and C.F. Barrett. 1993. Saltating particles over flat beds. *Journal of Fluid Mechanics* **251**, 661–685.

Namikas, S.L. 2002. Field evaluation of two traps for high-resolution aeolian transport measurements. *Journal of Coastal Research* **18**(1), 136–148.

Nickling, W.G. 1978. Eolian sediment transport during dust storms: Slims River Valley, Yukon Territory. *Canadian Journal of Earth Science* **15**, 1069–1084.

Nickling, W.G. and M. Ecclestone. 1981. The effects of soluble salts on the threshold shear velocity of fine sand. *Sedimentology* **28**, 505–510.

Nickling, W.G. 1983. Grain-size characteristics of sediment transported during dust storms. *Journal of Sedimentary Petrology* **53**, 1011–1024.

Nickling, W.G. 1984. The stabilizing role of bonding agents on the entrainment of sediment by wind. *Sedimentology* **31**, 111–117.

Nickling, W.G. 1988. The initiation of particle movement by wind. *Sedimentology* **35**(3), 499–512.

Nickling, W.G. 1989. Prediction of soil loss by wind. In: S. Rimwanich (ed.), *Land conservation for future generations*. Bangkok, Ministry of Agriculture, pp. 75–94.

Nickling, W.G. and J.A. Gillies. 1989. Emission of fine grained particles from desert soils. In: M. Leinen and M. Sarnthein (eds.), *Palaeoclimatology and palaeometeorology: modern and past patterns of global atmospheric transport*. Amsterdam, Kluwer, pp. 133–165.

Nickling, W.G. and C.L. McKenna Neuman. 1995. Development of deflation lag surfaces. *Sedimentology* **42**, 403–414.

Nickling, W.G. and C. McKenna Neuman 1997. Wind tunnel evaluation of a wedge-shaped aeolian sediment trap. *Geomorphology* **18**, 333–345.

Oke, T.R. 1978. *Boundary layer climates*. New York, Methuen.

Okin, G.S. 2005. Dependence of wind erosion and dust emission on surface heterogeneity: Stochastic modeling, *Journal of Geophysical Research* **110**(D11), doi:10.1029/2004JD005288.

Okin, G.S. and D.A. Gillette. 2001. Distribution of vegetation in wind-dominated landscapes: Implications for wind erosion modeling and landscape processes, *Journal of Geophysical Research* **106**(D9), 9673–9683.

Okin, G.S. and D.A. Gillette. 2004. Modeling wind erosion and dust emission on vegetated surfaces. In: R. Kelly and N.A. Drake (eds.), *Spatial Modeling of the Terrestrial Environment*. John Wiley, Hoboken, NJ, pp. 137–156.

Owen, P.R. 1964. Saltation of uniform grains in air. *Journal of Fluid Mechanics* **20**, 225–242.

Pluis, J.L. and B. de Winder. 1989. Spatial patterns of algae colonization of dune blowouts. *Catena* **16**, 499–506.

Pluis, J.L. and J.H. van Boxel. 1993. Wind velocity and algal crusts in dune blowouts. *Catena* **20**, 581–594.

Prandtl, L. 1935. The mechanics of viscous fluids. In: F. Durand (ed.), *Aerodynamic theory*. Volume III, Berlin, Springer, pp. 57–109.

Pye, K. 1980. Beach salcrete and eolian sand transport: evidence from North Queensland. *Journal of Sedimentary Petrology* **50**, 257–261.

Rajot, J.-L., S.C. Alfaro, L. Gomes, and A. Gaudichet. 2003. Soil crusting on sandy soils and its influence on wind erosion. *Catena* **53**(1), 1–16.

Rasmussen, K.R. and H.E. Mikkelsen. 1998. On the efficiency of vertical array aeolian field traps. *Sedimentology* **45**(4), 789–801.

Rasmussen, K.R. and M. Sørensen. 1999. Aeolian mass transport near the saltation threshold. *Earth Surface Processes and Landforms* **24**(5), 413–422.

Raupach, M.R. 1992. Drag and drag partition on rough surfaces, *Boundary-Layer Meteorology* **60**, 375–395.

Raupach, M.R., D.A. Gillette and J.F. Leys. 1993. The effect of roughness elements on wind erosion threshold, *Journal of Geophysical Research* **98**(D2), 3023–3029.

Raupach, M.R., D.E. Hughes and H.A. Cleugh 2006. Momentum absorption in rough-wall boundary layers with sparse roughness elements in random and clustered distributions, *Boundary-Layer Meteorology* **120**, 201–218, DOI 10.1007/s10546-006-9058-4.

Ravi, S., T. Zobeck, T. Over, G. Okin and P. D'odorico. 2006. On the effect of moisture bonding forces in air-dry soils on threshold friction velocity of wind erosion. *Sedimentology* **53**, 597–609.

Rice, M.A., B.B. Willetts and I.K. McEwan. 1995. An experimental study of multiple grain size ejecta produced by collisions of saltating grains with a flat bed. *Sedimentology* **42**(4), 695–706.

Rice, M.A., B.B. Willetts and I.K. McEwan. 1996a. Observation of collisions of saltating grains with a granular bed from high-speed cine-film. *Sedimentology* **43**, 21–31.

Rice, M.A., B.B. Willetts and I.K. McEwan. 1996b. Wind erosion of crusted soil sediments. *Earth Surface Processes and Landforms* **21**, 279–293.

Rice, M.A., C.E. Mullins and I.K. McEwan. 1997. An analysis of soil crust strength in relation to potential erosion by saltating particles. *Earth Surface Processes and Landforms* **22**(9), 859–884.

Rice, M.A., I.K. McEwan and C.E. Mullins. 1999. A conceptual model of wind erosion of soil surfaces by saltating particles. *Earth Surface Processes and Landforms* **24**, 383–392.

Rice, M.A. and I.K. McEwan. 2001. Crust strength: a wind tunnel study of the effect of impact by saltating particles on cohesive soil surfaces. *Earth Surface Processes and Landforms* **26**(7), 721–733.

Richards L.A. 1953. Modulus of rupture as an index of crusting of soil. *Proceedings Soil Science Society of America* **17**, 321–323.

Rumpel, D.A. 1985. Successive aeolian saltation: studies of idealized collisions. *Sedimentology* **32**, 267–280.

Sarre, R.D. 1987. Aeolian sand transport. *Progress in Physical Geography* **11**, 157–182.

Schlichting, H. 1936. Experimentle untersuchungen zum rauhigkeitsproblem. *Ingeniew-Archiv* **7**, 1–34. (English Translation: NACA Technical Memorandum 823, 1936).

Shao, Y. 2000. *Physics and Modelling of Wind Erosion*. Dordrecht, Kluwer Academic Publishers.

Shao Y. and H. Lu. 2000. A simple expression for wind erosion threshold friction velocity. *Journal of Geophysical Research* **105**(D17), 22437–22443.

Shao, Y. and M.R. Raupach. 1992. The overshoot and equilibration of saltation. *Journal of Geophysical Research* **97**(D18), 20559–20564.

Shao, Y., G.H. McTainsh, J.F. Leys and M.R. Raupach. 1993. Efficiency of Sediment Samplers for Wind Erosion Measurement. *Australian Journal of Soil Research* **31**, 519–532.

Shao, Y. and A. Li. 1999. Numerical modelling of saltation in the atmospheric surface layer. *Boundary-Layer Meteorology* **91**(2), 199–225.

Sharp, R.P. 1964. Wind-driven sand in Coachella Valley, California. *Bulletin of the Geological Society of America* **75**, 785–804.

Sherman, D.J., D.W.T. Jackson, S.L. Namikas and J. Wang. 1998. Wind-blown sand on beaches: an evaluation of models. *Geomorphology* **22**, 113–133.

Smalley, I.J. 1970. Cohesion of soil particles and the intrinsic resistance of simple soil. *Journal of Soil Science* **21**(1), 154–161.

Sørensen, M. 1991. An analytical model of wind-blown sand transport. *Acta Mechanica Supplementum*, 67–82.

Sørensen, M. 2004. On the rate of aeolian sand transport. *Geomorphology* **59**(1–4), 53–62.

Spies, P.J., I.K. McEwan and G.R. Butterfield. 1995. On wind velocity profile measurements taken in wind tunnels with saltating grains. *Sedimentology* **42**(3), 515–521.

Sterk, G., A.F.G. Jacobs and J.H. van Boxel. 1998. The effect of turbulent flow structures on saltation sand transport in the atmospheric boundary layer. *Earth Surface Processes and Landforms* **23**(10), 877–887.

Stockton, P.H. and D.A. Gillette. 1990. Field measurements of the sheltering effect of vegetation on erodible land surfaces. *Land Degradation and Rehabilitation* **2**, 77–86.

Stout, J.E. and T.M. Zobeck. 1997. Intermittent saltation. *Sedimentology* **44**, 959–970.

Stout, J.E. 1998. Effect of averaging time on the apparent threshold for aeolian transport. *Journal of Arid Environments* **39**(3), 395–401.

Svasek, J.N. and J.H.J. Terwindt. 1974. Measurements of sand transport by wind on a natural beach. *Sedimentology* **21**, 311–322.

Tennekes, H. and J.L. Lumley. 1972. *A first course in turbulence*. Cambridge: MIT Press.

Tsoar, H. and K. Pye. 1987. Dust transport and the question of desert loess formation. *Sedimentology* **34**, 139–154.

Tuller, M. and D. Or. 2005. Water films and scaling of soil characteristic curves at low water contents. *Water Resources Research* **41**(W09403), 1–6.

Ungar, J.E. and P.K. Haff. 1987. Steady-state saltation in air. *Sedimentology* **34**, 289–299.

van Boxel, J.H., G. Sterk and S.M. Arens. 2004. Sonic anemometers in aeolian sediment transport research. *Geomorphology* **59**(1–4), 131–147.

van den Anker, J.M., P.D. Jungerius and L.R. Mur. 1985. The role of algae in the stabilization of coastal dune blowouts. *Earth Surface Processes and Landforms* **10**, 189–192.

Van Pelt, R.S., Zobeck, T.M., Peters, P. and S.Visser. 2006. Wind tunnel testing and comparison of three saltation impact sensors. Abstracts, International Conference on Aeolian Research. July 24–28, 2006, Guelph, Ontario, Canada.

Walker, I.J. 2000. *Secondary airflow and sediment transport in the lee of transverse dunes*. Doctoral thesis, University of Guelph, Ontario, Canada, 256pp.

Walker, I.J. and W.G. Nickling. 2003. Simulation and measurement of surface shear stress over isolated and closely spaced transverse dunes in a wind tunnel. *Earth Surface Processes and Landforms* **28**(10), 1111–1124.

Walker, I.J. 2005. Physical and logistical considerations of using ultrasonic anemometers in aeolian sediment transport research. *Geomorphology* **68**(1–2), 57–76.

Wang, H.T., X.H. Zhang, Z. Dong and M. Ayrault. 2006. Experimental determination of saltating glass particle dispersion in a turbulent boundary layer. *Earth Surface Processes and Landforms* **31**, 1746–1762.

Werner, B.T. 1988. *A steady-state model of wind-blown sand transport*. The Office of Naval Technology, The Naval Weapons Center, 1.

West, N.E. 1990. Structure and function of microphytic soil crusts in wildland ecosystems of arid to semi-arid regions. *Advances in Ecological Research* **20**, 179–223.

White, B.R. 1979. Soil transport by winds on Mars. *Journal of Geophysical Research* **84**, 4643–4651.

White, B.R. and J.C. Schultz. 1977. Magnus effect in saltation. *Journal of Fluid Mechanics* **81**, 497–512.

Wiggs, G.F.S. 2001. Desert dune processes and dynamics. *Progress in Physical Geography* **25**(1), 53–79.

Wiggs, G.F.S., R.J. Atherton and A.J. Baird. 2002. The dynamic effects of moisture on the entrainment and transport of sand by wind. *Geomorphology* **59**(1–4), 13–30.

Wiggs, G.F.S., R.J. Atherton and A.J. Baird 2004. Thresholds of aeolian sand transport: establishing suitable values. *Sedimentology* **51**(1), 95–108.

Wiggs, G.F.S., I. Livingstone, D.S.G. Thomas and J.E. Bullard. 1996. Airflow and roughness characteristics over partially vegetated linear dunes in the southwest Kalahari, Desert. *Earth Surface Process and Landforms* **21**(1), 19–34.

Willetts, B.B. and M. Rice. 1985. Wind tunnel tracer experiments using dyed sand. *Proceedings of the International Workshop on the Physics of Blown Sand*.

Willetts, B.B. and M.A. Rice. 1986. Collision in aeolian transport: the saltation/creep link. In: W.G. Nickling (ed.), *Aeolian Geomorphology*. London, Allen & Unwin, pp. 1–18.

Willetts, B.B. and M.A. Rice. 1989. Collisions of quartz grains with a sand bed: The influence of incident angle. *Earth Surface Processes and Landforms* **14**, 719–730.

Williams, G. 1964. Some aspects of the eolian saltation load. *Sedimentology* **3**, 257–287.

Williams, J.J., G.R. Butterfield and D.G. Clark. 1990. Rates of aerodynamic entrainment in developing boundary layer. *Sedimentology* **37**, 1039–1048.

Wolfe, S. and W.G. Nickling. 1993. The protective role of sparse vegetation in wind erosion. Progress in Physical Geography **17**(1), 50–68.

Wolfe, S.A. and W.G. Nickling. 1996. Shear stress partitioning in sparsely vegetated desert canopies. *Earth Surface Processes and Landforms* **21**, 607–619.

Wu, H. and T. Stathopoulos. 1994. Further experiments on Irwin's surface wind sensor, *Journal of Wind Engineering and Industrial Aerodynamics* **53**, 441–452.

Wyatt, V.E. and W.G. Nickling. 1997. Drag and shear stress partitioning in spare desert creosote communities. *Canadian Journal of Earth Sciences* **34**(11), 1486–1498.

Yang, Y. and R. Davidson-Arnott. 2005. Rapid Measurement of Surface Moisture Content on a Beach. *Journal of Coastal Research* **21**(3), 447–452.

Zingg, A.W. 1953. Wind tunnel studies of the movement of sedimentary material. In Proceedings of the Fifth Hydraulic Conference. *Studies in Engineering, Bulletin* **34**, 111–135. Iowa City, University of Iowa.

Zou, X., Z. Wang, Z. Wang, Q. Hao, C. Zhang, Y. Liu and G. Dong. 2001. The distribution of velocity and energy of saltating sand grains in a wind tunnel. *Geomorphology* **36**(3–4), 155–165.

Chapter 18

Dune Morphology and Dynamics

Nicholas Lancaster

Introduction

Sand dunes form part of a hierarchical self-organized system of aeolian bedforms which comprises: (i) wind ripples (spacing 0.1–1 m), (ii) individual simple dunes or superimposed dunes on mega dunes (also called draa or compound and complex dunes) (spacing 50–500 m), and (iii) mega dunes (spacing >500 m). Most dunes occur in contiguous areas of aeolian deposits called ergs or sand seas (with an area of >100 km^2). Smaller areas of dunes are called dune fields. Major sand seas occur in the old world deserts of the Sahara, Arabia, central Asia, Australia, and southern Africa, where sand seas cover between 20 and 45% of the area classified as arid (Fig. 18.1). In North and South America there are no large sand seas, and dunes cover less than 1% of the arid zone. The majority of dunes are composed of quartz and feldspar grains of sand size, although dunes composed of gypsum, carbonate, and volcaniclastic sand, as well as clay pellets, also occur.

The formation of areas of dunes is determined by the production of sediment of a range of suitable particle sizes, the availability of this sediment for transport by wind, and the transport capacity of the wind (Kocurek and Lancaster, 1999). Most dunes are derived from material that has been transported by fluvial or littoral processes. Important sources include marine and lacustrine beaches, dry lake basins, river flood plains, and deltas. The availability of sediment (defined as the probability of entrainment of sand for transport) is determined by its moisture content, vegetation cover, crusting, and cohesion. The transport capacity of the wind (or the potential sand transport rate) is a cubic function of wind speed or surface shear stress above the transport threshold (see Chapter 17).

Dunes are created and modified by a series of interactions between granular material (sand) and shearing flow (the atmospheric boundary layer) as shown by Fig. 18.2. The resulting landforms are bedforms that are dynamically similar to those developed in subaqueous shearing flows (e.g. rivers, tidal currents). Their forms reflect the characteristics of the sediment (primarily its grain size) and the surface wind regime (especially its directional variability). In some areas vegetation may be a significant factor. As the bedform grows upwards into the boundary layer, the primary air flow is modified by interactions between the form and the flow which give rise to modifications of the local wind speed, shear stress, and turbulence intensity and create secondary flow circulations, especially in the lee of the dune. Many large dunes (megadunes or draa) are also characterized by superimposed bedforms that respond to changes in airflow and sediment transport on the megadune itself. Not all dunes are the products of contemporary processes and dynamics. In many areas, megadunes have a long and complex history in which the legacy of past climates and wind regimes is a significant factor in determining present-day dune morphology.

Dunes occur in self-organized patterns that develop over time as the response of sand surfaces to the wind regime (especially its directional variability) and the supply of sand (Werner, 1995). Development of these patterns is modulated by the effects of changes in climate and sea level on sediment supply, dune mobility,

N. Lancaster (✉)
Desert Research Institute, 2215 Raggio Parkway, Reno, NV 89512-1095, USA
e-mail: nick.lancaster@dri.edu

A.J. Parsons, A.D. Abrahams (eds.), *Geomorphology of Desert Environments*, 2nd ed., DOI 10.1007/978-1-4020-5719-9_18, © Springer Science+Business Media B.V. 2009

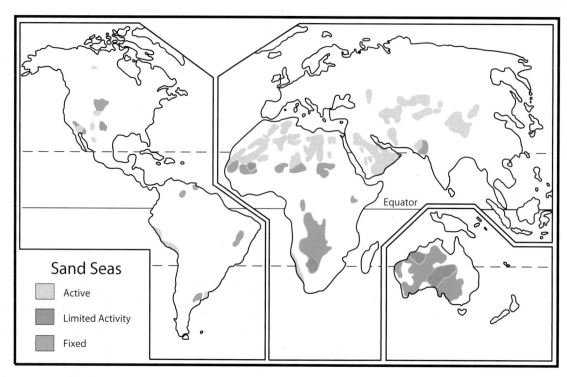

Fig. 18.1 Location of major desert sand seas and dunefields

and wind regime characteristics, often resulting in the formation of different generations of dunes. Characteristic features of dune patterns include close correlations between the height and spacing of dunes and systematic spatial variations in dune type, orientation, and sediment volume. The dune types discussed below represent the steady state attractors of the aeolian transport system and can evolve from a wide range of initial conditions. The orientation of dunes with respect to the wind regime is another aspect of the self-organizing nature of the system, in which dunes are oriented to maximize the gross sand transport normal to the crest (Rubin and Hunter, 1987).

The past two decades have seen major advances in our understanding of desert dune processes and dynamics at different temporal and spatial scales. These advances, together with parallel changes in the understanding of bedforms in subaqueous environments, have resulted in important changes in the paradigms that guide dune research. Key new paradigms include the gross bedform-normal concept for dune trends; the concept of dune generations to explain complex dune

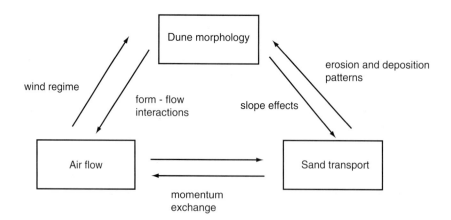

Fig. 18.2 A conceptual framework for process-form interactions on aeolian dunes

patterns; the sediment state model to explain episodic development of sand seas and dunefields; and the overarching principle of self-organization of dune systems and patterns.

Changes in the field of dune studies have been documented by new books on desert dunes and aeolian geomorphology (Lancaster, 1995a; Livingstone and Warren, 1996; Pye and Tsoar, 1990); and reviews by Wiggs (Wiggs, 2001), Walker (Walker and Nickling, 2002), and McKenna Neuman and Nickling (Nickling and McKenna Neuman, 1999) and (Livingstone et al., 2007). Progress in the field has been encouraged by the series of international conferences on aeolian research (the ICAR conferences), and the many special journal issues and books resulting from them (e.g. Goudie et al., 1999).

The following discussion of dune morphology and dynamics will proceed up the hierarchy of aeolian bedforms, with an emphasis on the fundamental processes operating at each level. Each element of the hierarchy responds to the dynamics of a component of the wind regime in an area and possesses a characteristic time period, termed the relaxation or reconstitution time (Allen, 1974), over which it will adjust to changed conditions. This increases from minutes in the case of wind ripples to millennia for draas. Change in bedforms involves the movement of sediment. Thus an increasing spatial scale is also involved at each level of the hierarchy.

Wind Ripples

Wind ripples (Fig. 18.3) are ubiquitous on all sand surfaces except those undergoing very rapid deposition and provide an excellent example of self-organization in aeolian systems (Anderson, 1990). They are the initial response of sand surfaces to sand transport by the wind, and form because flat sand surfaces over which transport by saltation and reptation takes place are dynamically unstable (Bagnold, 1941). The formation and movement of wind ripples are therefore closely linked to the processes of saltation and reptation.

Wind ripples trend perpendicular to the sand-transporting winds, although (Howard, 1977) has emphasized the effects of slope on ripple orientation. Because they can be reformed within minutes, wind

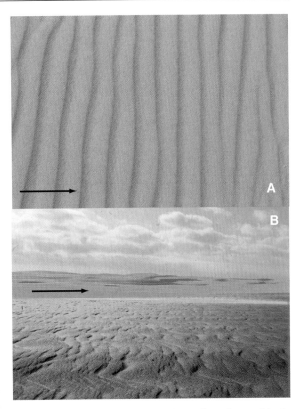

Fig. 18.3 Wind ripples: (**A**) ripples in medium–fine sand in the Gran Desierto sand sea, and (**B**) granule ripples on the Skeleton Coast, Namibia. Arrow indicates formative wind direction

ripples provide an almost instantaneous indication of local sediment transport and wind directions. Typical wind ripples have a wavelength between 13 and 300 mm (hundreds of grain diameters) and an amplitude of 0.6–14 mm (tens of grain diameters) (Anderson, 1990; Boulton, 1997). The ratio between ripple length and height is given by the ripple index, which can be used to compare ripples in different environments. Typical ripple indices for aeolian wind ripples range from 15 to 20. Much longer wavelength (0.5–2 m or more) ripples with an amplitude of 0.1 m or more are composed of coarse sand or granules (1–4 mm median grain size) and are termed 'granule ripples' (Fryberger et al., 1992; Sharp, 1963) and 'megaripples' by Greeley and Iversen (1985). Granule ripples are not distinct forms, and form one end of a continuum of wind ripple dimensions (Ellwood et al., 1975) (Fig. 18.4). Ripple wavelength is a function of both particle size and sorting and wind speed so that ripples in coarse sands have a greater spacing than those in fine sands (Sharp, 1963). For sands of a given size, ripple wavelength increases with wind

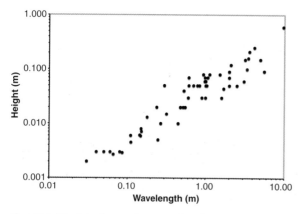

Fig. 18.4 Wind ripple morphometry. Note that there is a continuum of ripple wavelength from 'normal' wind ripples to granule ripples. Data from Sharp (1963), Walker (1981), and author's field observations

friction speed (Seppälä and Linde, 1978). Most wind ripples are asymmetric in cross-section with a slightly convex stoss slope at an angle of 1.4–7° and a lee slope that varies between 1.8 and 10°. The existence of very steep (20–30°) lee slope angles reported by Sharp (1963) is uncertain. In all cases the crest of the ripple is composed of grains that are coarse relative to the mean size of the surface sand.

Several models have been put forward to explain the formation and characteristic wavelength of ripples. Bagnold (1941) pointed to the close correspondence between calculated saltation path lengths and observed ripple wavelengths in wind tunnel experiments. He argued that chance irregularities in 'flat' sand surfaces give rise to variations in the intensity of saltation impacts (Fig. 18.5a) creating zones that would be preferentially eroded or protected. Grains from the zone of more intense impacts (A–B) would land downwind at a distance equal to the average saltation pathlength such that a zone of more intense saltation impacts would propagate downwind. In addition, variations in surface slope and saltation impact intensity cause variations in the reptation rate. Bagnold argued that interactions between the developing surface microtopography and the saltating and reptating grains would soon lead to a coincidence between the characteristic saltation pathlength and the ripple wavelength (Fig. 18.5b). Bagnold's concept of the formation of wind ripples was first challenged by Sharp (1963) who argued that grains in ripples are moved mostly by reptation or surface creep. Irregularities in the bed and interactions between grains moving at different speeds give rise to

local increases in bed elevation. These 'proto-ripples' begin as short-wavelength, low-amplitude forms and grow to their steady state dimensions by the growth of larger forms at the expense of smaller. Each developing ripple creates a 'shadow zone' in its lee (Fig. 18.5c), with a width proportional to ripple wavelength and impact angle. The size of the shadow zone determines the position of the next ripple downwind. Sharp argued that the controls of ripple wavelength are impact angle and ripple amplitude, both of which are dependent on grain size and wind speed, but he could see no obvious reason why ripple wavelength should be dependent on the mean saltation pathlength. His observations on ripple development to an equilibrium size and spacing have been confirmed experimentally by wind tunnel experiments (Seppälä and Linde, 1978; Walker, 1981) and by numerical model simulations of sediment surfaces (Anderson, 1990; Werner, 1988).

Anderson has provided a rigorous model for ripple development based on experimental data and numerical simulations of sand beds (Anderson, 1987; Anderson, 1990; Anderson and Bunas, 1993). Recent experimental and theoretical work on aeolian saltation (see Chapter 17) has demonstrated that saltating sand consists of two populations: (a) long trajectory, high impact energy 'successive saltations' and (b) short trajectory, low impact energy 'reptations'. There is a wide distribution of saltation trajectories with typical pathlengths that are much longer than ripple wavelengths, and a low range (1–2°) of impact angles. This suggests that the high impact energy grains do not contribute directly to ripple formation, as Bagnold hypothesized, but drive the reptation process. Using a simplified model of aeolian saltation, Anderson (1987, 1990) was able to show that a flat bed is unstable to infinitesimal variations in bed elevation, giving rise to spatial variations in the mass flux of reptating grains. Convergence and divergence of mass flux rates result in the growth of grain-scale perturbations on the bed, with the fastest growing perturbations having a wavelength 6–10 times the mean reptation distance (Fig. 18.5d). These perturbations subsequently develop into a self-organized pattern of ripples by coalescence and convergence of bedforms moving at different speeds, with the rate of change decreasing asymptotically with time. A quasi-stable wavelength emerges that is the effect of the sharply decreasing rate of ripple mergers with increasing ripple

Fig. 18.5 Models for wind ripple formation. (**A, B**) Variation in impact intensity over a perturbation in the bed (after Bagnold, 1941). Note higher impact intensity in zone A–B compared with B–C. (**C**) Alternation of impact and shadow zones on a developing wind ripple (after Sharp, 1963). (**D**) Growth and movement of developing bed perturbations that evolve to wind ripples (after Anderson, 1987)

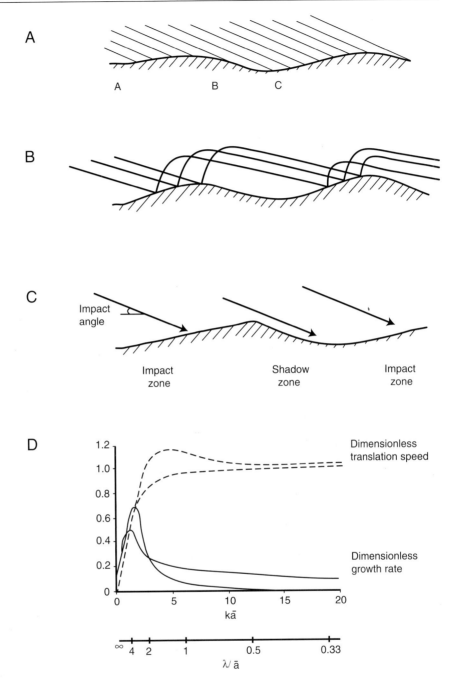

size (Werner, 1988). Numerical models of ripples suggest that their wavelength is partially dependent on the length of the shadow zone, and therefore the angle of incidence of the saltating grains, thus validating the model of Sharp (1963). Ripple height may depend in part on impact angle, but is also influenced by the higher wind shear stresses experienced on the ripple crest.

Dune Morphology and Morphometry

Dunes occur in a variety of morphologic types, each of which displays a range of sizes (height, width, and spacing). Most dune patterns are quite regular, as evidenced by distinct statistical populations of dunes defined by dune crest orientation, crest length and spacing

(Ewing et al., 2006) as well as the correlations between dune height and spacing (Fig. 18.6).

Many different classifications of dune types have been proposed (see Mainguet, 1983; 1984a, for a list of different schemes). They fall into two groups: (a) those based on the external morphology of the dunes (morphological classifications), and (b) those that imply some relationship of dune type to formative winds or sediment supply (morphodynamic classifications). Figure 18.7 provides a framework for classifying dunes based on morphology, sediment thickness, and other key parameters.

Crescentic Dunes

The simplest dune types and patterns are those that form in wind regimes characterized by a narrow range of wind directions in which dune crests are oriented tranverse to the sand transport direction. In the absence of vegetation, crescentic dunes are the dominant form. (Hunter et al., 1983) and (Tsoar, 1986) indicate that this dune type occurs where the directional variability of sand transporting winds is 15° or less about a mean value. Isolated crescentic dunes or barchans occur in areas of limited sand availability. As sand supply increases, barchans coalesce laterally to form crescentic or barchanoid ridges that consist of a series of connected crescents in plan view (McKee, 1966). Larger forms with superimposed dunes are termed compound crescentic dunes (e.g. Breed and Grow, 1979; Havholm and Kocurek, 1988; Lancaster, 1989b).

Barchans (Fig. 18.8a) are common in areas of limited sand supply and unidirectional winds (Andreotti et al., 2002), where they can migrate long distances with only minor changes of form (Gay, 1999; Hastenrath, 1987; Haynes, 1989; Long and Sharp, 1964). Recent field studies and modelling suggest that the morphological stability of these dunes may be overemphasized, and collisions leading to merging of faster-moving smaller dunes with larger dunes, as well as creation of new dunes are common (Elbelrhiti et al., 2005).

Barchans occur in two main areas: (a) on the margins of sand seas and dune fields (Sweet et al., 1988) and (b) in sand transport corridors linking sand source zones with depositional areas (Corbett, 1993; Embabi, 1982; Embabi and Ashour, 1993; Hersen

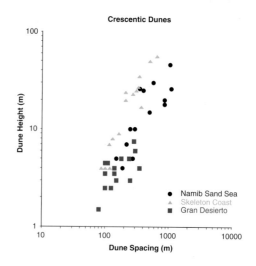

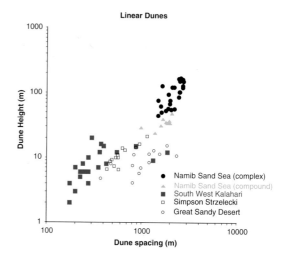

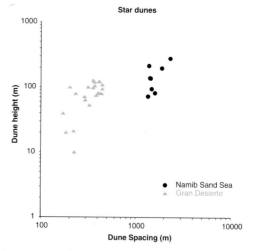

Fig. 18.6 Relations between dune height and spacing. Data from author's measurements and Wasson and Hyde (1983b)

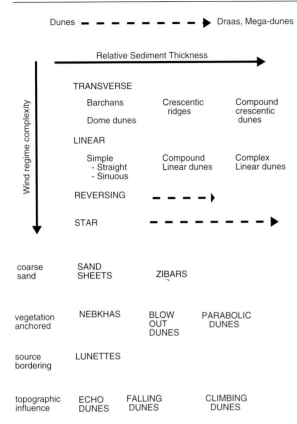

Fig. 18.7 A scheme for the classification of desert dunes. Modified from Lancaster (1995a)

are between 3 and 10 m high, with a spacing of 100–400 m. Typical areas of simple crescentic ridges occur at White Sands, New Mexico (McKee, 1966), Guererro Negro, Baja California (Inman et al., 1966), and in the Skeleton Coast dunefield in the northern Namib (Lancaster, 1982a).

Compound crescentic dunes (Fig. 18.8c) occupy some 20% of the area of sand seas world-wide. They usually consist of a main dune ridge 20–50 m high with a spacing of 800–1200 m. Smaller crescentic ridges 2–5 m high and 50–100 m apart are superimposed on their upper stoss slopes and crestal areas. Compound crescentic dunes are typified by those along the western edge of the Namib Sand Sea (Lancaster, 1989b), the Algodones Dunes of California (Havholm and Kocurek, 1988; Sweet et al., 1988), and the Liwa area of the United Arab Emirates (Stokes and Bray, 2005).

Linear Dunes

Linear dunes are characterized by their length (often more than 20 km) straightness, parallelism, and regular spacing, and high ratio of dune to interdune areas. Lancaster (1982b) estimated that 50% of all dunes are of linear form, with the percentage varying between 85 and 90% for areas of the Kalahari and Simpson–Strzelecki Deserts to 1–2% for the Ala Shan and Gran Desierto sand seas. Linear dunes are the dominant form in sand seas in the Southern Hemisphere and in the southern and western Sahara.

Many linear dunes consist of a lower gently sloping plinth, often partly vegetated, and an upper crestal area where sand movement is more active. Slip faces develop on the upper part of the dune, their orientation depending on the winds of the season. The average form of the dune may be symmetrical with an approximately triangular profile, but in each wind season its profile tends to an asymmetric form with a convex upper stoss slope and a well-developed lee face (Tsoar, 1985). Several varieties of linear dunes are recognized (Fig. 18.9). Simple linear dunes (Fig. 18.9a,b) are of two types: the long, narrow, straight, partly vegetated linear dunes of the Simpson (Wasson, 1983; Wasson et al., 1988) and Kalahari Deserts (Bullard and Nash, 1998; Bullard et al., 1995) (equivalent to the vegetated linear dunes of Tsoar [1989]), and the more

et al., 2004). Barchans are characterized by an ellipsoidal shape in plan view, with a concave slip face and 'horns' extending downwind (Sauerman et al., 2000). Dune height is typically about one-tenth of the dune width (Finkel, 1959; Hastenrath, 1967). Strongly elongated horns and asymmetric development of barchan plan shapes occur in some areas (e.g. Hastenrath, 1967; Lancaster, 1982a), and have been attributed to asymmetry in the wind regime or sand supply, but may be a product of merging and reconstitution of dunes as the pattern becomes self-organized (Hersen and Douady, 2005). In some areas, barchans of this type are transitional to linear dunes (Lancaster, 1980; Tsoar, 1984).

Crescentic dunes of simple and compound varieties occupy some 10% of the area of sand seas world-wide (Fryberger and Goudie, 1981) and occur in all desert regions. The patterns of most crescentic dunes are quite regular, as indicated by the close correlations that exist between dune height and spacing (Fig. 18.6). Most simple crescentic and barchanoid ridges (Fig. 18.8b)

Fig. 18.8 Crescentic dunes: (**A**) barchans on the upwind margin of the Skeleton Coast dunefield, Namibia; (**B**) barchans and simple crescentic dunes, Conception Bay, Namib Sand Sea; and

(**C, D**) compound crescentic dunes, Liwa area, UAE: note two orders of dune spacing, primary dunes and superimposed dunes

sinuous seif-type dunes found in Sinai [Tsoar, 1983], and in parts of the eastern Sahara. Compound linear dunes consist of two to four seif-like ridges on a broad plinth and are typified by those in the southern Namib sand sea (Lancaster, 1983). Large (50–150 m high, 1–2 km spacing) complex linear dunes (Fig. 18.9c) with a single sinuous main crestline with distinct star-form peaks and crescentic dunes on their flanks (Lancaster, 1983) occur in the Namib and parts of the Rub al Khali sand seas. Wide (1–2 km) complex linear dunes with crescentic dunes superimposed on their crests occur in the eastern Namib, parts of the Wahiba Sands (Warren, 1988; Warren and Allison, 1998), the Takla Makan (Wang et al., 2004) and the Akchar sand sea of Mauritania (Kocurek et al., 1991).

The origins of linear dunes and their relationship to formative wind directions have been the subject of considerable controversy (Lancaster, 1982b; Tsoar, 1989). A widely held view has been that linear dunes form parallel to prevailing or dominant wind directions (Folk, 1970; Glennie, 1970; Wilson, 1972).

Their parallelism and straightness are believed to result from roller vortices in which helicoidal flow sweeps sand from interdune areas to dunes (Hanna, 1969; Wilson, 1972). However, there are inconsistencies between the spacing of many dunes and the reported dimensions and stability of helical roll vortices (Lancaster, 1982b; Livingstone, 1986), which would have to be positioned at exactly the same place at successive wind episodes in order to allow dunes to grow and extend (Greeley and Iversen, 1985). The only observational evidence for helical roll vortices in dune areas comes from studies of tethered kites in the Simpson Desert, Australia (Tseo, 1990). There is, however, a substantial body of empirical evidence that indicates that linear dunes form in bidirectional wind regimes. Correlations between dune types and wind regimes (e.g. Fryberger, 1979), studies of internal sedimentary structures (Bristow et al., 2000; McKee, 1982; McKee and Tibbitts, 1964) and detailed process studies on linear dunes (Livingstone, 1986, 1988, 1993; Tsoar, 1983) support such a view.

Fig. 18.9 Linear dunes: (**A, B**) vegetated simple linear dunes, south-western Kalahari; and (**C, D**) complex linear dunes, Namib sand sea

Star Dunes

Star dunes (Fig. 18.10) are the largest dunes in many sand seas and may reach heights of more than 300 m. They contain a greater volume of sand than any other dune type (Wasson and Hyde, 1983b) and appear to occur in areas that represent depositional centres in many sand seas (Dong et al., 2004; Lancaster, 1983; Mainguet and Callot, 1978). Star and related reversing dunes comprise 9–12% of all dunes in the Namib, Gran Desierto, and central Asian sand seas. They are absent from the Australian deserts, from the Kalahari, and from India, but comprise 40% of dunes in the Grand Erg Oriental (Mainguet and Chemin, 1984).

Star dunes are characterized by a pyramidal shape, with three or four arms radiating from a central peak and multiple avalanche faces. Each arm has a sharp sinuous crest, with avalanche faces that alternate in aspect as wind directions change. The arms may not all be equally developed and many star dunes have dominant or primary arms on a preferred orientation. The upper parts of many star dunes are very steep with slopes at angles of 15–30°; the lower parts consist of a broad, gently sloping (5–10°) plinth or apron. Small crescentic or reversing dunes may be superimposed on the lower flank and upper plinth areas of star dunes.

Star dunes have been hypothesized to form at the centres of convection cells, at the nodes of stationary waves in oscillating flows, above rock hills, or at the nodal points of complex dune alignment patterns created by crossing or converging sand transport paths (Lancaster, 1989a). Comparisons between the distribution of star dunes and wind regimes suggest that they form in multidirectional or complex wind regimes (Fryberger, 1979). A strong association between the occurrence of star dunes and topographic barriers has also been noted (Breed and Grow, 1979). Topography may modify regional wind regimes to increase their directional variability, as in the Erg Fachi Bilma and Namib (Mainguet and Callot, 1978; McKee, 1982;

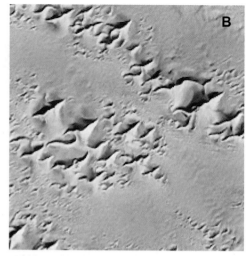

Fig. 18.10 Star dunes, Gran Desierto sand sea: (**A**) ground view; (**B**) Landsat image. Note preferred east-west orientation of major dune arms and small reversing dunes in middle distance

Lancaster, 1983), or create traps for sand transport, as at Kelso Dunes (Sharp, 1966), Dumont Dunes (Nielson and Kocurek, 1987) and Great Sand Dunes (Andrews, 1981).

Parabolic Dunes

Parabolic dunes (Fig. 18.11), common in many coastal dunefields and semi-arid to sub-humid areas, have a restricted distribution in arid region sand seas. The only major sand sea with significant areas of this dune type is in the Thar Desert of India (Kar, 1993; Singhvi and Kar, 2004; Verstappen, 1968). Small areas of parabolic

dunes occur in the south-western Kalahari (Eriksson et al., 1989), Saudi Arabia (Anton and Vincent, 1986), north-east Arizona (Hack, 1941), and at White Sands (McKee, 1966).

Parabolic dunes are characterized by a U shape with trailing partly vegetated parallel arms 1–2 km long and an unvegetated active 'nose' or dune front 10–70 m high that advances by avalanching. In the Thar Desert, compound parabolic dunes with shared arms occur in some areas and result from merging or shingling of several generations of dunes with different migration rates (Wasson et al., 1983). The conditions under which parabolic dunes form are not well known. They seem to be associated with the presence of a moderately developed vegetation cover, and with unidirectional wind

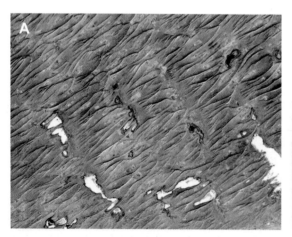

Fig. 18.11 Parabolic dunes: (**A**) Thar Desert, India (ASTER image); (**B**) White Sands, New Mexico (photograph by David Bustos, NPS)

regimes. Downwind, some parabolic dunes are transitional to crescentic dunes as vegetation cover decreases (Anton and Vincent, 1986).

Zibars and Sand Sheets

Not all aeolian sand accumulations are characterized by dunes. Low relief sand surfaces such as sand sheets are common in many sand seas and occupy from as little 5% of the area of the Namib sand sea to as much as 70% of the area of Gran Desierto (Lancaster et al., 1987). Extensive sand sheets also occur in the eastern Sahara (Breed et al., 1987; Maxwell and Haynes, 2001). Fryberger and Goudie (1981) estimate that 38% of aeolian deposits are of this type. Many sand sheets and interdune areas between linear and star dunes are organized into low rolling dunes, without slipfaces, known as zibars (Holm, 1960; Nielson and Kocurek, 1986; Warren, 1972) with a spacing of 50–400 m and a maximum relief of 10 m. Typically zibars are composed of coarse sand, and occur on the upwind margins of sand seas.

Sand sheets develop in conditions unfavourable to dune formation (Kocurek and Nielson, 1986). These may include a high water table, periodic flooding, surface cementation, coarse-grained sands, and presence of a vegetation cover all of which act to limit sand supply for dune building. Coarse grains armour the surface of sand sheets in the eastern Sahara (Breed et al., 1987). Sand sheets in the north-western Gran Desierto and the eastern Sahara are also composite features resulting from multiple generations of aeolian deposition separated by episodes of soil formation (Lancaster, 1993; Stokes et al., 1998). Those in the Gran Desierto have developed in conditions of a sparse vegetation cover which is insufficient to prevent sand transport taking place, but sufficient to cause divergence and convergence of airflow around individual plants in the manner suggested by Ash and Wasson (1983) and Fryberger et al. (1979) giving rise to localized deposition by wind ripples and shadow dunes.

Dune Processes and Dynamics

The initiation, development, and equilibrium morphology of all aeolian dunes are determined by a complex series of interactions between dune morphology, airflow, vegetation cover, and sediment transport rates. In turn, the developing bedforms exert a strong control on local transport rates through form-flow interactions and secondary flow circulations, leading to a dynamic equilibrium between dune morphology and local airflow. In multidirectional wind regimes, the nature of interactions between dune form and airflow change as winds vary direction seasonally, and lee side secondary flows become important. A conceptual framework for dune dynamics is provided by Fig. 18.2.

Dune Initiation

Dune initiation is poorly understood and little studied but is important in understanding how dunes and dune patterns develop. It involves localized deposition leading to bedform nucleation, which will then fix a pattern that can propagate downwind (Wilson, 1972). Reductions in the local sediment transport rate can occur through convergence of streamlines, by changes in surface roughness (e.g. vegetation cover, surface particle size), or by variations in microtopography (slope changes, relict bedforms).

There are few studies of the initiation and early development of dunes. Cooper, (1958), Jäkel (1980) and Kocurek et al. (1992) have described the development of barchans and transverse ridges from thin sand patches with no flow separation in their lee to small dunes with lee side flow separation, but only Kocurek et al. (1992) and Lancaster (1996) have documented the initiation of sand patches where changes in aerodynamic roughness or microtopography cause a reduction in near-surface wind speeds and transport rates.

Kocurek et al. (1992) recognized five stages of dune initiation and development (Fig. 18.12) with a progressive evolution of the lee face and bedform-induced secondary flow expansion and separation: (a) irregular patches of dry sand a few centimetres high, (b) 0.1–0.35-m-high protodunes with wind ripples on all surfaces, (c) 0.25–0.40-cm protodunes with grainfall on lee slopes, (d) 1–1.5-m-high barchans with grainflow, and (e) 1–2-m-high crescentic ridges. The developing dunes were characterized by a reverse asymmetry (steeper stoss slope) in stage (b) similar to that noted by Cooper (1958). The change from flow expansion to flow separation came as

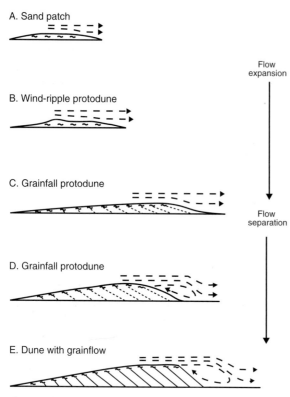

Flow expansion

Flow separation

Fig. 18.12 Stages of dune development at Padre Island, Texas (after Kocurek et al., 1992)

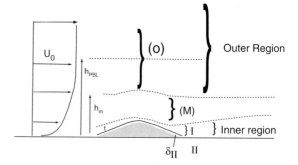

Fig. 18.13 Airflow over an isolated two-dimensional hill, showing components of the boudary layer development (after Nickling and McKenna Neuman, 1999). M – middle region

lee slopes exceeded 22°. Subsequent evolution and self-organization of the dunefield mainly takes place by repeated mergings, splittings, lateral linking, and cannibalization of dunes. Dune growth is at the expense of interdune areas and the pattern evolves so that the height: spacing ratio tends toward 1:20, a figure that is common in many areas of crescentic dunes.

Airflow Over Dunes

As dunes grow, they project into the atmospheric boundary layer so that they affect the airflow around and over them in a manner similar to isolated hills. Boundary layer theory and numerical modelling of flow over low hills provide a conceptual framework for understanding air flow over dunes, (e.g. Hunt et al., 1988; Jackson and Hunt, 1975; Jensen and Zeman, 1985; Mason and Sykes, 1979). Based on this framework, air flow over dunes can be divided into an outer inviscid region and an inner region which follows the topography of the dune (Fig. 18.13). The

inner layer is further divided into two sub-layers: (1) a very thin inner surface layer, in which the shear stress is constant, with a thickness equivalent to the surface roughness (1/30 of the grain diameter for a sand surface); and (2) a shear stress layer in which shear stress effects decrease with height above the surface. It is the shear stress generated in this part of the boundary layer that is responsible for sediment transport on dunes, yet its measurement in the field presents many problems (Wiggs, 2001). Estimates of the inner layer thickness range between 0.54 m (Lancaster et al., 1996) and 0.8 m (Wiggs, 2001). The basic features of this conceptual model have been confirmed by field studies of air flow over dunes and reproduced in numerical models (Parsons et al., 2004; Weng et al., 1991).

The Stoss or Windward Slope

Winds approaching the upwind toe of a dune stagnate slightly and are reduced in velocity (Wiggs, 1993) On the stoss, or windward slope of the dune, streamlines are compressed and winds accelerate up the slope. Data from field studies (Burkinshaw et al., 1993; Lancaster, 1985; Mulligan, 1988; Tsoar, 1985) have shown that the magnitude of the velocity increase is represented by the speed-up ratio Δs or amplification factor Az represented by U_2/U_1 where U_2 is the velocity at the dune crest and U_1 is the velocity at the same height above the surface at the upwind base of the dune. Typical values for Az range between 1.1 and 2.5 and vary with dune height and aspect ratio (Fig. 18.14), in good agreement with the Jackson and Hunt model. Field studies show that the rate of

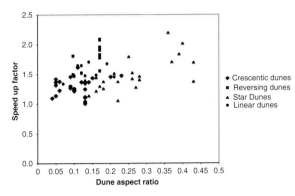

Fig. 18.14 Relations between velocity increase (amplification factor) and dune aspect ratio h/L. Data from author

velocity increase is not linear, and follows the surface of the dune closely (Fig. 18.15). Significant problems and uncertainties exist with the field measurement of wind shear velocity and shear stress on dunes in conditions of non-uniform flow and wind profiles that are not log-linear as a result of flow acceleration and development of internal boundary layers (Frank and Kocurek, 1996a). Wind-tunnel simulations and models indicate however that shear stress increases on dune slopes in a manner similar to that illustrated in Fig. 18.15. In addition to the effects of velocity

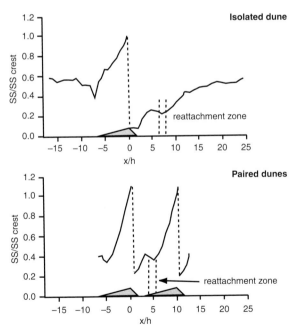

Fig. 18.15 Patterns of normalized shear stress (SS/SS crest) measured over wind tunnel models of isolated and paired transverse dunes (after Walker and Nickling, 2002)

amplification, curvature of wind streamlines may play an important role in the pattern of wind shear stress on dunes (Fig. 18.16). Concave upward curvature of streamlines at the toe of the dune enhances shear stress, whereas convex streamline curvature on the mid stoss slope and between the crest and the brink acts to decrease shear stress (Wiggs et al., 1996).

Lee Side Flow

In the lee of the crest of dunes, wind velocities and transport rates decrease rapidly as a result of flow expansion between the crest and brink of the lee or avalanche face and flow separation on the avalanche face itself. There is a complex pattern of flow separation, diversion, and re-attachment on the lee slopes of dunes, which is determined by the angle between the wind and the dune crest (angle of attack) and the dune aspect ratio (Walker and Nickling, 2002). Secondary flows, including lee-side flow diversion, are especially important where winds approach the dune obliquely, and are an important component of air flow on linear and many star dunes (Fig. 18.17). (Sweet and Kocurek, 1990) suggested that there are three types of flow in the lee of dunes: (a) separated, (b) attached, and (c) attached deflected. The nature of lee-side flow is controlled by the dune shape (aspect ratio), the incidence angle between the primary wind and the crestline, and the stability of the atmosphere (Fig. 18.18). High angles of attack on high aspect ratio (steep) dunes result in flow separation in the lee, which results in the development of an eddy in the lee of the dune. This may have the form of a roller vortex if flow is truly transverse. A conceptual model for airflow in the lee of flow-transverse dunes is provided by Walker and Nickling (Fig. 18.19) based on field and wind-tunnel studies (Frank and Kocurek, 1996b; Walker, 1999; Walker and Nickling, 2002). The model identifies a series of distinct regions of flow that vary in wind speed, shear, and turbulence intensity: a separation cell extending for 4–10 dune heights downwind (A), two wake regions (B, C) that merge at 8–10 h, and an internal boundary layer (D) that grows downwind of the point at which the separated flow reattaches to the surface. With isolated dunes, the internal boundary layer is identifiable at 8–10 h, and comes into equilibrium at around 25–30 h.

Fig. 18.16 Conceptual model of streamline convergence and divergence over a transverse dune (after Wiggs et al., 1996)

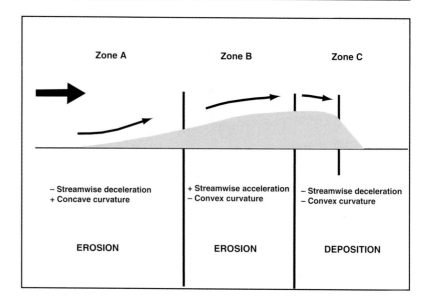

Zone A — Streamwise deceleration + Concave curvature — EROSION

Zone B + Streamwise acceleration − Convex curvature — EROSION

Zone C − Streamwise deceleration − Convex curvature — DEPOSITION

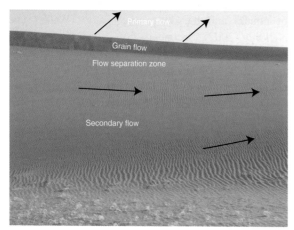

Fig. 18.17 Flow separation and diversion in the lee of a dune

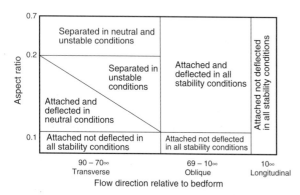

Fig. 18.18 Relations between lee-side flow speed, incidence angle, and aspect ratio for crescentic dunes (after Sweet and Kocurek, 1990)

When flow is oblique to the dune crest a helical vortex develops. The oblique flow is deflected along the lee slope parallel to the dune crest, with the degree of deflection being inversely proportional to the incidence angle between the crestline and the primary wind. Relatively high lee-side wind velocities are associated with low aspect ratio dunes and oblique primary winds. The degree of wind deflection is, in all cases, a cosine function of the incidence angle between the crestline and the primary wind.

When the angle between the dune and the wind is less than 40° the velocity of the deflected wind is greater than that at the crest and sand is transported along the lee side of the dune (Tsoar, 1983; Nielson and Kocurek, 1987; Lancaster, 1989a). Such a process is especially important on linear dunes, where it leads to dune extension, and on star dunes where it extends the arms of the dune. When winds are at more than 40° to the crestline the velocity of the deflected wind is reduced, giving rise to lee-side deposition. Changes in the local incidence angle between primary winds and a sinuous dune crest result in a spatially varying pattern of deposition and along-dune transport on the lee face. Deposition dominates where winds cross the crest line at angles approaching 90°, and erosion or along-dune transport occurs where incidence angles are <40° (Fig. 18.20).

Fig. 18.19 Conceptual model for flow in the lee of a transverse dune (after Walker and Nickling, 2002). Labelled regions represent: A – outer flow; B – overflow; C – upper wake; D – lower wake; E – separation cell; F – turbulent shear layer; G – turbulent stress maximum; H – turbulent shear zone; I – internal boundary layer

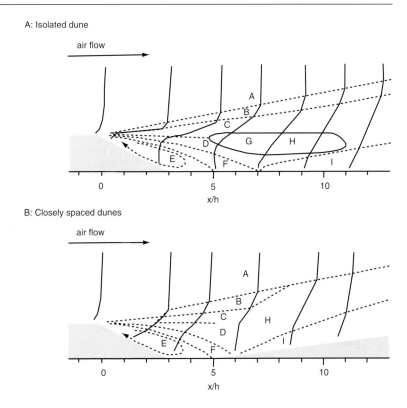

A: Isolated dune

B: Closely spaced dunes

Erosion and Deposition Patterns on Dunes

Flow acceleration, coupled with effects of stream line curvature, on the windward slopes of all dunes give rise to an exponential increase in sediment transport rates (Fig. 18.21) towards the dune crest (Lancaster et al., 1996; McKenna Neuman et al., 1997), resulting in erosion of the stoss slope. There are indications that the pattern of erosion and deposition varies with overall wind speed, so that areas near dune crests may experience erosion at lower wind speeds and deposition at wind speeds significantly above threshold (McKenna Neuman et al., 2000). In addition, numerical models suggest that the non-linear increase in sediment transport with height on a dune limits dune size and results in an equilibrium dune configuration (Momiji et al., 2000).

Downwind, the wind has to transport an increasing amount of sand eroded from the dune slope. This in turn requires that wind velocities and surface shear stress should change to increase transport rates proportionately. If the amount of sand in transport exceeds the capacity of the wind to transport it, deposition will occur, leading to adjustment of dune form (increasing dune steepness) and, hence, of local wind speed and transport rates. There is thus a high degree of interaction between the shape of the dune, the amount of change in wind velocities and sand transport rates and the rate and pattern of erosion or deposition.

Measurements of erosion and deposition on linear, reversing, and star dunes show that they consist of a crestal area where erosion and deposition rates are high and a plinth zone in which there is little surface change but considerable throughput of sand (Sharp, 1966; Lancaster, 1989b; Livingstone, 1989, 2003; Wiggs et al., 1995) (Fig. 18.22). On Namib linear dunes and Gran Desierto star dunes, over half the total amount of erosion and deposition takes place in the crestal zone (Lancaster, 1989a,b; Livingstone, 1989), as the position of the crestline varies seasonally over a distance of 3–15 m. The observed patterns of erosion and deposition follow changes in wind speed over the dunes and are directly related to the magnitude of velocity and transport rate amplification on dune slopes through the requirements of sediment continuity. This does not lead to a lowering of the dune crest because this zone is reworked from season to season, and there is very little net change in the position of the crestline. Time-series of erosion and

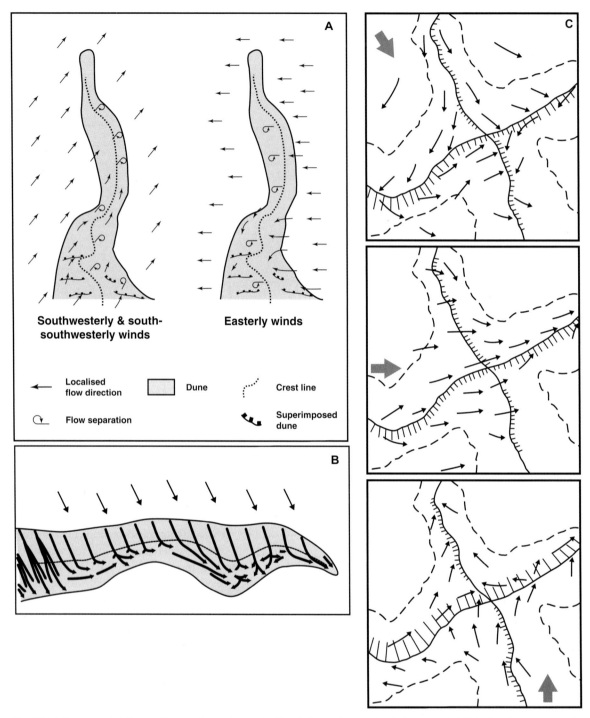

Fig. 18.20 Patterns of airflow on linear and star dunes: (**A**) Simple linear dune – after Bristow et al., 2000; (**B**) Linear dune – after Tsoar, 1983; (**C**) Star dune – after Lancaster, 1989a

Fig. 18.21 Changes in wind speed and sediment mass flux on the stoss slopes of flow transverse dunes: Silver Peak - reversing transverse dune (McKenna Neuman et al., 1997); Salton Sea - barchan dune (Lancaster et al., 1996)

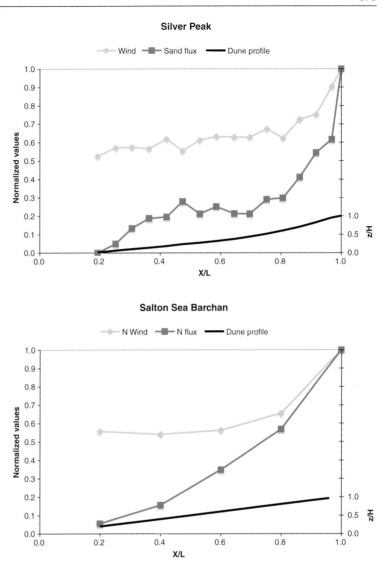

deposition show that the greatest amount of change occurs when winds change direction seasonally. This is the result of winds encountering a dune form that is out of equilibrium with a new wind direction. Field observations and models show that the crestal profiles of linear and star dunes tend toward a convex form similar to that of transverse dunes (Tsoar, 1985) and erosion and deposition rates near the crest decline as the dune comes into equilibrium with a new wind direction.

Despite their importance to the dynamics of modern dunes and the interpretation of the rock record (Howell and Mountney, 2001), there have been few studies of the depositional processes that occur on dune lee faces.

Prior studies consist of two modeling efforts (Anderson, 1988; Hunter, 1985), and three known field experiments (Hunter, 1985; McDonald and Anderson, 1995; Nickling et al., 2002). Following the terminology of (Hunter, 1977), grain flows occur when sediment deposited on the lee face by grain fall from the overshoot of saltating grains transported over the brink of the lee face builds up so that the lee slope is steepened above the angle of repose and fails, initiating a grain flow or avalanche.

McDonald and Anderson (1995) proposed a numerical model to account for the pattern of grainfall deposition developed on the lee of transverse dunes that suggested an exponential decay in deposition rates, rather

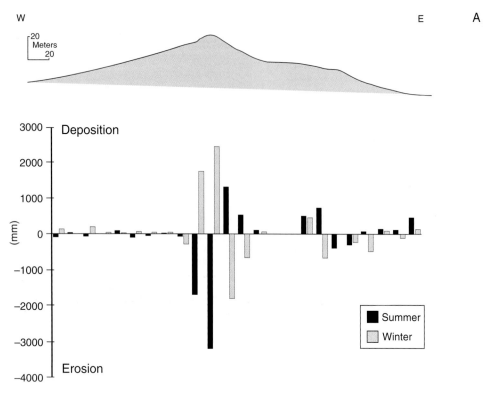

Fig. 18.22 Patterns of annual net erosion and deposition across a Namib complex linear dune. Primary winds is oblique to the dune from the south-south-west to the south-west (summer) or NE-E (winter). After Livingstone (1989)

than the power function suggested by Hunter (1985). Field testing of the model appears to confirm the exponential decay as well as the development of a depositional maximum on the upper lee slope. The experimental data, while providing a good fit to the numerical model, also indicates the importance of wake effects and suspension of grains in the lee of the dune. McDonald and Anderson also concluded that increased wind speeds resulted in more frequent, rather than larger avalanches.

Field experiments conducted by Nickling et al. (2002) clearly indicated that the decay of grainfall with horizontal distance is best described by an exponential function, with 55–95% of the grainfall being deposited within 1 m of the crest (Fig. 18.23a). However, comparison of field data with predictions from the model of Anderson (1988) showed significant lack of agreement with respect to the magnitude of deposition rates and the magnitude and location of the lee slope depositional maximum, which was an order of magnitude larger and located further down slope than indicated by the model. Avalanches appear to be

initiated downslope from this area, as confirmed by field observations in Namibia and experimental wind tunnel studies.

Wind tunnel experiments using a small but true-scale artificial flow-transverse sand dune (Cupp et al., 2004) indicate that grainfall decreases approximately exponentially with distance from the brink, confirming prior field studies. The grainfall flux to the lee slope scales with overall crest wind speed and therefore incoming saltation flux (Fig. 18.23a) and shows the development of a small depositional maximum just downslope from the brink. Significant quantities of grainfall are redistributed downslope by the reptation of grains on the surface of the lee slope, a previously unrecognized process. Reptation rates scale with grainfall rates (Fig. 18.23b) and therefore decline rapidly with distance from the brink. The spatial variations in reptation rates downslope are reflected in changes in the profile of the lee slope and suggest that reptation is primarily responsible for the formation of a large sediment wedge across the upper lee slope. This sediment wedge contributes to the failure of the

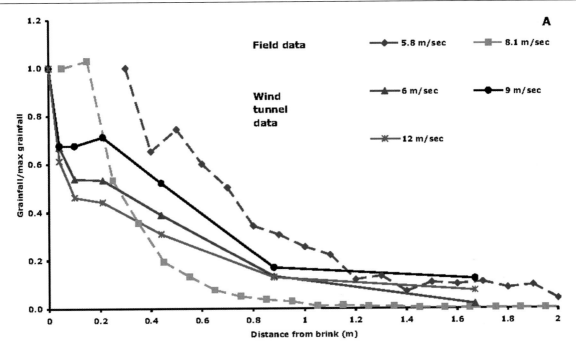

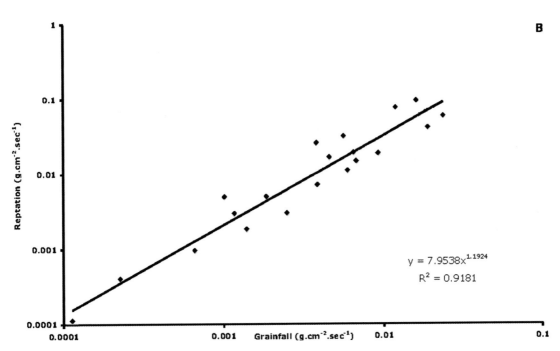

Fig. 18.23 (**A**) Depositional patterns in the lee of a transverse dune at varying incident wind speeds (field data from Nickling et al., 2002; and wind tunnel data from author); and (**B**) relations between rate of grainfall and rate of downslope reptation on experimental wind tunnel dune (author unpublished data)

upper lee slope, and the resulting redistribution of sand by grainflow. The experiments show that grainflow magnitude remained relatively constant with wind speed whereas grainflow frequency increased with wind speed.

Long Term Dune Dynamics

The majority of the process studies discussed above have a duration of days or weeks, and the measured rates of erosion and deposition are difficult to extrapolate to the longer time scales over which dunes and dune fields form. One approach is to model dune behaviour over time using information on the fundamentals of dune dynamics. This field is still in its infancy but recent studies show that this approach can produce realistic dune shapes and patterns using a number of different approaches (Partelli et al., 2006; Schwämmle and Herrmann, 2004; Werner, 1995).

Long term monitoring can provide insights into dune dynamics over annual to decadal timescales. Comparison of the position of dunes on time-series of aerial photographs or satellite images has provided a large data set on dune migration rates. An inverse relationship between barchan dune height and migration rate (Fig. 18.24) has been determined by using this approach by numerous investigators (Haff and Presti, 1995; Long and Sharp, 1964; Sweet et al., 1988; Finkel, 1959; Hastenrath, 1967; Marîn et al., 2005; Slattery, 1990). Such investigations are now facilitated by GIS and GPS methods which also enable changes in dune volume (including dissipation of dunes) to be determined (Bristow and Lancaster, 2004; Stokes et al., 1999).

Studies of the long-term dynamics of individual dunes are rare. Monitoring of surface topographic change on a large linear dune in the central Namib Desert since 1980 (Livingstone, 1989, 1993, 2003) has shown that the crest region of the dunes is the most active as it is reworked by winds from different directions according to season. The crestlines migrate over a lateral distance of as much as 14 m over a 12-month period but with little net change over periods of years. The crest also changed from a relatively high single crest in the 1980s to a slightly lower double crest form in the 1990s, and then back to single crest form by 2001, regaining much of its original height (Fig. 18.25). The lower, or plinth, areas of the

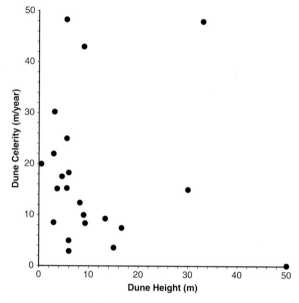

Fig. 18.24 Relations between crescentic dune migration rate (celerity) and dune height. Scatter in points in y direction is a function of the overall wind energy at the different localities. Data are mean values calculated from Finkel (1959), Long and Sharp (1964), Hastenrath (1967), Tsoar (1974), Embabi (1982), Endrody-Younga (1982), Haynes (1989), and Slattery (1990)

dune showed little change over the period of study. Livingstone (2003) attributed the changes in dune crest characteristics to changes in the relative magnitude and frequency of strong easterly winter winds, which increased in the late 1980s. The studies also suggested that these large dunes are not migrating laterally.

A longer-term (centuries to millennia) perspective on dune dynamics can now be gained through the application of high-resolution OSL dating of dunes, especially when used in combination with GPR studies of dune structures. Recent studies in the Namib Desert show that crescentic dunes superimposed on the southern edge of a N-S oriented linear dune (Fig. 18.26) have migrated to the west at an average rate of 0.12 m/yr over the past 1570 years (Bristow et al., 2005), while a large linear dune shows evidence of episodic development and lateral migration at rates of up to 0.13 m/yr over the past 6000 years (Bristow et al., 2007). Bray and Stokes adopted a similar approach to show that a 20 m high transverse dune in the Liwa area had migrated 250 m to the south over the past 320 years at an average rate of 0.78 m/yr, with more rapid migration of 0.91 m/year between 220 and 110 years ago (Stokes and Bray, 2005).

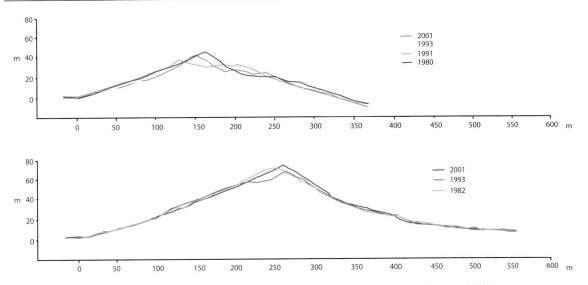

Fig. 18.25 Changes in the profile form of a Namib linear dune over a 20 year period (after Livingstone, 2003)

GPR studies can also provide information on the development and growth of dunes through information on sedimentary structures. Bristow et al. (2000) showed how linear dunes in the Namib Sand Sea go through a series of stages in which secondary flows and form-flow interactions become more important as the dune increases in size, leading ultimately to the establishment of superimposed bedforms (Fig. 18.27).

Controls of Dune Morphology

Recent work has recognized that the directional variability of the wind regime is a major determinant of dune type. Wind speed, grain size, and vegetation play subordinate roles. The effects of sand supply are uncertain (Rubin, 1984; Wasson and Hyde, 1983b).

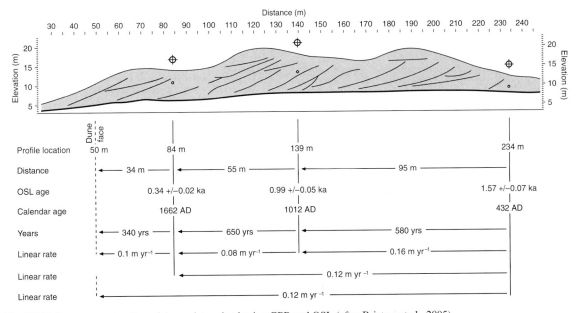

Fig. 18.26 Long term migration of dunes determined using GPR and OSL (after Bristow et al., 2005)

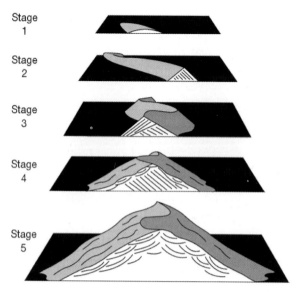

Fig. 18.27 Growth and extension of a linear dune (after Bristow et al., 2000)

Sediment Characteristics

There appears to be no evidence for a genetic relationship between the grain size and sorting character of sands and dune type, except that sand sheets and zibar are often composed of coarse, poorly sorted, often bimodal, sands (Bagnold, 1941; Kocurek and Nielson, 1986; Lancaster, 1982a; Maxwell and

Haynes, 1989; Nielson and Kocurek, 1986; Warren, 1972). The hypothesis of Wilson (1972) that dune spacing is related to sand particle size is not supported by empirical data (Wasson and Hyde, 1983a).

Wind Regimes

The association of dunes of different morphological types with wind regimes which have different characteristics, especially of directional variability, has been noted by many investigators. Fryberger (1979) compared the occurrence of each major dune type (crescentic, linear, star) on Landsat images of sand seas with data on local wind regimes, using the ratio between total (DP) and resultant sand flow (RDP) as an index of directional variability. High RDP/DP ratios characterize near unimodal wind regimes, whereas low ratios indicate complex wind regimes. Fryberger (1979) found that the directional variability or complexity of the wind regime increases from environments in which crescentic dunes are found to those where star dunes occur (Fig. 18.28). Crescentic dunes occur in areas where RDP/DP ratios exceed 0.50 (mean RDP/DP ratio 0.68) and frequently occur in unimodal wind regimes, often of high or moderate energy. Linear dunes develop in wind regimes with a much greater degree of directional variability and

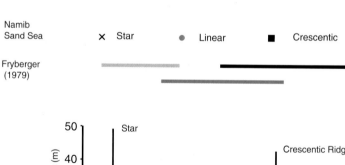

Fig. 18.28 Relations between dune type, wind regime variability (RDP/DP ratio), and sand thickness (RDP/DP data from Fryberger, 1979; Wasson and Hyde, 1983a; and Lancaster, 1989b). Note that linear dunes in the Namib occur in less variable wind regimes than those in Wasson and Hyde's sample

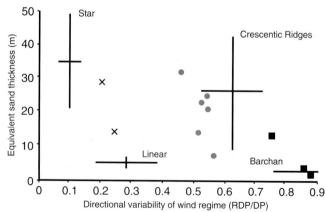

commonly form in wide unimodal or bimodal wind regimes with mean RDP/DP ratios of 0.45. Star dunes occur in areas of complex wind regimes with RDP/DP ratios less than 0.35 (mean = 0.19).

Comparison of the distribution of dunes of different morphological types in the Namib sand sea with the information on sand-moving wind regimes (Lancaster, 1989b) tends to confirm Fryberger's hypothesis that there is an increase in the directional variability of the sand-moving wind regime from crescentic to star dunes. However, the overall directional variability of wind regimes in the Namib sand sea is less than that in Fryberger's global sample (Fig. 18.28). Computer simulations of dune patterns using a cellular automaton approach further demonstrate that each of the major morphological types is independent of initial

conditions and forms an attractor in a complex system (Werner, 1995; Werner and Kocurek, 1997; Werner and Kocurek, 1999), in which wind regime variability (equivalent to variability in transport directions) is the main determinant of dune type and orientation (Fig. 18.29).

Process studies give some indications as to why dunes of different types should occur in different wind regimes. The primary response of sand surfaces to the wind is to form an asymmetric transverse dune with a convex stoss slope. This is the most common dune form in unidirectional wind regimes. In multidirectional wind regimes, profiles of the crestal areas of linear and star dunes tend toward this form in each wind season. As dunes grow, the cross-sectional area increases exponentially and their reconstitution time

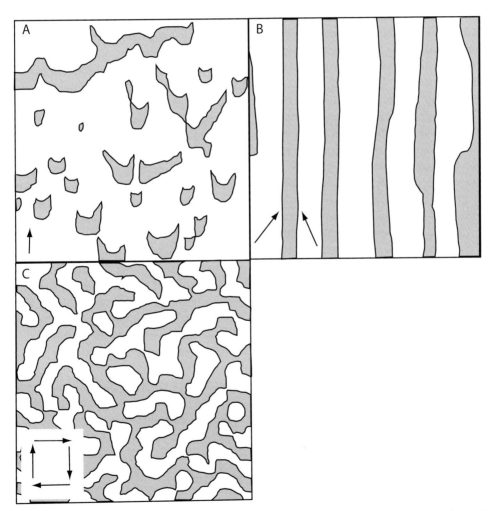

Fig. 18.29 Numerical simulation of dune morphological types and patterns (after Werner, 1995). Arrows show principal sand transporting wind directions used in simulations

increases by one or two orders of magnitude. Because of their size, they can no longer be remodelled in each season, so form–flow interactions become significant, and they develop a morphology that is controlled by more than one wind direction. There are many examples of dunes of different types occurring together in the same wind regime: the small dunes are almost always crescentic forms because they can be reformed entirely in each wind season; larger dunes often tend to be linear or star types.

The essential mechanism for linear dune formation is the deflection of winds that approach at an oblique angle to the crest to flow parallel to the lee side and transport sand along the dune. Thus any winds from a 180° sector centred on the dune will be diverted in this manner and cause the dune to elongate downwind. Linear dunes are not stable in a unidirectional wind regime because erosion and deposition are concentrated at the same localities, resulting in their eventual break-up (Tsoar, 1983). Thus a seasonal change in wind direction is necessary to maintain the dune and its triangular profile.

In bimodal wind regimes, deflection of oblique winds on the lee side will tend to elongate the form, producing a linear dune. The effects of each wind direction will be controlled by their direction relative to the dune. If a high percentage of winds are at optimal angles for lee-side deflection, then the dune will extend strongly, and such dunes will tend to be long and relatively low. If winds blow at higher angles to the dune, then longitudinal movements of sand will be replaced by deposition on lee-side avalanche faces. Sand will tend to stay on the dune and increase its height. A limiting case will occur when winds are perpendicular to the crest, producing a reversing dune. Such dunes tend to accrete vertically and develop major lee-side secondary flow cells that move sand toward the centre of the dune, leading to the formation of star dunes.

The development of star dunes is strongly influenced by the high degree of form–flow interaction, which occurs as a result of seasonal changes in wind direction, and the existence of a major lee-side secondary circulation. The pattern of winds on star dunes indicates that most of the resultant erosion and deposition involves the reworking of deposits laid down in the previous wind season. Sand, once transported to the dune, tends to stay there and add to its bulk. This is a result of the high deposition rates on lee faces and the fact that

wind velocity and sand transport rates decrease away from the central part of the dune. The prominent lee-side secondary flows also tend to move sand towards the centre of the dune and promote the development of the arms.

Sand Supply

The availability of sand for dune building has long been considered a factor influencing dune morphology (Hack, 1941; Wilson, 1972). Using a sample of dunes of different types from sand seas in all major desert areas, Wasson and Hyde (1983b) established that the mean equivalent or spread-out thickness (EST) of dune sand in a given area for all dune types was statistically identical and concluded that although sediment availability was a significant variable determining dune type, it was not the only one.

However, by plotting EST against Fryberger's RDP/DP ratio, a clear discrimination of dune types was achieved (Fig. 18.28), leading to the conclusion that barchans occur where there is very little sand and almost unidirectional winds; transverse dunes are located where sand is abundant and winds variable; linear dunes develop where sand supply is small, but winds are more variable; and star dunes form in complex wind regimes with abundant sand supply. Similar relationships are evident in the Namib sand sea, although the range of directional variability is less (Fig. 18.28).

However, EST is not a measure of sand supply, but of the volume of sand contained in the dunes and may be a reflection of dune type with the dune type being influenced by the other factors, especially the wind regime (Rubin, 1984). In the Namib sand sea it is possible to clearly discriminate between dune types on the basis of their relationships with wind regimes. The EST data merely suggest that there is more sand in complex linear and star dunes than in compound linear and all types of crescentic dunes.

Vegetation

The effects of vegetation on sand transport, which include direct protection of the surface, absorption of momentum, and partitioning of shear stress between the surface and plants are discussed in depth in Chapter 17. The effects of vegetation on dune morphology are, however less well known. Hack (1941) suggested that in north-eastern Arizona there was a transition from crescentic to parabolic dunes with increased vegetation cover, and that linear dunes occurred in areas with less sand and vegetation than parabolic dunes (Fig. 18.30). In the Negev Desert, Tsoar and Møller (1986) documented changes in the morphology of linear dunes as vegetation was removed, including development of sharp sinuous 'seif' dune crests and braided patterns.

Many dunes, even in hyperarid regions like the Namib, are vegetated to some extent, mostly in the plinth areas of linear and star dunes where relatively little surface change occurs (Thomas and Tsoar, 1990). In other areas, such as the Kalahari and Australian deserts, dune crests may be sparsely vegetated, while vegetation cover is sufficient to restrict sand transport on dune flanks. Partially vegetated linear dunes are widespread in the Australia deserts and in the Kalahri of southern Africa. Stratigraphic evidence and OSL dating (e.g. Stokes et al., 1997) shows that many linear dunes of this type were originally formed during the late Pleistocene and have since become stabilized by vegetation in more humid and/or less windy conditions, although localized Holocene activity has been documented. Studies of dune dynamics in relation to vegetation cover show that the crests of these dunes can be very active, with relatively large amounts of erosion and deposition being recorded, especially in periods of drought or after disturbance e.g. by fire (Bullard et al., 1997; Hesse and Simpson, 2006; Thomas and Leason, 2005; Wiggs et al., 1994, 1995). As pointed out by Bullard and others, inter-annual and decadal scale changes in rainfall, temperature, and wind strength give rise to significant temporal changes in the amount of vegetation cover and dune surface activity. Empirical studies (Lancaster and Baas, 1998; Wiggs et al., 1995) suggest that a vegetation cover of 14% is sufficient to restrict sand transport on dune surfaces. Thomas and Leason (2005) used Landsat image data to show that the proportion of the area of the SW Kalahari dune field with less than 14% vegetation cover ranged between 10 and 16% for dry years but was only 3–6% for wet years, suggesting that dune activity is likely much more extensive after periods of extended drought.

Controls on Dune Orientation

The controls on the orientation or alignment trend of dunes have long been a subject for speculation. Many workers have concluded that dunes are oriented relative to the resultant or vector sum of sand transport (Fryberger, 1979). Thus dunes can be classified as transverse (strike of crestline approximately normal to resultant), longitudinal (crest parallel to resultant), or oblique (15–75° to resultant direction) (Hunter et al., 1983; Mainguet and Callot, 1978). Others believe that dunes are oriented parallel to or normal to prevailing winds (Bagnold, 1953; Glennie, 1970), or that the trend of dunes is influenced by secondary flows (Tsoar, 1983). As discussed above, numerous explanations of why dunes are longitudinal or transverse to transport directions have been advanced, including wind regime characteristics, existence of helical vortices, and sand availability.

Field and laboratory experiments with wind ripples and sub-aqueous dunes (Rubin and Hunter, 1987; Rubin and Ikeda, 1990) suggest that all types of bedforms are oriented in the direction subject to the maximum gross bedform-normal transport across the crest (Fig. 18.31) leading to the conclusion that a wide variety of wind regimes can produce the same dune trend.

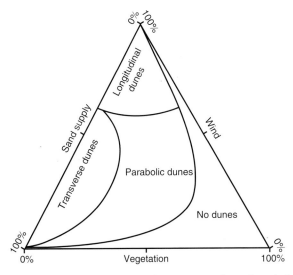

Fig. 18.30 Relations between dune type, sand supply, wind strength, and vegetation cover (after Hack, 1941)

Fig. 18.31 Gross bedform
normal (GBN) transport
direction and dune trends;
(**A**) relations between dune
trend and GBN direction
(after Lancaster, 1991, with
additional data); (**B**)
Explanation of Gross
Bedform Normal (after Rubin
and Ikeda, 1990)

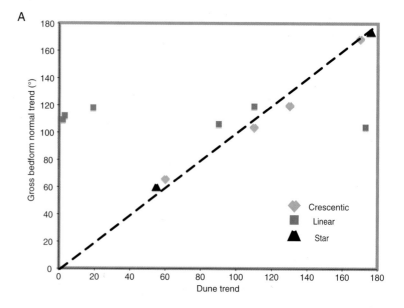

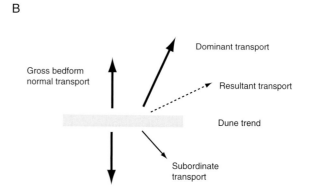

In this approach, transport from all directions contributes to bedform genesis and growth. By contrast, use of the resultant direction of transport as a control of bedform orientation implies that transport vectors from opposing directions cancel out each other and thus do not contribute to bedform growth. The type of dune that occurs is determined by the divergence angle between the dominant and subordinate transport vectors and the ratio between the two transport directions (Transport ratio). A trend parallel to or normal to the resultant direction of sand transport is purely coincidental.

Using a sample of dunes from sand seas in Namibia and the southwestern United States, Lancaster (1991) determined that there was a close agreement between observed and predicted gross-bedform normal orienta-

tions in the case of many barchans, crescentic dunes, simple linear dunes, and star dunes (Fig. 18.31), suggesting that all major dune types are oriented to maximize gross bedform-normal sediment transport and therefore are dynamically similar. Differences between the form and orientation of dunes therefore result from variations in the directional characteristics of the flow, especially the angle(s) between the major transport directions and the ratio(s) between the magnitudes of flows from different directions.

This approach can also be used to identify dune trends that are not in equilibrium with the modern sand-transporting wind regime, as demonstrated for Mauritania (Lancaster et al., 2002), and to suggest wind regimes that could have produced the trends of dunes constructed in the past.

Development of Compound and Complex Dunes

Many large dunes (complex and compound dunes, draas, mega dunes) exhibit superimposed dunes. There has been much debate over the orgins of the superimposed bedforms. Lancaster (1988) showed that crescentic dunes superimposed on linear dunes were clearly in equilibrium with secondary airflow patterns on the larger dune form. Similar observations were made for crescentic dunes (Havholm and Kocurek, 1988; Rubin and Hunter, 1982). The slopes of large dunes present an effectively planar surface on which sand transport takes place. Therefore, variations in sand transport rates on mega dunes in time or space will lead to the formation of superimposed dunes if the major dune is sufficiently large. This suggests that there is a minimum size for compound and complex dunes. Although the sample size is small, data on the width and spacing of simple, compound, and complex crescentic and linear dunes (Breed and Grow, 1979) indicate that the mean sizes of simple, compound, and complex dunes are statistically significantly different from one another, suggesting that a minimum dune size must be reached before superimposed dunes can develop. In the Namib sand sea, simple crescentic dunes have a spacing of less than 500 m, whereas compound dunes are all more than 500 m apart. There is, however, a continuum of the height and spacing of the major bedform from simple to compound and complex types. This suggests that, given sufficient sand supply and time, simple dunes may grow into compound and complex forms. This seems to be confirmed by the sedimentary structures of linear dunes (e.g. Bristow et al., 2000), which show the dominance of superimposed dunes once the dune reaches a certain size.

In some areas, however, the larger primary dunes are clearly a product of past wind regime and sediment supply conditions as demonstrated by OSL dating of the major form. Large dunes have considerable inertia and require significant periods of time to adjust to changed conditions of wind regime and sediment supply (Warren and Allison, 1998). The inertia of a dune can be represented by their reconstitution time, or the time required for the bedform to migrate its length in the direction of net transport. In the Namib, typical complex linear dunes in the northern parts of the sand sea have a width of approximately 600 m and migrate laterally at a rate of $0.1\,\mathrm{m\,y}^{-1}$ (Bristow et al., 2007). The minimum reconstitution time for these dunes is therefore 6000 years. Crescentic dunes superimposed on the flanks of the linear dunes have a mean spacing of 90 m and migrate at a rate of $3\,\mathrm{m\,y}^{-1}$, giving a reconstitution time of 30 years. Reconstitution time therefore increases by several orders of magnitude from simple to complex dunes. This implies that the morphology of simple dunes and superimposed dunes is governed principally by annual or seasonal patterns of wind speed and direction and by spatial changes in wind speed over draas. The lifespan of these dunes is about 10–100 years. Compound and complex dunes are relatively insensitive to seasonal changes in local air flow conditions and may persist for 1000–10000 years, as confirmed by OSL ages obtained from many large dunes.

Sand Seas

Sand seas (also called ergs) are dynamic sedimentary bodies that form part of regional-scale sand transport systems in which sand is moved by the wind from source zones to depositional sinks. Major desert sand seas occur in the Saharan and Arabian deserts, the Thar Desert of India, interior Australia, and southern Africa (Namib and Kalahari deserts) (Breed et al., 1979; Wilson, 1973). The source of sediment for sand sea accumulation is commonly external to the sand body, although internal sources, e.g. alluvial deposits in interdune areas, may be important (e.g. in Australia). Sediment for transport by wind may be derived from deflation of bedrock (Besler, 1980), but the primary source of sand-sized grains is sediments deposited by fluvial/alluvial, coastal, or lacustrine systems (e.g. Muhs, 2004; Muhs et al., 1995). In those sand seas for which the source is known, fluvial systems provide the major input of sand for wind transport, either directly through deflation of alluvial sediments, or indirectly through transfer of sediment in coastal sediment transport systems, as in the Namib, Gran Desierto, Sinai, Atacama, and Arabian sand seas (Lancaster, 1999).

Deposition of sand and the accumulation of sand seas occur downwind of source zones wherever sand transport rates are reduced as a result of changes in climate or topography. Compilations of wind data together with information from aerial photographs and satellite images (Fryberger and Ahlbrandt, 1979; Mainguet, 1984b; Wilson, 1971) show that long-distance transport of sand by the wind occurs in the Namib, Sahara, and Arabian Deserts (Fig. 18.32). However, Australian sand seas and many North American dunefields receive sand from local sources (Blount and Lancaster, 1990; Muhs, 2004; Muhs et al., 2003; Ramsey et al., 1999; Wasson et al., 1988). Some sand seas accumulate where sand transport pathways converge in the lee of topographic obstacles (e.g. Fachi-Bilma Erg) or in areas of low elevation (Wilson, 1971; Mainguet and Callot, 1978; Wasson

et al., 1988). Others occur where winds are checked by topographic barriers (e.g. Grand Erg Oriental, Kelso Dunes, Great Sand Dunes). Decelerating winds and reduced potential rates of sand transport may be the result of changes in regional circulation patterns that decrease wind speeds and/or increase their directional variability. Sand seas and dunefields in the western and southern Sahara, Saudi Arabia, the eastern Mojave and Sonoran Deserts, and the Namib occur in areas of low total or net sand transport compared with areas without sand sea development (Fryberger and Ahlbrandt, 1979; Fryberger et al., 1984; Lancaster, 1989b; Mainguet, 1984b). Many sand seas (e.g. the Akchar and Makteir sand seas in Mauritania

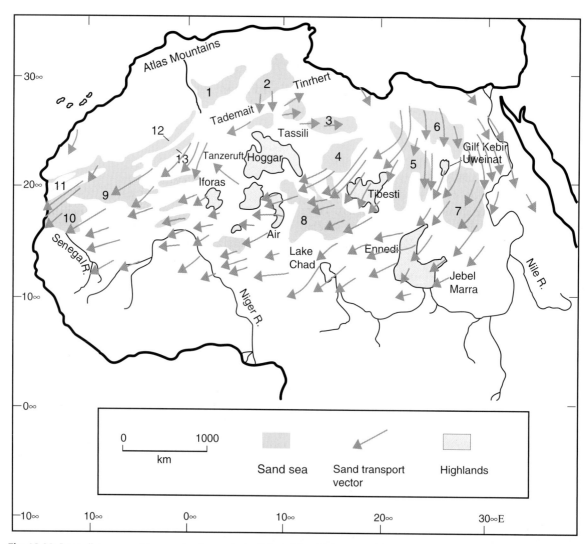

Fig. 18.32 Long distance sand transport pathways in the Sahara (after Mainguet and Chemin, 1983)

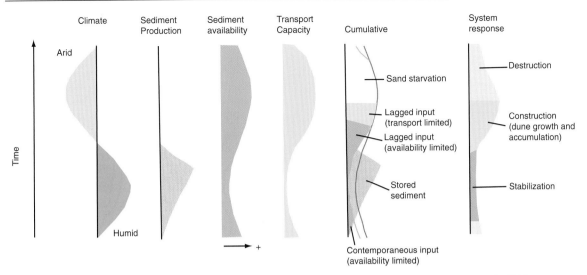

Fig. 18.33 Variation in sediment states and system response over time in a hypothetical example (after Kocurek, 1998)

and the Nafud of Saudi Arabia) are crossed by sand transport pathways. The same winds that transport sand to the sand sea can also remove it at the downwind end. These are the 'flow crossed' sand seas of Wilson (1971).

Sand seas are important depositional landforms and have accumulated over periods of 10^3 to 10^4 years, during which changes in climate and sea level have affected sediment supply, availability, and mobility. The relations between sediment supply, availability, and mobility at any point in time, as well as their variation through time, define a series of states of the system (Kocurek, 1998; Kocurek and Lancaster, 1999), as summarized in Fig. 18.33. They also determine the response of the system to external forcing factors so that a sand sea may be described as:

(1) Transport limited, in which actual rate of sand transport (Qa) equals the potential rate (Qp) and the system is limited only by the capacity of the wind to move sediment from source zones. Examples are the Namib sand seas and many coastal dune fields.
(2) Availability limited, in which Qa < Qp and the system response is controlled by vegetation cover, for example as in the Kalahari and Australia.
(3) Supply limited, in which Qa << Qp and the system is starved of sediment, as in many central and

northern sand Saharan sand seas, some of which are apparently eroding in present conditions.

Dune Patterns in Sand Seas

Many sand seas show a clear spatial patterning of dune types (Fig. 18.34), dune size and spacing, crest orientation, and sediment thickness (Breed et al., 1979; Ewing et al., 2006; Lancaster, 1999). Once thought to be largely the product of regional changes in wind regimes (e.g. Lancaster, 1983), sharp transitions between dune types and differences in their sedimentary characteristics (e.g. in the Simpson-Strzelecki, the Wahiba Sands, and the Gran Desierto) suggest that many sand seas are composed of different generations of dunes, each with a distinct morphology and sediment source (Lancaster, 1999). Dune generations (Fig. 18.35) can be recognized in many ways including: differences in dune morphology giving rise to statistically distinct populations of dunes with different crest orientation, spacing, and defect density (Kocurek and Ewing, 2005); variations in dune sediment composition, color, and particle size (Teller et al., 2002; Wasson, 1983); variations in vegetation cover and dune activity (Lancaster, 1992); and geomorphic relations between dunes of different types (e.g. crossing patterns, superposition of dunes) (Lancaster et al., 2002).

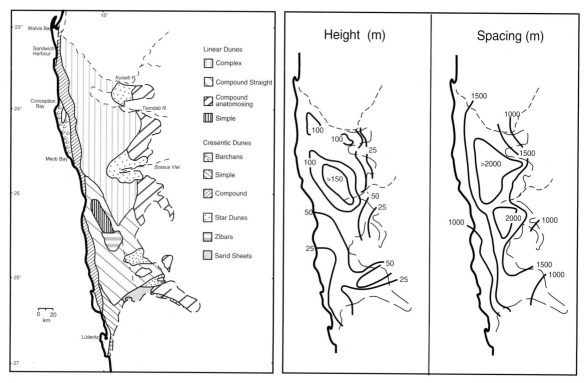

Fig. 18.34 Spatial variation dune types and morphology in the Namib Sand Sea (after Lancaster, 1989b)

Concepts of self-organization of complex systems (Werner, 2003; Werner and Kocurek, 1999), indicate that dune patterns should evolve through time as a result of interactions between dunes so that spacing increases asymptotically and defect density decreases. Further, each generation of dunes represents the product of a set of initial conditions of wind regime and sediment supply, together with evolution of the pattern

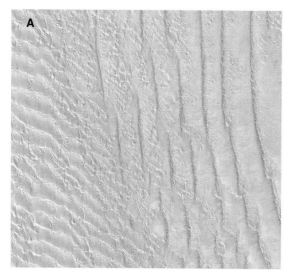

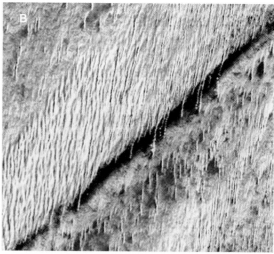

Fig. 18.35 Examples of dune generations: (**A**) juxtaposition and superposition of pale compound crescentic and redder complex linear dunes, Namib Sand Sea; and (**B**) superimposition of modern N-S oriented linear dunes on NE-SW oriented late Pleistocene red brown linear dunes, Azefal Sand Sea, Mauritania

over time by merging, lateral linking, migration of defects, and creation of terminations (Fig. 18.36) (Kocurek and Ewing, 2005). As a result, dune patterns (especially those of crescentic dunes) should evolve downwind, as demonstrated by dunes at White Sands and on the Skeleton Coast of Namibia. Change in dune patterns can only occur at terminations, so that dunes with few terminations per unit crest length (e.g. linear dunes) will tend to be more stable than those with large numbers of terminations (e.g. many crescentic dunes). These concepts are broadly supported by empirical data from several sand seas (Ewing et al., 2006), as discussed below.

The Azefal, Agneitir, and Akchar Sand Seas of western Mauritania show three distinct crossing trends of linear dunes (Fig. 18.37). Analyses of Landsat images, in conjunction with geomorphic and stratigraphic studies, and OSL (optically stimulated luminescence) dating of dunes demonstrate the existence of three main generations of dunes that were formed during the periods 25–15 ka (centered around the Last Glacial Maximum), 10–13 ka (spanning the Younger Dryas event), and after 5 ka (Lancaster et al., 2002). Modelling of the wind regimes that produced these dunes using the gross bedform normal approach shows that the wind regimes that occurred during each of these periods were significantly different, leading to the formation of dunes on three distinct superimposed trends-northeast, north-northeast, and north. Satellite images of the Azefal Sand Sea show clear evidence for reorientation of the termini of older dunes, supporting the model of Werner (1995) and Werner and Kocurek (1997) for change in dune patterns at termination sites. In this model, because linear dunes have few terminations per crest length, these dunes are remarkably stable features. The stability of linear dunes is further illustrated by the superposition of progressive generations of linear dunes at an angle to older trends, arguing for a rate of dune formation that is greater than the rate at which the older, larger dunes can be reoriented.

The Gran Desierto Sand Sea of northern Mexico exhibits a spatially diverse and complex pattern of dunes of several morphological types (Figs. 18.38) (Blount and Lancaster, 1990; Lancaster, 1995b). Analyses of dune patterns, coupled with OSL dating, shows that the sand sea has evolved by coalescence of several generations of dunes, each created during relatively short-lived periods of aeolian construction in the late Pleistocene and Holocene (Beveridge

et al., 2006). The characteristic star dunes represent a late Holocene modification (<3 ka) of linear dunes originally constructed between 26 and 12 ka; degraded crescentic dunes formed around 12 ka; while compound crescentic dunes on the easteren margin of the sand sea formed around 7 ka.

Conclusions and Future Research Directions

Major advances in our knowledge of the dynamics of dunes and sand seas have occurred in the past three decades. There is now a general understanding of the processes that form or at least maintain most major dune types. The importance of scale effects is recognized, so that the processes that form wind ripples or simple dunes need to be considered at different temporal and spatial scales from those that form compound and complex dunes and sand seas. The importance of past conditions of sediment supply, availability, and mobility, determined by climatic and sea level changes, in the formation of mega dunes and sand seas and dunefields is now well recognized.

This progress has come about as a result of several important methodological and conceptual advances. Increasing sophistication of instrumentation has enabled more precise measurement of airflow and sediment transport on dunes, although the limitations of this reductionist approach are being increasingly recognized (Livingstone et al., 2007). The application of increasingly precise optical dating techniques has provided new data on the age of dunes. When coupled with ground penetrating radar studies of dune sedimentary structures, this technique can provide unique information on dune dynamics on timescales of centuries to millennia. Numerical modelling of dunes and dune systems has been facilitated by vastly increased computing power, as well as better understanding of fundamental dune processes, although this area of study is still evolving very rapidly. Remote sensing images and GIS techniques now make it possible to analyze dune patterns in detail and trace patterns of sediment movement from source areas to depositional sinks.

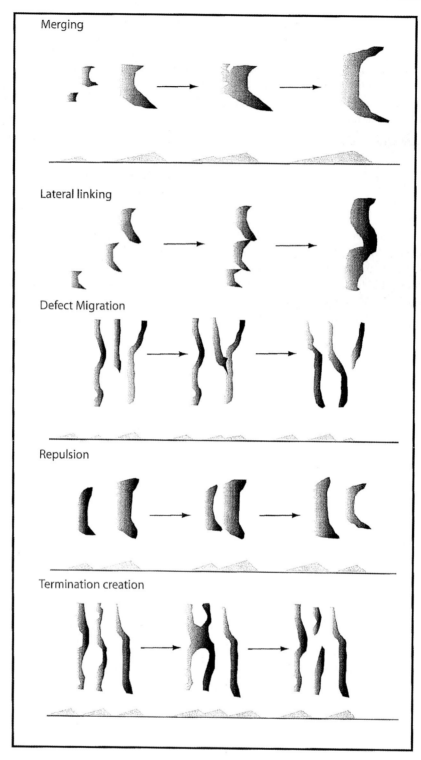

Fig. 18.36 Evolution of dune patterns (after Kocurek and Ewing, 2005)

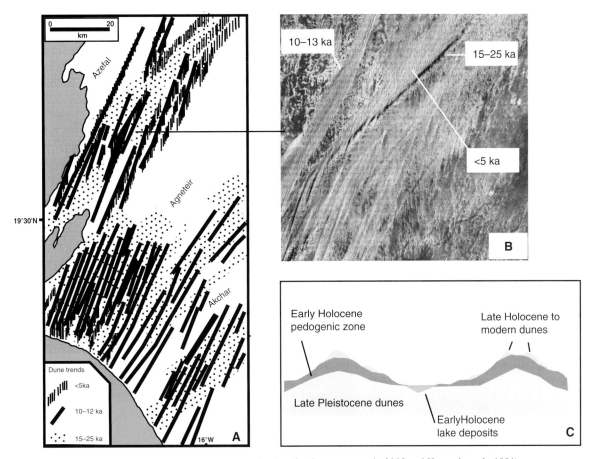

Fig. 18.37 Dune patterns and dune generations in Mauritania (after Lancaster et al., 2002 and Kocurek et al., 1991)

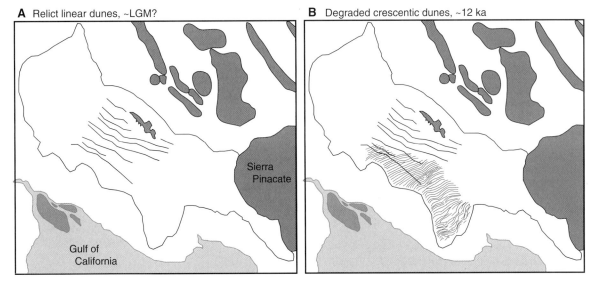

Fig. 18.38 Development of dune patterns in the Gran Desierto, Mexico by assembly and modification of different dune generations (after Beveridge et al., 2006)

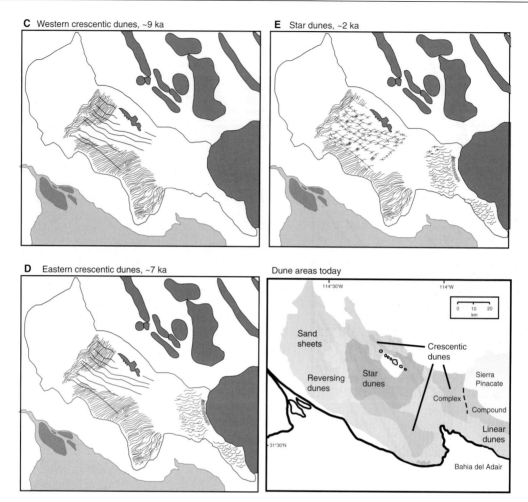

Fig. 18.38 (continued)

References

Allen, J.R.L., 1974. Reaction, relaxation and lag in natural sedimentary systems: general principles, examples and lessons. Earth Science Reviews, 10: 263–342.

Anderson, R.S., 1987. A theoretical model for aeolian impact ripples. Sedimentology, 34: 943–956.

Anderson, R.S., 1988. The pattern of grainfall deposition in the lee of aeolian dunes. Sedimentology, 35(2): 175–188.

Anderson, R.S., 1990. Eolian ripples as examples of self-organization in geomorphological systems. Earth Science Reviews, 29: 77–96.

Anderson, R.S. and Bunas, K.L., 1993. Grain size segregation and stratigraphy in aeolian ripples modeled with a cellular automaton. Nature, 365: 740–743.

Andreotti, B., Claudin, P., and Douady, S., 2002. Selection of dune shapes and velocities. Part I: Dynamics of sand, wind, and barchans. European Physics Journal, 28 B.

Andrews, S., 1981. Sedimentology of Great Sand Dunes, Colorado. In: F.P. Ethridge and R.M. Flores (Eds.), Recent and Ancient Non Marine Depositional Environments: models for exploration. The Society of Economic Paleontologists and Mineralogists, Tulsa, OK, pp. 279–291.

Anton, D. and Vincent, P., 1986. Parabolic dunes of the Jafurah Desert, Eastern Province, Saudi Arabia. Journal of Arid Environments, 11: 187–198.

Ash, J.E. and Wasson, R.J., 1983. Vegetation and sand mobility in the Australian desert dunefield. Zeitschrift fur Geomorphologie Supplement, 45: 7–25.

Bagnold, R.A., 1941. The Physics of Blown Sand and Desert Dunes. Chapman and Hall, London, 265pp.

Bagnold, R.A., 1953. The surface movement of blown sand in relation to meteorology, Desert Research, Proceedings of the International Symposium. Research Council of Israel, Jerusalem, pp. 89–93.

Besler, H., 1980. Die Dunen-Namib: entstehung und dynamik eines ergs. Stuttgarter Geographische Studien, 96: 241.

Beveridge, C., Kocurek, G., Ewing, R., Lancaster, N., Morthekai, P., Singhvi, A. and Mahan, S., 2006. Development of spatially diverse and complex dune-field patterns: Gran Desierto Dune Field, Sonora, Mexico: Sedimentology, 53: 1391–1409.

Blount, G. and Lancaster, N., 1990. Development of the Gran Desierto Sand Sea. Geology, 18: 724–728.

Boulton, J.W., 1997. Quantifying the morphology of aeolian impact ripples formed in a natural dune setting, University of Guelph, Guelph, Ontario, Canada, 172pp.

Breed, C.S., Fryberger, S.G., Andrews, S., McCauley, C., Lennartz, F., Geber, D. and Horstman, K., 1979. Regional studies of sand seas using LANDSAT (ERTS) imagery, In: McKee, E.D. (Ed.), A Study of Global Sand Seas: USGS Professional Paper, 1052, pp. 305–398.

Breed, C.S. and Grow, T., 1979. Morphology and distribution of dunes in sand seas observed by remote sensing. In: E.D. McKee (Editor), A Study of Global Sand Seas. Professional Paper. United States Geological Survey Professional Paper 1052, pp. 253-304.

Breed, C.S., McCauley, J.F. and Davis, P.A., 1987. Sand sheets of the eastern Sahara and ripple blankets on Mars. In: L.E. Frostick and I. Reid (Eds.), Desert Sediments: Ancient and Modern. Blackwell Scientific Publications, Oxford, London, Edinburgh, Boston, Palo Alto, Melbourne, pp. 337–359.

Bristow, C.S., Bailey, S.D. and Lancaster, N., 2000. Sedimentary structure of linear sand dunes. Nature, 406: 56–59.

Bristow, C.S., Duller, G.A.T. and Lancaster, N., 2005. Combining ground penetrating radar surveys and optical dating to determine dune migration in Namibia. Journal of the Geological Society (London), 162(2): 315–321.

Bristow, C.S., Duller, G.A.T. and Lancaster, N., 2007. Age and dynamics of linear dunes in the Namib Desert. Geology, 35:555–558.

Bristow, C.S. and Lancaster, N., 2004. Movement of a small slip-faceless dome dune in the Namib Sand Sea, Namibia. Geomorphology, 59: 189–196.

Bullard, J.E. and Nash, D.J., 1998. Linear dune pattern variability in the vicinity of dry valleys in the southwest Kalahari. Geomorphology, 23: 35–54.

Bullard, J.E., Thomas, D.S.G., Livingstone, I. and Wiggs, G.F.S., 1995. Analysis of linear sand dune morphological variability, southwestern Kalahari Desert. Geomorphology, 11: 189–203.

Bullard, J.E., Thomas, D.S.G., Livingstone, I. and Wiggs, G.F.S., 1997. Dunefield activity and interactions with climatic variability in the southwest Kalahari Desert. Earth Surface Processes and Landforms, 22(2): 165–174.

Burkinshaw, J.R., Illenberger, W.K. and Rust, I.C., 1993. Wind speed profiles over a reversing transverse dune. In: K. Pye (Ed.), The Dynamics and Environmental Context of Aeolian Sedimentary Systems. Geological Society, London, pp. 25–36.

Cooper, W.S., 1958. Coastal Sand Dunes of Oregon and Washington. Geological Society of America Memoir, 72: 167.

Corbett, I., 1993. The modern and ancient pattern of sandflow through the southern Namib deflation basin. International Association of Sedimentologists Special Publication, 16: 45–60.

Cupp, K., Lancaster, N. and Nickling, W.G., 2004. A Dune Simulation Wind Tunnel for Studies of Lee Face Processes.

Eos, Transactions American Geophysical Union, 85(47, Fall Meeting Supplement): Abstract P31B-0989.

Dong, Z., Wang, T. and Wang, X., 2004. Geomorphology of megadunes in the Badain Jaran Desert. Geomorphology, 60(1–2): 191–204.

Elbelrhiti, H., Claudin, P. and Andreotti, B., 2005. Field evidence for surface-wave-instability of sand dunes. Nature, 437: 720–723.

Ellwood, J.M., Evans, P.D. and Wilson, I.G., 1975. Small scale aeolian bedforms. Journal of Sedimentary Petrology, 45: 554–561.

Embabi, N.S., 1982. Barchans of the Kharga Depression. In: F. El Baz and T.A. Maxwell (Eds.), Desert Landfroms of Egypt: a basis for comparison with Mars. NASA, Washington D.C., pp. 141–156.

Embabi, N.S. and Ashour, M.M., 1993. Barchan dunes in Qatar. Journal of Arid Environments, 25; 49–69.

Endrody-Younga, S., 1982. Dispersion and translocation of dune specialist tenebrionids in the Namib area. Cimbebasia (A), 5: 257-271.

Eriksson, P.G., Nixon, N., Snyman, C.P. and Bothma, J.d.P., 1989. Ellipsoidal parabolic dune patches in the southern Kalahari Desert. Journal of Arid Environments, 16: 111–124.

Ewing, R.C., Kocurek, G. and Lake.L.W., 2006. Pattern analysis of dune-field parameters. Earth Surface Processes and Landforms, 31(9): 1176–1191.

Finkel, H.J., 1959. The barchans of Southern Peru. Journal of Geology, 67: 614–647.

Folk, R., 1970. Longitudinal dunes of the northwestern edge of the Simpson Desert, Northern Territory, Australia, 1. geomorphology and grain size relationships. Sedimentology, 16: 5–54.

Frank, A. and Kocurek, G., 1996a. Airflow up the stoss slope of sand dunes: limitations of current understanding. Geomorphology, 17(1–3): 47–54.

Frank, A. and Kocurek, G., 1996b. Towards a model for airflow on the lee side of aeolian dunes. Sedimentology, 43(3): 451–458.

Fryberger, S., Ahlbrandt, T. and Andrews, S., 1979. Origin, sedimentary features, and significance of low-angle eolian "sand sheet" deposits, Great Sand Dunes National Monument and vicinity, Colorado. Journal of Sedimentary Petrology, 49(3): 733–746.

Fryberger, S. and Goudie, A.S., 1981. Arid Geomorphology. Progress in Physical Geography, 5(3): 420–428.

Fryberger, S.G., 1979. Dune forms and wind regimes. In: E.D. McKee (Ed.), A Study of Global Sand Seas: United States Geological Survey, Professional Paper. U.S.G.S. Professional Paper, pp. 137–140.

Fryberger, S.G. and Ahlbrandt, T.S., 1979. Mechanisms for the formation of aeolian sand seas. Zeitschrift für Geomorphologie, 23: 440–460.

Fryberger, S.G., Al-Sari, A.M., Clisham, T.J., Rizoi, S.A.R. and Al-Hinai, K.G., 1984. Wind sedimentation in the Jafarah sand sea, Saudi Arabia. Sedimentology, 31(3): 413–431.

Fryberger, S.G., Hesp, P. and Hastings, K., 1992. Aeolian granule ripple deposits, Namibia. Sedimentology, 39: 319–331.

Gay, S.P., Jr., 1999. Observations regarding the movement of barchan sand dunes in the Nazca to Tanaca area of southern Peru. Geomorphology, 27(3–4): 279–294.

Glennie, K.W., 1970. Desert Sedimentary Environments. Developments in Sedimentology, 14. Elsevier, Amsterdam, 222pp.

Goudie, A.S., Livingstone, I. and Stokes, S. (Eds.), 1999. Aeolian Environments, Sediments, and Landforms. John Wiley & Sons, Chichester, 325pp.

Greeley, R. and Iversen, J.D., 1985. Wind as a Geological Process. Cambridge University Press, Cambridge, 333pp.

Hack, J.T., 1941. Dunes of the Western Navajo County. Geographical Review, 31(2): 240–263.

Haff, P.K. and Presti, D.E., 1995. Barchan dunes of the Salton Sea region, California. In: V.P. Tchakerian (Ed.), Desert Aeolian Processes. Chapman and Hall, New York, pp. 153–178.

Hanna, S.R., 1969. The formation of longitudinal sand dunes by large helical eddies in the atmosphere. Journal of Applied Meteorology, 8: 874–883.

Hastenrath, S., 1987. The barchan dunes of Southern Peru revisited. Zeitschrift fur Geomorphologie, 31(2): 167–178.

Hastenrath, S.L., 1967. The barchans of the Arequipa region, Southern Peru. Zeitschrift fur Geomorphologie, 11: 300–311.

Havholm, K.G. and Kocurek, G., 1988. A preliminary study of the dynamics of a modern draa, Algodones, southeastern California, USA. Sedimentology, 35: 649–669.

Haynes, C.V.J., 1989. Bagnold's barchan: a 57-yr record of dune movement in the eastern Sahara and implications for dune origin and palaeoclimate since Neolithic times. Quaternary Research, 32(2): 153–167.

Hersen, P., Anderson, K.H., Elbelrhiti, B., Andreotti, B., Claudin, P. and Douady, S., 2004. Corridors of barchan dunes: stability and size selection. Physics Review, E, 69: 011304.

Hersen, P. and Douady, S., 2005. Collision of barchan dunes as a mechanism of size regulation. Geophysical Research Letters, 34(21): L21403.

Hesse, P.P. and Simpson, R.L., 2006. Variable vegetation cover and episodic sand movement on longitudinal desert dunes. Geomorphology, 81: 276–291.

Holm, D.A., 1960. Desert geomorphology in the Arabian Peninsula. Science, 123: 1369–1379.

Howard, A.D., 1977. Effect of slope on the threshold of motion and its application to orientation of wind ripples. Geological Society of America Bulletin, 88: 853–856.

Howell, J. and Mountney, N., 2001. Aeoian grain flow architecture: hard data for reservoir models and implications for red bed sequence stratigraphy. Petroleum Geoscience, 7: 51–56.

Hunt, J.C.R., Leibovich, S. and Richards, K.J., 1988. Turbulent shear flows over low hills. Quarterly Journal of the Royal Meteorological Society, 114: 1435–1470.

Hunter, R.E., 1977. Basic types of stratification in small eolian dunes. Sedimentology, 24: 361–388.

Hunter, R.E., 1985. A kinematic model for the structure of leeside deposits. Sedimentology, 32: 409–422.

Hunter, R.E., Richmond, B.M. and Alpha, T.R., 1983. Storm-controlled oblique dunes of the Oregon Coast. Geological Society of America Bulletin, 94: 1450–1465.

Inman, D.L., Ewing, G.C. and Corliss, J.B., 1966. Coastal sand dunes of Guerrero Negro, Baja California, Mexico. Geological Society of America, Bulletin, 77: 787–802.

Jackson, P.S. and Hunt, J.C.R., 1975. Turbulent wind flow over a low hill. Quarterly Journal of the Royal Meteorological Society, 101: 929–955.

Jäkel, D., 1980. Die bildung von barchanen in Faya-Largeau/Rep. du Tchad. Zeitschrift für Geomorphologie, N.F., 24: 141–159.

Jensen, N.O. and Zeman, O., 1985. Perturbations to mean wind and turbulence in flow over topographic forms. In: O.E. Barndorff-Nielsen, J.T. Møller, K.R. Rasmussen and B.B. Willetts (Eds.), Proceedings of International Workshop on the Physics of Blown Sand. University of Aarhus, Aarhus, pp. 351–368.

Kar, A., 1993. Aeolian processes and bedforms in the Thar Desert. Journal of Arid Environments, 25: 83–96.

Kocurek, G., 1998. Aeolian System Response to External Forcing Factors – A Sequence Stratigraphic View of the Saharan Region. In: A.S. Alsharan, K.W. Glennie, G.L. Whittle and C.G.S.C. Kendall (Eds.), Quaternary Deserts and Climatic Change. Balkema, Rotterdam/Brookfield, pp. 327–338.

Kocurek, G. and Ewing, R.C., 2005. Aeolian dune field self-organization – implications for the formation of simple versus complex dune field patterns. Geomorphology, 72: 94–105.

Kocurek, G., Havholm, K.G., Deynoux, M. and Blakey, R.C., 1991. Amalgamated accumulations resulting from climatic and eustatic changes, Akchar Erg, Mauritania. Sedimentology, 38(4): 751–772.

Kocurek, G. and Lancaster, N., 1999. Aeolian Sediment States: Theory and Mojave Desert Kelso Dunefield example. Sedimentology, 46(3): 505–516.

Kocurek, G. and Nielson, J., 1986. Conditions favourable for the formation of warm-climate aeolian sand sheets. Sedimentology, 33: 795–816.

Kocurek, G., Townsley, M., Yeh, E., Havholm, K. and Sweet, M.L., 1992. Dune and dunefield development on Padre Island, Texas, with implications for interdune deposition and water-table-controlled accumulation. Journal of Sedimentary Petrology, 62(4): 622–635.

Lancaster, N., 1980. The formation of seif dunes from barchans - supporting evidence for Bagnold's hypothesis from the Namib Desert. Zeitschrift fur Geomorphologie, 24: 160–167.

Lancaster, N., 1982a. Dunes on the Skeleton Coast, SWA/Namibia: geomorphology and grain size relationships. Earth Surface Processes and Landforms, 7: 575–587.

Lancaster, N., 1982b. Linear dunes. Progress in Physical Geography, 6: 476–504.

Lancaster, N., 1983. Controls of dune morphology in the Namib sand sea. In: T.S. Ahlbrandt and M.E. Brookfield (Eds.), Eolian Sediments and Processes. Developments in Sedimentology. Elsevier, Amsterdam, pp. 261–289.

Lancaster, N., 1985. Variations in wind velocity and sand transport rates on the windward flanks of desert sand dunes. Sedimentology, 32: 581–593.

Lancaster, N., 1988. Controls of eolian dune size and spacing. Geology, 16: 972–975.

Lancaster, N., 1989a. Star Dunes. Progress in Physical Geography, 13(1): 67–92.

Lancaster, N., 1989b. The Namib Sand Sea: Dune forms, processes, and sediments. A.A. Balkema, Rotterdam, 200pp.

Lancaster, N., 1991. The orientation of dunes with respect to sand-transporting winds: a test of Rubin and Hunter's gross bedform-normal rule, NATO Advanced Research Workshop on sand, dust, and soil in their relation to aeolian and littoral processes. University of Aarhus, Sandbjerg, Denmark, pp. 47-49.

Lancaster, N., 1992. Relations between dune generations in the Gran Desierto, Mexico. Sedimentology, 39: 631–644.

Lancaster, N., 1993. Origins and sedimentary features of super-surfaces in the northwestern Gran Desierto Sand Sea. IAS Special Publication, 16: 71–86.

Lancaster, N., 1995a. Geomorphology of Desert Dunes. Routledge, London, 290pp.

Lancaster, N., 1995b. Origin of the Gran Desierto Sand Sea: Sonora, Mexico: Evidence from dune morphology and sediments. In: V.P. Tchakerian (Ed.), Desert Aeolian Processes. Chapman and Hall, New York, pp. 11–36.

Lancaster, N., 1996. Field studies of proto-dune initiation on the northern margin of the Namib Sand Sea. Earth Surface Processes and Landforms, 21: 947–954.

Lancaster, N., 1999. Sand Seas. In: A.S. Goudie, I. Livingstone and S. Stokes (Eds.), Aeolian Environments, Sediments, and Landforms. Wiley, Chichester, New York, pp. 49–70.

Lancaster, N. and Baas, A., 1998. Influence of vegetation cover on sand transport by wind: field studies at Owens Lake, California. Earth Surface Processes and Landforms, 23(1): 69–82.

Lancaster, N., Greeley, R. and Christensen, P.R., 1987. Dunes of the Gran Desierto Sand Sea, Sonora, Mexico. Earth Surface Processes and Landforms, 12: 277–288.

Lancaster, N., Kocurek, G., Singhvi, A.K., Pandey, V., Deynoux, M., Ghienne, J.-P. and Lo, K., 2002. Late Pleistocene and Holocene dune activity and wind regimes in the western Sahara of Mauritania. Geology, 30: 991–994.

Lancaster, N., Nickling, W.G., McKenna Neuman, C.K. and Wyatt, V.E., 1996. Sediment flux and airflow on the stoss slope of a barchan dune. Geomorphology, 17(1–3): 55–62.

Livingstone, I., 1986. Geomorphological significance of wind flow patterns over a Namib linear dune. In: W.G. Nickling (Ed.), Aeolian Geomorphology. Boston, Allen and Unwin, pp. 97–112.

Livingstone, I., 1988. New models for the formation of linear sand dunes. Geography, 73: 105–115.

Livingstone, I., 1989. Monitoring surface change on a Namib linear dune. Earth Surface Processes and Landforms, 14: 317–332.

Livingstone, I., 1993. A decade of surface change on a Namib linear dune. Earth Surface Processes and Landforms, 18(7): 661–664.

Livingstone, I., 2003. A twenty-one-year record of surface change on a Namib linear dune. Earth Surface Processes and Landforms, 28(9): 1025–1032.

Livingstone, I. and Warren, A., 1996. Aeolian Geomorphology: an introduction. Addison Wesley Longman, Harlow, 211pp.

Livingstone, I., Wiggs, G.F.S. and Weaver, C.M., 2007. Geomorphology of desert sand dunes: A review of recent progress. Earth Science Reviews, 80(3–4): 239–257.

Long, J.T. and Sharp, R.P., 1964. Barchan dune movement in Imperial Valley, California. Geological Society of America Bulletin, 75: 149–156.

Mainguet, M., 1983. Dunes vives, dunes fixées, dunes vêtues: une classification selon le bilan d'alimentation, le régime éolien et la dynamique des édifices sableux. Zeitschrift für Geomorphologie, Suppl. Bd. 45: 265–285.

Mainguet, M., 1984a. A classification of dunes based on aeolian dynamics and the sand budget. In: F. El-Baz (Ed.), Deserts and arid lands. Martinus Nijhoff, The Haguw, pp. 31–58.

Mainguet, M., 1984b. Space observations of Saharan aeolian dynamics. In: F. El Baz (Ed.), Deserts and Arid Lands. Nyhoff, The Hague, pp. 59–77.

Mainguet, M. and Callot, Y., 1978. L'erg de Fachi-Bilma (Tchad-Niger). Mémoires et Documents CNRS, 18: 178.

Mainguet, M. and Chemin, M.-C., 1984. Les dunes pyramidales du Grand Erg Oriental. Travaux de l'Institut de Géographie de Reims, 59–60: 49–60.

Marîn, L., Forman, S.L., Valdez, A. and Bunch, F., 2005. Twentieth century dune migration at the Great Sand Dunes National Park and Preserve, Colorado, relation to drought variability. Geomorphology, 70: 163–183.

Mason, P.J. and Sykes, R.I., 1979. Flow over an isolated hill of moderate slope. Quarterly Journal of the Royal Meteorological Society, 105: 383–395.

Maxwell, T. and Haynes, C., 2001. Sand sheet dynamics and Quaternary landscape evolution of the Selima Sand Sheet, southern Egypt. Quaternary Science Reviews, 20: 1623–1647.

Maxwell, T.A. and Haynes, C.V., Jr., 1989. Large-scale, low-amplitude bedfroms (chevrons) in the Selima sand sheet, Egypt. Science, 243: 1179–1182.

McDonald, R.R. and Anderson, R.S., 1995. Experimental verification of aeolian saltation and lee side deposition models. Sedimentology, 42(1): 39–56.

McKee, E., 1982. Sedimentary structures in dunes of the Namib Desert, South West Africa. Geological Society of America Special paper, 188: 60.

McKee, E. and Tibbitts, G.C., Jr., 1964. Primary structures of a seif dune and associated deposits in Libya. Journal of Sedimentary Petrology, 34(1): 5–17.

McKee, E.D., 1966. Structures of dunes at White Sands National Monument, New Mexico (and a comparison with structures of dunes from other selected areas). Sedimentology, 7(1): 1–69.

McKenna Neuman, C., Lancaster, N. and Nickling, W.G., 1997. Relations between dune morphology, airflow, and sediment flux on reversing dunes, Silver Peak, Nevada. Sedimentology, 44: 1103-1114.

McKenna Neuman, C., Lancaster, N. and Nickling, W.G., 2000. The effect of unsteady winds on sediment transport on the stoss slope of a transverse dune, Silver Peak, Nevada. Sedimentology, 47(1): 211–226.

Momiji, H., Carretero-Gonzalez, R., Bishop, S.R. and Warren, A., 2000. Simulation of the effect of wind speedup in the formation of transverse dune fields. Earth Surface Processes and Landforms, 25: 905–918.

Muhs, D.R., 2004. Mineralogical maturity in dunefields of North America, Africa, and Austrlia. Geomorphology, 59(1–2): 247–269.

Muhs, D.R., Bush, C.A., Cowherd, S.D. and Mahan, S., 1995. Source of sand for the Algodones Dunes. In: V.P. Tchakerian (Ed.), Desert Aeolian Processes. Chapman and Hall, New York, pp. 37–74.

Muhs, D.R., Reynolds, R.R., Been, J. and Skipp, G., 2003. Eolian sand transport pathways in the southwestern United States: importance of the Colorado River and local sources. Quaternary International, 104: 3–18.

Mulligan, K.R., 1988. Velocity Profiles measured on the windward slope of a transverse dune. Earth Surface Processes and Landforms, 13(7): 573–582.

Nickling, W.G. and McKenna Neuman, C., 1999. Recent investigations of airflow and sediment transport over desert dunes. In: A.S. Goudie, I. Livingstone and S. Stokes (Eds.), Aeolian Environments, Sediments and Landforms. Chichester, John Wiley & Sons.

Nickling, W.G., McKenna Neuman, C. and Lancaster, N., 2002. Grainfall Processes in the Lee of Transverse Dunes, Silver Peak, Nevada. Sedimentology, 49(1): 191–211.

Nielson, J. and Kocurek, G., 1986. Climbing zibars of the Algodones. Sedimentary Geology, 48: 1–15.

Nielson, J. and Kocurek, G., 1987. Surface processes, deposits, and development of star dunes: Dumont dune field, California. Geological Society of America Bulletin, 99:177–186.

Parsons, D.R., Walker, I.J. and Wiggs, G.F.S., 2004. Numerical modelling of flow structures over an idealised transverse dunes of varying geometry. Geomorphology, 59: 149–164.

Partelli, E.J.R., Schwämmle, V., Herrman, H.J., Monteiro, L.H.U. and Maia, L.P., 2006. Profile measurement and simulation of a transverse dune field in the Lencois Maranhenses. Geomorphology, 81: 29–42.

Pye, K. and Tsoar, H., 1990. Aeolian Sand and Sand Dunes. Unwin Hyman, London, 396pp.

Ramsey, M.S., Christensen, P.R., Lancaster, N. and Howard, D.A., 1999. Identification of sand sources and transport pathways at the Kelso Dunes, California using thermal infrared remote sensing. Geological Society of America Bulletin, 111: 646–662.

Rubin, D.M., 1984. Factors determining desert dune type (discussion). Nature, 309: 91–92.

Rubin, D.M. and Hunter, R.E., 1982. Bedform climbing in theory and nature. Sedimentology, 29: 121–138.

Rubin, D.M. and Hunter, R.E., 1987. Bedform alignment in directionally varying flows. Science, 237: 276–278.

Rubin, D.M. and Ikeda, H., 1990. Flume experiments on the alignment of transverse, oblique, and longitudinal dunes in directionally varying flows. Sedimentology, 37(4): 673–684.

Sauerman, G., Rognon, P., Poliakov, A. and Herrmann, H.J., 2000. The shape of the barchan dunes of Southern Morocco. Geomorphology, 36(1–2): 47–62.

Schwämmle, V. and Herrmann, H., 2004. Modelling transverse dunes. Earth Surface Processes and Landforms, 29(6): 769–784.

Seppälä, M. and Linde, K., 1978. Wind tunnel studies of ripple formation. Geografiska Annaler, 60(Series A): 29–42.

Sharp, R.P., 1963. Wind Ripples. Journal of Geology, 71: 617–636.

Sharp, R.P., 1966. Kelso Dunes, Mohave Desert, California. Geological Society of America Bulletin, 77: 1045–1074.

Singhvi, A.K. and Kar, A., 2004. The aeolian sedimentation record of the Thar Desert. Proceedings of the Indian Academy of Sciences (Earth Sciences), 113(3): 371–401.

Slattery, M.C., 1990. Barchan migration on the Kuiseb River Delta, Namibia. South African Geographical Journal, 72: 5–10.

Stokes, S. and Bray, H.E., 2005. Late Pleistocene eolian history of the Liwa region, Arabian Peninsula. Geological Society of America Bulletin, 117(11/12): 1466–1480.

Stokes, S., Goudie, A.S., Ballard, J., Gifford, C., Samieh, S., Embabi, N. and El-Rashidi, O.A., 1999. Accurate dune displacement and morphometric data using kinematic GPS. Zeistschrift für Geomorphologie Supplementbände, 11: 195–214.

Stokes, S., Maxwell, T.A., Haynes, C.V. and Horrocks, J., 1998. Latest Pleistocene and Holocene sand sheet construction in the Selima Sand Sheet, Eastern Sahara. In: A.S. Alsharan, K.W. Glennie, G.L. Whittle and C.G.S.C. Kendall (Eds.), Quaternary Deserts and Climatic Change. Balkema, Rotterdam/Brookfield, pp. 175–184.

Stokes, S., Thomas, D.S.G. and Shaw, P.A., 1997. New chronological evidence for the nature and timing of linear dune development in the southwest Kalahari Desert. Geomorphology, 20(1–2): 81–94.

Sweet, M.L. and Kocurek, G., 1990. An empirical model of aeolian dune lee-face airflow. Sedimentology, 37(6): 1023–1038.

Sweet, M.L., Nielson, J., Havholm, K. and Farralley, J., 1988. Algodones dune field of southeastern California: case history of a migrating modern dune field. Sedimentology, 35(6): 939–952.

Teller, J.T., Glennie, K.W., Lancaster, N. and Singhvi, A.K., 2002. Calcareous dunes of the United Arab emirates and Noah's Flood: the postglacial reflooding of the Persion (Arabian) Gulf. Quaternary International, 68–71: 297–308.

Thomas, D.S.G. and Leason, H.C., 2005. Dunefield activity response to climate variability in the southwest Kalahari. Geomorphology, 64(1–2): 117–132.

Thomas, D.S.G. and Tsoar, H., 1990. The geomorphological role of vegetation in desert dune systems. In: J.B. Thornes (Editor), Vegetation and Erosion. John Wiley & Sons Ltd., Chichester, pp. 471–489.

Tseo, G., 1990. Reconnaissance of the dynamic characteristics of an active Strzelecki Desert longitudinal dune, southcentral Australia. Zeitschrift für Geomorphologie N.F., 34(1): 19–35.

Tsoar, H., 1974. Desert dunes morphology and dynamics, El Arish (northern Sinai). Zeitschrift für Geomorphologie Supplementband, 20: 41–61.

Tsoar, H., 1983. Dynamic processes acting on a longitudinal (seif) dune. Sedimentology, 30: 567–578.

Tsoar, H., 1984. The formation of seif dunes from barchans – a discussion. Zeitschrift fur Geomorphologie, 28(1): 99–103.

Tsoar, H., 1985. Profile analysis of sand dunes and their steady state significance. Geografiska Annaler, 67A: 47–59.

Tsoar, H., 1986. Two-dimensional analysis of dune profile and the effect of grain size on sand dune morphology. In: F. El-Baz and M.H.A. Hassan (Eds.), Physics of Desertification. Martinus Nyhoff, Dordrecht, pp. 94–108.

Tsoar, H., 1989. Linear dunes – forms and formation. Progress in Physical Geography, 13(4): 507–528.

Tsoar, H. and Møller, J.T., 1986. The role of vegetation in the formation of linear sand dunes. In: W.G. Nickling (Ed.), Aeolian Geomorphology. Allen and Unwin, Boston, London, Sydney, pp. 75–95.

Verstappen, H.T., 1968. On the origin of longitudinal (seif) dunes. Zeitschrift für Geomorphologie N.F., 12: 200–220.

Walker, D.J., 1981. An experimental study of wind ripples. MSc Thesis, Massachusetts Institute of Technology.

Walker, I.J., 1999. Secondary airflow and sediment trasport in the lee of a reversing dune. Earth Surface Processes and Landforms, 24: 437–448.

Walker, I.J. and Nickling, W.G., 2002. Dynamics of secondary airflow and sediment transport over and the lee of transverse dunes. Progress in Physical Geography, 26(1): 47–75.

Wang, X., Dong, Z., Zhang, J. and Qu, J., 2004. Formation of the complex linear dunes of the central Taklimakan sand sea. Earth Surface Processes and Landforms, 29(6): 677–686.

Warren, A., 1972. Observations on dunes and bimodal sands in the Tenere desert. Sedimentology, 19: 37–44.

Warren, A., 1988. The dunes of the Wahiba Sands. In: R.W. Dutton (Ed.), Scientific Results of the Royal Geographical Society's Oman Wahiba Sands Project 1985–1987. Journal of Oman Studies, Special Report 3, Muscat, Oman, pp. 131–160.

Warren, A. and Allison, D., 1998. The palaeoenvironmental significance of dune size hierarchies. Palaeogeography, Palaeoclimatology, Palaeocology, 137: 289–303.

Wasson, R.J., 1983. Dune sediment types, sand colour, sediment provenance and hydrology in the Strzelecki-Simpson Dunefield, Australia. In: M.E. Brookfield and T.S. Ahlbrandt (Eds.), Eolian Sediments and Processes. Developments in Sedimentology. Elsevier, Amsterdam, Oxford, New York, Tokyo, pp. 165–195.

Wasson, R.J., Fitchett, K., Mackey, B. and Hyde, R., 1988. Large-scale patterns of dune type, spacing, and orientation in the Australian continental dunefield. Australian Geographer, 19: 89–104.

Wasson, R.J. and Hyde, R., 1983a. A test of granulometric control of desert dune geometry. Earth Surface Processes and Landforms, 8: 301–312.

Wasson, R.J. and Hyde, R., 1983b. Factors determining desert dune type. Nature, 304: 337–339.

Wasson, R.J, Rajaguru, S.N. Misra, V.N. Agrawal, D.P. Dhir, R.P., Singhvi, A.K., Kameswara Rao, K.., 1983. Geomorphology, late Quaternary stratigraphy and paleoclimatology of the Thar dunefield. Zeitschrift für Geomorphologie, Supplementband, 45: 117-151.

Weng, W.S., Hunt, J.C.R., Carruthers, D.J., Warren, A., Wiggs, G.F.S., Livingstone, A. and Castro, I., 1991. Air flow and sand transport over sand dunes. Acta Mechanica Supplement, 2: 1–22.

Werner, B.T., 1988. A steady-state model of wind-blown sand transport. Journal of Geology, 98(1): 1–17.

Werner, B.T., 1995. Eolian dunes: computer simulations and attractor interpretation. Geology, 23(12): 1107–1110.

Werner, B.T., 2003. Modeling Landforms as Self-Organized, Hierarchical Dynamical Systems. Predictions in Geomorphology, Geophysical Monograph, (135): 133–150.

Werner, B.T. and Kocurek, G., 1997. Bed-form dynamics: Does the tail wag the dog? Geology, 25(9): 771–774.

Werner, B.T. and Kocurek, G., 1999. Bedform spacing from defect dynamics. Geology, 27(8): 727–730.

Wiggs, G.F.S., 1993. Desert dune dynamics and the evaluation of shear velocity: an integrated approach. In: K. Pye (Ed.), The Dynamics and Environmental Context of Aeolian Sedimentary Systems. Geological Society, London, pp. 37–48.

Wiggs, G.F.S., 2001. Desert dune processes and dynamics. Progress in Physical Geography, 25(1): 53–79.

Wiggs, G.F.S., Livingstone, I., Thomas, D.S.G. and Bullard, J.E., 1994. Effect of vegetation removal on airflow patterns and dune dynamics in the southwestern Kalahari Desert. Land Degradation and Rehabilitation, 5: 13–24.

Wiggs, G.F.S., Livingstone, I. and Warren, A., 1996. The role of streamline curvature in sand dune dynamics:evidence from field and wind tunnel measurements. Geomorphology, 17(1–3): 29–46.

Wiggs, G.F.S., Thomas, D.S.G., Bullard, J.E. and Livingstone, I., 1995. Dune mobility and vegetation cover in the southwest Kalahari Desert. Earth Surface Processes and Landforms, 20(6): 515–530.

Wilson, I.G., 1971. Desert sandflow basins and a model for the development of ergs. Geographical Journal, 137(2): 180–199.

Wilson, I.G., 1972. Aeolian bedforms – their development and origins. Sedimentology, 19: 173–210.

Wilson, I.G., 1973. Ergs. Sedimentary Geology, 10: 77–106.

Chapter 19

Landforms, Landscapes, and Processes of Aeolian Erosion

Julie E. Laity

Introduction

Aeolian erosion develops through two principal processes: deflation (removal of loosened material and its transport as fine grains in atmospheric suspension) and abrasion (mechanical wear of coherent material). In a vegetation-free environment, the relative significance of each of these processes is a function of surface material properties, the availability of abrasive particles, and climate. The resulting landforms include ventifacts, ridge and swale systems, yardangs, desert depressions (pans), and inverted relief. Dust is an important by-product of some forms of erosional activity.

The significance of wind erosion as a geomorphological process has been long debated (Goudie 1989). In recent years, remote sensing images of terrestrial deserts and rover and satellite imagery of Mars (Greeley et al. 2002, Bridges et al. 2004) have stimulated new research, providing information on the extent of large-scale erosion systems, surface textures of Martian rocks, and the timing, frequency, and size of sand and dust storms associated with erosion. On Mars, ventifacts and yardangs are important proxies of climate, wind regime, sediment type and transportation, and thus consistent interpretations of geomorphic process are critical.

Despite recent advances, much remains to be learned about landforms of wind erosion. With few exceptions, detailed environmental analyses and short- or long-term process measurements are hampered by

remoteness. Consequently, landform ages, processes and rates of formation, and evolutionary history remain poorly understood.

Ventifacts

Published work on ventifacts dates back to the mid-19th century. They were first described by Blake (1855) from the Salton Sea region, California. Faceting of stones by wind abrasion was documented by Travers (1870), and further discussion on wind erosion was provided by Gilbert (1875). Evans (1911) proposed that "ventifact" encompass the multiple and compound terms (wind-grooved stones, wind-faceted stones, etc.) then in use. By 1931, Bryan had published a bibliography of 258 titles on ventifacts, including a useful discussion on terminology, and had provided an overview of prevailing theories concerning ventifact formation.

The study of ventifacts has spawned a number of controversies, particularly because many are fossil (relict of earlier surface and climatic conditions) and are thus not subject to field-based process studies. An area of early debate concerned whether faceted pebbles are shaped by mono-directional winds, winds from two opposing directions, or winds from variable directions. In addition, the factors that control the final shape of ventifacts were in contention. Woodworth (1894) suggested that a rock of moderate size will have a facet cut at right angles to the wind, with new facets formed as a result of accidental overturning or rotation. However, researchers in Europe proposed a final form primarily conditioned by (a) the shape of the original base (a square base would yield a pyramid of four faces), (b) winds that split along

J.E. Laity (✉)
Department of Geography, California State University,
Northridge, CA 91330, USA
e-mail: julie.laity@csun.edu

A.J. Parsons, A.D. Abrahams (eds.), *Geomorphology of Desert Environments*, 2nd ed.,
DOI 10.1007/978-1-4020-5719-9_19, © Springer Science+Business Media B.V. 2009

the ground and impinge upon the rock from variable directions, or (c) the original size and shape of the rock. Woodworth's ideas (1894) were later supported by Sharp (1949, 1964, 1980) and appear to represent the present majority opinion. In the literature of the past few decades, differences of opinion have appeared with regard to the relative importance of saltating and suspended grains in ventifact formation and the orientation of the abraded face relative to the wind.

The term ventifact, although widely applied, is ill-defined, describing as it does wind eroded forms of varying size, form, and material composition. Ventifacts range in size from small pebbles to large boulders. Although facets are often regarded as fundamental features of ventifacts (Sharp and Malin 1984), they are often absent from large boulders and rock outcrops, where the presence of pits, flutes, and grooves may be the principal indicators of wind erosion. On lava flows and bedrock outcrops, eroded semi-planar surfaces may develop and linear grooves form on playa surfaces subject to saltating sand grains. The material that forms ventifacts may not be stone at all: over short time periods, playa (Williams and Greeley 1981) or river bed ventifacts may form in cohesive sediments.

Ventifacts are valuable as evidence of past climatic conditions (Figs. 19.1 and 19.2). The presence of

Fig. 19.2 In this photograph taken in the Bristol Mountains, California, fossil ventifacts are located on the lower two-thirds of the hill, whereas actively-forming ventifacts are found near the crest and summit. Wind speed and sand transport increases towards the crest of a hill, and active sand is more common here

Fig. 19.1 Abraded granite boulder approximately 2.7 m high located on a hill crest near June Lake, California, east of the Sierra Nevada. Pitting covers much of the high-angle windward face, transitional to fluting as the wind approaches at an oblique angle on the left side of the ventifact. The fluted abrasion-hardened rind stands out in relief (detached by about 25 cm) relative to the pitted surface

wind-worn stones in ancient formations (Precambrian, Cambrian, Devonian, Permian, Jurassic, Triassic, Cretaceous, Tertiary, and Pleistocene) has been used as evidence of more arid conditions (Bryan 1931). Palaeoventifacts are indicative of intense wind activity, but they do not necessarily signify warm desert conditions, as they also form in temperate regions along ocean or lake margins, and in periglacial settings. Aspects of ventifact form (the orientation of facets, grooves, flutes, and pits) enable a determination of palaeowind direction, and the presence of ventifacts provides clues to the relative abundance of abrasive materials and vegetation cover in earlier environments. Although it is recognized that many ventifacts are no longer forming ("relict" or "fossil" forms), the actual age of most ventifacts is unknown.

Environments of Ventifact Formation

Ventifacts form in environments characterized by (a) a supply of an abrasive material, (b) lack of complete vegetative cover, (c) strong winds, (d) a topographic situation that allows the free sweep of wind or that locally accelerates wind, and (e) ground surface stability. Ventifacts are found in desert, periglacial, and coastal environments on Earth and are probably even more widespread on Mars. Many are fossil in nature, as a change in any of the aforementioned environmental factors will cause the diminishment or cessation of wind erosion. As the processes of ventifact formation in coastal and periglacial settings are similar to those in deserts, all environments will be briefly reviewed, and the research results incorporated into the discussion.

Ventifacts occur in numerous present-day and palaeo-periglacial environments, including North America (Blackwelder 1929, Powers 1936, McKenna-Neuman and Gilbert 1986, Bach 1995, Dorn 1995), Antarctica (Lindsay 1973, Miotke 1982, Pickard 1982, Hall 1989), Europe (Schlyter 1989, Christiansen 2004), Iceland (Antevs 1928, Greeley and Iversen 1985, Mountney and Russell 2004), Greenland, the Falkland Islands (Clark and Wilson 1992), and higher elevations in South America (Czajka 1972). These ventifacts range in size from small faceted forms to boulders of considerable size: palaeoventifacts along the eastern base of the Sierra Nevada, California exceed a metre in height (Blackwelder 1929) (Fig. 19.1) and Powers (1936) described 60-cm-long grooves in crystalline boulders strewn on the sandy Lake Wisconsin plain. Clark and Wilson (1992), Mountney and Russell (2004), McKenna-Neuman (1993) and Tremblay (1961) emphasize that most erosion occurs in the presence of sand (associated with sand sheets or sand dunes). In Antarctic dry valleys, very high wind speeds (up to 300 km h^{-1}) allow very coarse material, including medium sand, granules, and pebbles, to abrade rocks (Hall 1989), such that the yearly abrasion rates average a few millimetres and ventifacts form within a few decades or, at most, a few centuries (Miotke 1982). McKenna-Neuman (1990) reported aeolian transport of coarse granules up to 4 m in height in an arctic proglacial setting in Canada.

Ventifacts occur in coastal locations where winds are strong and drifts of sand provide an effective abrading agent (Travers 1870, King 1936, Wentworth and Dickey 1935, Bishop and Mildenhall 1994, Knight and Burningham 2003). Few ventifacts have been reported for moist temperate climates outside of the coastal

zone. In Nova Scotia, Canada, they occur as a result of highly permeable sediments that result in locally dry surface conditions (Hickox 1959).

Ventifacts appear to be fairly common in warm deserts and semi-arid environments. However, their spatial distribution is poorly known because such small features cannot be identified from aerial imagery. Our knowledge of ventifact sites thereby depends on fortuitous discovery by individuals who recognize these features and furthermore publish their observations. High altitude photography is sometimes useful in locating ventifacts, because their occurrence is often coincident with deposits of active and stabilized sand, with wind streaks, and with regions where the air is locally accelerated, either by topographic constrictions or by passing over a hill (Laity 1987).

In appearance, there is no evident difference between ventifacts formed in desert, coastal, or periglacial environments, suggesting that the processes of formation are broadly similar. Most researchers have assumed that the abrading agent in deserts is wind-blown sand (Fig. 19.3) (Blake 1855, Blackwelder 1929, Maxson 1940, Sharp 1964, 1980, Smith 1967, 1984, Greeley and Iversen 1986, Laity 1987, 1992, 1995, Bridges et al. 2004), but few have provided specific information as to grain size, shape, or lithology. Sediment analyses by Sharp (1964), Greeley and Iversen (1986), and Laity (1995) suggest that fine to medium-sized sand dominates in ventifact formation.

Fig. 19.3 Actively forming ventifacts surrounded by aeolian sand at the hillslope summit shown in Fig. 19.2. The rocks are dense basalts, with very little vesicularity. Multiple wind directions are represented by several facets, but there are few lineations developed

Morphology of Ventifacts

Lithology and Size of the Original Rock

Ventifacts develop in a wide range of rock types, ranging from exceptionally hard quartzite to softer limestone and marble. Most lithologies develop characteristic faceting, pitting, fluting, or grooves. Rocks such
as schist, gneiss, and ignimbrites are particularly prone
to linear erosion features and to etching (Sharp 1949,
Lancaster 1984). Dense, hard, and homogeneous rocks
(such as cherts or basalts of low vesicularity) are less
likely to develop pits, flutes, or grooves, as they lack
vesicles or soft inhomogeneities that abrasion can exploit (Fig. 19.3). Polishing and faceting of rock surfaces have been observed on all lithologies.

In deserts, ventifact preservation is affected by
weathering processes that modify or destroy surface
features. Limestones and dolomites are particularly
susceptible to solution, even in arid environments.
In a periglacial setting in Wyoming, Sharp (1949)
found no evidence of erosion in rocks of this type.
Laity (unpublished data) has observed rillenkarren
that cross perpendicular to grooves on fossil ventifacts
of the Panamint Valley, California. Coarse-grained
granites are particularly susceptible to granular disintegration and exfoliation in deserts, and tend to
preserve evidence of erosion less well than other materials (Lancaster 1984, Laity 1992). Likewise schists
and pegmatites of the Namib Desert (Selby 1977,
Lancaster 1984) and andesite in the Mojave Desert
(Laity 1992) were found to weather too rapidly to
preserve significant evidence of abrasion.

The ultimate form of the ventifact is strongly
controlled by the size of the original material. If the
rock is small, with a diameter of only a few centimetres, the ventifacts are more likely to be the classic
faceted types (with planar faces and smooth surfaces)
that lack lineations (Schoewe 1932, Wentworth and
Dickey 1935, King 1936, Needham 1937, Maxson 1940, Glennie 1970, Czajka 1972, Whitney and
Dietrich 1973, Babikir and Jackson 1985, Nero 1988).
Schoewe (1932) suggested that the ratio of the height
of a faceted ventifact to the height of the sand-laden
current does not exceed 1:8. Thus, if the zone of
saltation extends to 64 cm, the diameter of the largest
fragment faceted would be 8 cm. Maxson (1940) noted
that faceted ventifacts in Death Valley, California, did

Fig. 19.4 Wind-eroded rock formed in a bidirectional flow
regime on a hill crest in the Little Cowhole Mountains, Mojave
Desert, California. Wind directions are from the north and south,
producing two grooved facets and a central keel. The keel is perpendicular to the wind flow. Material surrounding the ventifacts
is active aeolian sand

not exceed a height of 8 cm, and that larger fragments
were striated and polished, but not faceted.

Given sufficient time for erosion, moderate-sized
boulders often develop a semi-planar face, or two, if
the wind is bidirectional (Fig. 19.4): these faces are invariably fluted or grooved on the windward exposure
(Powers 1936, Sharp 1949, Czajka 1972, Laity 1987).
Rocks with three facets are occasionally observed in
regions with three distinct wind directions. Owing to
their great mass, facets usually do not develop on boulders exceeding 1 m in diameter, although considerable bevelling may occur. Grooves, pitting, and fluting have been observed up to heights of several metres
on rocks (Fig. 19.1). Powers (1936) noted such features on a crystalline boulder 4.5 m in diameter. Outcrop ledges may show evidence of planation and be
pitted or grooved.

Erosional Forms

Ventifact surfaces are characterized by a wide range of
erosional features, that vary both in form and scale,
resulting from differences in rock type, face angle,
weathering history, sediment supply, and wind speed.
To date, there has been little examination of the causes
of these features.

Some degree of rock heterogeneity seems important
in the development of specific features. For example,

pits are best developed in materials that have vesicles or a softer matrix, such as coarse grained granites, volcanic tuffs, and basalts. Flutes and grooves also appear to develop from initial vesicles or weaker spots within rocks. In 1993, I placed 6 large blocks of modelling foam of different densities, but a homogeneous texture, in an area of active abrasion in the Little Cowhole Mountains. The blocks ranged from 50 g to 325 g in weight, and were placed at a 45° angle to the horizontal, more-or-less parallel to the slope. Thirteen years later, the foams have all been strongly abraded, with the low density foam almost completely eroded away. However, over this period, erosion was essentially uniform across the face of the targets and no lineations formed. By contrast, styrofoam blocks (that are heterogeneous in texture) and plaster targets with small air vesicles developed lineations within a few months.

Although the various erosional forms that occur on ventifacts are discussed separately below, it is important to note that some forms are transitional in nature. For example, a pit may be seen to be elongating into an incipient flute. In soft materials such as tuffs, a case-hardened fluted surface may be undermined by subsequent erosional episodes, to produce a surface that is deeply eroded and almost fretted. Additional images and discussion of ventifact features may be found in Viles and Bourke (2007).

Smoothing and Polishing of Rock Surfaces

Smoothing and polishing of rock surfaces is perhaps the most common feature reported for ventifacts. Polish occurs both on smooth facets and within flutes and grooves (Maxson 1940). In periglacial settings, ventifact surfaces may exhibit a high gloss that exceeds that of glacial polish (Tremblay 1961). Rock coatings may develop after abrasion ceases, including the development of a silica glaze on periglacial ventifacts and desert varnish in drylands (Dorn 1995).

Facets

King (1936) and Sharp (1949) used the term facet to describe a relatively plane surface cut at right angles to the wind, regardless of the original shape of the stone, and the term face to describe the original surface of the rock fragment. Facets commonly join along a sharp

ridge or keel (Figs. 19.3 and 19.4), and the number of keels (kante) is used to describe the stones as einkanter, zweikanter, dreikanter (one-, two-, three-ridged), etc. (Bryan 1931). Much effort has been expended on the morphological classification of ventifacts (Bryan 1931, King 1936, Czajka 1972), particularly for small faceted forms.

Multiple facets developed on ventifacts have been attributed to (a) the original shape of the stone, (b) splitting of the wind around the rock, (c) winds from different directions, and (d) shifting of ventifacts owing to undermining by wind scour, and overturning by wind, frost action, rain wash, and animal disturbance. In cold environments, frost shifting may be a particularly effective mechanism, as discussed by Sharp (1949) and Lindsay (1973).

Pits

A pitted surface is one indented by closed depressions, often of irregular shape (Fig. 19.5). According to Whitney (1978, 1979), McCauley et al. (1979), and Garvin (1982), the wind is capable of producing pits on the surface of dense, homogeneous stones. However, Sharp and Malin (1984) emphasized that pitting is not an inevitable by-product of aeolian erosion, as material such as chert and limestone may develop facets, flutes, and grooves, but no pits. It is probably easier for the wind to modify pre-existing pits (such as vesicles in basalts or tuffs) by enlargement or integration, or to

Fig. 19.5 Pits developed on high-angle windward face in tuff. Soft inclusions are preferentially eroded, and abrasion in tuff yields pits in a great range in shapes. In basalts, pits tend to be rounder and more regular in form

erode softer minerals, as is evident in pitting of coarse-grained granites (Fig. 19.1).

Pits occur on surfaces that are inclined at high angles to the wind (55–90°) and thereby indicate the windward side of boulders. As the angle between the face and the wind decreases, a transition from pits to deep flutes with overhanging ends occurs (Sharp 1949).

Flutes

In form, flutes are open at one end and closed at the other, and broadly U-shaped in cross-section (Figs. 19.6, 19.7, 19.8, and 19.9). They may appear as "arrowheads" that point in a downwind direction. Flutes form independently of material hardness, composition, or rock structure on surfaces that are nearly horizontal or inclined at low angles (less than 40°) to the wind: flutes become shorter and deeper as the inclination of the surface steepens (Maxson 1940, Sharp 1949). Figure 19.6 indicates how the scale of the flute may increase in height up the ventifact. Occasionally, smaller flutes will begin to develop within larger ones as a second cycle of erosion begins (Fig. 19.9).

Flute development by aeolian processes is not understood. Maxson (1940) proposed that flutes are cut beneath vortices of fine sand, and Whitney (1978) argued that vortex pits may coalesce into flute pits and pit chains. However, most flutes exhibit smooth interior surfaces that lack pits.

Fig. 19.7 (**a, b**) Intensely pitted and fluted basaltic ventifact, approximately 1.8 m in height, located near a hill crest in the Cady Mountains, Mojave Desert, California. The flutes and grooves radiate outward from a central area. The elevation of the ventifact, relative to the surrounding sand-covered plain is shown in Fig. 19.7b. Ventifacts occur across the entire slope, with feature scale and intensity increasing with altitude

Fig. 19.6 Ventifact in sandy terrain near Silver Lake, California. The windward face has been bevelled and fluted. The scale of the flutes increases from the base of the rock to the upper face

Scallops

U-shaped or scalloped erosional features are open at one end and closed at the other and have similar length to width ratios. They lack the "arrowhead" form of flutes. They appear to be rare on terrestrial ventifacts, but are occasionally observed on Martian rocks.

Grooves

Grooves are longer than flutes and open at both ends. They are best developed on surfaces gently inclined or parallel to the wind. However, they are sometimes seen on the vertical sides of rocks, particularly where the wind has accelerated between

Fig. 19.8 Fossil ventifact with exceptionally large flutes located on a hill that rises ~100 m above the Mojave River Sink, Mojave Desert, California. The largest flutes are 10–17 cm wide, up to 60 cm in length, and 7 cm in depth. The side illustrated here faces into the westerly winds. Part of the upper boulder surface has detached and fallen to the base of the ventifact. The fossil ventifacts in this area are heavily varnished, but those near the crest show signs of reactivation (varnish abrasion) and sand is present in some of the flutes

Fig. 19.9 Closeup of flutes shown in Fig. 19.8. A secondary cycle of new flute generation can be seen within some of the larger forms. Litre-sized water bottle for scale

adjacent large boulders. Like flutes, they may cut indiscriminately across mineral grains and rock structures.

Ventifact groove and flute trends are remarkably parallel on near-horizontal surfaces and reflect the flow direction of the highest velocity winds in an area (Fig. 19.4) (Laity 1987). On large curved surfaces facing the wind, flutes or grooves often radiate outward from a central pitted area. Sharp (1949) observed that

Fig. 19.10 Helical forms developed in marble. These features occur at scales up to 20 cm in length and several centimetres in width and depth. The abrading agent is sand, some of which can be seen trapped within the helical forms. Fine lineations (striae) also cross the lower left face of the ventifact

flutes and grooves are not mutually exclusive, for the surfaces of large grooves are often fluted on a small scale.

Tremblay (1961) characterized three scales of groove development: striae are fine lineations (Fig. 19.10), grooves are of intermediate scale, and channels attain several centimeters in depth (up to 13 cm deep in his study area). None of the lineations exceeds a metre in length, and most are considerably shorter. Whereas fine striae appear to cross a whole outcrop, in detail each comprises a succession of short scoop-like depressions a few centimetres long.

As in the case of flutes, there is little understanding of groove formation. According to Maxson (1940), grooves suggest vortices descending along the rock surface in the wind direction: once formed, the grooves may be modified by saltating grains at air velocities below those critical in the generation of vortices. Schoewe (1932) discovered that sand grains impinging at low angles on hard, smooth surfaces skid instead of rebounding directly into the air and proposed that this action may be related to the development of flutes and grooves.

Etching and fretting

Etching occurs when the composition of a rock mass is not homogeneous, and the wind selectively erodes

Fig. 19.11 Etching and development of incipient lineations and small-scale pits on a heterogeneous layered plaster target placed in the Little Cowhole Mountains for approximately 3 months. The target, placed on the hill crest, is in a bi-directional wind regime. The target is ~22 cm in length and 10 cm in width

less resistant strata or foliations. In nature, etching is often well developed on layered ignimbrites. Etching also develops in artificial rock targets with layers of different hardnesses (Fig. 19.11).

Fretting develops where there are harder inclusions within the rock material, with projecting points, knobs (Fig. 19.12), and ridges forming a particularly rough surface (Sharp and Malin 1984). The inclusions resist erosion, while the surrounding material is removed by abrasion. This may result in the formation of erosional "fingers" with a visible inclusion at the tip (sometimes referred to as dedos) (Fig. 19.13) or, in some cases, large rounded xenoliths that project out from the overall ventifact surface (Fig. 19.12).

Helical Forms

Helical forms are uncommon. They begin as shallow grooves, deepen and spiral in a downwind direction, and terminate in a sharp point (Fig. 19.10); range from several millimeters to several centimetres in width and depth; and maintain a consistent form as their scale increases. Helical forms may be found in association with flutes. Observations in both desert and periglacial settings in California suggest they occur where wind velocities are very high, such as within topographic saddles and near hill crests. They develop on the upper face of the ventifact and have been observed in marble, basalt, and granite.

Fig. 19.12 A knobby texture formed on the high-angle windward face of a granite boulder. The large xenolith inclusions are finer grained and more resistant to erosion than the matrix, standing out several centimeters in relief

Fig. 19.13 Marble ventifact from the Little Cowhole Mountains, Mojave Desert, California. The finger-like projections (sometimes called dedos) extending from the ventifact result from visible inclusions of resistant material

Abrasion-Hardened Surfaces

Abrasion may create a hardening of the surface that resists subsequent mechanical and chemical weathering. This process may require a silicic form of rock, as it has been observed in volcanic tuffs, granites, and andesites, but is less evident in basalts, and does not appear to be a factor in marbles. Any loss of the protective rind (fracturing or sloughing) allows renewed erosion of the underlying rock, and can result in an exposed abrasion carapace standing free of the main rock (Fig. 19.1).

Processes of Rock Destruction

Three processes have the potential for material erosion in an aeolian regime: abrasion, deflation, and rock wedging. Abrasion results principally from the impact of particles in saltation and is accomplished by sand-sized grains (about 60–2000 μm in diameter) (Greeley et al. 1984). Finer material in suspension is thought by some to erode rock mass at the microscopic level (Whitney and Dietrich 1973). Deflation is the removal of previously weathered material by strong winds and appears to be largely insignificant in ventifact formation. Whitney (1979) proposed erosion by pure air flow alone, but as no ventifacts have yet been recorded in a nonabrasive environment, and as particulate material is generally abundant, the efficacy of erosion by pure air need not be an issue. Rock breakdown by wedging occurs when grains moving at high velocity are packed into cracks, and succeeding particles impact grains that are already wedged. Hall (1989) observed this process in Antarctica, but it has yet to be recorded in warm deserts. Solution processes may play a role in enlarging or initiating pits or in forming rills, but these processes appear subordinate to wind erosion (McCauley et al. 1980).

In some environments the apparent absence of sand has led to a consideration of snow as an abrasive (Teichert 1939, Dietrich 1977b, Schlyter 1989). Although snow at $-44°C$ has a hardness similar to quartz, it has half the specific gravity, so that sand will have two to three times more kinetic energy upon impact (McKenna-Neuman 1993). Furthermore, field experiments do not support abrasion by snow. McKenna-Neuman and Gilbert (1986) showed that poles covered with eight coats of exterior enamel

were rapidly stripped and eroded by aeolian sands, but remained perfectly preserved when subject to blowing snow and ice. These observations agree with those of other researchers (Blackwelder 1929, Tremblay 1961, Miotke 1982, Nero 1988), and indicate that sand is the most important abrasonal agent in cold climates.

Nature of Abradant

Dust

Although most researchers attribute ventifact formation to a sandblast action, dust has also been invoked to explain polish and the formation of finely detailed features such as flutes. Higgins (1956) suggested that dust in suspension was the abrasive agent for small multifaceted pebbles embedded in bedrock, but later concluded that the stones were probably not ventifacts (Higgins, personal communication 1988). Sharp (1949) considered that polish on ventifacts may result from material finer than sand. In the Namib Desert, Lancaster (1984) suggested that dust particles are probably more important than sand in creating smooth polished rock surfaces, flutes, and grooves, as the effects of sand laden winds would be akin to the destructive effects of industrial sand blasting processes. Nonetheless, sand was abundant at his field site. Maxson (1940) and Whitney (1978) attributed flutes or lineations to cutting by particles fine enough to follow vortex currents, suggesting that most sand grains travelling by saltation probably pass through small vortices without much change of path.

Wind tunnel experiments by Dietrich (1977a) indicated that wind-blown materials, which are relatively soft or small in particle diameter, can produce minor amounts of erosion. For 455 days, dust was blown continuously against blocks of low hardness (halite and sylvite) and against synthetic periclase. At the end of the experiment, the microtexture was rougher, but the blocks had no measurable weight loss. As most natural materials are harder than those tested, and as dust storms are not common, averaging approximately 22 per year in warm deserts of Africa and the Middle East and less than 5 per year in North America (Middleton 1989), a very long period of time would be necessary to remove the mass essential for ventifact development.

A review by Breed et al. (1989) also posed several arguments for dust abrasion abrasion, asserting that sandblasting can produce facets, but cannot account for delicately-textured erosional markings (pits, flutes, grooves, or helical scores); that ventifacts within the saltation zone lack fine detail such as fluting on their windward faces; and that ventifacts rarely occur in or near sand dune fields, but are common on the stony surfaces of sand-poor regions. Furthermore, ventifacts were said to have more rock mass eroded from their lee areas than from their windward. As a further complication, McCauley et al. (1979) concluded from the negative flow observed to the lee of a ventifact subject to a bubble-generating device (simulating suspended particles), that pits and flutes on all sides of rocks could be explained by wind from a single direction. These points will be discussed further in the following section.

Sand

Despite laboratory evidence that dust may be capable of abrading very soft rock over extremely long time periods, it is likely that sand is the most effective agent of abrasion in ventifact formation. Areas where sand has been identified as an agent of erosion include coastal environments (King 1936, Knight and Burningham 2003), periglacial regions (Powers 1936, Tremblay 1961, Miotke 1982, McKenna-Neuman and Gilbert 1986, Nero 1988, McKenna-Neuman 1990, 1993), and desert regions (Sharp 1964, 1980, Sugden 1964, Selby 1977, Smith 1984, Laity 1992, 1994, 1995). To date, there is no field evidence of ventifacts forming in areas where dust is the sole agent.

The following arguments can be made for sand as the leading agent in aeolian erosion.

(a) Analytical models indicate that sand is a much more effective agent of erosion than dust. The mass of material lost per particle impact is directly proportional to the kinetic energy of impacting grains (Greeley et al. 1984, Anderson 1986). Dust not only has less mass, but is well coupled to the wind, being deflected around rocks and rarely impacting the surface directly. For suspended grains ($D \leq 0.31$ mm), Anderson (1986) showed that particle deflection by the air flow around the ventifact leads to a reduction in the delivery of kinetic energy to the surface. By contrast, saltating sand achieves enough mometum to be decoupled from the boundary layer velocity profile at peak acceleration near the top of its trajectory, with the sand velocity on the order of 50% or less of wind velocity at that height (Bridges et al. 2005). Although the velocity of sand is less than that of dust, the mass, which increases by the cube of particle size, is much greater. Indeed, abrasion experiments using particles ranging in size from 75 to 160 μm show that S_a (the susceptibility to abrasion by rocks) varies with particle diameter D to approximately the third power (Greeley et al. 1982, 1984). Thus, the mass of a 100 μm sand grain has 1,000 times the mass of a 10 μm dust particle. Considering the effects of both velocity and mass, the kinetic energy of sand upon impact is ~50–100 times that of dust. Additionally, large particles are more likely to impact the surface than small particles. Anderson (1986) predicted that the number of impacts from 10 μm particles would only be 10% of that of 100 μm grains. Taking into account both the kinetic energy upon impact and the potential number of impacts, it is likely that the total kinetic energy transferred to rocks by sand would be on the order of 1,000 times greater than that of dust (Laity and Bridges, 2008) . In the field, Sharp (1980) showed that the cutting rate increased coincident with the increased flux of windborne sand.

(b) Experiments also show that the amount of abrasion incurred by a ventifact impacted by particles less than 90 μm in diameter decreases much more rapidly with decreasing particle size and velocity than predicted by kinetic energy considerations or experiments using larger sized particles (Stewart et al. 1981). Two factors account for this. First, where clay particles are present in the entrained material, these particles are transferred to the surface of the target, sheltering it from subsequent abrasion by larger particles (a cushioning effect). Second, there is an apparent kinetic energy threshold for impact fracture, suggesting that the abrasion mechanisms change with kinetic energy.

(c) The present-day absence of sand from a region does not preclude it from consideration as the primary abrasive agent. Mainguet (1972) and Sharp and Malin (1984) emphasized that the movement of sand in aeolian corridors tends to be episodic, so that sites which have been traversed by large quantities of saltating grains now harbour only small accumulations. Also, many ventifacts are relict, and the abrasive sand that carved them has been deposited elsewhere.

(d) In many areas, the distribution of ventifacts is clearly associated with the presence of sand (King 1936, Tremblay 1961, Smith 1984, Laity 1992,

1995 and others). In the eastern Mojave Desert, fossil forms occur in association with deposits of stabilized sand or in aeolian corridors where sand has travelled to nearby sites of deposition. Actively forming ventifacts occur in rock outcrops presently being traversed by fine-grained sands (Figs. 19.3, 19.4, and 19.14). Ventifacts are not found in areas subject to dust influx alone. More than a metre of silt has been deposited in some areas of the Cima volcanic field (Dohrenwend 1987), yet despite long exposure to dust, there is no evidence of abrasion on rock surfaces.

(e) Actively forming ventifacts lying within the saltation zone clearly demonstrate fine details on their surfaces (Figs. 19.3, 19.4, and 19.13). Sharp (1964) described the formation of flutes on bricks in a wind-driven sand environment. In the Little Cowhole Mountains, Mojave Desert, fine sands are seasonally redistributed by the winds, alternately burying and exposing rocks (Laity 1995). All of the marble within the vicinity shows considerable mass removal and planation as well as sharply defined erosional markings (flutes, grooves, or helical forms). Some of the striations are only 1 mm across, separated in parallel arrays by sharp ridges. If sand is not responsible for the creation of these microfeatures, it could be argued

Fig. 19.14 Abrasion of ventifacts during the passage of a front in the Mojave Desert. Wind speeds reached 23 m s^{-1}, with higher gusts. Abrasion patterns on the hill slopes were complex, as sand was driven from near the base of the slope up towards the crest of the 100-m hill. As the sand moved up and over the hill, ventifacts were alternately exposed by wind scour and buried. Later in the day, the wind reversed and the opposite faces of ventifacts were abraded. Upslope wind acceleration results in maximum erosion near the crest. Note that whereas the windward face is exposed to abrasion, the lee face is commonly protected by a sand tail. The fossil ventifact shown in Figs. 19.8 and 19.9 is located further to the south along the same ridge, at a site presently less subject to sand passage

that they would be destroyed by the protracted impact of saltating grains.

Saltating sand grains are also carried across the Amboy (Greeley and Iversen 1986) and Pisgah (Laity 1987) lava fields of the Mojave Desert. The flutes and grooves that occur on the subhorizontal flow surfaces are developed by saltating grains on a descending path. Near the surface, where wind velocity approaches zero, dust would be an ineffective agent of abrasion (Anderson 1986).

(f) Polished surfaces and fine features can result from sand abrasion. Actively forming ventifacts of the Pisgah lava flow (basalt) and the Little Cowhole Mountains (marble) demonstrate a high degree of sheen and are macroscopically smooth. Under a microscope, there is abundant sand visible within the cavities and, at higher magnifications, scanning electron micrographs of flute interiors show considerable topographic roughness indicative of repeated chipping. Occasional large impact structures have fresh cleavage facets that show no loss in definition at high magnifications (Laity 1992). Thus, surfaces that appear very smooth to the eye can be very rough at the microscale. This evidence suggests that dust need not be invoked to explain polish or fine features.

(g) Most ventifacts show significant erosion on their windward faces and no erosion on their lee surfaces (Blackwelder 1929, Sharp 1964, McKenna-Neuman 1993). This relationship has been demonstrated unequivocally since 1993 at an instrumented site in the Little Cowhole Mountains, California. An automated weather station records the bidirectional winds, that seasonally move a reversing sand dune across a hillslope covered by marble ventifacts. The abrasion of painted poles, plaster and simulated sandstone targets, foam boards, and balsa wood posts has been related to wind direction throughout this period. Abrasion was recorded only on the windward face: no "lee side" abrasion from negative flow has occurred. Furthermore, in small-scale plaster "rocks," lineations have developed within the saltation zone, paralleling the direction of the highest velocity winds and with the same orientation as adjacent ventifacts.

Large boulders, that are not subject to movement and overturning, also exemplify this relationship. Blackwelder's (1929) study of ~1 m high boulders showed the windward surfaces to be strongly grooved and drilled, whereas on the leeward face "no such markings were found, but on the contrary the rocks are merely cracked and exfoliated in the ordinary manner" (p. 256). Similarly, McKenna-Neuman and Gilbert (1986) observed well-polished flutes and grooves on the windward face and a lee face typically covered in lichen, with no evidence of abrasion. Regardless of lithology, ventifacts in California show the anticipated relationship between maximum abrasion on the windward face, and a facet that slopes away from the wind. Furthermore, field observations during a strong wind event (up to $23\,ms^{-1}$) in the Mojave Desert (near Rasor Road) showed no lee side abrasion. Indeed, the deposition of sand on the lee face of some ventifacts acted to protect them from abrasion (Fig. 19.14).

(h) Pits are generally considered to be diagnostic of high-angle impact by sand grains on windward faces (Sharp 1949). Studies of pitted surfaces and wind flow in the Mojave Desert confirm this relationship. Pits are not developed equally on all sides of rocks, but are limited to high-angle windward surfaces (Smith 1984, Laity 1987, 1992).

In summary, laboratory experiments have shown that dust is capable of changing the surface micromorphology of rocks but, in contrast to sand abrasion, mass removal is generally too small to be measured and would be quantitatively insignificant, even over long time periods. To date, no actively forming ventifacts have been identified that develop in an environment dominated by dust alone. Dust has been invoked primarily as an agent in the formation of fossil ventifacts when a sand source is not evident, and to explain the polish and fine flutes and grooves that develop on ventifact surfaces. However, lack of sand in the immediate vicinity of ventifacts is not conclusive, as the movement of sand across landscapes is commonly episodic. Furthermore, ventifacts with polished surfaces and fine features are shown to occur in regions where sand dominates the flux of particles and dust storms are rare. Whereas polish appears to be macroscopically smooth, at the microscale the surface is very rough, with features indicative of repeated chipping and gouging by sand particles.

Formation of Ventifacts

The formation of ventifacts has been assessed analytically, by wind tunnel experiments, and by field exami-

nation. Early wind tunnel experiments are discussed in Schoewe (1932). Laboratory experiments have sought to assess ventifact formation by examining different abradants, target materials and slope angles, and wind conditions. Several studies have addressed abrasion rates.

There are several limitations to wind tunnel studies. They are typically conducted in tunnels that are too small to allow natural conditions of grain bounce and do not replicate natural surface conditions. The abradant varies greatly in grain size (Schoewe 1932, Kuenen 1960, Whitney and Dietrich 1973, Dietrich 1977a) and in angularity (Miotke 1982), and seldom represents the size, shape, or sorting characteristics of material abrading ventifacts in the field. Natural rocks are rarely used as the target material as they abrade too slowly. Nonetheless, wind tunnels provide the ability to control parameters that affect aeolian processes, giving insight into fundamental mechanisms. Bridges et al. (2004) conducted wind tunnel experiments using sandstone simulants and foams, with five shapes defined by the angle of the front face relative to the wind ($15°$, $30°$, $45°$, $60°$, and $90°$) and, as a ground truth calibration, placed similar targets at a ventifact site in the Little Cowhole Mountains, Mojave Desert (Fig. 19.11). Intermediate-angled faces exhibited the greatest angle changes, but not the greatest mass losses. Although wind tunnels yield useful information on short term abrasion and feature formation, wind abrasion in the field is more aerodynamically complex and is impossible to fully reproduce in the laboratory.

Process-oriented field studies were pioneered by Sharp (1964, 1980). A long-term field study (1993 to present) in the Little Cowhole Mountains has provided further insights into the formation of abrasional features in a regime of active sand transport (Laity 1995 and unpublished data, Greeley et al. 2002, Bridges et al. 2004). Other field studies have concentrated on aspects of ventifact morphology and on the use of ventifacts in palaeoclimatic interpretations.

Magnitude of Erosion

The magnitude of ventifact erosion is dependent on the susceptibility of the rock to erosion, S_a (as determined by density, hardness, fracture–mechanical properties, and shape of the rock) and on properties of the impacting particle, such as particle diameter D, density ρ_p, speed V, and angle of incidence α($90° =$ perpendicular impact) (Greeley et al. 1984, Anderson 1986). The mass of material lost per impact, A (Scattergood and Routbort 1983) is

$$A = S_\mathrm{a}\rho_\mathrm{p}(V\sin\alpha - V_\mathrm{o})^n(D - D_\mathrm{o})^m \qquad (19.1)$$

where V_o and D_o are the threshold particle speed and diameter that will initiate erosion. Abrasion experiments indicate values of $n = 2$ and $m = 3$ (Anderson 1986). The mass of material removed per impact is roughly proportional to the kinetic energy of the impact.

Height of Abrasion and Development of the Erosion Profile

Ventifacts develop within the curtain of saltating sand grains, and their shape and development is dependent on particle fluxes within this zone (Fig. 19.15). Saltating grains follow a path termed the saltation trajectory. Grain velocity increases throughout the

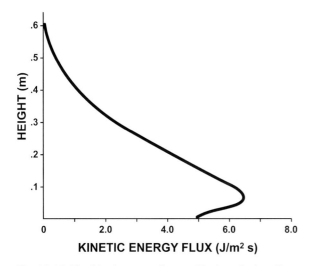

Fig. 19.15 The kinetic energy-flux profile for saltation (from Anderson 1986). The typical erosion profile, developed in fence posts, on rocks, or on yardangs, will be similar to this form, although the height of maximum kinetic energy flux (and hence erosion) above the ground varies according to surface conditions and slope angle. The faces of many ventifacts reflect the lower part of the curve, receding backwards at a nearly constant angle

path, reaching 50% of maximum velocity near the peak of the trajectory. In general, particles travelling at greater heights have higher velocities, owing to an increase in wind speed above the ground and the longer saltation paths that allow more time for them to be accelerated by the wind (Greeley et al. 1984). Calculated kinetic energy fluxes of saltating particles are greater where liftoff velocities are increased by grain bounce on elastic surfaces rather than mobile beds (Anderson 1986). In deserts, harder surfaces that promote grain bounce include extensive areas of stone pavement, boulder-strewn slopes, surfaces impregnated by late stage calcium carbonate, and exposed bedrock areas, including basalt flows. Such settings are subject to more vigorous erosion (Greeley and Iversen 1985, Laity 1987).

The range of effective sand abrasion rarely exceeds 1 m above the surface on level ground. Hobbs (1917) observed this limit on a variety of materials in Egypt, including cast-iron telegraph poles, thick adobe walls, and granite knobs, where a lower polished zone intersected the remaining weathered rock along a fairly sharp boundary. "Pedestal rocks" develop as a result of concentrated erosion near the rock base. Ventifacts 1 m in height show sand blast effects to their upper surfaces. On hillslopes, abrasion may occur to a height of several meters (Figs. 19.1 and 19.7).

Within this general zone of abrasion, erosion profiles develop with distinct maxima of mass removal (Sharp 1964, 1980, Wilshire et al. 1981, Anderson 1986). The pattern of erosion is similar for all materials, although the magnitude depends on material properties. The height of maximum erosion is influenced by such parameters as wind speed and the degree of grain bounce. Sharp (1964, 1980) recorded erosion maxima 0.10–0.12 m above a level ground surface in Lucite rods exposed to the wind for 15 years. The height of maximum abrasion shifts upward as wind velocity increases (Liu et al. 2003). Elevated heights also result from greater grain bounce on hard surfaces relative to softer ones, moving upwards from 0.28 m to 0.43 m (Wilshire et al. 1981). On hillslopes, the erosion maxima are also higher, owing to wind acceleration and other effects related to shifting sand base levels: balsa block array maxima on slopes in the Little Cowhole Mountains were at heights of 0.3–0.5 m (T. Boyle and J. Laity, unpublished data).

As a consequence of the increase in abrasion up to the maximum level, many ventifacts ultimately develop semi-planar faces, with the upper part of the abrasion face receding more rapidly than the lower part. Very large ventifacts may also exhibit a surface that slopes away from the prevailing wind, but owing to the greater height and mass of the rock, to a limited time of exposure to abrasion, and to variable rock resistance, such features are usually less well developed than those of smaller ventifacts.

Friction with the ground results in wind speed that is minimal near the surface. A sill of uneroded material at ground level developed on bricks and hydrocal blocks in an experimental plot (Sharp 1964) and on targets in wind tunnel experiments (Bridges et al. 2004). Owing to the complexity of interactions between topography, wind flow, and shifting sands, such a sill is commonly, but not always, observed in nature. On subhorizontal surfaces such as lava flows, saltating grains on a descending path erode flutes and grooves.

Characteristics of the Particle

As discussed earlier, the composition, shape, size and quantity of windblown materials all affect abrasion. Greeley et al. (1984) showed that there is little difference in abrasion by quartz and basalt particles, that ash is less efficient in erosion, and that aggregates may be plastered on to the target, forming a protective coating, and thereby lessening erosion. Laboratory experiments have shown that abrasion increases with particle diameter and with angularity, but field studies have rarely assessed these characteristics. Most abradants are moderately well sorted and in the size range of fine- to medium-aeolian sands (Greeley and Iversen 1986). The sharp-edged particles that develop in polar areas may be particularly abrasive (Miotke 1982).

Sharp (1949, p. 185) characterized the amount of sand necessary for ventifact formation as being "adequate but not too abundant". He demonstrated a direct correlation between rates of erosion at his experimental plot and the influx of grains (Sharp 1964, 1980). Laboratory experiments show that where blowing sand concentrations are great, rebound effects from the target rock interfere with incoming grains and lessen the abrasion (Suzuki and Takahashi 1981). In the field, too great a supply of sand results in burial of ventifacts (Fig. 19.14).

Susceptibility to Abrasion of the Target Materials

The erodibility of different materials has been assessed experimentally by controlling for impact velocity, impact angle, and impacting particle size and type (Greeley and Iversen 1985). The bond strength of the rock appears more important than its hardness in predicting abrasion (Dietrich 1977a, Suzuki and Takahashi 1981). Experiments by Greeley et al. (1982) indicate that glassy materials such as obsidian will erode quickly for surfaces perpendicular to the wind, whereas crystalline materials such as granite and basalt erode more quickly when surfaces are subparallel to the wind.

The role of the angle of incidence of the incoming grain to the target has been assessed primarily through wind tunnel experiments. For steep angles (~90°), the abrasion rate is lowered because rebounding grains hit incoming grains, slow them, and result in lower impact energies (Greeley et al. 1982). In the case of shallower angles (~15°), grains tend to skid along the surface, and hit at the lower velocities characteristic of grains at the end of their trajectories, so that the surface is more likely to be lowered than reduced in angle. Thus, very steep (90°) and very shallow (15°) targets more or less maintained their shape.

By contrast, slopes in the range of 30–60° tend to undergo changes in slope with abrasion. Within this range, the abrasion rate is greatest for higher angled slopes (Bridges et al. 2004). Schoewe (1932) showed that the rate of abrasion on faces sloping at 30° is about one-third as great as faces inclined at 60°. As a result, abrasion rates tend to lessen through time as the facet becomes more inclined. Preliminary field evidence (Bridges et al. 2004) of eroded targets tends to support the wind tunnel results and suggests that, given enough time, targets will evolve to an angle of ~30°.

The published results on slope angle appear to differ when the ventifacts become very small, although there is little information on this topic. Needham (1937) measured the facet angle for very small ventifacts (up to 1 cm in diameter) and found that most (63%) fell within the range 45–69°, 22% had high angle faces (70–89°), whereas very few (15%) had low angle faces (25–44°). No ventifacts were measured with facet angles less than 25°.

Rate of Abrasion

Natural rates of abrasion are difficult to determine and are probably highly variable through time owing to the many different controlling factors that influence ventifact formation. Wind velocities are not constant, but vary according to season, time of day, and the passage of fronts. Most of the abrasion occurs during periods of high velocity winds, which occur for only a small percentage of the time. Abrasion rates also change because, as discussed above, the rocks gradually wear, lowering the angle of incidence of the impacting grain and the height of the rock. If abrasion is episodic through time, weathering of the rock between erosional episodes may prepare it for abrasion and increase subsequent rates of surface wear. During periods of stability (no abrasion), weathering rinds may develop that cause surface hardening, necessitating higher than normal abrasion to renew erosion of the surface.

Time-dependent particle flux also determines abrasion rates. Even when sand is available, it may move intermittently through an area. A 15-year study of abrasion by Sharp (1980) showed an annual rate of wear 15 times greater during the last 3-year interval than in preceding years owing to an increased flux of windborne material derived from nearby fluvial flooding debris. Megascopically visible effects (polish, pitting, and incipient fluting) developed within 10 months during periods of intense erosion characterized by an abundance of windborne particulate material (Sharp 1980). Over short time periods (seasonally), ventifacts may be buried by sand and be protected. Over longer time periods, the availability of particles may decline through time owing to climatic change.

Rates of abrasion inferred from various ventifact sites range from approximately $0.01-1$ mm y^{-1} (Greeley et al. 1984) and the time taken to form ventifacts may range from hours to days to months along storm-exposed coastlines (Kuenen 1960), to dozens or hundreds of years (Sharp 1964, 1980), to thousands of years (Selby 1977). Knight and Burningham (2003) examined coastal ventifacts formed within the past century and estimated abrasion rates of $0.24-1.63$ mm y^{-1}. Abrasion is more rapid in Antarctica, primarily owing to higher wind speeds, with Miotke (1982) estimating rates of $5-20$ mm y^{-1}. Liu et al. (2003) demonstrated that the abrasion capacity of saltating sand (the ratio of abrasion rate to aeolian sand transport rate) increased logarithmically with wind velocity.

In addition to wind speed, abrasion rates are influenced by particle supply, particle diameter and density, periods of burial, target hardness, and target surface roughness. Long-term observations of modelling foam targets at the Little Cowhole Mountains clearly indicate that softer materials abrade more rapidly than harder ones. Additionally, targets that are initially pitted abrade more rapidly than those that have smooth surfaces (Bridges et al. 2004).

Topographic Influences on the Development and Spatial Distribution of Ventifacts

Local and regional topography affect wind velocity, sand flux, and the direction of wind flow, and thereby influence the location of ventifacts, the orientation of flutes, grooves and facets, and the magnitude of erosion. Two topographic situations that commonly affect ventifact development are (a) wind speed increase through topographic constrictions, and (b) wind acceleration up the windward flanks of hills. At large scales, wind acceleration through constrictions may involve passage through a valley, and at a small scale, through a saddle or dip in an outcrop. A series of eight 100-m transects, laid out at 2-km intervals in a topographic constriction in a structurally controlled valley in the Mojave Desert, California, showed that faceting and grooving affected 70–90% of all exposed cobbles and boulders. In the area downwind of the constriction velocity generally declines as streamlines spread out, and ventifacts are absent (Laity 1987).

Wind also accelerates as it moves up the windward flanks of hills or dunes (Ash and Wasson 1983, Lancaster 1985). The compression of streamlines in the boundary layer causes the wind to accelerate towards the crest of the slope and then to decelerate on the downwind side. This is important geomorphologically because the increase in wind speed over the surface of the hill produces increased sand transport as well as additional surface shear stress (Jackson and Hunt 1975, Lancaster 1985). Mason (1986) found that for a smooth, nearly circular hill rising 70 m above the surrounding terrain, mean velocity 8 m above the surface $u(8)$ was reduced to a minimum of 0.8 $u(8)$ at the base of the upstream face and flow over the summit increased to 2.0 $u(8)$. These observations suggest that flow is reduced on the uphill face, increases on the sides and summit of the hill, and separates on the lee slope. The actual acceleration of wind depends on the height of the hill and the angle of the incident flow. The largest shear stress values, pressure changes, form drags, and wind velocities are recorded when the near-surface wind approaches approximately normal to the topography. The effect of velocity speed-up increases with the height of the topographic obstacle.

The threshold velocity for the movement of sand may be reached as a consequence of this acceleration of wind near the crest. Therefore, an increase in sand transport at higher elevations is anticipated (Fig. 19.14). This effect is particularly marked because the rate of sand transport is proportional to the cube of the excess of wind velocity over the threshold velocity for sand movement (Bagnold 1941). Lancaster (1985) reported that with a 6.5 m s^{-1} wind, sand movement at the crest of a transverse dune is 19.5 times greater than at its base if the dune is 5 m high, and 121 times greater if the dune is 20 m high. Higher wind speeds keep sand active at some hill crests in the Mojave Desert. Fossil ventifacts may be found on the lower hillslopes and active-forming ventifacts near the summits (Figs. 19.2 and 19.3).

Wentworth and Dickey (1935), in their survey of ventifact localities in the United States, noted that many ventifacts occur on the surfaces or margins of mesas where pebbles are exposed to strong, persistent winds from adjacent areas of lower elevation. Field mapping of ventifacts in the Mojave Desert showed that upper hill slopes and crests are favourable sites for ventifact formation and that, in many cases, the intensity of abrasion, as measured by pit diameter and groove width and length, often increases with elevation up the slope (Laity 1987). Local topographically enhanced velocity increase may allow winds of moderate velocity to be effective in abrasion (McKenna-Neuman and Gilbert 1986). Where regional winds are strong, deep grooving and pitting may occur, particularly on large boulders that present high-angle faces to winds (Figs. 19.1, 19.7, and 19.8).

Use of Ventifacts in Determining an Aeolian Regional History and Palaeocirculation

Numerous investigators have shown that wind direction may be determined by reference to the position of the sharpest bounding edge of a facet, by pitting on the face, or by the direction of grooving and fluting (Maxson 1940, Selby 1977, McCauley et al. 1980, Laity 1987, Nero 1988). For faceted ventifacts, the keel is oriented in a large majority of cases at right angles to the wind (Maxson 1940): in the central Namib Desert 93% of ventifacts have facets indicating a dominant wind from the north-east (Selby 1977).

Fossil ventifacts provide an excellent record for palaeocirculation reconstruction. As small ventifacts can change their orientation through time, large stable boulders and outcrops are the best choice for mapping. The relict nature of ventifacts is often indicated by the weathered, dulled and partly exfoliated condition of the rock surfaces (Blackwelder 1929, Smith 1967, 1984) and by the presence of rock varnish (Dorn 1986) or lichens. Grooves and fluting may not cover the entire surface of the boulder, but rather occur in patches where weathering has failed to remove them (Powers 1936). The growth of vegetation surrounding ventifacts, the lack of any apparent wind-blown material, or the stabilization and incipient soil development of aeolian deposits also indicates the fossil nature of ventifacts.

Numerous studies have used ventifacts to infer wind direction and palaeoclimate (Powers 1936, Sharp 1949, Tremblay 1961, Nero 1988, Smith 1984, Laity 1992) In the east-central Mojave Desert, California, three principal flow directions – westerly, northerly, and southerly – were identified from the analysis of mapped grooves (Smith 1984, Laity 1992). Relative ages of ventifacts were assessed by field relationships and an examination of the surface micromorphology of ventifacts. The widespread cessation of abrasion is marked by weathering of the micro-impact structures in grooves and flutes, by the formation of rock varnish, and by the stabilization of sands in the immediate vicinity of ventifacts.

Problems in reconstructing surface palaeowinds by ventifact mapping occur when there has been more than one pulse of erosion and high velocity winds emanated from different directions in succeeding episodes. The most recent winds may erase the imprint of earlier activity, although in rare cases cross-grooves are discernible. Multiple episodes of erosion are best preserved in hilly or mountainous regions, because erosion by earlier winds may be preserved on topographically protected (leeward) slopes.

Yardangs

Yardangs are elongate wind-eroded ridges that develop at a range of scales, from microyardangs (centimetre-scale ridges), to meso-yardangs (metres in height and length), to megayardangs (also called ridge and swale systems) that are tens of metres high and kilometres in length. Although closely related, mega-yardangs and meso-yardangs differ in their aerodynamic form and scale and probably in the relative roles of abrasion and deflation. Mega-yardangs are best developed in more resistant materials (Figs. 19.16 and 19.17), whereas micro- and meso-yardangs commonly form in softer sediments.

Fig. 19.16 Space Shuttle image of wind-eroded ridges following, in an arc, the deflection of the trade winds around the Tibesti Mountains. The system crosses the Aorounga Crater, Chad, near the centre of the photograph. Source: Image Science and Analysis Laboratory, NASA-Johnson Space Center. 10 Jul. 2006. "Astronaut Photography of Earth – Display Record." http://eol.jsc.nasa.gov/scripts/sseop/photo.pl?mission=ISS014&roll=E&frame=6304

Mega-Yardangs

Mega-yardangs have been recorded in a number of locations on Earth (Goudie 2007) and are of interest owing to their significance in understanding the

aeolian history and geomorphic development of the Martian surface. In the Lut basin in southeastern Iran, yardangs up to 80 m high form very elongate ridges with flat to rounded summits, separated by troughs more than 100 m in width (Fig. 19.18) (McCauley et al. 1977). West of the Rio Ica, near Cerros Las Tres Pirámides, Peru are streamlined yardangs up to 1 km in length, developed in clastic Tertiary sediments. Near the town of Mangnai in central China, very large yardangs are eroded into lacustrine deposits at a scale comparable to some aeolian features observed on Mars. Martian yardangs include those of the Medusae Fossae Formation (MFF), spread across the Martian equator in the Amazonis and Elysium Planitiae regions (Bradley et al. 2002). In some areas of the MFF, yardangs reach 150 m in height: elsewhere, yardangs average 10–40 m in height. Jointing may play an important role in establishing yardang orientation in the easily-erodible MFF deposits.

One of the most spectacular mega-yardang systems is that lying on the southeastern flanks of the Tibesti massif of northern Africa (Figs. 19.16 and 19.17) (Grove 1960, Mainguet 1970). Mainguet et al. (1980) consider the Sahara a single vast aeolian unit, with wind being the most active geomorphic agent. The region is divided into sectors where either sand transport or deposition dominates. Sand seas or ergs represent the depositional sector. Zones of sand transport are characterized by landforms of erosion, including

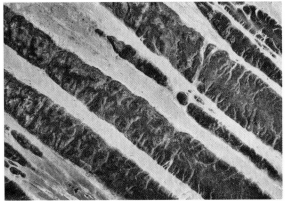

Fig. 19.17 Ridge and swale systems of the Sahara. Vast systems occur in a zone bordering the Tibesti Mountains to the east, south, and west. On aerial images the ridges appear dark owing to rock varnish, and the swales are lighter coloured owing to the presence of sand. In the Bembéché region of Chad, illustrated here, the ridges are very wide (up to 1 km) in proportion to their height (Cliché Institut National Géographique, Paris)

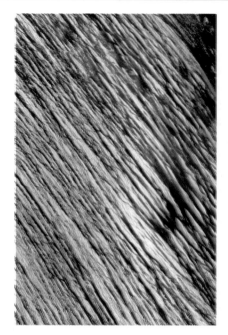

Fig. 19.18 The central portion of the Dasht-e Lut, Iran, is carved into yardangs that exhibit remarkable parallelism and a high length-to-width ratio. NASA Advanced Spaceborne Thermal Emission and Reflection Radiometer (ASTER) image, May 13, 2003

wind-abraded ridge and swale systems, yardangs, and zones of deflation.

Ridges and swales are best demonstrated by vast systems that occur in a zone bordering the Tibesti Mountains to the east, south, and west. On satellite and aerial images the ridges appear dark, owing to the well-developed patina of rock varnish, and the corridors show as lighter-coloured lineations. In orbital views, the systems appear to be continuous for hundreds of kilometres, sweeping in a broad arc around the mountains. On aerial photographs (1:50,000), the discontinuous nature of the systems is evident, with the largest ridges not longer than 4 km (Mainguet 1972). In the Bembéché region, which lies to the north-east of Faya-Largeau, the ridges are very wide (up to 1 km) in proportion to their height.

There are three factors that account for the development of these remarkable features: (a) extensive exposures of sandstone, (b) a dense network of joints that channelizes the wind, and (c) a monodirectional wind, charged with sand. The cover of Palaeozoic and Mesozoic sandstones is largely preserved on the southern flank of the mountains, whereas on the northern flank only isolated patches persist (Hagedorn 1980). The sandstone appears particularly susceptible to erosion, and forms of this type do not appear in

basalts, crystalline bedrock, schists or siltstones (Mainguet 1972). In diatomites, the ridges have a scale of development ten times less than that of sandstone.

The circum-Tibesti region is deformed and fractured along two major axes, NE–SW and NW–SE. Wind exploits the fracture system most closely aligned with its own direction, gradually enlarging corridors, the size of which is a function of the deviation between the wind direction and joint trends. Where the coincidence is best, the corridors are, with respect to the ridges, the least enlarged and the most regular in form. Fractures that run counter to the ridge trends are often exploited to form the fronts of ridges or yardings, these appearing to line up in a row.

The general direction of the wind is determined by trade wind circulation, except where it is diverted around major topographic obstacles such as the Tibesti Mountains. Northern Chad, to the east of the Tibesti, is characterized by the constancy of sand-laden winds that blow 8 months out of 12 from the north-east. Maximum velocities are reached during the daytime, when sand is transported at velocities of 6–8 m s^{-1} or greater. The ridges follow, in an arc, the deflection of the wind.

Topographic chanelling between the Tibesti and Ennedi Mountains enhances the Bodélé Low Level Jet (LLJ), which in turn increases the frequency of erosive surface winds (Fig. 19.19). To the lee of the Tibesti, there is a pronounced downslope flow, with a subsidence core at around 1.5 km overlying the Bodélé depression. The large scale topography increases the magnitude of the jet over the region and increases the frequency with which episodes of deflation and abrasion occur. Simulation models suggest that the Bodélé LLJ would have been even stronger during the Last Glacial Maximum (LGM) (Washington et al. 2006).

The amount of sand in the corridors is variable, with some being totally engulfed. Sand abrasion acts preferentially in the corridors and is ineffective on the middle and upper slopes of ridges, which are marked by a deep patina above the basal fringe area of eroded rock. There is no evidence that deflation plays a significant role. Wider swales may be occupied by barchans, whose axes parallel those of the ridge systems and the resultant winds (Mainguet 1972). The barchans are indicative of sand transport through the region and are clearly evident on satellite images.

The channelized topography has a wavelength that, in cross-section, appears to be relatively constant for any given group of ridges and swales (Mainguet 1970). The periodicity is determined to some degree by the

09:15 UTC (Terra MODIS)

12:15 UTC (Aqua MODIS)

Fig. 19.19 Northeastern Harmattan winds are funnelled and intensified as they pass between the Tibesti and Ennedi Mountains (*upper right corner*) in northern Chad, causing episodes of abrasion and deflation in the Bodélé Depression, one of the most active global dust source regions. The erodible lake sediments are lowering rapidly, with 4-m high yardangs having formed in a period of 1,200–2,400 years. These NASA Moderate Resolution Imaging Spectroradiometer (MODIS) images show a February 11, 2004 dust storm with a 3-h interval between images. In the lower image, the storm is approaching Lake Chad

fracture density, but topography and wind strength also appear to play a role, with the largest ridges found on the more elevated parts of the terrain, and the smaller forms in the lowest. Abrupt changes in scale may occur as an escarpment is surmounted by the wind (Mainguet 1972). In some areas, two families of ridges and corridors may develop, with the larger system being the most elevated and the longest, and the smaller superimposed within the corridors.

An interplay through time between fluvial and aeolian processes is evident. Lacustrine sediments occur in the largest aeolian corridors, apparently emplaced after the initial cutting of the ridges. Deposition was followed by a renewed phase of aeolian excavation. Lacustrine deposits, as well as ravines developed on the slopes of some ridges, suggest alternating wind and water dominance. Fluvial action may also have etched and enlarged many of the joints in the sandstone, thereby allowing wind channelization to occur. At the present time, wind action is dominant, and acts to erase most traces of fluvial activity.

Meso–yardangs

In contrast to ridge systems, yardangs are more streamlined and aerodynamic in form. They are often described as resembling an inverted ship's hull, although in many cases the yardangs are flat-topped. The windward face of the yardang is typically blunt-ended, steep and high (Fig. 19.20), whereas the leeward end declines in elevation and tapers to a point (Whitney 1984) (Fig. 19.21). It should be noted, however, that yardangs take on many different forms (Whitney 1983, Halimov and Fezer 1989).

Measurements of yardang length:width ratios commonly average 3:1 or greater. This elongate form minimizes the drag or resistance to the wind. Yardangs form parallel to one another, typically occurring as extensive fields (Fig. 19.18), and their long axes are oriented parallel to the strongest regional winds. They may occur as tight arrays, separated from one another by either U-shaped or flat-bottomed troughs, or as widely spaced, highly streamlined features on wind-bevelled plains.

History of Yardang Research

Yardangs have received attention as curiosities and geological oddities since the late 1800s. Stapff (1887) gave an account of "aerodynamic landforms" sculpted

Fig. 19.20 Windward face of yardang at Rogers Lake, Mojave Desert, California, showing a blunt end and erosional markings

Fig. 19.21 Yardangs at Rogers Lake, illustrating the sub-parallel nature of the long axes and a form that tapers to the lee

out of bedrock in the Kuiseb valley of the Namib Desert of south-west Africa, and yardangs were described by Walther (1891, 1912) in Egypt and by Kozlov (1899) east of Lop Nor. The term yardang was introduced by Hedin (1903, volume 1, p. 350) to describe a labyrinth of clay "terraces" in the Lop Nor (Nur) (Mongolian *nur* means lake) region at the eastern edge of the Taklimakan Desert in China, which the natives of the area called yardang. Hedin (volume 2, p. 139) attributed their formation to initial erosion by running water, and subsequent resculpturing by the wind. More recent interest in yardangs was fostered by the discovery of large yardang fields on Mars (Ward 1979) and by the availability of aerial photography and satellite imagery (Mainguet 1972, Ward 1979, Halimov and Fezer 1989).

Yardangs have not been studied in detail and much remains to be determined about their formation. There is little meteorological information available for most yardang fields. Although it is apparent that most form in association with monodirectional wind regimes, the velocity of winds within the corridors and the frequency of sand-transporting winds are not known. Nor is there any knowledge of the interplay between wind and topography in major yardang and ridge systems. Our understanding of wind flow around individual yardangs is based primarily on laboratory and theoretical determinations (Ward and Greeley 1984, Whitney 1985), rather than instrumentation. The long-term evolutionary development of yardang fields, the role of fluvial erosion and jointing in initial channelization of winds, and the rates of formation are poorly understood. The role of abrasion and deflation

in determining the ultimate form of yardangs also remains uncertain, and probably varies according to the rock type.

Factors Affecting the Distribution of Yardangs

Yardangs occur in desert regions on all continents, but they vary greatly in their scale, development, and spatial extent. Overall, they occupy only a very small part of the Earth's surface, as they require conditions of great aridity, nearly unidirectional winds, and, in some cases, a favourable material and some assistance from weathering to form. Nonetheless, aerial photography and satellite imagery show that yardang fields in some regions are of great areal extent. McCauley et al. (1977) and Goudie (2007) provide comprehensive discussions of the Earth's major yardang fields. In Africa, yardangs occur in the Arabian Peninsula, Egypt, Libya, Chad, Niger, the Namib Desert of southern Africa, and possibly along the coast of Mauritania. Asia has several yardang fields, including those of the northwestern Lut Desert, Iran, the Taklimakan Desert in the Tarim Basin (described by Hedin (1903)), and the Qaidam depression of Central Asia. Well-formed yardangs are found along the coastal desert of Peru in South America. They occur as minor groups in North America and Australia. In Europe, relict yardangs are present as a small field in the semiarid Ebro Depression of Spain (Gutiérrez-Elorza 2002). Satellite imagery suggests that yardangs are also widespread on Mars.

Yardangs develop where wind action dominates over fluvial processes, and are thereby limited to extremely arid deserts. In order for the wind to be effective, plant cover and soil development are generally minimal. In the Borkou region of Chad, for example, no vegetation occurs expect where subsurface oasis water is within the reach of plant roots. Many yardang fields develop where strong unidirectional winds occur throughout much of the year (Hobbs 1917, McCauley et al. 1977), but others, such as those of the Lut Desert, develop in regimes of seasonally opposing winds, where one wind is dominant and the other lighter and less frequent.

Yardangs form in a broad range of geologic materials, including sandstones, limestones, claystones, dolomites, granites, gneisses, schists, volcanic ignimbrites and basalts, and lacustrine sediments. The lithologies of many of the major yardang fields are summarized in Goudie (1989).

The role of topography in yardang development has not been fully explored. A number of yardang fields, including those of Lop Nur and the Lut Desert, occur in topographic depressions or are surrounded by mountain ranges that rise to considerable heights above the desert floor. Hagedorn (1980) observed that around the Tibesti Mountains there is an altitudinal zonation of geomorphic processes, with aeolian corrasion dominant at the lowest elevations, decreasing in intensity up to an altitude of 800 m.

Form and Scale Relationships

Wind tunnel experiments and measurement of mature, streamlined yardangs suggest an ideal length-to-width ratio of 4:1 (Ward and Greeley 1984), independent of scale. Yardang form is highly variable, however, owing to variations in lithology, wind strength and direction, and yardang age. The ideal proportions are probably approached only after a long period of erosion. In Peru, well-developed, streamlined yardangs have ratios ranging from 3:1 to 10:1 (McCauley et al. 1977). In the Qaidam Depression of Asia, some yardangs attain a length: width: height ratio of 10:2:1, apparently under conditions of high wind speeds that result from a long fetch (Halimov and Fezer 1989).

Although usually of dimensions measured in a few metres, some yardangs can attain heights of as much as 200 m and be several kilometres long (Mainguet 1968). Ridges in the Qaidam Depression attain 5 km in length (Halimov and Fezer 1989) and megayardangs up to 10 km in length are developed in Holocene basaltic flows in the Payun Matru Volcanic Field in the southern Andes Mountains, Argentina (Inbar and Risso 2001).

The inter-yardang spaces have been variously termed troughs, couloirs, corridors, swales, and boulevards. Where the inter-yarding space is narrow, troughs appear to be U-shaped; as they widen, their bottoms become flattened (Blackwelder 1934). In some cases, flattening may be due to a resistant stratum, such as a hard clay layer. Although attention is often focused on the yardang, most of the geomorphic activity appears to be concentrated in the troughs themselves. The troughs may be totally engulfed in sand, or be only partially sand covered (Mainguet 1972), show low transverse ridges of fine gravel (Blackwelder 1934) or ripple trains that diverge at the head and converge in the downwind direction around the yardang flanks (McCauley et al. 1977). Lag surface of pebbles or even mollusc shells may develop around yardangs and reduce erosion in the corridors (Mainguet 1972, Brookes 2001, Compton 2007). Corridors are commonly occupied by migrating barchan dunes whose major axes parallel those of the ridges and the resultant winds (Gabriel 1938, Hagedorn 1971, Mainguet et al. 1980). The rocks in the corridors often carry numerous marks of aeolian erosion, including longitudinal striations, and shallow erosional basins, metres or tens of metres in length, that occur either as isolated forms or groups (Mainguet 1972).

Processes of Yardang Field Formation

Yardangs are probably produced by a combination of abrasion and deflation and further modified by fluvial erosion, weathering, and mass movement. The significance of each of these processes in determining the ultimate form varies according to climate and yardang lithology and structure.

Abrasion

Abrasion by sand particles is probably the dominant process by which most yardangs form (Fig. 19.22). In many yardang fields, the passages are filled with aeolian sand (Grolier et al. 1980, Halimov and Fezer 1989) or gravel (Blackwelder 1934) that erodes the corridors and lower yardang slopes. Mainguet (1972) and Hagedorn (1971) emphasize the episodic nature of sand transport, so that yardang fields may be temporarily free of drifting sand even within a wind-corrasion landscape. The inter-yardang corridors are zones of transportation and erosion, not deposition.

In addition to influencing the overall form of the yardang, abrasion also affects the micromorphology, fluting and polishing the surface, and affecting its colour. The windward face commonly develops a well-developed re-entrant form. Polish and fluting typically occur to a height of one or two metres (Hobbs 1917, Hagedorn 1971, Grolier et al. 1980).

Fig. 19.22 Abrasional detail on the sides of a yardang at Rogers Lake

In sandstone yardangs in Africa, the upper yardang surface is rough, unshaped by the wind, and covered by a dark weathered crust, whereas the lower slopes are smoothed and lighter in colour (Hagedorn 1971, Mainguet 1972). In the southern Namib Desert, crystalline dolomite yardangs are streamlined up to 10–15 m in height (Corbett 1993). Actively forming yardangs are smoothly polished and show fluting, whereas relict forms have rough surfaces colonized by lichen and affected by solutional weathering.

Numerous small-scale features may retard abrasion, including salt crusts in the couloirs, armoring by mollusc shells in pan sediments (Compton 2007), and clay-rich drapes resulting from rainfall (Hörner 1932).

Deflation

Deflation removes relatively loose materials from the yardang surface, including unconsolidated sediments, or grains that are weathered from consolidated or crystalline materials. The dark patina of sandstone yardangs in the Sahara indicates that there is little active removal of material from the ridge summits, and indeed the varnish probably aids in cementing the grains and protecting them from deflation. Similarly, limestone yardangs in Egypt commonly have flat irregular tops that retain weathered surfaces that pre-date erosion (El-Baz et al. 1979).

Deflation may play a more important role in poorly indurated lacustrine material. Formed in fine-bedded white siltstones, yardangs 30–50 m high and up to 1.5 km long on the Pampa de la Averia, Peru, possess smooth, streamlined shapes from base to crest (McCauley et al. 1977). The aerodynamic form is the primary evidence of deflation. As yet, there have been no field studies to confirm that erosion and modification of yardangs take place by this process.

Researchers differ in their opinion as to the relative importance of abrasion and deflation in forming yardangs, even in a small field such as that at Rogers Lake, California (Figs. 19.20, 19.21 and 19.22). These yardangs are carved in moderately consolidated deposits containing beds of fine gravel, sand, silt, and clay (Ward and Greeley 1984). McCauley et al. (1977) felt that they attained their streamlined shape and smooth ridge crests principally by deflation, as evidenced by the lack of small-scale grooving and scouring, and the winnowed appearance of the rocks caused by quartz grains and clay swept away in suspension: abrasion played a minor role, causing some undercutting of the windward end and flanks, and contributing to trough lowering. On the other hand, Ward and Greeley (1984) considered abrasion to be the most important process in the development of the Rogers Lake yardangs, dominating trough formation and initial sculpting of the ridges. Thereafter, deflation increases in importance as it combines with abrasion to maintain the aerodynamic shape. Blackwelder (1934) also considered that the rounded form of low yardangs at Rogers Lake results from abrasion by saltating grains that flow both around and over the forms.

Fluvial Erosion

Running water acts in several ways to aid in yardang field development. At the outset, yardang fields may be initiated along stream courses that are enlarged and modified by the wind. Li Daoyuan (466–527 AD) referred to the formation of the yardangs of Lop Nor, China, as "rills cut by water is [*sic*] blown by wind subsequently" (Xia 1987). As climate changes or streams are diverted, yardang landscapes may be flooded. Hörner (1932) describes a condition at Lop-Nor where "millions of small yardang hillocks and groups of them stick up like islets, while the hollows between them are filled with water (p. 312)." Proximity to piedmont slopes and their periodic streamflow was considered to be critical to the dissection of yardangs in the Lut Desert (Krinsley 1970). Subsequently, the inter-yardang troughs may be fluvially eroded and the yardang slopes gullied. Although fluvial processes often initiate and abet yardang formation, the development of yardangs in hyperarid environments (including Mars) suggest that running water may not be essential to their formation.

The role of running water in yardang modification is most evident in yardangs composed of relatively soft deposits, such as lakebed silts and clays. At Rogers Lake, the concave-upward form of yardang flanks is thought to result from sheet-wash and gullying (Fig. 19.23): in more arid regions, such as coastal Peru, streamlined yardangs have broader crestlines and convex upward flanks. Intense aeolian abrasion destroys evidence of fluvial erosion at the windward ends of Rogers Lake yardangs, with the more stable mid-sections harbouring the largest gullies, and gully development hampered at the leeward by deposition of a porous mantle of aeolian sediment (Ward and Greeley 1984). Small alluvial cones deposited at the gully bases are destroyed by sand blasting (Blackwelder 1934). In the Lut Desert, Iran, winter rains

cause intense gullying, earthflows, and solution on the flanks and summits of 60-m-high yardangs formed in playa sediments. The upper slopes are well above the range of effective abrasion and deflation is too slow a process to remove evidence of fluvial dissection (Krinsley 1970).

In areas of the Sahara, the gullies observed on yardang flanks may be relict of earlier, wetter climates, formed when lacustrine sediments were deposited in the inter-yardang corridors. Today, erosion of the deposits and the filling of gullies by sand indicate that wind is the dominant process (Mainguet 1972).

Almost all yardangs are found in very arid conditions, as frequent fluvial erosion destroys the yardang form and vegetation limits wind erosion. An exception appears to be a cluster of approximately 50 relict yardangs in the Ebro Basin, Spain, that still retain their form despite an annual rainfall of ~400 mm and partially-vegetated flanks (Gutiérrez-Elorza 2002).

Mass Movement

Mass movement has received little attention as a modifying process, although the presence of slump blocks alongside yardangs appears to be quite common, particularly where the nose or base of the yardang has been undermined by abrasion or where the yardang is formed of strongly jointed materials such as sandstone (Hörner 1932).

Weathering

It is likely that weathering plays a role in preparing material for removal of the wind, although such processes are not well documented. As many yardangs are developed in or near to playa sediments, it is likely that salt weathering is particularly significant. Stein (in Hörner 1932) refers to "salt encrusted yardangs" in the Lop-Nor region. Grains weathered free from yardangs by salt weathering are subject to removal by deflation.

Models of Yardang Development

Yardang fields may be initiated in gullies, fractures, and tectonic features that are aligned parallel to the

Fig. 19.23 Fluvial erosion of yardang at Rogers Lake

direction of the prevailing wind and become enlarged by this wind (Hedin 1903, Blackwelder 1934, Ward and Greeley 1984). As air enters the passage, the streamlines are compressed and the wind accelerates. Sand that is carried through the passage abrades the bottom and sides of the trough, causing the slopes to become progressively steeper (Blackwelder 1934). During the early stages of formation, wind and occasional fluvial erosion erode the passages more rapidly than the yardangs, causing the passages to deepen and the yardangs to grow (Halimov and Fezer 1989). The effects of abrasion are most evident within a metre of the floor of the trough, decreasing in intensity on the upper slopes of the yardang. The low levels to which abrasion is effective account for the commonly ragged appearance of the crests of yardang ridges.

Blackwelder (1934) suggested that yardang troughs are initially elongate and somewhat sinuous in form, become wider with time, and finally breach the ridges at places of weakness. As abrasion is more intense at the yardang prows, the ridges become shorter and smaller, eventually developing into conical hills, mesas, and pyramids (Halimov and Fezer 1989). Blackwelder (1934) likened the processes that destroy the windward ends of yardangs to those which affect ventifacts or wind-abraded outcrops.

When the yardangs are lowered to approximately 1 m (Rogers Dry Lake) or 2–3 m (Qaidam Basin) in height, sand blasting affects the entire form, and the crests become more rounded and streamlined. In some cases, the hills vanish, leaving wide interspaces. Halimov and Fezer (1989) noted that only the streamlined whalebacks, which appear to be the most aerodynamically adjusted forms, survive for a long period. These progressive changes in yardang form are slowed by two processes: the development of a wind-resistant lag material on aeolian sands (Brookes 2001), and the cementation of yardang walls or crests by crusts variously composed of sand grains, loess, clay drapes, and salt.

Variations in the evolutionary form of yardangs occur because of material differences. Where the wind erodes horizontally layered rocks of varying resistances, tabular forms with protruding shelves or stair-step profiles are formed. On the other hand, inclined or vertically layered sediments will be eroded into ridges characterized by grooves and fins (Blackwelder 1934). Yardangs sometimes develop at angles to the structural grain of the rock. Joints and faults cross wind-shaped forms in the Sahara, and differential erosion in schists may result in a structural grain that runs obliquely to the form of the yardang (Hagedorn 1971).

Rate of formation of yardang fields

The rate of formation and modification of yardang fields is not well constrained, but several studies suggest that development takes less than 10,000 years, with much faster erosion in soft playa sediments than hard rocks. Yardangs at Rogers Lake, Mojave Desert, appear to be eroding by abrasion of the headward end at a rate of 2 cm y^{-1}, and by lateral erosion caused by both abrasion and deflation at 0.5 cm y^{-1} (Ward and Greeley 1984). The attachment of some yardangs to the original shoreline deposits suggests that this small yardang field is still evolving. Halimov and Fezer (1989) calculated that small yardangs in the Qaidam Basin formed in less than 1,500–2,000 years, based on dated pottery in the silty sediment. Yardangs in the Lop-Nor region, China, are also relatively young, and still retain vestiges of dead vegetation (copses) that are remanent of earlier times when the Tarim River flowed near the abandoned town of Lou-Lan. Hörner (1932) estimates that they formed within the past 1,500 years by the action of drifting sand. In the eastern Sahara, Haynes (1980) estimated that the development of mature yardangs, several meters in height, took place over several thousand years in soft playa muds, facilitated by a change to hyperarid conditions and a drop in the water table. In the Bodélé Depression, Chad, 4 m high yardangs in diatomites probably formed within 1,200 to 2,400 years (Washington et al. 2006).

There has been very little research on the rates of landform erosion in harder materials. Basaltic yardangs, 2–3 m in height, formed in less than 10,000 years in the Payun Matru Volcanic Field, Argentina (Inbar and Risso 2001). Rates of wind erosion in bedrock megayardangs, such as along the coast of Namibia or in the Sahara, are poorly constrained, although the "erasure" of large parts of the landscapes suggests that very long time periods are required. Millions of years may be involved, particularly in deserts that had their origin before the Pleistocene (Goudie 2007).

Wind Tunnel Simulations of Yardang Development

Ward and Greeley (1984) conducted wind tunnel experiments to simulate yardang development. Natural rock samples require an impractical duration of exposure to abrasion before significant erosion occurs, and therefore synthetic sediments were used, and soap bubbles substituted for sand. Several forms (such as mounds and cubes) were moulded and subjected to uniform winds. The samples evolved in a common sequential order, which resulted in final length-to-width ratios of about 4:1. Erosion proceeded from the windward corners, to the front slope and crestline, to the leeward corners, and finally to the leeward slope. Abrasion dominated at the windward end of the yardang, whereas deflation and reverse flow were more important near the middle and at the leeward end. Rates of erosion were greatest at the beginning of the experiment and diminished as the form became more streamlined.

Streamlined shapes are developed by modification of the original form. Where the original form is broad, erosion decreases its width, and elongation may occur by deposition of a tail. If the original form is more elongate, the dominant change will be a decrease in length.

Small-Scale Aeolian Grooving

Small-scale channelling is a feature less often mentioned in the literature. Worrall (1974) described channelling on nearly horizontal surfaces in the Faya district of Chad. The grooves are remarkably parallel and regularly spaced, but shallow (a few centimetres in depth and width) and discontinuous. They occur in homogeneous diatomite and on saline crusts and are formed by saltating sand grains. Grooving has also been observed by this author on clay surfaces of the Mojave River bed, traversed by blowing sand; on dry playa surfaces where saltating grains have created dust events; and on frozen sand dune surfaces at the Coral Pink Sand Dunes, Utah.

Desert Depressions

The role of wind erosion in the excavation of both small- and large-scale desert depressions is difficult to determine, as closed depressions may form by a number of different mechanisms, often acting in combination. These include (a) block faulting, (b) broad shallow warping, (c) crater lake development in volcanic areas, (d) chemical weathering and solution, (e) zoogenic processes, and (f) wind erosion.

The origin of very large enclosed basins, such as the noteworthy depressions of the Western Desert of Egypt (Siwa, Qattara, Baharia, Farafra, Dakhla, and Kargha), has been subject to several interpretations. Several of these basins have floors that lie below sea level (e.g., Kharga at 18 m BSL and Qattara at 143 m BSL). Depressions occur along the boundaries of northward-dipping strata and are bounded to the north by escarpments and to the south by gently rising valley floors. The depressions are often cited as textbook examples of deflation, with the depth of erosion limited by the groundwater table which forms a base level (Ball 1927). Knetsch and Yallouze (1955) invoked tectonic action in their models of depression formation. Said (1960) excluded the possibility of a tectonic origin, noting the absence of faults in the regions, which indicates that uplift was not accompanied by any significant tensional stresses.

The role of wind action is suggested by the conformity of depression locations with areas of thinner and more easily breached limestone capping. However, the sheer volume of the depressions, most notably the Qattara depression (20,000 km^2), suggests that the wind alone could not have excavated the material. Said (1981, 1983) showed that the southern part of the Western Desert experienced a wet climate through most of the Tertiary and Quaternary, punctuated by brief episodes of aridity. Albritton et al. (1990) proposed that the Qattara Depression was originally excavated as a stream valley, subsequently modified by karstic activity, and further deepened and extended by mass wasting, deflation, and fluviatile processes. Salt also played a role, by weathering and preparing materials for deflation. In a self-enhancing process, the creation of low areas alllows water and solutes to accumulate, and the crystallization of salts weakens sedimentary cements, allowing further deflation and salt accumulation in the lowered basins (Haynes 1982). Thus, the research to date suggests that the depressions probably had a polygenetic origin, with wind erosion playing a major role only during arid phases of the Quaternary.

Remote sensing has revealed that smaller depressions, termed pans, are widespread. They are especially well developed in southern Africa, on the High Plains of the U.S.A, and in western and southern Australia (Goudie and Wells 1995). Pans may develop in interdunal basins, as palaeodrainage depressions aligned along former river courses, and by excavation on the floors of former pluvial lakes (Goudie 1999).

Research into the origin of pans indicates that a single mode of genesis is unlikely, with several predisposing factors for their development. A dry climate is important, as is a vegetation-free surface that enhances wind flow and permits deflation. During periods of enhanced aridity, the water table lowers, allowing deepening of the basin (Haynes 1982, Holliday 1997, Langford 2003). Pan growth requires materials susceptible to deflation, and depressions form preferentially in poorly consolidated sediments, shales, and fine-grained sandstones. Space Shuttle photography indicates that pans are important source areas for dust storms (Middleton et al. 1986).

Several feedback mechanisms enhance deflation. Water that accumulates in the depressions evaporates to leave salts, which further retard vegetation growth and make sediments more susceptible to deflation by comminution of debris through salt weathering. (Haynes 1982, Goudie 1989). As the depressions enlarge, they become more attractive to grazing animals who, attracted by the lack of cover and availability of water and salt licks, further disrupt the surface and render it more prone to erosion (Goudie 1989).

Inverted Topography

Inverted relief develops when previously low areas of the landscape, such as river channels, are left high standing by erosion in a later phase of topographic development. This process is not commonly described in the aeolian erosion literature. Hörner (1932) remarks that in the yardang landscape of the Lop-Nor basin, China, former rivercourses have become inverted and are now marked by ridges and remnant hillocks. This inversion is attributed to the comparative resistance of the silty riverbed with respect to the other soft, young sediments in the region.

Integrated Landscapes of Aeolian Erosion and Landform Hierarchies

Although desert depressions, yardangs, and ventifacts may develop in isolation from one another, they are formed by similar processes and, as such, are often found in close proximity, often as part of an integrated aeolian system of erosion, transportation, and deposition. An example of such a landscape is found in the southern Namib Desert, where sand derived from the coast moves through an elongate deflation basin (20 km wide and 125 km long, including subbasins lowered to below sea level by salt weathering and deflation) to merge to the north with the Namib Sand Sea. (Corbett 1993). The system conveys sand in barchan trains from the coast to the sand sea, forming ventifacts, depressions, and yardangs en route. Changes in sea level dictate sediment availability which, in turn, affects the development of erosional features (Corbett 1993, Compton 2007).

Landscapes of aeolian erosion are also strongly associated with dust production. The role of saltating sand grains in producing dust is well established. Sand grains break up soil aggregates by ballistic impact, overcoming the otherwise strong cohesive forces associated with small particles (Gomes et al. 1990, Shao et al. 1993). This is nowhere more apparent than in the Bodélé depression, presently considered to be the world's greatest dust source (Fig. 19.19). An analysis of this landscape suggests that erosion is part of a system of processes that must be viewed over a palaeotimescale. Enhanced deflation by the Bodélé Low Level Jet (LLJ) during the Last Glacial Maximum (LGM) lowered the surface, creating a depression that was populated during wetter Holocene phases by diatoms associated with Mega-Lake Chad. This highly erodible material is conservatively lowering at a rate of 0.16–$0.31\,cm\,yr^{-1}$, allowing 4-m high yardangs to form within 1,200–2,400 years (Washington et al. 2006).

Aeolian bedforms arise from a two-way interaction between the surface and the airflow, involving a transfer of material and modification of form until a dynamic equilibrium is reached. Within the erosional landscape, hierarchies of landforms develop, from small scale aeolian grooving, to ventifacts of varying scale, through micro-yardangs, to normal yardangs, to megayardangs. As in other landform systems, in an aeolian erosional landscape there are large numbers of small features and very few large ones. Evans (2003) notes that lineation and streamlining are often associated with scale-specificity. On a local scale, yardangs within a field are all similar in size, but between different areas, their scale varies greatly, owing to differences in materials, time span, and process activity and intensity. Crest spacing is broadly similar within a field, such as the 20–40 m spacing in Holocene diatomites in Chad and the 1.6 km spacing of Palaeozoic sandstone megayardangs (Mainguet 1970, Evans 2003). For ventifacts, the scale hierarchy also exists: numerous ventifacts with small-scale features are found on the plains and lower hillslopes, but feature scale increases with altitude and sandblast activity and intensity, with the largest forms occurring on selected hilltop locations.

Yardangs have a microrelief that is quite different from the general form: in other words, small yardangs are generally not found to piggyback on larger structures. Erosional features such as flutes on yardangs are similar to those found on nearby ventifacts, emphasizing the similarity of process at the smaller scale.

Conclusions

Landforms of aeolian erosion vary in scale from vast systems of ridges several kilometers in length and up to one kilometre in width to small grooves only millimetres in amplitude. None the less, they exhibit many forms in common. Small faceted ventifacts exhibit shallow grooves; as ventifacts increase in size the grooves and flutes become more fully developed and dominate the form; outcrops in the immediate vicinity may be similarly abraded, as are the lower slopes of some yardangs; near-horizontal surfaces may be fluted and grooved; and ultimately small-scale aeolian grooving (Worrall 1974) may be transitional to yardangs and ridge and swale systems.

The two primary mechanisms of aeolian erosion are deflation and abrasion. Deflation contributes to the formation of large-scale features such as depressions and streamlined yardangs composed of easily erodible materials. Abrasion by the impact of sand grains is the primary process by which ventifacts form and is a major contributor to the development of yardangs and ridge systems in indurated material. Erosion by suspended

grains has not been documented in the field, with particles commonly swept around an obstacle rather than striking it directly.

Landforms of wind erosion have received much less attention than those of deposition and are still quite poorly understood. Fundamental questions as to the relative role of abrasion and deflation in the formation of yardangs, the mechanism of microfeature formation on yardangs and ventifacts, the interaction of yardang and ridge systems and wind flow, and the age, evolutionary history and rates of formation of erosional landforms remain await further process-oriented research. Such studies will provide an improved basis for understanding the climatic and geomorphic history of arid regions.

References

Albritton, C.C., J.E. Brooks, B. Issawi and A. Swedan 1990. Origin of the Qattara Depression, Egypt. *Bulletin of the Geological Society of America* **102**, 952–60.

Anderson, R.S. 1986. Erosion profiles due to particles entrained by wind: application of an aeolian sediment-transport model. *Bulletin of the Geological Society of America* **97**, 1270–8.

Antevs, E. 1928. Wind deserts in Iceland. *Geographical Review* **18**, 675–6.

Ash, J.E. and R.J. Wasson 1983. Vegetation and sand mobility in the Australian desert dunefield. *Zeitschrift für Geomorphologie Supplement Band* **45**, 7–25.

Babikir, A.A.A. and C.C.E. Jackson 1985. Ventifacts distribution in Qatar. *Earth Surfaces Processes and Landforms* **10**, 3–15.

Bach, A.J. 1995. Aeolian modifications of glacial moraines at Bishop Creek, Eastern California. In *Desert Aeolian Processes*. V. P. Tchakerian (ed.), 179–97, London: Chapman & Hall.

Bagnold, R.A. 1941. *The physics of blown sand and desert dunes*. London: Chapman & Hall.

Ball, J. 1927. Problems of the Libyan Desert. *Geographical Journal* **70**, 209–24.

Bishop, D.G. and D.C. Mildenhall 1994. The geological setting of ventifacts and wind-sculpted rocks at Mason Bay, Stewart Island, and their implications for late Quaternary paleoclimates. *New Zealand Journal of Geology and Geophysics* **37**, 169–80.

Blackwelder, E. 1929. Sandblast action in relation to the glaciers of the Sierra Nevada. *Journal of Geology* **37**, 256–60.

Blackwelder, E. 1934. Yardangs. *Bulletin of the Geological Society of America* **45**, 159–65.

Blake, W.P. 1855. On the grooving and polishing of hard rocks and minerals by dry sand. *American Journal of Science* **20**, 178–81.

Bradley, B.A., S.E.H. Sakimoto, H. Frey and J.R. Zimbelman 2002. Medusae Fossae Formation: New perspectives from Mars Global Surveyor. *Journal of Geophysical Research* **107**(E8), 5058, doi:10.1029/2001JE0001537.

Breed, C.S., J.F. McCauley and M.I. Whitney 1989. Wind erosion forms. In *Arid Zone Geomorphology*, D.S.G. Thomas (ed.), 284–307. New York: Wiley.

Bridges, N.T., J.E. Laity, R. Greeley, J. Phoreman and E.E. Eddlemon 2004. Insights on rock abrasion and ventifact formation from laboratory and field analog studies with applications to Mars. *Planetary and Space Science* **52**, 199–213.

Bridges, N.T., J. Phoreman, B.R. White, R. Greeley, E. Eddlemon, G. Wilson, and C. Meyer 2005. Trajectories and energy transfer of saltating particles onto rock surfaces: Application to abrasion and ventifact formation on Earth and Mars. *Journal of Geophysical Research* **110**, E12004, doi:10.1029/2004JE002388.

Brookes, I.A. 2001. Aeolian erosional lineations in the Libyan Desert, Dakhla Region, Egypt. *Geomorphology* **39**, 189–209.

Bryan, K. 1931. Wind-worn stones or ventifacts – a discussion and bibliography. Washington, DC: *National Research Council Circular* 98, Report of the Committee on Sedimentation, 1929–1930, 29–50.

Christiansen, H.H. 2004. Windpolished boulders and bedrock in the Scottish Highlands: evidence and implications of Late Devensian wind activity. *Boreas* **33**, 82–94.

Clark, R.C. and P. Wilson 1992. Occurrence and significance of ventifacts in the Falkland Islands, south Atlantic. *Geografiska Annaler* **74A**, 35–46.

Compton, J.S. 2007. Holocene evolution of the Anichab Pan on the south-west coast of Namibia. *Sedimentology* **54**, 55–70.

Corbett, I. 1993. The modern and ancient pattern of sandflow through the southern Namib deflation basin. In *Aeolian Sediments: Ancient and Modern*. K. Pye and N. Lancaster (eds), 45–60. *International Association of Sedimentologists Special Publication* **16**, Oxford: Blackwell Scientific Publications.

Czajka, W. 1972. Windschliffe als Landschaftmerkmal. *Zeitschrift für Geomorphologie* **16**, 27–53.

Dietrich, R.V. 1977a. Impact abrasion of harder by softer materials. *Journal of Geology* **85**, 242–6.

Dietrich, R.V. 1977b. Wind erosion by snow. *Journal of Glaciology* **18**(78), 148–9.

Dohrenwend, J.C. 1987. Basin and Range. In *Geomorphic Systems of North America*, W.L. Graf (ed.), 303–42. Boulder, CO: Geological Society of America, Centennial Special Volume 2.

Dorn, R. 1986. Rock varnish as an indicator of aeolian environmental change. In *Aeolian Geomorphology*, W.G. Nickling (ed.), 291–307. Boston: Allen & Unwin.

Dorn, R.I. 1995. Alterations of ventifact surfaces at the glacier/desert interface. In *Desert Aeolian Processes*. V.P. Tchakerian (ed.), 199–217, London: Chapman & Hall.

El-Baz, F., C.S. Breed, M.J. Grolier and J.F. McCauley 1979. Eolian features in the Western Desert of Egypt and some applications to Mars. *Journal of Geophysical Research* **84**, 8205–21.

Evans, J.W. 1911. Dreikanter. *Geological Magazine* **8**, 334–5.

Evans, I.S. 2003. Scale-specific landforms and aspects of the land surface. In *Concepts and Modelling in Geomorphology: International Perspectives*. I.S. Evans, R. Dikau, E. Tokunaga, H. Ohmori and M. Hirano (eds), 61–84. Tokyo: TERRAPUB.

Gabriel, A. 1938. The southern Lut and Iranian Baluchistan. *Geographical Journal* **92**, 193–210.

Garvin, J.B. 1982. Characteristics of rock populations in the western desert and comparison with Mars. In *Desert Landforms of Southwest Egypt: A Basis for Comparison with Mars*. F. El Baz and T.A. Maxwell (eds), 261–80. Washington, DC: National Aeronautics and Space Administration, Contractor Report 3611.

Gilbert, G.K. 1875. Report on the Geology of Portions of Nevada, Utah, California, and Arizona. *Geographical and Geological Surveys West of the 100th Meridian*, 3, 17–187.

Glennie, K.W. 1970. *Desert Sedimentary Environments*. Developments in Sedimentology, No. 14. Amsterdam: Elsevier.

Gomes, L., G. Bergametti, G. Coudé-Gaussen and P. Rognon 1990. Submicron desert dusts: a sandblasting process. *Journal of Geophysical Research* **95**(12),927–35.

Goudie, A.S. 1989. Wind erosion in deserts. *Proceedings of the Geologists' Association* **100**, 89–92.

Goudie, A.S. 1999. Wind erosional landforms; yardangs and pans. In *Aeolian Environments, Sediments and Landforms*. A.S. Goudie, I. Livingstone and S. Stokes (eds), 167–180. Chichester: John Wiley & Sons.

Goudie, A.S. 2007. Mega-Yardangs: A Global Analysis. *Geography Compass* **1**, 65–81.

Goudie, A.S. and G.L. Wells 1995. The nature, distribution and formation of pans in arid zones. *Earth-Science Reviews* **38**, 1–69.

Greeley, R. and J.D. Iversen 1985. *Wind as a Geological Process*. Cambridge: Cambridge University Press.

Greeley, R. and J.D. Iversen 1986. Aeolian processes and features at Amboy Lava field, California. In *Physics of Desertifcation*, F. El-Baz and M. Hassan (eds), 210–40. Dordrecht: Martinus Nijhoff.

Greeley, R., R.N. Leach, S.H. Williams, B.R. White 1982. Rate of wind abrasion on Mars. *Journal of Geophysical Research* **87**, B12, 10009–14.

Greeley, R., S.H. Williams, B.R. White, J.B. Pollack, et al. 1984. Wind abrasion on Earth and Mars. In *Models in Geomorphology*, M.J. Woldenberg (ed.), 373–422. Boston: Allen & Unwin.

Greeley, R., N. Bridges, N., R.O. Kuzmin and J. E. Laity 2002. Terrestrial analogs to windrelated features at the Viking and Pathfinder landing sites on Mars. *Journal of Geophysical Research* **107**, 5–1–5–21.

Grolier, M.J., J.F. McCauley, C.S. Breed and N.S. Embabi 1980. Yardangs of the Western Desert. *The Geographical Journal* **146**, 86–7.

Grove, A.T. 1960. Geomorphology of the Tibesti region with special reference to western Tibesti. *The Geographical Journal* **126**, 18–27.

Gutiérrez-Elorza, M., G. Desir and F. Gutiérrez-Santolalla 2002. Yardangs in the semiarid central sector of the Ebro Depression (NE Spain). *Geomorphology* **44**, 155–170.

Hagedorn, H. 1971. Untersuchungen über Relieftypen arider Räume an Beispielen aus dem Tibesti-Gebirge und seiner Umgebung. *Zeitschrift für Geomorphologie Supplement Band* **11**, 1–251.

Hagedorn, H. 1980. Geological and geomorphological observations on the northern slope of the Tibesti Mountains, central Sahara. In *The Geology of Libya*, Volume III, M.J. Salem and M.T. Busrewil (eds), 823–35. London: Academic Press.

Halimov, M. and F. Fezer 1989. Eight yardang types in central Asia. *Zeitschrift für Geomorphologie* **33**, 205–17.

Hall, K. 1989. Wind blown particles as weathering agents? An Antarctic example. *Geomorphology* **2**, 405–10.

Haynes, C.V. Jr. 1980. Geologic evidence of pluvial climates in the El Nabta area of the Western Desert, Egypt. In: *Prehistory of the Eastern Sahara*. F. Wendorf and R. Schild (eds), 353–71. New York: Academic Press.

Haynes, C.V. Jr. 1982. The Darb El-Arba'in desert: a product of Quaternary climatic change. In *Desert Landforms of Southwest Egypt: A Basis for Comparison with Mars*. F. El-Baz and T.A. Maxwell (eds), 91–117. Washington, DC: National Aeronautics and Space Administration, Contractor Report 3611.

Hedin, S. 1903. *Central Asia and Tibet*. New York: Greenwood Press.

Hickox, C.F. 1959. Formation of ventifacts in a moist, temperate climate. *Bulletin of the Geological Society of America* **70**, 1489–90.

Higgins, C.G., Jr 1956. Formation of small ventifacts. *Journal of Geology* **64**, 506–16.

Hobbs, W.H. 1917. The erosional and degradational processes of deserts, with especial reference to the origin of desert depressions. *Annals of the Association of American Geographers* **7**, 25–60.

Holliday, V.T. 1997. Origin and evolution of lunettes on the High Plains of Texas and New Mexico. *Quaternary Research* **47**, 54–69.

Hörner, N.G. 1932. Lop-nor. Topographical and Geological Summary. *Geografiska Annaler* **14**, 297–321.

Inbar, M. and C. Risso 2001. Holocene yardangs in volcanic terrains in the southern Andes, Argentina. *Earth Surface Processes and Landforms* **26**, 657–66.

Jackson, P.S. and J.C.R. Hunt 1975. Turbulent flow over a low hill. *Quarterly Journal of Royal Meteorological Society* **101**, 929–55.

King, L.C. 1936. Wind-faceted stones from Marlborough, New Zealand. *Journal of Geology* **44**, 201–13.

Knetsch, G. and M. Yallouze 1955. Remarks on the origin of the Egyptian oasis depressions. *Bulletin Societe Géographique Egypte* **28**, 21–33.

Knight, J.K. and H. Burningham 2003. Recent ventifact development on the central Oregon coast, western USA. *Earth Surface Processes and Landforms* **28**, 87–98.

Kozlov, R.K. 1899. *Otschet pomoshnika nachaunica expedici*. Moskva.

Krinsley, D.B. 1970. A geomorphological and palaeoclimatological study of the playas of Iran. *U.S. Geological Survey Final Report*, Contract No. PRO CP 70–800.

Kuenen, P.H. 1960. Experimental abrasion 4: eolian action. *Journal of Geology* **68**, 427–49.

Laity, J.E. 1987. Topographic effects on ventifact development, Mojave Desert, California. *Physical Geography* **8**, 113–32.

Laity, J.E. 1992. Ventifact evidence for Holocene wind patterns in the east-central Mojave Desert. *Zeitschrift für Geomorphologie, Supplement Band* **84**, 1–16.

Laity, J.E. 1994. Landforms of aeolian erosion. In *Geomorphology of Desert Environments*. A.D. Abrahams and A.J. Parsons (eds), 506–535. London: Chapman & Hall.

Laity, J.E. 1995. Wind abrasion and ventifact formation in California. In *Desert Aeolian Processes*. V.P. Tchakerian (ed), 295–321. London: Chapman & Hall.

Laity, J.E. and N.T. Bridges 2008. Ventifacts on Earth and Mars: Analytical, field, and laboratory studies supporting sand abrasion and windward feature development. *Geomorphology*, doi:10.1016/j.geomorph.2008.09.014

Lancaster, N. 1984. Characteristics and occurrence of wind erosion features in the Namib Desert. *Earth Surfaces Processes and Landforms* **9**, 469–78.

Lancaster, N. 1985. Variations in wind velocity and sand transport on the windward flanks of desert sand dunes. *Sedimentology* **32**, 581–93.

Langford, R.P. 2003. The Holocene history of the White Sands dune field and influences on eolian deflation and playa lakes. *Quaternary International* **104**, 31–39.

Lindsay, J.F. 1973. Ventifact evolution in Wright Valley, Antarctica. *Bulletin of the Geological Society of America* **84**, 1791–8.

Liu, L.-Y., Z.-B. Dong, S.-Y. Gao, P.-J. Shi and X.-Y. Li 2003. Wind tunnel measurements of adobe abrasion by blown sand: Profile characteristics in relation to wind velocity and sand flux. *Journal of Arid Environments* **53**, 351–63.

Mainguet, M. 1968. Le Bourkou – Aspects d'un modelé éolien. *Annales de Géographie* **77**, 296–322.

Mainguet, M. 1970. Un étonnant paysage: les cannelures gréseuses du Bembéché (N. du Tchad). Essai d'explication géomorphologique. *Annales de Géographie* **79**, 58–66.

Mainguet, M. 1972. *Le modelé des Grès*. Paris: Institute Geographie National.

Mainguet, M., L. Canon and M.C. Chemin 1980. Le Sahara: géomorphologie et paléogéomorphologie éoliennes. In *The Sahara and the Nile; Quaternary environments and Prehistoric Occupation in Northern Africa*, M.A.J. Williams and H. Faure (eds), 17–35. Rotterdam: Balkema.

Mason, P.J. 1986. Flow over the summit of an isolated hill. *Boundary-Layer Meteorology* **37**(4), 385–405.

Maxson, J.H. 1940. Fluting and faceting of rock fragments. *Journal of Geology* **48**, 717–51.

McCauley, J.F., C.S. Breed and M.J. Grolier 1977. Yardangs. In *Geomorphology in Arid Regions*. D.O. Doehring (ed.), 233–69. Boston: Allen & Unwin.

McCauley, J.F., C.S. Breed, M.J. Grolier and F. El-Baz 1980. Pitted rocks and other ventifacts in the western desert. In *Journey to the Gulf Kebir and Uweinat*, F. El-Baz (ed.), *Geographical Journal* **146**, 84–5.

McCauley, J.F., C.S. Breed, F. El-Baz, M.I. Whitney, et al. 1979. Pitted and fluted rocks in the Western Desert of Egypt: Viking comparisons. *Journal of Geophysical Research* **84**, 8222–32.

McKenna-Neuman, C. and R. Gilbert 1986. Aeolian processes and landforms in glaciofluvial environments of south-eastern Baffin Island, N.W.T., Canada. In *Aeolian Geomorphology*, W.G. Nickling (ed.), 213–35. Boston: Allen & Unwin.

McKenna-Neuman, C. 1990. Winter aeolian transport and niveo-aeolian deposition at Crater Lake, Pangnirtung Pass, N.W.T., Canada. *Permafrost and Periglacial Processes* **1**, 235–47.

McKenna-Neuman, C. 1993. A review of aeolian transport processes in cold environments. *Progress in Physical Geography* **17**, 137–55.

Middleton, N.J. 1989. Desert dust. In *Arid Zone Geomorphology*, D.S.G. Thomas (ed.), 262–83. New York: Wiley.

Middleton, N.J., A.S. Goudie and G.L. Wells 1986. The frequency and source areas of dust storms. In *Aeolian*

Geomorphology, W.G. Nickling (ed.), 237–59. Boston: Allen & Unwin.

Miotke, F. 1982. Formation and rate of formation of ventifacts in Victoria land. *Polar Geography and Geology* **6**, 98–113.

Mountney, N.P. and A.J. Russell 2004. Sedimentology of cold-climate aeolian sandsheet deposits in the Askja region of northeast Iceland. *Sedimentary Geology* **166**, 223–44.

Needham, C.E. 1937. Ventifacts from New Mexico. *Journal of Sedimentary Petrology* **7**, 31–3.

Nero, R.W. 1988. The ventifacts of the Athabasca sand dunes. *The Musk Ox* **36**, 44–50.

Pickard, J. 1982. Holocene winds of the Vestfold Hills, Antarctica. *New Zealand Journal of Geology and Geophysics* **25**, 353–58.

Powers, W.E. 1936. The evidences of wind abrasion. *Journal of Geology* **44**, 214–19.

Said, R. 1960. New light on the origin of the Qattara depression. *Bulletin Societe Géographique Egypt* **33**, 37–44.

Said, R. 1981. *The Geological Evolution of the River Nile*. Berlin: Springer.

Said, R. 1983. Remarks on the origin of the landscape of the eastern Sahara. *Journal of African Earth Sciences*, 153–8.

Scattergood, R.O. and J.L. Routbort 1983. Velocity exponent in solid-particle erosion. *Journal of the American Ceramic Society* **66**, C184–6.

Schlyter, P. 1989. Periglacial ventifact formation by dust or snow, some south Swedish examples. Second International Conference on Geomorphology, Symposium No. 5 '*Polar Geomorphology*', Bremen.

Schoewe, W.H. 1932. Experiments on the formation of wind-faceted pebbles. *American Journal of Science* **24**, 111–34.

Selby, M.J. 1977. Palaeowind directions in the central Namib Desert, as indicated by ventifacts, *Madoqua* **10**, 195–8.

Shao, Y., M.R. Raupach and P.A. Findlater 1993. The effect of saltation bombardment on the entrainment of dust by wind. *Journal of Geophysical Research* **98D**, 12719–26.

Sharp, R.P. 1949. Pleistocene ventifacts east of the Big Horn Mountains, Wyoming. *Journal of Geology* **57**, 173–95.

Sharp, R.P. 1964. Wind-driven sand in Coachella Valley, California. *Bulletin of the Geological Society of America* **75**, 785–804.

Sharp, R.P. 1980. Wind-driven sand in Coachella Valley, California: further data. *Bulletin of the Geological Society of America* **91**, 724–30.

Sharp, R.P. and M.C. Malin 1984. Surface geology from Viking landers on Mars: a second look. *Bulletin of the Geological Society of America* **95**, 1398–412.

Smith, H.T.U. 1967. Past versus present wind action in the Mojave Desert region, California. *Air Force Cambridge Research Laboratories*, 67–0683, 1–26.

Smith, R.S.U. 1984. Eolian geomorphology of the Devils Playground, Kelso Dunes and Silurian Valley, California. In *Western Geological Excursions. Vol. 1: Geological Society of America 97th Annual Meeting Field Trip Guidebook, Reno, Nevada*, J. Lintz (ed.), 239–51 Boulder, Colorado: Geological Society of America.

Stapff, F.M. 1887. Karte des unteren Khuisebtals. *Petermanns Geographische Mitteilungen* **33**, 202–14.

Stewart, G., D. Krinsley and J. Marshall 1981. An experimental study of the erosion of basalt, obsidian and quartz by fine sand, silt and clay. *National Aeronautics and Space Administration Technical Manual* 84211, 214–5.

Sugden, W. 1964. Origin of faceted pebbles in some recent desert sediments of southern Iraq. *Sedimentology* **3**, 65–74.

Suzuki, T. and K. Takahashi 1981. An experimental study of wind abrasion. *Journal of Geology* **89**, 23–36.

Teichert, C. 1939. Corrasion by wind-blown snow in polar regions. *American Journal of Science* **37**, 146–8.

Travers, W.T.L. 1870. On the sandworn stones of Evan's Bay. *Transactions of the New Zealand Institute* **2**, 247–8.

Tremblay, L.P. 1961. Wind striations in northern Alberta and Saskatchewan, Canada. *Bulletin of the Geological Society of America* **72**, 1561–4.

Viles, H.A. and Bourke, M.C. 2007. Aeolian Features. In *A Photographic Atlas of Rock Breakdown Features in Geomorphic Environments*, M. Bourke and H. Viles (eds.), 7–22, Tucson: Planetary Science Institute.

Walther, J. 1891. Die Denudation in der Wüste und ihre geologische Bedeutung. *Abhandlungen Sächsische Gesellschaft Wissenshaft* **16**, 345–570.

Walther, J. 1912. *Das Gesetz der Wüstenbildung in Gegenwart und Vorzeit*. Leipzig: Von Quelle und Meyer.

Ward, A.W. 1979. Yardangs on Mars: evidence of recent wind erosion. *Journal of Geophysical Research* **84**, B14 8147–63.

Ward, A.W. and R. Greeley 1984. Evolution of the yardangs at Rogers Lake, California. *Bulletin of the Geological Society of America* **95**, 829–37.

Washington, R., M.C. Todd, G. Lizcano, I. Tegen 2006. Links between topography, wind, deflation, lakes and dust: The case of the Bodélé Depression, Chad. *Geophysical Research Letters* **33**, L09401

Wentworth, C.K. and R.I. Dickey 1935. Ventifact localities in the United States. *Journal of Geology* **43**, 97–104.

Whitney, M.I. 1978. The role of vorticity in developing lineation by wind erosion. *Bulletin of the Geological Society of America* **89**, 1–18.

Whitney, M.I. 1979. Electron micrography of mineral surfaces subject to wind-blast erosion. *Bulletin of the Geological Society of America* **90**, 917–34.

Whitney, M.I. 1983. Eolian features shaped by aerodynamic and vorticity processes. In *Eolian Sediments and Processes*, M.E. Brookfield and T.S. Ahlbrandt (eds), 223–45. Amsterdam: Elsevier.

Whitney, M.I. 1984. Comments on 'Shapes of streamlined islands on Earth and Mars: experiments and analyses of the minimum-drag form'. *Geology* **12**, 570–1.

Whitney, M.I. 1985. Yardangs. *Journal of Geological Education* **33**, 93–6.

Whitney, M.I. and R.V. Dietrich 1973. Ventifact sculpture by windblown dust. *Bulletin of the Geological Society of America* **84**, 2561–82.

Williams, S. and R. Greeley 1981. Formation and evolution of playa ventifacts, Amboy, California. *National Aeronautics and Space Administration Technical Memorandum* 84211, 197–9.

Wilshire, H.G., J.D. Nakata and B. Hallet 1981. Field observations of the December 1977 wind storm, San Joaquin Valley, California. In *Desert Dust*, T.J. Péwé (ed.), 233–51. Geological Society of America Special Paper 186.

Woodworth, J.B. 1894. Post-glacial aeolian action in southern New England. *American Journal of Science* **47**, 63–71.

Worrall, G.A. 1974. Observations on some wind-formed features in the southern Sahara. *Zeitschrift für Geomorphologie* **18**, 291–302.

Xia, X.C. 1987. *A Scientific Expedition and Investigation to Lop Nor Area*. Scientific Press, Beijing.

Chapter 20

Dust

Joanna E. Bullard and Ian Livingstone

Introduction

In aeolian systems there is a fundamental difference between the behaviour of coarse sediments (sands) and fine sediments (dusts). Sand-sized material (63–2000 μm) travels predominantly by saltation, reptation and creep within the lowest levels of the atmospheric boundary layer (<3 m above the surface) and travels short distances. In contrast, dust-sized material, generally defined as <63 μm, is transported in suspension at a wide range of heights above the surface and can rapidly travel considerable distances. Dust plumes disperse as they travel away from source diffusing the concentration of sediment. Consequently, although there are clearly definable sources for dust emissions, these finer particles can be transported around the globe and their deposits can be both far removed from their origins and extensive. Every year up to three billion tonnes of dust are released into the atmosphere from the Earth's surface. The spatial and temporal patterns of these dust emissions are often closely controlled by desert geomorphology, and in turn have an impact both directly and indirectly on the desert landscape and further afield.

Some of the research on dust has been driven by environmental concerns which have included issues associated with health (as in the 'Dust Bowl' of the 1930s in the Great Plains of the USA) and with climate change because dust in the atmosphere affects radiation budgets and may, through iron fertilisation, enhance carbon dioxide uptake in the oceans. Mineral dust in the atmosphere can cause cooling (via scattering) or warming (via absorption) and its impact depends on factors such as mineralogy, chemical composition and particle size, each of which is strongly-correlated with dust source characteristics, as well as the position of the particles within atmospheric layers. There has also been considerable interest in the movement of dust in geomorphological systems. These impacts include, for example: deflation of dust-sized material leading to the formation or enhancement of topographic lows such as playas; ventifact and yardang formation; stone pavement formation; duricrust and rock varnish development. And the movement of dust has an impact on soils, for the deflation of dust-sized material leads to a loss of soil nutrients, usually concentrated at the surface, and a reduction in soil moisture-holding capacity at source, but can enhance soil nutrients in areas of deposition.

Characteristics of Dust Particles

Airborne dust is derived from a range of sources, both natural and anthropogenic, and the former constitutes 89% of global emissions (Satheesh and Moorthy 2005). It is becoming increasingly important to be able to distinguish rates of natural dust entrainment and its impacts from human-induced entrainment (such as mechanised agriculture and off-road vehicle use) or anthropogenically-generated dusts (such as industrial emissions), due to the impact of the latter on the physical and chemical properties of the atmosphere. This chapter focuses principally on

J.E. Bullard (✉)
Department of Geography, Loughborough University
of Technology, Loughborough, LE11 3TU, UK
e-mail: J.E.Bullard@lboro.ac.uk

A.J. Parsons, A.D. Abrahams (eds.), *Geomorphology of Desert Environments*, 2nd ed.,
DOI 10.1007/978-1-4020-5719-9_20, © Springer Science+Business Media B.V. 2009

natural dust. The dominant material comprising dust is SiO_2, usually in the form of quartz, but it can also include feldspars, calcite, halite and organic material such as pollen and diatoms.

Most sedimentological definitions of dust include all particles below a maximum size in the range 60–70 μm and consequently include the both the silt (4–63 μm) and clay (<4 μm) classes of the Udden-Wentworth grain-size scale as well as occasionally part of the very fine sand class (63–125 μm) (Wentworth 1922). In terms of aeolian geomorphology, the important distinction between 'sand' and 'dust' is the way in which the particles are transported. This difference in behaviour (and hence the precise size boundary) is governed by the balance between the forces holding particles aloft in the atmosphere and those pulling them towards the ground surface.

If the rate at which gravitational forces are pulling the particle towards the surface (its terminal velocity – w_t) is the same as, or less than the mean velocity at which the air parcel within which the particle is contained is moving upwards by atmospheric turbulence (the Lagrangian vertical velocity), then the particle is carried in suspension. In a thermally neutral atmospheric surface layer, the Lagrangian vertical velocity is κu_*, where κ is von Karman's constant (0.4) and u_* is the friction velocity (Hunt and Weber 1979). Particles travel in suspension when $| w_t | /\kappa u_* << 1$ and travel by saltation or creep when $| w_t | /\kappa u_* >> 1$. The critical size boundary between 'sand' and 'dust' (d_1) can be calculated by solving $| w_t | /(d_1) = \alpha_d \kappa u_*$. The value of α_d makes a significant difference to the position of the boundary (Fig. 20.1) but for dust entrainment studies is commonly 0.5, which means that once dust particles are lifted from the surface they are likely to remain suspended for some time (Shao et al. 1996, Shao 2000). This balance between upward and downward forces means that larger particles can be considered to be dust particles in stronger and more turbulent airflows whilst in weaker or less turbulent air flows the same sized particle might travel in saltation or by creep. It may also help to explain why particles far larger than 63 μm can occasionally be found in dust deposits located far from their source – for example the >75 μm quartz particle transported in a dust storm from Asia to Hawaii (Betzer et al. 1988).

In reality, when close to the sediment source, any body of sediment transported by the wind is likely to comprise a mix of saltating, creeping and suspended

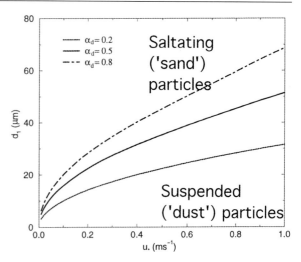

Fig. 20.1 The impact of variations in α_d on the size boundary between saltating (sand) particles and particles travelling in suspension (dust) (modified from Shao 2000)

particles and also to include particles travelling in a mode part way between pure suspension and pure saltation in what has been called modified suspension (e.g. Nalpanis 1985) where particle trajectories are influenced by both their inertia and settling velocity (see Nickling and McKenna Neuman, this volume).

With increasing distance from the source, coarser particles will be deposited and only true suspended particles will remain in the air mass. Pye (1987) suggested that particles in the size range 20–70 μm will travel in short-term suspension or by modified saltation (for example moving up to 30 km from source) whilst those <20 μm are likely to travel in long-term suspension (travelling thousands of km from source). Tegen and Fung (1994) estimate that the majority of contemporary dust emissions are in the size range 1–25 μm (Table 20.1).

The forces acting to transport or stabilise particles are discussed in chapter 17 and include cohesive, gravitational and aerodynamic forces. The relative importance of each of these is dependent upon the size of the particles: below 10 μm cohesive forces tend to dominate, for particles in the range 10–300 μm

Table 20.1 Estimates of global dust emissions by size (Tegen and Fung 1994)

Particle-size range	Global dust source strength
0.5–1 μm	390 Mt a^{-1}
1–25 μm	1960 Mt a^{-1}
25–50 μm	650 Mt a^{-1}

aerodynamic forces dominate, and $>300\,\mu m$ gravitational force dominates. For this reason the majority of wind-transported sediments are in the $10–300\,\mu m$ category. However, finer particles can be transported because the dominance of cohesive forces means that there is a tendency for fine particles to aggregate together rather than to act as individual grains. Aggregation can occur when fine silts and clays attach themselves to coarser sand-sized particles as individual adhesions or to form thin layers or coatings around them. In soils with a high clay content dust-sized particles are usually aggregated together. Under low wind speeds aggregates and dust-coated grains behave like coarser particles – the size of the aggregate. In higher wind speeds, aggregates break up, releasing the individual dust particles into suspension (Gomes et al. 1990). It is very difficult to quantify the degree of aggregation of particles, or the transport-stable size of aggregates for different wind speeds because methods for establishing particle size distributions tend to break-up the aggregates. An approximation can be obtained by examining the difference between minimally-dispersed and fully-dispersed sample treatments – for example Shao (2000) demonstrated a shift in modal particle size from $\approx 100\,\mu m$ to $\approx 4\,\mu m$ following fully-dispersed analysis of clay soil (Fig. 20.2).

In addition to size, the shape of dust particles can be important in terms of entrainment, transport and deposition, and is increasingly recognised as a key factor affecting the radiative properties of mineral dust particles in the atmosphere (e.g. Kalashnikova and Sokolik 2004, Meloni et al. 2005). The terminal velocity (w_t) of dust depends on the particle's mass and shape

and these factors can affect the size of particles that can be transported in suspension. Spherical particles are streamlined and consequently have higher settling velocities for a given mass than platy particles. Natural dust particles are rarely perfectly spherical and it is more common for particles to be platy or 'flattened' (Baba and Komar 1981, Le Roux 2002). The shape of a particle is linked to lithology but also to the mechanism by which it was formed – for example crushing or grinding results in blade-shaped particles (Assallay et al. 1998). For clay-sized sediments the effective particle shape is determined by aggregation. As the number of particles comprising the aggregate increases, its overall shape will become more irregular and there will be a concomitant decrease in density (Goossens 2005). Wind tunnel experiments have shown marked differences between measured and predicted rates of dust deposition depending on whether or not grain shape has been taken in to account. For example, using sub-angular to sub-rounded dust particles with a Corey Shape Factor of 0.5–0.9 (blocky to near-spherical), Goossens (2005) demonstrated that predicted mass deposition flux was closest to the measured rates when grain shape was taken into account, although the accuracy of the prediction varied considerably for individual sediment size fractions.

Mechanisms of Dust Production

Whilst the direct release of particles from fine-grained rocks might be a locally-important source of silt-sized material, the mean grain size of quartz in igneous and metamorphic rocks is around $700\,\mu m$ compared with

Fig. 20.2 The difference between fully-dispersed and minimally-dispersed particle-size analysis may indicate the presence of aggregation in clay soils (**a**) and results in a shift in modal grain size from $\approx 100\,\mu m$ to $\approx 4\,\mu m$. The two treatments have very little impact on the particle-size distributions of sandy soils (**b**) in which few aggregates would be expected (redrawn from Shao (2000) using unpublished data from McTainsh)

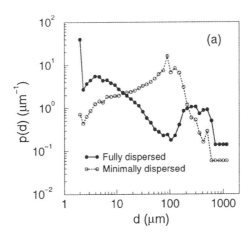
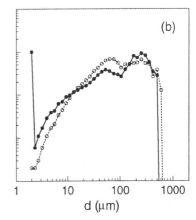

a mean size of detrital quartz of just 60 μm. Particles therefore need to undergo processes that can reduce their size by about 90% before they can be carried as suspended dusts. This size reduction can be achieved by the fracturing of whole particles into smaller grains, the removal of corners and protuberances from larger grains, usually as a result of physical stresses and impacts, or the gradual reduction of the grain size by rubbing or grinding against larger particles or more resistant surfaces. These processes are not mutually exclusive, for example the components of a fractured particle may subsequently undergo rounding or grinding (Fig. 20.3).

The main geomorphological processes proposed for silt-sized particle production are physical and chemical weathering, glacial grinding, fluvial comminution and aeolian abrasion, whilst chemical weathering is thought to be the dominant primary source of particles <4 μm in diameter (Pye 1987). Some of these mechanisms are far more important in deserts

than others, and it is these desert processes that we concentrate on here.

Weathering Processes

In desert environments salt weathering appears to be one of the most effective mechanisms of rock breakdown and has been demonstrated to produce both silt- and clay-sized particles. The process can be effective on both rock blocks and on sand grains. For example, field experiments by Goudie et al. (1997) in the Namib Desert found that on disintegration oolitic limestone blocks in contact with the ground surface produced fine particles with a bimodal size distribution. All the samples reported had a well-defined mode at 32–63 μm and a second, more variable mode in the medium-coarse sand range. The mean diameter of the ooids in the limestone was 689 μm (range 150–1500 μm) suggesting particle disintegration rather

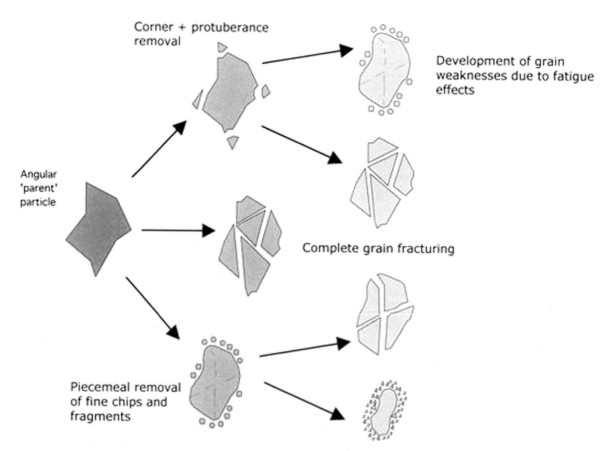

Fig. 20.3 Possible mechanisms of fine particle production from sand-sized material (modified from Wright 1993)

than simply the release of ooids by weathering. Viles and Goudie (2007) also reported a mean grain size of 32–63 μm for weathered schists in the Namib with >30% of the debris released from these rocks being finer than 63 μm diameter. The impact of salt weathering on quartz sand grains examined experimentally by Goudie et al. (1970) and Pye and Sperling (1983) has been shown to produce angular silt-sized particles with the shape of the resulting fine particles a function of both the lithology of the parent grain and also the weathering process.

Similarly, frost weathering has been widely demonstrated as capable of causing rock disintegration and the formation of silt-sized material in the laboratory (e.g. Lautridou and Ozouf 1982, Wright et al. 1998), and field observations in contemporary cold climates point to the importance of frost action in silt formation (e.g. Pye and Paine 1984). Whilst frost weathering is not generally considered to be as effective as salt weathering for fine particle production (Wright et al. 1998), the two processes can act in concert in cold, arid regions, however local environmental conditions are likely to determine whether their combined actions increase or decrease the rate of weathering (Pye 1987).

Aeolian Abrasion

Aeolian transport of sand involves both grain-to-ground impacts and mid-air grain-to-grain collisions which together provide considerable potential for particle breakage and fracturing and consequently for the production of fine material (Kuenen 1960, Whalley et al. 1982, 1987, Wright et al. 1998). Controlled experiments on selected size fractions of sediments show positive relationships between rates of aeolian abrasion and increasing particle size, angularity and surface roughness. Angular particles initially yield high amounts of fine material as corners and protuberances are removed, followed by a decrease in the production of fine particles as the grains become more rounded (Kuenen 1960, Whalley et al. 1987, Wright et al. 1998). However the extent to which the samples used are representative of sediments in desert environments or these experiments simulate the impacts of natural saltation is debatable. For example, Wright (2001a) noted that aeolian abrasion of artificial or freshly-crushed quartz grains with a very angular

initial particle shape resulted in higher fine particle production than is likely in natural conditions where particles are generally sub-angular to sub-rounded (Goudie and Watson 1991) and Bullard et al. (2007) found no relationship between particle size or sorting and fine particle production during experiments using natural dune sands.

Other Dust Sources

As discussed above, clay minerals <10 μm may aggregate to form larger sand-sized particles or may adhere to sand grains forming a surface coating. These aggregates can be broken down during aeolian activity. Gillette and Walker (1977), for example, found that sandblasting by saltation removed clay platelets from the surfaces of quartz grains and separated aggregates. There has been some debate about the extent to which clay coatings can be removed by aeolian abrasion (Walker 1979, Gomes et al. 1990, Wopfner and Twidale 2001). Experiments by Evans and Tokar (2000) to determine the robustness of iron-rich coatings on natural sands showed that after one week of dry abrasion in a rock tumbler about 50% of the grains lost their clay coating entirely and the remainder of the particles retained only 5–20% of their coatings (by surface area). These results did not correspond to field data, however, as clay coatings were found to persist for more than two and a half years in a coastal environment (Evans and Tokar 2000). Other experiments (Bullard et al. 2007) indicated that for sub-rounded to sub-angular grains, the quantity of clay coating on a grain surface is the most important determinant of the amount of material <63 μm produced by aeolian abrasion. The removal of iron-rich clay coatings has received attention in the literature due to the impact of the process on the release of bio-available iron oxides to the atmosphere. Whilst it is difficult to study in the field, the removal of iron-rich red-coloured clay coatings can cause visible changes in soils over very short time periods (Shao et al. 1993a, McTainsh 1985) suggesting that this may be an important process.

Dust-sized particles can also be generated by the breakdown of organic materials. For example diatomaceous material deposited during wet phases can form extensive deposits when lakes desiccate – flakes of the diatomite entrained by the wind undergo a 'self-abrasion' process in which they break up into fine dust-sized particles on collision with each other or the ground surface (Tegen et al. 2006). This process is similar to the break-up of particles by aeolian abrasion as the diatomite flakes are single particles rather than aggregates.

Relative Importance of the Mechanisms

The relative importance of each of the mechanisms described above is a function of macro- and micro-climatic conditions, geomorphological setting and lithology (Bullard et al. 2007, Viles and Goudie 2000, Smith et al. 2000, Wright 2001b, 2007). Wright et al. (1998) compared the relative efficiency of five mechanisms of fine-particle production during a range of experiments (Table 20.2), however the magnitude and frequency characteristics of the different processes make it difficult to draw conclusions. For example, the amount of material produced by weathering is low but the fact that it is a ubiquitous process may mean weathering is potentially the most important geomorphological process in silt formation (Wright et al. 1998). Fluvial tumbling and aeolian abrasion are efficient producers of silt but fluvial activity is usually spatially and temporally confined and aeolian activity is also sporadic in most desert environments. It is likely that over long periods of time, several different mechanisms of fine particle production will take place: for example in the production of the silts comprising desert loess deposits (Wright 2001b, Fig. 20.4). In contemporary environments, the presence of specific agents of weathering may determine the relative

importance of different mechanisms, for example salt weathering is an important contemporary process operating in the Namib due to the presence of salts and frequent fog in the environment. Similarly, dune sands that have developed a coating of clay-sized materials may be a significant source of dust during periods of sustained aeolian activity. The effectiveness of a particular dust-production process and the precise particle-size distribution of the resulting material is also related to the character of the source materials and the presence of defects and susceptibilities in the particles (Assallay et al. 1998, Kumar et al. 2006).

Global Dust Emissions

Globally most atmospheric dust is derived from the deflation of surface material in arid and semi-arid areas including North Africa, the Middle East and Asia, in the northern hemisphere, and Australia and southern Africa, in the southern hemisphere. Within these dryland zones, ground meteorological observations and satellite remote sensing data have enabled the identification of the main global dust sources and transport paths.

Remote Sensing of Dust Sources, Transport and Sinks

Detailed analyses of global dust sources by Prospero et al. (2002) using data from the Total Ozone Monitoring Spectrometer (TOMS) mounted on the Nimbus 7 satellite launched in September 1978 (Herman et al. 1997) demonstrated that all major contemporary dust sources are identified with topographic lows in areas with an annual rainfall of usually less than 250 mm (Fig. 20.5). These areas are predominantly located within internally draining basins with seasonally active rivers, streams and playas. The implication is that fine alluvial material deposited within these channels and on the flood plains following flood events and playa deposits developed during more humid conditions are the major global sources of dust. This is the cornerstone of the 'inland-basins hypothesis' (Bullard

Table 20.2 A comparison of the relative effectiveness of silt-producing mechanisms (calculation of the theoretical maximum amount of silt produced from 1 kg of the original sample) (Wright et al. 1998)

Run type	Amount of fine (<63 μm) material produced (g kg^{-1})	Run duration
Glacial grinding	47.4	24 h
Aeolian abrasion	287	96 h
Fluvial tumbling: spheres	900	32 h
Salt weathering: Na$_2$SO$_4$	41.6	40 cycles
Frost weathering	0.44	360 cycles

Fig. 20.4 Possible mechanisms for the production of fine particles in a hot desert environment (after Wright 2001b)

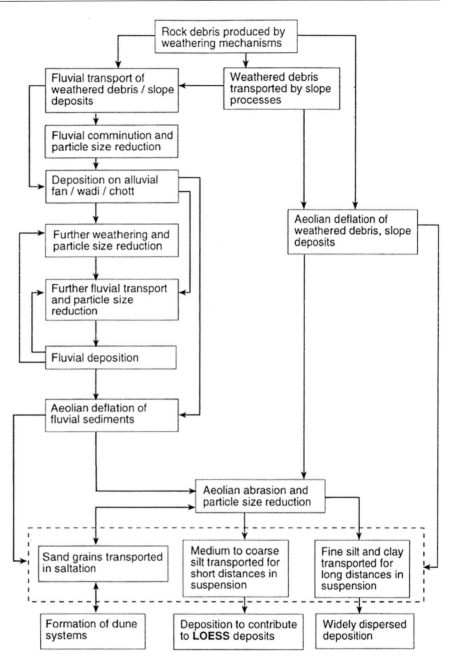

and McTainsh 2003). In some cases, deep alluvial deposits laid down during the Pleistocene are also dust sources (Prospero et al. 2002).

Satellite-based sensors have revolutionised the mapping of dust emissions and have proved an excellent source of data for ascertaining spatial patterns of dust emissions at the global scale, seasonal temporal trends, and likely zones of deposition (Prospero et al. 2002, Washington et al. 2003). No single satellite data source can identify all dust emissions. The frequently used Total Ozone Mapping Spectrometer Aerosol Index (TOMS-AI), for example, can be affected by cloud, dust mineralogy and chemical composition, and is more sensitive to middle and upper troposphere aerosols than lower level (<1–2 km) aerosols. This sensitivity difference can result in an under-representation of dust events at some locations (Washington et al. 2003), and of some types, such as haboob-style dust storms that are confined to the lower 1 km of the atmosphere (Miller et al. 2008).

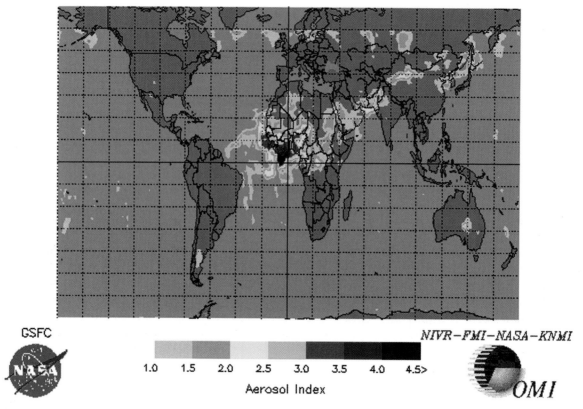

Fig. 20.5 A composite image showing the Ozone Monitoring Instrument aerosol index (TOMS-AI) on 22 February 2008 (reproduced courtesy of the Ozone Processing Team at NASA's Goddard Space Flight Center)

However, when a range of complementary satellite sensors is used, such as TOMS, MODIS (Moderate Resolution Imaging Spectroradiometer) (e.g. Kaufman et al. 2005), GOES (Geostationary Operational Environmental Satellite), SeaWIFS (Sea-viewing Wide Field of View Sensor) and LiDAR (Laser Imaging Detection and Ranging) (e.g. Berthier et al. 2006), each of which has strengths and limitations, then characterisation of dust emissions can be very precise (e.g. Rivera et al. 2006, Miller et al. 2008).

Global Dust Models

As dust is increasingly recognised as an important component of the atmosphere capable of driving or affecting global climate change, a number of attempts have been made to model global emissions. Models vary but generally include a wind erosion or erodibility component coupled with a global climate model. 'Bulk mobilization' models (Zender et al. 2003a) assume a

simple relationship between dust emission and wind speed for particular sizes of emitted dust (e.g. Tegen and Fung 1994, Perlwitz et al. 2001) whilst more complex models include detailed specification of the physical processes acting in the emission zones, such as saltation mass flux (e.g. Shao 2001). Different models predict varying amounts of dust with the majority suggesting total global dust emissions in the range 1000–2000 Tg a^{-1} (Table 20.3) but considerably higher and lower estimates have been reported due largely to the different underlying assumptions incorporated into the models (Zender et al. 2004). Table 20.3 also demonstrates the impact of different model parameters on the relative proportions of dust attributed to different regions of the globe, with for example estimates for Australia ranging from 37–148 Tg a^{-1} and for Arabia from 43–415 Tg a^{-1}.

Researchers attempting to model global and continental-scale dust emissions increasingly recognise the need, not only to acknowledge surface heterogeneity in their calculations (Sokolik et al. 2001), but also to obtain high-quality land erodibility and

Table 20.3 Comparison of the regional annual mean dust flux (Tg a^{-1}) from selected global dust models. Numbers in brackets indicate the percentages for the annual mean global emission flux predicted by each model (from Tanaka and Chiba 2006)

	Africa		Asia			America			
	North	South	Arabia	Central	East	North	South	Australia	Global
Ginoux et al. (2004)	1430 (69.0)	63 (3.4)	221 (11.8)	140 (7.5)	214 (11.4)	2 (0.1)	44 (2.3)	106 (5.7)	1877
Luo et al. (2003)	1114 (67.0)		119 (7.2)		54 (3.2)			132 (8.0)	1654
Miller et al. (2004)	517 (51.0)		43 (4.2)	163 (16.0)	50 (4.9)	53 (5.2)		148 (15.0)	1019
Tanaka and Chiba (2006)	1087 (57.9)	63 (3.4)	221 (11.8)	140 (7.5)	214 (11.4)	2 (0.1)	44 (2.3)	106 (5.7)	1988
Werner et al. (2002)	693 (65.0)		101 (9.5)	96 (9.0)				52 (4.9)	1060
Zender et al. (2003a)	980 (66.0)		415 (28)			8 (0.5)	35 (2.3)	37 (2.5)	1490

dust concentration data at suitable resolutions. Research using global models has demonstrated that it is unrealistic to assume that soil erodibility is uniform and factors such as topography, geomorphology and hydrology have to be accounted for to explain global patterns of dust emissions. For example, Zender et al. (2003b) found that a model based on the assumption of uniform soil erodibility could explain none of the spatial distribution of dust emissions in Australia and was also a poor predictor of spatial patterns in North Africa and the Arabian Peninsula. However the inclusion of geomorphic erodibility improved their model considerably.

Sub-Continental Scale Dust Emissions

Although different models predict different absolute mean annual dust emissions, Table 20.3 and Fig. 20.4 indicate clear conclusions that can be drawn about the relative importance of different geographical areas. The largest global dust source is the Sahara in north Africa, followed by Arabia and southwest and central Asia. At present, Australia is the largest contributor of dust in the southern hemisphere and southern Africa and the Americas all contribute significantly lower quantities of dust to the global system. The TOMS-AI can be used to identify dust 'hot spots' within these zones more precisely, to the extent of identifying specific geomorphological regions (Table 20.4). For example the 37,000 km^2 Mkgadikgadi basin ephemeral lake complex is the largest dust source in Botswana, the south-central area of the 1.14 million km^2 Lake Eyre Basin is the most significant dust source in Australia and the Salar de Uyuni, possibly the world's largest salt flat extending 10,582 km^2 across the Bolivian altiplano in South America, is also a clearly definable dust source.

Whilst Table 20.4 emphasises the importance of topographic basins in areas of low rainfall as major sources of dust, these are rarely the only two parameters contributing to high dust emissions. For example, the Earth's single largest source of dust is the Bodélé Depression, estimated to emit 640–780 Tg a^{-1} (Ginoux et al. 2001). The Bodélé Depression, on the southern edge of the Sahara Desert, is part of the bed of Lake Chad. Currently the lake itself is approximately 1350 km^2, having shrunk from 25,000 km^2 in 1963 and up to 400,000 km^2 at its maximum extent in the mid-Holocene (Lake Mega-Chad) (Coe and Foley 2001, Ghienne et al. 2002). The recession of the water has left an extensive area of desiccated, highly erodible alluvial and diatomaceous material. The status of the Bodélé Depression as the dustiest place on Earth (Giles 2005) results from the coincidence of several

Table 20.4 Maximum mean aerosol index (AI) values for major global dust sources determined by TOMS (Goudie and Middleton 2006). The higher the value of the index, the greater the proportion of absorbing aerosols in the atmosphere

Location	AI value	Average annual rainfall (mm)
Bodélé Depression, south central Sahara	>30	17
West Sahara in Mali and Mauritania	>24	5–100
Arabia (Southern Oman/Saudi border)	>21	<100
Eastern Sahara (Libya)	>15	22
Southwest Asia (Makran coast)	>12	98
Taklamakan/Tarim Basin	>11	<25
Etosha Pan (Namibia)	>11	435–530
Lake Eyre Basin (Australia)	>11	150–200
Mkgadikgadi Basin (Botswana)	>8	460
Salar de Uyuni (Bolivia)	>7	178
Great Basin of the USA	>5	400

variables that combine to give consistent high winds and sediment supply. First, the area is a topographic low in which considerable quantities of sediment have collected. These sediments include thick layers of diatomite formed from the shells of freshwater diatoms which lived in the lake and which correspond with the areas of maximum dust deflation. High levels of dust production occur during the transport of saltating diatomite flakes which disintegrate during collision with one another and on impact with the ground surface (Giles 2005). The topographic characteristics of the area also promote strong winds with low directional variability as the winds are funnelled by the Tibesti Mountains to the north and by the Ennedi Mountains (Koren and Kaufman 2004, Washington and Todd 2005). Washington et al. (2006) suggested that enhancement of these topographically-generated winds (the Bodélé Low Level Jet) during drier, windier conditions in the past may have helped to create the depression, thus promoting the topographic and environmental conditions that make the area such an intense contemporary dust source.

At the local scale ($<10\,\mathrm{km}^2$) there have been many high-resolution studies of the susceptibility of different surfaces to aeolian erosion. These studies have provided detailed information concerning critical threshold wind velocities for dust entrainment and the influences of particle size, surface roughness, soil moisture content and biocrust development on soil erodibility (see Nickling and McKenna Neuman, this volume). The most significant gap in our current understanding of dust emissions is at the meso-scale or sub-drainage basin scale (10^4–$10^6\,\mathrm{km}^2$). Most inland basins are patchworks of soil and land types that yield both different quantities and different types of dust. For example, using measured erosion rates, Breshears et al. (2003) projected annual erosion mass flux to be $4.5\,\mathrm{g\,m^{-2}\,a^{-1}}$ from semi-arid grassland compared with $14.3\,\mathrm{g\,m^{-2}\,a^{-1}}$ in semi-arid shrubland. In southern Nevada and California, Reheis and Kihl (1995) found that playa and alluvial surfaces produced the same amount of dust per unit area (9–$14\,\mathrm{g\,m^{-2}\,a^{-1}}$) but the greater areal extent of alluvial surfaces meant they contributed a higher total volume of dust. The composition of dust from the two sources also differs: the playas producing dust much richer in soluble salts and carbonate compared with the majority of alluvial sources.

Whilst the spatial distribution of dust sources is becoming well known, and some dust sources are very

consistent, the majority of dust sources are intermittent in terms of temporal productivity. The causes of this intermittency are variable, and a key current area of research lies in determining what causes the 'switching on' or 'switching off' of different dust sources. For some dust events, the triggers of dust emissions are relatively easy to discern. For example, in a supply-limited system such as in parts of the Mojave Desert, USA, if the supply of fine materials to dust source areas is not maintained (for example by regular input from flood events) the magnitude and frequency of dust events diminishes (e.g. Clarke and Rendell 1998). Where currently active ephemeral streams are major sources of dust, a temporal relationship between fluvial events and dust sources can be discerned. However this relationship can be complicated if the associated increase in soil moisture content affects other parts of the system, such as vegetation cover, soil salt content or surface crust development (Offer and Goossens 2001). McTainsh et al. (1999) reported that under conditions of low rainfall (150 mm in 1994) in the Channel Country of Australia, dust flux was higher from sand dunes than from clay pan surfaces. However subsequent increases in rainfall caused improved vegetation cover on the dunes which reduced dust emissions. Higher rainfall (370 mm in 1997) maintained the non-erodible vegetation cover on the dunes, but resulted in a pulse of sediment being delivered by flood waters to an adjacent clay pan facilitating a series of dust emission events. Similarly, Bryant (2003) and Mahowald et al. (2003) explored the interactions amongst ephemeral lakes, flood inundation, land use and dust emissions in the Etosha Basin, Namibia and found that although there was some indication that dust emissions increased following post-flood desiccation of the pan surface, the relationship between the two variables was not a simple one. Even where sediments are present they are not always available for deflation, for example due to vegetation protecting the land surface from erosion. Bullard et al. (2008) found an increase in the frequency of dust emissions from the Simpson desert, Australia, following widespread fires that removed vegetation cover, and Rostagno (2007) recorded rapid removal of 90% of the clay and silt fractions from soils during dust storms originating from rangelands following fires.

Whilst flood events in dryland areas can be an important source of dust-sized material, the delivery of coarser, sand-sized material to dust-emission areas can be important in controlling the extent to which the fine material can be raised into the atmosphere. Several field studies have highlighted the importance of sand-sized sediment in releasing dust into the atmosphere – the impact of the sand particles on the fine, often consolidated or compacted clays and silts, causes dust-sized material to be ejected into the airstream. (e.g. Nickling et al. 1999, Shao et al. 1993b, Nickling and Gillies 1993). This process of 'sandblasting' not only triggers the release of dust but can, by aeolian abrasion of the coarse saltators, simultaneously result in dust production. Grini and Zender (2004) found that predictions of dust emission were significantly improved when saltation and sandblasting were factored into a dust emission model.

Meteorological Conditions Promoting Dust Events

Dust emissions have a distinct spatial distribution and, as discussed above, are temporally discontinuous. Whilst variables such as sediment supply are important, there are also certain meteorological conditions that promote the entrainment and transport of dust. Nickling and McKenna Neuman (this volume) discuss thresholds of entrainment for fine particles; this section will focus on broader meteorological parameters.

Dust storms are defined as 'the result of turbulent winds raising large quantities of dust into the air and reducing visibility to less than 1000 m' (McTainsh and Pitblado 1987, p. 416). Dust storms vary in extent and duration, and can include both local and long-distance dust transport. There are four main types of weather systems commonly associated with dust storm generation.

Dust Storms Associated with Cold Fronts

The passage of frontal depressions is one of the most widespread causes of dust storms (Pye 1987, Livingstone and Warren 1996), especially in the Middle East, USA and Australia. These low-pressure fronts have steep pressure gradients that generate strong winds and are often accompanied by an increase in relative humidity and a fall in temperature. As a cold front moves displacing warm air, increasing instability and vertical air movements increase turbulence which is conducive

to the raising of dust. The passage of a cold front is commonly marked by a sharp change in wind direction. Dust is raised along the cold front and can be lifted to high levels in the atmosphere and transported long distances. Widespread dust haze, such as that associated with the harmattan in north Africa can be caused by frontal activity. At lower levels (typically <2000 m) cold fronts can generate belt dust storms hundreds of kilometres long and tens of kilometres wide that can travel at speeds of over 20–30 ms^{-1}.

One of the best documented recent dust storms associated with a cold front occurred in Australia on 22–23 October 2002 (Chan et al. 2005, Leslie and Speer 2006, McTainsh et al. 2005, Shao et al. 2007). The storm was generated by a cold front moving eastwards from the Great Australian Bight. On 22 October 2002 moderate-strong pre-frontal north westerly winds affected parts of southeast Australia reducing visibility in some areas to <500 m (McTainsh et al. 2005). As the front travelled across the continent, large quantities of dust were entrained from rangelands and the central deserts and at its maximum extent the dust plume was 2400 km long (north-south), 400 km across and 1.5–2.5 km in height (Fig. 20.6). The estimated total dust load was 3.35–4.85 Mt and dust in the atmosphere increased

Fig. 20.6 SeaWIFS satellite image showing the dust transported by pre-frontal north westerly winds in southeast Australia on 22 October 2002

Fig. 20.7 Dust storm at Big Spring, Texas, USA, on 16 June 1997. Photo taken by and reproduced by kind permission of Weinan Chen

Goossens 2001), across the Arabian peninsula (Miller et al. 2008), in the southwest USA (Brazel and Nickling 1986), and in Australia (Leslie and Speer 2006). Chen and Fryrear (2002) measured wind speeds of $1.8-2\,\mathrm{ms}^{-1}$ rising to $2.5-5.5\,\mathrm{ms}^{-1}$ at the leading edge and $13\,\mathrm{ms}^{-1}$ at the rear of a haboob dust storm at Big Spring, Texas on 16 June 1997 (all velocities measured at 10 m above the ground surface) (Fig. 20.7). Dust concentration (integrated at 27 heights up to 1.567 m) was $84,960\,\mathrm{kg\,km}^{-1}\,\mathrm{h}^{-1}$ (of which 21% was in the size range 10–20 μm). The mean diameter of the dust particles entrained by the event decreased with height above the surface from 33.4 μm (0.41 m) to 23 μm (1.567 m) and the dust became more poorly sorted with height (Chen and Fryrear 2002).

Dust Storms Associated with Major Depressions

In the northern hemisphere spring, off the coast of north Africa there is an extreme contrast between the surface temperatures of the sea and the desert air. This contrast leads to the generation of large, synoptic-scale depressions which travel eastwards over the Sahara and south-central Mediterranean. Passage of these depressions is associated in most cases with strong, hot, dry winds which entrain sediments and can lead to severe dust storms as well as widespread dust haze across the region.

the average concentration of particles $<10\,\mathrm{\mu m}$ (PM_{10} load) in Brisbane and Mackay on the east coast to 161 and $475\,\mathrm{\mu g\ m}^{-3}$, respectively, (greatly exceeding the recommended $50\,\mathrm{\mu g\ m}^{-3}$ air quality standard) (Chan et al. 2005). Average maximum wind speeds associated with the event were $32.4\,\mathrm{km\,h}^{-1}$ ($9.0\,\mathrm{ms}^{-1}$) which is lower than some comparable events. The magnitude of the storm is thought to have been exacerbated by six months of preceding drought conditions and high temperatures that reduced vegetation cover. Whilst the desert regions were the strongest dust source, fine particles were also entrained from grazing and farmlands (Shao et al. 2007).

Dust Storms Associated with Thundery Conditions

Dust storms are often associated with strong downdrafts of cooled air that descend steeply from cumulonimbus clouds and thunderstorm cells. These downdrafts can cause wind speeds as high as $50\,\mathrm{ms}^{-1}$ (Goudie and Middleton 2006) which raise dust causing localised dust storms along the storm line. This type of event, known in north Africa as a 'haboob', is typically characterised by a dense wall of dust which moves across the landscape. Haboob-type storms in north Africa (particularly Sudan) are associated with northward movement of the intertropical front bringing moisture inland from the central Atlantic and are most common from May to September, although they can occur at any time. They also occur in Israel (Offer and

Dust Storms Associated with Mountain Winds

Air that has undergone orographic lifting cools and dries as a result of condensation of water vapour. This cool, dry air becomes dynamically-heated as it descends from lower to higher atmospheric pressures creating strong, warm, dry, turbulent airflow known as föhn (or foehn) winds. Winds associated with mountain ranges are common but do not all trigger dust storms due to sediment supply limitations. Common dust-raising föhn winds include the Santa Ana winds of southern California and Berg winds of southern Africa. Soderberg and Compton (2007) report a berg wind event in Cape Town that occurred on 19–20 May 2002 in which visibility was reduced by wind-blown dust. On the 20 May, PM_{10} levels at Khayelitsha averaged

$101 \, \mu g \, m^{-3}$ over the 24 period (classed as 'very high') and hourly records at stations including Table View exceeded $200 \, \mu g^{-3}$ (City of Cape Town Air Quality Monitoring Network 2002). In this area dust storms are an important source of fine particles and nutrients (e.g. K, Al, Ca) for soil and vegetation (Soderberg and Compton 2007).

Small-Scale Dust Events

One of the most localised types of dust-raising event is the 'dust devil'. Dust devils form as air at the ground surface is heated and starts to rise rapidly through overlying cool air. The air column may start to rotate, gaining increasing wind speeds which enable entrainment of sediments from the surface that then rise with the hot air through the air column. Dust devils form on surfaces of low gradient that are susceptible to heating, such as bare soils, although vegetation can be present (Mattsson et al. 1983) and their development can be exacerbated by local topography. For example Hess and Spillane (1990) reported dust devils forming parallel to topographic ridges. The requirement for surface heating means that dust devils are most frequent during the hottest part of the day (late morning or early afternoon) and during spring and summer seasons (Wigner and Peterson 1987, Oke et al. 2007a). Dust devils range from <0.5 m to >100 m in diameter, although most are <10 m across. They are usually at least five times higher than they are wide (Hess and Spillane 1990) and exceptionally can reach heights of over 1000 m on Earth, although over 80% lie in the height range 3–300 m (Balme and Greeley 2006). The height of a dust devil is determined not only by atmospheric conditions but also by the characteristics of the dust particles incorporated within it. Oke et al. (2007b) found that over 80% of particles transported in dust devils in Australia (known locally as willy-willies) were $\leq 63 \, \mu m$ in diameter and suggest that this preferential entrainment of fine particles could have implications for soil surface texture, vegetation sustainability and hence geomorphology. Until recently there had been little geomorphological research on dust devils when compared with that concerning other dust-event phenomena, but the discovery that dust devils on Mars are significant enough in terms of magnitude and frequency to affect surface albedo and possibly regional-scale atmospheric circulation (e.g. Michaels 2006) has led to more studies

on Earth which might act as useful analogues for other planets (e.g. Towner et al. 2004).

Dust Deposition

Dust Deposition in the Oceans

Of course, once entrained dust can travel long distances in suspension. Grousset et al. (2003), for example, reported that dust deflated from the Taklamakan Desert in western China had travelled over 20,000 km east across the Pacific Ocean, North America and the Atlantic to be deposited in western Europe. Similarly, Tanaka (2005) reported a two-day dust event recorded at meteorological stations in Japan that had its origins in dust storms in north Africa and the Middle East several days earlier, and he concluded that the Sahara and Arabian deserts were potentially important sources of dust in east Asia.

As a consequence of this propensity for long-distance travel, much of the dust generated in the desert areas is exported and deposited in the oceans. The deep-sea drilling programmes have provided much evidence in the form of sedimentary cores from the ocean floors of export of aeolian material from deserts, early examples of the recognition of aeolian dust in ocean cores including the work of Chester et al. (1971), Duce (1980) and Prospero et al. (1981). This dust has been seen as an important palaeoenvironmental proxy: accumulation rates indicating dust fluxes and particle sizes indicating wind strength. There is good evidence, for instance, of increased dust input into ocean cores at the time of the last glacial maximum around 18,000 years ago, notably off the west coast of north Africa (Sarnthein and Koopman 1980, Tetzlaff and Peters 1986) and into the Pacific off the east coast of Australia (Hesse and McTainsh 1999). In both cases these cores point to increased dust production in the deserts, probably associated with lowered precipitation rates, but in some instances they may also reflect greater wind energy, although Hesse and McTainsh (1999) suggested that this was not the case in Australia.

Dust deposited in oceans has negligible geomorphological impact but there are a range of consequences of this deposition. There may for instance be a significant link between aeolian dust and biological productivity, for example in the Pacific (Yuan and Zhang 2006), Indian (Piketh et al. 2000) and southern Oceans although there are still considerable uncertainties concerning this relationship (Jickells et al. 2005). Additionally, the geochemical properties of deposited dust have been seen as a potential tracer for dust sources that could find application in palaeoenvironmental studies of dust transport and deposition (e.g. Moreno et al. 2006). Dust deposited in the oceans can therefore be used as a tracer for global atmospheric circulation (Biscaye 1997, Prospero 1999, Grousset and Biscaye 2005, Tanaka 2005) and this potential has meant that dust transport and deposition has been the subject of modelling (e.g. Tanaka and Chiba 2006).

Contemporary Terrestrial Dust Deposition Records

At least part of the difficulty of establishing the importance of aeolian transport of dust has been the lack of effective monitoring. While dust flux may be monitored remotely (see the section on remote sensing above), it is much more difficult to measure the amounts of dust being deposited. Where attempts have been made, the characteristics and mass of dust trapped, and hence how representative the sample is of actual dust deposition, is highly dependent upon the type of trap used. A number of recent studies have compared the performance of different methods for ascertaining dust flux and rates of dust deposition. There is a fundamental difference between horizontal transport flux and vertical deposition of dust and most traps measure effectively one or the other, but not both. Relationships between the two variables can be determined (e.g. Goossens 2008) but these vary with grain size. Goossens and Offer (2000) evaluated the efficiency of five horizontal dust flux samplers and found this varied with wind speed. The semi-isokinetic, self-orienting Big Spring Number Eight (BSNE) sampler (Fryrear 1986), which is widely used in field studies, was recommended because its trap efficiency is largely independent of wind speed, however it is better suited to coarser particles than very fine particles and was not tested at high wind speeds. A wide range of studies have explored the relative efficiencies of dust deposition samplers (Goossens 2005, Goossens 2007, Goossens and Rajot, 2008, Sow et al. 2006), the big practical difficulty to be overcome being developing equipment that can prevent the re-entrainment of dust deposited on the trap. Common approaches include

using vessels (which vary widely in terms of their aerodynamics) containing marbles or water to prevent re-entrainment, or traps with high roughness values such as astroturf (O'Hara et al. 2006).

Dust deposition varies both spatially and temporally and the types of traps used should be tailored to the research questions being asked, but will also be dependent on available resources and how frequently data can be retrieved. Recently, a number of projects have been established to trap airborne dust and estimate deposition rates both spatially and over long time scales. The US Geological Survey (USGS), for instance, has established the CLIM-MET network of meteorological stations in the south-west USA which include a number of instruments for monitoring both dust transport and deposition. From their data they have been able to piece together a picture of dust sources, fluxes and deposition rates in the Mojave Desert (e.g. Reynolds et al. 2006). (Data from the CLIM-MET project are publically available from: http://esp.cr.usgs.gov/info/sw/clim-met/index.html). In North Africa O'Hara et al. (2006) monitored dust deposition in Libya over an annual cycle and found values in the range 38.6–$311.0\,\mathrm{g\,m^{-2}\,a^{-1}}$, with a mean of $129.1\,\mathrm{g\,m^{-2}\,a^{-1}}$. Dust activity in the McMurdo valleys (Antarctica) showed high inter-annual variability ($<1\,\mathrm{g\,m^{-2}\,a^{-1}}$ to $>200\,\mathrm{g\,m^{-2}\,a^{-1}}$) from 1999 to 2005 with rates of dust deposition strongly influenced by the availability of material (Lancaster 2002). These studies have all used different methods for quantifying dust deposition, but one possibility is that sampler comparison studies, such as those cited above, could provide a framework for comparing studies conducted using different techniques.

Factors Promoting Dust Deposition

The nature of the transport of dust in the wind means that it is often carried large distances and spread over large areas, as a consequence of which it only rarely builds distinguishable deposits and arguably never builds distinctive landforms. Frequently this transport is away from the terrestrial source and out over the oceans where it may be incorporated into marine sediments. On land it is often spread so thinly that it is indistinguishable from other surface materials. Particularly in areas where weathering rates are sufficient, the dust will be quickly incorporated

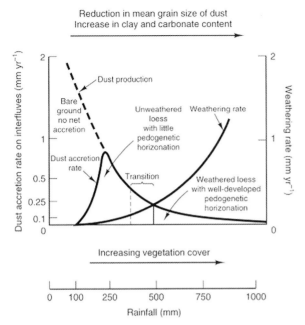

Fig. 20.8 Schematic model of dust accretion in relation to mean annual rainfall (from Pye and Tsoar 1987)

into soils. If deposited where there is little vegetation or other surface materials to trap the dust, it will be re-entrained. There are therefore rather particular conditions that need to be satisfied before aeolian dust deposits accumulate sufficiently to be recognisable. Where the terrestrial dust deposits are recognisable they are often called 'loess'.

Pye and Tsoar (1987) summarised the conditions for terrestrial dust deposition by plotting dust accretion as a function of rainfall (Fig. 20.8). Where rainfall is low there may be plenty of dust generated by all the mechanisms discussed above (and represented by the dashed line on Fig. 20.8) but it is too easily re-entrained to accumulate. Increasing vegetation cover generated by greater rainfall acts as an effective trap to the dust that does fall to the surface and helps to prevent re-entrainment. Increasing precipitation is also often the mechanism for deposition by washing the dust out of the atmosphere. However, more moisture also intensifies the weathering rate and this means that deposited dust is quickly incorporated into soils. Although airborne desert dust does reach temperate areas, it does so in small quantities and is soon 'lost' into the soil. It may also be redistributed by runoff processes generated by the rainfall. Consequently, the schematic model of Pye and Tsoar showed that dust most readily accu-

mulated in areas of around 250 mm a^{-1}. Drier than this and the dust is easily re-entrained; wetter and it is 'lost' into the soil.

Airborne dust does also accumulate in combination with other surface materials. McFadden et al. (1987) suggested that dust was being trapped by coarser material in the Mojave Desert, in their example on the surface of a lava flow. Once it reached the surface they argued it was washed into the interstices between fragments of the lava to create a stone pavement with finer material concentrated below the surface under an armour of coarser material. Stone pavements have been described as forming by a number of mechanisms, including aeolian deflation (Cooke 1970, Chapter 19) but this mechanism of trapping of fine material carried by suspension in the air has been reported from a number of locations (e.g. Gerson and Amit 1987, Goossens 1995, Li et al. 2005).

The trapping of wind-blown dusts by rough surfaces can also lead to distinctive localised deposits. For example, in northern Chile dust streaks have formed on uneven salar surfaces (Stoertz and Ericksen 1974). Figure 20.9 shows dust deposits extending from the base of a hillside on the southwest of the Salar de Pintados. Gullies in the hillside funnel the wind which entrains fine sediments and deposits them forming streaks with an average length of 1.5 km, although the longest extend over 3 km on to the salar and have a

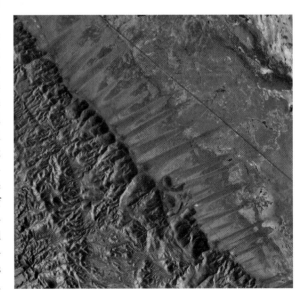

Fig. 20.9 Dust streaks on the surface of the Salar de Pintados, north Chile. Aerial photograph extract reproduced by kind permission of Servicio Aero Fotogrametrico, Chile

maximum width of 350 m. The dust particles, which are predominantly silt-sized aggregates, are trapped in pockets on the salar surface and the streaks are clearly visible because the dusts are dark grey in colour and contrast with the pale salt crust of the salar.

Loess

Where recognisable deposits of wind-blown dust accumulate they are termed 'loess'. There is a huge literature reflecting the disputes over the recognition of loess. For a long time loess was defined by its size, and the best known of the world's loess deposits were associated with glacial outwash – often specifically linked to greater glacial extents at the time of the last global glaciation – and the source of dust-sized material in the world's deserts was contested. Even as recently as 1990, Smalley and Derbyshire felt able to argue that 'most of what was called desert loess can now be seen as mountain loess' (Smalley and Derbyshire 1990, p.301). Their argument was that loess was a Quaternary phenomenon because only glaciers were able to grind rock to the size fraction required for loess deposition. Their concession was that in some low-latitude mountain environments it might be possible for cold-climate conditions to produce loess-sized material by periglacial weathering processes. As Yaalon (1991) pointed out in his rebuttal to Smalley and Derbyshire, other sedimentary deposits are not distinguished by their place of origin; we do not talk, for instance, of igneous sand dunes or plutonic alluvium. But in addition, as we have seen, deserts do provide an excellent environment for generating silt-sized material to be carried by the wind, and Yaalon (1991) could not see how loess at the margins of deserts could be called 'mountain loess'. Even in 1995, Smalley was reluctant to admit that deserts might provide the material for loess. He pointed to the mountains as the source for the loess of central Asia and north China. He did admit that Africa provided him with something of a problem and failed to mention loess deposits in Arabia or Australia. Assallay et al. (1996) suggested that the particles in the Libyan deposits were too large for them to be termed 'loess' although Coudé-Gaussen (1987) attributed large particle size elsewhere in north Africa to proximity to the source. Sarnthein and Koopman (1980) reported that there were deposits of terrigenous quartz silt in the size range characteris-

tic of loess in the Atlantic Ocean off the West African coast. There is now no doubt that deserts produce the basic material for loess deposits.

Notwithstanding the lengthy debates about the origin, provenance, mineralogy and size of loess, loess can most straightforwardly be defined as 'a terrestrial clastic sediment, composed predominantly of silt-size particles, which is formed essentially by the accumulation of wind-blown dust' (Pye 1995, p. 653). The mineralogy reflects the mineralogy of wind-blown dust discussed above: it is predominantly quartz with components of feldspar, mica, carbonates and clay minerals. Despite some arguments that true loess must be exclusively silt-sized, the modal size is nonetheless 20–40 μm, and many loess deposits at desert margins include a significant sand component. Arbitrarily Pye (1987) termed loess with more than 20% sand-sized material as 'sandy loess', an example being the peri-desert loess in Tunisia which has a mode of 63 μm (Coudé-Gaussen and Rognon 1988). In part this may be because much loess associated with deserts is relatively close to its source. Smith et al. (2002) noted studies (Morales 1979) in which dusts blowing out of the Sahara were described as having modal particle sizes predominantly less than 10 μm.

The conditions that lead to loess accumulation (summarised by Smith et al. 2002) are: first, atmospheric and ground conditions in the source area that are conducive to deflation of material that can be carried in suspension; second, a specific combination of atmospheric and ground conditions that encourage the preferential deposition of silts from all the particle sizes being carried by the wind; and third, prolonged duration of these conditions, during which time there is minimal reworking of deposits or additions of material from other sources via different transport mechanisms.

Given these conditions and the model of Pye and Tsoar (Fig. 20.8) we might expect that material is being entrained and exported from the arid cores of the deserts, and deposited once moisture and vegetation cover are great enough to trap the material in the semi-arid areas at the desert margins. These deposits are commonly termed 'peri-desert loess'. Until the 1980s these peri-desert loesses were frequently omitted from or poorly represented on world maps of loess deposits partly because the deposits were not recognised as wind-blown dust. In Australia, for example, Butler (e.g. 1956) felt unable to use the term 'loess' for deposits that he recognised as wind blown and instead used the local term 'parna' for what he described as 'wind-blown clays'. Consequently 'these so-called parna soils were relegated to the position of a Southern Hemisphere curiosity, rather than representing early examples of desert loess' (Hesse and McTainsh 2003), a situation addressed by Haberlah (2007).

The recognition in the 1970s and 1980s that deserts provided efficient machines for generating suspendible material and the desert margins were ideal environments to trap that material led to a flurry of reports of peri-desert loess from around the world. The thickest loess deposits are in China, where up to 300 m of loess has been deposited over the past 2.5 million years and at least some of these deposits are thought to have a desert origin. Other peri-desert loesses are much thinner than the Chinese loess or than their high-latitude counterparts but they are now reported from the margins of all the world's major deserts (Table 20.5). At least part of the debate about peri-desert loess has been concerned with whether the deposited material was generated in the desert or whether the desert was merely an intermediate stopping-off point on the way from origin to loess deposit. Either way, of course, the desert environment has played an important part in the creation of peri-desert loess deposits (Figs. 20.10 and 20.11).

The largest accumulations of loess on the planet are in China (Fig. 20.12) where they cover 440,000 km^2 (Liu 1985) and appear to have built up over at least the last 22 million years according to Guo and co-workers (Guo et al. 2002, Hao et al. 2008). One view has been that the three northwestern inland basins (the

Junggar Basin, the Tarim Basin and the Qaidam Basin) have been important source areas for the loess. In contrast, Sun (2002) argued that the gobi (stony desert) in southern Mongolia and the adjoining gobi and sand deserts (the Badain Jaran Desert, Tengger Desert, Ulan Buh Desert, Hobq Desert and Mu Us Desert) in China, rather than the three inland basins, were the dominant source areas. However, although Sun regarded these gobi and sand deserts as the main source regions, he suggested that they served as dust and silt holding areas rather than dominant producers and that the Gobi Altay, Hangayn and Qilian mountains may have been the initial sources. It appears that the material of the right size is created by a variety of processes in the mountains but is entrained by the wind from the desert basins and then deposited as loess on the Chinese loess plateau. The considerable thickness of the Chinese loess may therefore be the consequence of a particular combination of rapid uplift of the Tibetan Plateau and surrounding mountain ranges, high rates of sediment production and supply to adjacent basins, a strong northwesterly and westerly wind regime, and the existence of effective dust traps downwind of the source regions (Pye 1995). The Libyan deserts have also been regarded as 'mixing points' for dust before being re-entrained (O'Hara et al. 2006).

Elsewhere conditions at desert margins do not favour the development of such thick loess deposits. For example, Breuning-Madsen and Awadzi (2005) argued that deposition rates of Harmattan dust in Ghana simply were not great enough to develop loess deposits. There may be a number of reasons for this, but clearly for thick loess deposits to develop there needs to be an effective trap downwind and nearby a considerable source of dust. In particular, Pye (1995) and Mason et al. (1999) have drawn attention to the importance of topographic effects on the distribution of loess. Loess tends to accumulate in basins and at the foot of mountains, both because of aerodynamic effects and because of post-depositional reworking by water and gravity.

Table 20.5 Examples of peri-desert loess locations

Location	Reference
Sahara (northern margin) ~ Matmata Plateau, Tunisia	Coudé-Gaussen and Rognon 1988, Dearing et al. 2001
Sahara (northern margin) ~ Jebel Gharbi, Libya	Giraudi 2005
Sahara (southern margin) Nigeria	McTainsh 1987
Namib Desert, Namibia	Blümel 1982, Brunotte and Sander 2000
Arabia UAE	Goudie et al. 2000, Edgell 2006
Australia	Butler 1956, Hesse et al. 2003
Peru	Eitel et al. 2005
Yemen	Nettleton and Chadwick 1996
Negev Desert, Israel	Ginzbourg and Yaalon 1963, Yaalon and Dan 1974, Yaalon and Ganor 1975

Loess in the Palaeoenvironmental Record

The importance of the deposition of airborne dust in the palaeoclimatic record was recognised by the establishment of the DIRTMAP project (e.g. Kohfeld

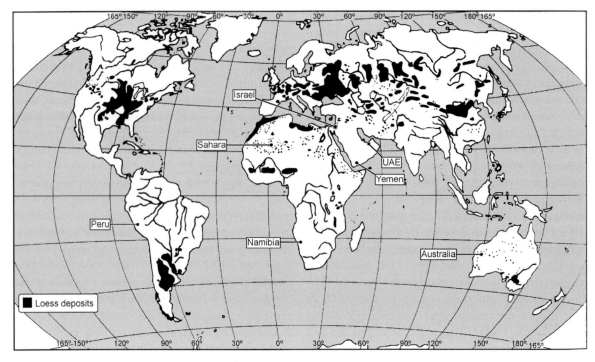

Fig. 20.10 World map showing major loess locations (after Livingstone and Warren 1996) and the major deserts

Fig. 20.11 Loess deposits in (**a**) China (courtesy Frank Eckardt) and (**b**) Tunisia

and Harrison 2001) which created a database of records of dust deposition both from contemporary monitoring and in the sedimentary record. Where loess has been deposited it has usually accumulated over a considerable period of time and therefore provides a very important terrestrial record of past environments (Sun et al. 2000). Rates of loess deposition appear to have varied quite considerably. Derbyshire (2003) suggested that 'some marine and terrestrial sedimentary records suggest that rates of aeolian deposition during the Last Glacial Maximum (LGM) may have

been up to 10 times greater than those of the present (see summary by Kohfeld and Harrison 2001)', and Hesse and McTainsh (2003) reported that the dust flux from Australia to the oceans was at least three times greater at the LGM than in the Holocene. Pye (1987) estimated that loess was accumulating at 0.5–$3.0\,\mathrm{mm}\,\mathrm{a}^{-1}$ at the LGM. Global climate changes have affected not only the dust fluxes and rates of dust deposition, but also changed the location of sources and sinks. In China, for instance, Derbyshire (2003) noted that during the LGM, China's dryland margins

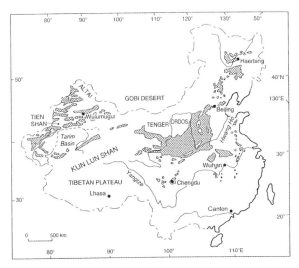

Fig. 20.12 Map of loess deposits in China (from Goudie and Middleton, 2006)

This record of dust deposition has become particularly important especially as luminescence dating techniques have developed since the 1980s. Luminescence dating allows the dating of minerals such as quartz and feldspar, both important constituents of loess, and has therefore removed the reliance on finds of organic material required for radiocarbon dating. There is now a plethora of luminescence dates that have been provided for the loess in China, but dated sections have also been provided for other peri-desert areas, such as the Matmata Plateau in Tunisia (Dearing et al. 2001) and bordering the northern Atacama Desert in Peru (Eitel et al. 2005). Some of the issues associated with luminescence dating of loess were discussed by Singhvi et al. (2001).

As noted above, peri-desert loess is frequently coarser than its temperate and higher latitude counterparts, but its particle size is far from homogeneous. Chinese dust is typically bimodal representing a far-travelled, poorly sorted, fine component and a more localised, well-sorted component (Sun et al. 2004). The variability of particle size has been seen as a useful indicator of past environmental conditions.

advanced several hundred kilometres south and east of those of the Holocene climatic optimum, affecting the winter monsoon intensity and influencing loess accumulation rates (Ding et al. 1999, Porter 2001).

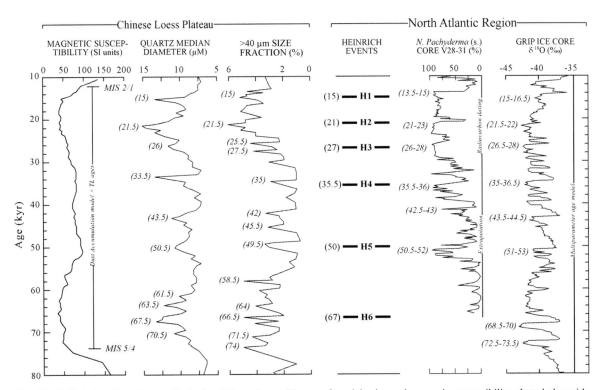

Fig. 20.13 Variation through a profile in the Chinese Loess Plateau of particle size and magnetic susceptibility plotted alongside variations in oxygen isotope ratios for the North Atlantic (from Porter and An 1995)

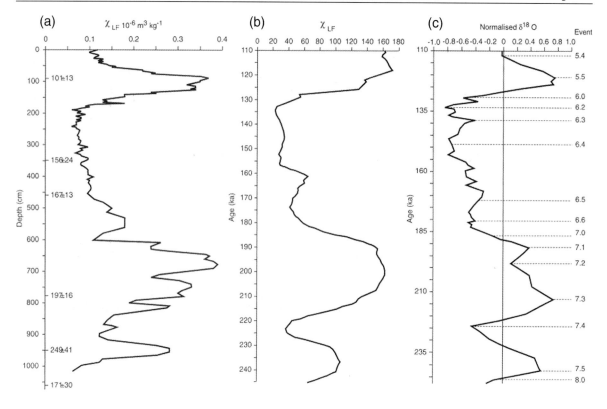

Fig. 20.14 Magnetic susceptibility records from the loess on the Matmata Plateau plotted against depth (**a**) with a similar record from the Chinese loess (Kukla et al. 1988) (**b**), compared with the oxygen isotope record (**c**) (Martinson et al. 1987) (from Dearing et al. 2001)

Porter and An (1995), for example, suggested that particle size (particularly as measured by the percentage >40 μm) was a proxy for the rate of deposition of the Chinese loess and therefore coarser particles indicated times when the capacity of winds to carry dust increased. They linked this measure of particle-size variability to oxygen isotope stages and were able to propose that windier periods occurred during colder periods in the marine record (linked to Heinrich iceberg-discharge events in the north Atlantic) (Fig. 20.13). Similar links have been made by, for example, Nugteren et al. (2004) and Sun and Huang (2006), who related particle-size variation in the northwestern Chinese Loess Plateau through the last interglacial to half-precessional cycles in insolation. Feng and Wang (2006), however, suggested that some caution was required because a number of factors contributed to the particle size distribution and it might not be possible to use particle size as a direct proxy for climate, particularly the strength of the winter monsoon.

Much of the effort in loess studies has concentrated on the recognition of palaeosols in the loess sequences. Generally the palaeosol has been seen as an indicator of surface stability and zero or reduced dust input, probably associated with greater humidity, whereas un-weathered loess without apparent soil development has been viewed as evidence of increased dust input and associated with periods of aridity. Often the palaeosols have been recognised using measures of magnetic susceptibility, increased susceptibility being taken as an indication of secondary mineralisation by pedogenic processes. However, Kemp (1999, 2001) was keen to urge caution when using palaeosols as palaeoenvironmental indicators particularly because of problems associated with reworking or syn-depositional alteration of palaeosols. Frequently, continuous deposition has been the basis for developing the Quaternary climate record from the Chinese loess. The assumption has been that loess sequences represent the whole record and that either the loess is accumulating or the ground surface is stable and a soil is developing. The reality is likely to be much more complicated with episodes of erosion when loess has been removed and the possibility of episodes when loess deposition is contemporaneous with soil formation. For example, work by Stevens et al. (2006), based on luminescence dates from sample points closely-spaced within a section, suggested that Chinese loess deposition has been episodic rather than continuous. Notwithstanding Kemp's concerns, there are a large number of studies of loess palaeosol sequences using mineral magnetic measures from the Chinese loess (see the reviews by Porter 2001 and Liu et al. 2007). There are few from other peri-desert loess areas although Dearing et al. (2001) provided magnetic data in combination with luminescence dates to place development of the Matmata Plateau loess in Tunisia alongside the oxygen isotope record in stages 8–5 (Fig. 20.14).

Conclusion

Plainly the past couple of decades have seen a very considerable increase in the interest in mineral dust generated in deserts. Although desert geomorphologists have long been aware of dust generation in deserts, the monitoring and accurate recording of dust entrainment, transport and deposition has been a relatively recent phenomenon. In part this has been made possible by improving technology. Unlike other geomorphological processes, the movement of dust often does not leave clear landform evidence of erosion and frequently the deposition of even quite large amounts of sediment is spread as thin mantles over large areas. Remote sensing has helped greatly in monitoring the transport of dust, such that we now have a much clearer impression of the global budget for dust transport. In addition the technology used in field studies has been developed to allow good measurements of dust fluxes at ground level to be ascertained. Added to these studies, the recent past has seen a widespread recognition that the dust generated within the world's deserts is frequently exported by the wind and deposited at their margins, either as recognisable loess deposits or as dust-sized additions to semi-arid soils. Some of this work on dust fluxes has been driven by a desire to comprehend more fully the role of suspended dust (of which deserts are one

source) in the global atmospheric system, but there has also been a general recognition that we need to develop a fuller understanding of the aeolian transport of dust if we are to explain the movement of clastic material in deserts.

References

Assallay, A.M., Rogers, C.D.F. and Smalley, I.J. 1996. Engineering properties of loess in Libya. *Journal of Arid Environments* **32**, 373–386.

Assallay, A.M., Rogers, C.D.F., Smalley, I.J. and Jefferson, I.F. 1998. Silt: 2-62 mu m, 9-4 phi. *Earth-Science Reviews* **45**, 61–88.

Baba, J. and Komar, P.D. 1981. Measurements and analysis of settling velocities of natural quartz sand grains. *Journal of Sedimentary Petrology* **51**, 631–640.

Balme, M. and Greeley, R. 2006. Dust devils on Earth and Mars. *Reviews of Geophysics* **44** (3), RG3003, doi:10.1029/2005RG000188.

Berthier, S., Chazette, P., Couvert, P., Pelon, J., Dulac, F., Thieuleux, F., Moulin, C. and Pain, T. 2006. Desert dust aerosol columnar properties over ocean and continental Africa from LIDAR in-Space Technology Experiment (LITE) and Meteosat synergy. *Journal of Geophysical Research – Atmospheres* **111** (D21), D21202.

Betzer, P.R., Carder, K.L., Duce, R.A., Merrill, J.T., Tindale, N.W., Uematsu, M., Costello, D.K., Young, R.W., Feely, R.A., Breland, J.A., Berstein, R.E. and Greco, A.M. 1988. Long-range transport of giant mineral aerosol particles. *Nature* **336**, 568–571.

Biscaye, P.E. 1997. Asian provenance of glacial dust (stage 2) in the Greenland Ice Sheet Project 2 Ice Core, Summit, Greenland. *Journal of Geophysical Research – Oceans* **102**, 26765.

Brazel, A.J. and Nickling, W.G. 1986. The relationship of weather types of dust storm generation in Arizona (1955–1980). *Journal of Climatology* **6**, 255–275.

Breshears, D.D., Whicker, J.J., Johansen, M.P. and Pinder, J.E. 2003. Wind and water erosion and transport in semi-arid shurbland, grassland and forest ecosystems: quantifying dominance of horizontal wind-driven transport. *Earth Surface Processes and Landforms* **28**, 1189–1209.

Breuning-Madsen, H. and Awadzi, T.W. 2005. Harmattan dust deposition and particle size in Ghana. *Catena* **63**, 23–38.

Bryant, R.G. 2003. Monitoring hydrological controls on dust emissions: preliminary observations from Etosha Pan, Namibia. *Geographical Journal* **169**, 131–141.

Bullard, J.E., Baddock, M.C., McTainsh, G.H. and Leys, J.F. 2008. Sub-basin scale dust source geomorphology detected using MODIS. *Geophysical Research Letters*, **35**, 15, L15404.

Bullard, J.E. and McTainsh, G.H. 2003. Aeolian-fluvial interactions in dryland environments: scales, concepts and Australia case study. *Progress in Physical Geography* **27**, 471–501.

Bullard, J.E., McTainsh, G.H. and Pudmenzky, C. 2007. Factors affecting the rate and nature of fine particle production by aeolian abrasion. *Sedimentology* **54**, 1169–1182.

Butler, B.E. 1956. Parna – an aeolian clay. *Australian Journal of Science* **18**, 145–151.

Chan, Y.-C., McTainsh, G., Leys, J., McGowan, H. and Tews, K. 2005. Influence of the 23 October 2002 dust storm on the air quality of four Australian cities. *Water, Air and Soil Pollution* **164**, 329–348.

Chen, W. and Fryrear, D.W. 2002. Sedimentary characteristics of a haboob dust storm. *Atmospheric Research* **61**, 75–85.

Chester, R., Elderfield, H. and Griffin, J.J. 1971. Dust transported in north-east and south-east trade winds in Atlantic Ocean. *Nature* **233**, 474–476.

City of Cape Town Air Quality Monitoring Network. 2002. http://www.capetown.gov.za/airqual/episodes/default. asp?yr=2002 Last accessed 5 December 2006.

Clarke, M.L. and Rendell, H.M. 1998. Climate change impacts on sand supply and the formation of desert sand dunes in the southwest USA. *Journal of Arid Environments* **39**, 517–531.

Coe, M.T. and Foley, J.A. 2001. Human and natural impacts on the water resources of the Lake Chad basin. *Journal of Geophysical Research* **106**, 3349–3356.

Cooke, R.U. 1970. Stone pavements in deserts. *Annals of the American Association of Geographers* **60**, 560–577.

Coudé-Gaussen, G. 1987. The perisaharan loess: sedimentological characterization and paleoclimatical significance. *GeoJournal* **15**, 177–183.

Coudé-Gaussen, G. and Rognon, P. 1988. The Upper Pleistocene loess of southern Tunisia: a statement. *Earth Surface Processes and Landforms* **13**, 137–151.

Dearing, J.A., Livingstone, I.P., Bateman, M.D. and White, K.H. 2001. Palaeoclimate records from OIS 8.0-5.4 recorded in loess-palaeosol sequences on the Matmata Plateau, southern Tunisia, based on mineral magnetism and new luminescence dating. *Quaternary International* **76–77**, 43–56.

Derbyshire, E. 2003. Loess, and the Dust Indicators and Records of Terrestrial and Marine Palaeoenvironments (DIRTMAP) database. *Quaternary Science Reviews* **22**, 1813–1819.

Ding, Z.L., Sun, J.M., Rutter, N.W., Rokosh, D. and Liu, T.S. 1999. Changes in sand content of loess deposits along a north-south transect of the Chinese Loess Plateau and the implications for desert variations. *Quaternary Research* **52**, 56–62.

Duce, R.A. 1980. Long-range atmospheric transport of soil dust from Asia to the tropical north Pacific – temporal variability. *Science* **209**, 1522.

Edgell, H.S. 2006. *Arabian Deserts: nature, origin and evolution*, Berlin: Springer.

Eitel, B., Hecht, S., Machtle, B. and Schukraft, G. 2005. Geoarchaeological evidence from desert loess in the nazca-palpa region, southern Peru: Palaeoenvironmental changes and their impact on pre-columbian cultures. *Archaeometry* **47**, 137–158.

Evans, J.E. and Tokar Jr, F.J. 2000. Use of SEM/EDS and X-ray diffraction analyses for sand transport studies, Lake Eyrie, Ohio. *Journal of Coastal Research* **16**, 926–933.

Feng, Z.D. and Wang, H.B. 2006. Geographic variations in particle size distribution of the last interglacial pedocomplex S1 across the Chinese Loess Plateau: their chronological and pedogenic implications. *Catena* **65**, 315–328.

Fryrear, D.W. 1986. A field dust sampler. *Journal of Soil and Water Conservation* **41**, 117–120.

Gerson, R. and Amit, R. 1987. Rates and modes of dust accretion and deposition in an arid region-the Negev, Israel. In: Desert Sediments: Ancient and Modern, Spec. Publ. Geol. Soc.

Ghienne, J.-F., Schuster, M., Bernard, A., Duringer, P. and Brunet, M. 2002. The Holocene giant Lake Chad revealed by digital elevation models. *Quaternary International* **87**, 81–85.

Giles, J. 2005. The dustiest place on earth. *Nature* **434**, 816–819.

Gillette, D.A. and Walker, T.R. 1977. Characteristics of airborne particles produced by wind erosion of sandy soil, High Plains of West Texas. *Soil Science* **123**, 97–110.

Ginoux, P., Chin, M., Tegen, I., Prospero, J.M., Holben, B., Dubovik, O. and Lin, S.J. 2001. Sources and distributions of dust aerosols simulated with the GOCART model. *Journal of Geophysical Research – Atmospheres* **106** (D17), 20255–20273.

Ginoux, P., Prospero, J.M., Torres, O., and Chin, M. 2004. Long-term simulation of global dust distribution with the GOCART model: correlation with the North Atlantic Oscillation. *Environmental Modelling and Software*, **19**, 113–128.

Ginzbourg, D., Yaalon, D.M. 1963. Petrography and origin of the loess in the Be'er Sheva basin. *Israel Journal of Earth Sciences* **12**, 68–70.

Giraudi, C. 2005. Eolian sand in peridesert northwestern Libya and implications for Late Pleistocene and Holocene Sahara Expansions. *Palaeogeography Palaeoclimatology Palaeoecology* **218**, 159–171.

Gomes, L., Bergametti, G., Coudé-Gaussen, G. and Rognon, P. 1990. Submicron desert dusts: a sandblasting process. *Journal of Geophysical Research* **95**, 13927–13935.

Goossens, D. 1995. Field experiments of aeolian dust accumulation on rock fragment substrata *Sedimentology* **42**, 391–402.

Goossens, D. 2005. Quantification of the dry aeolian deposition of dust on horizontal surfaces: an experimental comparison of theory and measurements. *Sedimentology* **52**, 859–873.

Goossens, D. 2007. Bias in grain size distribution of deposited atmospheric dust due to the collection of particles in sediment catchers. *Catena* **70**, 16–24.

Goossens, D. 2008. Relationships between horizontal transport flux and vertical deposition flux during dry deposition of atmospheric dust particles. *Journal of Geophysical Research-Earth Surface* **113** (F2), art.no. F02S13.

Goossens, D. and Offer, Z.Y. 2000. Wind tunnel and field calibration of six aeolian dust samplers. *Atmospheric Environment* **34**, 1043–1057.

Goossens, D. and Rajot, J.L. 2008. Techniques to measure the dry aeolian deposition of dust in arid and semi-arid landscapes: a comparative study in west Niger. *Earth Surface Processes and Landforms* **33**, 178–195.

Goudie, A.S., Cooke, R.U. and Evans, I. 1970. Experimental investigation of rock weathering by salts. *Area* **4**, 42–48.

Goudie, A.S. and Middleton, N.J. 2006. *Desert Dust in the Global System*. Berlin: Springer.

Goudie, A.S. and Watson, A. 1991. The shape of desert sand dune grains. *Journal of Arid Environments* **4**, 185–190.

Goudie, A.S., Viles, H.A. and Parker, A.G. 1997. Monitoring of rapid salt weathering in the central Namib desert using limestone blocks. *Journal of Arid Environments* **37**, 581–598.

Goudie, A.S., Parker, A.G., Bull, P.A., White, K. and Al-Farraj, A. 2000. Desert loess in Ras Al Khaimah, United Arab Emirates. *Journal of Arid Environments* **46**, 123–135.

Grini, A. and Zender, C.S. 2004. Roles of saltation, sandblasting, and wind speed variability on mineral dust aerosol size distribution during the Puerto Rican Dust Experiment (PRIDE). *Journal of Geophysical Research* **109**, D07202, doi:10.1029/2003JD004233.

Grousset, F.E. and Biscaye, P.E. 2005. Tracing dust sources and transport patterns using Sr, Nd and Pb isotopes. *Chemical Geology* **222**, 149–167.

Grousset, F.E., Ginoux, P., Bory, A. and Biscaye, P.E. 2003. Case study of a Chinese dust plume reaching the French Alps. *Geophysical Research Letters* **30**, Art. No. 1277.

Guo, Z.T., Ruddiman, W.F., Hao, Q.Z., Wu, H.B., Qiao, Y.S., Zhu, R.X., Peng, S.Z., Wei, J.J., Yuan, B.Y., Liu, T.S. 2002. Onset of Asian desertification by 22 Myr ago inferred from loess deposits in China. *Nature* **416**, 159–163.

Haberlah, D. 2007. A call for Australian loess. *Area* **39** 224–229.

Hao, Q.Z., Oldfield, F., Bloemendal, J., Guo, Z.T. 2008. The magnetic properties of loess and paleosol samples from the Chinese Loess Plateau spanning the last 22 million years. *Palaeogeography Palaeoclimatology Palaeoecology* **260**, 389–404.

Herman, J.R., Bhartia, P.K., Torres, O., Hsu, C., Seftor, C. and Celarier, E. 1997. Global distribution of UV-absorbing aerosols from Nimbus-7/TOMS data. *Journal of Geophysical Research* **102**, 16,911–16,922.

Hesse, P.P. and McTainsh, G.H. 1999. Last glacial maximum to Early Holocene wind strength in the mid-latitudes of the Southern Hemisphere from aeolian dust in the Tasman Sea. *Quaternary Research* **52**, 343–349.

Hesse, P.P. and McTainsh, G.H. 2003. Australian dust deposits: modern processes and the Quaternary record. *Quaternary Science Reviews* **22**, 2007–2035.

Hesse, P.P., Humphreys, G.S., Smith, B.L., Campbell, J. and Peterson, E.K. 2003. Age of loess deposits in the Central Tablelands of New South Wales. *Australian Journal of Soil Research* **41**, 1115–1131.

Hess, G.D. and Spillane, K.T. 1990. Characteristics of dust devils in Australia. *Journal of Applied Meteorology* **27**, 305–317.

Hunt, J.C.R. and Weber, A.H. 1979. A Lagrangian statistical analysis of diffusion from a ground-level source in a turbulent boundary layer. *Quarterly Journal of the Royal Meteorological Society* **105**, 423–443.

Jickells, T.D., An, Z.S., Andersen, K.K., Baker, A.R., Bergametti, G., Brooks, N., Cao, J.J., Boyd, P.W., Duce, R.A., Hunter, K.A., Kawahata, H., Kubilay, N., laRoche, J., Liss, P.S., Mahowald, N., Prospero, J.M., Ridgwell, A.J., Tegen, I. and Torres, R. 2005. Global iron connections between desert dust, ocean biogeochemistry and climate. *Science* **308**, doi.10.1126.

Kalashnikova, O.V. and Sokolik, I.N. 2004. Modeling the radiative properties of nonspherical soil-derived mineral aerosols. *Journal of Quantitative Spectroscopy and Radiative Transfer* **87**, 137–166.

Kaufman, Y.J., Koren, I., Remer, L.A., Tanre, D., Ginoux, P. and Fan, S. 2005. Dust transport and deposition observed from the Terra-Moderate Resolution Imaging Spectroradiometer (MODIS) spacecraft over the Atlantic Ocean, *Journal of Geophysical Research – Atmospheres* **110** (D10), D10S12.

Kemp, R.A. 1999. Micromorphology of loess-paleosol sequences: a record of paleoenvironmental change. *Catena* **35**, 179–196.

Kemp, R.A. 2001. Pedogenic modification of loess: significance for palaeoclimatic reconstructions. *Earth-Science Reviews* **54**, 145–156.

Kohfeld, K.E. and Harrison, S.P. 2001. DIRTMAP: the geological record of dust. *Earth-Science Reviews* **54**, 81–114.

Koren, I. and Kaufman, Y.J. 2004. Direct wind measurements of Saharan dust events from Terra and Aqua satellites. *Geophysical Research Letters* **31**, L06122, doi:10.1029/2003GL019338.

Kuenen, P.H. 1960. Experimental abrasion 4: aeolian action. *Journal of Geology* **68**, 427–449.

Kukla, G., Heller, F., Ming, L.X., Chun, X.T., Sheng, L.T. and Sheng, A.Z. 1988. Pleistocene climaes in China dated by magnetic susceptibility. *Geology* **16**, 811–814.

Kumar, R., Jefferson, I.F., O'Hara-Dhand, K. and Smalley, I.J. 2006. Controls on quartz silt formation by crystalline defects. *Naturwissenschaften* **93**, 185–188.

Lancaster, N. 2002. Flux of eolian sediment in the McMurdo Dry Valleys, Antarctica: a preliminary assessment. *Arctic, Antarctic and Alpine Research* **34**, 3, 318–323.

Lautridou, J.-P. and Ozouf, J.C. 1982. Experimental frost shattering: fifteen years of research at the Centre de Geomorphologie du CNRS. *Progress in Physical Geography* **6**, 215–232.

Le Roux, J.P. 2002. Shape entropy and settling velocity of natural grains. *Journal of Sedimentary Research* **73**, 363–366.

Leslie, L.M. and Speer, M.S. 2006. Modelling dust transport over central eastern Australia. *Meteorological Applications* **13**, 141–167.

Liu, T.S. 1985. *Loess and the Environment*. Beijing: China Ocean Press.

Liu, Q.S., Deng, C.L., Torrent, J. and Zhu, R.X. 2007. Review of recent developments in mineral magnetism of the Chinese loess. *Quaternary Science* **26** 368–385.

Li, X.Y., Liu, L.Y., Gao, S.Y., Shi, P.J., Zou, X.Y., Hasi, E. and Yan, P. 2005. Aeolian dust accumulation by rock fragment substrata: influence of number and textural composition of pebble layers on dust accumulation, *Soil and Tillage Research* **84**, 139–144.

Livingstone, I. and Warren, A. 1996. *Aeolian Geomorphology: an introduction*. London: Longman.

Luo, C., Mahowald, N.M. and del Corral, J. 2003. Sensitivity study of meteorological parameters on mineral aerosol mobilization, transport and distribution. *Journal of Geophysical Research* **108** (D15), 4447. Doi:10.1029/2003JD003483.

McFadden, L.D., Wells, S.G. and Jercinovich, M.J. 1987. Influences of eolian and pedogenic processes on the origin and evolution of desert pavements. *Geology* **15**, 504–508.

McTainsh, G.H. 1985. Dust processes in Australia and West Africa: a comparison. *Search* **16**, 104–106.

McTainsh, G. 1987. Desert loess in northern Nigeria. *Zeitschrift für Geomorphologie* **31**, 145–165.

McTainsh, G.H. and Pitblado, J.R. 1987. Dust storms and related phenomena measured from meteorological records in Australia. *Earth Surface Processes and Landforms* **12**, 415–424.

McTainsh, G.H., Leys, J.F. and Nickling, W.G. 1999. Wind erodibility of arid lands in the Channel Country of western Queensland, Australia. *Zeitschrift für Geomorphologie NF Supplementband* **116**, 113–130.

McTainsh, G., Chan, Y.-C, McGowan, H., Leys, J. and Tews, K. 2005. The 23rd October 2002 dust storm in eastern Australia: characteristics and meteorological conditions. *Atmospheric Environment* **39**, 1227–1236.

Mahowald, N.M., Bryant, R.G., del Corral, J. and Steinberger, L. 2003. Ephemeral lakes and desert dust sources. *Geophysical Research Letters* **30**, art.no. 1074.

Martinson, D.G., Pisias, N.G., Hays, J.D., Imbrie, J., Moore, Jr. T.C. and Shackleton, N.J. 1987. Age dating and the orbital theory of the ice ages: development of a high-resolution 0 to 3,000,000-year chronostratigraphy. *Quaternary Research* **27**, 1–29.

Mason, J.A., Nater, E.A., Zanner, C.W. and Bell, J.C. 1999. A new model of topographic effects on the distribution of loess. *Geomorphology* **28**, 223–236.

Mattsson, J.O., Nihlén, T. and Yue, W. 1993. Observations of dust devils in a semi-aird district of southern Tunisia. *Weather* **48**, 359–363.

Meloni, D., di Sarra, A., Di Iorio, T. and Fiocco, G. 2005. Influence of the vertical profile of Saharan dust on the visible direct radiative forcing. *Journal of Quantitative Spectroscopy and Radiative Transfer* **93**, 397–413.

Michaels, T.I. 2006. Numerical modeling of Mars dust devils: albedo track generation. *Geophysical Research Letters* **33**, art no. L19S08.

Miller, R.L., Tegen, I. and Perlwitz, J. 2004. Surface radiative forcing by soil dust aerosols and the hydrologic cycle. *Journal of Geophysical Research* **106** (D16), 18193. Doi:10.1029/2003JD004085.

Miller, S.D., Kuciauskas, A.P., Liu, M., Ji, Q., Reid, J.S., Breed, D.W., Walker, A.L., and Al Mandoos, A. 2008. Haboob dust storms of the southern Arabian Peninsula. *Journal of Geophysical Research* **113** (D1), art.no. D01202.

Morales, C. (ed.) 1979. *Saharan Dust: mobilization, transport, deposition.* New York: Wiley.

Moreno, T., Querol, X., Castillo, S., Alastuey, A., Cuevas, E., Herrmann, L., Mounkaila, M., Elvira, J. and Gibbons, W. 2006. Geochemical variations in aeolian mineral particles from the Sahara-Sahel Dust Corridor. *Chemosphere* **65**, 261–270.

Nalpanis, P. 1985. Saltating and suspended particles over flat and sloping surfaces. II. Experiments and numerical simulations. In: Barndorff-Nielsen, O.E., Møller, J.T., Romer-Rasnussen, K., Willetts, B.B. (eds.) *Proceedings of an International Workshop on the Physics of Blown Sand.* Department of Theoretical Statistics, Institute of Mathematics, University of Aarhus, Memoir 8, 1, 37–66.

Nettleton, W.D. and Chadwick, O.A. 1996. Late Quaternary, re-deposited loess-soil developmental sequences, south Yemen. *Geoderma* **70**, 21–36.

Nickling, W.G. and Gillies, J.A. 1993. Dust emission and transport in Mali, Africa. *Sedimentology* **40**, 859–868.

Nickling, W.G., McTainsh, G.H. and Leys, J.F. 1999. Dust emissions from the Channel Country of western Queensland, Australia. *Zeitschrift für Geomorphologie NF Supplementband* **116**, 1–17.

Nugteren, G., Vandenberghe, J., Van Huissteden, J.K. and An, Z. 2004. A Quaternary climate record based on grain size analysis from the Luochuan loess section on the Chinese Loess Plateau, China. *Global and Planetary Change* **41**, 167–183.

Offer, Z.Y. and Goossens, D. 2001. Ten years of aeolian dust dynamics in a desert region (Negev Desert, Israel): analysis of airborne dust concentration, dust accumulation and high-magnitude dust events. *Journal of Arid Environments* **47**, 211–249.

O'Hara, S.L., Clarke, M.L. and Eltrash, M.S. 2006. Field measurements of desert dust deposition, *Atmospheric Environment* **40**, 3881–3897.

Oke, A.M.C., Dunkerley, D. and Tapper, N.J. 2007a. Willy-willies in the Australian landscape: the role of key meteorological variables and surface conditions in defining frequency and spatial characteristics. *Journal of Arid Environments* **71**, 201–215.

Oke, A.M.C., Dunkerley, D. and Tapper, N.J. 2007b. Willy-willies in the Australian landscape: sediment transport characteristics. *Journal of Arid Environments* **71**, 216–228.

Perlwitz, J., Tegen, I. and Miller, R.L. 2001. Interactive soil dust aerosol model in the GISS GCM: 1. Sensitivity of the soil dust cycle to radiative properties of soil dust aerosols. *Journal of Geophysical Research* **106** (18), 167–18,192.

Piketh, S.J., Tyson, P.D., Steffen, W. 2000. Aeolian transport from southern Africa and iron fertilization of marine biota in the South Indian Ocean. *South African Journal of Science* **96**, 244–246.

Porter, S.C. 2001. Chinese loess record of monsoon climate during the last glacial-interglacial cycle. *Earth-Science Reviews* **54**, 115–128.

Porter, S.C. and An, Z. 1995. Correlation between climate events in the North Atlantic and China during the last glaciation. *Nature* **375**, 305–308.

Prospero, J.M. 1999. Long-term measurements of the transport of African mineral dust to the southeastern United States: implications for regional air quality. *Journal of Geophysical Research – Atmospheres* **104**, 15917.

Prospero, J.M., Glaccum, R.A. and Nees, R.T. 1981. Atmospheric transport of soil dust from Africa to South-America. *Nature* **289**, 570–572.

Prospero, J.M., Ginoux, P., Torres, O., Nicholson, S.E. and Gill, T.E. 2002. Environmental characterization of global sources of atmospheric soil dust identified with the Nimbus-7 Total Ozone Mapping Spectrometer (TOMS) absorbing aerosol product. *Reviews of Geophysics* **40**, doi: 10.1029/2000RG000095.

Pye, K. 1987. *Aeolian dust and dust deposits.* London: Academic Press.

Pye, K. 1995. The nature, origin and accumulation of loess. *Quaternary Science Reviews* **14**, 653–667.

Pye, K. and Paine, A.D.M. 1984. Nature and source of aeolian deposits near the summit of Ben Arkle, northwest Scotland. *Geologie en Mijnbouw* **63**, 13–18.

Pye, K. and Sperling, C.H.B. 1983. Experimental investigation of silt formation by static breakage processes: the effect of temperature, moisture and salt on quartz dune sand and granitic regolith. *Sedimentology* **30**, 49–62.

Pye, K. and Tsoar, H. 1987. The mechanics and geological implications of dust transport and deposition in deserts with particular reference to loess formation and dune sand diagenesis in the northern Negev, Israel, In: Frostick, L. and Reid I. (eds.) *Desert Sediments: Ancient and Modern. Geological Society Special Publication* **35**, 139–156.

Reheis, M.C. and Kihl, R. 1995. Dust deposition in southern Nevada and California, 1984–1989: relations to climate, source area and source lithology. *Journal of Geophysical Research* **100** (D5), 8893–8918.

Reynolds, R.L., Reheis, M., Yount, J. and Lamothe, P. 2006. Composition of aeolian dust in natural traps on isolated surfaces of the central Mojave Desert – Insights to mixing, sources, and nutrient inputs. *Journal of Arid Environments* **66**, 42–61.

Rivera Rivera, N.I., Bleiweiss, M.P., Hand, J.L. and Gill, T.E. 2006. Characterization of dust storm sources in southwestern US and northwestern Mexico using remote sensing imagery. 14th Conference on Satellite Meteorology and Oceonography, American Meteorological Society, February 2006, 17pp.

Rostagno, C.M. 2007. Fire and wind erosion in the northeastern portion of Patagonia. Abstract. Multidisplinary workshop on southern South American dust, Puerto Madryn, Argentina, October 3–5, 2007.

Sarnthein, M. and Koopman, B. 1980. Late Quaternary deep-sea record of northwest African dust supply and wind circulation. *Palaeoecology of Africa* **12**, 239–253.

Satheesh, S.K. and Moorthy, K.K. 2005. Radiative effects of natural aerosols: a review. *Atmospheric Environment* **39**, 2089–2110.

Shao, Y. 2000. *Physics and Modelling of Wind Erosion.* Dordrecht: Kluwer Academic.

Shao, Y. 2001. A model for mineral dust erosion. *Journal of Geophysical Research* **106**, 20, 239–20,254.

Shao, Y., McTainsh, G.H., Leys, J.F. and Raupach, M.R. 1993a. Efficiency of sediment samplers for wind erosion measurement. *Australian Journal of Soil Research* **31**, 519–532.

Shao, Y., Raupach, M.R. and Findlater, P.A. 1993b. Effect of saltation bombardment on the entrainment of dust by wind. *Journal of Geophysical Research* **98**, 12719–12726.

Shao, Y., Raupach, M.R. and Leys, J.F. 1996. A model for predicting aeolian sand drift and dust entrainment on scales from paddock to region. *Australian Journal of Soil Research* **34**, 309–342.

Shao, Y., Leys, J.F., McTainsh, G.H. and Tews, K. 2007. Numerical simulation of a dust event in Australia. *Journal of Geophysical Research-Atmospheres* **112**, D8, D08207.

Singhvi, A.K., Bluszcz, A., Bateman, M.D. and Rao, M.S. 2001. Luminescence dating of loess-palaeosol sequences and coversands: methodological aspects and palaeoclimatic implications. *Earth-Science Reviews* **54**, 193–211.

Smalley, I. 1995. Making the material: the formation of silt-sized primary mineral particles for loess deposits. *Quaternary Science Reviews* **14**, 645–651.

Smalley, I. and Derbyshire, E. 1990. The definition of 'ice-sheet' and 'mountain' loess. *Area* **22**, 300–301.

Smith, B.J., Warke, P.A. and Moses, C.A. 2000. Limestone weathering in a contemporary arid environment: a case study from southern Tunisia. *Earth Surface Processes and Landforms* **25**, 1343–1354.

Smith, B.J., Wright, J. and Whalley, W.B. 2002. Sources of nonglacial, loess-size quartz silt and the origins of "desert loess". *Earth-Science Reviews* **59**, 1–26.

Sokolik, I.N., Winker, D.M., Bergametti, G., Gillette, D.A., Carmichael, G., Kaufman, Y.J., Gomes, L., Schuetz, L. and Penner, J.E. 2001. Introduction to special section: outstanding problems in quantifying the radiative impacts of mineral dust. *Journal of Geophysical Research – Atmospheres* **106** (D16), 18015–18027.

Soderberg, K. and Compton, J.S. 2007. Dust as a nutrient source for fynbos ecosystems, South Africa. *Ecosystems* **10**, 550–561.

Sow, M., Goossens, D. and Rajot, J.L. 2006. Calibration of the MDCO dust collector and of four versions of the inverted Frisbee dust deposition sampler. *Geomorphology* **82**, 360–375.

Stevens, T., Armitage, S.J., Lu, H.Y. and Thomas, D.S.G. 2006. Sedimentation and diagenesis of Chinese loess: implications for the preservation of continuous, high-resolution climate records. *Geology* **34**, 849–852.

Stoertz, G. and Ericksen, G. 1974. Geology of salars in northern Chile. *USGS Professional Paper*, **811**, 65pp.

Sun, D.H., Bloemendal, J., Rea, D.K., An, Z.S., Vandenberghe, J., Lu, H.Y., Su, R.X. and Liu, T.S. 2004. Bimodal grain-size distribution of Chinese loess, and its palaeoclimatic implications. *Catena* **55**, 325–340.

Sun, J.M. 2002. Provenance of loess material and formation of loess deposits on the Chinese Loess Plateau. *Earth and Planetary Science Letters* **203**, 845–859.

Sun, J. M. and Huang, X.G. 2006. Half-precessional cycles recorded in Chinese loess: response to low-latitude insolation forcing during the Last Interglaciation. *Quaternary Science Reviews* **25**, 1065–1072.

Sun, J., Kohfeld, K.E. and Harrison, S.P. 2000. *Records of aeolian dust deposition on the Chinese Loess Plateau during the Late Quaternary.* Jena, Germany: Max Planck Institute for Biogeochemistry.

Tanaka, T.Y. 2005. Possible transcontinental dust transport from North Africa and the Middle East to East Asia. *Atmospheric Environment* **39**, 3901.

Tanaka, T.Y. and Chiba, M. 2006. A numerical study of the contributions of dust source regions to the global dust budget. *Global Planetary Change* **52**, 88–104.

Tegen, I. and Fung, I. 1994. Modeling of mineral dust in the atmosphere: sources, transport and optical thickness. *Journal of Geophysical Research – Atmospheres* **99** (D16), 22897–22914.

Tegen, I., Heinold, B., Todd, M., Helmert, J., Washington, R. and Dubovik, O. 2006. Modelling soil dust aerosol in the Bodélé depression during the BoDEx campaign. *Atmospheric Chemistry and Physics Discussions* **6**, 4171–4211.

Tetzlaff, G. and Peters, M. 1986. Deep-sea sediments in the eastern equatorial Atlantic of the African coast and meteorological flow patterns over the Sahel. *Geologische Rundschau* **75**, 71–79.

Towner, M.C., Ringrose, R.J., Patel, M.R. Balme, M., Metzger, S.M., Greeley, R. and Zarnecki, J.C. 2004. A close encounter with a terrestrial dust devil. *Lunar and Planetary Science* **XXXV**, 1259.

Viles, H.A. and Goudie, A.S. 2000. Weathering, geomorphology and climatic variability in the central Namib desert. In: McLaren, S.J. and Kniveton, D.R. (eds.) *Linking Climate Change to Land Surface Change.* Amsterdam: Kluwer, 65–82.

Viles, H.A. and Goudie, A.S. 2007. Rapid salt weathering in the coastal Namib Desert: implications for landscape development. *Geomorphology* **85**, 49–62.

Walker, T.R. 1979. Red colour in eolian sand. In: E.D. McKee (ed.) *A Study of Global Sand Seas*. USGS Professional Paper **1052**, 62–81.

Washington, R. and Todd, M.C. 2005. Atmospheric controls on mineral dust emission from the Bodélé Depression, Chad: the role of the low level jet. *Geophysical Research Letters* **32**, L17701, doi:10.1029/2005GL023597.

Washington, R., Todd, M., Middleton, N.J. and Goudie, A.S. 2003. Dust-storm source areas determined by the Total Ozone Monitoring Spectrometer and surface observations. *Annals of the Association of American Geographers* **93**, 297–313.

Washington, R., Todd, M.C., Lizcano, G., Tegen, I., Flamant, C., Koren, I., Ginoux, P., Engelstaedter, S., Britsow, C.S., Zender, C.S., Goudie, A.S., Warren, A. and Prospero, J.M. 2006. Links between topography, wind, deflation, lakes and dust; the case of the Bodélé Depression, Chad. *Geophysical Research Letters* **33**, L09401, doi:10.1029/2006GL025827.

Wentworth, C.K. 1922. A scale of grade and class terms for clastic sediments. *Journal of Geology* **30**, 377–392.

Werner, M., Tegen, I., Harrison, S.P., Kohfeld, K.E., Prentice, I.C., Balkanski, Y., Rodhe, H. and Roelandt, C. 2002. Seasonal and interannual variability of the mineral dust cycle under present and glacial climate conditions. *Journal of Geophysical Research* **107** (D24), 4744. Doi: 10.1029/2002JD002365.

Whalley, W.B., Marshall, J.R. and Smith, B.J. 1982. Origin of desert loess from some experimental observations. *Nature* **300**, 433–435.

Whalley, W.B., Smith, B.J., McAlister, J.J. and Edwards, A.J. 1987. Aeolian abrasion of quartz particles and the production of silt-size fragments: preliminary results. In: L. Frostick and I. Reid (eds.) *Desert Sediments: ancient and modern*. Geological Society Special Publication **35**, 129–138.

Wigner, K.A. and Peterson, R.E. 1987. Synoptic climatology of blowing dust on the Texas South Plains. *Journal of Arid Environments* **13**, 199–209.

Wopfner, H. and Twidale, C.R. 2001. Australian desert dunes: wind rift or depositional origin? *Australian Journal of Earth Sciences* **48**, 239–244.

Wright, J.S. 1993. Non-glacial origins of loess-sized quartz silt. Unpublished PhD thesis, 1993, The Queen's University of Belfast, 502 pp.

Wright, J.S. 2001a. Making loess-sized quartz silt: data from laboratory simulations and implications for sediment transport pathways and the formation of 'desert' loess deposits associated with the Sahara. *Quaternary International* **76/77**, 7–19.

Wright, J.S. 2001b. 'Desert' loess versus 'glacial' loess: quartz silt formation, source areas and sediment pathways in the formation of loess deposits. *Geomorphology* **36**, 231–256.

Wright, J.S. 2007. An overview of the role of weathering in the production of quartz silt. *Sedimentary Geology* **202**, 337–351.

Wright, J.S., Smith, B.J. and Whalley, W.B. 1998. Mechanisms of loess-sized quartz silt production and their relative effectiveness: laboratory simulations. *Geomorphology* **23**, 15–34.

Yaalon, D.H. 1991. Mountain loess is not a suitable term. *Area* **23**, 255–256.

Yaalon, D.H. and Dan, J. 1974. Accumulation and distribution of loess-derived deposits in the semi-arid desert-fringe areas of Israel. *Zeitschrift für Geomorphologie Supplementband* **20**, 27–32.

Yaalon, D.H. and Ganor, E. 1975. Rate of aeolian dust accretion in the Mediterranean and desert fringe environments of Israel. 19th International Congress on Sedimentology Theme 2, 169–174.

Yuan, W. and Zhang, J. 2006. High correlations between Asian dust events and biological productivity in the western North Pacific. *Geophysical Research Letters* **33**, L07603.

Zender, C.S., Bian, H. and Newman, D. 2003a. Mineral Dust Entrainment and Deposition (DEAD) model: description and 1990s dust climatology. *Journal of Geophysical Research* **108** (D17), 4543, doi: L10.1029/2002/JD002775.

Zender, C.S., Newman, D. and Torres, O. 2003b. Spatial heterogeneity in aeolian erodibility: uniform, topographic, geomorphic and hydrologic hypoetheses. *Journal of Geophysical Research* **108**, doi: 10.1029/ 2002JD003039.

Zender, C.S., Miller, R.L. and Tegen, I. 2004. Quantifying mineral dust mass budgets: terminology, constraints and current estimates. *Eos* **85** (48), 30 November 2004.

Part VIII
Climatic Change

Chapter 21

Rock Varnish and its Use to Study Climatic Change in Geomorphic Settings

Ronald I. Dorn

Introduction

The dusky brown to black coating of rock varnish dominates bare rock surfaces of many desert landforms (Oberlander, 1994). Thicknesses less than even 0.020 mm (or 20 micrometres, μm) are enough to darken light-colored rock types (Fig. 21.1). The gradual pace of change on many desert landforms permits the slow accretion of rock varnish at rates of a few micrometres per thousand years (Dorn, 1998; Liu and Broecker, 2000). Just about any rock type will accumulate varnish, in so long as rock-surface erosion is slow enough to permit varnish accretion.

Rock varnish formation starts when bacteria oxidize and concentrate manganese (Mn) and iron (Fe) (Dorn and Oberlander, 1981; Krumbein and Jens, 1981; Palmer et al., 1985; Hungate et al., 1987; Jones, 1991; Drake et al., 1993; Dorn, 1998; Krinsley, 1998). Although random collection of rock varnish reveals an impressive array of dozens of different types of microorganisms and their molecular by-products (Nagy et al., 1991; Dragovich, 1993; Perry et al., 2002; Kuhlman et al., 2005; Kuhlman et al., 2006), the bacteria that actually initiate varnish grow very infrequently (Dorn, 1998). Hence, rock varnish accumulates very slowly. If all of the organisms purported to cause varnish formation (Taylor-George et al., 1983; Eppard et al., 1996; Perry et al., 2003) actually contributed, rates of formation would reach millimetres per thousand years.

Simple mathematics helps to visualize constraints on varnish formation. With both sides of an oxide-encrusted sheath of bacteria being ca. 0.3 μm thick and with clay minerals making up about 70% of varnish (Potter and Rossman, 1977; Potter and Rossman, 1979c; Krinsley et al., 1995; Dorn, 1998), only a few varnish-producing bacteria need to grow every thousand years to produce varnish. Five Mn-enhancing bacteria per hundred years is enough to generate a very fast-accumulating varnish in deserts. This means that the budding bacteria (Hirsch, 1974) making varnish (Dorn and Oberlander, 1982) can wait in a dormant state until a gentle wetting event takes place, and clear visual *in situ* evidence reveals they then enhance the Mn-Fe that glues clay minerals to rock surfaces (Dorn, 1998; Krinsley, 1998).

The details of varnish accumulation are important to understand in utilizing varnish as a desert geomorphic tool able to decipher climatic change. During wetter conditions, Mn enhancement is favored over Fe (Dorn, 1990; Jones, 1991; Cremaschi, 1996; Broecker and Liu, 2001). Mn fits into the crystalline structure of interstratified clay minerals that come to rest as dust on rock surfaces. The hexagonal arrangement of oxygen in clay acts as a template (Potter, 1979: 174–175) for the nanometre-scale bits of Mn that derive from bacterial sheaths (Fig. 21.2). The net effect is to form the type of Mn-mineral found in rock varnish, typically birnessite (Potter and Rossman, 1979a,b; McKeown and Post, 2001). This process of nanometre-scale varnish accretion is summarized in Fig. 21.2.

The slow and steady accumulation of rock varnish forms the basis of its use as a powerful tool to understand desert geomorphology. Mn-rich microlaminations (Fig. 21.3) result from wetter times in a

R.I. Dorn (✉)
School of Geographical Sciences, Arizona State University, Tempe, AZ 85287, USA
e-mail: ronald.dorn@asu.edu

A.J. Parsons, A.D. Abrahams (eds.), *Geomorphology of Desert Environments*, 2nd ed., DOI 10.1007/978-1-4020-5719-9_21, © Springer Science+Business Media B.V. 2009

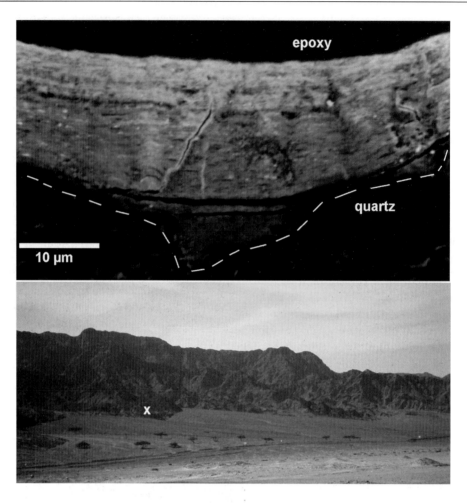

Fig. 21.1 Rock varnish dramatically darkens the appearance of desert landforms. This backscattered (BSE) electron microscope image derives from the "x" on the corresponding ground photo-graph of varnished slopes, Sinai Peninsula. The greater brightness of rock varnish in the BSE image reflects the higher atomic number of the manganese and iron in rock varnish

desert, because these conditions are more favorable for the growth of the Mn-enhancing bacteria. When the climate dries with fewer gentle winter-season wetting events, varnish does not stop growing. Clay minerals still deposit in dust. Mn-enhancing bacteria still grow. However, the ratio of Mn to Fe drops with the greater aridity, due to more alkaline materials deposited on desert surfaces (Dorn, 1990; Jones, 1991; Cremaschi, 1996; Broecker and Liu, 2001; Lee and Bland, 2003). The net effect is that drier periods generate varnish microlaminations with a higher percentage of clays and lower abundance of Mn (Fig. 21.3).

The next section of this chapter presents a revolution in varnish palaeoclimatic research brought about in just the past few years.

Microlamination Revolution

Orange and black varnish microlaminations (VML) were first reported three decades ago (Perry and Adams, 1978), although Charles B. Hunt showed me thin section images with these microlaminations dated to 1958. Others then started to explore the potential for VML to serve as a palaeoclimatic indicator with Mn-rich layers indicating wetter conditions (Dorn, 1984; Dorn, 1990; Cremaschi, 1992, 1996; Diaz et al., 2002; Lee and Bland, 2003; Thiagarajan and Lee, 2004). This prior research, however, represented only a tiny fraction of the work needed to connect VML to specific palaeoclimatic events.

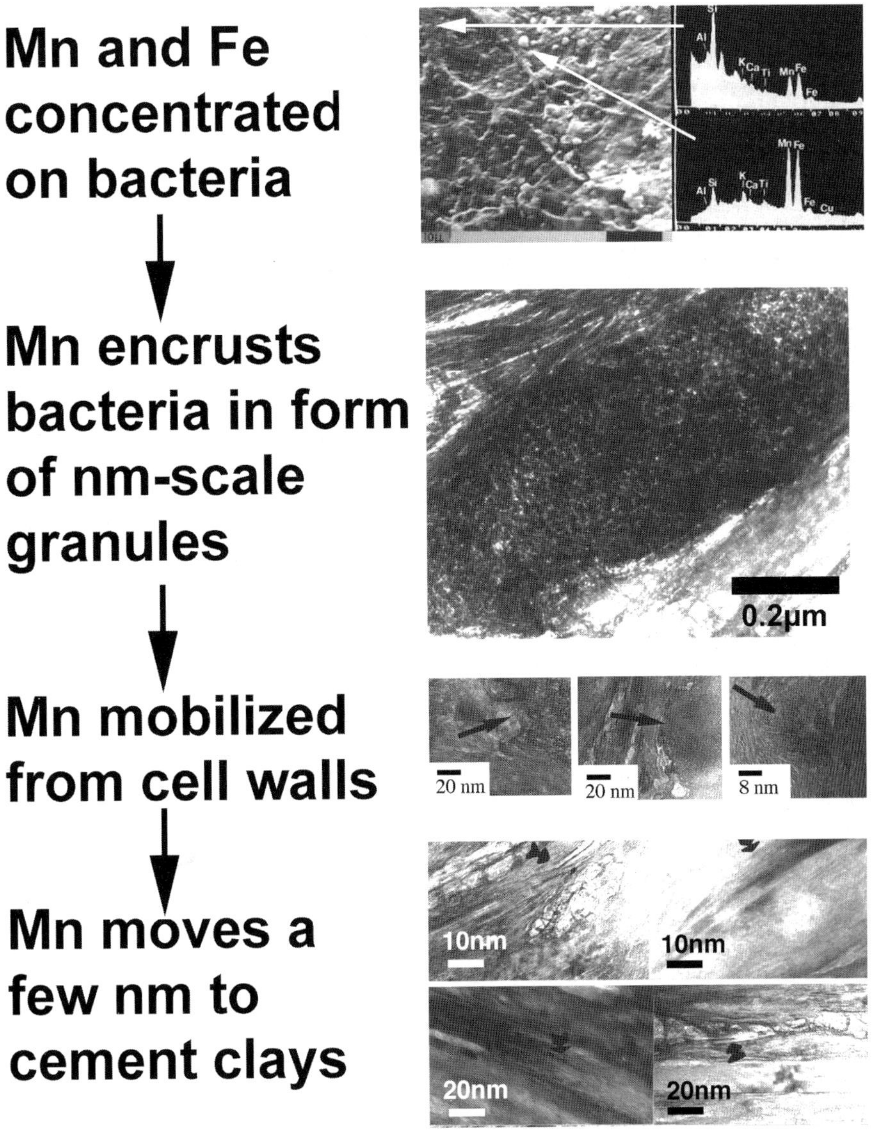

Mn and Fe concentrated on bacteria

↓

Mn encrusts bacteria in form of nm-scale granules

↓

Mn mobilized from cell walls

↓

Mn moves a few nm to cement clays

Fig. 21.2 Varnish formation starts with the oxidation and concentration of Mn (and Fe) by bacteria. Gentle wetting events provide enough carbonic acid to dissolve nanometre-scale fragments of Mn. Ubiquitous desert dust supplies interstratified clay minerals. The Mn-bacterial fragments fit into the weathered edges of clays, cementing clays much like mortar cements brick. The upper image is from secondary electrons, and the others are high resolution transmission electron microscopy (HRTEM). A more detailed discussion of alternative hypotheses of varnish formation can be found elsewhere (Dorn, 2007)

Tanzhuo Liu started to unravel the puzzle of VML in the early-1990s as a part of his dissertation applied to desert geomorphology (Liu, 1994; Liu and Dorn, 1996). "TL" to his colleagues then took a post-doctoral position at Lamont-Doherty Earth Observatory and continued to analyze VML from rock varnishes around the world (Liu and Broecker, 2000; Liu et al., 2000; Zhou et al., 2000; Broecker and Liu, 2001; Liu, 2003; Liu and Broecker, 2007, 2008a, 2008b).

Focusing on geomorphic surfaces of known age, Liu analyzed ultrathin cross sections of tens of thousands of rock-surface microbasins filled with rock varnish. These sedimentary deposits were then evaluated through light microscopy, backscattered-electron microscopy, and microchemical mapping methods.

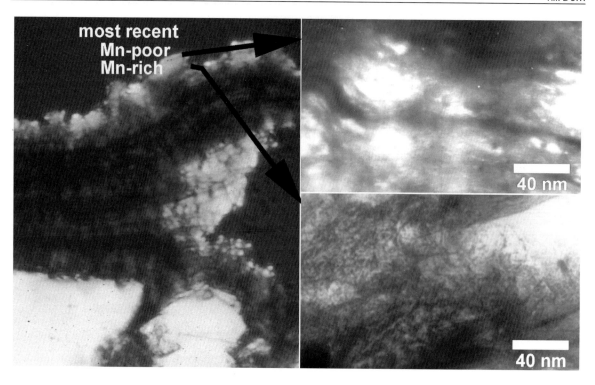

Fig. 21.3 Mn-rich and Fe-rich microlaminae of rock varnish from the Coso Range of eastern California. The left image is a light microscope image of alternating Mn-rich and Fe-rich layers. The upper right image is a HRTEM view of the most recent Mn-poor layer dominated by laminated clay minerals. In contrast, the most recent Mn-rich layer directly underneath still displays the granular texture of bacterial casts that are remobilizing into adjacent clay minerals while in nanoscale disequilibria

Just as detailed analysis of varves led to a revolution in understanding of deglaciation chronology (De Geer, 1930), and painstaking analysis of grains in sediment cores in the North Atlantic led to a revolution in understanding of iceberg armadas related to global cold snaps (Heinrich, 1988), Liu's work over the past decade has led to a similar revolution in rock varnish palaeoclimatic research.

Liu subjected his VML method to a blind test administered by Richard Marston, editor of *Geomorphology* (Liu, 2003; Marston, 2003; Phillips, 2003). Liu and Fred Phillips analyzed samples from the Mojave Desert for VML and [36]Cl ages, respectively. They then sent their results separately to Marston, who published a summary of the blind test:

> "This issue contains two articles that together constitute a blind test of the utility of rock varnish microstratigraphy as an indicator of the age of a Quaternary basalt flow in the Mohave Desert. This test should be of special interest to those who have followed the debate over whether varnish microstratigraphy provides a reliable dating tool, a debate that has reached disturbing levels of acrimony in the literature. Fred Phillips (New Mexico Tech) utilized cosmogenic [36]Cl dating, and Liu (Lamont-Doherty Earth Observatory, Columbia University) utilized rock varnish microstratigraphy to obtain the ages of five different flows, two of which had been dated in previous work and three of which had never been dated. The manuscripts were submitted and reviewed with neither author aware of the results of the other. Once the manuscripts were revised and accepted, the results were shared so each author could compare and contrast results obtained by the two methods. In four of the five cases, dates obtained by the two methods were in close agreement. Independent dates obtained by Phillips and Liu on the Cima "I" flow did not agree as well, but this may be attributed to the two authors having sampled at slightly different sites, which may have in fact been from flows of contrasting age. Results of the blind test provide convincing evidence that varnish microstratigraphy is a valid dating tool to estimate surface exposure ages." (Marston, 2003: p. 197)

The analysis of VML at sites of known age led to development of calibrations. The best-developed cali-

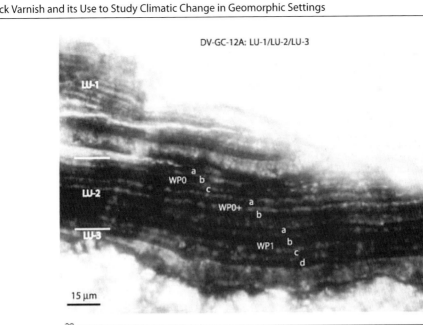

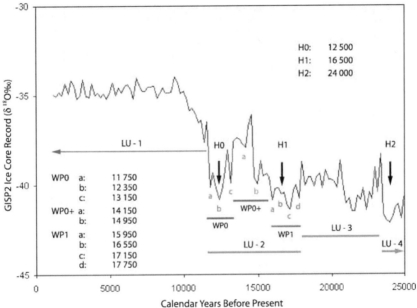

Fig. 21.4 Example of the rock varnish record of the latest Pleistocene wet events in Death Valley, California, and its possible correlation with the GISP2 ice core record in Greenland (Bond et al., 1999). WP = wet event in Pleistocene. LU = Layering Unit; H = Heinrich event. WP0 corresponds with the Younger Dryas (courtesy of Tanzhuo Liu from http://www.vmldatinglab.com/)

brations are for arid regions of the western USA, although Liu's research includes the deserts of Patagonia, China, Australia and elsewhere (Liu, 2008). The late Pleistocene (Figs. 21.4 and 21.5) and Holocene (Figs. 21.6 and 21.7) calibrations for the interior western USA form the basis of exemplars presented in the next section.

Examples of VML Use in Desert Geomorphology

Desert geomorphic research requiring an understanding of both chronology and climatic change can benefit from the VML method. The potential is limited only by

Heinrich Events   Late Pleistocene
(calendar ka)   Microlaminations

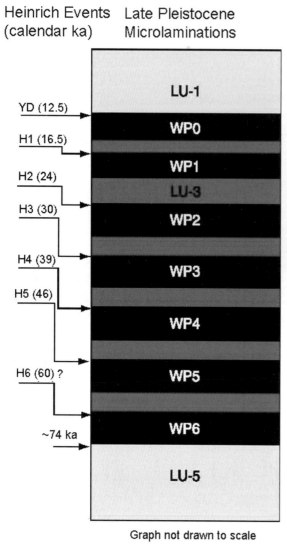

Fig. 21.5 Late Pleistocene calibration of varnish (adapted from Liu, 2003)

Fig. 21.6 Example of a Holocene rock varnish microstratigraphy from Searles Lake in the Mojave Desert, western USA (maximum varnish thickness is 220 microns). The *dark layers* in this image represent wet events that are correlated in time with Holocene millennial-scale cooling events in the North Atlantic region. Preliminary radiometric age calibration indicates that the topmost *dark layer* was deposited during the Little Ice Age around 300–650 calendar years, and the basal *dark layers* formed during a wet/cold period of early Holocene around 10,300 calendar years. Since the climate events recorded in the varnish are contemporaneous on a regional scale, these *dark layers* can be used as a tephrachronology-like dating tool to yield age estimates for varnished features (courtesy of Tanzhuo Liu from http://www.vmldatinglab.com/)

the creativity of the researcher. This section illustrates only a few possibilities.

VML in Desert Weathering Processes

Because rock varnish formation starts after the subaerial exposure of a rock surface, VML can yield unique insights into weathering that results in spalling—exposing new rock faces. One example comes from lightning as a weathering process.

Lightning strikes act as agents of rock weathering and erosion (Karfunkel et al., 2001). In a personal ob-servation of the process, I observed a strike opening preexisting fractures in the host ignimbrite of the Su-perstition Mountains, Arizona. The strike also made a thin fulgurite film from desert dust—fusing dust to the newly opened fracture. Fulgurite crusts are noted else-where (Julien, 1901; Libby, 1986). The net result was a crust that formed on top of pre-existing fissuresol (Villa et al., 1995) (Fig. 21.8).

A preliminary study then ensued to examine the fea-siblity of using VML to study the role of lightning strikes in desert weathering. The crest of the Super-stition Mountains, Sonoran Desert, Arizona served as

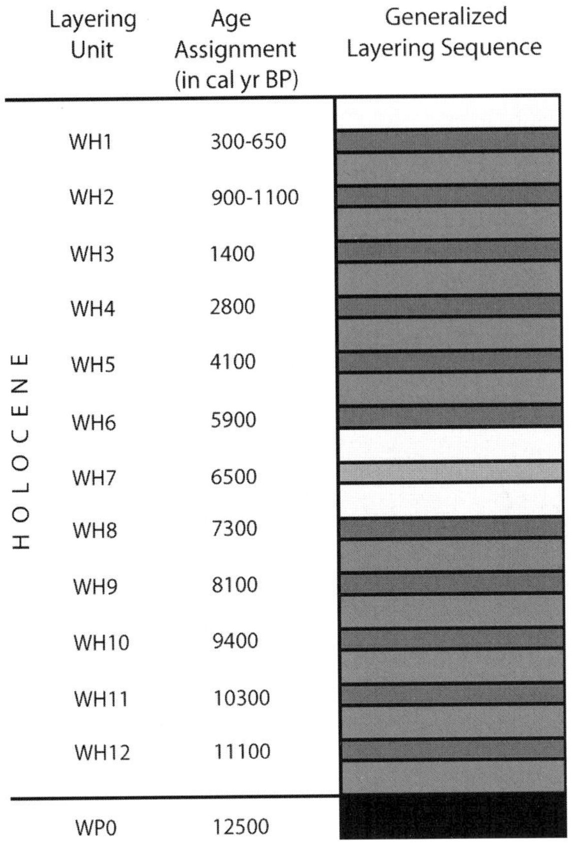

Layering Unit	Age Assignment (in cal yr BP)	Generalized Layering Sequence
WH1	300-650	
WH2	900-1100	
WH3	1400	
WH4	2800	
WH5	4100	
WH6	5900	
WH7	6500	
WH8	7300	
WH9	8100	
WH10	9400	
WH11	10300	
WH12	11100	
WP0	12500	

(Layering Units WH1–WH12 are labelled within the HOLOCENE bracket)

Fig. 21.7 Holocene calibration of varnish microlaminations (adapted from Liu, 2008)

the pilot study area. This rhyolite caldera experiences summer thunderstorms each July through September. Of the 912 spalled fissuresols examined along a 0.45 km transect, 14 displayed evidence of micro-fulgurites. The climatic geomorphology question is whether the frequency of lightning-induced erosion may have been different in the early Holocene, a period of more summer precipitation (Van Devender et al., 1987).

Only Holocene VML sequences (cf. Fig. 21.8) formed on top of these micro-fulgurites. A simple histogram of different VML events failed to reveal any clear indication of temporal clustering (Fig. 21.8). Certainly, a larger number of fused-dust fulgurites will be needed to further test the palaeo-weathering of lightning. However, having the ability to group weathering events in temporal clusters opens the door to new sets of questions related to desert processes that result in detachment.

Another type of desert weathering problem relates to quantifying rates of *in situ* weathering. Consider

the classic phenomenon of weathering rinds. Dryland researchers often measure rind thicknesses, having to assume that rind erosion has a minimal effect (Colman, 1982; Pinter et al., 1994)—an assumption untenable in periglacial settings (Etienne, 2002). The formation of VML on desert surfaces permits the study of rind spalling as a collaborative process in weathering-rind development in deserts.

The keys to solving the conundrum of sorting the effects of spalling from weathering rests in knowing both the age of the desert landform and knowing the last spalling event. Both conditions were met for the Tioga-3 glacial moraine of Bishop creek in the semi-arid Owens Valley of eastern California (Gordon and Dorn, 2005). The age of the landform was established by [36]Cl measurements of morainal boulders (Phillips et al., 1996). Boulder spalling ages were determined by VML. The amount of *in situ* weathering was measured directly underneath the varnish VML by counting pores in backscattered electron microscope imagery (Dorn, 1995b).

At this semi-arid locale, less than 10% sampled granodiorite boulder surfaces remain uneroded (the WP1 samples in Fig. 21.9). These uneroded spots hosted the thickest weathering rinds. In contrast, about half of the boulder surfaces experienced weathering-rind detachment during Holocene, and these have the thinnest rinds (the LU-1 samples in Fig. 21.9). While the conclusion of this study (Gordon and Dorn, 2005) is certainly not surprising, VML analyses now permit research strategies promoting dialogue between *in situ* weathering and weathering-limited detachment.

VML in Desert Soil Processes

Desert pavements are living entities, constantly interacting with adjacent biophysical processes. With each successive pavement-altering process, there is potential for some original clast surface to erode and reset the VML signal. For example, particle alteration might be from spalling (Amit et al., 1993), biodisturbance (McAullife, 1994), water flow (Wainwright et al., 1999), dust interaction (Mabbutt, 1979; Gossens, 2005), and other processes (Dixon, 1994). An important part of pavement studies, therefore, rests in quantifying clast disturbance rates—a problem that has made harder by an inability to gather basic data.

Fig. 21.8 VML formation on fulgurites from the Superstition Mountains crest, central Arizona. (**a**) An observed lightning strike created a sequence of fused dust on Mn-rich fissuresol varnish, on iron film fissuresol, then on laminar calcrete. (**b**) VML patterns on top of fourteen fused-dust fulgurites from the Superstition Mountains crest do not reveal a clear temporal pattern

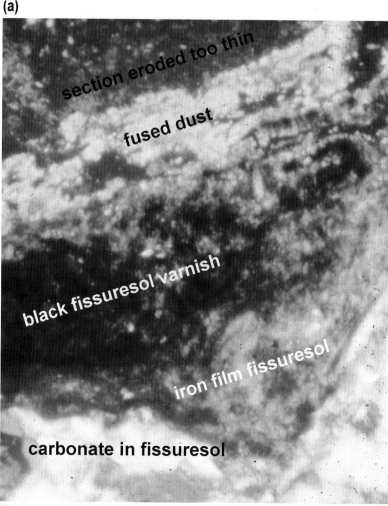

(a)

section eroded too thin

fused dust

black fissuresol varnish

iron film fissuresol

carbonate in fissuresol

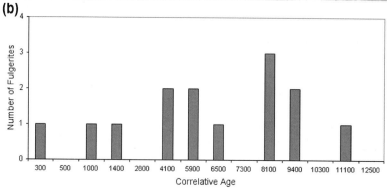

(b)

It has certainly been possible to obtain reasonable ages for the onset of a pavement through cosmogenic nuclide dating of large boulders carried by debris flow (Nishiizumi et al., 1993), as well as finding datable materials in the sediment hosting the pavement (Reheis et al., 1996). VML analysis now provides an oppor-

tunity to obtain detailed information about pavement disturbance over timescales of thousands of years.

A conventional radiocarbon age on woody material at Hanaupah Canyon alluvial fan in Death Valley starts the clock of pavement modification at one site on the fan's late Pleistocene surface. The 24 ka conventional

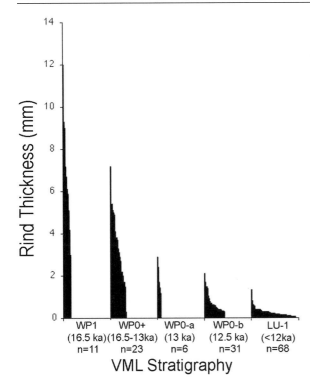

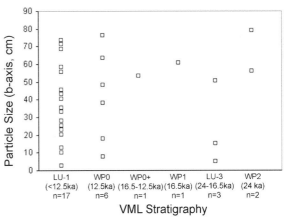

Fig. 21.10 Clast exposure on a 26 ka-old desert pavement in Death Valley, as measured by the late Pleistocene VML calibration (Fig. 21.6)

Fig. 21.9 Distribution of rind thickness under rock varnish thin sections collected from the Tioga-3 glacial moraine of Bishop Creek, Owens Valley, eastern California. Rind thickness was measured by the microporosity in BSE thin sections. VML assignments are based on the late Pleistocene calibration curve, with the goal of providing only broad time ranges. These data show that rinds are constantly eroding, "resetting" the VML clock, even as they thicken over time

^{14}C measurement (Hooke and Dorn, 1992) has a calendar calibration of ca 26,800 years. Thus, undiminished particles should have a VML stratigraphy of WP2 in Fig. 21.5.

In a 64 m^2 plot, randomly selected pavement clasts were examined to assess suitability for VML analysis. A sample size of 30 was reached after evaluating 242 samples. This relatively high ratio of suitable-to-unsuitable samples was possible because the Death Valley alluvial fans are an optimal area for VML dating (Liu and Dorn, 1996). No constraints were placed on particle sizes. The only discriminating issue in clast selection was whether the varnish was appropriate for VML, using criteria established by Liu (1994, 2008).

Only two clasts had a VML sequence as old as the conventional radiocarbon age for the deposit (Fig. 21.10). More than half of the pavement clasts hosted varnishes developed within the past 12.5 ka, as revealed by the presence of only the LU-1 VML

Holocene unit (Fig. 21.10). Eleven clasts experienced enough disturbance to expose surfaces anew in the latest Pleistocene. There does appear to be a slight particle-size effect in surface disturbance, where the most stable pavement surfaces tend to be on the largest clasts (Fig. 21.10). However, VML analysis reveals that it is possible for small pavement clasts to remain stable for long periods.

A different issue related to pavement stability concerns the occurrence of old plant scars. Desert pavements can host perennial plants for a time; and then when those plants die, the reworked fan and aeolian deposits undergo a series of changes that eventually generates a new pavement (Peterson, 1981; Peterson et al., 1995; McAuliffe and McDonald, 2006). Reformation of pavement within ancient plant scars might record ancient droughts (McAuliffe and McDonald, 2006).

To assess the potential of VML to detect old drought periods, thirty old plant scars were sampled on the Shadow Mountain Fan, McDowell Mountains, Scottsdale. These fossil plant scars are noticeable from their circular to oval shape, from the smaller clast sizes inside subtle depressions, and from the VML signal itself. Away from these plant scars, the oldest VML sequence on the fan yields a WP2 or last glacial maxima minimum age. Inside the plant scars, however, the VML sequence is always Holocene. Although the sample size and single site location of this pilot study is limited, pavement clast VML ages do appear to reveal two distinct age clusters (Fig. 21.11). One cluster is in the mid-Holocene, and a second is at the transition

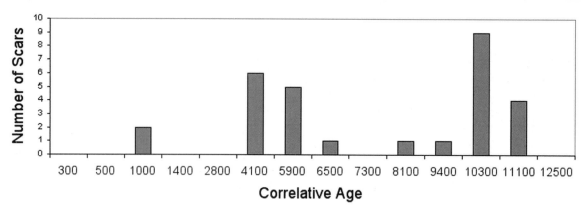

Fig. 21.11 Clast exposure in fossil plant scars from desert pavements, southern McDowell Mountains, Sonoran Desert, Arizona

between the Pleistocene and Holocene. Because the VML signal would reflect an unknown amount of time between plant death and pavement reformation in the scar, the VML signal would necessarily have to be a minimum age for the loss of vegetative cover.

Researchers no longer have to settle for indirect deduction about the ages of pavement-altering events. The following sort of intuitive guess work at inter-site correlations is all too common: "the light to moderate coatings of rock varnish on surfaces of pavement clasts within depressions are very similar to the degree of varnish formation on clasts of gravelly alluvial fans deposited 8,000–12,000 yr ago at a site in the Mojave Desert" (McAuliffe and McDonald, 2006: 213). These and many other researchers have been forced to utilize the appearance of varnish because of the lack of a better tool. Now, the VML revolution means that studies of pavement evolution can be grounded in clast-specific, site-specific data.

VML in Hillslope Processes

Desert slopes can be areas of historic activity or they can host varnishes indicative of thousands of years of stone stability. Liu exemplifies VML's potential to study hillslopes through case studies of colluvial boulder stripes at Yucca Mountain, Nevada, fault escarpments in the western Great Basin, and of a debris-flow cone in Death Valley (Liu, 2008).

Desert landslides represent another fertile area for VML research, especially as urban expansion continues to abut steep mountains. Larger landslides are better dated with cosmogenic nuclides, because previously unexposed rocks are brought to the surface

(Hermanns et al., 2001; Ballantyne and Stone, 2004). However, smaller landslides on the order of $100\,m^2$ are more difficult to study, because many of the clasts have a prior exposure history on the slope prior to mass wasting.

An example of the potential of VML to study desert landslides comes from metropolitan Phoenix, Arizona. Smaller landslides pepper the urban area and pose a hazard to development that continues to move up against steep mountains. A key hazard question is whether these smaller deposits are fossil artifacts of a wetter Pleistocene, as is the case for the larger events in the region (Douglass et al., 2005).

South Mountain, central Arizona, offers an opportunity to study the problem. The largest urban park, this metamorphic core complex (Reynolds, 1985) presents steep flanks on many margins. For most of the Sonoran Desert, the boundary between grussified and relatively fresh granite rests deep underneath the surface. The granitic portions of this range, however, pose a landslide hazard because the subsurface weathering front is within a few metres of the surface. This permeability contrast generated a sturzstrom in the region (Douglass et al., 2005) and is associated with smaller granitic landslides in the area.

A pilot study of VML on six landslides abutting development at South Mountain reveals that their ages all rest in the Holocene. Using Liu's Holocene chronology (Fig. 21.7), their ages range from more recent than the WH1 layer (less than 300 years) to the early Holocene WH11 layer (ca. 10,300 years). Figure 21.12 illustrates one of the more recent landslide events. The ongoing nature of landslides throughout the arid Holocene, therefore, suggests potential for contemporary mass wasting.

Fig. 21.12 Small landslide in Phoenix, Arizona, on the border margin of South Mountain Park. VML analysis of the source area of the landslide suggests that the landslide is slightly older than 1100 years, the start of the WH2 layer

A major question in desert geomorphology is whether the terminal-Pleistocene to Holocene transition increased rates of hillslope erosion. This "transition to a drier climate" hypothesis is favoured by many investigators who argue that the reduction in vegetation cover accompanied this desiccation, leading to flash floods that transported sediment out onto alluvial fans during the early Holocene (Huntington, 1907; Eckis, 1928; Bull, 1991; Harvey and Wells, 1994; Reheis et al., 1996). One way to test this hypothesis is to analyze VML on debris flow levées that represent discrete erosional events.

The hillslopes of the Panamint Range above Hanaupah Canyon alluvial fan host an extensive number of debris-flow levées. Thirty such levées were sampled sequentially along a south-to-north transect below 1000 m.

The VML patterns on levée boulders do not support the "transition to a drier climate" hypothesis in this field area (Fig. 21.13). Overwhelming erosion in the early Holocene should have produced an overwhelming bias towards Holocene-aged (LU-1) VML. Instead, more than half of the sampled debris-flow levées formed during the late Pleistocene (Fig. 21.13). Still, this is just the first study of its kind. VML analyses of debris-flow levées on other hillslopes may reveal a different pattern.

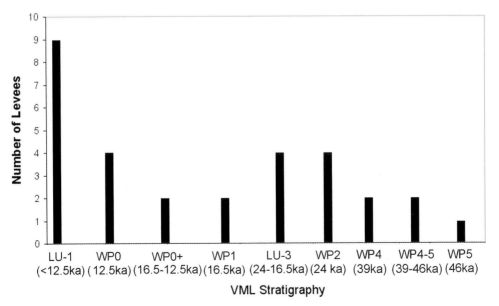

Fig. 21.13 Frequency of debris-flow levées on the lower slopes of the Panamint Range, Death Valley, above Hanaupah Canyon alluvial fan

VML Palaeolake Events

Palaeolake shorelines have served as important calibration sites for the VML method (Liu and Broecker, 2000; Liu et al., 2000; Broecker and Liu, 2001; Liu, 2003, 2008; Liu and Broecker, 2007, 2008a, 2008b). Bonneville, Coyote, Lahontan, Manix, Manly, Panamint, Searles, Silver, and Sumner western USA palaeolakes all yielded VML consistent with prior age control. Considerable potential, therefore, rests in VML providing insight into the chronometry of palaeolakes lacking other age control.

VML in Study of Hydraulic Activity

Large spheroidally-weathered granitic boulders characterize the debris slopes of inselbergs abutting pediments throughout the Mojave and Sonoran Deserts. A climatic-change interpretation holds that these inselberg slopes represent fossil landforms, relics of subsurface weathering during wetter periods of the Cenozoic (Oberlander, 1972, 1974, 1989; Twidale, 1982). Although the overarching landform might be a by-product of a regional climatic desiccation throughout the Cenzoic, there exists clear field evidence that the fines on these slopes are in adjustment with hydraulic processes (Abrahams et al., 1985; Parsons and Abrahams, 1987). VML analysis offers potential to reconcile these disparate perspectives.

Teutonia Peak inselberg at Cima Dome (Davis, 1905) and the Apple Valley, Mojave Desert (Abrahams et al., 1985; Parsons and Abrahams, 1987), debris slopes offer two venues to analyze the stability of inselberg boulder slopes. The goal rests in understanding the stability of the boulders themselves. There are complexities, however, in working with granitic boulders. One difficulty in boulder sampling rests in the potential for subjectivity, because varnish development varies considerably over a single boulder. Another problem in boulder sampling rests in the processes of boulder erosion; spalling of fissuresols (Villa et al., 1995) is a common erosional process, and these surfaces are not suitable for VML analysis. Thus, sampling had to be random and had to avoid fissuresol surfaces. This was accomplished by running transects from the piedmont junction up these slope.

The centre each intersected boulder was sampled, or—if a fissuresol—the closest suitable locale.

Ten boulders from each debris slopes suggest that ongoing boulder erosion has been taking place throughout the late Quaternary. No sampled boulder surface had a VML sequence even reaching back even to the Eemian interglacial or the LU-5 VML microstratigraphy. More than half of the sampled boulder surfaces at each site hosted only a LU-1 VML sequence, indicating Holocene erosion (Fig. 21.14). These data reveal that ongoing boulder erosion supplies sediment to hydraulic systems on inselberg debris slopes. Although these preliminary data do not falsify a climatic change origin for pediment-inselberg granitic complexes, they do indicate active late Quaternary erosion of inselberg boulders.

One of the first uses of VML was the study of alluvial-fans in Death Valley (Liu and Dorn, 1996). In addition to providing chronometric insight into specific depositional events on fan surfaces, VML analyses help decipher the connection between climatic events and fan deposition. If dry periods generate fan building through flash floods mobilizing unvegetated sediment, then Mn-poor dry-period VML should initiate the VML sequence. If wet periods generate fan building through fluvial transport, then Mn-rich black layers should rest at the lowest layers of the varnish.

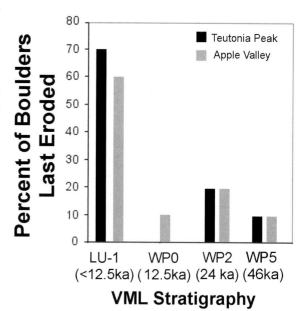

Fig. 21.14 Frequency of boulder erosion on inselberg debris slopes backing pediments in the Apple Valley and Cima Dome, Mojave Desert

Why did it Take so Long?

There are several basic reasons why it took a quarter century to go from the first publication of VML (Perry and Adams, 1978) to Liu's calibration (Figs. 21.5 and 21.7).

First, varnish ultrathin sections are difficult to make. Normal geologic thin sections are too thick, and further thinning to the host glass slide creates a triangular geometry that essentially hides many VML. Section zones too thin result in seeing no VML or in seeing a seemingly random pattern generated by areas of slightly greater resistance. Section zones too thick result in seeing too few VML. Having made a few hundred sections myself, the remainder of this chapter could be spent on anecdotes of how to make mistakes in section preparation and how these mistakes can generate errors in interpretation.

Second, patterns in sedimentary deposits may not necessarily be obvious with analysis of just a few samples. Particularly rapid accretion will form extra layers that combine in varnishes with slower accumulation rates. Variable rates of accretion will, thus, disguise patterns that emerge only after the study of many different sampling locales. It took Liu a decade of careful laboratory work to extract the complete VML pattern from thousands of rock varnish samples. Thus, a general problem comes from researchers who collect a few samples, as Oberlander (1994: 118) correctly emphasized: "researchers should be warned against generalizing too confidently from studies of single localities". Simply put, Liu is the only researcher who was dedicated enough to test the potential of detailed regional patterns of varnish VML accretion.

Third, researchers often select the wrong type of varnish, a problem discussed in detail elsewhere (Dorn, 1998: 214–224). The only type of varnish appropriate for VML analysis forms on subaerial surfaces exposed to only rain and dust deposition.

Fourth, some researchers fail to sample pure rock varnish. Instead, they collect samples that have mixtures of rock varnish with other types of rock coatings, such as heavy metal skin, iron film, oxalate crust, phosphate skin, or silica glaze. Certainly, rock varnish exists in the analyzed samples, as revealed by enhancements of Mn. Yet, published details in many papers reveal clear evidence that these other rock coatings actually interfinger with varnish in the analyzed samples, thus invalidating many conclusions.

Fifth, the occurrence of microcolonial fungi and other acid-producing lithobionts dissolves varnish, erasing microlaminations in the surrounding volume. Even if these organisms are not readily apparent on contemporary surfaces, they may have dissolved VML in the past. The selection of uneroded sequences, then, becomes a vital step analogous to avoiding bioturbation in the study of soils and sediments.

The above difficulties directly imply that the VML method requires that a researcher spend months to years in training. The large number of iterative cycles of field collection, thin section preparation, and laboratory analyses means that this powerful method should be undertaken by someone willing to devote years of constant research to the topic. The development of this expertise is analogous to musicians or surgeons who study for years to master a combination of method and art. Although he was hesitant at first, with the encouragement of colleagues TL has opened a dating laboratory that can accept samples from outside users (Liu, 2008). Hopefully, this VML dating laboratory will serve to encourage a group of future varnish scientists willing to put in the years needed to develop this expertise.

Other Climatic Change Signals

As with any sedimentary deposit, there are a host of ways whereby climatic-change signals might be extracted from rock varnish. The micromorphology of varnish, as seen in secondary electrons, provides a record of palaeodust (Dorn, 1986). Unconformities inside varnish can record aeolian abrasion events (Dorn, 1995a). Interdigitation of other types of rock coatings, such as oxalates, can indicate climatic change favouring a different type of accretion (Dorn, 1998). The occurrence of different types of organic matter might similarly reveal information of past environments (Dorn and DeNiro, 1985; Nagy et al., 1991; Perry and Kolb, 2003). Researchers focused on the search for extraterrestrial life concern themselves with the importance of fossil organisms that might be found entrapped by rock varnish (Probst et al., 2002; Allen et al., 2004; Perry et al., 2006). Even the occurrence of varnish on palaeosurfaces has been used to indicate palaeoaridity (Dorn and Dickinson, 1989; Marchant et al., 1993).

In the past, all varnish palaeoenvironmental methods such as those listed in the previous paragraph ranged from experimental to highly speculative. Some evidence existed to support their use. Yet the detailed and repetitive research needed to move the rock varnish science forward had not been conducted—until now.

The varnish microlaminations method has moved forward into a class of methods that can be considered reliable. VML is based on analyses of thousands sedimentary deposits (varnish microbasins). The VML method has been thoroughly calibrated at sites with numerical age control from multiple methodologies (e.g. radiocarbon, uranium-series, cosmogenic nuclide, and others). Critically, the VML method has been subjected to a rigorous blind test. For desert geomorphologists interested in learning about the chronology and palaeoclimatogy of desert landforms, there even exists a commercial laboratory to assist data acquisition. The time is ripe to make dramatic advances at the interface of desert geomorphic processes and climatic change.

Acknowledgments Thanks to Arizona State University for providing sabbatical support, to the late James Clark for his brilliance on the microprobe, and to T. Liu for permission to use graphics from his website http://www.vmldatinglab.com.

References

Abrahams, A. D., Parsons, A. J., and Hirsch, P. J., 1985. Hillslope gradient-particle size relations: Evidence of the formation of debris slopes by hydraulic processes in the Mojave Desert. *Journal of Geology* **93**: 347–357.

Allen, C., Probst, L. W., Flood, B. E., Longazo, T. G., Scheble, R. T., and Westall, F., 2004. Meridiani Planum hematite deposit and the search for evidence of life on Mars – iron mineralization of microorganisms in rock varnish. *Icarus* **171**: 20–30.

Amit, R., Gerson, R., and Yaalon, D. H., 1993. Stages and rate of the gravel shattering process by salts in desert Reg soils. *Geoderma* **57**: 295–324.

Ballantyne, C. K., and Stone, J. O., 2004. The Beinn Alligin rock avalanche, NW Scotland: cosmogenic Be-10 dating, interpretation and significance. *Holocene* **14**: 448–453.

Bond, G. C., Showers, W., Elliot, M., Evans, M., Lotti, R., Hajdas, I., Bonani, G., and Johnson, S., 1999. The North Atlantic's 1–2 kyr climate rhythm, relation to Heinrich events, Dansgaard/Oeschger cycles and the little ice age. In: *Mechanisms of Global Climate Change at Millennial Time Scales*, P. U. Clark, R. S. Webb and L. D. Keigwin (eds.), American Geophysical Union, Washington, D.C., pp. 35–58.

Broecker, W. S., and Liu, T., 2001. Rock varnish: recorder of desert wetness? *GSA Today* **11**(8): 4–10.

Bull, W. B., 1991. *Geomorphic responses to climatic change*. Oxford University Press, Oxford, 326pp.

Colman, S. M., 1982. Chemical weathering of basalts and andesites: evidence from weathering rinds. *U.S. Geological Survey Professional Paper* **1246**: 51.

Cremaschi, M., 1992. Genesi e significato paleoambientale della patina del deserto e suo ruolo nello studio dell'arte rupestre. Il caso del Fezzan meridionale (Sahara Libico). In: *Arte e Culture del Sahara Preistorico*, M. Lupaciollu (ed.), Quasar, Rome, pp. 77–87.

Cremaschi, M., 1996. The desert varnish in the Messak Sattafet (Fezzan, Libryan Sahara), age, archaeological context and paleo-environmental implication. *Geoarchaeology* **11**: 393–421.

Davis, W. M., 1905. The geographical cycle in arid climate. *Journal of Geology* **13**: 381–407.

De Geer, G., 1930. The Finiglacial Subepoch in Sweden, Finland and the New World. *Geografiska Annaler* **12**: 101–111.

Diaz, T. A., Bailley, T. L., and Orndorff, R. L., 2002. SEM analysis of vertical and lateral variations in desert varnish chemistry from the Lahontan Mountains, Nevada. *Geological Society of America Abstracts with Programs* **May 7–9 Meeting**: <///gsa.confex.com/gsa/2002RM/finalprogram/abstract_33974.htm>.

Dixon, J. C., 1994. Aridic soils, patterned ground, and desert pavements. In: *Geomorphology of Desert Environments*, A. D. Abrahams and A. J. Parsons (eds.), Chapman, London, pp. 64–81.

Dorn, R. I., 1984. Cause and implications of rock varnish microchemical laminations. *Nature* **310**: 767–770.

Dorn, R. I., 1986. Rock varnish as an indicator of aeolian environmental change. In: *Aeolian Geomorphology*, W. G. Nickling (ed.), Allen & Unwin, London, pp. 291–307.

Dorn, R. I., 1990. Quaternary alkalinity fluctuations recorded in rock varnish microlaminations on western U.S.A. volcanics. *Palaeogeography, Palaeoclimatology, Palaeoecology* **76**: 291–310.

Dorn, R. I., 1995a. Alterations of ventifact surfaces at the glacier/desert interface. In: *Desert aeolian processes*, V. Tchakerian (ed.), Chapman & Hall, London, pp. 199–217.

Dorn, R. I., 1995b. Digital processing of back-scatter electron imagery: A microscopic approach to quantifying chemical weathering. *Geological Society of America Bulletin* **107**: 725–741.

Dorn, R. I., 1998. *Rock coatings*. Elsevier, Amsterdam, 429p.

Dorn, R. I., 2007. Rock varnish. In: *Geochemical Sediments and Landscapes*, D. J. Nash and S. J. McLaren (eds.), Blackwell, London, pp. in press – Chapter 8, pp. 246–297.

Dorn, R. I., and DeNiro, M. J., 1985. Stable carbon isotope ratios of rock varnish organic matter: A new paleoenvironmental indicator. *Science* **227**: 1472–1474.

Dorn, R. I., and Dickinson, W. R., 1989. First paleoenvironmental interpretation of a pre-Quaternary rock varnish site, Davidson Canyon, south Arizona. *Geology* **17**: 1029–1031.

Dorn, R. I., and Oberlander, T. M., 1981. Microbial origin of desert varnish. *Science* **213**: 1245–1247.

Dorn, R. I., and Oberlander, T. M., 1982. Rock varnish. *Progress in Physical Geography* **6**: 317–367.

Douglass, J., Dorn, R. I., and Gootee, B., 2005. A large land-slide on the urban fringe of metropolitan Phoenix, Arizona. *Geomorphology* **65**: 321–336.

Dragovich, D., 1993. Distribution and chemical composition of microcolonial fungi and rock coatings from arid Australia. *Physical Geography* **14**: 323–341.

Drake, N. A., Heydeman, M. T., and White, K. H., 1993. Distribution and formation of rock varnish in southern Tunisia. *Earth Surface Processes and Landforms* **18**: 31–41.

Eckis, R., 1928. Alluvial fans in the Cucamonga district, southern California. *Journal of Geology* **36**: 111–141.

Eppard, M., Krumbein, W. E., Koch, C., Rhiel, E., Staley, J. T., and Stackebrandt, E., 1996. Morphological, physiological, and molecular characterization of actinomycetes isolated from dry soil, rocks, and monument surfaces. *Archives of Microbiology* **166**: 12–22.

Etienne, S., 2002. The role of biological weathering in periglacial areas: a study of weathering rinds in south Iceland. *Geomorphology* **47**: 75–86.

Gordon, S. J., and Dorn, R. I., 2005. In situ weathering rind erosion. *Geomorphology* **67**: 97–113.

Gossens, D., 2005. Effect of rock fragment embedding on the aeolian deposition of dust on stone-covered surfaces. *Earth Surface Processes and Landforms* **30**: 443–460.

Harvey, A. M., and Wells, S. G., 1994. Late Pleistocene and Holocene changes in hillslope sediment supply to alluvial fan systems: Zzyzx, California. In: *Environmental Change in Drylands: Biogeographical and Geomorphological Perspectives*, A. C. Millington and K. Pye (eds.), Wiley & Sons, London, pp. 67–84.

Heinrich, H., 1988. Origin and consequences of cyclic ice rafting in the Northeast Atlantic Ocean during the past 130,000 years. *Quaternary Research* **29**: 143–152.

Hermanns, R. L., Niedermann, S., Garcia, A. V., Gomez, J. S., and Strecker, M. R., 2001. Neotectonics and catastrophic failure of mountain fronts in the southern intra-Andean Puna Plateau, Argentina. *Geology* **29**: 619–622.

Hirsch, P., 1974. The budding bacteria. *Annual Review Microbiology* **28**: 391–444.

Hooke, R. L., and Dorn, R. I., 1992. Segmentation of alluvial fans in Death Valley, California: New insights from surface-exposure dating and laboratory modelling. *Earth Surface Processes and Landforms* **17**: 557–574.

Hungate, B., Danin, A., Pellerin, N. B., Stemmler, J., Kjellander, P., Adams, J. B., and Staley, J. T., 1987. Characterization of manganese-oxidizing (MnII–>MnIV) bacteria from Negev Desert rock varnish: implications in desert varnish formation. *Canadian Journal Microbiology* **33**: 939–943.

Huntington, E., 1907. Some characteristics of the glacial period in non-glaciated regions. *Geological Society of America Bulletin* **18**: 351–388.

Jones, C. E., 1991. Characteristics and origin of rock varnish from the hyperarid coastal deserts of northern Peru. *Quaternary Research* **35**: 116–129.

Julien, A. A., 1901. A study of the structure of fulgurites. *Journal of Geology* **9**: 673–693.

Karfunkel, J., Addad, J., Banko, A. G., Hadrian, W., and Hoover, D. B., 2001. Electromechanical disintegration - an important weathering process. *Zeitschrift fur Geomorphologie* **45**: 345–357.

Krinsley, D., 1998. Models of rock varnish formation constrained by high resolution transmission electron microscopy. *Sedimentology* **45**: 711–725.

Krinsley, D. H., Dorn, R. I., and Tovey, N. K., 1995. Nanometer-scale layering in rock varnish: implications for genesis and paleoenvironmental interpretation. *Journal of Geology* **103**: 106–113.

Krumbein, W. E., and Jens, K., 1981. Biogenic rock varnishes of the Negev Desert (Israel): An ecological study of iron and manganese transformation by cyanobacteria and fungi. *Oecologia* **50**: 25–38.

Kuhlman, K. R., Allenbach, L. B., Ball, C. L., Fusco, W. G., La_Duc, M. T., Kuhlman, G. M., Anderson, R. C., Stuecker, T., Erickson, I. K., Benardini, J., and Crawford, R. L., 2005. Enumeration, isolation, and characterization of ultraviolet (UV-C) resistant bacteria from rock varnish in the Whipple Mountains, California. *Icarus* **174**: 585–595.

Kuhlman, K. R., Fusco, W. G., Duc, M. T. L., Allenbach, L. B., Ball, C. L., Kuhlman, G. M., Anderson, R. C., Erickson, K., Stuecker, T., Benardini, J., Strap, J. L., and Crawford, R. L., 2006. Diversity of microorganisms within rock varnish in the Whipple Mountains, California. *Applied and Environmental Microbiology* **72**: 1708–1715.

Lee, M. R., and Bland, P. A., 2003. Dating climatic change in hot deserts using desert varnish on meteorite finds. *Earth and Planetary Science Letters* **206**: 187–198.

Libby, C. A., 1986. Fulgurite in the Sierra Nevada. *California Geology* **39**(11): 262.

Liu, T., 1994. Visual microlaminations in rock varnish: a new paleoenvironmental and geomorphic tool in drylands, Ph.D. thesis, 173 pp., Arizona State University, Tempe.

Liu, T., 2003. Blind testing of rock varnish microstratigraphy as a chronometric indicator: results on late Quaternary lava flows in the Mojave Desert, California. *Geomorphology* **53**: 209–234.

Liu, T., 2008. VML Dating Lab, *http://www.vmldatinglab.com/* <accessed November 14, 2008>.

Liu, T., and Broecker, W. S., 2000. How fast does rock varnish grow? *Geology* **28**: 183–186.

Liu, T., and Broecker, W., 2007. Holocene rock varnish microstratigraphy and its chronometric application in drylands of western USA. *Geomorphology* **84**: 1–21.

Liu, T., and Broecker, W. S., 2008a. Rock varnish microlamination dating of late Quaternary geomorphic features in the drylands of western USA. *Geomorphology* **93**: 501–523.

Liu, T., and Broecker, W. S., 2008b. Rock varnish evidence for latest Pleistocene millennial-scale wet events in the drylands of western United States. *Geology* **36**: 403–406.

Liu, T., Broecker, W. S., Bell, J. W., and Mandeville, C., 2000. Terminal Pleistocene wet event recorded in rock varnish from the Las Vegas Valley, southern Nevada. *Palaeogeography, Palaeoclimatology, Palaeoecology* **161**: 423–433.

Liu, T., and Dorn, R. I., 1996. Understanding spatial variability in environmental changes in drylands with rock varnish microlaminations. *Annals of the Association of American Geographers* **86**: 187–212.

Mabbutt, J. A., 1979. Pavements and patterned ground in the Australian stony deserts. *Stuttgarter Geographische Studien* **93**: 107–123.

Marchant, D. R., Schisher, C., Lux, D., West, D., and Denton, G., 1993. Pliocene paleoclimate and East Antarctic ice-sheet history from surficial ash deposits. *Science* **260**: 667–670.

Marston, R. A., 2003. Editorial note. *Geomorphology* **53**: 197.

McAuliffe, J. R., and McDonald, E. V., 2006. Holocene environmental change and vegetation contraction in the Sonoran Desert. *Quaternary Research* **65**: 204–215.

McAullife, J. R., 1994. Landscape evolution, soil formation, and ecological patterns and processes in Sonoran Desert Bajadas. *Ecological Monographs* **64**(2): 111–148.

McKeown, D. A., and Post, J. E., 2001. Characterization of manganese oxide mineralogy in rock varnish and dendrites using X-ray absorption spectroscopy. *American Mineralogist* **86**: 701–713.

Nagy, B., Nagy, L. A., Rigali, M. J., Jones, W. D., Krinsley, D. H., and Sinclair, N., 1991. Rock varnish in the Sonoran Desert: microbiologically mediated accumulation of manganiferous sediments. *Sedimentology* **38**: 1153–1171.

Nishiizumi, K., Kohl, C., Arnold, J., Dorn, R., Klein, J., Fink, D., Middleton, R., and Lal, D., 1993. Role of in situ cosmogenic nuclides [10]Be and [26]Al in the study of diverse geomorphic processes. *Earth Surface Processes and Landforms* **18**: 407–425.

Oberlander, T. M., 1972. Morphogenesis of granite boulder slopes in the Mojave Desert, California. *Journal of Geology* **80**: 1–20.

Oberlander, T. M., 1974. Landscape inheritance and the pediment problem in the Mojave Desert of Southern California. *American Journal of Science* **274**: 849–875.

Oberlander, T. M., 1989. Slope and pediment systems. In: *Arid Zone Geomorphology*, D. S. G. Thomas (ed.), Belhaven Press, London, pp. 56–84.

Oberlander, T. M., 1994. Rock varnish in deserts. In: *Geomorphology of Desert Environments*, A. Abrahams and A. Parsons (eds.), Chapman and Hall, London, pp. 106–119.

Palmer, F. E., Staley, J. T., Murray, R. G. E., Counsell, T., and Adams, J. B., 1985. Identification of manganese-oxidizing bacteria from desert varnish. *Geomicrobiology Journal* **4**: 343–360.

Parsons, A. J., and Abrahams, A. D., 1987. Gradient-particle size relations on quartz monzonite debris slopes in the Mojave Desert. *Journal of Geology* **1987**: 423–452.

Perry, R. S., and Adams, J., 1978. Desert varnish: evidence of cyclic deposition of manganese. *Nature* **276**: 489–491.

Perry, R. S., Dodsworth, J., Staley, J. T., and Gillespie, A., 2002. Molecular analyses of microbial communities in rock coatings and soils from Death Valley, California. *Astrobiology* **2**(4): 539.

Perry, R. S., Engel, M., Botta, O., and Staley, J. T., 2003. Amino acid analyses of desert varnish from the Sonoran and Mojave deserts. *Geomicrobiology Journal* **20**: 427–438.

Perry, R. S., and Kolb, V. M., 2003. Biological and organic constituents of desert varnish: Review and new hypotheses. In: *Instruments, methods, and missions for Astrobiology VII*, vol. 5163, R. B. Hoover and A. Y. Rozanov (eds.), SPIE, Bellingham, pp. 202–217.

Perry, R. S., Lynne, B. Y., Sephton, M. A., Kolb, V. M., Perry, C. C., and Staley, J. T., 2006. Baking black opal in the desert sun: The importance of silica in desert varnish. *Geology* **34**: 737–540.

Peterson, F., 1981. Landforms of the Basin and Range Province, defined for soil survey. *Nevada Agricultural Experiment Station Technical Bulletin* **28**: 52.

Peterson, F. F., Bell, J. W., Dorn, R. I., Ramelli, A. R., and Ku, T. L., 1995. Late Quaternary geomorphology and soils in Crater Flat, Yucca Mountain area, southern Nevada. *Geological Society of America Bulletin* **107**: 379–395.

Phillips, F. M., 2003. Cosmogenic 36Cl ages of Quaternary basalt flows in the Mojave Desert, California, USA. *Geomorphology* **53**: 199–208.

Phillips, F. M., Zreda, M. G., Plummer, M. A., Benson, L. V., Elmore, D., and Sharma, P., 1996. Chronology for fluctuations in Late Pleistocene Sierra Nevada glaciers and lakes. *Science* **274**: 749–751.

Pinter, N., Keller, E. A., and West, R. B., 1994. Relative dating of terraces of the Owens River, Northern Owens Valley, California, and correlation with moraines of the Sierra Nevada. *Quaternary Research* **42**: 266–276.

Potter, R. M., 1979. The tetravalent manganese oxides: clarification of their structural variations and relationships and characterization of their occurrence in the terrestrial weathering environment as desert varnish and other manganese oxides. Ph.D. thesis, California Institute of Technology, Pasadena, 245 pp.

Potter, R. M., and Rossman, G. R., 1977. Desert varnish: The importance of clay minerals. *Science* **196**: 1446–1448.

Potter, R. M., and Rossman, G. R., 1979a. The manganese- and iron-oxide mineralogy of desert varnish. *Chemical Geology* **25**: 79–94.

Potter, R. M., and Rossman, G. R., 1979b. Mineralogy of manganese dendrites and coatings. *American Mineralogist* **64**: 1219–1226.

Potter, R. M., and Rossman, G. R., 1979c. The tetravalent manganese oxides: identification, hydration, and structural relationships by infrared spectroscopy. *American Mineralogist* **64**: 1199–1218.

Probst, L. W., Allen, C. C., Thomas-Keprta, K. L., Clemett, S. J., Longazo, T. G., Nelman-Gonzalez, M. A., and Sams, C., 2002. Desert varnish - preservation of biofabrics and implications for Mars. *Lunar and Planetary Science* **33**: 1764.pdf.

Reheis, M. C., Slate, J. L., Throckmorton, C. K., McGeehin, J. P., SarnaWojcicki, A. M., and Dengler, L., 1996. Late Qaternary sedimentation on the Leidy Creek fan, Nevada-California: Geomorphic responses to climate change. *Basin Research* **8**: 279–299.

Reynolds, S. J., 1985. Geology of the South Mountains, central Arizona. *Arizona Bureau of Geology and Mineral Technology Bulletin* **195**: 1–61.

Taylor-George, S., Palmer, F. E., Staley, J. T., Borns, D. J., Curtiss, B., and Adams, J. B., 1983. Fungi and bacteria involved in desert varnish formation. *Microbial Ecology* **9**: 227–245.

Thiagarajan, N., and Lee, C. A., 2004. Trace-element evidence for the origin of desert varnish by direct aqueous atmospheric deposition. *Earth and Planetary Science Letters* **224**: 131–141.

Twidale, C. R., 1982. *Granite landforms*. Amsterdam, Elsevier, pp. 312.

Van Devender, T. R., Thompson, R. S., and Betancourt, J. L., 1987. *Vegetation history of the deserts of southwestern North America; the nature and timing of the late Wisconsin-Holocene transition.* Geological Society of America, Boulder, Colo, pp. 323–352.

Villa, N., Dorn, R. I., and Clark, J., 1995. Fine material in rock fractures: aeolian dust or weathering? In: *Desert aeolian processes*, V. Tchakerian (ed.), Chapman & Hall, London, pp. 219–231.

Wainwright, J., Parsons, A. J., and Abrahams, A. D., 1999. Field and computer simulation experiments on the formation of desert pavement. *Earth Surface Processes and Landforms* **24**: 1025–1037.

Zhou, B. G., Liu, T., and Zhang, Y. M., 2000. Rock varnish microlaminations from northern Tianshan, Xinjiang and their paleoclimatic implications. *Chinese Science Bulletin* **45**: 372–376.

Chapter 22

Hillslopes as Evidence of Climatic Change

Karl-Heinz Schmidt

Introduction

Geomorphic systems disclose great differences in their sensitivity to climatic change, and the various relief units carry the imprints of past processes to dissimilar degrees. Fluvial systems are highly susceptible to climatically induced changes in process. Hillslopes, on the other hand, are generally regarded as being rather resistant to such changes. In addition to the sensitivity of relief units to climatic change, another point of crucial geomorphic interest is how long the legacies of past processes are preserved in the form elements. Unfortunately, sensitivity to change and the length of time of preservation of past changes are usually inversely correlated. This means that we have either a detailed record of short-term changes for a limited period of time or a relatively inaccurate record of only major changes for a longer timespan. Where a detailed sedimentary record of past climatic changes has survived, related landform records may not be complete, because climatic changes are generally more frequent than landform changes.

The critical factors rendering a process change geomorphologically significant are the intensity and magnitude of the climatic change as well as its direction and the duration of the new climatic regime. These factors are counteracted by the resistance of the geomorphic system to external change. It is this resistance to change which predominantly controls the relaxation time of a system. Relief units with very short relaxation times will only document legacies of former processes for the very recent past, and relief units with very long relaxation times will only document long-lasting and very intensive changes.

Resistance to Change

With special regard to hillslopes the resistance of a relief unit to process change is dependent on a number of internal system variables relating to form, process, and lithological characteristics (Littmann and Schmidt 1989, Selby 1993).

(a) Slope angle: A high slope angle leads to a low resistance to gravitational and water erosion on the slope units compared with pediment and planation surfaces or lithologically controlled stripped plateau surfaces. On free faces no legacies of past processes are preserved.

(b) Lines of process concentration: If the relief unit is affected by lines of process concentration (e.g. valley floors, erosional gullies, slope rills) which channel water and material throughput, it becomes more susceptible to change. If a scarp is breached by a major river or if a river runs at its foot (Fig. 22.1), resistance to change is reduced by the closer connection to base level fluctuations and flood effects.

(c) Lithological characteristics of the surface material and substrate: This type of control is especially important in arid regions, where weathering and erosion processes operate in a highly selective manner in contrast to more humid regions which have greater moisture availability and an increase of chemical weathering. Outcrops of resistant

K.-H. Schmidt (✉)
Department of Geoscience, Universität Halle,
Von-Seckendorff-Platz 4, 06120 Halle, Germany
e-mail: karl-heinz.schmidt@geo.uni-halle.de

A.J. Parsons, A.D. Abrahams (eds.), *Geomorphology of Desert Environments*, 2nd ed.,
DOI 10.1007/978-1-4020-5719-9_22, © Springer Science+Business Media B.V. 2009

Fig. 22.1 Valley-side scarp in the Canyonlands section of the Colorado Plateau, Utah. The Colorado River and a tributary flow at the base of the scarp. The caprock consists of the sandstones of the Glen Canyon Group, and the lower slope is formed in the redbeds of the Chinle Formation (slope type 3). The lower slope is directly affected by fluvial processes and base level fluctuations. No legacies of past slope processes have been preserved

bedrock layers on slopes, hillslope surfaces consisting of consolidated talus or gravels, and old pediment surfaces show a high resistance to erosion, whereas unconsolidated hillslope debris, particularly when consisting of fine-grained colluvial deposits, is easily rearranged by changing process regimes.

Taken overall, the effectiveness of a past geomorphic process E associated with a past climate in shaping a hillslope is proportional to the product of the duration of the former process T_1 and its intensity I_1 divided by the resistance of the hillslope unit R.

$$E \propto \frac{T_1 I_1}{R} \qquad (22.1)$$

On the other hand, the degree of preservation P of a past process on a hillslope is dependent on its resistance to change R divided by the duration T_2 and intensity I_2 of the present process.

$$P \propto \frac{R}{T_2 I_2} \qquad (22.2)$$

If the resistance of a hillslope unit is considered to remain constant over time, the response of the system is mainly controlled by the duration and intensity of the processes involved. Only if the product of the duration and intensity of a past process is greater than the product of the duration and intensity of subsequent

processes will the effects of the past process remain visible on the landform. With this general background, we will examine the suitability of desert hillslopes for the elucidation of past climatic changes.

Indicator Potential of Hillslopes

The form of present-day slopes has evolved mainly over the past million years or less and the climatic changes over this period have been on a scale sufficient to have had a considerable effect on surface processes, particularly in the subtropical desert margins, which are most susceptible to environmental changes and tend to states of instability (Schmidt and Meitz 1996). The ages of hillslopes strongly depend on the strength of the material of which they are composed and the related denudational activity (Parsons 1988, p. 121). Consequently, the ages of hillslopes are decisively controlled by their resistance to change (see Equations 22.1 and 22.2). When hillslopes or parts of them are old enough to have experienced the climatic changes of the Pleistocene, it is necessary to investigate

(a) whether the climatic changes were strong and long enough to shift form–process relationships from one state of equilibrium to a significanttly different one (a change in principle), or

(b) whether they only enhanced or attenuated existing form-process relationships (a change in degree), or

(c) whether they were too weak and short-lived to have caused any effect on the general trend of hillslope development (no morphological change).

Only in the first case will truly independent climatogenetic landform elements in Büdel's (1982) sense have come into existence.

Slope systems are controlled by lateral backwearing, sometimes in combination with downwearing and slope angle reduction where processes of slope decline are involved. On bipartite cuesta scarps parallel retreat is the dominant mechanism (first described by Powell 1875). Slope denudation implies an inherent mechanism of surface destruction along the entire slope reach. This surface-destruction mechanism is a major drawback to the preservation of forms created by past climates.

The ability of hillslopes to bear clear testimony to distinct climatic phases is complicated by the fact that process change may not simply be expressed in a characteristic form assemblage of a slope system. The work of varying processes can also result in a slope configuration, which represents a typical expression of a particular sequence of different climatic periods. The morphology of these slopes is due neither to present-day nor to past processes alone but has been shaped by a temporal succession of processes. Descriptions of hillslopes characterized by this type of development have already been presented by Oberlander (1972) for the Mojave Desert and by Moss (1977) for southern Arizona. The term 'morphogenetic sequence' has been proposed by Mensching (1974) for the evolution of landforms, which owe their present shape to a combination of different processes. He described a number of examples from North African arid and semi-arid areas.

As will be shown below, most legacies of former slope processes are found on bipartite slopes, where a hard caprock overlies an easily erodible formation. These hillslopes are generally called cuesta scarps. Homolithic slopes do not tend to furnish similar information. The indicator potential of the bipartite slopes can be explained by their specific composition. They normally consist of an upper steep slope in the resistant caprock and a lower, moderately inclined slope in less resistant beds (see Fig. 22.5). There has been much confusion in the terminology of the different form elements of cuesta scarps (cf. Oberlander 1997). Terms such as subtalus slope (Koons 1955), footslope (Oberlander 1977), rampart (Howard and Kochel 1988) or substrate ramp (Oberlander 1997) have been utilized for the part of the scarp developed in soft rocks below the caprock. The more neutral term lower scarp slope (Schmidt 1987a, Schmidt and Meitz 2000a,b) will be used in this chapter. The caprock provides resistant talus which protects portions of the lower scarp slope from the slope's general fate of being destroyed by the backwearing processes. Mainly for this reason most studies dealing with evidence of past processes on desert hillslopes have chosen cuesta scarps as their research subject; as will this chapter. This type of scarp may serve as a useful indicator for major climatic fluctuations (Gerson 1982, Schmidt 1989b, 1996, Gutiérrez et al. 2006).

The resistance to change and consequently the inertia is greater for hillslope systems than for rivers, floodplains, and alluvial fans. Fluvial systems are able to store a detailed record, but this record also includes some 'noisy' information. For the Colorado Plateau in the south-western United States the Holocene and late Pleistocene depositional and erosional history of the river systems is well known from the alluvial record (e.g. Hack 1942, Euler et al. 1979, Wells et al. 1982, Graf et al. 1987, Hereford 1987). There is, however, no comparable detailed evidence of past climatic oscillations on cuesta scarps (Schmidt and Meitz 2000a,b). Apparently the relaxation times are longer and thresholds for formative change are higher for slopes. The base-level fluctuations were neither long enough nor strong enough to be propagated upslope (Schmidt 1988, Littmann and Schmidt 1989). On the hillslopes only former climates of long duration and characterized by highly effective processes have left their imprint (see Equation 22.1). This makes these hillslopes a valuable indicator for the low frequency climatic changes of the Pleistocene. But not all desert cuesta scarps bear the imprint of former processes. A strong influence is exerted by the lithological and structural attributes of the caprock and substrate and other controlling factors (Gerson and Grossman 1987, Schmidt 1989b, Schmidt and Meitz 1996). These factors will be discussed in a subsequent section.

Stability or Activity in Dry Periods

The question whether desert landforms remain more stable in dry (interpluvial) than in more humid (pluvial) conditions has for long been a matter of controversy. For desert washes (arroyos) in the American Southwest the discussion about the influence of different hydrologic regimes including human interference has been summarized by Cooke and Reeves (1976), Graf (1983) and Karlstrom (1988). Conclusions on this important subject remain very mixed. For cuesta scarps in the same region this controversy has initially been highlighted in the papers of Ahnert (1960) and Schumm and Chorley (1966). Ahnert (1960) argued that erosional processes, particularly slump block production and sapping processes supported by subterranean wash, were much more active during the pluvials, and that present-day landform modifications are only of minor importance. He derived a model of four main scarp profiles from the rock sequence in Monument Valley, which were formed by varying combinations of pluvial and interpluvial processes (Ahnert 1960). The greater moisture availability in the humid periods may well have enhanced groundwater flow and seepage above impervious beds. Howard and Kochel (1988) and Laity and Malin (1985) concluded from their detailed field observations in Navajo Sandstone alcoves that present-day sapping activity and alcove development is much less than during a past pluvial climate. With respect to talus production, Schumm and Chorley (1966) maintained that the frequently observed lack of basal talus on the lower scarp slopes is not a consequence of slope stability but a result of effective weathering and a high rate of talus removal. Ahnert's (1960) climato-cyclic interpretation contrasts sharply with this equilibrium view of Schumm and Chorley (1966). Schmidt and Meitz (2000a,b) in their comprehensive study of scarp slope types on the Colorado Plateau conclude that changes in humidity may induce opposing tendencies of reactions in different slope types corresponding to their intrinsic lithological attributes. This means that there is no general trend of activation or stabilization with changing rainfall input (see below).

Presently inactive landslide material, slump blocks, and major rockfall accumulations on cuesta slopes have frequently been interpreted as indicators of decreasing geomorphological activity from the last humid period to the present dry phase. Figure 22.2 shows the distribution of mean annual precipitation on the central Colorado Plateau, the area to which most of the discussion in this section refers. The gravitational deposits have been regarded as relict landforms by many field geologists, as depositional features inherited from past pluvial processes. Symptomatic of this kind of interpretation with regard to landslides is Hunt et al. (1953, p. 171) statement: 'They appear to have survived from a more humid climate when conditions were more conducive for weathering and recession of cliffs.' Landslides and large-scale rotational slumps as evidence of humid phase activity have been described from the Vermilion Cliffs and Echo Cliffs, Arizona (Strahler 1940, Phoenix 1963), from the Red House Cliffs, Utah (Mullens 1960), and the White Canyon area, Utah (Thaden et al. 1964). All of these sites lie in areas with a present-day precipitation of about 200 mm and are characterized by massive sandstones (Glen Canyon Group), overlying swelling bentonitic Chinle shale. These landslides are enormous in size, being up to 2 km long and 800 m wide (Thaden et al. 1964,

p. 75). The absence of youthful slumps is taken as an indicator of present stability. With similar arguments Reiche (1937) reported landslides from the Black Mesa area, Arizona, where Mesa Verde Sandstone overlies Mancos Shale. Reiche, however, also reported two major young slumping events (Toreva Blocks) from 1870 and 1927.

Schumm and Chorley (1966) collected eyewitness accounts of active scarp retreat processes from the National Parks and National Monuments Services on the Colorado Plateau. These accounts show that rockfalls of different magnitudes have frequently occurred in recent years. This conclusion is supported by descriptions of field geologists for different parts of the Colorado Plateau. Ford et al. (1976) gave a detailed account of rockfalls in the Grand Canyon area. Cliff retreat is also a presently active process in the Montezuma Canyon area, southeastern Utah, with isolated rockfalls occurring occasionally in the Dakota and Salt Wash sandstones (Huff and Lesure 1965). In the Slick Rock district, southwestern Colorado a number of slides were observed during fieldwork in the years 1953–1956. The slides involved the Burro Canyon formation and the Salt Wash sandstone and occurred after heavy autumn rain (Shawe et al. 1968). Apparently present-day intensive precipitation events are well capable of triggering mass movements.

Schumm and Chorley (1964) described a major rockfall event in Mesa Verde sandstones in the Chaco Canyon National Monument, New Mexico (Fig. 22.4), dating from 1941. They concluded (1966, p. 19): 'Although such contemporary occurrences of this magnitude may be rare, the fact that slumps have occurred during the past 100 years illustrates the need for caution in attributing all occurrences of an erosional phenomena [sic] to past climates greatly different from the present'. This quotation implies that the morphological features of cuesta scarps can be to some extent explained by presently active processes and that, if there have been changes at all, they may have been only in the relative rates of processes. The continuity of large-scale cliff retreat processes has also been stressed by Davidson (1967) in the Circle Cliffs area, southeastern Utah and by Shawe et al. (1968) in the Slick Rock district, Colorado.

Any conclusion concerning activity or stability of geomorphic processes must try to avoid some potential hazards of misinterpretation. Stability or activity is highly dependent on the altitudinal location and on

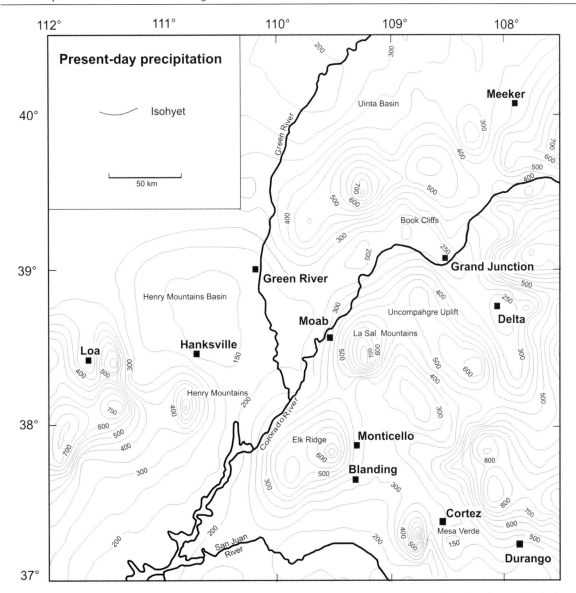

Fig. 22.2 Distribution of mean annual precipitation on the Colorado Plateau in southwestern Colorado and southeastern Utah. Some of the localities mentioned in the text are shown

the lithological compostion of the slope (Schmidt and Meitz 2000a,b). It is also extremely hazardous to judge from a one- or two-year field investigation or even from a 100-year photographic record (Baars 1971, Shoemaker and Stephens 1975) whether a specific type of landform in a desert environment is stable or not. A period of 100 years covers not more than 1% of the Holocene. Geomorphological events on a hillslope are not only highly sporadic in time but also unevenly distributed in space in contrast to fluvial action. Gravitational events may be ordered according

to the magnitude of the masses of material involved. They range from the fall of small grains through the fall of individual stones, slabs, and blocks, to rockfalls, rockslides, slumps, and landslides. Masses ranging from 10^{-3} g through 10^8 g to 10^{12} g are transported by the different processes. Obviously the authors who deduce present-day morphological stability from their observations refer to giant landslides with lengths of several hundred metres or large slump blocks with lengths of 50 metres or more (e.g. Reiche 1937, Strahler 1940, Mullens 1960). Gravitational processes of smaller magnitude such as falls of individual rock fragments, minor rockfalls of a few cubic metres, and slab failures along joints or exfoliation structures have also been observed by authors, who otherwise deny the present-day activity of cuesta scarps (Strahler 1940). 'After severe summer rains and during early spring the Vermilion and Echo Cliffs sometimes resound with noise of falling rocks. The blocks that fall are sometimes 15 or 20 feet in diameter' (Phoenix 1963, p. 40). The activity of processes has to be evaluated in the light of magnitude and frequency concepts. Moreover, on the scarp slopes the gravitational processes are spatially and temporally discontinuous. The variability in space and time increases with increasing process magnitudes. It is difficult to decide whether a particular type of high-magnitude process is really extinct or whether we just meet a cuesta slope in the interval between two such rare events depending on the specific recurrence intervals (Viles and Goudie 2003).

When there are no signs of present-day high-magnitude events (rockfalls, slump blocks, or landslides) on a particular cuesta scarp section, this does not mean that there is a general state of morphodynamic stability. So the cuesta scarp section under consideration may well experience events of smaller magnitude, while at other scarp sections events of greater magnitude may happen at the same time (Fig. 22.3a,b). Only the chronological integration of processes of different magnitude and frequency and the spatial integration of complete scarp sections will lead to a reliable evaluation of process activity in desert regions.

A space-integrating argument in support of the view of present-day morphological activity is the current high rate of suspended sediment transport of the Colorado River and its tributaries (Smith et al. 1960, Iorns et al. 1965, Schmidt 1985, Graf 1985). The overall denudation rate lies between 70 and 350 $m^3 km^{-2} y^{-1}$ (or mm per 1000 years) (Schmidt 1985), but it is difficult to determine how much material is eroded from the slopes and how much originates from the alluvial valley fills which have been dissected since the end of the 19th century (Graf 1983). Erosion measurements on lower scarp slope rocks yield additional evidence for high process intensity. Colbert (1966) determined a denudation rate of 5.7 $mm y^{-1}$ for steeper parts of the lower scarp slope of the Echo Cliffs, northern Arizona, in the bentonitic shales of the Petrified Forest member of the Chinle Formation. Surface lowering

in the easily erodible Mancos Shale at the foot of the Book Cliffs, southeastern Colorado, varies between 0.2 and 2.2 $mm y^{-1}$ and is a function of inclination and land use (Lusby 1979). In the same area a rate of denudation of 2.7 mm was measured in a single event (Hadley and Lusby 1967). Intensive lowering on Mancos Shale slopes (0.15–1.5 $mm y^{-1}$) was also determined by stake exposure measurements (Schumm 1964). The high erosion rates in the lower slope rocks are also indicative of the general erosional activity of the Mancos Shale slopes (cf. Fig. 22.8b).

In Northern Africa, too, different views have been expressed concerning activity or stability of gravitational processes on cuesta scarps. The cuesta scarps in Nubian Sandstone on the margins of the Mourzouk basin have been investigated by Barth and Blume (1975) and Grunert (1983). Grunert stated that the landslides on the western flank of the basin are 'fossil' landforms dating from more humid periods in the middle Quaternary with annual precipitation amounts in excess of 400 mm (presently about 10 mm)! The landslides are presently destroyed by fluvial erosion. Barth and Blume (1975) take a broader view, pointing out that there are cuesta sections with extremely intensive denudational processes and others without any recent activity. The variations are explained by non-climatic controlling factors such as differences in resistance between caprock and lower slope rocks, differences in the thickness ratio, and differences in the elevation above base level. Apparently these controlling factors influence the resistance to change of the cuesta scarps.

In a study of limestone scarps in the northwestern Sahara, Smith (1978) concluded that karstic and sapping processes and, consequently, cliff retreat are now stagnant. This view is again expressed in another case study on Cenomanian–Turonian limestone scarps in south-eastern Morocco (Smith 1987). He deduced that the landforms are undergoing superficial modifications by weathering, but that the landscape as a whole has been essentially stable since mid to late Quaternary times (Tensiftien: about 1.3×10^5 years BP). He contended that undermining of the caprock is largely the product of spring sapping. The present author has also investigated the limestone cuesta scarps in Southern Morocco (Schmidt 1987b, 1989a). Certainly karst processes and sapping by karst water were much more efficient during humid periods than today. But rockfalls are not only caused by spring sapping and related

Fig. 22.3 (**a**) West Mitten
Butte in Monument Valley,
Arizona. Vertical cliff is
formed in de Chelly
sandstone with resistant
Triassic layers on top, the
lower slope is developed in
the Organ Rock siltstones of
the Cutler Formation. Note
the intercalation of more
resistant strata, which form
ledges on the lower scarp
slope. The slope belongs to
slope type 3 (see Fig. 22.5).
The lower slope is almost
talus-free, which might lead
the observer to assume
general geomorphic stability.
(**b**) Complex scarp in Capitol
Reef National Park in
southeastern Utah. The upper
part consists of a vertical cliff
in the Wingate Sandstone
underlain by Chinle redbeds
forming a type 3 slope. A
fresh rockfall with large
blocks covers the lower slope
extending donwards to the
convex slope (type 2) in the
Petrified Forest Member of
the Chinle Formation.
Apparently, rockfalls can be
triggered without pluvial
sapping. This document
might lead the observer to
assume general geomorphic
activity. However, a reliable
evaluation of process activity
or stability must use a time
and space integrating
survey

alcove formation. Slope dissection and steepening are much more important for the destabilization of the caprock, and these mechanisms are also active in the present dry period. Limestone talus lies on and below the pediment level, which was formed in the last pluvial of the Pleistocene (Soltanien) (Schmidt 1989a) (Fig. 22.4).

In the upper Colorado River basin, where there is a great variety of different climatic regimes with rainfall amounts generally in the range between 150 and 600 mm, the tributaries with the lowest precipitation and specific yield have the highest rates of mechanical denudation (Schmidt 1985), which shows that solid load production is most active in dry environments. Barth and Blume (1973) investigated cuesta scarps in several localities in the dry regions of the United States with different degrees of moisture availability from 165 mm annual precipitation with ten arid months on the central Colorado Plateau to 425 mm of annual precipitation with seven arid months in the Black Hills

Fig. 22.4 Cuesta scarp (Hamada de Meski, southern Morocco) capped by Cenomanian/Turonian limestones, which are underlain by soft continental redbeds. Climato-cyclic talus flatirons have developed on the scarp slope and pediment flatirons in the foreland. Different generations of flatirons can be discerned. The backslopes of the flatirons are firmly consolidated with carbonate cement. The talus flatiron on the right is assumed to be of Soltanien (Würm) age, the pediment flatiron on the left is assumed to be of Tensiftien (Riss) age. Note the coarse debris in the slope rill below the younger level

in Wyoming. The latter conditions might be representative of the situation in the last pluvial on the Colorado Plateau (Barth and Blume 1973, Schmidt and Meitz 2000b). They demonstrated that the cuesta scarps with the strongest morphological activity are found in the drier areas with precipitation amounts lower than 300 mm. In these areas the landform elements are interpreted as being in equilibrium with present-day processes and the authors rejected Ahnert's (1960) view of greater process activity in pluvial periods.

The preceding discussion shows that the question of alternating climates and of stable or unstable periods is difficult to resolve (i) when the climatic changes only cause minor alterations (changes in degree) in the process regime, (ii) when different climatically induced triggering processes – spring sapping under more humid conditions or lower slope dissection under more arid conditions – result in the same morphological response (rockfalls), and (iii) when different lithologies are involved, which show dissimilar reactions to increasing/decreasing humidity. The stability of erosion-controlled slopes is generally strengthened by increasing humidity (denser vegetation cover), whereas lower slopes controlled by gravitational processes become generally more

susceptible to denudational activity when humidity increases.

Slope Types and Their Reactions to Climatic Change

Cuesta scarp slopes have developed in various lithological contexts and in different altitudinal belts on the Colorado Plateau in the southwestern United States. Cuesta scarps are found in elevations ranging from 1300 to 3000 m with corresponding mean annual precipitation amounts between less than 150 and more than 600 mm (Fig. 22.2). There is also a wide lithological variety of caprocks and soft lower slope substrates composing the cuesta scarps (Barth and Blume 1973, Schumm and Chorley 1966, Schmidt 1988). In the semi-arid parts (<250 mm) in the lower elevations (1300–1800 m) four different slope types were identified (concave, convex, straight and vertical) (Schmidt 1987a, 1988, Schmidt and Meitz 1996). Figure 22.5 shows the principal slope types with a brief description of their main characteristics. The classification in this altitudinal belt refers essentially to the form and process features of the lower scarp slope, which are mainly dependent on the lithological and structural attributes of the soft lower slope substrates. The form variations of the upper scarp slopes are not as evident because in the semi-arid belt, with its selective weathering and erosion, all the resistant caprocks tend to form nearly vertical cliffs. The form and process response to changing climatic conditions was systematically investigated in the more humid higher elevations of the Colorado Plateau (southern flank of Uinta Basin, Uncompahgre Uplift, Elk Ridge, Mesa Verde, for locations see Fig. 22.2). Specific attributes of the scarp slopes such as vegetation and soil cover, dissection density, activity of mass movements, talus production and general morphometry were documented (Schmidt and Meitz 1996, 2000a,b). The data collection and interpretation showed how slope attributes change with increasing altitude and precipitation under the present climatic regime (Fig. 22.6). This information can be used as an indicator of morphoclimatic changes during the cold (humid) phases of the late Quaternary in a space-time substitution. Altitudinal shifts of the hygric environment during the last glacial period

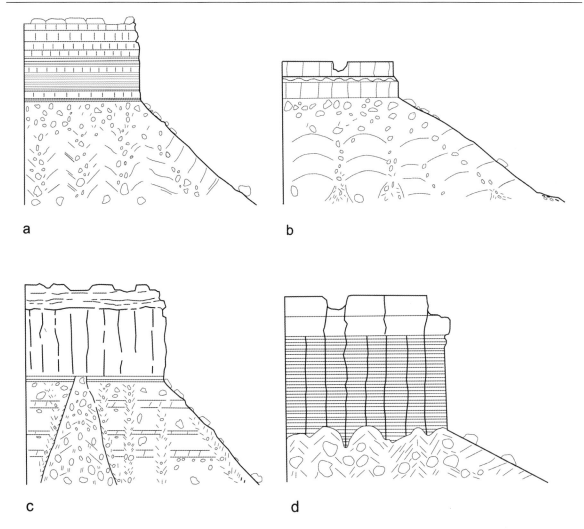

Fig. 22.5 Litologically controlled different types of scarp slopes in the arid to semi-arid altitudinal belt (Schmidt 1988, Schmidt and Meitz 2000b). (**a**) Slope type 1: concave lower slope, average inclination 35°, deeply dissected by slope rills, debris concentrated in the rills, pediment at scarp foot, developed in impermeable homogeneous shales (type 1 is shown in Fig. 22.8b, developed in the impermeable Mancos Shale), (**b**) Slope type 2: convex slopes with lower inclinations (30°), in swelling bentonitic clays, influenced by piping, only moderately dissected, debris concentrated in the upper parts of the lower slope and at the scarp foot (type 2 is shown in the central section of the complex slope in Fig. 22.3b, developed in the Petrified Forest Member of the Chinle Formation), (**c**) Slope type 3: slopes composed of segments of different inclinations in heterogeneous lower slope rocks with intercalations of more resistant beds, the resistant beds impede deep dissection, parts of the slope covered by cones of landslide material, in the overall profile the slopes are straight with inclinations close to 40° (type 3 is shown in Figs. 22.1 and 22.3a,b). (**d**) Slope type 4: vertical cliffs witrh talus slopes at their base in evenly bedded gypsiferous fine grained sandstones, siltstones and shales with vertical joints (type 4 is shown in the lower part of Fig. 22.3b, developed in the Moenkopi Formation)

resulted in associated shifts of vegetation and morphoclimatic belts, and lower elevations were affected by present-day high-altitude process combinations.

The specific responses of the principal slope types: (1) concave; (2) convex; (3) straight are demonstrated in Fig. 22.6 for north-facing slopes. Slope type 4 (vertical) showed no systematic altitudinal variations. Stratigraphic examples for the individual slope types are indicated in the captions of Fig. 22.6. In the arid to semi-arid altitudinal belt slope type 1 is deeply dissected with talus concentrated in the slope rills (cf. Fig. 22.5a). It has a concave profile with an average inclination of about 35°. The establishment of woodland vegetation in the more humid belts leads to slope

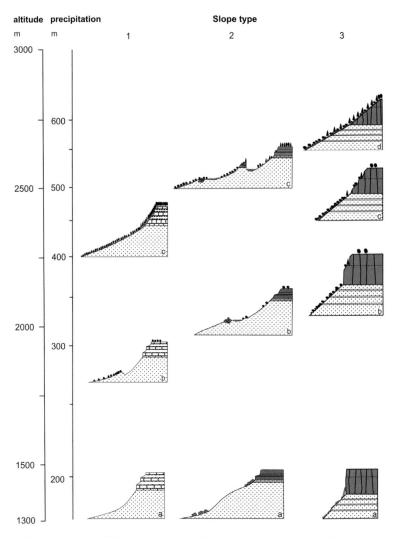

Fig. 22.6 Altitudinal and hygric zonation of the geomorphological attributes of slope types 1-3 on north-facing slopes (on south-facing slopes the altitudinal belts are shifted upwards).

1) concave slope (e.g. Mesaverde Sandstone/Mancos Shale cuesta scarp)

   (a) arid steep, densely and deeply dissected slope (see Fig. 22.5a)
   (b) arid to semi-arid dissected slope with talus flatirons (<350 mm)
   (c) smooth, only moderately dissected slope in altitudes with more than 400 mm rainfall

2) convex slope (e.g. Dakota Sandstone/Brushy Basin Shale cuesta scarp)

   (a) arid zone convex slope (see Fig. 22.5b)
   (b) semi-arid zone with inactive mass movement accumulations in the lower parts of the slope
   (c) active mass movements, irregular slope profile with depressions (>500 mm)

3) Straight slope type (e.g. Glen Canyon Group/Chinle Formation cuesta scarp)

   (a) arid zone straight lower slope with vertical cliff in the upper slope (see. Fig. 22.5c, see also Fig. 22.3a)
   (b) beginning stabilization of lower slope and beginning segmantation of the cliff (precipitation about 400 mm)
   (c) development of small platforms on the cliff, denser vegetation cover, reduction of dissection and lower slope inclination (precipitation between 400 and 550 mm)
   (d) moderately inclined straight slope in caprock and lower slope rocks, densely vegetated (rainfall >550 mm)

stabilization, reduced dissection and inclination and the development of a smooth profile with a continuous debris cover (profile c for slope type 1 in Fig. 22.6), Talus and pediment flatirons have originated on the Mancos-Shale slope as a consequence of alterations between cold(humid) and warm(dry) climatic phases in the intermediate altitudinal belt (profile b for slope type 1 in Fig. 22.6, see also Figs. 22.8b and 22.9a). In the semi-arid altitudinal belt slope type 2 is moderately dissected and has a convex profile with an average inclination of about 30° (Fig. 22.5b). With increasing humidity the convex slope profile disappears and mass movement activity and associated irregular slope forms become more and more characteristic for the scarp slope (profile b and c for slope type 2 in Fig. 22.6). The lower limit of inactive landslides is indicative of how far humid phase mass movement processes extended downslope. On slope type 3 scarp form variations are not only found on the lower, but they are also visible on the upper scarp slope. On the lower slope in the arid to semi-arid altitudinal belt dissection is less intense than on slope type 1 and talus is concentrated on the divides between the slope rills. The slope has a straight profile with inclinations close to 40° (see Fig. 22.3a for an examples from the Monument Valley rock sequence). Inclination, dissection and the amount and size of talus decrease in more humid zones. In the drier elevations the sandstones form vertical cliffs, which with increasing altitude and humidity change to horizontally segmented and compartmental cliffs and eventually to a straight slope (profiles b, c and d for slope type 3 in Fig. 22.6). Due to their relatively high resistance to change, the duration of the last cold (humid) phase was not long enough to shift the sandstone slopes of the individual altitudinal belts from their given states of form-process equilibrium to completely different ones.

Only hillslope form elements that give evidence of significantly different processes of weathering and erosion will help to elucidate the effects of climatic change and the succession of different climatic regimes. The slopes must be able to preserve forms created by past processes, i.e. the form elements must be firm and stable enough to survive climatically induced process change for a longer period of time. There are cuesta scarp hillslopes, which display multiphase assemblages of climato-genetic form elements. On slope type 2 inactive mass movement accumulations are found in the presently stable dry altitudinal belt. On slope type 1 talus and pediment flatirons give clear evidence of climatic change (Fig. 22.6).

Talus and Pediment Flatirons

Talus and pediment flatirons on the slopes of dryland cuesta scarps have a triangular to trapezoidal shape with their tops directed towards the scarp (Fig. 22.4). Talus flatirons are located on the scarp slope and pediment flatirons in the transition zone between scarp and foreland. In their distal parts, pediment flatirons may merge into river or lake basin terraces. Both the talus and pediment flatirons were parts of formerly active continuous slope systems, from which they received their debris cover. They have been detached from the presently active scarp by small rills running perpendicular to the general slope inclination, by slope steepening, and by cliff retreat. The inner slope of the flatirons is inclined towards the cuesta scarp and consists of the soft bedrock material of the lower portion of the scarp. The outer slope, which is inclined towards the scarp foreland with slope angles from 2 to 30°, is covered by a debris mantle consisting of resistant caprock talus protecting the soft lower slope rock from being eroded (Figs. 22.4 and 22.8b). The talus is generally not more than 5 m thick and has an average thickness of 2 m. In the proximal parts, however, the talus may contain blocks with a-axis diameters of more than 2 m which are remnants of rockfall debris.

Models of Flatiron Formation

Talus and pediment flatirons have attracted growing attention in recent years. They have been described from a variety of arid and semi-arid environments including the southwestern United States (Koons 1955, Blume and Barth 1972, Barth and Blume 1973, Schmidt 1988, 1996, Schmidt and Meitz 1996, 2000a,b), Spain (Gutiérrez et al. 1998, 2006, Sancho et al. 1988), the Saharan countries (Ergenzinger 1972, Grunert 1983), southern Morocco (Joly 1962, Dongus 1980, Schmidt 1987b, 1989a), Saudi Arabia (Barth 1976), Syria (Sakaguchi 1986), Cyprus (Everard 1963), and

Israel (Gerson 1982, Gerson and Grossman 1987). In these publications different interpretations have been offered for the processes and climatic controls involved in the formation of these landforms. The debate has centred on the question of whether the flatirons are the product of climatically controlled distinct processes or whether they develop independently from climatic change. The respective conceptions of flatiron formation are called the climato-cyclic model and the non-cyclic model (Schmidt 1989b, 1996). Only in the former case can the flatirons be used as 'keys to major climatic fluctuations' (Gerson 1982, p. 123).

The non-cyclic model, which was first proposed by Koons (1955), is characterized by the sequence of events as depicted in Fig. 22.7a. No climatic change is needed for this sequence to occur. The successive stages of non-cyclic flatiron development from talus-free slopes to fresh talus cones and finally to complete separation of cones can be mapped in the field in close spatial proximity (Fig. 22.7b). On the Colorado Plateau non-cyclic flatirons are most frequently found on slope type 3 in the arid to semi-arid altitudinal belt (Schmidt 1988, 1989b).

The climato-cyclic model explains talus and pediment flatirons as consequences of climatically dependent alternations of significantly different processes on scarp slopes (Gerson 1982, Gerson and Grossman 1987, Gutiérrez et al. 1998, 2006, Schmidt 1989a,b, 1996, Schmidt and Meitz 2000b). In the more humid periods (pluvials) scarp slopes are controlled by aquatic debris transport and slopewash, which results in the formation of smooth, undissected concave slope profiles with some gravitational activity in the proximal parts. When these conditions are replaced by more arid climatic regimes (interpluvials), dissection of the smooth debris slopes begins, bedrock is exposed in the rills, and gullying and rock collapse are the dominant processes. Parts of the formerly continuous slope are separated from the active scarp, and talus and pediment flatirons begin to form, with the backslopes of the flatirons representing the remnants of the smooth debris slopes (Figs. 22.4, 22.8a,b and 22.9a). In the current dry period, dissection of the smooth debris slopes and talus removal are ubiquitously observed processes on scarps in Northern Africa and the Middle East (Gerson 1982, Schmidt 1987b) as well as in the southwestern US on slope type 1 in the lower elevations (Figs. 22.8b and 22.9a) (Schmidt 1996).

Lithological and Structural Prerequisites for the Formation of Talus and Pediment Flatirons

Apart from climatic influence a number of lithological and structural controls are essential to the formation of the two separate types of flatirons (Schmidt 1989b). Scarps with **non-cyclic talus flatirons** need to be fed with rockfall material which covers the complete lower slope with a talus cone. This continuous talus cone can only be supplied by thick caprocks and on scarp slopes where the thickness ratio between soft rock and caprock is not much greater than unity. The angle of the lower slope must be high enough to destabilize large parts of the caprock, and dissection of the lower scarp slope must not be so deep that the rockfall talus

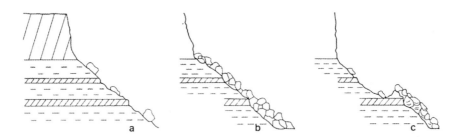

Fig. 22.7a The non-cyclic model of talus flatiron development (from Schmidt 1989b). (**a**) The caprock is undermined and destabilized by dissection and steepening of the lower slope. (**b**) A rockfall is triggered, which covers the complete lower slope with a talus cone, at the same time reducing the inclination of the lower slope. (**c**) the flanks of the talus cone are undermined by slope rills, the apex of the cone is detached from the scarp by tributaries to the major slope rills, and the talus cone becomes isolated from the scarp and a talus flatiron originates. After this succession of events the lower slope is again dissected and steepened, until a critical state of caprock destabilization is reached and a new rockfall is triggered

Fig. 22.7b Cuesta scarp with Glen Canyon sandstones overlying redbeds and bentonitic claystones (base of the scarp) of the Chinle Formation, Comb Ridge, Southeastern Utah (slope type 3). Different stages of non-cyclic talus flatiron development can be observed along this reach of the Comb Ridge. From right to left: talus flatiron separated from the scarp slope, flanks under-cut by slope rills, Chinle redbeds are exposed on the flanks, apex detached from the caprock; on the left a talus cone in parts still connected with the vertical cliff; between these talus cones and on the extreme left scarp sections relatively free of talus. Note the coarse-grained talus (>3 m) on the flatiron backslopes

is trapped in the large slope gullies. All of these conditions are met on slope type 3 on the Colorado Plateau (cf. Fig. 22.7b).

As the **climato-cyclic talus flatirons** attest to past climates and climatic change, the conditions necessary for their development are of particular interest.

(a) In the more humid phase, a smooth concave debris-flow-controlled slope profile must be formed as the initial stage of flatiron formation. This is only possible in thick lower slope rocks of relatively uniform resistance. In North Africa the continental redbeds below the Cretaceous limestones belong to this category of rock type (Fig. 22.4), as does the Mancos Shale

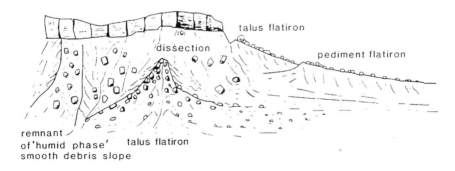

Fig. 22.8a The climato-cyclic model of talus and pediment flatiron development (from Schmidt 1989b, 1996). A smooth concave debris-flow-controlled slope is formed in the more humid phase (pluvial). The slope is then dissected and steepened in the more arid phase (interpluvial). Parts of the debris slope are detached from the active scarp by dissection and scarp retreat, and talus and pediment flatirons originate. The flatiron backslopes are remnants of the humid phase debris slope

Fig. 22.8b Scarp slope (slope type 1) on the eastern Book Cliffs, Colorado. Mancos Shale crops out on the lower scarp slope. Caprock is the Sego Sandstone of the Mesaverde Group. Remnants of the smooth, concave cold(humid) phase debris slope have been preserved. Parts of the debris covered slope are stll connected to the upper slope (left side of the photo). But most parts have been detached from the active slope, and talus flatirons have developed. The flatiron backslopes are covered with a continuous talus mantle and support the growth of dwarf junipers. The presently active warm(dry) phase steep Mancos Shale slope is deeply dissected and almost free of vegetation. Talus is concentrated in the slope rills

Fig. 22.9a Multi-level flatiron sequence on the slope of the eastern Book Cliffs, Colorado and Utah. The higher pediment flatiron in the right centre stands 50 m above the lower level. This lower level flatiron can be seen in the left centre close to the foot of the Book Cliffs. Its backslope is covered by dwarf forest. A schematic diagram of the sequence is shown in Fig. 22.9b

in the southwestern United States. Where the Mancos Shale crops out in the lower slope (slope type 1), cyclic flatirons are found on the Colorado Plateau (Schmidt 1996), such as on the slopes of the Mesa Verde near Cortez, Colorado and on the Book Cliffs, near Grand Junction, Colorado (Figs. 22.8b and 22.9a) (for locations see Fig. 22.2).

On lower scarp slopes with intercalated resistant layers (slope type 3, Figs. 22.3a and 22.5c), the resistance to change is too great for smooth debris slopes to form, and, consequently, for cyclic flatirons to develop (Schmidt and Meitz 1996). Intensity and duration of the past process were not able to gain ascendancy over lithological form resistance (Equation 22.1). On the other hand, free faces, as in the Moenkopi or Summerville Formations (Fig. 22.5d) with their short relaxation times, are also not characterized by flatiron forms (Schmidt and Meitz 2000b).

(b) The outer slopes of the talus and pediment flatirons need a highly protective talus cover to survive periods of cyclic duration. Only very resistant caprocks such as limestones, quartzites, or well-cemented sandstones and conglomerates can supply the lower slope with a sufficiently protective talus cover. The protective effect is enhanced when the talus becomes cemented by carbonates. Limestone and cemented sandstones and conglomerates can supply the slope debris with a lateral input of water containing carbonates in solution. In southern Morocco, for instance, the caprocks either guarantee the supply of carbonate cement (limestones, cemented conglomerates) or are very resistant (limestones, quartzites) (Schmidt 1987b, 1989b). Also the caprocks described from Israel (Gerson 1982) are very resistant (flints, limestones). On the Colorado Plateau, however, many of the caprocks are poorly cemented sandstones such as the Navajo, Entrada or de Chelly sandstones, but some have a stronger carbonate cement, such as parts of the Mesaverde Sandstones (Figs. 22.8b and 22.9a). If the flatirons carry a resistant backslope cover, their resistance to change and their tendency to be preserved will increase. The intensity and duration of the subsequent processes are not competent to obliterate the legacies of the past processes (Equation 22.2).

(c) The thickness ratio between soft rock and caprock is another important controlling factor. When the caprock is relatively thin and the thickness ratio in excess of 5, the mass of caprock material is not sufficient to cover and protect the complete lower slope with a talus apron, especially when the caprock is not very resistant. On the other hand, when the thickness ratio is less than unity, talus production exceeds talus transport and removal, the lower slope 'drowns' in debris and no smooth concave debris slope – the initial stage of flatiron formation – can develop. Additionally, great differences in resistance between caprock and soft rock promote cyclic flatiron formation.

(d) On complex scarps, where more than one caprock–soft-rock sequence is involved in scarp composition, the intercalated caprocks on the slopes impede the formation of a continuous concave debris slope and, hence, the development of talus or pediment flatirons (cf. Fig. 22.3b).

(e) Undercutting rivers or washes at the foot of the scarp will steepen the basal part of the lower slope and preclude basal concavities from being formed (Fig. 22.1). Also deeply incised canyons in the immediate scarp foreland with the resulting high vertical distance to base level make the evolution of cyclic flatirons impossible by decreasing the slope's resistance to change.

Climatic Thresholds

In those areas where the smooth debris slopes are being destroyed, ubiquitous dissection is taking place, and a new generation of cyclic flatirons is being formed. The annual precipitation lies between 75 and 175 mm in southern Morocco (Schmidt 1987b) and between 150 and well below 100 mm in Israel (Gerson and Grossman 1987). In both areas high precipitation intensities with intensive gullying have been recorded. The question arises: what is the threshold precipitation at which slope dissection gives way to smooth debris slope formation, and vice versa? Gerson and Grossman (1987) specifiy that a change to a moderately arid (150–250 mm) or semi-arid (250–400 mm) climate will cause the onset of 'debris-flow-controlled' processes. In a study of a cuesta scarp in the northwestern Tibesti area, Ergenzinger (1972) came to the conclusion that annual precipitation amounts in excess of 200 mm are needed for the formation of a smooth slopes controlled by sheetwash.

Investigations of Cenomanian–Turonian limestone scarps in southern Morocco showed that, although flatiron development is extremely active on the southern margin of the High Atlas, no slope dissection with flatiron formation occurs in its northern foreland (Schmidt 1989b). In this area mean annual precipitation exceeds 200 mm. In southern Morocco the 200-mm isohyet seems to represent a critical threshold value for the formation of talus relics on scarps with limestone caprocks underlain by soft continental redbeds. Apparently the smooth undissected concave lower scarp slopes, which are more densely vegetated than the slopes south of the High Atlas, are in equilibrium with the processes operating under the current climatic regime. Also no flatirons are formed in the Middle Atlas where precipitation exceeds 400 mm. Caution is needed in spatially extrapolating this threshold precipitation value of 200 mm, which is defined only for a specific set of lithological, structural, environmental, and biotic conditions.

On the Colorado Plateau smooth concave slopes are presently being formed on slope type 1 in areas with precipitation amounts exceeding 400 mm (profile c for slope type 1 in Fig. 22.6). They are dissected on cuesta scarp slopes where mean annual precipitation is 350 mm or less (profiles a. b for slope type 1 in Fig. 22.6). A detailed evaluation of the critical conditions and thresholds initiating variations in slope morphometry and process attributes for diverse slope types with different lithological and structural characteristics has been elaborated by Schmidt and Meitz (1996, 2000a,b) for the Colorado Plateau (see Fig. 22.6).

Timespans Required for the Formation of Climato-Cyclic Flatiron Forms and the Interpretation of Flatiron Sequences

The resistance to change and the relaxation times of hillslopes (see above) strongly control the timespans needed for the climato-cyclic talus relics to develop. Short-term climatic changes of a few hundred years or less and individual events are generally not recorded on hillslopes (Gerson 1982, Gerson and Grossman 1987, Littmann and Schmidt 1989). Gerson (1982) gave a tentative minimum estimate of 25 000–50 000 years for a smooth debris slope to develop after dry-phase dissection. And several thousand years are required to convert a talus slope into a slope controlled by gullying and dissectional activity (Gerson and Grossman 1987). Dissection of smooth debris slopes during the late Holocene dry period is an example (Littmann and Schmidt 1989). On the other hand, Gutiérrez et al. (2006) derive intra-Holocene ages for their youngest ^{14}C dated flatiron sequences in northeastern Spain.

There is a general paucity of chronostratigraphic information for coarse-grained hillslope deposits, since they are mostly free of fossils, organic carbon remains, or archaeological evidence. With the present shortage of absolute age determinations in most cases only some inferences about relative chronologies are possible. According to the model of climato-cyclic flatiron formation, the latest continuous smooth debris slope formed during the last wet period of at least several thousands years' duration. In Israel a talus relic has been dated by Mousterian artefacts to have originated during the early to middle Würm, which was a relatively wet period in the eastern Mediterranean region (Gerson and Grossman 1987). In southern Morocco many of the backslopes of the youngest flatiron generation merge into Soltanien (Würmian) terrace surfaces (Schmidt 1987b, Littmann and Schmidt 1989). Judging from the late Quaternary climatic record for areas where climato-cyclic flatirons have developed (northern fringe of the Sahara, Middle East, eastern Mediterranean area), periods of greater moisture availability coincide with certain stages of the last glaciation (Gerson 1982, Schmidt 1987b). Littmann (1989) showed that the Atlas region remained wetter than today until the late glacial. More arid climates with slope dissection seem to have prevailed in interglacial times (such as the Holocene).

On the slopes and in the foreland of many cuesta scarps, sequences of talus and pediment flatirons are found, each flatiron backslope preserving the latest stage of a pluvial period (Figs. 22.4, 22.8b and 22.9a). The inter-flatiron erosional gaps are the product of scarp recession during the dissection-controlled interpluvial and the subsequent debris-flow-controlled pluvial periods. They represent the denudational activity of a complete interpluvial–pluvial cycle. Multicycle flatiron generations are best preserved on scarp spurs and on the flanks of erosional outliers, where the intensity of slope denudational processes is reduced. In southern Morocco up to four successive generations have been found on the limestone scarps (Schmidt 1987b). Six levels were mapped in the Makhtesh Ramon area in Israel (Gerson and Grossman 1987). The number of flatiron generations is an indication of the number of late and middle Quaternary pluvial periods, and the distance between two successive flatirons (Fig. 22.9b) can be used for estimating scarp retreat during an interpluvial–pluvial cycle (Gerson 1982, Gutiérrez et al. 1998, Sancho et al. 1988, Schmidt 1987b, 1989a,b). To estimate rates of retreat, the lengths of the cycles must be known, which, in many cases, is still a matter of speculation. In Morocco the cycle Tensiftien (Riss) to Soltanien (Würm) lasted about 10^5 years, and the cycle Amirien (Mindel) to Tensiftien about 2×10^5 years. With these rough figures rates of retreat of about 0.5 m per 1000 years were calculated for the limestone scarps in southern Morocco (Schmidt 1987b, 1989a). In the southwestern United States rates of retreat of 3.5 m per 1000 years were estimated for the Mesaverde Sandstone/Mancos Shale cuesta scarp of the Book Cliffs for the Illinoian-Wisconsin interval (Schmidt 1996). Since the last pluvial, the scarp has retreated a mean distance of about 10 m (Fig. 22.9a,b). Progress in dating

techniques will certainly help to fit the cyclic flatiron sequences into the climatic pattern of the Quaternary and make it possible to use them for palaeoclimatic interpretations in a more detailed way than presently possible (Gutiérrez et al. 2006). One of the most promising methods is surface exposure dating (Dorn and Philips 1991). Yet in the special case of the Book Cliffs the sandstones of the Mesaverde group on the backslope of the flatirons were too easily weathered to allow reliable datings (Schmidt and Meitz 1997).

Conclusion

Relief elements on desert hillslopes may serve as valuable indicators of the low-frequency climatic variations during the Quaternary because, owing to their relatively high resistance to change and their usually long relaxation times, they record changes of long duration and high process effectiveness. Particularly bipartite slopes (cuesta scarps) may bear clear testimony of climatic change. Yet many of the interpretations of climatic change derived from hillslope form and process evidence are based on inference. Surveys of sediment yield (Schmidt 1985) and comparative investigations of cuesta scarps in areas of varying moisture availability (Barth and Blume 1973, Schmidt and

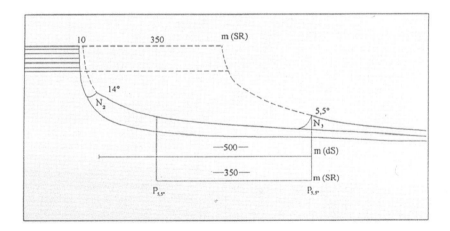

Fig. 22.9b Schematic diagram of the flatiron sequence shown in Fig. 22.9a. Sequences of flatirons can be used to estimate amounts of scarp retreat. Amounts of scarp retreat can be obtained by measuring the distance from the top of the older (*higher*) flatiron to the point of equal inclination (P 5,5°) on the next younger (*lower*) flatiron backslope. The amount of retreat (SR) in the example is 350 m. Measuring the distance between

the tops of successive flatirons or the distance to scarp slope (dS) will overestimate the amount of scarp recession, because the inner slopes of the flatirons have also retreated after their formation. Assuming that the cycle from end-Illinoian to end-Wisconsin lasted about 10^5 years, a rate of retreat of about 3.5 m per 1000 years is obtained. Since the last cold(humid) phase (Wisconsin) the scarp has retreated about 10 m (Schmidt 1996)

Meitz 1996, 2000a,b) strongly indicate that in the more arid regions of the southwestern United States denudational processes are in a state of high morphodynamic activity. On the other hand, landslide-controlled slopes are more active during humid periods. On a worldwide scale, the understanding of different process combinations under changing climatic conditions remains fragmentary. With regard to gravitational processes on cuesta scarps, a promising field of research seems to be the association of alluvial chronologies with rockfalls and slope deposits to decide whether events have been more frequent during specific periods in the past.

Process alternations, which show not only changes of process rates but of process type, are of major interest for palaeoclimatic interpretations. Landform assemblages which give evidence of geomorphic processes of different character are talus and pediment flatirons. Present possibilities of dating these landform successions are still unsatisfactory, though progress has been made (Gutiérrez et al. 2006). But surface exposure dating may permit the dating of landforms in deserts previously thought to be undatable (Dorn and Philips 1991, Beck 1994) and may help attain greater precision in the palaeoclimatic interpretation of hillslope form sequences. Climate, however, is only one of several factors controlling hillslope processes, and results obtained in one region cannot readily be transferred to other regions in different structural and lithological settings.

Acknowledgments Research in southern Morocco and on the Colorado Plateau was financially supported by the Deutsche Forschungsgemeinschaft (German Science Foundation) in a number of research grants from 1978 to 1998. Without this funding field and laboratory work would not have been possible.

References

Ahnert, F. 1960. The influence of Pleistocene climates upon the geomorphology of cuesta scarps on the Colorado Plateau. *Annals of the Association of American Geographers* **50**, 139–56.

Baars, D.L. 1971. Rates of change, 1871–1968. In *Geology of Canyonlands and Cataract Canyon*, D.L. Baars and C.M. Molenaar (eds.), 89–99. Durango, CO: Four Corners Geological Society.

Barth, H.K. 1976. Pedimentgenerationen und Reliefentwicklung im Schichtstufenland Saudi-Arabiens. *Zeitschrift für Geomorphologie Supplement Band* **24**, 111–9.

Barth, H.K. and H. Blume 1973. Zur Morphodynamik und Morphogenese von Schichtkamm- und Schichtstufenrelief in den Trockengebieten der Vereinigten Staaten. *Tübinger Geographische Studien* **53**, 1–102.

Barth, H.K. and H. Blume 1975. Die Schichtstufen in der Umrahmung des Murzuk-Beckens (Iibysche Zentralsahara). *Zeitschrift für Geomorphologie Supplement Band* **23**, 118–29.

Beck, C. (ed.) 1994. Dating in Exposed and Surface Contexts. Albuquerque.

Blume, H. and H.K. Barth 1972. Rampenstufen und Schuttrampen als Abtragungsformen in ariden Schichtstufenlandschaften. *Erdkunde* **26**, 108–16.

Büdel, J. 1982. *Climatic geomorphology*. Princeton, NJ: Princeton University Press.

Colbert, E.H. 1966. Rates of erosion in the Chinle Formation – ten years later. *Plateau* **38**, 68–74.

Cooke, R.U. and R.W. Reeves 1976. *Arroyos and environmental change in the American South-West*. Oxford: Oxford University Press.

Davidson, E.S. 1967. Geology of the Circle Cliffs area, Garfield and Kane Counties, Utah. *U.S. Geological Survey Bulletin* **1229**, 1–140.

Dongus, H. 1980. Rampenstufen und Fußflächenrampen. *Tübinger Geographische Studien* **80**, 73–8.

Dorn, R.I. and F.M. Philips 1991. Surface exposure dating: review and critical evaluation. *Physical Geography* **12**, 303–33.

Euler, R.C., G.C. Gumerman, T.N.V. Karlstrom, J.S. Dean, et al. 1979. The Colorado Plateaus: cultural dynamics and palaeoenvironment. *Science* **205**, 1089–101.

Ergenzinger, P. 1972. Reliefentwicklung an der Schichtstufe des Massif d'Abo (Nordwesttibesti). *Zeitschrift für Geomorphologie Supplement Band* **15**, 93–112.

Everard, C.E. 1963. Contrasts in the form and evolution of hillside slopes in central Cyprus. *Transactions of the Institute of British Geographers* **32**, 31–47.

Ford, T.D., P.W. Huntoon, W.J. Breed and G.H. Billingsley 1976. Rock movement and mass wastage in the Grand Canyon. In *Geology of the Grand Canyon*, W.J. Breed and E. Roat (eds.), 116–28. Flagstaff, AZ: Museum of Northern Arizona.

Gerson, R. 1982. Talus relicts in deserts: a key to major climatic fluctuations. *Israel Journal of Earth Sciences* **31**, 123–32.

Gerson, R. and S. Grossman 1987. Geomorphic activity on escarpments and associated fluvial systems in hot deserts. In *Climate, history, periodicity, predictability*, R. Rampino, J.E. Sanders, W.S. Newman and L.K. Königsson (eds.), 300–22. New York: Van Nostrand Reinhold.

Graf, W.L. 1983. The arroyo problem – palaeohydrology and palaeohydraulics in the short term. In *Background to palaeohydrology*, K.J. Gregory (ed.), 279–302, Chichester: Wiley.

Graf, W.L. 1985. The Colorado River, Instability and Basin Management. Resource Publications in Geography, Washington.

Graf, W.L., R. Hereford, J. Laity and R.A. Young 1987. Colorado Plateau. In *Geomorphic systems of North America*, W.L. Graf (ed.), 259–302. Geological Society of America, Centennial Special Volume 2.

Grunert, J. 1983. Geomorphologie der Schichtstufen am Westrand des Murzuk-Beckens (Zentrale Sahara). *Relief, Boden, Paläoklima* **2**, 1–269.

Gutiérrez, M., F. Gutiérrez and G. Desir 2006. Considerations on the chronological and causal relationships between talus flatirons and paleoclimatic changes in central and northeastern Spain. Geomorphology **73**, 50–63.

Gutiérrez, M., C. Sancho, T. Arauzo and J.L. Pena 1998. Scarp retreat rates in semiarid environments from talus flatirons (Ebro Basin, NE Spain). Geomorphology **25**, 11–21.

Hack, J.T. 1942. The changing physical environment of the Hopi Indians of Arizona. *Peabody Museum Paper* **35**, 1–86.

Hadley, F.H. and G.C. Lusby 1967. Runoff and hillslope erosion resulting from high-intensity thunderstorms near Mack, Western Colorado. *Water Resources Research* **3**, 139–43.

Hereford, R. 1987. Modern alluvial history of the Paria River drainage basin. Quaternary Research **25**, 293–311.

Howard, A.D. and R.C. Kochel 1988. Introduction to cuesta landforms and sapping processes on the Colorado Plateau. In *Sapping features on the Colorado Plateau, a comparative planetary geology field guide*, A.D. Howard, R.C. Kochel and H.E. Holt (eds.), 6–56. National Aeronautics and Space Administration Special Publication 491.

Huff, L.C. and F.G. Lesure 1965. Geology and uranium deposits of Montezuma Canyon area, San Juan County, Utah. *U.S. Geological Survey Bulletin* 1190.

Hunt, C.B., P. Averitt and R.L. Miller 1953. Geology and geography of the Henry Mountains region, Utah. *U.S. Geological Survey Professional Paper* 228.

Iorns, W.V., C.H. Hembree and G.L. Oakland 1965. Water resources of the upper Colorado River basin, technical report. *U.S. Geological Survey Professional Paper* 441.

Joly, F. 1962. Etudes sur le relief du Sud-est marocain. *Travaux de l'Institut Scientifique Chérifien, Série Géologie et Géographie Physique* **10**, 1–578.

Karlstrom, T.N.V. 1988. Alluvial chronology and hydrologic change of Black Mesas and nearby regions. In *The Anasazi in a changing environment*, G.J. Gumerman (ed.), 45–91. Cambridge: Cambridge University Press.

Koons, D. 1955. Cliff retreat in the southwestern United States. *American Journal of Science* **253**, 44–52.

Laity, J.E. and M.C Malin 1985. Sapping processes and the development of theater-headed valley networks on the Colorado Plateau. *Geological Society of America Bulletin* **96**, 203–217.

Littmann, T. 1989. Spatial patterns and frequency distribution of late Quaternary water budget tendencies in Africa. *Catena* **16**, 163–88.

Littmann, T. and K.-H. Schmidt 1989. The response of different relief units to climatic change in arid environments (southern Morocco). *Catena* **16**, 343–55.

Lusby, G.C. 1979. Effects of converting sagebrush cover to grass on the hydrology of small watersheds at Boco Mountain, Colorado. *U.S. Geological Survey Water Supply Paper* 1532–J.

Mensching, H. 1974. Landforms as a dynamic expression of climatic factors in the Sahara and Sahel – a critical discussion. *Zeitschrift für Geomorphologie Supplement Band* **20**, 168–77.

Moss, J.R. 1977. The formation of pediments: scarp backwearing or surface downwasting. In *Geomorphology in arid regions*, D.O. Doehring (ed.), 51–78. London: Allen & Unwin.

Mullens, T.E. 1960. Geology of the Clay Hills area, San Juan County, Utah. *U.S. Geological Survey Bulletin* 1087–H, 259–336.

Oberlander, T.M. 1972. Morphogenesis of granitic boulder slopes in the Mojave Desert, California. *Journal of Geology* **80**, 1–20.

Oberlander, T.M. 1977. Origin of segmented cliffs in massive sandstones in Southeastern Utah. In *Geomorphology in arid regions*, D.O. Doehring (ed.), 79–114. London: Allen & Unwin.

Oberlander, T.M. 1997. Slope and pediment systems. In *Arid zone geomorphology*, D.S.G. Thomas (ed.), 135–163. London: Wiley.

Parsons, A.J. 1988. *Hillslope form*. London: Routledge.

Phoenix, D.A. 1963. Geology of the Lees Ferry area, Coconino County, Arizona. *U.S. Geological Survey Bulletin* **1137**, 1–86.

Powell, J.W. 1875. *Exploration of the Colorado River of the West and its tributaries*. Washington, US Government Printing Office.

Reiche, P. 1937. The Toreva Block – a distinctive landslide type. *Journal of Geology* **45**, 538–48.

Sakaguchi, Y. 1986. Pediment – a glacial cycle topography? *Bulletin of the Department of Geography, University of Tokyo* **18**, 1–19.

Sancho, C., M. Gutierrez, J.L. Pena and F. Burillo 1988. A quantitative approach to scarp retreat starting from triangular slope facets, central Ebro basin, Spain. *Catena Supplement* **13**, 139–46.

Schmidt, K.-H. 1985. Regional variation of mechanical and chemical denudation, Upper Colorado River basin, USA. *Earth Surface Processes and Landforms* **10**, 497–508.

Schmidt, K.-H. 1987a. Factors influencing structural landforms dynamics on the Colorado Plateau – about the necessity of calibrating theoretical models by empirical data. *Catena Supplement* **10**, 51–66.

Schmidt, K.-H. 1987b. Das Schichtstufenrelief der präsaharischen Senke, Süd-Marokko. *Zeitschrift für Geomorphologie Supplement Band* **66**, 23–36.

Schmidt, K.-H. 1988. Die Reliefentwicklung des Colorado Plateaus. *Berliner Geographische Abhandlungen* **49**, 1–183.

Schmidt, K.-H. 1989a. Stufenhangabtragung und geomorphologische Entwicklung der Hamada de Meski, Südostmarokko. *Zeitschrift für Geomorphologie Supplement Band* **74**, 33–44.

Schmidt, K.-H. 1989b. Talus and pediment flatirons – erosional and depositional features on dryland cuesta scarps. *Catena Supplement* **14**, 107–18.

Schmidt, K.-H. 1996. Talus and pediment flatirons – indicators of climatic change on scarp slopes on the Colorado Plateau, USA. *Zeitschrift für Geomorphologie Supplement Band* **103**, 135–158.

Schmidt, K.-H. and P. Meitz 1996. Cuesta scarp forms and processes in different altitudinal belts of the Colorado Plateau as indicators of climatic change. In *Advances in Hillslope Processes*, Anderson, M.G. and S.M. Brooks (eds.), 1079–1097, Chichester (Wiley).

Schmidt, K.-H. and P. Meitz 1997. Surface Exposure Dating of talus and pediment flatirons on the Colorado Plateau. In *PRIME Lab 1994-1997 Progress Report: 19–21*, Sharma, P. (ed.), Purdue University, West Lafayette.

Schmidt, K.-H. and P. Meitz 2000a. Effects of increasing humidity on slope morphology, studied on cuesta scarps on the Colorado Plateau, USA. – IAHS Publications **261**, 165–181.

Schmidt, K.-H. and P. Meitz 2000b. Schichtstufenhänge auf dem Colorado Plateau, USA – lithologische Steuerung und klimatische Höhenstufendifferenzierung. *DIE ERDE* **131**, 181–204.

Schumm, S.A. 1964. Seasonal variations of erosion rates and processes on hillslopes in western Colorado. *Zeitschrift für Geomorphologie Supplement Band* **5**, 215–38.

Schumm, S.A. and R.J. Chorley 1964. The fall of Threatening Rock. *American Journal of Science* **262**, 1041–54.

Schumm, S.A. and R.J. Chorley 1966. Talus weathering and scarp recession in the Colorado Plateau. *Zeitschrift für Geomorphologie* **10**, 11–36.

Selby, M.J. 1993. Hillslope Materials and Processes. Oxford University Press.

Shawe, D.R., G.C. Simmons and N.L. Archbold 1968. Stratigraphy of Slick Rock District and vicinity, San Miguel and Dolores Counties, Colorado. *U.S. Geological Survey Professional Paper* 576-A.

Shoemaker, E.M. and H.G. Stephens 1975. First photographs of the Canyonlands. *In Canyonlands country*, J.E. Fassett (ed.), 111–22. Durango, CO: Four Corners Geology Society.

Smith, B.J. 1978. The origin and geomorphic implications of cliff foot recesses and tafoni on limestone hamadas in the north-west Sahara. *Zeitschrift für Geomorphologie* **22**, 21–43.

Smith, B.J. 1987. An integrated approach to the weathering of limestone in an arid area and its role in landscape evolution: a case study from Southeast Morocco. In *International Geomorphology, part II*, V. Gardiner (ed.), 637–57. Chichester: Wiley.

Smith, W.O., C.P. Vetter and G.B. Cummings 1960. Comprehensive survey of sedimentation in Lake Mead, 1948–49. *U.S. Geological Survey Professional Paper* 295.

Strahler, A.N. 1940. Landslides of the Vermilion and Echo Cliffs, Northern Arizona. *Journal of Geomorphology* **3**, 285–301.

Thaden, R.E., A.F. Trites and T.L. Finell 1964. Geology and ore deposits of the White Canyon area, San Juan and Garfield Counties, Utah. *U.S. Geological Survey Bulletin* **1125**, 1–166.

Viles, H.A. and A.S. Goudie 2003. Interannual, decaded and multi-decadal scale climatic variability and geomorphology. *Earth Science Reviews* **61**, 105–181.

Wells, S.G., T.F. Bullard and L.N. Smith 1982. Origin and evolution of deserts in the Basin and Range and the Colorado Plateau provinces of western North America. *Striae* **17**, 101–11.

Chapter 23

River Landforms and Sediments: Evidence of Climatic Change

Ian Reid

Introduction

Significant changes in the style of superimposed fluvial deposits have long been used by sedimentologists as an indication of broad changes in climate. Certainly, in traversing the globe from one environmental extreme to another – from humid to arid regions – it is possible to discern substantial differences in the character of rivers, both from the point of view of the forms that they adopt and the sediments that they carry (Schumm 1977; Wolman and Gerson 1978). However, rivers react to a number of large-scale stimuli, and it is often difficult to determine whether a change in character reflects tectonic or climatic influences, or both (Frostick and Reid 1989a). Attempts to attribute cause have inevitably and ingeniously simplified the setting by choosing systems in areas where most factors are presumed to have remained more-or-less constant whilst the putative controlling factor has varied monotonically. So, for example, location on a stable craton may allow an assessment of the impacts of climate change on river systems without the additional complications that arise from tectonic instability (Schumm 1968; Rust and Nanson 1986). However, while this approach may be useful in deducing the response of rivers to recent shifts in climate – say, those of the late Pleistocene and Holocene – and may be instructive in indicating the direction and magnitude of likely changes to be expected in river systems during periods of environmental change, the sedimentary legacies are often too similar to those assumed to have arisen from tectonic influences to ascribe changes in sedimentary style confidently to either set of causes when stepping further back in time (Frostick et al. 1992).

In fact, rivers are comparatively insensitive to changes in climate unless these changes are substantial. Moreover, it is likely that there is considerable hysteresis in the relation between climate change and river metamorphosis. Intuitively, it might be argued that the rate of change in a parameter such as channel sinuosity would be greater during a shift towards aridity, with the tendency to increased flashiness in the flood regime and, therefore, greater potential erosional damage, than it would in a corresponding period of increasing humidity, when channel change might depend upon sediment accretion at a time of progressively decreasing rates of sediment yield (see, e.g., Hereford 1984, 1986).

Nanson and Tooth (1999) implied that they were surprised by the declamation (made by the author in the 1st edition of this book) that rivers are insensitive to comparatively small degrees of climate change, though, later, they conceded that the response rate might depend on local circumstances, indicating that 'dryland rivers which are confined within bed-rock gorges, stabilised by effective riparian vegetation, or restrained by indurated or very clay-rich alluvium, are probably relatively insensitive to all but extreme environmental changes, whereas rivers within erodible silty or sandy alluvium will probably respond much more readily to even modest changes' (Nanson et al. 2002).

The sensitivity of rivers to climate change is also conditioned, in part, by the wide range in flood magnitude that is experienced under a singular climatic regime and by the erosional or depositional legacy of large events. Baker and his colleagues (Baker 1977;

I. Reid (✉)
Department of Geography, Loughborough University
Loughborough, LE11 3TU, UK
e-mail: ian.reid@lboro.ac.uk

A.J. Parsons, A.D. Abrahams (eds.), *Geomorphology of Desert Environments*, 2nd ed.,
DOI 10.1007/978-1-4020-5719-9_23, © Springer Science+Business Media B.V. 2009

Baker et al. 1983; Patton et al. 1982) have drawn attention to the erosional implications of very large floods having recurrence intervals in the order of 10^3 years in the Northern Territories of Australia and western Texas. Nanson (1986) has described what he calls flood-plain stripping in the coastal valleys of New South Wales. Here, he has deduced that substantial quantities of accumulated sediment are removed periodically during the passage of rare high-magnitude floods, leaving channel dimensions that bear little relation to the more frequent floods ordinarily thought to be channel-formative and a flood-plain topography that is difficult to interpret in the context of observed modern flows. Under a drier regime, Schumm and Lichty (1963) collated historical records to show that the Cimarron River of south-western Kansas was widened dramatically from an average of 15 to 200 m by a single high-magnitude flood in 1914. During the next 25 years, channel width increased to an average value of 365 m with the passage of other damaging floods, after which it declined progressively to 183 m over a period of 21 years, in sympathy with lower flood discharges. But the fact that at-a-point channel width can be shown to have had a 24-fold range during a period when climate did not change dramatically suggests that parameters such as these – so important as tools in palaeohydrology – may have only limited value in diagnosing climate change, at least in arid zones.

The damaging impact of large floods highlights another problem in using channel dimensional parameters as a diagnostic of climate change. Schumm and Lichty's (1963) study of the Cimarron River suggested a relaxation period (the lag between infliction of the flood damage and a return to pre-flood conditions) of much more than 50 years in this semi-arid environment. This compares with a relaxation period of less than 20 years in a temperate humid environment where flood frequency and the prospects of channel reparation are higher (Gupta and Fox 1974). Wolman and Gerson (1978) have suggested from rather more casual observations made in Sinai that the lack of sediment-trapping riparian vegetation and the extreme infrequence of floods in arid and hyper-arid zones removes almost entirely the potential for channel restoration. In this type-environment, the relaxation period is, to all intents and purposes, infinite. In these circumstances, and with the inevitability of poor or non-existent hydrological records in such drylands,

it may be difficult to judge whether or not a channel and its sediments are anachronistic. In other words, it may be difficult to establish whether they represent the prevailing climate or one that is now past, especially if the differences involve a comparatively narrow range of annual effective precipitation.

Having cautioned that diagnosing climate change from channel and sediment characteristics can be hazardous where extremes of weather inflict damage that is only slowly restored to reflect the prevailing climatic average, it has to be pointed out that rivers may not react entirely predictably to changing climate – at least as far as can be ascertained, given that shifts in climate can rarely be defined precisely in terms of the consequences for runoff. One problem is that each drainage basin is a unique permutation of all those factors that encourage or discourage runoff and sediment transfer. Factors which act as regulators of change may be more important in one basin than in another. Despite this complexity and the fact that the infrequence of runoff in deserts has deterred the collection of data relating to flood flows and sediment transport – so that understanding of fluvial processes in such environments is undernourished – certain patterns of climate-induced behavioural change have emerged.

Quaternary Climate Change and Continental Drainage

The changing pattern of climate in middle and low latitudes, particularly for the last 30 ka, has now been successfully documented (e.g. Street and Grove 1979; Nanson et al. 1992). Because of the equivocal nature of river responses to subtle shifts in the character of water catchments, fluvial landforms and depositional legacies have usually been used to corroborate evidence derived from other sources – principally datable, lacustrine, strand-line deposits – rather than acting as primary sources of information. However, set against evaporation, changes in river regime are, and have been, an important determinant of lake levels. Nowhere has this been more significant than in dryland lake basins such as those of the East African Rift (Butzer et al. 1972; Gasse and Street 1978), the Levant (Begin et al. 1974), Australasia (Bowler et al. 1976), western North America (Benson 1978) and, increasingly, Central Asia (Kroonenberg et al. 1997).

The water-balance curve of Lake Chad can be taken as a typical example of the changing relationship between runoff and evaporation in the period running up to and since the last glacial maximum (Fig. 23.1; Servant and Servant-Vildary 1980). Particularly noteworthy in this curve is the trend towards extreme desiccation between 20 and about 13 ka BP. This pattern was widespread throughout low and middle latitude drylands and obviously had important implications for river systems. At a comparatively local scale, low base-levels in internal drainage systems led to fluvial incision. This is not particularly well documented, if only because, although present lake levels are comparatively low, they are nevertheless higher than their late glacial minima. This means that evidence is either sub-lacustrine or buried by aggradation. So, for example, the surface of the Dead Sea is thought, on the basis of the geochemistry of evaporites, to have lain at about −700 m (relative to global ocean surface), some 300 m lower than at present (Katz et al. 1977), and hydrographic surveys have shown sub-lacustrine valleys running offshore as projections of present lines of drainage (Neev and Emery 1967). A subsurface seismic examination

of small deltas on the shores of Lake Turkana in northern Kenya has revealed palaeochannels some 20 m below the present alluvial feeders. These buried palaeochannels have gradients about 1.5 times those of their modern counterparts and were given an age of about 17 ka BP, albeit on the flimsiest of arguments – that is, through an assessment of the thickness of the alluvial wedge in the context of contemporary sedimentation rates and questionable presumptions of uniformitarianism (Frostick and Reid 1986).

On a broader geographical canvas, the comparative aridity of the last glacial maximum had considerable consequence for the nature and behaviour of the so-called exotic perennial rivers that traverse the world's drylands. Perhaps best known (in circumstances where information is always sparse, either because of the poor sedimentary preservation potential of many desert environments or because information is being synthesised from a huge continental area) is that of the Nile (Adamson et al. 1980; Adamson 1982). There is inevitable controversy over the development of the integrated drainage that exists today. De Heinzelin and Paepe (1965) were of the view that the present river system is a product of the Upper Pleistocene, and that,

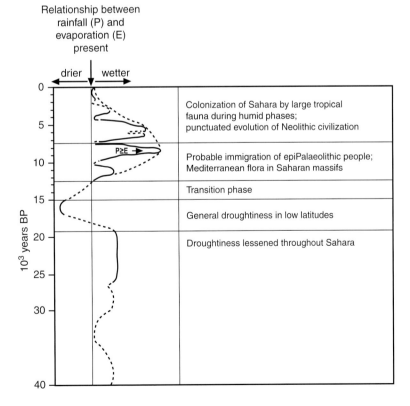

Fig. 23.1 Climate change in the Sahel over the last 40,000 years as inferred from strand-line and other evidence in the Lake Megachad Basin (after Servant and Servant-Vildary 1980)

previously, there was a series of separate and unconnected basins. Butzer and Hansen (1968) acknowledge that the Blue Nile and other rivers rising in the highlands of Ethiopia had probably contributed to the trunk stream since the beginning of the Pleistocene, while Williams and Williams (1980) suggest that the Nile has had Ethiopian headwaters since Tertiary times. The history of the White Nile is even less certain. Kendall (1969) showed, from evidence contained in bottom sediments, that Lake Victoria had no outlet during an undefined period prior to 12 ka BP, and Williams and Adamson (1974) have speculated that the White Nile was, at best, highly seasonal if not completely ephemeral during the hyper-desiccation of the terminal Pleistocene arid phase. The picture that emerges merely reinforces a view that the Nile system is peculiar. Climate change and the ruptured tectonic setting of the White Nile headwaters give us the clues we need to understand why the present river network defies the dendritic laws of drainage that help to describe large river systems on other relatively stable cratons.

The other notable element in the water-balance of drylands in the recent past was the shift towards more humid conditions during the terminal Pleistocene and early Holocene (about 12–5 ka BP; Fig. 23.1). The impact on rivers was as dramatic as was the preceding phase of desiccation, and has to be understood if only to interpret the fluvial relics that litter the present arid landscape.

In the Sahel of West Africa, Talbot (1980) has drawn attention to the now defunct northern tributaries of the Niger, indicating that the catchment area contributing water to the trunk stream was perhaps as much as one-and-a-half times that of the present-day (Fig. 23.2). Indeed, he infers, from a broad consideration of valley configuration, that the Dallol Bosso (now largely relic) may have been the trunk stream and that the upper Niger was one of its right-bank tributaries. Further east, the catchment area contributing to Lake Chad was enormous during this humid phase of the terminal Pleistocene/early Holocene, with drainage focused on the thermally subsided Cretaceous Niger Rift, at the southern end of which now lies the shrunken rump of the present lake. In fact, Lake Megachad, spreading into the Bodele Depression, occupied an area about six times that of present Lake Victoria and not far short of the area occupied by the modern Caspian Sea (Servant and Servant-Vildary 1980; Fig. 23.2). Apart from the now relic northern feeders of Megachad that, at this time, carried seasonal discharge down from the Tibesti Massif, the other element of the regional drainage that is of considerable interest (in part because it explains patterns of sedimentation beyond the present Chad catchment and outside the area of modern

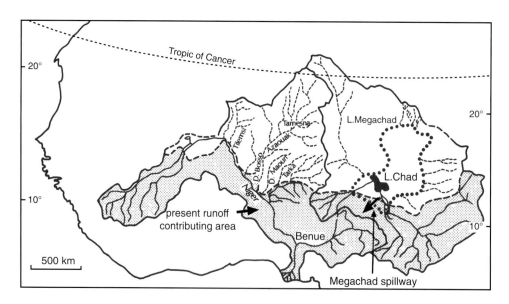

Fig. 23.2 Water catchment of the Niger and Benue Rivers *c.* 9,000 BP and the present area contributing runoff to the Niger delta and to Lake Chad (stippled). The maximum extent of palaeo-Lake Megachad and its early Holocene overspill to the Benue River are shown together with the largely relic drainage of the southern Sahara (after Talbot 1980)

desert) lies to the south-west of the present remnant lake. For a brief period around 9 ka BP, when the water-balance favoured rainfall over evaporation, Megachad spilled into the Benue River and, from there, waters originating in the heart of the present Sahara reached the South Atlantic via the Niger delta (Servant and Servant-Vildary 1980; Grove 1985; Figs. 23.1 and 23.2).

Neither was this the only large-scale integration of drainage on the African continent at this time. In East Africa, a change in regional water-balance led to a rise in lake-levels within closed rift basins and in overspill from one to the other. In fact, it was at this time that the Nile catchment expanded enormously relative to its present extent, if only for a brief period. Water from Lakes Chiamo and Abaya overspilled by way of the Sagan River into Chew Bahir, which, in turn, overspilled through the Bakate Corridor to a much expanded Lake Turkana (Grove et al. 1975; Frostick and Reid 1989b; Fig. 23.3). To the south of Lake Turkana, swollen Lake Logipi spilled from the Suguta Valley westward into the Kerio River and thence to Lake Turkana (Truckle 1976), while Turkana, already collecting the seasonal rains of a portion of the

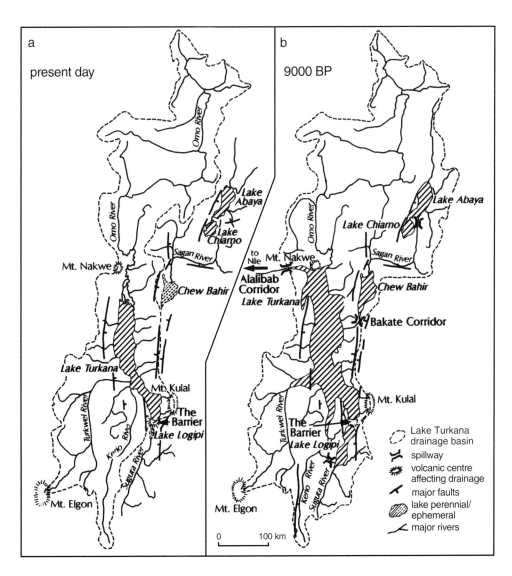

Fig. 23.3 Water catchment of Lake Turkana in the northern section of the Gregory Rift of East Africa: (**a**) during the present phase of relative desiccation; and (**b**) during the early Holocene humid phase about 9,000 BP, showing the swollen lakes and the connecting spillways from basin to basin and to the River Nile (after Frostick and Reid 1989b; Harvey and Grove 1982)

Ethiopian highlands through the Omo River, as it does today, spilled to the north-west through the Alalibab Corridor. From here, the waters crossed the Lotogipi Swamps to enter the Pibor River, and then the River Sobat to join the White Nile (Harvey and Grove 1982; Fig. 23.3b).

Given this temporary increase in contributing catchment, together with the overspill of Lake Victoria via the Falls at Jinja and the prevalence of relatively humid conditions across Equatoria, it is not surprising that the White Nile was three to four times the width of the present river (Adamson et al. 1982) and that Fairbridge (1963) was able to deduce a palaeodischarge three times that typical of the present-day. Williams et al. (2000) have documented the late Pleistocene/early Holocene relic meanders of the White Nile in the vicinity of El Geteina township, about 150 km upstream from Khartoum (Fig. 23.4). The interplay of the swollen sinuous river with the dunes of a large regional sand sea – legacy of the arid terminal Pleistocene – is reminiscent of the present-day Channel Country of central Australia (Nanson et al. 1995), although the White Nile, drawing not only on the Ethiopian highlands through the Rivers Omo and Sagan, but on the Mt Elgon massif through the Turkwel River and the drainage to and through Lake Victoria had sufficient discharge to maintain a very large, singular, meandering channel, unlike Cooper Creek.

Neither was this pattern of overspill unique to the early Holocene. For instance, Frostick and Reid (1980) have drawn attention to a previous overspill from Chew Bahir to Lake Turkana, dated tentatively as older than 100 ka. This was responsible, in part, for deepening the Bakate Corridor (re-used so conveniently in the early Holocene) and for spreading the last, coarse-grained unit of the hominid-bearing Plio-Pleistocene Koobi Fora Formation in the East Turkana Basin, the fluvio-lacustrine sediments of which have revealed so much about early human phylogeny.

Fluvial Landform Reconstruction

Understanding this climate-driven sequence of events has been of considerable importance in arriving at an interpretation of the underfit or relict nature of drainage in present-day drylands. It has also provided the information and confidence necessary to reconstruct large erosional and aggradational landforms even where evidence is flimsy.

An interesting example comes from the hyperarid core of the Sahara, a region which has been used by the US National Aeronautics and Space Administration as an Earthbound analogue of the Martian surface (El-Baz and Maxwell 1982). The Gilf Kebir plateau and surrounding area of SW Egypt – one objective of Bagnold's celebrated desert expeditions in the 1930s (Bagnold 1931) – is currently dominated by aeolian processes that form, among other features, the giant Selima Sandsheet (Breed et al. 1987). However, the area has not always been hyperarid. Peel (1939) advanced the view that the region bore the vestiges of a Tertiary landscape developed by fluvial processes, and this has been corroborated recently by surface-penetrating spaceborne imaging radar which has revealed broad alluvial valleys underlying features such as the Selima Sandsheet (McCauley et al. 1982b). Armed with this information, McCauley et al. (1982a) have recon-

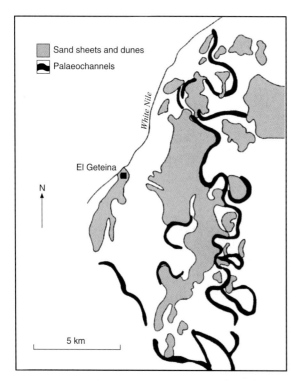

Fig. 23.4 Late Pleistocene-early Holocene relic channel meanders of the once-swollen lower White Nile east of El Geteina township, which is sited on the right-bank of the current diminutive river. The palaeochannels inter-finger with an extensive late Pleistocene sand sheet (after Williams et al. 2000)

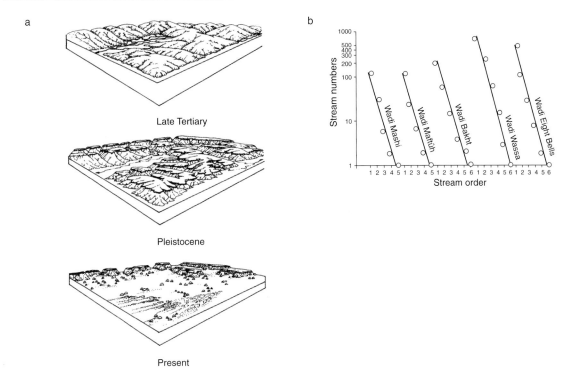

a

Late Tertiary

Pleistocene

Present

b

Fig. 23.5 (**a**) Cartoons showing a reconstruction of the humid Tertiary, semi-arid Pleistocene and modern landscapes of the Gilf Kebir plateau, which lies in the currently hyper-arid core of the Sahara in south-western Egypt. (**b**) Stream-ordering of reconstructed Tertiary drainage systems in the Gilf Kebir plateau (after McCauley et al. 1982a)

structed the Gilf Kebir landscape (Fig. 23.5a). Using the argument that vestigial knolls littering the present landscape represent mass-wasted remnants of interfluves, they have drawn a best approximation of the former drainage net. This reconstruction has then been 'tested' using stream-ordering rules. The regression coefficients that describe the relationship between number of streams and stream order are indistinguishable from those derived for modern drainage nets in environments less arid than those that prevail in the present-day Gilf Kebir (Fig. 23.5b).

Another case where knowledge of the climate-driven sequence of events during the late Pleistocene and early Holocene has allowed a confident interpretation of large fluvial landforms is that of the Gezira in Sudan (Williams and Adamson 1973; Adamson et al. 1982; Fig. 23.6). The Gezira is a large plain that lies between the White and Blue Niles. Superficially, it consists of clay. However, Williams and Adamson were able to establish conclusively that the plain is a relict low-angle fan of the Blue Nile, and that it is crossed by a number of defunct sand-filled distribu-

taries. It is now clear that the distributaries were a product of the terminal Pleistocene desiccation phase, when flows from the Blue Nile were probably even more seasonal and somewhat more diminutive that at present, but when a change in land-cover would have meant higher and coarser specific sediment yield from the Ethiopian hinterland. Further upstream from the fan itself, the rivers Rahad, Dinder and Blue Nile have been shown to have been elements in a complex of interconnected relatively shallow sandy channels (Fig. 23.6b_1) separated only by the silty-clay of the flood-plain that accumulated from frequent overbank spills. The more humid phase of the terminal Pleistocene and early Holocene was marked by incision of the Blue Nile, an abandonment of the Gezira fan, and clarification of the drainage pattern into a trunk stream and two substantial right-bank tributaries (Fig. 23.6b_2). As this occurred, silty-clay continued to accrete on the flood plain, but the change in soil-water relations brought about by fluvial incision and decreasing frequency of overbank flooding is nicely recorded by the mollusc population (Fig. 23.6c).

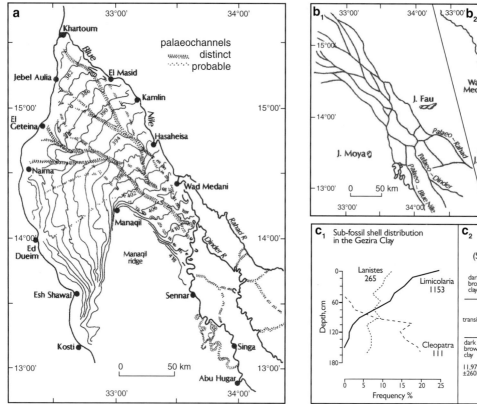

Fig. 23.6 (**a**) Late Pleistocene low-angle alluvial fan of the Blue Nile, the Gezira, showing sandy distributary channels and contours. (**b₁**) Reconstructed channel complex of the Blue Nile, Rahad and Dinder Rivers at the head of the Gezira fan during the arid-phase of the terminal Pleistocene. (**b₂**) Present-day river planform, following early Holocene incision of the Nile. (**c₁**) Changes in the abundance of three genera of molluscs with depth in the late Pleistocene/Holocene flood-plain muds of the Blue Nile at the head of the Gezira fan. *Cleopatra* is aquatic; *Lanistes* is amphibious; *Limicolaria* is terrestrial. (**c₂**) Soil profile on flood-plain of the Blue Nile, showing ^{14}C date derived from shell material (after Adamson et al. 1982, mollusc data originally from Tothill 1946)

Aquatic genera such as *Cleopatra*, abundant during the early Holocene humid phase, rapidly died out as flood-plain and slack-water pools became less common or non-existent; amphibious genera such as *Lanistes* were able to survive because the self-mulching properties of the crusted mud maintained a suitably moist subsoil environment for survival between successive flood inundations; terrestrial genera such as *Limicolaria* would have found conditions too wet during the early part of the Holocene humid phase, but obviously enjoyed conditions more and more as the flood-plains became drier. Adamson et al. (1982) considered that this reorganization of the Blue Nile drainage from multi-thread to single-thread stream-type had occurred somewhat before 5 ka BP.

Climate and Channel Planform

Changes in channel planform have long been thought of as indicative of climate change, differences in the meander wavelengths of valleys and rivers that flow through them being ascribed to changes in runoff volumes during the Quaternary. However, it was Dury (e.g. 1955; 1964; 1976) who honed ideas about cause and effect, and who showed that the meander geometry of single-thread rivers is functionally related, among other things, to water discharge, at least in humid environments.

Dryland rivers tend to be wider and less sinuous than counterparts in humid regions (Schumm 1960; Wolman and Gerson 1978). Schumm (1961) ascribes this difference, in large part, to the nature of the material deposited from the transport load: the lower the clay-silt content, the lower the bank cohesion and, consequently, the higher the width-depth ratio which, in turn, discourages the secondary currents associated with encouragement of meandering (Reid and Frostick 1997).

Whether subtle changes in climate that are effective over short spans of time are able to bring about channel change is open to question. Graf (1981; 1984; Fig. 23.7) provides some insight through a collation of historical records for the Gila River and its tributaries in the North American South-west. The conclusions that can be drawn from these studies is that channel shifting is a frequent occurrence but that channel sinuosity has remained largely constant throughout a period characterized by changing rainfall intensity, if not by changes in annual total rainfall (Cooke 1974). In fact, sinuosity of the Gila River deviated by no more than ±3 percent from the mean value of 1.13 over a 104-year period of record.

Another archival study was conducted by Burkham (1972; 1976) along the same river but upstream (Fig. 23.8). Here, over the same period, a succession of large, damaging floods was shown to have led to spasmodic straightening of the channel through the erosion of floodplain bottomland. Subsequent years saw the build-up of bars and their attachment to remnants of the pre-existing flood plain so that sinuosity was eventually restored to pre-damage levels. Of incidental interest, but

having important implications for sedimentation, is the fact that the increase in flow resistance and the decrease in channel gradient that accompanied the restoration process reduced the celerity of flood waves by about 60% as sinuosity increased from 1 to 1.2.

The lesson to be learnt from these two studies is that channels within a single semi-arid province (here, even the same river) can experience quite different planform changes over the same period. Burkham's study illustrates (as does Schumm and Lichty's study of channel width on the Cimarron River mentioned above) that any changes which might result from climatic alteration have to be viewed in the context of the dramatic changes brought about by individual high-magnitude floods. This suggests that rivers are comparatively insensitive as indicators of more subtle shifts in climate, responding too readily in arid environments to unusual weather, as much as reflecting the longer-term weather patterns that form an average condition – that is, climate.

Having noted this, for the same dryland province and for the same historical period, Hereford (1984, 1986) has provided evidence that climate change can encourage a shift in river sedimentation and channel cross-sectional dimensions through a change in flood behaviour. He obtained rainfall, tree-ring (a surrogate for rainfall) and runoff records of the period from the 1920s through 1981 for the Little Colorado and Paria Rivers, left- and right-bank tributaries of the Colorado River that drain catchments lying in northern Arizona and southern Utah. He was able to show that the discharge record of a sub-period up to 1940 was punctuated with floods of comparatively high magnitude (Fig. 23.9). These had swept the incised valleys, widening and deepening the channels and, in places, reaching bedrock. Historical photographs revealed broad, sandy channels, free of vegetation and with steep cut-banks. From the 1940s onward, a reduction in storm rainfall intensities produced a change in runoff and a commensurate shift in flood peak discharges. This change appears to have encouraged alluviation, the development of a flood plain and consequent narrowing of the active low-flow channels by about one-third. Hereford (1986) used the spread of salt cedar (*Tamarisk chinensis* Lour.) to date floodplain development, examining growth rings to establish the age of trees that had colonized the floodplain. From this evidence, he identified a sub-horizontal substrate

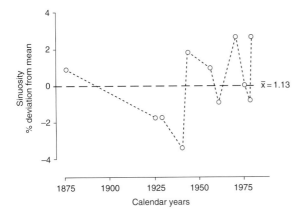

Fig. 23.7 Percentage deviation of channel sinuosity from the 1868–1980 period mean for the Gila River between Monument Hill and Powers Butte, near Phoenix, Arizona. (Data from Graf 1981)

Fig. 23.8 (a) Historical changes in the bottomland of the Gila River in Safford Valley, Arizona as a result of high magnitude floods in 1891, 1905–17, 1941 and 1967. (b) Changes in Gila River channel sinuosity in Safford Valley following flood-plain change and channel straightening by successive large flood events. (c) Changes in the average celerity of Gila River flood waves ranging in peak discharge between 283 and 566 m^3 s^{-1} due to changes in channel sinuosity and, hence, flow resistance. (**a** and **b** after Burkham 1972, **c** after Burkham 1976)

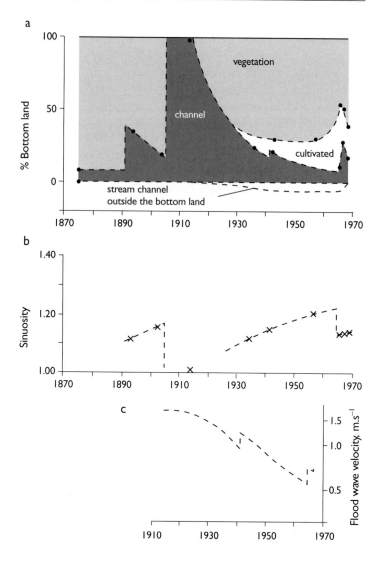

in the alluvial microstratigraphy (the 'germination zone' in Fig. 23.9), so confirming that the progressive confinement of the channels by floodplain building was coincident with the period of reduced flood flows. The reaches studied by Hereford are confined and, because of this, the rivers had limited scope to reflect the changes in discharge and sedimentation by changes in planform. However, these case studies show that alluvial architecture and channel dimensions can indeed respond fairly rapidly to changes in the nature of storm rainfall and the non-linear response of runoff.

These comments form a backcloth against which to judge the climatic implications of those studies that have revealed systematic changes in channel form over much longer periods.

Maizels (1987) provided an interesting example from the western Sharqiya of Oman (Fig. 23.10). The palaeochannels that she described hold added fascination because preferential cementation of the coarse-grained channel sediments has allowed them to resist the erosion and ablation that has affected the surrounding non-channel sediments. As a result, the channels lie perched above the deflated plain by up to 30 m. In the most widespread sedimentary unit that was laid down in front of the Eastern Oman Mountains, Fan 1, Maizels identified as many as five generations of palaeochannels. The oldest are most sinuous, with index values that are highly variable but generally exceed 1.5. The youngest are almost straight with a typical sinuosity index of 1.03. As for

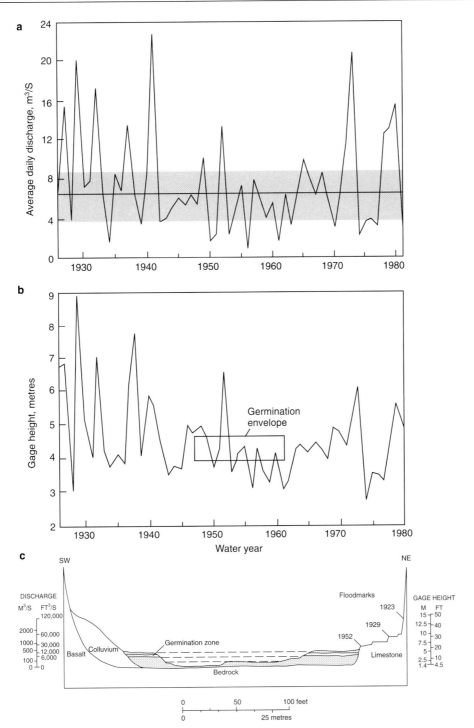

Fig. 23.9 Time-series of flow and a cross-section of the Little Colorado River, Arizona. (**a**) Annual average daily discharge 1926–81, showing the effects of greater flood magnitudes up to the early 1940s and reduced flood magnitudes especially in the period 1942–1963. Horizontal line is the median and the stipple encloses the inter-quartile range. (**b**) River stage of the annual flood at Grand Falls gauging station 1926–80. The 'Germina-tion envelope' includes the root bowl elevations and ages of the colonial salt cedars. (**c**) Cross-section 200 m below the Grand Falls gauging station showing progressive decline in elevation of identified flood marks, the scoured pre-1940s channel, and the post-1943 flood plain growth with tell-tale salt cedar 'germina-tion zone' (after Hereford 1984)

Fig. 23.10 Plio-Pleistocene gravel-bed channel remnants of western Sharqiya, Oman, exhumed and left as bas-relief by erosion and deflation of the surrounding non-channel sediments. Sinuosity and average width decreases progressively from the oldest to the youngest generation of channels. Reproduced by permission of the Geological Society of London from 'Plio-Pleistocene raised channel systems of the western Sharqiya (Wahiba), Oman' by J.K. Maizels in *Geological Society of London Special Publication* 35, 1987

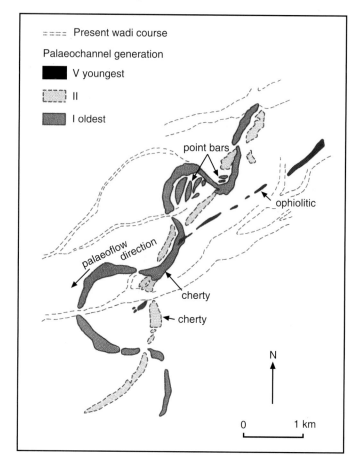

channel width, again, there appears to be a progressive decrease with age from greater than 100 m to about 50 m. These changes may reflect a change in flow regime as much as a change in peak flood flows, since Maizels' exhaustive palaeohydraulic calculations show insignificant differences in peak channel discharge between the oldest and youngest generations of channels. She rationalizes that the most sinuous channels would need at least seasonal flows to establish an equilibrium meandering form, which implies that the almost straight, youngest channels were formed under flashier, (?) ephemeral hydrographs.

In fact, Wolman and Gerson's (1978) compilation of data that compares flows in arid and humid regions may offer corroboration for the apparent contradiction between undifferentiated values of calculated peak flow and the very obvious change in channel planform. They have shown that flood peak discharges are indistinguishable from one environment to another, at least where water catchments range in area up to 100 km², even though the exponent relating total annual runoff to

catchment area in humid environments is almost double its arid zone equivalent.

There are no dates attached to the Sharqiya palaeochannels, though they may be of Pliocene to Early Pleistocene age. In any case, there are no other independent lines of evidence for climate change, so the channels and their sediments become a valuable, if imprecise, guide to changing climate, as they do in even older desert settings (e.g. the reconstruction of Triassic landscape in a rift basin of the North Sea; Frostick et al. 1992).

The importance of dating sediments in order to provide a framework for palaeoenvironmental interpretation of channel form is nowhere more self-evident than on the Riverine Plain of south-eastern Australia (Fig. 23.11). Schumm (1968) had offered a skilfully argued interpretation of channel adjustments to climate change. Three generations of channels had previously been identified: the oldest had been called the Prior Streams by Butler (1950), while another group of palaeochannels, younger than the Prior Streams, had

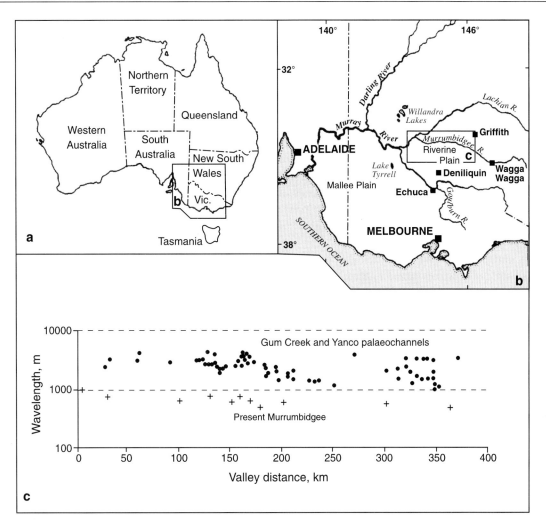

Fig. 23.11 (**a**) Map of Australia showing the location of the Riverine Plain. (**b**) Modern rivers of the Riverine Plain of southeast Australia. (**c**) Meander wave-lengths of the Gum Creek (35–25 ka) and Yanco (20–13 ka) palaeochannels and of the modern Murrumbidgee River (after Page et al. 1996)

been designated as the Ancestral Rivers (Pels 1964); both of these had been superseded by the modern channels, amongst which, the Murrumbidgee River was selected by Schumm for comparison with its predecessors. The present-day Murrumbidgee is fairly sinuous, values of the sinuosity index lying in the region of two, and it was deemed to be a suspended-load type channel. This is matched more or less by the Ancestral Rivers, although they had channel widths typically double those of the Murrumbidgee and estimated bankfull discharges that were some five times those of the modern river. The Prior Streams were deemed by Schumm to have had a completely different planform,

with a sinuosity index generally lying just above unity and channel widths as much as eight times that of the Murrumbidgee.

Schumm (1968) drew on his previous analysis of channel form on the Great Plains of N America (Schumm 1963) to point to the important role of sediment in part-determining channel character. Schumm (1960) had originally speculated that the inverse relation between channel width-depth ratio and the silt-clay fraction of the channel perimeter reflected the cohesion of clays and the ability of clay-rich deposits to maintain near-vertical sidewalls. In his 1968 paper, he also acknowledged the need to take account of the calibre of sediment presented to the river by hillslope processes, and the adjustment of river form needed to maintain a measure of transport efficiency and avoid choking. In effect, he proposed that a shift in sediment transport towards bedload (represented by the sand that was thought to dominate the bed material of the Prior Streams) had been accommodated by a straightening and, hence, steepening of the channel and by channel widening. The widening was facilitated by the lack of shear strength inherent in the sand that now formed a large fraction of the channel deposits. It was postulated that the Prior Streams were a response to aridification and the increased mobilization of coarse-grained sediment.

This interpretation was questioned by Bowler (1978), but it has been overturned by Page et al. (1996) who had the advantage of further field analyses of channel sediments and dimensions and, more significantly, a series of thermoluminescence dates to determine channel ages. They had also the advantages of the rapid increase in understanding of late Quaternary climate change that marked the last third of the 20th century, to which they themselves were contributing much in the context of drylands. Page et al. were able to identify four generations of palaeochannel: (1) the Coleambally system, dating from 100 to 80 ka and corresponding with some of Schumm's Prior Streams, represents a high degree of fluvial activity; Schumm had speculated that the channels were much younger and formed under arid conditions during the last glacial maximum; (2) the Kerarbury system, dating between 55 and 35 ka and corresponding to Schumm's northern Prior Streams, was active during what is now regarded as a sub-pluvial phase in inland and southern Australia; (3) the Gum Creek system, dating from 35 to 25 ka and corresponding with Schumm's Ancestral Rivers, reflects enhanced fluvial activity; and (4) the Yanco system, dating from 20 to 13 ka and grouped by Schumm with the Ancestral Rivers, is characterized by large lateral accretion features such as scroll bars and ox-bows. Significantly, there is a dearth of palaeochannels dated between 80 and 55 ka. Page et al. have drawn attention to this as an interval that includes Oxygen Isotope Stage 4, a period known to be characterized by widespread aridity.

The four palaeochannel systems indicate substantially larger discharges than occur in the modern Murrumbidgee River. This is reflected in the channel geometry (Fig. 23.11c). For example, the Gum Creek and Yanco systems have channel widths averaging 225 m and meander wavelengths averaging 3000 m, compared with the modern Murrumbidgee, where equivalents are 60 and 550 m. Page et al. have suggested a cooler climate with reduced evaporation and an increase in seasonal snow-pack melt in the south-eastern highlands that form the headwaters as being responsible for the episodic increase in river discharge during the late Pleistocene. They further suggest that sediment availability was governed by reductions in vegetation and an increase in periglacial activity and that this change contributed to the coarsening of river sediment character. The amelioration of climate during the Holocene led to disappearance of the headwater glaciers, an altitudinal rise in the tree-line and a lessening of coarse sediment supply. The river response to a reduction in both water discharge and sediment flux is seen in the high sinuosity and modest width of the modern Murrumbidgee.

Some corroboration of this Holocene trend in channel planform comes from the Gran Chaco of northern Argentina, where channel widths have decreased in response to a reduction in runoff volume even though meander geometry has remained essentially unchanged (Baker 1978; Fig. 23.12). However, the unfolding story of fluvial form and sediments on the Riverine Plain indicates that caution is required in using rivers as diagnostic indicators of climatic aridification and humidification.

Another study from the Australian continent prompts geomophologists to be circumspect about the use of channel form and sediments as an indicator of climate change. The sand-bed, anastomosing channels of Cooper Creek, a major feeder of the endorheic Lake Eyre Basin, appear to be incised in a braid-plain which had previously been interpreted as a relic of a former flow regime. Nanson et al. (1986) have argued, albeit on the self-confessed basis of a paucity of hydrological records, that the braid-plain is a contemporary feature and that it is utilized as a floodway for very high discharges. The incised, anastomosed channels are there to accommodate the flows of more moderate floods. The braids and braid bars are unusual in that they consist of sand-sized clay-rich aggregates. However, the contemporaneity of two distinct channel types over such a large area and the suggestion that neither is anachronistic counsels the need for caution

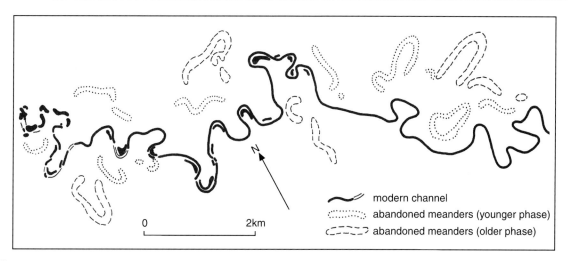

Fig. 23.12 *Osage-type* underfit modern channels and remnants of wider precursors on the Gran Chaco of Argentina (after Baker 1978)

in making climatic inferences from apparent changes in channel form and deposits both elsewhere and for earlier times.

This fascinating confusion is compounded further by another example of juxtaposed, contemporary, but contrasting channel types, in this case from the Northern Plains of central Australia. Tooth and Nanson (2004) have examined two ephemeral channels that, for several tens of kilometres, run sub-parallel courses within a few kilometres of each other along the same topographical depression. The longitudinal slopes of the two channels do not differ measurably, but the Plenty River has a bed of medium to coarse sand and has few tributaries, while the Marshall River transports coarse sand and granules and has several tributaries. The Plenty River has a single-thread channel, contrasting surprisingly with the Marshall River, which consists of numerous narrow anabranching channels (Fig. 23.13). Tooth and Nanson explain the juxtaposition of contrasting channel types by the frequency of tributary flows and the provision of perched groundwater that has encouraged tree growth in the Marshall River and the consequent establishment of mid-channel bars and islands as sediment has accumulated in the lee of the trunks. Maintenance of sediment-transporting competence has required adjustment of flow resistance by streamlining the islands to form elongate ridges that now separate the sub-parallel anabranched channels. Regardless of whether this analysis is correct, there remains the intriguing fact that two completely

contrasting channel-types lie alongside each other. Yet, there are no differences in climate between the two catchments, suggesting that a change in channel form from one to the other could not be used to indicate unequivocally climate change in a stratigraphic sequence.

The Arroyo Problem

Having examined the possibilities and problems of matching changes in the fluvial system to climate change and *vice versa* on the grand scale, the moment has come to turn the spotlight on the intense debate that surrounded, and still surrounds, arroyo development in the American South-west.

Briefly, the end of the 19th century witnessed an apparently sudden, geographically widespread, progressive and deep incision of previously flat-floored alluvial bottomland. A comprehensive account is given by Cooke and Reeves (1976) who indicate that while the exact timing of initial incision may vary, most of the trenching was accomplished between 1865 and 1915. Today, there is some suggestion that the process has been reversed and that net aggradation is occurring (Leopold 1976), despite reports that individual arroyos continue to grow by head-cutting (Malde and Scott 1977). In fact, rates of aggradation will be much lower than rates of incision, and Leopold et al. (1966) have advised that long-term monitoring of processes

a - Plenty River

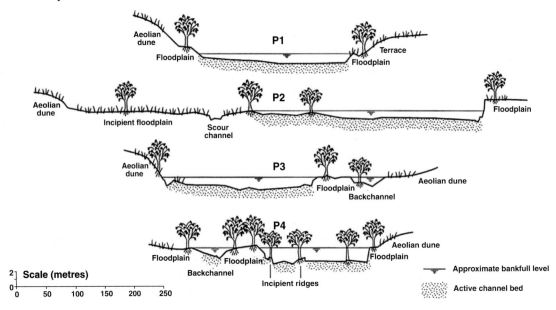

b - Marshall River

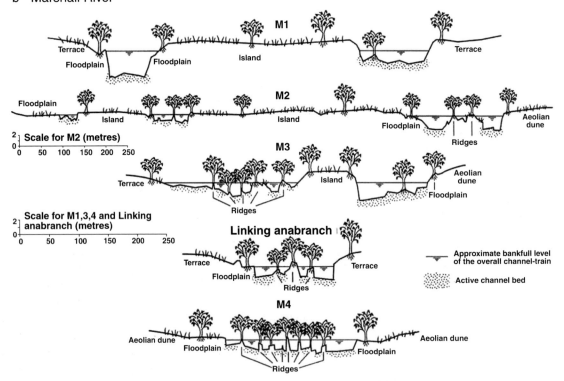

Fig. 23.13 Cross sections of two ephemeral channels that run sub-parallel courses less than 2.5 km apart on the Northern Plains of central Australia. (**a**) The Plenty River, which is a single-thread channel, largely devoid of trees, with a bed material composed of sand. (**b**) The Marshall River, which has multiple narrow anabranches separated by tree-induced ridges and islands and a bed material of coarse sand and granules (after Tooth and Nanson 2004)

will be needed in order to achieve an understanding of the current direction that arroyo development is taking.

The controversy surrounding the initiation of arroyos was fuelled by the fact that not only were there some shifts in climate that might indirectly have brought about trenching, but that, simultaneously, landuse of the area was fairly abruptly changed through the introduction of cattle by colonial non-indigenes. Whatever the cause, it is generally agreed that runoff was encouraged through a depletion of vegetative ground cover. Bailey (1935), among others, blamed overgrazing and rejected the case for climate change. Devevan (1967) sat on the fulcrum and argued that higher intensity summer storms had fallen on drought-weakened and overgrazed vegetation. Bryan (1941), in reviewing a longer period of time leading up to the events of the late 19th century and including the period of Pueblo Culture, was inclined to argue for increased runoff as a product of climate change and a loss of vegetation. Leopold (1951) approached the changing climate problem from a more subtle point of view. He was able to show that the number of light raindays (thought to be vital to sustaining the herb and grass community as a partial ground cover) had been significantly lower in the last part of the 19th century (Fig. 23.14), and speculated that this reduction allowed the less frequent but higher magnitude runoff events to ravage valley bottomlands, so carving out the arroyos. His analyses of rainfall in New Mexico were later corroborated by Cooke's (1974) collation and analyses of records obtained at stations in Arizona and the suggested implications for runoff were supported by Hereford's (1984, 1986) assessment of flood flows up to and including the early 1940s in the Paria and Little Colorado Rivers (Fig. 23.9).

Leopold's rainfall analysis suggests a mid 20th century trend in storm intensity that would bring a return of rainfall patterns that characterized the period of arroyo cutting. In this circumstance, it should be possible to see signs of another epicycle of erosion. However, other factors have come to bear that complicate the picture. Graf (1978; Fig. 23.15a,b) has traced the spread of riparian tamarisk in the Colorado Plateau since its introduction from Europe in the late 19th century. In addition, he has drawn attention to the role of both riparian and channel vegetation in increasing flow resistance, showing that the amount of biomass on the floodway distinguishes entrenched and unentrenched channels, at least in upland Colorado

(Graf 1979; Fig. 23.15c). These and other factors have complicated landform response to what are, after all, subtle changes in climatic variables, so that the cause of arroyo initiation still remains elusive.

Indeed, some believe that climate change is unnecessary as an explanation for arroyo initiation (Patton and Schumm 1981). Schumm and Hadley (1957; Fig. 23.16) have proposed a cycle of trenching and alluviation that they suggest accounts for the discontinuous nature of many arroyos. Transmission losses during flash flooding are thought to be responsible for encouraging localised deposition along a channel, which in turn, produces equally localised steepening of the channel gradient and eventually encourages incision and knickpoint recession. The reduction of gradient downstream of the knick, together with erosional displacement of material from the knickpoint brings about alluviation of the previously entrenched lower reaches, while arroyo development progresses upstream. In the fullness of time, renewed incision occurs at steeper sections in the alluvium that has been deposited in the trunk stream and a new knick moves progressively upstream in a new cycle.

A great deal of scientific enquiry has gone into the arroyo problem, yet the results remain equivocal. Perhaps as important an upshot of what at one time seemed a pressing problem in need of a solution has been the investment in obtaining a general understanding of fluvial processes, especially those of the world's drylands.

Dryland River Deposits

Dryland ephemeral rivers differ from perennial counterparts in the rate at which they convey sediment and, to some degree, in the character of their deposits (Reid and Frostick 1987, 1997). The reasons for such differences lie as much outside as within the river channel and are associated with differences in the degree of protection of the ground by vegetation, the relative (in)stability of topsoil structure, presence/absence of biopore control of soil hydraulic conductivity and the incidence of surface ponding and overland flow, and so on – that is, all those factors that influence the availability and transfer of material from hillslopes. Within the river, the hydraulic feedback that is associated with the type and quantity of material with which it is presented leads not only to differences in channel plan and

Fig. 23.14 Frequency of non-summer (Oct.–June) rainfalls by storm size and the average rainfall per rainy day at Santa Fe, New Mexico, covering the major period of arroyo incision (1865–1915) and subsequent developments (after Leopold et al. 1966)

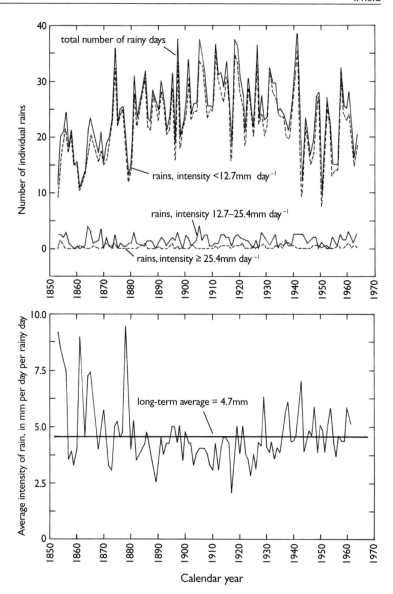

cross-sectional form, as discussed above, but also in the architecture of the alluvium.

Alexandrov et al. (2007) have used the fact that the northern Negev enjoys (or suffers) from seasonal differences in the nature of storm rainfall. This not only highlights the reasons for high specific annual yields of suspended sediment in drylands, but shows how a single variable – storm type – impacts on sediment delivery in the same catchment, so removing the complication of comparing different catchments, each with its peculiar, even if similar, lithological, soil and landuse characteristics. In autumn and spring, convectively-enhanced cellular storms deliver high intensity rainfall whereas in winter, storms are frontal in origin and deliver ubiquitous rainfall of low intensity. Over a 10-year period, cellular convective storms produced 21% of the runoff but yielded 56% of the suspended sediment, while frontal storms contributed 79% and yielded 44%, respectively. This study helps point to what consequences there might be for fine-grained sedimentation in response to subtle shifts in rainfall intensity as climate changes.

For coarser-grained sediment, Laronne and Reid (1993) have demonstrated that bedload flux in ephemeral gravel-bed channels is several orders of magnitude higher that in perennial channels at similar

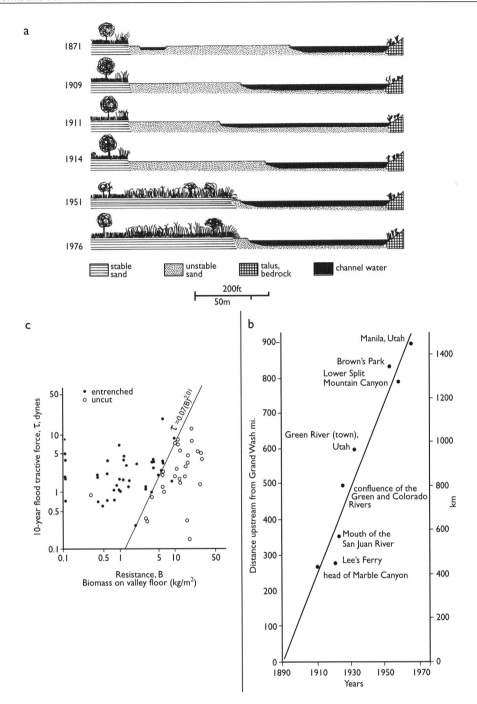

Fig. 23.15 (**a**) Sketches of Green River cross-section at Bowknot Bend, Utah, derived from historical photographs and field survey before and after the arrival of tamarisk (salt cedar) in the 1930s, showing the dramatic stabilization of alluvium, constriction of the low-water channel and increased flow resistance of the floodway. (**b**) Spread of tamarisk up the Colorado and Green Rivers – a rate of colonization averaging $20\,\text{km a}^{-1}$ – after its introduction to North America from Europe in the 1880s. (**c**) Separation of entrenched and un-entrenched valley bottomlands in upland Colorado as a function of biomass-induced flow resistance, allowing for the estimated average tractive force of the local 10-year flood at peak discharge. The line is a fitted discriminator and not a line derived by regression analysis (**a** and **b** after Graf 1978; **c** after Graf 1979)

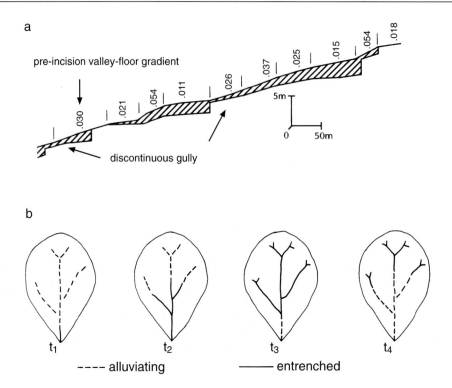

Fig. 23.16 (a) Discontinuous gullies in Dam 17 Wash, near Cuba, New Mexico, showing knick-points that have moved up-channel from over-steepened sections of the pre-incision alluvium. (b) Cycle of trenching and alluviation in semi-arid drainage basins (after Schumm and Hadley 1957)

levels of boundary shear stress or unit stream power (Fig. 23.17). The essential reason for the differences has been shown to relate to the lack of armour layer development in ephemeral channels. This has been related functionally to the abundant supply of sediment in drylands, with their poorly vegetated slopes, and the ready replacement of smaller grain-sizes that are selectively removed from the channel bed by flood flows. Reid et al. (1999) have used the differences in bedload flux between perennial and ephemeral rivers to speculate on what effects a major change in climate might have on coarse-grained sediment yield. They have shown that the efficiency of bedload transport (effectively the percentage of stream power used in transport) is inversely related to the channel-bed armour ratio (the relation between a parameter of the grain-size distributions of the surface and sub-surface layers). One ephemeral stream is shown to be, on average, 345 times more efficient than the most armoured perennial stream for which there is comparable data. They also show that the specific bedload

sediment yield of one upland ephemeral channel is about 18 times that of an equivalent perennial channel having a wooded catchment of about the same size.

How does this information about sedimentary processes become useful in understanding sedimentary sequences? Talbot (1980) has given a schematic and regionally generalized sequence of river sediments that has arisen from shifts in climate in the western Sahel during the terminal Pleistocene and Holocene (Fig. 23.18). The sediment column is dominated by sand (which is derived in part from the reworking of southward driven Saharan dunes), but the period of maximum regional humidity that centres on 9 ka BP is characterized by a change towards suspended-load type systems and indicates greater vegetation cover and soil development. These comparatively large changes in sediment calibre reflect what, for this region, have been end-members of the climatic spectrum. However, curiously, there is no suggestion that the more subtle shifts in climate that are reflected in contemporaneous

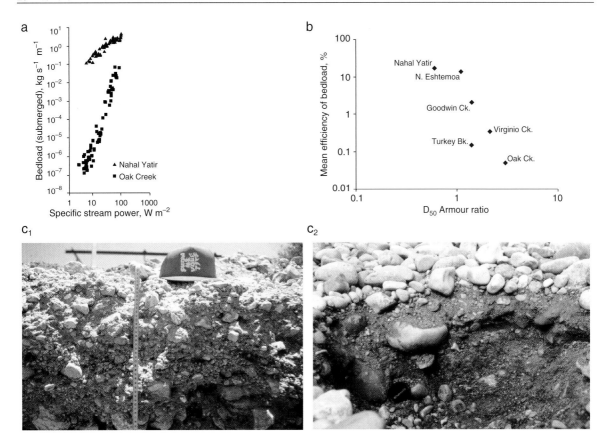

Fig. 23.17 (**a**) Bedload sediment flux as a function of unit stream power in the unarmoured ephemeral gravel-bed Nahal Yatir, Israel, and the armoured perennial gravel-bed Oak Creek, Oregon, showing the much higher transports rates in the Yatir as a function of the environmentally-controlled absence of armour development. (**b**) Bagnold's (1973) index of percentage bedload transport efficiency, $E = 100 i_b/(\omega/\tan\alpha)$, where i_b is the bedload flux (submerged weight) per unit channel width, ω is specific stream power, and $\tan\alpha$ is assumed constant at 0.63, as a function of the armour ratio of the channel bed sediments expressed as the dividend of the median grain sizes of the surface and subsurface layers in two ephemeral (Nahal Yatir, Nahal Eshtemoa), one seasonal (Goodwin Creek) and three perennial (Virginio Creek, Turkey Brook, Oak Creek) gravel-bed channels where bedload was established with bed-slot samplers. (**c**) Sections through the gravel beds of the unarmoured ephemeral Nahal Yatir (c_1), the hat resting on the channel-bed surface, and the armoured perennial River Wharfe, England (c_2) (**a** after Laronne and Reid 1993, **b** after Reid et al. 1999)

lesser risings of palaeo-Lake Chad are translated into deposit character.

Indeed, Love (1982) points to the equivocal relations between fluvial deposition and climate using the mid and late Holocene channel-fills that are exposed by present-day incision within Chaco Canyon, New Mexico (Fig. 23.19). These palaeochannels range in size, and, therefore, in their probable role and importance as part of the contemporary drainage system. But they also range widely in the calibre and style of sedimentation and include palaeochannels typical of ephemeral, flash-flood, 'bedload-type' streams (Fig. 23.19b, lower left), those of 'mixed-load type' (upper left) as well as those of 'suspended-load type' (lower right). Yet, tree-ring and other sources of data show no dramatic shifts in climate during the period covering this episode of canyon filling. On the other hand, there are indications of phases of greater-than and less-than normal rainfall that may be reminiscent of the pattern established for the period of arroyo incision and possible refilling during the late-19th/20th century. Although Love remains sceptical that these climatic wobbles can be used to explain differences in channel and channel-fill character, it suggests that a thorough examination of present-day

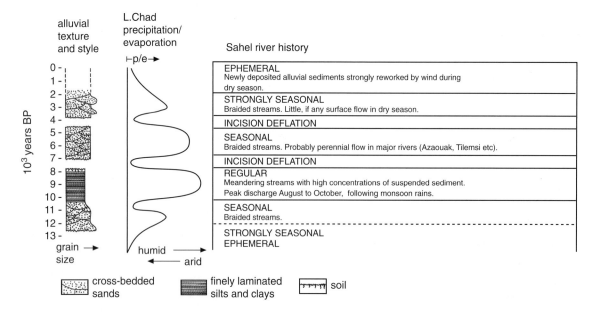

Fig. 23.18 Schematic generalized sequence of sedimentation by West Sahel rivers during the terminal Pleistocene and Holocene set against the palaeoclimate curve derived from strand-line and other evidence in the Lake Chad Basin (after Talbot 1980)

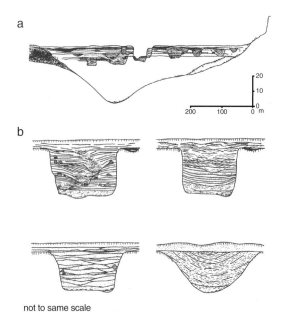

Fig. 23.19 Styles of late Holocene channel-fills exposed in section by present-day arroyo development in Chaco Canyon, New Mexico (after Love 1982)

arroyo fills might be useful in at least indicating the range of alluvial architecture that might be expected following known, subtle shifts in climate.

Having issued a cautionary note, it is possible to state that desert stream sediments may well carry telltale signs of relatively small changes in climate. Grossman and Gerson (1987) have drawn a distinction between Pleistocene and Holocene river terraces in the southern Negev (Fig. 23.20). The older deposits are built partly by debris flows and display a commensurately disorganized fabric, a wide range of clast size (up to 1.5 m), and a poor degree of stratification. The younger terrace deposits are fluviatile in origin, displaying a high degree of stratification, an ordered fabric, and so on. There are also differences in the reg soils that developed on the two sets of terraces. Those of Pleistocene origin show greater illuvial horizonation and a deeper profile. The inference is that the older terraces reflect a moderately arid to semi-arid climate in which rainfall was occasionally sufficient to encourage mass-wasting of talus slopes in headwater reaches and so provide material of the right consistency for debris flows. At the same time, the rainfall regime, while not generous, nevertheless encouraged soil development. The Holocene terraces reflect the onset of the extreme aridity that characterizes the region at present. Debris flows were not a feature of the sediment transfer system and soil development was inhibited by the infrequency of rainfall.

Fig. 23.20 Differentiation of the largely fluviatile sediments of the Holocene terraces of the Timna Valley, southern Negev, laid down by flash floods that punctuated extremely arid conditions and those terrace sediments of Pleistocene age which consist, in part, of debris-flow deposits and indicate a slightly ameliorated climate that produced moderately arid to semi-arid conditions. Reproduced by permission of the Geological Society of London from 'Fluviatile deposits and morphology of alluvial surfaces as indicators of Quaternary environmental changes in the southern Negev, Israel' by S. Grossman and R. Gerson in *Geological Society of London Special Publication* 35, 1987

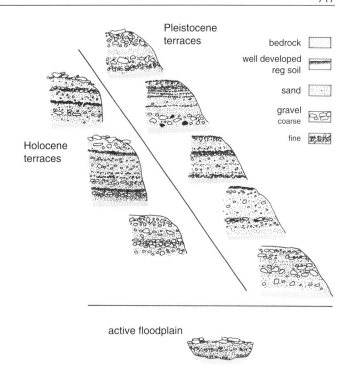

Maizels (1987; Fig. 23.21) offers another glimpse of the sedimentary legacy of changes in climate, albeit that it is the sediments and not independent lines of evidence that provide the clues, so that there is always the danger of circular argument. Notwithstanding this problem, the lowest unit of the Wadi Andam Fan complex (Fan I) exhibits a cut-and-fill structure that matches the meandering nature of the channels that drained from the Eastern Oman Mountains during the wetter conditions of the early Pleistocene. The beds of Fan II are, in contrast, thinner and more tabular, reflecting the flashier flood regime of the braided streams that developed in the region during arid phases of the mid to late Pleistocene.

The examples discussed so far are, by and large, headwater or upper catchment deposits of small to moderate-sized drainage systems. However, the deposits of larger rivers also reflect shifts in climate. The patterns of change are not dissimilar to those already described except that the character of the sediment is governed by processes of comminution and selective entrainment that operate over the long transport distances that are involved. Adamson's (1982; Fig. 23.22) compilation of hydrological and sedimentological

information for the expanded Nile basin and adjacent areas provides a thoroughly researched example. In particular, the fine sediments deposited along the main Nile and in its delta are equated with the ameliorated climate of the period that preceded the last glacial maximum and of the early Holocene, while coarser sands are a product of the terminal Pleistocene and late Holocene arid phases. A similar pattern of events is drawn from analyses of the sediments of the Niger Delta (Pastouret et al. 1978) and those of the deep-sea fan that lies off the mouth of the Amazon (Damuth and Fairbridge 1970).

In the Channel Country of central Australia, Rust and Nanson (1986) found a regionally extensive sand sheet underlying the mud-pellet braid-plain of the modern Cooper Creek, which they date tentatively as between 200 and 50 ka. They relate the sand to high-sinuosity meandering channels visible on air photography where the mud-pellet braid plain is undeveloped and infer a wetter climate and more sustained flow regime. There is no explanation for the switch in sediment character to the peculiar clay-pellet clasts of the modern braid plain, but it is taken to signify aridification of the catchment.

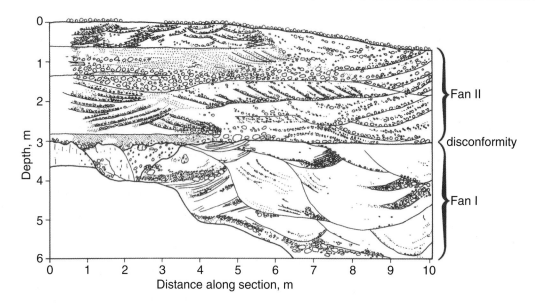

Fig. 23.21 Deposits of the Plio-Pleistocene Wadi Andam Fan Complex of western Sharqiya, Oman. Fan I was laid down by sinuous streams that probably had at least a seasonal discharge regime and indicate a wetter climate than Fan II, which was a braid-plain built by ephemeral flash floods. Reproduced by permission of the Geological Society of London from 'Plio-Pleistocene raised channel systems of the western Sharqiya (Wahiba), Oman' by J.K. Maizels in *Geological Society of London Special Publication* 35, 1987

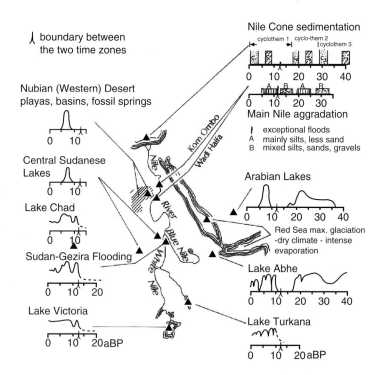

Fig. 23.22 Summary of hydrological and sedimentological data for the greater Nile Basin and adjacent areas during the late Pleistocene and Holocene (after Adamson 1982)

Conclusions

Desert rivers undoubtedly respond to those gross changes in climate that bring a shift in environment from arid to semi-arid or even sub-humid conditions and *vice versa*. Perhaps this is not surprising given that Langbein and Schumm's (1958) celebrated (if contentious) curve relating present-day specific sediment yield to effective rainfall is at its steepest for the world's drylands. At finer resolution, it is still debatable whether river response reflects lesser changes in climate, if only because the alterations which occur in ground cover and soil that affect the runoff process are stepped in character and do not vary continuously with increasing or decreasing rainfall. In any case, several studies in drylands suggest that storm type and rainfall intensity rather than rainfall amounts are as important in determining runoff response and, hence, sediment transport. Besides this, the impact of occasional large floods undoubtedly tends to mask the gradual alteration of channel form that might be expected to accompany diminutive changes in climate.

There are well-documented examples where both river planform and sedimentary deposits suggest that considerable uncertainty lies in attempting to interpret them in terms of climate change. As a consequence, considerable circumspection is required and other lines of evidence should be sought as corroboration. However, despite the difficulties that are experienced in calibrating the fluvial measuring-stick in terms of climate, river character is a valuable, and sometimes singular, guide to climate-induced environmental change, especially in deserts where biogenic evidence is often absent.

Acknowledgments   I am grateful to Mark Szegner of the Drawing Office, Department of Geography, Loughborough University for the considerable effort involved in converting to digital format the diapositives of drawings brought forward from the first edition as well as for composing new illustrations.

References

Adamson, D.A. 1982. The integrated Nile. In *A land between two Niles*. M.A.J. Williams and D.A. Adamson (eds.), Balkema, Rotterdam, 221–234.

Adamson, D.A., F. Gasse, F.A. Street and M.A.J. Williams 1980. Late Quaternary history of the Nile. *Nature* 288, 50–55.

Adamson, D.A., M.A.J. Williams and R. Gillespie 1982. Palaeogeography of the Gezira and the lower Blue and White Nile valleys. In *A land between two Niles*. M.A.J. Williams and D.A. Adamson (eds.), Balkema, Rotterdam, 165–219.

Alexandrov, Y., J.B. Laronne and I. Reid 2007. Intra-event and inter-seasonal behaviour of suspended sediment in flash floods of the semi-arid northern Negev, Israel. *Geomorphology* 85, 85–97.

Bagnold, R.A. 1931. Journeys in the Libyan Desert 1929 and 1930. *Geographical Journal* 78, 13–39.

Bagnold, R.A. 1973. The nature of saltation and of 'bedload' transport in water. Proceedings of the Royal Society of London A332, 473–504.

Bailey, R.W. 1935. Epicycles of erosion in the valleys of the Colorado Plateau province. *Journal of Geology* 43, 337–355.

Baker, V.R. 1977. Stream-channel response to floods, with examples from central Texas. *Geological Society America Bulletin* 88, 1057–1071.

Baker, V.R. 1978. Adjustment of fluvial systems to climate and source terrain in tropical and subtropical environments. In *Fluvial sedimentology*. A.D. Miall (ed.), Canadian Society Petroleum Geologists Memoir, Calgary 5, 211–230.

Baker, V.R., R.C. Kochel, P.C. Patton and G. Pickup 1983. Palaeohydrologic analysis of Holocene flood slack-water sediments. In *Modern and ancient fluvial systems*. J.D. Collinson and J. Lewin (eds.), Special Publication International Association Sedimentologists Blackwell, Oxford 6, 229–239.

Begin, Z.B., A. Ehrlich and Y. Nathan 1974. Lake Lisan the Pleistocene precursor of the Dead Sea. *Bulletin Geological Survey of Israel* 63.

Benson, L.V. 1978. Fluctuations in the level of pluvial Lake Lohontan during the last 40 000 years. *Quaternary Research* 9, 300–318.

Bowler, J.M. 1978. Quaternary climate and tectonics in the evolution of the Riverine Plain, southeastern Australia. In *Landform evolution in Australia*. J.L. Davies and M.A.J. Williams (eds.), Australian National University Press, Canberra.

Bowler, J.M., G.S. Hope, J.N. Jennings, G. Singh and P. Walker 1976. Late Quaternary climates of Australia and New Guinea. *Quaternary Research* 6, 359–394.

Breed, C.S., J.F. McCauley and P.A. Davis 1987. Sand sheets of the eastern Sahara and ripple blankets on Mars. In *Desert sediments: ancient and modern*. L.E. Frostick and I. Reid (eds.), Special Publication Geological Society, London, 35, 219–242.

Bryan, K. 1941. Pre-Columbian agriculture in the Southwest, as conditioned by periods of alluviation. *Annals American Association Geographers* 31, 219–242.

Burkham, D.E. 1972. Channel changes of the Gila River in Safford Valley, Arizona 1846–1970. *Professional Paper United States Geological Survey* 655-G.

Burkham, D.E. 1976. Effects of changes in an alluvial channel on the timing, magnitude and transformation of flood waves, southeastern Arizona. *Professional Paper United States Geological Survey* 655-K.

Butler, B.E. 1950. A theory of prior streams as a causal factor of soil occurrence in the Riverine Plain of southeastern Australia. *Australian Journal Agricultural Research* 1, 231–252.

Butzer, K.W. and C.L. Hansen 1968. *Desert and river in Nubia: geomorphology and prehistoric environments at the Aswan Reservoir*. University of Wisconsin Press, Madison.

Butzer, K.W., G. Ll. Isaac, J.L. Richardson and C. Washbourn-Kamau 1972. Radiocarbon dating of East African lake levels. *Science* 175, 1069–1076.

Cooke, R.U. 1974. The rainfall context of arroyo initiation in southern Arizona. *Zeitschrift fur Geomorphologie N.F. Supplement Band* 21, 63–75.

Cooke, R.U. and R.W. Reeves 1976. *Arroyos and environmental change in the American South-West.* Clarendon Press, Oxford.

Damuth, J.E. and R.W. Fairbridge 1970. Equatorial Atlantic deep-sea arkosic sands and ice-age aridity in tropical South America. *Geological Society America Bulletin* 81, 189–206.

De Heinzelin, J. and R. Paepe 1965. The geological history of the Nile Valley in Sudanese Nubia. In *Contributions to the prehistory of Nubia.* F. Wendorf (ed.), Southern Methodist University Press, Dallas, 29–56.

Denevan, W.M. 1967. Livestock numbers in Nineteenth-Century New Mexico, and the problem of gullying in the Southwest. *Annals American Association Geographers* 57, 691–703.

Dury, G.H. 1955. Bedwidth and wave-length in meandering valleys. *Nature* 176, 31.

Dury, G.H. 1964. Principles of underfit streams. *Professional Paper United States Geological Survey* 452-A.

Dury, G.H. 1976. Discharge prediction, present and former, from channel dimensions. *Journal of Hydrology* 30, 219–245.

El-Baz, F. and T.A. Maxwell (eds.). 1982. Desert landforms of Southwest Egypt, a basis for comparison with Mars. National Aeronautics and Space Administration Contractors' Report CR-3611, 372pp.

Fairbridge, R.W. 1963. Nile sedimentation above Wadi Halfa during the last 20 000 years. *Kush* 11, 96–107.

Frostick, L.E. and I. Reid 1986. Evolution and sedimentary character of lake deltas fed by ephemeral rivers in the Turkana basin, northern Kenya. In *Sedimentation in the African rifts.* L.E. Frostick, R.W. Renaut, I. Reid and J.-J. Tiercelin (eds.), Special Publication Geological Society, London, 25, 113–125.

Frostick, L.E. and I. Reid 1980. Sorting mechanisms in coarse-grained alluvial sediments: fresh evidence from a basalt plateau gravel, Kenya. *Journal Geological Society London* 137, 431–441.

Frostick, L.E. and I. Reid 1989a. Climatic versus tectonic controls of fan sequences: lessons from the Dead Sea, Israel. *Journal Geological Society London* 146, 527–538.

Frostick, L.E. and I. Reid 1989b. Is structure the main control of river drainage and sedimentation in rifts? *Journal African Earth Sciences* 8, 165–182.

Frostick, L.E., T.K. Linsey and I. Reid 1992. Tectonic and climatic control of Triassic sedimentation in the Beryl Basin, northern North Sea. *Journal Geological Society London* 149, 13–26.

Gasse, F. and A. Street. 1978. Late Quaternary lake-level fluctuations and environments of the Northern Rift Valley and Afar region (Ethiopia and Djibouti). *Palaeogeography, Palaeoclimatology and Palaeoecology* 24, 279–325.

Graf, W.L. 1978. Fluvial adjustments to the spread of tamarisk in the Colorado Plateau region. *Geological Society America Bulletin* 89, 1491–1501.

Graf, W.L. 1979. The development of montane arroyos and gullies. *Earth Surface Processes* 4, 1–14.

Graf, W.L. 1981. Channel instability in a braided, sand bed river. *Water Resources Research* 17, 1087–1094.

Graf, W.L. 1984. A probabilistic approach to the spatial assessment of river channel instability. *Water Resources Research* 20, 953–962.

Grossman, S. and R. Gerson 1987. Fluviatile deposits and morphology of alluvial surfaces as indicators of Quaternary environmental changes in the southern Negev, Israel. In *Desert sediments: ancient and modern.* L.E. Frostick and I. Reid (eds.) Special Publication Geological Society, London 35, 17–29.

Grove, A.T. 1985. *The Niger and its neighbours.* Balkema, Rotterdam.

Grove, A.T., F.A. Street and A.S. Goudie 1975. Former lake levels and climatic change in the rift valley of southern Ethiopia. *Geographical Journal* 141, 171–202.

Gupta, A. and H. Fox 1974. Effects of high-magnitude floods on channel form: a case study in Maryland Piedmont. *Water Resources Research* 10, 499–509.

Harvey, C.P.D. and A.T. Grove 1982. A prehistoric source of the Nile. *Geographical Journal* 148, 327–336.

Hereford, R. 1984. Climate and ephemeral-stream processes: Twentieth-century geomorphology and alluvial stratigraphy of the Little Colorado River, Arizona. *Geological Society of America Bulletin* 95, 654–668.

Hereford, R. 1986. Modern alluvial history of the Paria River drainage basin, southern Utah. *Quaternary Research* 25, 293–311.

Katz, A., Y. Kolodny and A. Nissenbaum 1977. The geochemical evolution of the Pleistocene Lake Lisan-Dead Sea system. *Geochimica et Cosmochimica Acta* 41, 1609–1626.

Kendall, R.L. 1969. An ecological history of the Lake Victoria basin. *Ecological Monographs* 39, 121–176.

Kroonenberg, S.B., G.V. Rusakov and A.A. Svitoch 1997. The wandering of the Volga delta: a response to rapid Caspian sea-level change. *Sedimentary Geology* 107, 189–209.

Langbein, W.B. and S.A. Schumm 1958. Yield of sediment in relation to mean annual precipitation. *Transactions American Geophysical Union* 39, 1076–1084.

Laronne, J.B. and I. Reid 1993. Very high rates of bedload sediment transport by ephemeral desert rivers. *Nature* 366, 148–150.

Leopold, L.B. 1951. Rainfall frequency, an aspect of climatic variation. *Transactions American Geophysical Union* 32, 347–357.

Leopold, L.B. 1976. Reversal of erosion cycle and climatic change. *Quaternary Research* 6, 557–562.

Leopold, L.B., W.W. Emmett and R.M. Myrick 1966. Channel and hillslope processes in a semiarid area New Mexico. *Professional Paper United States Geological Survey* 352-G.

Love, D.W. 1982. Quaternary fluvial geomorphic adjustments in Chaco Canyon, New Mexico. In *Adjustments of the fluvial system.* D.D. Rhodes and G.P. Williams (eds.), George, Allen and Unwin, London, 277–308.

Maizels, J.K. 1987. Plio-Pleistocene raised channel systems of the western Sharqiya (Wahiba), Oman. In *Desert sediments: ancient and modern.* L.E. Frostick and I. Reid (eds.), Special Publication Geological Society, London, 35, 31–50.

Malde, H.E. and A.G. Scott 1977. Observations of contemporary arroyo cutting near Santa Fe, New Mexico, USA. *Earth Surface Processes* 2, 39–54.

McCauley, J.F., C.S. Breed and M.J. Grolier 1982a. The interplay of fluvial, mass-wasting and aeolian processes in the eastern Gilf Kebir region. In *Desert landforms of southwest*

Egypt: a basis for comparison with Mars. F. El-Baz and T.A. Maxwell (eds.), National Aeronautics and Space Administration, Washington, DC, CR-3611, 207–239.

McCauley, J.F., G.G. Schaber, C.S. Breed, M.J. Grolier, C.V. Haynes, B. Issawi, C. Elachi and R. Blom 1982b. Subsurface valleys and geoarcheology of the eastern Sahara revealed by Shuttle Radar. *Science* 218, 1004–1020.

Nanson, G.C. 1986. Episodes of vertical accretion and catastrophic stripping: a model of disequilibrium flood-plain development. *Geological Society America Bulletin* 97, 1467–1475.

Nanson, G.C., X.Y. Chen and D.M. Price 1995. Aeolian and fluvial evidence of changing climate and wind patterns during the past 100 ka in the western Simpson Desert, Australia. *Palaeogeography, Palaeoclimatology, Palaeoecology* 113, 87–102.

Nanson, G.C., D.M. Price and S.A. Short. 1992. Wetting and drying of Australia over the past 300 ka. *Geology* 20, 791–794.

Nanson, G.C., B.R. Rust and G. Taylor 1986. Coexistent mud braids and anastomosing channels in an arid-zone river: Cooper Creek, central Australia. *Geology* 14, 175–178.

Nanson, G.C. and S. Tooth 1999. Arid-zone rivers as indicators of climate change. In *Paleoenvironmental reconstruction in arid lands*, A.K. Gupta and E. Derbyshire (eds.), Oxford and IBH Publishing Co., New Delhi, 175–216.

Nanson, G.C., S. Tooth and A.D. Knighton 2002. A global perspective on dryland rivers: perceptions, misconceptions and distinctions. In *Dryland rivers: hydrology and geomorphology of semi-arid channels*, L.J. Bull and M.J. Kirkby (eds.), John Wiley and Sons, Chichester, 17–54.

Neev, D. and K.O. Emery 1967. The Dead Sea. *Bulletin Geological Survey of Israel* 41.

Page, K., G. Nanson and D. Price 1996. Chronology of Murrumbidgee River palaeochannels on the Riverine Plain, southeastern Australia. *Journal of Quaternary Science* 11, 311–326.

Pastouret, L., H. Chamley, G. Delibrias, J.-C. Duplessy and J. Thiede 1978. Late Quaternary climatic changes in western tropical Africa deduced from deep-sea sedimentation off the Niger Delta. *Oceanologica Acta*, 1, 217–232.

Patton, P.C., V.R. Baker and R.C. Kochel 1982. Slack-water deposits: a geomorphic technique for interpretation of fluvial paleohydrology. In *Adjustments of the fluvial system*. D.D. Rhodes and G.P. Williams (eds.), George, Allen and Unwin, London, 225–253.

Patton, P.C. and S.A. Schumm 1981. Ephemeral-stream processes: implications for studies of Quaternary valley fills. *Quaternary Research* 15, 24–43.

Peel, R.F. 1939. The Gilf Kebir. *Geographical Journal* 93, 295–307.

Pels, S. 1964. The present and ancestral Murray River system. *Australian Geographical Studies* 2, 111–119.

Reid, I. and L.E. Frostick 1987. Flow dynamics and suspended sediment properties in arid zone flash floods. *Hydrological Processes* 1, 239–253.

Reid, I. and L.E. Frostick 1997. Channel form, flows and sediments in deserts. In *Arid zone geomorphology*, 2nd Edition. D.S.G. Thomas (ed.), John Wiley and Sons, Chichester, 205–229.

Reid, I., J.B. Laronne and D.M. Powell 1999. Impact of major climate change on coarse-grained river sedimentation: a speculative assessment based on measured flux. In *Fluvial processes and environmental change*. A.G. Brown and T.A. Quine (eds.), John Wiley and Sons, Chichester, 105–115.

Rust, B.R. and G.C. Nanson 1986. Contemporary and palaeo channel patterns and the late Quaternary stratigraphy of Cooper Creek, south-west Queensland, Australia. *Earth Surface Processes and Landforms* 11, 581–590.

Schumm, S.A. 1960. The shape of alluvial channels in relation to sediment type. *Professional Paper United States Geological Survey* 352-B.

Schumm, S.A. 1961. Effect of sediment characteristics on erosion and deposition in ephemeral stream channels. *Professional Paper United States Geological Survey* 352-C.

Schumm, S.A. 1963. Sinuosity of alluvial rivers on the Great Plains. *Geological Society America Bulletin* 74, 1089–1100.

Schumm, S.A. 1968. River adjustment to altered hydrologic regimen – Murrumbidgee River and paleochannels, Australia. *Professional Paper United States Geological Survey* 298.

Schumm, S.A. 1977. *The fluvial system.* Wiley-Interscience, New York.

Schumm, S.A. and R. F. Hadley 1957. Arroyos and the semiarid cycle of erosion. *American Journal Science* 255, 161–174.

Schumm, S.A. and R.W. Lichty 1963. Channel widening and flood-plain construction along Cimarron River in southwestern Kansas. *Professional Paper United States Geological Survey* 352-D.

Servant, M. and S. Servant-Vildary 1980. L'Environment quaternaire de basin du Tchad. In *The Sahara and the Nile.* M.A.J. Williams and H. Fauré (eds.), Balkema, Rotterdam, 133–162.

Street, F.A. and A.T. Grove 1979. Global maps of lake-level fluctuations since 30 000 yr BP. *Quaternary Research* 12, 83–118.

Talbot, M.R. 1980. Environmental responses to climatic change in the West African Sahel over the past 20 000 years. In *The Sahara and the Nile.* M.A.J. Williams and H. Fauré (eds.), Balkema, Rotterdam, 37–62.

Tooth, S. and G.C. Nanson 2004. Forms and processes of two highly contrasting rivers in arid central Australia, and the implications for channel-pattern discrimination and prediction. *Geological Society of America Bulletin* 116, 802–816.

Tothill, J.D. 1946. The origin of the Sudan Gezira clay plain. *Sudan Notes and Records* 27, 153–183.

Truckle, P.H. 1976. Geology and late Cainozoic lake sediments of the Suguta Trough, Kenya. *Nature* 263, 380–383.

Williams M.A.J. and D.A. Adamson 1973. The physiography of the central Sudan. *Geographical Journal* 139, 498–508.

Williams, M.A.J. and D.A. Adamson 1974. Late Pleistocene desiccation along the White Nile. *Nature* 248, 584–586.

Williams, M.A.J., D. Adamson, B. Cock and R. McEvedy 2000. Late Quaternary environments in the White Nile region, Sudan. *Global and Planetary Change* 26, 305–316.

Williams, M.A.J. and F.M. Williams 1980. Evolution of the Nile basin. In *The Sahara and the Nile.* M.A.J. Williams and H. Fauré (eds.), Balkema, Rotterdam, 207–224.

Wolman, M.G. and R. Gerson 1978. Relative scales of time and effectiveness of climate in watershed geomorphology. *Earth Surface Processes* 3, 189–208.

Chapter 24

The Role of Climatic Change in Alluvial Fan Development

Ronald I. Dorn

The Persistence of Climatic Change in Alluvial-Fan Studies

Alluvial fans develop at the base of drainages where feeder channels release their solid load (Blair and McPherson, 2009; Leeder et al., 1998; Harvey et al., 2005). A classic fan-shape forms where there is a well-defined topographic apex. Multiple feeder channels, however, often blur the fan-shape resulting in a merged bajada. Alluvial fans can be found in almost all terrestrial settings. These include alpine (Beaudoin and King, 1994), humid tropical (Iriondo, 1994; Thomas, 2003), humid mid-latitude (Bettis, 2003; Mills, 2005), Mediterranean (Robustelli et al., 2005; Thorndrycraft and Benito, 2006), periglacial (Lehmkuhl and Haselein, 2000), and different paraglacial settings (Ballantyne, 2002). The geographical focus of this chapter, however, rests on alluvial fans in regions that are currently deserts or that experienced episodes of aridity in the Quaternary.

The research literature contains a host of different ways of thinking about and conducting research on desert alluvial fans (Table 24.1). Despite the wealth of research hypotheses and perspectives, many researchers keep returning to climate change as a vital forcing factor on desert fan evolution. Although some reject climatic change as important (De Chant et al., 1999; Webb and Fielding, 1999; Rubustelli et al., 2005), the following sorts of judgments commonly pepper the literature on fans found in arid and semi-arid regions:

R.I. Dorn (✉)
School of Geographical Sciences, Arizona State University, Tempe, AZ 85287, USA
e-mail: ronald.dorn@asu.edu

Hence, climate is an exclusive controlling factor of the transition from periods of geomorphodynamic activity to periods of stability (Gunster and Skowronek, 2001: 27).

The field evidence indicates that the Tabernas fan/lake system responded to regional tectonics, but that the fan sediment sequences were primarily climatically driven (Harvey et al., 2003: 160).

It is probably no coincidence that the first major episode of fan sedimentation occurred in MIS 5, the longest and more severe episode of cold and arid climates during the Pleistocene... (Pope and Wilkinson, 2005: 148).

Even along Dead Sea, climatic changes appear to be more important in fan development than base level or tectonic changes (Bowman, 1988; Klinger et al., 2003).

A persistent return to the importance of variable climate may result, in some small part, to the history of geomorphic thought where climatic change remains a major theme (Tricart and Cailleux, 1973; Besler, 1976; Mabbutt, 1977; Büdel, 1982; Hagedorn and Rapp, 1989; Derbyshire, 1993; Twidale and Lageat, 1994; Wendland, 1996; Elorza, 2005). Even if the tradition of climatic geomorphology shapes thought, it is the newly gathered evidence that drives researchers towards climate as an allocyclic process along with tectonic and base-level fluctuations (Roberts, 1995; Bettis, 2003; Harvey et al., 2005). The next section, however, argues that there are substantial obstacles to scientific investigations of the role of climatic change in desert alluvial-fan research.

Limitations of a Climatic Change Focus

This section makes three arguments that climatic change studies of desert alluvial fans should be viewed with considerable methodological skepticism. Sedimentology cannot be used to match fan depositional

Table 24.1 Examples of different research foci on desert alluvial fan research

Focus	Synopsis
Accommodation space	Different tectonic (Viseras et al., 2003), sea-level (Robustelli et al., 2005), base-level (Harvey, 1984; Calvache et al., 1997), basin width and sediment supply (Weissmann et al., 2005), accommodation space (Posamentier and Vail, 1988; Muto and Steel, 2000) conditions alter fan dynamics.
Catastrophism	Catastrophic changes dramatically alter fans (Beaty, 1974) where sediment-generating events can derive from fire (Wohl and Pearthree, 1991; Moody and Martin, 2001), anthropogenic landscape use (Eriksson et al., 2000; Gomez et al., 2003; Gómez-Villar et al., 2006), release of glacial damned lakes (Benn et al., 2006), rock avalanches (Blair, 1999), or high magnitude floods (Beaty, 1974; Kale et al., 2000; Baker, 2006).
Complex response	A variable response to the same external stimuli (Schumm, 1977) has been used interpreting alluvial fans experiencing different responses to similar conditions of climate, land cover and sediment supply (Harvey, 1997; Kochel et al., 1997; Coulthard et al., 2002).
Coupling	Coupling fosters linkage of processes at different spatial, temporal scales (Brunsden, 1993; Allen, 2005). As applied to alluvial fans (Harvey, 2002a), coupling analyses explain fan events over short and long time scales and small and large drainage basins.
Dynamic Equilibrium	Alluvial fans may represent a dynamic equilibrium in transportation of course debris from range to basin (Denny, 1965; Denny, 1967), but a dynamic equilibrium that may require millennial (Davies and Korup, 2006) or longer (Tricart and Cailleux, 1973) time scales.
Hazards	Fan hazard studies include process geomorphology (Chawner, 1935; Schick et al., 1999; Field, 2001), historical geomorphology (Kochel et al., 1997; Crosta and Frattini, 2004), Quaternary studies (Keefer et al., 2003; House, 2005), as well as engineering and policy issues (Committee on Alluvial Fan Flooding, 1996).
Megafans	The causes of and processes on megafans may involve periods of aridity (Krohling and Iriondo, 1999; Leier et al., 2005), and arid drainages may require different conditions to produce megafans (Rodgers and Gunatilaka, 2002; Arzani, 2005) than in other climates.
Modeling	Modeling (Schumm et al., 1987; Coulthard et al., 2002) helps understand sediment waves (Tucker and Slingerland, 1997), high-frequency variations in sediment supply (Hardy and Gawthorpe, 2002; Davies and Korup, 2006), landscape evolution (Coulthard et al., 2002; Clevis et al., 2003), how fan morphology affects groundwater recharge (Blainey and Pelletier, 2008) and understanding linkages between specific geomorphic processes and corresponding forms (Weaver and Schumm, 1974).
Morphometry	Rich understanding of fan and landscape change develops from morphometry studies (Hooke and Rohrer, 1977; Kostaschuk et al., 1986; Jansson et al., 1993; Calvache et al., 1997; Harvey et al., 1999a; Viseras et al., 2003; Staley et al., 2006; Volker et al., 2007; Wasklewicz et al., 2008), including links to steady-state (Hooke, 1968), allometry (Bull, 1975; Crosta and Frattini, 2004) and other larger concepts.
Process studies	Process research forms the core of fan theory development (Hooke, 1967; Kostaschuk et al., 1986; Blair, 1987; Wohl and Pearthree, 1991; Blair and McPherson, 1994; Blair, 1999; Schick et al., 1999; Al Farraj and Harvey, 2004; Crosta and Frattini, 2004; Benn et al., 2006; Griffiths et al., 2006).
Remote Sensing	Digital image processing of satellite (White, 1993; Farr and Chadwick, 1996; Robinson, 2002; Garcia-Melendez et al., 2003) and ground-based imagery (Crouvi et al., 2006) generates valuable perspectives on mapping and fan processes.
Sedimentology	Sedimentary and stratigraphic analyses (Robinson, 2002) yields insight about processes (Blair and McPherson, 1994; Robinson, 2002; Harvey et al., 2005), high magnitude events (Lafortune et al., 2006), low-magnitude changes in a basin (Calvache et al., 1997; Robinson, 2002), fan fossilization (Stokes et al., 2007), and sometimes potential sources (Krzyszkowski and Zielinkski, 2002; Harvey et al., 2003).
Tectonics	Tectonic setting permits most fan development (Singh et al., 2001; Hartley et al., 2005). Although some disregard tectonics as important in certain settings (Klinger et al., 2003; Colombo, 2005), tectonism can alter relief, generate headward erosion, alter stratigraphy, change fan gradients, drop base levels, and change accommodation space (Kesel and Spicer, 1985; Owen et al., 1997; Clevis et al., 2003; Guerra-Merchan et al., 2004; Pope and Wilkinson, 2005; Rubustelli et al., 2005; Quigley et al., 2007; Sancho et al., 2008).

records with climatic changes. Dating methods are simply not up to the task of correlating geomorphic events with climatic changes, and even a new method that directly connects climatic change with aggradational events can only suggest millennial-scale correlations. Lastly, controlled experiments are not possible with field-based studies.

Sedimentology Limitations

Processes leading to alluvial-fan deposits "differ remarkably little between humid and arid environments, or between arctic and subtropical environments" (Harvey et al., 2005: 3), a conclusion reached in many different studies (Brierley et al., 1993; Ibbeken et al., 1998; Ballantyne, 2002; Krzyszkowski and Zielinkski, 2002; Lafortune et al., 2006). Although researchers sometimes connect climatic changes to sedimentological changes using independent chronometric control (Calvache et al., 1997; Singh et al., 2001), there is a danger that sediment-based analyses alone could suffer from circular reasoning in inferring the importance of climatic change (Jain and Tandon, 2003).

> No matter how detailed the geomorphological and stratigraphic examination that is undertaken of alluvial fan systems, no estimate of age obtained by these methods can ever be deemed reliable except in the grossest possible terms. It is only with the application of chronometric dating that a reliable temporal framework can be constructed, and only with such a framework can the triggers of fan-forming processes be independently assessed (Pope and Wilkinson, 2005: 149).

Unlike lake shorelines, periglacial features and glacial moraines whose existence directly connects to climatic events, alluvial-fan studies cannot currently infer climatic change through sedimentological, stratigraphic, or geomorphological analyses.

Are Dating Methods Up to the Task?

The Target is Decadal, Century and Millennial Climatic Changes

With the growth of increasingly precise proxy records, the last few decades has seen a substantial trans-

formation in palaeoclimatology. Whereas much of the alluvial-fan research in the twentieth century focused on correlations with Milankovitch-scale (Shaffer et al., 1996) oscillations (Wells et al., 1987; Bull, 1991; Reheis et al., 1996), twenty-first century geomorphic research must articulate to records of millennial, century and even decadal climatic change (Starkel, 1999; Birks and Ammann, 2000; Viles and Goudie, 2003; Thomas, 2004; Anderson, 2005). There exists a clearly identified need for research on fluvial responses to allogenic forcing over sub-Milankovitch time scales of 10^2–10^3 years (Blum and Tornqvist, 2000: 2).

The vast preponderance of newer palaeoclimatological research now emphasizes sub-Milankovitch high frequency and high magnitude climatic events (Fig. 24.1). Examples of millennial events include iceberg surges generating Heinrich Events (Vidal et al., 1997) and Dansgaard-Oeschger cycles with an asymmetry of decadal warming (Taylor et al., 1997) and then longer cooling (Bond et al., 1997; Curry and Oppo, 1997).

Climatic variability exists at all timescales, and processes that drive climatic changes are closely coupled. Accordingly, there has been increased attention to ever

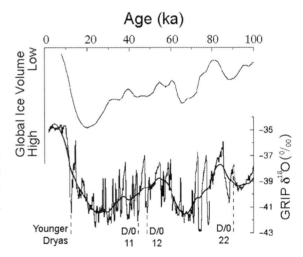

Fig. 24.1 Researchers advocating the importance of climatic change on desert fans originally argued for the role of (A) Milankovitch-scale oscillations reflecting global ice volume. A new generation of palaeoclimatic studies emphasize sub-Milankovitch sudden climate shifts such as the (B) changing location of global moisture recorded in Greenland Ice Core Project (GRIP) records. Vertical lines identify examples of rapid climatic change GRIP events of the Younger Dryas and Dansgaard-Oeschger (D/O) cycles 11, 12 and 22. Diagram is adapted from (Masson-Delmotte et al., 2005)

shorter climate-change time scales. These fluctuations include the El Niño Southern Oscillation over a sub-decade time scale (Philander, 1999), the North Atlantic Oscillation over a decadal scale (Wanner et al., 2001), the Pacific Decadal Oscillation over a bi-decade scale (Houghton et al., 2001), the Atlantic Multidecal Oscillation over a seventy-year scale (Enfield et al., 2000) and others. Sub-Milankovitch oscillations show up in a variety of high resolution biotic and geological proxy records (Arz et al., 1998; Proctor et al., 2000; House and Baker, 2001; Benson et al., 2002; Madsen et al., 2003; Rohling and Palike, 2005; Ellwood and Gose, 2006; Henderson, 2006; Schaefer et al., 2006), but not in desert-fan research for reasons made clear in the next section.

Fan Dating Methods Have Trouble Distinguishing High Frequency Events

At issue here is whether methods used to date alluvial-fan events are up to the task of a correlation with millennial-scale changes, let alone century or decadal oscillations. This challenge was issued a decade ago with the argument that climatic change hypotheses for desert alluvial fans are not testable, because even the most precise time intervals for a fan event could be assigned to wet, dry, or transition climatic events (Dorn, 1996). The problem was repeated again: "no detailed assessment of the response of the clastic sedimentary depositional environment to such abrupt, high amplitude changes... is available." (Fard, 2001) The same difficulty was explained a bit later for northeast Queensland:

> The switch from fan building to fanhead trenching constitutes an 'abrupt' change in the behaviour of the fluvial system, but we have little idea of the transitional time from one mode of flow to the other. It is likely that such changes were (a) diachronous between basins, and (b) in response to more than one threshold-crossing event. But both these sources of variation may have occurred within 10^1–10^2 years, and when viewed across a 10^3 year time period may appear synchronous (Thomas, 2004: 112).

The challenge was issued yet again in the context of modelling and field-based studies:

> we wish to demonstrate that persistent alluvial fanhead morphology may result from rare, large sediment inputs not necessarily related to climatic or tectonic perturbations. This possibility has largely been ignored when us-

ing alluvial fans as indicators of past climatic or tectonic regimes (Davies and Korup, 2006).

Proponents of fan-climate correlations have yet to address these critiques.

The mainstay of alluvial-fan climatic change research in the arid western USA has been soil-stratigraphic studies punctuated with occasional tephrachronology and radiometric (e.g. U-series, ^{14}C, cosmogenic) data (Wells et al., 1987; Bull, 1991; Harvey et al., 1999a; Harvey et al., 1999b; McDonald et al., 2003; Western Earth Surface Processes Team, 2004; Harkins et al., 2005; Knott et al., 2005). These studies yield fan depositional events where the highest precision generates broad age ranges for correlated surfaces that span 10^3 to 10^5 years. Even lower precision derives from such morphologic evolution dating strategies as scarp diffusion (Hsu and Pelletier, 2004). This is not to infer that geomorphic or traditional soil-stratigraphy strategies have no value. They certainly do (Huggett, 1998), but not to test scientific hypotheses of alluvial-fan development related to climatic change. No method has yet enabled a correlation of fan surfaces, based on soils and geomorphic parameters, with sub-Milankovitch climatic fluctuations.

Radiocarbon measurement is certainly precise enough to discriminate millennial-scale climatic events, especially with the use of accelerator mass spectrometry (Keefer et al., 2003). Aside from tremendous problem of a general lack of availability of suitable material on arid fans, there are concerns about whether precise measurements are truly accurate. Worries occur over the effect of groundwater (Bird et al., 2002), whether extant models of pedogenic carbonate accumulation are appropriate (Stadelman, 1994; Wang et al., 1996; Alonso-Zarza et al., 1998), over contamination by old carbon sources (Chitale, 1986; Falloon and Smith, 2000; Six et al., 2002), contamination by younger carbon (Ljungdahl and Eriksson, 1985), and the importance of experienced and rigorous lab processing (Gillespie et al., 1992).

Alluvial-fan climate change researchers outside of the USA often employ optically stimulated luminescence (OSL) (White et al., 1996; Krohling and Iriondo, 1999; Roberts et al., 1999; Eriksson et al., 2000; Singh et al., 2001; Suresh et al., 2002; Stokes et al., 2003; McLaren et al., 2004; Robinson et al., 2005; Gardner et al., 2006; Suresh et al., 2007; Spencer et al., 2008; Zazo et al., 2008), while OSL has seen very limited application in the arid southwestern USA (Hanson, 2005; DeLong and Arnold, 2007; Mahan et al., 2007; Sohn et al., 2007). With the best available protocol it is difficult, but possible, for OSL to obtain reliable and precise enough ages to discriminate millennial-scale oscillations (Olley et al., 2004).

Considerable recent attention has been paid to cosmogenic nuclide dating on alluvial fans (Liu et al., 1996; Phillips et al., 1998; Keefer et al., 2003; Matmon et al., 2005; Evenstar et al., 2006; DeLong and Arnold, 2007; Dühnforth et al., 2007; Frankel et al., 2007). By itself, cosmogenic nuclides do not yield accurate enough results for millennial-scale correlations. There exists 5% error in counting, 50% error in chemical and blank corrections, and 10–15% in production rates (Brown et al., 2003; Benn et al., 2006). This is all before other uncertainties are considered, such as boulder erosion rates, changes in the geomorphic position of sampled surfaces, potential sampling bias that often goes unidentified, prior exposure history, and periodic cover by snow or soil—all leading to potential offsets from reported exposure dates. It is the rare researcher (Robinson, 2002; Brown et al., 2003; Benn et al., 2006) who actually presents real uncertainties associated with cosmogenic results. This is not to say that cosmogenic nuclide data on fan sediment lacks value. Far from it. New and creative strategies for unraveling complex signals are under development (Robinson, 2002). The simple point here is that the magnitude of identified and often unidentified errors simply makes it impossible at the present time to link cosmogenic ages on fan sediment to sub-Milankovitch climatic events.

New Strategy Linking Fan Events with Climatic Change

The ideal research method would be one where the deposit can be directly correlated with a climatic event. One such method has just passed from the experimental realm, varnish microlaminations (VML) developed by Tanzhuo Liu. Liu subjected his VML method to a successful blind test administered by Richard Marston, editor of *Geomorphology* (Liu, 2003; Marston, 2003; Phillips, 2003). Both general and specific aspects of ex-

tracting palaeoclimatic information from varnish layering have also been replicated (Dorn, 1984, 1990; Cremaschi, 1992; Cremaschi, 1996; Diaz et al., 2002; Lee and Bland, 2003; Thiagarajan and Lee, 2004). The method was originally applied to Milankovitch-scale correlations of alluvial fan units (Fig. 24.2) (Liu and Dorn, 1996). However, since this original exploration VML dating had another decade of development based on scrutiny of more than 10,000 varnish microsedimentary basins.

The latest technical advances in VML dating now permit the resolution of twelve millennial-scale events during the Holocene, at least in the southwestern USA (Liu and Broecker, 2007). Such high resolution permits the assignment of specific ages to deposits such as found on a well-photographed debris-flow fan on the east side of Death Valley (Fig. 24.3) (Liu, 2008). Seven analyzed fan units were correlated with "relatively wet periods during the Holocene" (Liu and Broecker, 2007).

VML directly links climatic change with aggradational events on alluvial-fan surfaces, but at its best VML can only resolve millennial-scale correlations. Two difficulties still remain if this methodology is to fulfill its potential of testing the importance of climatic change in alluvial-fan development. First, there must be a clear linkage between climatic thresholds needed to change VML and the millennial-scale climatic events altering alluvial fans. In other words, the drainage-basin/fan under examination may not necessarily respond to the same climatic forcings as the varnish. Second, if century and decade-scale wet phases were vital in generating fan surfaces during a millennial-scale dry period, even this finest-scale methodology could misidentify the fan/climate correlation. For example, it might be extreme events in decadal dry phases during a millennial wet phase that actually generated the alluvial-fan deposits in Fig. 24.3, and VML would not be able to identify such high resolution patterns.

The best strategy available today rests in utilizing several high resolution methods together. For example VML might be used in tandem with OSL, much in the way that several fan researchers utilize as many different methods as possible to identify systemic uncertainties with a single method that might otherwise go undetected (e.g., Roberts et al., 1999; Poisson and Avouac, 2004; Owen et al., 2006).

Fig. 24.2 Six Springs Alluvial Fan, Death Valley, where fan mapping corresponds with the varnish microlamination (VML) sequence. In broad Milankovitch-scale terms, VML layering units (LU) 2, 4, 6 and 8 correspond with marine oxygen isotope stages 2, 4, 6 and 8. At this course scale of resolution, there is no clear relationship between Death Valley fan aggradation and Milankovitch-scale climatic change (Dorn, 1996; Liu and Dorn, 1996)

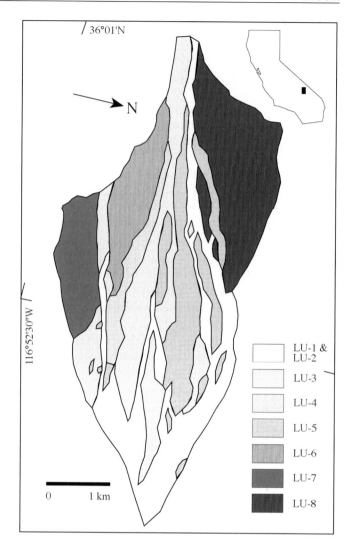

Intrinsic Lags

Beyond the chronometric limitations of a fan event/climate event correlation, there are well known inherent challenges offered by the geomorphic system (Brunsden, 2001; Harvey, 2002a; Viles and Goudie, 2003; Oguchi and Oguchi, 2004; Thomas, 2004). A few of these correlation challenges include time lags in how fast a climatic change impacts the ability of the geomorphic system to adjust with changes of slope and drainage systems, mediation by slow or fast (e.g. wildfire) shifts in vegetation cover, available sediment stores and lags in its exhaustion, the nature of climatic-vegetation discontinuities at the onset of a change, magnitude of the change, rate of change at the discontinuity, and complex

responses. When we do not understand drainage basin sediment production, storage or transport, it becomes extraordinarily difficult to connect climatic change with alluvial-fan deposits over sub-Milankovitch timescales.

Lack of a Control

Scientific research requires independent controls to isolate the effect of a variable. Different types of modelling (e.g. Clevis et al., 2003) do sometimes permit researchers to isolate the impact of climatic change. A quasi-steady approximation "suggests that environmental variables (e.g. climate, lithology) play

Fig. 24.3 A small debris-flow fan approximately 0.05 km² reveals distinct Holocene VML. Optical thin sections of varnish on two of the older deposits are exemplified here, revealing VML ages of ~11,100–12,500 cal BP and ~9400 cal BP (These images are courtesy of T. Liu.) The seven distinct fan depositional units appear to be correlated with millennial-scale wetter periods during the Holocene (Liu and Broecker, 2008)

a less significant role in overall fan morphology than do basic sedimentary and flow processes." (De Chant et al., 1999: 651). A cellular automaton model, in contrast, "shows that the sediment discharge upstream of the alluvial fan closely follows the climate signal" (Coulthard et al., 2002: 280). Models can explore the implication of climatically forced sediment waves moving down channel networks (Tucker and Slingerland, 1997) or the implication of climatic changes for fan progradation and aggradation (Hardy and Gawthorpe, 2002).

Controlled studies evaluating the role of climatic change, however, are simply not possible in field-based research. Unlike tectonic and base-level allocyclic processes, where it is possible to reasonably assume no tectonic or base-level change influences to compare

with field sites impacted by these variables (Harvey, 2002b; Harkins et al., 2005), all desert alluvial fans have experienced climatic oscillations. This makes it impossible to craft an experimental design isolating just climate change. Thus, modelling research will inevitably play an increasingly important role in the future of the scientific study of the role of climatic change on desert fans.

20th Century USA Research and the Transition-To-Drier-Climate Model

The notion has been around a long time. A drier climate leads to a sparser cover of woody vegetation.

As infiltration capacity decreases, the location of the channel head moves upslope and excavates weathered material. Alluvium then moves down channels towards alluvial fans. This transition-to-drier-climate model (Fig. 24.4) had its birth in the southwestern USA (Huntington, 1907; Eckis, 1928; Melton, 1965; Knox, 1983; Wells et al., 1987; Bull, 1991). The hypothesis of regional desiccation as the key process forcing fan aggradation continues to dominate the conclusions of southwestern USA research (Throckmorton and Reheis, 1993; Dorn, 1994; Bull, 1996; Harvey and Wells, 1994; Harvey et al., 1999b; Monger and Buck, 1999; Baker et al., 2000; McDonald et al., 2003; Western Earth Surface Processes Team, 2004; Mahan et al., 2007; Sohn et al., 2007). The following conclusions are typical of the regional literature:

> Fan deposition was probably triggered by a change from relatively moist to arid conditions causing a decrease in vegetation cover and increases in flash floods and sediment yield. We think that this scenario applies to most of the other valleys in the southern Basin and Range (Reheis et al., 1996: 279).
>
> Thus, it appears that the initiation of hillslope erosion, fan building, and valley deposition was associated with a climatic shift from moister to drier conditions and a significant change in the nature of uplands vegetation (Miller et al., 2001: 385).

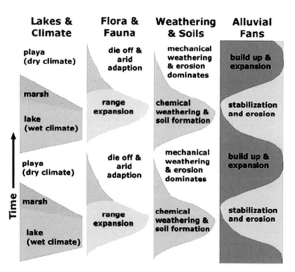

Fig. 24.4 The transition-to-a-drier-climate model has been adopted by the U.S. Geological Survey's Western Earth Surface Processes Team (2004) that explains: "when dry conditions return, the plant cover would eventually become reduced, and episodic desert storms would strip away soil (formed during the proceeding wet period), contributing to the influx of greater amounts of sediments downstream onto alluvial fans"

The transition-to-drier-climate model has certainly been used outside of the USA, in India (Kale and Rajaguru, 1987), Israel (Bull and Schick, 1979; Klinger et al., 2003), and elsewhere. In South Australia incision into a ∼45–40 ka surface (Dorn, 1994: 607), perhaps correlated with a ∼40–30 ka paleosol (Quigley et al., 2007), led to subsequent inset aggradation thought to derive from a transition "from more humid to more arid continental climatic conditions" (Quigley et al., 2007). In Argentina, an abrupt desiccation led to a condition where "the large rivers of the province built alluvial fans in their lower tracts". (Carignano, 1999: 130)

Although there has been international use of the concept, the transition-to-dry hypothesis dominates USA desert alluvial-fan thinking. One signal that a strict mindset exists is when available information is stretched to fit a desired conclusion. For example, very precise pronouncements sometimes emerge from what are truly very broad age ranges: "global climate changes that caused synchronous pulses of alluviation in the Mojave Desert at about 125, 55, and 10 (12 and 8) ka" (Bull, 1996: 217). No such pin-point millennial precision actually supports such a sweeping regional generalization. Another signal of a fixed paradigm comes when one's own evidence is passed over in reaching a conclusion. Despite presenting evidence of ongoing aggradation during transitions from drier to wetter conditions, during wetter conditions, and during a particular high magnitude event, cumulative evidence is still interpreted according to the acceptable southwestern USA paradigm: "[t]he pulse of sediment at the Pleistocene-Holocene transition is consistent with other depositional events identified elsewhere and with geomorphic models (Bull, 1991; McDonald et al., 2003)." (Nichols et al., 2006: 8).

In summary, Southwestern USA researchers have largely restricted themselves to a narrow theoretical framework and methodology that has very little potential to resolve correlation difficulties outlined in the second section of this chapter.

Diversity of Thought Outside of the USA

Alluvial-fan research on arid and semi-arid fans has burgeoned in the last few decades outside of the USA. Furthermore, this non-USA literature that explores connections between dryland fans and climatic change has shown a far greater theoretical flexibility, hosting a large number of alternatives to "transition-to-drier" thinking.

One alternative model is that of paraglacial processes (Ballantyne, 2002) generating large fans found in deserts that can, in some cases, be traced back directly to moraines (Krzyszkowski and Zielinkski, 2002). Many of the large desert fan complexes in central Asia appear to be paraglacial in origin (Rost, 2000; Stokes et al., 2003; Owen et al., 2006). In the southwestern USA, some researchers in the 20th (Huntington, 1907; Dorn, 1994) and 21st centuries (Weissmann et al., 2002; Benn et al., 2006; Dühnforth et al., 2007) have invoked glacial processes as the source of nearby desert-fan sediment. Similarly, enhanced snowfall or periglacial processes might increase sediment flow to fan heads (Dorn, 1994; Carignano, 1999).

A number of researchers argue for enhanced alluvial-fan aggradation during drier periods. The case is made for Calabria, Italy (Sorriso-Valvo, 1988) and in southern Greece (Pope and Millington, 2000; Pope and Wilkinson, 2005). Drier conditions are also thought to generate fan aggradation in the northeastern Tibetan Plateau (Hetzel et al., 2004), western India (Chamyal et al., 2003), and coastal Ecuador (Iriondo, 1994).

Other study sites produce evidence that fan aggradation occurs during the transition period from drier to wetter conditions (Roberts and Barker, 1993). OSL dating in Death Valley, California indicates that the "25 ka Q2d alluvial-fan deposits correlate to a globally and regionally low-effective moisture that is followed by a relatively rapid increase in moisture" (Sohn et al., 2007: 57), and cosmogenic ages from Death Valley suggest fan aggradation also occurred at the dry-wet transition between 63 and 70 ka (Frankel et al., 2007). In western India, it is thought that:

> A sudden change from dry to wet climate can lead to a sudden increase in the discharge resulting in gravel or sand bedload streams with high aggradation rates in the presence of high sediment availability (from the preceding dry phase)...On the other hand, a climatic transition from wet to dry will eventually lead to decimation of the fluvial activity and a simultaneous increase in aeolian activity (Jain and Tandon, 2003: 2231).

A similar argument was made for Tanzania where: in the northeast Irangi Hills, the shift from a dry to a wet climate [deposition occurred]... during this "window" of high erosivity formed by increasing rainfall combined with incomplete vegetation cover. (Eriksson et al., 2000: 123). Commensurately, many argue that a transition from dry to wetter conditions can be one cause of fan-head incision (Owen et al., 1997; Nott et al., 2001; Brown et al., 2003; Jain and Tandon, 2003; Bowman et al., 2004; Poisson and Avouac, 2004) explained as follows:

> [A climate change from cold and dry to warm and humid] encouraged the expansion of vegetation cover over basin slopes, thereby reducing the volume of sediment supplied by each basin. With streams transferring less sediment, an increase in the discharge (Q) to sediment load (Qs) ratio (or more water per unit of sediment) resulted in major entrenchment of the fanhead and proximal fan surfaces (Pope and Millington, 2000: 611).

Another common model, mostly rejected in southwestern USA research, invokes enhanced aggradation during wetter periods. Wet-phase fan aggradation is thought to occur during high lake periods in the Qaidam Basin (Owen et al., 2006), in the Gobi-Altay, Mongolia (Fitz et al., 2003), Australia (Nanson et al., 1992; Kershaw and Nanson, 1993), western India (Bhandari et al., 2005), the northern United Arab Emerites (Al Farraj and Harvey, 2004), Arabia (Glennie and Singhvi, 2002), Oman (Mazels, 1990; Rodgers and Gunatilaka, 2002), southern Spain (Zazo et al., 2008), and Jordan (McLaren et al., 2004). Only a few western USA studies have argued against the transition-to-dry model, suggesting that wetter conditions may generate southwestern USA fan aggradation during the late Pleistocene (Dorn, 1988; Hanson, 2005; DeLong and Arnold, 2007; Dühnforth et al., 2007; DeLong et al., 2008) and Holocene (Liu and Broecker, 2007).

Although the limitation of fan chronometry cannot currently test these different climatic models of fan aggradation, this methodological restriction may not always be the case. The rich and diverse international theory, based on detailed case study analysis, will offer future chronometricians ample opportunity to evaluate these and future theoretical options.

Coupling may also offer potential to sort out the seemingly overwhelming problem of linking intrinsic geomorphic lags and dating uncertainties. Coupling is a fluvial geomorphology concept that conceptually links processes at different spatial and temporal scales (Brunsden, 1993; Allen, 2005). As applied to alluvial fans, coupling could potentially explain fan aggradation during all of the aforementioned climatic conditions: dry intervals, wet phases, as well as transitions to and from aridity (Harvey, 2002a: 189).

Increasing Importance of High Magnitude Floods

There exists a growing momentum in international alluvial-fan research favouring the importance of high magnitude, low frequency floods—regardless of whether the general climatic state is in an arid, humid or transitional phase. Certainly, geomorphologists have long recognized the importance of large, but infrequent floods in deserts (Beaty, 1974; Schick, 1974; Wolman and Gerson, 1978; Talbot and Williams, 1979; Frostick and Reid, 1989; Pickup, 1991; Ely et al., 1993; Schick and Lekach, 1993). However, the last decade has seen a great expansion of interest in and evidence for high magnitude storms as being vital to the interpretation of desert alluvial fans.

Areas first recognized as being heavily impacted by ENSO have been the focus of some of this research (Grosjean et al., 1997). In northern Chile large floods appear to be the dominant cause of sedimentation on fans during the late Pleistocene and Holocene (Mather and Hartley, 2005). Aggradational events in southern Peru are "evidently associated with extremely heavy El Nino-induced precipitation" (Keefer et al., 2003: 70), and these events do appear to be generated by "Mega-Niños" with "higher amplitude climatic perturbations than any in the Peruvian historical record except for the AD 1607–1608 event." (Keefer et al., 2003: 74). The alluvial record in Ecuador indicates changing periodicity of ENSO aggradation over the last 15,000 years (Rodbell et al., 1999). Similarly in Argentina:

> Holocene sedimentary accumulations which are present over a large region, could have been controlled by one specific climatic favor, the activity of the El Niño Southern Oscillation (ENSO). The dynamics of this oscillation suggest. . .that very intense and randomly distributed rainfall could cause floods that are locally very important (Colombo, 2005: 81).

The importance of ENSO events has also been recognized in other deserts (Ely, 1997).

Arid southern Asia also yields research pointing to the key role of large storms. Western India studies suggests that such events may have been impor-

tant in the late Pleistocene ~10–14 ka (Jain and Tandon, 2003) and during marine oxygen isotope stages 3 and 5 (Juyal et al., 2000). "Development of alluvial fans [in Pinjaur Dun] requires intense but infrequent precipitation to create flash-flood discharge needed for transporting sediments from the drainage basin to the fan site. . ." (Suresh et al., 2002: 1273).

In southwestern Asia, in Iran's Abadeh Basin "episodic thundershowers, in an arid-semi-arid climate, resulted in periodic high magnitude runoff and created flash floods towards the feeder channel at the fan apex" (Arzani, 2005: 58). In Syria, there is field evidence of "[a]brupt increases in storm activity, steep talus slopes sensitive to erosion, and the hillslopes directly connected to the alluvial fan over very short distances together accounted for the rapid geomorphic response." (Oguchi and Oguchi, 2004: 138).

The Australian literature has long recognized the importance of extreme events (Pickup, 1991). Even in a situation where the Milankovitch-scale changes suggest a correlation of fan aggradation during a transition-to-drier climate, there is clear recognition of the importance of high magnitude events:

> Despite the general drying of the climate [after approximately 27 ka], the occurrence of major debris flows during this period suggests that extreme rainfall events must have occurred. These high rainfall events would have resulted in dramatic erosion of soils and regolith on slopes covered by the sparser vegetation communities (Nott et al., 2001: 881).

In southeastern Australia, high-energy flood events took place just before the last glacial maxima (Gardner et al., 2006). The Wilkatana alluvial fan displays evidence of "very large magnitude flood events" at millennial-scale intervals (Quigley et al., 2007) during the Holocene.

Century-scale (Thorndrycraft and Benito, 2006) and millennial-scale (Gunster and Skowronek, 2001) increases in large magnitude floods are also recognized as important in Spain. Millennial-scale increases in flooding is viewed as important to understand fans in the hyper-arid intermontane basins of central Asia (Owen et al., 2006), in central Turkey (Kashima, 2002), and in Italy where "infrequent but intense rainfalls" played a key role in mobilizing slope debris (Zanchetta et al., 2004). Recent southwestern USA fan research also indicates importance of high magnitude floods, at least during the Holocene (Griffiths and Webb, 2004;

Lave and Burbank, 2004; Anderson, 2005; Griffiths et al., 2006).

Information on event frequency and magnitude are obviously critical for developing accurate hazard assessment and mitigation strategies (Soeters and van Westen, 1996). This is certainly the case in the Sonoran Desert of Arizona where large infrequent floods have led to channel avulsions on alluvial fans (Field, 2001). Current strategies to assess hazards generate probabilistic flood hazard maps with a constancy in climate (Pelletier et al., 2005), an assumption that may not be valid. Sub-Milankovitch oscillations may change the frequency of high magnitude events in the southwest (Ely, 1997) and elsewhere (Viles and Goudie, 2003).

Given this uncertainty, a study of VML in south-central Arizona in metropolitan Phoenix has been focusing on understanding whether fan avulsion events took place during dry or wet phases of the Holocene. This study utilizes the revolution in the VML technique for analyzing Holocene desert surfaces (Liu and Broecker, 2007). Old channel avulsions are represented by abandoned alluvial-fan segments that occur throughout the metropolitan Phoenix region (Committee on Alluvial Fan Flooding, 1996). In all 42 abandoned Holocene fan surfaces exiting three ranges hosting development on the urban fringe (Fig. 24.5) have been sampled for VML.

There does appear to be a weak association between fan avulsion events and millennial-scale climate change in this northern portion of the Sonoran Desert (Fig. 24.6). Avulsions leading to fan-surface abandonment appear to have occurred three-quarters of the time during wetter periods of the Holocene. Thus, there may be a need to adjust current probabilistic strategies (Pelletier et al., 2005) for assessment of fan hazards in this sprawling urban centre.

As indicated previously, there are inherent limitations in the use of VML results to connect these central Arizona fan events with climatic change. First, the avulsion events—likely from large floods but not necessarily (Field, 2001)—could have taken place during decadal or century droughts that reduced vegetation cover, all nested within a millennial-scale wet phase. Second, single chronometric tools should always be eschewed, especially when attempting correlations with sub-Milankovitch-scale events. The use of multiple chronometric tools such as OSL in tandem with VML will help identify definitive clustering of fan-altering flood events. Third, intrinsic geomorphic

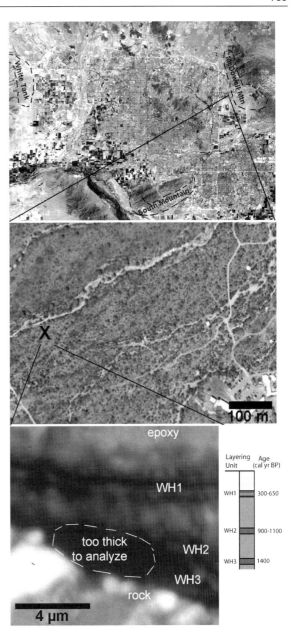

Fig. 24.5 VML analysis is used to study fan avulsion events on the edges of a sprawling desert metropolis. Three study areas are on the southern (South Mountain piedmont), western (western White Tank Mountains piedmont), and northeastern (western McDowell Mountains piedmont) fringes of metropolitan Phoenix, Arizona, as identified on a 2005 Landsat image ~75 km across. The middle aerial photograph from the McDowell Piedmont identifies the collection location of the lower image of an ultra-thin section from this site. Annotations correspond with the VML Holocene calibration (Liu and Broecker, 2007)

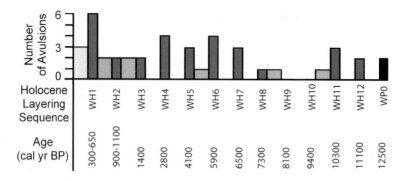

Fig. 24.6 Number of different fan avulsions on the fringe of metropolitan Phoenix, Arizona, associated with different millennial-scale climatic intervals in the Holocene—as recorded by varnish microlaminations (Liu and Broecker, 2007). For ex- ample, WH1 is the wet Holocene period 1. Six identified avul- sion events took place during this Little Ice Age interval. Three avulsions took place in the time since WH1, and two avulsions took place in the dry interval between WH1 and WH2

adjustments likely differ from one drainage basin to another, setting the geomorphic table at different times with differential time lags. Fourth, the climatic con- ditions needed to cause a change in varnish layering may not necessarily match the threshold needed to alter alluvial fans. Lastly, just because these results for central Arizona match a debris fan in Death Valley (Liu and Broecker, 2007) should not in anyway be used to infer a regional pattern. Similar studies in other areas may identify broad correlations between extreme fan-flooding events and drier periods of the Holocene. Thus, even using the highest resolution dating method available to desert alluvial-fan researchers, linkages between fan development and climatic change are tenuous.

Summary

The connection between desert alluvial fans and cli- matic change runs very deep in desert geomorphology. Initial thoughts in the early 20th century connected fan evolution to wetter glacial periods and to times when climate change reduced protective vegetation cover. The mid-20th century saw desert fan research focused in the southwestern USA, and a climatic-change paradigm evolved for this region. The vast majority of southwestern USA alluvial-fan research carried out in the last two decades promotes the hypothesis that alluvial-fan surfaces found throughout the region were produced during transitions between wetter and drier Milankovitch-scale climatic intervals.

International desert alluvial-fan studies have bur- geoned in the last decade and now carry the lead of in- novation in method and theory building. This extensive literature includes hypotheses that fan building may have taken place during dry periods, wet phases, transi- tions from dry to wet conditions, as well as transitions from wet to dry times.

Despite the persistent focus by desert geomorpholo- gists on linkages between climatic change and alluvial- fan development, there are major limitations to this entire subfield of desert geomorphology. First, fan de- posits are not diagnostic of any particular climatic con- dition, and thus sedimentology and stratigraphy cannot be used as an indicator of climatic change without in- dependent chronometric support. Second, dating meth- ods are not capable of correlating geomorphic events to sub-Milankovitch climatic changes with any degree of certainty. The highest resolution methods available to desert geomorphologists can only suggest group- ings of events with millennial-scale climatic periods. The ability to connect fan-building events to decadal or century-scale oscillations is highly speculative at best. Third, controlled research designs rest at the founda- tion of our science, yet controlled experiments are not possible in field research because all study sites have experienced climatic change. Thus, modelling studies that control climate must play an increasingly impor- tant role in the future. Fourth, high magnitude storms have taken on increased importance in the past decade in alluvial-fan research, and there exists no mechanism to falsify the hypothesis that "catastrophic" meteoro- logical storms truly form desert fans, regardless of the climatic period of the event. Fifth, perhaps the single largest obstacle to understanding the role of climatic change rests in the uncertainty of time scales of internal geomorphic adjustments to climatic changes (Brunsden, 2001; Harvey, 2002a; Viles and Goudie, 2003; Oguchi and Oguchi, 2004; Thomas, 2004). For these reasons, understanding the role of climatic change on desert alluvial-fan development remains an incredibly challenging task.

Acknowledgments Thanks to Arizona State University for providing sabbatical support, to T. Liu for permission to adapt graphics for use in Fig. 24.3, and to reviews from colleagues and students. The views stated here, however, are those of the author.

References

Al Farraj, A., and Harvey, A. M., 2004. Late Quaternary interactions between aeolian and fluvial processes: A case study in the northern UAE. *Journal of Arid Environments* **56**: 235–248.

Allen, P., 2005. Striking a chord. *Nature* **434**: 961.

Alonso-Zarza, A. M., Silva, P. G., Goy, J. L., and Zazo, C., 1998. Fan-surface dynamics and biogenic calcrete development: Interactions during ultimate phases of fan evolution in the semi-arid SE Spain (Murcia). *Geomorphology* **24**: 147–167.

Anderson, D. E., 2005. Holocene fluvial geomorphology of the Amargosa River through Amargosa Canyon, California. *Earth-Science Reviews* **73**: 291–307.

Arz, H. W., Patzold, J., and Wefer, G., 1998. Correlated millennial-scale changes in surface hydrography and terrigenous sediment yield inferred from last-glacial marine deposits off northeastern Brazil. *Quaternary Research* **50**: 157–166.

Arzani, N., 2005. The fluvial megafan of Abarkoh Basin (Central Iran): an exmaple of flash-flood sedimentation in arid lands. In: *Alluvial fans: Geomorphology, sedimentology, dynamics*, A. M. Harvey, A. E. Mather and M. Stokes (eds.), Geological Society Special Publications 251, London, pp. 41–59.

Baker, R. G., Fredlund, G. G., Mandel, R. D., and Bettis III, E. A., 2000. Holocene environments of the central Great Plains: multi-proxy evidence from alluvial sequences, southeastern Nebraska. *Quaternary International* **67**: 75–88.

Baker, V. R., 2006. Palaeoflood hydrology in a global context. *Catena* **66**: 161–168.

Ballantyne, C. K., 2002. Paraglacial geomorphology. *Quaternary Science Reviews* **21**: 1935–2017.

Beaty, C. B., 1974. Debris flows, alluvial fans, and a revitalized catastrophism. *Zeitschrift für Geomorphology N.F. Supplement Band* **21**: 39–51.

Beaudoin, A. B., and King, R. H., 1994. Holocene palaeoenvironmental record preserved in a paraglacial alluvial fan, Sunwapta Pass, Jasper National Park, Alberta, Canada. *Catena* **22**: 227–248.

Benn, D. I., Owen, L. A., Finkel, R. C., and Clemmens, S., 2006. Pleistocene lake outburst foods and fan formation along the eastern Sierra Nevada,California:implications for the interpretation of intermontane lacustrine records. *Quaternary Science Reviews* **25**: 2729–2748.

Benson, L., Kashgarian, M., Rye, R., Lund, S., Paillet, F., Smoot, J., Kester, C., Mensing, S., Meko, D., and Lindstrom, S., 2002. Holocene multidecadal and multicentennial droughts affecting northern California and Nevada. *Quaternary Science Reviews* **21**: 659–682.

Besler, H., 1976. Climatic conditions and climatic geomorphology with respect to zoning in central Namib (Southwest Africa). *Geographische Zeitschrift* **64**: 155.

Bettis III, E. A., 2003. Patterns in Holocene colluvium and alluvial fans across the prairie-forest transition in the midcontinent USA. *Geoarchaeology* **18**: 779–797.

Bhandari, S., Maurya, D. M., and Chamyal, L. S., 2005. Late Pleistocene alluvial plain sedimentation in Lower Narmada Valley, Western India: Palaeoenvironmental implications. *Journal of Asian Earth Sciences* **24**: 433–444.

Bird, M. I., Turneya, C. S. M., Fifield, L. K., Jones, R., Aylifee, L. K., Palmer, A., Cresswell, R., and Robertson, S., 2002. Radiocarbon analysis of the early archaeological site of Nauwalabila I, Arnhem Land, Australia: implications for sample suitability and stratigraphic integrity. *Quaternary Science Reviews* **21**: 1061–1075.

Birks, H. H., and Ammann, B., 2000. Two terrestrial records of rapid climatic change during the glacial-Holocene transition (14,000–9,000 calendar years BP) from Europe. *Proceedings of the National Academy of Sciences of The United States of America* **97**(4): 1390–1394.

Blair, T. C., 1987. Sedimentary processes, vertical stratification sequences, and geomorphology of the Roaring River alluvial fan, Rocky Mountain National Park. *Journal of Sedimentary Petrology* **57**: 1–18.

Blair, T. C., 1999. Alluvial fan and catchment initiation by rock avalanching, Owens Valley, California. *Geomorphology* **28**: 201–221.

Blainey, J. B., and Pelletier, J. D., 2008. Infiltration on alluvial fans in arid environments: Influence of fan morphology. *Journal of Geophysical Research – Earth Surface* **113**: Article number F03008.

Blair, T.C., and McPherson, J.G., 1994. Alluvial Fan processes and Forms. In: *Geomorphology of Desert Environments*. A.J. Parsons and A.D. Abrahams (eds.), Chapman and Hall, London, pp. 344–402.

Blair, T. C., and McPherson, J. G., 2009. Alluvial fan processes and forms. In: *Geomorphology of Desert Environments (2nd Edn)*, A. D. Abrahams and A. J. Parsons (eds.)

Blum, M. D., and Tornqvist, T. E., 2000. Fluvial responses to climate and sea-level change: a review and look forward. *Sedimentology* **47**: 2–48.

Bond, G., Showers, W., Cheseby, M., Lotti, R., Almasi, P., deMenocal, P., Priore, P., Cullen, H., Hajdas, I., and Bonani, G., 1997. A pervasive millennial-scale cycle in North Atlantic Holocene and glacial climates. *Science* **278**: 1257–1266.

Bowman, D., 1988. The declining but non-rejuvinating base level – Lisan Lake, the Dead Sea area, Israel. *Earth Surface Processes and Landforms* **13**: 239–249.

Bowman, D., Korjenkov, A., Porat, N., and Czassny, B., 2004. Morphological response to Quaternary deformation at an intermontane basin piedmont, the northern Tien Shan, Kyrghyzstan. *Geomorphology* **63**: 1–24.

Brierley, G. J., Liu, K., and Crook, K. A. W., 1993. Sedimentology of coarse- grained alluvial fans in the Markham Valley, Papua New Guinea. *Sedimentary Geology* **86**: 297–324.

Brown, E. T., Bendick, R., Bourles, D. L., Gaur, V., Molnar, P., Raisbeck, G., and Yiou, F., 2003. Early Holocene climate recorded in geomorphological features in Western Tibet. *Palaeogeography, Palaeoclimatology, Palaeoecology* **199**: 141–151.

Brunsden, D., 1993. Barriers to geomorphological change. In: *Landscape Sensitivity*, D. S. G. Thomas and R. J. Allison (eds.), Wiley, Chichester, pp. 7–12.

Brunsden, D., 2001. A critical assessment of the sensitivity concept in geomorphology. *Catena* **42**: 83–98.

Büdel, J., 1982. *Climatic Geomorphology. Translated by Lenore Fischer and Detlef Busche*. Princeton University Press, Princeton, New Jersey, 443p.

Bull, W. B., 1975. Allometric change of landforms. *Geological Society of America Bulletin* **86**: 1489–1498.

Bull, W. B., 1991. *Geomorphic responses to climatic change*. Oxford University Press, Oxford, 326p.

Bull, W. B., 1996. Global climate change and active tectonics: effective tools for teaching and research. *Geomorphology* **16**: 217–232.

Bull, W. B., and Schick, A. P., 1979. Impact of climatic change on an arid watershed: Nahal Yael, southern Israel. *Quaternary Research* **11**: 153–171.

Calvache, M. L., Viseras, C., and Fernandez, J., 1997. Controls on fan development – evidence from fan morphometry and sedimentology; Sierra Nevada, SE Spain. *Geomorphology* **21**: 69–84.

Carignano, C. A., 1999. Late Pleistocene to recent climate change in Cordoba Province, Argentina: Geomorphological evidence. *Quaternary International* **57/58**: 117–134.

Chamyal, L. S., Maurya, D. M., and Raj, R., 2003. Fluvial systems of the drylands of western India: a synthesis of Late Quaternary environmnetal and tectonic changes. *Quaternary International* **104**: 69–86.

Chawner, W. D., 1935. Alluvial fan flooding. *Geographical Review* **25**: 255–263.

Chitale, J. D., 1986. Study of Petrography and Internal Structures in Calcretes of West Texas and New Mexico (Microtextures, Caliche). Ph.D. Dissertation, Thesis thesis, 120pp., Texas Tech University, Lubbock. ·

Clevis, Q., de Boer, P., and Wachter, M., 2003. Numerical modelling of drainage basin evolution and three-dimensional alluvial fan stratigraphy. *Sedimentary Geology* **163**: 85–110.

Colombo, F., 2005. Quaternary telescopic-like alluvial fans, Andean Ranges, Argentina. In: *Alluvial fans: Geomorphology, sedimentology, dynamics*, A. M. Harvey, A. E. Mather and M. Stokes (eds.), Geological Society Special Publications 251, London, pp. 69–84.

Committee on Alluvial Fan Flooding, N. R. C., 1996. *Alluvial fan flooding*. National Academy of Science Press, Washington D.C., 172p.

Coulthard, T. J., Macklin, M. G., and Kirkby, M. J., 2002. A cellular model of Holocene upland river basin and alluvial fan evolution. *Earth Surface Processes and Landforms* **27**: 269–288.

Cremaschi, M., 1992. Genesi e significato paleoambientale della patina del deserto e suo ruolo nello studio dell'arte rupestre. Il caso del Fezzan meridionale (Sahara Libico). In: *Arte e Culture del Sahara Preistorico*, M. Lupaciollu (ed.), Quasar, Rome, pp. 77–87.

Cremaschi, M., 1996. The desert varnish in the Messak Sattafet (Fezzan, Libryan Sahara), age, archaeological context and paleo-environmental implication. *Geoarchaeology* **11**: 393–421.

Crosta, G. B., and Frattini, P., 2004. Controls on modern alluvial fan processes in the central Alps, northern Italy. *Earth Surface Processes and Landforms* **29**: 267–293.

Crouvi, O., Ben-Dor, E., Beyth, M., Avigad, D., and Amit, R., 2006. Quantitative mapping of arid alluvial fan surfaces using field spectrometer and hyperspectral remote sensing. *Remote Sensing of the Environment* **104**: 103–117

Curry, W. B., and Oppo, D. O., 1997. Synchronous high frequency oscillations in tropical sea-surface temperatures and North Atlantic deep water production during the last glacial cycle. *Paleooceanography* **12**: 1–14.

Davies, T. R. H., and Korup, O., 2006. Persistent alluvial fanhead trenching resulting from large, infrequent sediment inputs. *Earth Surface Processes and Landforms* **32**: 725–742.

De Chant, L. J., Pease, P. P., and Tchakerian, V. P., 1999. Modelling alluvial fan morphology. *Earth Surface Processes and Landforms* **24**: 641–652.

DeLong, S. B., and Arnold, L. J., 2007. Dating alluvial deposits with optically stimulated luminescence, AMS C-14 and cosmogenic techniques, western Transverse Ranges, California, USA. *Quaternary Geochronology* **2**: 129–136.

DeLong, S. B., Pelletier, J. D., and Arnold, L. J., 2008. Climate change triggered sedimentation and progressive tectonic uplift in a coupled piedmont-axial system: Cuyama Valley, California, USA. *Earth Surface Processes and Landforms* **33**: 1033–1046.

Denny, C. S., 1965. Alluvial fans in the Death Valley region of California and Nevada. *U.S. Geological Survey Professional Paper* **466**.

Denny, C. S., 1967. Fans and pediments. *American Journal Science* **265**: 81–105.

Derbyshire, E. D., 1993. *Climatic Geomorphology*. Macmillan, London, 296p.

Diaz, T. A., Bailley, T. L., and Orndorff, R. L., 2002. SEM analysis of vertical and lateral variations in desert varnish chemistry from the Lahontan Mountains, Nevada. *Geological Society of America Abstracts with Programs* **May 7–9 Meeting**: <///gsa.confex.com/gsa/2002RM/finalprogram/abstract_33974.htm>.

Dorn, R. I., 1984. Cause and implications of rock varnish microchemical laminations. *Nature* **310**: 767–770.

Dorn, R. I., 1988. A rock varnish interpretation of alluvial-fan development in Death Valley, California. *National Geographic Research* **4**: 56–73.

Dorn, R. I., 1990. Quaternary alkalinity fluctuations recorded in rock varnish microlaminations on western U.S.A. volcanics. *Palaeogeography, Palaeoclimatology, Palaeoecology* **76**: 291–310.

Dorn, R. I., 1994. Alluvial fans an an indicator of climatic change. In: *Geomorphology of Desert Environments*, A. D. Abrahams and A. J. Parsons (eds.), Chapman & Hall, London, pp. 593–615.

Dorn, R. I., 1996. Climatic hypotheses of alluvial-fan evolution in Death Valley are not testable. In: *The Scientific Nature of*

Geomorphology, B. L. Rhoads and C. E. Thorn (eds.), Wiley, New York, pp. 191–220.

Dühnforth, M., Densmore, A. L., Ivy-Ochs, S., Allen, P. A., and Kubik, P. W., 2007. Timing and patterns of debris flow deposition on Shepherd and Symmes creek fans, Owens Valley, California, deduced from cosmogenic [10]Be. *Journal of Geophysical Research* **112**: Art. No. F03S15.

Eckis, R., 1928. Alluvial fans in the Cucamonga district, southern California. *Journal of Geology* **36**: 111–141.

Ellwood, B. B., and Gose, W. A., 2006. Heinrich H1 and 8200 yr B.P. climate events recorded in Hall's Cave, Texas. *Geology* **34**: 753–756.

Elorza, M. G., 2005. *Climatic geomorphology*. Elsevier, Amsterdam, 774p.

Ely, L., 1997. Response of extreme floods in the southwestern United States to climatic variations in the late Holocene. *Geomorphology* **19**: 175–201.

Ely, L. L., Enzel, Y., Baker, V. R., and Cayan, D. R., 1993. A 5000-year-record of extreme floods and climate change in the southwestern United States. *Science* **262**: 410–412.

Enfield, D. B., Mestas-Nunez, A. M., and Trimble, P. J., 2000. The Atlantic multidecadal oscillation and its relation to rainfall and river flows in the continental US. *Geophysical Research letters* **28**: 2077–2080.

Eriksson, M. G., Olley, J. M., and Payton, R. W., 2000. Soil erosion history in central Tanzania based on OSL dating of colluvial and alluvial hillslope deposits. *Geomorphology* **36**: 107–128.

Evenstar, L., Hartley, A., Rice, C., Stuart, F., Mather, A. E., and Guillermo, C., 2006. Miocene-Pliocene climate change in the Peru-Chile Desert. *6th International Symposium on Andean Geodynamics (ISAG 2005, Barcelona)* **Extended Abstracts**: 258–260.

Falloon, P. D., and Smith, P., 2000. Modeling refractory soil organic matter. *Biology and fertility of soils* **30**: 388–398.

Fard, A. M., 2001. Recognition of abrupt climatic changes in clastic sedimentary environments: an introduction. *Global and Planetary Change* **28**: ix–xii.

Farr, T. G., and Chadwick, O. A., 1996. Geomorphic processes and remote sensing signatures of alluvial fans in the Kun Lun mountains, China. *Journal of Geophysical Research-Planets* **101**(E10): 23091–23100.

Field, J., 2001. Channel avulsion on alluvial fans in southern Arizona. *Geomorphology* **37**: 93–104.

Fitz, J.-F., Bourles, D., Brown, E. T., Carretier, S., Chery, J., Enhtuvshin, B., Galsan, P., Finkel, R. C., Hanks, T. C., Kendrick, K. J., Philip, H., Raisbeck, G., Schlupp, A., Schwartz, D. P., and Yiou, F., 2003. Late Pleistocene to Holocene slip rates for the Gurvan Bulag thrust fault (Gobi-Altay, Mongolia) estimated with [10]Be dates. *Journal of Geophysical Research* **108**(B3): 2162, doi:10.1029/2001JB000553, 2003.

Frankel, K. L., Brantley, K. S., Dolan, J. F., Finkel, R. C., Klinger, R. E., Knott, J. R., Machette, M. N., Owen, L. A., Phillips, F. M., Slate, J. L., and Wernicke, B. P., 2007. Cosmogenic Be-10 and Cl-36 geochronology of offset alluvial fans along the northern Death Valley fault zone: Implications for transient strain in the eastern California shear zone. *Journal of Geophysical Research – Solid Earth* **112**: Art. No. B06407.

Frostick, L. E., and Reid, I., 1989. Climatic versus tectonic control of fan sequences – lessons from the Dead Sea, Israel. *Journal of the Geological Society, London* **146**: 527–538.

Garcia-Melendez, E., Goy, J. L., and Zazo, C., 2003. Neotectonics and Plio-Quaternary landscape development within the eastern Huercal-Overa Basin (Betic Cordilleras, southeast Spain). *Geomorphology* **50**: 111–133.

Gardner, T. W., Webb, J., Davis, A. G., Cassel, E. J., Pezzia, C., Merritts, D. J., and Smith, B., 2006. Late Pleistocene landscape response to climate change: eolian and alluvial fan deposition Cape Liptrap, southeastern Australia. *Quaternary Science Reviews* **25**: 1552–1569.

Gillespie, R., Prosser, I. P., Dlugokencky, E., Sparks, R. J., Wallace, G., and Chappell, J. M. A., 1992. AMS Dating of Alluvial Sediments on the Southern Tablelands of New-South-Wales, Australia. *Radiocarbon* **34**: 29–36.

Glennie, K. W., and Singhvi, A. K., 2002. Event stratigraphy, paleoenvironment and chronology of SE Arabian deserts. *Quaternary Science Reviews* **21**: 853–869.

Gomez, B., Banbury, K., Marden, M., Trustrum, N. A., Peacock, D. H., and Hoskin, P. J., 2003. Gully erosion and sediment production: Te Weroroa Stream, New Zealand. *Water Resources Research* **39**(7): 1187.

Gómez-Villar, A., Álvarez-Martínez, J., and García-Ruiz, J. M., 2006. Factors influencing the presence or absence of tributary-junction fans in the Iberian Range, Spain. *Geomorphology* **81**: 252–264.

Griffiths, P. G., Hereford, R., and Webb, R. H., 2006. Sediment yield and runoff frequency of small drainage basins in the Mojave Desert, USA. *Geomorphology* **74**: 232–244.

Griffiths, P. G., and Webb, R. H., 2004. Frequency and initiation of debris flows in Grand Canyon, Arizona. *Journal of Geophysical Research* **109**: F04002.

Grosjean, M., Nunez, L., Castajena, I., and Messerli, B., 1997. Mid-Holocene climate and culture changes in the Atacama Desert, northern Chile. *Quaternary Research* **48**: 239–246.

Guerra-Merchan, A., Serrano, F., and Ramallo, D., 2004. Geomorphic and sedimentary Plio–Pleistocene evolution of the Nerja area (northern Alboran basin, Spain). *Geomorphology* **60**: 89–105.

Gunster, N., and Skowronek, A., 2001. Sediment- soil sequences in the Granada Basin as evidence for long-and short-term climatic changes during the Pliocene and Quaternary in the Western Mediterranean. *Quaternary International* **78**: 17–32.

Hagedorn, H., and Rapp, A., 1989. *Geomorphology and Geoecology: Climatic Geomorphology. Proceedings of the Second International Conference on Geomorphology and Geoecology*. Zeitschrift für Geomorphologie Supplementbände Band 84, Frankfurt, 201p.

Hanson, P. R., 2005. Alluvial fan response to climate change along the lower Colorado River. *Geological society of America Abstracts with Program, Salt Lake City Meeting* **37**(7): 110.

Hardy, S., and Gawthorpe, R., 2002. Normal fault control on bedrock channel incision and sediment supply: Insights from numerical modeling. *Journal of Geophysical Research* **107**(B10): 2246.

Harkins, N. W., Anastasio, D. J., and Pazzaglia, F. J., 2005. Tectonic geomorphology of the Red Rock fault, insights into segmentation and landscape evolution of a developing range front normal fault. *Journal of Structural Geology* **27**: 1925–1939.

Hartley, A., Mather, A. E., Jolley, E., and Turner, P., 2005. Climatic controls on alluvial-fan activity, Coastal Cordillera, northern Chile. In: *Alluvial fans: Geomorphology, sedimentology, dynamics*, A. M. Harvey, A. E. Mather and M. Stokes (eds.), Geological Society Special Publications 251, London, pp. 95–115.

Harvey, A. M., 1984. Aggradation and dissection sequences on Spanish alluvial fans: influence on morphological development. *Catena* **11**: 289–304.

Harvey, A. M., 1997. The role of alluvial fans in arid zone fluvial systems. In: *Arid Zone Geomorphology*, D. S. G. Thomas (ed.), Wiley, Chichester, pp. 231–260.

Harvey, A. M., 2002a. Effective timescales of coupling within fluvial systems. *Geomorphology* **44**: 175–201.

Harvey, A. M., 2002b. The role of base-level change in the dissection of alluvial fans: case studies from southeast Spain and Nevada. *Geomorphology* **45**: 67–87.

Harvey, A. M., Foster, G., Hannam, J., and Mather, A., E., 2003. The Tabernas alluvial fan and lake system, southeast Spain: applications of mineral magnetic and pedogenic iron oxide analyses towards clarifying the Quaternary sediment sequences. *Geomorphology* **50**: 151–171.

Harvey, A. M., Mather, A. E., and Stokes, M., 2005. Alluvial fans: geomorphology, sedimentology, dynamics – introduction. A review of alluvial-fan research. In: *Alluvial fans: Geomorphology, sedimentology, dynamics*, A. M. Harvey, A. E. Mather and M. Stokes (eds.), Geological Society Special Publications 251, London, pp. 1–7.

Harvey, A. M., Silva, P. G., Mather, A. E., Goy, J. L., Stokes, M., and Zazo, C., 1999a. The impact of Quaternary sea-level and climatic change on coastal alluvial fans in the Cabo de Gata ranges, southeast Spain. *Geomorphology* **28**: 1–22.

Harvey, A. M., and Wells, S. G., 1994. Late Pleistocene and Holocene changes in hillslope sediment supply to alluvial fan systems: Zzyzx, California. In: *Environmental Change in Drylands: Biogeographical and Geomorphological Perspectives*, A. C. Millington and K. Pye (eds.), Wiley & Sons, London, pp. 67–84.

Harvey, A. M., Wigand, P. E., and Wells, S. G., 1999b. Response of alluvial fan systems to the late Pleistocene to Holocene climatic transition: contrasts between the margins of pluvial Lakes Lahontan and Mojave, Nevada and California, USA. *Catena* **36**: 255–281.

Henderson, G. M., 2006. Caving in to new chronologies. *Science* **313**: 620–622.

Hetzel, R., Tao, M., Stokes, S., Niedermann, S., Ivy-Ochs, S., Gao, B., Strecker, M. R., and Kubik, P. W., 2004. Late Pleistocene//Holocene slip rate of the Zhangye thrust (Qilian Shan, China) and implications for the active growth of the northeastern Tibetan Plateau. *Tectonics* **23**: TC6006.

Hooke, R. L., 1967. Processes on arid region alluvial fans. *Journal Geology* **75**: 438–460.

Hooke, R. L., 1968. Steady-state relationships on arid-region alluvial fans in closed basins. *American Journal Science* **266**: 609–629.

Hooke, R. L., and Rohrer, W. L., 1977. Relative erodibility of source-area rock types, as determined from second order variations in alluvial fan size. *Geological Society of America Bulletin* **88**: 1177–1182.

Houghton, J. T., Ding, Y., Griggs, D. J., Noguer, M., van der Linden, P. J., Dai, X., Maskell, K., and Johnson, C. A., Climate change 2001. *The scientific basis*. Cambridge University Press, Cambridge, p. 881.

House, P. K., 2005. Using geology to improve flood hazard management on alluvial fans – An example from Laughlin, Nevada. *Journal of the American Water Resources Association* **41**: 1431–1447.

House, P. K., and Baker, V. R., 2001. Paleohydrology of flash floods in small desert watersheds in western Arizona. *Water Resources Research* **37**: 1825–1839.

Hsu, L., and Pelletier, J. D., 2004. Correlation and dating of Quaternary alluvial-fan surfaces using scarp diffusion. *Geomorphology* **60**: 319–335.

Huggett, R. J., 1998. Soil chronosequences, soil development, and soil evolution: a critical review. *Catena* **32**: 155–172.

Huntington, E., 1907. Some characteristics of the glacial period in non-glaciated regions. *Geological Society of America Bulletin* **18**: 351–388.

Ibbeken, H., Warnke, D. A., and Diepenbroek, M., 1998. Granulometric study of the Hanaupah fan, Death Valley, California. *Earth Surface Processes and Landforms* **23**: 481–492.

Iriondo, M., 1994. The Quaternary of Ecuador. *Quaternary International* **21**: 101–112.

Jain, M., and Tandon, S. J., 2003. Fluvial response to Late Quaternary climate changes, western India. *Quaternary Science Reviews* **22**: 2223–2235.

Jansson, P., Jacobson, D., and Hooke, R. L., 1993. Fan and playa areas in southern California and adjacent parts of Nevada. *Earth Surface Processes and Landforms* **18**: 109–119.

Juyal, N., Rachna, R., Maurya, D. M., Chamyal, L. S., and Singhvi, A. K., 2000. Chronology of Late Pleistocene evironmental changes in the lower Mahi basin, Western India. *Journal of Quaternary Science* **15**: 501–508.

Kale, V. S., and Rajaguru, S. N., 1987. Late Quaternary alluvial history of the Northwestern Deccan upland region. *Nature* **325**: 612–614.

Kale, V. S., Singhvi, A. K., Mishra, P. K., and Banerjee, D., 2000. Sedimentary records and luminescence chronology of Late Holocene palaeofloods in the Luni River, Thar Desert, northwest India. *Catena* **40**: 337–358.

Kashima, K., 2002. Environmental and climatic changes during the last 20,000 years at Lake Tuz, central Turkey. *Catena* **48**: 3–20.

Keefer, D. K., Moseley, M. E., and deFrance, S. D., 2003. A 38 000-year record of floods and debris flows in the Ilo region of southern Peru and its relation to El Niño events and great earthquakes. *Palaeogeography, Palaeoclimatology, Palaeoecology* **194**: 41–77.

Kershaw, A. P., and Nanson, G. C., 1993. The last full glacial cycle in the Australian region. *Global and Planetary Change* **7**: 1–9.

Kesel, R. H., and Spicer, B. E., 1985. Geomorphologic relationships and ages of soils on alluvial fans in the Rio General Valley, Costa Rica. *Catena* **12**: 149–166.

Klinger, Y., Avouac, J. P., Bourles, D., and Tisnerat, N., 2003. Alluvial deposition and lake-level fluctuations forced by Late Quaternary climate change: the Dead Sea case example. *Sedimentary Geology* **162**: 119–139.

Knott, J. R., Sarna-Wojcicki, A. M., Machette, M. N., and Klinger, R. E., 2005. Upper Neogene stratigraphy and tectonics of Death Valley – A review. *Earth Science Reviews* **73**: 245–270.

Knox, J. C., 1983. Responses of river systems to Holocene climates. In: *Late Quaternary Environments of the United States. Vol. 2. The Holocene*, H. E. J. Wright (ed.), University Minnesota Press, Minneapolis, pp. 26–41.

Kochel, R. G., Miller, J. R., and Ritter, J. B., 1997. Geomorphic response to minor cyclic climate changes, San Diego County, California. *Geomorphology* **19**: 277–302.

Kostaschuk, R. A., MacDonald, G. M., and Putnam, P. E., 1986. Depositional process and alluvial fan-drainage basin morphometric relationships near Banff, Alberta, Canada. *Earth Surface Processes and Landforms* **11**: 471–484.

Krohling, D. M., and Iriondo, M., 1999. Upper Quaternary Palaeoclimates of the Mar Chiquita area, North Pampa, Argentina. *Quaternary International* **57/58**: 149–163.

Krzyszkowski, D., and Zielinkski, T., 2002. The Pleistocene end moraine fans: controls on the sedimentation and location. *Sedimentary Geology* **149**: 73–92.

Lafortune, V., Filion, L., and Hétu, B., 2006. Impacts of Holocene climatic variations on alluvial fan activity below snowpatches in subarctic Québec. *Geomorphology* **76**: 375–391.

Lave, J., and Burbank, D., 2004. Denudation processes and rates in the Transverse Ranges, southern California: Erosional response of a transitional landscape to external and anthropogenic forcing. *Journal of Geophysical Research* **109**: F01006.

Lee, M. R., and Bland, P. A., 2003. Dating climatic change in hot deserts using desert varnish on meteorite finds. *Earth and Planetary Science Letters* **206**: 187–198.

Leeder, M. R., Harris, T., and Kirkby, M. J., 1998. Sediment supply and climate change: implications for basin stratigraphy. *Basin Research* **10**: 7–18.

Lehmkuhl, F., and Haselein, F., 2000. Quaternary paleoenvironmental change on the Tibetan Plateau and adjacent areas (Western China and Western Mongolia). *Quaternary International* **65**: 121–145.

Leier, A. L., DeCelles, P. G., and Pelletier, J. D., 2005. Mountains, monsoons, and megafans. *Geology* **33**: 289–292.

Liu, B., Phillips, F. M., Pohl, M. M., and Sharma, P., 1996. An alluvial surface chronology based on cosmogenic ^{36}Cl dating, Ajo Mountains (Organ Pipe Cactus National Monument), Southern Arizona. *Quaternary Research* **45**: 30–37.

Liu, T., 2003. Blind testing of rock varnish microstratigraphy as a chronometric indicator: results on late Quaternary lava flows in the Mojave Desert, California. *Geomorphology* **53**: 209–234.

Liu, T., 2008. VML Dating Lab. *http://www.vmldatinglab.com/*: accessed June 24, 2006.

Liu, T., and Broecker, W., 2007. Holocene rock varnish microstratigraphy and its chronometric application in drylands of western USA. *Geomorphology* **84**: 1–17.

Liu, T., and Broecker, W.S., 2008. Rock varnish microlamination dating of late Quaternary geomorphic features in the drylands of western USA. *Geomorphology* **93**: 501–523.

Liu, T., and Dorn, R. I., 1996. Understanding spatial variability in environmental changes in drylands with rock varnish microlaminations. *Annals of the Association of American Geographers* **86**: 187–212.

Ljungdahl, L., and Eriksson, K., 1985. Ecology of microbial cellulose degradation. *Advances in Microbial Ecology* **8**: 237–299.

Mabbutt, J. C., 1977. *Desert landforms*. Australian National University Press, Canberra, 340p.

Madsen, D. B., Chen, F., Oviatt, G., Zhu, Y., Brantingham, P. J., Elston, R. G., and Bettinger, R. L., 2003. Late Pleistocene/Holocene wetland events recorded in southeast Tengger Desert lake sediments, NW China. *Chinese Science Bulltin* **48**: 1423–1429.

Mahan, S. A., Miller, D. M., Menges, C. M., and Yount, J. C, 2007. Late Quaternary stratigraphy and luminescence geochronology of the northeastern Mojave Desert. *Quaternary International* **166**: 61–78.

Marston, R. A., 2003. Editorial note. *Geomorphology* **53**: 197.

Masson-Delmotte, V., Jouzel, J., Landais, A., Stievenard, M., Johnsen, S. J., White, J. W. C., Werner, M., Sveinbjornsdottir, A., and Fuhrer, K., 2005. GRIP deuterium excess reveals rapid and orbital-scale changes in Greenland moisture origin. *Science* **209**: 118–121.

Mather, A. E., and Hartley, A., 2005. Flow events on a hyper-arid alluvial fan: Quebrada Tambores, Salar de Atacama, northern Chile. In: *Alluvial fans: Geomorphology, sedimentology, dynamics*, A. M. Harvey, A. E. Mather and M. Stokes (eds.), Geological Society Special Publications 251, London, pp. 9–29.

Matmon, A., Schwartz, D. P., Finkel, R., Clemmens, S., and Hanks, T. C., 2005. Dating offset fans along the Mojave section of the San Andreas fault using cosmogenic ^{26}Al and ^{10}Be. *Geological Society of America Bulletin* **117**: 795–807.

Mazels, J. L., 1990. Long-term palaeochannel evolution during episodic growth of an exhumed alluvial fan, Oman. In: *Alluvial Fans: A Field Approach*, A. H. Rachocki and M. Church (eds.), Wiley, Chichester, pp. 271–304.

McDonald, E. V., McFadden, L. D., and Wells, S. G., 2003. Regional response of alluvial fans to the Pleistocene-Holocene climatic transition, Mojave Desert, California. *Geological Society of America Special Paper* **368**: 189–205.

McLaren, S. J., Gilbertson, D. D., Grattan, J. P., Hunt, C. O., Duller, G. A. T., and Barker, G. A., 2004. Quaternary palaeogeomorphologic evolution of the Wadi Faynan area, southern Jordan. *Palaeogeography, Palaeoclimatology, Palaeoecology* **205**: 131–154.

Melton, M. A., 1965. The geomorphic and paleoclimatic significance of alluvial deposits in southern Arizona. *Journal Geology* **73**: 1–38.

Miller, J. R., Germanoski, D., Waltman, K., Rausch, R., and Chambers, J., 2001. Influence of late Holocene hillslope processes and landforms on modern channel dynamics in upland watersheds of central Nevada. *Geomorphology* **38**: 373–391.

Mills, H. H., 2005. Relative-age dating of transported regolith and application to study of landform evolution in the Appalachians. *Geomorphology* **67**: 63–96.

Monger, H. C., and Buck, B. J., 1999. Stable isotopes and soil-geomorphology as indicators of Holocene climate change, northern Chihuahuan Desert. *Journal of Arid Environments* **43**: 357–373.

Moody, J. A., and Martin, D. A., 2001. Initial hydrologic and geomorphic response following a wildfire in the Colorado

Front Range. *Earth Surface Processes and Landforms* **26**: 1049–1070.

Muto, T., and Steel, R. J., 2000. The accomodation concept in sequence stratigraphy: some dimensional problems and possible redefinition. *Sedimentary Geology* **130**: 1–10.

Nanson, G. C., Price, D. M., and Short, S. A., 1992. Wetting and drying of Australia over the past 300 ka. *Geology* **20**: 791–794.

Nichols, K. K., Bierman, P. R., Foniri, W. R., Gillespie, R., Caffee, M., and Finkel, R., 2006. Dates and rates of arid region geomorphic processes. *GSA Today* **16**(8): 4–11.

Nott, J. F., Thomas, M. F., and Price, D. M., 2001. Alluvial fans, landslides and Late Quaternary climaticchange in the wet tropics of northeast Queensland. *Australian Journal of Earth Sciences* **48**: 875–882.

Oguchi, T., and Oguchi, C. T., 2004. Late Quaternary rapid talus dissection and debris flow deposition on an alluvial fan in Syria. *Catena* **55**: 125–140.

Olley, J. M., Pietsch, T., and Roberts, R. G., 2004. Optical dating of Holocene sediments from a variety of geomorphic settings using single grains of quartz. *Geomorphology* **60**: 337–358.

Owen, L. A., Finkel, R. C., Haizhou, M., and Barnard, P. L., 2006. Late Quaternary landscape evolution in the Kunlun Mountains and Qaidam Basin, Northern Tibet: A framework for examining the links between glaciation,lake level changes and alluvial fan formation. *Quaternary International* **154–155**: 73–86.

Owen, L. A., Windley, B. F., Cunningham, W. D., Badamgarav, J., and Dorjnamjaa, D., 1997. Quaternary alluvial fans in the Gobi of southern Mongolia: evidence for neotectonics and climate change. *Journal of Quaternary Science* **12**: 239–252.

Pelletier, J. D., Mayer, L., Pearthree, P. A., House, P. K., Demsey, K. A., Klawon, J. E., and Vincent, K. R., 2005. An integrated approach to flood hazard assessment on alluvial fans using numerical modeling, field mapping, and remote sensing. *Geological Society of America Bulletin* **117**: 1167–1180.

Philander, S. G., 1999. El Niño and La Nina predictable climate fluctuations. *Reports on Progress in Physics* **62**: 123–142.

Phillips, F. M., 2003. Cosmogenic ^{36}Cl ages of Quaternary basalt flows in the Mojave Desert, California, USA. *Geomorphology* **53**: 199–208.

Phillips, W. M., McDonald, E. V., and Poths, J., 1998. Dating soils and alluvium with cosmogenic Ne-21 depth profiles: Case studies from the Pajarito Plateau, New Mexico, USA. *Earth and Planetary Science Letters* **160**: 209–223.

Pickup, G., 1991. Event frequency and landscape stability on the floodplain systems of arid central Australia. *Quaternary Science Reviews* **10**: 463–473.

Poisson, B., and Avouac, J. P., 2004. Holocene hydrological changes inferred from alluvial stream entrenchment in North Tian Shan (northwestern China). *Journal of Geology* **112**: 231–249.

Pope, R. J. J., and Millington, A. C., 2000. Unravelling the patterns of alluvial fan development using mineral magnetic analysis: Examples from the Sparta Basin, Lakonia, southern Greece. *Earth Surface Processes and Landforms* **25**: 601–615.

Pope, R. J. J., and Wilkinson, K. N., 2005. Reconciling the roles of climate and tectonics in Late Quaternary fan development on the Spartan piedmont, Greece. In: *Alluvial fans: Geo-*

morphology, sedimentology, dynamics, A. M. Harvey, A. E. Mather and M. Stokes (eds.), Geological Society Special Publications 251, London, pp. 133–152.

Posamentier, H. W., and Vail, P. R., 1988. Eustatic control on clasticdeposition: II. Sequence and systems tracts models. In: *Sea Level Changes: An Integrated Approac*, C. K. Wilgus, B. S. Hastings, C. Kendall, H. W. Posamentier, C. A. Ross and J. C. Van Wagoner (eds.), Society of Economic Paleontologists and Mineralogists Special Publication 42, Tulsa, pp. 125–154.

Proctor, C. J., Baker, A., Barnes, W. L., and Gilmour, M. A., 2000. A thousand year speleothem proxy record of North Atlantic climate from Scotland. *Climate Dynamics* **16**: 815–820.

Quigley, M. C., Sandiford, M., and Cupper, M. L., 2007. Distinguishing tectonic from climatic controls on range-front sedimentation. *Basin Research* **19**: 491–505.

Reheis, M. C., Slate, J. L., Throckmorton, C. K., McGeehin, J., P., SarnaWojcicki, A. M., and Dengler, L., 1996. Late Qaternary sedimentation on the Leidy Creek fan, Nevada-California: Geomorphic responses to climate change. *Basin Research* **8**: 279–299.

Roberts, N., 1995. Climatic forcing of alluvial fan regimes during the Late Quaternary in Konya basin, south central Turkey. In: *Mediterranean Quaternary River Environment*, J. Lewin, M. G. Macklin and J. Woodward (eds.), Balkema, Rotterdam, pp. 205–217.

Roberts, N., and Barker, P., 1993. Landscape stability and biogeomorphic response to past and future climatic shifts in intertropical Africa. In: *Landscape sensitivity*, D. S. G. Thomas and R. J. Allison (eds.), Wiley, London, pp. 65–82.

Roberts, N., Black, S., Boyer, P., Eastwood, W. J., Griffiths, H. I., Lamb, H. F., Leng, M. J., Parish, R., Reed, J. M., Twigg, D., and Yigitbasioflu, H., 1999. Chronology and stratigraphy of Late Quaternary sediments in the Konya Basin, Turkey: Results from the KOPAL Project. *Quaternary Science Reviews* **18**: 611–630.

Robinson, R. A. J., Spencer, J. Q. G., Strecker, M. R., Richter, A., and Alonso, R. N., 2005. Luminescence dating of alluvial fans in intramontane basins of NW Argentina. In: *Alluvial fans: Geomorphology, sedimentology, dynamics*, A. M. Harvey, A. E. Mather and M. Stokes (eds.), Geological Society Special Publications 251, London, pp. 153–168.

Robinson, S. E., 2002. Cosmogenic nuclides, remote sensing, and field studies applied to desert piedmonts, Dissertation thesis, 387pp., Arizona State University, Tempe.

Robustelli, G., Muto, F., Scarciglia, F., Spina, V., and Critelli, S., 2005. Eustatic and tectonic control on Late Quaternary alluvial fans along the Tyrrhenian Sea coast of Calabria (South Italy). *Quaternary Science Reviews* **24**: 2101–2119.

Rodbell, D. T., Seltzer, G. O., Anderson, D. M., Abbott, M. B., Enfield, D. B., and Newman, J. H., 1999. 15000-year record of El Niño-driven alluviation in southwetsern Ecuador. *Science* **283**: 516–520.

Rodgers, D. W., and Gunatilaka, A., 2002. Bajada formation by monsoonal erosion of a subaerial forebuldge, Sultanate of Oman. *Sedimentary Geology* **154**: 127–147.

Rohling, E. J., and Palike, H., 2005. Centennial-scale climate cooling with a sudden cold event around 8,200 years ago. *Nature* **434**: 975–979.

Rost, K. T., 2000. Pleistocene paleoenvironmental changes in the high mountain ranges of central China and adjacent regions. *Quaternary International* **65/66**: 147–160.

Rubustelli, G., Muto, F., Scarciglia, F., Spina, V., and Critelli, S., 2005. Eustatic and tectonic control on Late Quaternary alluvial fans along the Tyrrhenian Sea coast of Calabria (South Italy). *Quaternary Science Reviews* **24**: 2101–2119.

Sancho, C., Pena, J. L., Rivelli, F., Rhodes, E., and Munoz, A., 2008. Geomorphological evolution of the Tilcara alluvial fan (Jujuy Province, NW Argentina): Tectonic implications and palaeoenvironmental considerations. *Journal of South American Earth Sciences* **26**: 68–77.

Schaefer, J. M., Denton, G. H., Barrell, D. J. A., Ivy-Ochs, S., Kubik, P. W., Andersen, B. G., Phillips, F. M., Lowell, T. V., and Schluchter, C., 2006. Near-synchronous interhemisheric termination of the last glacial maximum in mid-latitudes. *Science* **213**: 1510–1513.

Schick, A. P., 1974. Formation and obliteration of desert stream terraces – a conceptual analysis. *Zeitschrift fur Geomorphology N.F. Supplement Band* **21**: 81–105.

Schick, A. P., Grodek, T., and Wolman, M. G., 1999. Hydrologic processes and geomorphic constraints on urbanization of alluvial fan slopes. *Geomorphology* **31**: 325–335.

Schick, A. P., and Lekach, J., 1993. An evaluation of two ten-year sediment budgets, Nahal yael, Israel. *Physical Geography* **14**: 226–238.

Schumm, S. A., 1977. *The fluvial system.* Wiley, New York, 338p.

Schumm, S. A., Mosley, M. P., and Weaver, W. E., 1987. *Experimental fluvial geomorphology.* Wiley & Sons, New York, 413p.

Shaffer, J. A., Cerveny, R. S., and Dorn, R. I., 1996. Radiation windows as indicators of an astronomical influence on the Devil's Hole chronology. *Geology* **24**: 1017–1020.

Singh, A. K., Parkash, B., Mohindra, R., Thomas, J. V., and Singhvi, A. K., 2001. Quaternary alluvial fan sedimentation in the Dehradun Valley Piggyback Basin, NW Himalaya: tectonic and palaeoclimatic implications. *Basin Research* **13**: 449–471.

Six, J., Conant, R. T., Paul, E. A., and Paustian, K., 2002. Stabilization mechanisms of soil organic matter: Implications for C-saturation of soils. *Plant and Soil* **241**: 155–176.

Soeters, R., and van Westen, C. J., 1996. Slope instability, recognition, analysis and zonation. In: *Landslides: Investigation and Mitigation*, vol. 247, A. K. Turner and R. L. Schuster (eds.), National Research Council, Transportation Research Board, Washington D.C., pp. 129–177.

Sohn, M. F, Mahan, S. A., Knott, J. R., and Bowman, D. D, 2007. Luminescence ages for alluvial-fan deposits in Southern Death Valley: Implications for climate-driven sedimentation along a tectonically active mountain front. *Quaternary International* **166**: 49–60.

Sorriso-Valvo, M., 1988. Landslide-related fans in Calabria. *Catena Supplement* **13**: 109–121.

Spencer, J. Q. G., and Robinson, R. A. J., 2008. Dating intramontane alluvial deposits from NW Argentina using luminescence techniques: Problems and potential. *Geomorphology* **93**: 144–155.

Stadelman, S., 1994. Genesis and post-formational systematics of carbonate accumulations in Quaternary soils of the South-western United States, Ph.D. Dissertation thesis, 124pp., Texas Tech University, Lubbock.

Staley, D. M., Wasklewicz, T. A., and Blaszczynski, J. S., 2006. Surficial patterns of debris flow deposition on alluvial fans in Death Valley, CA using airborne laser swath mapping data. *Geomorphology* **74**: 152–163.

Starkel, L., 1999. Space and time scales in geomorphology. *Zeitschrift für Geomorphologie Supplementband* **115**: 19–33.

Stokes, M., Nash, D. J., and Harvey, A. M., 2007. Calcrete 'fossilisation' of alluvial fans in SE Spain: The roles of groundwater, pedogenic processes and fan dynamics in calcrete development. *Geomorphology* **85**: 63–84.

Stokes, S., Hetzel, R., Bailey, R. M., and Mingzin, T., 2003. Combined IRSL-OSL single aliquot regeneration (SAR) equivalent dose (De) estimates from source proximal Chinese loess. *Quaternary Science Reviews* **22**: 975–983.

Suresh, N., Bagati, T. N., Thakur, V. C., Kumar, R., and Sangode, S. J., 2002. Optically stimulated luminescence dating of alluvial fan deposits of Pinjaur Dun, NW Sub Himalaya. *Current Science* **82**: 1267–1274.

Suresh, N., Bagati, T. N., Kumar, R., and Thakur, V. C., 2007. Evolution of Quaternary alluvial fans and terraces in the intramontane Pinjaur Dun, Sub-Himalaya, NW India: interaction between tectonics and climate change. *Sedimentology* **54**: 809–833.

Talbot, M. R., and Williams, M. A. J., 1979. Cyclic alluvial fan sedimentation on the flanks of fixed dunes, Janjari, central Niger. *Catena* **6**: 43–62.

Taylor, K. C., Mayewski, P. A., Alley, R. B., Brook, E. J., Gow, J., Grootes, P. M., Meese, D. A., Saltzmann, E. S., Severinghaus, J. P., Twickler, M. S., White, J. W. C., Whitlow, S., and Zelinsky, G. A., 1997. The Holocene-Younger Dryas transition recorded at Summit, Greeland. *Science* **278**: 825–827.

Thiagarajan, N., and Lee, C. A., 2004. Trace-element evidence for the origin of desert varnish by direct aqueous atmospheric deposition. *Earth and Planetary Science Letters* **224**: 131–141.

Thomas, M. F., 2003. Late Quaternary sediment fluxes from tropical watersheds. *Sedimentary Geology* **162**: 63–81.

Thomas, M. F., 2004. Landscape sensitivity to rapid environmental change – a Quaternary perspective with examples from tropical areas. *Catena* **55**: 107–124.

Thorndrycraft, V. R., and Benito, G., 2006. Late Holocene fluvial chronology of Spain: The role of climatic variability and human impact. *Catena* **66**: 34–41.

Throckmorton, C. K., and Reheis, M. C., 1993. Late Pleistocene and Holocene environmental changes in Fish Lake Valley, Nevada-California: Geomorphic response of alluvial fans to climate change. *U.S. Geological Survey Open File Report* **93–620**: 1–82.

Tricart, J., and Cailleux, A., 1973. *Introduction to Climatic Geomorphology. Translated from the French by Conrad J. Kiewiet de Jonge.* St. Martin's Press, New York, 295p.

Tucker, G. E., and Slingerland, R., 1997. Drainage basin responses to climate change. *Water Resources Research* **33**: 2031–2047.

Twidale, C. R., and Lageat, Y., 1994. Climatic geomorphology. A Critique. *Progress in Physical Geography* **18**: 319–334.

Vidal, L., Laberyie, L., Cortijo, E., Arnold, M., Duplessy, J. C., Michel, E., Becque, S., and vanWearing, T., 1997. Evidence for changes in the North Atlantic deep water linked to meltwater surges during the Heinrich events. *Earth and Planetary Science Letters* **146**: 13–27.

Viles, H. A., and Goudie, A. S., 2003. Interannual, decadal and multidecadal scale climatic variability and geomorphology. *Earth-Science Reviews* **61**: 105–131.

Viseras, C., Calvache, M. L., Soria, J. M., and Fernandez, J., 2003. Differential features of alluvial fans controlled by tectonic or eustatic accommodation space. Examples from the Betic Cordillera, Spain. *Geomorphology* **50**: 181–202.

Volker, H. X., Wasklewicz, T. A., and Ellis, M. A., 2007. A topographic fingerprint to distinguish alluvial fan formative processes. *Geomorphology* **88**: 34–45.

Wang, Y., McDonald, E., Amundson, R., McFadden, L., and Chadwick, O., 1996. An isotopic study of soils in chronological sequences of alluvial deposits, Providence Mountains, California. *Geological Society of America Bulletin* **108**: 379–391.

Wanner, H., Bronniman, S., Casty, C., Gyalistras, D., Luterbacher, J., Schmutz, C., Stephenson, D. B., and Xoplaki, E., 2001. North Atlantic Oscillation: concepts and studies. *Surveys in Geophysics* **22**: 321–382.

Wasklewicz, T. A., Mihir, M., and Whitworth, J., 2008. Surface variability of alluvial fans generated by disparate processes, Eastern Death Valley, CA. *Profesional Geographer* **60**: 207–223.

Weaver, W. E., and Schumm, S. A., 1974. Fan-head trenching: an example of a geomorphic threshold. *Geological Society America Abstracts with Program* **6**: 481.

Webb, J. A., and Fielding, C. R., 1999. Debris flow and sheetflood fans of the northern Prince Charles Mountains, East Antarctica. In: *Varieties of Fluvial Form*, A. J. Miller and A. Gupta (eds.), Wiley, Chichester, pp. 317–341.

Weissmann, G. S., Bennett, G. L., and Lansdale, A. L., 2005. Factors controlling sequence development on Quaternary fluvial fans, San Joaquin Basin, California, USA. In: *Alluvial fans: Geomorphology, sedimentology, dynamics*, A. M. Harvey, A. E. Mather and M. Stokes (eds.), Geological Society Special Publications 251, London, pp. 169–186.

Weissmann, G. S., Mount, J. F., and Fogg, G. E., 2002. Glacially driven cycles in accumulation space and sequence stratigraphy of a stream-dominated alluvial fan, San Joaquin Valley, California, USA. *Journal of Sedimentary Research* **72**: 270–281.

Wells, S. G., McFadden, L. D., and Dohrenwend, J. C., 1987. Influence of late Quaternary climatic changes on geomorphic and pedogenic processes on a desert piedmont, eastern Mojave Desert, California. *Quaternary Research* **27**: 130–146.

Wendland, W. M., 1996. Climatic changes: Impacts on geomorphic processes. *Engineering Geology* **45**: 347–358.

Western Earth Surface Processes Team, U. S. G. S., 2004. Stream Channel Development in the Changing Mojave Climate. *http://deserts.wr.usgs.gov/mojave/* **Last Updated 1–14–2004**: Accessed 7/25/06.

White, K., 1993. Image processing of Thematic Mapper data for discriminating piedmont surficial materials in the Tunisian Southern Atlas. *International Journal of Remote Sensing* **14**: 961–977.

White, K., Drake, N., Millington, A., and Stokes, S., 1996. Constraining the timing of alluvial fan response to Late Quaternary climatic changes, southern Tunisia. *Geomorphology* **17**: 295–304.

Wohl, E. E., and Pearthree, P. P., 1991. Debris flows as geomorphic agents in the Huachuca Mountains of Southeastern Arizona. *Geomorphology* **4**: 273–292.

Wolman, M. G., and Gerson, R., 1978. Relative scales of time and effectiveness of climate in watershed geomorphology. *Earth Surface Processes and Landforms* **3**: 189–208.

Zanchetta, G., Sulpizio, R., and Di Vito, M. A., 2004. The role of volcanic activity and climate in alluvial fan growth at volcanic areas: an example from southern Campania (Italy). *Sedimentary Geology* **168**: 249–280.

Zazo, C., Mercier, N., Lario, J., Roquero, E., Goy, J., Silva, P., Cabero, A., Borja, F., Dabrio, C., Bardaji, T., Soler, V., Garcia-Blazquez, A., and de Luque, L., 2008. Palaeoenvironmental evolution of the Barbate-Trafalgar coast (Cadiz) during the last 140 ka: Climate, sea-level interactions and tectonics. *Geomorphology* **100**: 212–222.

Chapter 25

Evidence for Climate Change From Desert Basin Palaeolakes

Dorothy Sack

Introduction

Lakes have long been recognized as being rich store-houses of environmental information. A lake basin collects water, but also sediment, much of which has been weathered and transported via fluvial processes from the near and far reaches of its drainage basin. The amount of water held in a lake is recorded on the landscape in coastal erosional and depositional landforms created at the water's edge. The sediments deposited on the bottom of the lake can be clastic, geochemical, or biogenic, and include materials derived within the standing water body itself, such as through coastal erosion, chemical precipitation, or biogenic concentration, as well as those delivered to the lake from the surrounding drainage basin. In most cases only a small percentage of a lake's sediment load is delivered from outside of the drainage basin as aeolian fallout. Because, under natural conditions, climate is the main determinant of the amount of water in a lake and because it influences some important characteristics of the lacustrine sediments and biota, changing climatic conditions are represented in the suites of abandoned shorelines and accumulations of sediments left by the lake over time (Fig. 25.1). This archival property makes the geomorphic and sedimentologic evidence of present and past lakes valuable as environmental and palaeoenvironmental indicators. Such evidence from late Quaternary palaeolakes, in fact, ranks among of the most complete and accessible sources of palaeoclimatic proxy

Fig. 25.1 An abandoned gravel shoreline in the Great Basin, USA, partially overlain by pelagic lacustrine deposits of calcium carbonate (marl)

data currently available for the late Pleistocene and Holocene.

Earth scientists have conducted comprehensive studies of the geomorphic and sediment evidence of late Pleistocene and Holocene palaeolakes, and have made palaeoclimatic interpretations from them, since the late 19th century (Russell 1885, Gilbert 1890). Limited by poor age control, and to some extent by interest in other topics when the Davisian cycle of erosion paradigm was popular (Davis 1899), the number of palaeolake studies waned during the first half of the 20th century. About mid-century, palaeolake research began a slow but steady growth under the process geomorphology paradigm and as numerical dating techniques became established and increasingly refined. Eventually the growth in palaeolake research began to accelerate, along with interest in earth-system science, starting about 1980. Since approximately the mid-1990s, the number of palaeolake researchers and publications has grown dramatically reflecting

D. Sack (✉)
Department of Geography, Ohio University, Athens, OH 45701, USA
e-mail: sack@ohio.edu

A.J. Parsons, A.D. Abrahams (eds.), *Geomorphology of Desert Environments*, 2nd ed.,
DOI 10.1007/978-1-4020-5719-9_25, © Springer Science+Business Media B.V. 2009

increasing social and scientific concern with human impacts on the environment and global climate change.

Today, palaeolake investigations contribute to climate studies in many ways. Researchers work on accurately reconstructing details of the timing and extent of palaeolake-level fluctuations (e.g. Fornari et al. 2001, Godsey et al. 2005), estimating the local and regional values of climatic variables and circulation attributes that could have led to those fluctuations (e.g. Benson 1993, Bookhagen et al. 2001, Stone 2006, Dühnforth et al. 2006), searching for spatial and temporal similarities and differences in the behavior of multiple palaeolakes (e.g. Benson et al. 1995, Krider 1998, Mensing 2001, Zhang et al. 2004, Balch et al. 2005), and comparing the palaeoclimatic signal determined from palaeolakes with climate signals derived from other sources (e.g. Benson et al. 1998, Broecker et al. 1998, Lin et al. 1998, Stager et al. 2002, Balch et al. 2005). These efforts provide information on the amount and rate of natural climate variability experienced during the late Quaternary, and therefore on what might be possible in the future. They supply a record of past climatic conditions that emerging models of global climate change should be able to successfully hindcast. Furthermore, comparing fluctuations in various palaeolakes around the globe with oscillations present in such data sources as the marine oxygen-isotope record, the Greenland ice cores, and the earth's orbital parameters helps scientists understand the mechanisms, sensitivities, and teleconnections of the natural climate system. Clearly, reconstructing the timing and extent of palaeolake fluctuations is the scientific foundation that makes these applications possible.

Desert Basin Palaeolakes

Lakes form wherever there is an adequate basin of containment and enough surplus water to accumulate in it. Topographic depressions that function as lake basins may be derived from a wide variety of sources. They originate through tectonic, volcanic, fluvial, aeolian, mass wasting, glacial, meteoritic, or other processes (Hutchinson 1957). Most lakes in humid climates receive so much inflow that the level of the standing water body permanently attains, and continually spills out over, the lowest point along the boundary of the containment basin. This low point is called the sill or threshold, and in humid regions the overflowing stream is typically part of an integrated, throughflowing, fluvial drainage system. Such open-basin, or externally drained, lakes have the elevation of their water level controlled by the elevation of the threshold. An increase of flow into an open-basin lake is handled by an increase in discharge out of the lake. Although the cross-sectional depth of the stream flowing out over the threshold will vary to some extent with discharge, much of the variation in volume of water is accounted for instead by the other two fluvial discharge variables, cross-sectional width and velocity of flow. As a result, the water level of open-basin lakes tends to be maintained very near the elevation of the threshold. Although this can lead to strongly developed coastal landforms within that narrow vertical zone, threshold control largely prohibits changes in the amount of water delivered to the lake from being sensitively recorded in distinct, multiple shorelines. A detailed record of changing conditions of effective moisture is thus lost. Alternatively, successively lower shorelines sometimes form in open lake basins as a result of fluvial erosion of the threshold and irrespective of vacillations in the regional effective moisture.

In many arid regions, topographic basins are commonly not connected with each other by throughflowing surface drainage, and this is primarily due to climatic factors (Langbein 1961). As in other climatic regions, topographic basins that may pond water can be formed by a variety of processes. In desert environments, however, once a large basin exists it is unlikely that sufficient surface water will be generated to completely fill the containment basin, spill over, and contribute to an integrated surface drainage system that reaches ultimate base level. Some desert basins contain perennial lakes while many others currently support only ephemeral or intermittent lakes (playas or playa lakes) (Mifflin and Wheat 1979, Smith and Street-Perrott 1983, Williams and Bedinger 1984). Perennial lakes in desert basins tend to be closed-basin, or subthreshold, lakes rather than open-basin lakes. As a result, they may exist for long periods of time because they are not destroying their own basin closure by fluvial erosion at the threshold. More importantly, by not being threshold controlled, the lakes are free to fluctuate in level in response to changes in effective moisture leaving telltale coastal landforms at a variety of water stillstand levels. The largest desert lakes in existence today are those like the Salton Sea in California that are supplied by exotic streams which originate in distant regions, those like Lake Eyre in Australia with very large drainage basins, lakes that lie in drainage basins which have some terrain outside of the arid climatic regime, such as the Dead Sea in Israel, and hemiarid lakes like Pyramid Lake in Nevada, which are those fed by adjacent nonarid highlands (Fig. 25.2) (Currey 1994, Wilkins and Currey 1997). Note that these categories are not all mutually exclusive.

When an arid region undergoes a climate change to circumstances of greater available moisture, existing perennial lakes expand while new closed-basin lakes become established in basins that previously held only playas. Because of late Pleistocene and Holocene climate fluctuations, many now-desert basins of the middle and subtropical latitudes display considerable geomorphic and sedimentological evidence of having contained larger lakes during various times of greater effective wetness in the late Quaternary. These are

sometimes referred to as pluvial lakes, but that term is discouraged since it implies that the climate responsible for them was only rainier than present with no change in temperature or other influential variables. Regardless, the return to arid conditions, with its concomitant sparse vegetation and limited weathering rates, has left much of the palaeolake evidence well preserved, visible, exposed, and accessible to scientific study (Fig. 25.3). Researchers have enumerated about 100 late Quaternary palaeolakes in the Basin and Range province of the western US alone (Williams and Bedinger 1984), with Lakes Bonneville and Lahontan being the largest (Fig. 25.2). Considerable scientific attention has also been directed toward palaeolakes on the Altiplano of Bolivia, Peru, Chile, and Argentina (Valero-Garcés et al. 1999, Fornari et al. 2001, D'Agostino et al. 2002); Megalake Chad in North Africa (Leblanc et al. 2006a); Lake Lisan and others in the Jordan-Dead Sea Rift Valley (Stein 2001, Bartov et al. 2002, Hazan et al. 2005, Migowski

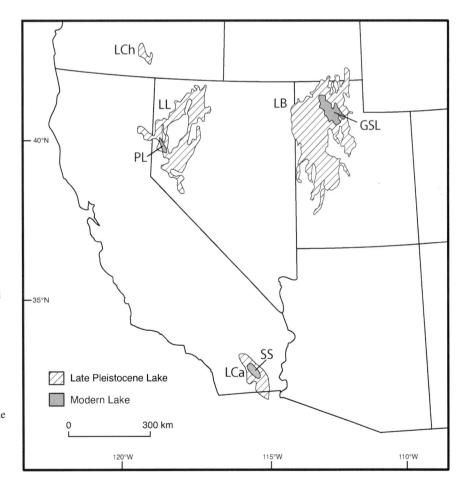

Fig. 25.2 Location of selected late Pleistocene and modern lakes of western North America (those discussed in the text). SS = Salton Sea, GSL = Great Salt Lake, LB = Lake Bonneville, PL = Pyramid Lake, LL = Lake Lahontan, LCh = Lake Chewaucan. The Salton Sea is shown here in relation to Lake Cahuilla (LCa), which occupied that basin in the late Pleistocene

Fig. 25.3 An impressive set of relict shorelines created by late Pleistocene Lake Bonneville, Utah

et al. 2006); predecessors of Lake Eyre in Australia (Croke et al. 1998, Nanson et al. 1998, Stone 2006); and Megalake Tengger and others in northwestern China and Mongolia (Qin and Huang 1998, Peck et al. 2002, Zhang et al. 2004, Gao et al. 2006, Jiang et al. 2006).

Geomorphology of Desert Basin Palaeolakes

Relict coastal landforms still visible in the arid, sub-aerial landscape typically constitute the most obvious and compelling evidence that a sizeable lake once existed in a now-desert basin (Avouac et al. 1996). It is only through the identification, correlation, and mapping of preserved shoreline segments that the spatial extent of the water body can be accurately reconstructed and its elevation and surface area determined (Migowski et al. 2006). Palaeolake surface area, as discussed later in this chapter, is a fundamental variable for assessing palaeoclimatic conditions. In addition to the highest water level attained by the lake, it is often desirable to delineate the extent of prominent lower shorelines, which may also mark important climatically induced stillstands or oscillations of the water plane. In some cases these lower shorelines are only visible in stratigraphic exposures because of burial by later lacustrine or subaerial deposits.

Identifying, correlating, and mapping segments of a given shoreline, even a prominent one, can be challenging. Although when it was formed the shoreline demarcated the complete perimeter of the lake, reworking or burial due to subsequent lacustrine processes and post-lacustrine attack by especially fluvial, alluvial fan, and aeolian processes obliterate geomorphic evidence of numerous shoreline segments (Fig. 25.4) (Sack 1995). Simple contour tracing is rarely an option for shoreline mapping because of local postlake geomorphic re-arrangement of the landscape. In addition, palaeolake shorelines, which were horizontal when created, can be warped by hydroisostatic rebound and offset in places by tectonism (Lambert et al. 1998, Adams and Wesnousky 1999). Hydroisostatic rebound is caused by a reduction in water load and elevates a shoreline from its original position by a distance that depends on the amount of unloading. The shallowest water, and therefore the smallest amount of unloading and rebound, will occur near the margin of a lake basin. Both isostatic and tectonic processes have impacted the impressive relict shorelines of late Pleistocene Lake Bonneville, for example, with maximum differential rebound of 74 m for the highest shoreline (Figs. 25.5 and 25.6) (Currey 1982). Another problem in shoreline mapping stems from the fact that desert piedmont

Fig. 25.4 Portion of a 1:20,000-scale vertical aerial photograph (AAH-14W-99) showing preserved and eroded shorelines and sediments of Lake Bonneville

escarpments can be made by tectonic, mass wasting, fluvial, aeolian, coastal, or other origins, and it is sometimes difficult to identify a bluff with certainty as a relict coastal feature (Gilbert 1890, Knott et al. 2002, Hooke 2004).

Geomorphology is also key to distinguishing those aspects of a palaeolake chronology that reflect climatic conditions from those that result from hydrographic conditions. Climatically induced changes in effective moisture cause fluctuations in lake level but so do basin geomorphic factors. It is critical, for example, to identify any periods of threshold control that might have occurred during the lake's existence. The stabilizing effect of external drainage on the level of a lake has already been noted. If a period of exterior drainage goes unrecognized, the threshold-controlled shoreline will be attributed to a prolonged stability in effective moisture, which probably did not occur. Conversely, during an open-basin phase of a lake, positive or negative changes in the elevation of the threshold due to volcanism, tectonism, fluvial erosion, or mass movement can cause the lake level to rise or fall without a climate change. The geomorphic event of threshold failure, and not a climatic event, caused Lake Bonneville to drop 104 m from the Bonneville to the Provo shoreline in less than a year (Fig. 25.7) (Gilbert 1890, Jarrett and Malde 1987, Burr and Currey 1988).

Geomorphic effects of isostasy produce changes in shoreline position that could be misinterpreted as climatic responses. The highest shoreline of Lake Bonneville was formed as a result of an extended period of threshold control. While at that level, hydroisostatic loading caused the central portion of this large lake basin to subside relative to the outlet (Currey 1980, Currey et al. 1983, Burr and Currey 1988). As a result, in basin interior locations the lake left a record of apparent transgressions as the water plane impinged on terrain that was subsiding from a subaerial to a subaqueous position.

Subbasins dynamics is yet another geomorphic element that can play an important role in controlling the level of a palaeolake (Fornari et al. 2001, Sack 2002, Brown et al. 2003). Many palaeolakes, including for example Lakes Bonneville, Lahontan, and Chewaucan in western North America (Fig. 25.2) (Eardley et al. 1957, Allison 1982, Benson and Thompson 1987, Sack 2002), Lake Chillingollah in Australia (Stone 2006), and Lake Lisan in Israel (Bartov et al. 2002), consisted of a collection of Subbasins separated from each other by interior thresholds (Fig. 25.8). Each Subbasins had a unique interval of integration with the main palaeolake determined by its local hydrologic balance and the elevation of the dividing interior threshold. Some Subbasins contained isolated, independent palaeolakes before and after their integration period with the larger water body (Allison 1982, Sack 2002). During the transgressive phase of a lake cycle, the water level will naturally rise at different rates in different Subbasins. When it reaches the elevation of the lowest interior threshold of the primary Subbasins it will flow over that divide into an adjoining closed Subbasins. Unless there is significant erosion or slope failure at the interior threshold, the water level in the main basin will be held approximately constant while the water body in the Subbasins undergoing integration rises to equilibrate with it. A shoreline will form in the main basin as the result of the stillstand, whereas the water level can rise too quickly to leave shoreline evidence in the adjacent, filling Subbasins (Sack 1990). The Subbasins nature of a palaeolake must be thoroughly investigated so that stillstands and rapid rises in lake level caused by Subbasins integrations and isolations are not given climatic interpretations. Most large palaeolakes consisted of Subbasins and underwent complex Subbasins dynamics.

Fig. 25.5 Theoretical curves of post-Bonneville deformation (Gilbert 1890, Plate L). The map depicts the general pattern of isostatic rebound of the highest shoreline of Lake Bonneville, called the Bonneville shoreline. Units are in feet above Great Salt Lake. The location of Lake Bonneville appears in Fig. 2

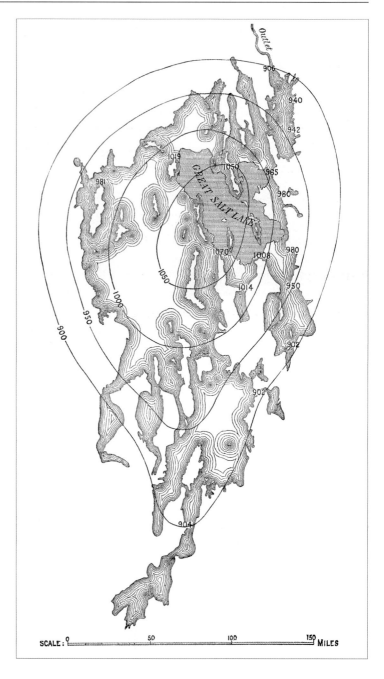

Drainage basin dynamics also lead to geomorphically induced fluctuations in the level of a terminal lake. An increase or decrease in the drainage basin area resulting from tectonism, volcanism, mass movement, or stream capture alters stream flow, which causes closed-basin lake level fluctuations without a change in climate. The Holocene successor to Lake Bonneville, Utah's Great Salt Lake (Fig. 25.2), receives its greatest inflow from the Bear River. This large river may have been diverted into the Bonneville basin between the last two major lake cycles (Bouchard et al. 1998, Hart et al. 2004), which occurred during marine oxygen-isotope stages (MIS) 2 and 6. The addition of the Bear River as a tributary could have contributed to the MIS 2 Bonneville basin palaeolake rising approximately 42 m higher than the MIS 6 palaeolake and attaining open-basin status (Currey 1982, McCoy 1987).

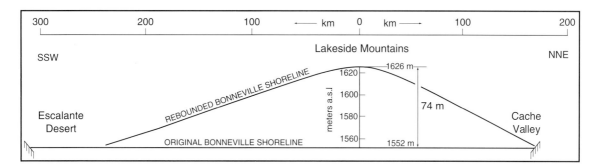

Fig. 25.6 Transect from south-southwest to north-northeast across the Bonneville basin showing the modern (rebounded) elevation of the Bonneville shoreline in relation to the elevation at which it was created (after Currey 1990, p. 203). This shoreline marks the maximum extent of Lake Bonneville. The deepest part of the lake basin lies near the center of the basin adjacent to the Lakeside Mountains, therefore preserved shoreline remnants located there display the greatest amount of hydroisostatic rebound

Geomorphic Techniques

Geomorphologists study desert-basin palaeolakes with a variety of field and laboratory techniques. Field geomorphic, and related sedimentary, observations and measurements remain fundamental to palaeolake studies, as does morphostratigraphy, which uses the form of sediment packages in stratigraphic exposure to interpret landforms subsequently buried by other sediments (Fig. 25.9). Stereoscopic interpretation of aerial photographs remains a valuable asset for shoreline mapping (Nanson et al. 1998). Air photo mapping requires close inspection of shoreline landforms on large- and intermediate-scale air photos. This process also aids in the identification of stage-specific geomorphic signatures, which may reflect important aspects of the palaeolake history, and in the identification of well-developed or well-exposed sites for detailed field investigation. Digital elevation models (DEMs) help researchers reconstruct shorelines and contend with the

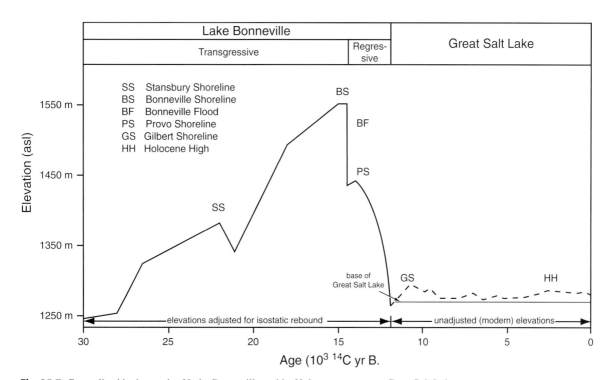

Fig. 25.7 Generalized hydrograph of Lake Bonneville and its Holocene successor, Great Salt Lake

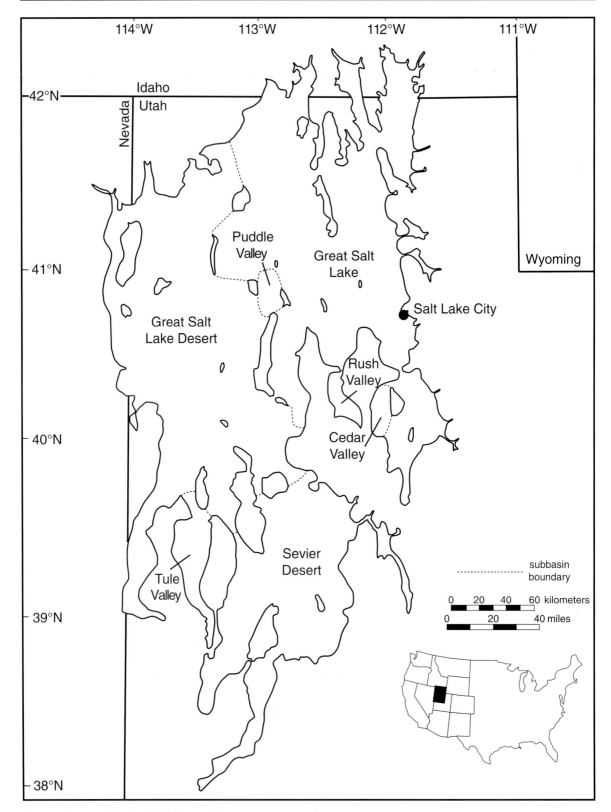

Fig. 25.8 Major subbasins of Lake Bonneville

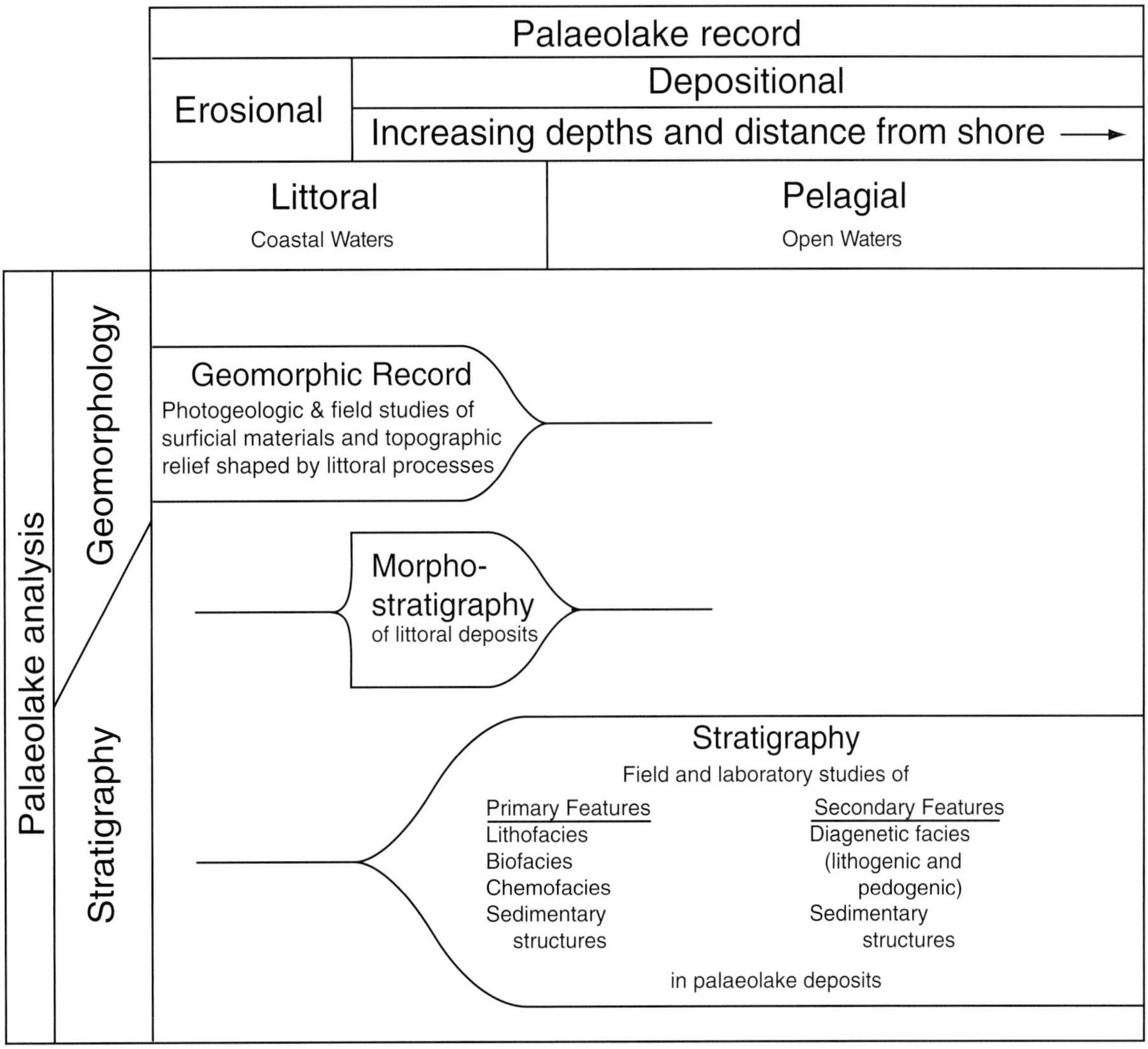

Fig. 25.9 Morphostratigraphy and its relationship to geomorphology and stratigraphy in palaeolake studies (after Currey 1990, p. 200). Sedimentology is a fundamental constituent of all three forms of analysis, being largely inapplicable only for the geomorphic study of erosional landforms

problem of discontinuous preservation of the shoreline perimeter (DeVogel et al. 2004). Palaeolake investigators use DEMs, radar topographic data, and various sources of satellite imagery to identify shorelines and thresholds particularly in regions with poor accessibility (Komatsu et al. 2001, Schuster et al. 2003, Leblanc et al. 2006a,b, Ghoneim and El Baz 2007). Palaeolake landforms submerged under present water bodies have been identified with depth-profiling techniques (Ricketts et al. 2001), whereas data handling, visualization, and virtual filling of palaeolakes are accomplished with GIS and computer modelling (DeVogel et al. 2004).

Obtaining accurate measurements of shoreline elevation and reliable numeric age determinations are necessary for shoreline segment correlation, for assessing amounts and rates of hydroisostatic and neotectonic offsets, and for constructing detailed palaeolake time vs. water level graphs. At present, the most accurate elevations are acquired in the field with electronic total stations, provided sufficient bench marks are available. Differential global positioning systems are also quite useful for field measurements of shoreline elevation (Hoelzmann et al. 2001). Algorithms are constructed to determine the original elevation of a shoreline that has been subjected to hydroisostatic rebound (Currey and Oviatt 1985). In most cases this rebound-free, or derebounded, elevation is what should be plotted on hydrographs of large palaeolakes (Fig. 25.7).

Portraying the time factor on a palaeolake hydrograph depends on finding datable material in unequivocal stratigraphic context with respect to a known water level. This is often quite challenging in desert palaeolake basins (Geyh et al. 1999). Accelerator mass spectrometer (AMS) radiocarbon dating, uranium-series (U-series) dating, amino acid dating, and occasionally optically stimulated luminescence (OSL) and tephrochronology, offer the most precise age determinations. Numeric ages based on radiocarbon, however, which are the most commonly employed, suffer from some unknowns, depending on the type of material dated. Carbonate samples, as from shells and tufa, might be contaminated with younger carbon or with older carbon derived from fossil water, which is called the reservoir effect (Benson 1993). Some organisms, moreover, tend to use ^{14}C-deficient carbon in making their shells so that they do not reflect the ^{14}C balance of the water body (Pigati et al. 2004). Relative age estimates and correlations between shoreline segments have been determined with such techniques as degree of shoreline development or degradation (Wilkins and Currey 1997, Hooke 1999), rock varnish accumulation (Liu et al. 2000), soil development (Adams and Wesnousky 1999, Stone 2006), and cross-cutting by features of known age.

Clearly, geomorphology is a major source of information concerning palaeolakes and their fluctuations, but that data source also has its limitations. The geomorphic record is naturally weak for small palaeolakes with limited fetch, consists of negative evidence for rapid changes in lake level, may be reworked by later water-level oscillations or buried by later lacustrine sediments, and becomes increasingly obliterated by subaerial processes with increasing time since exposure (Sack 1995). Fortunately, the record of accumulated palaeolake sediment supplies additional insights into the nature of the palaeolake and its regional environment.

and depositional coastal landforms; materials that have been deposited within the lake provide other significant palaeoenvironmental information. Indeed, a continuous record of a palaeolake's sedimentation history will exist in the deepest part of the basin if the lake did not experience complete desiccation during which sediment was lost through deflation (Magee et al. 1995, Nanson et al. 1998). Completeness of the sediment record decreases with proximity to the shoreline.

Palaeolake sediment records are studied primarily from sediment cores and outcrops, but seismic profiles (Valero-Garcés et al. 1996) and ground penetrating radar images have also been used. Researchers have developed an impressive array of approaches for gleaning palaeoenvironmental data from lacustrine sediments. A vertical increase or decrease in the grain size of lithic fragments and sediment density suggests a falling or rising water level, respectively (e.g. Davies 2006). Pollen reflects the climate of the drainage basin (e.g. Mensing 2001, Zhang et al. 2002). Salinity, relative water depth, and/or subsurface versus surface sources of lake water are commonly investigated with diatom and ostracod assemblages, isotope geochemistry, elemental chemistry, carbonate content, amount of organic carbon, total inorganic carbon, concentration of magnetic minerals, and sedimentary structures (e.g. Benson et al. 1998, Ricketts et al. 2001, Balch et al. 2005, Flower et al. 2006). Multiple proxies from the same interval, however, sometimes indicate conflicting climatic signals (Dearing 1997, Grosjean et al. 2003). In addition, none of these analyses reveal exact lake size, although when multiple cores are retrieved from across a palaeolake basin, their correlation discloses some spatial characteristics of the palaeolake. The sediment record from cores, on the other hand, contains a much greater abundance of materials that indicate age than do coastal landforms. AMS radiocarbon, U-series, amino acid, and tephrochronology analyses are the most commonly employed means of determining the age of a fluctuation interpreted from cores. These, plus various thermoluminescence techniques, provide reliable age determinations from outcrop samples.

Sedimentology and Stratigraphy of Desert Basin Palaeolakes

The size of a palaeolake is directly indicated only by the position of its shoreline as marked by erosional

Palaeoclimatic Reconstruction

Once a palaeolake chronology has been reliably reconstructed and the influences of geomorphically induced versus climatically induced water-level changes have been identified, it is still difficult to deduce specific palaeoclimatic variables from what is essentially a palaeohydrologic record. Even if only climatic variables were involved, there are multiple climatic scenarios that could have resulted in a rise in lake level. Some of these include increased average annual precipitation, decreased average annual evapotranspiration, increased cool season precipitation, decreased warm season evapotranspiration, or combined changes in precipitation and evapotranspiration on an average annual or seasonal basis. Evapotranspiration itself, moreover, responds to a variety of atmospheric variables, such as temperature, cloudiness, and windiness. Because an increase in precipitation would result in more vegetation, an increase in evapotranspiration should accompany a precipitation increase as well (Mifflin and Wheat 1979, Qin and Huang 1998). The most likely scenario incorporates changes in both precipitation and evapotranspiration.

An important link between the geomorphic evidence of lake size and the specific climatic variables responsible for it consists of the z ratio of lake surface area, A_L, to tributary basin area, A_B, with the latter consisting of the drainage basin area excluding A_L (Snyder and Langbein 1962, Mifflin and Wheat 1979, Street-Perrott and Harrison 1985):

$$z = A_L/A_B. \qquad (25.1)$$

For a closed-basin lake that has insignificant groundwater transfer and annual input of water equal to annual output, the water balance is represented by the equation:

$$R + A_L P_L = A_L E_L, \qquad (25.2)$$

where R is annual tributary runoff into the lake, P_L is direct precipitation onto the lake, and E_L is evaporation from the lake. Runoff, however, can be expressed in terms of tributary basin precipitation, P_B, and tributary basin evaporation, E_B:

$$R = A_B(P_B - E_B). \qquad (25.3)$$

By substituting for R, Equation (25.2) becomes:

$$A_B(P_B - E_B) + A_L(P_L) = A_L(E_L) \qquad (25.4a)$$

$$A_B(P_B - E_B) = A_L(E_L - P_L) \qquad (25.4b)$$

$$A_L/A_B = (P_B - E_B)/(E_L - P_L). \qquad (25.4c)$$

Therefore,

$$z = A_L/A_B = (P_B - E_B)/(E_L - P_L). \qquad (25.5)$$

Through the z ratio, the geomorphic evidence of lake area and basin area is related directly to the palaeoclimatic variables of precipitation and evaporation (Snyder and Langbein 1962, Mifflin and Wheat 1979, Street-Perrott and Harrison 1985). Variables A_L and A_B are reconstructed from relict shoreline evidence of late Pleistocene palaeolakes, but determining values for the evaporation and precipitation variables represents a greater challenge. A common approach consists of using modern relationships among the study region's temperature, precipitation, evapotranspiration, and runoff to estimate values for the palaeoclimatic variables, within limits set by such proxies as pollen, and solving iterations of the formula until the z ratio calculated from climatic variables approaches the z ratio obtained from the geomorphic data (Mifflin and Wheat 1979, Barber and Finney 2000, Menking et al. 2004). Although a unique solution does not exist, sensitivity tests reveal the most likely ranges of palaeoclimatic variables that would have resulted in a palaeolake of the observed extent (Bookhagen et al. 2001, Jones et al. 2007).

Any discrepancies between geomorphically derived values of the z ratio and values derived by estimating the palaeoclimatic variables may be due to (a) errors in determining A_L and A_B, (b) imprecision in the relationships established for modern data (Kotwicki and Allan 1998), or (c) the possibly inaccurate assumption that the relationships among the palaeoclimatic variables can be adequately modelled by the relationships among the modern climatic variables (Mifflin and Wheat 1979, Benson and Thompson 1987). The first type of error is minimized by careful mapping and fieldwork and by employing in calculations only those lake basins that have well preserved geomorphic evidence which can be mapped with a high degree of confidence. In this regard, late Quaternary palaeolake basins in arid regions probably have the greatest potential. The second source of error will no doubt decrease as geomorphologists and climatologists continue to investigate the modern empirical relationships among the relevant climatic and hydrologic variables. Validity of the assumption that the empirical relationships remain uniform over time can be checked by comparing palaeoclimatic inferences drawn from palaeolake data with inferences drawn from other sources of palaeoenvironmental evidence. Potential error in reconstructing palaeoclimatic variables from palaeolake data can be minimized by using information from multiple proxies and/or many basins to investigate regional trends instead of focusing on single data sources, and recent research has moved in these directions. Further accuracy should derive from balancing energy (Bergonzini et al. 1997) and isotopic budgets (Jones et al. 2007) in addition to the hydrologic budgets, and this approach holds considerable promise for palaeoclimatic research based on desert basin palaeolakes.

Conclusions

Specific values of palaeotemperature and palaeoprecipitation cannot yet be determined with certainty from the palaeohydrologic evidence of desert basin palaeolakes, but approaches continue to become more sophisticated and multivariate. In the meantime, with increasing concern for global climate change and human impacts on the environment, a growing body of palaeolake research focuses on characterizing the relative amplitude, duration, and chronology of past changes in regional effective moisture, explaining these in terms of altered atmospheric circulation patterns, especially shifts in the jet streams and in monsoonal circulation, and correlating palaeolake fluctuations with climatic events represented in the marine oxygen-isotope record and Greenland ice cores. This broader, i.e., more global, perspective remains grounded in the science of reconstructing in detail the fluctuation chronology of individual palaeolakes. Palaeolake researchers accomplish this using all of the tools at their disposal – with fundamental geomorphic, morphostratigraphic, and sedimentologic/stratigraphic methods and technologically evolving techniques.

References

Adams KD, Wesnousky SG (1999) The Lake Lahontan highstand: Age, surficial characteristics, soil development, and regional shoreline correlation. Geomorphology 30:357–392

Allison IS (1982) Geology of pluvial Lake Chewaucan, Lake County, Oregon. Or State Univ Stud Geol No. 11

Avouac J-P, Dobremez J-F, Bourjot L (1996) Palaeoclimatic interpretation of a topographic profile across middle Holocene regressive shorelines of Longmu Co (western Tibet). Palaeogeogr Palaeoclim Palaeoecol 120:93–104

Balch DP, Cohen AS, Schnurrenberger DW, Haskell BJ, Valero Garces V, Beck JW, Cheng H, Edwards RL (2005) Ecosystem and paleohydrological response to Quaternary climate change in the Bonneville basin, Utah. Palaeogeogr Palaeoclim Palaeoecol 221:99–122

Barber VA, Finney BP (2000) Late Quaternary paleoclimatic reconstructions for interior Alaska based on paleolake-level data and hydrologic models. J Paleolimn 24:29–41

Bartov Y, Stein M, Enzel Y, Agnon A, Reches Z (2002) Lake levels and sequence stratigraphy of Lake Lisan, the late Pleistocene precursor of the Dead Sea. Quat Res 57:9–21

Benson LV (1993) Factors affecting ^{14}C ages of lacustrine carbonates: Timing and duration of the last highstand lake in the Lahontan basin. Quat Res 39:163–174

Benson LV, Thompson RS (1987) Lake-level variation in the Lahontan basin for the past 50,000 years. Quat Res 28:69–85

Benson LV, Kashgarian M, Rubin M (1995) Carbonate deposition, Pyramid Lake subbasin, Nevada: 2. Lake levels and polar jet stream positions reconstructed from radiocarbon ages and elevations of carbonates (tufas) deposited in the Lahontan basin. Palaeogeogr Palaeoclim Palaeoecol 117:1–30

Benson LV, Lund SP, Burdett JW, Kashgarian M, Rose TP, Smoot JP, Schwartz M (1998) Correlation of Late-Pleistocene lake-level oscillations in Mono Lake, California, with North Atlantic climatic events. Quat Res 49:1–10

Bergonzini L, Chalié F, Gasse F (1997) Paleoevaporation and paleoprecipitation in the Tanganyika basin at 18,000 years B.P. inferred from hydrologic and vegetation proxies. Quat Res 47:295–305

Bookhagen B, Haselton K, Trauth MH (2001) Hydrological modelling of a Pleistocene landslide-dammed lake in the Santa Maria basin, NW Argentina. Palaeogeogr Palaeoclim Palaeoecol 169:113–127

Bouchard DP, Kaufman DS, Hochberg A, Quade J (1998) Quaternary history of the Thatcher basin, Idaho, reconstructed from the ^{87}Sr/^{86}SR and amino acid composition of lacustrine fossils: Implications for the diversion of the Bear River into the Bonneville basin. Palaeogeogr Palaeoclim Palaeoecol 141:95–114

Broecker WS, Peteet D, Hajdas I, Lin J, Clark E (1998) Antiphasing between rainfall in Africa's Rift Valley and North America's Great Basin. Quat Res 50:12–20

Brown ET, Bendick R, Bourlès DL, Gaur V, Molnar P, Raisbeck GM, Yiou F (2003) Early Holocene climate recorded in geomorphological features in western Tibet. Palaeogeogr Palaeoclim Palaeoecol 199:141–151

Burr TN, Currey DR (1988) The Stockton Bar. Utah Geol Miner Surv Misc Publ 88–1:66–73

Croke J, Magee J, Price D (1998) Major episodes of Quaternary activity in the lower Neales River, northwest of Lake Eyre, central Australia. Quat Res 124:1–15

Currey DR (1980) Coastal geomorphology of Great Salt Lake and vicinity. Utah Geol Miner Surv Bull 116:69–82

Currey DR (1982) Lake Bonneville: Selected features of relevance to neotectonic analysis. US Geol Surv Open-File Rep 82–1070

Currey DR (1990) Quaternary palaeolakes in the evolution of semidesert basins, with special emphasis on Lake Bonneville and the Great Basin, U.S.A. Palaeogeogr Palaeoclim Palaeoecol 76:189–214

Currey DR (1994) Hemiarid lake basins: Hydrographic patterns. In: Abrahmas AD, Parsons AJ (eds) Geomorphology of desert enviornments. London, Chapman and Hall

Currey DR, Oviatt CG (1985) Durations, average rates, and probable causes of Lake Bonneville expansions, stillstands, and contractions during the last deep-lake cycle, 32,000 to 10,000 years ago. In: Kay PA, Diaz HF (eds) Problems of and prospects for predicting Great Salt Lake levels. University of Utah, Salt Lake City

Currey DR, Oviatt CG, Plyler GB (1983) Lake Bonneville stratigraphy, geomorphology, and isostatic deformation in west-central Utah. Utah Geol Miner Surv Spec Stud 62: 63–82

D'Agostino K, Seltzer G, Baker P, Fritz S, Dunbar R (2002) Late-Quaternary lowstands of Lake Titicaca: Evidence from high-resolution seismic data. Palaeogeogr Palaeoclim Palaeoecol 179:97–111

Davies CP (2006) Holocene paleoclimates of southern Arabia from lacustrine deposits of the Dhamar highlands, Yemen. Quat Res 66:454–464

Davis WM (1899) The geographical cycle. Geogr J 14:481–504

Dearing JA (1997) Sedimentary indicators of lake-level changes in the humid temperate zone: A critical review. J Paleolimn 18:1–14

DeVogel SB, Magee JW, Manley WF, Miller GH (2004) A GIS-based reconstruction of late Quaternary paleohydrology: Lake Eyre, arid central Australia. Palaeogeogr Palaeoclim Palaeoecol 204:1–13

Dühnforth M, Bergner AGN, Trauth MH (2006) Early Holocene water budget of the Nakuru-Elementeita basin, central Kenya rift. J Paleolimn 36:281–294

Eardley AJ, Gvosdetsky V, Marsell RE (1957) Hydrology of Lake Bonneville and sediments and soils of its basin. Geol Soc Am Bull 68:1141–1201

Flower RJ, Stickley C, Rose NL, Peglar S, Fathi AA, Appleby PG (2006) Environmental changes at the desert margin: An assessment of recent paleolimnological records in Lake Qarun, middle Egypt. J Paleolimn 35:1–24

Fornari M, Risacher F, Féraud G (2001) Dating of paleolakes in the central Altiplano of Bolivia. Palaeogeogr Palaeoclim Palaeoecol 172:269–282

Gao Q, Tao Z, Li B, Jin J, Zou X, Zhang Y, Dong G (2006) Palaeomonsoon variability in the southern fringe of the Badain Jaran Desert, China, since 130 ka BP. Earth Surf Process Landf 31:265–283

Geyh MA, Grosjean M, Núñez L, Schotterer U (1999) Radiocarbon reservoir effect and the timing of the late-glacial/early Holocene humid phase in the Atacama Desert (northern Chile). Quat Res 52:143–153

Ghoneim E, El Baz F (2007) The application of radar topographic data to mapping of a mega-paleodrainage in the eastern Sahara. J Arid Environ 69:658–675

Gilbert GK (1890) Lake Bonneville. US Geol Surv Monogr 1, 438p

Godsey HS, Currey DR, Chan MA (2005) New evidence for an extended occupation of the Provo shoreline and implications for regional climate change, Pleistocene Lake Bonneville, Utah, USA. Quat Res 63:212–223

Grosjean M, Cartajena I, Geyh MA, Nuñez L (2003) From proxy data to paleoclimatic interpretation: The mid-Holocene paradox of the Atacama Desert, northern Chile. Palaeogeogr Palaeoclim Palaeoecol 194:247–258

Hart WS, Quade J, Madsen DB, Kaufman DS, Oviatt CG (2004) The $^{87}Sr/^{86}Sr$ ratios of lacustrine carbonates and lake-level history of the Bonneville paleolake system. Geol Soc Am Bull 116:1107–1119

Hazan N, Stein M, Agnon A, Marco S, Nadel D, Negendank JFW, Schwab MJ, Neev D (2005) The late Quaternary limnological history of Lake Kinneret (Sea of Galilee), Israel. Quat Res 63:60–77

Hoelzmann P, Keding B, Berke H, Kröpelin S, Kruse H-J (2001) Environmental change and archaeology: Lake evolution and human occupation in the eastern Sahara during the Holocene. Palaeogeogr Palaeoclim Palaeoecol 169: 193–217

Hooke RLeB (1999) Lake Manly(?)shorelines in the eastern Mojave Desert, California. Quat Res 52:328–336

Hooke RLeB (2004) Letter to the editor. Quat Res 61:339–343

Hutchinson GE (1957) A treatise on limnology, vol 1, Geography, physics, and chemistry. Wiley, New York

Jarrett RD, Malde HE (1987) Paleodischarge of the late Pleistocene Bonneville flood, Snake River, Idaho, computed from new evidence. Geol Soc Am Bull 99:127–134

Jiang W, Guo Z, Xun X, Wu H, Chu G, Yuan B, Hatté C, Guiot J (2006) Reconstruction of climate and vegetation changes of Lake Bayanchagan (Inner Mongolia): Holocene variability of the East Asian monsoon. Quat Res 65:411–420

Jones MD, Roberts CN, Leng MJ (2007) Quantifying climatic change through the last glacial-interglacial transition based on lake isotope palaeohydrology from central Turkey. Quat Res 67:463–473

Knott JR, Tinsley III JC, Wells SG (2002) Are the benches at Mormon Point, Death Valley, California, USA, scarps or strandlines? Quat Res 58:352–360

Komatsu G, Brantingham PJ, Olsen JW, Baker VR (2001) Paleoshoreline geomorphology of Böön Tsagaan Nuur, Tsagaan Nuur and Orog Nuur: The Valley of Lakes, Mongolia. Geomorphology 39:83–98

Kotwicki V, Allan R (1998) La Niña de Australia – Contemporary and palaeo-hydrology of Lake Eyre. Palaeogeogr Palaeoclim Palaeoecol 144:265–280

Krider PR (1998) Paleoclimatic significance of late Quaternary lacustrine and alluvial stratigraphy, Animas Valley, New Mexico. Quat Res 50:283–289

Lambert A, James TS, Thorleifson LH (1998) Combining geomorphological and geodetic data to determine postglacial tilting in Manitoba. J Paleolimn 19:365–376

Langbein WB (1961) Salinity and hydrology of closed lakes. US Geol Surv Circ 52

Leblanc M, Favreau G, Maley J, Nazoumou Y, Leduc C, Stagnitti F, van Oevelen PJ, Delclaux F, Lemoalle J (2006a) Reconstruction of Megalake Chad using shuttle radar topographic mission data. Palaeogeogr Palaeoclim Palaeoecol 239:16–27

Leblanc MJ, Leduc C, Stagnitti F, van Oevelen PJ, Jones C, Mofor LA, Razack M, Favreau G (2006b) Evidence for

Megalake Chad, north-central Africa, during the late Quaternary from satellite data. Palaeogeogr Palaeoclim Palaeoecol 230:230–242

Lin JC, Broecker WS, Hemming SR (1998) A reassessment of U-Th and ^{14}C ages for late-glacial high-frequency hydrological events at Searles Lake, California. Quat Res 49:11–23

Liu T, Broecker WS, Bell JW, Mandeville CW (2000) Terminal Pleistocene wet event recorded in rock varnish from Las Vegas Valley, southern Nevada. Palaeogeogr Palaeoclim Palaeoecol 161:423–433

Magee JW, Bowler JM, Miller GH, Williams DLG (1995) Stratigraphy, sedimentology, chronology and palaeohydrology of Quaternary lacustrine deposits at Madigan Gulf, Lake Eyre, South Australia. Palaeogeogr Palaeoclim Palaeoecol 113:3–42

McCoy WD (1987) Quaternary aminostratigraphy of the Bonneville basin, western United States. Geol Soc Am Bull 98:99–112

Menking KM, Anderson RY, Shafike NG, Syed KH, Allen BD (2004) Wetter or colder during the last glacial maximum? Revisiting the pluvial lake question in southwestern North America. Quat Res 62:280–288

Mensing S (2001) Late-glacial and early Holocene vegetation and climate change near Owens Lake, eastern California. Quat Res 55:57–65

Mifflin MD, Wheat MM (1979) Pluvial lakes and estimated pluvial climates of Nevada. Nev Bur Mines Geol Bull 94, 57p

Migowski C, Stein M, Prasad S, Negendank JFW, Agnon A (2006) Holocene climate variability and cultural evolution in the Near East from the Dead Sea sedimentary record. Quat Res 66:421–431

Nanson GC, Callen RA, Price DM (1998) Hydroclimatic interpretation of Quaternary shorelines on South Australian playas. Palaeogeogr Palaeoclim Palaeoecol 144:281–305

Peck JA, Khosbayar P, Fowell SJ, Pearce RB, Ariunbileg A, Hansen BCS, Soninkhishig N (2002) Mid to late Holocene climate change in north central Mongolia as recorded in the sediments of Lake Telmen. Palaeogeogr Palaeoecol 183:135–153

Pigati JS, Quade J, Shahanan TM, Haynes CV Jr (2004) Radiocarbon dating of minute gastropods and new constraints on the timing of late Quaternary spring-discharge deposits in southern Arizona, USA. Palaeogeogr Palaeoclim Palaeoecol 204:33–45

Qin B, Huang Q (1998) Evaluation of the climatic change impacts on the inland lake – A case study of Lake Qinghai, China. Clim Chang 39:695–714

Ricketts RD, Johnson TC, Brown ET, Rasmussen KA, Romanovsky VV (2001) The Holocene paleolimnology of Lake Issyk-Kul, Kyrgyzstan: Trace element and stable isotope composition of ostracodes. Palaeogeogr Palaeoclim Palaeoecol 176:207–227

Russell IC (1885) Geologic history of Lake Lahontan. US Geol Surv Monogr 11

Sack D (1990) Quaternary geology of Tule Valley, west-central Utah. Utah Geol Miner Surv Map 124, 26p

Sack D (1995) The shoreline preservation index as a relative-age dating tool for late Pleistocene shorelines: An example from the Bonneville basin, U.S.A. Earth Surf Process Landf 20:363–377

Sack D (2002) Fluvial linkages in Lake Bonneville subbasin integration. Smithson Inst Contrib Earth Sci 33:129–144

Schuster M, Duringer P, Ghienne J-F, Vignaud P, Beauvilain A, Mackaye HT, Brunet M (2003) Coastal conglomerates around the Hadjer El Khamis inselbergs (western Chad, central Africa): New evidence for Lake Mega-Chad episodes. Earth Surf Process Landf 28:1059–1069

Smith GI, Street-Perrott FA (1983) Pluvial lakes of the western United States. In: Porter SC (ed) Late-Quaternary environments of the United States, vol 1, The late Pleistocene. University of Minnesota Press, Minneapolis

Snyder CT, Langbein WB (1962) The Pleistocene lake in Spring Valley, Nevada, and its climatic implications. J Geophys Res 67:2385–2394

Stager JC, Mayewski PA, Meeker LD (2002) Cooling cycles, Heinrich event 1, and the desiccation of Lake Victoria. Palaeogeogr Palaeoclim Palaeoecol 183:169–178

Stein M (2001) The sedimentary and geochemical record of Neogene-Quaternary water bodies in the Dead Sea basin – Inferences for the regional paleoclimatic history. J Paleolimn 26:271–282

Stone T (2006) Last glacial cycle hydrological change at Lake Tyrrell, southeast Autstralia. Quat Res 66:176–181

Street-Perrott FA, Harrison SP (1985) Lake levels and climate reconstruction. In: Hecht AD (ed) Paleoclimate analysis and modeling. Wiley, New York

Valero-Garcés B, Grosjean M, Schwalb A, Geyh MA, Messeli B, Kelts K (1996) Limnogeology of Laguna Miscanti: Evidence for mid to late Holocene moisture changes in the Atacama Altiplano. J Paleolimn 16:1–21

Valero-Garcés BL, Grosjean M, Kelts K, Schreier H, Messerli B (1999) Holocene lacustrine deposition in the Atacama Altiplano: Facies models, climate and tectonic forcing. Palaeogeogr Palaeoclim Palaeoecol 151:101–125

Wilkins DE, Currey DR (1997) Timing and extent of late Quaternary paleolakes in the Trans-Pecos closed basin, west Texas and south-central New Mexico. Quat Res 47:306–315

Williams TR, Bedinger MS (1984) Selected geologic and hydrologic characteristics of the Basin and Range province, western United States. US Geol Surv Misc Investig Map I-1522D

Zhang H, Wünnemann B, Ma Y, Peng J, Pachur H-J, Li J, Qi Y, Chen G, Fang H, Feng Z (2002) Lake level and climate changes between 42,000 and 18,000 ^{14}C yr B.P. in the Tengger Desert, northwestern China. Quat Res 58:62–72

Zhang HC, Peng JL, Ma YZ, Chen GJ, Feng Z-D, Li B, Fan HF, Chang FQ, Lei GL, Wünnemann B (2004) Late Quaternary palaeolake levels in Tengger Desert, NW China. Palaeogeogr Palaeoclim Palaeoecol 211:45–58

Chapter 26

Palaeoclimatic Interpretations From Desert Dunes and Sediments

Vatche P. Tchakerian

Introduction

During the late Quaternary, the world's major deserts experienced dramatic changes in the nature and frequency of aeolian processes (Fig. 26.1). Sand seas (ergs) cover 5% of the global land surface and reveal evidence of repeated phases of dune formation (Thomas et al. 2005). This paper presents a review of dune-building episodes during late Quaternary time and their palaeoclimatic significance. The emphasis of the paper is on African and North American sand seas. Although beyond the scope of this paper, a more detailed synthesis and chronologies of global sand seas is presented by Tchakerian (1999), Goudie (2002), Munyikwa (2005) and Lancaster (2007).

In the Sahara-Sahel drylands of North Africa, peak dune deposition is believed to have ocurred between 20 and 12 ka (Sarnthein 1978), during the Last Glacial Maximum (LGS). On the other hand, aeolian activity was drastically reduced between 11 and 5 ka, as the region experienced a period of humid conditions characterised by increased vegetation cover, high lake stands, incised fluvial channels, and a rise in neolithic cultures (Williams 1982). Owing to the geologic and climatic setting of the North African sand seas, the major dune constructional episodes are believed to have been nearly synchronous between regions, although significant variations in the timing of the dune-building episodes did occur (Tchakerian 1999, Lancaster 2007). Since 5 ka, aridity has slowly returned to the Sahara-Sahel, as the subtropical high pressure cells assumed their current position, with today's hyper-arid central Saharan core region well established by 2 ka. Re-activation of some of the stabilized Sahelian dormant/relict dune systems has been the direct result of increased population pressure in the region, owing to the fact that stabilized dunes provide a richer plant cover for grazing amd firewood gathering, and are easier to cultivate.

In the North American arid zone (with the exception of some mid-Holocene aeolian depositional periods in the Nebraska Sand Hills in the Great Plains), major dune-building episodes in the Mojave Desert and the Great Basin, appear to be highly episodic and discontinuous in time and space, and controlled primarily by sediment production, availability and transport capacity systems (e.g. Kocurek 1998), rather than by hemispheric or regional changes in atmospheric pressure fields, such as in the Sahara-Sahel region. Although highly episodic, the period between 25 and 7 ka witnessed multiple pulses of aeolian deposition as recognized from the dune fields and sand ramps of the Mojave Desert (Lancaster and Tchakerian 2003).

Quaternary Aeolian Activity From Continental Deserts

Africa

In the ergs of the African continent, the last Wisconsin cool substage most likely coincided with an arid episode as initially suggested by Grove and Warren (1968). This tropical aridity is believed to have been the result of high-latitude glaciations

V.P. Tchakerian (✉)
Department of Geography, Texas A & M University, College Station, TX 77843, USA
e-mail: v-tchakerian@tamu.edu

A.J. Parsons, A.D. Abrahams (eds.), *Geomorphology of Desert Environments*, 2nd ed., DOI 10.1007/978-1-4020-5719-9_26, © Springer Science+Business Media B.V. 2009

Fig. 26.1 Global distribution of sand seas for (**a**) the present and (**b**) the Last Glacial Maximum (LGM), 18 ka (after Sarnthein 1978). H denotes humid conditions

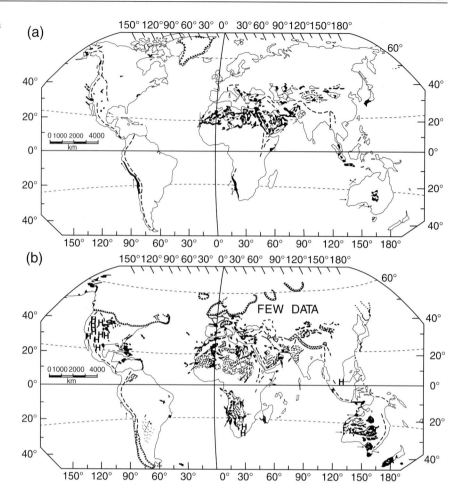

(Williams 1975, Sarnthein 1978). In North Africa, the evidence for late Wisconsin aridity comes in the form of extensive belts of fixed, fossil, degraded dunes, which now extend from the Sahara south to latitude 10–12° N in the Sahel over a latitudinal distance of about 5000 km (Grove and Warren 1968, Sarnthein 1978, Rognon 1987, Thomas 1989). The majority of the Sahara-Sahel dune systems are believed to have been formed between 20 and 12 ka, a time interval characterised by drier and windier than present climatic conditions, with the driest period occurring between 14 and 12 ka (Rognon and Williams 1977, Bowler 1978, Williams and Faure 1980, Alimen 1982, Goudie 1983, Grove 1985, Thomas 1987, 1989).

Based on detailed geomorphic and stratigraphic relations, together with luminescence ages from the different aeolian units, Kocurek et al. (1991) found evidence for the formation of large linear dunes between

20 and 13 ka in the Akchar Erg in Mauritania. The dunes were subsequently stabilized by vegetation and paleosols between 11 and 4.5 ka. In the western Sahara of Mauritania, OSL (Optical Simulated Luminescence) dating, combined with sedimentological studies, revealed three main generations of linear dunes formed during periods 25–15 ka (centered around the LGM), 13–10 ka (spanning the Younger Dryas cooling episode), and after 5 ka (Lancaster et al. 2002). The above sequence of widespread aeolian activity beginning at or near the LGM and continuing until about 12 ka, followed by humid conditions and dune stabilization from 11 to 5 ka, and a return to more arid and widespread aeolian activity since 5 ka, seems to have occurred in large parts of the central and southern Sahara-Sahel drylands (Swezey 2001). Figure 26.2 is a chronological synthesis of the main dune-building phases and/or activity from the continental deserts of Africa, Asia and Australia.

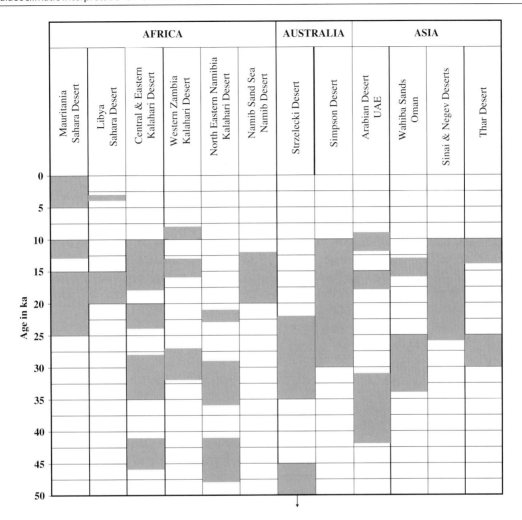

Fig. 26.2 A chronological synthesis of late Quaternary major dune-building and/or aeolian activity from the tropical and sub-tropical continental deserts of Africa, Asia and Australia. Most dates are from luminescence methods. Some radiacarbon and other proxy data also included. More references are found in the text as well as in Tchakerian (1999), Munyikwa (2005) and Lancaster (2007). A much longer luminescence chronology is also provided by Lancaster (2007)

In the Selima sand sea of Libya in the eastern Sahara Desert, Stokes et al. (1998), based on sedimentary and OSL dating studies, propose two major episodes of dune construction of 20–15 and 4–3 ka, similar in chronology to the Mauritanian ergs, with the exception of the missing 13–10 ka Younger Dryas event. On the other hand, in some of the northern Saharan ergs, aeolian activity was diminished owing to the incursions of moisture from the quasi-permanent low pressure cells associated with the colder conditions in Europe as well as the presence of glaciers in the higher latitudes (Nicholson and Flohn 1980, Rognon 1987). Based on atmospheric general circulation models, Nicholson

and Flohn (1980) proposed that the northern sections of the Sahara located in the Maghreb countries of North Africa, experienced relatively wet conditions between 20 and 14 ka. The Grand Erg Occidental and Erg Chech were inactive during the LGM and resumed their dune-building activities only some time after 10 ka (Nicholson and Flohn 1980, Rognon 1987). Thus the majority of the northern Saharan ergs are believed to have been inactive during most of the late Pleistocene, while in the central and southern Sahara and the Sahel, aridity was more widespread with major dune-building episodes between 20 and 12 ka.

Additional evidence for glacial age aridity comes from ocean cores, in which significant concentrations of aeolian quartz grains have been found reflecting the enhanced trade-wind transport of dusts from arid regions (Sarnthein and Koopman 1980). Pokras and Mix (1985) presented evidence for high concentrations of aeolian biosiliceous particles in deep sea cores from the tropical Atlantic and attribute its presence to the deflation of diatomaceous deposits from dry lake beds beginning at about 20 ka. Quartz-rich layers in marine sediments from the eastern Atlantic are believed to have been deposited during the LGM, as a result of increased aeolian activity and dust deflation from the western Sahara Desert (Pokras and Mix 1985). The proponents of cool-stage aridity in the tropics have also utilized information from meteoric groundwater. In the Sahara, Sonntag et al. (1980) indicated that very little discharge occurred from 20 to 14 ka, and suggest a period of substantially reduced hydrological activity during the last cold stage of the Wisconsin.

In the Mega-Kalahari sand sea of central southern Africa, Stokes et al. (1997) and Thomas et al. (1997), using detailed luminscence dating methods, have identified multiple dune-building episodes during the late Pleistocene (Fig. 26.2). The Mega-Kalahari erg consists primarily of three major linear dune complexes, interspersed with interdune corridors and pans (dry lakes): northern, eastern and southern. In the southern Mega-Kalahari, two late-Pleistocene dune-building episodes have been identified by optically stimulated luminescence (OSL) dating, the first between 30 and 23 ka, and the second between 16 and 10 ka (Stokes et al. 1997). A comprehensive study, based on OSL chronology on quartz grains of the central and eastern Mega-Kalahari erg, revealed five late Pleistocene major arid intervals with linear dune construction: 115–95, 46–41, 35–28, 24–20 and 18–10 (Stokes et al. 1997, 1998, Thomas et al. 2000). On the other hand, some of the dune-building activity in parts of the Mega-Kalahari in western Zambia indicate dune building phases from 32 to 27, 16 to 13, and 10 to 8 ka (O'Connor and Thomas, 1999). Linear dunes in the north-eastern Namibian portion of the Mega-Kalahari also exhibit three stages of dune construction: 48–41, 36–29, and 23–21 ka, with limited aeolian activity since 21 ka, and hence no dune phases at the LGM (Thomas and Shaw 2002). The arid intervals lasted between 5 and 20 ka, separated by humid phases lasting between 20 and 40 ka. Linear dune emplacements are believed to be linked to changes in southeast Atlantic

sea surface temperatures (SST), which causes changes in the NE-SW summer rainfall gradient, with periods of low SST's corresponding to periods of enhanced aridity and hence dune construction (Thomas and Shaw 2002, Munyikwa 2005). Episodic aeolian activity throughout the Mega-Kalahari has continued from 9 ka to the present, especially around 5 ka and also between 2 and 1 ka (Thomas et al. 1997, O'Connor and Thomas, 1999). Holocene aeolian activity has been confined largely to the reworking of Pleistocene linear dune sands and the formation of lunettes (clay dunes) on the lee side of the numerous pans that dot the region (Goudie 2002). Remobilization of the stabilized dunes in the Kalahari Desert (and the Sahel) under various land degradation scenarios owing to global warming, can have detrimental effect on the pastoral and agricultural activities in the region (Thomas et al. 2005).

The Namib erg, a sand sea that covers about 35,000 km^2, is dominated by a core of inland linear dune complexes associated with bi-directional wind regimes, mostly SSW-SW and NE-E (Goudie 2002). According to Lancaster (1989), the linear dune complexes constitute about 75% of all the aeolian sands in the erg, followed by a belt of highly mobile coastal crescentic (mostly barchans) dunes (14%) and star dunes (9%). Owing to the paucity of luminescence dates from the Namib erg, palaeoenvironmental reconstruction is primarily through proxy records. Uranium series dating from speleothems in presently flooded caves and cenotes (solution hollows in carbonates), indicate dry conditions from 15 to 11 ka and between 9 and 7 ka (Brook et al. 1999). Palaeohydrologic evidence suggests periods of increased river discharge and groundwater flow, as well as the presence of shallow lakes and ponds in interdune corridors centered on 8–12, 20–24 and 26–32 ka (Lancaster 2002), thus implying arid conditions potentially favorable to dune construction between 20 and 12 ka. Using GPR (ground-penetrating radar for identifying changes in dune sedimentary structures) and OSL dating on linear dunes in the northern Namib erg, Bristow et al. (2007) recently report two phases of major dune-building: the first between 5.7 and 5.2 ka and the second between 2.4 and 1 ka, thus lending support to the return of more widespread and episodic aeolian activitiy after 5 ka.

The available data from the African deserts suggest that late glacial aridity and aeolian depositional phases were not synchronous between the two desert regions, north (Sahara-Sahel) and south (Kalahari, Namib) of the equator. In parts of the Kalahari Desert, wetter conditions are believed to have prevailed up until c. 10 ka with no major dune development during the LGM (Thomas and Shaw 2002).

Australia

Dunes cover over 40% of the Australian continent, with linear dunes as the most widespread aeolian landform type (Goudie 2002). The majority of the linear dunes are found in the Simpson and Strzelecki deserts. Luminescence, radiocarbon and stratigraphic data from the Simpson desert indicate multiple dune-building phases between 33 and 9 ka (Nanson et al. 1995, Twidale et al. 2001). According to Wasson (1984), peak dune formation in the Simpson Desert occurred between 20 and 16 ka, with significant reduction in aeolian activity and dune formation after 13 ka. In the Strzelecki Desert, dune-building episodes are concentrated between 65 and 45 ka and again between 35 and 22 ka (Lomax et al. 2003). Based on thermoluminescence and radiocarbon dating, Gardner et al. (1987) proposed that the linear dunes in the southwestern Australian deserts were formed between 30 and 10 ka. The combined chronologies from the Australian arid zone indicate that major dune constructional episodes occurred between 35 and 10 ka (Fig. 26.2).

Wyrwoll and Milton (1976) first proposed that aridity during glacial stages was widespread, especially in northwestern Australia, owing largely to increased anticyclonic activity. Additional evidence for increased aeolian activity during this arid episode of the last glacial maximum comes from the submerged discordant dunes in Fitzroy Sound, dunes formed when sea level was lower (Jennings 1975). Renewed aeolian activity is evidenced by the reworking of stabilized linear dune sediments in parts of the Australian interior between 3 and 1 ka, when lake levels and temperatures fell from their high levels during the early Holocene (Wasson 1984).

Asia

Geomorphic, chrono-stratigraphic and luminescence data from the fossil dunes in the Thar Desert of India indicate that the major dune-building episodes took place between 30 and 25 ka and 14 to 10 ka (no dune construction during the LGM), with some reactivation of stabilized dunes in the Holocene (Singhvi and Kar 2004). Data from the dune complexes in the UAE suggest periods of dune construction as indicated by OSL ages of 42–31, 18–15 and 12–9 ka (Glennie and Singhvi 2002). In the Wahiba sand sea of Oman, the periods from 34 to 25, and 16 to 13 ka represent major times of sand accumulation and dune formation, including some of the spectacular, 50–100 m high, south-north oriented complex linear dune systems (Preusser et al. 2005). The above dune complexes in the Wahiba sand sea are in turn underlain by older dune deposits and aeolianites (carbonate cemented aeolian sands). There is evidence from Lake Lisan sediments in the Dead Sea and from the northern Sinai Desert for hyperaridity and increased rates of aeolian sedimentation from 26 to 15 ka (Gerson 1982). Based on radiocarbon dates from playa sediments, two dune-building episodes from 21 to 16 ka and 12 to 10 ka are recognized in the north-western Negev Desert of Israel (Magaritz and Enzel 1990). In the Al-Jafr basin of southern Jordan, there are indications of dune deposition after 26 ka (Huckreide and Weissmann 1968). Playa sediments from the Iranian Plateau also show evidence for increased concentrations of aeolian sands between 22 and 12 ka (Wright 1966). Once again (with the exception of the Thar Desert dunes), intensified periods of aeolian activity and dune-building seem to correspond with the LGM, with glacial episodes in the high-latitudes and arid (with major aeolian depositional phases) episodes in the low-latitude deserts of Africa, Australia and Asia (Fig. 26.2).

North America

Great Plains

The majority of the evidence for late Quaternary aeolian activity comes from stabilized (dormant and relict) sand dunes and sand sheets in the Central Plains of the USA (Fig. 26.3). In the Sandhills of Nebraska (the largest stabilized sand sea in the western hemisphere), recent studies based on detailed radioacarbon and OSL dating, indicate that the dunes formed in two major depositional phases during the Holocene: the first from

Fig. 26.3 A chronological synthesis of late Quaternary major dune-building and/or sand sheet and sand ramp activity from the Central Plains, Great Basin and the Mojave Desert, USA. Most dates are from luminescence methods. Some radiacarbon and other proxy data also included. More references are found in the text as well as in Tchakerian (1999) and Lancaster (2007)

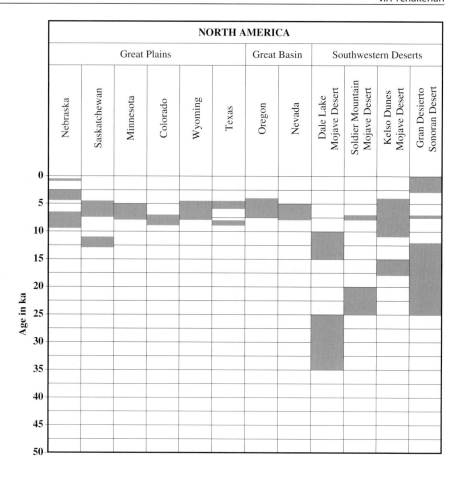

6 to 5.7 ka and the second, from 4 to 2 ka, with three smaller reactivation phases during the last 800 years (Stokes and Swinehart 1997). Based on 95 OSL ages from dunes and loess from the Nebraska Sandhills, Miao et al. (2007) propose three major dune-building periods in the Holocene: 9.6–6.5, 4.5–2.3 and 1.0–0.7 ka. Ahlbrandt and Fryberger (1980) reported radiocarbon dates for regional dune formation from 8.5 to 7 and from 5 to 3 ka, with the latter constituting the major dune-building phase. Additional studies from outside the Nebraska Sandhills also suggest an early to mid-Holocene phase for dune deposition. In the Canadian High Plains of Saskatchewan, Wolfe et al. (2006) report two major peiods of dune activity: the first between 13 and 11 ka (following deglacation and present drainage development), and the second between 7.5 and 4.7 ka. Grigal et al. (1976) found two periods of dune formation between 8 and 5 ka from ^{14}C dates on lacustrine organics in north-central Minnesota. In the same vicinity, from the analysis of pollen and lacus-

trine sediments, Dean et al. (1984) found two periods of aridity between 8.5 and 4 ka. Based on radiocarbon and thermoluminescence age estimates for buried A horizons in aeolian sands from northern Colorado, Forman and Maat (1990) reported heightened dune reactivation and deposition some-time between 9 and 7 ka. From southwestern Wyoming (Kilpecker Dunes), Mayer and Mahan (2004) suggest a number of aeolian depositional events between 8.2 and 6.6 ka. Using geomorphic and stratigraphic analysis of a 25-m-thick sequence of dune and interdune deposits at Clear Creek in south-central Wyoming, Gaylord (1990), found at least four episodes of increased aeolian sedimentation between 7.5 and 4.5 ka, with pronounced hyperaridity from 7.5 to 7 ka and 5.9 to 4.5 ka. Based on geomorphic, stratigraphic, pedologic, and archaeological evidence from draws, dunes, lunettes, and playa sediments from the southern High Plains in Texas, Holliday (1989) reported widespread aeolian activity beginning c. 9 ka with dune deposition between 6 and

4.5 ka. It is apparent that the central High Plains experienced major aeolian depositional episodes between 9 and 3 ka, particularly between 7 and 5 ka (Fig. 26.3).

Since 3 ka, there is supporting evidence for discontinuous and episodic accumulation or re-activation of sand dunes from the Great Plains, in essentially modern climatic conditions. These dune reactivations are most likely occurring in response to periods of extended droughts (more intense than the Dust Bowl events of the 1930s), such as 20 to 30 year quasi-periodic cycles that extend back thousands of years (Muhs et al. 1996). The late Holocene dune constructional episodes most likely took advantage "pirating" of the Pleistocene sand dunes and sand sheets deposited during the LGM or earlier. The current vegetation stablized dunes and sand sheets in the Great Plains (such as the Nebraska Sandhills) can be reactivated if the plant cover is significantly reduced because of extended drought or anthropogenic activities. According to Muhs and Maat (1993), wind speeds in the Great Plains typically exceed the threshold wind velocity 30–60% of the time. The sand dunes and sand sheets are thus "poised" for renewed aeolian activity if an when the present vegetation cover is removed or significantly disturbed.

South-Western Deserts

Compared with other continental deserts of the world, aeolian deposits form only a minor component of the total surficial deposits in the North American arid zone (Tchakerian 1997). With the exception of the Great Plains, the Mojave Desert in California and Nevada, and the Gran Desierto del Altar in Sonora, Mexico, are the only areas with significant aeolian depositional landforms.

The few detailed studies from the Great Basin also indicate dune-building episodes during early to mid-Holocene times (Fig. 26.3), similar to those in the Great Plains (Tchakerian 1999). In Catlow Valley, south-east Oregon, two dune episodes overlie lake deposits between 7.5 and 4 ka (Mehringer and Wigand 1986). In the dissected badlands at Corn Creek Flat, 30 km north-west of Las Vegas, Nevada, Quade (1986) found evidence of increased aeolian sedimentation between 8 and 5 ka accompanied by a drop in the water table of at least 25 m. Renewed aeolian activity in the form of coppice dunes of up to 4 m high is believed to have started within the last

1000 years (Quade 1986). Additionally, renewed late Holocene activity commencing around 3 ka is also documented in the Tucson Mountains in southern Arizona based on ^{14}C dates from archaeological remains embedded in sand dunes (Brakenridge and Schuster 1986).

The aeolian sediments in the Mojave Desert consist primarily of small dune fields, sand sheets and sand ramps (amalgamated deposits consisting primarily of aeolian sands, as well as colluvial and alluvial deposits on the slopes of desert mountains). Because sand ramps contain multiple aeolian depositional units separated by paleosols formed in periods of geomorphic stability, they offer a unique opportunity for detailed geomorphic, stratigraphic and luminscence dating, thus their use in palaeoenvironmental reconstruction. Whitney et al. (1985) were one of the first to conclude from limited studies of sand ramps at four localities in the northern Amargosa Desert in Nevada, that the aeolian sediments were deposited during dry and windy Pleistocene climatic episodes as long ago as 750 ka. They noted the presence of up to 10 buried soils with Bishop Tuff, K/Ar dated at 740 ka, near the base of two of the four sand ramps.

Sand ramps are found astride topographically well-defined sand-transport corridors that follow the region's geologic and tectonic setting (Zimbelman et al. 1995), and have been the focus of detailed geomorphic research (Lancaster and Tchakerian 1996). Most sand ramps are relict features and are not accumulating today, with boulder-to-gravel size talus mantles and incised stream channels that expose the underlying sequence of sedimentary units (Lancaster and Tchakerian 1996). Chronological dating based on optical stimulated luminescence (OSL) and infrared stimulated luminescence (IRSL) dating of aeolian deposits in sand ramps indicate a clustering of depositional phases between 30 to 20 ka and 15 to 7 ka (Fig. 26.3). An 18 m thick sand ramp at Dale Lake in the southeastern Mojave Desert, has been the subject of many studies (Tchakerian 1991, Rendell et al. 1994, Lancaster and Tchakerian 1996, 2003, Clarke and Rendell 1998), and based on detailed analysis of geomorphic, granulometric, soil-stratigraphic, sedimentological and luminescence dating, a long period of aeolian accumulation, with intervening periods of stability and soil formation, has been proposed. A total of 6 sedimentary units are recognized and two main periods of dune emplacement suggested: the first between >35 and 25 ka and the second between

15 and 10 ka. In the central Mojave Desert, the stratigraphy and sediments of the Soldier Mountain sand ramp (about 30 m thick) have been exposed by a combination of mining activity and gully incision, enabling the use of detailed luminescence dating. OSL and IRSL ages indicate that the Soldier Mountain sand ramp accumulated between 20–25 ka and 7–8 ka (Lancaster and Tchakerian 2003). One of the largest and best studied dunefields in the Mojave Desert is the Kelso Dunes, located about 50 km downwind from its initial source, the Mojave River Wash (via Afton Canyon) and Pleistocene Lake Manix. The Kelso dune complex features many juxtaposed areas of dunes of distinctly different morphologies, type, size, spacing, alingment and age (Lancaster and Tchakerian 2003). The distinct mosaic of the dune complex suggests the independent development of the aeolian units, each representing a depositional episode either from new sediment sources or from the reworking of existing dunes. Luminescence ages for dune sediments at the Kelso Dunes indicate numerous aeolian constructional episodes, especially between 18 and 15 ka, and from 11 to 4 ka, with extensive reworking and reactivation of dunes during the last 1500 years (Lancaster and Tchakerian 2003).

Investigations of the late Quaternary history of Lake Manix in the Mojave Desert indicate that the lake spilled and rapidly cut Afton Canyon sometime between 18 and 16 ka (Meek 1999), hence the paucity of luminescence ages greater than 20 ka in the Kelso Dunes. It is highly probable that the exposed lake sediments were ultimately mobilized by the regional winds and ended up in the Kelso Dunes, thus the clustering of ages within the 18–15 ka range.

Additional evidence for Quaternary aeolian activity in the Mojave Desert comes from alluvial stratigraphy, and stages of soil development on dated lava flows of the Cima volcanic field. Studies by Wells et al. (1987) on the late Quaternary geomorphic history of the Silver Lake in the eastern Mojave Desert have documented two aeolian depositional episodes based on stratigraphic relationships between alluvial fan deposits and ^{14}C dated high shoreline stands. These two episodes are believed to have occurred sometime between 10.5 and 8 ka. Indirect evidence for late Pleistocene aeolian activity is also recognized from buried soils within accretionary loess mantles on K/Ar dated basalt flows from the Cima volcanic field in the eastern Mojave Desert (McFadden et al. 1986). The youngest basalt flow, dated at 16 ka, is overlain by several aeolian units (McFadden et al. 1986).

In the Sonoran Desert of Mexico, the dune systems in the Gran Desierto del Altar have been the focus of several studies (Lancaster 1995). The dune complex, its sediments largely derived from the Colorado River and its delta, has a spatially heterogenous pattern consisting of multiple generations of linear, crescentic, and star dunes, with extensive sand sheets surrounding the various dunes. Recent geomorphic, sedimentologic, stratigraphic and luminescence dating by Beveridge et al. (2006), indicates a continuous pattern of dune construction as well as reworking of older sands from 27 ka to the present, with peak dune-building periods between 25 and 12 ka (mostly the linear dunes), around 7 ka (the eastern crescentic dunes) and during the last 3 ka (the western crescentic dunes and the star dunes).

The foregoing discussion reveals a sharp contrast in the timing of major dune-building episodes between North America, where these episodes exhibit a nearly continuous but episodic sedimentation pattern beginning around 35 ka, with an early to mid-Holocene peak, and the deserts of Africa, Asia, and Australia (with the exception of sections of the Thar Desert and some of the southern African ergs of the Mega-Kalahari and nearby areas), where dune deposition seems to have peaked between 18 and 12 ka (around the LGM), with a distinct humid period between 11 and 5 ka. In the next section the relationship between general atmospheric conditions and dune-building episodes will be discussed.

Late Quaternary Palaeoclimates and Aeolian Episodes

Africa, Asia, and Australia

For the majority of the tropical and subtropical deserts of Africa, Asia, and Australia, the currently accepted paradigm equates the extension of the ergs to the development of high-latitude glaciations during the late Quaternary, with most of the major dune systems being formed on or just after the onset of the LGM at ~20 ka (Williams 1975, Sarnthein 1978). With some exceptions from the Mega-Kalahari dunes and the Thar

Desert, the above pardigm seems to still support the hemispheric nature of aeolian activity with peak dune-building episodes concentrated between 18 and 12 ka (Munyikwa 2005). The general atmospheric characteristics associated with such a scenario are given by Nicholson and Flohn (1980) and include the migration of the climatic belts towards the Equator and southwards, the increase in meridional (pole to Equator) atmospheric temperature gradients, the lessening of thermal contrasts between the two hemispheres, and an increase in wind speeds of the trades and mid-latitude westerlies. During the LGM, the equatorial trough (including the ITCZ) and the subtropical high pressure belts were displaced equatorwards and are believed to have been instrumental in the development of the extensive belts of dunes in Africa. The intrusion of the ITCZ south of the Equator may explain why some of the the southern African ergs in the Mega-Kalahari were inactive or were contracting, and thus out of phase with their northern counterparts in the Sahara-Sahel.

The reduction in temperature and precipitation patterns in conjunction with changes in wind regime also affected the overall vegetation. According to Talbot (1984), in addition to an increase in the wind regime, precipitation had to have been 25–50% less than present values for the establishment of dunes in the Sahel between 20 and 13 ka. A prolonged period of aridity and reduced moisture most likely led to a reduction of vegetation cover and density on dunes (especially on linear dunes).

According to Thomas and Shaw (1991), the use of vegetated fossil dunes as indicators of former arid conditions should be used with caution. Linear dunes are believed to be preferred sites for vegetation establishment (especially along the windward slopes) because of their migratory or extending forms (Thomas and Shaw 1991). On the other hand, transverse dunes (such as barchans) are too migratory to allow any form of vegetation foothold, while complex dunes exhibit vertical accretion and are usually found within ergs in hyperarid climates, and not conducive for vegetation establishment. Additionally, vegetated linear dunes are preferred sites for plant anchorage and growth owing to the unique moisture-retaining capabilities of their sands, especially within interdune corridors and lower windward slopes. For example, aeolian transport has been documented from vegetated linear dunes with up to 35% plant cover (Ash and Wasson 1983), especially at the dune crests, where wind speeds tend to be accel-

erated and sediment transported at a much higher rate (Mulligan 1987). Throughout much of the Australian desert dunefields (where vegetation densities are below 10%), low wind velocities are thought to be responsible for the immobility of the sand bodies rather than the anchoring role of vegetation (Ash and Wasson 1983, Bowler and Wasson 1984, Wasson 1984). Thus enhanced aeolian activity and dune-building phases at the glacial maximum are thought to have resulted from strengthening of the anticyclonic winds, steeper meridional temperature gradients, and increased continentality owing to lower sea levels, as well as from lowered temperatures and precipitation, and to a lesser degree, a decrease in vegetation cover and density.

For the late Pleistocene, the picture that emerges is one of increased aeolian sedimentation in the southern Sahara and southwards (Sahel), with humid and stable conditions in the northern Sahara and the Maghreb (Rognon 1987). In the southern African ergs, aeolian activity around the LGM seems to have been limited to the central Mega-Kalahari, with relatively humid conditions prevalent in some parts of the Mega-Kalahari and adjacent deserts (Thomas 1989). For example, there was no dune construction in parts of the northwestern Mega-Kalahari in Zambia and Namibia during the LGM (Thomas and Shaw 2002).

The understanding of larger spatial and/or temporal scales of dune construction, driven primarily by hemispheric scale atmospheric processes, need to be balanced by the complex interplay that occurs at small spatial scales (such as one basin) and/or temporal scales (such as decadal) which ultimately control the rates of sediment supply, storage, entrainment, and transport capacity, as well as basin geomorphology and the nature and cover of vegetation.

During the late Pleistocene and the early Holocene (after the cessation or diminution of major aeolian activities around the LGM), and because of pronounced thermal differences between oceans and lands (owing to significant variations in solar insolation), a strong monsoonal flow developed from 12 to 6 ka with the majority of the Saharan lakes exhibiting high water stands (Street-Perrott et al. 1985, Zubakov and Borzenkova 1990). Lake Megachad reached its highest level sometime between 9 and 6 ka (Maley 1977, Nicholson and Flohn 1980, Servant and Servant-Vildary 1980). In the Malian Sahara, two lacustrine phases are believed to have occurred, the first between 9.5 and 6.4 ka, and the second between

5.4 and 4 ka, with the period in between characterized by a dry episode (Petit-Maire and Riser 1983). Also, pollen samples from Holocene lacustrine sediments in north-west Sudan indicate a period of wetter conditions and high lakes between 9.5 and 4.5 ka (Ritchie and Haynes 1987). The lacustrine sediments are overlain by aeolian sands and are believed to represent the beginning of the arid conditions which still dominate the region (Ritchie and Haynes 1987). The early to middle Holocene moist periods are thought to have been instrumental for the development of the famous Neolithic cultures and rock art of the Sahara, and arguably the rise of Egyptian civilization in the Nile Valley (Williams 1982, Zubakov and Borzenkova 1990, Kuper and Kröpelin 2006). During the same period, lake levels were generally lower in the south-western basins of the United States (Smith and Street-Perrott 1983, Benson et al. 1990). Current arid conditions in the Sahara are thought to have begun sometime after 4 ka (Rognon 1987).

North America

An earlier paradigm proposed that dune deposition in the basins of the southwestern deserts of the United States was at it peak during early to middle Holocene time, a period of aridity and drought from about 8 to 5 ka (Smith 1967). This period was characterized by higher temperatures, drying of basin lakes, reduction in effective moisture, and a concomittant increase in aeolian activity (Van Devender et al. 1987, Wells et al. 1987, Spaulding 1991). Antevs (1955) first proposed such a climatic interval, which he labelled the Altithermal (it is currently referred to as the Holocene Climatic Optimum).

Currently, it is generally understood that the various aeolian sediments presently found within the basin and range type deserts of the south-western United States have accumulated primarily in response to the lowering of lake basins and a consequent increase in the availability of fine sediment, as well as to changes in hillslope, fluvial and alluvial geomorphic systems (Lancaster and Tchakerian 2003). Lake-level fluctuations can provide extensive sandy sediments for subsequent deflation. Sediments are also supplied or replenished by ephemeral desert streams. Additional sources for aeolian dune construction include fan-deltaic and beach deposits formed in and around lakes, and from alluvial fans and wadis and/or arroyos. These environmental changes are attributed to significant climatic oscillations (Wells et al. 1987). Additionally, there is increasing evidence that winds were stronger in intensity and more persistent during the glacial maximum (Kutzbach 1987). As sediment transport varies with the cube of the shear velocity, any increase in the speed and persistence of the wind (especially over longer timespans) can lead to an increase in sediment transport and thus dune deposition. This increased aeolian component, combined with falling lake levels, contributed to the formation of the sand dunes and sheets in the deserts of the American South-west (Wells et al. 1987, Lancaster and Tchakerian 2003).

In the Mojave Desert, between ~35 and 9 ka, a series of full and intermittant lakes occupied the drainage basin of the Mojave River, the largest fluvial system in the Mojave Desert (Fig. 26.4). The Mojave River originates in the San Bernardino Mountains of southern California and currently ends in Silver Lake. Three basins and their lakes have received Mojave River flows: Lake Mojave (including present day Soda and Silver lakes), Cronese Basin, and Manix Basin (including Afton, Troy, Coyote and Harper sub-basins). Sediments from the above fluctuating and dessicating lake basins (and their adjoining alluvial and fluvial sub-basins) are believed to be the primary source for the sand dunes and sand ramps of the Mojave Desert (Tchakerian and Lancaster 2002). The relations between dune-building episodes and lake levels can be seen in Fig. 26.5. Aeolian depositional events seem to take place during both high and intermittent lake levels, as well as during periods of lake dessication. Dune construction appears to be mostly episodic, discontinuous or in discrete pulses controlled largely by the availability and mobility of sediment. A time-lagged, temporally and spatially disjointed, Sediment State System (Fig. 26.6) has been proposed by Kocurek (1998) and Kocurek and Lancaster (1999) that explains the rather complex, but inter-related processes associated with lacustrine, hillslope, fluvial, and aeolian sediment production, storage, transport and deposition. These processes ultimately control the rates, fluxes and the timing of the aeolian depositional episodes for the Mojave Desert (and in similar tectonically controlled geographic drylands). High sediment production occurs during sub-humid and/or semi-arid climatic periods,

Fig. 26.4 The Mojave River system and its associated Pleistocene lakes and the Kelso Dunes. See text for further discussion

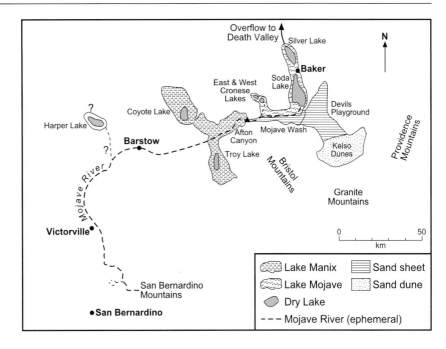

Fig. 26.5 Comparison of the history of dune-building episodes and palaeolake fluctuations from the Mojave Desert. Aeolian constructional periods based on luminescence ages. Arrow indicates the draining of Lake Manix. See text for further discussion and references

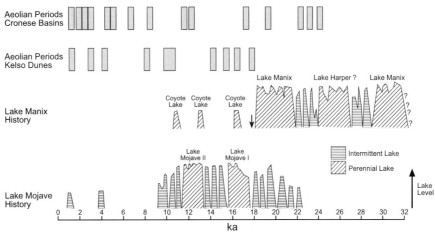

while at the same time, aeolian sediment mobility and availability remain low because of higher precipitation and vegetation cover. This climatic interval is characterized by abundant sediment supply and storage in various geomorphic environments, particularly around lake basins and adjoining lower piedmont slopes. During more humid and/or wetter periods (e.g. high to intermediate lake-levels), increased vegetation cover inhibits or dramatically reduces sediment entrainment, deflation, and transport, leading to the stabilization of most aeolian sands, and the formation of palaeosols (e.g. on the sand ramps). During more arid periods, sediment supply from fluvial and lacustrine sources

is drastically curtailed, vegetation cover is sparse, and the stored sediment in lake basins and adjoining geomorphic environments (fan-deltas, distal fluvial deposits) is mobilized or re-activated.

On the other hand, during the more arid and perhaps windier Climatic Optimum (Altithermal), between 7 and 5 ka, dune construction in the Mojave Desert was more localized and not as widespread, since most sediments had already been mobilized and new sediment supply was at negligible levels. Arid and hyper-arid sub-cycles tend not to be conducive for the deposition of new aeolian deposits, particularly if not enough time has elapsed to generate, store, and

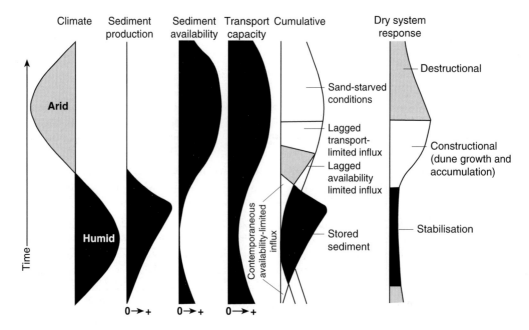

Fig. 26.6 The Sediment State System. A process-response model based on sediment production, sediment availability (supply) and transport capacity of wind as first proposed by Ko-

curek 1998) for Saharan sand seas and modified subsequently by Kocurek and Lancaster (1999) for the Kelso Dunes in the Mojave Desert, California, USA

mobilize fine sediments for deflation and subsequent deposition. In contrast, the Pleistocene to Holocene climatic transition (12–10 ka) was characterized by fluctuating environmental conditions and witnessed warmer temperatures, greater effective moisture (the latter both from winter precipitation and summer monsoonal rains), and brief high lake stands, owing primarily to the northward displacement of the jet stream and its associated atmospheric phenomena, and sporadic aeolian activity (Spaulding and Graumlich 1986, Kutzbach 1987, COHMAP Members 1988, Tchakerian and Lancaster 2002). Intermittant aeolian activity and brief lake level fluctuations have been established for the past 800 years among the basins of the Mojave River. Between 1850 and 1994, fourteen extreme floods on the Mojave River produced documemted high lake stands, albeit lasting for only a few months in Silver Lake and the Cronese Basin (Wells et al. 2003). These lacustrine sediments (after desiccation), along with fluvial sands and silts from the Mojave River floodwaters, were eventually entrained all along the Mojave River Wash/Devil's Playground Sands/Kelso Dunes sand transport corridor (Clarke and Rendell 1998). These events of the last 150 years provide additional support to the Sediment State model

of Kocurek and Lancaster (1999) in that most aeolian deposition is highly episodic and discontinuous, and primarily driven by sediment supply, availability, storage, and the transport capacity of wind.

During the LGM, atmospheric conditions were dramatically different in the deserts of the American South-west. At the same time that the tropical and subtropical deserts were experiencing significant dune-building episodes and low-lake stands, aeolian activity in the basins of the south-western deserts of the United States was highly episodic, with fluctuating climatic conditions and desert basin lakes at high levels (Benson et al. 1990, Tchakerian and Lancaster 2002). According to general circulation models (COHMAP Members. 1988), in the south-western deserts, temperatures were 2°C to 3°C lower than at present, while precipitation and potential evapotranspiration were higher. The changes reflected increased winter rains (associated with the southward shift of the storm track and jet), reduced evaporation (associated with lower temperatures), and stronger and more persistent winds, the latter associated with the steep gradient in atmospheric pressure between glaciers in the mountains, such as the Sierra Nevada, and warmer land areas farther south, such as the Mojave Desert. Regional studies indicate that the majority of closed-basin lakes in the south-western deserts of the United States were high to intermittant from about 25 to 11 ka (Wells et al. 2003).

In contrast to the American South-west and the Great Plains, late Quaternary aeolian activity in most other deserts (some of the southern African ergs excluded) appears to have peaked between 20 and 12 ka (Fig. 26.1). The contemporaneity of earlier Holocene aeolian activity in different parts of western North America is probably the result of the re-establishment of arid conditions, previously displaced by a cooler moisture atmospheric circulation system around the margins of continental ice sheets farther north. In Eurasia, the continental deserts were farther removed from the effects of smaller ice sheets and deglaciation episodes to the north and responded instead to fluctuations in subtropical anticyclones and SST's. When these anticyclones weakened and SST's increased (warmed), for example between 8 and 5 ka, intrusive westerly and/or tropical airflows introduced moisture to such areas as the Sahara Desert (COHMAP Members 1988).

Discussion

The concept of synchronous hemispheric tropical aridity (northern glaciations = tropical aridity) and dune episodes (Williams 1975, Sarnthein 1978), needs to be refined to accommodate the complex temporal and spatial nature of dune construction. At larger spatial and/or temporal scales, dune formation in the Sahara-Sahel region and in the southern African ergs is influenced by the intensity and persistence of trade wind circulation, as well as changes in Atlantic sea surface temperatures (SST's), with periods of low SST's corresponding to periods of enhanced aridity. Changes in the strength of the African monsoon has also been invoked to explain some of the disparities between dune construction in the Thar Desert of India with those in Africa (Lancaster 2007). In contrast to the Sahara-Sahel, major dune-building episodes in the southwestern deserts of North America is influenced by the nature of the polar jet and its accompanied atmospheric phenomena (e.g. increased wind and storm frequencies) and their role in the presence or absence of pluvial lakes and major flood events on desert streams. Continental or regional scale changes in atmospheric phenomena and pressure

fields are not alone sufficient for ultimately triggering dune deposition in desert basins, although major climatic transitions can produce significant changes in geomorphic processes.

At small spatial and/or temporal scales, dune-building episodes need not be correlated with aridity owing to the fact that aeolian deposition is most likely controlled by rates of sediment supply and storage, local and regional topographic and geomorphic controls, the frequency, magnitude, persistence and direction of the prevailing winds (which strongly influence sediment entrainment and transport capacity), and vegetation (as proposed by the Sediment State model). The complex loops within and among the above variables seem to ultimately control whether dune construction will take place. For example, a decrease in the vegetation cover over dunes and stabilized sand sheets, can lead to increased sediment mobility. High wind speeds combined with lower vegetation densities (and reduced roughness) and ample sediment supply (from dry or intermittant lake beds and surrounding piedmont areas) were most likely responsible for the accumulation of sand dunes, sand sheets and sand ramps in the basins of the American Southwest.

Because of the complex nature of aeolian depositional processes, a global systems approach need not necessarily be the correct paradigm through which to investigate the relationships between major dune-building episodes and global aridity associated with climatic changes in the deserts of the world. A more regional and systematic analysis of the aeolian deposits found in the basins of the major deserts should be accomplished before any generalizations can be made as to the global synchroneity of major atmospheric and climatic events and their likely effects on dune-building episodes.

With further refinements and fine tuning in luminescence dating methods (and other dating tools), coupled with new theoretical concepts (such as the Sediment State model), and field techniques (such as GPR), the temporal and spatial patterns of global dune deposition and the various geomorphic, climatic and ecological factors that control the complex nature of sediment supply, availability and mobility, will be better elucidated in the near future. Reactivation of the world's stabilized and vegetated dunes by 21st century global warming would represent one of the most daunting environmental challenges for society (Thomas et al. 2005).

References

Ahlbrandt, T.S. and S.G. Fryberger 1980. Eolian deposits in the Nebraska Sand Hills. *U.S. Geological Survey Professional Paper* 1120.

Alimen, H.M. 1982. Le Sahara – grande zone desertiques Nord Africaine. In *The geological story of the world's deserts*, T.L. Smiley (ed.), 35–51. Uppsala: University of Uppsala Press.

Antevs, E.A. 1955. Geologic–climatic dating in the West. *American Antiquity* **20**, 317–35.

Ash, J.E. and R.J. Wasson 1983. Vegetation and sand mobility in the Australian desert dunefield. *Zeitschrift für Geomorphologie Supplement Band* **45**, 7–25.

Benson, L.V., D.R. Currey, R.I. Dorn, K.R. Lajoie, et al. 1990. Chronology of expansion and contraction of four Great Basin systems during the past 35,000 years. *Palaeogeography, Palaeoclimatology, Palaeoecology* **78**, 241–86.

Beveridge, C., G. Kocurek, R.C. Ewing, N. Lancaster, P. Morthekai, A.K. Singhvi and S.A. Mahan 2006. Development of spatially diverse and complex dune-field patterns: Gran Desierto Dune Field, Sonora, Mexico. *Sedimentology* **53**, 1391–1409.

Bowler, J.M. 1978. Glacial age eolian events at high and low latitudes: a southern hemisphere perspective. In *Antarctic glacial history and world paleoenvironments*, E.M. van Zinderen Bakker (ed.), 149–72. Rotterdam: Balkema.

Bowler, J.M. and R.J. Wasson 1984. Glacial age environments of inland Australia. In *Late Cainozoic paleoclimates of the Southern Hemisphere*, J.C. Vogel (ed.), 183–208. Rotterdam: Balkema.

Brakenridge, G.R. and J. Schuster 1986. Late Quaternary geology and geomorphology in relation to archaeological site locations, southern Arizona. *Journal of Arid Environments* **10**, 225–39.

Bristow, C.S., G.A.T. Duller and N. Lancaster 2007. Age and dynamics of linear dunes in the Namib Desert. *Geology* **35**, 555–58.

Brook, G.A., E. Marais and J.B. Cowart 1999. Evidence of wetter and drier conditions in Namibia from tufas and submerged cave speleothems. *Cimbebasia* **15**, 29–39.

Clarke, M.L. and H.M. Rendell 1998. Climate change impacts on sand supply and the formation of desert sand dunes in the south-west USA. *Journal of Arid Environments* **39**, 517–31.

COHMAP Members 1988. Climatic changes of the last 18,000 years: observations and model simulations. *Science* **241**, 1043–52.

Dean, W.E., J.P. Bradbury, R.Y. Anderson and C.W. Barnosky 1984. The variability of Holocene climate change: evidence from varved lake sediments. *Science* **226**, 1191–4.

Forman, S.L. and Maat, P. 1990. Stratigraphic evidence for late Quaternary dune activity near Hudson on the Piedmont of northern Colorado. *Geology* **18**, 745–8.

Gardner, G.J., A.J. Mortlock, D.M. Price, M.L. Readheas, et al. 1987. Thermoluminescence and radiocarbon dating of Australian desert dunes. *Australian Journal of Earth Sciences* **34**, 343–57.

Gaylord, D.R. 1990. Holocene paleoclimatic fluctuations revealed from dune and interdune strata in Wyoming. *Journal of Arid Environments* **18**, 123–38.

Gerson, R. 1982. The Middle East: landform of a planetary desert through environmental changes. In *The geological story of the world's deserts*, T.L. Smiley (ed.), 52–78. Uppsala: University of Uppsala Press.

Glennie, K.W. and Singhvi, A.K. 2002. Event stratigraphy, palaeoenvironment and chronology of SE Arabian deserts. *Quaternary Science Reviews* **21**, 853–69.

Goudie, A.S. 1983. The arid Earth. In *Mega-geomorphology*, R. Gardner and H. Scoging (eds.), 152–71. Oxford: Clarendon Press.

Goudie, A.S. 2002. *Great warm deserts of the world*. Oxford University Press.

Grigal, D.F., R.C. Severson and G.E. Goltz 1976. Evidence of eolian activity in north-central Minnesota 8,000 to 5,000 yr. ago. *Bulletin of the Geological Society of America* **87**, 1251–4.

Grove, A.T. 1985. The physical evolution of the river basins. In *The Niger and its neighbours*, A.T. Grove (ed.), 2–60. Rotterdam: Balkema.

Grove, A.T. and A. Warren 1968. Quaternary landforms and climate on the south side of the Sahara. *Geographical Journal* **134**, 194–208.

Holliday, V.T. 1989. Middle Holocene drought on the southern High Plains. *Quaternary Research* **31**, 74–82.

Huckreide, R. and G. Weissmann 1968. Der JungPleistozane pluvialsee von Elgafr und Weitere daten zum Quartär Jordaniens. *Geologische Paleontologie* **2**, 13.

Jennings, J.N. 1975. Desert dunes and estuarine fill in the Fitzroy Estuary, north-western Australia. *Catena* **2**, 215–62.

Kocurek, G. 1998. Aeolian system response to external forcing factors – a sequence stratigraphic view of the Saharan region. In *Quaternary Deserts and Climate Change*, A.S. Alsharhan, K.W. Glennie. G.L. Whittle and C.G. Kendall (eds.), 327–38. Rotterdam: Balkema.

Kocurek, G. and N. Lancaster 1999. Aeolian sediment states: Theory and Mojave Desert Kelso dunefield example. *Sedimentology* **46**, 505–16.

Kocurek, G., K.G. Havholm, M. Deynoux, R.C. Blakey 1991. Amalgamated accumulations resulting from climatic and eustatic changes, Akchar Erg, Mauritania. *Sedimentology* **38**, 751–72.

Kuper R. and Kröpelin, S. 2006. Climate-controlled Holocene occupation in the Sahara: Motor of Africa's evolution. *Science* **313**, 803–7.

Kutzbach, J.E. 1987. Model simulations of the climatic patterns during the glaciation of North America. In *North America and adjacent oceans during the last deglaciation, Volume K-3, Geology of North America*, W.F. Ruddiman and H.E. Jr Wright (eds.), 425–46. Boulder, CO: Geological Society of America.

Lancaster, N. 1989. *The Namib Sand Sea: Dune forms, processes and sediments*. Rotterdam: Balkema.

Lancaster, N. 1995. Origin of the Gran Desierto Sand Sea, Sonora, Mexico: Evidence from dune morphology and sedimentology. In *Desert aeolian processes*, V.P. Tchakerian (ed.), 11–35. London: Chapman and Hall.

Lancaster, N. 2002. How dry was dry? Late Pleistocene palaeoclimates in the Namib Desert. *Quaternary Science Reviews* **21**, 769–82.

Lancaster, N. 2007. Low latitude dune fields. In *Encyclopedia of Quaternary Science*, S.A. Elias (ed.), 626–42. Amsterdam: Elsevier.

Lancaster, N. and V.P. Tchakerian 1996. Geomorphology and sediments of sand ramps in the Mojave Desert. *Geomorphology* **17**, 151–66.

Lancaster, N. and V.P. Tchakerian 2003. Later Quaternary eolian dynamics, Mojave Desert, California. In *Paleoenvironments and Paleohydrology of the Mojave and Southern Great Basin Deserts*, Y. Enzel, S.G. Wells and N. Lancaster (eds.), 231–249. Boulder, CO: Geological Society of America Special Paper 368.

Lancaster, N., G. Kocurek, A. Singhvi, A. Pandey, V. Deynoux, J.F. Ghienne and K. Lo 2002. Late Pleistocene and Holocene dune activity and wind regimes in the western Sahara Desert of Mauritania. *Geology* **30**, 991–94.

Lomax J., A. Hilgers, H. Wopner, R. Grun and C.R. Twidale 2003. The onset of dune formation in the Strzelecki Desert, South Australia. *Quaternary Science Reviews* **22**, 1067–76.

Magaritz, M. and Y. Enzel 1990. Standing-water deposits as indicators of late Quaternary dune migration in the northwestern Negev, Israel. *Climatic Change* **16**, 307–18.

Maley, J. 1977. Paleoclimates of Central Sahara during the early Holocene. *Nature* **269**, 573–7.

Mayer, J.H. and S.A. Mahan 2004. Late Quaternary stratigraphy and geochronology of the western Killpecker Dunes, Wyoming, USA. *Quaternary Research* **61**, 72–84.

McFadden, J.C., S.G. Wells and J.C. Dohrenwend 1986. Cumulic soils formed in eolian parent materials on flows of the Cima volcanic field, Mojave Desert, California. *Catena* **13**, 361–89.

Meek, N. 1999. New discoveries about the Late Wisconsinan history of the Mojave River system. In *Tracks along the Mojave: A field guide from Cajon Pass to the Calico Mountains and Coyote Lake*, R.E. Reynolds and J. Reynolds (eds.), 113–17. San Bernardino, CA: San Bernardino County Museum Publication.

Mehringer, P.J. Jr and P.E. Wigand 1986. Holocene history of Skull Creek dunes, Catlow Valley, southeastern Oregon, U.S.A. *Journal of Arid Environments* **11**, 117–38.

Miao X.D., J.A. Mason, J.B. Swinehart, D.B. Loope, P.R. Hanson, R.J. Goble and X.D. Liu 2007. A 10,000 year record of dune activity, dust storms, and severe drought in the central Great Plains. *Geology* **35**, 119–22.

Muhs, D.R. and P.B. Maat 1993. The potential response of aeolian sands to greenhouse warming and precipitation reduction on the Great Plains of the USA. *Journal of Arid Environments* **25**, 905–18.

Muhs, D.R., T.W. Stafford, S.D. Cowherd, S.A. Mahan, R. Kihl, P.B. Maat, C.A. Bush, amd K. Nehring 1996. Origin of the late Quaternary dunefields of northeast Colorado. *Geomorphology* **17**, 129–50.

Mulligan, K.R. 1987. Velocity profiles measured on the windward slope of a transverse dune. *Earth Surface Processes and Landforms* **13**, 573–82.

Munyikwa, K. 2005. Synchrony of Southern Hemisphere Late Pleistocene arid episodes: A review of luminescence chronologies from arid aeolian landscapes south of the Equator. *Quaternary Science Reviews* **24**, 2555–2583.

Nanson, G.C., X.Y. Chen and D.M. Price 1995. Aeolian and fluvial evidence of changing climate and wind patterns during the past 100 ka in the western Simpson Desert, Australia. *Palaeogeography, Palaeoclimatology, Palaeoecology* **113**, 87–102.

Nicholson, S.E. and H. Flohn 1980. African environmental and climatic changes and the general atmospheric circulation in late Pleistocene and Holocene. *Climatic Change* **2**, 313–48.

O'Connor, P.W. and D.S.G. Thomas 1999. The timing of and environmental significance of the Late Quaternary linear dune development in western Zambia. *Quaternary Research* **52**, 44–52.

Petit-Maire, N. and J. Riser (eds.) 1983. *Sahara ou Sahel? Quaternaire Recent du Bassin de Taoudenni, Mali*. Marseille: Imprimerie Lamy.

Pokras, E.M. and A.C. Mix 1985. Eolian evidence for spatial variability of late Quaternary climates in tropical Africa. *Quaternary Research* **24**, 137–49.

Preusser, F., D. Radies, F. Driehorst and A. Matter 2005. Late Quaternary history of the coastal Wahiba Sands, Sultanate of Oman, *Journal of Quaternary Science* **20**, 395–405.

Quade, J. 1986. Late Quaternary environmental changes in the upper Las Vegas Valley, Nevada. *Quaternary Research* **26**, 340–57.

Rendell, H.M., N. Lancaster and V.P. Tchakerian 1994. Luminescence dating of late Quaternary aeolian deposits at Dale Lake and Cronese Mountains, Mojave Desert, California. *Quaternary Science Reviews* **13**, 417–22.

Ritchie, J.C. and C.V. Haynes 1987. Holocene vegetation zonation in the eastern Sahara. *Nature* **330**, 645–7.

Rognon, P. 1987. Late Quaternary climatic reconstruction for the Maghreb (North Africa). *Palaeogeography, Palaeoclimatology, Palaeoecology* **58**, 11–34.

Rognon, P. and M.A.J. Williams 1977. Late Quaternary climatic change in Australia and North Africa: a preliminary interpretation. *Palaeogeography, Palaeoclimatology, Palaeocology* **21**, 285–327.

Sarnthein, M. 1978. Sand deserts during glacial maximum and climatic optimum. *Nature* **272**, 43–6.

Sarnthein, M. and B. Koopman 1980. Late Quaternary deep-sea record of northwest Africa: dust supply and wind direction. *Paleoecology of Africa* **12**, 239–53.

Servant, M. and S. Servant-Vildary 1980. L'environnement quaternaire du bassin du Tchad. In *The Sahara and the Nile*, M.A.J. Williams and H. Faure (eds.), 133–62. Rotterdam: Balkema.

Singhvi, A.R. and A. Kar 2004. The aeolian sedimentation record of the Thar Desert. *Proceedings of the Indian Academy of Sciences* **113**, 371–401.

Smith, G.I. and F.A. Street-Perrott 1983. Pluvial lakes of the western United States. In *Late Quaternary environments of the United States I. The late Pleistocene*, S.C. Porter (ed.), 190–211. Minneapolis: University of Minnesota Press.

Smith, H.T.U. 1967. Past versus present wind action in the Mojave Desert region, California, *U.S. Air Force Cambridge Research Labs Publication* AFLCRL-67-0683.

Sonntag, C., U. Thorweibe, J. Rudolph, E.P. Lohnert, et al. 1980. Isotope identification of Saharian groundwater. *Paleoecology of Africa* **12**, 159–71.

Spaulding, W.G. 1991. A middle Holocene vegetation record from the Mojave Desert of North America and its paleoclimatic significance. *Quaternary Research* **35**, 427–37.

Spaulding, W.G. and L.J. Graumlich 1986. The last pluvial climate episodes in the deserts of southwestern North America. *Nature* **320**, 441–4.

Stokes, S. and J.B. Swinehart 1997. Middle and late Holocene dune reactivation in the Nebraska Sand Hills, USA. *The Holocene* **7**, 273–82.

Stokes, S., D.S.G. Thomas and R. Washington 1997. Multiple episodes of aridity in southern Africa since the last interglacial period. *Nature* **388**, 154-58.

Stokes, S., T.A. Maxwell, C.V. Haynes and J. Horrocks 1998. Latest Pleistocene and Holocene sand sheet construction in the Selima Sand Sheet, Eastern Sahara. In *Quaternary Deserts and Climate Change*. A.S. Alsharhan, K.W. Glennie. G.L. Whittle and C.G. Kendall (eds.), 175–84. Rotterdam: Balkema.

Street-Perrott, F.A., N. Roberts and S. Metcalfe 1985. Geomorphic implications of late Quaternary hydrological and climatic changes in the Northern Hemisphere tropics. *In Environmental change and tropical geomorphology*, I. Douglas and T. Spencer (eds.), 165–83. London: Allen & Unwin.

Swezey, C. 2001. Eolian sediment responses to late Quaternary climate changes: temporal and spatial patterns in the Sahara. *Palaeogeography, Palaeoclimatology, Palaeoecology* **167**, 119–55.

Talbot, M.R. 1984. Late Pleistocene dune building and rainfall in the Sahel. *Paleoecology of Africa* **16**, 203–14.

Tchakerian, V.P. 1991. Late Quaternary aeolian geomorphology of the Dale Lake Sand Sheet, southern Mojave Desert, California. *Physical Geography* **12**, 347–69.

Tchakerian, V.P. 1997. North America. In *Arid zone geomorphology*, D.S.G. Thomas (ed.), 523–41. New York: John Wiley & Sons.

Tchakerian, V.P. 1999. Dune palaeoenvironments. In *Aeolian environments, sediments and landforms*, A.S. Goudie, I. Livingstone and S. Stokes (eds.), 261–92. London: John Wiley & Sons.

Tchakerian, V.P. and N. Lancaster 2002. Late Quaternary arid/humid cycles in the Mojave Desert and western Great Basin of North America. *Quaternary Science Reviews* **21**, 799–810.

Thomas, D.S.G. 1987. Discrimination of depositional environments using sedimentary characteristics in the Mega Kalahari, central southern Africa. In *Desert sediments: ancient and modern*, L. Frostick and I. Reid (eds.), 293–306. Oxford: Blackwell, Geological Society of London Special Publication 35.

Thomas, D.S.G. 1989. Reconstructing ancient arid environments. In *Arid zone geomorphology*, D.S.G. Thomas (ed.), 311–34. New York: John Wiley & Sons.

Thomas, D.S.G. and P.A. Shaw 1991. 'Relict' desert dune systems: interpretations and problems. *Journal of Arid Environments* **20**, 1–14.

Thomas, D.S.G. and P.A. Shaw 2002. Late Quaternary environmental change in central southern Africa: New data, synthesis, issues and prospects. *Quaternary Science Reviews* **21**, 783–97.

Thomas, D.S.G., S. Stokes and P.A. Shaw 1997. Holocene aeolian activity in the southwestern Kalahari Desert, southern Africa: Significance and relationships to late-Pleistocene dune-building events. *The Holocene* **7**, 273–81.

Thomas, D.S.G., M. Knight and G.F.S. Wiggs 2005. Remobilization of southern African desert dune systems by twenty-first century warming. *Nature* **435**, 1218–21.

Thomas, D.S.G., P.W. O'Connor, M.D. Bateman, P.A. Shaw, S. Stokes and D.J. Nash 2000. Dune activity as a record of late Quaternary aridity in the northern Kalahari. *Palaeogeography, Palaeoclimatology, Palaeoecology* **156**, 243–59.

Twidale, C.R., J.R. Prescott, J.A. Bourne and F.M. Williams 2001. Age of desert dunes near Birdsville, southwest Queensland. *Quaternary Science Reviews* **20**, 1355–64.

Van Devender, T.R., R.S. Thompson and J.L. Betancourt 1987. Vegetation history of the deserts of southwestern North America: the nature and timing of the late Wisconsin–Holocene transition. In *North America and adjacent oceans during the last deglaciation, Volume K-3, Geology of North America*, W.R. Ruddiman and H.E. Wright Jr (eds.), 323–52. Boulder, CO: Geological Society of America.

Wasson, R.J. 1984. Late Quaternary paleoenvironments in the desert dunefields of Australia. In *Late Cainozoic paleoclimates of the Southern Hemisphere*, J.C. Vogel (ed.), 419–32. Rotterdam: Balkema.

Wells, S.G., L.D. McFaddena and J.C. Dohrenwend 1987. Influence of the late Quaternary climatic changes on geomorphic and pedogenic processes on a desert piedmont, eastern Mojave Desert, California. *Quaternary Research* 27, 130–46.

Wells, S.G., W.J. Brown, Y. Enzel, R.Y. Anderson and L.D. McFadden 2003. Late Quaternary geology and paleohydrology of pluvial Lake Mojave, southern California. In *Paleoenvironments and Paleohydrology of the Mojave and Southern Great Basin Deserts*, Y. Enzel, S.G. Wells and N. Lancaster (eds.), 79–114. Boulder, CO: Geological Society of America Special Paper 368.

Whitney, J.W., W.C. Swadley and R.R. Shroba 1985. Middle Quaternary sand ramps in the southern Great Basin, California and nevada. *Geological Society of America Abstracts with Programs* **17**, A750.

Williams, M.A.J. 1975. Late Quaternary tropical aridity synchronous in both hemispheres? *Nature* **253**, 617–8.

Williams, M.A.J. 1982. Quaternary environments in northern Africa. In *A land between two Niles*, M.A.J. Williams and D.A. Adamson (eds.), 43–63. Rotterdam: Balkema.

Williams, M.A.J. and H. Faure (eds.) 1980. *The Sahara and the Nile*. Rotterdam: Balkema.

Wolfe, S.A., J. Ollerhead, D.J. Huntley and Q.B. Lian 2006. Holocene dune activity and environmental change in the prairie parkland and boreal forest, central Saskatchewan, Canada. *The Holocene* **16**, 17–29.

Wright, H.E., Jr 1966. Stratigraphy of lake sediments and the precision of the paleoclimatic record. In *World climate from 8,000 to 0 BC*, S.E. Soyer (ed.), 157–73. London: Royal Meteorological Society.

Wyrwoll, K.H. and D. Milton 1976. Widespread late Quaternary aridity in Western Australia. *Nature* **264**, 429–30.

Zubakov, V.A. and I.I. Borzenkova 1990. *Global paleoclimate of the late Cenozoic*, Developments in Paleontology and Stratigraphy 12. Amsterdam: Elsevier.

Zimbelman, J.R., S.H. Williams and V.P. Tchakerian 1995. Sand transport paths in the Mojave Desert, southwestern United States. In *Desert aeolian processes*, V.P. Tchakerian (ed.), 101–30. London: Chapman & Hall.

Chapter 27

Early Humans in Dryland Environments: A Geoarchaeological Perspective

Sue J. McLaren and Tim Reynolds

Introduction

Geoarchaeology is a particular focus of archaeology; some academics argue that it is a subdivision of the natural sciences (e.g. Rapp and Hill 1998); others suggest it is a combination of the sciences and humanities, using the same logic, principles and dating techniques (Huckleberry 2000). If amalgamation is the case, Gladfelter (1981) argues that geoarchaeology should be defined in a generalised and hence inclusive way. So according to Gladfelter (1981: p. 343), geoarchaeology 'deals with earth history within the time frame of human history' or more specifically applies the theory and methods used in Earth Sciences in the study of the human past (Gifford and Rapp 1985; Rapp and Hill 1998). Hence, geoarchaeologists are concerned with reconstructing past environments, using a range of different proxies, in areas of archaeological interest. Geoarchaeologists apply skills from the fields of geomorphology, pedology, sedimentology, hydrology, biology, geochemistry, remote sensing, geographical information systems and geology to study the evidence of past human-landscape interactions. Such studies involve a range of field and laboratory skills which include surveying techniques, the use of sedimentological, botanical, palaeontological, palaeogeomorphological, pedological, and geochemical techniques for palaeoenvironmental reconstructions, documentation of the site in terms of natural and cultural processes and activities, as well as developing chronological frameworks. Geoarchaeology therefore uses multi-disciplinary (Yerkes 2006) and multi-proxy approaches. As a consequence, geoarchaeologists do not have to be archaeologists, indeed much geoarchaeological research involves people from a variety of disciplines. This diversity of backgrounds has led to a wide assortment of definitions of geoarchaeology (e.g. see Butzer 1971, 1982; Dincauze 2000; Renfrew 1976, 1983; Taylor 2003; Yerkes 2006).

Understanding the landscape, its structure and form, the processes operating on the surface and how it has changed over time is key to understanding how humans utilised and/or were constrained by their environment. This is of particular importance in drylands as they are often perceived to be impenetrable barriers to human occupation (Smith et al. 2005); yet there is significant evidence for people in deserts around the world for long periods through the Quaternary. Thus the role of geoarchaeology in dryland regions is to examine the timing and nature of human presence against a background of changing (wetter/drier) environments and to identify the importance of deserts both as potential blockades to human dispersal and habitation during drier phases, and as corridors when it was wetter (Gamble 1993). There is much debate about the environmental conditions that existed during times of human evolution and occupation in deserts. In terms of human evolution, Sherratt (1997) argues that biological changes are mainly caused by changes in the environment and states that 'it is no coincidence that successive species of hominid made their appearance during the Quaternary Period, with its rapid pace and massive scale of environmental alteration' (p. 283). Human occupation of arid lands can only be truly dated to the Holocene, prior to this period the archaeology of arid regions is often the archaeology

S.J. McLaren (✉)
Department of Geography, University of Leicester, Leicester LE1 7RH, UK
e-mail: sjm11@le.ac.uk

A.J. Parsons, A.D. Abrahams (eds.), *Geomorphology of Desert Environments*, 2nd ed., DOI 10.1007/978-1-4020-5719-9_27, © Springer Science+Business Media B.V. 2009

of areas that were formerly occupied when conditions were wetter or more humid.

This book has shown that geomorphological processes, landforms and their associated sediments in drylands are complicated and it is often difficult to determine whether (i) the presence/absence of, and/or (ii) a change in the character of, a geomorphological feature actually reflects a climatic control. Equally, it is necessary to be cautious about what can be learnt about the history of climate change in deserts from the archaeological record as there is not always a direct and contemporaneous link between regional climate, climate change and human activity. Evidence for past human occupation of modern-day deserts has commonly been related to changes in available moisture sources (Mandel 1999). However, one does not always need to invoke wetter conditions for people to live in deserts. People can exist in drylands as long as they have sufficient food and water resources. Although climatically an area may be described as arid (or even hyper-arid), there are many conditions under which people and animals could concentrate and thrive. Examples include: - where allogenic (exotic) rivers (e.g. the Nile) run through desert regions; areas with access to springs and groundwater discharge zones (e.g. in the Fazzān during Garamantean times, Libya); or coastal shorelines such as the Atacama and Namib Deserts. In addition, although people may not be able to live permanently in deserts, the land may be suitable for periodic exploration or migration through. However, it may be very difficult to discriminate from some archaeological remains whether people were occupying a site long-term or briefly residing in temporary camps.

Many of the theories that attempt to link adaptations in humans with climate change are problematical as they often rely on correlations between small scale, localised information derived from sediments containing hominin remains and global scale climate variations preserved in ocean cores (Behrensmeyer 2006). Working in East Africa, Trauth et al. (2005) argue that marine records may not always reflect contemporaneous environmental changes on land. However, Thomas et al. (2003) have found sound correlations between pollen records and sea surface temperatures from south east Atlantic cores (Ning et al. 2000; Little et al. 1997) and multi-proxy land-based evidence for environmental changes in the late Quaternary at Tsodilo Hills, Botswana.

It is important to remember that not only did temporal changes in the environment have an impact on humans but equally humans have affected and changed the landscape over time, making palaeoenvironmental interpretations much more difficult. For example, during the mid-Holocene in Wadi Faynan, southern Jordan, it is not a simple task to identify whether vegetation changes have resulted from increasing aridification or have been a consequence of human disturbance through the removal of trees and vegetation for fuelwood, land clearance and/or overgrazing (McLaren et al. 2004).

Preserved archaeological remains are highly variable in their usefulness as palaeoenvironmental indicators. Lithics, for example, provide little/no information about the environment/climate at the time they were being utilised; on the other hand, preserved remains of human palaeodiets (allowing isotopic analyses); farming and water harvesting techniques; and rock art all provide varying amounts of information that can be used to support the evidence from geoscientific proxies. However, each preserved stratigraphic sequence only represents relatively small segments of the time-space framework of hominin evolution (Behrensmeyer 2006). In addition, many of the archaeological artefacts present problems in terms of being able to provide an absolute chronological framework for the environment in which they are found and there are still numerous geoarchaeological sites that remain poorly dated. Inadequately dated hominin remains/artefacts have resulted in problems with trying to correlate this evidence with data on climate change.

Nonetheless, the archaeological evidence from drylands provides a valuable counterpoint to the biased record of locations with long-term and intensive occupation. Understanding the processes involved in the development of the landscape and how it changes over time in drylands are, therefore, equally important in integrating the evidence these regions can provide into an effective synthesis.

Issues of Scale Between Geoscientists and Archaeologists

It is clear that inter-disciplinary collaborations are necessary as human artefacts are preserved in and on the earth's surface. Researchers from different disciplines undertake studies at 'effective scales' appropriate to their research aims and objectives (Marquardt 1992). The problem with interpreting geoarchaeological evidence is that human behavioural and geomorphic processes operate at different spatial and temporal scales (Dean 1993). Archaeologists tend to study at a 'human scale' (Stein 1993) with sites ranging from a few metres to a few kilometres and on time scales of hundreds to a few thousands of years. Archaeologists in general will tend to study smaller scale events in greater detail when compared to geographers; and as such their research is somewhat limited in terms of understanding large scale changes in palaeoclimates that the humans may have experienced. Quaternary scientists and geoscientists tend to conduct palaeoenvironmental reconstruction studies on timescales from a few thousand to hundreds of thousands of years and on spatial scales that can vary from micro-scale geochemical or diagenetic studies, to synoptic interpretations on the scale of hundreds of kilometres using satellite imagery.

As a result of the different approaches used problems of scale arise both in terms of description and interpretation of data (Stein 1993). Therefore geoarchaeologists need to have a good comprehension of the range of scales involved in the overall research project and be able to accommodate them in their analyses and interpretations. Geoarchaeologists need to think about scale at three broad spatial, as well as temporal, levels.

1. *The Archaeological Site*: Both the spatial and temporal scale of study within the archaeological locale is generally small and at a high resolution. Here the research questions tend to focus on site specific environmental issues. It is important to remember that all sites are artificial constructs determined by the archaeologist as much as by past activity. As such the composition of an archaeological site depends upon assumptions that ought to be tested through taphonomic study including, for example, the reconstruction of the site's local setting, the location of access points and activity foci. The nature of the sediments, their origin, accumulation and erosion all determine human perceptions of a 'site'. Such factors/features all have an important bearing on what a 'site' becomes and how it is later interpreted.

At the archaeological site scale, geoscientists interests include, local sedimentation and erosion patterns, stratigraphy, identification of soils and geochemical sediments, diagenetic processes, collection of biological materials for palaeoenvironmental inter-

pretations and sampling for numerical dating purposes. As an example, Hunt et al. (2004) studied Holocene fluvial sequences that were associated with Neolithic artefacts in the desertic area around Wadi Faynan in southern Jordan. The sediments contained pollen, plant macrofossils and molluscs for palaeoenvironmental analyses and organic matter for radiocarbon dating. The research found evidence for meandering perennial rivers prior to 6000 cal. BP. Through the Holocene the landscape was increasingly affected by the impacts of both an aridification of the climate and early farming practices, with evidence of significant land degradation by the Chalcolithic period (c. 6,000 years ago).

Wind erosion, transportation and deposition in deserts can leave cultural artefacts exposed as lags (e.g. Wilkinson et al. 2006) or buried by sand (e.g. Hurtak 1986). River channel change and flash floods in ephemeral channels can disturb and wash away materials (e.g. Waters and Ravesloot 2001, 2003). Elsewhere slow rates of, or episodic, sediment accumulation may mean that the age of a deposit or palaeosurface and a period of occupation may not be related (Renfrew 1976). Thus, geoarchaeologists need to study the processes of aggradation before and during phases of human activity to gain an understanding of what the environment was and if there was any change over the period of time of interest. It is also important to study the sediment accumulation patterns after the evidence of human activity because of the potential affects of later burial and loss of evidence of the archaeology at the surface. Equally, the processes of erosion need to be considered as they have implications for the destruction of the archaeological evidence either completely or in terms of altering the palaeoenvironmental context of the artefacts through disturbance and movement.

2. *The Surrounding Landscape*: Depositional and erosional forces that operate at the scale of the area directly surrounding the archaeological site are of concern to geoarchaeologists, as such factors influence the size and shape of the site and how it appears in the modern day landscape (Linse 1993). There is huge variation in the nature of archaeological sites because of the different locations that they occur within a landscape (Johnson et al. 2002). According to Barker and Gilbertson (2000) although archaeological artefacts in deserts are often the most visible within the landscape they are also the most likely to be destroyed by geomorphological agents. But not all archaeological sites

are visible on the ground surface, having been buried by sediments such as migrating sand dunes or building up of alluvial fan sediments, incorporated in soils during wetter phases, or eroded away creating biases or loss of information in the archaeological data surveyed/sampled. It is important to be able to determine whether the observed distribution of artefacts or sites is a function of prehistoric behaviour rather than site preservation and visibility (Ravesloot and Waters 2004). Often due to the lack of visibility at the start of an investigation there are challenges in terms of devising a good sampling strategy from a poorly known target data source. Geoarchaeologists are needed to evaluate the sediments, landforms and the landscape to enable the development of sampling strategies to avoid any biases and to ensure that the sediments/deposits associated with the archaeology and that will provide the palaeoenvironmental history, are analysed, described and sampled prior to the commencement of the archaeological excavation. In the landscape, landforms, sediments and their stratigraphy, soils and their profiles, as well as their physical and chemical properties, plus any fauna and flora they contain all form the essential context for the archaeological remains and palaeoenvironmental reconstructions.

3. *The Regional Setting*: The regional scale largely falls into the realm of the geoscientist. Archaeologists will often have a different (smaller) view on what is regarded as a regional setting by geographers (Linse 1993). People live(d) within dryland landscapes over a much wider area than is likely to be preserved by the archaeological remains. Hence it is important to understand a region's natural physical features that may have affected decision making. For example, migrational or communication routes may have been influenced by the location of coastlines, mountain passes or rivers; availability of water, variations in microclimate, location of food sources or suitable materials to create tools and weapons would all be sites potentially explored or exploited by humans. Bolten et al. (2006), for example, have found that hunter-gatherer sites in the Western Desert of Egypt varied over time as a function of the physical relief of the environment under changing climatic conditions. During the wetter early Holocene, people appeared to prefer vegetated highland and plain areas as there was plenty of available water. However, during the more arid mid Holocene, people concentrated in depressions and around ephemeral rivers. Without an understanding of the regional setting it would be easy to misinterpret the available archaeological evidence.

4. *Chronological scales*: The older the environmental and archaeological evidence is and/or the broader the time period being covered, the lower the resolution of the interpretations (Stein 1993) because there is an exponential decay of both information (Thornes 1983) and understanding as we go further back in time. The use of numerical dating techniques (especially when calibrated using a range of techniques) allows ages to be established and correlations between the various approaches to be made.

There are significantly different approaches to how archaeologists and geoscientists handle time within the archaeological site. The latter prefer to study vertical sections and conduct their studies from the oldest to the youngest events; whereas archaeologists do the opposite and work down from younger to older using horizontal layers. This latter approach can lead to confusion as horizontal layers may incorporate a number of sediment types that are of different ages and reflect different palaeoenvironments. It is crucial that geoarchaeologists understand these potential chronostratigraphic problems.

Scope of the Chapter

The majority of books on geoarchaeology provide the background information and skills of physical geographers and geoscientists, for archaeologists and budding geoarchaeologists not already trained in these fields. This chapter on the other hand aims to provide brief synopses of what is currently understood about the early occupation of humans in modern-day drylands; and considers the environments that these humans lived in. Such research is largely a result of the collaborative research conducted by physical geographers, geoscientists and archaeologists – the essence of this field of geoarchaeology.

Late Quaternary Human Activity and Palaeoclimates

Africa

Humans as a group are members of the Primates order and as such are atypical in their current geographical range. Most extant primates are to be found within 30° of the equator. The lineage leading to modern humans is believed to have split off some time around 5–6 Ma ago. The earliest fossil evidence to support this date comes from Chad in what is now arid land but was at that time a warm gallery forest surrounding palaeolake Chad (Brunet et al. 1995, 2002; Vignaud et al. 2002). Perilacustrine and lacustrine facies containing abundant aquatic remnants were found in association with hominid remains (Vignaud et al. 2002). The bulk of subsequent evidence for pre-*Homo* hominins comes from eastern and southern Africa. The changes in climate, towards cooler and drier phases breaking up the extensive African forests, have been used to suggest why the human lineage split off from that of the other African great apes which remain tied to forests. Trauth et al. (2005) studied lake sediments in ten East African rift basins and found evidence for rapid variations between arid and humid periods between 1 and 3 Ma, which they have argued may have provided the stress required to initiate speciation.

Evidence from cores of deep marine sediments indicates that over the last 3 Ma there has been an overall general trend towards cooler climates (e.g. deMenocal 1995). Contemporaneous land-based data (such as faunal remains, pollen and isotope analyses on soil carbonates) in tropical Africa show gradual increases in aridity and seasonality as well as expansion of open habitats (Behrensmeyer et al. 1997; Cerling 1992; O'Connell et al. 1999; Reed 1997). Sepulchre et al. (2006) argue that tectonic uplift in East Africa led to aridification as a result of a reorganisation of atmospheric circulation. With the change in climate towards more arid conditions, the human lineage shows an increasing ground-based ecology so that even though the fossil evidence is ambiguous on the complete locomotor pattern of early hominins it is clear that bipedal walking similar to our own is present by 4 Ma at Laetoli, Tanzania (Leakey and Harris 1987). The hominin responsible for the footprint trails (two adults and a single child) is thought to be *Australopithecus afarensis*. This diverse group has human-like feet but fingers that are curved and still reflect a tree-dwelling ancestry. It is likely that this group actually contains a mixture of traits showing that the human and ape lineages were experimenting with different locomotor habits against a climatic background of increasing aridity and reduction in forest habitats (Stringer and Andrews 2005). The evidence from Chad is a curious outlier at present

and it remains a fact that the majority of our evidence for earliest hominins is limited to east and South Africa. Much more research is needed in Chad and surrounding dryland areas if the assumptions derived from sub-Saharan Africa are to be adequately tested. Thus, arid-zone geoarchaeology still has much to contribute to one of the most important debates on human evolution.

In terms of human evolution, a big question is the origin of our own Genus and its spread out from Africa. It may be that our Genus, *Homo*, is the direct result of further desiccation of climate and reduction of forest cover. The contemporary hominins adapting to the spreading grasslands and incorporating a greater proportion of meat into their diet which then enables larger brain sizes, greater sociability and information storage. This change begins c. 2.4 Ma ago and produces *Homo habilis* and/or *Homo rudolphensis*. There is a debate as to how much variability should be allowed in a palaeospecies and thus whether one or more species can be identified. Some authors even suggest *Homo habilis* could be considered an australopithecine! (Chamberlain and Wood 1987; Groves 1989; Wood 1987, 1992). The geoarchaeological record thus appears to suggest a correlation between major climate changes and significant human evolution but further research is clearly needed.

Out of Africa

A crucial part of the story from drylands is the phasing of human spread out from sub-Saharan Africa into North Africa, the Middle East and beyond. Can the timing of the crossings (there were surely more than one) be related to changes in climate, changes in the type of human present, or to the culture and economy of the hominin itself? Evidence for past environments is extremely patchy and poorly dated in the Early and Mid Quaternary and relies largely on evidence from stone artefacts, faunal remains, preserved sediments, dating of volcanics and palaeomagnetism studies. The earliest acceptable date for a human presence outside Africa comes from Dmanisi, Georgia at 1.8 Ma where a number of individuals have been discovered (Lordkipanidze et al. 2007; Vekua et al. 2002) along with animal fossils, embedded in tuffaceous sands. These individuals could have derived from an earlier *Homo habilis/rudolphensis* ancestor or be part of an emerging new group labelled *Homo ergaster* (which first appears in Africa at c. 1.9 Ma) or less likely *Homo erectus* which appears at 1.7 Ma (Stringer and Andrews 2005). What is clear is that hominins are spreading alongside herds of herbivores and social carnivores at a time when the climate allows them to travel across a grassland habitat. A contemporary movement sees hominins spread to the coastal districts of North Africa (Ternifine and Ain Hanech, Algeria, Sidi Abderrahman quarry and Sale both in Morocco) (Sahnouni et al. 2002; Biberson 1961; Day 1986; Isaac 1982). At Ain Hanech, the artefacts used for processing the carcasses of animals by early hominids have been found along a shallow river embankment (Sahnouni et al. 2002). At Ternifine human remains have been found associated with a fauna that included camels, elephants, hippotamus, equids, antelope and rhinoceros. Carnivores were also present. The associated stone tools comprised both pebble tools and handaxes (Arambourg 1963). There is as yet no evidence from within the Sahara itself for these population movements, perhaps indicating that it was too arid here. Further south in the Sahel, the human fossils from Yayo, Chad might be associated (Coppens 1966) but it is believed that wetter conditions prevailed and corridors of grassland associated with lakes and rivers allowed the spread. It is possible that a coastal route was followed but there is even less evidence for this possibly as a result of sea level changes.

Currently, there is no evidence of occupation in neighbouring drylands for this spread but the site of 'Ubeidiya, Israel, suggests that a route along the rift system was likely (Bar-Yosef 1994; Tchernov 1988). Routes of spread around both the northern and southern edges of the Arabian peninsula are possible (Forster and Matsumura 2005; Beyin 2006). Hominins are present in the Far East by 1.6 Ma (China) and either 1.8 or 1.4 Ma (Java, Indonesia) and 0.8 Ma in Island Southeast Asia at Mata Menge, Flores, Indonesia (Anton and Swisher 2004; Dennell and Roebroeks 2005; Hyodo et al. 2002; Morwood et al. 1998; Semah et al. 2002; Swisher et al. 1994). Where hominin fossils have been found they are associated with grassland and open habitat bovids such as buffalo and gaur (Aziz 2001).

The evidence for human presence during this time is mostly limited to stone artefacts; associated faunal or human bones are very rare. In many dryland areas including the Sahara, as a result of erosion of the surrounding sediments, lithics have been left exposed as deflated surface scatters (Reynolds 2006). In such situations, the archaeological site provides no palaeoenvironmental context and the lithic artefacts act as poor chronological tools. The presence and absence of specific artefact types could be a sampling phenomenon based on visibility, degree of exposure, weathering, and ability of researchers to recognise the material. An additional problem recently has been the selection of well-made artefact types to trade in the antiquities market removing the typological indicators from lithic scatters and leaving spreads of unclassifiable flaking debris (Reynolds 2006). The earliest stone tools are pebble-based pieces and flakes associated with low gradient fluvial, paludal and lacustrine environments, from the Awash Depression in Ethiopia at 2.6 Ma (Quade et al. 2004). Calcic vertisols capping the top of aggradation successions indicate that over the last 3.4 Ma the palaeoclimate has been semi-arid and strongly seasonal (Quade et al. 2004). In addition, isotopic analyses of pedogenic carbonates indicate a gradual vegetation change from dominantly forested to increasing levels of grassland. Subsequent assemblages comprise pebble-based choppers and chopping tools and flakes. These are the first identifiable industrial complex and known as Oldowan after the type site at Olduvai Gorge and dates between 1.8 and 1.2 Ma (Isaac 1984). The environmental setting of the hominid locality was an ephemeral river channel containing *Corbicula* (a freshwater clam that needs shallow flowing water), fish, crocodile and hippopotamus (Blumenschine et al. 2003). Large choppers and chopping tools are found at Ain Hanech in Algeria and 'Ubeidiya, Israel beyond the Saharan region (Sahnouni et al. 2002; Bar-Yosef 1989a). The Oldowan is replaced by the Acheulean techno-complex from c. 1.6 Ma. The Acheulean is characterised by the presence of handaxes or bifaces (e.g. see Roche and Kibunjia 1994) that come in a variety of shapes and sizes and can be made from pebbles or large flakes. Use patterns on them and experimental work suggests a variety of functions for them from animal butchery (Jones 1980) to digging up tubers and hence they provide little palaeoenvironmental information on their own. A number of different sites have yielded handaxes across the Sahara (Clark 2000; Haour and Winton 2003; Reynolds 2006; Tillet 1985). It is a problem for geoarchaeology that these pieces are relatively ubiquitous and preserve well but yield remarkably little chronological patterning although there have been some attempts using associations of different shapes and technologies with coastal terraces in northwestern Africa (Alimen 1978; Biberson 1967; Camps 1974). The Acheulean techno-complex lasts through until c. 100,000 years and spans parts of both the Lower and Middle Palaeolithic and so, again, can rarely be used to provide useful dating or palaeoenvironmental evidence. Detailed analysis of tool-use patterns, distribution (both intra- and inter-site), and exploitation systems of raw materials are needed across the Saharan region before a more subtle and useful understanding of human presence and absence (and any associated palaeoenvironmental evidence for climatic conditions) can be made for this period of time. As it is, the presence of Oldowan and Acheulean stone tools shows human presence at different intervals as environmental circumstances favour the spread of grasslands (Quade et al. 2004). Evidence for hominin ecology during these periods suggests a dependency upon meat and there has been some debate as to whether it was obtained by scavenging or hunting (e.g. Blumenschine 1987; Dominguez-Rodrigo 2002). This has significance for understanding the trans-Saharan routes for human movements for if scavenging was the main means of obtaining meat it requires sufficient water for game and predators to be present to supply scavenging opportunities. However, the presence of water to support a healthy animal population for food on its own is not an indicator of palaeoclimate because the water source may be from palaeo-groundwater sources rather than as a result of increased/higher rainfall. Evidence for the Acheulean suggests big game hunting was a part of hominin behaviour for *Homo erectus*. It remains to be seen if there is more specific variation in tool form and use during the different phases of occupation that can be documented. Within the Acheulean techno-complex a new technology known as Levallois (after a site in northern France) comes in at around 500,000 BP. This technique involved greater control of cores when producing flakes and can be used to produce a desired shape of blank on the core in advance of its removal. Levallois technique spreads within the Acheulean techno-complex through time but is not an effective chronological indicator as some Acheulean assemblages do not use the technique even when it has been well-documented in a region for a long period beforehand. It also spans both the Acheulean

techno-complex and succeeding Mousterian/Middle Stone Age industries. Hominins associated with the Acheulean include *Homo ergaster*, *Homo erectus*, *Homo antecessor* and *Homo heidelbergensis*. The industry is a poor marker of which group is present. One important aspect of the stone tool assemblages of this period that can inform upon is the routes taken through the Sahara by studying the provenance of raw materials used. This can show distances and directions travelled by groups but sourcing materials is complex and currently rare in the region. The importance of suitable rock for lithic production may force humans (and leave evidence of their activities) to move in to areas that would otherwise be regarded as inhospitable. However, the few studies undertaken to date suggest use of local raw materials.

As we move in to the Late Quaternary there is better preservation of sedimentological and geomorphological evidence of palaeoenvironments as well as an increase in the accuracy and reliability of a range of different dating techniques. The next grouping of industries are Middle Palaeolithic or Middle Stone Age assemblages and there are three main forms known within the Saharan region (Wendorf and Schild 1992). As a background to this variability, there are at least five wet/dry cycles during the span of the Middle Palaeolithic documented in the Western Desert of Egypt, the last being associated with Marine Isotope Stages (MIS) 5 and the earliest may be MIS 7 (Wendorf et al. 1993). The first industrial form is a continuation of the Acheulean techno-complex which may or may not include a Levallois element. The Acheulean techno-complex is known from lake-side sites in the Fazzān in Libya (Reynolds 2006; Drake et al. 2006; McLaren et al. 2006) and other sites in the Egyptian Western Desert and probably dates between 500,000 and 100,000 BP, but there are no radiometric dates as yet.

The second industry is the Aterian, a flake-based industry where many of the flake tools are tanged and they seem to indicate the extensive use of hafting technology (Clark 1993). The use of hafts introduces a complexity into the use of space and ethnographic evidence might suggest that maintaining the haft itself is the most complex component in planning the use of stone tools (Hayden 1979). To replace a haft requires materials and the means to manipulate them (including fire to soften mastics). Hafting also suggests use of wooden materials implying a need for stands of woody plants which might have implications for the environment and climate present at the time. There are very few excavated Aterian sites so understanding this industry and the palaeoenvironments at the time are difficult. It is considered by some to be the first industry produced by anatomically modern humans in the region (Barton et al. 2005; Bouzouggar et al. 2002). Large game hunting is a significant component of such human groups and their ecology. A butchery site adjacent to a palaeolake has been well described (Bir Tarfawi, in the Egyptian Western Desert) where gazelle formed the main prey species but there were also occasional kills of antelope, buffalo, giraffe, camel, rhinoceros and equids (Wendorf et al. 1977). In this area, Wendorf et al. (1977) found fossil springs, playas and basins that provided evidence of plentiful water supplies. Bifacial foliate points that may have evolved out from the Acheulean but suggest a function as spear heads or projectile points. This pattern, the fossils and sediments all show that a diverse fauna was available and environmental conditions were significantly different from today. The artefact assemblage also included grinding stones which may have been used for processing plant foods.

The third Middle Palaeolithic industry is the Mousterian (named after the type site Le Moustier, southwestern France). This is a flake-based industry characterised by the presence of large numbers of scrapers, denticulated and notched tools with some points, burins and sometimes bifaces. Occasionally the assemblage may also contain stone balls, an element more frequent in sub-Saharan industries (Willoughby 1985). The main difference between this group of industries and those of the Aterian is the absence of tanged tools and unifacial rather than bifacial points. It is thought that the Aterian is younger than the late Acheulean and Mousterian and it has been dated to between 90,000 and 60,000 BP in the Acacus, Libya (Cremaschi et al. 2000; di Lernia 1999). The Aterian has also been dated to between 80,000 and 40,000 BP in the Jebel Ghari (Jebel Nafusah) in northwestern Libya where it is associated with active springs (Garcea and Giraudi 2006). In the Nile Valley three forms of Middle Palaeolithic have been described within a broad 'Mousterian' heading; an early Middle Palaeolithic which has a generalised flake tool component and some handaxes. The artefacts are often heavily worn and come from surface sites and they are most frequent in the desert east of Wadi Halfa

(Van Peer 1993). A mid-Middle Palaeolithic occurs in local wadi deposits suggesting humid conditions such as those dated by thermoluminescence techniques to 60,000 BP at Wadi Kubbaniya (Schild 1987). These industries have many denticulated tools and varying frequencies of handaxes. They have been termed the Nubian Mousterian (Vermeersch et al. 1990). The late Middle Palaeolithic is associated with drier conditions in Nilotic sediments and sand dunes. There are clear quarry sites for extracting raw materials in this phase (Vermeersch and Paulissen 1993) and there is some transport of materials within an oasis area (Hawkins and Kleindienst 2002). Flake tools predominate and Levallois technique is common, although there remains considerable intra and inter site variability (Van Peer 1993). There remain significant problems with correlating data from the Nile Valley with what are mostly surface assemblages in the western desert and beyond.

In the Fazzān area, there is another industry based upon the production of massive blades on locally-outcropping quartzite which is undated but may belong to the Middle Palaeolithic (Reynolds 2006). Blade-rich industries with and without handaxes have been tentatively dated to marine isotope stage (MIS) 5, in the Mediterranean and the Middle East, when conditions are thought to have been wetter (e.g. Linvat and Kronfeld 1985; Moeyersons et al. 2002; Vaks et al. 2007).

After the Middle Palaeolithic (∼30 ka BP) climate deteriorated and palaeolakes dried up (Street and Grove 1979) and there appears to be a substantial gap in human presence. There is little evidence for a full Upper Palaeolithic or Late Stone Age spanning the period 30,000–10,000 and populations may only have begun re-entering the Saharan region after the Last Glacial Maximum (LGM) at 18,000 BP. According to Wendorf et al. (1976) the Western Desert of Egypt was 'apparently devoid of surface water and any sign of life' during the late Pleistocene (p. 113). However, Barton et al. (2005) provide well dated (luminescence and AMS radiocarbon techniques) palaeoenvironmental evidence along with human occupation layers from Kehf el Hammar Cave in Morocco (Barton et al. 2005; Roche 1976) that indicate that Late Upper Palaeolithic occupation of the area occurred during a period of severe aridification. The nearness of the coast may have played a part in accounting for the presence of humans at this time. There is also archaeological material of this age from Algeria (Saxon et al. 1974) and Libya (Garcea and Giraudi 2006).

The first clear Upper Palaeolithic industry is the Iberomaurusian which is a set of blade and bladelet industries using a hafted, backed blade industry and it continues until the start of the Holocene. The Iberomaurusian is found around the coastal districts of northwestern Africa and is known as the Eastern Oranian at Haua Fteah, Libya (Barton et al. 2005; Garcea and Giraudi 2006; McBurney 1967). Haua Fteah is a large limestone cave that contains evidence of repeated human habitation. Palaeoclimatic evidence comes from studying the preserved sediments, fossils (bovines indicating warmer/drier environments and Barbary sheep reflecting colder climates) and oxygen isotope analyses of marine shells (McBurney 1967). At the site of Kehf el Hamar, Morocco, the Eastern Oranian is associated with a cooler and more arid environment than the present day. At the LGM dustiness increases and active sand dunes move further south (Mainguet et al. 1980). There are major drops in lake levels in Africa at 17–18,000 BP and also at 13–14,000 BP (Street and Grove 1979). In contrast, Maley (2000) has found evidence for high rainfall in the mountains of central Sahara during the LGM, which has been linked to activity of the Subtropical Jet Stream. Through studying lacustrine and fluvial deposits and analysing pollen, Maley (2000) argues for increased levels of precipitation and decreased evaporation rates (due to cooler climates) between 20–15.5 ka and 15–12.5 ka. Little is known of the archaeology of these mountainous regions during the LGM and so it is unclear if they formed refuges for human occupation in an otherwise hostile lowland Saharan environment. There is a significant issue as to when modern humans are present in the area and the nature of their adaptations. The complex, hafted Aterian may be seen as the first evidence of the behavioural sophistication seen in other facets of modern human behaviour but there is, at present, no sign of the increased use of pigments and colouring materials, increased blade production, or the manufacture and use of art objects associated with the 'modern human revolution' although these are present in southern Africa (d'Errico et al. 2005). The Acheulean has been found in association with *Homo erectus*, *Homo antecessor*, *Homo heidelbergensis* and a Mousterian form of it with *Homo neanderthalensis*.

Towards the end of the Pleistocene and into the Holocene, the Sahara became wetter with the development of extensive palaeolakes (Brooks et al. 2003). Modern humans are present towards the end of the Pleistocene when blade and micro-blade-based industries appear associated with well made hearths, perforated animal teeth and other art objects. The late Pleistocene has industries that are generally Iberomaurusian or Early Oranian grade into true microlithic industries at about 10,000 BP following a drier episode (McBurney 1967; Smith 1993). The early Holocene was relatively wet (11–9,000 BP) but with a number of fluctuations. Roberts (1989) argues that large areas of the Sahara Desert were not present during the early Holocene. The humans present at the end of the Pleistocene were still hunting species such as deer, gazelle, hartebeest, giraffe, ostrich and hippotamus. They were also exploiting aquatic resources such as fish, waterfowl, reptiles and gathering shellfish (Sutton 1974, 1977). The productivity of grain species was relatively high and predictable which allowed the development of semi-sedentism and associated rapid population growth (Connah 2005). By 9,800 BP these hunter-gatherers are settled along the available lakeshores, oases and in the highlands (Reynolds 2006). They are also collecting wild grain species. The use of ceramics appears between 9,800 and 9,000 BP with different patterns of decoration suggesting local traditions emerging (Smith 1980; Vernet 1993; Wendorf and Hassan 1990). Mobility would not have been restricted to land, a boat dating to 8,000 BP has been discovered at Dufuna, Chad (Breunig et al. 1996) and the role of the lakes and rivers in human adaptation at this time should not be underestimated. Pastoralism based upon cattle followed shortly afterwards as conditions moved towards cyclical droughts between 9 and 7,000 BP (Banks 1984; Wendorf and Schild 2005). From 8 to 5,000 BP the environment became increasingly drier (Brooks et al. 2003) and reliance upon livestock increased with longer journeys and seasonally-based aggregation sites being used when resources could support longer stays and higher population numbers. Cattle provided a means of coping with decreased resources not simply as a meat supply but through 'secondary products' such as blood and milk. Cattle keeping also increased the division of labour and disperses population across grazing lands and can reduce localised population pressure on resources

(Banks 1984; Ehret 2002). Groups were able to exploit the seasonal growth of grasses around the lakes that existed and travelled between the lakes and other sources of water on the edges of uplands where springs were present. Each lake would grow and shrink (Fig. 27.1) according to climatic fluctuations (Brooks et al. 2003) and human exploitation of each, and its surrounding area, would vary accordingly. Some lakes would resist drier periods more effectively as they had greater catchments, groundwater feeding, etc and so the pattern is not necessarily consistent across the Sahara with time. Aquatic resources continued to be exploited where possible. Ground stone technology also appears commonly, perhaps to allow exploitation of the stands of grasses. Whether pottery and ground stone technology was invented elsewhere and introduced into the region or developed *in situ* at a variety of times and places is unclear although the current

evidence might suggest separate invention in a limited number of places (such as the Nile Valley for the Eastern Sahara, and in the Central Sahara and Sahel area), and introduction from there, although it is likely that mobile groups were in contact across the region as a means of minimising the risks of food supply (Barich 1993). Domesticated sheep and goats first appear in North Africa having come across from Western Asia around 7,400 BP (Smith 2005). Settlement mounds form showing increasingly intense occupation at favoured locations. A cattle cult of burial mounds with cattle burials emerges and these monuments are then later supplanted by human burial mounds. (di Lernia 2004). Such investment in architectural monuments could be an indicator of territoriality but also as 'way marks' for a highly mobile population. Although these cultures are termed Neolithic there is little evidence for domestic plants. All the industries from the LGM onwards could be termed Late Stone Age (including the Iberomaurusian and the Eastern Oranian). These industries begin much earlier in sub-Saharan Africa and along the Mediterranean coast but do not appear to make many inroads into the Sahara.

Rock art is widely spread across the Sahara and includes both engravings and paintings. Rock art spans at least the last 7,000 years but some researchers have argued that it may be as old as pre-Holocene (Barnett 2006; Lutz and Lutz 1995; Mori 1998). There are common themes to the art but with regional variations. Animals no longer found in the region, such as elephant, giraffe, ostrich, hippopotamus and rhinoceros are commonly depicted (Figs. 27.2 and 27.3). There are occasional hunting scenes, and later images include domestic scenes. Rock art forms an invaluable resource often depicting aspects of human activity that would otherwise be missed from the material record.

The archaeological record shows how tightly linked human presence is to the appropriate environmental conditions. It is only when water to support stands of vegetation and livestock is available that people are present in the region. The exception to this is the establishment of settlements that can exploit local springs and use irrigation and wells to supplement water supplies, such as occur in the Wadi al Agial, Fazzān (Mattingly 2003). Here there is evidence for agriculture developing from c. 3,000 BP. Desiccated and charred remains of emmer and bread wheat, barley, dates, grapes and figs (Van Der Veen 2006). There are questions as to

Fig. 27.1 Palaeolake deposits capped with calcrete and now displaying inverted relief, Fazzān, southern Libya

Fig. 27.2 Engraving of a rhinoceros, Fazzān, southern Libya

Fig. 27.3 Engraving of a giraffe, Fazzān, southern Libya

how far the chronological sequence documented above is synchronous across the Sahara and indeed, whether all phases are to be found in all places. For example, a Neolithic presence involving sheep and goat breeding only becomes established between 4,000 and 2,000

BC in the Egyptian eastern desert (Dittman 1993). Cattle-based pastoralism remains controversial – is it part of a package of adaptations introduced from the Near East or does domestication take place *in situ*? Work is still continuing on this subject but a preliminary DNA study suggests African domestic Bos has an equal time depth to those from elsewhere (Troy et al. 2001).

As populations, farming and technological practices all expanded in the Holocene, it becomes increasingly difficult to distinguish between natural environmental changes and those as a result of human activities. It is likely that the Saharan region saw a mixed set of economies with widespread pastoralism, some cattle-based and some sheep/goat-based alongside some settled farmers. Systems of exchange would link the different populations although raids by the mobile groups upon the settled farms would also be possible. A series of trans-Saharan trade routes was probably established during this time taking sub-Saharan products to exchange with Mediterranean ones and return. This trade was an element in the early state formation which appears to develop *in situ* in the Fazzān with the rise of the Garamantes. This development may have coincided with a wetter phase between the fourth and second centuries BC (di Lernia and Merighi 2006) although Mattingly (2006) places the origins of this society earlier (900 BC). The Garamantes were a state society using draught animals, wheeled transport, written language and metallurgy. They were sedentary farmers and lived in towns and villages (Fig. 27.4). Garamantian society lasted from c. 900 BC until AD 500 (Mattingly 2003, 2006). Evidence of a state society living

Fig. 27.4 Mud buildings of the Garamantian (900 BC–AD 500) capital at Jarma, Fazzān, southern Libya

in the arid interior of the Sahara at this time is not evidence for wetter conditions but rather of exploitation of spring and groundwater resources. The Garamantes extracted water from the spring lines of the local sandstone hills using an irrigation system based upon foggara. These comprise a linear series of shafts dug down to the water table and then linked below ground to form a 'watermain'. These foggara take water from the hills out into the wadis to support agriculture (Wilson 2006). Foggara seem to have been developed in Egypt in the first millennium BC and spread across the Sahara along established trade routes. The Garamantian evidence occurs in a central location that can control this trans-Saharan trade (Mattingly 2006) and Wilson suggests the spread of this form of irrigation along these trade routes (Wilson 2006).

The above overview provides a model for human activities in the Sahara, how they relate to climate change and outlines a number of research issues. It then becomes a question as to how far the Saharan pattern can be used as a model more widely for arid zones.

The Middle East

The nearest comparable arid area lies in the Middle East and is adjacent to an extension of the African Rift system. As such it was occupied relatively early and does show marked similarities with the history of hominin use of the Saharan region. Stone artefacts are the most common archaeological finds, often as surface lags (so provide little palaeoenvironmental information), but their existence is used to indicate human presence. The earliest site, 'Ubeidiya (Bar-Yosef 1989a; Bar-Yosef and Belfer-Cohen 2001; Tchernov 1988), is a lakeside location with both Oldowan and Acheulean industries stratified within fluviatile and lacustrine deposits. The Jordan Valley has a number of undated sites but based upon geomorphology and associated volcanic materials are thought to be 'early' Acheulean (Muhesen 1985; Muhesen et al. 1988). A site at Gesher Benot Ya'aqov in Israel has been correlated to the Matuyama-Brunhes chron boundary at 780 ka and contains artefacts in fluvial conglomerates, organic-rich calcareous muds and coquinas that accumulated along the shorelines of the palaeo-Hula lake (Goren-Inbar et al. 2000).

There are likely gaps in human presence reflecting environmental change but by 500,000 years BP Acheulean sites are distributed along most of the major water courses and also appear at springs. Once again, there is a clear association of the location of human activities and water sources (McLaren et al. 2004). In Wadi Faynan in southern Jordan, the evidence for human utilisation in the Pleistocene consists entirely of surface collections of struck stone. This material could span a minimum period of 100,000 years and could quite reasonably be considered to span a far greater period (McLaren et al. 2007). Late Acheulean is found across the Middle East and well characterised in Israel (both in open air and cave sites) and Jordan (Copeland 1988; Gilead 1970; Hublin 2000). The Acheulean found in the desert wadis of Jordan is associated with a fauna including camel, hartebeest, wild boar, rhinoceros, elephant, equids and aurochs. Rech et al. (2007) have identified deflated Acheulian sites in the al–Jafr basin in Jordan; they account for the high density of Lower Palaeolithic material in the area as being a result of perennial springs at the time. There is little in the way of later Palaeolithic tools which Rech et al. (2007) suggest is a result of the drying up of this water source. Surface scatters of bifaces occur in the Arabian Peninsula, Iraq and Iran (Petraglia 2003, 2005). The Aterian does not occur in this region but the Mousterian is widespread from Gaza into the eastern deserts of Jordan, the Arabian Peninsula and Mesopotamia (Bar-Yosef 1998). There is evidence in Israel for both Neanderthals and modern humans using Mousterian technology and their faunal remains show similar lifestyles (Bar-Yosef 1992; Henry 1995). The Israeli evidence appears to show modern humans being replaced by Neanderthals as conditions change towards cooler environments after the last interglacial (MIS 5) with Neanderthals being replaced by modern humans at c. 40–35,000 years ago (Valladas et al. 1987; 1988; Bar-Yosef 1989b).

There has been much debate concerning the influence of climate on migration of modern humans out of Africa. Vaks et al. (2007), studying speleothems in caves in the Negev Desert in southern Israel, have found evidence for major wet episodes between 140 and 110 ka. Vaks et al. (2007) argue that this climate change caused the Sahara to decrease in area which created a window for early modern human dispersal into the Levant region. Evidence for modern humans has been found in the Carmel area of Israel and dates to 130–100 ka (Vanhaeren et al. 2006).

Details of late Quaternary environmental changes can be found through studies of the lacustrine Lisan marls in and around the Dead Sea (McLaren et al. 2004). Lake Lisan existed between 63,000 and 23,000 years ago (Abed and Yaghan 2000). During the LGM between 23,000 and 15,000 years ago, the Lisan marls were replaced by the formation of gypsum, which represents the driest phase when conditions were brackish. There was a return to freshwater conditions between 15,000 and 12,000 years ago with the formation of the Damya Lake (Abed and Yaghan 2000).

In the area of the Rift Valley human occupation seems to continue with an Upper Palaeolithic presence that may have evolved *in situ* out of the last Mousterian industries that are rich in levallois points with basal trimming, Emiran points. The Upper Palaeolithic is a blade-rich complex with a greater diversity of tool forms, larger and more complex sites and more sites overall (Bar-Yosef 1998; Coinman 1998), and modern humans are the sole hominin present. The greater continuity of human presence in this area compared to the Sahara, reflects its greater diversity in a smaller geographical area so that although populations may find environmental conditions challenging, there is always sufficient water, flora and fauna to permit some survivors. The larger area of the Sahara and its geomorphology would allow human populations to become isolated and locally extinct more frequently. Another major contrast to the situation in the Sahara is the human response to the difficult conditions produced at the LGM. Whereas in the Sahara populations disappeared, human groups using a Kebaran stone tool industry intensified their economy upon stands of wild barley and emmer wheat, acorns and legumes and at the same time diversified the range of species that they exploited to include not just gazelle and deer but fox, hare, fish and birds (Watkins 2005). This allowed populations to remain in the region and, as conditions began to ameliorate after the LGM, those populations increased in size and complexity. There was greater investment in certain sites. Some localities such as those with access to reliable water sources such as at Ohalo II situated adjacent to Lake Lisan in Israel (Nadel and Hershkovitz 1991; Nadel and Werker 1999) were occupied year round within a

system that exploited both upland and lowland sites seasonally. It appears likely that a proportion of the population was becoming sedentary (Watkins 2005). These conditions provided the basis for the *in situ* development of farming and the domestication of both plants possible as early as 10,000 BC (cereals) and animals c. 8,800 BC (sheep/goats and cattle) (Rodrigue 2005; Zeder and Hesse 2000). Hunting remained important with gazelle and ibex providing significant meat supplies and upland-lowland diversification of the economy took place. Settlements became larger and were occupied for longer. There was an increasing use of grinding technology to exploit plant foods and storage facilities to keep them. During the Natufian (12,000–9,600 BC) trade links with the Red Sea, the Mediterranean and southern Turkey were established and populations became part of a regional exploitation pattern based on farming and the production of surplus (Watkins 2005). Hunted food remained a significant factor, however. Between 9,600 and 6,900 BC pre-pottery Neolithic societies were established and ceramics were introduced circa 6,900–6,000 BC. Crops were introduced at different times at different places depending upon local factors including the availability of a reliable water supply. Once sites were inhabited for longer periods water supplies were improved through irrigation schemes (from c. 6,000 BC) and full-time farming emerged (Matthews 2005; Watkins 2005). At this time defended settlements appeared and conflict over land was significant (Watkins 2005). The presence of domesticated plants and animals during the early Holocene in Israel, Jordan, Iraq, Iran and Turkey contrasts the Sahara where such innovations appear to be introduced from outside the area and are later. Southwest Asia had the advantage of a diversified topography and range of potential domesticates (e.g. chickpeas, barley, einkorn, emmer wheat, peas, beans and lentils) (Bar-Yosef and Meadows 1995). The cereals need a rainfall of more than 250 mm annually and so cultivation began in these areas first but as domestication proceeded it spreads outside this zone associated with water retention (Fig. 27.5) and irrigation schemes (Barker et al. 2007; Matthews 2005). The intensification continued with the rise of early cities and writing in the 4th millennium BC. A number of city states based on irrigated agriculture, surplus production and long distance trade were established (e.g. see Safar et al. 1981). Again through the Holocene it

Fig. 27.5 Ancient field systems in the Wadi Faynan, southern Jordan

becomes increasingly difficult to distinguish between natural environmental changes and those as a result of human activities. The arid zone became more peripheral to these cultures as widespread localised settlements using both agriculture and pastoralism became centralised into more complex systems. The arid zone then was left to seasonal nomads.

Southern Africa

The evidence for human occupation of the Kalahari and the southwestern African arid zone remains very poor and there are only a few open air sites of Middle and Late Stone Age dates (Deacon and Lancaster 1988; Hiscock and O'Connor 2005). Late Quaternary climatic changes in southern Africa are both spatially and temporally complex (Thomas and Shaw 2002). Southern Africa is, as yet, not well explored archaeologically and no clear patterns that can be compared to those of the Sahara or the Middle East yet exist. However, there are claims for a continuous occupation of neighbouring hills (Robbins et al. 1994) and numerous rock art sites to the east of the sand flats in the Kalahari. Apollo 11 Cave in the Huns Mountains in Namibia, has been used for about 70,000 years and contains ancient paintings that date back to between 29,000 and 25,000 years ago (Phillipson 2005). In rock shelters and caves in the Tsodilo Hills, Botswana, artefact assemblages of both Middle Stone Age and Late Stone Age have been found and the area contains more than 3,500 rock paintings (Robbins et al. 1994).

Occupation from the Late Stone Age is documented with small backed tools present, and in the lake areas harpoon-using fishing cultures are documented (Sadr 1997; Reid 2005). Evidence for lake stands around Tsodilo Hills, indicating regionally wetter conditions, occurred between 40 and 32 ka but they became more seasonal after 36 ka (Thomas et al. 2003). Sand dune activity has been dated to between 36 and 28 ka and this was followed by another wetter phase from 27 to 12 ka (possibly drying out between 22 and 19 ka). Contrary to the Sahara, evidence suggests that conditions in some parts of the Mega-Kalahari were less arid around the LGM probably due to the movement of the ITCZ south of the equator (Nicholson and Flohn 1980). It appears that the region saw seasonal exploitation of wet season stands of vegetation by pastoralists who were themselves a possible extension of farming groups from the east of the area. The Holocene around Tsodilo Hills has seen little geomorphic activity (Thomas et al. 2003). Evidence suggests that over the last two millennia people were herding cattle in the Kalahari (Sadr 2005). The Kalahari area itself could support limited hunting and gathering activities. This limited account might suggest some similarities with the hunter-gatherer-fisher cultures of the Late Stone Age from the Lake Chad district (Holl 2005) but not enough research has been conducted to allow meaningful comparisons.

Australia

In many area of Australia there is a good correlation between human occupation and evidence of wetter phases from both geomorphological and palaeontological evidence. Archaeological evidence demonstrates that humans arrived in Australia between 60,000 and 45,000 years ago (Bowler 1998; Bowler and Magee 2000; Bowler et al. 2003; Gillespie and Roberts 2000; Gillespie et al. 2006; Grün et al. 2000). During this time, the Australian continent was experiencing a climate a little cooler and drier than present according to Flood (1995) Lourandos (1996) and Mulvaney and Kamminga (1999) or similar to present (Miller et al. 2006). Lomax et al. (2003) found evidence for dune building between 65 and 45 ka in the Strzelecki Desert. Evidence for a wide range of small to medium sized game being hunted, including reptiles, marsupials and aquatic organisms (Hiscock and Wallis 2005), has been found. There is little to suggest that populations at this time were able to exploit the more arid central zones and they may, in fact, have been coastally-focused hunter-gatherer-fishers. At the time when humans did start to explore the interior of Australia, conditions were probably wetter and there was no single uniform pattern of colonisation (Hiscock and Wallis 2005). During the Last Glacial Maximum, when conditions are generally regarded to have been drier (Hesse et al. 2004), there are many areas that do not contain any evidence of occupation, indicating local or regional abandonment (Bellwood and Hiscock 2005). At 9,000 BC it became hotter and wetter (Hesse et al. 2004) and populations expanded but the land mass was significantly reduced as Tasmania and New Guinea separated from the continental mainland due to rising sea levels. By c. 8,000 BC populations were present even in the central area (Bellwood and Hiscock 2005). At 2,500–1,000 BC a series of standardised tool forms are employed but with regional variants (Hiscock 1994). These tool forms are often microlithic and assumed to be hafted. This may reflect a greater need for mobility in the changing environmental conditions as at this time rainfall decreased and summer rains became less predictable (Hesse et al. 2004). There were frequent droughts and dune formations were re-initiated. The domestic dog was also introduced at c. 2,000 BC (Corbett 1995) and so new hunting patterns may also have been a factor in the human response to changing climate (Bellwood and Hiscock 2005). It is thought that populations grew c. three times over this period. The changes in adaptation occurred at different times across the continent from a start date of c. 4,000 BC. From about 2,000 years ago rainfall began to increase again and human populations continued to grow and long-distance trade was established. Stone axes, shell pendants and medicinal plants were exchanged over a wide network. The exploitation systems used are tied seasonally to larger bodies of water and water courses with groups breaking up as resources become limited to forage more widely across the landscape. At Lake Eyre in central Australia stone tool accumulations are found around spring mounds as populations responded to drought (Bellwood and Hiscock 2005). The arid zone in Australia remained marginal and was rarely exploited by groups of any significant size.

North America

When humans first arrived in the Americas is a controversial subject, but most researchers argue that there is little convincing evidence for humans being present prior to at least 18,000 years ago (Dillehay 2000). Between 13.7 and 11.4 ka Lake Mojave levels were high (Wells et al. 2003). At the end of the Pleistocene established big-game hunting hunter-gatherers faced increasing aridity and droughts. After 9,000 BC the climate warmed up and wetlands decreased. Coinciding with these climatic changes, populations adopted ground-stone technology probably reflecting a move to exploitation of plant foods, especially nuts and seeds (Browman et al. 2005; Plog 1997). In the Great Basin (parts of Oregon, Nevada, Utah and California) the late Pleistocene saw many pluvial lakes (Fig. 27.6) and run-off fed rivers and lakes (Meek 1999; Tchakerian and Lancaster 2002). By 8.7 ka there was a total drying out of Lake Mojave (Wells et al. 2003). As aridity increased salt flats and small lakes replaced the larger lakes and populations broadened their exploitation patterns to include waterfowl and other birds. They also ate pickleweed seeds (Rhode et al. 2006). As the salt lakes increased in size, population use of the area became more seasonal and exploitation territories increased (Basgall 2000; Bettinger 1999). In the Southwest, population levels were low and groups moved through a series of small scale, short term, settlements. Humans exploited deer, pronghorn antelope, mountain sheep and the occasional bison (Cordell 1997). As

Fig. 27.6 Palaeolake sediments of former late Pleistocene Lake Manix, Mojave Desert, California

conditions became more arid, rabbits and rodents become part of the diet (Arnold et al. 2004). At about 2,000 BC maize was introduced into New Mexico and Arizona from the south to supplement the existing plant foods of pinyon nut, walnuts, agaves, yuccas, cactus fruits and grasses (Browman et al. 2005). The grinding stone technology to process maize was, therefore, already established. The first sites to use maize are in areas where natural stands can be supported and incorporated into seasonal movements of the population (Bayman et al. 1996; Hildebrand and Hagstrum 1995). Through time settlements began to include storage and roasting pits, stone vessels, figurines and traded coastal products (such as seashell beads) and they increased in size (Bayman et al. 1996). At 1,400 BC irrigation systems using canals and walls were introduced (Hurt 1996; Mabry 1999). It is debated how far this economy developed indigenously or was imported as a whole from the south (Browman et al. 2005). Is it a sign of population movements or simply the spread of technology?

Additional evidence of southern links includes the use of squash/pumpkin by 1,400 BC and common beans in New Mexico by the start of the Christian era (Tagg 1996). By AD 100 ceramics are in use, water diversion systems, soil moisture retention and ground temperature control techniques are all developed. New forms of maize are introduced and domestic turkeys kept (Hurt 1996). By AD 700 villages of c. 1–2,000 people exist and irrigation is extensive. At AD 1150 a more centralised system with larger settlements including public architecture is found and more peripheral settlements decline. Long distance trade with Mexico and the Pacific coast is expanded. Wild foods remain an important part of the economy, however. The system begins to collapse at c. AD 1450 as result of possible local climate change or the over-use of fields resulting in salinisation.

In the northern part of the American Southwest a more diversified topography produces vegetation zones based on altitude and human exploitation patterns reflect this for much of the Holocene. Populations could switch between foraging and farming as conditions suited as the shifting system left large areas fallow. The use of pottery and an increased intensification of farming appears at AD 200–400 (Wills 2001). Storage facilities increase, grinding equipment is more common and larger in scale and buildings are bigger (e.g. see Dean et al. 1994). Settlements are still dispersed. AD

900 sees an increased population, concentrated settlements and 'great houses'. How this increased population was supported in an arid zone is debated but recent evidence for increased water flow through the area with high silt loads refreshing soils has been discovered and a lake may have been present west of the main complex at Chaco Canyon (Force et al. 2002). There are a number of defensive sites constructed to protect and control water in the period AD 1150–1300 and food production is increased (e.g. see Varien et al. 2000). The proportion of wild foods exploited remains important. There is evidence for increased violence between groups at this stage (Kuckelman et al. 2002) and severe drought between AD 1276 and 1299 sees incursions by other hunter-gatherer groups, continued irrigation using terraced fields, and diversion dams and more reliance on wild foods (Van West 1996).

South America

Despite the Atacama Desert being hyperarid, populations have been able to exist due to food and water resources both along the coastline and concentrated in deep fluvially-incised ravines. During wetter phases, evidence for human occupation has been found around palaeolakes, in palaeowetlands as well as in caves (Núñez et al. 2002). Evidence for some of the earliest humans in South America is found predominantly in coastal locations (Santoro et al. 2005). The early-Mid Holocene Palaeoindian Paijan archaeological complex found on the coastal districts of Peru and Chile was centred upon coastal resources but exploited some terrestrial resources in what is now the driest desert in the world (Betancourt et al. 2000). Populations become established along the coast towards the end of the Pleistocene and exploit shellfish, fish and some terrestrial game by 10,200 BC (Núñez 1983; Lynch 1999) and such sites are widespread by 9,500–7,000 BC. Technology used to fish includes both hook and line fishing and the use of nets (Keefer et al. 1998; Sandweiss et al. 1989, 1998). There are also links with populations in the highlands with obsidian being exchanged presumably for coastal products (Llagostera 1992; Moseley 1975; Sandweiss et al. 1998). Grosjean et al. (1997), working in the Atacama basin in northern Chile, have found evidence of thirty debris flows containing twenty intercalated archaeological campsites. Radiocarbon dating of these hearths reveals evidence of short-term extreme flood events during the hyper-arid mid Holocene period. Many lakes in the surrounding area (e.g. Lake Titicaca and Laguna Miscanti) were at extremely low levels and others had dried out (e.g. the Tauca palaeolakes) at this time (Wirrmann and Mourguiart 1995; Markgraf 1993; Grosjean and Nûñez 1994; Valero et al. 1996). The wealth of the marine resources allows groups to develop complex settlements with public areas and architecture. Burials of mummies is developed 6,000–1,700 BC (Arriaza and Standen 2002). Stable isotope analysis of human bones suggests that at this time as much as 90% of the diet was based on marine resources such as shellfish, sea lions, beached whales and fish (Moseley and Heckenberger 2005). From c. 5,000 BC the use of plant foods increases, settlements become larger and more permanent. These plant foods include manioc, quinoa and later, sweet potato, potatoes, and squash (Browman et al. 2005; Moseley 2001). The use of terrestrial game also increases. Rising population numbers are increasing their reliance upon land resources and systems to trap water where valleys open out begin to be used. At 1,800 BC such irrigation systems are becoming widespread and parts of the desert are farmed. These developments are more common in the north on the Peruvian coast where more large rivers flow into the sea than in the south where such rivers are further apart and rarer (Moseley and Heckenberger 2005). The Moche and Nazca cultures develop between 200 BC and AD 650 and again use irrigation systems, complex relationships with highland populations and marine resources in an interlinked economy that provides support at difficult times of resource failure or drought (Aveni 1991; Bawden 1996). These cultures are developed to state level but die out in the face of increasing aridity, competition from highland states (Fig. 27.7) and the failure of their irrigation systems. Small coastal populations continue to exploit the marine resources.

Summary

Studies of palaeoenvironmental reconstruction are intrinsic to archaeological research (Branch et al. 2005). Within the field of dryland geoarchaeology there

Fig. 27.7 View across Tiwanaku, on the former lake shore of Lake Titicaca, Bolivia, occupied between AD 500 and 950

is still a lack of detailed paired environmental and archaeological information in many locations (Veth et al. 2000); many sites lack good geochronological frameworks and much more research is needed.

The earliest members of the human lineage originated in Africa in an environment that was gradually becoming cooler and drier over time. Evidence exists in the form of sediments, human and animal bones, stone artefacts, footprints and other signs of activity. Establishing the ages for these palaeoenvironmental and archaeological events comes largely from lithostratigraphy, biostratigraphy, potassium-argon dating, argon-argon dating, palaeomagnetism and correlation of tuffs. It has been argued that environmental change played an important role in the evolution, movement and extinction of different species of hominins. It is known that early humans migrated out of sub-Saharan Africa into North Africa, the Middle East and even further afield.

Archaeological evidence from Africa, the Middle East and Australia indicates that anatomically modern humans occupied deserts about the same time (60,000–40,000 years ago). Following on, there were numerous wet and dry phases in the late Pleistocene which affected human migration and occupation patterns. However, there was always a close association between the location of human activities and water sources. During the LGM there were relatively low and concentrated levels of human activity in the Sahara, Middle East and Australia due to dry conditions. In some parts of southern Africa, as a result of the intrusion of the ITCZ south of the equator, climate was a little wetter, elsewhere remained arid. In North America, the southward shift of the jet and storm track resulted in cooler and wetter conditions around the LGM. It is uncertain if humans were present at this time. The early Holocene in Africa, the Middle East and Australia was a much wetter phase and there is abundant evidence of human activity, rapid expansions in populations, farming and technology. In North America, there is evidence of aeolian activity and lowering of lake levels at the start of the Holocene. Population levels in desert regions were low and many groups migratory. In South American drylands people were concentrated along coastlines in the early Holocene and were less affected by climate changes. In the late Holocene, the landscapes of many desert areas around the world became increasingly affected by aridification and land degradation caused by significant human activities.

Thus during wetter phases in drylands, increased exploration and exploitation occurred due to greater availability and predictability of resources (Hiscock and Wallis 2005). Human *adaptation* to drylands is limited to occasional exploitation strategies within the Holocene. That does not mean that the geoarchaeology of drylands is unimportant – the evidence from them informs upon major issues in human evolution, human exploitation of natural resources, the origins of agriculture, the early stages in state formation and palaeoenvironmental changes in desert landscapes.

References

Abed, A.M. and R. Yaghan. 2000. On the palaeoclimate of Jordan during the last glacial maximum. *Palaeogeography Palaeoclimatology Palaeoecology* **160**, 23–33.

Alimen, H. 1978. *L'evolution de l'Acheuleen au Sahara Nord-Occidental (Saoura, Ougarta, Tabellala).* Paris, Meudon.

Anton, S.C. and C.C. Swisher. 2004. Early Dispersals of *Homo* from Africa. *Annual Review of Anthropology* **33**, 271–296.

Arambourg, C. 1963. Le gisement de Ternifine. *Archives de l'Institut Paleontologie Humaine* **31**, 1–190.

Arnold, J.E., M.R. Walsh, and S.E. Hollimon. 2004. The archaeology of California. *Journal of Archaeological Research* **12**, 1–73.

Arriaza, B. and V. Standen. 2002. *Death, Mummies and Ancestral Rites: The Chinchorro Culture,* Arica, Serie Patrimonio Cultural I Region de Tarapacá, Universidad de Tarapacá.

Aveni, A. (ed.) 1991. *The Lines of Nazca.* Philadelphia: American Philosophical Society.

Aziz, F. 2001. New insight on the Pleistocene fauna of Sangiran and other hominid sites in Java, In *Sangiran: Man, Culture and Environment in Pleistocene Time*s, Simanjuntak, T., B. Prasetyo, and R. Handini (eds.), 260–271. Jakarta: Yayosan Obor Indonesia.

Banks, K.M. 1984. *Climates, Cultures and Cattle. The Holocene Archaeology of the Eastern Sahara.* Dallas: Department of Anthropology, Southern Methodist University.

Barich, B.E. 1993. Culture and environment between the Sahara and the Nile in the Early and Mid-Holocene. In *Environmental Change and Human Culture in the Nile Basin and Northern Africa until the Second Millennium BC,* Krzyzaniak, L., M. Kobusiewicz, and J. Alexander (eds.), 171–183. Poznan: Archaeological Museum.

Barker, G. and D.D. Gilbertson. 2000. Living at the margin: themes in the archaeology of drylands. In *The Archaeology of Drylands: Living at the margin*, Barker, G., D.D. Gilbertson (eds.), 3–18. London: Routledge, One World Archaeology 39.

Barker, G., D. Gilbertson, and D. Mattingly (eds.) 2007. *Archaeology and Desertification. The Degradation and Well-Being of the Wadi Faynan Landscape in Southern Jordan.* Oxbow: Council for British Research in Levant.

Barnett, T. 2006. Libyan Rock Art as a Cultural Heritage Resource. In *The Libyan Desert. Natural Resources and Cultural Heritage*, Mattingly, D., S. McLaren, E. Savage, Y. al-Fasatwi, and K. Gadgood (eds.), 95–110. London: Society for Libyan Studies Monograph No. 6.

Barton, R.N.E., A. Bouzouggar, S. Collcutt, R. Gale, T.F.G. Higham, L.T. Humphrey, S. Parfitt, E. Rhodes, C.B. Stringer, and F. Malek. 2005. The Late Upper Palaeolithic Occupation of the Moroccan Northwest Maghreb During the Last Glacial Maximum. *African Archaeological Review* **22**, 77–100.

Bar-Yosef, O. 1989a. The excavations at 'Ubeidiya in retrospect: an eclectic view. In *Investigations in South Levantine Prehistory*, Bar-Yosef, O. and B. Vandermeersch (eds.), 101–112. Oxford: British Archaeological Reports.

Bar-Yosef, O. 1989b. Geochronology of the Levantine Middle Palaeolithic. In The *Human Revolution: Behavioural Biological Perspectives on the Origins of Modern Humans*, Mellars, P. and C. Stringer (eds.), 589–610. Edinburgh: Edinburgh University Press.

Bar-Yosef, O. 1992. Middle Palaeolithic human adaptations in the Mediterranean Levant. In *The Evolution and Dispersal of Modern Humans in Asia*, Akazawa, T., K. Aoki, and T. Kimura (eds.), 198–215. Tokyo: Hokusen-sha.

Bar-Yosef, O. 1994. The lower paleolithic of the Near East. *Journal of World Prehistory* **8**, 211–265.

Bar-Yosef, O. 1998. The Natufian Culture in the Levant, Threshold to the Origins of Agriculture. *Evolutionary Anthropology*, **6**, 159–177.

Bar-Yosef, O. and A. Belfer-Cohen. 2001. From Africa to Eurasia – early dispersals. *Quaternary International* **75**, 19–28.

Bar-Yosef, O. and R. Meadows. 1995. The origins of agriculture in the Near East. In *Last Hunters First Farmers: New Perspectives on the Transition to Agriculture*, Price, D. and A.-B. Gebauer (eds.), 39–94. Santa Fe: School of American Research.

Basgall, M.E. 2000. The structure of archaeological landscapes in the north-central Mojave Desert. In *Archaeological Passages: A Volume in Honor of Claude Nelson Warren*, Scheider, J.S., R.M.II. Yohe, and J.K. Gardner (eds.), 123–138. Hermet: Western Center for Archaeology and Palaeontology Publications in Archaeology No. 1.

Bawden, G. 1996. *The Moche.* Oxford: Blackwell.

Bayman, J.M., R.H. Hevley, B. Johnson, K.J. Reinhard, and R. Ryan. 1996. Analytical perspectives on a protohistoric cache of ceramic jars from the lower Colorado Desert. *Journal of California and Great Basin Anthropology* **18**, 131–154.

Behrensmeyer, A.K. 2006. Climate change and human evolution. *Science* **311**, 476–478.

Behrensmeyer, A.K., N.E. Todd, R. Potts, and G.B. McBinn. 1997. Late Pliocene faunal turnover in the Turkana Basin, Kenya and Ethiopia. *Science* **278**, 1589–1594.

Bellwood, P. and P. Hiscock. 2005. Australia and the Austronesians. In *The Human Past. World Prehistory and the Development of Human Societies*, Scarre, C. (ed.), 264–305. London: Thames and Hudson.

Betancourt, J., C. Latorre, J. Rech, J. Quade, and K. Rylander. 2000. A 22,000 year record of monsoonal precipitation from northern Chile's Atacama Desert. *Science* **289**, 1546–1550.

Bettinger, R.L. 1999. From traveller to processor: regional trajectories of hunter-gatherer sedentism in the Inyo-Mono region, California. In *Settlement Pattern Studies in the Americas, Fifty Years since Viru*. Billman, B.R. and G.M. Feinman (eds.), 39–55. Washington: Smithsonian Institution Press.

Beyin, A. 2006. The Bab al Mandab versus the Nile-Levant: an appraisal of the two dispersal routes for early modern humans out of Africa. *African Archaeological Review* **23**, 5–30.

Biberson, P. 1961. *Le Paleolithique inferieur du Maroc Atlantique*. Publications du Service des Antiquities du Maroc 17.

Biberson, P. 1967. Some aspects of the Lower Palaeolithic of Northwest Africa. In *Background to Evolution in Africa*, Bishop, W.W. and J.D. Clark (eds.), 447–475. Chicago: University of Chicago Press.

Blumenschine, R.J. 1987. Characteristics of an early hominid scavenging niche. *Current Anthropology* **28**, 383–407.

Blumenschine, R.J., C.R. Peters, F.T. Masao, R.J. Clarke, A.L. Deino, R.L. Hay, C.C. Swisher, I.G. Stanistreet, G.M. Ashley, L.J. McHenry, N.E. Sikes, N.J. van der Merwe, J.C. Tactikos, A.E. Cushing, D.M. Deocampo, J.K. Njau, and J.I. Ebert. 2003. Late Pliocene *Homo* and Hominid land use from western Olduvai Gorge, Tanzania. *Science* **299**, 1217–1221.

Bolten, A., O. Bubenzer, and F. Darius. 2006. A digital elevation model as a base for the reconstruction of Holocene Land-use potential in arid regions. *Geoarchaeology* **21**, 751–762.

Bouzouggar, A., J.K. Kozlowski, and M. Otte. 2002. Etude des ensembles lithiques ateriens de la grotte d'El Aliya a Tanger (Maroc). *L'Anthropologie* **106**, 207–248.

Bowler, J.M. 1998. Willandra Lakes revisited: environmental framework for human occupation. *Archaeology in Oceania* **33**, 120–155.

Bowler, J.M. and J.W. Magee. 2000. Redating Australia's oldest human remains: a sceptic's view. *Journal of Human Evolution* **38**, 719–726.

Bowler, J.M., H. Johnston, J.M. Olley, J.R. Prescott, R.G. Roberts, W. Shawcross, and N.A. Spooner. 2003. New ages for human occupation and climatic change at Lake Mungo, Australia. *Nature* **421**, 837–840.

Branch, N., M. Canti, P. Clark, and C. Turney (2005) *Environmental Archaeology: Theoretical and Practical Approaches*. London: Arnold.

Breunig, P., K. Neumann, and W. Van Neer. 1996. New Research on the Holocene Settlement and Environment of the Chad Basin in Nigeria. *The African Archaeological Review* **13**, 111–145.

Brooks, N., N. Drake, S. McLaren, and K. White. 2003. Studies in Geography, Geomorphology, Environment and Climate. In *The archaeology of the Fazzān. Volume 1: – Synthesis*, Mattingly, D.J. (ed.), 37–75. Tripoli: Socialist People's Libyan Arab Jamahariya, The Department of Antiquities, Tripoli and The Society of Libyan Studies.

Browman, D.L., G.J. Fritz, and P.J. Watson. 2005. Origins of Food-producing Economies in the Americas. In *The Human Past. World Prehistory and the Development of Human Societies*, Scarre, C. (ed.), 306–349. London: Thames and Hudson.

Brunet, M., A. Beauvilain, Y. Coppens, E. Heintz, A.H.E. Moutaye, and D. Pilbeam. 1995. The first australopithecine 2,500 kilometres west of the Rift Valley (Chad). *Nature* **378**, 273–275.

Brunet, M., F. Guy, D. Pilbeam, H.T. Mackaye, A. Likius, D. Ahiunta, A. Beauvilain, C. Blondel, H. Bocherons, J.-R. Boisserie, L. de Bonis, Y. Coppens, J. Dejax, C. Denys, P. Duringer, V. Eisenmann, G. Fanone, P. Fronty, D. Geraads, T. Lehmann, F. Lihoreau, A. Louchart, A. Mahamats, G. Merceron, G. Mouchelon, O. Otero, P. Pelaez Campomanes, M. Ponce de Leon, J.-C. Rage, M. Sapanet, M. Schuster, J. Sudre, P. Tassy, X. Valentin, P. Vignaud, L. Viriot, A. Zazzo, and C. Zollikofer. 2002. A new hominid from the Upper Miocene of Chad, Central Africa. *Nature* **418**, 145–151.

Butzer, K.W. 1971. *Environment and Archaeology*, Aldine Chicago.

Butzer, K.W. 1982. *Archaeology as Human Ecology*. Cambridge: Cambridge University Press.

Camps, G. 1974. *Les Civilisations Prehistoriques de l'Afrique du Nord et du Sahara*. Paris: Doin.

Cerling, T.E. 1992. Development of grasslands and savannas in East Africa during the Neogene. *Palaeogeography Palaeoclimatology Palaeoecology* **97**, 241–247.

Chamberlain, A.T. and B.A. Wood. 1987. Early hominid phylogeny. *Journal of Human Evolution* **16**, 119–133.

Clark, J.D. 1993. The Aterian of the Central Sahara. In *Environmental Change and Human Culture in the Nile Basin and Northern Africa Until the Second Millennium B.C.*, Krzyzaniak, L., M. Kobusiewicz, and J. Alexander (eds.), 49–67. Poznan: Poznan Archaeological Museum.

Clark, J.D. 2000. Human Populations and cultural adaptations in the Sahara and Nile during prehistoric times. In *Recent Research Into the Stone Age of Northeastern Africa*, Krzyzaniak, L., K. Kroeper, and M. Kobusiewicz (eds.), 527–582. Poznan: Poznan Archaeological Museum.

Coinman, N.R. 1998. The Upper Palaeolithic of Jordan. In *The Prehistoric Archaeology of Jordan*, Henry, D.O. (ed.), 39–63. Oxford: British Archaeological Reports International series **705**.

Connah, G. 2005. Holocene Africa. In *The Human Past. World Prehistory and the Development of Human Societies*, Scarre, C. (ed.), 350–391. London: Thames and Hudson.

Copeland, L. 1988. Environment, Chronology and Lower – Middle Paleolithic Occupations of the Azraq Basin, Jordan. *Paleorient* **14.2**, 66–75.

Coppens, Y. 1966. An Early Hominid from Chad. *Current Anthropology* **7**(5), 584–585.

Corbett, L. 1995. *The Dingo in Australia and Asia*. Natural History Series, Wales, UK: University of South Wales Press.

Cordell, L. 1997. *Archaeology of the Southwest*. San Diego: Academic Press.

Cremaschi, M, S. di Lernia, and E.A.A. Garcea. 2000. First chronological indications on the Aterian in the Libyan Sahara. In *Recent Research Into the Stone Age of Northeastern Africa*, Krzyzaniak, L., K. Kroeper, and M. Kobusiewicz (eds.), 229–237. Poznan: Poznan Archaeological Museum.

Day, M.H. 1986. *Guide to fossil man* (4th edition). London: Cassell.

Deacon, J. and N. Lancaster. 1988. *Late Quaternary Palaeoenvironments of Southern Africa*. Oxford: Clarendon Press.

Dean, J.S. 1993. Geoarchaeological perspectives on the past: chronological considerations. In *Effects of Scale on Archaeological and Geoscientific Perspectives*, Stein, J.K. and A.R. Linse (eds.), 59–66. Boulder: Geological Society of America Special Paper **283**.

Dean, J.S., W.H. Doelle, and J.D. Orcutt. 1994. Adaptive stress: environment and demography. In *Themes in Southwest Prehistory*, Gumerman, G. (ed.), 53–86. Santa Fe: School of American Research Press.

Dennell, R. and W. Roebroeks. 2005. An Asian perspective on early human dispersal from Africa. *Nature* **438**, 1099–1104.

deMenocal, P.B. 1995. Plio-Pleistocene African climate. *Science* **270**, 53–59.

d'Errico, F., C. Henshilwood, M. Vanhaeren, and K. van Niekerk. 2005. *Nassarius kraussianus* shell beads from Blombos Cave. *Journal of Human Evolution* **48**, 3–24.

di Lernia, S. (ed.) 1999. *The Uan Afuda Cave*. Hunter-Gatherer Societies of Central Sahara. Firenze: Arid Zone Archaeology Monograph1.

di Lernia, S. 2004. Aridity, Cattle, and Rites. Social responses to rapid environmental changes in the Saharan Pastoral societies, 6500–5000 yr BP. IGCP 490 and ICSU Environmental catastrophes in Mauritania, the desert and the coast. Field Conference Atar, Mauritania. Accessed via http://atlasconferences.com/c/a/m/u/24.htm on 25th October 2006.

di Lernia, S. and F. Merighi. 2006. Transitions in the Later Prehistory of the Libyan Sahara, as seen from the Acacus Mountains. In *The Libyan Desert. Natural Resources and Cultural Heritage*, Mattingly, D., S. McLaren, E. Savage, Y. al-Fasatwi, and K. Gadgood (eds.), 111–121. London: Society for Libyan Studies Monograph No. 6.

Dillehay, T.D. 2000. *The Settlement of the Americas: A New Prehistory*. New York: Basic Books.

Dincauze, D. 2000. *Environmental Archaeology: Principles and Practice*. New York: Cambridge University Press.

Dittman, A. 1993. Environmental and climatic change in the northern part of the Eastern Desert during Middle Palaeolithic and Neolithic times. In *Environmental Change and Human Culture in the Nile Basin and Northern Africa Until the Second Millennium B.C.*, Krzyzaniak, L., M. Kobusiewicz, and J. Alexander (eds.), 145–152. Poznan: Poznan Archaeological Museum.

Dominguez-Rodrigo, M. 2002. Hunting and scavenging in early humans: the state of the debate. *Journal of World Prehistory* **16**, 1–56.

Drake, N., K. White, and S.J. McLaren. 2006. Quaternary Climate Change in the Jarma Region of Fazzan, Libya. In *Natural Resources and Cultural Heritage of Libya (Socialist People's Libyan Arab Jamahariya)*, Mattingly, D., S. McLaren, E. Savage, Y. al-Fasatwi, and K. Gadgood (eds.), 133–144. London: Society of Libyan Studies Monograph No. 6. Lanes Ltd.

Ehret, C. 2002. *The Civilizations of Africa: A History to 1800*. Oxford: James Currey.

Flood, J. 1995. *Archaeology of the Dreamtime* (3rd edition). Sydney: Angus and Robertson.

Force, E.R., R.G. Vivian, T.C. Windes, and J.S. Dean. 2002. *Relation of 'Bonito' Paleo-Channels and Base-Level Variations to Anasazi Occupation, Chaco Canyon, New Mexico*. Arizona State Museum Archaeological Series 194. Tucson: University of Arizona.

Forster, P. and S. Matsumura. 2005. Did early humans go north or south? *Science* **308**, 965–966.

Gamble, C. 1993. *Timewalkers: The Prehistory of Global Colonization*. Pheonix Mill: Alan Sutton Publishing.

Garcea, E.A.A. and C. Giraudi. 2006. Late Quaternary human settlement patterning in the Jebel Gharbi. *Journal of Human Evolution* **51**, 411–421.

Gillespie, R., B.W. Brook, and A. Baynes. 2006. Short overlap of humans and megafauna in Pleistocene Australia. *Alcheringa* **1**, 163–186.

Gifford, J.A. and G. Jr. Rapp. 1985. The early development of archaeological geology in North America. In *Geologists and Ideas: a History of North American Geology*, Drake, E.T. and W.M. Jordan (eds.), 409–421. Boulder: Geological Society of America Centennial Special Volume **1**.

Gillespie, R. and R.G. Roberts. 2000. On the reliability of age estimates for human remains at Lake Mungo. *Journal of Human Evolution* **38**, 727–732.

Gladfelter, B.G. 1981. Developments and directions in geoarchaeology. *Advances in Archaeological Methods and Theory* **4**, 344–364.

Gilead, I. 1970. Handaxe Industries in Israel and the Near East. *World Archaeology* **2.1**, 1–11.

Goren-Inbar, N., C.S. Feibel, K.L. Verosub, Y. Melamed, M.E. Kislev, E. Tchernov, and I. Saragusti. 2000. Pleistocene milestones on the Out-Of –Africa corridor at Gesher Benot Ya'aqov, Israel. *Science* **289**, 944–947.

Grosjean, M. and L. Nũnez. 1994. Late glacial, early and middle Holocene environments, human occupation, and resource use in the Atacama (Northern Chile). *Geoarchaeology* **9**, 271–286.

Grosjean, M., L. Nũnez, I. Cartajena, and B. Messerli. 1997. Mid-Holocene climate and culture change in the Atacama Desert, Northern Chile. *Quaternary Research* **48**, 239–246.

Groves, C.P. 1989. *A theory of human and primate evolution*. Oxford: Oxford University Press.

Grün, R., N.A. Spooner, A. Thorne, G. Mortimer, J.J. Simpson, M.T. McCulloch, L. Taylor, and D. Curnoe. 2000. Age of the Lake Mungo 3 skeleton – reply to Bowler and Magee and to Gillespie and Roberts. *Journal of Human Evolution* **38**, 733–742.

Haour, A. and V. Winton. 2003. A Palaeolithic Cleaver from the Sahel: Freak or Fact? *Antiquity* Online Vol. **77** (No.297): *Antiquity*, %20Project%20Gallery%20Haour%20 %26%20Winton.htm accessed on 25th October 2006.

Hawkins, A.L. and M.R. Kleindienst. 2002. Lithic raw material usages during the Middle Stone Age at Dakhleh Oasis, Egypt. *Geoarchaeology* **17**, 601–624.

Hayden, B. 1979. *Palaeolithic Reflections. Lithic technology and ethnographic excavations among Australian Aborigines*. Canberra: Australian Institute of Aboriginal Studies.

Henry, D.O. (ed.) 1995. *Prehistoric Cultural Ecology and Evolution*. New York: Plenum Press.

Hesse, P.P., J.W. Magee, and S. van der Kaars. 2004. Late Quaternary climates of the Australian arid zone: a review. *Quaternary International* **118–19**, 87–102.

Hildebrand, J.A. and M.B. Hagstrum. 1995. Observing subsistence change in native southern California: the late Prehistoric Kumeyaay. *Research in Economic Anthropology* **16**, 85–127.

Hiscock, P. 1994. Technological responses to risk in Holocene Australia. *Journal of World Prehistory* **8**, 267–292.

Hiscock, P. and S. O'Connor. 2005. Arid paradises or dangerous landscapes: a review of explanations for Palaeolithic assemblage change in arid Australia and Africa. In *Desert Peoples: archaeological perspectives*, Veth, P., M. Smith, and P. Hiscock (eds.), 58–78. Oxford: Blackwell.

Hiscock, P. and L.A. Wallis. 2005. Pleistocene settlement of deserts from an Australian perspective. In *Desert Peoples: archaeological perspectives*, Veth, P., M. Smith, and P. Hiscock (eds.), 34–57. Oxford: Blackwell.

Holl, A.F.C. 2005. Holocene 'Aquatic' Adaptations in North Tropical Africa. In *African Archaeology*, Stahl, A.B. (ed.), 174–186. Oxford: Blackwell.

Hublin, J.J. 2000. Modern/Non Modern Human Interactions: A Mediterranean Perspective. In *The Geography of Neanderthals and Modern Humans in Europe and the Greater Mediterranean*, Bar-Yosef, O. and D. Pilbeam (eds.), 157–182. Harvard: Peabody Museum Bulletin 8.

Huckleberry, G. 2000. Interdisciplinary and Specialized Geoarchaeology: a Post-Cold War Perspective. *Geoarchaeology* **15**, 523–536.

Hunt, C.O., H.A. Elrishi, D.D. Gilbertson, J. Grattan, S. McLaren, F.B. Pyatt, G. Rushworth, and G. Barker. 2004. Early Holocene environments in the Wadi Faynan, Jordan. *The Holocene* **14**, 921–930.

Hurt, R.D. 1996. *Indian Agriculture in America: Prehistory to the Present*. Lawrence, KS: Kansas Press.

Hurtak, J.J. 1986. Subsurface morphology and geoarchaeology revealed by spaceborne and airborne radar. http://www.afs.org/html/geoarchaeology.html accessed 27/10/06.

Hyodo, M., H. Nakaya, A. Urabe, H. Saegusa, S. Shunrong, Y. Jiyun, and J. Xuepin. 2002. Paleomagnetic dates of hominid remains from Yuanmou, China and other Asian sites. *Journal of Human Evolution* **43**, 27–41.

Isaac, G. 1982. The earliest traces. In *The Cambridge History of Africa. Volume 1: From the earliest times to c.500B.C.*, Clark, J.D. (ed.), 157–247. Cambridge: Cambridge University Press.

Isaac, G. 1984. The archaeology of human origins: studies of the Lower Pleistocene in East Africa 1971–1981. *Advances in World Archaeology* **3**, 1–87.

Johnson, D.L., R.D. Mandel, J.D. Leach, and M. Petraglia. 2002. Introduction. *Geoarchaeology* **17**, 3–6.

Jones, P.R. 1980. Experimental butchery with modern stone tools and its relevance for Palaeolithic archaeology. *World Archaeology* **12**, 153–175.

Keefer, D.K., S.D. DeFrance, M.E. Moseley, III J.B. Richardson, D.R. Satterlee, and A. Day-Lewis. 1998. Early maritime economy and El Niño events at Quebrada Tacahuay, Peru. *Science* **281**, 1833–1835.

Kuckelman, K.A., R.R. Lightfoot, and D. Martin. 2002. The bioarchaeology and taphonomy of violence at Castle Rock and Sand Canyon pueblos, Southwestern Colorado. *American Antiquity* **67**, 486–513.

Leakey, M.D. and J.M. Harris (eds.) 1987. *Laetoli: a Pliocene site in northern Tanzania*. Oxford: Clarendon.

Linse, A.R. 1993. Geoarchaeological scale and archaeological interpretation: examples from the central Jornada Mogollon. In *Effects of Scale on Archaeological and Geoscientific Perspectives*, Stein, J.K. and A.R. Linse (eds.), 11–28. Boulder: Geological Society of America Special Paper **283**.

Linvat, A. and J. Kronfeld. 1985. Palaeoclimatic implications of U-series dates for lake sediments and travertines in the Arava Rift Valley, Israel. *Quaternary Research* **24**, 164–172.

Little, M.G., R.R. Schneider, D. Kroon, B. Price, T. Bickert, and G.Wefer. 1997. Rapid palaeooceanographic changes in the Benguela upwelling system for the last 160,000 years as indicated by abundances of planktonic foraminifera. *Palaeogeography, Palaeoclimatology, Palaeoecology* **130**, 135–161.

Llagostera, A. 1992. Fishermen on the Pacific Coast of South America. *Andean Past* **3**, 87–109.

Lomax, J., A. Hilgers, H. Wopfner, R. Grün, C.R. Twidale, and U. Radtke. 2003. The onset of dune formation in the Strzelecki Desert, South Australia. *Quaternary Science Reviews* **22**, 1067–1076.

Lordkipanidze, D., T. Jashashvili, A. Vekua, M.S. Ponce de León, C.P.E. Zollikofer, G.P. Rightmire, H. Pontzer, R. Ferring, O. Oms, M. Tappen, M. Bukhsianidze, J. Agusti, R. Kahlke, G. Kiladze, B. Martinez-Navarro, A. Mouskhelishvili, M. Nioradze, and L. Rook. 2007. Postcranial evidence from early Homo from Dmanisi, Georgia. *Nature* **449**, 305–310.

Lourandos, H. 1996. *Continent of Hunter-gatherers: New Perspectives in Australian Prehistory*. Cambridge: Cambridge University Press.

Lutz, R. and G. Lutz. 1995. *The Secrets of the Desert: the Rock art of Messak Sattafet and Messak Mellet, Libya*. Innsbruck.

Lynch, T.F. 1999. The earliest South American lifeways. In *The Cambridge History of the Native Peoples of the Americas. Volume 3, part 1: South America*, Salomon, F. and S.B. Schwartz (eds.), 188–263. Cambridge: Cambridge University Press.

Mabry, J.B. 1999. Las Capas and early irrigation farming. *Archaeology Southwest* **11**, 6.

Mainguet, M., L. Canon, and M.C. Chemin. 1980. Le Sahara: géomorphologie et paléogeomorphologie éoliennes. In *The Sahara and the Nile*, Williams, M.A.J. and H. Faure (eds.), 17–35. Rotterdam: Balkema.

Maley, J. 2000. Last Glacial Maximum lacustrine and fluviatile formations in the Tibesti and other Saharan mountains, and large-scale climatic teleconnections linked to the activity of the Subtropical Jet Stream. *Global and Planetary Change* **26**, 121–136.

Mandel, R.D. 1999. Introduction. *Geoarchaeology* **14**, 727–728.

Markgraf, V. 1993. Climatic history of Central and South America since 18,000 yr B.P.: comparison of pollen records and model simulations. In *Global Climates Since the Last Glacial Maximum*, Wright Jr., H.E., J.E. Kutzbach, T. Webb III, W.F. Ruddiman, F.A. Street-Perrott, and P.J. Bartlein. (eds.), 357–385, Minneapolis: University of Minnesota Press.

Marquardt, W.H. 1992. Dialectical archaeology. *Archaeological Method and Theory* **4**, 101–140.

Matthews, R. 2005. The rise of civilization in southwest Asia. In *The Human Past. World Prehistory and the Development of*

Human Societies, Scarre, C. (ed.), 432–471. London: Thames and Hudson.

Mattingly, D. (ed.) 2003. *The Archaeology of the Fazzān. Volume 1, Synthesis*. London and Tripoli: Society of Libyan Studies and Department of Antiquities, Libya.

Mattingly, D. 2006. The Garamantes: the First Libyan State. In *The Libyan Desert. Natural Resources and Cultural Heritage*, Mattingly, D., S. McLaren, E. Savage, Y. al-Fasatwi, and K. Gadgood (eds.), 189–204. London: Society for Libyan Studies Monograph No. 6.

McBurney, C. 1967. The Haua Fteah (Cyrenaica) and the stone age of the south-east Mediterranean. Cambridge: Cambridge University Press.

McLaren, S.J., D.D. Gilbertson, J.P. Grattan, C.O. Hunt, G.A.T. Duller, and G. Barker. 2004. Quaternary palaeogeomorphologic evolution of the Wadi Faynan area, southern Jordan. *Palaeogeography Palaeoclimatology Palaeoecology* **205**, 131–154.

McLaren, S., N. Drake, and K. White. 2006. Late Quaternary environmental changes in the Fazzān, southern Libya: evidence from sediments and duricrusts. In *Natural Resources and Cultural Heritage of Libya (Socialist People's Libyan Arab Jamahariya)*, Mattingly, D., S. McLaren, E. Savage, Y. al-Fasatwi, K. Gadgood (eds.), 157–166. London: Society of Libyan Studies Monograph No. 6., Lanes Ltd.

McLaren, S., T. Reynolds, D.D. Gilbertson, J. Grattan, C. Hunt, G. Barker, and G. Duller. 2007. Pleistocene environments and human settlement. In *Archaeology and Desertification. The Degradation and Well-Being of the Wadi Faynan Landscape in Southern Jordan*, Barker, G., D. Gilbertson, and D. Mattingly (eds.), Oxbow: Council for British Research in Levant, 177–198.

Meek, N. 1999. New discoveries about the late Wisconsinan history of the Mojave river system. In *Tracks along the Mojave. A field guide from Cajon Pass to the Calico Mountains and Coyote Lake*, Reynolds, R.E. and J. Reynolds (eds.), 113–117. San Bernardino: San Bernardino County Museum.

Miller, G.H., J.W. Magee, M.L. Fogel, and M.K. Gagan. 2006. Detecting human impacts on floors, fauna, and summer monsoon of Pleistocene Australia. *Climate of the Past Discussions* **2**, 535–562.

Moeyersons, J., P.M. Vermeersch, and P. Van Peer. 2002. Dry cave deposits and their palaeoenvironmental significance during the last 115 ka, Sodeim Cave, Red Sea Mountains, Egypt. *Quaternary Science Reviews* **21**, 837–851.

Mori, F. 1998. *The Great Civilisations of the Ancient Sahara. Neolithisation and the Earliest Evidence of Anthropomorphic Religion*. Rome: L'Erma di Bretschneider.

Morwood, M., P.B. O'Sullivan, F. Aziz, and A. Raza. 1998. Fission-track ages of stone tools and fossils on the east Indonesian island of Flores. *Nature* **392**, 173–176.

Moseley, M.E. 1975. *The Maritime Foundations of Andean Civilisation*. Menlo Park: Cummings Publications.

Moseley, M.E. 2001. *The Incas and Their Ancestors: The Archaeology of Peru*. London and New York: Thames Hudson.

Moseley, M.E. and M.J. Heckenberger. 2005. From Village to Empire in South America. In *The Human Past. World Prehistory and the Development of Human Societies*, Scarre, C. (ed.), 640–677. London: Thames and Hudson.

Muhesen, S. 1985. *L'Acheuleen recent evolve de Syrie*. Oxford: British Archaeological Reports International Series **248**.

Muhesen, S., T. Akazawa, and A. Abdul-Salam. 1988. Prospections prehistoriques dans le region d'Afrin (Syrie). *Paleorient* **14**(2), 145–153.

Mulvaney, J. and J. Kamminga. 1999. *Prehistory of Australia*. Washington DC: Smithsonian Institution Press.

Nadel, D. and I. Hershkovitz. 1991. New Subsistence data and human remains from the earliest Levantine Epipalaeolithic. *Current Anthropology* **32**, 631–635.

Nadel, D. and E. Werker. 1999. The oldest ever brush hut plant remains from Ohalo II, Jordan Valley, Israel (19,000 BP). *Antiquity* **73**, 755–764.

Nicholson, S.E. and H. Flohn. 1980. African environmental and climatic changes and the general atmospheric circulation in Late Pleistocene and Holocene. *Climatic Change* **2**, 313–348.

Ning, S., L.M. Dupont, H.-J. Beug, and R. Schnieder. 2000. Correlation between vegetation in southwestern Africa and oceanic upwelling in the past 21,000 years. *Quaternary Research* **54**, 72–80.

Núñez, L. 1983. Paleoindian and Archaic Cultural Period in the arid and semi-arid region of northern Chile. *Advances in World Archaeology* **2**, 161–222.

Núñez, L, M. Grosjean, and I. Catajena. 2002. Human occupations and climate change in the Puna de Atacama, Chile. *Science* **298**, 821–824.

O'Connell, J.F., K. Hawkes, and N.G. Blurton Jones. 1999. Grandmothering and the evolution of *Homo erectus*. *Journal of Human Evolution* **36**, 461–485.

Petraglia, M.D. 2003. The Lower Palaeolithic of the Arabian Peninsula: occupations, adaptations, and dispersals. *Journal of World Prehistory* **17**, 141–179.

Petraglia, M.D. 2005. Hominin responses to Pleistocene environmental change in Arabia and South Asia. In *Early-Middle Pleistocene Transitions: The Land-Ocean Evidence*, Head, M.J. and P.L. Gibbard 305–319. London: Geological Society, Special Publications **247**.

Phillipson, D.W. 2005. *African Archaeology*. Cambridge: Cambridge University Press.

Plog, S. 1997. *Ancient Peoples of the American Southwest*. London: Thames and Hudson.

Quade, J., N. Levin, S. Semaw, D. Stout, P. Renne, M. Rogers, and S. Simpson. 2004. Paleoenvironments of the earliest stone toolmakers, Gona, Ethiopia. *Bulletin of the Geological Society of America* **116**, 1529–1544.

Rapp, G. Jr. and C.L. Hill. 1998. *Geoarchaeology: The Earth-Science Approach to Archaeological Interpretation*. New Haven: Yale University Press.

Ravesloot, J.C. and M.R. Waters. 2004. Geoarchaeology and archaeological site patterning on the middle Gila River, Arizona. *Journal of Field Archaeology* **29**, 203–214.

Rech, J.A., L.A. Quintero, P.J. Wilke, and E.R. Winer. 2007. The lower Palaeolithic landscape of 'Ayoun Qedim, al-Jafr Basin, Jordan. *Geoarchaeology* **22**, 261–275.

Reed, K.E. 1997. Early hominid evolution and ecological changes through the African Plio-Pleistocene. *Journal of Human Evolution* **32**, 289–322.

Reid, A. 2005. Interaction, Marginalization and the Archaeology of the Kalahari. In *African Archaeology*, Brower Stahl, A. (ed.), 353–377. Oxford: Blackwell.

Renfrew, C. 1976. Archaeology and the earth sciences. In *Geoarchaeology*, Davidson D.A. and M.L. Shackley (eds.), 1–5. London: Duckworth.

Renfrew, C. 1983. Geography, Archaeology and Environment. 1. Archaeology. The *Geographical Journal* **149**, 316–322.

Reynolds, T. 2006. The importance of Saharan lithic assemblages. In *The Libyan Desert. Natural Resources and Cultural Heritage*, Mattingly, D., S. McLaren, E. Savage, Y. al-Fasatwi, and K. Gadgood (eds.), 81–90. London: Society for Libyan Studies Monograph No. 6.

Rhode, D., D.B. Madsen, and K.T. Jones. 2006. Antiquity of early Holocene small-seed consumption and processing at Danger Cave. *Antiquity* **80**, 328–339.

Robbins, L.H., M.L. Murphy, K.M. Stewart, A.C. Campbell, and G.A. Brook. 1994. Barbed Bone Points, Palaeoenvironment, and the Antiquity of Fish Exploitation in the Kalahari Desert, Botswana. *Journal of Field Archaeology* **21**, 257–264.

Roberts, N. 1989. *The Holocene: An Environmental History*. Oxford: Blackwell.

Roche, J. 1976. *Cadre chronologique de l'Epipaleolithique marocain. Actes du IXe Congres de l'UISPP: Chronologie et synchronisme dans la prehistoire circum-mediterraneenne*, 153–167.

Roche, J. and M. Kibunjia. 1994. Les sites archéologiques Plio-Pléistocènes de la formation de Nachukui, West Turkana, Kenya. *Comptes Rendus de l'Académie des Sciences, Paris* **318**(Série II), 1145–1151.

Rodrigue, C.M. 2005. James Blaut's critique of diffusionism through a Neolithic lens: early animal domestication in the Near East. *Antipode* **37**, 981–989.

Sadr, K. 1997. Kalahari Archaeology and the 'Bushman Debate'. *Current Anthropology* **38**, 104–112.

Sadr, K. 2005. Hunter-gatherers and herders of the Kalahari during the late Holocene. In *Desert Peoples: archaeological perspectives*, Veth, P., M. Smith, and P. Hiscock (eds.), 206–221. Oxford: Blackwell.

Safar, F., M.A. Mustafa, and S. Lloyd. 1981. *Eridu*. Baghdad: State Org of Antiquities and Heritage.

Sahnouni, M., D. Hadjouis, J. van der Made, Ael-K. Derradji, A. Canals, M. Medig, H. Belahrech, Z. Harichane, and M. Rabhi. 2002. Further research at the Oldowan site of Ain Hanech, North-eastern Algeria. *Journal of Human Evolution* **43**(6), 925–937.

Sandweiss, D., J. III, Richardson, E. Reitz, J. Hsu, and R. Feldman. 1989. Early maritime adaptations in the Andes: preliminary studies at the Ring Site, Peru. In *Ecology, Settlement and History in the Osmore Drainage, Peru*, Rice, D.C. Stanish, and P. Scarr (eds.), 35–84. Oxford: British Archaeology International Series 545.

Sandweiss, D., H. McInnis, R. Burger, A. Cano, B. Ojeda, R. Paredes, M. Sandweiss, and M. Glascock. 1998. Quebrada Jaguay: Early South American maritime adaptations. *Science* **281**, 1830–1832.

Santoro, C.M., B.T. Arriaza, V.G. Standen, and P.A. Marquet. 2005. People of the coastal Atacama Desert living between sand dunes and waves of the Pacific Ocean. In *Desert Peoples: archaeological perspectives*, Veth, P., M. Smith, and P. Hiscock (eds.), 58–78. Oxford: Blackwell.

Saxon, E.C., A.E. Close, C. Cluzel, V. Morse, and N.J. Shackleton. 1974. Results of recent excavations at Tamar Hat. *Libyca* **XXII**, 49–91.

Schild, R. 1987. Unchanging contrast? The late Pleistocene Nile and eastern Sahara. In *Prehistory of arid North Africa Essays in honour of Fred Wendorf*, Close, A.E. (ed.), 13–28. Dallas: Southern Methodist University.

Semah, F., A.-M. Semah, and T. Simanjuntak. 2002. More than a Million Years of Human Occupation in Insular Southeast Asia. The Early Archaeology of Eastern and Central Java. In *Under the Canopy: The Archaeology of Tropical Rainforests*, Mercader, J. (ed.), 161–190. New Brunswick: Rutgers University Press.

Sepulchre, P., G. Ramstein, F. Fluteau, M. Schuster, J.J. Tiercelin, and M. Brunet. 2006. Tectonic uplift and Eastern African aridification. *Science* **313**, 1419–1423.

Sherratt, A. 1997. Climatic cycles and behavioural revolutions: the emergence of modern humans and the beginning of farming. *Antiquity* **71**, 271–287.

Smith, A.B. 1980. The Neolithic Tradition in the Sahara. In *The Sahara and the Nile. Quaternary environments and prehistoric occupation in northern Africa*, Williams, M.A. and H. Faure (eds.), 451–465. Rotterdam: Balkema.

Smith, A.B. 1993. Terminal Palaeolithic industries of Sahara: a discussion of new data. In *Environmental Change and Human Culture in the Nile Basin and Northern Africa Until the Second Millennium B.C.*, Krzyzaniak, L., M. Kobusiewicz, and J. Alexander (eds.), 69–75. Poznan: Poznan Archaeological Museum.

Smith, A.B. 2005. Desert solitude: the evolution of ideologies among pastoralists and hunter-gatherers in arid North Africa. In *Desert Peoples: archaeological perspectives*, Veth, P., M. Smith, and P. Hiscock (eds.), 261–275. Oxford: Blackwell.

Smith, M., P. Veth, P. Hiscock, and L.A. Wallis. 2005. Global deserts in perspective. In *Desert Peoples: archaeological perspectives*, Veth, P., M. Smith, and P. Hiscock (eds.), 1–14. Oxford: Blackwell.

Stein, J.K. 1993. Scale in Archaeology, Geosciences and Geoarchaeology. In *Effects of Scale on Archaeological and Geoscientific Perspectives*, Stein, J.K. and A.R. Linse (eds.), 1–10. Boulder: Geological Society of America Special Paper **283**.

Street, A.F. and A.T. Grove. 1979. Global maps of lake-level fluctuations since 30,000 yr B.P. *Quaternary Research* **12**, 83–118.

Stringer, C.B. and P. Andrews. 2005. *The Complete World of Human Evolution*. London and New York: Thames and Hudson.

Sutton, J.E.G. 1974. The aquatic civilisation of middle Africa. *Journal of African History* **15**, 527–546.

Sutton, J.E.G. 1977. The African aqualithic. *Antiquity* **51**, 25–34.

Swisher, C.C., G.H. Curtis, T. Jacob, A.G. Getty, A. Suprijo, and A. Widiasmoro. 1994. Age of the earliest known hominids in Java, Indonesia. *Science* **263**, 1119–1121.

Tagg, M.D. 1996. Early cultigens from Fresnal Shelter, Southeast New Mexico. *American Antiquity* **61**, 311–324.

Taylor, R.E. 2003. Commentary: the "two cultures" in American anthropological archaeology. *The Review of Archaeology* **24**, 1–11.

Tchakerian, V.P. and N. Lancaster. 2002. Late Quaternary arid/humid cycles in the Mojave Desert and western Great

Basin of North America. *Quaternary Science Reviews* **21**, 799–810.

Tchernov, E. 1988. The age of 'Ubeidiya formation (Jordan Valley, Israel) and the earliest Hominids in the Levant. *Paleorient* **14.2**, 63–65.

Tillet, T. 1985. The palaeolithic and its environment in the northern part of the Chad basin. *The African Archaeological Review* **3**, 163–177.

Thornes, J.B. 1983. Geography, archaeology and recursive ignorance. The *Geographical Journal* **149**, 326–333.

Thomas, D.S.G. and P.A. Shaw. 2002. Late Quaternary environmental change in central southern Africa: new data, synthesis, issues and prospects. *Quaternary Science Reviews* **21**, 783–797.

Thomas, D.S.G., G. Brook, P. Shaw, M. Bateman, K. Haberyan, C. Appleton, D. Nash, S.J. McLaren, and F. Davies. 2003. Late Pleistocene wetting and drying in the NW Kalahari: an integrated study from the Tsodilo Hills, Botswana. *Quaternary International* **104**, 53–67.

Trauth, M.H., M.A. Maslin, A. Deino, and M.R. Strecker. 2005. Late Cenozoic moisture history of East Africa. *Science* **309**, 2051–2053.

Troy, C.S., D.E. MacHugh, J.F. Bailey, D.A. Magee, R.T. Loftus, P. Cunningham, A.T. Chamberlain, B.C. Sykes, and D.G. Bradley. 2001. Genetic evidence for Near Eastern origins of European cattle. *Nature* **410**, 1088–1091.

Vaks, A., M. Bar-Matthews, A. Ayalon, A. Matthews, L. Halicz, and A. Frumkin. 2007. Desert speleothems reveal climatic window for African exodus of early modern humans. *Geology* **35**, 831–834.

Valero-Garces, B., M. Grossjean, A. Scwalb, M.A. Geyh, B. Messerli, and K. Kelts. 1996. Late Holocene environmental change in the Atacama Altiplano: limnogeology of Laguna Miscanti, Chile. *Journal of Palaeolimnology* **16**, 1–121.

Valladas, J., L. Joron, G. Valladas, B. Arensburg, O. Bar-Yosef, A. Belfer-Cohen, P. Goldberg, H. Laville, L. Meignen, Y. Rak, E. Tchernov, A.M. Tillier, and B. Vandermeersch. 1987. Thermoluminescence dates for the Neanderthal burial site at Kebara in Israel. *Nature* **330**, 159–160.

Valladas, H., J.L. Reyss, J.L. Joron, G. Valladas, O. Bar-Yosef, and B. Vandermeersch. 1988. Thermoluminescence dating of Mousterian Troto-Cro-Magnon' remains from Israel and the origin of modern man. *Nature* **331**, 614–616.

Vanhaeren, M.F. d'Errico, C. Stringer, S.L. James, J.A. Todd, and H.K. Mienis. 2006. Middle Palaeolithic shell beads in Israel and Algeria. *Science* **23**, 1785–1788.

Van Peer, P. 1993. Levallois variability and the Middle Palaeolithic of the lower Nile Valley and the eastern Sahara. In *Late Prehistory of the Nile Basin and the Sahara. Studies in African Archaeology*, Krzyzaniak, L. and M. Kobusiewicz (eds.), 129–143. Poznan: Poznan Archaeological Museum.

Van West, C. 1996. Agricultural potential and carrying capacity in southwestern Colorado A.D. 901 to 13000. In *The Prehistoric Pueblo World A.D. 1150–1350*, Adler, M. (ed.), 214–227. Tucson: University of Arizona Press.

Varien, M.D., C.R. Van West, and G.S. Patterson. 2000. Competition, cooperation, and conflict: agricultural production and community catchments in the central Mesa Verde region. *Kiva* **66**, 45–66.

Veen M. van der. 2006. Food and Farming in the Libyan Sahara. In *The Libyan Desert. Natural Resources and Cultural Heritage*, Mattingly, D., S. McLaren, E. Savage, Y. al-Fasatwi, and K. Gadgood (eds.), 171–178. London: Society for Libyan Studies Monograph No. 6.

Vekua, A., D. Lordkipanidze, G.P. Rightmire, J. Agusti, R. Ferring, G. Maisuradze, A. Moustkhelishivili, M. Nioradze, P. de Leon, M. Tappen, M. Tvalchrelidze, and L.P.E. Zollikofer. 2002. A new skull of early *Homo* from Dmanisi, Georgia. *Science* **297**, 85–89.

Vermeersch, P.M. and E. Paulissen. 1993. Palaeolithic chert quarrying and mining in Egypt. In *Environmental Change and Human Culture in the Nile Basin and Northern Africa Until the Second Millennium BC*. Krzyzaniak, L., M. Kobusiewicz, and J. Alexander (eds.), 337–349. Poznan: Poznan Archaeological Museum.

Vermeersch, P.M., E. Paulissen, and P. van Peer. 1990. Le Paleolithique dans la Vallee du Nil egyptien. *L'Anthropologie* **94**, 435–458.

Vernet, R. 1993. Le Neolithique recent dans le sud-ouest du Sahara. In *Environmental Change and Human Culture in the Nile Basin and Northern Africa Until the Second Millennium B.C.* Krzyzaniak, L., M. Kobusiewicz, and J. Alexander (eds.), 91–101. Poznan: Poznan Archaeological Museum.

Veth, P.M., S. O'Connor, and L.A. Wallis. 2000. Perspectives on ecological approaches in Australian archaeology. *Australian Archaeology* **50**, 54–66.

Vignaud, P., P. Duringer, H.T. Mackaye, A. Likius, C. Louis de Bonis, V. Eisenmann, M.-E. Etienne, D. Geraads, F. Guy, N. Lopes-Martinez, C. Mourer-Chauviré, O.Otero, J.-C. Rage, M. Zazzo, and M. Brunet. 2002. Geology and palaeontology of the Upper Miocene Toros-Menalla hominid locality, Chad. *Nature* **418**, 152–155.

Waters, M.R. and J.C. Ravesloot. 2001. Landscape change and the cultural evolution of the Hohokam along the middle Gila River and other river valleys in south-central Arizona. *American Antiquity* **66**, 285–299.

Waters, M.R. and J.C. Ravesloot. 2003. Disaster or catastrophe: human adaptation to high- and low-frequency landscape processes – a reply to Ensor, Ensor and Devries. *American Antiquity* **68**, 400–405.

Watkins, T. 2005. From Foragers to Complex Societies in Southwest Asia. In *The Human Past. World Prehistory and the Development of Human Societies*, Scarre, C. (ed.), 200–233. London: Thames and Hudson.

Wells, S.G., W.J. Brown, Y. Enzel, R.Y. Anderson, and L.D. McFadden. 2003. Late Quaternary Geology and paleohydrology of pluvial Lake Mojave, south California. In Paleohydrology of the Mojave and southern Great Basin Deserts. Boulder, Colorado: Geological Society of America Special Paper 368, 79–114.

Wendorf, F. and F.A. Hassan. 1990. Holocene ecology and prehistory in the Egyptian Sahara. In *The Sahara and the Nile. Quaternary Environments and Prehistoric Occupation in Northern Africa*, Williams, M.A. and H. Faure (eds.), 407–419. Rotterdam: Balkema.

Wendorf, F. and R. Schild. 1992. The Middle Palaeolithic of North Africa. A status report. In *New Light on the Northeast African Past*, Klees, F. and R. Kuper (eds.), 39–78. Current Prehistoric Research, Cologne: Heinrich Barth Institut.

Wendorf, F. and R. Schild. 2005. Are the early Holocene cattle in the eastern Sahara domestic or wild? *Evolutionary Anthropology: Issues, News and Reviews* **3**, 118–128.

Wendorf, F., R. Schild, and A.E. Close. 1993. Middle Palaeolithic occupations at Bir Tarfawi and Bir Sahara East, Western Desert of Egypt. In *Environmental Change and Human Culture in the Nile Basin and Northern Africa Until the Second Millennium B.C.*, Krzyzaniak, L., M. Kobusiewicz, and J. Alexander (eds.), 103–111. Poznan: Poznan Archaeological Museum.

Wendorf, F., R. Schild, R. Said, C.V. Haynes, A. Gautier, and P. Kobusiewicz. 1976. The Prehistory of the Egyptian Sahara. *Science* **193**, 103–116.

Wendorf, F., A.E. Close, R. Schild, P. Kobusiewicz, R. Wieckowska, R. Said, C.V. Haynes, A. Gautier, and A. Hadidi. 1977. Late Pleistocene and recent climatic changes in the Egyptian Sahara. *The Geographical Journal* **143**, 211–234.

Wilkinson, K.N., A.R. Beck, and G. Philip. 2006. Satellite imagery as a resource in the prospection for archaeological sites in Syria. *Geoarchaeology* **21**, 735–750.

Willoughby, P.R. 1985. Spheroids and battered stones in the African Early Stone Age. *World Archaeology* **17**, 44–60.

Wills, W.H. 2001. Pithouse architecture and the economics of household formation in the prehistoric American Southwest. *Human Ecology* **29**, 477–499.

Wilson, A. 2006. The Spread of Foggara-based Irrigation in the Ancient Sahara. In *The Libyan Desert. Natural Resources and Cultural Heritage*, Mattingly, D., S. McLaren, E. Savage, Y. al-Fasatwi, and K. Gadgood (eds.), 205–216. London: Society for Libyan Studies Monograph No. 6.

Wirrmann, D. and Ph. Mourguiart. 1995. Late Quaternary spatio-temporal limnological variations in the Altiplano of Bolivia and Peru. *Quaternary Research* **43**, 344–354.

Wood, B.A. 1987. Who is the 'real' *Homo habilis*? *Nature* **327**, 187–188.

Wood, B.A. 1992. Origin and evolution of the genus *Homo. Nature* **355**, 783–790.

Yerkes, R.W. 2006. Book review – Branch, N., M. Canti, P. Clark, and C. Turney (2005) Environmental Archaeology: Theoretical and Practical Approaches. *Geoarchaeology* **21**, 878–880.

Zeder, M.A. and B. Hesse. 2000. The initial domestication of goats (*Capra hicus*) in the Zagros Mountains 10,000 years ago. *Science* **287**, 2254–2257.

Chapter 28

Cenozoic Climates in Deserts

M.A.J. Williams

Introduction

Deserts are superb repositories of geological, geomorphic and archaeological evidence. The very aridity to which they owe their existence has enabled them to preserve a remarkably good record of past depositional and erosional events. The fossil river valleys of the Sahara, the great salt lakes of Australia, China, and Patagonia, the dissected volcanic mountains of the Arabian peninsula and the Afar Desert – all are legacies of former tectonic, volcanic, and climatic episodes which ultimately gave rise to the deserts we see today. Each desert reflects its own individual geological inheritance and geomorphic history; each is unique in its assemblage of landforms; each ideally deserves detailed and separate study in its own right (Pesce 1968, McKee 1978, Rognon 1989).

Two contrasting themes permeate the study of desert landscapes: longevity and change. In many parts of the desert world, morphogenesis is virtually inactive, and relief is presently being conserved. In other more limited areas, recurrent dust storms and ever shifting dunes convey an impression of a dynamic and changing landscape. This latter impression is reinforced by the presence of now vegetated and stable dunefields located well beyond the present-day confines of active desert dunes, as well as by the occurrence of relict river and lake basins deep within the desert – silent witness to previously wetter climates. The paradox here is that deserts can be simultaneously both young and old, conserving as well as destroying relief,

morphologically stable as well as geomorphically dynamic. In the deserts of Western Australia, for instance, late Quaternary dunefields adjoin valleys that have changed very little since the final separation of Australia from Antarctica ~50 Ma ago. The frequent juxtaposition of very old elements of the landscape with others that are very young reflects the polygenetic nature of desert landscapes (Mabbutt 1977, Frostick and Reid 1987) .

The aim of this chapter is to examine the role of past tectonic, volcanic, and climatic events in shaping the landscapes of our major deserts, focusing upon the Cenozoic legacy in particular, while noting that many aspects of desert geomorphology cannot be fully understood without an appreciation of much older tectonic and sedimentary events, many of them extending well back into Phanerozoic times, and some even into the Precambrian. We begin with a summary of the causes of present-day aridity, followed by a discussion of the problems and assumptions involved in reconstructing Cenozoic climatic changes in deserts. We then consider the geomorphic history of the North African and Australian deserts, and conclude with an evaluation of the impact of Cenozoic climatic fluctuations upon the evolution of desert landscapes worldwide.

Causes of Aridity

We will now examine some of the causes of present-day aridity and some of the factors which control the distribution of our existing deserts. Deserts are regions of rare and unreliable rainfall. They are not restricted to any particular latitude, but are especially extensive

M.A.J. Williams (✉)
Geographical and Environmental Studies, University of Adelaide, Adelaide, SA 5005, Australia
e-mail: martin.williams@adelaide.edu.au

A.J. Parsons, A.D. Abrahams (eds.), *Geomorphology of Desert Environments*, 2nd ed.,
DOI 10.1007/978-1-4020-5719-9_28, © Springer Science+Business Media B.V. 2009

astride the two tropics, in latitudes characterized by more or less permanent high pressure cells and hot dry subsiding air. The two polar deserts are under the influence of semi-permanent anticyclones and of cold dry subsiding air. As air subsides and becomes compressed, it also becomes warmer, so that its relative humidity is decreased even though the absolute amount of water vapour held in desert air may be substantial and may become evident in the cold hours before dawn in the form of the evanescent desert dew. The anticyclonic hot deserts result from the global Hadley circulation, and have very little to do with the regional distribution of land and sea. Oceans in strictly tropical and high polar latitudes also receive minimal precipitation and are simply the oceanic extensions of the terrestrial deserts.

A second and very common cause of aridity on land is distance inland. Except for the equatorial zone, precipitation usually decreases rapidly away from the coast. The great temperate deserts of the continental interior of Asia are a prime example of such deserts, and the effects of continentality reinforce those of latitude in the hot tropical deserts of North Africa, Arabia, and Australia.

Two additional factors accentuate the aridity resulting from latitude and increasing distance from the nearest source of moist maritime air. One is the rain-shadow effect, and the other is the presence offshore of cold upwelling water or a cold ocean current. When warm moist maritime air reaches the land it frequently encounters mountain ranges, as in the case of the Andes, the Rockies, the Himalayas, the Ethiopian uplands, and the Eastern Highlands of Australia. The moist air rises and is cooled adiabatically, rapidly attaining vapour saturation, and shedding its precipitated water as rain or snow – hence such local names as Sierra Nevada and Snowy Mountains. If the coastal ranges are reasonably elevated relative to the interior, as is true of the Americas and Australia, there will be a pronounced rain-shadow effect inland of the ranges, with desiccation of the previously moist air accentuated as it flows downhill, becoming warmer and drier in the process. In extreme cases, the air may shed its moisture on mountains 2000–4000 m high, before descending to valleys lying close to or even below sea level, as in the Afar Depression and the Dead Sea Rift.

Deserts such as the Atacama and the Namib are flanked offshore by cold upwelling water. In fact, the western borders of all the Trade Wind or tropical deserts are washed by the cool waters of ocean currents generated by the gyres which flow clockwise in the Northern Hemisphere and anticlockwise in the Southern Hemisphere. The result is predictable: cool moist air from the ocean reaches land which is warmer than the adjacent cool ocean, so that the relative humidity of the air is reduced, and its capacity to absorb moisture rather than to shed it is increased. Such air masses therefore usually function as a desiccator rather than as a welcome source of moisture, leaving the plants and animals dependent on coastal fog for their survival. Erkowit, situated high in the Red Sea Hills of the eastern Sudan, is a good example of a mist oasis, receiving much of its precipitation from fog blowing off the Red Sea in winter while all around it is sweltering lowland desert.

Aridity should not be confused with desertification. Within the timescale of the last two million years the deserts have occupied essentially their present locations on the globe (Fig. 28.1). They are where they are for sound and relatively immutable geographical reasons. A combination of at least five major factors accounts for the distribution of modern deserts and for their low and erratic precipitation (Williams 2002a). These are the prevalence of dry subsiding air over the deserts (itself linked to latitude and to global atmospheric circulation), a vast land area, low inland relief and high coastal ranges, and the presence of cool ocean water close offshore. An additional factor is the presence aloft of a subtropical jetstream, the existence and course of which are partly controlled by the presence or absence of extensive areas of high elevation, such as the Tibetan Plateau immediately north of the Himalayas, which exerts a strong influence on the easterly jetstream which flows from Tibet across the Arabian peninsula towards Somalia (Rognon and Williams 1977, Flohn 1980), accentuating the aridity in those regions.

The previous discussion now requires some qualification, elaboration, and amendment. All of the 'relatively immutable' geographical controls over aridity alluded to in the previous paragraph are true only for the very late Cenozoic. Prior to that, lithospheric plate movements created a different and constantly changing distribution of land and sea, of warm and cold ocean currents, and of high and low terrestrial relief. We will discuss the global climatic repercussions of Cenozoic plate tectonic movements in the final section of this chapter.

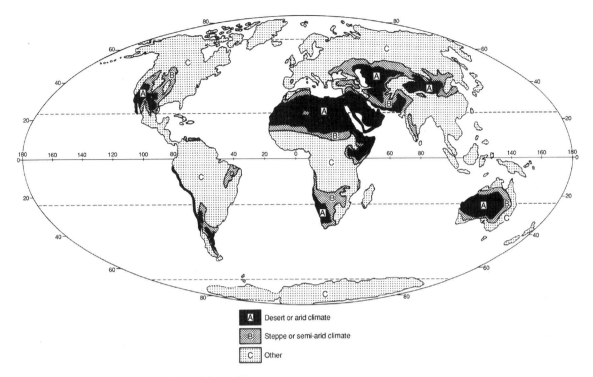

Fig. 28.1 The semi-arid and arid regions of the world

Even within the limited time frame of the Quaternary (c. 1.8 Ma), the sphere of influence of the deserts has waxed and waned from times of maximum expansion, such as during the very late Pleistocene ∼ 20 ka ago, to times when Nile perch, crocodiles, and hippos swam in the lake-studded early Holocene Sahara (Kuper and Kröpelin 2006). We will examine the causes and consequences of these Quaternary climatic fluctuations later in this chapter.

Finally, the role of prehistoric and historic human activities cannot be ignored. Trivial during Lower and Middle Palaeolithic times, when prehistoric world population was sparse, there are discernible signs of Upper Palaeolithic impact in many continents, especially through the widespread use of fire. The result was an increase in savanna at the expense of forest, and of grassland at the expense of woodland. With the onset of plant and animal domestication at the start of the Holocene, and the associated rapid increase in world population from roughly 10 million people at the start of the Neolithic (c. 10 ka) to roughly 100 million by 5 ka, increasing exponentially to a billion (10^9) by 2 ka (May 1978), the stage was set for unprecedented deforestation using polished stone axes

and adzes initially, later to be replaced with bronze, iron, and steel.

Desertification is the general term used for a change to desert-like conditions in areas previously beyond the climatic limits of existing deserts (Williams and Balling 1996, Williams 2002b). It is brought about by a combination of poor land management (overgrazing, indiscriminate clearing of the vegetation cover, increased soil salinization) and prolonged local or regional intervals of drought. The result is a reduction in plant biomass, animal-carrying capacity, and overall ecological resilience and diversity, with occasionally catastrophic impacts on local pastoral and farming communities Despite eloquent but unsubstantiated claims to the contrary, there is no credible evidence that the deserts depicted on Fig. 28.1 are the result of human action. Sweeping and apparently plausible assertions that the modern deserts are wholly or partly man-made need to be placed in context. The expanding desert syndrome erupts after every severe drought, but the re-establishment of a modern plant cover along the desert margins during the wetter intervals between droughts is seldom mentioned.

Reconstructing Cenozoic Climates in Deserts

Reconstruction of Cenozoic climatic changes in deserts is based on a growing body of evidence from both land and sea, including sediments, plant and animal fossils, isotope geochemistry, landforms, soils, instrumental records, prehistoric archaeology and historical archives. The length of the climatic record attainable from the various independent lines of evidence and the time resolution possible with each of them vary widely, and cover a time range from 10^{-2} to 10^7 years (Williams et al. 1998). It is therefore critical that the scope and limitations of each type of proxy evidence are thoroughly appreciated. Six very different recent studies drawn from central Asia, North America, the Pyrenees, Tanzania, Indonesia and the Arctic demonstrate the subtleties involved in reconstructing Cenozoic climates and illustrate vividly how our present understanding is still in a state of flux.

Uplift of the Tibetan plateau as a result of the collision of India and Asia ~ 45 Ma ago caused a major change in the distribution of land and sea and was followed by severe desiccation of the region to the north and east of the plateau (the present-day Taklimakan, Badain Jaran and Gobi deserts of China, Inner Mongolia and Mongolia, respectively). The inception or intensification of the Asian winter and summer monsoons has also been attributed to these tectonic events. However, Dupont-Nivet et al. (2007) have developed a fine-resolution magnetostratigraphic chronology for the Eocene-Oligocene transition (33–34 Ma ago) in the Xining basin at the northeastern edge of the Tibetan plateau. Widespread sedimentation in playa lakes persisted during the Eocene and ended abruptly at the Eocene-Ologocene transition, coincident with Cenozoic global cooling at 33–34 Ma associated with the inception of permanent Antarctic ice sheets at that time. The timing of uplift in Tibet is poorly constrained and probably time-transgressive, prompting Dupont-Nivet et al. (2007) to conclude that the desiccation evident at the Eocene-Ologocene transition in this part of Asia is more likely to have been a result of global cooling than of regional tectonic events, although these undoubtedly helped to accentuate aridity.

Zanazzi et al. (2007) arrived at very different conclusions relating to the climatic changes associated with the Eocene-Oligocene transition in the northern

Great Plains of central North America. They analysed the stable carbon and oxygen isotopic composition of fossil tooth enamel and fossil bone to obtain a 400 ka record at 40 ka resolution and found that the mean annual temperature fell by $8.2 \pm 3.1°C$ during that time, attributing this drop to an inferred major decrease in the concentration of atmospheric carbon dioxide. They found no evidence of any increase in aridity across the transition, in contrast to the palaeoclimatic inferences of Dupont-Nivet et al. (2007) for central Asia.

The importance for the hydrological cycle of Cenozoic changes in the concentration of atmospheric carbon dioxide is also evident from a recent study from the Pyrenees. A major increase in the concentration of atmospheric carbon dioxide (pCO_2) at the Paleocene-Eocene transition ~ 55 Ma ago ushered in the warmest 100 ka of the entire Cenozoic. In the Spanish Pyrenees a vast alluvial fan deposited coarse gravels over an area of 500–2000 km^2 within a few thousand years at the start of the Eocene (Schmitz and Pujalte 2007). The gravels imply intense but highly seasonal rainfall at this time, with carbonate nodules within the fan deposits indicating seasonally dry periods. The authors concluded that regional hydrological cycles were highly sensitive to changes in pCO_2.

A further matter for debate in interpreting Cenozoic climates is the issue of long-term Eocene cooling and whether or not it was synchronous from high to low latitudes. Pearson et al. (2007) have re-interpreted previous studies of the oxygen isotopes of deep-sea tropical foraminifera shells using new drill core samples from Tanzania. They were able to demonstrate that in contrast to the exceptionally well-preserved Tanzanian samples, previously analysed deep-sea plankton shells had been diagenetically recrystallised, so that the apparent tropical sea-surface cooling trend inferred by earlier workers was an artefact of diagenetic overprinting. The unaltered Tanzanian foram shells showed a stable warm tropical climate throughout the Eocene, indicating that low latitudes were unresponsive to the long-term Eocene cooling evident in higher latitudes.

The time when continental ice first appeared in Greenland also remains controversial. Eldrett et al. (2007) have recovered macroscopic dropstones from ice-rafted debris from Eocene and Oligocene sediments that were laid down in the Norwegian-Greenland Sea between 38 and 30 Ma ago, with East Greenland as the most likely source. These findings, if correct, show that glaciers were present on Greenland

~20 Ma earlier than previously demonstrated, but at a time when global temperatures and pCO$_2$ were both relatively high. The timing and the causes of northern high latitude ice accumulation thus remain contoversial.

The late Pliocene increase in aridity evident in East Africa and Ethiopia 3–4 Ma ago (Feakins et al. 2005) has been attributed to closure of the Panama Isthmus and northward diversion of the warm equatorial water which until then had flowed westwards from the Atlantic into the Pacific Ocean. The presence of warm moist air over the North Atlantic coupled to a decrease in insolation linked to increased orbital eccentricity and a decrease in tilt of the earth's axis (leading to cooler high latitude northern summers and milder northern winters) was a prerequisite for widespread and persistent snow accumulation over North America (Williams et al. 1998). The rapid accumulation of ice over North America at 3.5–2.5 Ma was accompanied by global cooling and intertropical aridification.

The causes and consequences of the late Pliocene build up of ice over North America and ensuing tropical climatic desiccation are better constrained than the Paleogene glacial events described earlier but still offer scope for differing interpretations. Cane and Molnar (2001) have proposed that closure of the Indonesian seaway 3–4 Ma ago as a result of northward displacement of New Guinea in the early Pliocene may have triggered a change in the source of water flowing through Indonesia into the Indian Ocean from previously warm South Pacific water to cooler North Pacific water. The concomitant decrease in sea-surface temperatures in the Indian Ocean could have reduced rainfall over East Africa. However, *pace* Cane and Molnar (2001), it seems unlikely that closure of the Indonesian seaway was the sole cause of late Pliocene desiccation in East Africa since the region derives its moisture from both the South Atlantic and the Indian Ocean. We return to the impact of late Neogene cooling and desiccation later in this chapter.

A further *caveat* concerns the processes involved in interpreting past climates. For instance, consider the steps involved in determining the climatic factors responsible for lake level fluctuations. Ideally, we would start with the present-day fluctuations in lake level and relate these to monitored changes in precipitation, evaporation, runoff, inflow, outflow, seepage losses and groundwater inputs. After allowing for the influence of extreme events and their erosional or depositional

legacy, the next step would be to use proxy data (e.g., diatoms, trace element geochemistry) to reconstruct past changes in water chemistry (salinity, alkalinity), depth and temperature. A stratigraphically consistent chronology, preferably based on several independent methods (e.g., AMS ^{14}C, OSL, U-series) is essential at each step in the historical reconstruction. Not all methods yield equally precise ages but all must be accurate. To then convert a history of lake level fluctuations to a set of climatic factors is fraught with additional uncertainties. Even assuming that tectonic or non-climatic factors may be ruled out, it is not always possible to decide whether changes in precipitation seasonality outweigh the influence of changes in annual abundance, and whether reduced evaporation offsets any rainfall decrease.

Much past discussion of climatic change in deserts has relied upon geomorphic evidence but great care is needed in using landforms to infer climate, partly because desert landforms are often polygenic and have evolved at different times under ever changing climates. The major erosional landforms in deserts tend to be far older than the depositional landforms. Erosional landforms include desert mountains such as the Hoggar and Tibesti massifs of the Sahara, stony tablelands and plateaux such as the Mesozoic plateaux of central Australia, and denuded lowlands such as the shield deserts of Mauritania and Western Australia, which are located on tectonically stable Archaean and Proterozoic formations with a long history of subaerial erosion.

Depositional landforms include desert dunes, alluvial fans, and floodplains of varying complexity and size, and lacustrine features such as playa lakes. There is often a close relationship between alluvial, lacustrine, and aeolian features, typified by source-bordering dunes emanating from alluvial point-bar sands or by the clay dunes (lunettes) developed on the downwind margins of certain now dry or saline lakes in Algeria, Australia, and Siberia. The distinction between erosional and depositional landforms is often somewhat arbitrary, and will depend upon the temporal and spatial scale in which one is interested. A Holocene diatomite etched by wind erosion is evidence of lacustrine deposition during a less arid interval, as well as evidence of the efficiency of wind erosion during an ensuing drier phase. The focus may be on the deposit (lacustrine diatomite) or on the landform (a series of yardangs eroded from the lake sediments). Alternatively, the emphasis may be on the origin of the lake basin (tectonic or deflationary?) in which the diatomites occur.

Many desert mountains and plateaux are flanked by rectilinear or gently concave rock-cut surfaces or pediments, which are usually mantled with a layer of sediment up to several metres thick. Spectacular stepped flights of pediments are a feature of some of the piedmont slopes of the Atlas, Aurès, Hoggar, and Tibesti mountains in the Sahara and of the Helan Shan in Inner Mongolia. In their distal sectors, the pediments may coalesce and become buried beneath alluvial fans and playa lakes. The larger fans and lake deposits often contain a fragmented but otherwise valuable record of late Tertiary and Quaternary depositional and hydrological fluctuations which may yield useful palaeoclimatic information. Before using the erosional and depositional landforms preserved in our deserts as evidence of past climatic changes, it is essential to distinguish between the effects of climate on the one hand, and those of tectonic and volcanic processes as well as the more subtle controls exercised by lithology and rock structure.

Geomorphic evidence of climatic change is seldom unequivocal and leads too readily to circular argument. Independent verification (or refutation) is vital if palaeoclimatology is to retain credibility. The appropriate question to ask is not whether or not there has been climatic change, but whether or not a particular climate is demanded by a particular suite of landforms. Since most erosional desert landforms can form in a variety of different ways, although the end-product, such as a pediment, may for all practical purposes have the same final appearance, such features will not be usefully diagnostic of past climatic events. More helpful, because more sensitive, is the depositional legacy of former wetter and drier episodes, although here too great caution is needed when invoking climatic change.

Consider, for example, some late Cenozoic lakes in semi-arid Ethiopia. A lake may be created by lava damming a river and may vanish as a result of seismic disruption. Pliocene Lake Gadeb in the semi-arid south-eastern uplands of Ethiopia originated when a lava flow blocked the course of the ancestral Webi Shebele channel some 2.76 m.y. ago (Williams et al. 1979, Eberz et al. 1988). Earlier still, a vast lake occupied what is now the middle Awash valley of the southern Afar Desert and dried out shortly after 3.8–4.0 Ma during an intense phase of tectonic and volcanic activity (Williams et al. 1986). A final and more recent example will suffice to show the climatically ambiguous nature of some lake records. During the height of the prolonged drought which began in 1968 in Ethiopia, at a time when many Ethiopian rift lakes were shrinking, Lake Besaka, a graben lake lying at the foot of Fantale volcano, started to rise at a rate of $0.5\,\mathrm{m\,y^{-1}}$ between 1974 and 1975 (Williams et al. 1981b). A resurgence of hot spring activity along the fault scarps bordering the lake may be the primary cause – possibly a precursor to a future eruption from the caldera, which last erupted about AD 1850. Within two years the brackish lake with its fauna of brine shrimps and flamingoes had become a freshwater lake with a new fauna of freshwater fish and crocodiles.

Only by rigorous multidisciplinary study is it possible to differentiate between desert lakes formerly fed solely by surface runoff (whether local or far-travelled), or solely by groundwater, or by a combination of runoff and groundwater – a combination which will very likely have varied through time (Gasse 1975, Williams et al. 1987, 1991b, Street 1980, Abell and Williams 1989). With these considerations in mind, it is instructive to review the geomorphic history of our two largest tropical deserts: the Sahara and the arid inland of Australia.

Geomorphic History of the Sahara

Precambrian Tectonic Legacy

The geomorphic history of the Sahara begins in the Precambrian, and the influence of the Precambrian tectonic and structural legacy is pervasive across North Africa and the Arabian Peninsula (Black and Girod 1970, Clifford and Gass 1970, Adamson and Williams 1980, 1987, Williams 1984a, Bowen and Jux 1987, El-Gaby and Greiling 1988, F.M. Williams et al. 2004, Thurmond et al. 2004, Guiraud et al. 2005). The Precambrian cover rocks are comparatively undeformed and unaltered, and appear to be of Middle and Upper Proterozoic age. They overlie the weakly to highly metamorphosed and strongly folded and faulted Archaean and Lower Proterozoic formations with a marked erosional unconformity and often form rugged plateaus, mesas, and monoclinal ridges of quartz arenite and conglomerate with intercalated dolerites and basalts. Other parts of Pangaea had a similar history of episodic Archaean and Lower Proterozoic metamorphism followed by prolonged intervals of widespread erosion and late Precambrian sedimentation. For example, the present-day landscapes of the Vindhyan Hills in north central India and the Arnhem Land Plateau in northern Australia differ from some of those in the western Sahara only in now being covered in savanna woodland.

Much of the Precambrian landscape is concealed beneath younger sedimentary and volcanic rocks in places up to 10 km thick, so that the areal extent of out-cropping Precambrian formations is confined to about 15% of the Sahara. Despite such limited outcrop, the influence of the ancient basement rocks of the Sahara is out of all proportion to their present surface distribution. In recent decades we have come to appreciate more and more that the patterns of Phanerozoic faulting, rifting, and volcanism, the emplacement of ring-complexes, and the distribution of major depocentres in arid northern Africa and Arabia are a direct reflection of the pervading influence of geological structures that developed during the 650–550 Ma Pan-African orogenic event as well as during the 2000–1250 Ma Eburnean orogeny. Later workers have shown how these early orogenic events have controlled the subsequent location of Phanerozoic uplands and sedimentary basins in the Sahara (Thurmond et al. 2004, Guiraud et al. 2005). Periodic reactivation of some of the major Precambrian structural trends has determined the location of Cenozoic faults, rifts, and volcanoes in northern Africa .

Palaeozoic and Mesozoic Sedimentation

The Pan-African orogeny which reached its climax towards 550 Ma ago was followed by prolonged and widespread erosion which reduced much of the Sahara to a gently undulating plain. Renewed uplift towards 450 Ma heralded the late Ordovician glaciations of the Sahara (Beuf et al. 1971). Around the present Hoggar mountains some of the erosional evidence of these Ordovician ice caps, such as the glacially striated pavements, is so well preserved that it almost appears to be Pleistocene. Melting of the Saharan ice caps was followed by the rapid and world-wide early Silurian rise in sea level, comparable to the glacio-eustatic sea-level rise which followed melting of the late Pleistocene

ice caps. Accumulation of Carboniferous mudstones, limestones and sandstones across the Sahara eventually gave way to Hercynian uplift in the western Sahara and Atlas associated with opening of the North Atlantic. Uplift and folding of the previously horizontal Palaeozoic sedimentary rocks triggered the Triassic and later erosion during which the woodlands of the northern Sahara were rapidly buried beneath the sandstones and mudstones laid down by Mesozoic rivers flowing from the south and east. Silicified tree-trunks, some still in upright positions of growth, are vivid indications of the speed and effectiveness of the Mesozoic fluvial aggradation. A hundred million years later, the prehistoric peoples of the Pleistocene Sahara were to use fragments of this silicified wood for stone tool-making.

The early Mesozoic opening of the South Atlantic and the ensuing separation of the African and South American lithospheric plates resulted in a series of Mesozoic marine transgressions. Much of the Sahara was relatively flat and low-lying at that time, and flanked by warm and shallow epi-continental seas. One such marine incursion flooded much of the central Sahara during the Upper Cenomanian and Lower Turonian, although the Hoggar uplands remained above sea level, as did much of the West African craton.

These Mesozoic marine and non-marine sedimentary formations comprise some of the greatest present-day aquifers in North Africa, and include the Nubian Sandstone Formation of the eastern Sahara and the 'Continental intercalaire' (Kilian 1931) of the central Sahara. The more or less horizontal Mesozoic formations are scattered across the Sahara today in the form of extensive sandstone and limestone plateaux or hamada. Australia had a similar Mesozoic history, and there is little to distinguish the great stony tablelands and associated gibber plains of the present Australian desert from the vast gravel-strewn hamada and stony reg surfaces of the Sahara.

A classic example of a Saharan sandstone plateau or hamada is the Gilf Kebir in the now hyperarid desert of south-eastern Libya and south-western Egypt. The margins of the Gilf Kebir as well as those of other great sandstone plateaux in southern Libya are highly crenulated and deeply dissected by now inactive river valleys. Peel (1966) and many later observers have emphasized the former efficiency of such fluvial erosion (Griffin 2006), but perhaps the most eloquent testimony to these pluvial episodes – apart from the dry valleys themselves – are the great galleries of Upper Palaeolithic and Neolithic rock paintings and engravings of now vanished herds of elephants, giraffes, and domesticated cattle (Muzzolini 1995, Coulson and Campbell 2001). More recently, Breed and her colleagues have made brilliant use of shuttle-imaging radar to identify Pleistocene and older river channels in the eastern Sahara (Breed et al. 1987). These now defunct watercourses flowed at least intermittently during the early to middle Pleistocene and at intervals thereafter, including the early to middle Holocene. During the early and middle Pleistocene, small bands of *Homo erectus* roamed the Sahara equipped with their all-purpose Acheulian toolkit of bifacially worked hand axes, cleavers and scrapers (Breed et al. 1987, McHugh et al. 1988). By Holocene times, the Palaeolithic hunters and gatherers had been replaced by Neolithic pastoralists who grazed their cattle throughout the Sahara (Williams and Faure 1980, Clark and Brandt 1984, McHugh et al. 1989).

Cenozoic Deep Weathering, Uplift, and Erosion

Withdrawal of the shallow, equatorial Cretaceous seas from the Sahara was followed by a very long interval of intense early Tertiary weathering and leaching of the forested lowlands of the tectonically quiescent southern Sahara (Faure 1962). Near-surface solution and redeposition of iron and silica during the Palaeocene and Eocene gave rise to the resistant caprocks of ironstone or silicified rock which now protect many of the Mesozoic and Tertiary plateau summits from erosion. The present-day geomorphic outcome is a process of slow and episodic scarp retreat by undercutting of the less resistant mudstones and softer sandstones during wetter phases, followed by collapse of the resistant caprocks. The undercutting is effective even today, and is aided by seepage at the cliff base, by salt weathering and by chemical weathering and deflation. Similar processes of scarp retreat have been invoked to explain boulder-mantled debris slopes in the semi-arid northwest of South Australia.

Post-Eocene uplift triggered a widespread phase of mid-Tertiary erosion within major massifs such as

Tibesti and the Hoggar, as well as in more isolated ring-complexes such as Jebel Arkenu and Jebel 'Uweinat in south-east Libya or Adrar Bous in central Niger. The mid-Tertiary drainage system appears to have been a highly efficient and well-integrated system which kept pace with the various epeirogenic uplifts across the Sahara. The Nile cut down through Nubian Sandstone capping the Sabaloka ring complex to form the Sabaloka gorge north of Khartoum – one of the many instances of superimposed Cenozoic drainage in the Sahara (Grove 1980, Williams and Williams 1980, Thurmond et al. 2004). The early Tertiary mantle of deeply weathered rock was virtually removed from the uplands of the southern Sahara, leaving a bare and rugged landscape of gaunt rocky pinnacles and boulder-mantled slopes. Episodic deep weathering followed by episodic erosion and exhumation of the weathering front became the geomorphic norm of the later Cenozoic (Dresch 1959, Thorp 1969, Williams 1971). There seems little doubt that mid-Tertiary tectonic movements performed a dominant role in the initial pulse of erosion but the late Cenozoic climatic oscillations became increasingly important erosional pacemakers thereafter (Williams et al. 1987).

The sandy colluvial–alluvial debris eroded from the Saharan uplands was carried away from the mountains by the Tertiary and early Quaternary rivers to be in part deposited in late-Cenozoic marine deltas such as those of the Nile, the Niger, and the Senegal. However, a considerable proportion of the sediment began to accumulate in the closed interior basins created during the course of late Mesozoic and Cenozoic faulting, rifting and epeirogenic movements.

It was the unconsolidated Tertiary sediments laid down in large subsiding sedimentary basins such as the Kufra-Sirte basin in Libya, or the Chad basin, which provided the source material for the late Tertiary and Quaternary desert dunes. Miocene tectonic uplift in East Africa may have contributed to the desiccation in this region from about 8 Ma onwards (Sepulchre et al. 2006). In the Chad basin, Servant (1973) identified wind-blown sands in a number of very late Tertiary stratigraphic sections. He concluded that the onset of aridity and the first appearance of desert dunes in this part of the southern Sahara was a late Tertiary phenomenon. Using fossil and sedimentary evidence, Schuster et al. (2006) have since confirmed that the onset of recurrent desert conditions in the Chad basin began at least 7 Ma ago. Further north, in the Hoggar, some elements of the late Tertiary flora were already physiologically well adapted to aridity (Maley 1980a, b, 1996). If we accept the sedimentological evidence of Servant (1973) and of Schuster et al. (2006) and the palynological evidence of Maley (1996), then it follows that the onset of climatic desiccation and the ensuing disruption of the integrated mid-Tertiary Saharan drainage network (Griffin 2006) was a feature of the very late Tertiary, long pre-dating the arrival of *Homo sapiens*. Before pursuing this topic in greater detail, we need to retrace our steps and examine the impact of Cenozoic volcanism, faulting, and rifting upon the Saharan landscape.

Cenozoic Volcanism, Uplift, and Rifting

The basin and swell topography of the present-day Sahara (Fig. 28.2) is very largely a function of the Tertiary epeirogenic movements discussed earlier. An additional and extremely important factor was the massive extrusion of lava which accompanied and accentuated the late Cenozoic uplift of all the existing uplands, including the Ethopian Highlands, Jebel Marra, Tibesti, the Hoggar, and the Aïr.

An estimated 8000 km^3 of volcanic rock was erupted during the formation of Jebel Marra, and about half that amount (3000 km^3) during the eruptions which gave rise to Tibesti (Francis et al. 1973). The volcanic eruptions which helped to create Jebel Marra (elevation: 3088 m), Tibesti (3415 m) and the Hoggar (2918 m) were associated with significant uplift of the basement rocks which now lie beneath the late Cenozoic volcanic rocks (Guiraud et al. 2005). How much of this uplift took place during as opposed to before the eruptions is not known, but of the uplift itself there seems no doubt. Relative to the basement rocks on the adjacent plains, the basement rocks of Tibesti and the Hoggar have risen about 1000 m and those of Jebel Marra about 500 m (Bordet 1952, Vincent 1963, Vail 1972a,b). Much of the volcanic activity is of Miocene age and younger, and significant erosion of the crystalline basement rocks immediately preceded the volcanic activity and its erosional aftermath (Rognon 1967, Williams et al. 1980).

The timing of uplift in the Aïr massif is less well established (Williams et al. 1987), but the presence of possible Cretaceous sedimentary rocks at 1400 m on Tamgak granite plateau and very high (c. 1500 m) on the Goundai ring-complex (Raulais 1951, Vogt and Black 1963) suggests a possible 1000–2000 m of post-Cretaceous uplift. Given the late Cenozoic updoming of the Hoggar (Bordet 1952, Rognon 1967) and Jebel Marra (Vail 1972a,b) it seems likely that uplift of the Aïr is also of late Cenozoic age. The Mesozoic sedimentary formations south and west of Agadès all dip westwards away from the main Aïr massif and are again consistent with post-Cretaceous updoming of the massif.

Uplift of the Ethiopian Highlands is also a late Cenozoic event (Pik et al. 2003). Pollen grains characteristic of tropical lowland rainforests are abundant in Miocene lignites intercalated between basalt flows dated by K/Ar content at 8 Ma (Yemane et al. 1985). The basalts crop out at roughly 2000 m elevation near Gondar in the north-western Ethiopian uplands, suggesting roughly 2000 m of uplift in the past 8 m.y., or an average of 0.25 mm y^{-1}. The great depth and steepness of the Blue Nile gorge and the Tekazze gorge in central western Ethiopia is consistent with late Cenozoic uplift, and Faure has argued for an acceleration of uplift from Tertiary to Quaternary based on uplift rates of coral reefs along the Red Sea (Faure 1975). Since there is no necessary link between coastal uplift and uplift of the Ethiopian Highlands, the question is best left open. One thing is certain, however, and that is the enormous volume of rock eroded and removed from Ethiopia since extrusion of the Upper Oligocene and Miocene Trap Series basalts. The Blue Nile, the Tekazze, and their tributaries have removed about 100 000–200 000 km^3 of rock from the north-western Ethiopian uplands in the last 10–20 m.y., which is also equivalent to the volume of the Nile cone in the eastern Mediterranean (McDougall et al. 1975, Williams and Williams 1980). A comparison of modern rates of erosion in upland Ethiopia and mean geological rates of denudation persuaded Williams and Williams (1980) that uplift of the Ethiopian Highlands was episodic, with prolonged intervals of relative stability between shorter intervals of rapid uplift. The abundance of montane forest podocarpus and juniper pollen in Pliocene sediments, which are now at low elevations in the lower Awash valley of the west-central Afar Desert (Bonnefille et al. 2004) also indicates that uplift of the Ethiopian uplands was accompanied by down-faulting

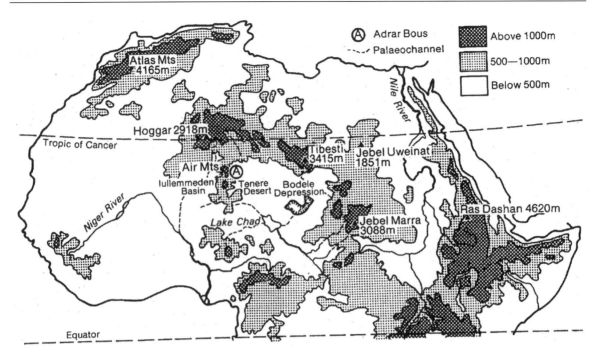

Fig. 28.2 Relief map of northern Africa, showing distribution of major Saharan uplands and lowlands. Also shown is the locatioan of Adrar Bous ring-complex in the heart of the Sahara, immediately east of the northern Aïr massif, in the Tenere Desert of Niger (after Williams et al. 1987, Fig. 1)

and subsidence along the adjacent margins of the eastern Ethiopian escarpment (Adamson and Williams 1987).

The pattern of rifting in Ethiopia has created a very particular type of desert in the Afar Depression (Tiercelin 1987, Williams et al. 1986, Adamson and Williams 1987). The Afar is a lava desert and is one of the hottest and most forbidding deserts on Earth. Flanked by the mighty Ethiopian escarpment on the west, it descends over 150 m below sea level at Lake Asal, a salt lake now separated from the adjacent Red Sea by a puny Pleistocene lava dam.

Neogene fluvio-lacustrine sediments in southern Afar have yielded one of the longest, richest, and most complete records of fossil hominid evolution anywhere in Africa (Clark et al. 1984). Associated with these australopithecine hominid fossils are abundant superbly preserved Pliocene fossils of elephants, pigs, bovids, hippos, crocodiles, and sundry non-aquatic carnivores. The fauna is a savanna fauna, and most taxa belong to now extinct species. The presence of such fossils in parts of the Afar that do not now support much life, together with the presence of thick lacus-

trine deposits within the Neogene formations again raises the issue of possible late-Cenozoic climatic desiccation, a topic to which we now turn.

Late Cenozoic Desiccation

Tertiary volcanism in the central and southern Sahara was preceded and accompanied by prolonged deep weathering. In central Niger kaolinitic and bauxitic weathering profiles up to 45 m thick are developed on rocks of Eocene to Precambrian age. Uplift in the mid-Tertiary resulted in a change from previously biogenic and chemical sedimentation in this region to dominantly clastic sedimentation (Faure 1962, Greigert and Pougnet 1967, p. 157). Rejuvenated rivers flowing down from the great watersheds of Tibesti, the Hoggar, and the Aïr deposited the fluvial gravels, sands, and clays of the 'Continental terminal' extensively around their parent uplands. Williams and co-workers have concluded that in Niger and adjacent areas 'the origin of the Sahara as a continental desert. . . may be said to stem from the Miocene Alpine orogeny and the subsequent stripping of the Eocene deep weathering profile' (Williams et al. 1987, p. 109). Apart from tectonic uplift, what other factors were responsible for this dramatic change from a landscape of lowland equatorial rainforest to one of bare rocky inselbergs and desert dunes?

A major influence, not mentioned so far, was the post-Palaeozoic northward drift of the African lithospheric plate. Triassic Africa was part of the Gondwana supercontinent, as were South America, Antarctica, Australia, and India. With the Jurassic and earlier separation of Gondwana into the two continents of West Gondwana (Africa and South America) and East Gondwana (Australia, Antarctica, and India) the stage was set for further break-up of these two large continents during the Cretaceous (Owen 1983).

The Cretaceous equator in Africa ran from southern Nigeria through central Chad and the northern Sudan into Arabia. During the late Mesozoic and Cenozoic, the African plate moved northward with a slow clockwise rotation (Habicht 1979, Owen 1983, F.M. Williams et al. 2004). Early Cenozoic Africa was south of its present position by only a few degrees of latitude and came into contact with Europe as a result of a slight clockwise rotation during the Miocene and Pliocene. One outcome of the ensuing crustal deformation was the uplift of the Atlas, noted earlier, which was also coeval with volcanism and updoming of the Hoggar, Tibesti, Aïr and Jebel Marra uplands, creating the major elements of the topography depicted on Fig. 28.2.

A further outcome of Africa's Mesozoic and Cenozoic northward drift and rotational movement was a corresponding southward shift of the equatorial rainforest. This zone once ran obliquely across the Sahara from Egypt and the northern Sudan south-westwards towards southern Nigeria. Aridity set in earlier in Morocco, Algeria, and Tunisia than in Egypt and the Sudan, as is evident from the abundance of Mesozoic and younger evaporite formations in the north-western Sahara, which by then had already reached dry tropical latitudes (Coque 1962, Conrad 1969, Williams 1984a).

Three additional influences contributed to the late Cenozoic desiccation of the Sahara. These were the uplift of the Tibetan plateau, the build-up of continental ice in Antarctica and the Northern Hemisphere, and cooling of the world's oceans. We consider the possible causes of these phenomena in the final section of this chapter; our concern here is purely with their effects upon the Sahara.

Late Cenozoic uplift of the Himalayas and of the vast Tibetan plateau was associated with the intensification (and perhaps the inception) of the easterly jet stream which today brings dry subsiding air to the deserts of Arabia and northern Africa. A major change in the flora and fauna of the Potwar plateau in the Siwalik foothills of Pakistan between 7.3 and 7.0 Ma may also be related to Himalayan uplift and is consistent with intensification of the Indian summer monsoon, if not with its origin at that time (Quade et al. 1989, Cerling et al. 1997).

Accumulation of continental ice in Antarctica may seem somewhat remote from Saharan desiccation but was in fact of critical importance. Mountain glaciers were present on Antarctica early in the Oligocene, and a large ice cap was well established by 10 Ma (Shackleton and Kennett 1975). Continental ice was slower to form in the Northern Hemisphere but was present in high northern latitudes by 3 Ma, and possibly by 5 Ma or even well before then, with a rapid increase in the rate of ice accumulation towards 2.5 to 2.4 Ma (Shackleton and Opdyke 1977, Shackleton et al. 1984). As temperatures declined over the poles, and sea surface temperatures at high latitudes grew colder, the temperature and pressure gradients between the Equator and the poles increased. There was a corresponding increase in Trade Wind velocities, and hence in the ability of these winds to mobilize and transport the alluvial sands of the Saharan depocentres and to fashion them into desert dunes. Higher wind velocities were also a feature of glacial maxima during the Pleistocene and were responsible for transporting Saharan desert dust far across the Atlantic (Parkin and Shackleton 1973, Parkin 1974, Williams 1975, Sarnthein 1978, Sarnthein et al. 1981). During the Last Glacial Maximum centred on 21 ± 2 ka (Mix et al. 2001) Australian desert dust was also blown as far as central Antarctica (Petit et al. 1981).

Late Cenozoic cooling of the ocean surface was also responsible for reducing intertropical precipitation. Galloway (1965) noted that two-thirds of global precipitation now falls between latitudes 40°N and 40°S and depends upon effective evaporation from the warm tropical seas. The ocean surface cooling, which was linked to global cooling associated with high-latitude continental ice build up and enhanced cold bottom-water circulation, would help to reduce evaporation from the tropical seas, thereby reducing rainfall across North Africa.

The late Cenozoic desiccation which created the largest desert in the world was therefore a result of a number of factors. Northward drift of the African plate ultimately helped to disrupt the warm Tethys Sea with its abundant supply of moist maritime air. Northern Africa moved away from wet equatorial latitudes into the dry subtropics. Growth of the great continental ice sheets and cooling of the oceans saw a decrease in precipitation and an increase in the strength of the Trade Winds. At the start of the Oligocene there was a sharp drop in sea surface temperatures, with eventual global repercussions.

Sudano-Guinean woodland covered much of the Sahara during the Oligocene and early Miocene, having replaced the equatorial rainforest of Palaeocene and Eocene times. During the late Miocene and early Pliocene a xeric flora, well adapted to aridity, began to replace the earlier woodland, so that many elements of the present Saharan flora were already present during the late Pliocene, when aridity became even more severe across the Sahara and the Horn of Africa (Bonnefille 1976, 1980, 1983, Maley 1980b).

The combination of a reduction in plant cover and a trend towards more erratic rainfall had a profound impact on the late Cenozoic rivers of the Sahara (Griffin 2006). Big rivers capable of carving large valleys became seasonal or ephemeral. Integrated drainage systems became segmented and disorganized. Wind mobilized the sandy alluvium into active dunefields. Dunes formed barriers across river channels no longer competent to remove them. Dust storms left the desert topsoils depleted in clay, silt, and organic matter. The Sahara was now a true wilderness, as the Arabic word implies.

Quaternary Climatic Fluctuations

Mid-Pliocene closure of the Panama isthmus towards 3.2 Ma paved the way for the rapid accumulation of continental ice sheets in high northern latitudes during the late Pliocene (Schnitker 1980, Loubere and Moss 1986, Prentice and Denton 1988). Oxygen isotope evidence from deep-sea cores indicates that the onset of major Northern Hemisphere continental glaciations at 2.4 ± 0.1 Ma (Shackleton et al. 1984) also coincided with cooling in high southern latitudes (Kennett and Hodell 1986). The 2.3–2.5 Ma tempera-

ture drop is also evident in the south-eastern uplands of Ethiopia (Bonnefille 1983) and the dry northern interior of China, with the beginning of widespread loess accumulation in the Loess Plateau of central China dated to 2.4 Ma (Heller and Liu 1982). In the north-western Mediterranean region, the presence of a Mediterranean vegetation adapted to winter rains and summer drought is already evident at 3.2 Ma, but it is not developed in its modern form until about 2.3 Ma (Suc 1984).

Magnetic susceptibility measurements of deep-sea cores from the Arabian Sea and the eastern tropical Atlantic also reveal a change in the length of astronomically controlled climatic cycles at this time. Before 2.4 Ma, the dominant cycles are the 23-ka and 19-ka precession cycles, but after 2.4 Ma, the 41-ka obliquity cycle becomes dominant (Bloemendal and deMenocal 1989). Although the boundary between Pliocene and Pleistocene is now defined by the International Union of Geological Sciences as 1.8 Ma, this somewhat arbitrary date should not obscure the fact that the continental glaciations characteristic of the Quaternary were ushered in by the dramatic fall in global temperatures towards 2.4 Ma.

There is now widespread recognition that the magnitude and frequency of the late Pliocene and Quaternary glaciations were strongly influenced by orbital perturbations. This recognition is thanks to the work of the brilliant Yugoslav astronomer and mathematician, Milutin Milankovitch, who was the first to persuade Quaternary scientists of the climatic importance of the Earth's orbital variations (Milankovitch 1920, 1930, Chappell 1974, Imbrie and Imbrie 1979, Berger 1981, Williams et al. 1998). The major cycles identified by Milankovitch, and for which he calculated the changes in insolation received on Earth at different seasons and latitudes for the successive stages of each cycle, are as follows. The 100-ka orbital eccentricity cycle is determined by the changing elliptical path of the Earth around the Sun. The changing tilt of the Earth's rotational axis gives the 41-ka obliquity cycle. The precession of the equinoxes varies with the changing distance between Earth and Sun and gives the 23-ka cycle.

Although the correlations between orbital perturbations, ice volume fluctuations, and glacio-eustatic sea level fluctuations are statistically significant and now well accepted (Hays et al. 1976), the relative influence of the various cycles has varied during the course of Quaternary time.

Scrutiny of oxygen isotope records from deep-sea cores spanning the full duration of the Quaternary persuaded Williams et al. (1981a) that the early Pleistocene from 1.8 to 0.9 Ma was subject to high-frequency but low-amplitude fluctuations in the oxygen isotope differences between glacial and interglacial maxima. In contrast, the last 0.9 Ma were characterized by high-amplitude but low-frequency fluctuations in oxygen isotopic composition (Williams et al. 1981a). Since changes in the oxygen isotopic composition of benthic Foraminifera are very broadly a reflection of changes in global ice volume (Shackleton 1977, 1987), the changing pattern of glaciation during the Quaternary will also be reflected in variations in the severity and frequency of cycles of glacial aridity in the tropics.

Ruddiman and Raymo (1988) demonstrated that the 100-ka orbital eccentricity cycle was dominant in North Atlantic cores during the last 0.78 Ma (i.e., during the Brunhes magnetic chron). Before then, during the Matuyama chron from 2.60 to 0.78 Ma, the 41-ka orbital cycle was dominant, reflected in more frequent but lower amplitude climatic fluctuations.

Given the powerful influence exerted by the North Atlantic upon both ice volume and precipitation in the Northern Hemisphere, certain palaeoclimatic inferences may be drawn with respect to the Pleistocene Sahara. We have long known that the last glacial maximum in the Sahara was a time of accentuated aridity, with reactivation (or advance) of desert dunes up to 500–1000 km beyond their present southern limits (Grove 1958, Grove and Warren 1968, Talbot 1980, Williams 1975, 1985). During these times of glacial aridity and desert expansion, vast plumes of Saharan and Arabian desert dust were mobilized and blown far out to sea. Over the past 0.6 m.y., maximum concentrations of Saharan desert dust in equatorial Atlantic deep-sea cores coincide with times of low sea-surface temperature or glacial maxima (Parmenter and Folger 1974, Bowles 1975). A similar pattern of glacial aridity is evident in the Red Sea and Gulf of Aden. Planktonic Foraminifera from deep-sea cores collected in this region reveal through their changing isotopic composition that during the last 250 000 years, at least, glacial maxima were times of extreme aridity, with much increased sea-surface salinity reflecting even higher local rates of evaporation than today (Deuser et al. 1976).

It would be misleading to portray all arid phases as coinciding with glacial maxima and all humid phases with peak interglacial times. The reality is more complex. Lake levels in Lake Chad (Servant 1973) and Lake Abhe (Gasse 1975) were high for at least 20 000 years before 18 ka when they fell rapidly. Lake Abhe remained dry until 12 ka, and Lake Chad intermittently dry until then, after which they both rose again rapidly, reaching peak levels towards 9 ka. Since about 4.5 ka these lakes have remained relatively low, with occasional short-lived transgressions. Very schematically, we could consider the interval of high lake levels from 30 to 18 ka as representing a humid glacial phase, the interval of low lake levels from 18 to 12 ka as an arid glacial phase; the interval of early Holocene high lake levels as a humid interglacial phase; and the interval of late Holocene relatively low lake levels as a dry interglacial phase. Even this fourfold subdivision is a caricature of reality, and does not take into account local hydrological and geomorphic controls over precipitation, runoff, evaporation, and groundwater inflow and seepage (Fontes et al. 1985, Abell and Williams 1989).

Whatever their ultimate causes (Kutzbach and Street-Perrott 1985, Gasse et al. 1990, Street-Perrott and Perrott 1990), the consequences of the alternating wetter and drier Quaternary climatic phases are very evident throughout the Sahara. For instance, at the isolated ring-complex of Adrar Bous in the geographical heart of the Sahara (Fig. 28.2), the geomorphic expressions of these past climatic fluctuations include active and stable dunes, lake strandlines and partially deflated lacustrine diatomites, alluvial fans, alluvial terraces, and partly buried palaeochannels and former backswamps (Fig. 28.3). The stratigraphic evidence is equally informative and extends well back into the middle Pleistocene (Williams et al. 1987). Phases of rapid erosion with associated deposition of coarse sands and gravels alternated with longer intervals of minimal erosion, fine-grained sedimentation in low-energy environments, and soil development in and around the mountain. The presence at Adrar Bous of late Pleistocene and Holocene freshwater snails and gastropods allows us to use palaeoecological and isotopic evidence (Williams et al. 1987) to test the inferences drawn from sedimentology and geomorphology, and additional evidence is yielded by prehistoric stone tool assemblages, hearths, graves,

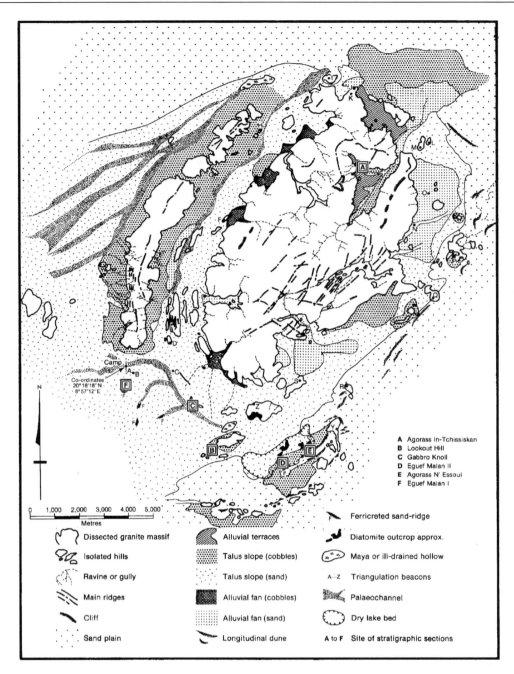

Fig. 28.3 Geomorphic map of Adrar Bous ring-complex in the Tenere Desert of Niger, showing the major landforms and lo-cation of the stratigraphic sections discussed in the text (after Williams et al. 1987, Fig. 3)

and middens, the last with burnt remains of locally consumed fauna (Clark et al. 1973, Smith 1980).

Similar Quaternary interdisciplinary or multidisci-plinary studies are now the norm in different parts of the Sahara, so that despite its vastness and periodic difficulties of access, the Sahara has provided the best dated and most comprehensive evidence of late Quaternary environmental fluctuations so far available from any of our deserts (Maley 2000, Hoelzmann et al. 2004).

Geomorphic History of the Australian Desert

The following account is distilled from Bowler (1973, 1976, 1998), Shackleton and Kennett (1975), Williams (1984b,c, 1991, 2000, 2001), Chen (1989), Williams et al. (1991a, 2001), Alley et al. (1995), Ayliffe et al. (1998), Bowler and Price (1998), Magee (1998), Bowler et al. (1998, 2001), Veevers (2000), McGowran et al. (2004), Fujioka et al. (2005), Martin (2006), Prideaux et al. (2007), Revel-Rolland et al. (2006), van der Kaars et al. (2006) and Sniderman et al. (2007).

Precambrian Tectonic Legacy

Until about 2500 Ma, much of what is now western and northern Australia consisted of a granitic basement of Archaean rocks. Localized faulting and rifting created a vast regional sediment trap in northern Australia towards 2500 Ma. Over the next few hundred million years, intercalated marine and non-marine gravels, sands, and muds, together with volcanic rocks, accumulated within this depocentre, attaining a thickness of over 14 000 m. Major regional metamorphism ensued towards 1870–1800 Ma, during which the Lower Proterozoic sedimentary and volcanic formations, together with some of the adjacent basement rocks, were folded, faulted, and metamorphosed to form the Lower Proterozoic metasediments which crop out today as rocky strike-ridges in northern Australia. This period of regional metamorphism was followed by uplift, erosion, and by several minor episodes of granite intrusion and volcanic activity. For several tens of millions of years thereafter, prolonged erosion, interrupted by episodic faulting and uplift, ultimately created a Precambrian landscape of gently undulating relief with sporadic hills and shallow valleys. Thereafter, the region has been tectonically stable apart from slow epeirogenic uplift during the late Phanerozoic.

Several hundred metres of horizontal sands, interbedded lavas, and minor basal gravels were laid down across the early Precambrian land surface towards 1690–1650 Ma. These Middle and Upper Proterozoic sandstones are the older Precambrian cover rocks in northern Australia and today form rugged sandstone plateaux such as the Arnhem Land plateau with its joint-controlled gorges and steep erosional cliffs. These Precambrian cover rocks or plateau sandstones have protected the underlying Lower Proterozoic metasediments from subsequent erosion. Slow scarp retreat has gradually exhumed the original Precambrian topography, so that the present relief of bevelled strike ridges and undulating rock-cut surfaces is, in parts, a resurrected and, except for the modern vegetation cover, a virtually unmodified Lower Proterozoic landscape.

Palaeozoic and Mesozoic Sedimentation

Palaeozoic sandstones and limestones crop out in the less elevated parts of northern Australia but are conspicuously absent from the stable cratonic areas of western Australia as well as from the summits of the northern Precambrian plateaux.

During the early Cretaceous, towards 135–100 Ma, Australia consisted of three large islands separated by shallow seas. The two western islands were the Precambrian shield region of western Australia and the Precambrian and Palaeozoic uplands of northern and central Australia, indicating that these upland areas have had a long history of subaerial erosion.

Some of the present northern Precambrian plateaux, such as the Arnhem Land plateau, have a thin cover of Cretaceous sediments. These Cretaceous mudstones and sandstones also mantle the Lower Proterozoic and Archaean rocks north and west of Arnhem Land, where they are up to 200 m thick. Scarp retreat of the plateau has probably been accelerated by the uplift which followed the 110 Ma Aptian marine transgression, and may be a Cenozoic rather than a Cretaceous phenomenon, since the Cretaceous sediments are uniformly and deeply weathered wherever they occur. In central Australia, the horizontal Mesozoic formations often have a caprock of silcrete. Further north, in what are now the seasonally wet tropics, the caprock consists of ferruginous ironstone (laterite or ferricrete) or even of bauxite. Such deep weathering implies prolonged and efficient leaching under conditions of slow or ineffective mechanical erosion, and normally requires low gradients, high effective precipitation and a dense vegetation cover.

Cenozoic Deep Weathering, Uplift, and Erosion

By late Cretaceous time, the sea had retreated from most of Australia, and in the centre and north there were several prolonged intervals of deep weathering of the newly emerged land surface. This prolonged weathering under conditions of tectonic stability was interrupted by minor intervals of mostly gentle uplift with local faulting and folding, notably in the middle to late Miocene.

At least three post-early Cretaceous erosion surfaces have been identified by different workers in central and northern Australia on the basis of elevation, laterite-capped remnants, and regional slope. Assuming that block-faulting can be ruled out, and assuming that the weathering mantle did indeed develop on a gently sloping surface, both of which at present are unproven assumptions, then the presence of such stepped surfaces is entirely consistent with episodic Cenozoic uplift and associated vertical and lateral erosion. The geomorphic evidence accords well with such an interpretation. Rejuvenation of the drainage in northern Australia has led to vertical incision and the development of rugged relief consisting of steep valleys, rocky strike-ridges, boulder-mantled granite hills, and steep-sided sandstone plateaux.

Late Cenozoic Desiccation

Until about 130 Ma, Australia, Antarctica, and Greater India were all part of East Gondwana. Very early in the Cretaceous a rift developed between Greater India and Australia–Antarctica, and the present Indian Ocean began to form. Sea-floor spreading was very rapid between 80 and 53 Ma, attaining rates up to 175 mm y^{-1}. The initial separation of Australia from Antarctica at 90 Ma was at first very slow and remained so until about 30 Ma. During the last 30 m.y., Australia has drifted north at a mean rate of roughly 50–70 mm y^{-1}. Concomitant rifting along the southern continental margin and the eastern edge of continental Australia led to separation of the Campbell Plateau, New Zealand, and the Lord Howe rise from Australia, and formation of the Tasman Sea, and, ultimately, of the Southern Ocean.

As Australia drifted north, the Eastern Highlands were the centre of sustained volcanism, although the locus of volcanic activity shifted south as Australia moved north. The Eastern Highlands were uplands well before the Tertiary, and sporadic uplift continued during the late Cretaceous and Cenozoic, even if the detailed pattern of uplift is still unclear. The west-flowing drainage of eastern Australia is thus very old, and great river basins such as the ancestral Murray–Darling were already in existence during the late Cretaceous.

In western Australia, early Cretaceous rifting, volcanism, uplift and later subsidence accompanied the separation of Greater India from Australia. Late Palaeocene to Eocene rifting also accompanied the separation of Australia from Antarctica. As a result of these tectonic upheavals, the once integrated network of drainage in western Australia became progressively disrupted from 130 to 40 Ma (early Cretaceous to late Eocene), and the rivers became less and less active during the Oligocene and early Miocene. By mid-Miocene times (15–10 Ma), a conspicuous network of linear salt lakes occupied the western half of the Australian continent, a witness to early Cenozoic tectonic disruption and later Cenozoic climatic desiccation.

As Australia moved north away from Antarctica, it came increasingly under the influence of tropical climatic systems and dry, subsiding, anticyclonic air masses. The long-term reduction in precipitation and runoff is reflected in the late Cenozoic trend towards the endoreic and areic drainage systems which are to-day characteristic of roughly two-thirds of the continent. Cooling of the Southern Ocean accentuated the trend towards aridity evident in the changing flora and fauna of inland Australia. Figure 28.4 shows a fluctuating decline in Southern Ocean sea-surface temperatures from roughly 19°C in the early Eocene to only 7°C in the Oligocene, associated with the progressive accumulation of ice in Antarctica discussed earlier in this chapter. The global cooling and ensuing intertropical desiccation which accompanied the late Pliocene expansion of continental ice sheets in North America added the final gloss to the late Cenozoic drying out of Australia's great inland rivers and lakes, which reached its full climax in the second half of the Quaternary (Figs. 28.5, 28.6, 28.7).

Fig. 28.4 Cenozoic sea
surface temperatures in the
Southern Ocean deduced
from changes in oxygen
isotopic composition of
planktonic Foraminifera at
DSDP sites 277, 279, and 281
(after Shackleton and
Kennett 1975, Fig. 2)

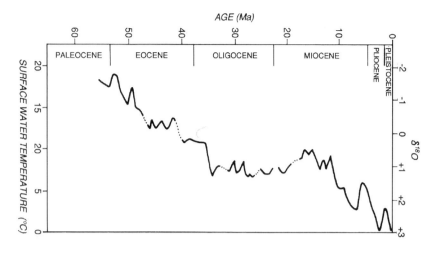

Fig. 28.5 Cenozoic events in
Australia in global context
(after McGowran et al. 2004,
Fig. 2)

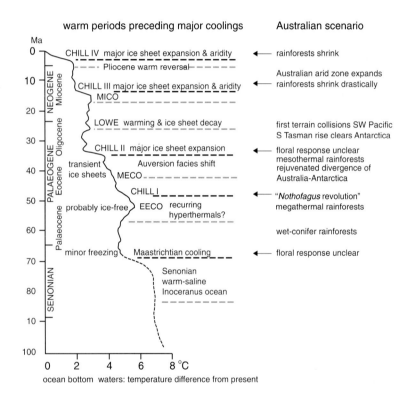

Quaternary Climatic Fluctuations

Recent palaeomagnetic, geomorphological, and geochemical investigations in the arid Amadeus Basin of central Australia have demonstrated that the onset of aridity did not become apparent until about the Jaramillo subchron (1.1 Ma) below the Brunhes–Matuyama palaeomagnetic boundary (Chen 1989).

Before that time, from at least 5 Ma until about 0.9 Ma, the pattern of sedimentation in the Amadeus Basin was dominantly one of fluviolacustrine clay accumulation with intermittent drier intervals. Thereafter, quartz dune sands, groundwater evaporites, and wind-blown gypsum sands indicate an alternation between aridity and less arid intervals characterized by high regional groundwater levels and weak pedogenesis on stabilized dune surfaces (Chen 1989).

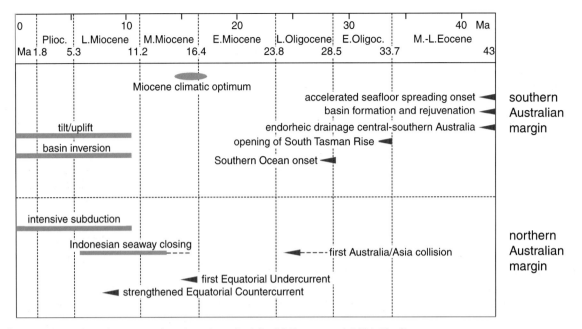

Fig. 28.6 Cenozoic environments of southern Australia (after McGowran et al. 2004, Fig. 9)

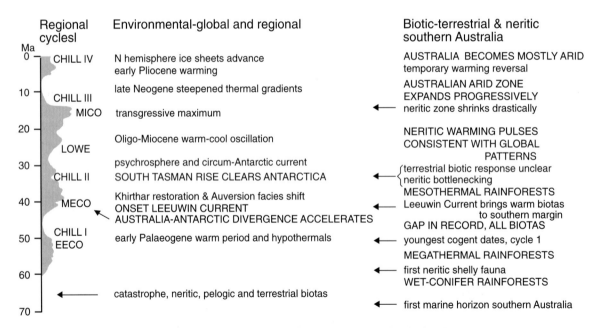

Fig. 28.7 Cenozoic events in northern and southern Australia (after McGowran et al. 2004, Fig. 28)

It is tempting to equate the change in regional climate evident in the change in depositional regime in central Australia to a change in global atmospheric circulation linked to the astronomically modulated cycles of changing global ice volumes and ocean temperatures. As a working hypothesis, we propose that aridity in both the Sahara and Australian deserts was more severe during the last 0.9 Ma than previously, and was associated with a change from the low-amplitude, high-frequency 41,000-year obliquity cycle, which dominated the late Pliocene and first half of the Pleistocene, to the high-amplitude but low-frequency 100,000-year orbital eccentricity cycle, which has dominated the last 0.9 Ma of the Pleistocene.

Desert dunes and sand plains occupy some 40% of mainland Australia and have long been a focus of attention. The relationship between dune orientation and present-day wind direction is well established. During and after the LGM dunes were active over 40% of the continent, including NE Tasmania. Preliminary OSL age estimates for dunes in the Strzelecki and Tirari deserts show well separated peaks at ~14 ka, 20 ka (LGM), and 34 ka, with minor peaks or shoulders at ~10.5 ka and 42 ka (Rhodes et al., 2004). The apparent trough between the 20 ka and 14 ka peaks suggests that few deposits have been preserved from that period. Either little dune building occurred then or subsequent reworking has removed the evidence. Dune sediments are best preserved if succeeded by a humid phase but have less chance of preservation if followed by an arid phase. Swezey (2001, 2003) makes a similar point in relation to the Saharan dunes. OSL ages for 20 aeolian dune samples from other locations in Australia show peaks at ~21 ka, 36 ka and 68 ka and a pronounced shoulder at ~43 ka. Combining all the data (54 age estimates) suggest that periods of dune-building and aeolian activity across Australia centred at ~14 ka, 20 ka, 35 ka, 42 ka and 70 ka. The event with the most samples is that at 20 ka when glaciers were at their maximum global extent and sea level at its lowest (LGM). Earlier TL ages (37 samples) by Nanson et al. (1992) lack the resolution of the new OSL data but do show broad peaks at ~30–40 ka and ~70 ka.

Hesse and McTainsh (2003) have investigated modern dust storms and Quaternary aeolian dust flux in marine cores. The marine record shows a threefold increase in dust flux during the LGM relative to the Holocene in temperate and tropical Australia. The causes appear to be weakened Australian monsoon rains (tropical north) and drier westerly circulation (temperate south). The Lake Eyre Basin/Simpson Desert and the Murray-Darling basin are the major present-day sources of dust. During 33–16 ka, there was enhanced aeolian dust flux to the east and south over the southern half of the continent. The northern limit of the dust plume was 350 km or 3° N of the present limit during 22–18 ka (Hesse et al., 2004). Williams and Nitschke (2005) have demonstrated that aeolian dust was a significant component of late Pleistocene valley-fills in and around the Flinders Ranges in semi-arid South Australia. These have been dated using paired OSL and AMS ^{14}C samples to ~33 ka near the base and ~17 ka near the top of the sequence. Drawing upon over 200 ^{14}C, TL and OSL ages, Bowler (1998) and Bowler and Price, (1998) have established that aeolian dust (or *Wüstenquarz*) began to accumulate in the lunettes on the eastern side of Pleistocene Lake Mungo and adjacent lakes from ~35 ka until ~16 ka with a peak centred around the LGM. Clay dunes and gypseous lunettes were active on the downwind margins of seasonally fluctuating lakes in many parts of the SE and SW immediately before and during 21–19 ka. Major deflation of dry lake beds coincided broadly with the time of extreme aridity centred on the LGM (e.g., Lake Eyre: Magee and Miller, 1998).

Gingele and De Deckker (2005) have recorded aeolian dust in two cores that span the last 170 ka located on the continental margin of South Australia immediately south of Kangaroo Island in the Murray Canyons area. During periods of minimum insolation at this latitude, strong westerly winds blew dust from the continental interior, with peaks at ~20 ka, ~45 ka and ~70–74 ka. These periods are coeval with times of lake desiccation, dune building and sparse vegetation cover in the centre and south of Australia (Croke et al., 1996). They also note a conspicuous double ∂^{13}C minimum at 13 ka and at ~12–10.5 ka that reflects a major hydrographic change south of the Polar Front, during which Antarctic Intermediate Water was transferred to lower latitudes. These events may coincide with the Antarctic Cold Reversal and with the Younger Dryas.

In the lead up to the cold and arid Last Glacial Maximum there was a progressive weakening of the Australian summer monsoon. Johnson et al. (1999) analysed carbon isotopes in fossil emu eggshell from around Lake Eyre in central Australia. They found significant changes in the proportions of C4 grasses over the last 65 000 years. The data imply that the Australian monsoon was most effective between 65 and 45 ka, least effective during the LGM, and moderately effective during the Holocene. They noted that the effectiveness of the summer monsoon decreased at about the time that the megafauna became extinct and at about the time that humans arrived on the continent.

The evidence from lakes, lunettes and wetlands provides further insights into the late Quaternary evolution of the Australian desert. During the LGM the tropical northern lakes were mostly dry, except for Lake Carpentaria. Lake Eyre in central Australia was totally dry and its bed was actively lowered by wind erosion (Magee et al., 1995; Croke et al., 1996; Magee and Miller, 1998). Southeastern Australia shows cooler conditions but significant regional variations in hydrology. Lakes in the southwest and southeast were mostly dry or saline at this time. In the now arid to semi-arid Flinders Ranges of South Australia a number of permanent wetlands occupied sheltered valleys within the mountains from at least 33 ka until somewhat after 16 ka (Williams et al., 2001). The persistence of such a wetland reflects the interactive influence of a number of factors: lower temperature and lower evaporation; low CO_2 levels and fewer trees, leading to rising groundwater levels; increased dust flux, trapped by the grass cover; reduced hillslope runoff and increased infiltration into the slope mantles, contributing to enhanced base flow; weak summer monsoon and fewer erosive events; lower cloud base and enhanced gentle winter rains (Chor et al., 2003).

Miller et al. (1997) used the temperature-dependent amino acid racemization reaction in radiocarbon-dated emu eggshells from the continental interior to reconstruct subtropical temperatures at low elevations over the last 45 ka. They concluded that millennial-scale average temperatures were at least 9°C lower between 45 and 16 ka than after 16 ka. There was a sharp change at 16 ka followed by rapid warming.

Impact of Key Tectonic and Climatic Events on Deserts

Throughout this chapter, we have been at pains to emphasize that many of the geomorphic attributes of our present-day deserts have a very ancient geological pedigree. We have stressed the role of Precambrian and early Phanerozoic tectonic inheritance in controlling the disposition of Mesozoic and Cenozoic uplands and lowlands in the Sahara. In Australia, we have seen that certain elements of the modern landscape are exhumed surfaces developed well before the final onset of Tertiary and Quaternary climatic fluctuations, and that some of these surfaces have been remarkably little altered since they were formed during the Precambrian. It is now appropriate to review some of the key tectonic and climatic events of the Mesozoic and Cenozoic which, directly or indirectly, had an important influence on the evolution of our modern deserts (Figs. 28.5, 28.6, 28.7).

Both hemispheres played a decisive role in the events which culminated in the late-Cenozoic desiccation of what are now the great tropical and temperate deserts of the world. Since Cenozoic intertropical desiccation was intimately associated with the cooling of Antarctica, we begin with a discussion of tectonic and climatic events in the Southern Hemisphere.

Initial separation of Australia from Antarctica started at 90 Ma and was fully effective by 45–50 Ma (Fig. 28.5). Later opening of the Drake Passage between South America and Antarctica towards 30–25 Ma resulted in the establishment of a circum-Antarctic ocean current (Fig. 28.5). Antarctica was now thermally isolated from warmer ocean waters to the north, and rapid cooling ensued. In the Southern Ocean, the changing isotopic composition of both planktonic and benthic Foraminifera is evidence of a dramatic cooling of deep ocean water as well as surface water. Cumulative ice build-up in Antarctica saw the creation of mountain glaciers followed by the growth of a major ice cap, first in East Antarctica and later in West Antarctica (Fig. 28.6). Australia, meanwhile, was moving north into dry subtropical latitudes. Within Australia, forest gave way to woodland, and woodland gave way to savanna. The net effects for Australia were climatic desiccation, progressive disruption of the drainage network, expansion of the desert, and successive plant and animal extinctions (Fig. 28.7). We turn now to the Northern Hemisphere.

Collision between Greater India and Asia extends back to about 45 Ma, and by 20–15 Ma the resulting underthrusting of Greater India beneath Asia had resulted in early uplift of the Himalayas. This uplift continued at least intermittently during the Miocene, Pliocene and Quaternary and continues to this day. Development of the Indian monsoon during the late

Miocene (or earlier) was one climatic outcome of this uplift. Another was the genesis of the easterly jet stream emanating from the Tibetan Plateau.

In East Africa, uplift and rifting created the Neogene depocentres with their unrivalled record of Pliocene and Pleistocene hominid evolution. The emergence of *Australopithecus afarensis*, a bipedal but small-brained early Pliocene hominid, may well be linked to the 6–5-Ma Messinian salinity crisis which led to the genetic isolation of Africa from Eurasia. During this terminal Miocene event, the Mediterranean Sea dried out completely, refilled and dried out again on about a dozen occasions, resulting in the accumulation of evaporites and anhydrite deposits up to a kilometre thick, representing roughly 6% of the total oceanic salt supply. The precursors to this 'salinity crisis' were the Miocene and earlier shrinking of the Tethys Sea (with an associated change to a more seasonal rainfall regime with lower annual precipitation) and the more immediate late Miocene glacio-eustatic drop in sea level caused by final accumulation of the West Antarctic ice cap. It was the 40-m fall in global sea level in the very late Miocene that exposed the Gibraltar sill separating the Atlantic from the Mediterranean, so that the Mediterranean became a closed basin in which evaporation greatly exceeded inputs from coastal rivers and local rainfall. The result was extreme aridity along the North African littoral, together with fluvial downcutting by the late Miocene Nile to carve a gorge over 1000 km long and up to 2 km deep at its northern end.

Late Pliocene cooling and desiccation towards 2.5 Ma (Fig. 28.5) caused the tropical lakes of the Sahara and East Africa to shrink, and in upland areas such as Ethiopia was associated with expansion of grassland at the expense of montane forest. It may be no coincidence that the first stone tools made to a replicated pattern make their first appearance at about this time, together with an increase in meat-eating by our hominid ancestors.

Northern polar cooling towards 15 Ma followed break-up of Laurasia. Antarctic cooling and ice build-up at about this time triggered a major change in ocean bottom water circulation (Fig. 28.6). Pliocene closure of the Panama isthmus was indirectly responsible for enhanced snowfall in high northern latitudes, reflected in Arctic ice accumulation by 3.5 Ma and a rapid expansion of North American ice caps at 2.5 Ma. Initiation of the Laurentide ice sheet ushered in the late

Pliocene and Quaternary glacial–interglacial cycles, with consequent glacio-eustatic sea-level oscillations and a global pattern of spatially and temporally alternating morphogenetic systems. Glacial maxima were times of desert expansion and forest retreat. During at least the last 0.7 Ma, each cycle was roughly 100 ka long and modulated by the 100-ka orbital eccentricity cycle.

The net effects of the late Cenozoic changes in tectonics, climate, and ocean circulation were therefore a global increase in latitudinal temperature gradients and in the seasonal incidence of rainfall, final evolution of the incipient deserts into regions of scanty and unreliable rainfall, and the emergence in Africa of upright-walking, stone toolmaking ancestral humans who later discovered how to make and use fire, and ultimately occupied every continent except Antarctica.

The story of the Cenozoic tectonic and climatic events which culminated in the slow emergence of the world's great deserts is thus also, in a very real sense, the story of the global environmental changes which were associated with the origins of the first humans in Africa some 2.5 Ma ago. From that time on, humans began to modify their habitats, sometimes contributing unwittingly to desertification or local expansion of the desert environment. Notwithstanding our destructive proclivities, the great deserts would be where they are even if humans had never appeared on the face of this Earth. We must needs learn to live with our Cenozoic inheritance.

Acknowledgments   The fieldwork on which this review is based was supported financially by the Australian Research Council, the National Science Foundation, and the Royal Geographical Society, to all of whom I am deeply grateful.

References

Abell, P.I. and M.A.J. Williams 1989. Oxygen and carbon isotope ratios in gastropod shells as indicators of palaeoenvironments in the Afar region of Ethiopia. *Palaeogeography, Palaeoclimatology, Palaeoecology* **74**, 265–78.

Adamson, D.A. and F. Williams 1980. Structural geology, tectonics and the control of drainage in the Nile Basin. In *The Sahara and the Nile*, M.A.J. Williams and H. Faure (eds), 225–52. Rotterdam: Balkema.

Adamson, D.A. and M.A.J. Williams 1987. Geological setting of Pliocene rifting and deposition in the Afar Depression of Ethiopia. *Journal of Human Evolution* **16**, 597–610.

Alley, N.F. and Lindsay, J.M. 1995. Tertiary. In *The geology of South Australia, Volume 2. The Phanerozoic*, J.F. Drexel and W.V. Preiss (eds), Bulletin 54, 151–217. South Australia: Geological Survey.

Ayliffe, L.K., Marianelli, P.C., Moriarty, K.C., Wells, R.T., McCulloch, M.T., Mortimer, G.E. and Hellstrom, J.C. 1998. 500 ka precipitation record from southeastern Australia: Evidence for interglacial relative aridity. *Geology* **26**, 147–50.

Berger, A.L. 1981. The astronomical theory of palaeoclimates. In *Climatic variations and variability: facts and theories*, A. Berger (ed), 501–2. Dordrecht: D. Reidel.

Beuf, S., B. Bijou-Duval, O. De Charpal, P. Rognon et al. 1971. Les Grés du Paléozoique inférieur au Sahara. *Publication de l'Institut français du Pétrole*. Paris: Technip.

Black, R. and M. Girod 1970. Late Palaeozoic to Recent igneous activity in West Africa and its relationship to basement structure. In *African magmatism and tectonics*, T.N. Clifford and I.G. Gass (eds), 185–210. Edinburgh: Oliver and Boyd.

Bloemendal, J. and P. deMenocal 1989. Evidence for a change in the periodicity of tropical climate cycles at 2.4 Myr from whole-core magnetic susceptibility measurements. *Nature* **342**, 897–900.

Bonnefille, R. 1976. Implications of pollen from the Koobi Fora Formation, East Rudolf, Kenya. *Nature* **264**, 403–7.

Bonnefille, R. 1980. Vegetation history of savanna in East Africa during the Plio-Pleistocene. *IV International Palynological Conference* **3**, 78–89.

Bonnefille, R. 1983. Evidence for a cooler and drier climate in the Ethiopian uplands towards 2.4 myr ago. *Nature* **303**, 487–91.

Bonnefille, R., Potts, R., Chalié, F., Jolly, D. and Peyron, O. 2004. High-resolution vegetation and climate change associated with Pliocene *Australopithecus afarensis*. *Proceedings of the National Academy of Sciences* **101**, 12125–29.

Bordet, P. 1952. Les appareils volcaniques récents de l'Ahaggar. *Monographic Régionale. Algérie* 1: No. 11, 19ᵉ Congrès de Géologie Internationale, Alger.

Bowen, R. and U. Jux 1987. *Afro Arabian geology. A kinematic view*. London: Chapman & Hall.

Bowler, J.M. 1973. Clay dunes: their occurrence, formation and environmental significance. *Earth-Science Reviews* **9**, 315–38.

Bowler, J.M. 1976. Aridity in Australia: age, origins and expression in aeolian landforms and sediments. *Earth-Science Reviews* **12**, 279–310.

Bowler, J.M. 1998. Willandra Lakes revisited: environmental framework for human occupation. *Archaeology in Oceania* **33**, 120–155.

Bowler, J.M. and Price, D.M. 1998. Luminescence dates and stratigraphic analyses at Lake Mungo: review and new perspectives. *Archaeology in Oceania* **33**, 156–168.

Bowler, J.M., Duller, G.A.T., Perret, N., Prescott, J.R. and Wyrwoll, K.-H. 1998. Hydrological changes in monsoonal climates of the last glacial cycle: stratigraphy and luminescence dating of Lake Woods, N.T., Australia. *Palaeoclimates* **3**, 179–207.

Bowler, J.M., Wyrwoll, K.-H. and Lu, Y. 2001. Variations in the northwest Australian summer monsoon over the last 300,000 years: the paleohydrological record of the Gregory (Mulan) Lakes System. *Quaternary International* **83–85**, 63–80.

Bowles, F.A. 1975. Palaeoclimatic significance of quartz/illite variations in cores from the eastern equatorial North Atlantic. *Quaternary Research* **5**, 225–35.

Breed, C.S., J.F. McCauley and P.A. Davis 1987. Sand sheets of the eastern Sahara and ripple blankets on Mars. In *Desert sediments: ancient and modern*, L. Frostick and I. Reid (eds), 337–59. Geological Society Special Publication No. 35.

Cane, M.A. and Molnar, P. 2001. Closing of the Indonesian seaway as a precursor to east African aridification around 3–4 million years ago. *Nature* **411**, 157–62.

Cerling, T.E., Harris, J.M., MacFadden, B.J., Leakey, M.G., Quade, J., Eisenmann, V. and Ehlerlinger, J.R. 1997. Global vegetation change through the Miocene/Pliocene boundary. *Nature* **389**, 153–8.

Chappell, J. 1974. Relationships between sealevels, ¹⁸O variations and orbital perturbations, during the past 250,000 years. *Nature* **252**, 199–202.

Chen, X.Y. 1989. *Lake Amadeus, central Australia: modern processes and evolution*. Unpublished Ph.D. thesis, Australian National University, Canberra.

Chor, C., Nitschke, N. and Williams, M. 2003. Ice, wind and water: Late Quaternary valley-fills and aeolian dust deposits in arid South Australia. *Proceedings of the Cooperative Research Centre for Landscape, Environment and Mineral Exploration (CRC LEME) Regional Regolith Symposia*, edited by I.C. Roach, Adelaide, November 13–14, 2003, pp.70–73.

Clark, J.D. and S.A. Brandt (eds) 1984. *From hunters to farmers. The causes and consequences of food production in Africa*. Berkeley: University of California Press.

Clark, J.D., M.A.J. Williams and A.B. Smith 1973. The geomorphology and archaeology of Adrar Bous, Central Sahara: a preliminary report. *Quaternaria* **17**, 245–97.

Clark, J.D., B. Asfaw, G. Assefa, J.W.K. Harris, et al. 1984. Palaeoanthropological discoveries in the Middle Awash Valley, Ethiopia. *Nature* **307**, 423–8.

Clifford, T.N. and I.G. Gass 1970. *African magmatism and tectonics*. Edinburgh: Oliver & Boyd.

Conrad, G. 1969. *L'évolution continentale post-hercynienne du Sahara algérien*. (Saoura, Erg Chech-Tanezrouft, Ahnet-Mouydir). Paris: CNRS.

Coque, R. 1962. *La Tunisie présaharienne, étude géomorphologique*. Paris: A. Colin.

Coulson, D. and Campbell, A. 2001. *African Rock Art: Paintings and Engravings on Stone*. New York: H.A. Abrams.

Croke, J., Magee, J. and Price, D. 1996. Major episodes of Quaternary activity in the lower Neales River, northwest of Lake Eyre, central Australia. *Palaeogeography, Palaeoclimatology, Palaeoecology* **124**, 1–15.

Dresch, J. 1959. Notes sur la géomorphologie de l'Air. *Bulletin de l'Association de Géographes français* **280–81**, 2–20.

Deuser, W.G., E.H. Ross and L.S. Waterman 1976. Glacial and pluvial periods: their relationship revealed by Pleistocene sediments of the Red Sea and Gulf of Aden. *Science* **191**, 1168–70.

Dupont-Nivet, G., Krijgsman, W., Langereis, C.G., Abels, H.A., Dai, S. and Fang, X. 2007.Tibetan plateau aridification linked to global cooling at the Eocene-Oligocene transition. *Nature* **445**, 635–38.

Eberz, G.W., F.M. Williams and M.A.J. Williams 1988. Plio-Pleistocene volcanism and sedimentary facies changes at

Gadeb prehistoric site, Ethiopia. *Geologische Rundschau* **77**, 513–27.

Eldrett, J.S., Harding, I.C., Wilson, P.A., Butler, E. and Roberts, A.P. 2007. Continental ice in Greenland during the Eocene and Oligocene. *Nature* **446**, 176–79.

El-Gaby, S. and R.O. Greiling (eds) 1988. *The Pan-African belt of northeast Africa and adjacent areas.* Braunschweig: Vieweg & Sohn.

Fabre, J. 1974. Le Sahara: un musée géologique. *La Recherche* **42**(5), 140–52.

Faure, H. 1962. Reconnaissance géologique des formation-post-paléozoiques du Niger oriental. *Mémoire du Bureau de Recherches Géologiques et Minières* (Dakar) (1966) No. 47.

Faure, H. 1975. Recent crustal movements along the Red Sea and Gulf of Aden coasts in Afar (Ethiopia and TFAI). *Tectonophysics* **29**, 479–86.

Feakins, S.J., deMenocal, P.B. and Eglinton, T.I. 2005. Biomarker records of late Neogene changes in northeast African vegetation. *Geology* **33**, 977–80.

Flohn, H. 1980. The role of the elevated heat source of the Tibetan Highlands for the large-scale atmospheric circulation (with some remarks on paleoclimatic changes). In *Proceedings of Symposium on Qinghai-Xizang (Tibet) Plateau (abstracts), Beijing*, China, May 25 – June 1, 1980. Beijing: Academia Sinica.

Fontes, J.C., F. Gasse, Y. Callot and J.-C. Plaziat, et al. 1985. Freshwater to marine-like environments from Holocene lakes in northern Sahara. *Nature* **317**, 608–10.

Francis, P.W., R.S. Thorpe and F. Ahmed 1973. Setting and significance of Tertiary-Recent volcanism in the Darfur Province of Western Sudan. *Nature Physical Science* **243**, 30–2.

Frostick, L.E. and I. Reid (eds) 1987. *Desert sediments: ancient and modern.* Geological Society Special Publication No. 35.

Fujioka, T., Chappell, J., Honda, M., Yatsevich, I., Fifield, K. and Fabel, D. 2005. Global cooling initiated stony deserts in central Australia 2–4 Ma, dated by cosmogenic ^{21}Ne-^{10}Be. *Geology* **33**, 993–996.

Galloway, R.W. 1965. A note on world precipitation during the last glaciation. *Eiszeitalter and Gegenwart* **16**, 76–7.

Gasse, F. 1975. L'évolution des lacs de l'Afar Central (Ethiopie et TFAI) du Plio-Pléistocène à l'Actuel. D.Sc. thesis, University of Paris VI.

Gasse, F., R. Téhet, A. Durand, E. Gibert, et al. 1990. The arid–humid transition in the Sahara and the Sahel during the last deglaciation. *Nature* **346**, 141–6.

Gingele, F.X and De Deckker, P. 2005. Late Quaternary fluctuations of palaeoproductivity in the Murray Canyons area, South Australian continental margin. *Palaeogeography, Palaeoclimatology, Palaeoecology* **220**, 361–373.

Greigert, J. and R. Pougnet 1967. Essai de description des formations géologiques de la République du Niger. *Mémoires du Bureau de Recherches Géologiques et Minières* (Dakar) No. 48, 1–236.

Griffin, D.L. 2006. The late Neogene Sahabi rivers of the Sahara and their climatic and environmental implications for the Chad Basin. *Journal of the Geological Society, London* **163**, 905–21.

Grove, A.T. 1958. The ancient erg of Hausaland and similar formations on the south side of the Sahara. *Geographical Journal* **124**, 528–33.

Grove, A.T. 1980. Geomorphic evolution of the Sahara and the Nile. In *The Sahara and the Nile*, M.A.J. Williams and H. Faure (eds), 7–16. Rotterdam: Balkema.

Grove, A.T. and A. Warren 1968. Quaternary landforms and climate on the south side of the Sahara. *Geographical Journal* **134**, 194–208.

Guiraud. R., Bosworth, W., Thierry, J. and Delplanque, A. 2005. Phanerozoic geological evolution of Northern and Central Africa: An overview. *Journal of African Earth Sciences* **43**, 83–143.

Habicht, J.K.A. 1979. Paleoclimate, paleomagnetism, and continental drift. *AAPG Studies in Geology No. 9.* Tulsa, Oklahoma: American Association of Petroleum Geologists.

Hays, J.D., J. Imbrie and N.J. Shackleton 1976. Variations in the earth's orbit: pacemaker of the ice ages. *Science* **194**, 1121–32.

Heller, F. and T.-S. Liu 1982. Magnetostratigraphical dating of loess deposits in China. *Nature* **300**, 431–3.

Hesse, P.P. and McTainsh, G.H. 2003. Australian dust deposits: modern processes and the Quaternary record. *Quaternary Science Reviews* **22**, 2007–2035.

Hesse, P.P., Magee, J.W. and van der Kaars, S. 2004. Late Quaternary climates of the Australian arid zone: a review. *Quaternary International* **118–119**, 87–102.

Hoelzmann, P., Gasse, F., Dupont, L. M., Salzmann, U., Staubwasser, M., Leuschner, D. C., Sirocko, F., 2004. Palaeoenvironmental changes in the arid and subarid belt (Sahara-Sahel-Arabian Peninsula) from 150 kyr to present. In *volume 6: Past Climate Variability through Europe and Africa* R.W. Battarbee, R.W., Gasse, F., Stickley, C.E. (eds.),. Springer, Dordrecht, p. 219–256.

Imbrie, J. and K.P. Imbrie 1979. *Ice ages: solving the mystery.* London: Macmillan.

Johnson, B.J., Miller, G.H., Fogel, M.L., Gagan, M.K. and Chivas, A.R. 1999. 65,000 years of vegetation change in Central Australia and the Australian summer monsoon. *Science* **284**, 1150–1152.

Kennett, J.P. and D.A. Hodell 1986. Major events in Neogene oxygen isotopic records. *South African Journal of Science* **82**, 497–8.

Kilian, C. 1931. Des principaux complexes continentaux du Sahara. *Compte rendu de la Société géologique de France, 1928–31*, 109–11.

Kuper, R. and Kröpelin, S. 2006. Climate-controlled Holocene occupation in the Sahara: Motor of Africa's evolution. *Science* **313**, 803–07.

Kutzbach, J.E. and F.A. Street-Perrott 1985. Milankovitch forcing of fluctuations in the level of tropical lakes from 18 to 0 kyr BP. *Nature* **317**, 130–4.

Loubere, P. and K. Moss 1986. Late Pliocene climatic change and the onset of Northern Hemisphere glaciation as recorded in the northeast Atlantic Ocean. *Bulletin of the Geological Society of America* **97**, 818–28.

Mabbutt, J.A. 1977. *Desert landforms.* Canberra: Australian National University Press.

Magee, J.M. 1998. *Late Quaternary environments and palaeohydrology of Lake Eyre, arid central Australia.* Unpublished PhD thesis, Australian National University.

Magee, J.W., Bowler, J.M., Miller, G.H. and Williams, D.L.G. 1995. Stratigraphy, sedimentology, chronology and palaeohydrology of Quaternary lacustrine deposits at Madigan

Gulf, Lake Eyre, South Australia. *Palaeogeography, Palaeo-climatology, Palaeoecology* **113**, 3–42.

Magee, J.W. and Miller, G.H. 1998. Lake Eyre palaeohydrology from 60 ka to the present: beach ridges and glacial maximum aridity. *Palaeogeography, Palaeoclimatology, Palaeoecology* **144**, 307–329.

Maley, J. 1980a. Etudes palynologiques dans le bassin du Tchad et paléoclimatologie de l'Afrique Nord-tropicale de 30 000 ans à l'époque actuelle. D.Sc. thesis, University of Montpellier.

Maley, J. 1980b. Les changements climatiques de la fin du Tertiaire en Afrique: leur conséquence sur l'apparition du Sahara et de sa végétation. In *The Sahara and the Nile*, M.A.J. Williams and H. Faure (eds), 63–86. Rotterdam: Balkema.

Maley, J. 1996. The African rain forest – main characteristics of changes in vegetation and climate from the Upper Cretaceous to the Quaternary. *Proceedings of the Royal Society of Edinburgh* **104B**, 31–73.

Maley, J. 2000. Last Glacial Maximum lacustrine and fluviatile Formations in the Tibesti and other Saharan mountains, and large-scale climatic teleconnections linked to the activity of the Subtropical Jet Stream. *Global and Planetary Change* **26**, 121–36.

Martin, H.A. 2006. Cenozoic climatic changes and the development of the arid vegetation of Australia. *Journal of Arid Environments* **66**, 533–563.

May, R.M. 1978. Human reproduction reconsidered. *Nature* **272**, 491–5.

McDougall, I., W.H. Morton and M.A.J. Williams 1975. Age and rates of denudation of Trap Series basalts at Blue Nile Gorge, Ethiopia. *Nature* **254**, 207–9.

McGowran, B., Holdgate, G.R., Li, Q. and Gallagher, S.J. 2004. Cenozoic stratigraphic succession in southeastern Australia. *Australian Journal of Earth Sciences* **51**, 459–496.

McHugh, W.P., C.S. Breed, G.G. Schaber, J.F. McCauley, et al. 1988. Acheulian sites along the 'radar rivers', southern Egyptian Sahara. *Journal of Field Archaeology* **15**, 361–79.

McHugh, W.P., G.G. Schaber, C.S. Breed and J.F. McCauley 1989. Neolithic adaptation and the Holocene functioning of Tertiary palaeodrainages in southern Egypt and northern Sudan. *Antiquity* **63**, 320–36.

McKee, E.D. (ed) 1978. A study of global sand seas. *United States Geological Survey, Professional Paper* 1052.

Milankovitch, M. 1920. *Théorie mathématique des phénomènes thermiques produits par la radiation solaire*. Paris: Gaultier-Villars.

Milankovitch, M. 1930. Mathematische Klimalehre und astonomische Theorie der Klimaschwankungen. In *Handbuch der Klimatologie*, Volume I(A), W. Köppen and R. Geiger (eds), 1–176. Berlin: Gebruder Borntraeger.

Miller, G.H., Magee, J.W. and Jull, A.J.T. 1997. Low-latitude glacial cooling in the Southern hemisphere from amino-acid racemization in emu eggshells. *Nature* **385**, 241–244.

Mix, A. C., Bard, E., and Schneider, R. 2001. Environmental processes of the ice age: land, oceans, glaciers (EPILOG). *Quaternary Science Reviews* **20**, 627–657.

Muzzolini, A. 1995. *Les Images Rupestres du Sahara*. Toulouse: Muzzolini.

Nanson, G.C., Price, D.M. and Short, S.A. 1992. Wetting and drying of Australia over the past 300 ka. *Geology* **20**, 791–794.

Owen, H.G. 1983. *Atlas of continental displacement, 200 million years to the present*. Cambridge: Cambridge University Press.

Parkin, D.W. 1974. Trade-winds during the glacial cycles. *Proceedings of the Royal Society of London A* **337**, 73–100.

Parkin, D.W. and N. Shackleton 1973. Trade-winds and temperature correlations down a deep-sea core off the Saharan coast. *Nature* **245**, 455–7.

Parmenter, C. and D.W. Folger 1974. Eolian biogenic detritus in deep sea sediments: a possible index of equatorial Ice Age aridity. *Science* **185**, 695–8.

Pearson, P.N., van Dongen, B.E., Nicholas, C.J., Pancost, R.D., Schouten, S., Singano, J.M. and Wade, B.S. 2007. Stable warm tropical climate through the Eocene Epoch. *Geology* **35**, 211–4.

Peel, R.F. 1966. The landscape in aridity. *Transactions of the Institute of British Geographers* **38**, 1–23.

Pesce, A. 1968. *Gemini space photographs of Libya and Tibesti. A geological and geographical analysis*. Tripoli: Petroleum Exploration Society of Libya.

Petit, J.R., M. Briat and A. Royer 1981. Ice age aerosol content from East Antarctic ice core samples and past wind strength. *Nature* **293**, 391–4.

Pik, R., Marty, B., Carignan, J., and Lave, J., 2003, Stability of the Upper Nile drainage network (Ethiopia) deduced from (U – Th)/He thermochronometry: Implications for uplift and erosion of the Afar plume dome. *Earth and Planetary Science Letters* **215**, 73–88.

Prentice, M.L. and G.H. Denton 1988. The deep-sea oxygen isotope record, the global ice sheet system and hominid evolution. In *Evolutionary history of the 'robust' australopithecines*, F.E. Grine (ed), 383–403. New York: Aldine de Gruyter.

Prideaux, G.J., Long, J.A., Ayliffe, L.K., Hellstrom, J.C., Pillans, B., Boles, W.E., Hutchinson, M.N., Roberts, R.G., Cupper, M.L., Arnold, L.J., Devine, P.D. and Warburton, N.M. 2007. An arid-adapted middle Pleistocene vertebrate fauna from south-central Australia. *Nature* **445**, 422–25.

Quade, J., T.E. Cerling and J.R. Bowman 1989. Development of the Asian monsoon revealed by marked ecological shift during the latest Miocene in northern Pakistan. *Nature* **342**, 163–6.

Raulais, M. 1951. Du Crétacé probable sur les hauts reliefs sahariens. *Compte rendu sommaire des séances de la Société géologique de France, 1951*, 22–3.

Revel-Rolland, M., De Deckker, P., Delmonte, B., Hesse, P.P., Magee, J.M., Basile-Doelsch, I., Grousset, F. and Bosch, D. 2006. Eastern Australia: A possible source of dust in East Antarctica interglacial ice. *Earth and Planetary Science Letters* **249**, 1–13.

Rhodes, E., Fitzsimmons, K., Magee, J., Chappell, J., Miller, G. and Spooner, N.G. (2004). The history of aridity in Australia: preliminary chronological data. In *Regolith 2004. Cooperative Research Centre for Landscape Environment and Mineral Exploration*, I.C.Roach (ed), pp. 299–302.

Rognon, P. 1967. *Le massif de l'Atakor et ses bordures (Sahara central): étude géomorphologique*. Paris: CNRS and CRZA.

Rognon, P. 1989. *Biographie d'un désert*. Paris: Plon.

Rognon, P. and M.A.J. Williams 1977. Late Quaternary climatic changes in Australia and North Africa:

a preliminary interpretation. *Palaeogeography, Palaeoclimatology, Palaeoecology* **21**, 285–327.

Ruddiman, W.F. and M.E. Raymo 1988. Northern Hemisphere climate régimes during the past 3 Ma: possible tectonic connections. *Philosophical Transactions of the Royal Society B* **318**, 411–30.

Sarnthein, M. 1978. Sand deserts during glacial maximum and climatic optimum. *Nature* **272**, 43–5.

Sarnthein, M., G. Tetzlaff, B. Koopmann, K. Wolter and U. Pflaumann, 1981. Glacial and interglacial wind regimes over the eastern subtropical Atlantic and north-west Africa. *Nature* **293**, 193–6.

Schmitz, B. and Pujalte, V. 2007. Abrupt increase in seasonal extreme precipitation at the Paleocene-Eocene boundary. *Geology* **35**, 215–8.

Schnitker, D. 1980. Global paleoceanography and its deep water linkage to the Antarctic glaciation. *Earth-Science Reviews* **16**, 1–20.

Schuster, M., Duringer, P., Ghienne, J.-F., Vignaud, P., Mackaye, H.T., Likius, A. and Brunet, M. 2006. The age of the Sahara desert. *Science* **311**, 821.

Sepulchre, P., Ramstein, G., Fluteau, F., Schuster, M., Tiercelin, J.-J. and Brunet, M. 2006. Tectonic uplift and Eastern African aridification. *Science* **313**, 1419–23.

Servant, M. 1973. *Séquences continentales et variations climatiques: Evolution du bassin du Tchad au Cénozoique supéricur.* D.Sc. thesis, University of Paris.

Shackleton, N.J. 1977. The oxygen isotope stratigraphic record of the late Pleistocene. *Philosophical Transactions of the Royal Society* **280**, 169–79.

Shackleton, N.J. 1987. Oxygen isotopes, ice volume and sea level. *Quaternary Science Reviews* **6**, 183–90.

Shackleton, N.J. and J.P. Kennett 1975. Paleotemperature history of the Cenozoic and the initiation of Antarctic glaciation: oxygen and carbon isotope analyses in DSDP sites 277, 279 and 281. In *Initial Reports of the deep sea drilling project No. 29*, J.P. Kennett, R.E. Houtz, P.B. Andrews, A.R. Edwards, et al. (eds), 743–55. Washington DC: U.S. Government Printing Office.

Shackleton, N.J. and N.D. Opdyke 1977. Oxygen isotope and palaeomagnetic evidence for early Northern Hemisphere glaciation. *Nature* **270**, 216–9.

Shackleton, N.J., J. Backman, H. Zimmerman, D.V. Kent, et al. 1984. Oxygen isotope calibration of the onset of ice-rafting and history of glaciation in the North Atlantic region. *Nature* **307**, 620–3.

Smith, A.B. 1980. The Neolithic tradition in the Sahara. In *The Sahara and the Nile*, M.A.J. Williams and H. Faure (eds), 451–65. Rotterdam: Balkema.

Sniderman, J.M. K., Pillans, B., O'Sullivan, P.B. and Kershaw, A.P. 2007. Climate and vegetation in southeastern Australia respond to Southern Hemisphere insolation forcing in the late Pliocene-early Pleistocene. *Geology* **35**, 41–44.

Street, F.A. 1980. The relative importance of climate and local hydrogeological factors in influencing lake-level fluctuations. *Palaeoecology of Africa* **12**, 137–58.

Street-Perrott, F.A. and R.A. Perrott 1990. Abrupt climatic fluctuations in the tropics: the influence of Atlantic Ocean circulation. *Nature* **343**, 607–12.

Suc, J.-P. 1984. Origin and evolution of the Mediterranean vegetation and climate in Europe. *Nature* **307**, 429–32.

Swezey, C. (2001). Eolian sediment response to late Quaternary climate changes: temporal and spatial patterns in the Sahara. *Palaeogeography, Palaeoclimatology, Palaeoecology* **167**, 119–155.

Swezey, C. (2003). The role of climate in the creation and destruction of continental stratigraphic records: An example from the northern margin of the Sahara Desert. *SEPM Special Publication* No. **77**, 207–225.

Talbot, M.R. 1980. Environmental responses to climatic change in the West African Sahel over the past 20,000 years. In *The Sahara and the Nile*, M.A.J. Williams and H. Faure (eds), 37–62. Rotterdam: Balkema.

Thorp, M.B. 1969. Some aspects of the geomorphology of the Aïr Mountains, southern Sahara. *Transactions of the Institute of British Geographers* **47**, 25–46.

Thurmond, A.K., Stern, R.J., Abdelsalam, M.G., Nielsen, K.C., Abdeen, M.M. and Hinz, E. 2004. The Nubian Swell. *Journal of African Earth Sciences* **39**, 401–7.

Tiercelin, J.J. 1987. The Pliocene Hadar Formation, Afar depression of Ethiopia. In *Desert sediments: ancient and modern*, L.E. Frostick and I. Reid (eds), 221–40. Oxford: Blackwell Scientific, Geological Society Special Publication No. 35.

Vail, J.R. 1972a. Jebel Marra, a dormant volcano in Darfur Province, western Sudan. *Bulletin Volcanologique* **36**, 251–65.

Vail, J.R. 1972b. Geological reconnaissance in the Zalingei and Jebel Marra areas of western Darfur Province, Sudan. *Bulletin of the Geological Survey, Sudan* **19**, 1–50.

Van der Kaars, S., De Deckker, P. and Gingele, F. X. 2006. A 100 000-year record of annual and seasonal rainfall and temperature for northwestern Australia based on a pollen record obtained offshore. *Journal of Quaternary Science* **21**, 879–889.

Veevers, J.J. (ed) 2000. *Billion-year earth history of Australia and neighbours in Gonwanaland.* Sydney: GEMOC Press.

Vincent, P. 1963. Les volcans tertiaires et quaternaires du Tibesti occidental et central (Sahara du Tchad). *Mémoires, Bureau de Recherches Géologiques et Minières* **23**, 1–307.

Vogt, J. and R. Black 1963. Remarques sur la géomorphologie de l'Aïr. *Bulletin, Bureau de Recherches Géologiques et Miniéres* (Dakar) **1**, 1–29.

Williams, M.A.J. 1971. Geomorphology and Quaternary geology of Adrar Bous. *Geographical Journal* **137**, 449–55.

Williams, M.A.J. 1975. Late Pleistocene tropical aridity synchronous in both hemispheres? *Nature* **253**, 617–8.

Williams, M.A.J. 1984a. Geology. In *Key environments. Sahara Desert*, J.L. Cloudsley-Thompson (ed.), 31–9. Oxford: Pergamon.

Williams, M.A.J. 1984b. Cenozoic evolution of arid Australia. In *Arid Australia*, H.G. Cogger and E.E. Cameron (eds), 59–78. Sydney: Australian Museum.

Williams, M.A.J. 1985. Pleistocene aridity in tropical Africa, Australia and Asia. In *Environmental change and tropical geomorphology*, I. Douglas and T. Spencer (eds), 219–33. London: Allen & Unwin.

Williams, M.A.J. 1991. Evolution of the landscape. In *Monsoonal Australia. Landscape, ecology and man in the northern lowlands*, M.G. Ridpath, C.D. Haynes and M.A.J. Williams (eds), 207–21. Rotterdam: Balkema.

Williams, M.A.J. (2000). Quaternary Australia: extremes in the Last Glacial-Interglacial cycle. In *Billion-year earth history of Australia and neighbours in Gondwanaland*, J.J. Veevers (ed),. 55–59. Sydney: GEMOC Press.

Williams, M.A.J. (2001). Morphoclimatic maps at 18 ka, 9 ka, & 0 ka. In *Atlas of Billion-year earth history of Australia and neighbours in Gondwanaland*, J.J. Veevers (ed), 45–48. Sydney: GEMOC Press.

Williams, M. (2002a). Deserts. In *Encyclopedia of Global Environmental Change (ISBN-0-471-97796-9) Volume 1: The earth system: physical and chemical dimensions of global environmental change*, M.C. MacCracken and J.S. Perry (eds), Chichester: Wiley, pp. 332–343.

Williams, M.A.J. (2002b). Desertification. In *Encyclopedia of Global Environmental Change (ISBN 0-471-97796-9) Volume 3: Causes and consequences of global environmental change*, I. Douglas (ed), 282–290. Chichester: Wiley.

Williams, M.A.J. and H. Faure (eds) 1980. *The Sahara and the Nile*. Rotterdam: Balkema.

Williams, M.A.J. and F.M. Williams 1980. Evolution of the Nile Basin. In *The Sahara and the Nile*, M.A.J. Williams and H. Faure (eds), 207–24. Rotterdam: Balkema.

Williams, M.A.J., D.A. Adamson, G.H. Curtis, & F. Gasse, et al. 1979. Plio-Pleistocene environments at Gadeb prehistoric site, Ethiopia. *Nature* **282**, 29–33.

Williams, M.A.J., D.A. Adamson, F.M. Williams, W.H. Morton, et al. 1980. Jebel Marra volcano: a link between the Nile Valley, the Sahara and Central Africa. In *The Sahara and the Nile*, M.A.J. Williams and H. Faure (eds), 305–37. Rotterdam: Balkema.

Williams, D.F., W.S. Moore and R.H. Fillon 1981a. Role of glacial Arctic Ocean ice sheets in Pleistocene oxygen isotope and sea level records. *Earth and Planetary Science Letters* **56**, 157–66.

Williams, M.A.J., F.M. Williams and P. Bishop 1981b. Late Quaternary history of Lake Besaka, Ethiopia. *Palaeoecology of Africa* **13**, 93–104.

Williams, M.A.J., G. Assefa and D.A. Adamson 1986. Depositional context of Plio-Pleistocene hominid-bearing formations in the Middle Awash Valley, southern Afar Rift, Ethiopia. In *Sedimentation in the African Rifts*, L. Frostick, R. Renaut, I. Reid and J.J. Tiercelin (eds), 233–43. Oxford: Blackwell Scientific, Geological Society Special Publication No. 25.

Williams, M.A.J., P.I. Abell and B.W. Sparks 1987. Quaternary landforms, sediments, depositional environments and gastropod isotope ratios at Adrar Bous, Tenere Desert of Niger, south-central Sahara. In *Desert sediments: ancient and modern*, L. Frostick and I. Reid (eds), 105–25. Geological Society Special Publication No. 35.

Williams, M.A.J., P. De Deckker and A.P. Kershaw (eds) 1991a. *The Cainozoic in Australia: a re-appraisal of the evidence*. Geological Society of Australia, Special Publication No. 18.

Williams, M.A.J., D.A. Adamson, P. De Deckker and M.R. Talbot 1991b. Episodic fluviatile, lacustrine and aeolian sedimentation in a late Quaternary desert margin system, central western New South Wales. In *The Cainozoic in Australia: a re-appraisal of the evidence*, M.A.J. Williams, P. De Deckker and A.P. Kershaw (eds), 258–87. Geological Society of Australia, Special Publication No. 18.

Williams, M.A.J. and Balling, R.C. Jr 1996. *Interactions of Desertification and Climate*. Arnold, London, 270 pp., including foreword, ten colour plates and index, xi–xii, 259–270.

Williams, M., Dunkerley, D., De Deckker, P., Kershaw, P. and Chappell, J. 1998. *Quaternary Environments*. 2nd Edition. London: Arnold.

Williams, M., Prescott, J.R., Chappell, J., Adamson, D., Cock, B., Walker, K. and Gell, P. 2001. The enigma of a late Pleistocene wetland in the Flinders Ranges, South Australia. *Quaternary International* **83-85**, 129–144.

Williams, F.M., Williams, M.A.J. and Aumento, F. (2004). Tensional fissures and crustal extension rates in the northern part of the Main Ethiopian Rift. *Journal of African Earth Sciences* **38**, 183–97.

Williams, M.A.J. and Nitschke, N. 2005. Influence of windblown dust on landscape evolution in the Flinders Ranges, South Australia. *South Australian Geographical Journal* **104**, 25-36.

Yemane, K., R. Bonnefille and H. Faure 1985. Palaeoclimatic and tectonic implications of Neogene microflora from the north-western Ethiopian highlands. *Nature* **318**, 653–6.

Zanazzi, A., Kohn, M.J, MacFadden, B.J. and Terry, D.O. 2007. Large temperature drop across the Eocene-Oligocene transition in central North America. *Nature* **445**, 639–42.

Index